Reference Series in Phytochemistry

This series provides a platform for essential information on plant metabolites and phytochemicals, their chemistry, properties, applications, and methods.

By the strictest definition, phytochemicals are chemicals derived from plants. However, the term is often also used to describe the large number of secondary metabolic compounds found in and derived from plants. These metabolites exhibit a number of nutritional and protective functions for human wellbeing and are used e.g. as colorants, fragrances and flavorings, amino acids, pharmaceuticals, hormones, vitamins and agrochemicals.

The series offers extensive information on various topics and aspects of phytochemicals, including their potential use in natural medicine, their ecological role, role as chemo-preventers and, in the context of plant defense, their importance for pathogen adaptation and disease resistance. The respective volumes also provide information on methods, e.g. for metabolomics, genetic engineering of pathways, molecular farming, and obtaining metabolites from lower organisms and marine organisms besides higher plants. Accordingly, they will be of great interest to readers in various fields, from chemistry, biology and biotechnology, to pharmacognosy, pharmacology, botany and medicine.

The Reference Series in Phytochemistry is indexed in Scopus.

Sylvester Chibueze Izah •
Matthew Chidozie Ogwu •
Muhammad Akram
Editors

Herbal Medicine Phytochemistry

Applications and Trends

Volume 1

With 245 Figures and 197 Tables

Editors
Sylvester Chibueze Izah
Department of Microbiology
Bayelsa Medical University
Yenagoa, Bayelsa, Nigeria

Matthew Chidozie Ogwu
Department of Sustainable Development
Appalachian State University
Boone, NC, USA

Muhammad Akram
Government College
University of Faisalabad
Islamabad, Pakistan

ISSN 2511-834X ISSN 2511-8358 (electronic)
Reference Series in Phytochemistry
ISBN 978-3-031-43198-2 ISBN 978-3-031-43199-9 (eBook)
https://doi.org/10.1007/978-3-031-43199-9

This Springer imprint is published by the registered company Springer Nature Switzerland AG.
The registered company address is: Gewerbestrasse 11, 6330 Cham, Switzerland

Preface

The advancement of evidence-based herbal medicine practice relies on understanding the chemical composition of medicinal plants and their bioactive components. Utilization techniques in herbal medicine draw from the traditional wisdom, knowledge, understanding, and practices of local healers. To ensure the safety and effectiveness of herbal medicines, ongoing quality assessment, and research are essential to help bridge the gap between traditional knowledge and modern scientific and societal methods. Herbal medicine phytochemistry identifies bioactive plant compounds, their mechanism of action, standardized herbal products, and the validation of traditional remedies. Challenges include the need for scientific validation, quality control, regulatory compliance, etc. Opportunities exist in the discovery of new medicines as well as the integration of herbal medicine into mainstream conventional healthcare. Quality assessment such as the precise identification of plant and phytochemicals and advanced chemical analysis can boost the cultural and ethical aspects and provide background for establishing relevant regulatory frameworks. Herbal medicine phytochemistry research is vital to integrate herbal medicine into modern healthcare effectively.

This book, entitled *Herbal Medicine Phytochemistry: Applications and Trends*, presents a roadmap through a broad interdisciplinary collection of reviews written by researchers, intellectuals, experts, practitioners, and professionals. The book focuses on various aspects of phytochemistry and herbal medicine. It delves into the historical context and the classification of phytochemicals in plants with herbal value as well as their utilization patterns. The book explores the diversity of medicinal plants used to treat viral, parasitic, bacterial, and viral diseases, emphasizing their nutritional profiles, bioactive components, and therapeutic potential. It also looks into specific plants like *Cola acuminata*, *Rosmarinus officinalis*, *Garcinia kola*, *Citrus aurantifolia*, *Citrus aurantium*, *Aframomum melegueta*, *Solanum torvum*, and others. It discusses herbal medicine and the use of plants in managing and treating various health conditions, including cardiovascular, respiratory, neurological, gastrointestinal, skin, cancer, arthritis, reproductive issues, eye, wound healing potential, and pregnancy. The utilization approach and practices in herbal medicine were also explored to highlight the significance of chemopreventive practices, trends in orthodox medicine practices, and the role of xenobiotics in traditional medicine. It further discusses quality control strategies, utilization methods in Africa and Asia, the value of herbal medicine in sustainable development, and its socioeconomic impact. The synergy of artificial intelligence in traditional medicine and its role in addressing antimicrobial resistance is also discussed. The book looks into herbal medicine's quality assessment and research needs, emphasizing topics like chemical and microbial contamination of herbal remedies, regulations, policies, and sustainability. It also covers the application of biotechnology in herbal medicine; the use of big data in herbal medicine; research needs related to medicinal plants; conservation and sustainable use of these plants; sustainable supply chain management in the herbal medicine industry; and the intersection of business, sustainability, and herbal medicine.

The book will be useful to students (undergraduates and postgraduates), academicians, researchers, experts, and practitioners in the fields of natural, life, health, clinical, and biomedical sciences, such as pharmaceuticals, herbal medicine, pharmacology, pharmacognosy, phytochemistry, industrial, food and public health scientists, human nutrition and dietetics, plant biology, and biotechnology/microbiology clinicians.

Bayelsa Medical University, Yenagoa, Bayelsa, Nigeria — Sylvester Chibueze Izah (Ph.D.)
Appalachian State University, Boone, USA — Matthew Chidozie Ogwu (Ph.D.)
Government College University Faisalabad, Eastern Medicine, Islamabad, Pakistan — Muhammad Akram (Ph.D.)
May 2024

Acknowledgments

Numerous individuals made significant contributions to ensure the successful completion of this book project. In particular, we thank the editors of the Reference Series in Phytochemistry, notably Prof. Kishan G. Ramawat, for his invaluable guidance and advice. We are equally thankful to the dedicated editorial team at Springer, including Ms. Lydia Mueller, Dr. Sofia Costa, Dr. Sylvia Blago, Ms. Johanna Klute, and others, for their unwavering commitment to the success of this book. Our heartfelt thanks also go to all the contributors who played a vital role in different work sections. We are grateful to our spouse and children for their patience, sacrifice, and understanding. To our mentors, we sincerely appreciate your shared knowledge, which has contributed to enriching humanity through this book.

Contents

Volume 1

Volume 2

About the Editors

Dr. Sylvester Chibueze Izah
Bayelsa Medical University
Yenagoa, Bayelsa State, Nigeria

Dr. Sylvester Izah holds a Ph.D. in Public Health, and Applied Microbiology and Environmental Health, an M. Sc. in Applied Microbiology, and a B.Sc. in Biological Sciences from Niger Delta University, Nigeria. He is a licensed Environmental Health Specialist in Nigeria. He brings over ten years of experience from his work with an environmental consultancy firm specializing in soil, water, and atmospheric assessments. During the early stages of his career, Dr. Izah conducted studies on vegetation cover and biodiversity, focusing on species diversity, composition, and utilization. Presently, he serves as a lecturer at Bayelsa Medical University in Yenagoa, Nigeria, where he holds the position of Assistant Director of Academic Planning, Research, and Innovations. Dr. Izah's extensive research portfolio encompasses water, air, and soil quality, applied microbiology, biotechnology, public and environmental health, risk assessment, bioenergy, toxicology, medicinal plants, herbal medicine, and biodiversity. He has an impressive track record of over 300 peer-reviewed publications, including journal articles, book chapters, and edited books. Dr. Izah's Google Scholar total citation, h-index, and i10-Index are >6330, 44, and 169, respectively. His current research interests are centered around Sustainable Human-Environmental Health Interactions, covering areas such as environmental sustainability trends, toxicology, hygiene, sanitation, food science, public-environmental health assessment, medicinal plants, herbal medicine, bioactive and natural compounds, vegetation, and wildlife. Dr. Izah has also collaborated on research projects with colleagues worldwide and actively contributes as an editorial and review board member for several esteemed journals.

Prof. Matthew Chidozie Ogwu
Appalachian State University
Boone, NC, USA

Dr. Ogwu is an assistant professor of Integrated Ecology and Sustainable Development in the Goodnight Family Sustainable Development Department at Appalachian State University, USA. He obtained his Ph.D. in Biological Sciences with a research focus in the Molecular Biology of Soil Microbial and Geographical Ecology from Seoul National University. He holds an M.Sc. (Distinction) and a B.Sc. (First Class) in Plant Diversity and Conservation and Plant Biology and Biotechnology respectively from the University of Benin, Nigeria. After receiving his Ph.D., he worked as a researcher at the Centre for Floristic Research of the Apennines, Italy, where he contributed to numerous European Union-funded research including the famous NATURA 2000 project for sustainable ecosystem and biodiversity management. Before that, Dr. Ogwu was a lecturer at the University of Benin, Nigeria, where he taught courses in plant biology and biotechnology, biometry, economic botany, cytogenetics, etc., and supervised and co-supervised honors and graduate students' theses. Dr. Ogwu is an interdisciplinary academic with transdisciplinary skills pertinent to the assessment of coupled human and natural as well as socio-ecological systems and has numerous awards, research grants, and scholarships to his name. His research and teaching revolve around sustainable biodiversity and ecosystem management by assessing the realm of coupled systems for sustainable environmental development as well as addressing sustainable development issues arising from various scales of biogeographical interactions. Some of his research interests include soil microbial and geographical ecology, molecular and genomic characterization of economic plant species for breeding, conservation, and sustainable utilization as well as biodiversity issues and policies, climate change science, gut and fecal microbiota, bioinformatics, biosystems and bioeconomy, sustainable agriculture and food production systems, ethnoscience (including traditional biotechnological practices), land use patterns, waste management, microbial diversity, environmental and social systems integrity, landscape preservation, and pollution ecology. He is spearheading some convergence works (One Health and Eco Health) in the Human Environmental and Agricultural Laboratory (HEAL lab) at Appalachian State University. Dr. Ogwu has published over 100 peer-reviewed works including several in high-impact journals and has presented his results at different international conferences. Dr. Ogwu has attended numerous professional courses and serves on the board of and as a reviewer for many peer-reviewed international journals. He continues to volunteer his time and skills to promote sustainable community development, especially in the Global South. Dr. Ogwu is fluent in several languages and loves to travel and meet new people.

Dr. Muhammad Akram
Government College University Faisalabad
Faisalabad, Pakistan

Dr. Muhammad Akram is an associate professor in the Department of Eastern Medicine, Government College University Faisalabad, Pakistan, where he serves as the chairman of his Department. He received his Ph.D. from Hamdard University, Karachi, Pakistan, in 2013. Dr. Akram was chairman of the Department of Eastern Medicine and Surgery, University of Poonch, Rawalakot Azad Kashmir from 2015 to 2017. He received many honors and awards during his career. He serves as an editor and invited reviewer of several national and international journals. He has numerous publications and presentations to his credit, and he is an active member of several professional societies. Dr. Akram has authored and co-authored 16 books, 74 book chapters, and over 350 journal articles. Dr. Akram's research interests include hyperuricemia, xanthine oxidase inhibition by some selected medicinal plants, enzyme inhibition, traditional medicine, phytochemistry, poisonous plants, bioactivity, and phytopharmaceutical evaluation of herbal drugs and their natural products, biochemistry, and bioinformatics.

Contributors

Maduamaka Cyriacus Abajue Department of Animal and Environmental Biology, Faculty of Science, University of Port Harcourt, Port Harcourt, Nigeria

Sherein Saied. Abdelgayed Department of Pathology, Faculty of Veterinary Medicine, Cairo University, Giza, Egypt

Sara Taha Abdelkhalek Hubei Insect Resources Utilization and Sustainable Pest Management Key Laboratory, College of Plant Science and Technology, Huazhong Agricultural University, Wuhan, China

Department of Entomology, Faculty of Science, Ain Shams University, Cairo, Egypt

Rukayat Abiola Abdulsalam Department of Biotechnology and Food Science, Faculty of Applied Sciences, Durban University of Technology, Durban, South Africa

Godwin T. W. Achana Department of Geography, SDD University of Business and Integrated Development Studies, Wa, Ghana

Munir K. Adegoke Wahab Department of Wildlife and Ecotourism Management, Faculty of Renewable Natural Resources Management, College of Agriculture, Osun State University, Ejigbo, Nigeria

Department of Forest Resources Management, Faculty of Renewable Natural Resources Management, College of Agriculture, Osun State University, Ejigbo, Nigeria

Bukola Omotomilola Adetola Department of Ecotourism and Wildlife Management, School of Agriculture and Agricultural Technology, Federal University of Technology, Akure, Ondo State, Nigeria

Sayan Adhikary Microbial Engineering & Algal Biotechnology Laboratory, Department of Biosciences, JIS University, Agarpara, Kolkata, India

Hassaan Ahmad Department of Agronomy, University of Agriculture, Faisalabad, Pakistan

Muhammad Akram Department of Eastern Medicine, Faculty of Medical Sciences, Government College University, Faisalabad, Pakistan

Ahmad Ali University Department of Life Sciences, University of Mumbai, Mumbai, India

Timinipre Amabie Department of Computer Science, Faculty of Science, Bayelsa Medical University, Yenagoa, Bayelsa State, Nigeria

Arinze Favour Anyiam Department of Medical Laboratory Science, Thomas Adewumi University, Oko, Nigeria

Onyinye Cecilia Arinze-Anyiam Department of Medical Laboratory Science, Thomas Adewumi University, Oko, Nigeria

Christiana Eleojo Aruwa Department of Biotechnology and Food Science, Faculty of Applied Sciences, Durban University of Technology, Durban, South Africa

Oladimeji Taiwo Babatunde Department of Chemical Sciences, Crown-Hill University, Eiyenkorin, Kwara State, Nigeria

Falak Bamne University Department of Life Sciences, University of Mumbai, Mumbai, India

Dishani Banerjee Microbial Engineering & Algal Biotechnology Laboratory, Department of Biosciences, JIS University, Kolkata, India

Sayani Banerjee Microbial Engineering & Algal Biotechnology Laboratory, Department of Biosciences, JIS University, Kolkata, India

Hasadiah Okon Bassey Department of Biological Sciences, Akwa Ibom State Polytechnic, Ikot Osurua, Nigeria

Poumita Bhattacherjee Microbial Engineering & Algal Biotechnology Laboratory, Department of Biosciences, JIS University, Agarpara, Kolkata, India

Rebecca Bockarie Sierra Leone Agricultural Research Institute, Kenema, Sierra Leone

Saket Singh Chandel Dr. C.V. Raman Institute of Pharmacy, Dr. C. V. Raman University, Bilaspur, India

Debjani Choudhury Department of Plant Pathology, School of Agriculture, Lovely Professional University, Phagwara, Punjab, India

Koko Sunday Daniel Department of Forestry and Wildlife, University of Uyo, Uyo, Nigeria

Kratika Daniel Oriental College of Pharmacy, Indore, India

Moyuri Das Department of Zoology, Cooch Behar Panchanan Barma University, Cooch Behar, West Bengal, India

Nibedita Datta Microbial Engineering & Algal Biotechnology Laboratory, Department of Biosciences, JIS University, Kolkata, India

Abantika De Microbial Engineering & Algal Biotechnology Laboratory, Department of Biosciences, JIS University, Kolkata, India

Shrestha Debnath Microbial Engineering & Algal Biotechnology Laboratory, Department of Biosciences, JIS University, Agarpara, Kolkata, India

Priyanka Devi Department of Agronomy, School of Agriculture, Lovely Professional University, Phagwara, India

Shipa Rani Dey Department of Agronomy, School of Agriculture, Lovely Professional University, Phagwara, India

Afamefuna Dunkwu-Okafor Department of Microbiology, Faculty of Life Sciences, University of Benin, Benin City, Edo State, Nigeria

J. Ekowati Department of Pharmaceutical Science, Faculty of Pharmacy, Airlangga University, Surabaya, Indonesia

Asma Elyas Department of Forest Management, Faculty of Forestry, University of Khartoum, Khartoum, Sudan

Ngozi Georgewill Emaikwu Department of Biotechnology, Federal University of Technology, Owerri, Nigeria

Kingsley Erhons Enerijiofi Department of Biological Sciences, College of Basic and Applied Sciences, Glorious Vision University, Ogwa, Edo State, Nigeria

Syeda Nishat Fathima Jayamukhi College of Pharmacy, Narsampet, Warangal, Andhra Pradesh, India

Moses Fayiah Department of Forestry, School of Natural Resources Management, Njala University, Njala, Sierra Leone

Muloma Seibatu Fayiah Department of Biological Sciences School of Environmental Sciences, Njala University, Njala, Sierra Leone

Natalia Shania Francis Department of Allied Health Sciences, Faculty of Science, Universiti Tunku Abdul Rahman (UTAR), Kampar, Malaysia

Dipankar Ghosh Microbial Engineering & Algal Biotechnology Laboratory, Department of Biosciences, JIS University, Agarpara, Kolkata, India

Bina Gidwani Department of Pharmaceutical Quality Assurance, Columbia Institute of Pharmacy, Tekari, Raipur, Chhattisgarh, India

Avinash Pratap Gupta Department of Environmental Science, Sant Gahira Guru University, Sarguja, Chhattisgarh, Ambikapur, India

Saurabh Gupta Deaprtment of Microbiology, Mata Gujri College, Fatehgarh Sahib, Punjab, India

Milan Hait Department of Chemistry, Dr. C. V. Raman University, Bilaspur, Chhattisgarh, India

Beckley Ikhajiagbe Department of Plant Biology and Biotechnology, Faculty of Life Sciences, University of Benin, Benin City, Nigeria

Omiagocho ThankGod Isaac Department of Biochemistry, Joseph Sarwuan Tarkaa University, Makurdi, Nigeria

Laboratory for Innovative Drugs Development, Moscow Institute of Physics and Technology, Moscow, Russia

Austin-Asomeji Iyingiala Department of Community Medicine, Faculty of Clinical Sciences, College of Medical Sciences, Rivers State University, Port Harcourt, Rivers State, Nigeria

Adams Ovie Iyiola Department of Fisheries and Aquatic Resources Management, Faculty of Renewable Natural Resources Management, College of Agriculture and Renewable Resources, Ejigbo Campus, Osun State University, Ejigbo, Osun State, Nigeria

Sylvester Chibueze Izah Department of Microbiology, Faculty of Science, Bayelsa Medical University, Yenagoa, Bayelsa State, Nigeria

Daniel Etim Jacob Department of Forestry and Wildlife, University of Uyo, Uyo, Nigeria

Sachin Kumar Jain IPS Academy College of Pharmacy, Oriental College of Pharmacy and Research, Oriental University, Indore, Madhya Pradesh, India

Manoj Kumar Jhariya Department of Farm Forestry, Sant Gahira Guru Vishwavidyalaya, Ambikapur, India

Mei -Xiang Jin Hubei Insect Resources Utilization and Sustainable Pest Management Key Laboratory, College of Plant Science and Technology, Huazhong Agricultural University, Wuhan, China

Marcella Tari Joshua Department of Medical Laboratory Science, Faculty of Health Sciences, Bayelsa Medical University, Yenagoa, Nigeria

Jogen Chandra Kalita Department of Zoology, Guwahati University, Assam, India

Mohamed Koiva Kallon Institute of Environmental Management and Quality Control, School of Environmental Sciences, Njala University, Njala, Sierra Leone

Ashish M. Kandalkar Vidyaniketan College of Pharmacy, Anjangaon, India

Nirmala Karam Department of Plant Pathology, School of Agriculture, Lovely Professional University, Phagwara, Punjab, India

Nand Kumar Kashyap Department of Chemistry, Dr. C. V. Raman University, Bilaspur, Chhattisgarh, India

Tabassum Khan Department of Pharmaceutical Chemistry & Quality Assurance, SVKM's Dr. Bhanuben Nanavati College of Pharmacy, Mumbai, India

Aman Khokhar Department of Agronomy, School of Agriculture, Lovely Professional University, Phagwara, India

Ayotunde Samuel Kolawole Department of Fisheries and Aquatic Resources Management, Faculty of Renewable Natural Resources Management, College of Agriculture, Ejigbo Campus, Osun State University, Ejigbo, Osun State, Nigeria

Enoch Akwasi Kosoe Department of Environment and Resource Studies, SDD University of Business and Integrated Development Studies, Upper West Region, Ghana

Amit Kumar Kharvel Subharti College of Pharmacy, Swami Vivekanand Subharti University, Meerut, Uttar Pradesh, India

Prasann Kumar Department of Agronomy, School of Agriculture, Lovely Professional University, Phagwara, Punjab, India

Ravindra Kumar Department of Agriculture, Mata Gujri College, Fatehgarh Sahib, Punjab, India

Sokindra Kumar Department of Pharmacology, Kharvel Subharti College of Pharmacy, Swami Vivekanand Subharti University, Meerut, Uttar Pradesh, India

Annaletchumy Loganathan Department of Allied Health Sciences, Faculty of Science, Universiti Tunku Abdul Rahman (UTAR), Kampar, Malaysia

Sudipta Kumar Maiti Faculty, Department of Botany, Raja N.L.Khan Women's College (Autonomous), Midnapore, West Bengal, India

Anumita Mallick Faculty, Department of Nutrition, Belda College, Belda, West Bengal, India

Hamza Maqsood Department of Agronomy, University of Agriculture, Faisalabad, Pakistan

Tamal Mazumder Department of Zoology, North Bengal St. Xavier's College, Rajganj, WB, India

Bharat Mishra R.G.S College of Pharmacy, Lucknow, India

Ganesh Prasad Mishra Department of Pharmacology, Kharvel Subharti College of Pharmacy, Swami Vivekanand Subharti University, Meerut, Uttar Pradesh, India

Munira Momin Department of Pharmaceutics, SVKM's Dr. Bhanuben Nanavati College of Pharmacy, Mumbai, India

Tapas Nag Department of Anatomy, All India Institute of Medical Sciences, New Delhi, India

Imaobong Ufot Nelson Department of Forestry and Wildlife, University of Uyo, Uyo, Nigeria

Clement Takon Ngun Cross River University of Technology, Calabar, Nigeria

Norhayati Magister of Pharmaceutical Science, Faculty of Pharmacy, Airlangga University, Surabaya, Indonesia

Ebiuwa Gladys Obahiagbon Glasgow School for Business and Society, Department of Management and Human Resource Management, Glasgow Caledonian University, London, UK

Chioma Peggy Obasi National Roots and Crops Research Institute, Umudike, Nigeria

Ejeatuluchukwu Obi Toxicology Unit, Department of Pharmacology and Therapeutics, College of Health Sciences, Nnamdi Azikiwe University, Awka, Nigeria

Nkereuwem Udoakah Obongodot Department of Plant Biology and Biotechnology, Faculty of Life Sciences, University of Benin, Benin City, Nigeria

Tamaraukepreye Catherine Odubo Department of Microbiology, Faculty of Science, Bayelsa Medical University, Yenagoa, Bayelsa State, Nigeria

Odangowei Inetiminebi Ogidi Department of Biochemistry, Faculty of Basic Medical Sciences, Bayelsa Medical University, Yenagoa, Bayelsa State, Nigeria

Olalekan Bukunmi Ogunro Reproductive and Endocrinology, Toxicology, and Bioinformatics Research Laboratory, Department of Biological Sciences, KolaDaisi University, Ibadan, Oyo State, Nigeria

Happiness Isioma Ogwu Department of Microbiology, Faculty of Life Sciences, University of Benin, Benin City, Edo State, Nigeria

Matthew Chidozie Ogwu Goodnight, Family Department of Sustainable Development, Appalachian State University, Boone, NC, USA

Joseph Etim Okon Department of Botany, Faculty of Biological Sciences, Akwa Ibom State University, Uyo, Akwa Ibom State, Nigeria

Okon Godwin Okon Department of Botany, Faculty of Biological Sciences, Akwa Ibom State University, Uyo, Akwa Ibom State, Nigeria

Ichehoke Austine Omakor Department of Plant Science and Biotechnology, Faculty of Science, Dennis Osadebay University, Anwai, Asaba, Nigeria

Moses Edwin Osawaru Department of Plant Biology and Biotechnology, Faculty of Life Sciences, University of Benin, Benin City, Edo State, Nigeria

Kurotimipa Frank Ovuru Department of Disease Control and Immunization, Bayelsa State Primary Health Care Board, Yenagoa, Bayelsa State, Nigeria

Piyush Pandey Department of Environmental Science, Sant Gahira Guru University, Sarguja, Chhattisgarh, Ambikapur, India

Priyanka Pandey Departments of Pharmaceutics, NKBR College of Pharmacy, Meerut, Uttar Pradesh, India

Ravindra Kumar Pandey Columbia Institute of Pharmacy, Tekari, Raipur, Chhattisgarh, India

Rupesh Kumar Pandey Department of Pharmacology, Kharvel Subharti College of Pharmacy, Swami Vivekanand Subharti University, Meerut, Uttar Pradesh, India

Astha Pathak Department of Pharmacy, Sant Gahira Guru University, Sarguja, Chhattisgarh, Ambikapur, India

Manish Pathak Departments of Pharmaceutical Chemistry, Kharvel Subharti College of Pharmacy, Swami Vivekanand Subharti University, Meerut, Uttar Pradesh, India

Anjolaolowa Mary Popoola Nigeria Centre for Disease Control and Prevention, Abuja, Nigeria

Omoniyi Michael Popoola Department of Fisheries and Aquaculture Technology, Federal University of Technology, Akure, Nigeria

Mrinalini Prasad Department of Botany, St. John's College, Agra, India

Rizki Rahmadi Pratama Master Program of Pharmaceutical Science, Faculty of Pharmacy, Universitas Airlangga, Surabaya, Indonesia

Abhay Punia Department of Zoology, DAV University, Jalandhar, Punjab, India

Morufu Olalekan Raimi Department of Community Medicine, Environmental Health Unit, Faculty of Clinical Sciences, Niger Delta University, Wilberforce Island, Bayelsa State, Nigeria

Abhishek Raj Pt. DeendayalUpadhyay College of Horticulture & Forestry, Dr. Rajendra Prasad Central Agriculture University, Samastipur, Bihar, India

Rajiv Ranjan Department of Botany, Dayalbagh Educational Institute, Agra, India

Glory Richard Department of Community Medicine, Faculty of Clinical Sciences, Niger Delta University, Wilberforce Island, Bayelsa State, Nigeria

Department of Clinical Sciences, Faculty of Clinical Sciences, Niger Delta University, Wilberforce Island, Bayelsa State, Nigeria

Torisa Roy Immunology & Cell Biology Laboratory, Department of Zoology, Cooch Behar Panchanan Barma University, Cooch Behar, WB, India

Saheed Sabiu Department of Biotechnology and Food Science, Faculty of Applied Sciences, Durban University of Technology, Durban, South Africa

Salimatu Saccoh Institute of Food Sciences, School of Agriculture and Food Science, Njala University, Njala, Sierra Leone

Lalit Saini Department of Agronomy, School of Agronomy, Lovely Professional University, Phagwara, Punjab, India

Muhammad Farrukh Saleem Department of Agronomy, University of Agriculture, Faisalabad, Pakistan

Saoban Sunkanmi Salimon School of Biological and Medical Physics, Moscow Institute of Physics and Technology, Dolgoprudny, Russia

Muhammad Sarwar Department of Agronomy, University of Agriculture, Faisalabad, Pakistan

Subham Saurabh Department of Agronomy, School of Agriculture, Lovely Professional University, Phagwara, Punjab, India

Wisdom Ebiye Sawyer Department of Community Medicine, Faculty of Clincal Sciences, Niger Delta University, Wilberforce Island, Bayelsa State, Nigeria

Nikhat Shaikh Regional Research Institute of Unani Medicine, Mumbai, India

Khushbu Sharma Department of Agronomy, School of Agriculture, Lovely Professional University, Phagwara, Punjab, India

Monika Sharma Department of Agronomy, School of Agronomy, Lovely Professional University, Phagwara, Punjab, India

Urvashi Sharma Faculty of Pharmacy, Medi-Caps University, Indore, India

Jin -Hua Shi Hubei Insect Resources Utilization and Sustainable Pest Management Key Laboratory, College of Plant Science and Technology, Huazhong Agricultural University, Wuhan, China

Irawati Sholikhah Department of Chemistry, Faculty of Science and Technology, Universitas Airlangga, Surabaya, Indonesia

Shiv Shankar Shukla Department of Pharmaceutical Quality Assurance, Columbia Institute of Pharmacy, Tekari, Raipur, Chhattisgarh, India

Lubhan Singh Department of Pharmacology, Kharvel Subharti College of Pharmacy, Swami Vivekanand Subharti University, Meerut, Uttar Pradesh, India

Nitu Singh R.G.S College of Pharmacy, Lucknow, India

Vijay Singh Department of Botany, Mata Gujri College, Fatehgarh Sahib, Punjab, India

Saket Singh Chandel Dr. C.V. Raman Institute of Pharmacy, Dr. C.V. Raman University, Bilaspur, Chhattisgarh, India

Nalini Singh Chauhan P.G Department of Zoology, Kanya Maha Vidyalya, Jalandhar, Punjab, India

Anita Thakur Department of Chemistry, RIMT University, Mandi Gobindgarh, Punjab, India

Kelly Ebelakpo Torru Department of Family Medicine, Faculty of Clinical Sciences, Bayelsa Medical University, Yenagoa, Nigeria

Najeeb Ullah Agricultural Research Station, Office of VP for Research & Graduate Studies, Qatar University, Doha, Qatar

M. M. Vaishnav Department of Chemistry, Government G. B. College, Korba, India

T. Venkatachalam JKKMMRFs-Amnai JKK Sampoorani Ammal College of Pharmacy B. Komarapalayam, Namakkal, Tamil Nadu, India

Man -Qun. Wang Hubei Insect Resources Utilization and Sustainable Pest Management Key Laboratory, College of Plant Science and Technology, Huazhong Agricultural University, Wuhan, China

Hunny Waswani Department of Botany, Dayalbagh Educational Institute, Agra, India

Retno Widyowati Department of Pharmaceutical Science, Faculty of Pharmacy, Universitas Airlangga, Surabaya, Indonesia

Michael Ndubuisi Wogu Department of Animal and Environmental Biology, Faculty of Science, University of Port Harcourt, Port Harcourt, Nigeria

Neelima Yadav Dr. C.V. Raman Institute of Pharmacy, Dr. C.V. Raman University, Bilaspur, Chhattisgarh, India

Baturh Yarkwan Department of Biochemistry, Joseph Sarwuan Tarkaa University, Makurdi, Nigeria

Hadida Yasmin Immunology & Cell Biology Laboratory, Department of Zoology, Cooch Behar Panchanan Barma University, Cooch Behar, WB, India

Babalola Ola Yusuf Department of Biotechnology and Food Science, Faculty of Applied Sciences, Durban University of Technology, Durban, South Africa

Zaharadeen Muhammad Yusuf Department of Biological Sciences, Al-Qalam University, Katsina, Nigeria

An Introduction to Herbal Medicine Phytochemistry: Applications and Trends

Background

The advancement of evidence-based herbal medicine hinges on a comprehensive grasp of the chemical makeup of medicinal plants and their bioactive elements, with phytochemistry as an indispensable element in this endeavor. The utilization techniques and approaches in herbal medicine, often involving the preparation of plant extracts and remedies with a historical lineage, are deeply rooted in traditional wisdom and the expertise of local healers. To guarantee the safety, effectiveness, and uniformity of herbal medicines, continual quality assessment and research efforts are imperative, especially as integrating modern scientific methodologies gains increasing significance within this field.

This book has three main components: phytochemistry and herbal medicine, utilization approach and practices in herbal medicine, and quality assessment and research needs of herbal medicine (Fig. 1).

Phytochemistry and Herbal Medicine

Phytochemistry is the field of chemistry that looks into plant compounds' molecular structures. It involves the breakdown, separation, characterization, and identification of the numerous chemical constituents present in plants, encompassing a wide range of compounds like terpenes, alkaloids, flavonoids, phenolic compounds, etc. Phytochemistry plays a pivotal role in comprehending the chemical composition of plants and their possible applications, particularly in medicine. The major facets of phytochemistry are summarized in Fig. 2.

Botanical medicine, also known as herbal medicine or phytotherapy, typically involves utilizing plant-based substances for their therapeutic properties. This traditional healing approach has served as a primary treatment method across various cultures worldwide for millennia, encompassing a diverse array of civilizations. The main elements of herbal medicine are summarized in Fig. 3.

Herbal medicine and phytochemistry are closely intertwined fields. Phytochemistry plays a pivotal role in identifying and isolating bioactive plant compounds,

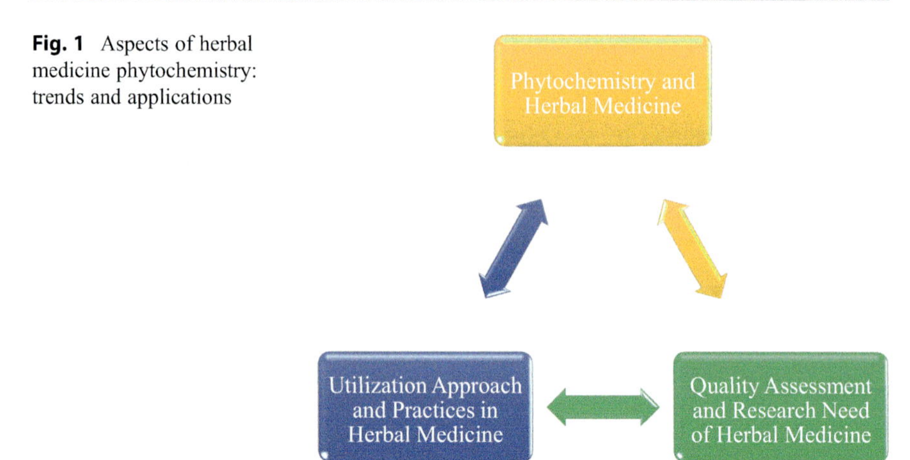

Fig. 1 Aspects of herbal medicine phytochemistry: trends and applications

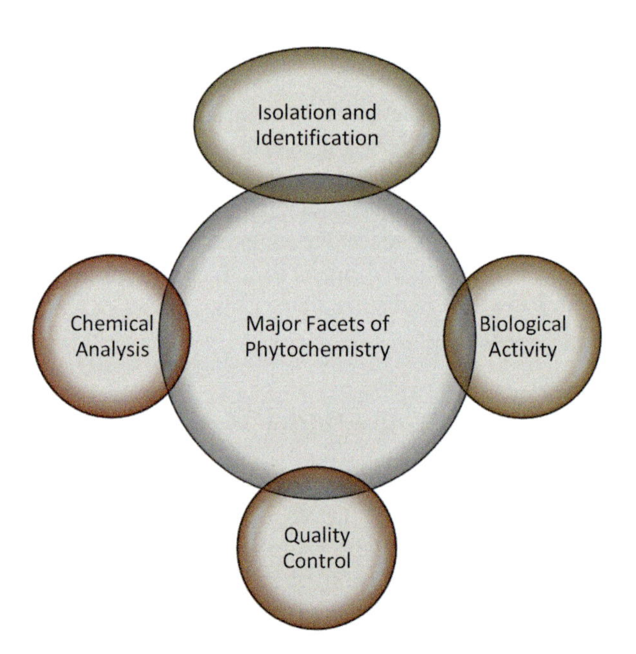

Fig. 2 The major facets of phytochemistry

which are subsequently employed in herbal medicine. This scientific understanding not only promotes the development of standardized herbal products, ensuring consistency in dosage and quality, but also substantiates the legitimacy of traditional herbal remedies by unveiling their chemical composition and pharmacological effects.

Both phytochemistry and herbal medicine share common challenges and potential prospects. Obstacles include the need for rigorous scientific validation, quality control of the products, and addressing regulatory issues. Nonetheless, they also

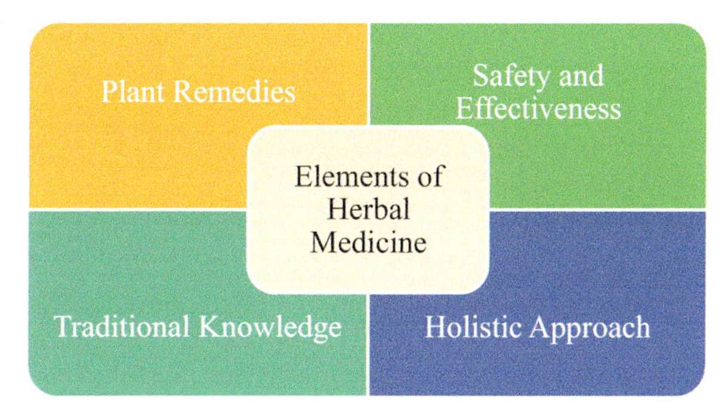

Fig. 3 The main elements of herbal medicine

offer promising opportunities, such as the potential for discovering novel medications, more potent natural remedies, and the integration of complementary and alternative medicine into mainstream healthcare.

This part of the book has 33 chapters which are summarized as follows:

i. Historical Perspectives and Overview of the Value of Herbal Medicine: Herbal medicine has a rich historical context and continues to play a crucial role in healthcare, offering a holistic approach to healing by harnessing the therapeutic potential of plant-based remedies.
ii. Plant Food for Human Health: Case Study of Indigenous Vegetables in Akwa Ibom State, Nigeria: Indigenous vegetables in Akwa Ibom State, Nigeria, hold substantial nutritional value, emphasizing the importance of these local plant foods in promoting human health and well-being.
iii. Classification of Phytochemicals in Plants with Herbal Value: The classification of phytochemicals within plants with herbal value provides insights into the diverse compounds responsible for their medicinal properties, facilitating a deeper understanding of their therapeutic potential.
iv. Nutritional Profile, Bioactive Components, and Therapeutic Potential of Edible Flowers of Chhattisgarh, India: Edible flowers in Chhattisgarh, India, are not only nutritionally rich but also contain bioactive components that offer therapeutic benefits, making them a valuable resource for both food and health.
v. Phytochemical Constituents of *Rosmarinus officinalis* Linn. and Their Associated Role in the Management of Alzheimer's Disease: *R. officinalis* harbors phytochemical constituents that may contribute to the management of Alzheimer's disease, shedding light on its potential role in neurological health.
vi. Plants Used in the Management and Treatment of Male Reproductive Health Issues: Case Study of Benin People of Southern Nigeria: The Benin people in Southern Nigeria rely on specific plants for the management and treatment of male reproductive health issues, illustrating the significance of traditional herbal remedies in addressing such concerns.

vii. *Cola acuminata*: Phytochemical Constituents, Nutritional Characteristics, Scientific Validated Pharmacological Properties, Ethnomedicinal Uses, Safety Considerations, and Commercial Values: *Cola acuminata* exhibits a diverse range of phytochemical constituents, nutritional properties, validated pharmacological benefits, cultural significance, safety considerations, and commercial value, making it a multi-faceted plant of interest.

viii. Diversity of Medicinal Plants Used in the Treatment and Management of Viral Diseases Transmitted by Mosquitoes in the Tropics: Medicinal plants in tropical regions serve as a diverse resource for treating viral diseases transmitted by mosquitoes, offering potential solutions to public health challenges.

ix. Medicinal Plants in the Tropics Used in the Treatment and Management of Parasitic Diseases Transmitted by Mosquitoes: Administration, Challenges, and Strategic Options for Management: Medicinal plants in tropical areas are integral in combating parasitic diseases transmitted by mosquitoes, though challenges in administration require strategic management approaches.

x. Efficacy of Ethnoherbal Medicines with Anti-inflammatory and Wound Healing Potentiality: A Case of West Bengal, India: Ethnoherbal medicines in West Bengal, India, demonstrate efficacy in anti-inflammatory and wound healing applications, highlighting their valuable role in traditional healthcare.

xi. Classification Methods and Diversity of Medicinal Plants: The categorization of medicinal plants sheds light on the rich diversity of plant species used in traditional medicine worldwide. Also, the classification methods provide insights into the vast array of medicinal plants employed for diverse health purposes.

xii. Plants Used in the Management and Treatment of Cardiovascular Diseases: Case Study of the Benin People of Southern Nigeria: The Benin people in Southern Nigeria use herbs to manage and treat cardiovascular diseases, thereby offering valuable insights into traditional remedies for heart health. This practice also highlights the potential of indigenous knowledge in cardiovascular disease management, thereby bridging the gap between traditional and modern medicine.

xiii. Mechanistic Approaches of Herbal Medicine in the Treatment of Arthritis: The mechanism by which herbal medicine effectively treats arthritis, thereby contributing to the understanding of alternative therapies for debilitating conditions. The elucidation of the mechanisms may offer a scientific foundation for the use of herbal remedies in arthritis treatment.

xiv. Phytochemicals and Overview of the Evolving Landscape in Management of Osteoarthritis: The exploration of phytochemicals in osteoarthritis management provides an overview of how bioactive compounds found in plants are shaping the evolving landscape of treatment options.

xv. Assessment of the Phytochemical Constituents and Metabolites in the Medicinal Plants and Herbal Medicine Used in the Treatment and Management of Respiratory Diseases: The assessment scrutinizes the phytochemical constituents and metabolites in medicinal plants and herbal remedies, thereby focusing on their application in the treatment and management of respiratory diseases. This could contribute to the understanding of the therapeutic properties of herbal remedies for respiratory conditions.

xvi. Assessment of the Phytochemical Constituents and Metabolites in Medicinal Plants and Herbal Remedies Used in the Treatment and Management of Reproductive Diseases: Polycystic Ovary Syndrome: The phytochemical constituents and metabolites in medicinal plants and herbal remedies could be essential in the treatment of reproductive diseases such as polycystic ovary syndrome (PCOS). Findings could help address a critical area of herbal medicine application in women's healthcare.

xvii. Assessment of the Phytochemical Constituents and Metabolites in the Medicinal Plants and Herbal Medicine Used in the Treatment and Management of Skin Diseases: Phytochemical constituents and metabolites found in medicinal plants and herbal medicine emphasize their significance in the treatment and management of skin diseases. This chapter sheds light on the potential of herbal remedies in dermatological care.

xviii. Metabolites and Phytochemicals in Medicinal Plants Used in the Management and Treatment of Neurological Diseases: The role of metabolites and phytochemicals in medicinal plants and their application in managing and treating neurological diseases, providing insights into alternative therapeutic approaches for brain-related conditions. This contributes to the growing body of knowledge about herbal remedies for neurological health.

xix. Diabetes Treatment and Prevention Using Herbal Medicine: The chapter centers on using herbal medicine to treat and prevent diabetes, offering a holistic and natural approach to managing this prevalent metabolic disorder. This chapter underscores the potential of herbal remedies as a complementary strategy for diabetes care and prevention.

xx. Therapeutic Applications of Herbal Medicines for the Prevention and Management of Cancer: This chapter explores the potential of herbal remedies in preventing and managing cancer, focusing on their therapeutic applications and outcomes.

xxi. Herbs and Herbal Formulations for the Management and Prevention of Gastrointestinal Diseases: This chapter looks into the use of medicinal plants and herbal formulations to address gastrointestinal disorders, thereby shedding light on their role in disease prevention and management.

xxii. Herbal Medicine and Pregnancy: This chapter addresses herbal medicine during pregnancy, providing insights into its safety and efficacy, and emphasizing the well-being of expectant mothers.

xxiii. Herbal Medicine and Rheumatic Disorders Management and Prevention: This chapter discusses how herbal remedies can be employed to manage and prevent rheumatic disorders, highlighting their potential to improve the quality of life for affected individuals.

xxiv. Physiological and Biochemical Outcomes of Herbal Medicine Use in the Treatment of Hypertension: This chapter offers an overview of herbal medicine's physiological and biochemical effects on hypertension, shedding light on its role as an alternative treatment.

xxv. Evidence from the Use of Herbal Medicines in the Management and Prevention of Common Eye Diseases: This chapter provides evidence-based insights into utilizing herbal medicines for managing and preventing common eye diseases, emphasizing their potential benefits.

xxvi. Assessment of the Phytochemical Constituents and Metabolites of Some Medicinal Plants and Herbal Remedies Used in the Treatment and Management of Injuries: This chapter assesses the phytochemical constituents and metabolites of medicinal plants and herbal remedies used in injury treatment and management.

xxvii. Plants Used in the Management and Treatment of Female Reproductive Health Issues: Case Study from Southern Nigeria: This chapter offers a case study of plants used in the treatment of female reproductive health issues, focusing on the practices in Southern Nigeria.

xxviii. *Citrus aurantium*: Phytochemistry, Therapeutic Potential, Safety Considerations, and Research Needs: This chapter presents an overview of *Citrus aurantium*, discussing its phytochemical composition, therapeutic potential, safety considerations, and areas for further research.

xxix. Medicinal Spice, *Aframomum melegueta*: An Overview of the Phytochemical Constituents, Nutritional Characteristics, and Ethnomedicinal Values for Sustainability: This chapter provides an overview of *A. melegueta*, highlighting its phytochemical constituents, nutritional characteristics, and cultural significance for sustainability.

xxx. "Turkey berry (*Solanum torvum* Sw. [Solanaceae]): An Overview of the Phytochemical Constituents, Nutritional Characteristics, and Ethnomedicinal Values for Sustainability: This chapter discusses *S. torvum* and its phytochemical constituents, nutritional qualities, and ethnomedicinal significance in the context of sustainability.

xxxi. *Garcinia kola* Heckel. (Clusiaceae): An Overview of the Cultural, Medicinal, and Dietary Significance for Sustainability: This chapter explores the cultural, medicinal, and dietary importance of *G. kola*, emphasizing its role in sustainable practices.

xxxii. *Vernonia amygdalina* Delile (Asteraceae): An Overview of the Phytochemical Constituents, Nutritional Characteristics, and Ethnomedicinal Values for Sustainability: This chapter provides an overview of *V. amygdalina*, focusing on its phytochemical constituents, nutritional attributes, and ethnomedicinal value within the context of sustainability.

xxxiii. *Citrus aurantifolia*: Phytochemical Constituents, Food Preservative Potentials, and Pharmacological Value: This chapter focuses on *C. aurantifolia*, detailing its phytochemical components, potential for use as a food preservative, and its pharmacological value.

Utilization Approach and Practices in Herbal Medicine

Herbal medicine has a rich history of transcending cultures, emphasizing plant diversity, traditional knowledge, and holistic care. It has influenced medicine throughout history, with roots in traditional systems, European herbalism, and a twentieth-century resurgence. Today, herbal medicine finds uses in supplements, integrative healthcare, and personalized medicine. Despite its growth, it faces challenges including quality control, safety concerns, the need for scientific evidence, and ethical issues related to cultural appropriation (Fig. 4). Therefore, herbal medicine's holistic, plant-based approach remains significant, requiring continued research and responsible practices to ensure its safety and effectiveness in contemporary healthcare.

This part of the book has 16 chapters which are summarized as follows:

i. Chemopreventive Practices in Traditional Medicine: This chapter focuses on traditional medicine incorporated in a wide range of practices aimed at preventing disease, including dietary modifications and the use of specific medicinal plants, often rooted in historical knowledge and cultural traditions.
ii. Chemopreventive Strategies in Herbal Medicine Practice: Current Aspects, Challenges, Prospects, and Sustainable Future Outlook: This chapter focuses on herbal medicine practices especially chemopreventive strategies, with a focus on identifying specific medicinal plants and phytochemicals that exhibit cancer prevention properties, which can contribute to mainstream healthcare but face challenges in terms of standardization and regulatory compliance. Furthermore, the sustainable future outlook for herbal medicine in chemoprevention lies in strengthening scientific validation, addressing quality control

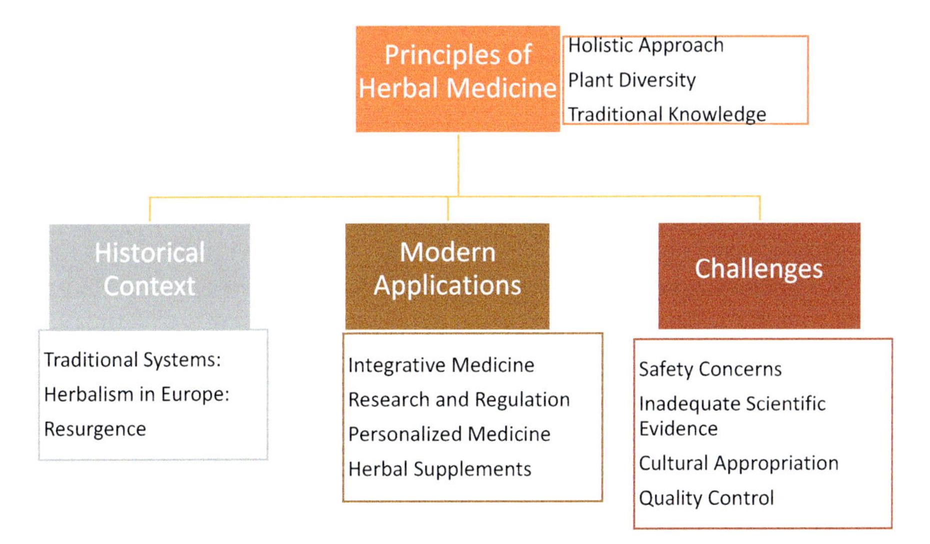

Fig. 4 Overview of the utilization approach and practices in herbal medicine

concerns, and fostering greater collaboration between traditional knowledge and modern research.

iii. Sustainable Current Trends and Future Directions in Orthodox Medicine Practice in Sierra Leone: In Sierra Leone, orthodox medicine practices are experiencing a shift toward more sustainable and community-oriented healthcare delivery, emphasizing accessibility, affordability, and cultural relevance. The future of orthodox medicine in Sierra Leone holds the promise of improved healthcare infrastructure, strengthened primary care, and integration with traditional healing practices to provide holistic health services.

iv. Herbal Medicine: A Case Study of Kherias, Lodhas, and Mundas Tribes in Southwest Bengal of India: The Kherias, Lodhas, and Mundas tribes in Southwest Bengal have preserved and continue to practice herbal medicine, employing a rich repertoire of plant-based remedies passed down through generations. This case study highlights the cultural significance and knowledge preservation of indigenous herbal medicine and underscores the need for their inclusion in discussions on healthcare sustainability.

v. Herbal Medicine—Exploring Its Scope Across Belief Systems of Indian Medicine: Herbal medicine in India transcends various belief systems, including Ayurveda, Siddha, and Unani, demonstrating a holistic approach to health and well-being that combines ancient knowledge with contemporary healthcare practices. This exploration sheds light on the versatility of herbal medicine across belief systems, fostering a more comprehensive understanding of its potential benefits.

vi. Xenobiotics in Traditional Medicine Practices and Quality Control Strategies: Traditional medicine practices sometimes involve the use of xenobiotics, foreign substances that require rigorous quality control measures to ensure safety and efficacy. Addressing the presence of xenobiotics in traditional medicine underscores the importance of quality control and regulation in these practices to safeguard public health.

vii. Herbal Medicine for Health Management and Disease Prevention: Herbal medicine offers an array of preventive and health management solutions, demonstrating its potential to reduce the burden of chronic diseases and improve overall well-being. The utilization of herbal medicine in health management presents an opportunity for a more holistic and cost-effective approach to healthcare.

viii. Utilization Methods and Practices of Herbal Medicine in Africa: Across Africa, diverse utilization methods and practices of herbal medicine are integral to healthcare, driven by cultural traditions and local knowledge. Understanding the rich tapestry of herbal medicine practices in Africa is vital for enhancing healthcare accessibility and addressing health disparities on the continent.

ix. Herbal Medicine Methods and Practices in Nigeria: In Nigeria, herbal medicine methods and practices have been deeply rooted in the culture and tradition, often involving the use of locally sourced plants and knowledge passed down through generations. These methods include the preparation of decoctions, infusions, and the use of specific plant parts for various ailments, showcasing the rich diversity of herbal remedies within the country.

x. Value of Herbal Medicine to Sustainable Development: Herbal medicine plays a vital role in sustainable development by offering affordable and locally available healthcare solutions, reducing the strain on modern healthcare systems and pharmaceutical industries. Additionally, the cultivation and sustainable harvesting of medicinal plants for herbal medicine contribute to biodiversity conservation and support the livelihoods of many rural communities.

xi. Herbal Medicine and Sustainable Development Challenges and Opportunities: Challenges related to herbal medicine and sustainable development include the need for quality control, standardization, and regulations to ensure safe and effective herbal remedies. However, opportunities lie in the potential for herbal medicine to provide alternative, ecofriendly healthcare solutions and promote sustainable practices in plant cultivation and harvesting.

xii. Current Trends on Phytochemicals Toward Herbal Medicine Development: Current trends in phytochemical research focus on identifying and isolating bioactive compounds from plants, understanding their mechanisms of action, and exploring their potential in developing new herbal medicines. The use of advanced analytical techniques and bioinformatics is enhancing our knowledge of phytochemicals, driving innovation in herbal medicine development.

xiii. Proximate Analysis of Herbal Drugs: Methods, Relevance, and Quality Control Aspects: Proximate analysis is essential for determining the nutritional and chemical composition of herbal medicine, ensuring its quality, safety, and efficacy. These methods are highly relevant in quality control, enabling the detection of contaminants and variations in herbal drug preparations, which is crucial for consumer safety and regulatory compliance.

xiv. Socioeconomic Values of Herbal Medicine: Herbal medicine contributes significantly to the socioeconomic well-being of communities involved in its production and trade, offering employment opportunities and income generation. Moreover, it serves as an accessible and cost-effective healthcare option for many people, particularly in regions with limited access to modern medicine.

xv. Harmonizing Tradition and Technology: The Synergy of Artificial Intelligence in Traditional Medicine: The integration of artificial intelligence with traditional medicine practices holds promise in enhancing the diagnosis, treatment, and management of health conditions based on extensive data analysis and predictive algorithms. This harmonization of tradition and technology showcases the potential for artificial intelligence to support the preservation and modernization of traditional medicine.

xvi. Antimicrobial Resistance and the Role of Herbal Medicine: Challenges, Opportunities, and Future Prospects: Herbal medicine presents opportunities to address antimicrobial resistance by providing alternative treatment options and potentially reducing the overuse of antibiotics. However, challenges include the need for scientific validation, quality control, and regulatory frameworks to harness the full potential of herbal medicine in combating this global health threat.

Quality Assessment and Research Needs of Herbal Medicine

Assessing the quality of herbal medicine is integral to ensuring the safety and reliability of these natural products. This process involves several key considerations, including precise plant identification to reduce the risk of adverse effects and contamination (Fig. 5). Advanced chemical analysis techniques, such as spectroscopy and chromatography, are essential for determining the chemical composition of herbal extracts and standardizing formulations. Moreover, establishing quality control protocols and guidelines is crucial to maintaining uniformity, safety, and potency in herbal products (Fig. 5).

There is a growing acceptance of herbal medicine in healthcare, but it requires further research to advance and integrate it into modern medical practices. This necessitates comprehensive studies to validate the therapeutic benefits and safety profiles of herbal remedies, addressing the concerns of both patients and healthcare professionals. Investigating the phytochemical composition of medicinal plants is vital to understanding the bioactive components underlying their therapeutic effects, facilitating the development of evidence-based herbal treatments. Furthermore, research must uncover potential interactions between herbal and pharmaceutical drugs to ensure patient safety and effective treatment approaches (Fig. 5). Acknowledging herbal medicine's cultural and ethical aspects is essential to preserve

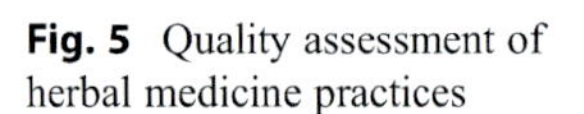
Fig. 5 Quality assessment of herbal medicine practices

indigenous practices and prevent the exploitation of traditional knowledge. Strict regulatory frameworks, encompassing labelling laws and quality requirements, are needed to safeguard consumer confidence and public health. Additionally, more research is needed to determine how herbal medicine can be seamlessly integrated into conventional healthcare practices, considering its potential in complementary and alternative medicine approaches.

Research and quality assessment in herbal medicine (Fig. 6) are imperative for ensuring safety and effectiveness. First, accurate botanical identification is crucial, as using the wrong plant species can be harmful or ineffective; modern techniques like DNA barcoding and microscopy aid in precise identification. Moreover, the growing conditions of medicinal plants, influenced by factors like farming practices, climate, and soil quality, significantly affect their chemical composition and therapeutic potential, necessitating careful consideration for consistent quality. Good agricultural and collection practice provides a sustainable cultivation and harvesting framework, emphasizing quality control throughout production. Detection of authentication and adulteration is essential, as periodic substitutions with subpar or hazardous components can occur; analytical techniques like high-performance liquid chromatography and mass spectrometry help identify adulterants. Additionally, standardization of extraction, processing, and herbal product combination methods is essential for ensuring uniform quality and therapeutic efficacy (Fig. 6).

This part of the book has 19 chapters which are summarized as follows:

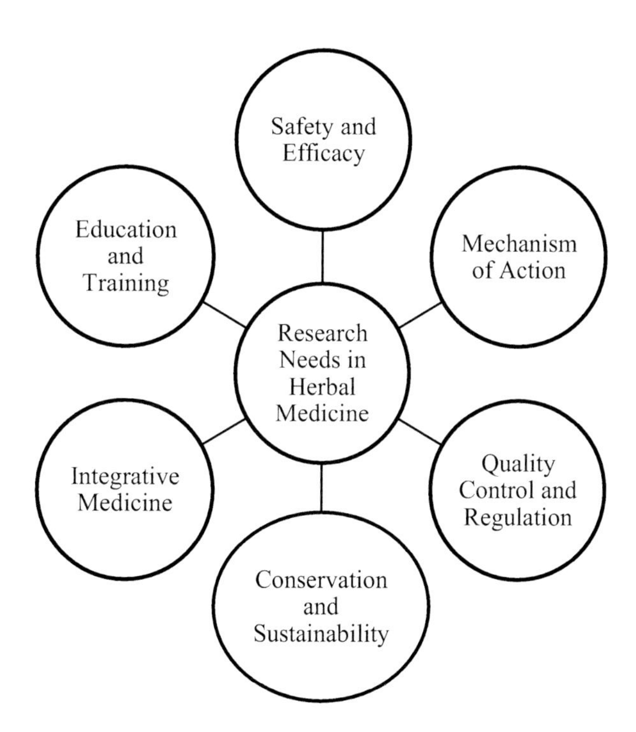

Fig. 6 Research needs in herbal medicine

i. Trace Metals Contamination of Herbal Remedies: Trace metals contamination in herbal remedies poses health risks due to potential toxic effects. Stringent quality control measures and standards are necessary to ensure consumer safety. Regulations and policies should address trace metal limits in herbal medicines, monitor their presence, and enforce compliance to safeguard public health.
ii. Regulations and Policies for Herbal Medicine and Practitioners: Establishing comprehensive regulations and licensing for herbal medicine practitioners is essential to ensure patient safety and standardize the practice. Striking a balance between traditional knowledge and modern regulations is key to promoting the integration of herbal medicine into mainstream healthcare systems.
iii. Toward Sustainability in the Source of Raw Materials for Herbal Remedies: Sustainable harvesting and cultivation practices are vital to conserve medicinal plant species and maintain a consistent supply of herbal remedies. Developing guidelines and incentives for sustainable sourcing can help preserve biodiversity and support local communities.
iv. Microbial Contaminants of Herbal Remedies: Health Risks and Sustainable Quality Control Strategies: Microbial contamination in herbal remedies can jeopardize consumer health. Quality control strategies, such as microbial testing and Good Manufacturing Practices, are essential for product safety. Sustainable sourcing, processing, and storage practices can mitigate microbial contamination risks in herbal medicines.
v. Adoption and Application of Biotechnology in Herbal Medicine Practices: The integration of biotechnology offers opportunities to enhance the quality, efficacy, and safety of herbal medicines. Regulation and ethical considerations must guide its responsible application. Biotechnological methods like tissue culture and genetic profiling can aid in the cultivation and conservation of valuable medicinal plant species.
vi. Challenges and Future of Nanotechnology in Global Herbal Medicine Practices: Nanotechnology holds promise for improving drug delivery and enhancing the therapeutic potential of herbal medicines, but challenges related to safety and regulation must be addressed. Collaborative efforts among scientists, regulators, and herbal practitioners are crucial to unlock the full potential of nanotechnology in herbal medicine.
vii. Eco-metabolomic Studies of Medicinal Plants and Herbal Medicine: Eco-metabolomic studies provide insights into the interactions between medicinal plants and their environments, aiding in sustainable cultivation and conservation efforts. This ecological perspective can guide the responsible use of medicinal plants and the development of herbal medicine industry practices that support environmental health.
viii. Physiological Ecology of Medicinal Plants: Implications for Phytochemical Constituents: Understanding the physiological ecology of medicinal plants helps elucidate how environmental factors influence the production of bioactive compounds, contributing to the standardization of herbal remedies.

Conservation strategies should consider the ecological requirements of medicinal plants to ensure a stable supply of quality raw materials.

ix. Big Data Application in Herbal Medicine – the Need for a Consolidated Database: The accumulation of big data can revolutionize herbal medicine research and practice, but a consolidated, accessible database is essential to harness this potential fully. Data sharing and standardized data collection methods can facilitate collaborative research and evidence-based decision-making in herbal medicine.

x. Challenges in the Storage of Herbal Medicine Products and Strategies for Sustainable Management: Proper storage of herbal medicine products is crucial to maintaining their efficacy and safety. Sustainable packaging and storage solutions can reduce environmental impacts. Strategies should address issues like humidity control and temperature stability, ensuring the long-term viability of herbal remedies.

xi. Herbal Medicine Formulation, Standardization, and Commercialization Challenges and Sustainable Strategies for Improvement: Herbal medicine formulation and standardization are necessary to ensure consistent product quality and efficacy. Sustainable practices can enhance market competitiveness. Collaboration between traditional healers, scientists, and industry stakeholders can lead to improved formulations and sustainable commercialization of herbal remedies.

xii. Research Needs of Medicinal Plants Used in the Management and Treatment of Some Diseases Caused by Microorganisms: Research is crucial to identify and develop herbal remedies for diseases caused by microorganisms. Collaboration between researchers and traditional healers can expedite discovery and validation. Addressing research needs can lead to the development of effective, evidence-based herbal treatments for infectious diseases.

xiii. Conservation and Sustainable Uses of Medicinal Plants Phytochemicals: Medicinal plants play a crucial role in traditional healthcare, with phytochemicals as their active compounds. Conservation efforts are essential to ensure the sustainable use of these plants, preserving their phytochemical richness and cultural significance.

xiv. Threats and Conservation Strategies of Common Edible Vegetables that Possess Pharmacological Potentials in Nigeria: Common edible vegetables in Nigeria possess valuable pharmacological properties, but they face threats from habitat loss and overharvesting. To safeguard these resources, conservation strategies must be developed to balance nutritional needs with sustainability.

xv. Sustainable Supply Chain Management in the Herbal Medicine Industry: Herbal medicine relies on various plant resources, making sustainable supply chain management crucial to ensure a consistent and ethical source of ingredients, while also supporting the livelihoods of local communities involved in cultivation and harvesting.

xvi. Consumer Perception and Demand for Sustainable Herbal Medicine Products and Market: Consumer demand for herbal medicine products with sustainability and ethical sourcing practices is on the rise. Understanding consumer perceptions and preferences is key to creating a thriving market for sustainable herbal remedies.

xvii. Indigenous Knowledge and Phytochemistry: Deciphering the Healing Power of Herbal Medicine: Indigenous knowledge plays a vital role in understanding the healing potential of herbal medicine. By combining traditional wisdom with modern phytochemistry, we can unlock the therapeutic properties of these natural remedies.

xviii. The Nexus of Business, Sustainability, and Herbal Medicine: The intersection of business, sustainability, and herbal medicine is an exciting frontier. It involves innovative approaches to commercializing herbal products while promoting environmentally friendly practices and ethical standards in the industry.

xix. Healing Trails: Integrating Medicinal Plant Walks into Recreational Development: Integrating medicinal plant walks into recreational development creates a unique experience for nature enthusiasts. These trails offer a blend of education, wellness, and conservation by introducing people to the world of medicinal plants while ensuring their sustainable use.

Conclusion

Herbal Medicine Phytochemistry: Applications and Trends offers a comprehensive exploration of the dynamic and ever-evolving world of herbal medicine and phytochemistry. Throughout this book, the rich tapestry of natural compounds found in plants, their intricate chemistry, and their diverse applications in healthcare, wellness, and beyond are presented. This is paramount considering the remarkable resurgence of interest in traditional herbal remedies, driven by the pursuit of natural alternatives, sustainability, and a profound understanding of the phytochemical treasures hidden within the plant kingdom, ancient cultures, and societies.

The book will take readers on a journey from the historical roots of herbal medicine to its modern renaissance, showcasing the valuable contributions of phytochemistry in unlocking the secrets of these medicinal plants. From the laboratories to the fields, it has revealed the diverse applications of herbal medicine, encompassing its use in traditional healing, mainstream healthcare, dietary supplements, and even cosmetics. It explores the transformative power of phytochemistry in identifying, isolating, and harnessing bioactive compounds that have the potential to combat diseases, alleviate symptoms, and enhance well-being. The innovative chapters featured in this book have illuminated the promising future of phytochemicals in modern medicine, paving the way for novel drug development, the creation of herbal formulations, and the enhancement of healthcare practices. It underscores the importance of responsible sourcing, sustainability, and quality control in the herbal medicine industry, highlighting the need to preserve nature's biodiversity and protect the heritage of indigenous knowledge. It has emphasized the imperative for rigorous research, safety evaluations, and standardized formulations to ensure that herbal medicine remains a safe and effective choice for individuals seeking natural solutions.

In a world where the demand for holistic and sustainable approaches to health is growing, *Herbal Medicine Phytochemistry: Applications and Trends* serves as an invaluable resource, shedding light on the boundless potential of phytochemistry and its impact on our lives. The book is a testament to the enduring significance of herbal medicine and the pivotal role of phytochemistry in the ongoing quest for wellness and the harmonious coexistence of humanity with the natural world. It is noteworthy that the world of herbal medicine and phytochemistry is a constantly evolving landscape. It is our hope that this book will provide readers, professional, practitioners, researchers, students, and enthusiasts with a deeper understanding of these fields and inspire curiosity, further research, policy establish, and responsible engagement. With this knowledge, we can look forward to a future where the healing powers of nature are harnessed, celebrated, and integrated into the healthcare systems of tomorrow.

Department of Microbiology, Bayelsa Medical University, Yenagoa, Bayelsa State, Nigeria — Sylvester Chibueze Izah
Goodnight Family Department of Sustainable Development, Appalachian State University, 212 Living Learning Center, Boone, NC, USA — Matthew Chidozie Ogwu
Chairperson, Department of Eastern Medicine, Faculty of Medical Sciences, Government College University Faisalabad, Faisalabad, Pakistan — Muhammad Akram

Part I

Introduction

Historical Perspectives and Overview of the Value of Herbal Medicine

1

Sylvester Chibueze Izah, Odangowei Inetiminebi Ogidi, Matthew Chidozie Ogwu, Saoban Sunkanmi Salimon, Zaharadeen Muhammad Yusuf, Muhammad Akram, Morufu Olalekan Raimi, and Austin-Asomeji Iyingiala

Contents

S. C. Izah (✉)
Department of Microbiology, Faculty of Science, Bayelsa Medical University, Yenagoa, Bayelsa State, Nigeria
e-mail: sylvester.izah@bmu.edu.ng

O. I. Ogidi
Department of Biochemistry, Faculty of Basic Medical Sciences, Bayelsa Medical University, Yenagoa, Bayelsa State, Nigeria

M. C. Ogwu (✉)
Goodnight, Family Department of Sustainable Development, Appalachian State University, Boone, NC, USA
e-mail: ogwumc@appstate.edu

S. S. Salimon
School of Biological and Medical Physics, Moscow Institute of Physics and Technology, Dolgoprudny, Russia

Z. M. Yusuf
Department of Biological Sciences, Al-Qalam University, Katsina, Nigeria

M. Akram
Department of Eastern Medicine, Faculty of Medical Sciences, Government College University, Faisalabad, Pakistan

M. O. Raimi
Department of Community Medicine, Environmental Health Unit, Faculty of Clinical Sciences, Niger Delta University, Wilberforce Island, Bayelsa State, Nigeria

A.-A. Iyingiala
Department of Community Medicine, Faculty of Clinical Sciences, College of Medical Sciences, Rivers State University, Port Harcourt, Rivers State, Nigeria
e-mail: iyingiala.austin-asomeji@ust.edu.ng

S. C. Izah et al. (eds.), *Herbal Medicine Phytochemistry*, Reference Series in Phytochemistry, https://doi.org/10.1007/978-3-031-43199-9_1

Abstract

Herbal medicine practices are as old as human history. Plants have been used to treat and manage different disease conditions in different parts of the world since time immemorial. Currently, about 80% of the global population uses herbal medicine directly or indirectly to manage or treat disease conditions. The use of plants to treat diseases depends on the local knowledge and cultural practices of the indigenous people of the area and requires a certain level of inborn skills or formal/informal training for the efficient and effective utilization of the herbal formulations. In herbal medicine, plants are used for spiritual processes including libation, sacrifice, appeasing of the gods, invocation, divination, food preparation, and as medicine. Whole plant or their parts are formulated into herbal medicine that is dependent on the bioactive constituents of the plant and their parts. Some diseases that have been successfully managed using herbal medicine include memory loss, cardiovascular issues, arthritis, osteoarthritis, digestive and respiratory diseases, reproductive problems, skin diseases, neurological issues, diabetes, diverse types of cancers, hypertension, and many others. Herbal medicines may be utilized in powder or liquid form and depending on the ailment may be ingested or externally applied on the affected parts of the body, but dosages are often unspecified or unclear. Herbal medicine is at a critical development stage, and there is a need for research and regulatory policies to support and standardize the processes involved in the exploration of raw materials, production of herbal remedies, branding, and marketing as well as sustainable utilization. Through research and regulatory policy, there will be greater clarity about herbal medicine processing, procedures, delivery routes, dosages, and compatibility, which will position the agelong process to be competitive, integrated into, or practiced alongside conventional medicine.

Keywords

Herbal medicine · Pharmacopeia · Traditional disease management · Physiological condition · Medicinal plants · Phytomedical research · Herbal medicine governance · Tradomedicine

Abbreviations

%	Percent or Percentage
AD	Anno Domini
AHM	Arab Herbal medicine
AM	Arabic medicine
AMM	Arabic materia medica
AMPK	5′ AMP-activated protein kinase
BC	Before Christ
BCE	Before the Christian Era
CAM	Complementary and alternative medicine
CAMHI	Conference of African Ministers of Health
COVID-19	Coronavirus disease 2019
FDA	Food and Drug Administration
HIV/AIDS	Human immunodeficiency virus acquired immunodeficiency syndrome
HPLC	High-performance liquid chromatography,
IIS	*C. elegans* insulin/IGF-1 signaling (IIS) pathway
JNK	c-Jun N-terminal kinases
MAPK	Mitogen-activated protein kinase
NFκB	Nuclear factor kappa-light-chain-enhancer
OAU	Organization of African Unity
ORID	Other related infectious diseases
SARS-CoV-2	Severe acute respiratory syndrome *coronavirus 2*
SATM	Standardized African traditional medicines
TB	Tuberculosis
TCM	Traditional Chinese Medicine
U.S.	United States
USA	United States of America
USD	United States Dollars
USG	United States Government
WHM	Western Herbal Medicine
WHO	World Health Organization

1 Introduction

Medicinal plants are the source and foundation for the effectiveness and efficacy of both traditional and modern medicine. In both industrialized and developing countries, medicinal plants are crucial in the management of diseases and are the foundation for many maladies that have been treated for thousands of years [85]. Traditional medicine is becoming more popular in both developed and developing nations due to its natural origins and perceived lack of negative side effects. Herbal medicine has gained increased attention at the policy level, as evidenced by the

Astana Declaration on Primary Healthcare [12], which promotes the impartial integration of both herbal medicine and scientific knowledge in basic healthcare. The declaration calls for greater political commitment to organizations and individuals that contribute to the provision of primary healthcare to everyone around the world. Herbal medicine is vital to primary healthcare because of the knowledge and capacity-building considerations. Nonetheless, it is impossible to not overstate the therapeutic significance of traditional medicine.

According to the World Health Organization (WHO), around 80% of the people in the world are reliant on herbal medicine for their basic healthcare needs ([85]; WHO, [208]). Based on the most recent estimates, two-thirds of the global populace are heavily reliant on unorthodox medical treatments because they are more widely accessible, more affordable, and more in line with patients' ideologies, placate the fears of negative effects from chemosynthetic medicines, satisfy the inclination for individualized healthcare, and enable widespread public access to personal and relevant health records [1, 37, 40]. As a result, several more individuals have embraced the use of indigenous plants. Chemically synthesized drugs are widely spread over the past century and have profoundly reformed medical services in several countries across the globe. Nevertheless, a larger fraction of the developing nation population continues to depend on natural remedies and traditional alternative medicine practitioners for their primary care. At least 90% of the African populace and 70% of the Indian descendants count on herbal medicine to aid and/or satisfy their healthcare needs [30, 49, 120]. Over 40% of all healthcare services provided in China are traditional medicine, whereas over 90% of general hospitals already established departments for traditional health medicine practice [54, 211, 216]. Furthermore, the use of traditional medicine is not restricted to third-world nations only; the last 200 years have witnessed an upsurge in the people's interest in natural remedies in developed nations, along with an expansion of the use of ethnobotanicals. Herbal medicine is at the heart of several healthcare systems across the globe – Western Herbal Medicine (WHM) is practiced in New Zealand, Canada, Australia, Great Britain, and Western Europe; Traditional Chinese Medicine (TCM) in China; Ayurveda in India; and Kampo in Japan. In addition, a variety of over-the-counter medicines approved by the European Union's Traditional Herbal Medicinal Product Regulation are widely available in pharmacies [62].

This chapter aims to provide a historical background to herbal medicine resources, practices and techniques, products, values, and challenges to the promotion of the agelong medical practice and highlight areas in need of attention to help increase the competitiveness of the practice and help integrate it into or ensure it is sustainably practiced alongside conventional medicine. It is organized into five parts – history of herbal medicine, utilization of herbal medicine, herbal medicine techniques, trends in herbal medicine, and herbal medicine research needs. This is the oldest form of medicine known to humans all over the world and is proven to be effective in the treatment and management of diverse diseases and ill health conditions. This work highlights the need for globally standardized procedures and research support to help clarify processing methods, routes of administration, doses, and compatibility issues.

2 Historical Perspectives of Herbal Medicine

It is difficult to clearly distinguish the history of herbal medicine into different eras because human societies evolved different herbal medicine practices and systems in isolation which contributed to the various forms of medicine available today. However, here we adopt a system that presents the views and historical practices of herbal medicine in some ancient human societies around different parts of the world, and these are arranged in no particular order.

2.1 Ancient Babylonians

The ancient Babylonians used plants as medical remedies as far back as 60,000 years ago. Written accounts of herbal medicine are at least 5000 years old in China and 2500 years old in Asia Minor and Greece [160]. There are several herbal medical systems, and each one's beliefs and practices are shaped by the area where it initially emerged [211]. Traditional Chinese medicine, which has been practiced for centuries, is the country's unique system [202]. The world's first known herbal book was published over 2000 years ago (*Devine Farmer's Classic of Herbalism*) and was written in China. Many herbal pharmacopeias and other monographs on certain plants are also available [200].

2.2 Ancient Hindustani Doctors

Ancient Hindu doctors and saints established Ayurveda, a medical system that has been practiced in India for over 5000 years. In its material medica, more than 1500 plants and 10,000 formulas are fully described. In contrast to Western medicine, the Indian government has acknowledged Ayurveda as a full healthcare system [53]. The Japanese herbal medication Kampo medicine has roughly 148 formulas and is over 1500 years old [205]. The majority of people continue to practice Ayurvedic medicine in India, Kampo medicine in Japan, Traditional Chinese Medicine, and Unani medicine in the Middle East and South Asia. The demand for herbal medicine and other botanicals by Western cultures has been continuously rising in the recent herbal renaissance period. Acupuncture and herbal treatment, which are both parts of TCM, are now more popular than ever in Western nations [21]. Together with herbal medications, other herbal items including cosmetics, perfumes, teas, health foods, and nutraceuticals are also widely used and make up a sizable component of the worldwide herbal market.

In actuality, it likely predates contemporary *Homo sapiens*. In Neanderthal graves in Iraq going back 60,000 years, archaeologists have discovered pollen and flower pieces from several different medicinal plants. Among them were species of *Ephedra*, *Centaurea*, *Senecio*, *Althea*, and *Achillea*. This can be evidence of the widespread usage of different herbal treatments at that very early time. According to legend, China has utilized *Cannabis* plant (*C. sativa* L.) for more than 8000 years [41].

The opium produced from the poppy (*Papaver somniferum* L.), which was first grown in Mesopotamia around 5400 years ago [19], has been used for medical purposes continuously ever since. Two fragments of the birch fungus *Piptoporus betulinus* (Bull.) Karst was found inside the mummified human body known as the "Iceman" that was found in the Italian Alps in 1991. According to scientists, this 5300-year-old person may have been utilizing the fungus as a medicine to cure intestinal parasites [24]. It is without dispute that herbal therapy has a long history.

2.3 Chinese Medicine

Chinese people have employed Traditional Chinese Medicine since ancient times. While components from animals and minerals have been employed, plants are the main source of cures. About 500 of the more than 12,000 items used by traditional healers are regularly used [103]. Botanical products are only used after being processed, which could include stir-frying or soaking in vinegar or wine, for example. In clinical practice, a complex and frequently individualized treatment may be prescribed after a conventional diagnosis. In China, conventional Chinese medicine is still widely practiced. Traditional medicine is regularly used by more than 50% of people, with rural areas having the highest rates of use. In China, there are over 5000 traditional medicines that makeup about one-fifth of the country's pharmaceutical sector [103]. Numerous herbal treatments made their way from China to the Japanese traditional medical systems. In the ninth century, the earliest pharmacopoeia of Japanese traditional medicine included a classification of native Japanese herbs [173].

2.4 Arab Herbalists

It is generally known that in the Middle Ages, Arab herbalists, pharmacologists, chemists, and doctors modified and enhanced classical Hippocratic-Greek medical knowledge. In addition, the vast majority of Arabs are Muslims, and Islamic philosophy and Arabic culture are intimately intertwined. As a result, Arabic medicine (AM), Arabic materia medica (AMM), and Arabic herbal medicine (AHM) are also known as Greco-Arab or Islamic medicine. Arabs in the Baghdad area were the first people in history to distinguish between medicine and pharmaceutical science in the eighth century. The Arab world is where the first pharmacies were opened (Baghdad, 754 CE). Certain medication formulations may still be found in pharmacopoeias today, and the forms utilized at the time are still used in treatment [116]. Mesopotamia produced the oldest records of herbs, which were recorded on clay tablets and written in cuneiform (dating back to 2600 BCE). The Ebers Papyrus, which dates back to 1500 BCE, is the most famous Egyptian medical record. It lists 700 herbal remedies (mostly derived from plants) and their dosage forms, including gargles, snuffs, poultices, infusions, pills, and ointments, as well as their carriers, including beer, milk, wine, and honey [171].

AHM has been practicing the use of organic and inorganic natural medicines for the prevention and cure of illnesses since the eighth century, including camel urine [123]. It is interesting to note that camel urine therapy had a significant cytotoxic impact on mouse bone marrow cells, according to pharmacological investigations [115]. Just 200–250 plant species are still used in traditional Arab medicine to treat different ailments, despite the Middle East being home to more than 2600 plant species, more than 700 of which are known for their usage as medical herbs or botanical insecticides [6].

From Alexandria to Sallum in Egypt, the western Mediterranean coastal area has 230 species of plants from 48 families; 89% of them were useful for medicine, 62% were common, around 24.9% were unusual, and 15% were rare [14]. The Mediterranean area and/or the worldwide market still sell or trades 236 plant species, 30 animal species, 29 organic compounds, and 9 materials of other or mixed sources that are still used to cure human ailments [6]. Around 250–290 plant species are still in use in the Mediterranean area, according to ethnopharmacologists' surveys of the region's plant species [101, 172]. A total of 129 plant species are used in AM in Israel to treat a range of illnesses. About 40 species of these plants are used to cure skin conditions, 27 species for digestive issues, 22 species for liver issues, 16 species for respiratory conditions and coughing, 22 species for different types of cancer, and 9 species for weight reduction and cholesterol lowering [15]. Yet throughout the Arab Empire, Islamic doctors employed more than 1400 different types of herbal remedies (632–1258). The late nineteenth and early twentieth centuries, up to the implementation of the Food and Drugs Act of 1906, were likely the height of herbal medicine use in the USA. Hundreds of patent herbal preparations, many of which included a significant amount of alcohol, were easily accessible on the US market before this regulation. Some of these, such as Swift's Syphilitic Specific, later shortened to SSS Tonic [191], and Lydia E. Pinkham's Vegetable Compound [192], continued to be sold and are still well remembered by many people today. Yet, the majority of these items and the absurd advertising that supported them were put out of business by the 1906 Act.

2.5 Aztec Medicine

This is a system of herbal medicine that evolved within Aztec societies where human health is managed holistically using spiritual and material bodies (plants and animal products and by-products). It began among the Nahuatl-speaking indigenous people of Mexico. Today, Mexico is the nation with the second-highest number of registered medicinal plants after China and has a long history of using medicinal plants [7, 50]. [36]. Indigenous traditional medicine in Mesoamerican societies began to emerge about 1500 B.C., beginning with the Olmec, the mother culture, and continuing with the Toltecs, Teotihuacans, and Mayans [170]. This came to an end, however, with the collapse and invasion of the Aztec empire in 1521 by Spanish colonizers, who lost a great deal of knowledge about the use of plants as medicines. After the conquest, however, it was possible to reconstruct the Aztecs' medical history. The *de la Cruz-Badiano Codex* (Código de la Cruz-Badiano), which was produced in 1552 by Nahuatl native Martin de la Cruz and translated by Juan Badiano, is the earliest book on medicinal plants used by Native Americans on the American continent [68]. Religion, astronomy, divination, and the polarity of cold and warm served as the foundation for Aztec medical doctrine. The estimated number of Indigenous people in Mexico now is 25.5 million (or 21.5%) [69]. The majority of herbal medicine used to treat common ailments is used by these Indigenous groups [29].

2.6 Indigenous or Native American Societies

Presently, 4.9% of Canada's population, or 1.7 million people, and 1.7%, or 5.6 million people in the USA, identify as Indigenous [110, 184]. The Indigenous people who resided on the land and were assimilated into several nations in Canada before the advent of European invaders, such as the Inuit and Metis, are known as First Nations [184]. Native Canadians' traditional medical practices are in danger of disappearing since they were historically replaced by the modern medical system and lost the trust of the populace as a result of conquering. As a result, governments, medical professionals, and European Canadian missionaries gradually banned indigenous healing practices [39]. For instance, with the Indian Act of 1876, Federal Indian policies increased and codified the monitoring and control of Native Americans' lives [52, 98].

Children from these villages were taken away by force in the 1960s and put in foster care to be adopted by non-Native families in Canada. Intergenerational trauma resulted from the assumption that the parents of these children did not have enough residences and educational opportunities to care for them [39, 88]. As a result, it has been challenging for the Indigenous community to transmit their ancestors' traditional medical knowledge. Nonetheless, this persecution and cultural marginalization have been reversing in recent decades. Indigenous peoples of Canada have used their traditional knowledge of medicinal plants for thousands of years, and it has been handed down verbally down the generations. Particularly significant in this sense for Indigenous peoples like the Metis, Cree, Dene, Sekani, Innu, Ojibwa, Chippewa, Abekani, and others are the boreal forest regions of Canada [195].

According to the most recent population census, the majority of the Indigenous people in the USA are American Indians, Alaskan Natives, and Hawaiian Natives, with a combined population estimated at five million (1.5%) [194]. Before the coming of European colonizers, Native Americans in the USA engaged in health practices that, during therapeutic sessions supported by goods like herbs, emphasized the connections between individuals (family and social groups) and the environment [215]. The loss of lands and other resources experienced by the American Indians and Alaskan Natives in particular during colonialism influenced their way of life, particularly the repression of their spiritual and healing practices, which is still happening today [215]. With the loss of their rights came the relocation of American Indians and Alaskan natives to reservations, separation from their holy places and medically significant flora and fauna. Congress was given extensive constitutional authority to negotiate with these groups in 1787. The American Indian Religious Freedom Act was enacted by the US Congress in 1978. However, many American Indians and Alaskan Natives are not recognized by the federal government [215].

Before the coming of the European conquistadors, traditional Native American medicine was used for generations. Yet these techniques were outlawed, and this information was given a lower value. Because of this, little is known about how these activities affect human health, and it is challenging to interpret both archaeological and human remains. The 1990 Native American Graves Protection and Repatriation Act, in particular, restricts research on the latter [193]. However, the contemporary tribal tribes are attempting to revive these historical rituals [215]. In the USA, the study of traditional indigenous medicine made from wild plants dates back more than a century [156]. The mesophytic woods of Appalachia are thought to include 1100 plant species of therapeutic use [25]. Eastern Tennessee, Western North Carolina, and Southeastern Virginia make up these woods; this area has a wide range of temperatures that favor the development of a wide diversity of plants. Although, traditional medicine and the use of medicinal plants can be traced back to Mesopotamia and Egypt, the foundation of European medicine was established by physicians and philosophers during Greek antiquity and the time of the Roman Empire, who began establishing a written consensus about what was considered effective medical knowledge. The oldest known fragment of a Greek herbal is preserved in the ninth book of Theophrastus of Eresos' (c. 370–287 BC) *Enquiry into Plants*, which is about the juices and therapeutic characteristics of plants [65]. Hippocrates of Kos (c. 460–370 BC), Galen (129–c. 199/200 or 216–217 AD), and Pedanios Dioscorides (first century AD) were the most illustrious and significant ancient writers who successfully merged botanical and medicinal knowledge. The most significant herbals ever published are considered to be Dioscorides' *De Materia Medica* and Galen's first alphabetical compilation of simple medications (*De simplicium medicamentorum facultatibus libri XI*). These writings had an impact on Mediterranean and European medicinal plant usage up to the eighteenth century via the constant copying and distribution of their content [11, 196].

Dioscorides discussed the therapeutic use of almost 900 herbal species, 600 medicinal plant species, 35 animals, and 90 mineral remedies [166]. A century later, Galen recorded roughly 850 simple medicines, most of which were similar to those named by Dioscorides but had far fewer medical use [100]. Each plant taxon was frequently referred to by a "code" made up of a list of the vernacular names and was known by various languages. This rigorous cross-referencing of names and uses served as a protocol for academic rigor to ensure accurate plant identification between regions, allowing herbal texts to be applicable in a wider geographical and cultural context and fostering the growth and consolidation of scientific knowledge.

2.7 Greek-Roman-European Herbalism

Greek herbals have a rather well-documented history, including their mutual effect and phylogeny. Lost copies may even be found by looking for missing connections [181]. Dioscorides, for example, cites and alludes to a variety of various writers and books, some of which have survived (such as works attributed to Theophrastus and Hippocrates) and others of which are now lost or have only partial copies left. Translations into Syriac served as a crucial intermediary step that permitted the transfer of Greek and Byzantine medical knowledge into Arabic and to the medieval Islamic world [169]. In addition to preserving classical medical knowledge, the Arab culture advanced it and made it possible for it to be transmitted back to the Occident via al-Andalus and the School of Salerno (about 1000–1300 AD) (711–1492; [196]). The analysis of the so-called Syriac Galen Palimpsest is likely to provide new and important insights on the transfer of Greco-Roman medical knowledge to the Arabic world. The earliest copy of Galen's writings on herbal remedies is now this manuscript, which is significant because it offers other readings from the more recent Greek versions that may be used to spot interpolations and other textual interpretations [18].

The middle and Northern European herbals were also affected by the traditional Mediterranean materia medica in terms of their organization, literary style, and substance. Back in the Medieval Ages according to monastic law, Benedictine monks were required to maintain therapeutic herb gardens, such as those at the abbeys at Montecassino, Italy, and St. Gall, Switzerland [190]. Classic Greco-Roman herbals and medicinal books underwent rigorous examination, commentary, and translation into contemporary languages throughout the Renaissance, garnering significant print runs. Renaissance herbals, however, also included the consensus of northern and central European medical folklore and were accompanied by woodcut drawings enforcing the commentators' botanical identifications.

The *Gart der Gesundheit* is recognized as the first thorough German herbal, authored by Johann Wonnecke von Kaub (ca. 1430–1504), physician to the city of Frankfurt and illustrated by Dutch artist Erhard Reuwich (1445–1505). [121]. The 435 monographs in the *Gart der Gesundheit* are based on texts from the Mediterranean region, including Pliny's *Natural History* from the first century AD, *The Canon of Medicine* by Avicenna (Ibn Sina, ca. 980–1037), the *Pseudo-Serapion* (Aggregator) by Ibn Wafid from the eleventh century in Toledo, and the *Circa Instans* by Matthaeus Platearius from the twelfth century [114].

The printing press' development led to a surge in the creation of European herbals and a race among writers, who weren't always free of nationalist emotions. The documentation of traditional folk knowledge on central and northern European materia medica was combined with the translation, commentary, and integration of ancient Greco-Roman knowledge by Peter Schoffer (1425–1503), Jan Stanko (1430–1493), Jean Ruelle (1474–1537), Otto Brunfels (1488–1534), Adam Lonitzer (1528–1586), Leonhard Fuchs (1501–1566), Hieronymus Bock (1498–1554), and William Turner (1508–1568). Nevertheless, a thorough examination of the problems of which plant species and applications were first recorded and in which herbal and how much materials were used is still absent.

The Greco-Roman and Arabian medical treatises were a major source of inspiration for the first formal pharmacopoeias. In addition to Hippocrates, Pliny, Dioscorides, and Galen, Matthaeus Platearius (twelfth century), Rhazes (865–925), and Avicenna were all significant figures (ca. 980–1037; [196]). Three texts written after 1000 AD at the School of Salerno, namely, the Antidotarium Nicolai (ca. 1100), the Antidotarium Nicolai Myrepsi (thirteenth century), and the Antidotarium or Grabadin of Pseudo-Mesue (ca. thirteenth century), are considered to be the forerunners of the official city pharmacopoeias due to the format and presentation used [196]. Constantinus Africanus (eleventh century), a Carthage herbal trader who collected Arabic medical works in North Africa and translated them into Latin, contributed a significant amount to the corpus of work created at the School of Salerno [121].

Roman era records of early imports of spices and herbal medicines from the Orient well-documented both the New World and Southeast Asia, showing the significance of exotic goods for European pharmacopoeias [198]. Prior to the creation of the European Pharmacopoeia, there were several factors that prevented attempts to synchronize pharmacopoeias, official or not, including the notion that locally sourced medicines would have a disproportionately high level of positive effects and the economic case for supporting locally produced goods [196]. The quest for effective treatments, often known as the trial and error method, during the outbreak of new epidemics (such as syphilis) and the testing of imported "strange" plant species supplied the laboratory for experimentation at the same time. After the Conquest, when Christopher Columbus and his ship returned from the Caribbean with their cargo, the syphilis pandemic was transferred from the Americas to Europe. The core wood of *Guaiacum officinale* L., a remedy used by the native people of the West Indies to treat syphilis, was imported to Europe and advertised against the disease, but later proved to be ineffective. This was because it was believed that the origin of specific diseases would be connected with the source of effective medicines [111]. Matthioli was the first to identify mercury's particular usefulness against syphilis, even though Arabian pharmacists had brought the substance to Europe as a therapy for skin conditions [111].

The Swiss physician Paracelsus (1493–1541) attempted to depart from the traditional medical theories of the time, particularly those of Galen and Avicenna, and promoted high hygienic standards for surgery. Paracelsus is best known for developing the fundamental toxicology principle that "the dose makes the poison" [111]. Yet even Paracelsus, an alchemist, was susceptible to esoteric and symbolic thinking. The "doctrine of signatures" was put out by Paracelsus in De natura rerum in 1537 as a holistic and harmonic theory that could foretell the medicinal properties of plants based on characteristics like form, color, smell, and taste [121]. While the theory of signatures' fundamental premise is founded in folk medical practices all over the globe, Paracelsus systematized it in conceptual contrast to Galen's humoral pathology [121]. The German physician Samuel Hahnemann (1755–1843) developed homoeopathy, which still adheres to the idea of signatures today, although conventional herbal therapy in Europe is likely losing ground in this regard.

Since there were fewer early herbals and because they had more time to exert their causal influence on the spread of medical knowledge and popular herbal knowledge, the earlier herbals were generally more important. Due to their large print runs, which are estimated to have exceeded 30,000 for the older editions of *I Discorsi* alone, others, like Matthioli's Renaissance commentary on Dioscorides' Materia Medica (*I Discorsi*), originally published in 1544, also had a significant influence [99]. The work of Dioscorides was included into Matthioli's herbal, which also included 600 new medicinal species. It was also published in Italian and translated into more vernacular languages, including French, German, Czech, and Arabic, with a total of almost 60 distinct editions (Barberi, 1967–1970). *I Discorsi* was founded on Matthioli's experience as a medical doctor and botanist who was acquainted with the Mediterranean flora, making Matthioli's translation simple to understand [185].

2.8 Alkebu-Lan or African Herbalism

Alkebu-lan medicine may be the oldest form of herbal medicine practice and system from which every other system evolved considering the history and origin of humans. The lack of scripts and high secrecy associated with the practice in most parts of Africa makes it difficult to trace most of these practices, but remnants still exist to date in the form of African traditional medicine (ATM) on which a large proportion of the population depends on for their primary health care because it is cheap, easily accessible, and highly relatable [132, 134, 138, 139, 147, 152]. In large part, the system is connected to how plants are used both as food and medicine and their conservation [128–131, 133, 135–137, 140–142, 149–151]. According to Ozioma and Chimwe [153], ATM involves spiritualism, divination, and herbalism and permeates the culture, religious beliefs, attitudes, knowledge, and understanding systems of African communities. Practices include the use of materials (plants and or animal parts) with incantations, sacrifice, and exorcisms. The system was challenged and almost pushed out by the European and Arab colonizers who seek to increase the spread of their religion and modern medicine practice while the locals where challenged with economic (poverty) and other developmental issues. Enslaved Africans who were taken to different parts of the world may have contributed to the growth and development of the practice in those parts. Although enslaved they maintained small plots containing medicinal plants that are used in the preparation of herbal tinctures, decoctions, teas, incense, etc. for the treatment and management of diverse ailments. According to Okaiyeto and Oguntibeju [143], some factors contributing to the preservation of ATM include:

1. Accessibility and cost-effectiveness.
2. The perception that it is natural and safe.
3. Connection to their gods, dead relatives, and nature.
4. Superior efficacy to modern medicine.

5. Ability to self-medicate.
6. High levels of confidentiality associated with ATM.
7. Fear of erroneous diagnosis in the hand of modern medicine practitioners.
8. Long waiting periods are often associated with modern medicine in Africa.
9. Advertisement of herbal products, practices, and practitioners.

3 Utilization of Herbal Medicine

Globally, the interest and use of medicinal herbs have seen exponential growth with an estimated market value of USD 152 billion and a forecast of USD 350 billion by 2030 [47]. In Europe, Germany has the largest market for herbal medications with medical claims and those that are typically prescribed in pharmacies [62]. According to German manufacturer prices, the market costs €1.33 billion annually, which is proportionate to 20.7% of the entire European market. Other significant nations in conformity with the provision and sales of herbal medicines include Spain (6.3%), Poland (7.2%), Italy (17.1%), Russia (9.2%), and France (13.0%) [62]. A survey by Rashrash et al. [164] revealed that more than a third of adults in the USA utilize alternative medicine for the management of their health. The most frequently utilized alternative medicine, except for vitamins and minerals, was herbal products [16]. In most developed countries, over-the-counter sales of herbal medicine products can take in the form of plant oils, herbal supplements, or teas (such as chamomile, ginger tea, peppermint tea, green tea, etc.) sold in drug stores alongside conventional medicine. Traditional medical practices vary considerably from one nation to another and area to region due to influences from such things as culture, history, philosophy, and individual attitudes. Usually, their theory and practical application differ markedly from those of mainstream medicine [210].

Traditional medicine has been utilized in Nigeria to treat a variety of illnesses. In a study by Abubakar et al. [3], it was found that 85% of the respondents used traditional medicine to treat a variety of health issues, including diabetes, typhoid fever, pneumonia, pile, cancer, fever, measles, diarrhea, and cough. Accessibility, affordability, perceived safety, and therapeutic recommendation potential for treating numerous ailments are believed to be the major driving factors for the surge in the utilization of herbal medicine [124]. As such, the upkeep of human health has become the primary focus of the widespread usage of herbal medicine [4, 5, 70]. Past studies have revealed that Nigerians of various socioeconomic status engage in the use of herbal medicine [45]. Botanical treatments persisted in the form of tinctures, fluid extracts, and concentrated extracts. One might argue that such galenical medicines served as the cornerstones of period medicine. Products like the fluid extract of Ergot U.S.P. and the tincture of *Digitalis* U.S.P. were quite popular. With the advent of sulfanilamide in the middle of the 1930s, the first suppression of plant treatments started. The race was on as synthetic organic chemists created hundreds of novel compounds with potential therapeutic qualities in response to this new German antibacterial medication. Old plant medications were quickly superseded by new, often very helpful, synthetic pharmaceuticals for two reasons [167]. They could be marketed at a premium price to pay the expenses of the research needed in generating them while still allowing the maker to make a sizable profit since they were unique chemical entities covered by patent protection. Also, since they were unique chemical entities, it was simple to standardize their action and achieve predictable potency. In contrast, many of the traditional botanicals were likewise efficient but were abandoned since they could not be patented and did not guarantee profit margins. They were challenging to standardize since they often ascribed their action to many constituents, many of which remained unknown. Almost impossible was action uniformity. The issue is typified by Digitalis (*Digitalis purpurea*). The herb is prized as a treatment for congestive heart failure because it contains both short-acting and long-acting glycosides with rapid and slow onset, making it basically superior to any single glycoside isolated from it. However, it was difficult to accurately standardize the leaf in any way. Researchers eventually gave up after trying a variety of animals with little success, including guinea pigs, goldfish, chick hearts, frogs, cats, water fleas, and pigeons. As a result, digitalis leaf is no longer used in America. Eventually, almost all herbal remedies vanished from pharmacy shelves, and by the 1960s, American medications, unlike those used in the majority of other nations, were almost entirely synthetic. According to local custom, herbal remedies are utilized in several EU nations. For instance, according to the Allensbach Study from 2017, 70% of Germans have tried "natural medicines," with herbal treatments being the most popular kind. Without a prescription, by mail order, or in pharmacies, herbal medications worth 1.36 billion market data were sold in Germany alone in 2015 [67]. In addition to properly approved medications, other goods have been registered as traditional medicines due to their extensive history of usage. This presents some significant issues for ethnopharmacology, particularly in terms of how to establish a solid evidence foundation for the use of such items and define what exactly qualifies as a tradition of usage as a medicine. In the 1970s and 1980s, sophisticated European nations, particularly Germany, started to publish scientific and clinical papers demonstrating the medicinal and financial value of herbal treatments, which had never been completely abandoned in that region [175]. The oldest and maybe most diverse treatment approach is African Traditional Medicine. With its great biological variety and cultural diversity, which is reflected in regional variations in healing methods, Africa is regarded as the birthplace of humanity [10, 56]. Sadly, the systems of medications are still not well documented. Traditional African Medicine encompasses the body and mind holistically in all of its manifestations. Before administering medications, especially medicinal plants to address the symptoms, the traditional healer often determines the psychological cause of a disease and treats it [57].

The main natural resources ever employed by humanity for maintaining preventive, curative, and rehabilitative health in Africa include plants, minerals, and animals. These materials have been employed by traditional healers for more than 10,000 years, much like any other continent. In traditional African medicine, herbalists, midwives, and diviners are all involved. The diagnosis of illnesses, which in certain circumstances are said to be brought on by ancestor spirits and other forces, is the responsibility of diviners. Native plants are often used by traditional midwives to facilitate delivery. A herb trading market in Durban reportedly draws between 700,000 and 900,000 merchants annually from South Africa, Zimbabwe, and Mozambique since herbalists are so well-liked on the continent. There are smaller herb markets in almost every town [58].

4 Herbal Medicine Applications and Techniques

Plants contain a wide range of compounds, including minerals (molybdenum, potassium, calcium, sodium, zinc, magnesium, and so on) and phytonutrients which include sugars (cellulose, starch, rhamnose, sucrose, ribose, fructose, maltose, etc.), carotenoids (flavonoids, lignans, lutein, carotene, plant sterols, isothiocyanates, lycopene, and others), and polyphenols (cinnamic acid, vanillic acid, curcumin, ellagic acid, tannins, etc.) all of which play integral roles in the prevention of diseases and health management [59, 86, 95]. A majority of these compounds possess antioxidant properties, enhance immunity, prevent cancer, suppress inflammation, protect the brain, and effectively reduce the progression of aging [66, 95, 182, 199]. Plant components are used directly as chemotherapeutics; thus, ethnobotanicals are crucial in pharmacognosy research, drug discovery, and development. They can also be used as building blocks in the design and production of drugs or serve as models for pharmacodynamic substances [104]. Morphine (named after Morpheus – the Greek god of dreams) was the first physiologically effective pure compound, isolated from Papaver somniferum (opium) by Friedrich Serturner at the start of the nineteenth century [13]. This stimulated an avalanche of research on medicinal plants and led to the discovery of numerous lifesaving bioactive compounds of natural origin. Some known examples include artemisinin (*Artemisia annua*), aspirin (*Salix* spp.), atropine (*Atropa belladonna*), cannabinoids (*Cannabis sativa*), cocaine (*Erythroxylum coca*), codeine (*Papaver somniferum*), digoxin (*Digitalis purpurea*), galanthamine (*Galanthus nivalis*), quinine (*Cinchona officinalis*), and reserpine (*Rauwolfia* spp.), among others [13]. This epoch of drug discovery was further broadened to include microorganisms by Fleming's discovery of penicillin which established the scientific and economic platform for modern medicine and the biopharmaceutical industry [38].

According to recent aging studies, bioactive compounds have exhibited enormous promise for managing even the most severe ailments and may one day replace conventional medications [197]. Nevertheless, the effects of many plant products on humans are largely a mystery, despite the multitudes of research in this area. The best active ingredients and doses can be chosen to produce the optimal beneficial effects from herbal medicines by carefully analyzing the mechanisms by which they affect the aging process. It is better to examine in vivo the efficacy of biologically active substances and plant extracts with geroprotective characteristics. The animal model systems employed for this purpose include but are not limited to zebra fish (*Danio rerio*), mice (*Mus musculus*), fruit fly (Drosophila melanogaster), nematode

(*Caenorhabditis elegans*), and rats (Rattus norvegicus) [64, 118]. These models make it easier to thoroughly research how newly created medications affect the aging process. However, the model organism *C. elegans* is gaining increasing popularity in these studies because of its numerous advantages. They include a short life span; easy lab maintenance; transparent body for real-time imaging; advanced genetic, genomic, and molecular tools and manipulations; high genetic homology with humans; etc. [189]. Havermann et al. [61] revealed that the bioactive compound baicalein extended life span and improved stress resistance via the activation of Keap-1/Nrf-2 signaling pathway using C. elegans. In addition, luteolin and chrysin demonstrated life span extension via the activation of AMPK pathway in both *D. melanogaster* and *C. elegans* [97]. Rosmarinic acid, a natural polyphenol, has been shown to increase the average life span of *C. elegans* by upregulating the S pathway via *ins-18* and *daf-16*, the MAPK pathway via *skn-1* and *sek-1*, and stress resistance and antioxidant genes like *ctl-1*, *sod-3*, and *sod-5* [106]. Curcumin has also been reported to increase life span in *C. elegans*, which is dependent on the functions of *age-1*, *skn-1*, *sir-2.1*, *sek-1*, *unc-43*, *osr-1*, and *mek-1*, which are related to the S, MAPK, and JNK signaling pathways [105].

Furthermore, natural products remain the fulcrum of more than 60% of cancer therapeutics that are available or undergoing investigation. Over 70% of the globally approved 177 anticancer drugs are derivatives of natural products, and several of them have been optimized using combinatorial chemistry. Camptothecin (*Camptotheca acuminata*), a precursor for irinotecan and topotecan; combretastatin derived from *Combretum erythrophyllum*; paclitaxel, isolated from *Taxus brevifolia*; vincristine from *Catharanthus roseus*; noscapine from *Papaver somniferum*; and flavopiridol, a structural isomer of rohitukine derived from *Dysoxylum binectariferum* are some examples of plant-derived medications that have been employed for cancer therapeutics [31, 32, 158, 178]. In addition, alpinumi isoflavone and 4′-methoxylicoflavanone isolated from the stem bark of *Erythrina suberosa* exhibited cytotoxicity against human leukemia (HL-60 cells) through the suppression of the membrane potential in the mitochondria and activation of apoptotic proteins [96]. The chemoprotective and anticancer activity of curcumin has also been well established both in vivo and in vitro. Some of its known mechanisms include apoptotic stimulation via the inhibition of cyclin D1, cyclo-oxygenase-2, and other relevant NFκB target genes [188]; autophagy [27], mitotic arrest [55], and antioxidant and anti-inflammatory activities [48].

In addition, the inherent capacity of herbal medicine to fight COVID-19 became an important research focus during the global pandemic. Due to their potent suppression of SARS-CoV-2, traditional Chinese medications garnered a lot of interest. For instance, Qingfei Paidu decoction demonstrated an outstanding clinical efficacy of >90% in treating COVID-19 patients at all phases [119, 163, 201]. Shuanghuanglian, also popular in TCM, suppresses SARS-CoV-2 Mpro replication in a dose-dependent manner [187]. Natural products such as graveospene A, deguelin, and erianin have been utilized mono-therapeutically for the management of lung cancer in vitro, while baicalein, resveratrol, and ginkgolic acid were employed during the fight against SARS-CoV-2 [217]. Moreover, the combination of some Food and Drug Administration (FDA)-approved drugs with biologically active compounds has been reported to mitigate the pernicious effect of SARS-CoV-2 [217]. For instance, the combination of nelfinavir and cepharanthine, remdesivir and linoleic acid, cisplatin, and curcumin have all been reported [217]. It is noteworthy that more than 40% of pharmaceuticals approved by the FDA have come from natural sources or their derivatives [157].

Possibly, due to the phytochemical content of plants, many of them have been reported to be used as an alternative agent for the treatment of several diseases. Hence, they can be used to treat diseases caused by microbes possibly due to their antimicrobial potentials [42–44, 71–73, 77–83, 89–93]. It can be used to control vectors or insects such as mosquitoes [17, 74–76, 176, 177, 218]. They can be used as preservatives. Other diseases and conditions that can be managed with plants include memory loss, cardiovascular diseases, arthritis, osteoarthritis, digestive, respiratory, reproductive, skin, neurological diseases, diabetes, cancer, hypertension, gastrointestinal, and conditions such as pregnancy, etc. Two main factors that support the continued interest in traditional medicine in the African healthcare system include affordability and accessibility. Contemporary medical care is often too expensive, or not available. Secondly, although having an almost worldwide distribution, certain diseases like malaria and HIV/AIDS, disproportionately affect Africans more, and the reliability of traditional medicine makes them try it out [126, 127]. Medicinal plants are often the community's primary readily available source of healthcare in many regions of Africa. Also, they are often the patients' first choice. For the vast majority of these individuals, traditional healers provide information, counseling, and therapy to patients and their families in a personal way while also being aware of their patient's surroundings [10, 56, 58]. Africa is endowed with abundant biodiversity [144, 145, 161, 162, 174]; it is thought to have between 40 and 45,000 plant species with the potential to become developed, of which 5000 species are utilized medicinally. This is not unexpected given that Africa has a tropical and subtropical climate and that plants naturally collect significant secondary metabolites as a method of surviving in a harsh environment [109].

Written records of herbal medicine go back 5000 years in Egypt. The Edwin Smith, Ebers, and Kahun papyri include a wealth of information on Ancient Egyptian medicine. The Edwin Smith Papyrus and the Ebers Papyrus, both published in 1930 by Breathed and Bryan, respectively, are from the seventeenth and sixteenth centuries BCE. It is thought that these writings came from previous sources. These include spells and recipes for a wide range of illnesses and symptoms. They go through illness diagnosis and provide anatomical facts. They go into great depth into the anatomy, physiology, and medicine of the Ancient Egyptians. A gynecological literature called the Kahun Papyrus [51] discusses issues such the reproductive system, conception, pregnancy testing, childbirth, and contraception. Crocodile dung, honey, and sour milk are some of the substances that are recommended for contraception [168]. Pygeum (*Prunus africana*), a traditional African remedy for mild-to-moderate benign *prostatic hyperplasia*, has gained widespread acceptance outside of Africa. It has been offered in Europe since the 1970s. Pygeum barks are gathered annually in Madagascar and Cameroon in quantities of 600 metric tons and 2000 metric tons, respectively. The bark is brewed into tea in Africa [125].

It is offered everywhere in the globe in the forms of powders, tinctures, and tablets; often, it is mixed with other herbs thought to be beneficial for prostate issues. Customers experience less inflammation and cholesterol buildup, as well as easier urination. The significance of this therapeutic method in Africa may be shown by comparing the numbers of traditional healers and physicians. There is one traditional healer for every 700–1200 individuals in the Venda region of South Africa, compared to 1 doctor for every 17,400 people. For every 110 individuals, Swaziland has one traditional healer. The percentage is the same in Benin City, Nigeria. In urban Kenya, there is 1 traditional healer for every 833 people.

The chiefs of state and government of the former Organization of African Unity (OAU) in Africa acknowledged that 85% of the continent's people rely on it for their healthcare requirements. The OAU proclaimed 2001 to be the Decade of Traditional Medicine. Following this historic pledge by African leaders, the Plan of Action and implementation mechanism that were endorsed by the AU summit heads of state and government in Maputo in 2003 were adopted by the First AU Session of the Conference of African Ministers of Health (CAMH1), held in April 2003 in Tripoli, Libya. The Plan of Action's primary goal is for all Member States to recognize, embrace, advance, and institutionalize traditional medicine within the region's public healthcare system by 2010. [58]. Additionally, the Maputo Declaration on Malaria, HIV/AIDS and Other Related Infectious Diseases (ORID) of July 2003 resolved to keep supporting the Plan of Action for the AU Decade of African Traditional Medicine (2001–2010), particularly research in the area of treatment for HIV/AIDS, tuberculosis (TB), malaria, and ORID. The Lusaka Summit proclaimed the years 2001–2010 to be the OAU Decade for African Traditional Medicine in July of the same year. The 11 priority areas, which have been developed into strategic activities, are institutional arrangements, information, education, and communication, resource mobilization, research and training, cultivation and conservation of medicinal plants, protection of traditional medical knowledge, local production of standardized African traditional medicines (SATMs), partnerships, and evaluation. The African Union (AU) Member States have been carrying out the plan of action of the AU Decade of African Traditional Medicine and the priority interventions of the WHO regional strategy since 2001, namely, policy formulation, capacity building, research promotion, development of local production, including cultivation of medicinal plants, and protection of traditional medical knowledge and intellectual property rights [168].

5 Trends in Herbal Medicine

Now, several nations throughout the globe are still reacting to the Beijing statement from November 8, 2008. The World Health Organization assembly urged nations to take charge of their citizens' health by creating national policies, rules, and standards that would guarantee the proper, safe, and efficient use of traditional medicine [212]. Since traditional medicine has helped millions of people throughout the globe get cheap treatment, it has shown the ability to meet the WHO's goal for universal healthcare [26]. Studies demonstrate that patients who see a general practitioner with extra training in alternative medicine have lower healthcare expenses and death rates than those who do not support this claim [94]. Less use of prescription medications and hospital stays was said to have reduced costs. Hence, every government in the world has made the significance of complementary and alternative medicine (CAM), which mostly employs plant items as pharmaceuticals, a top priority. These medications are prescribed over the counter, utilized as over-the-counter home treatments, and serve as pharmaceutical industry raw ingredients. They make up a significant share of the global medication industry and help millions of people throughout the globe get basic healthcare. This section will examine the use, regulation, integration, and quality control trends for herbal medicine on a worldwide scale. Almost 100 million individuals in Europe use traditional medicine-related goods and procedures, with 5% utilizing it as an alternative to mainstream treatment and the same amount as a supplement to it [148]. According to reports from the UK, 40% of doctors reportedly recommend patients to alternative healthcare providers. Complementary and Alternative Medicine (CAM) is used by the populations of France, Canada, and Australia at rates of 49, 76, and 46%, respectively. In Asia, China is reported to have produced 83.1 billion US dollars worth of Chinese herbal medicine in 2012, while the Republic of Korea increased its yearly spending on traditional medicines from 4.4 billion US dollars in 2004 to 7.4 billion US dollars in 2009 [148].

Regarding regulation, conventional Western medicine is completely controlled by government agencies, as is the use of Traditional Chinese Medicine. Almost 1146 species had been listed by 2005 in the well-established Chinese Pharmacopoeia, which was created to assure the quality of Chinese material medicine. These species may be examined using over 479 different HPLC techniques, 45 TLC methods, and 47 GC methods [33, 34]. The Ministry of Standardization and Methodology in Japan has accepted and standardized over 148 formulas from various manufacturers [60]. In addition, over 210 formulations are available over the counter and are utilized in medical institutions. In Hong Kong, traditional healers are consulted by around 60% of the population. For instance, local researchers in Singapore employ software to examine multi-herb mixtures [203]. The usage of straightforward preparations (cut and dried) and those in other dosage forms is regulated by law (an example is existing statutes in Singapore). Only items that adhere to the necessary safety and quality criteria may be listed and authorized for production, importation, supply, or sale. In the USA, it was predicted that 14.8 billion dollars were spent directly on natural goods in 2008. According to further South American figures, traditional medicines are used by 71% and 48% of the populations of Chile and Colombia, respectively. In several of these nations, medical schools educate about traditional remedies. To assure the identification, quality, purity, potency, and consistency of herbal medications, the US FDA suggests a variety of tests. Chemical assays, distinctive markers, and chromatographic fingerprints are used in many testing for drug substances and products. Raw material process controls and drug substance validation are also conducted. Moreover, in the USA, the creation of herbal monographs has benefited greatly from microscopy [9].

Several African nations have not yet included these types of medications in their primary healthcare systems. According to Ghana, traditional medicine has been incorporated into South Africa's and Nigeria's current healthcare systems [209]. Others have partly incorporated it, such as Tanzania, Ethiopia, and Rwanda. More nations, such as Uganda, Chad, and Gabon, are creating integration policies. According to the World Health Organization, traditional medicines are the primary source of healthcare for 70–90% of Africa's rural population [210]. In a 10-year assessment, the World Health Organization found that the majority of African communities continue to rely on herbal therapy [213]. According to research by Chatora et al. [28], the majority of basic healthcare on the African continent is still provided by traditional medical practices and products. To substantiate Chatora's research report from 2003, Abdullahi [2] references the increased usage of traditional medicines in Africa despite the problems it faces. This indicates that, despite the changing circumstances, the need for herbal treatments has grown over time due to the growing population. The growth in chronic and noncommunicable illnesses, as well as their availability and cost, have all been linked to an increase in the use of traditional medicine, particularly herbal remedies, across the world [26, 102]. One reason for the widespread usage, for instance, is that middle- and low-income nations in Africa lack access to contemporary pharmaceuticals. According to a survey conducted in 36 low- and middle-income countries by Cameron et al. [23], a significant portion of the population cannot afford traditional medications. The evidence from nine African nations may support this.

The World Health Organization has created a traditional medicine plan for the years 2014–2023 to improve the integration of herbal medicines into member nations' primary healthcare systems. The main objectives of the approach are to assess, control, incorporate, and maximize the potential of traditional medicine for people's health and welfare [214]. By 2023, the plan will have identified some of the problems harming traditional medicines and strategies to work with the relevant member states and other stakeholders to resolve them. Its continuous use, expanding economic significance on a worldwide scale, advancements in research and development, intellectual property rights, and integration with healthcare systems are among the problems [214]. Evaluation of both safety and efficacy, quality control, regulatory and safety monitoring, education, and training of health authorities and control agencies are some of the areas in which research is required [211]. Australia is an island continent that has been inhabited by Aboriginal people for thousands of years. The many ecofloristic zones, including wet tropics, savannahs, evergreen forests, shrublands, grasslands, and wetlands, have taught over 500 different clan groupings or countries to coexist peacefully [87, 117]. While there are many distinct tribal groups in these very diverse places, what unites them all is their close and deep connection to nature and their utilization of its resources [107]. Native plants are an important part of their culture and are used to manufacture a variety of products, including food, medicine, narcotics, stimulants, adornments, ceremonial artifacts, weapons, clothing, shelter, tools, and artwork [35, 186]. Using local flora, animals, and abiotic resources over thousands of years, this native wisdom has developed [155]. Aboriginal lore serves as a means of knowledge preservation and transmission in the absence of written language. Songs, storytelling, dance, and other forms of art have been used to pass down practices and stories from one generation to the next. There are still hundreds of Aboriginal people who speak their native language and maintain the knowledge, songs, and rituals of their ancestors despite the disruption of oral traditions and lore practices by colonial contact, with English being the second or third language [113, 159]. Several Aboriginal clan groupings continue to practice Aboriginal ethnomedicine, although the amount to which it is employed varies greatly across rural and urban areas of Australia and between communities [146].

Aboriginal cultures still use numerous plants such as bush food and bush medicine, particularly in isolated locations. This makes Aboriginal plant knowledge the oldest surviving pharmacopoeia [154]. Australia has a total of 1511 plant species that have been documented as being used medicinally [154, 204]. The Aboriginal groups in the Northern Territory, New South Wales, South Australia, and Western Australia are responsible for the majority of the medicinal plant knowledge that has been documented in the literature [179]. Just a few publications on Aboriginal medicinal plants of Cape York Peninsula have been made, and there hasn't been much written on Native medicinal plants of Queensland [122, 180]. Several Native medicinal plants continue to be either unreported in the literature or unavailable to the general population [179]. Like many indigenous knowledge throughout the globe, Australian Aboriginal knowledge has been underexplored in a "Western scientific" sense and is at grave risk of disappearing entirely [84]. According to Stack [183], because the European immigrants brought their ailments and utilized their traditional treatments, Aboriginal traditional medicine did not play a role in their lives 200 years ago. Nonetheless, there exist documents that demonstrate that certain doctors and botanists collaborated with Native healers and conducted experiments with local plants for therapeutic reasons throughout the early years of European arrival. For instance, Denis Considen, the first assistant surgeon to Surgeon-General John White in the first fleet (1788–1794), claimed to have discovered indigenous medicinal plants before any other European physicians; however, his methods of discovery are unknown, and it is unclear whether he enlisted the help of indigenous informants [108]. Macpherson and Considen [108] noted the effectiveness of various indigenous medicinal plants, such as native sarsaparilla (*Smilax glyciphylla*) as an anti-scorbutic and myrtle (*Eugenia australis*) and yellow gum (*Xanthorrhoea hastilis*) for dysentery [108]. In addition, local sarsaparilla was reportedly thought to be more pleasant than Jamaican or Central American varieties, according to Macpherson and Considen [108]. He said that Sydney herbalists had used it often as a trade item before 1927. Macadamia nuts were one of the plants that both the Aboriginal people and the European settlers traded regularly (*Macadamia* sp.). Notwithstanding the disruption of Native medical traditions by Europeans, Aboriginal people have made some significant contributions to global medical knowledge. For instance, Pearn [159] claims that Native childcare ethnomedicine is among "the world's oldest pediatric traditions" [159]. Native ethnomedical traditions are still a vibrant part of many Aboriginal communities today. Many groups in Australia employ various medicinal herbs depending on their local flora and habitat [146].

For instance, Southern Australia uses the fruits of the native plant *Solanum laciniatum*, whereas Eastern Australia once utilized *Solanum aviculare* (Kangaroo Apple) [183]. For joint swelling, both species were used as poultices [183].

The alkaloid solasodine, found in both of these *Solanum* species, serves as a precursor to cortisone and other steroids used in the creation of oral contraceptives (often known as "the pill") [112]. Due to their ability to biosynthesize this important phytochemical, these plants have been introduced to Russia and Eastern Europe, where they are now grown extensively [183]. A similar study was conducted on the congeneric *Duboisia myoporoides* when it was discovered that the native Australian Aboriginal narcotic plant pituri (*Duboisia hopwoodii*) was particularly useful. Aboriginals were chewing "pituri," a plant, in a manner like that of tobacco or East Indian betel, according to Joseph Banks, the first European botanist to visit the East Coast of Australia (in 1770) [46]. Ferdinand von Mueller, a botanist for the Victorian Government, recognized the plant as *Duboisia hopwoodii* a century later (in 1872) [165]. The related *Duboisia myoporoides* were discovered to generate hyoscine, now known as scopolamine, an alkaloid that is a very efficient remedy for motion sickness as reported by Von Mueller [63].

6 Research Needs in Herbal Medicine

There is a significant amount of work that needs to be done in the field of herbal medicines; however, this could be counterbalanced by the potential advantages of using these products, which are demonstrated by their consumption by significant populations, particularly in nations that are still developing. Be that as it may, it is necessary to research to ensure the identification of the specific compounds and bioactive elements contained within these herbal formulations [78, 79]. This will allow for the formulations to be purified and put to use treating specific diseases, as opposed to treating the disease with the whole plant [72, 73]. In addition, research is required to determine whether or not the use of extracted bioactive substances is preferable to the utilization of entire herbs. There is a significant need for additional studies to be conducted on the standardization procedures used to prepare herbal products. The evaluation of herbal medicine does not have a study approach that is considered adequate or generally acknowledged. It is necessary to have precise and repeatable procedures for the preparation of these medications to guarantee the distribution of the same herbal products despite differences in brand names and manufacturers. It is necessary to research to guarantee the safety of herbal goods in general. Further research is necessary to determine the processing methods, routes of administration, dosages, and compatibility with other herbal remedies or conventional treatments.

7 Conclusion

Herbal remedies have been used traditionally as medicine from the beginning of recorded human history. Plants have been used by humans in many parts of the world as a form of treatment and management for a wide variety of diseases and conditions. According to the World Health Organization (WHO) [206, 207], more than 80% of people all over the world utilize herbs as a kind of treatment when they are sick. Education is a crucial component in the usage of herbs since the capacity to utilize herbs as a therapy for diseases is based on the amount of information possessed by the native people of the area. Plants play a vital part in a wide variety of traditional religious rites, such as libation, sacrifice, the appeasing of the gods, invocation, and divination. In addition, plants are used in the production of food and the delivery of medicine. Plant parts are typically used in the manufacture of herbal treatments since these plant parts are the active elements in herbal medications. Some of the diseases and conditions that have been successfully managed by using herbal medicine include memory loss, cardiovascular disease, arthritis, osteoarthritis, digestive illness, respiratory illness, reproductive illness, skin illness, neurological illness, diabetes, cancer, hypertension, gastrointestinal illness, conditions such as pregnancy, and prenatal and postnatal procedures. Herbal therapy has also been shown to be effective in the treatment of a variety of other diseases and disorders. Herbal drugs are often administered either in powdered or liquid form, with the powdered form being the more prevalent of the two. Some of these products are intended to be used internally, while others can be put on the skin or rubbed into the part of the body that is problematic. There is still a significant amount of work that has to be done in the field of herbal medicines; however, this can be counterbalanced by the potential advantages that can be gained through the utilization of these items. Despite this, it is very necessary to conduct a study to guarantee the accurate identification of the particular chemical components as well as the physiologically active components that are found in these herbal treatments. There is a need for research to be conducted so that it can be determined whether or not the use of herbs in their unprocessed form, as opposed to the use of bioactive substances that have been extracted from the herbs, is more beneficial. There is a major need for additional studies to be conducted on the standardization procedures that are utilized during the manufacturing of herbal goods. This research should be conducted as soon as possible. To ascertain the processing methods, routes of administration, doses, and compatibility with other herbal medicines or conventional treatments, additional research is required.

References

1. Abaya ST, Orga, C, Raimi AG, Raimi MO, Kakwi, DJ (2023) Assessment of the implementation of electronic records management system in Bayelsa state. Lippincott® preprints. Preprint. https://doi.org/10.1097/preprints.22643812.v1
2. Abdullahi AA (2011) Trends and challenges of traditional medicine in Africa. Afri J Trad Comp Altern Med 8(5):115–123
3. Abubakar US, Osodi FA, Aliyu I, Jamila GA, Saidu SR, Fatima SS, Sani SI, Ahmad SA, Tsoken BG (2016) The use of traditional medicine among Bayero university community, Kano. J Med Plants Stud 4(6):23–25
4. Aigberua AO, Izah SC (2019) pH variation, mineral composition and selected trace metal concentration in some liquid herbal products sold in Nigeria. Int J Res Stud Biosci 7(1):14–21
5. Aigberua AO, Izah SC (2019) Macro nutrient and selected heavy metals in powered herbal medicine sold in Nigeria. Int J Med Plants Nat Prod 5(1):23–29

6. Al-Harbi MM, Qureshi S, Ahmed MM, Raza M, Baig MZA, Shah AH (1996) Effect of camel urine on the cytological and biochemical changes induced by cyclophosphamide in mice. J Ethnopharmacol 52(3):129–137
7. Alonso-Castro AJ, Dominguez F, Maldonado-Miranda JJ, Pérez LJC, Carranza-Álvarez C et al (2017) Use of medicinal plants by health professionals in Mexico. J Ethnopharmacol 198:81–86
8. Allensbach Survey (2017) http://www.ifd-allensbach.de/uploads/tx_studies/7528_Naturheilmittel_2010.pdf. Last accessed 12 Jan 2017
9. American herbal Pharmacopoeia AHP (2011) Botanical Pharmacognosy microscopic characterization of botanical medicines. CRC press, New York
10. Aone Mokaila (2001) http://www.blackherbals.com/atcNewsletter913.pdf
11. Arber A (1953) Herbals, their origin and evolution. A chapter in the history of botany 1470–1670. Cambridge University Press, Cambridge, MA
12. Astana (2019) Global conference on primary health care. https://www.who.int/docs/default-source/primary-health/declaration/gcphc-declaration.pdf?ua=1. Accessed 28 Feb 2023
13. Atanasov AG, Waltenberger B, Pferschy-Wenzig EM, Linder T, Wawrosch C, Uhrin P, Temml V, Wang L, Schwaiger S, Heiss EH, Rollinger JM (2015) Discovery and resupply of pharmacologically active plant-derived natural products: a review. Biotechnol Adv 33(8): 1582–1614
14. Azaizeh H, Saad B, Cooper E, Said O (2010) Traditional Arabic and Islamic medicine, a re-emerging health aid. Evid Based Complement Alternat Med 7(4):419–424
15. Azaizeh H, Saad B, Khalil K, Said O (2006) The state of the art of traditional Arab herbalmedicine in the eastern region of the Mediterranean: a review. Evid Based Complement Alternat Med 3(2):229–235
16. Barnes PM, Bloom B, Nahin R (2007) Complementary and alternative medicine use among adults and children: United States. CDC National Health Statistics Report
17. Bassey SE, Izah SC (2017) Nigerian plants with insecticidal potentials against various stages of mosquito development. ASIO J Med Health Sci Res 2(1):07–18
18. Bhayro S, Hawley R, Kessel G, Pormann PE (2013) The syriac galen palimpsest: progress, prospects and problems. J Semit Stud 58:131–148
19. Booth M (1998) Opium. St Martin's Press. New York. p 15
20. BreatHed JH (1930) The Edwin Smith surgical papyrus. University of Chicago, University of Chicago Press
21. Brendler T, Eloff LN, Gurib-Fakim A, Phillips LD (2010) African herbal Pharmacopeia. AAMPS Publishing, Mauritius
22. Bryan PW (1930) The Papyrus Ebers (Geoffrey Bles: London) Crystalinks website. Ancient Egyptian Medicine – Smith Papyrus – Ebers Papyrus. http://www.crystalinks.com/egyptmedicine.htm. Accessed 11 Oct 2022
23. Cameron A, Even M, Ross Degnan D, Bail D, Laing R (2008) Medicine prices availability and affordability in 36 developing middle income countries, vol 373. A secondary analysis, Geneva WHO, p 240
24. Capasso L (1998) 5300 years ago the ice man used natural laxatives and antibiotics. Lancet 352:1864
25. Cavender A (2006) Folk medical uses of plant foods in southern Appalachia, United States. J Ethnopharmacol 108:74–84
26. Chan M (2013) Speech by WHO director general Dr. Margaret Chan at the international conference on traditional medicine for South East Asian countries, New Delhi India, pp 12–15
27. Chatterjee SJ, Pandey S (2011) Chemo-resistant melanoma sensitized by tamoxifen to low dose curcumin treatment through induction of apoptosis and autophagy. Cancer Biol Ther 11(2):216–228
28. Chatora R (2003) An overview of the traditional medicine situation in the African region. Afr Health Monit 4:4–7
29. Chavarría A, Espinosa G (1970) Cruz-Badiano codex and the importance of the Mexican medicinal plants. J Pharm Technol Res Manag 7:15–22

30. Che CT, George V, Ijinu TP, Pushpangadan P, Andrae-Marobela K (2017) Traditional medicine. In Pharmacognosy. Academic Press, pp 15–30
31. Che E, Gao Y, Wan L, Zhang Y, Han N, Bai J, Li J, Sha Z, Wang S (2015) Paclitaxel/gelatin coated magnetic mesoporous silica nanoparticles: preparation and antitumor efficacy in vivo. Microporous Mesoporous Mater 204:226–234
32. Chen X, Dang TTT, Facchini PJ (2015) Noscapine comes of age. Phytochemistry 111:7–13
33. Chinese Pharmacopeia (1992) The state pharmacopoeia commission of People's Republic of China. Beijing, China
34. Chinese Pharmacopeia (2005) Pharmacopoeia of the People's Republic of China. Chemical Industry Press, Beijing, pp 7–117
35. Clarke PA (2007) Aboriginal people and their plants. Rosenberg Publishing Pty Ltd, Dural Delivery Centre, NSW
36. CONABIO (2020) Medicinal plants; National Commission for the knowledge and use of biodiversity: Mexico City, Mexico. https://www.biodiversidad.gob.mx/diversidad/medicinal/plantas. Accessed 30 Jan 2023
37. Daniyal M, Akram M, Zainab R, Munir N, Shah SM, Liu B, Wang W, Riaz M, Jabeen F (2019) Progress and prospects in the management of psoriasis and developments in phyto-therapeutic modalities. Dermatol Ther 32(3):e12866
38. David B, Wolfender JL, Dias DA (2015) The pharmaceutical industry and natural products: historical status and new trends. Phytochem Rev 14:299–315
39. Dickason OP, Newbigging W (2010) A concise history of Canada's first nations, 2nd edn. Oxford University Press, New York, pp 1–180
40. Erezina AE, Raimi AG, Emmanuel OO, Raimi MO, Abaya ST, Kakwi DJ (2023) Level of professional awareness among health record officers in Bayelsa state and their implications for patient care, health systems, and health policy. Lippincott® preprints. Preprint. https://doi.org/10.1097/preprints.22637689.v1
41. Emboden W (1979) Narcotic plants. Macmillan Publishing Co, New York, p 50
42. Enaregha EB, Izah SC, Okiriya Q (2021) Antibacterial properties of *Tetrapleura tetraptera* pod against some pathogens. Res Rev Insights 5:1–5. https://doi.org/10.15761/RRI.1000165
43. Epidi JO, Izah SC, Ohimain EI, Epidi TT (2016) Phytochemical, antibacterial and synergistic potency of tissues of *Vitex grandifolia*. Biotechnol Res 2(2):69–76
44. Epidi JO, Izah SC, Ohimain EI (2016) Antibacterial and synergistic efficacy of extracts of *Alstonia boonei* tissues. Br J Appl Res 1(1):0021–0026
45. Ezeome ER, Anarado AN (2007) Use of complementary and alternative medicine by cancer patients at the University of Nigeria Teaching Hospital, Enugu, Nigeria. BMC Complement Altern Med 7:28
46. Foley P (2006) *Duboisia myoporoides*: the medical career of a native Australian plant. Hist Rec Aust Sci 17(1):31–69
47. Fortune Business Insights (2021) The global herbal medicine market is projected to grow from $165.66 billion in 2022 to $347.50 billion by 2029, at a CAGR of 11.16% in forecast period. https://www.fortunebusinessinsights.com/herbal-medicine-market-106320. Accessed 28 Feb 2023
48. Fridlender M, Kapulnik Y, Koltai H (2015) Plant derived substances with anti-cancer activity: from folklore to practice. Front Plant Sci 6:799
49. Galabuzi C, Agea J, Fungo B, Kamoga R (2010) Traditional medicine as an alternative form of health care system: a preliminary case study of Nangabo sub-county, Central Uganda. Afr J Tradit Complement Altern Med 7(1). https://doi.org/10.4314/ajtcam.v7i1.57224
50. Geck MS, Cristians S, Berger-González M, Casu L, Heinrich M, Leonti M (2020) Traditional herbal medicine in Mesoamerica: toward its evidence base for improving universal health coverage. Front Pharmacol 11:1160
51. Ghalioungui P (1975) Les plus anciennes femmes-médecins de l'histoire. BIFAO 75:159–164

52. Government of Canada (2022) Crown-indigenous relations and northern affairs Canada. A history of treaty-making in Canada. https://www.rcaanc-cirnac.gc.ca/eng/1314977704533/1544620451420. Accessed 10 Jan 2023
53. Goyal M, Sasmal D, Nagori BP (2012) Ayurveda the ancient science of healing: an insight. In: Vallisuta O, Olimat SM (eds) Drug discovery research in Pharmacognosy. InTech, Croatia, pp 1–10
54. Gu S, Pei J (2017) Innovating Chinese herbal medicine: from traditional health practice to scientific drug discovery. Front Pharmacol 8:381
55. Gupta KK, Bharne SS, Rathinasamy K, Naik NR, Panda D (2006) Dietary antioxidant curcumin inhibits microtubule assembly through tubulin binding. FEBS J 273(23):5320–5332
56. Gurib-Fakim A (2006) Medicinal plants: traditions of yesterday and drugs of tomorrow. Mol Asp Med 27(1):1–93
57. Gurib-Fakim A, Brendler T, Phillips LD, Eloff LN (2010) Green gold – success stories using southern African plant species. AAMPS Publishing, Mauritius
58. Gurib-Fakim A, Mahomoodally MF (2013) African flora as potential sources of medicinal plants: towards the chemotherapy of major parasitic and other infectious diseases – a review. Jordan J Biol Sci 6:77–84
59. Hartmann T (2007) From waste products to ecochemicals: fifty years research of plant secondary metabolism. Phytochemical 68:2831–2846
60. Haruki Y, Kampo Y (2004) Medicine and chemotherapy (published in Japanese) Nippon Kagaku Ryoho Gakkai Zasshi 52: 547–555
61. Havermann S, Humpf HU, Wätjen W (2016) Baicalein modulates stress-resistance and life span in C. elegans via SKN-1 but not DAF-16. Fitoterapia 113:123–127
62. Heinrich M, Sharma SK, Suetterle U, Bhamra SK (2023) Herbal medicine use in the UK and Germany and pharmacy practice-a commentary. Res Soc Adm Pharm 19(3):535–540
63. Hines HJG (1947) T.L. Bancroft memorial lecture. Proc R Soc Qld 57:75–78
64. Holtze S, Gorshkova E, Braude S, Cellerino A, Dammann P, Hildebrandt TB et al (2021) Alternative animal models of aging research. Front Mol Biosci 8:660959
65. Hort A (1977) Theophrastus enquiry into plants – and minor works on Odours and weather signs, vol II. Harvard University Press, Cambridge, MA
66. Huang X, Li N, Pu Y, Zhang T, Wang B (2019) Neuroprotective effects of ginseng phytochemicals: recent perspectives. Molecules 24(16):2939
67. IMS Health OTC®Report (EVP) (2015). https://www.imshealth.com/files/web/Germany/Marktbericht/IMS_Pharmamarktbericht_Juli_2015.pdf. Accessed 12 Jan 2017
68. Instituto Nacional de Antropología e Historia (INAH) (1552) Martín de la Cruz-Badiano 1552. https://www.codices.inah.gob.mx/pc/contenido.php?id=12. Accessed 30 Jan 2023
69. Instituto Nacional de Estadística y Geografía Population Census (2015) https://www.inegi.org.mx/programas/intercensal/2015/#Tabulados. Accessed 12 Jan 2020
70. Izah SC, Aigberua AO, Richard G (2022) Concentration, source, and health risk of trace metals in some liquid herbal medicine sold in Nigeria. Biol Trace Elem Res 200:3009–3302
71. Izah SC, Chandel SS, Etim NG, Epidi JO, Venkatachalam T, Devaliya R (2019) Potency of unripe and ripe express extracts of long pepper (Capsicum frutescens var. baccatum) against some common pathogens. Int J Pharm Phytopharmacol vRes 9(2):56–70
72. Izah SC, Etim NG, Ilerhunmwuwa IA, Silas G (2019) Evaluation of crude and ethanolic extracts of Capsicum frutescens var. minima fruit against some common bacterial pathogens. Int J Complement Altern Med 12(3):105–108
73. Izah SC, Etim NG, Ilerhunmwuwa IA, Ibibo TD, Udumo JJ (2019) Activities of express extracts of Costus afer Ker–Gawl. [Family COSTACEAE] against selected bacterial isolates. Int J Pharm Phytopharmacol Res 9(4):39–44
74. Izah SC, Chandel SS, Epidi JO, Venkatachalam T, Devaliya R (2019) Biocontrol of Anopheles gambiae larvae using fresh ripe and unripe fruit extracts of Capsicum frutescens var. baccatum. Int J Green Pharm 13(4):338–342

75. Izah SC, Youkparigha FO (2019) Larvicidal activity of fresh aqueous and Ethanolic extracts of Cymbopogon citratus (DC) Stapf on malaria vector, Anopheles gambiae. BAOJ Biotech 5:040
76. Izah SC (2019) Activities of crude, acetone and ethanolic extracts of *Capsicum frutescens* var minima fruit against larvae of anopheles gambiae. J Environ Treat Tech 7(2):196–200
77. Izah SC, Uhunmwangho EJ, Dunga KE, Kigigha LT (2018) Synergy of methanolic leave and stem-back extract of *Anacardium occidentale L.* (cashew) against some enteric and superficial bacteria pathogens. MOJ Toxicol 4(3):209–211
78. Izah SC, Uhunmwangho EJ, Dunga KE (2018) Studies on the synergistic effectiveness of methanolic extract of leaves and roots of *Carica papaya L.* (papaya) against some bacteria pathogens. Int J Complement Altern Med 11(6):375–378
79. Izah SC, Uhunmwangho EJ, Etim NG (2018) Antibacterial and synergistic potency of methanolic leaf extracts of *Vernonia amygdalina L.* and *Ocimum gratissimum L.* J Basic Pharmacol Toxicol 2(1):8–12
80. Izah SC, Zige DV, Alagoa KJ, Uhunmwangho EJ, Iyamu AO (2018) Antibacterial efficacy of aqueous extract of *Myristica fragrans* (common nutmeg). EC Pharmacol Toxicol 6(4):291–295
81. Izah SC, Uhunmwangho EJ, Eledo BO (2018) Medicinal potentials of *Buchholzia coriacea* (wonderful kola). J Med Plant Res 8(5):27–43
82. Izah SC, Aseibai ER (2018) Antibacterial and synergistic activities of Methanolic leaves extract of lemon grass (Cymbopogon citratus) and rhizomes of ginger (Zingiber officinale) against Escherichia coli, Staphylococcus aureus and Bacillus subtilis. Acta Sci Microbiol 1(6): 26–30
83. Izah SC (2018) Some determinant factors of antimicrobial susceptibility pattern of plant extracts. Res Rev Insight 2(3):1–4
84. Jamie JF (2020) Macquarie-Yaegl partnership: community capability strengthening through Western and indigenous science. Aust J Chem 74(1):28–33
85. Jamshidi-Kia F, Lorigooini Z, Amini-Khoei H (2018) Medicinal plants: past history and future perspective. J Herbmed Pharmacol 7(1):1
86. Jenke-Kodama H, Müller R, Dittmann E (2008) Evolutionary mechanisms underlying secondary metabolite diversity. Progress Drug Res 119:121–140
87. Keith DA (2017) Australian vegetation. Cambridge University Press, Cambridge
88. Kirmayer L, Simpson C, Cargo M (2003) Healing traditions: culture, community and mental health promotion with Canadian aboriginal peoples. Australas Psychiatry 11:S15–S23
89. Kigigha LT, Izah SC, Uhunmwangho EJ (2018) Assessment of hot water and ethanolic leaf extracts of *Cymbopogon citratus* Stapf (lemon grass) against selected bacteria pathogens. Ann Microbiol Infect Dis 1(3):1–5
90. Kigigha LT, Selekere RE, Izah SC (2018) Antibacterial and synergistic efficacy of acetone extracts of Garcinia kola (bitter kola) and Buchholzia coriacea (wonderful kola). J Basic Pharmacol Toxicol 2(1):13–17
91. Kigigha LT, Biye SE, Izah SC (2016) Phytochemical and antibacterial activities of *Musanga cecropioides* tissues against *Escherichia coli, Pseudomonas aeruginosa Staphylococcus aureus, Proteus* and *bacillus* species. Int J Appl Res Technol 5(1):100–107
92. Kigigha LT, Izah SC, Okitah LB (2016) Antibacterial activity of palm wine against *pseudomonas, bacillus, staphylococcus, Escherichia,* and *Proteus* spp. point. J Bot Microbiol Res 2(1):046–052
93. Kigigha LT, Apreala A, Izah SC (2016) Effect of cooking on the climbing pepper (*Piper nigrum*) on antibacterial activity. J Environ Treat Tech 4(1):6–9
94. Koreeman P, Baars EW (2012) Patients whose general practitioner (GP) knows complementary medicine tend to have lower costs and live longer. Eur J Health Econ 13(6):769–776
95. Kumar N, Goel N (2019) Phenolic acids: natural versatile molecules with promising therapeutic applications. Biotechnol Rep 24:e00370
96. Kumar S, Pathania AS, Saxena AK, Vishwakarma RA, Ali A, Bhushan S (2013) The anticancer potential of flavonoids isolated from the stem bark of Erythrina suberosa through

induction of apoptosis and inhibition of STAT signaling pathway in human leukemia HL-60 cells. Chem Biol Interact 205(2):128–137

97. Lashmanova E, Zemskaya N, Proshkina E, Kudryavtseva A, Volosnikova M, Marusich E et al (2017) The evaluation of geroprotective effects of selected flavonoids in *Drosophila melanogaster* and *Caenorhabditis elegans*. Front Pharmacol 8:884
98. Lawrence B (2012) Survivance, identity, and the Indian act. In: Burnett K, Read G (eds) Aboriginal history a reader, 1st edn. Oxford University Press, Don Mills, pp 195–206
99. Leonti M, Cabras S, Weckerle CS, Solinas MN, Casu L (2010) The causal dependence of present plant knowledge on herbals – contemporary medicinal plant use in Campania (Italy) compared to Matthioli (1568). J Ethnopharmacol 130:379–391
100. Leonti M, Staub PO, Cabras S, Castellanos ME, Casu L (2015) From cumulative cultural transmission to evidence-based medicine: evolution of medicinal plant knowledge in southern Italy. Front Pharmacol 6:207
101. Lev E, Amar Z (2002) Ethnopharmacological survey of traditional drugs sold in the kingdom of Jordan. J Ethnopharmacol 82(2–3):131–145
102. Lewal OA, Banjo AD (2007) A survey of usage of arthropods in traditional medicine in south West Nigeria. J Entomol 4(2):104–112
103. Li L (2000) Opportunity and challenge of traditional Chinese medicine in face of the entrance to WTO (World Trade Organization). Chin Inform Trad Chin Med 7:7–8. (in Chinese)
104. Li JWH, Vederas JC (2009) Drug discovery and natural products: end of an era or an endless frontier? Science 325:161–165
105. Liao VHC, Yu CW, Chu YJ, Li WH, Hsieh YC, Wang TT (2011) Curcumin-mediated lifespan extension in Caenorhabditis elegans. Mech Ageing Dev 132(10):480–487
106. Lin C, Xiao J, Xi Y, Zhang X, Zhong Q, Zheng H et al (2019) Rosmarinic acid improved antioxidant properties and healthspan via the IIS and MAPK pathways in Caenorhabditis elegans. Biofactors 45(5):774–787
107. Maclean K, Woodward E, Jarvis D, Turpin G, Rowland D, Rist P (2022) Decolonising knowledge co-production: examining the role of positionality and partnerships to support indigenous-led bush product enterprises in northern Australia. Sustain Sci 17:333–350
108. MacPherson J, Considen (1927) Dennis Considen, assistant surgeon of the first fleet. Med J Aust 2(23):770–773
109. Manach C, Scalbert A, Morand C, R'em'esy C, Jim'enez L (2004) Polyphenols: food sources and bioavailability. Am J Clin Nutr 79(5):727–747
110. Mangla A, Agarwal N (2022) Clinical practice issues in American Indians and Alaska natives. [updated 2022 Nov 30]. In: StatPearls [internet]. Treasure Island: StatPearls Publishing; 2023 Jan https://www.ncbi.nlm.nih.gov/books/NBK570601/
111. Mann RD (1984) Modern drug use: an enquiry on historical principles. MTP Press Limited, Lancaster
112. Manosroi J, Manosroi A, Sripalakit P (2005) Extraction of solasodine from dry fruits and leaves of *Solanum laciniatum* Ait and the synthesis of-16-dehydropregnenolone acetate from solasodine by phase-transfer catalysis. Acta Hortic 679:105–111
113. Marmion D, Obata K, Troy J (2014) Community, identity, wellbeing: the report of the second National Indigenous Languages Survey. Australian Institute of Aboriginal and Torres Strait Islander Studies, Canberra
114. Mayer JG (2011) Die Wahrheit uber den Gart der gesundheit (1485) und sein Weiterleben in den Krauterbuchern der fruhen Neuzeit. In: Anagnostou S, Egmond F, Friedrich C (eds) A passion for plants: *materia medica* and botany in scientific networks from the 16th to 18th centuries, Quellen und Studien zur geschichte der Pharmazie, vol 95, Stuttgart, pp 119–128
115. Medicine in Europe and the United States (MEUS) (2013) The medieval world. http://science.jrank.org/pages/10139/Medicinein-Europe-United-States-Medieval-World.html
116. Medicine in the Medieval Islamic world (MMIW) (2013) http://en.wikipedia.org/wiki/Medicine in the medieval Islamic world

117. Metcalfe D, Bui E (2016) Land: vegetation. In: Australia state of the environment 2016, Australian Government Department of the Environment and energy, Canberra, 2016. Accessed from: https://soe.environment.gov.au/theme/land/topic/2016/veget ation-0. https://doi.org/10.4226/94/58b6585f94911
118. Mitchell SJ, Scheibye-Knudsen M, Longo DL, de Cabo R (2015) Animal models of aging research: implications for human aging and age-related diseases. Annu Rev Anim Biosci 3(1): 283–303
119. Morufu OR, Aziba-anyam GR, Teddy CA (2021) 'Silent Pandemic': evidence-based environmental and public health practices to respond to the Covid-19 crisis. IntechOpen. https://doi.org/10.5772/intechopen.100204. ISBN 978-1-83969-144-7. https://www.intechopen.com/online-first/silent-pandemic-evidence-based-environmental-and-public-health-practices-to-respond-to-the-covid-19-Published: December 1st 2021; ISBN: 978-1-83969-144-7; Print ISBN: 978–1–83969-143-0; eBook (PDF) ISBN: 978-1-83969-145-4. Copyright year: 2021
120. Mukherjee PK (2001) Evaluation of Indian traditional medicine. Drug Inf J 35(2):623–632
121. Muller I (2011) Gart der Gesundheit. Botanik im Buchdruck von den Anfangen bis 1800. Verlag der Franckeschen Stiftungen zu Halle, Schweinfurt
122. Ndi CP, Sykes MJ, Claudie DJ, McKinnon RA, Semple SJ, Simpson BS (2016) Antiproliferative aporphine alkaloids from *Litsea glutinosa* and ethnopharmacological relevance to Kuuku I'yu traditional medicine. Aust J Chem 69(2):145–151
123. Newman DJ, Cragg GM, Snader KM (2000) The influence of natural products upon drug discovery. Nat Prod Rep 17(3):215–234
124. Nworu CS, Udeogaranya PO, Okafor CK, Adikwu AO, Akah PA (2015) Perception, usage and knowledge of herbal medicines by students and academic staff of University of Nigeria: a survey. Eur J Integr Med 7(3):218–227
125. Ogidi OI (2020) A review on the use of herbal remedies and clinical therapeutics for the management of Covid-19 pandemic. ASIO J Pharm Herb Med Res 6(2):68–77
126. Ogidi OI (2023) Sustainable utilization of important medicinal plants in Africa in: sustainable utilization and conservation of Africa's biological resources and environment, Sustainable development and biodiversity. Springer Nature Singapore Pte Ltd. https://doi.org/10.1007/978-981-19-6974-4_12
127. Ogidi OI, Enenebeaku UE (2023) Medicinal potentials of aloe Vera (*Aloe barbadensis* miller): technologies for the production of therapeutics in: sustainable utilization and conservation of Africa's biological resources and environment, sustainable development and biodiversity. Springer Nature Singapore Pte Ltd. https://doi.org/10.1007/978-981-19-6974-4_11
128. Ogwu MC (2009) The significance and contribution of conservation education in Nigeria: an appraisal and a call for improvement. In: Babalola F (ed) Proceedings of the maiden seminar of Nigerian Tropical Biology Association. Nigerian Tropical Biology Association and Tropical Biology Association, pp 109–114
129. Ogwu MC (2010) Conserving biodiversity even in poverty: the African experience. In: Babalola F (ed) Seminar proceedings of Nigerian Tropical Biology Association. Nigerian Tropical Biology Association and Tropical Biology Association, pp 112–117
130. Ogwu MC (2019) Towards sustainable development in Africa: the challenge of urbanization and climate change adaptation. In: Cobbinah PB, Addaney M (eds) The geography of climate change adaptation in urban Africa. Springer Nature, Switzerland, pp 29–55. https://doi.org/10.1007/978-3-030-04873-0_2
131. Ogwu MC (2020) Value of *Amaranthus* [L.] species in Nigeria. In: Waisundara V (ed) Nutritional value of Amaranth. IntechOpen, UK, pp 1– 21. https://doi.org/10.5772/intechopen.86990
132. Ogwu MC (2023) Local food crops in Africa: sustainable utilization, threats, and traditional storage strategies. In: Izah SC, Ogwu MC (eds) Sustainable utilization and conservation of Africa's biological resources and environment, Sustainable development and biodiversity, vol 888. Springer, Singapore, pp 353–376. https://doi.org/10.1007/978-981-19-6974-4_13

133. Ogwu MC, Osawaru ME (2022) Traditional methods of plant conservation for sustainable utilization and development. In: Izah SC (ed) Biodiversity in Africa: potentials, threats and conservation, Sustainable development and biodiversity, vol 29. Springer, Singapore, pp 451–472. https://doi.org/10.1007/978-981-19-3326-4_17
134. Ogwu MC, Ahana CM, Osawaru ME (2018) Sustainable food production in Nigeria: a case study for Bambara groundnut (*Vigna subterranean* (L.) Verdc. Fabaceae). J Energy Nat Resour Manag 1:68–77
135. Ogwu MC, Chime AO, Oseh OM (2018) Ethnobotanical survey of tomato in some cultivated regions in southern Nigeria. Maldives Natl Res J 6(1):19–29
136. Ogwu MC, Izah SC, Iyiola AO (2022) An overview of the potentials, threats and conservation of biodiversity in Africa. In: Izah SC (ed) Biodiversity in Africa: potentials, Threats and conservation, Sustainable development and biodiversity, vol 29. Springer, Singapore, pp 3–20. https://doi.org/10.1007/978-981-19-3326-4_1
137. Ogwu MC, Osawaru ME, Ahana CM (2014) Challenges in conserving and utilizing plant genetic resources (PGR). Int J Genet Mol Biol 6(2):16–22. https://doi.org/10.5897/IJGMB2013.0083
138. Ogwu MC, Osawaru ME, Chime AO (2015) A contemporary approach to the history of plant taxonomy. In: Amusa TO, Babalola FD (eds). Proceedings of the fifth biodiversity conference of Nigeria Tropical Biology Association. Nigerian Tropical Biology Association and Tropical Biology Association. pp. 110–117
139. Ogwu MC, Osawaru ME, Obahiagbon GE (2017) Ethnobotanical survey of medicinal plants used for traditional reproductive care by Usen people of Edo state, Nigeria. Malaya J Biosci 4 (1):17–29
140. Ogwu MC, Osawaru ME, Owie MO (2023) Effects of storage at room temperature on the food components of three cocoyam species (*Colocasia esculenta, Xanthosoma atrovirens, and X. sagittifolium*). Food Stud 13(2):59–83. https://doi.org/10.18848/2160-1933/CGP/v13i02/59-83
141. Ogwu MC, Osawaru ME, Aiwansoba RO, Iroh RN (2016) Status and prospects of vegetables in Africa. In: Borokini IT, Babalola FD (eds) Conference proceedings of the joint biodiversity conservation conference of Nigeria Tropical Biology Association and Nigeria chapter of Society for conservation biology on MDGs to SDGs: toward sustainable biodiversity conservation in Nigeria. University of Ilorin, Nigeria, pp 47–57
142. Ogwu MC, Osawaru ME, Amodu E, Osamo F (2023) Comparative morphology, anatomy and chemotaxonomy of two *Cissus* Linn. Species. Rev Bras Bot:1. https://doi.org/10.1007/s40415-023-00881-0
143. Okaiyeto K, Oguntibeju OO (2021) African herbal medicines: adverse effects and cytotoxic potentials with different therapeutic applications. Int J Environ Res Public Health 18(11):5988. https://doi.org/10.3390/ijerph18115988
144. Okosodo EF, Ogidi OI (2023) Biodiversity conservation strategies and sustainability. In: Sustainable utilization and conservation of Africa's biological resources and environment, sustainable development and biodiversity. Springer Nature Singapore Pte Ltd
145. Olalekan RM, Omidiji AO, Williams EA, Christianah MB, Modupe O (2019) The roles of all tiers of government and development partners in environmental conservation of natural resource: a case study in Nigeria. MOJ Ecol Environ Sci 4(3):114–121. https://doi.org/10.15406/mojes.2019.04.00142
146. Oliver SJ (2013) The role of traditional medicine practice in primary health care within aboriginal Australia: a review of the literature. J Ethnobiol Ethnomed 9:46
147. Omoigui DO, Osawaru ME, Aiwansoba RO, Ogwu MC (2016) Morphological and phytodermological evaluation of Oka-Uselu (maize- *Zea mays* L.). Appl Trop Agric 21(3):96–101
148. Onyambu MO, Gikonyo NK, Nyambaka HN, Thoithi GN (2019) A review of trends in herbal drugs standardization, regulation and integration to the National Healthcare Systems in Kenya and the globe. Int J Pharmacogn Chinese Med 3(3):1–13
149. Osawaru ME, Ogwu MC (2014) Conservation and utilization of plant genetic resources. In: Omokhafe K, Odewale J (eds) Proceedings of 38th annual conference of The Genetics Society of Nigeria. Empress Prints Nigeria Limited, pp 105–119

150. Osawaru ME, Ogwu MC (2014) Ethnobotany and germplasm collection of two genera of cocoyam (*Colocasia* [Schott] and *Xanthosoma* [Schott], Araceae) in Edo state Nigeria. Sci Technol Arts Res J 3(3):23–28. https://doi.org/10.4314/star.v3i3.4
151. Osawaru ME, Ogwu MC (2020) Survey of plant and plant products in local markets within Benin City and environs. In: Filho LW, Ogugu N, Ayal D, Adelake L, da Silva I (eds) African handbook of climate change adaptation. Springer Nature, Switzerland, pp 1–24. https://doi.org/10.1007/978-3-030-42091-8_159-1
152. Osawaru ME, Ogwu MC, Omoigui ID, Aiwansoba RO, Kevin A (2016) Ethnobotanical survey of vegetables eaten by Akwa Ibom people residing in Benin City, Nigeria. Uni Benin J Sci Technol 4(1):70–93
153. Ozioma JE-O, Chinwe OAN (2019) Herbal medicines in African traditional medicine. IntechOpen. https://doi.org/10.5772/intechopen.80348
154. Packer J, Gaikwad J, Harrington D, Ranganathan S, Jamie J, Vemulpad S (2012) Medicinal plants of New South Wales, Australia. In: Medicinal plants, genetic resources, chromosome engineering, and crop improvement, vol 6. CRC Press, Boca Raton, pp 259–296
155. Packer J, Turpin G, Ens E, Venkataya B, Hunter J et al (2019) Building partnerships for linking biomedical science with traditional knowledge of customary medicines: a case study with two Australian indigenous communities. J Ethnobiol Ethnomed 15(1):69
156. Parker H (1907) Folk-Lore of the North Carolina mountaineers. J Am Folk 20:241
157. Patridge E, Gareiss P, Kinch MS, Hoyer D (2016) An analysis of FDA-approved drugs: natural products and their derivatives. Drug Discov Today 21(2):204–207
158. Pawar AP, Vinugala D, Bothiraja C (2014) Nanocochleates derived from nanoliposomes for paclitaxel oral use: preparation, characterization, in vitro anticancer testing, bioavailability and biodistribution study in rats. Biomed Pharmacother 3502:1–9
159. Pearn J (2005) The world's longest surviving paediatric practices: some themes of aboriginal medical ethnobotany in Australia. J Paediatr Child Health 41:284–290
160. Qazi MA, Molvi K (2016) Herbal medicine: a comprehensive review. J Pharm Res 8(2):1–5
161. Raimi MO, Abiola OS, Atoyebi B, Okon GO, Popoola AT, Amuda-KA OL, Austin-AI & Mercy T (2022) The challenges and conservation strategies of biodiversity: the role of government and non-governmental organization for action and results on the ground. In: Chibueze Izah S (ed) Biodiversity in Africa: potentials, threats, and conservation. Sustainable development and biodiversity, vol 29. Springer, Singapore. https://doi.org/10.1007/978-981-19-3326-4_18
162. Raimi MO, Austin-AI OHS, Abiola OS, Abinotami WE, Ruth EE, Nimisingha DS, Walter BO (2022) Leaving no one behind: impact of soil pollution on biodiversity in the global south: a global call for action. In: Chibueze Izah S (ed) Biodiversity in Africa: potentials, threats and conservation. Sustainable development and biodiversity, vol 29. Springer, Singapore. https://doi.org/10.1007/978-981-19-3326-4_8
163. Raimi MO, Mcfubara KG, Abisoye OS, Ifeanyichukwu EC, Henry SO, Raimi GA (2021) Responding to the call through translating science into impact: building an evidence-based approaches to effectively curb public health emergencies [COVID-19 crisis]. Global J Epidemiol Infect Dis 1(1). https://doi.org/10.31586/gjeid.2021.010102. Retrieved from https://www.scipublications.com/journal/index.php/gjeid/article/view/72
164. Rashrash M, Schommer JC, Brown LM (2017) Prevalence and predictors of herbal medicine use among adults in the United States. J Patient Exp 4(3):108–113
165. Ratsch A, Steadman KJ, Bogossian F (2010) The pituri story: a review of the historical literature surrounding traditional Australian aboriginal use of nicotine in Central Australia. J Ethnobiol Ethnomed 6:26
166. Riddle JM (1985) Dioscorides on pharmacy and medicine. University of Texas Press, Austin
167. Robbers JE, Speedie MK, Tyler VE (1996) Pharmacognosy and Pharmacobiotechnology. Williams & Wilkins, Baltimore, pp 1–14
168. Rosalie D, Patricia LZ (2008) Egyptian mummies and modern science. Cambridge University Press, Cambridge
169. Russell GA (2010) Chapter 6: after Galen late antiquity and the Islamic world. Handb Clin Neurol 95:61–77

170. Russell PL (2015) The essential history of Mexico. From pre-conquest to present, 1st edn. Routledge Taylor & Francis Group, New York, pp 1–93
171. Saad B, Azaizeh H, Said O (2005) Tradition and perspectives of Arab herbal medicine: a review. Evid Based Complement Alternat Med 2(4):475–479
172. Said O, Khalil K, Fulder S, Azaizeh H (2002) Ethnopharmacological survey of medicinal herbs in Israel, the Golan Heights and the West Bank region. J Ethnopharmacol 83(3):251–265
173. Saito H (2000) Regulation of herbal medicines in Japan. Pharmacol Regul 41:515–519
174. Saliu AO, Komolafe OO, Bamidele CO, Raimi MO (2023) The value of biodiversity to sustainable development in Africa. In: Izah SC, Ogwu MC (eds) Sustainable utilization and conservation of Africa's biological resources and environment, Sustainable development and biodiversity, vol 888. Springer, Singapore. https://doi.org/10.1007/978-981-19-6974-4_10
175. Schulz V, HaÈnsel R, Tyler VE (1998) Rational phytotherapy, 3rd edn. Springer Berlin, p 306
176. Seiyaboh EI, Seiyaboh Z, Izah SC (2020) Environmental control of mosquitoes: a case study of the effect of *Mangifera Indica* root-bark extracts (family Anacardiaceae) on the larvae of *Anopheles gambiae*. Ann Ecol Environ Sci 4(1):33–38
177. Seiyaboh EI, Odubo TC, Izah SC (2020) Larvicidal activity of *Tetrapleura tetraptera* (Schum and Thonn) Taubert (Mimosaceae) extracts against *Anopheles gambiae*. Int J Adv Res Microbiol Immunol 2(1):20–25
178. Shah U, Shah R, Acharya S, Acharya N (2013) Novel anticancer agents from plant sources. Chin J Nat Med 11(1):16–23
179. Simpson B, Claudie D, Smith NM, McKinnon R, Semple S (2013) Learning from both sides: experiences and opportunities in the investigation of Australian aboriginal medicinal plants. J Pharm Pharm Sci 16(2):259–271
180. Simpson BS, Claudie DJ, Smith NM, Gerber JP, McKinnon RA, Semple SJ (2011) Flavonoids from the leaves and stems of *Dodonaea polyandra*: a northern Kaanju medicinal plant. Phytochemistry 72(14–15):1883–1888
181. Singer C (1927) The herbals in antiquity and its transmission to later ages. J Hell Stud 47:1–52
182. Soldati L, Di Renzo L, Jirillo E, Ascierto PA, Marincola FM, De Lorenzo A (2018) The influence of diet on anti-cancer immune responsiveness. J Transl Med 16(1):1–18
183. Stack EM (1989) Aboriginal pharmacopoeia. Northern Territory Library Service 3(10):1–7
184. Statistics Canada (2022) Statistics on indigenous peoples. https://www.statcan.gc.ca/en/subjects-start/indigenous_peoples. Accessed 10 Jan 2023
185. Staub PO, Casu L, Leonti M (2016) Back to the roots: a quantitative survey of herbal drugs in Dioscorides' De Materia Medica (*ex* Matthioli, 1568). Phytomedicine 23:1043–1052
186. Stump D (2018) Routledge handbook of landscape and food. Landscape 19(1):80–81
187. Su HX, Yao S, Zhao WF, Li MJ, Liu J, Shang WJ, Xie H, Ke CQ, Hu HC, Gao MN, Xu YC (2020) Anti-SARS-CoV-2 activities in vitro of Shuanghuanglian preparations and bioactive ingredients. Acta Pharmacol Sin 41(9):1167–1177
188. Thangapazham RL, Sharad S, Maheshwari RK (2013) Skin regenerative potentials of curcumin. Biofactors 39(1):141–149
189. Tissenbaum HA (2015) Using C. elegans for aging research. Invertebr Reprod Dev 59(sup1): 59–63
190. Tschirch A (1910) Handbuch der Pharmakognosie. Allgemeine Pharmakognosie, Erster Band II. Abteilung. Chr. Herm. Tauchnitz, Leipzig
191. Tyler VE (1984) Three proprietaries and their claims as American 'Indian' remedies. Pharm Hist 26:146–149
192. Tyler VE, Was Lydia E (1995) Pinkham's vegetable compound an effective remedy? Pharm Hist 37:24–28
193. U.S. Department of the Interior (2022) Bureau of land management. Native American Graves Protection & Repatriation Act. https://www.blm.gov/NAGPRA. Accessed 22 Jan 2023
194. United States Government (2022) United States census bureau. https://www.census.gov/quickfacts/fact/table/US/PST045221. Accessed 22 Jan 2023

195. Uprety Y, Asselin H, Dhakal A, Julien N (2012) Traditional use of medicinal plants in the boreal forest of Canada: review and perspectives. J Ethnobiol Ethnomed 8:7
196. Urdang G (1951) The development of pharmacopoeias; a review with special reference to the pharmacopoea Internationalis. Bull World Health Organ 4:577–603
197. Urits I, Borchart M, Hasegawa M, Kochanski J, Orhurhu V, Viswanath O (2019) An update of current cannabis-based pharmaceuticals in pain medicine. Pain Ther 8:41–51
198. Van der Veen M, Morales J (2015) The Roman and Islamic spice trade: new archaeological evidence. J Ethnopharmacol 167:54–63
199. Velmurugan BK, Rathinasamy B, Lohanathan BP, Thiyagarajan V, Weng CF (2018) Neuroprotective role of phytochemicals. Molecules 23(10):2485
200. Wachtel-Galor S, Benzie IFF (2011) Herbal medicine. In: Benzie I, Wachtel-Galor S (eds) Herbal medicine: biomolecular and clinical aspects, 2nd edn. CRC Press/Taylor & Francis, Boca Raton
201. Wang Z, Yang L (2021) Chinese herbal medicine: fighting SARS-CoV-2 infection on all fronts. J Ethnopharmacol 270:113869
202. Wang S, Li Y (2005) Traditional Chinese medicine. In: Devinsky O, Schachter S, Pacia S (eds) Complementary and alternative therapies for epilepsy. Demos Medical Publishing, New York, pp 177–182
203. Wang JF, Cai CZ, Kong CY, Cao ZW, Chen YZ (2005) A computer method for validating traditional Chinese medicine herbal prescriptions. Am J Chin Med 33(2):281–297
204. Wangchuk P, Tobgay T (2015) Contributions of medicinal plants to the gross National Happiness and biodiscovery in Bhutan. J Ethnobiol Ethnomed 11:48
205. Watanabe K, Matsuura K, Gao P, Hottenbacher L, Tokunaga H et al (2011) Traditional Japanese Kampo medicine: clinical research between modernity and traditional medicine–the state of research and methodological suggestions for the future. Evid Based Complement Alternat Med 2011:1–19
206. World Health Organization (WHO) (2005) National Policy on traditional medicine and regulation of herbal medicines. Report of WHO global survey, Geneva
207. World Health Organization (2002) WHO traditional medicine strategy 2002–2005, Geneva
208. World Health Organization (2018) Antimicrobial resistance and primary health care. World Health Organization, Geneva
209. WHO (2001) Legal status of traditional medicine and complementary/alternative medicine. A worldwide review, WHO, Geneva
210. WHO (2002) Regulatory situation of herbal medicines; a worldwide review. World health organization, Geneva
211. WHO (2005) National Policy on traditional medicine and regulation of herbal medicines. World Health Organization, Geneva
212. WHO (2009) Traditional medicine in 62nd world health assembly 18–22nd May 2009, resolutions and decisions (WHA 62/2009/REC/1)
213. WHO (2011) Progress report on decade of traditional medicine in the African region, Brazzaville, WHO, Africa region (AFRO) AFR/RC61/PR/2
214. WHO (2013) World health organization, traditional medicine strategy 2014–2023
215. Wurtzburg SJ (2016) Traditional medicine: native American tribes. In: Boslaugh SE (ed) The SAGE encyclopedia of pharmacology and society, 1st edn. SAGE Publications, Inc, Thousand Oaks, pp 1412–1415
216. Xu J, Yang Y (2009) Traditional Chinese medicine in the Chinese health care system. Health Policy 90(2–3):133–139
217. Yang L, Wang Z (2021) Natural products, alone or in combination with FDA-approved drugs, to treat COVID-19 and lung cancer. Biomedicine 9(6):689
218. Youkparigha FO, Izah SC (2019) Larvicidal efficacy of aqueous extracts of *Zingiber officinale* Roscoe (ginger) against malaria vector, *Anopheles gambiae* (Diptera: Culicidae). Int J Environ Agric Sci 3:020

Part II

Phytochemistry and Herbal Medicine

Plant Food for Human Health: Case Study of Indigenous Vegetables in Akwa Ibom State, Nigeria

2

Nkereuwem Udoakah Obongodot and Matthew Chidozie Ogwu

Contents

N. U. Obongodot
Department of Plant Biology and Biotechnology, Faculty of Life Sciences, University of Benin, Benin City, Nigeria

M. C. Ogwu (✉)
Goodnight Family Department of Sustainable Development, Appalachian State University, Boone, NC, USA
e-mail: ogwumc@appstate.edu

S. C. Izah et al. (eds.), *Herbal Medicine Phytochemistry*, Reference Series in Phytochemistry, https://doi.org/10.1007/978-3-031-43199-9_2

Abstract

The people of a particular culture and locale utilize plant resources within their environment for diverse purposes, e.g., as food. Plants that are utilized as food also contribute to the health and well-being of the populace. This chapter assesses the relationship between the people of Akwa Ibom State, Nigeria, and the crop plants found in their locality that contribute to health security. People from Akwa Ibom State rely mostly on vegetables for their diet and as medicines, and these plants play numerous roles in their culture. A total of 88 plants from 15 higher plant families were found to be used as food and medicine by the Akwa Ibom people. This chapter will describe these plants as well as their cultural practices, conservation status, and traditional knowledge and utilization patterns. These plants are either used solely or in combination with other plant resources to prepare different local relishes that possess medicinal value for the treatment and management of diverse ailments. These plants include *Cucurbita maxima, Gongronema latifolium, Abelmoschus esculentus, Piper guineense, Amaranthus hybridus, Justicia schimperi, Ocimum gratissimum, Telfairia occidentalis, Gnetum africana, Crassocephalum crepidioides, Heinsia crinita, Talinum triangulare, Vernonia amygdalina, Lasianthera africana, Colocasia esculentus, Cucumis sativus,* and *Ipomoea batatas*. Plants have been used as food and medicine for decades by all age groups in Akwa Ibom State, Nigeria, although some of them are considered underutilized. However, their utilization as food and medicine is on the increase in the present day, and they will most likely continue to be relevant in the future. Nonetheless, practical measures like value addition, better packaging, and conservation strategies are necessary.

Keywords

Ethnobotany · Indigenous vegetables · Akwa Ibom · Nigeria · Plant Conservation · Plant medicine · Food as medicine

1 Introduction

Ethnobotany is concerned with the dynamic relationship between people of a particular culture and their interactions with the diverse plant resources in their locality in terms of how they are utilized, what they are used for, when they are used, as well as their general relevance to life within the community or society. The field of ethnobotany evolved from the study of plants used by primitive societies [6, 42, 74, 76]. Subsequently, many other workers [43, 93] improved the concepts to include an interdisciplinary and holistic approach that focuses on the realm of the human-plant relationship. This realm of interaction between plants and human defined by economic, sociocultures, and environmental relationships is not limited to the use of plants for food, clothing, and shelter but also include their use for religious ceremonies, ornamentation, and health care [72, 80]. For instance, a previous survey revealed some vegetables eaten by Akwa Ibom people residing in Benin City, Nigeria, which highlighted ten vegetables including Ikong-ubong (*Telfairia occidentalis*), Mmong-mmongikong (*Talinum triangulare*), Atama (*Heinsia crinita*), Afang (*Gnetum africanum*), and Editan (*Lasianthera africana*), either consumed alone or in combination with other plant-based food resources or used in traditional medicines [99]. An ethnobotanical survey of plants is useful in identifying plants that serve in the alleviation of ailments [79, 80]. These plants work either in combination with one or more plants or are used solely. These plants have served and are still serving as leads in the production of orthodox medicines.

Indigenous vegetables are edible plant species peculiar to a particular locality or region with their diversity and distribution depending on geographical location and edaphic and climatic conditions [44, 46, 73, 75, 80]. Some of these plants are naturalized species and are well adapted to specific local conditions. According to Ogwu et al. [76, 77], these plants are grown for their leaves, succulent stems, young shoots, fruits, and/or combination of these plant parts. They may not require formal cultivation because they grow easily in the wild [69]. As protective foods, these indigenous vegetables are widely consumed in tropical regions including Nigeria as important sources of proteins, vitamins, and essential minerals [99]. The knowledge of indigenous plants used as food and medicines is often embedded in indigenous cultures but is slowly eroding with modernization. The decline of cultural diversity and the gradual erosion of human knowledge of medicinal plant species as well as their distribution, management, and methods of extracting medicinal chemicals is a contemporary challenge that is likely to increase in the future unless they are linked to their use as food [23, 24, 81].

This chapter aims to highlight some important vegetables with medicinal properties that are used by the people of Akwa Ibom State, Nigeria, and share their culinary uses and value in the treatment and management of diverse ailments. The chapters begin with an introduction to indigenous vegetables, Nigeria and Akwa Ibom State, and climax by presenting a list of indigenous vegetables associated with people from Akwa Ibom State and discussing their botany, traditional knowledge, and practices as well as their ethnomedicinal usage. The list of indigenous vegetables associated with Akwa Ibom people was sourced from personal knowledge of these vegetables, the work of Etukudo [31], and a few scholarly articles [95, 99]. The essence of the work is to give proper documentation of some of these species for easy access and to make recommendations on the need to conserve these vegetables as they are seen as underutilized crops.

2 Akwa Ibom State, Southern Nigeria

Akwa Ibom State (latitudes 4°32 and 5°53 North and longitudes 7°25 and 8°25 E) is flanked on the east, west, and north by Cross River, Abia, and Rivers states, respectively in the south-southern axis of Nigeria on the sandy deltaic coastal plain of the Guinea Coast (Fig. 1). To the south of the state is the Atlantic Ocean.

Akwa Ibom has a total land area of 8412 km^2 with altitudes of 45–70 m.a.s.l. and includes the Qua Iboe River basin and the lower Imo River basin [28, 99, 103]. The state is located within Nigeria's forest zone and has a tropical rainy climate. It has 31 local government areas composed mainly of the Ibibio, Eket, Annang, Oron, and Okobo ethnic groups. Ibibio is the largest ethnic group in Akwa Ibom and speaks the Ibibio language with diverse dialects [117]. The majority of the populace engages in farming, fishing, trading, hunting, wood-carving, raffia works, blacksmithing, pottery, ironworks, tailoring, and crafts creation. Native diets are mainly plant-based and are extolled for their medicinal values including efere afang, Edikang ikong,

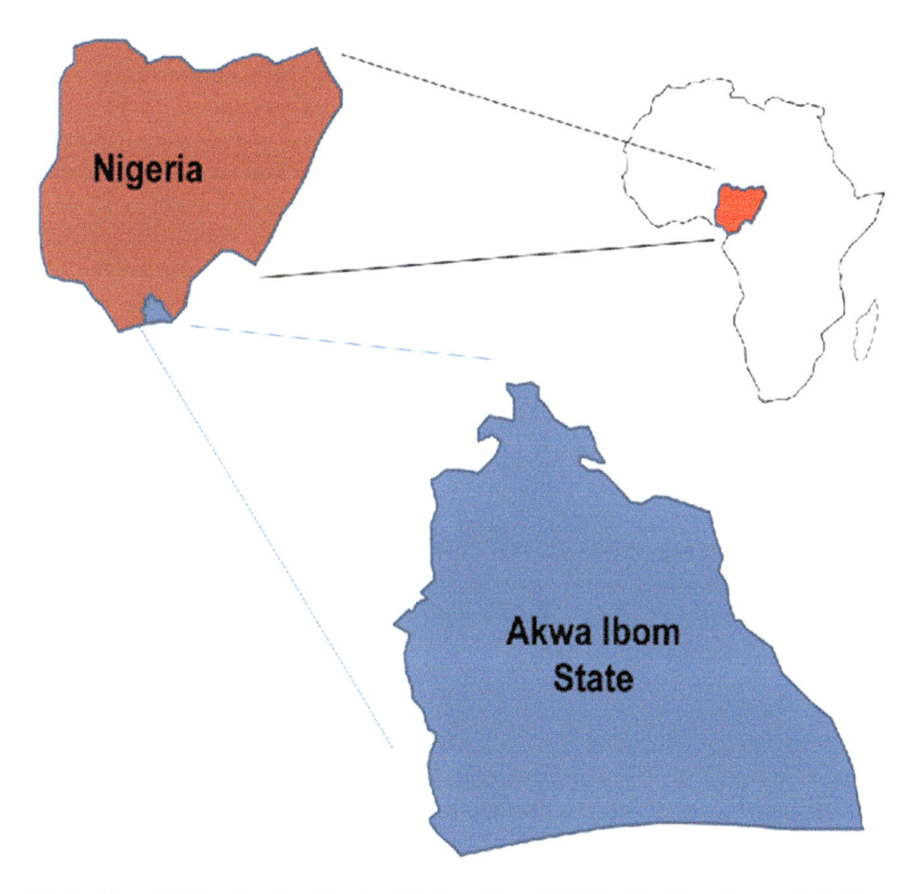

Fig. 1 Map of Africa showing Nigeria with Akwa Ibom highlighted in blue. (Source: Ref. [9])

efere etidot (bitter leaf soup), efere atama, efere editan, efere ikon, efere otong, afiaefere, eferendekiyak, efere etighi, efere ndukpauyo, efere mmongmmongikong, and eferenyama. Most of these soups go with coarse starchy food such as garri, pounded yam, wheat flour, and fufu.

3 Indigenous Vegetables Used as Food and Medicine in Akwa Ibom

The term vegetable is not strictly botanical but arbitrary and largely based on the mode of plant utilization rather than on their plant morphology [37]. For instance, Vainio and Bianchini defined vegetables as all edible plant parts including stem, stalk, root, tuber, bulb, leaves, flowers, fruits, or seeds, whereas Remison [106] suggested that plants are grown for their useable parts like leaves or young shoots, fruits, or a combination of usable plant parts like leaves and fruits. Vegetables can be fruity or leafy and may be soft and fleshy plant parts eaten raw or cooked, mature or immature, or both. Some plants used as vegetables also have horticultural values. Asaolu et al. [12] reported that vegetables are widely consumed in the processed, semi-processed, or raw form as part of a main dish or salad because of the taste they add to food. Many plants used as vegetables in Akwa Ibom are native to the Flora of Nigeria and considered resilient, adaptive, and tolerant to adverse climates. Knowledge of these plants has been transmitted from one generation to the next orally and without adequate documentation. The practice is still prevalent among rural and tribal communities within the state albeit unsustainable because of the potential loss of information about important species and best practices such as the method of propagation, mode of preparation, and preservation. These indigenous vegetables are cheap, flavory, and colorful and add aesthetic appeal to local diets [62]. They make up an important part of daily diets because of their carbohydrate, protein, mineral, vitamins, fiber, and other nutritional contents [32, 65]. In Akwa Ibom, despite their low calorific value, they are eaten either in the form of leaves, seeds, fruits, flowers or pods, roots, stems, and tubers and are a source of many nutrients including potassium, folic acid, vitamin A, and vitamin C [85, 91]. The low caloric content of these local vegetables is because plants produce food in the leaves but many do not store food in the leaves [32, 45, 73, 83]. Fibers in vegetables are known to promote digestion and prevent constipation [46, 64, 96–98]. The nutrients in vegetables can be absorbed as regulatory and protective nutrients, as well as for bodybuilding [12]. During the rainy season, vegetables are abundant in the wild, market, and home gardens and are relatively cheaper unlike in the dry season when there is general scarcity leading to higher prices as demands cannot be met [94]. A few of these vegetables are domesticated and marketed, thus serving as a source of household income.

Akwa Ibom State indigenes are well known for their local delicacies with high food and medicinal values. These delicacies are prepared using traditional methods. The scientific and common names of some of these vegetables are presented in Table 1, while their plant family distribution is presented

Table 1 Indigenous vegetables common to Akwa Ibom people

Family	Species	Names in Akwa Ibom
Gnetaceae	*Gnetum Africana*	Afang
Rubiaceae	*Heinsia crinita*	Atama
Portulaceae	*Talinum triangulare*	Mmongm mongikong
Cucurbitaceae	*Telfairia occidentalis*	Nkong
Icacinaceae	*Lasianthera Africana*	Editan
Asteraceae	*Vernonia amygdalina*	Etidod
Asclepiadaceae	*Gongronema latifolium*	Utasi
Malvaceae	*Abelmoschus esculentus*	Etighi
Lamiaceae	*Ocimum gratissimum*	Nton
Piperaceae	*Piper guineense*	Odusa
Amaranthaceae	*Amaranthus hybridus*	Iyanafia
Acanthaceae	*Justicia schimperi*	Mmemme
Araceae	*Colocasia esculentus*	Eka ikpori
Asteraceae	*Crassocephalum crepidioides*	Mkpafid
Convolvulaceae	*Ipomoea batatas*	Ediam makara
Cucurbitaceae	*Cucumis sativus*	Okokon
Cucurbitaceae	*Cucurbita maxima*	Ndise
Euphorbiaceae	*Microdesmis puberula*	Ntanebid, Ntabid
Zingiberaceae	*Aframomum melegueta*	Ntuen-ibok
Amaryllidaceae	*Allium sativum*	Etebe-owo inua
Liliaceae	*Aloe vera* (and *A. barbadensis*).	Akokafid
Gentianaceae	*Anthocleista djalonensis*	Ibu
Acanthaceae	*Acanthus montanus*	Mbara ekpe
Amaranthaceae	*Achyranthes aspera*	Udok mbiok, Udok mbiet
Amaranthaceae	*Alternanthera bettzickiana*	Nkpok isip essien
Amaranthaceae	*Cyathula prostrata*	Nkibe ubuk
Amaryllidaceae	*Allium cepa*	Ayim
Caricaceae	*Carica papaya*	Okpod, popo
Polygalaceae	*Carpolobia lutea*	Ikpafum
Anacardiaceae	*Lannea acida*	Ayara nsukakara
Anacardiaceae	*Lannea nigritana*	Odok eto
Anacardiaceae	*Anacardium occidentale*	Cashew
Anacardiaceae	*Mangifera indica*	Mango
Annonaceae	*Uvaria chamae*	Nkarika ekpo, atama nkana
Annonaceae	*Monodora myristica*	Enwun
Araceae	*Caladium bicolor*	Ikpon ekpo, udia edi
Araceae	*Xanthosoma sagittifolium*	Ikpon mbakara
Asteraceae	*Tridax procumbens*	Ayara utimense
Asteraceae	*Emilia sonchifolia*	Utime nse, usio mmon
Asteraceae	*Crassocephalum crepidioides*	Mkpafit
Boraginaceae	*Heliotropium indicum*	Otukeyin eka, esin ono
Bignoniaceae	*Spathodea campanulata*	Esenim
Bignoniaceae	*Newbouldia laevis*	Itumo, oboti

(continued)

Table 1 (continued)

Family	Species	Names in Akwa Ibom
Bignoniaceae	*Kigelia Africana*	Ntabinim
Bromeliaceae	*Ananas comosus*	Eyop mbakara
Burseraceae	*Dacryodes klaineana*	Eben ikot
Clusiaceae	*Allanblackia floribunda*	Udiaebion, ekporo-enin
Combretaceae	*Combretum micranthum*	Asaka
Combretaceae	*Terminalia ivorensis*	Nkot ebene
Convulvulaceae	*Ipomoea quamoclit*	Ediam ikanikot
Convulvulaceae	*Ipomoea pileata*	Mkpafiafian
Costaceae	*Costus afer*	Mbritem
Crassulaceae	*Bryophyllum pinnatum*	Ndodop afiaiy
Cucurbitaceae	*Citrullus colocynthis*	Ikon, ikpan
Cucurbitaceae	*Momordica balsamina*	Mbiadon edon
Dioscoreaceae	*Dioscorea dumetorum*	Enem, edidia iwa
Euphorbiaceae	*Euphorbia hirta*	Etinkene ekpo
Euphorbiaceae	*Jatropha curcas*	Ukim eyio
Euphorbiaceae	*Manihot esculenta*	Iwa
Euphorbiaceae	*Manniophyton fulvum*	Ekonikon
Euphorbiaceae	*Mallotus oppositifolius*	Uman nwariwa
Fabaceae	*Afzelia Africana*	Eyin mbukpo
Fabaceae	*Afzelia bella*	Eyin mbukpo
Fabaceae	*Albizia lebbeck*	Ubam
Fabaceae	*Baphia nitida*	Afuo
Fabaceae	*Cajanus cajan*	Nkoti
Fabaceae	*Cassia alata*	Adaya okon
Fabaceae	*Glycine max*	Nkoti eto
Fabaceae	*Lonchocarpus cyanescens*	Awa
Fabaceae	*Parkia biglobosa*	Ukon uyayak
Fabaceae	*Pentaclethra macrophylla*	Ukana
Fabaceae	*Pterocarpus erinaceus*	Ukpa
Fabaceae	*Pterocarpus santalinoides*	Nkpa-inyan
Fabaceae	*Tetrapleura tetraptera*	Uyayak
Hypericaceae	*Harungana madagascariensis*	Oton
Irvinginiaceae	*Irvingia gabonensis*	Uyo

in Fig. 2. Almost 80 plants from 15 higher plant families are presented here based on oral historical records and the work of Etukudo [31]. Cucurbitaceae and Asteraceae were the plant families with the highest representations. Earlier, Osawaru et al. [99] conducted an ethnobotanical survey of vegetables eaten by Akwa Ibom residents in Benin City, Nigeria, and reported a total of ten vegetables. Similarly, Ajibesin et al. [9] surveyed plants in Akwa Ibom State used for skin disease and other related ailments and documented 183 medicinal plant species representing 153 genera and 59 families.

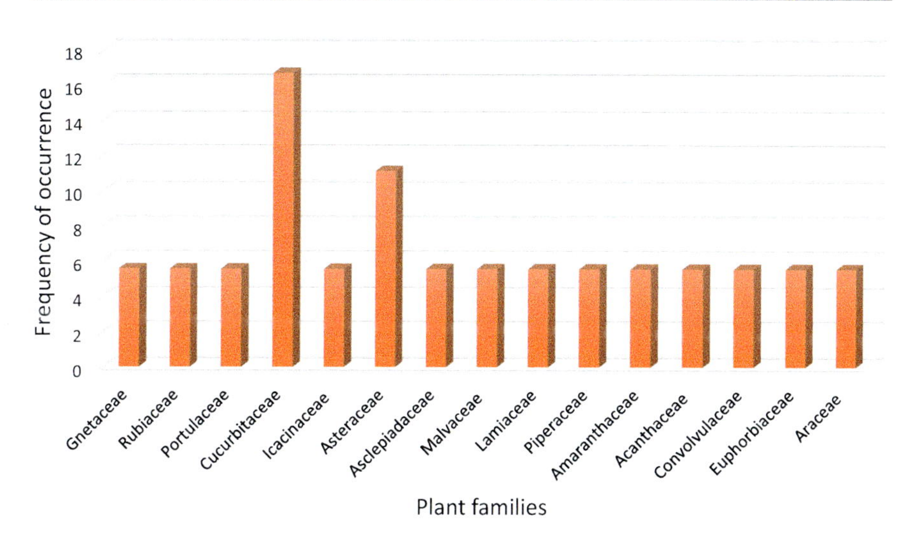

Fig. 2 Plant family distribution of indigenous plants used as food and medicine by Akwa Ibom people, Nigeria

The utilization mode and pattern, parts used, and cultivation status and sources of these indigenous vegetables used for food and health by Akwa Ibom people are presented in Table 2.

4 Botanical Description, Knowledge, and Practices Associated with the Vegetables Indigenous Used as Food and Medicine by Akwa Ibom People

The description and local utilization and knowledge and practices associated with some indigenous vegetables used as food and medicine in Akwa Ibom are presented below.

4.1 *Gnetum africanum*

Description, utilization, knowledge, and practices. It is a creeping vine. The leaves are collected as forest products. The leaves are used in the preparation of the local delicacy "Efere afang." The different recipes on how the dish is prepared mark the cultural identity of the Akwa Ibom people. Soup is one of the main meals for any kind of ceremony. The stem is used in preparations for medical concoctions to ease childbirth. The seeds and leaves are used in the treatment of piles, sore throat, and high blood pressure. It also serves as a purgative and the supple stem is sometimes used as rope. The mineral element composition of *G. africanum* is as follows: Ca, 130 ppm; Mg, 89.0 ppm; Fe, 76.1 ppm; and Zn, 1.3 ppm, while cadmium and lead were not detected [48].

Table 2 Status, utilization, and conservation status of the indigenous vegetables

Species	Status	Utilization (food and/or medicine)	Part used
Gnetum Africana	Wild; cultivated	Food, medicine	Leaves, stem, and seeds
Heinsia crinita	Wild; cultivated	Food, medicine	Leaves
Talinum triangulare	Wild; cultivated	Food, medicine	Leaves, stems
Telfairia occidentalis	Cultivated	Food, medicine	Leaves, seeds
Lasianthera Africana	Cultivated; wild	Food, medicine	Leaves
Vernonia amygdalina	Wild; cultivated	Food, medicine	Leaves
Gongronema latifolium	Wild; cultivated	Food, medicine	Leaves
Abelmoschus esculentus	Cultivated	Food, medicine	Leaves, seeds, pods
Ocimum gratissimum	Cultivated	Food, medicine	Leaves
Piper guineense	Cultivated	Food, medicine	Leaves
Amaranthus hybridus	Cultivated	Food, medicine	Leaves
Justicia schimperi	Cultivated	Food, medicine	Leaves
Colocasia esculentus	Cultivated, wild	Food	Whole plant
Crassocephalum crepidioides	Cultivated	Food, medicine	Leaves
Ipomoea batatas	Cultivated	Food, medicine	Leaves, roots
Cucumis sativus	Cultivated	Food, medicine	Whole plants
Cucurbita maxima	Cultivated	Food, medicine	Leaves, seeds, roots
Microdesmis puberula	Cultivated	Food, medicine	Whole plant
Aframomum melegueta	Cultivated	Medicine	Leaves, seeds
Allium sativum	Cultivated	Food, medicine	Bulb
Aloe vera (and *A. barbadensis*)	Cultivated	Medicine	Leaves, juice
Anthocleista djalonensis	Wild	Medicine	Bark, leaves
Acanthus montanus	Wild	Medicine	Stem, leaves
Achyranthes aspera	Wild	Medicine	Seeds, leaves
Alternanthera bettzickiana	Wild, cultivated	Medicine	Leaves, young shoot
Cyathula prostrata	Cultivated	Medicine	Leaves
Allium cepa	Cultivated	Food, medicine	Bulb, leaves
Carica papaya	Cultivated	Food, medicine	Leaves, seeds, fruits
Carpolobia lutea	Cultivated	Food, medicine	Leaves, fruits, stembark
Lannea acida	Semi-cultivated, wild	Medicine	Bark, leaves
Lannea nigritana	Cultivated, wild	Food, medicine	Seeds, fruit, roots
Anacardium occidentale	Cultivated, wild	Food, medicine	Bark, leaves, fruits
Mangifera indica	Cultivated	Food, medicine	Fruits, leaves, stem

(continued)

Table 2 (continued)

Species	Status	Utilization (food and/or medicine)	Part used
Uvaria chamae	Wild	Medicine	Root
Monodora myristica	Cultivated	Food, Medicine	Seeds
Caladium bicolor	Wild	Medicine	Leaves, rhizome, corm
Tridax procumbens	Wild	Medicine	Leaves
Emilia sonchifolia	Wild	Medicine	Leaves, young shoot
Crassocephalum crepidioides	Wild	Medicine	Leaves, roots
Heliotropium indicum	Wild	Medicine	Leaves
Spathodea campanulata	Cultivated	Medicine	Seeds, roots, bark
Newbouldia laevis	Cultivated	Medicine	Stem- bark
Kigelia africana	Cultivated	Medicine	Seeds, fruits
Ananas comosus	Cultivated	Food, medicine	Fruit
Dacryodes klaineana	Cultivated	Food, medicine	Fruit, leaves
Allanblackia floribunda	Semi-cultivated, wild	Medicine	Seed, bark
Combretum micranthum G	Cultivated	Medicine	Seed, leaves, roots
Terminalia ivorensis	Cultivated, wild	Medicinal	Bark, root
Ipomoea quamoclit	Cultivated	Medicine	Leaves
Ipomoea pileata	Cultivated	Medicine	Leaves
Costus afer	Wild, cultivated	Medicine	Leaves, fruits
Bryophyllum pinnatum	Cultivated	Medicine	Leaves
Citrullus colocynthis	Cultivated	Food, medicine	Seed, fruit
Momordica charantia	Cultivated	Food, medicine	Fruit, young shoots, leaves
Dioscorea dumetorum	Cultivated	Food, medicine	Root, tuber
Euphorbia hirta	Wild	Medicine	Young leaves and shoot
Jatropha curcas	Cultivated	Food	Seeds
Manihot esculenta	Cultivated	Food, medicine	Tuberous roots, young leaves
Manniophyton fulvum	Wild, semi-cultivated	Medicine	Seed, root, leaves, stem
Mallotus oppositifolius	Wild, semi-cultivated	Medicine	Leaves, stem bark
Afzelia africana	Semi-cultivated	Medicine	Leaves
Afzelia bella	Cultivated	Medicine	Leaves, bark
Baphia nitida	Cultivated	Medicine	Seeds
Cajanus cajan	Cultivated	Food, medicine	Seeds, shoot, leaves
Cassia alata	Wild, cultivated	Medicine	Leaves, seedpods
Glycine max	Cultivated	Food, medicine	Seeds

(continued)

Table 2 (continued)

Species	Status	Utilization (food and/or medicine)	Part used
Lonchocarpus cyanescens	Cultivated	Medicine	Leaves, roots
Parkia biglobosa	Cultivated	Food, medicine	Seed, Leaves
Pentaclethra macrophylla	Cultivated	Food, medicine	Stem, fruit, seed
Pterocarpus erinaceus	Cultivated	Medicine	Stem bark
Pterocarpus santalinoides	Cultivated	Medicine	Seed, leaves, root
Tetrapleura tetraptera	Cultivated	Food, medicine	Seed
Harungana madagascariensis	Cultivated	Food, medicine	Fruit, bark
Irvingia gabonensis	Wild, cultivated	Food, Medicine	Seed

Cultivation requirements and management within Akwa Ibom: The vegetables are often grown in the wild although some are cultivated in the home gardens. The stem cuttings are usually planted in the soil and properly watered till the plant germinates. As they grow, the plant requires some support such as a tree to grow into.

4.2 *Heinsia crinita*

Description, utilization, knowledge, and practices: It is usually cultivated or wild. The vegetable is used in the preparation of the local dish "Efere Abak." It is also used in pediatric care for the treatment of measles. The mineral element composition of *H. crinita* is reported as follows: K, 84.41 mg/g; Ca, 24.20 mg/g; Mg, 11.31 mg/g; and P, 9.14 mg/g [89].

Cultivation requirements and management within Akwa Ibom: The vegetable is propagated by seed and stem cuttings on loamy soil. The cultivated crops grow into a shrub or trees and are available all year round.

4.3 *Talinum triangulare*

Description, utilization, knowledge, and practices: The leaves and stems are used as a softener when cooking fibrous vegetables such as Afang (*G. africanum*), Atama (*H. crinita*), and fluted pumpkin (*T. occidentalis*). It is also used for its high medicinal value, usually in the treatment of diuretics and stomach problems [67].

Cultivation requirements and management within Akwa Ibom: Waterleaf is propagated by seed but also vegetatively by cutting 15–20 cm of the matured stem. It is a fast-growing plant that once established can easily reseed itself because it is self-pollinating and produces flowers early year-round. It is an ephemeral crop that is due for harvest between 35 and 45 days after planting (Udoh and Etim 2008). Some are wild or weedy.

4.4 *Telfairia occidentalis*

Description, utilization, knowledge, and practices. It is a vine plant often grown in a homestead garden. Due to its creeping nature, it is usually staked into a framework of bamboo, or a shelf is made for it to climb on when twining [88]. *Telfairia occidentalis* shoots and leaves are used in preparing Edikang ikong soup. Much importance is placed on the crop for its edible seeds which are rich in protein and fat and can be eaten as a whole by boiling. The sliced leaves can be mixed with coconut water and salt used for the treatment of convulsion. The leaves are also used in the management of anemia because of the high iron content, while the roots are used as rodenticides [35, 68].

Cultivation requirements and management within Akwa Ibom: Seeds are planted directly in the soil, typically in groups of three to increase output in a case of a failed germination. Although dependent on soil type, the fluted gourd can grow and subsequently produce many flushes of fruit over long periods [7]. The seeds cannot be stored for more than 3 days once they are extracted from the fruit.

4.5 *Lasianthera africana*

Description, utilization, knowledge, and practices. It is a perennial, glabrous shrub. The leaves are highly valued for their medicinal properties including the ability to relieve stomach complaints with or without diarrhea, as vermifugal, and for cold as a wash against headache. Leaf sap is considered anti-psoric and is crushed in warm water, which raises an abundant froth that is used to bathe feverish infants. Pulped leaf or pulped bark is made up into a dressing to bind over fractures. Tests for alkaloids have given mild response for the leaf, moderate for the bark, and strong for the root, and tannin is present in all these parts. Notwithstanding the presence of alkaloids and tannin, certain tribes like the Akwa Ibom people put the leaf in soup. The leaves are used as antacid, analgesic, antispasmodic, laxative, antipyretic, antiulcerogenic, antidiabetic, and antimalarial because of their bacteriostatic, anti-ulcer, fungicidal, antidiabetic, and antiplasmodial properties [29, 49, 50, 86, 87]. The plant also has social roles in ceremonies as well as in food preparation. The mineral element composition of *L. Africana* is as follows: potassium, 56.24 mg/g; calcium, 31.43 mg/g; magnesium 31.99 mg/g; and phosphorus, 5.00 mg/g [89].

Cultivation requirements and management within Akwa Ibom: It is cultivated through the propagation of adult stem cuttings [84].

4.6 *Vernonia amygdalina*

Description, utilization, knowledge, and practices. It is a shrub and can grow into a small tree. The leaves have a characteristically bitter taste and the extracts are used medicinally in the treatment of cough, fever, and pile and as a laxative. The vegetable is used in the preparation of the local relish such as efere etidod, and efere ikon. The leaves are washed before use to get rid of their bitter taste. The stems are used as a chewing stick for oral hygiene as well as the management of dental problems. They can be found as weeds in the wild and cultivated in the home garden. The active ingredient (vernonioside B) possesses antiparasitic, antitumoral, and antibacterial agents [114].

Cultivation requirements and management within Akwa Ibom: It is cultivated through the propagation of adult stem cuttings.

4.7 *Gongronema latifolium*

Description, utilization, knowledge, and practices: It is a climbing shrub usually used as a spice and vegetable [119]. In soups, it imparts its beautiful and agreeable bitter taste to the food. It is also eaten alone raw or added to other meals like plantain porridge, yam porridge, etc. The leaves are also incorporated into salads. The twigs and branches are used as chewing sticks [31]. Abbiv [1] records that the leaves are rubbed on joints to enable children to walk early. It is a cough, diabetes, malaria, and hypertension remedy [30]. Leaf extract or cold infusion of pounded leaf mixed with lime or pineapple juice is used as a blood tonic and is also known to expel intestinal worms and help with upset stomach and for the treatment of typhoid fever [30].

Cultivation requirements and management within Akwa Ibom: *G. latifolium* is propagated by soft stem or hard stem cuttings.

4.8 *Abelmoschus esculentus*

Description, utilization, knowledge, and practices. It is a tall erect annual plant with edible leaves, pods, seeds, and flowers. The leaves, flowers, and young shoots are eaten as vegetables. It contains mucilage which gives the soup a slimy, sticky texture. It can be sliced, dried, powdered, and preserved for future use. This serves as a flavoring and as a soup thickener. The seeds are eaten as snacks after roasting.

Cultivation requirements and management within Akwa Ibom: *A. esculentus* are cultivated through seedlings. Usually, the seeds are soaked overnight to have uniform germination. Two to three seeds are sowed per hole at 3 cm depth. The crop is usually watered regularly.

4.9 *Ocimum gratissimum*

Description, utilization, knowledge, and practices: It is a perennial shrub. The leaves have a strong fragrance with a strong aroma. The leaves are used as a seasoning in the preparation of pepper soup. Medicinally, it is used in the treatment of high fever and piles. The leaves may be taken for catarrh medication. A paste of the leaves is applied topically against ringworm and skin diseases. Crushed leaf extract together with leaf extracts is used to treat diabetes and pile. More so, most of the respondents claim that the scent from the leaves of the plant is also used to protect against snakes.

Cultivation requirements and management within Akwa Ibom: *O. gratissimum* is propagated by seeds and soft stem cutting and is usually available all year round.

4.10 *Piper guineense*

Description, utilization, knowledge, and practices: It is a climbing shrub commonly used as a spice in the preparation of meals especially soups like efereetek, edikangikong, afiaefere. This crop is used to facilitate the swallowing of coarse-textured starchy food. The vegetable is also used by Akwa Ibom people to add flavours to their local dishes. Ethnomedicinally, it is used in the regulation of the menstrual cycle. The fruits and leaves are used as a spice for preparing soup for postpartum women.

Cultivation requirements and management within Akwa Ibom: The crop is propagated by seeds.

4.11 *Amaranthus hybridus*

Description, utilization, knowledge, and practices: It is an erect annual plant with leaves used as vegetables in cooking porridge and vegetable sauce and also in the local delicacy ekpang nkukwo [125]. The crushed leaves are useful as an antidote for snake and scorpion bites. The plants are also used by the tribal people to feed animals.

Cultivation requirements and management within Akwa Ibom: Propagation is by seeds and they germinate readily if the soil is warm.

4.12 *Justicia schimperi*

Description, utilization, knowledge, and practices: It is an annual or perennial herb. The leaves are edible and are used in cooking soups, plantain porridge, and many other delicacies. The leaves are cooked in a palm fruit filtrate and fed to babies.

It is used as an enema to treat umbilical problems in infants. The leaves are also used in treating chest and heart problems (Inyang [31]).

Cultivation requirements and management within Akwa Ibom: It is usually propagated by seeds and germinates readily with the availability of rain.

4.13 *Colocasia esculentum*

Description, utilization, knowledge, and practices: It is a perennial plant commonly grown for its starchy edible underground tuber that is consumed after boiling, frying, or roasting with soup or stew. It can be cooked as cocoyam porridge or grated and used to prepare ayan ekpang which is eaten with a mucilaginous soup rich in meat, periwinkle, and fish [31]. The corm and leaves are used for the preparation of the local delicacy ekpang nkukwo. They are mostly cultivated in home gardens around the homestead, while some can be found as weeds. The tender petioles are leaves that are also eaten when young. The matured leaves are used as wrapping materials. Consumption of the plant (corm and leaves) is believed to have effect on constipation, diarrhea, skin rash, injuries and neurological disorders.

Cultivation requirements and management within Akwa Ibom: The top of the corm or the whole cormel is used for propagation usually on ridges when the rains are reliable [106].

4.14 *Crassocephalum crepidioides*

Description, utilization, knowledge, and practices: It is an annual herb with soft stems. The leaves and tender shoots are edible as vegetables or spinach, especially by elderly people on very auspicious occasions. The leaves when used as soup cure indigestion and give general comfort and well-being to the gastrointestinal system [31].

Cultivation requirements and management within Akwa Ibom: It is propagated by seeds or through stem cuttings.

4.15 *Ipomoea batatas*

Description, utilization, knowledge, and practices: It is a herbaceous perennial creeping vine. The tender leaves and young shoots are eaten as vegetables or spinach. They are also used to wrap grated cocoyam mash during the preparation of ekpang nkukwo, especially in the dry season when the usual wrapping leaves are scarce and costly. Leaves are antidiabetic and antiscorbutic.

Cultivation requirements and management within Akwa Ibom: Cultivation is done by stew cuttings about 30 cm long or portions of tubers. Planting is usually done in ridges or mold. The crops are planted when the rains are well established [106].

4.16 *Cucumis sativus*

Description, utilization, knowledge, and practices: It is a trailing or climbing monoecious annual herb with an extensive and largely superficial root system [106]. The green fruits are eaten fresh in salad and are rich in vitamin B, iron, and calcium and believed to have diuretic and purgative effects. The leaves are eaten as vegetables either raw in a salad or cooked [20]. The fruit contains 95% water. Proteins, fat, carbohydrates, fiber, and ash make up the chemical composition of *Cucumis sativus*.

Cultivation requirements and management within Akwa Ibom: Propagation is by seeds. The seeds are sown directly in ridges or molds or prepared planting holes. Plants seeded directly may be thinned to two or three plants per stand after establishment. The plant is sometimes stalked especially for some creeping cultivars [106].

4.17 *Cucurbita maxima*

Description, utilization, knowledge, and practices. It is an annual herb with thick climbing or creeping stems [107]. The leaves, tender shoots, and flowers of the plants are used as vegetables or pot-herb cooked in soups and is believed to help relive intestinal infections, stomach discomfort, and kidney issues. They are also used to prepare yams, cocoyam, sweet yam, and water porridge. The pulp is eaten after the whole fruit had been cooked. It can be eaten salted with stew or soup or as a vegetable.

Cultivation requirements and management within Akwa Ibom: It is propagated through direct seedlings in homestead gardens and grows well on organic matter [107].

4.18 *Microdesmis puberula*

Description, utilization, knowledge, and practices: *Microdesmis puberula* is a shrub or miniature tree. Edible leaves are used to prepare afang soup. The leaves may be pounded and mixed with alligator pepper seeds or ginger and applied on sprains, burns, and bruises for healing [31]. The twigs are used as chewing sticks to clean teeth and strengthen gums. It is sometimes taken orally or applied as an enema to treat diarrhea and is prescribed for pregnant women and young children [128].

Cultivation requirements and management within Akwa Ibom: It is propagated by seeds and can germinate massively in fallow land.

4.19 *Aframomum melegueta*

Description, utilization, knowledge, and practices: It is a perennial herb. The seeds are used as a traditional spice (ground or whole) for native ceremonies to welcome guests and ward off evil spirits. The seeds have a hot, pungent black-peppery flavor with a hint of citrus [113]. It is used traditionally in threes or sevens. It is used in almost if not all medicaments. The seeds are revulsive and carminative. They are used to strengthen drinks among other uses. The seeds are edible and are a stimulant.

Cultivation requirements and management within Akwa Ibom: It is propagated by seeds.

4.20 *Allium sativum*

Description, utilization, knowledge, and practices: It is a perennial herb. It is used to flavor soups, salads, and sausages. It is a classic ingredient of many cuisines and has a pungent, spicy flavor. The leaves are also edible. It is used in the treatment of several health-related issues like dysuria, and bronchitis [31]. They can also be used as antibiotics and antiseptics [102].

Cultivation requirements and management within Akwa Ibom: Propagation is by asexual reproduction. The cloves are planted in the soil and probably watered.

4.21 Aloe vera (*A. barbadensis*)

Description, utilization, knowledge, and practices: It is a perennial herb up to 160 cm tall, usually with or without a stem. It is a well-known medicinal plant with dermatological and other health benefits in Akwa Ibom. It is used in skin care, to alleviate stress, for heartburns, indigestion, etc. among other uses.

Cultivation requirements and management within Akwa Ibom: It is propagated vegetatively.

4.22 *Anthocleista djalonensis*

Description, utilization, knowledge, and practices: It is a tree growing in the wild. It is widely used as a strong purgative and diuretic. The bark is usually used to cure painful menstruation and gonorrhea. It is used as an abortifacient [26]. It is used in the treatment of several other ailments.

Cultivation requirements and management within Akwa Ibom: Propagation is by seeds.

4.23 *Acanthus montanus*

Description, utilization, knowledge, and practices: It is a prickly semi-woody herb, nearly 2 m high. The plant is used as an ornamental because of its beautiful flowers and peculiar leaves. The leaves are taken by pregnant women to stop internal lower abdominal heat. Similarly, leafy twigs are used to cure several problems such as upset stomach, painful menstruation, cough, and whooping cough [31].

Cultivation requirements and management within Akwa Ibom: It is vegetatively propagated.

4.24 *Achyranthes aspera*

Description, utilization, knowledge, and practices: *Achyranthes aspera* is a perennial, erect herb used in the treatment of dropsy, piles, and boils in children as well as of asthma, bleeding, bronchitis, cold, cough, colic, debility, dog bite, dysentery, headache, ear complications, leukoderma, pneumonia, renal complications, scorpion bite, snake bite, and skin diseases [51]. It is also used as herbal medicine, especially in obstetrics and gynecology to treat abortion, induction of labor, and postpartum bleeding [56].

Cultivation requirements and management within Akwa Ibom: It grows in the wild as an invasive species.

4.25 *Alternanthera bettzickiana*

Description, utilization, knowledge, and practices: *Alternanthera bettzickiana* is an herbaceous perennial herb. It is used in amenity planting [31]. The leaves are cooked as a vegetable and used in the management of anemia in children, as well as hepatitis, tight chest, and asthma, and to improve general health and well-being [60].

Cultivation requirements and management within Akwa Ibom: It is propagated by seeds or cuttings.

4.26 *Cyathula prostrata*

Description, utilization, knowledge, and practices: It is an annual to perennial branched herb/shrub often harvested for local use, especially as a medicine but also as food. It is used to treat many ailments including dysentery, wounds, and eye trouble [20].

Cultivation requirements and management within Akwa Ibom: It is propagated by seeds and stem cuttings.

4.27 *Allium cepa*

Description, utilization, knowledge, and practices: It is a perennial plant grown as an annual. It is used as a spice, flavoring agent, food plant, and vegetable. The immature and matured bulbs may be eaten fresh or cooked and eaten as a vegetable. *Allium cepa* is added to soups and sauces as a seasoning and flavoring agent. Onions are eaten raw to ease cough [31]. It is used as hypotensive, diuretics, and depurative.

Cultivation requirements and management within Akwa Ibom: It is propagated by sowing the bulb directly into the soil.

4.28 *Carica papaya*

Description, utilization, knowledge, and practices: Large herbaceous perennial plant with a soft single stem and sparely arranged leaves at the top of the trunk, whereas the lower trunk is scarred where leaves and fruits are born [112]. The fruits are edible. It is also used medically for weak digestion. Pawpaw juice is used in the treatment of eczema and ulcer [31]. Traditionally, the leaves are used in the treatment of malaria as well as an abortifacient and purgative agent or smoked to help remedy asthma (Titanji et al. 2008).

Cultivation requirements and management within Akwa Ibom: This plant is cultivated by seed.

4.29 *Carpolobia lutea*

Description, utilization, knowledge, and practices: It is a shrub or small tree with sweet edible fruits. The soft stem is used as a chewing stick. It is used to cure stomachaches and bone fractures and to boost sexual performance [70]. The root is used in the treatment and management of reproductive issues including as an aphrodisiac, the facilitation of childbirth, and treatment of sterility as well as other ailments like headache and worm infestation [63].

Cultivation requirements and management within Akwa Ibom: It is propagated by seed.

4.30 *Lannea acida*

Description, utilization, knowledge, and practices: It is a multipurpose shrub cultivated for its medicinal and edible values. Its berry-like fruits occur in large clusters and may be consumed fresh or dried [66]. The plant parts are used in the treatment of various ailments including skin injuries and inflammations, general body pain, gastrointestinal problems, fever and malaria, gynecological and pregnancy disorders, hemorrhoids, skin diseases, and infections [39].

Cultivation requirements and management within Akwa Ibom: Propagation is done by seeds.

4.31 *Lannea nigritana*

Description, utilization, knowledge, and practices: It is a small shrub with brownish flowers and fruits. It is used for amenity planting. The leaves are used to treat burns hence its Ibibio name (Uma-ikan). The bark is used by Akwa Ibom people to treat intestinal pains and dysentery. The leaves are administered as an enema for treating abdominal complaints [19].

Cultivation requirements and management within Akwa Ibom: Propagation is done by seeds.

4.32 *Anacardium occidentale*

Description, utilization, knowledge, and practices: It is a shrub cultivated for diverse uses including as medicine and as a source of commodities (raw material). The fruits are eaten raw or cooked and usually have an astringency that leaves the mouth feeling furry. The leaves are used as a febrifuge and in the treatment of malaria [10]. The fruit is anti-scorbutic, astringent, and diuretic [109]. The nutshell is used locally to draw tattoos. The sap or bark is considered to be a contraceptive [15]. The root is used as a purgative.

Cultivation requirements and management within Akwa Ibom: Propagation is done by seeds and usually requires very little management practices.

4.33 *Mangifera indica*

Description, utilization, knowledge, and practices: *M. indica* is a large, evergreen tree grown for its edible fruits. It is one of the oldest cultivated plants. The fruit is consumed for its fresh spicy flavored taste. The bark is used either through decoction or infusion is used to treat jaundice, rheumatism, diarrhea, and dysentery. The flowers are used to repel mosquitoes [15]. The leaves have astringent properties and are infused to reduce blood pressure and treat asthma, cough, and diabetes [15].

Cultivation requirements and management within Akwa Ibom: Propagation is done by seeds and usually requires very little management practices.

4.34 *Uvaria chamae*

Description, utilization, knowledge, and practices: It is a climbing shrub or small tree. The plant is harvested from the wild for edible fruits, while the medicinally important roots are sold in local markets. The root purgative, respiratory, and antipyretic benefits are used by Akwa Ibom people to treat dysentery, nasal congestions, and catarrh [31].

Cultivation requirements and management within Akwa Ibom: It grows in the wild.

4.35 *Monodora myristica*

Description, utilization, knowledge, and practices: it is a perennial edible plant. The seeds are used as a spice because of the nutmeg-like odor and taste. The seed has anti-sickling properties and is consumed to maintain good health [122]. The stem and bark are used in the treatment of stomachaches, fever pains, and eye disease [57].

Cultivation requirements and management within Akwa Ibom: It is cultivated by seeds.

4.36 *Caladium bicolor*

Description, utilization, knowledge, and practices: *Caladium bicolor* also called fancy-leafed caladium, elephant's ear, and heart of Jesus because of its leaves and ornamental value. In Akwa Ibom, it grows in the wild as tubers and produces berry fruits with several small ovoid seeds. Fresh leaves are used in the treatment of angina, while dried powdery leaves are placed on sores for them to heal and as facial treatments [17].

Cultivation requirements and management within Akwa Ibom: It grows in the wild.

4.37 *Tridax procumbens*

Description, utilization, knowledge, and practices: *Tridax procumbens* is a prostrate herbaceous plant that is covered with erect stiff hairs [120]. The leaves are opposite, simple, and thick and with dense hairs. The extracts of *Tridax procumbens* leaves have antimicrobial activity and stimulate wound healing [115]. In Akwa Ibom, the leaves are crushed and used in the treatment of skin spots [31].

Cultivation requirements and management within Akwa Ibom: It grows in the wild.

4.38 *Emilia sonchifolia*

Description, utilization, knowledge, and practices: *Emilia sonchifolia* is a branching, annual herb with lyrate-pinnatilobed leaves [52, 66]. In Akwa Ibom, the plant leaves are crushed to release the juice which is then applied to wound, measles, and rashes.

Cultivation requirements and management within Akwa Ibom: It is propagated wild by seed.

4.39 *Crassocephalum crepidioides*

Description, utilization, knowledge, and practices: It is an erect annual, slightly succulent herb. The fleshy, mucilaginous leaves and stems may be eaten despite the presence of plant toxins [5, 25, 38]. The leaves and stems help with indigestion and act as a laxative, purgative, and remedy for lever problems [13]. In Akwa Ibom it is also used to treat wounds, boils, and burns.

Cultivation requirements and management within Akwa Ibom: It is propagated wild by seed.

4.40 *Heliotropium indicum*

Description, utilization, knowledge, and practices: *Heliotropium indicum* is a robust annual herb that occurs as a weed [101, 105]. Traditionally, the plant leaves are used for wound healing, antidote, bone fracture, boils, febrifuge, eye infections, menstrual disorders, nerve disorders, kidney problems, and antiseptic purposes [116].

Cultivation requirements and management within Akwa Ibom: It is propagated wild by seed.

4.41 *Spathodea campanulata*

Description, utilization, knowledge, and practices: It is a tree that grows between 7 m and 25 m (23–82 ft) tall and is valued for its timber used as firewood, fodder, and fence [33, 41, 120]. The sap sometimes stains yellow on fingers and clothes. The seeds are used as a poison to kill pests [21]. The stem, bark, and leaves of *Spathodea campanulata* may be crushed and soaked in water or gin and used for the treatment of skin eruption, skin lesions, burns, ulcers, and bruises [8].

Cultivation requirements and management within Akwa Ibom: It is propagated wild by seed.

4.42 *Newbouldia laevis*

Description, utilization, knowledge, and practices: Newbouldia laevis is a fast-growing evergreen shrub or small tree [19]. In Akwa Ibom, the stem, bark, and root are used to prepare concoctions for treating boils, skin lesions, and spots.

Cultivation requirements and management within Akwa Ibom: It is propagated by seed.

4.43 *Kigelia africana*

Description, utilization, knowledge, and practices: *Kigelia africana* is a tree grows that bears poisonous fruits and can grow up to 60 cm. The wood of *Kigelia africana* is pale brown or yellowish, undifferentiated, and not prone to cracking. [108]. The stem and bark are used in Akwa Ibom in the treatment of wounds and sore.

Cultivation requirements and management within Akwa Ibom: It is propagated by seed.

4.44 *Ananas comosus*

Description, utilization, knowledge, and practices: Ananas comosus is a herbaceous, monocot biennial, or perennial tropical plant that produces fleshy, edible fruits. In Akwa Ibom, it is commonly known as Eyop mbakara. Mature pineapple is used in traditional medicine by mixing the fruit peel with other materials to treat rashes [8] or stomachache.

Cultivation requirements and management within Akwa Ibom: It is propagated by seed.

4.45 *Dacryodes klaineana*

Description, utilization, knowledge, and practices: *Dacryodes klaineana* is an evergreen perennial tree known as Eben Ikot in Akwa Ibom. The leaves and roots are used in the treatment of skin spots and rashes.

Cultivation requirements and management within Akwa Ibom: It is propagated by seed.

4.46 *Allanblackia floribunda*

Description, utilization, knowledge, and practices: *Allanblackia floribunda* is an evergreen tree, which grows up to 30 m [55]. The seeds are eaten fresh or made into jams and jellies. Stem decoction and leaves are used to treat skin infections.

Cultivation requirements and management within Akwa Ibom: It is propagated by seed.

4.47 *Combretum micranthum*

Description, utilization, knowledge, and practices: *Combretum micranthum* is a shrub is known as Asaka in Akwa Ibom. The seeds are edible. Infusions of the leaves are used in the treatment of skin lesions and spots [8].

Cultivation requirements and management within Akwa Ibom: It is propagated by seed.

4.48 *Terminalia ivorensis*

Description, utilization, knowledge, and practices: *Terminalia ivorensis* is known as Nkot ebene in Akwa Ibom. The plant is useful in the treatment of ulcers, cuts, wound sores, body pains, high temperatures, hemorrhoids, and diuresis, as well as malaria and yellow fever [20, 59].

Cultivation requirements and management within Akwa Ibom: It is propagated by seed.

4.49 *Ipomoea quamoclit*

Description, utilization, knowledge, and practices: Ipomoea quamoclit is a herbaceous vine from the family Convolvulaceae known as Ediam ikanikot in Akwa Ibom. The leaves are Poultice and are used traditionally for the treatment of boils and small open injuries.

Cultivation requirements and management within Akwa Ibom: It is propagated by seed.

4.50 *Ipomoea pileata*

Description, utilization, knowledge, and practices: *Ipomoea pileate* is an annual or perennial twining herb that grows from a thick taproot. In Akwa Ibom, it is commonly known as Mkpafiafian. The leaves are crushed in water and applied to skin spots.

Cultivation requirements and management within Akwa Ibom: It is propagated by seed.

4.51 *Costus afer*

Description, utilization, knowledge, and practices: *Costus afer* is a tall perennial semi-woody herb. In Akwa Ibom, it is generally known as Mbritem and is mostly used for several ailments including arthritis, rheumatism, eye treatments, nasopharyngeal affections, stomach pains and laxatives, sexual stimulants, antidepressants, venereal diseases, febrifuges, leprosy, dropsy, swellings, oedema, and gout stem.

Cultivation requirements and management within Akwa Ibom: It is propagated by seed.

4.52 *Bryophyllum pinnatum*

Description, utilization, knowledge, and practices: *B. pinnatum* is commonly called the resurrection plant or cathedral bells. In Akwa Ibom, it is generally known as Ndodop Afiaiyo. The leaf extract is used for the treatment of kidney stones, ulcers, bacterial and viral infections, asthma, and ulcers. It has analgesic properties, and recent research shows that it has antimutagenic and anticancer properties [127].

Cultivation requirements and management within Akwa Ibom: Propagation is done by seeds. In addition, plantlets arise from the edges of their leaves and take root when it drops on the ground. It requires little care, but during the hot seasons, it should be shielded from direct sunlight or else the leaves will be burned.

4.53 *Citrullus colocynthis*

Description, utilization, knowledge, and practices: *C. colocynthis* is a perennial creeping herbaceous vine commonly found in desert areas. It looks like a watermelon, and common names include bitter apple, desert gourd, bitter apple, vine of Sodom, etc. In Akwa Ibom, it is called Ikon. The seed oil has antioxidant, anticancer, antidiabetic, and insecticidal properties. The fruit is bitter and pungent and used as a purgative, anthelminthic, and carminative cure for tumors, ulcers, asthma, leukoderma, etc. The seeds are crushed and used to clear skin spots and abscesses [2].

Cultivation requirements and management within Akwa Ibom: Cultivation is by seeds. It requires good light and good soil-based compost. It requires care only at the initial stage because it thrives in pots and nursery beds at the initial stage.

4.54 *Momordica charantia*

Description, utilization, knowledge, and practices: *M. balsamina* is commonly called an African pumpkin. In Akwa Ibom, it is called Mbiadon Edon. It is a tendril-bearing, wild climber with small lobed leaves, monoecious flowers, and ovoid ellipsoid fruits which are softly warted and fleshy. The leaves, fruits, seeds, and bark contain resins, alkaloids, flavonoids, glycosides, steroids, terpenes, cardiac glycosides, and saponins that are known to have medicinal properties including antiviral, antiplasmodial, antidiarrheal, antiseptic, antibacterial, antiviral, antimicrobial, analgesic, and hepatoprotective properties [40]

Cultivation requirements and management within Akwa Ibom: Cultivation is by seeds. Care is required at the initial stage because it thrives better in a nursery.

4.55 *Dioscorea dumetorum*

Description, utilization, knowledge, and practices: *D. dumetorum* is commonly called bitter yam. In Akwa Ibom, it is called Anim. It has distinct trifoliate leaves each having 3–7 veins per leaflet. The vine has ridged ascending prickles and a characteristic twining to the left. The vine grows annually from the underground tubers which are deeply lobed and occur in clusters just below the surface of the soil.

Bitter yam lowers blood sugar level, boosts fertility, reduces inflammation, enhances brain function, and prevents cancer [4].

Cultivation requirements and management within Akwa Ibom: Propagation is by seed or cutting off the tuber, setts, or bulbils. It requires staking for maximum yield.

4.56 *Euphorbia hirta*

Description, utilization, knowledge, and practices: *E. hirta* is commonly referred to as an asthma plant and is a hairy herb In Akwa Ibom, it is called Etinkene ekpo. It exudes white latex as soon as it is cut. It is used as a traditional medicine to treat asthma, skin ailments, and hypertension [110].

Cultivation requirements and management within Akwa Ibom: It goes in the wild.

4.57 *Jatropha curcas*

Description, utilization, knowledge, and practices: *J. curcas* is a semi-evergreen shrub or small tree. In Akwa Ibom, it is called Ukim Eyio. It is used for the treatment of rheumatism, toothache, jaundice, dysentery, and gum bleeding and inhibits HIV by inducing cytopathic effects with low cytotoxicity, etc. [100].

Cultivation requirements and management within Akwa Ibom: It can be cultivated by both seeds or stem cuttings and it needs little management to survive.

4.58 *Manihot esculenta*

Description, utilization, knowledge, and practices: *M. esculenta* is commonly called cassava.

In Akwa Ibom, it is called Iwa.

It is a perennial shrub plant with conspicuous, almost palmate alternate leaves which are deeply parted into 5–9 lobes. It produces elongated tubers. The stems can grow up to 5 m high. It is used to induce labor, treats dehydration in people with diarrhea, and is effective for tiredness and sepsis [11].

Cultivation requirements and management within Akwa Ibom: This is mostly propagated using stem cuttings. To enhance productivity, the stem cuttings must be planted in the right position on flat ground for soft soil and on mounds for hard soil.

4.59 *Manniophyton fulvum*

Description, utilization, knowledge, and practices: *M. fulvum* is a straggling bush or liane and in Akwa Ibom is called Ekonikon. Its stem can be about 30 m long to

10 cm thick. It is used for the treatment of whooping cough, leprosy, dysentery, piles, and painful menses [20].

Cultivation requirements and management within Akwa Ibom: It grows in the wild, although it can be propagated by seed and requires little management.

4.60 *Mallotus oppositifolius*

Description, utilization, knowledge, and practices: *M. oppositifolius* is a shrub or small tree with pale greenish-gray bark which is slightly rough and scaly and covered with mixed simple and stellate hairs [104]. In Akwa Ibom, it is commonly called Uman nwariwa. The leaves are useful for the treatment of epilepsy, diarrhea, inflammation, dysentery, convulsion, pain, etc. [3].

Cultivation requirements and management within Akwa Ibom: Propagation is by seeds, although vegetative propagation may be possible given its easy growth. They have a self-supporting growth form and hence do not need much management for survival. Similarly, it grows in the wild.

4.61 *Afzelia africana*

Description, utilization, knowledge, and practices: *A. africana* is a tree in Akwa Ibom called Eyin Mbukpo. The leaves are alternate, petiolated, or paripinnate, up to 30 cm long with 7–17 pairs of leaflets; it has sweet-smelling flowers and produces oblong straight flattened dehiscent pod fruits which are black-brown [27, 34, 92]. The leaves are used for pain relief and treating digestive problems such as constipation and vomiting and internal bleeding.

Cultivation requirements and management within Akwa Ibom: It is propagated by seeds and vegetative techniques through budding. Since the seeds are readily preyed upon by animals, care should be taken for the right depth while planting.

4.62 *Acacia ataxacantha*

Description, utilization, knowledge, and practices: *A. ataxacantha* is a woody, shrub or small tree growing up to 10 m in height. In Akwa Ibom, it is called Mbara Okpok and is used in the treatment of abscesses, backache, cough, dental caries, toothache, pneumonia, malaria, sores, wounds, stomach problems, etc. [61].

Cultivation requirements and management within Akwa Ibom: Propagation is by seeds as well as stem cuttings.

4.63 *Afzelia bella*

Description, utilization, knowledge, and practices: *A. bella* is a tall tree. The leaves produce sweet-scented flowers. The ground bark is used for the treatment of topical skin infections. Bark decoctions and macerations are taken to treat intestinal parasites, diarrhea, menstruation problems, rheumatism, and hemorrhoids and can be used as a tonic. Leaves are administered against constipation [71]. It is commonly called pod mahogany and referred to as Enyin Mbukpo in Akwa Ibom.

Cultivation requirements and management within Akwa Ibom: Propagation is via seeds which must be sown no deeper than 2 cm with the hilum facing downward. Little management is required.

4.64 *Baphia nitida*

Description, utilization, knowledge, and practices: *B. nitida* is commonly called camwood and Afuo in Akwa Ibom. It is a shrubby, leguminous, hardwooded tree with evergreen leaves. It can grow to about 4–5 m tall. It has characteristic smooth, green leaves which are oval shaped measuring about 10–15 cm long. It produces flowers that are white and pea-like and have a yellow center. The leaves have inflammatory, antidiarrheal, and analgesic effects. The extract of camwood can be formed into a soft soap-like material that promotes healthy skin [111].

Cultivation requirements and management within Akwa Ibom: Propagation is by seeds and cuttings (from young parts). It does not require special management [22].

4.65 *Cajanus cajan*

Description, utilization, knowledge, and practices: *Cajanus cajan* is called Nkoti in Akwa Ibom and is an erect, branched, hairy shrub. The leaves are used for treating diabetes, sores, skin irritations, hepatitis, measles, jaundice, dysentery, stabilizing menstrual period, expelling bladder stones, etc. [126].

Cultivation requirements and management within Akwa Ibom: Propagation is via seeds in holes 2 m apart. The management required constant weeding for the first 2 months.

4.66 *Cassia alata*

Description, utilization, knowledge, and practices: *Cassia alata* is called Adaya Okon in Akwa Ibom. It is an annual shrub or occasionally a biannual herb with zygomorphic flowers. Seed extracts are used to cure skin diseases, while the leaves are used to treat constipation and fresh leaves are applied to the skin to treat fungus [90].

Cultivation requirements and management within Akwa Ibom: Propagation is by seeds.

4.67 *Glycine max*

Description, utilization, knowledge, and practices: *Glycine max* is called Nkoti eto in Akwa Ibom. It is an erect, bushy, hairy annual legume. Seeds are used for the prevention and treatment of cardiovascular diseases, diabetes, cancer, obesity, cholesterol, etc. [14].

Cultivation requirements and management within Akwa Ibom: Cultivation is by seeds and requires little management.

4.68 *Lonchocarpus cyanescens*

Description, utilization, knowledge, and practices: *Lonchocarpus cyanescens* is a shrub. In Akwa Ibom, it is called Awa. All the plant parts are medicinal and can be used in the treatment of diarrhea, leprosy, yaws, arthritis, rheumatism, genital stimulants, dysentery, etc. [36].

Cultivation requirements and management within Akwa Ibom: Propagation is by seeds and it requires little management to thrive.

4.69 *Parkia biglobosa*

Description, utilization, knowledge, and practices: *Parkia biglobosa* is called Ukon Uyayak in Akwa Ibom and is a perennial deciduous tree. It is useful for the treatment of hypertension, wound healing, antimalaria, management of bacterial infection, etc. [53].

Cultivation requirements and management within Akwa Ibom: Propagation is by seeds and it doesn't require special management.

4.70 *Pentaclethra macrophylla*

Description, utilization, knowledge, and practices: *Pentaclethra macrophylla* is called Ukana in Akwa Ibom. It is a large size tree with log bipinnate compound leaves. It has antibacterial and antimicrobial properties, and the seeds are used to treat infertility.

Cultivation requirements and management within Akwa Ibom: Cultivation is by seeds and little management is required to make it grow.

4.71 *Pterocarpus erinaceus*

Description, utilization, knowledge, and practices: *Pterocarpus erinaceus* is commonly called bar wood and also known as Ukpa in Akwa Ibom. The leaves are 3–7-cm-long petioles with hairy pinnae. It has a golden yellow flower that grows up to 13 mm long. The fruit is straw colored, circular, flattened, and indehiscent. It has a seeded pod of 4–7 cm in diameter. These pods contain seeds. It is used to treat inflammatory disease, malaria, anemia, ulcer, rheumatism, dermatitis, etc.

Cultivation requirements and management within Akwa Ibom: It is propagated by planting nursery-raised seedlings or rooted cuttings. Careful management is required for its survival before sowing which requires dipping the seeds in water or sulfuric acid for some time. After germination, frequent pruning is required [123].

4.72 *Pterocarpus santalinoides*

Description, utilization, knowledge, and practices: *Pterocarpus santalinoides* is a tree with slash yellowish-white exuding drops of red gum when slashed. It produces fruits that are orbicular with the flattened indehiscent pod which is hairy and pal brown in color. The pods have seeds [58]. It is commonly called Mututi in English and Nkpa-inyan in Akwa Ibom. It is used for the treatment of gastroenteritis.

Cultivation requirements and management within Akwa Ibom: Cultivation is by seeds.

4.73 *Tetrapleura tetraptera*

Description, utilization, knowledge, and practices: *Tetrapleura tetraptera* is commonly called aridan or gum tree in English and Uyayak in Akwa Ibom. It is a single-stemmed, robust, perennial tree [54]. It is used for the treatment of diabetes, hypertension, asthma, epilepsy, schistosomiasis, arthritis, etc.

Cultivation requirements and management within Akwa Ibom: Cultivation is by seeds that require scarification for germination due to their hard coats.

4.74 *Harungana madagascariensis*

Description, utilization, knowledge, and practices: *Harungana madagascariensis* is called Oton in Akwa Ibom and is a small bushy tree. It produces almond-scented white or cream flowers and globular berry-like small fruits [16]. It is used in the treatment of chest pain, urogenital infections, ringworm, eczema, malaria, etc.

Cultivation requirements and management within Akwa Ibom: Propagation is by seeds and it requires thinning after planting to enhance growth.

4.75 *Irvingia gabonensis*

Description, utilization, knowledge, and practices: *Irvingia gabonensis* is commonly called Uyo in Akwa Ibom. It is useful for weight loss and treatment of high cholesterol and diabetes [18].

Cultivation requirements and management within Akwa Ibom: Cultivation is by seeds.

5 Conclusion

The plants documented here are regularly used by Akwa Ibom people as food and medicine despite the lack of any elaborate institutional effort to manage, protect, and conserve the genetic resources. The food and medicinal benefits of these plants are linked to the phytochemical composition and non-secondary metabolites that are bioactive at different levels because their ingestion in diverse forms influence biological processes. This is unsustainable because of the ongoing biodiversity loss, climate change, and sociocultural transitions away from traditional past and calls for formal management strategies. Knowledge about the food and medicinal values of these plants are mostly held by community elders, village heads, chiefs, head of households, academics, researchers, and traditional medical practitioners. There is a need to undertake the collection, characterization, and conservation of the plant resources recorded in here to protect them from ongoing changes and ensure future generations can benefit from these resources. By documenting the relevant information connected to the utilization and cultivation of these plant resources as well as encouraging their cultivation on large scales, employing the use of plant breeding programs for better qualities and employing the use of conservation methods for all varieties, these resources and their inherent values can be saved

References

1. Abbiv DK (1990) Useful plants of Ghana: west African uses of wild and cultivated plants. Intermediate Technology Publications and The Royal Botanic Gardens, Kew
2. Abdalbasit A, Robert LJ (2022) Antioxidant, antimicrobial, and antidiabetic activities of Citrullus colocynthis seed oil. In: Abdalbasit AM (ed) Multiple biological activities of unconventional seed oils. Academic Press. pp 139–146. https://doi.org/10.1016/B978-0-12-824135-6.00005-2
3. Adetunji TL et al (2022) Mallotus Oppositiofolius (Geisler) Mull: the first review of its botany, ethnomedicinal uses, phytochemistry and biological activities. S Afr J Bot 147(2):245
4. Adigoun A, Akintola F, Adoukonou S, Hubert FC, Tchougourou A, Agassounon TM, Ahanhanzo C (2019) Diversity, distribution and ethnobotanical importance of cultivated and wild African trifoliate yam (*Dioscorea dmetorum*) in Benin. Genet Resour Crop Evol 66:659–683
5. Agbogidi OM (2010) Ethno-botanical survey of non-timber forest products in Sapele local government of Delta State. Niger Afr J Plant Sci 4(6):183–189
6. Ahana MC, Osawaru ME, Ogwu MC (2022) Status and potentials of the genetic resources of Cocoyam (*Xanthosoma* Schott., Araceae) in Nigeria. In: Proceedings of Cukurova 8th international scientific researches conference, Adana Turkey. 15–17 April, 2022, pp 1–11
7. Aiyelaagbe IO, Kintomo AA (2002) Nitrogen response of fluted pumpkin (*Telfairia occidentalis* Hook. F) grown solely or intercropped with Banana. Nutr Cycl Agroecosyst 64: 231–235
8. Ajibesin KK, Ekpo BA, Bala DN, Essien EE, Adesanya SA (2008) Ethnobotanical survey of Akwa Ibom State of Nigeria. Journal of ethnopharmacology 115(3):387–408. https://doi.org/10.1016/j.jep.2007.10.021
9. Ajibesin KK, Ekpo BA, Bala DN, Essien EE, Adesanya SA (2008) Ethnobotanical survey of Akwa Ibom state of Nigeria. J Ethnopharmacol 115(3):387–408

10. Aliyu OM (2017) Analysis of absolute nuclear DNA content reveals a small genome and intra-specific variation in Cashew (Anacardium occidentale L.), Anacardiaceae" Silvae Genetica, 63:285–292. https://doi.org/10.1515/sg-2014-0036
11. Alves AAC (2002) Cassava botany and physiology. In: Hillocks RJ, Thresh JM, Bellotti AC (eds) Cassava: biology, production and utilization. CABI Publishing, New York, pp 67–89
12. Asaolu SS, Adefemi OS, Oyahkilome IG, Ajibulu KE, Asaolu MF (2012) Proximate and mineral consumption of Nigeria leafy vegetables. J Food Res 1:214–218
13. Ayodele AE (2007) The medicinally important leafy vegetables of South Western Nigeria. [Online] http://www.siu.edu/webl/leaflets/ayodele.ht
14. Bachheti RK, Worku LA, Gonfa YH, Zebeaman M, Deepti, Pandey DP, Bachheti A (2022) Prevention and treatment of cardiovascular diseases with plant phytochemicals: a review. Evid Based Complement Alternat Med 2022:5741198. https://doi.org/10.1155/2022/5741198
15. Barwick M (2004) Tropical & subtropical trees. A worldwide encyclopaedic guide. Thames & Hudson, London, 64 p
16. Beentje HJ (1994) Kenya trees, shrubs and lianas. National Museums of Kenya
17. Biswas MK, Mridha SA, Rashid MA, Sharmin T (2013) Membrane stabilizing and antimicrobial activities of *Caladium bicolor* and *Chenopodium album*. IOSR J Pharm Biol Sci 6(5): 62–65
18. Brennan D (2021) What Are the Benefits of African Mango? https://www.medicinenet.com/what_are_the_benefits_of_african_mango/article.htm
19. Burkil HM (2004)Brief descriptions and details of the uses of over 4,000 plants. A superb, if terse, resource, it is also available on http://www.aluka.org/
20. Burkill HM (1985) The useful plants of west tropical Africa, families A – D, vol 1, 2nd edn. Royal Botanic Garden, Kew. 960p
21. CABI (2021) *Spathodea campanulata*. In: Invasive species compendium. CAB International, Wallingford. https://www.cabi.org/isc/datasheet/51139#tosummaryOfInvasiveness
22. Cardon D, Jansen PCM (2005) *Baphia nitida* Lodd. Plant Resources of Tropical Africa, Wageningen
23. Chime AO, Aiwansoba RO, Danagogo SJ, Egharevba II, Osawaru ME, Ogwu MC (2016) Effects of lining with leaves of *Triclisia dictyophylla* on the fungal composition of *Cola nitida* during storage. J Ind Res Technol 5(2):128–136
24. Chime AO, Aiwansoba RO, Ogwu MC (2018) Pathological status of plant germplasm and sustainable crop production and conservation. J Energy Nat Resour Manag 1:17–21
25. Dairo FAS, Adanlawo IG (2007) Nutritional quality of *Crasocephalum crepidioides* and Seneciobiafrae. Pak J Nutr 6(1):35–39. https://doi.org/10.3923/pjn
26. De Ruijter A (2007) Anthocleista djalonensis. A. Chev. Record from Protobase. Schmelzer G H. & Gurib-Fakin A. PROTA (Plant Resourses of Tropical Africa). [Online]
27. Donkpegan ASL, Hardy OJ, Lejeune Ph, Oumorou M, Daïnou K, Doucet JL (2014) On a species complex, Afzelia, in African forests of economic and ecological interest. A review. Biotechnol Agron Soc Environ 18(2):233–246
28. Effah G, Ekpenyong B, Babatunde G, Ajayi I, Dairo D (2019) Compliance with malaria rapid diagnostic test results and correlates among clinicians in Uyo, Akwa Ibom State, Nigeria: 2018. Open J Epidemiol 9:259–288
29. Ekanem A (2006) Antidiabetic activity of ethanolic leaf extract and fractions of *Lasianthera africana* on alloxan diabetic rats. M.Sc. thesis, University of Uyo, Nigeria. 67p
30. Etetim EN, Okokon J, Useh MF (2008) Pharmacological screening and evaluation of anti-plasmodial activity of *Gongronema latifolium* against *Plasmodium berghei* infection in mice. Niger J Health Biomed Sci 7(2):51–55
31. Etukudo I (2003) Ethnobotany: conventional and traditional use of plants, 1st edn. The Verdict Press, Akwa Ibom State, pp 4–108
32. Fasuyi AO (2006) Nutritional potential of some tropical vegetable meal: chemical characterisation and functional properties. Afr J Biotechnol 5:49–53

33. Fongod AGN, Ngoh LM, Veranso MC (2014) (2014). Ethnobotany, indigenous knowledge and unconscious preservation of the environment: an evaluation of indigenous knowledge in south and southwest regions of Cameroon. Int J Biodivers Conserv 6:85–99. https://doi.org/10.5897/ijbc2013.0637
34. Gérard J, Louppe D (2011) Afzelia africana Sm. ex Pers. In: Lemmens RHMJ, Louppe D, Oteng-Amoako AA (eds) PROTA (Plant resources of tropical Africa/Ressources végétales de l'Afrique tropicale). Wageningen
35. Gill LS (1992) Ethnomedical uses of plants in Nigeria. Uniben Press, University of Benin, Benin City, pp 228–229
36. Gillow J (2010) Textiles of the Islamic World. Thames & Hudson
37. Gregory EW (2015) Vegetable production and practices. CABI, Boston, pp 1–2
38. Grubben GJH (2004) Vegetables, volume 2 of plant resources of tropical Africa. PROTA. ISBN 90-5782-147-8
39. Gruca M, Cámara-Leret R, Macía MJ, Balslev H (2014) New categories for traditional medicine in the economic botany data collection standard. J Ethnopharmacol 155:1388–1392
40. Gulab Singh T et al (2009) *Mormordica balsamina*: a medicinal neutraceutical plant for health care management. Curr Pharm Biotechnol 10(7):667–682
41. Hargreaves D, Hargreaves B (1964) Tropical trees of Hawaii. Hargreaves, Kailua
42. Harshberger JW (1896) The purpose of ethnobotany. Bot Gaz 21:146–158
43. Idu M (2009) Current trends in ethnobotany. Trop J Pharm Res 8(4):12–24
44. Ikhajiagbe B Ogwu MC, Ogochukwu OF, Odozi EB, Adekunle IJ, Omage ZE (2021) The place of neglected and underutilized legumes in human nutrition and protein security. Crit Rev Food Sci Nutr. https://doi.org/10.1080/10408398.2020.1871319
45. Ikhajiagbe B, Atoe R, Ogwu MC, Loveniers P-J (2021) Changes in *Telfaria occidentalis* leaf morphology, quality and phytochemical composition under different local preservation regimes. VEGETOS Int J Plant Res Biotechnol 31(1):29–36. https://doi.org/10.1007/s42535-021-00188-z
46. Imaobong U, Promise E (2013) Assessment of proximate compositions of twelve edible vegetables in Nigeria. Int J Modern Chem 4(2):79–89
47. Imarhiagbe O, Ogwu MC (2022) Sacred groves in the global south: a panacea for sustainable biodiversity conservation. In: Chibueze Izah S (ed) Biodiversity in Africa: potentials, threats and conservation. Sustainable development and biodiversity, vol 29. Springer, Singapore, pp 525–546. https://doi.org/10.1007/978-981-19-3326-4_20
48. Isong EU, Adewusi SAR, Nkanga EU, Umoh EE, Offiong EE (1998) Nutritional and phytogeriatological studies of three varieties of *Gnetum africanum* (afang). Food Chem 64:489
49. Itah AY (1996) Screening of plant's parts for fungicidal properties. Trans Nig Soc Bio Conserv 4(1):26–40
50. Itah AY (1997) Bactericidal and bacteriostatic effect of edible leafy vegetable extract on growth of canned food borne bacteria. Trans Nig Soc Bio Conserv 6:103–111
51. Jain SP, Singh J (2010) Traditional medicinal practices among the tribal people of Raigarh (Chatisgarh), India. Indian J Nat Prod Resour 1(1):109–115
52. Jeffrey C (1986) Notes on Compositae: IV. The Senecioneae in East Tropical Africa. Kew Bull 41(4):873
53. Karou S, Tchacondo T, Tchibozo MD, Abdoul-Rahaman S, Anani K, Kouduvo K et al (2011) Ethnobotanical study of botanical plats used in the management of diabetes mellitus and hypertension in the central region of Togo. Pharm Biol 49:1286
54. Katende AB et al (1995) Useful trees and shrubs for Uganda. Identification, propagation and management for Agricultural and Pastoral communities, Regional Soil Conservation Unit (RSCU), Swedish International Development Authority (SIDA)
55. Keay RWJ (1989) Trees of Nigerian. A reverse version of Nigeria trees. Clarendo Press, Oxford. 281 pp
56. Khan MTJ, Ahmad K, Alvi MN, Noor-Ul-Amin B, Mansoor M, Asif Saeed FZ, Khan M, Jamshaid S (2010) Achyranthes aspera. Pak J Zool 42(1):93–97

57. Koudou J, Etou OAW, Akikokou K, Abenna AA, Gbeassor M, BessiereJM (2007) Chemical composition and hypertensive effects ofessential oil of Monodora myristica (Gaerth). J Boil Sci 7:937–942. https://doi.org/10.3923/jbs.2007.937.942
58. Lemmens RHMJ (2008) Pterocarpus santalinoides DC. In: Louppe D, Oteng-Amoako AA, Brink M (eds) Plant resources of tropical Africa. Backhuys Publishers, Leiden Netherlands/ CTA, Wageningen Netherlands. 704p
59. Lincoln WA (1986) World woods in colour. Stobard Davies, Hertford. ISBN 0-85442-028-2
60. Manan M, Saleem U, Akash MSH, Qasim M, Hayat M, Raza Z, Ahmad B (2020) Antiarthritic potential of comprehensively standardized extract of Alternanthera bettzickiana: in vitro and in vivo studies. ACS Omega 5(31):19478–19496. https://doi.org/10.1021/acsomega.0c01670
61. Maroyi A (2018) Review of ethnopharmacology and phytochemistry of *Acacia ataxacantha*, vol 17, p 2301
62. Mepba HD, Eboh L, Banigo DEL (2002) Effects of processing treatments on the nutritive composition and consumer acceptance of some Nigerian edible leafy vegetables. Afr J Food Agric Nutr Dev 11:12–24
63. Mitaine-Offer A, Miyamoto T, Khan IA, Delaude C, Lacaille-Dubois MA. (2002) Three new triterpene saponins from two species of Carpolobia. J Nat Prod 65:533–537. https://doi.org/10.1021/np010546e
64. Mohammed MI, Sharif N (2011) Mineral composition of some leafy vegetables consumed in Kano, Nigeria. J Basic Appl Sci 19:208–211
65. Mosha TC, Gaga HE (1999) Nutritive value and of blanching on trypsin and chymotrypsin inhibitor activities of selected leafy vegetables. Plant Foods Human Nutr 54:271–283
66. Natural Resources Conservation Service (NRCS) (2008) PLANTS profile, *Emilia sonchifolia*. The PLANTS Database. United States Department of Agriculture. Washington, USA
67. Nkang, N.M., Omonona, B.T and Ibana, S.E. (2006). Modelling water demand and use behavior of dry season waterleaf (*Talinum triangulare*) cultivators in Calabar, Nigeria: A discrete choice approach. J Agric Soc Sci, 2(4):242–248
68. Nkang A, Omokaro D, Egbe A, Amanke G (2002) Nutritive value of *Telfairia occidentalis*. Afr J Biotechnol 2(3):33–39
69. Nnamani CV, Oselebe HO, Okporie EO (2010) Aspect of ethnobotany of traditional leafy vegetables utilized as human food in rural tropical communities. Animal Research International 7(1):1110–1115
70. Nwidu LL, Airhihen B, Ahmadu A (2016) Anti-inflammatory and anti-nociceptive activities of stem-bark extracts and fractions of Carpolobia lutea (Polygalaceae). J Basic Clin Pharm 8 (1):25–32. https://doi.org/10.4103/0976-0105.195097
71. Ofeimum J et al (2020) In vitro antioxidant and antimicrobial activities of methanol leaf extract and fractions of *Afzellia bellaHarms* (Fabaceae). Ethiop Pharm J 36(1):19
72. Ogwu MC (2019) Towards sustainable development in Africa: the challenge of urbanization and climate change adaptation. In: Cobbinah PB, Addaney M (eds) The geography of climate change adaptation in urban Africa. Springer Nature, Cham, pp 29–55. https://doi.org/10.1007/978-3-030-04873-0_2
73. Ogwu MC (2020) Value of *Amaranthus* [L.] species in Nigeria. In: Waisundara V (ed) Nutritional value of Amaranth. IntechOpen, London, pp 1–21. https://doi.org/10.5772/intechopen.86990
74. Ogwu MC, Osawaru OP (2022) State of the genetic resources of West African Okra (*Abelmoschus caillei* [A. Chev.] Stevels.): a taxon with industrial potentials. In: Ilkim M, Amanzholova A (eds) Proceedings of the II Başkent International Conference on Multidisciplinary Studies, Ankara, Turkey, pp 93–102
75. Ogwu MC, Osawaru ME (2022) Traditional methods of plant conservation for sustainable utilization and development. In: Chibueze Izah S (ed) Biodiversity in Africa: potentials, threats and conservation. Sustainable development and biodiversity, vol 29. Springer, Singapore, pp 451–472. https://doi.org/10.1007/978-981-19-3326-4_17

76. Ogwu MC, Osawaru ME, Atsenokhai EI (2016) Chemical and microbial evaluation of some uncommon indigenous fruits and nuts. Borneo Sci 37(1):54–71
77. Ogwu MC, Osawaru ME, Aiwansoba RO, Iroh RN (2016) Status and prospects of vegetables in Africa. In: Borokini IT, Babalola FD (eds) Conference proceedings of the joint biodiversity conservation conference of Nigeria Tropical Biology Association and Nigeria chapter of Society for Conservation Biology on MDGs to SDGs: toward sustainable biodiversity conservation in Nigeria. University of Ilorin, Nigeria, pp 47–57
78. Ogwu MC, Osawaru ME, Aiwansoba RO, Iroh RN (2016) Ethnobotany and collection of West African Okra [*Abelmoschus caillei* (A. Chev.) Stevels] germplasm in some communities in Edo and Delta States, Southern Nigeria. Borneo J Resour Sci Technol 6(1):25–36. https://doi.org/10.33736/bjrst.212.2016
79. Ogwu MC, Osawaru ME, Obahiagbon GE (2017) Ethnobotanical survey of medicinal plants used for traditional reproductive care by Usen people of Edo State, Nigeria. Malaya J Biosci 4 (1):17–29
80. Ogwu MC, Chime AO, Oseh OM (2018) Ethnobotanical survey of tomato in some cultivated regions in Southern Nigeria. Mal Natl Res J 6(1):19–29
81. Ogwu MC, Chime AO, Aiwansoba RO, Emere AO (2019) Effects of storage methods and duration on the microbial composition and load of tomato (*Solanum lycopersicum* [L.], Solanaceae) fruits. Bitlis Eren Univ J Sci Technol 9(1):1–7. https://dergipark.org.tr/download/article-file/746056
82. Ogwu MC, Izah SC, Iyiola AO (2022) An overview of the potentials, threats and conservation of biodiversity in Africa. In: Chibueze Izah S (ed) Biodiversity in Africa: potentials, threats and conservation. Sustainable development and biodiversity, vol 29. Springer, Singapore, pp 3–20. https://doi.org/10.1007/978-981-19-3326-4_1
83. Ogwu MC, Osawaru ME, Owie MO (2023) Effects of storage at room temperature on the food components of three cocoyam species (*Colocasia esculenta, Xanthosoma atrovirens, and X. sagittifolium*). Food Stud Interdiscip J 13(2):59–83. https://doi.org/10.18848/2160-1933/CGP/v13i02/59-83
84. Okafor JC (1981) Woody plants of nutritional importance in traditional farming systems of the Nigerian humid tropics. Ph.D. thesis, University of Ibadan, Ibadan, Nigeria, 264p
85. Okigbo BN (1986) Broadening the food base in Africa. The potential of traditional food plants. Food Nutr 12:4–17
86. Okokon JE, Antia BS, Essiet GA (2007) Evaluation of in vivoantiplasmodial activity of ethanolic leaf extract of *Lasianthera africana*. Res J Pharmacol 1(2):30–33
87. Okokon JE, Antia BS, Umoh BB (2009) Antiulcerogenic activity of ethanolic leaf extract of *Lasianthera africana*. Afr J Tradit Complement Alternat Med 6(2):150–154
88. Okoli BE, Mgbeogu CM (1983) Fluted pumpkin, *Telfairia occidentalis*. West African vegetable crop. Biol Sci 37(2):145–149
89. Okon OG, James US (2014) Proximate and mineral composition of some traditional vegetables in Akwa Ibom State, Nigeria. Int J Sci Res Publ 4(8):5–8
90. Oladele AT, Elujoba AA, Oyelami AO (2012) Clinical Studies of Three Herbal Soaps in the Management of Superficial Fungal Infections. Res J Med Plants 6:56–64
91. Olaiya C, Adebisi J (2010) Phytoevaluation of the nutritional values of ten green leafy vegetables in South Western Nigeria. Int J Nutr Wellness 9(2):1937–8297
92. Orwa C, Mutua A, Kindt R, Jamnadass R, Anthony S (2009) Agroforestree database: a tree reference and selection guide version 4.0. World Agroforestry Centre, Kenya
93. Osawaru ME, Ogwu MC (2014) Ethnobotany and germplasm collection of two genera of cocoyam (*Colocasia* [Schott] and *Xanthosoma* [Schott], Araceae) in Edo State Nigeria. Sci Technol Arts Res J 3(3):23–28
94. Osawaru ME, Ogwu MC (2020) Survey of plant and plant products in local markets within Benin City and environs. In: Filho LW, Ogugu N, Ayal D, Adelake L, da Silva I (eds) African handbook of climate change adaptation. Springer Nature, Cham, pp 1–24. https://doi.org/10.1007/978-3-030-42091-8_159-1

95. Osawaru ME, Ogwu MC, Ahana CM (2013) Current status of plant diversity and conservation in Nigeria. Niger J Life Sci 3(1):168–178
96. Osawaru ME, Ogwu MC, Ogbeifun NS, Chime AO (2013) Microflora diversity of the phylloplane of wild Okra (*Corchorus olitorius* L. Jute). Bayero J Pure Appl Sci 6(2):136–142
97. Osawaru ME, Ogwu MC, Imarhiagbe O (2013) Biochemical characterization of some Nigerian *Corchorus* L. species. Bayero J Pure Appl Sci 6(2):69–75
98. Osawaru ME, Ogwu MC, Imarhiagbe O (2013) Agro-morphological characterization of some Nigerian *Corchorus* (L.) species. Biol Environ Sci J Tropics 10(4):148–158
99. Osawaru ME, Ogwu MC, Omoigui ID, Aiwansoba RO, Kevin A (2016) Ethnobotanical survey of vegetables eaten by Akwa Ibom people residing in Benin City, Nigeria. Univ Benin J Sci Technol 4(1):70–93
100. Oskoueian E, Oskoueian A, Shakeri M, Jahromi MF (2021) Benefits and challenges of Jatropha meal as novel biofeed for animal production. Vet Sci 8(9):179. https://doi.org/10.3390/vetsci8090179
101. Osungunn MO, Adedeji KA (2005) Phytochemical and antimicrobial screening of methanol extract of *Heliotropium indicum* leaf. J Microbiol Antimicrob 3(8):213–216
102. Pamplona-Roger GD (1998) Encyclopedia of medicinal herbs 1. Education and Health Library, Editorial Safeliz SL. Aravaca, Madrid, pp 230–377
103. Petters SW, Iwok ER, Uya OE (1994) Akwa Ibom State: the land of Promise – a compendium. Gabumo Press, Lagos, pp 19–244
104. Radcliffe-Smith A (1996) Euphorbiaceae. Flora Zambesiaca 9(4)
105. Rahman MA, Mia M, Shahid I (2001) Pharmacological and phytochemical screen activities of roots of *Heliotropium indicum* Linn. Pharmacol On Line 1(1):185
106. Remison SU (2005) Arable and vegetable crops of the tropics. Gift Prints Associates, Edo State, pp 112–202
107. Reznicek AA, Voss EG, Walters BS (2011) Michigan Flora. University of Michigan. Web, http://michiganflora.net/species.aspx?id=883. Accessed 12 Mar 2021
108. Roodt V (1992) Kigelia africana in the Shell field guide to the common trees of the Okavango Delta and Moremi game reserve. Shell Oil Botswana, Gaborone
109. Salehi B, Gültekin-Özgüven M, Kırkın C, Özçelik B, Morais-Braga MFB, Carneiro JNP, Bezerra CF, Silva TGD, Coutinho HDM, Amina B, Armstrong L, Selamoglu Z, Sevindik M, Yousaf Z, Sharifi-Rad J, Muddathir AM, Devkota HP, Martorell M, Jugran AK, Martins N, ... Cho WC (2019) Anacardium plants: chemical, nutritional composition and biotechnological applications. Biomolecules 9(9):465. https://doi.org/10.3390/biom9090465
110. Sharma NK, Dey S, Prasad R (2007) In vitro antioxidant potential evaluation of *Euphorbia hirta* L. Pharmacologyonline 1:91–98
111. Soladoye MO (1985) A revision of Baphia (Leguminosae-Papilionoideae). Kew Bull 40(2): 291
112. Srivastava AK, Singh VK (2016) Carica Papaya – a herbal medicine. Int J Res Stud Biosci 4(11):19–25
113. Sudeep HV, Aman K, Jestin TV, Shyamprasad K (2022) Aframomum melegueta seed extract with standardized content of 6-paradol reduces visceral fat and enhances energy expenditure in overweight adults – a randomized double-blind, placebo-controlled clinical study. Drug Des Devel Ther 16:3777–3791. https://doi.org/10.2147/DDDT.S367350
114. Swerdlow JL (2000) Natures medicine-plants that heals. National Geographic Society, Washington, DC
115. Taddel A, Rosas Romero AJ (2000) Bioactivity studies of extracts from *Tridax procumbens*. Phytomedicine 7(3):235–238
116. Togola A, Diallo D, Dembélé S, Barsett H, Paulsen BS (2005) Ethnopharmacological survey of different uses of seven medicinal plants from Mali, (West Africa) in the regions Doila, Kolokani and Siby. J Ethnobiol Ethnomed 1:7
117. Udo EA (1984) Who are the Ibibio? FEP Publishers, Onitsha, p 107

118. Udoh EJ, Etim NA (2008) Measurement of farm-level efficiency of water-leaf (Talinum triangulare) production among city farmers in Akwa Ibom state, Nigeria. J Sustain Devel Agric Environ 3(2):47–54
119. Ugochukwu NH, Babady NE (2003) Antihyperglycemic effect of aqueous and ethanolic extracts of *Gongronema latifolium* leaves on glucose and glycogen metabolism in livers of normal and streptozotocin-induced diabetic rats. Life Sci 73(15):1925–1938
120. USDA (2015) *Spathodea campanulata*. The PLANTS database (plants.usda.gov). National Plant Data Team, Greensboro
121. USDA, NRCS (2015) *Tridax procumbens*. The PLANTS database (plants.usda.gov). National Plant Data Team, Greensboro
122. Uwakwe AA, Nwaoguikpe RN (2008) In vitro antisickling effects of Xylopia aethiopica and Monodora myristica. J Med Plants Res 2(6):119–124
123. van Houten H (1997) ICRAF annual report 1996. ICRAF, Nairobi, 340 p
124. Voorhoeve AG (1965) Liberian high forest trees. Centre for Agricultural Publications and Documentation, Wageningenb, p 59
125. Vwioko DE, Okoekhian I, Ogwu MC (2018) Stress analysis of *Amaranthus hybridus* L. and *Lycopersicon esculentum* mill. Exposed to Sulphur and nitrogen dioxide. Pertanika J Trop Agric Sci 41(3):1169–1191
126. Wu N, Fu K, Fu YJ, Zu YG, Chang FR, Chen YH et al (2009) Antioxidant activities of extracts and main components of pigeon pea leaves. Molecules 14:1032
127. Yamagishi T, Haruna M, Yan XZ, Chang JJ, Lee KH (1989) Antitumor agents, 110. Bryophilin B, a novel potent cytotoxic bufadienoide from *Bryophyllum pinnatum*. J Nat Prod 52:1071
128. Zamble A, Yao D, Martin-Nizard F, Sahpez S, Offoumou M, Duriez P, Brunet C, Baileul F (2006) Vasoactivity and antioxidant properties of M. Keayana roots. J Ethnopharmacol 104: 263–269

Classification Methods and Diversity of Medicinal Plants

3

Okon Godwin Okon, Joseph Etim Okon, and Hasadiah Okon Bassey

Contents

O. G. Okon (✉) · J. E. Okon
Department of Botany, Faculty of Biological Sciences, Akwa Ibom State University, Uyo, Akwa Ibom State, Nigeria
e-mail: okonokon@aksu.edu.ng; josephokon@aksu.edu.ng

H. O. Bassey
Department of Biological Sciences, Akwa Ibom State Polytechnic, Ikot Osurua, Nigeria
e-mail: hasadiah.bassey@akwaibompoly.edu.ng

S. C. Izah et al. (eds.), *Herbal Medicine Phytochemistry*, Reference Series in Phytochemistry, https://doi.org/10.1007/978-3-031-43199-9_15

Abstract

This chapter discusses the diversity and classification of medicinal plants. Medicinal plants occupy a place of prime importance due to their complementary applications in modern medicines, and their use as food supplements, nutraceuticals, pharmaceutical intermediates, folk medicines, and secondary metabolites they provide for novel drugs. Therefore, an understanding of the diversity and classification of medicinal plants is useful for practitioners of ethnomedicine, botanists, and the common man for home remedies. Medicinal plants are mostly either angiosperms or gymnosperms with about 4171 and 706 plant species, respectively. Among angiospermic medicinal plants, dicotyledons have the most common representation of medicinal plants. Medicinal plants may be classified based on the following; parts used, habitat, habit, therapeutic, and ayurvedic formulations. The majority of medicinal plants are herbaceous, and about 51.3% grow mainly in the wild. Tropical rain forest provides the highest percentage of species (54.6%), followed by home gardens (25.4%) and unused land (19.5%). The medicinal use of about 72 (38.5%) plants has not before been reported. Many medicinal plants and their related indigenous values have not been adequately documented despite numerous investigations on medicinal plant products. The bulk of plant parts utilized in phytomedicine are the leaves (145 species), followed by the roots (119 species), and the stem (53 species). Herbs make up the highest percentage of species in terms of habit (44%), followed by trees (26%), shrubs (16%), lianas (5%), and others. Also, the sustainable use of medicinal plants should be ensured, and thorough documentation of the traditional knowledge linked with their usage should also be made and passed down across generations.

Keywords

Ayurvedic · Bioactive · Classification · Diversity · Herbs · Medicinal plants

Abbreviations

BGCI	Botanic Garden Conservation International
IUCN	International Union for Conservation of Nature
AKSU	Akwa Ibom State University
Ag	Agriculture
Aq	Aquaculture
FV	Food value
CO	Construction
UR	Urbanization
NT	Near threatened
NYBA	Not yet assessed
Fd	Food
Md	Medicine

Om	Ornamental
Tb	Timber
UV	Use-value

1 Introduction

A medicinal plant can be any plant in which one or more of its components contains bioactive substances, which can be utilized therapeutically or as precursors to the manufacture of valuable pharmaceuticals. Since time immemorial, natural plant products have occupied a place of prime importance in the field of medicine. They continue to be vital to many who do not have access to modern medicines or cannot afford synthetic drugs. Medicinal plants, for example, herbs and trees contain chemical compounds known to modern and ancient civilizations for their bioactive properties. Until the development of chemistry, particularly, the synthesis of secondary metabolites in the nineteenth century, medicinal plants and herbs have been the sole source of bioactive compounds capable of healing human ailments. Researchers believe that if medicinal herbs are taken in the appropriate doses and forms, they will be more effective than modern pharmaceutical drugs (synthetic drugs).

Plant products contain secondary metabolites that heal a variety of ailments and conditions in humans and animals. Over 75% of the world's population relies mainly on plants and their products for primary healthcare. Approximately 30% of all plant species were used for medical purposes at some point in history [1].

In developed nations like the United States, it has been estimated that up to 25% of all medications are made from plants. Fast-developing nations like China and Singapore make up as much as 80% of the total. As a result, countries like China use medicinal herbs for economic purposes considerably more frequently than the rest of the globe. Indigenous systems of medicine are employed to treat rural people, and these nations contribute two-thirds of the botanicals used in contemporary medical practices. Over 80,000 of the 250,000 higher plant species that exist on Earth are used for medicinal purposes [2].

Medicinal plants are plants that have the potency in the treatment of illnesses and the prevention of specific problems and diseases that impact both humans and animals. There are other varieties of herbal treatments that may differ from region to region, producing a similar pattern of "size" and "shapes." Natural plants have excellent secondary metabolites or chemical properties from their roots down to leaves even in their saps. Leaves of some herbs, for example, Karpooravalli (*Coleus ambonicus*), Bitter leaf (*Vernonia* species), Podina (*Mentha arvensis*), finger root (*Uvaria chamae*), Neem (*Azadirachta indica*), Thudhuvalai (*Solanum trilobatum*), Fern (*Diplazium sammatii*), Basil (*Ocimum sanctum*) and some edible species of fungi (mushroom), etc., are used by humans today. Some leaves have secondary metabolites that treat skin conditions, colds, headaches, constipation, and indigestion, among other conditions. Therefore, plants used for their specific health benefits to both human and animal health are considered medical plants [3, 4].

A significant number of plants are rarely used whole; at least one of its parts (leaf, stem, root, rhizome, bark, seeds, buds and leaves, etc.) can be utilized medicinally. The value of other parts of the same plant can vary. Plants that produce secondary metabolites can also be utilized as food, condiments, or even for purifying drinks. Since the beginning of time, the theory of signatures, which was systematized in the sixteenth century, has been essential and significant in the distinction by analogy of plants useful for human healing. From the seventeenth century, however, it has been expanded, contested, and canvassed, and by the eighteenth century, the elite community has completely abandoned it. According to data on plant use around the world, 14–28% of plants are included for their therapeutic properties. According to twenty-first-century surveys, 3–5% of patients in Western nations, 80% of local inhabitants in poor nations, and 85% of populations in the Southern Sahara chose medicinal plants as their primary form of treatment [2, 5].

There are still many people who practice traditional medical approaches. According to estimates, 80% of the world's population, or about four million people, cannot afford the goods produced by the Western pharmaceutical sector. They have to rely on the use of conventional medications, many of which have botanical origins. The inventory of medicinal plants, which includes over 20,000 kinds of plants, provides extensive documentation of this fact. Despite the overwhelming influence, reliance, and technological advancements in synthetic medications, a sizable portion of the global population still favours drugs derived from plants due to their accessibility and lower cost.

This chapter seeks to discuss the methods of classification of medicinal plants. We will also give details on their diversity, usage, formulation, habits, nutrition, habitats, nature, size, life span, as well as their therapeutic values.

1.1 Medicinal Plant Species and Their Therapeutic Value

Different plant groups have different medicinal plants that are synonymous to these groups depending on the number of species that belong to the said group. A breakdown of the different plant groups as well as the total number of species is given in Table 1. These plants and their bioactive components have been exploited

Table 1 Plant species with therapeutic value under different plant groups

Different plant groups	Number of plant species
Angiosperms:	
Monocotyledons	676
Dicotyledons	3495
Gymnosperms:	
Thallophytes	250
Bryophytes	54
Pteridophytes	402
Total	**4877**

Source: [6]

Table 2 Plant species, drugs, and their uses

Plant Species	Drugs	Uses
Catharanthus roseus	Vinblastine	Anti-cancer
Tradescantia spathacea	Taurine	Anti-inflammatory
Catharanthus roseus	Ajmalacine	Anti-cancer, hypotensive
Erythroxylum coca	Cocaine	Local anaesthetic
Rauvolfia serpentina	Rescinnamine	Tranquilizer
Rauvolfia serpentine	Reserpine	Tranquilizer
Papaver somniferum	Codeine	Anti-tussive
Cinchona species	Quinine	Antimalarial, amoebic dysentery
Erythroxylum coca	Cocaine	Tropical anaesthetic
Papaver somniferum	Morphine	Painkiller
Colchicum autumnale	Colchiceine, Amide	Anti-tumour
Xanthium strumarium	Tomentocin	Anti-sedative
Papaver somniferum	Codeine	Anti-cough
Digitalis species	Cardiac glycosides	Congestive heart failure
Artemesia annua	Artemisinin	Anti-malarial
Plumbago indica	Plumbagin	Anti-bacterial, antifungal
Gossypium species	Gossypol	Anti-spermatogenic
Allium sativum	Allicin	Anti-fungal
Xanthium strumarium	Xanthinin	Hair growth
Ricinus communis	Ricin	Anti-fungal
Digitalis thevetia	Digitoxin, Digoxin	Cardiotonic
Thevetia species	Nerrifolin	Cardiotonic
Podophyllum emodi	Podophyllin	Anti-cancer
Aspilia Africana	Aspicine	Wound healing
Musa parasidiaca	Musadine	Cuts and sores
Sesamum indicum	Sesamin	Anti-anaemic
Tithonia diversifolia	Tithonine	Skin diseases
Senna occidentalis	Carfeine	Pain relief
Tinospora cordifolia	Tinosporone	Anti-rheumatism
Eryngium foetidium	Arialgeine	Analgesic and inflammation
Colchicum autumnale	Colchicine	Anti-tumour, anti-gout

Sources: [7–10]

by pharmaceutical companies for many years now to produce drugs with several benefits, some of which are listed in Table 2.

2 Methods of Classifying Medicinal Plants

It has been estimated that there are 250,000 species of higher plants on earth; more than 80,000 species have at least some known medicinal uses. There are around 5000 species with specific medicinal benefits. They are classified according to: parts used, habit, habitat, therapeutic value, and Ayurvedic formulation [3, 9, 10].

2.1 Some Medicinal Plants Based on Parts Used

The following are some medicinal plants used based on their parts:

Whole plant: *Boerhaavia diffusa, Phyllanthus neruri, P. amarus, Euphorbia hirta*
Rhizome: *Curcuma longa*, *Zingiber officinale*
Roots: *Uvaria chamae, Daucus carota, Nuaclea latifolia*
Stem: *Tinospora cordifolia, Acorus calamus*
Bark: *Saraca asoca, Manifera indica, Azadiractita indica, Persea americana*
Leaf: *Indigofera tinctoria, Lawsonia inermis, Aloe vera, Sesamum indicum, Centalla asiatica*
Flower: *Biophytum sensitvum, Mimusops elenji*
Fruit: *Aframumum melequeta*, *Solanum* species, *Psidum gaujava, Cucumis melo*
Seed: *Datura stramonium, Cola nitida, Cola acuminata, Persea americana*

2.2 Some Medicinal Plant Based on Habit

The following are some medicinal plants used based on their habits:

Grass: *Cynodon dactylon*
Sedge: *Cyperus rotundus*
Herbs: *Vernonia cineria, Emilia sonchifolia*
Shrubs: *Solanum* species, *Capsicum* species
Climbers: *Asparagus racemosus, Dioscorea* species, *Phildendron domesticum, Cleome ciliate, Smilax anceps*
Creeper: *Commelina erecta, Commelina begalensis, Ipomoea batatas, Ipomoea cordatotriloba*
Trees: *Azadiractita indica, Mangifera indica, Dacroydes edulis, Croton* species.
Lianas: *Heinsia bussei, Tetracarpidium conophorum*

2.3 Some Medicinal Plant Based on Habitat

The following are some medicinal plants used based on their habitats:

Tropical: *Andrographis paniculata*
Sub-tropical: *Mentha arvensis*
Temperate: *Atropa belladona*

2.4 Some Medicinal Plant Based on Therapeutic Values

The following are some medicinal plants used based on their therapeutic values:

Anti-malaria: *Cinchona officinalis, Artemisia annua*
Anti-cancer: *Catharanthus roseas, Taxus baccata*
Anti-ulcer: *Azadirachta indica, Viscum album*
Anti-diabetic: *Catharanthus roseus, Musa acuminata, Dioscorea bulbifera*
Anti-cholesterol: *Allium sativum*
Anti-inflammatory: *Curcuma domestica, Desmodium gangeticum, Eryngium foetidum*
Anti-viral: *Acacia* species
Anti-bacterial: *Plumbajo indica*
Anti-fungal: *Allium sativum*
Anti-diaharrhoeal: *Psidium guajava, Curcuma domestica*
Anti-hypertensive: *Allium sativum, Persea appericara*
Anti-cardiotonic: *Digitalis* species, *Telfairia occidentalis, Peristrophe bicalyculata*

2.5 Medicinal Plant Based on Ayurvedic Formulation

This type of medicine is practiced in some parts of the world specifically in India, for example, *Desmodium gangeticum, Solanum indicum, Gmelina arborea, Premna spinosus, Ipomoea maxima, Emilia sonchifolia, Cardiospermum halicacabum, Ficus racemose, Ficus microcarpa, Phyllanthus emblica, Terminalia bellerica.*

3 Diversity in Medicinal Plants

Plantae as a Kingdom has existed for about 410 million years as green algae evolve from water to land, medicinal plants included. The land is a rich resource of medicinal plants and was comparatively not colonized. However, terrestrial habitats give more light and carbon dioxide needed by plant for growth and survival.

Medicinal plants are multicellular and mostly photosynthetic and may be terrestrial or aquatic and some are epiphytic or ephemeral. Medicinal plants can be found almost in all habitats on Earth.

According to research carried out at the Botanical Garden in the main campus of Akwa Ibom State University at Ikot Akpaden, Mkpat Enin Local Government Area of Akwa Ibom State – Nigeria by Okon et al. [11]. The Garden has a rich floral diversity of medicinal plants that have been conserved for the past ten years both in situ and ex situ mainly for research purposes. According to Sikolia and Omondi [12], the preservation of plants in botanical gardens dates back to the beginning of time. According to Botanic Garden Conservation International (BGCI), a botanic garden is an establishment that is accessible to the general public for assessment and evaluation, has a respectable level of permanence, and conducts scientific or technical research on plants (including species of wild origin) in its collections that are used for medicinal, ornamental, cultural, or other purposes. The Botanical Garden of Akwa Ibom State University (AKSU), Nigeria, was the site of a study by Okon et al. [11] that indicated the percentage habits of medicinal plants found in the highlighted

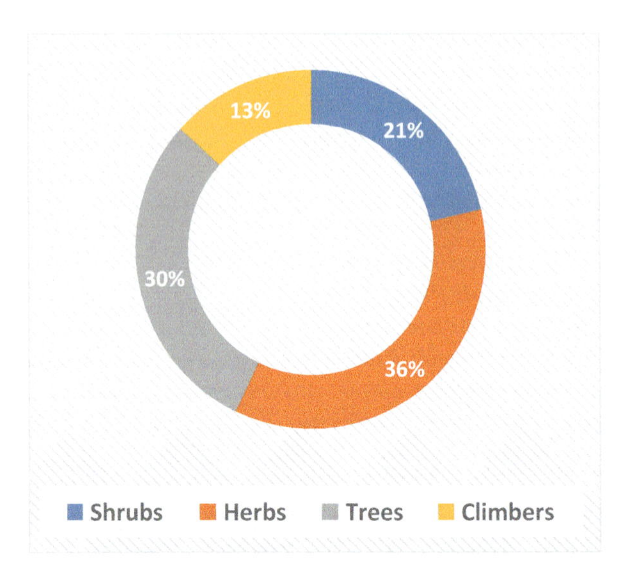

Fig. 1 Percentage distribution of medicinal plants habits in AKSU Botanical Garden, Nigeria [11]

area of the AKSU botanical garden examined documented diversity of medicinal plants based on their habits as follows: herbs as highest (35.71%), trees follow with 29.59%, then shrubs with 21.43%, while climbers were the least with 13.27% (Fig. 1).

Medicinal plant diversity can also be attributed to how well protected they are. To determine the condition of their conservation, the total number of plant species discovered in the AKSU Botanical Garden were cross-checked on the websites of two conservation organizations: the International Union for Conservation of Nature (IUCN) and Natural Resources/Nature Serve. Research indicates that 42.86% of the medicinal plants identified at the AKSU Botanical Garden are assessed as least concern/secure and face minimal risk, while 52.04% are rated as data deficient/ unranked. Near threatened species were recorded at 4.08% and 1.02% susceptible [11].

3.1 Medicinal Plant Diversity Based on Their Habitat

Based on the different characteristics and morphology of plants, there are diverse types of plants. Medicinal plants are grouped into the following categories according to their habitat [13, 14].

3.1.1 Hydrophytes

The word hydrophytes means plants that grow near water or in. These plants have weak vascular tissue (xylem and phloem), soft stems, and poor root systems. Here the majority of the tissue has air spaces and is spongy.

Hydrophytic Plants Include the Following Characteristic Features

a. Submerged plants, for example, *Elodea*, *Hydrilla*, *Sphgnum*, *Zostera*, *Vallisneria,* and *Potamogeton*, etc.
b. Fixed-floating and free-floating, for example, *Eichhornia, Utricularia, Wolffia, Salvinia, Egeria, Lemna, Ceratophyllum, Pistia*, *Trapa,* and *Eichornia,* etc.
c. Amphibious, that is, partly submerged, for example, *Sagittaria, Ranunculus aquatilis,* and *Alisma plantago*, etc.). Two angiosperms that are also marine, for example, *Thalassia* and *Zostera*.

3.1.2 Hygrophytes

Hygrophytic plants require wet, dark, and moist surroundings to develop and grow. Their fragile, spongy stems and root systems limited growth. Their leaves are completely formed and have stomata. Examples of common plants are several types of grasses, ferns, and begonias [6].

3.1.3 Halophytes

Halophytes are plants that can thrive in salty environments like soil or water. They can adapt or withstand high salt concentrations, such as those of NaCl, MgCl2, and MgSO4, quite well. They possess peculiar, negatively geotropic breathing roots known as pneumatophores, such as *Rhizophora racemosa, Rhizophora mangle Ceriops,* and *Avicennia* etc., which are some common examples.

3.1.4 Mesophytes

Mesophytes are mostly angiosperms that thrive in environments with little or moderate water supply. They always grow bigger and faster. They also have a strong root and leaf system. Both woody and herbaceous stems are possible. Some of them grow mesophytically in the summer and xerophytically in the winter, much like deciduous trees.

3.1.5 Xerophytes

Most xerophytic plants are medicinal. Examples include *Ziziphus*, *Nerium, Acacia, Euphorbia, Amaranthus, Argemone,* etc. These plants do well in xeric, arid, or places with insufficient water availability. Due of their ability to store water in their stems, many xerophytes are succulents like *Opuntia*, leaves like *Bryophyllum* species, and *Agave* or roots, for example, *Asparagus* species [7].

3.1.6 Epiphytic Plants

These plants can be found growing as an epiphyte on trees or on the trunks or branches of other plants, such as orchids or lichen like *Mangifera indica, Persia americana, Citrus* species, *Kola* species, and *Pterocarpus* species, etc. Epiphytic plants are thought of as the parasitic space occupants on other plants. The interaction between an orchid, a commensal, and a tree, the host, is an illustration of commensalism where the host is unharmed [13].

3.1.7 Parasitic Plants

Parasitic plants such as *Cuscuta, Nuytsia floribuda, Monotropa uniflora, Castilleja,* and *Striga* are parasites that grow on other plants on the roots of *Sorghum bicolor* (jowar). They produce a specialized structure mainly for feeding called haustoria that form a functional cap between the two plants.

3.2 Medicinal Plant Diversity Based on Habit

Medicinal plants in the group of angiosperms are categorized into four different groups depending on their form, shape, and size [14, 15].

3.2.1 Herbaceous Plants

These are plants with short, green, and soft stems. Usually with a brief or short life span like wheat or gram. Many herbaceous plants have a much-reduced underground stem portion, but the aerial branch with flowers at the apex develops from the underground part during reproduction. Such a stem is recognized as a scape, for example, *Allium cepa* (onion) and *Allium sativum* (garlic) in the family Amaryllidaceae [7].

3.2.2 Shrubby Plants

Shrubby plants are more advanced and significantly woody with branches than herbaceous plants. They lack an axis but generally have multiple stems, common examples are Ixora dwarf, black kodia, *Schefflera*, henna, roses, and China roses [16].

3.2.3 Arborescent (Trees)

They are usually taller than bushes with thick, rigid, and woody. Arborescent have a noticeable trunk, for example, *Irvingia* species, *Khaya* species, *Hura crepitans*, *Garcinia* species, and *Musanga cecropioides*.

3.2.4 Culms

The nodes and internodes in culms are very apparent. Most plants in these groups have hollow internodes. A typical example is *Bambusa vulgaris* (bamboo), which are grasses but cannot be classified as herbs, shrubs, or trees, etc.

3.2.5 Flowering Vine

These are very weak stem plants, normally supported by another plant. Common examples are Wisteria, Kiwifruit, ivy, and morning glory, etc.

3.3 Medicinal Angiosperm Diversity Based on the Nature of Stem

Some medicinal plants under angiosperm can be grouped based on the type of stem they possess [14, 15].

3.3.1 Erect Stem

Plants in this group develop upright/upward with hard or strong stems. Based on their strong stems, most trees, shrubs, and many herbaceous plants can stand upright on the ground without any support from another plant. Such plants include *Terminalia catappa*, *Terminalia ivorensis*, *terminelia superba*, *Irvingia* species, *Khaya* species, *Hura crepitans*, *Garcinia* species and *Musanga cecropioides, Hibiscus rosa-sinensis, Zea mays,* and *Lysimacchia vulgaris,* etc.

3.3.2 Creepers

Here, the plant's stem dangles and creeps on the ground, and the roots grow from the nodes (i.e., leaves come from nodes) of the culm from which the branches emerge. Consider strawberries and grasses (Cynodon and Oxalis). Adventitious roots are created by the nodes and the length of the stem.

3.3.3 Trailers

Trailers resemble creepers, they do not grow adventitious roots at nodes. There are two different kinds of trailers: Prostrate/procumbent – the stem creeps on the ground, as in the case of *Evolvulus, Tribulus,* or Decumbent – when the prostrate stem extends its tip, as in the case of *Portulaca* and *Linderbergia.*

3.3.4 Climbers

These plants have long, weak stems and are equipped with roots, petioles, spines, and tendrils that allow them to climb or attach themselves to other plants or objects for support. They are of three different types: **Rootlet Climbers** – Roots are produced at the nodes to assist in climbing the object, for example, *Tecoma*, *Pothos,* and *Piper betal* (pan). **Hook Climbers** – *Bougainvillea*, *Duranta,* and *Carrisa* species where the thorns are the modification of an axillary vegetative bud which helps in climbing. Unlike in *Bignonia*, the terminal leaflet is modified into a hook which helps in climbing, **Tendril Climbers** – The tendrils here are thread-like in structure which helps the plant in climbing, for example, *Lathyrus sativus*, *Pisum sativus*, *Clematis*, *Nepenthes*, *Smilex*, *Gloriosa*, *Antigonon*, *Vitis,* and *Passiflora,* etc. [16].

3.4 Medicinal Plant Diversity Based on Size

The medicinal angiosperm plants come in a wide range of sizes. The smallest angiosperm medicinal plant is the rootless aquatic plant *Wolffia*. It has a diameter of roughly 0.1 mm. Aquatic *Lemna* has a diameter of roughly 0.1 cm. With a height of more than 100 metres, the *Eucalyptus regnans* tree is the tallest angiosperm medicinal plant. Some eucalyptus trees can reach heights of 130 metres. The largest plant species is the banyan tree (*Ficus bengalensis*). It can cover a space of between two and five acres and contains more than 200 prop roots.

3.5 Medicinal Plant Diversity Based on Life Span

Based on life span, angiosperms are categorized into the following four groups [14].

3.5.1 Ephemerals

These types of medicinal plants reach the end of their life span relatively quickly before the start of the true dry season. These plants, such as *Argemon Mexicana, Cassia tora*, and *Solanum xanthocarpum*, among others, are not genuine xeric and are frequently referred to as drought evaders, escapers from drought, or drought avoiders.

3.5.2 Annuals

As soon as they have finished their life cycle within a year, medicinal plants like *Oryza sativa* (rice), *Triticum aestivum* (wheat), and Cicer arietinum (gramme), among others, die.

3.5.3 Biennials (Biannuals)

These classes of medicinal plants finished their entire life cycle in two years. Only the first year after planting do they show signs of vegetative development; the following year, they begin to produce flowers, fruits, and seeds as well as attain full maturity. These plants are typically herbs, such as carrot, turnip, and radish plants.

3.5.4 Perennials

Perennial medicinal plants can live for more than 200 years. The Kolkata Botanical Garden is home to a massive banyan tree (*Ficus bengalensis*). A *Ficus religiosa* that can be found in Gaya is about 2500 years old. After reaching full maturity, a large number of medicinal perennials bloom and bear fruit during a particular time of the year. They are polycarpic and include *Mangifera indica*, *Acacia* species, and *Cocos nucifera*. Some perennial medicinal plants, such as *Agave* (Agarvaceae) and *Bambusa vulgaris* (Poaceae), are monocarpic, meaning they only bear fruit once in their lives. All annuals and biennials are monocarpic [16].

3.6 Medicinal Plant Diversity Based on Nutrition

Based on their method of nutrition, medicinal plants in this category are separated into the following groups [14].

3.6.1 Autophytes/Autotrophs

They produce their own food. They are divided into two groups: chemotrophs, which use chemical energy to manufacture their food, and phototrophs, which use sunlight to perform photosynthesis to generate their food.

3.6.2 Heterotrophs

Here, plants are unable to produce their own food and must rely on external sources. Heterotrophs can be insectivorous, saprophytic, parasitic, or symbiotic. Other plant varieties include Polygamous plants, for example, *Mangifera indica*, Stolon plants, for example, *Ajuga*, *Stachys,* and *Mentha*, seedless vascular plants, sucker plants, for example, red raspberry, lilac, and *Forsythia*, air layering plants, for example, *Forsythia*, jasmine, *Hamamelis,* and *Philodendron,* cutting plants. There are different kinds of medicinal plants, and each one needs a certain setting in order to grow and propagate. The essential elements are sunlight, nutrients, water, air, soil, and temperature that plants rely on for their life, germination, growth, development, and full maturity [14–16].

3.7 Diversity in Frequency of Occurrence and Relative Density of Medicinal Plants

The frequency of occurrence and relative density of some medicinal plants found/grown in Uyo Local Government Area, Nigeria; *Cleome ciliata, Diplazium sammatii, Hibiscus rosa-sinensis, Justicia insularis,* and *Sesamum indicum* each had the highest frequency of occurrence of 60% each while medicinal plants such as *Bombax buonopozense, Glyphaea brevis, Heinsia bussei, Talinum triangulare,* and *Vernonia polysphaera* had the least frequency value of 20% each. In density value, *Cleome ciliata* dominated (8000 plants/ha) while the lowest value was associated with *B. buonopozense* (40 plants/ha) as reported in Table 3 by Okon [17].

In Ikot Ekpene Local Government Area, Nigeria (Table 4), *Justicia insularis, Lasienthera africanum, Microdesmis puberula,* and *T. triangulare* had the highest frequency of occurrence of 60% each while species like *B. buonopozense, Curcubita maxima, D. samamatii, Pterocarpus mildbraedii, Vernonia amygdalina,* and *V. polysphaera* had the lowest frequencies of occurrence (20%). *D. sammatii* and *Pterocarpus mildbraedii* had the highest and lowest density values of 1200 plants/ha and 40 plants/ha, respectively. The frequency occurrence of medicinal plants in Eket Local Government Area as shown in Table 5 revealed that *Glyphaea brevis* had the highest frequency of occurrence (80%) while the least frequency was associated with species like *Amaranthus hybridus* (20%), *C. ciliata* (20%), *D. sammatii* (20%), and *M. puberula* (20%). *S. indicum* had the highest density value of 4200 plants/ha while *B. buonopozense* had the least density value of 40 plants/ha. Also, the results revealed that *Gnetum africanum* was near threatened. Threats that affected the medicinal plants were as a result of agricultural activities, aquaculture, food value, construction, and urbanization (Tables 3, 4 and 5) [17].

3.8 Diversity in Use-Value Index of Medicinal Plants in Akwa Ibom State, Nigeria

The use-value index was determined based on the different categories for a particular medicinal plant. Results obtained showed a high index for the following medicinal

Table 3 Frequency of occurrence and relative density of medicinal plants in Uyo, Nigeria

Vegetables	Habitats	Frequency (%)	Density (Plts/ha.)	Rel. Freq.	Rel. Density	Conservation Status	Threats
Amaranthus hybridus	Herb	40	220	6.67	1.37	NYBA	Fv, Ag
Bombax buonopozense	Tree	20	40	3.33	0.25	NYBA	Ag, UR, CO
Cleome ciliata	Herb	60	8000	10.00	49.81	NYBA	CO
Cucurbita maxima	Herb	0	0	0	0	NYBA	–
Diplazium sammatii	Herb	60	1600	10.00	9.96	NYBA	–
Glyphaea brevis	Shrub	20	80	3.33	0.49	NYBA	–
Gnetum africanum	Herb	0	0	0	0	NT	Fv, Ag, Aq
Heinsia bussei	Herb	20	60	3.33	0.37	NYBA	–
Hibiscus rosa-sinensis	Shrub	60	180	10.00	1.12	NYBA	UR, CO
Justicia insularis	Herb	60	1000	10.00	6.23	NYBA	UR, CO
Lansianthera africamum	Shrub	40	240	6.67	1.49	NYBA	Fv, UR, CO
Microdesmis puberula	Shrub	40	240	6.67	1.49	NYBA	CO, UR
Piper guineense	Herb	40	60	6.67	0.37	NYBA	Fv, Ag
Pterocarpus mildbraedii	Tree	0	0	0	0	NYBA	–
Sesamum indicum	Herb	60	2200	10.00	13.69	NYBA	Fv, CO, UR
Sterculia tragacantha	Tree	40	80	6.67	0.49	NYBA	CO, UR
Talinum triangulare	Herb	20	800	3.33	4.98	NYBA	Fv, Ag, UR
Telfairia occidentalis	Herb	0	0	0	0	NYBA	Fv, UR, CO
Vernonia amygdalina	Shrub	0	0	0	0	NYBA	Fv, UR, CO
Vernonia polysphaera	Herb	20	1260	3.33	7.85	NYBA	CO, UR

Ag agriculture, *Aq* aquaculture, *FV* food value, *CO* construction, *UR* urbanization, *NT* near threatened, *NYBA* not yet assessed
Source: [17]

Table 4 Frequency of occurrence and relative density of medicinal plants in Ikot Ekpene, Nigeria

Vegetables	Habitats	Frequency (%)	Density (Plts/ha.)	Rel. Freq.	Rel. Density	Conservation Status	Threats
A. hybridus	Herb	40	280	7.14	4.93	NYBA	Fv, Ag
B. buonopozense	Tree	20	60	3.57	1.06	NYBA	Ag, UR, CO
C. ciliata	Herb	40	400	7.14	7.04	NYBA	CO
C. maxima	Herb	20	300	3.57	5.28	NYBA	–
D. sammatii	Herb	20	1200	3.57	21.13	NYBA	–
G. brevis	Shrub	40	300	7.14	5.28	NYBA	–
G. africanum	Herb	0	0	0	0	NT	Fv, Ag, Aq
H. bussei	Herb	0	0	0	0	NYBA	–
H. rosa-sinensis	Shrub	40	300	7.14	5.28	NYBA	UR, CO
J. insularis	Herb	60	1000	10.71	17.61	NYBA	UR, CO
L. africamum	Shrub	60	360	10.71	6.34	NYBA	Fv, UR, CO
M. puberula	Shrub	60	460	10.71	8.09	NYBA	CO, UR
P. guineense	Herb	40	60	7.14	1.06	NYBA	Fv, Ag
P. mildbraedii	Tree	20	40	3.57	0.70	NYBA	–
S. indicum	Herb	0	0	0	0	NYBA	Fv, CO, UR
S. tragacantha	Tree	0	0	0	0	NYBA	CO, UR
T. triangulare	Herb	60	600	10.71	10.56	NYBA	Fv, Ag, UR
T. occidentalis	Herb	0	0	0	0	NYBA	Fv, UR, CO
V. amygdalina	Shrub	20	220	3.57	3.87	NYBA	Fv, UR, CO
V. polysphaera	Herb	20	100	3.57	1.76	NYBA	CO, UR

Ag agriculture, *Aq* aquaculture, *FV* food value, *CO* construction, *UR* urbanization, *NT* near threatened, *NYBA* not yet assessed
Source: [17]

Table 5 Frequency of occurrence and relative density of medicinal plants in Eket, Nigeria

Vegetables	Habitats	Frequency (%)	Density (Plts/ha.)	Rel. Freq.	Rel. Density	Conservation Status	Threats
A. hybridus	Herb	20	140	3.33	1.52	NYBA	Fv, Ag
B. buonopozense	Tree	40	40	6.67	0.43	NYBA	Ag, UR, CO
C. ciliata	Herb	20	420	3.33	4.55	NYBA	CO
C. maxima	Herb	40	260	6.67	2.81	NYBA	–
D. sammatii	Herb	20	1200	3.33	12.98	NYBA	–
G. brevis	Shrub	80	480	13.33	5.19	NYBA	–
G. africanum	Herb	0	0	0	0	NT	Fv, Ag, Aq
H. bussei	Herb	60	140	10.00	1.52	NYBA	–
H. rosa-sinensis	Shrub	40	180	6.67	1.95	NYBA	UR, CO
J. insularis	Herb	40	600	6.67	6.49	NYBA	UR, CO
L. africamum	Shrub	0	0	0	0	NYBA	Fv, UR, CO
M. puberula	Shrub	20	200	3.33	2.16	NYBA	CO, UR
P. guineense	Herb	0	0	0	0	NYBA	Fv, Ag
P. mildbraedii	Tree	40	120	6.67	1.29	NYBA	–
S. indicum	Herb	40	4200	6.67	45.45	NYBA	Fv, CO, UR
S. tragacantha	Tree	0	0	0	0	NYBA	CO, UR
T. triangulare	Herb	40	400	6.67	4.33	NYBA	Fv, Ag, UR
T. occidentalis	Herb	0	0	0	0	NYBA	Fv, UR, CO
V. amygdalina	Shrub	40	220	6.67	2.38	NYBA	Fv, UR, CO
V. polysphaera	Herb	60	640	10.00	6.93	NYBA	CO, UR

Ag agriculture, *Aq* aquaculture, *FV* food value, *CO* construction, *UR* urbanization, *NT* near threatened, *NYBA* not yet assessed
Source: [17]

plants: *T. triangulare, V. amygdalina, P. guineense, P. mildbraedii, L. africamum, J. insularis, G. fricanum, C. maxima, A. hybridus,* and *T. occidentalis* while the least use-value was recorded in *M. puberula, B. buonopozense, S. tragacantha, G. brevis, H. bussei* and *V. polysphaera, C. ciliata,* and *D. sammatii.* Also, all the ten medicinal plants selected for analysis were observed to be available in the wild except *H. rosa-sinensis* (Table 6) [17].

3.9 Medicinal Plants Diversity Among Tree Canopies

Most temperate moist forests in different regions are home to a variety of flora that are pivotal in sustaining the livelihoods of the local plant communities. The abundance and distribution of plant life in these forests, particularly medicinal species, is comparatively underreported (with scant or no supporting documentation). Zubair et al. [18] identified a total of 45 species from 34 different groups in their study site in North-Eastern Pakistan. The majority of plants were identified as being medicinal (45%), followed by food (26%) species that also included some medicinal components. Based on richness and abundance, tree canopy cover has an impact on the general growth of ethnomedicinal plants. In comparison to open spaces and closed canopy, the site with partial canopy showed the highest diversity, richness, and dominance. These results are crucial in revealing the abundance of medicinal flora variety in the Balakot temperate forest in the north-east and the potential for preserving the native plant communities' means of subsistence through public–private collaborations in several sectors.

It has been observed that the diversity of vegetation has a direct impact on the biogeochemical cycles, ecosystem processes, and vegetation structure of a particular ecosystem [19, 20]. Statistics on plant biodiversity in mountainous areas offer vital information and insight on the suitability of the particular habitat, ecosystem productivity, and successional pathway prediction [21–23].

The engine room and hotspots of plant diversity, metabolic processes, ecological zones, and contacts between creatures and the atmosphere are thought to be found under forest canopies [24]. The maintenance of biodiversity and meeting multiple environmental needs are both dependent on the canopy cover of the forest stand [25, 26]. Nearly 40% of the world's current plant and animal life is supported by various forest canopies, of which 10% are recognized as canopy authority [27]. Different strata of medicinal plant species rely on forest canopies for sunlight, nutrients, and assistance with reproduction. According to research, the density and quantity of understory vegetation are directly impacted by disturbance in the various layers. In a forest's understory vegetation, medicinal plants play a crucial function [28, 29]. Holscher and Adnan [30] evaluated the variety and abundance of medicinal plants in old growth forest and degraded forest, reporting that the density of medicinal plants in old growth forest was significantly higher than that in the latter.

Furthermore, it was noted that the quantity of quantitative vegetation surveys in the Himalayas is extremely little and insufficient to draw a precise picture of the diversity of understory vegetation [30, 31]. For the purpose of developing strategies

Table 6 Plant diversity/categories and use-value index of medicinal plants

Scientific name	Family	Common name	Local name	Harvest type	Uses	No. of categories Used	No. of respondents	∑	Rank/%	UV
M. puberula	Euphorbiaceae	–	Ntabid	Wild	Md	1	5	5	12/10	0.02
B. buonopozense	Bombacaceae	Red silt cotton tree	Ukim	Wild/ cultivated	Tb	1	2	2	15/4	0.02
S. tragacantha	Sterculiaceae	African tragacanth	Udod Eto	Wild/ cultivated	Md	1	4	4	13/8	0.02
S. indicum	Pedaliaceae	Sesame	Etehedeh	Wild/ cultivated	Fd, Md	2	14	28	9/28	0.04
G. brevis	Tiliaceae	–	Ndodido	Wild/ cultivated	Md	1	6	6	11/12	0.02
H. rosa-sinensis	Malvaceae	Red hibiscus	Frawa	Cultivated	Md, Om	2	14	28	9/28	0.04
H. bussei	Rubiaceae	–	Atama Idim	Wild	Md	1	2	2	15/4	0.02
V. polysphaera	Compositae/ Asteraceae	Bitter leaf	Asio-isong	Wild/ cultivated	Md	1	10	10	10/20	0.02
C. Ciliata	Capparidaceae	Consumption weed	Mkpat unen	Wild	Md	1	3	3	14/6	0.02
D. sammatii	Athyriaceae	Fern	Nyama Idim	Wild	Om	1	1	1	16/2	0.02
T. triangulare	Portulaceae	Waterleaf	Mmon-mmon ikon	Cultivated	Fd, Md, Om	3	42	126	6/84	0.06

V. amygdalina	Compositae/ Asteraceae	Bitter leaf	Etidod	Cultivated	Fd, Md	2	49	98	2/98	0.04
P. guineense	Piperaceae	Guinea black pepper	Odusa	Wild/ cultivated	Fd Md	2	50	100	1/100	0.04
P. mildbraedii	Papilionaceae	White Camwood	Mkpa	Wild/ cultivated	Fd, Md, Tb	3	48	144	3/98	0.06
L. africamum	Icacinaceae	–	Editan	Wild/ cultivated	Fd, Md	2	50	100	1/100	0.04
J. insularis	Acanthaceae	Hunter's weed	Mmeme	Wild/ cultivated	Fd, Md	2	50	100	1/100	0.04
G. africanum	Gnetaceae	African salad	Afang	Wild/ cultivated	Fd, Md, Om	2	46	92	4/92	0.04
C. maxima	Cucurbitaceae	Melon pumpkin	Ndise	Wild/ cultivated	Fd, md	2	37	74	7/74	0.04
A. hybridus	Amarantheceae	Amaranth	Inyan-afia	Wild/ cultivated	Fd, Md	2	43	86	5/86	0.04
T. occidentalis	Cucurbitaceae	Fluted pumpkin	Ikon ubon	Cultivated	Fd, Md	2	49	98	2/98	0.04

Notes: *Fd* food, *Md* medicine, *Om* ornamental, *Tb* timber, *UV* use-value
Source: [6, 17].

and employing sustainable management techniques, it is critical to document the variety of vegetation, particularly the situation of medicinal plants in the area. It is important that these untamed and priceless resources are well-maintained because a large section of the forest communities rely on them for economics, food, construction, and medical uses, as well as the restoration of ecosystem functions and biodiversity. Similar to this, there are not many studies on how various forest canopies affect the variety of medicinal plants. In numerous studies, the biodiversity of diverse medicinal plants in the chosen forest is portrayed, as well as the impact of different canopies on the variety and abundance of the medicinal plants. Despite the fact that there are many studies emphasizing the importance and diversity of medicinal plants in many places, this research offers a new dimension to our understanding of how differing forest canopies affect the variety, distribution, and abundance of medicinal plants.

3.10 Medicinal Plant Cover and Abundance Between the Various Forest Canopies

Research revealed that the comparison of the different forest canopies reported that the partial canopy had 65% medicinal plants as the maximum number, followed by the closed canopy with 56% and 41% for open spaces.

3.11 Types of Medicinal Plant Canopy

3.11.1 Closed Canopy

The closed canopy recorded about 56% of medicinal plant species identified and classified as extremely important medicinal plants, in which the most dominant one was *Valeriana wallichii* (Frequency (F) = 100%; Density (D) = 173). *Perilla frutescens* was the second medicinal plant species that was significantly present (F = 70%; D = 113). *Geranium wallichianum* (F = 60%; D = 56), a threatened and endemic medicinal plant (Zubair et al. 2021). Infrequent medicinal plants species were also recorded, the most treasured of which was *Podophyllum hexandrum* Royle (F = 10%; D = 1). Some edible plants of about 25% were noted and the most prominent was *Potentilla indica* (Andrews) T. Wolf having the maximum F and D (F = 100%; D = 320). Other medicinal plants such as *Viburnum grandiflorum* Wall. (F = 10% D = 2) and *Dryopteris filix-mas* (L.) Schott (F = 20% D = 18) were not quite as abundant (Table 7).

3.11.2 Open Spaces

Zubair et al. [18] reported the frequency (F) and Density (D) of the vegetation in the open space to be little to no different when compared to the closed canopy. The difference was that the open spaces were the representation of grasses and forage with 47% in the majority that were utilized by the grazing community. The most abundant of medicinal plants were *Chrysopogon gryllus* (F = 70%; D = 171), *Trifolium repens* L. (F = 60%; D = 185), *Gladiolus communis* (F = 50%;

Table 7 Species and classification in closed canopy in Naga forest

Species	Classification
Perilla frutescens	Medicinal
Adiantum stenochlamys	Medicinal
Viburnum grandiflorum	Fruit
Chrysopogon montanus	Animal forage
Lavatera cachemiriana	Medicinal
Podophyllum hexandrum	Medicinal
Geranium wallichianum	Medicinal
Arisaema triphyllum	Medicinal
Dryopteris filix-mas	Vegetable
Sonchus asper	Animal forage
Chrysopogon gryllus	Animal forage
Potentilla indica	Fruit
Gentiana decumbens	Medicinal
Geranium nepalense	Medicinal
Dryopteris austriaca	Vegetable
Valeriana wallichii	Medicinal

Source: [18]

D = 37), *Primula denticulata* (F = 30%; D = 11), *Cyperus rotundus* (F = 30; D = 46). Plants species showing medicinal characteristics (some chemical constituents) were also represented in this paper, having about 41% of the species, *Hypericum perforatum* (F = 30%; D = 41) being the most prominent and important followed by *Plantago ovata* (F = 20% D = 21) and *Rumex nepalensis* (F = 20%; D = 13), others were quite rare, having minimum frequencies and densities. Edible vegetables were not very dominant and hence were sparsely scattered around different canopies. Among the edibles, only *Potentilla indica* (F = 40% D = 56) had considerable representation (Table 8).

3.11.3 Partial Canopy

Partial canopy cover depicted a different trend away from open and closed spaces and recorded the greatest number of medicinal plant species. The maximum number of medicinal plants was 65%, the most dominant species being the *Mentha spicata* (F = 50%; D = 41) followed by *Polygonum amplexicule* (F = 30%; D = 55) and *Skimmia laureola* (F = 20%; D = 58), while the other 52% of the medicinal plant species were quite rare and were not uniformly distributed in the canopy [18]. Among the rare species, the most important medicinal plant species present in the canopy were *Paeonaia emodi* (F = 10%; D = 35), *Berberis lycium Royle* (F = 10%; D = 1), *Asparagus racemosus* (F = 10%; D = 14), and *Geranium wallichianumi* (F = 10%; D = 40). As compared to the closed canopy-covered site, this area had quite a few varieties of grass species, the most dominant being the *Chrysopogon gryllus* (F = 100%; D = 150) and *Trifolium repens* (F = 40%; D = 244). In this site, a few edible species were represented. A quite popular edible plant, *Potentilla indica* (F = 80%; D = 299), was in abundance when compared to other edible vegetables while other important edible plants such as *Dryopteris filix-mas* (F = 10%; D = 18), *Phyllanthus niruri* (F = 10%; D = 14), and *Solanum nigrum* (F = 10%; D = 16) were quite rare as assessed in the canopy (Table 9).

Table 8 Species and classification in open spaces in Naga forest

Species	Classification
Hyoscyamus niger	Medicinal
Skimmia laureola	Medicinal
Cyperus rotundus	Animal forage
Cynodon dactylon	Animal forage
Rubus niveus	Fruit
Adiantum stenochlamys	Medicinal
Primula denticulate	Animal forage
Artemisia absinthium	Medicinal
Potentilla collettiana	Fruit
Rumex nepalensis	Animal forage
Gladiolus communis	Aromatic
Chrysopogon gryllus	Animal forage
Plantago ovata	Medicinal
Hypericum perforatum	Medicinal
Trifolium repens	Animal forage
Potentilla indica	Fruit

Source: [18]

Table 9 Species and classification in partial Canopy in Naga forest

Species	Classification
Asparagus racemosus	Vegetable
Adiantum stenochlamys	Medicinal
Polygonum amplexicaule	Medicinal
Dryopteris austriaca	Vegetable
Paeonaia emodi	Medicinal
Geranium wallichianum	Medicinal
Arisaema triphyllum	Medicinal
Miscanthus nepalensis	Animal forage
Rumex nepalensis	Medicinal
Berberis lyceum	Medicinal
Plantago major	Medicinal
Potentilla indica	Fruit
Chrysopogon gryllus	Animal forage
Skimmia laureola	Medicinal
Dryopteris filix-mas	Vegetable
Trifolium repens	Animal forage
Mentha spicata	Medicinal

Source: [18]

4 Conclusion

This chapter has extensively discussed the diversity and different methods of medicinal plant classification. Medicinal plants are plants that have the potency in the treatment of ailments and prevention of certain diseases and conditions which affect humans and animals alike. It is estimated that out of the 350,000 higher plant species on earth, more than 80,000 have medicinal properties. They are classified according to: part use, habit, habitat, and therapeutic value as well as Ayurvedic formulation. Plantae as a Kingdom has existed for about 410 million years as green algae evolve from water to land, medicinal plants included. The land is a rich resource of medicinal plants and was comparatively not colonized. This chapter expounds on the diversity and classification of these medicinal plants to aid understanding and usage of these plants. This chapter has also offered a new dimension to our understanding of how differing forest canopies affect the variety, distribution, and abundance of medicinal plants. Research has revealed that from the comparison of the different forest canopies, the forest ecosystem with the partial canopy had 65% medicinal plants as the maximum number, followed by the closed canopy with 56% and 41% for open spaces. Thus, for the purpose of developing strategies and employing sustainable management techniques in the conservation of medicinal plants, it is critical to document the variety of vegetation, particularly the situation or status of medicinal plants in the area. It is important that these untamed and priceless resources are well-maintained and conserved because a large section of humans living in forest communities rely on them for economics, food, construction and medical uses, as well as the restoration of ecosystem functions and biodiversity.

Acknowledgment The authors wish to thank Mr. Felix Udo and AK18 students of the Department of Botany, Akwa Ibom State University, Ikot Akpaden, Mkpat Enin for their role in the survey of the plants used in some parts of this chapter.

References

1. Salehi B, Kumar NV, Sener B, Sharifi-Rad M, Kılıç M, Mahady GB, Vlaisavljevic S, Iriti M, Kobarfard F, Setzer WN (2018) Medicinal plants used in the treatment of human immunodeficiency virus. Int J Mol Sci 19:1459
2. Hu R, Lin C, Xu W, Liu Y, Long C (2020) Ethnobotanical study on medicinal plants used by Mulam people in Guangxi, China. J Ethnobiol Ethnomed 16:1–50
3. Nelly A, Annick DD, Frederic D (2008) Plants used as remedies antirheumatic and antineuralgic in the traditional medicine of Lebanon. J Ethnopharmacol 120:315–334
4. Raskin I, Ribnicky DM, Komarnytsky S, Ilic N, Poulev A, Borisjuk N, Brinker A, Moreno DA, Ripoll C, Yakoby N (2002) Plants and human health in the twenty-first century. Trends Biotechnol 20:522–531
5. Amenu E (2007) Use and management of medicinal plants by indigenous people of Ejaji area (Chelya Woreda) West Shoa, Ethiopia: an ethnobotanical approach. Master's Thesis, University in Addis, Ababa, Ethiopia
6. Bassey M (2013) Introductory pteridology. Modern Business Press Ltd, Uyo, p 96
7. Etukudo I (2003) Ethnobotany: conventional and traditional uses of plants. The Verdict Press, Uyo, p 191

8. Sofowora EA (2008) Medicinal plants and traditional medicine in Africa, 3rd edn. Spectrum Books Ltd., Ibadan, p 436
9. Kurian JC (2014) Healing wonders of plants, vol 1. Zambia Adventist Press, Zambia, p 193
10. Kurian JC (2014) Healing wonders of plants, vol 2. Zambia Adventist Press, Zambia, p 200
11. Okon OG, Joseph EO, Sunday MS, Lovina IU, Felix EU (2021) Checklist, conservation status and health status assessment via total photosynthetic pigment contents of plants found at the Akwa Ibom State University Botanical Garden, Nigeria. J Biodivers Environ Sci 19(6):20–29
12. Sikolia SF, Omondi S (2017) Checklist of plants in the university botanic garden of Maseno and their significances to the society. J Pharm Biol Sci 2(1):27–49
13. Etukudo I (2000) Forest: our divine treasure. Dorand Publishers, Uyo, p 194
14. Raafat HAE, Mohamed SZ, Wafaa MK, Abdel RAM (2008) Diversity and distribution of medicinal plants in North Sinai, Egypt. Afr J Environ Sci Technol 2(7):157–171
15. Ebukanson GJ, Bassey ME (1992) About seed plants. Baraka Press and Publisher Limited, p 150
16. Paul S, Manssan K (2007) Medicinal plant diversity and uses in the Sango Bay Area, Southern Uganda. J Ethnopharmacol 113(3):521–540
17. Okon JE (2017) Nutritional qualities, phytochemical constituents and use value index of some wild edible leafy vegetables in Akwa Ibom State. Unpublished Ph.D. Thesis. University of Uyo, Uyo, Nigeria, p 235
18. Zubair M, Jamil A, Hussain SB, Ul Haq A, Hussain A, Zahid DM, Hashem A, Alqarawi AA, Abd Allah EF (2021) Diversity of medicinal plants among different tree canopies. Sustainability 13:2640
19. Roy A, Kushwaha SPS (2018) Landscape level plant diversity characterization in Indian Himalayan region. In: Berra S, Das AP (eds) Plant diversity in the Himalaya hotspot region. Bishen Singh Mahendra Pal Singh, Uttarakhand
20. Ruiz-Jaén MC, Aide TM (2015) Vegetation structure, species diversity, and ecosystem processes as measures of restoration success. Forest Ecol Manag 218:159–173
21. Grêt-Regamey A, Brunner SH, Kienast F (2012) Mountain ecosystem services: who cares? Mt Res Dev 32
22. Rawal RS, Rawal R, Rawat B, Negi VS, Pathak R (2018) Plant species diversity and rarity patterns along altitude range covering treeline ecotone in Uttarakhand: conservation implications. Trop Ecol 59:225–239
23. Tali BA, Khuroo AA, Nawchoo IA, Ganie AH (2019) Prioritizing conservation of medicinal flora in the Himalayan biodiversity hotspot: an integrated ecological and socioeconomic approach. Environ Conserv 46:147–154
24. Bagaram MD, Giuliarelli G, Chirici F, Giannetti F, Barbati AUAV (2018) Remote sensing for biodiversity monitoring: are forest canopy gaps good covariates? Remote Sens 10:1397
25. Nadkami NM, Mewin MC, Niedert J (2001) Forest canopies: plant diversity. Encycl Biodivers 3:27–40
26. Keller HW (2004) Tree canopy biodiversity: student research experiences in Great Smoky Mountains National Park. Syst Geogr Plants 74:47–65
27. Ozanne CM, Anhuf D, Boulter SL, Keller M, Kitching RL, Körner C, Stork NE (2003) Biodiversity meets the atmosphere: a global view of forest canopies. Science 301:183–186
28. Fazal HINA, Ahmad N, Rashid ABDUR, Farooq S (2010) A checklist of phanerogamic flora of Haripur Hazara, Khyber Pakhtunkhwa, Pakistan. Pak J Bot 42:1511–1522
29. Kumar V (2015) Ethno-medicinal plants in five forest ranges in Dang district, South Gujarat, India. Indian J Trop Biodiv 23:1–9
30. Barkatullah IM, Rauf A, Hadda TB, Mubarak MS, Patel S (2015) Quantitative ethnobotanical survey of medicinal flora thriving in Malakand Pass Hills, Khyber Pakhtunkhwa, Pakistan. J Ethnopharmacol 169:335–346
31. Shaheen SY, Bibi M, Hussain H, Iqbal Saira I, Safdar LS (2017) A review on *Geranium wallichianum* D-Don ex-sweet: an endangered medicinal herb from Himalaya region. Med Aromat Plants 6:2167–2412

Classification of Phytochemicals in Plants with Herbal Value

4

Sara Taha Abdelkhalek, Jin -Hua Shi, Mei -Xiang Jin, Sherein Saied. Abdelgayed, and Man -Qun. Wang

Contents

S. T. Abdelkhalek
Hubei Insect Resources Utilization and Sustainable Pest Management Key Laboratory, College of Plant Science and Technology, Huazhong Agricultural University, Wuhan, China

Department of Entomology, Faculty of Science, Ain Shams University, Cairo, Egypt

J. -H. Shi · M. -X. Jin · M. -Q. Wang (✉)
Hubei Insect Resources Utilization and Sustainable Pest Management Key Laboratory, College of Plant Science and Technology, Huazhong Agricultural University, Wuhan, China
e-mail: shijinhua@webmail.hzau.edu.cn; meixiangJin@webmail.hzau.edu.cn; mqwang@mail.hzau.edu.cn

S. S. Abdelgayed
Department of Pathology, Faculty of Veterinary Medicine, Cairo University, Giza, Egypt
e-mail: Sherein.abdelgayed@cu.edu.eg

S. C. Izah et al. (eds.), *Herbal Medicine Phytochemistry*, Reference Series in Phytochemistry, https://doi.org/10.1007/978-3-031-43199-9_12

Abstract

In the past decade, there has been an increased concern about the effects of medicinal plants. Traditional medicinal herbs from diverse habitats and locations can be evaluated as novel treatment and prevention methods for injuries and diseases. Natural products, especially secondary metabolites in medicinal herbs, including those utilized in conventional and ethnic health care systems, provide prospective components for developing novel drug candidates. Phytochemicals have many potential roles as they can protect plants from enemies and act as antimicrobial, anti-inflammatory, antidiabetic, and chemopreventive agents. Their identification and classification are usually according to the chemical formula, such as flavonoids, terpenoids, alkaloids, saponins, phytosterols, carotenoids, essential oils, nonessential amino acids, and aromatic and aliphatic acids. Each group has characteristics, including anticancer, anthelmintic, and antigenotoxic. Additionally, they can offer direct/indirect protection against pathogens or hazardous conditions. Due to their potency and cost-effectiveness, phytochemicals have recently received considerable interest in this area. The impact of medicinal plant utilization is international and has been developing in many countries. Notably, as a potential source of alternative treatments, traditional medicine has attracted attention worldwide. This chapter will focus on classifying phytochemicals (primary and secondary metabolites), identifying some active secondary metabolites (such as flavonoids, alkaloids, terpenoids, phytosterols, and phenolic compounds), studying their potentiality in the treatment of some disorders, and the modern research advances in herbal medicine field.

Keywords

Phytochemicals · Herbal medicine · Flavonoids · Alkaloids · Cancer treatment · Antimicrobial properties · Artificial phytochemicals production

Abbreviations

5-FU	5-fluorouracil
ABTS	2,2′-azinobis-(3-ethylbenzothiazoline-6-sulfonic acid)
ACE	Angiotensin-converting enzyme
ATP	Adenosine triphosphate
COX	Cyclooxygenase
DPPH	α, α-diphenyl-β-picrylhydrazyl
ECa-09	Esophageal squamous carcinoma cell in the human
ESCC	Esophageal squamous cell carcinoma
HL-60	Promyelocytic leukemia cell line
HNSCC	Head and neck squamous cell carcinoma

IC_{50}	Half-maximal inhibitory concentration
JAK/STAT	The Janus kinase/signal transducers and activators of the transcription pathway
NF-B	Nuclear factor-B
Nrf-2	Nuclear factor-erythroid factor 2-related factor 2

1 Introduction

Phytochemicals (secondary metabolites) are the chemical substances synthesized by the plant through various chemical mechanisms. Most phytochemicals support plants' ability to compete with other plants and provide defense against diseases, herbivores, or abiotic challenges, such as high UV radiation doses. Usually, humans and animals can detect or smell and sometimes taste the secreted phytochemicals. Moreover, phytochemicals contain various chemical constituents, including alkaloids, flavonoids, saponins, steroids, terpenoids, and many other active groups. Plant metabolites can be classified depending on their physiological function in the plant system as follows: (1) Primary metabolites: They are biomolecules that directly affect the metabolism and the development of different species, and they are broadly distributed in plants, thus being appropriate for use as an edible human source. In addition, these primary components (carbohydrates, proteins, lipids, vitamins, and nucleic acid constituents) are produced through a complex system of metabolic processes to fulfill energy demands and serve as precursors for some physiological processes and biochemical synthesis of vital components (secondary metabolites) [1]. (2) Secondary metabolites are derived from the primary metabolites, in particular developmental stages of the plant. These sophisticated compounds have diverse purposes in the plant system (defense against enemies) and potential biological activity, such as medicinal components [2].

The identification and nomenclature of plant species and understanding their relationships to other species are fundamental for many researchers. Moreover, the botanical identification and taxonomic positions of studied plants, especially medicinal herbs, are crucial in the scientific investigation of their therapeutic usage or the foundation of crude pharmaceuticals. Numerous medicinal plants belonging to diverse plant families include Fabaceae, Apiaceae, Apocynaceae, Lamiaceae, Malvaceae, Mimosaceae, Papaveraceae, Phytolaccaceae, Asteraceae, Boraginaceae, Brassicaceae, Caryophyllaceae, Cesalpinaceae, and Cucurbitaceae [3].

Attributed to their significant biological activity, phytochemicals have been used for decades in traditional medicine, reflecting the critical therapeutic benefits of these chemical components [4]. Additionally, different tissues and organs of medicinal herbs may have unique therapeutic characteristics at distinct stages of development [5]. They are involved in valuable industries like pharmaceuticals, cosmetics, and fine chemicals [6].

In this chapter, the classification of some phytochemicals, including main active groups (terpenoids, flavonoids, alkaloids, phytosterols, and phenolic components) are highlighted and their general biosynthesis pathway, their vital activity in the treatment of some diseases, artificial production of secondary metabolites, and some recent advances in their research are presented.

2 Classification of Phytochemicals

The secondary metabolites are very specialized and can be prevalent in several plant species. The complicated combinations of plant secondary metabolites provide distinctive chemical characteristics among the plant classes, which is an essential classification tool for taxonomists [7]. A summary of some active compounds of different medicinal plant species and their potential treatment effects are presented in Table 1.

2.1 Flavonoids

Flavonoids belong to the polyphenol family of plant secondary metabolites, with more than 6,000 structures, widely distributed in various parts of the plant due to the multiple potential benefits provided by the experts that have received significant attention [56]. Like most metabolites described, flavonoids are essential for adapting plants to their environment, helping to cope with biotic and abiotic stresses, and having critical pharmacological activities. The biosynthesis of flavonoids was studied by using many plant species, including *Arabidopsis thaliana*, *Petunia hybrida*, and *Zea mays*. The synthesis of flavonoids begins with the phenylalanine pathway; multiple enzymes are involved, which makes flavonoids one of the rich families in the plant kingdom [57–59]. The basic structure of most flavonoids contains two phenyl rings (A and B), connected by a heterocyclic pyrene ring with an oxygen atom (C-ring) with the general formula of C6-C3-C6 (Fig. 1). Usually, flavonoids are water-soluble metabolites and classified following the position of ring B into isoflavone and other types, while according to the degree of saturation of the central heterocyclic C-ring can be subdivided into seven categories, including flavanone, flavone, flavanonol, flavanol, flavonol, anthocyanidins, and chalcones (Fig. 2) [60].

The most significant flavonoids are quercetin, quercitrin, and kaempferol, which are expressed in about 70% of all plant species. Several flavonoids play a vital role in anticancer and improving cardiovascular health. For example, hesperidin (Hsp) is a vital flavonoid with high anticancer activity. Also, quercetin is an antioxidant flavonoid, improving blood vessel health and reducing cardiovascular disease risk [61, 62].

2.1.1 Isoflavone

The distribution of isoflavone in plants is limited. It is mainly existed in soybean and legumes and known as phytoestrogens. It has a unique structure to other flavonoids; the position of the B ring for isoflavone is in the third place of the C ring, while the B ring is in the second place of the C ring for other flavonoids. Isoflavone has potential

Table 1 Overview of some bioactive secondary metabolites in herbal medicine

Scientific name	Family of the plant	Active chemical compound	Medical importance	Reference
Cajanus cajan	Fabaceae	Hydroalcoholic, cyanidin-3-monoglucoside, longistylin C	Antioxidant, anti-inflammatory activities, anticancer	[8, 9]
Coriandrum sativum	Apiaceae	Linalool, γ-terpinene, and α-pinene	Hepatoprotective	[10]
Calotropis procera	Apocynaceae	Group of cardiac glycosides (calotropin, calactin, and calotoxin)	Anti-ATPase, antidiarrheal, anticonvulsant, and antiviral	[11–13]
Psychotria nervosa Sw.	Rubiaceae	Emetine (emetine dihydrochloride) and cephaeline	Anti-dengue virus infection	[14–17]
Hippeastrum puniceum (Lam.) Kuntz	Amaryllidaceae	isoquinoline alkaloids (9-O-demethyllycoramine, 9-demethyl-2α-hydroxyhomolycorine, lycorine and tazettine)	Antioxidant, antifungal, antiparasitic, and acetylcholinesterase inhibitory activity	[18]
Hymenocallis caribaea (L.) Herb	Amaryllidaceae	Narciclasine and prancristatin	Antineoplastic compounds	[15, 19, 20]
Thespesia populnea (L.) Sol. ex Corrêa	Malvaceae	Thespesenone and dehydrooxoperezinone-6-methyl ether	Antihepatotoxic compounds	[21–23]
Lagerstroemia speciose (L.) Pers.	Lythraceae	Corosolic acid and ellagitannins	Diabetes treatment, antihyperlipidemic, and antioxidant activities	[24]
Persea americana Mill.	Lauraceae	Peptone, b-galactoside, and glycosylated abscisic acid	Hypertension, stomachache, bronchitis, diarrhea, and diabetes	[25]
Cocos nucifera L.	Arecaceae	Folate (vitamin B_9)	Reduce the risk of breast cancer and anemia during pregnancy	[26, 27]
Acorus calamus L.	Acoraceae	β-asarone, eugenol, asaronaldehyde, and acorin	Antidiabetic, anti-obesity, antihypertensive, anti-inflammatory, and anticonvulsant activities	[28–31]
Alstonia scholaris L. R. Br.	Apocynaceae	Ditamine, echitamine, echitenine, alschomine, isoalschomine, tubotaiwine, and N^b-oxide	Anti-ulcer, antirheumatic, carminative, aphrodisiac, and antiperiodic	[32]

(continued)

Table 1 (continued)

Scientific name	Family of the plant	Active chemical compound	Medical importance	Reference
Phyllanthus epiphyllanthus	Phyllanthaceae	Michellamine B (novel alkaloid), gallotannins, and triterpenes	Inhibition of human immunodeficiency virus (HIV), hepatitis B virus, and antibacterial activity against typhoid fever bacteria	[15, 33–35]
Citrus limon L.	Rutaceae	Eriodictyol	Prevention of diabetic retinopathy, antioxidant, anti-inflammatory, and analgesic effects	[36, 37]
Euphorbia hypericifolia L.	Euphorbiaceae	Isomotiol, espinendiol A, ursolic acid, juglangenin A, and teuviscin A	Anticancer activity	[38–41]
Valeriana officinalis L.	Caprifoliaceae	Isovaleric acid, gamma-aminobutyric acid, valerenic acid, and valerine	Relieves mild nervous tension and helps in sleep	[42–44]
Eucalyptus obliqua L'Hér.	Myrtaceae	1,8-cineole, catechins, isorhamnetin, luteolin, kaempferol, phloretin, and quercetin	Anticancer, antioxidant, antiseptic, stimulant, antimalarial, anthelmintic action, anti-inflammatory, and antihistaminic	[45–48]
Panax ginseng C. A.Mey.	Araliaceae	Ginsenosides	Improve phagocytosis, interferon production, and vasodilation, and affects the hypoglycemic activity	[49–51]
Stachytarpheta jamaicensis (L) Vahl	Verbenaceae	Gamma-aminobutyric acid and dopamine	Inhibition of the chicken pox virus	[15, 52]
Moringa oleifera Lam.	Moringaceae	Quercetin, kaempferol glycosides, myricetin, and epicatechin	Hypolipidemic, hypoglycemic, antioxidant, anti-inflammatory, and anticancer properties	[53–55]

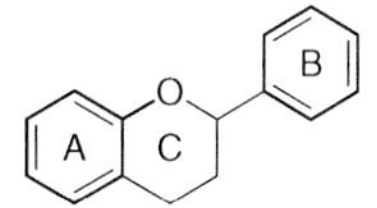

Fig. 1 Basic structure of flavonoids

antibacterial and antioxidant activities [64, 65]. Besides, it is protective against acute lung injuries, cardiovascular diseases, and breast cancer [66–68].

2.1.2 Flavanone

Flavanone is the direct precursor of most flavonoid synthesis. It has a saturated and oxidized C ring. For instance, naringenin and hesperetin are analogue compounds for flavanones and excipients richer in citrus, lemon, grapefruit, and fruit peels. Naringenin is potentially antiviral and antitumoral, preventing cardiovascular disease [69–71]. Hesperetin and its derivatives positively improve acute lung injury, diabetes, and Alzheimer's disease [72–75].

2.1.3 Flavone

Flavone is one of the largest subgroups of flavonoids. Compared to flavanones, besides having a keto group at position four on the C ring, there is also a double bond between the C-2 and C-3 of the C ring. Flavone is present in practically all plant tissue. It has commercial and pharmaceutical value as neuroprotective [76, 77]. For example, apigenin protects neuronal cells from injury by inhibiting microglia cells [78]. In addition, some studies showed that apigenin is safe for humans, which makes it a suitable target for future drug development [79, 80].

2.1.4 Flavanol

Flavanol has a saturated and oxidized C ring and a hydroxyl group in C-3 or C-4 of the C ring. It is mainly found in cocoa, fruits, cereals, and vegetables [81]. Flavanols are important bioactive compounds in *Theobroma cacao*. Researchers have proven its efficacy in anti-inflammatory, neuroprotective, and cardiovascular diseases [82–85]. Recently, Hidalgo et al. [86] investigated the therapeutic activity of epicatechin and epigallocatechin-3-gallate (the main active compound in green tea) flavanols toward nonalcoholic fatty liver disease. Their findings reflected several beneficial properties besides its ability to prevent or treat nonalcoholic fatty liver disease, such as antihyperglycemic, antihypertensive, antithrombotic, anti-inflammatory, and antifibrotic effects. Moreover, following the US Food and Drug Administration classification, they reported its safety on humans as harmless.

2.1.5 Flavonol

Flavonol has a hydroxyl group in C-3 of the C ring, like flavanol. However, there is also a double bond between the second and third carbons in the C ring. Kaempferol and quercetin are representative compounds of flavonols. Both have anxiolytic, antiviral, and anti-inflammatory effects [62, 87]. Tian et al. [88] studied the antioxidant (through the activities of phagocytosis, DPPH, and ABTS radical scavenging)

Flavonols

Kaemferol

Flavonols

Catechin

Anthocyanins

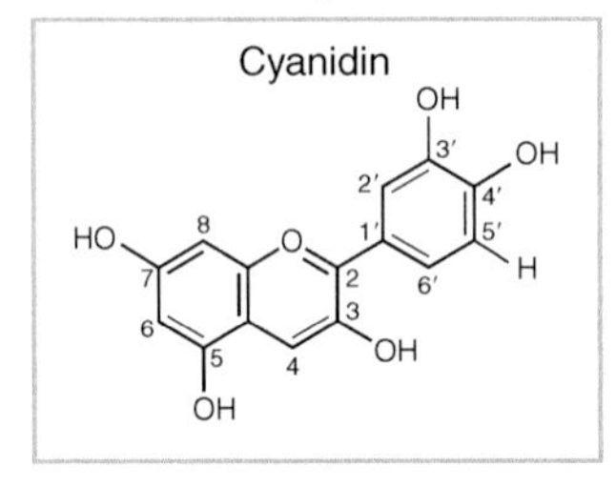

Flavonols

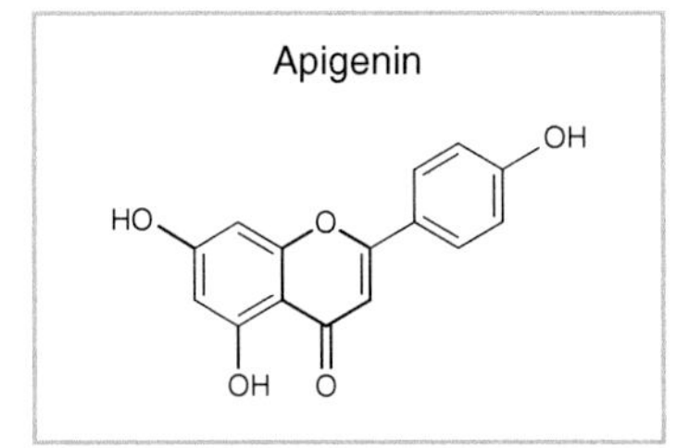

Isoflavones

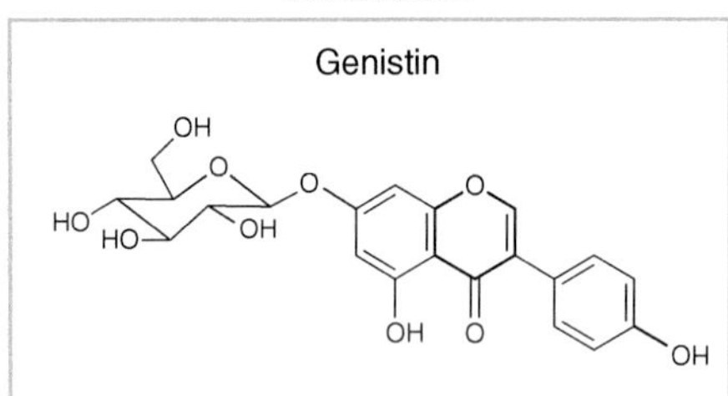

Flavanones

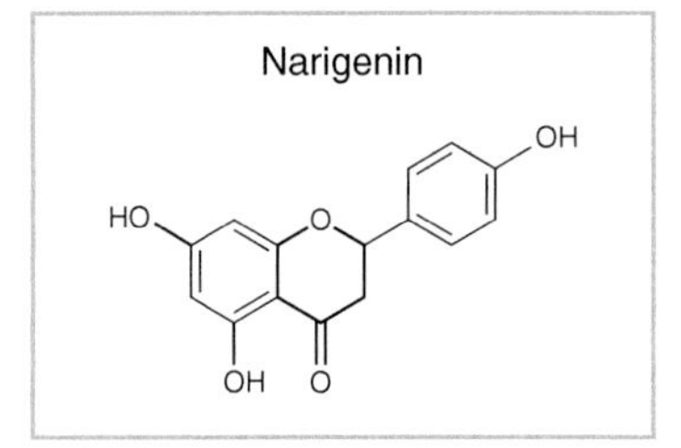

Flavanonol

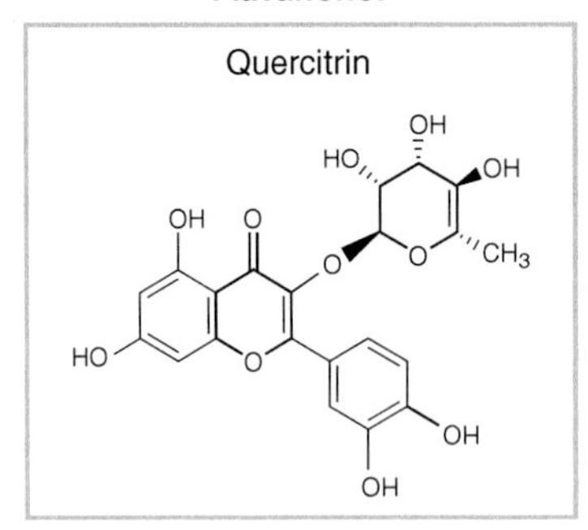

Chalcones

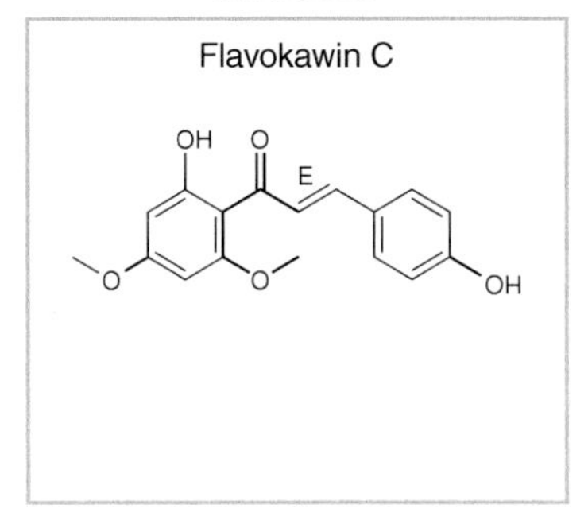

Fig. 2 Examples of different flavonoid categories. (Modified after Bešlo et al. [63])

and anti-inflammatory properties of some flavonols (kaempferol, luteolin, quercetin, and apigenin). The results showed that quercetin is a promising anti-inflammatory and antioxidant substance potentially as a therapeutic adjunct for inflammatory disorders and oxidative stress. Furthermore, preliminary findings from this study indicated that antioxidant activity is directly correlated with the number of phenolic hydroxyl groups. A comparison of anti-inflammatory and antioxidant activities revealed that compounds with enol groups were better than those without.

2.1.6 Flavanonol

Flavanonol has a hydroxyl group at position C-3 and a keto group at C-4. Its compounds can exist in free or combined form and are mainly abundant in citrus. Additionally, flavanonols are found in various medicinal herbs families, like Rutaceae, Leguminosae, and Rosaceae [89]. Flavanonol has significant medical value in vascular protection; it has been found that taxifolin has some therapeutic properties, including preventing the damage of vascular structure in diabetic patients, improving capillary microcirculation, and improving blood flow in the retina [90].

2.1.7 Anthocyanidins

Anthocyanidins are water-soluble pigments widely distributed in flowers and can also be detected in leaves, roots, and fruits (strawberry, chokeberry, and elderberry) [91–95]. The soluble pigment helps plants show different colors (red, orange, blue, and green). Consequently, approximately all angiosperm species contain anthocyanins [96, 97]. The structure is characterized by the absence of ketone in the C-4 position of the central heterocyclic ring and the positive charge of the first oxygen atom in the C ring. In flora, anthocyanins are the most prevalent anthocyanidins (containing glycosylated flavylium ion). Up to 300 anthocyanidins have been identified [98], whereas there are approximately 8,000 possible anthocyanins, such as the various anthocyanidin subtypes and the glycosylated [99].

Polyphenolics, especially anthocyanins, have recently become more relevant in treating chronic diseases like cardiovascular diseases [100]. Pergola et al. [101] investigated that the blackberry extract (about 88% of the cyanidin-3-glucoside out of the total anthocyanin content) inhibited the deleterious cardiovascular activity of nitric acid. Moreover, Graf et al. [102] studied the antiatherogenic activity of anthocyanins in the grape-bilberry extract. The results demonstrated that a dose of 1.55 mg/L decreased the total content of cholesterol and triglyceride levels in treated models. Additionally, anthocyanins reflected a physiological disruption of many groups of fatty acids, including a reduction in saturated fatty acids and an elevation in the long-chain fatty acids in plasma. Consequently, these findings showed the antiatherosclerosis activity of anthocyanins-rich grape-bilberry juice.

2.1.8 Chalcones

Chalcones are open-chain flavonoids (alpha, beta-unsaturated ketones) lacking the structural formula's complete C ring. The basic structure of chalcones includes two conformations: cis and trans isomers [103]. It is widely distributed in medicinal herbs and can be modified into diversified forms (prenylated chalcones (3, 4, 7), licochalcones, and dihydrochalcones). Chalcones and derivatives have been reported for pharmacological activities in the aspect of antiviral, cardiovascular disease prevention, antimicrobial, antioxidant, analgesic, and antidiabetic functions [104–108]. Many researchers reported the pharmaceutical properties of chalcones. Tang et al. [109], Acharjee et al. [110], and Attarde et al. [111] showed the antidiabetic activity of chalcone, piperonal chalcones derivative, and 3-(4-hydroxyphenyl)-1-phenylprop-2-en-1-one (chemically synthesized chalcone) exhibited significant suppression of the activity of α-glucosidase and α-amylase.

2.2 Alkaloids

Alkaloids are a prominent structurally diverse family of heterocyclic substances containing nitrogen in the heterocyclic ring and obtained from amino acids [112]. They exhibit a variety of significant physiological impacts on humans and other mammals. Morphine, strychnine, quinine, ephedrine, and nicotine are prominent alkaloids. Alkaloids are mainly found in plants and are particularly prevalent in certain flowering plant families [113]. Around 20% of higher plant species are estimated to contain alkaloids, with several thousand distinct varieties, besides some other alkaloid-rich families, including Ranuculaceae, Amaryllidaceae, and Solanaceae [114]. They are primarily associated with plant defense against herbivores (feeding deterrence) and pathogens (antibacterial and antifungal activities). Traditional and recent applications of alkaloids range from 25–75% in pharmaceuticals, demonstrating their enormous medicinal potential [115, 116]. The fundamental quality of alkaloids is no longer the requirement for an alkaloid, and the chemical reactivity of nitrogen atoms permits at least four classes of nitrogenous substances.

Moreover, several synthesized substances with equivalent structures are also referred to as alkaloids [117]; however, others can produce salts with organic acids like oxalic and acetic acids. Certain botanical alkaloids, such as solanine in the Solanaceae family, exist in a glycosidic form. The biosynthetic pathway of the alkaloid is involved, especially in the decarboxylation of substances [118]. For example, aspirin is the most common antiplatelet (pain treatment) medicine, prepared mainly from salicin alkaloids originating from the willow plant (*Salix alba* L.) [119].

According to their biochemical precursor and heterocyclic ring arrangement, alkaloids have been categorized into several groups, including indole, piperidine, tropane, purine, pyrrolizidine, imidazole, quinolizidine, isoquinoline, and pyrrolidine [112, 120]. Fig. 3 shows the structural formula of some potent alkaloid substances.

Some pure isolated alkaloids and synthetic analogues (berberine, serotonin, dopamine, and gamma-aminobutyric acid) are employed as fundamental therapeutic agents globally due to their antispasmodic, antimalarial, antibacterial, and analgesic activities [121, 122]. Alkaloids can prevent the beginning of several degenerative disorders by scavenging free radicals or interacting with the oxidation reaction catalyst. Numerous studies have evaluated the vast spectrum of pharmacological properties of alkaloids

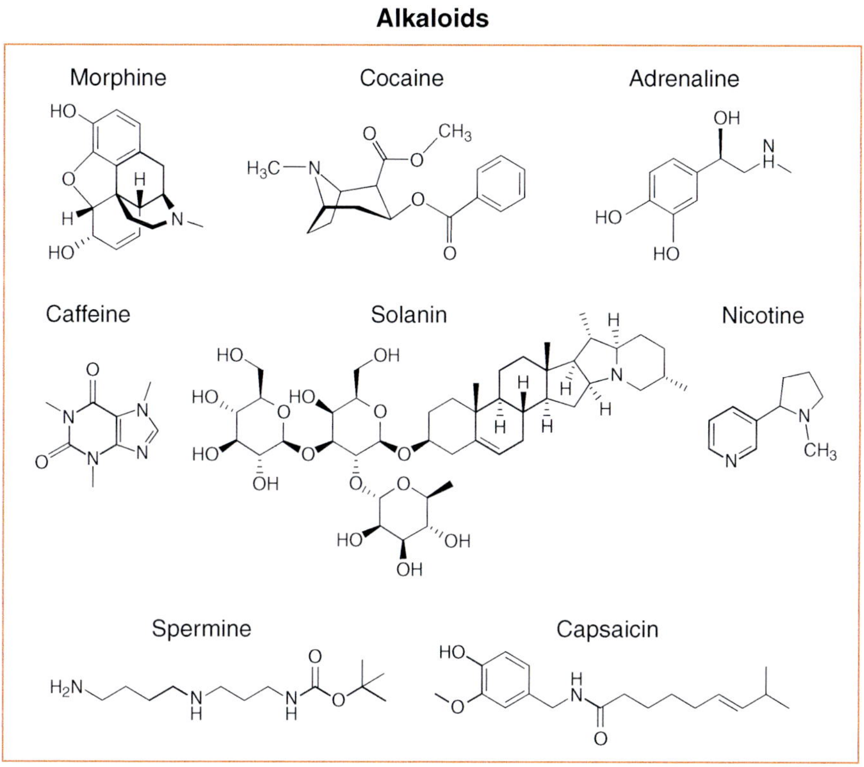

Alkaloids

Morphine Cocaine Adrenaline

Caffeine Solanin Nicotine

Spermine Capsaicin

Fig. 3 Examples of some vital alkaloid compounds

from diverse plants [120]. For instance, the active ingredients of *Papaver somniferum* L., the opium poppy (morphine), and its methyl ether derivatives (codeine) are comparatively nonaddictive analgesics [123]. Moreover, some alkaloids stimulate the cardiovascular or respiratory systems. Uzor [124] reported that quinine, chloroquine, amodiaquine, mefloquine, artemisinin, and artemether are the most vital antimalarial alkaloidal drugs. In addition, Kouam et al. [125] isolated three novel antimalarial indolosesquiterpene alkaloids from the bark extract of *Polyalthia oliveri*. Several studies [112, 126–131] investigated the activity of some active alkaloid constituents, such as atropine (from *Atropa belladonna*) acts as an antidote; ephedrine (from *Ephedra sinica*) can be used as antiasthmatics; and emetine (from *Carapichea ipecacuanha*) as antiprotozoal. Furthermore, noscapine (from *Papaver somniferum*) is used as an antitussive and used as a product in the following names Bequitusin, Degoran, and Tussisedal. In addition to pharmaceutical drugs, some alkaloids have been used as antiplatelet substances. Rutaecarpine, an alkaloid derived from *Evodia rutaecarpa*, was demonstrated as an antiplatelet, and its derivatives (3-methylenedioxyrutaecarpine, 3-chlororutaecarpine, and 3-hydroxyrutaecarpine) reflecting a significant enhancement of its role by interacting with several mediators of coagulation factors [132, 133].

2.3 Terpenoids

Terpenoids are the largest class of phytochemicals (plant-specialized metabolites) with several significant pharmacological effects [134]. They are isoprene-based (5-carbon) metabolites commonly known as isoprenoids. Terpenes can be divided into six categories based on the number of isoprene units in their chemical structure: monoterpene, sesquiterpene, diterpenes, triterpenes, tetraterpenes, and polyterpenoids (Fig. 4) [135]. Terpenoids exist mainly in flowering plants and are responsible for the distinct aroma, flavors, scents, and colors of many species, such as eucalyptus, cinnamon, tomatoes, sunflower, ginger, and clove [136].

Hundreds of terpenoid compounds possess pharmaceutical characteristics; for example, most terpenes in *cannabis* sp. (tetrahydrocannabinol) have diverse pharmacologic effects, including anti-inflammatory, sedative, immunomodulatory, and neuromodulatory properties [137].

2.3.1 Monoterpenoids

The chemical structure of monoterpenes and their derivatives contains two isoprene units. They incorporate three subclasses, acyclic monoterpenes, monocyclic monoterpenes, and bicyclic monoterpenes. For example, the *Cymbopogon citratus*

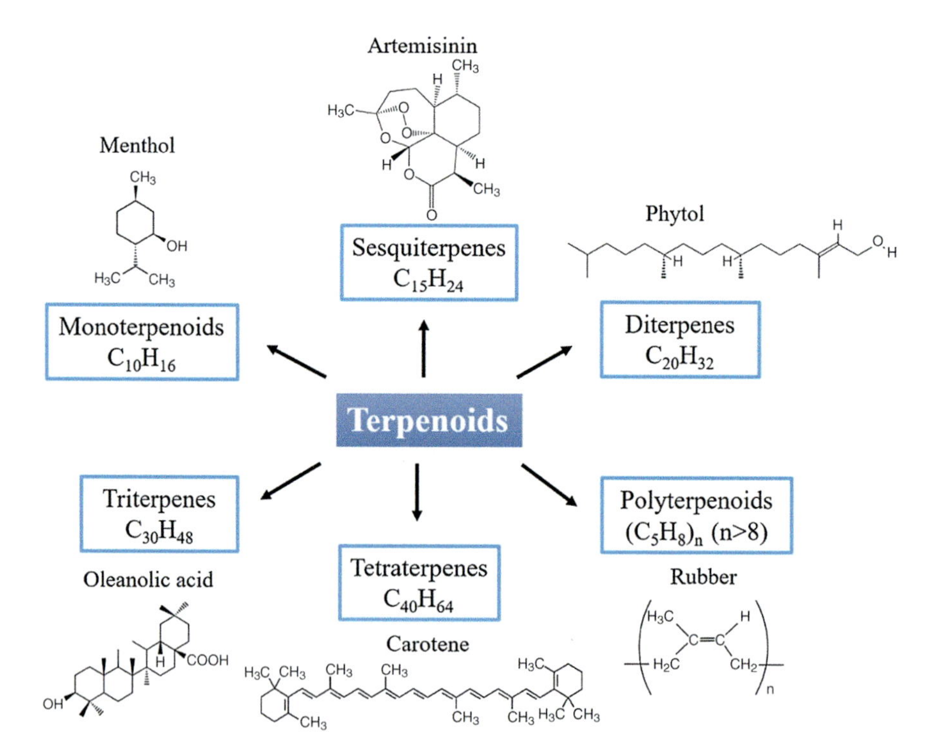

Fig. 4 Chemical structure of some representative terpenoid substances

(lemongrass) produces a large quantity of citronellol (acyclic monoterpene). Santos et al. [138] investigated that citronellol significantly affects bacterial, fungal, allergic, spastic, and diabetic resistance. More acyclic monoterpenes showed antibacterial, anti-inflammatory, and antianxiety properties, such as linalool and its derivatives [139, 140]. In addition, menthol is a monocyclic monoterpene with a prevalent analgesic activity in pain control [141]. α-and β-pinene are representative bicyclic monoterpenoid substances; however, they share the same molecular formula of C10H16, their chemical structures are distinct, and both exhibit anticancer and antimicrobial bioactivity [142–144].

2.3.2 Sesquiterpenes

Sesquiterpene compounds are the most common kind of terpene with three isoprene units. It is subdivided into straight sesquiterpenes (e.g., farnesol and nerolidol) and annular sesquiterpenes based on the stability of the carbon rings (e.g., artemisinin and guaiol). Almost 50% of sesquiterpene compounds contain lactones [145]. It has significant biological and physiological functions. Recent research demonstrated that sesquiterpene lactones could contribute to inflammation and cancer therapies [146, 147].

2.3.3 Diterpenes

Diterpene compounds comprise four isoprene units with one or more carbon rings within their molecular structure. Saha et al. [148] reported that diterpenes had a broad spectrum of microbial inhibition, including bacteria, viruses, and fungi. In addition, several studies reflected that it possesses anticancer, antidiabetic, and antitumor properties [149–153]. Diterpene compounds can not only inhibit the growth and spread of cancer cells but also promote the apoptosis of these cells. Thus, they are widely used to treat liver, lung, breast cancers, and other malignant tumors [151, 154, 155].

2.3.4 Triterpenes

Triterpene chemicals refer to six isoprene units–containing terpenes. The most common triterpene molecules are tetracyclic and pentacyclic, which have four- and five-carbon rings, respectively. Triterpenes showed potential medical activities; for example, oleanolic acid might combat diabetes by suppressing the function of α-glucosidase [156]. Moreover, Sohag et al. [157] revealed that lupeol (pentacyclic triterpene) safeguard against cardiovascular, renal, liver, and skin disorders. Sun et al. [158] summarized the effect of *Centella asiatica* triterpenes in treating common diseases and the underlying mechanisms, such as Alzheimer's disease, acne, chronic and recurrent liver injuries, pelvic inflammatory disease, and rheumatoid arthritis.

2.3.5 Tetraterpenes

Tetraterpenes are one of the isoprenoids containing eight isoprene units with molecular formula C40H64 – they involve many pigmentary substances (carotenoids) [159] and nonpigmentary compounds (poduran) [160]. Carotene is the first isolated tetraterpene discovered in carrots. Furthermore, beta-carotene has significant antioxidant effects that help in the treatment of many chronic diseases, such as optical diseases and particular cancers [161]. Zerres and stahl [162] reported that carotenoids are characterized by UV-absorbing properties and can be used as photoprotectants. They revealed that carotenoids provide comprehensive photoprotection after supplementation or consuming carotenoid-rich meals. Additionally, lycopene offers therapeutic potential to treat human malignancies [163].

2.3.6 Polyterpenoids

Polyterpenoid molecules have more than eight isoprene units and have a greater molecular weight than other terpenes. These can be extracted from some citrus oils or tree sap, and their yields fluctuate according to the growing seasons (e.g., rosin and its derivatives) [164]. Few investigations have been carried out on the pharmacological characteristics of polyterpenoids. Rubber is a natural polyterpenoid compound with between 1500 and 15000 isopentenyl units. de Barros et al. [165] and Mendonca et al. [166] showed that it influences wound healing and oxytocin-sustained release.

2.4 Phenolic Components

Phenolic acids are a large family of secondary aromatic metabolites. They are extensively distributed in their seeds, roots, stems, and leaves and are associated with xylans, pectin, and lignin. Similarly, to flavonoids, phenolic acids are essential members of the phenolic family. They comprise the most frequently distributed nonflavonoid phenolic molecules in plants and exist in various forms, like free, conjugated soluble, and insoluble bound. Phenolic compounds are the derivatives of benzoic and cinnamic acids [167]. The most common benzoic acid derivatives are p-hydroxybenzoic acid, salicylic acid, gallic acid, and ellagic acid, whereas the most prevalent derivatives of cinnamic acid are p-coumaric acid, caffeic acid, and ferulic acid [168]. Phenolic acids compounds can be classified into two fundamental subclasses hydroxybenzoic and hydroxycinnamic acids (Fig. 5).

In hydroxybenzoic acid, one of the hydrogens on the benzene ring is substituted by the carboxyl group (—COOH). The majority of hydroxybenzoic acid compounds exist as free acids, while a few occur as esters or glycosides. Hydroxycinnamic acids comprise a three-carbon chain connected to the benzene ring (C6H5CHCHCOOH) with at least one hydrogen atom that an OH group can replace [169]. In drug development technology, phenolic components have a significant interest worldwide due to their diverse chemical structures and biological activities. They play a crucial function in antihypertensive, antidiarrheal [170], neuroprotective, antidepressant, anti-inflammatory, anticancer, and antihyperglycemic drugs [171]. In addition, phenolic acids can suppress the activity of acetylcholinesterase and butyrylcholinesterase, which play a crucial role in treating Alzheimer's disease [172].

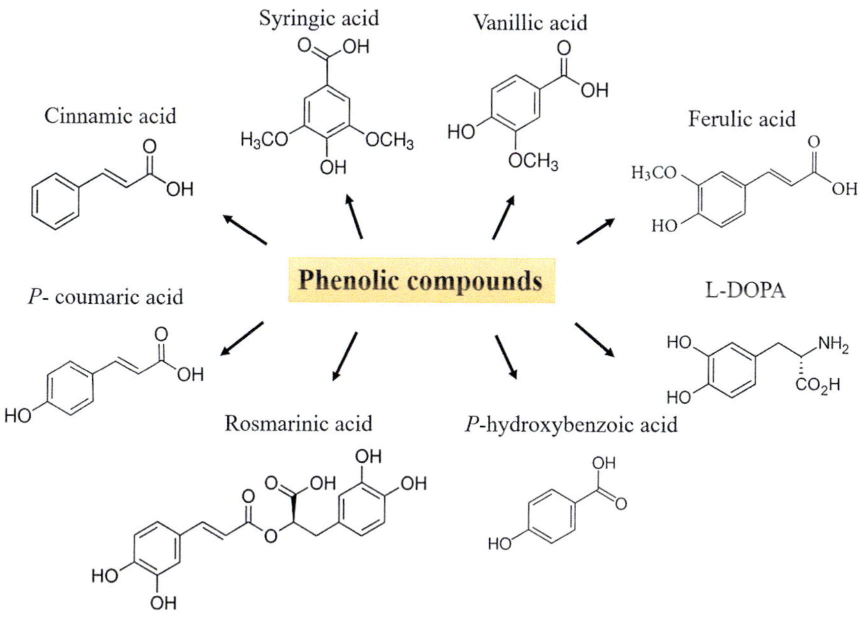

Fig. 5 Examples of some phenolic compounds

2.4.1 Hydroxybenzoic Acids

Hydroxybenzoic acid comprises a benzene ring, a carboxyl group, and additional hydroxyls, methyl (CHn), or methoxy (CHnO) groups attached to the benzene ring. The primary components of hydroxybenzoic acids include gallic, p-hydroxybenzoic, protocatechuic, and vanillic-hydroxybenzoic acid components produced as a resultant product of aromatic amino acids through the shikimic acid pathway [173]. Here, chloroformate intermediate is used to convert shikimic acid to L-phenylalanine. Consequently, L-phenylalanine is translated into p-hydroxybenzoic acid (a precursor to other phenolic acid metabolites) [174]. Moreover, the hydroxylation and methylation of its benzene ring produces other hydroxycinnamic acids (including ferulic and caffeic acids) or hydroxybenzoic acids (such as protocatechuic and p-hydroxybenzoic acids). It is hypothesized that hydroxybenzoic acid is synthesized by hydroxycinnamic acid with a similar structure via CoA-dependent (oxidative) or CoA-independent (nonoxidative) or a mixture of both mechanisms [169].

P-hydroxybenzoic acid, its derivatives, and vanillic acid are the major phenolic acids in many plant species, including cereals, flaxseeds, chia, and sunflower seeds [175–177]. Most of the hydroxycinnamic acid derivatives in chokeberry are chlorogenic and neochlorogenic acids. At the same time, strawberries and red raspberries involve predominant types of hydroxybenzoic acids, such as p-coumaric acid, p-hydroxybenzoic acid, and ellagic acid, respectively [168]. Hydroxybenzoic acid has contributed to significant medical research advances. Adefegha et al. [178] reported that some hydroxybenzoic acids have antidiabetic properties. For instance, gallic and protocatechuic acids inhibit the activity of type 2 diabetes–related enzymes (amylase and glucosidase). Furthermore, Yi et al. [179] investigated that the last mentioned acids (extracted from muscadine grapes) exhibited anticancer activities against cervical cancer.

2.4.2 Hydroxycinnamic Acids

From a structural perspective, hydroxycinnamic acids contain functional groups that can contribute to the coordination, like phenolic hydroxyl and carboxyl groups [180]. In addition, some compounds have phenolic hydroxyl groups and carboxyl groups of hydroxycinnamic acid and have one or two methoxy groups adjacent to phenolic hydroxyl groups on the benzene ring that have multiple coordination sites (which is crucial to the formation of complex and diverse compounds) [181]. Hydroxycinnamic acid is produced by lignin's metabolic pathway, which provides mechanical support for the plant cell wall [182]. It is catalyzed by phenylalanine ammonia-lyase, which deaminases L-phenylalanine to generate (E)-cinnamic acid. Then it undergoes further enzymatic transformations to create some related metabolites [183].

Recently, hydroxycinnamic acids' antioxidant, anti-inflammatory, and antibacterial properties have been investigated and employed in various pharmaceutical domains [184]. Many studies focused on vasodilation, antioxidant activity, and inflammation characteristics of hydroxycinnamic acid. Recent research showed that hydroxycinnamic acid compounds could reduce the severity of cardiovascular diseases [185, 186], hypertension [187, 188], depression [189], and diabetes [190]. Moreover, Tresserra-Rimbau et al. [179] and Adriouch et al. [180] studied that the high-intake percentages of hydroxycinnamic acids in daily meals significantly reduced the risks of cardiovascular disease in two countries (Spain and France).

2.5 Phytosterols

Phytosterols are cholesterol-like molecules found in various plants, with high concentrations occurring in the cell membranes of vegetables, fruits, and other plants. Phytosterol is a 3-hydroxy compound of cyclopentaphthene. The structure of phytosterols is similar to cholesterol; the main difference is that sterols are usually connected with a methyl or ethyl group at C-24, and common sterols have unsaturated double bonds between C-22 and C-23. Their structure comprises a steroid skeleton characterized by the saturated bond between C-5 and C-6 of sterol moiety [191]. Their aliphatic side chains are connected to C-17, while the hydroxyl groups are attached to the C-3 atom. They can be divided into sterols and stanols, representing unsaturated and saturated molecules. Phytosterols exist in four forms in plants, in the form of fatty acid esters, free alcohols, sterol glycosides, and acylated sterol glycosides [192]. They mainly exist in free and esterified forms (in edible oils), are found in plants, and cannot be synthesized in the human body. People mostly get phytosterols by ingesting dietary fiber, such as vegetable oil, beans, nuts, vegetables, and fruits [193]. Phytosterols show the characteristics of directly inhibiting tumor growth, including reducing cell cycle progression, inhibiting tumor metastasis, and inducing apoptosis, such as reducing the risk of oesophagal and ovarian cancers [194]. Campesterol, β-sitosterol, ergosterol, and stigmasterol are four representative compounds widely studied in treating diseases (Fig. 6).

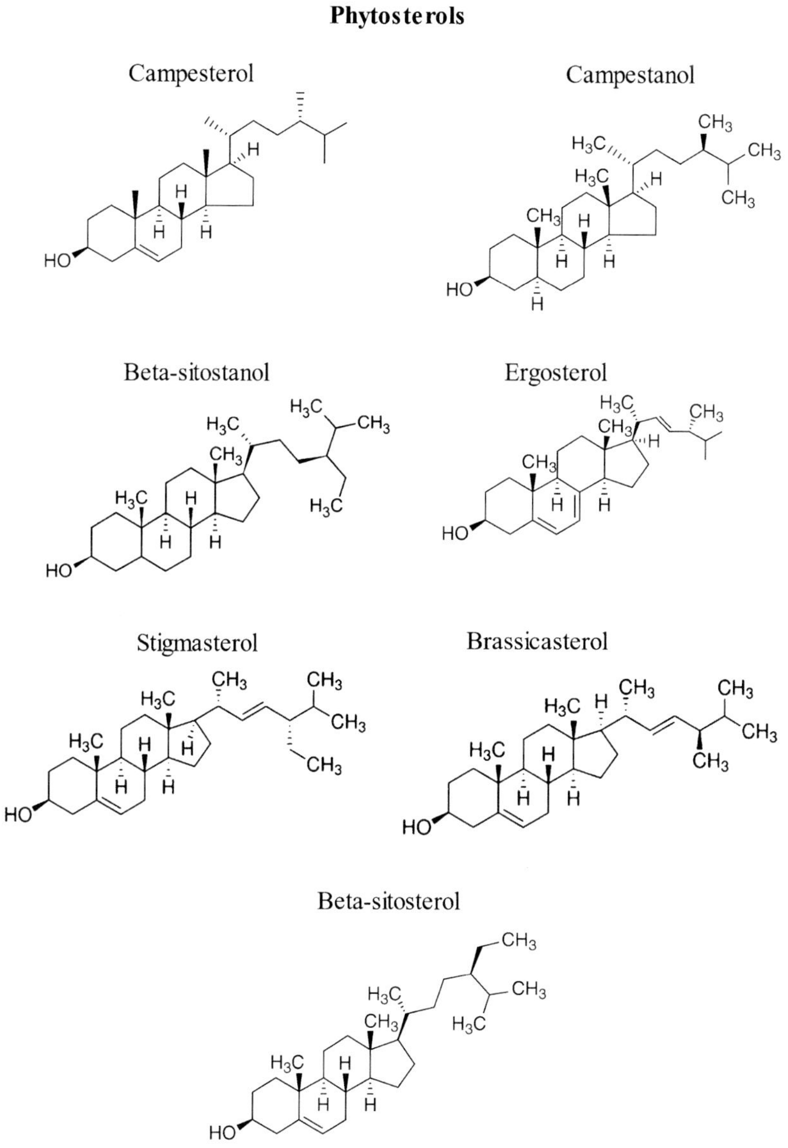

Phytosterols

Campesterol

Campestanol

Beta-sitostanol

Ergosterol

Stigmasterol

Brassicasterol

Beta-sitosterol

Fig. 6 Examples of some phytosterol compounds

Campesterol is abundant in canola and corn oils [195]. Campesterol has played an important role in inhibiting the proliferation of ovarian cancer cells and has the potential as a new antiovarian cancer drug. It was isolated from *Strychnos innocua* (Delile) and has been found to potentially affect antibacterial activity [196].

β-sitosterol has an unsaturated double bond on C5-C6, easily oxidized by active oxygen. Because of this structural characteristic, oxyphytosterols are one of the most common phytosterols in plants. Oxyphytosterols can play an important role as anti-inflammatory and antivirus agents. β-sitosterol can interfere with various cell signaling pathways, including cell cycle, proliferation, anti-inflammatory, hepatoprotective, and antioxidation [197]. Moreover, von Holtez et al. [198] discovered that low concentrations of β-sitosterol in cancer treatment dimensioned the development of cancer cells and even caused cancer cell death.

Ergosterol is the major steroidal component of the cell membrane of filamentous fungi and is not present or a minor component in most higher plants. It also exists in yeast cell walls and the mitochondrial membrane [193]. Its content has been widely used as an estimate of fungal biomass in soil and the aquatic environment because it was found that there was a strong correlation between ergosterol content and dry fungal biomass [194].

Stigmasterol is mainly found in soybeans. In addition to reducing cholesterol activity, stigmasterol has anti-inflammatory effects. Gabay et al. [199] showed that stigmasterol extracted from the bark of *Butea monosperma* (Lam.) Taub. has hypoglycemic, antiperoxidative, and thyroid-inhibiting effects.

3 Role of Secondary Metabolites in Diseases Treatment

Phytochemicals have been increasingly studied in disease treatment globally. In this section, the role of some phytochemicals in treating some disorders, including hepatic, renal, and cardiovascular disorders, cancer treatment, antimicrobial, anti-inflammatory activities, and neurological disorders, are discussed.

3.1 Hepatic Disorders

Nowadays, alternative therapy methods for several ailments are being embraced that use phytochemicals derived from natural resources. Despite significant developments in contemporary medicine, a persistent issue has been the lack of suitable and effective hepatoprotective medication [200]. Recent research has focused mostly on phytochemical screening to identify distinct types of biological activity. Recent years have seen a tremendous increase in interest in phytotherapy research as researchers look for new pharmacological cures. Numerous phytochemical substances, especially from medicinal herbs, are incorporated into the treatment of liver problems, including *Hedyotis corymbos*, *Rosa damascene*, *Casuarina equisetifolia*, *Cajanus cajan*, *Trichosanthes dioica*, *Glycosmis pentaphylla*, *Justicia gendarussa*, *Tinospora crispa*, *Bixa orellana*, *Moringa oleifera*, *Premna esculenta*,

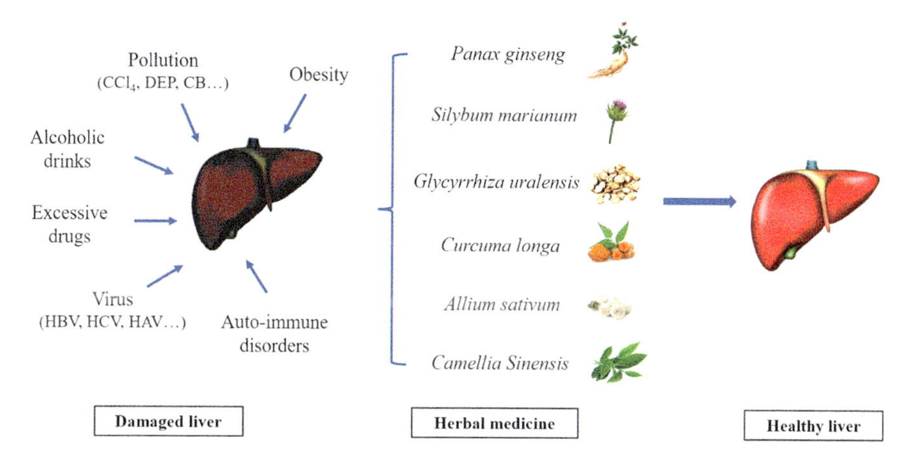

Fig. 7 Some vital phytochemicals support the hepatic disorders treatment. (modified after Das et al. [202])

Dendrophthoe pentandra, *Argemone mexicana*, *Leea macrophylla*, *Physalis minima*, *Synedrella nodiflora*, *Hylocereus polyrhizus*, *Caesalpinia bonduc*, *Piper chaba*, and *Bombax ceiba* (Fig. 7) [201].

Aguirre et al. [203] investigated some polyphenol components, like resveratrol (trans-3,4′,5-trihydroxystilbene) and stilbenoid involved in the enhancement of glycemic control and glucose tolerance, decrease in lipid levels, as well as anti-lipogenic, anti-inflammatory, and antioxidant properties. In addition, thymohydroquinone exhibited hepatoprotective characteristics toward drug-induced hepatotoxicity [204].

3.2 Renal Disorders

Renal damage and chronic kidney disorders are severe medical cases globally; the average number of patients increased by 8–16%. Besides Chinese herbal medicine metabolites, bioflavonoids, resveratrol, quercetin, and curcumin were demonstrated to be potential in chronic kidney disorders [205]. Green tea contains various polyphenolic substances, among them is epigallocatechin-3-gallate. Many reports showed the reactivity of epigallocatechin-3-gallate in renal damage treatment by delaying lupus nephritis via boosting the antioxidant Nrf2 pathway and diminishing the NLRP3 inflammasome activation. Additionally, ursolic acid (pentacyclic triterpenoid) is frequently found in fruit peels, herbs, and spices. Kunkel et al. [206] investigated that ursolic acid has a protective activity against chronic kidney disorders by suppressing the activity of STAT3 and the NF-B mechanism, hence decreasing the inflammatory activity. Moreover, allicin (organosulfur compound) enhances renal function by modifying the AT1, Nrf2/Keap1, and eNOS pathways and lowering CKD-mediated systemic hypertension [207].

3.3 Cardiac Disorders

According to current epidemiological forecasts, the globe is on course to experience a vascular typhoon of cardiovascular disease burden. In emerging nations, coronary heart disease raised by 120% in women and 137% in men in 2020 [208]. Therefore, researchers searched for natural plant-derived cardiovascular medicines. Soya bean isoflavones reported a powerful ability to manage the risk factors for cardiovascular disorders. Key isoflavones found in soya beans are present as glycosides (e.g., genistein, daidzein, and glycitin) and support cardiac function through various methods [209]. Many epidemiological, clinical, and experimental investigations revealed a positive correlation between green tea consumption and cardiovascular health, as green tea contains a high yield of catechins [210]. In addition, an epidemiological study suggests dietary carotenoids may lower the incidence of ischemic stroke, myocardial infarction, and coronary heart disease. Several phytochemical components, such as glucoraphanin, cucurbitacins, diosgenin, sulforaphane, and tocopherols, provide cardiovascular protection, reduce oxidative stress, enhance lipid profiles, and lower the blood pressure [211, 212].

3.4 Cancer Treatment

The number of cancer patients is still rising yearly, despite research into many ways to prevent and treat the disease. Reregulating cellular processes has been the focus of cancer treatment. Many clinical trials have examined cancer treatments using radiation, chemotherapy, antibody therapy, and immunotherapy. Because they cause deleterious effects on healthy cells, radiation and chemotherapy have serious adverse effects. Immunotherapy and antibody therapy exhibit exact cancer-targeting abilities, but they have a small therapeutic window and are sometimes expensive [213]. About 25% of medications utilized in cancer therapeutic settings come from plants (anticancer activity). For example, Lestari et al. [214] reported the anticancer activity of curcumin (1,7-bis(4-hydroxy-3-methoxyphenyl)-1,6-heptadiene-3,5-dione) (isolated from *Curcuma longa*) in colorectal cancer treatment.

Additionally, curcumin can be investigated as an antiangiogenic, anti-inflammatory, and antioxidant drug. Polyphenol flavonoid resveratrol (3,4,50-hydroxystilbene) can stop the growth of *H. pylori* and the division of gastric NSCLC cells preventing/treating gastric cancer [215]. Recent research revealed that the secondary metabolite hypericin from the plant *Hypericum* L. inhibited the overexpression of the ABC transporter in the HL-60 subclone leukemia cells via accelerating mitoxantrone-induced cell death [216]. Furthermore, β-carotene has been demonstrated to enhance the suppression activity of the anticancer drug (5-FU) on the development of tumors by oesophagal carcinoma (Eca109) [217].

3.5 Antimicrobial Activity

In the past 50 years, the development of vaccines and antibiotics has successfully eliminated infectious diseases such as *Mycobacterium tuberculosis* and smallpox, saving many lives. Plant secondary metabolites, such as flavonoids, phenolic, alkaloids, organic acids, and essential oils, have a significant role in antimicrobial properties. *Saussurea gossypiphora* contains multiple secondary metabolites, including apigenin and luteolin, which have antimicrobial activity against *Escherichia coli* and *Staphylococcus aureus* [218]. Other examples of heterologous expression of phytochemical pathways have shown it is possible to develop plants into pharmaceutical production platforms. *Crocosmia* spp. is an ornamental plant found to produce montbretin A, a potent inhibitor of human pancreatic amylase, which is a promising treatment for type II diabetes pending clinical validation.

Moreover, rosmarinic acid extracted from *Origanum vulgare* L. has significant antibacterial activity against *Helicobacter pylori* associated with an ulcer [219]. Flavonoids have multitarget characteristics, which can target bacterial plasma membranes to inhibit drug-resistant bacteria through different sterilization mechanisms. Isoprenylation of flavonoids plays a crucial role in antibacterial activity. Both α-mangostin and isobavachalcone target phospholipids, which destroy the membrane homeostasis and combine with the outer membrane permeabilizer against gram-positive bacteria. In addition, the antibacterial activity of phenolic secondary metabolites is attributed to the acidic characteristics of hydroxyl groups. The compounds change the permeability of cells, interfere with the enzymes involved in productivity, interrupt protein synthesis, and eventually lead to cell death [220]. Alkaloids affect the integrity of the cell wall of *Candida albicans*, leading to mitochondrial dysfunction, which in turn leads to the upregulation of oxidative stress. Similarly, essential oils have antimicrobial mechanisms that affect the ATP concentration and peptidoglycan of *Escherichia coli* and *Staphylococcus aureus*, resulting in cell wall damage [221].

3.6 Anti-Inflammatory Activity

Inflammation is the common process in the body's defensive response to hazardous stimuli, such as biological factors (bacteria, fungi, parasites, and viruses), physical factors (ultraviolet rays, extreme temperature, radioactive substances, and mechanical damage), chemical elements (a poisonous gas), and others. The uncontrolled inflammatory reaction is one of the main causes of allergies, cardiovascular diseases, and autoimmune diseases. Inflammation can cause various conditions, such as asthma, arteriosclerosis, and arthritis, which greatly threaten health [222]. Treating some inflammation-related diseases depends on steroidal and nonsteroidal anti-inflammatory drugs but treating such drugs will have several side effects. Therefore, using natural herbal medicines in treatment is particularly important. How to develop drugs with anti-inflammatory activity by using natural herbal medicine has become the hotspot of research. Natural herbal medicines can synthesize various anti-inflammatory compounds, including flavonoids, terpenoids, alkaloids, and essential oils.

Turmeric has anti-inflammatory and anti-arteriosclerosis effects. In India, turmeric is a traditional medicine used to treat rheumatic diseases [223]. It is reported that the genus *Ipomoea* and *Alstonia* species in Egypt have anti-inflammatory activity with little side effects. Furthermore, *Ipomoea pescaprae* extract has a therapeutic impact on dermatitis caused by jellyfish sting and oedema caused by ethyl phenylpropionate in experimental animals [224]. Studies illustrated this activity is due to the separation of some lipid and phenolic compounds.

In addition, licorice is the most frequently used traditional Chinese medicine; it can produce a variety of bioactive secondary metabolites, such as flavonoids, polysaccharides, and triterpenes. They have anti-inflammatory activity and are important as antibacterial, antiviral, and hepatoprotective substances [225]. Stilbenoids are found in tea, grapes, nuts, and berries. Resveratrol, pterostilbene, and piceatannol are well-known stilbenoids with indisputable anti-inflammatory activity in vitro and in vivo. Stilbene compounds are a group of plant phytoalexin polyphenols that can protect plants against pathogenic bacteria [226]. In folk medicine, these compounds are extensively used to treat skin inflammation, hepatitis, and stomachache. At the same time, they also have anticancer, neuroprotective, and antiviral properties. Recent studies reported that phenylpropanoids in essential oils have anti-inflammatory activities. Phenylpropionic acids are organic substances produced by plants that play an important role in preventing and treating trauma and infection. They are widely used in the medical field due to their corresponding pharmacological characteristics [227]. For instance, *Cinnamomum cassia* is commonly used to treat gastritis, anti-inflammation, and dyspepsia [228].

Moreover, alkaloids are widely distributed in *Menispermum dauricum* DC., *Tripterygium wilfordii*, and *Begonia kunmingshanensis* [229]. Wei et al. [230] stated that alkaloids have strong anti-inflammatory activity and are commonly used in treating ankylosing spondylitis, systemic lupus erythematosus, and other diseases. Berberine, sinomenine, dauricine, tetrandrine, stephanine, and lycorine in isoquinoline alkaloids have good anti-inflammatory effects. Piperidine alkaloids have anti-inflammatory, anti-arrhythmia, and anticancer effects. In addition to anti-inflammation, terpenoid alkaloids have antipyretic, analgesic, and antihypertensive effects [231].

3.7 Neurological Disorders

Neurological disorders affect the central nervous system and its peripheral nerves. The symptoms of neurological disorders include headaches, memory loss, slowness of movement, speech disorders, limb weakness, or pain. The causes of the disease are complicated; some neurological diseases have obvious heritability. Furthermore, environmental factors and age also affect nervous system health. The active ingredients in plants are very important in preventing and treating neurological diseases. Herbal medicines can delay or reduce the symptoms of neurological disorders. For example, tea polyphenols are a well-known active ingredient in tea. It can inhibit acetylcholinesterase and butyrylcholinesterase in improving Alzheimer's disease. In addition, tea polyphenols can modulate and shape gut microbiota to boost immunity and enhance sleep quality by the gut-brain axis. Phenolic substances extracted from grapes have a good antioxidant effect, which can reduce the formation of free radicals and prevent damage to protective cells, helping to alleviate neurological disorders and brain aging. Moreover, numerous research reported that cannabinoids (isolated from *Cannabis Sativa*) have sedative, anti-inflammatory, and analgesic effects. Thus, it can potentially relieve various neurological diseases [232, 233].

4 Artificial Phytochemicals Production

Herbal medicines have been utilized since antiquity, and several therapeutic plant secondary metabolites are employed directly as medications and raw resources for semisynthetic alterations. Their structural diversity, which frequently limits the cost-effective chemical synthesis and relatively low extracted content of many plant species, mandates the field-cultivated materials. The novel biotechnological fabrication techniques of these chemicals offer a variety of benefits, including predictable, steady, and longtime sustainable production, adaptability, and more straightforward isolation and purification [234]. Many researchers use genetic engineering to manipulate plant metabolism to manufacture artificial phytochemicals. This entails the introduction of genes that encode enzymes involved in synthesizing specific bioactive components in plants. For instance, scientists have successfully modified plants to generate secondary Taxol® (paclitaxel) metabolite. It is an isoprenoid extracted mainly from *Taxus brevifolia* plant and widely used in cancer treatment [235–237]. Moreover, the artificial synthesis of phytochemicals involves using genetically engineered microbes or "cell factories" engineered to manufacture certain compounds, including carotenoids, flavonoids, and terpenoids. The extracted chemicals are subsequently utilized as dietary supplements or food additives to enhance their nutritional value.

Plant synthetic biology attempts to avoid several challenges by overexpressing bioactive chemicals for large-scale synthesis and purification. The complete characterization of the biosynthetic process for the target molecule is the critical step in synthetic biology. Recognizing the interactions between produced intermediates and identifying the necessary components of a biosynthetic route based on conventional classes of enzymes can assist in completing several fragmented molecular mechanisms [238]. Moreover, contemporary sequencing technology has progressed, so whole genome sequencing and phylogenetic analyses of nonmodel species are now feasible [239]. For example, Natt et al. [240] identified a biosynthetic mechanism for producing and extracting colchicine (alkaloid) from *Gloriosa superba*. The study was conducted through previously derived RNA sequencing data of many plant species following metagenomics to ascertain eight critical genes in the pathway and to reassemble a 16-gene scheme in *Nicotiana benthamiana* with overexpression to generate N-formyldemecolcine (the colchicine precursor).

5 Recent Advances in Herbal Medicine

To understand the most recent advances in herbal medicine achievements, in this part, groups of studies conducted on the potentiality of plant secondary metabolites activity in disorders treatment are discussed. Several natural products derived from Chinese herbal medicine display anticancer properties, such as antiproliferative, proapoptotic, antimetastatic, and antiangiogenic effects, as well as the ability to regulate autophagy, modify multidrug resistance, balance immunity, and augment chemotherapy in vitro and in vivo. For instance, the therapeutic activity of curcumin, epigallocatechin gallate, berberine, artemisinins, ginsenosides, ursolic acid, silibinin, emodin, triptolide, cucurbitacins, tanshinones, ordonin, shikonin, gambogic acid, artesunate, wogonin, β-elemene, and cepharanthine has been discussed in more than 100 published research articles [241]. Curcumin has been found to possess diverse pharmacological effects, such as anticancer, anti-inflammatory, and anti-xidative activities, as supported by clinical data and comprehensive research [242, 243].

Curcumin and its derivatives have demonstrated promising potential as therapeutic agents for various malignant conditions, including cancer. Several research studies have demonstrated that curcumin and its derivatives can impede the growth of tumors in different human body regions, such as the head, neck, skin, ovarian, and gastric cancers[244–246]. Liu et al. [243] reported that the natural polyphenol curcumin (extracted from turmeric rhizome) suppressed the Janus kinase/signal transducers and activators of the transcription (JAK/STAT) pathway in cultured oesophagal squamous cell carcinoma (ESCC) cells. The attachment of cytokines to receptors activates JAKs, which phosphorylates STATs. STATs are translocated into the nucleus after being dimerized and phosphorylated to regulate gene expression. These genes, including cyclins and antiapoptosis proteins, are crucial for cell proliferation and survival [242]. The results showed that curcumin inhibits JAK2 activation, downregulating STAT3 signaling, suppressing cell proliferation and colony formation, cell cycle arrest, and apoptosis. In addition, curcumin substantially reduced tumor growth in xenografts derived from ESCC patients. These findings indicated that curcumin is a powerful agent for preventing ESCCs harboring constitutively active STAT3 proteins. In addition, curcumin substantially reduced tumor growth in xenografts derived from ESCC patients. These findings indicated that curcumin is a powerful agent for preventing ESCCs harboring constitutively active STAT3 proteins. A study by Sivanantham et al. [247] reflected the anticancer activity of curcumin and its role in treating head and neck squamous cell carcinoma (HNSCC). They combined curcumin with three anticancer drugs, including docetaxel, doxorubicin, 5-fluorouracil, and diammine dichloroplatinum (II) (cisplatin), and evaluated their joint influence on the HNSCC cell line NT8e. The outcomes showed that the combined administration of 5- fluorouracil or doxorubicin with curcumin significantly inhibited NT8e cancer cell proliferation and enhanced apoptosis. NT8e cells treated with 5-fluorouracil or doxorubicin in combination with curcumin exhibited apoptosis via inhibition of Bcl-2 and elevation of Bax, caspase-3, and poly-ADP ribose polymerase. The researchers undergo some confirmation experiments to ensure the results are obtained through DAPI staining and decreased red/green fluorescence by JC-1 observations of apoptotic cell features, such as membrane blebbing, nuclear condensation, and cell contraction.

Some phenolic substances collaborate with nonsteroidal anti-inflammatory drugs to block pro-inflammatory mediators' functioning or gene expression, such as cyclooxygenase (COX). Diverse phenolic components can also act on transcription factors, like nuclear factor-B (NF-B) and nuclear factor-erythroid factor 2-related factor 2 (Nrf-2), to upregulate or downregulate antioxidant response pathway constituents. Moreover, they may suppress enzymes involved in developing human diseases and have been utilized for treating various common human disorders, such as hypertension, metabolic issues, incendiary infections, and neurological disorders [247]. Hypertension is a prevalent and frequently progressive condition associated with a high risk of heart failure and complications [248]. According to estimates, up to 25% of adults struggle with hypertension globally [249]. Hypertension is a serious and becoming more prevalent global health issue. Multiple investigations have demonstrated that polyphenol-rich foods effectively prevent and treat hypertension, particularly through angiotensin-converting enzyme (ACE) inhibition [250]. Patten et al. [235] recently discovered about 74 plant families with substantial ACE inhibitory activity. Similarly, Han et al. [251] proved that some cocoa polyphenols, including catechins, flavonol glycosides, anthocyanins, and procyanidins, are bioavailable substances with antihypertensive properties by inhibiting ACE [252].

In addition, an in vitro study carried out by Bhandari et al. [253] investigated the antidiabetic efficacy of two soluble active compounds(−)-3-O-galloylepicatechin and (−)-3-O-galloylcatechin found in the ethyl acetate Bergenia ciliate extract served as catalysts for the dose-dependent suppression of porcine pancreatic amylase and rat intestinal maltase activation. The half-maximal inhibitory concentration (IC_{50}) value is the compound concentration required for inhibiting enzyme activity by 50%. The in vivo and in vitro suppressive enzyme activity of these compounds against α-glucosidase and α-amylase indicate that they have outstanding prospects for development as a treatment for type 2 diabetes [254].

Aging can be described as an accumulation of multiple detrimental changes in cells and tissues, which increases the risk of disease and mortality with advancing age. The free radical and oxidative stress theory [255] represents one of the most commonly accepted hypotheses for the aging mechanism. Despite normal conditions, oxidative damage occurs; however, as antioxidative and restoration processes become less effective with age, the rate of oxidative destruction rises [256, 257]. Plasma antioxidant capacity is correlated with antioxidant dietary consumption; it has been observed that a diet rich in antioxidants can mitigate the negative effects of aging and behavior. Multiple investigations suggest that a combination of antioxidant/anti-inflammatory polyphenolic substances found in fruits and vegetables may be effective antiaging compounds [258]. Anthocyanins, a category of flavonoids, are exceptionally abundant in brightly colored fruits, like berries, Concord grapes, and grape seeds. Anthocyanins have potent anti-inflammatory and antioxidant effects and inhibit lipid peroxidation and COX-1 and -2, which are inflammatory mediators. The antioxidant activity of formulations of flavonoid-rich fruits and vegetables, including spinach, strawberries, and blueberries, is very high. According to the results of Shukitt-Hale et al. [259], supplemental nutrition with spinach, strawberry, or blueberry extracts in a control diet was similarly effective in restoring age-related deficits in the brain and behavioral function of geriatric rats. Moreover, tea catechins have potent antiaging characteristics, and drinking green tea abundant in these catechins could potentially delay the advent of aging [260].

6 Conclusion

Natural plant secondary metabolites (phytochemicals) have unique chemical structures with a wide range of diversity in medicinal and biological characteristics. They significantly treated some diseases and chronic disorders (cancers, cardiovascular, hyperglycemic, and type 2 diabetes). Moreover, they are involved in many pharmaceutical industries, like cosmetics and food supplements. The classification and the taxonomic position of each group and plant species are crucial in understanding each phytochemical's distinct function and characteristics (physical and chemical properties, synthesis pathways, and extraction methods). Furthermore, nature may have more based on the number of secondary metabolites identified and extracted. Considering the developments in synthesis technology as well as the invention of more advanced isolation and analysis approaches, it should be possible to identify a significant amount of these additional phytochemicals. Moreover, artificial phytochemical manufactoring also offers the potential to increase the effectiveness and safety of natural phytochemicals. Scientists can confirm the absence of impurities and pollutants by synthesizing molecules in a sterile atmosphere. In addition, they can alter the structure of a substance to increase its efficacy or decrease its toxicity. Recently, many researchers reflected on the potency of many isolated active metabolites in treating some vital disorders and may replace commercial and chemically synthesized drugs, especially those offered for tumor treatment.

References

1. Dowd CJ, Kelley B (2011) Purification process design and the influence of product and technology platforms
2. Shitan N (2016) Secondary metabolites in plants: transport and self-tolerance mechanisms. Biosci Biotechnol Biochem 80:1283–1293
3. Yazaki K (2005) Transporters of secondary metabolites. Curr Opin Plant Biol 8:301–307
4. Jamwal K, Bhattacharya S, Puri S (2018) Plant growth regulator mediated consequences of secondary metabolites in medicinal plants. J Appl Res Med Aromat plants 9:26–38

5. Bartwal A, Mall R, Lohani P, Guru SK, Arora S (2013) Role of secondary metabolites and brassinosteroids in plant defense against environmental stresses. J Plant Growth Regul 32: 216–232
6. Liu Z, Wang H, Xie J, Lv J, Zhang G, Hu L, Luo S, Li L, Yu J (2021) The roles of cruciferae glucosinolates in disease and pest resistance. Plan Theory 10:1097
7. Thrane U (2001) Development in the taxonomy of fusarium species based on secondary metabolites. In Proceedings of the fusarium: Paul E. Nelson memorial symposium, APS Press, pp 29–49
8. Pal D, Mishra P, Sachan N, Ghosh AK (2011) Biological activities and medicinal properties of Cajanus Cajan (L) Millsp. J Adv Pharm Technol Res 2:207
9. Tungmunnithum D, Hano C (2020) Cosmetic potential of Cajanus Cajan (L.) Millsp: botanical data, traditional uses, phytochemistry and biological activities. Cosmetics 7:84
10. Mahmoud MF, Ali N, Mahdi I, Mouhtady O, Mostafa I, El-Shazly AM, Abdelfattah MAO, Hasan RA, Sobeh M (2023) Coriander essential oil attenuates dexamethasone-induced acute liver injury through potentiating Nrf2/HO-1 and ameliorating apoptotic signaling. J Funct Foods 103:105484
11. Roy S, Sehgal R, Padhy BM, Kumar VL (2005) Antioxidant and Protective Effect of Latex of Calotropis Procera against Alloxan-Induced Diabetes in Rats. J Ethnopharmacol 102:470–473
12. Al-Rowaily SL, Abd-ElGawad AM, Assaeed AM, Elgamal AM, El Gendy AE-NG, Mohamed TA, Dar BA, Mohamed TK, Elshamy AI (2020) Essential oil of calotropis procera: comparative chemical profiles, antimicrobial activity, and allelopathic potential on weeds. Molecules 25:5203
13. Al-Thobaiti SA, Konozy EHE (2022) Purification, partial characterization, and evaluation of the antiulcer activity of calotropis procera leaf lectin. Protein Pept Lett 29:775–787
14. Panda H (2002) Medicinal plants cultivation & their uses. Asia Pacific Business Press Inc., ISBN 8178330962
15. Cohall D, Carrington S (2012) A comparison of the chemical constituents of barbadian medicinal plants within their respective plant families with established drug compounds and phytochemicals used to treat communicable and non-communicable diseases. West indian Med J 61
16. Rangan C (2012) Diuretics, ipecac, and laxatives. Med. Toxicol. Drug Abus. Hoboken, NJ John Wiley Sons, Inc 200–232
17. Debnath B, Singh WS, Das M, Goswami S, Singh MK, Maiti D, Manna K (2018) Role of plant alkaloids on human health: a review of biological activities. Mater today Chem 9:56–72
18. Soprani LC, Andrade JP de, Santos VD dos, Alves-Araújo A, Bastida J, Silva CAG, Silveira D, Borges WDeS, Jamal CM (2021) Chemical evaluation and anticholinesterase activity of Hippeastrum Puniceum (Lam.) Kuntz Bulbs (Amaryllidaceae). Brazilian J Pharm Sci, 57
19. Shen C-Y, Xu X-L, Yang L-J, Jiang J-G (2019) Identification of narciclasine from lycoris radiata (L'Her.) herb. and its inhibitory effect on lps-induced inflammatory responses in macrophages. Food Chem Toxicol 125:605–613
20. Kornienko A, Evidente A (2008) Chemistry, biology, and medicinal potential of narciclasine and its congeners. Chem Rev 108:1982–2014
21. Patil VS, Harish DR, Vetrivel U, Deshpande SH, Khanal P, Hegde HV, Roy S, Jalalpure SS (2022) Pharmacoinformatics analysis reveals flavonoids and diterpenoids from andrographis paniculata and thespesia populnea to target hepatocellular carcinoma induced by hepatitis B virus. Appl Sci 12:10691
22. Saravanakumar A, Venkateshwaran K, Vanitha J, Ganesh M, Vasudevan M, Sivakumar T (2009) Evaluation of antibacterial activity, phenol and flavonoid contents of thespesia populnea flower extracts. Pak J Pharm Sci:22
23. Gritto MJ, Nandagopalan V, Doss A (2015) GC-MS analysis of bioactive compounds in methanolic extract ofthespesia populnea (L.) SOL. Ex Correa
24. Miura T, Takagi S, Ishida T (2012) Management of diabetes and its complications with banaba (Lagerstroemia Speciosa L.) and corosolic acid. Evid-Based Compl Altern Med 2012

25. Yasir M, Das S, Kharya MD (2010) The phytochemical and pharmacological profile of persea americana mill. Pharmacogn Rev 4:77
26. Zhang SM, Willett WC, Selhub J, Manson JE, Colditz GA, Hankinson SE (2003) A prospective study of plasma total cysteine and risk of breast cancer. Cancer Epidemiol Biomark Prev 12:1188–1193
27. Goh YI, Koren G (2008) Folic acid in pregnancy and fetal outcomes. J Obstet Gynaecol (Lahore) 28:3–13
28. Ahmed F, Chandra J, Urooj A, Rangappa KS (2009) In vitro antioxidant and anticholinesterase activity of acorus calamus and nardostachys jatamansi rhizomes. J Pharm Res 2:830–883
29. Prisilla DH, Balamurugan R, Shah HR (2012) Antidiabetic activity of methanol extract of acorus calamus in STZ induced diabetic rats. Asian Pac J Trop Biomed 2:S941–S946
30. Sharma V, Singh I, Chaudhary P (2014) Acorus calamus (the healing plant): a review on its medicinal potential, Micropropagation and conservation. Nat Prod Res 28:1454–1466
31. Velichkova K, Sirakov I, Stoyanova S, Zhelyazkov G, Staykov Y, Slavov T (2019) Effect of Acorus Calamus L. extract on growth performance and blood parameters of common carp (Cyprinus Carpio L.) cultivated in a recirculation system. J Cent Eur Agric 20:585–591
32. Channa S, Dar A, Ahmed S (2005) Evaluation of alstonia scholaris leaves for broncho-vasodilatory activity. J Ethnopharmacol 97:469–476
33. del Barrio G, Parra F (2011) 14 antiviral activities of phyllanthus orbicularis, an endemic cuban species. Phyll Species Sci Eval Med Appl 219
34. Dabanka CP (2013) Antibacterial activity of phyllanthusamarus (schumand thonn) extract against salmonella typhicausative agent of typhoid fever
35. Eichelbaum SR (2016) Screening of plants for antibacterial properties: growth inhibition of staphylococcus aureus by artemisia tridentata
36. Bucolo C, Leggio GM, Drago F, Salomone S (2012) Eriodictyol prevents early retinal and plasma abnormalities in streptozotocin-induced diabetic rats. Biochem Pharmacol 84:88–92
37. Islam A, Islam MS, Rahman MK, Uddin MN, Akanda MR (2020) The pharmacological and biological roles of eriodictyol. Arch Pharm Res 43:582–592
38. Singh H, Kapoor VK, Piozzi F, Passannanti S, Paternostro M (1978) Isomotiol, a new triterpene from strychnos potatorum. Phytochemistry 17:154–155. https://doi.org/10.1016/S0031-9422(00)89704-6
39. Srivastava SK, Jain DC (1989) Triterpenoid saponins from plants of araliaceae. Phytochemistry 28:644–647. https://doi.org/10.1016/0031-9422(89)80074-3
40. Zhang Y-W, Lin H, Bao Y-L, Wu Y, Yu C-L, Huang Y-X, Li Y-X (2012) A new triterpenoid and other constituents from the stem bark of juglans mandshurica. Biochem Syst Ecol 44:136–140. https://doi.org/10.1016/j.bse.2012.04.015
41. Li Z-Y, Qi F-M, Zhi D-J, Hu Q-L, Liu Y-H, Zhang Z-X, Fei D-Q (2017) A novel spirocyclic triterpenoid and a new taraxerane triterpenoid from teucrium viscidum. Org Chem Front 4:42–46. https://doi.org/10.1039/C6QO00460A
42. Shahidi F, Naczk M (2003) Phenolics in food and nutraceuticals. CRC press. ISBN 0203508734
43. Patočka J, Jakl J (2010) Biomedically relevant chemical constituents of valeriana officinalis. J Appl Biomed 8:11–18
44. Fernández S, Wasowski C, Paladini AC, Marder M (2004) Sedative and sleep-enhancing properties of linarin, a flavonoid-isolated from valeriana officinalis. Pharmacol Biochem Behav 77:399–404
45. Kokate CK, Purohit AP, Gokhale SB (2003) Text book of pharmacognosy. Pune Nirali Prakashan 8:1–624
46. Nagpal N, Shah G, Arora NM, Shri R, Arya Y et al (2010) Int J Pharm Sci Res 1:28–36
47. Hardel DK, Laxmidhar S (2011) A review on phytochemical and pharmacological of eucalyptus globulus: a multipurpose tree. Int J Res Ayurveda Pharm 2:1527–1530
48. Dixit A, Rohilla A, Singh V (2012) Eucalyptus globulus: a new perspective in therapeutics. Int J Pharm Chem Sci 1:1678–1683

49. Park HW, Kim OT, Hyun DY, Kim YB, Kim JU, Kim YC, Bang KH, Cha SW, Choi JE (2013) Overexpression of farnesyl diphosphate synthase by introducing CaFPS gene in panax ginseng CA Mey. Korean J Med Crop Sci 21:32–38
50. Shellie RA, Marriott PJ, Huie CW (2003) Comprehensive two-dimensional gas chromatography (GC× GC) and GC× GC-quadrupole MS analysis of asian and american ginseng. J Sep Sci 26:1185–1192
51. Bahukhandi A, Upadhyay S, Bisht K (2021) Panax Ginseng ca Meyer. In: Naturally Occurring Chemicals Against Alzheimer's Disease. Elsevier, Amsterdam, pp 217–223
52. Neamsuvan O, Bunmee P (2016) A survey of herbal weeds for treating skin disorders from southern Thailand: Songkhla and Krabi province. J Ethnopharmacol 193:574–585
53. Ahmad J, Khan I, Johnson SK, Alam I, ud Din Z (2018) Effect of Incorporating Stevia and Moringa in Cookies on Postprandial Glycemia, Appetite, Palatability, and Gastrointestinal Well-Being. J Am Coll Nutr 37:133–139
54. Jain PG, Patil SD, Haswani NG, Girase MV, Surana SJ (2010) Hypolipidemic activity of Moringa Oleifera Lam., moringaceae, on high fat diet induced hyperlipidemia in albino rats. Rev Bras 20:969–973
55. Ma ZF, Ahmad J, Zhang H, Khan I, Muhammad S (2020) Evaluation of phytochemical and medicinal properties of moringa (Moringa Oleifera) as a potential functional food. South African J Bot 129:40–46
56. Panche AN, Diwan AD, Chandra SR (2016) Flavonoids: an overview. J Nutr Sci 5:e47
57. Marrs KA, Alfenito MR, Lloyd AM, Walbot V (1995) A glutathione s-transferase involved in vacuolar transfer encoded by the maize gene bronze-2. Nature 375:397–400. https://doi.org/10.1038/375397a0
58. Winkel-Shirley B (2001) It takes a garden. how work on diverse plant species has contributed to an understanding of flavonoid metabolism. Plant Physiol 127:1399–1404. https://doi.org/10.1104/pp.010675
59. Martens S, Preuß A, Matern U (2010) Multifunctional flavonoid dioxygenases: flavonol and anthocyanin biosynthesis in Arabidopsis Thaliana L. Phytochemistry 71:1040–1049. https://doi.org/10.1016/j.phytochem.2010.04.016
60. Tariq H, Asif S, Andleeb A, Hano C, Abbasi BH (2023) Flavonoid production: current trends in plant metabolic engineering and de novo microbial production. Meta 13:124
61. Cao Y-L, Lin J-H, Hammes H-P, Zhang C (2022) Flavonoids in treatment of chronic kidney disease. Molecules 27:2365
62. Jan R, Khan M, Asaf S, Asif S, Kim K-M (2022) Bioactivity and therapeutic potential of kaempferol and quercetin: new insights for plant and human health. Plan Theory 11:2623
63. Bešlo D, Došlić G, Agić D, Rastija V, Šperanda M, Gantner V, Lučić B (2022) Polyphenols in ruminant nutrition and their effects on reproduction. Antioxidants 11:970
64. Dastidar SG, Manna A, Kumar KA, Mazumdar K, Dutta NK, Chakrabarty AN, Motohashi N, Shirataki Y (2004) Studies on the antibacterial potentiality of isoflavones. Int J Antimicrob Agents 23:99–102
65. Wang T, Liu Y, Li X, Xu Q, Feng Y, Yang S (2018) Isoflavones from green vegetable soya beans and their antimicrobial and antioxidant activities. J Sci Food Agric 98:2043–2047
66. Chin-Dusting JPF, Fisher LJ, Lewis TV, Piekarska A, Nestel PJ, Husband A (2001) The vascular activity of some isoflavone metabolites: implications for a cardioprotective role. Br J Pharmacol 133:595–605
67. Aboushanab SA, Ali H, Narala VR, Ragab RF, Kovaleva EG (2021) Potential therapeutic interventions of plant–derived isoflavones against acute lung injury. Int Immunopharmacol 101:108204
68. Shah U, Patel A, Patel S, Patel M, Patel A, Patel S, Patel S, Maheshwari R, Mtewa AG, Gandhi K (2022, 2063–2079) Role of natural and synthetic flavonoids as potential aromatase inhibitors in breast cancer: structure-activity relationship perspective. Anti-Cancer Agents Med Chem (Formerly Curr Med Chem Agents) 22

69. Morimoto R, Matsubara C, Hanada A, Omoe Y, Ogata T, Isegawa Y (2022) Effect of structural differences in naringenin, prenylated naringenin, and their derivatives on the anti-influenza virus activity and cellular uptake of their flavanones. Pharmaceuticals 15:1480
70. Moghaddam RH, Samimi Z, Moradi SZ, Little PJ, Xu S, Farzaei MH (2020) Naringenin and naringin in cardiovascular disease prevention: a preclinical review. Eur J Pharmacol 887: 173535
71. Nouri Z, Fakhri S, El-Senduny FF, Sanadgol N, Abd-ElGhani GE, Farzaei MH, Chen J-T (2019) On the neuroprotective effects of naringenin: pharmacological targets, signaling pathways, molecular mechanisms, and clinical perspective. Biomol Ther 9:690
72. Yang H, Wang Y, Xu S, Ren J, Tang L, Gong J, Lin Y, Fang H, Su D (2022) Hesperetin, a promising treatment option for diabetes and related complications: a literature review. J Agric Food Chem 70:8582–8592
73. Ye J, Guan M, Lu Y, Zhang D, Li C, Li Y, Zhou C (2019) Protective effects of hesperetin on lipopolysaccharide-induced acute lung injury by targeting MD2. Eur J Pharmacol 852:151–158
74. Zheng Y, Zhang Y, Li Z, Shi W, Ji Y, Guo Y-H, Huang C, Sun G, Li J (2021) Design and synthesis of 7-O-1, 2, 3-triazole hesperetin derivatives to relieve inflammation of acute liver injury in mice. Eur J Med Chem 213:113162
75. Wu M, Zhu X, Zhang Y, Wang M, Liu T, Han J, Li J, Li Z (2021) Biological evaluation of 7-O-amide hesperetin derivatives as multitarget-directed ligands for the treatment of alzheimer's disease. Chem Biol Interact 334:109350
76. Woodman OL, Meeker WF, Boujaoude M (2005) Vasorelaxant and antioxidant activity of flavonols and flavones: structure-activity relationships. J Cardiovasc Pharmacol 46:302–309
77. Dajas F, Juan Andres A-C, Florencia A, Carolina E, Felicia R-M (2013) Neuroprotective actions of flavones and flavonols: mechanisms and relationship to flavonoid structural features. Cent Nerv Syst Agents Med Chem (Formerly Curr Med Chem Nerv Syst Agents) 13:30–35
78. Ha SK, Lee P, Park JA, Oh HR, Lee SY, Park J-H, Lee EH, Ryu JH, Lee KR, Kim SY (2008) Apigenin inhibits the production of NO and PGE2 in microglia and inhibits neuronal cell death in a middle cerebral artery occlusion-induced focal ischemia mice model. Neurochem Int 52: 878–886
79. Ross JA, Kasum CM (2002) Dietary flavonoids: bioavailability, metabolic effects, and safety. Annu Rev Nutr 22:19–34
80. Salehi B, Venditti A, Sharifi-Rad M, Kręgiel D, Sharifi-Rad J, Durazzo A, Lucarini M, Santini A, Souto EB, Novellino E (2019) The therapeutic potential of apigenin. Int J Mol Sci 20:1305
81. Manach C, Scalbert A, Morand C, Rémésy C, Jiménez L (2004) Polyphenols: food sources and bioavailability. Am J Clin Nutr 79:727–747
82. Pearson DA, Paglieroni TG, Rein D, Wun T, Schramm DD, Wang JF, Holt RR, Gosselin R, Schmitz HH, Keen CL (2002) The effects of flavanol-rich cocoa and aspirin on ex vivo platelet function. Thromb Res 106:191–197
83. Selmi C, Mao TK, Keen CL, Schmitz HH, Gershwin ME (2006) The anti-inflammatory properties of cocoa flavanols. J Cardiovasc Pharmacol 47:S163–S171
84. Nehlig A (2013) The neuroprotective effects of cocoa flavanol and its influence on cognitive performance. Br J Clin Pharmacol 75:716–727
85. Lalonde R, Strazielle C (2022) Cocoa flavanols and the aging brain. Curr Aging Sci 15
86. Hidalgo I, Ortiz-Flores M, Villarreal F, Fonseca-Coronado S, Ceballos G, Meaney E, Nájera N (2022) Is it possible to treat nonalcoholic liver disease using a flavanol-based nutraceutical approach? basic and clinical data. J Basic Clin Physiol Pharmacol 33:703–714
87. Khazdair MR, Anaeigoudari A, Agbor GA (2021) Anti-viral and anti-inflammatory effects of kaempferol and quercetin and COVID-2019: a scoping review. Asian Pac J Trop Biomed 11: 327

88. Tian C, Liu X, Chang Y, Wang R, Lv T, Cui C, Liu M (2021) Investigation of the anti-inflammatory and antioxidant activities of luteolin, kaempferol, apigenin and quercetin. South African J Bot 137:257–264
89. Desam NR, Al-Rajab AJ (2022) Herbal biomolecules: anticancer agents. In: Herbal biomolecules in healthcare applications. Elsevier, Amsterdam, pp 435–474
90. Raj U, Aier I, Varadwaj PK (2017) Taxifolin: a wonder molecule in making multiple drug targets. Ann Pharmacol Pharm 2:1083
91. Alvarez-Suarez JM, Giampieri F, Tulipani S, Casoli T, Di Stefano G, González-Paramás AM, Santos-Buelga C, Busco F, Quiles JL, Cordero MD (2014) One-month strawberry-rich anthocyanin supplementation ameliorates cardiovascular risk, oxidative stress markers and platelet activation in humans. J Nutr Biochem 25:289–294
92. Curtis PJ, Kroon PA, Hollands WJ, Walls R, Jenkins G, Kay CD, Cassidy A (2009) Cardiovascular disease risk biomarkers and liver and kidney function are not altered in postmenopausal women after ingesting an elderberry extract rich in anthocyanins for 12 weeks. J Nutr 139:2266–2271
93. Zapolska-Downar D, Bryk D, Małecki M, Hajdukiewicz K, Sitkiewicz D (2012) Aronia melanocarpa fruit extract exhibits anti-inflammatory activity in human aortic endothelial cells. Eur J Nutr 51:563–572
94. Naruszewicz M, Łaniewska I, Millo B, Dłużniewski M (2007) Combination therapy of statin with flavonoids rich extract from chokeberry fruits enhanced reduction in cardiovascular risk markers in patients after myocardial infraction (MI). Atherosclerosis 194:e179–e184
95. Kokotkiewicz A, Jaremicz Z, Luczkiewicz M (2010) Aronia plants: a review of traditional use, biological activities, and perspectives for modern medicine. J Med Food 13:255–269
96. Kumar R, Khurana A, Sharma AK (2013) Role of plant hormones and their interplay in development and ripening of fleshy fruits. J Exp Bot 65:4561–4575
97. Bhatnagar A, Singh S, Khurana JP, Burman N (2020) HY5-COP1: the central module of light signaling pathway. J Plant Biochem Biotechnol 29:590–610
98. Doughty J, Aljabri M, Scott RJ (2014) Flavonoids and the regulation of seed size in arabidopsis. Biochem Soc Trans 42:364–369
99. Mouradov A, Spangenberg G (2014) Flavonoids: a metabolic network mediating plants adaptation to their real estate. Front Plant Sci 5:620
100. Vauzour D, Rodriguez-Mateos A, Corona G, Oruna-Concha MJ, Spencer JPE (2010) Polyphenols and human health: prevention of disease and mechanisms of action. Nutrients 2:1106–1131
101. Pergola C, Rossi A, Dugo P, Cuzzocrea S, Sautebin L (2006) Inhibition of nitric oxide biosynthesis by anthocyanin fraction of blackberry extract. Nitric Oxide 15:30–39
102. Graf D, Seifert S, Jaudszus A, Bub A, Watzl B (2013) Anthocyanin-rich juice lowers serum cholesterol, leptin, and resistin and improves plasma fatty acid composition in fischer rats. PLoS One 8:e66690
103. Zhuang C, Zhang W, Sheng C, Zhang W, Xing C, Miao Z (2017) Chalcone: a privileged structure in medicinal chemistry. Chem Rev 117:7762–7810
104. Elkhalifa D, Al-Hashimi I, Al Moustafa A-E, Khalil A (2021) A comprehensive review on the antiviral activities of chalcones. J Drug Target 29:403–419
105. Al-Saheb R, Makharza S, Al-Battah F, Abu-El-Halawa R, Kaimari T, Abu Abed OS (2020) Synthesis of new pyrazolone and pyrazole-based adamantyl chalcones and antimicrobial activity. Biosci Rep 40
106. Higgs J, Wasowski C, Marcos A, Jukič M, Pavan CH, Gobec S, de Tezanos Pinto F, Colettis N, Marder M (2019) Chalcone derivatives: synthesis, in vitro and in vivo evaluation of their anti-anxiety, anti-depression and analgesic effects. Heliyon 5:e01376
107. Rammohan A, Bhaskar BV, Venkateswarlu N, Gu W, Zyryanov GV (2020) Design, synthesis docking and biological evaluation of chalcones as promising antidiabetic agents. Bioorg Chem 95:103527

108. Rawat P, Singh RN, Ranjan A, Gautam A, Trivedi S, Kumar M (2021) Study of antimicrobial and antioxidant activities of pyrrole-chalcones. J Mol Struct 1228:129483
109. Tang C, Zhu L, Chen Y, Qin R, Mei Z, Xu J, Yang G (2014) Synthesis and biological evaluation of oleanolic acid derivative–chalcone conjugates as α-glucosidase inhibitors. RSC Adv 4:10862–10874
110. Acharjee S, Maity TK, Samanta S, Mana S, Chakraborty T, Singha T, Mondal A (2018) Antihyperglycemic activity of chalcone based novel 1-{3-[3-(Substituted Phenyl) Prop-2-Enoyl] Phenyl} thioureas. Synth Commun 48:3015–3024
111. Attarde M, Vora A, Varghese A, Kachwala Y (2014) Synthesis and evaluation of chalcone derivatives for its alpha amylase inhibitory activity. Org Chem An Indian J 10:192–204
112. Kaur R, Arora S (2015) Alkaloids-important therapeutic secondary metabolites of plant origin. J Crit Rev 2:1–8
113. T. Editors of Encyclopaedia Alkaloid|definition, structure, & classification | britannica https://www.britannica.com/science/alkaloid. Accessed 29 Mar 2023
114. Amirkia V, Heinrich M (2014) Alkaloids as drug leads–a predictive structural and biodiversity-based analysis. Phytochem Lett 10:xlviii–liii
115. Perviz S, Khan H, Pervaiz A (2016) Plant alkaloids as an emerging therapeutic alternative for the treatment of depression. Front Pharmacol 7:28
116. Khan H (2017) Anti-inflammatory potential of alkaloids as a promising therapeutic modality. Lett Drug Des Discov 14:240–249
117. Khan H, Mubarak MS, Amin S (2017) Antifungal potential of alkaloids as an emerging therapeutic target. Curr Drug Targets 18:1825–1835
118. Grynkiewicz G, Gadzikowska M (2008) Tropane alkaloids as medicinally useful natural products and their synthetic derivatives as new drugs. Pharmacol Rep 60:439
119. Ain Q-U, Khan H, Mubarak MS, Pervaiz A (2016) Plant alkaloids as antiplatelet agent: drugs of the future in the light of recent developments. Front Pharmacol 7:292
120. Roy A (2017) A review on the alkaloids an important therapeutic compound from plants. IJPB 3:1–9
121. Starý F (1994) The natural guide to medicinal herbs and plants. Barnes & Noble, ISBN 0880298286
122. Roberts MF (2013) Alkaloids: biochemistry, ecology, and medicinal applications; Springer Science & Business Media, ISBN 1475729057
123. Sadiq IS, Balogun JB, Ajayi SS (2016) A review of natural products chemistry-their distribution, effects and usage to man. Dutse J Pure Appl Sci:265–276
124. Uzor PF (2020) Alkaloids from plants with antimalarial activity: a review of recent studies. Evid Based Complement Altern Med, 2020
125. Kouam SF, Ngouonpe AW, Lamshöft M, Talontsi FM, Bauer JO, Strohmann C, Ngadjui BT, Laatsch H, Spiteller M (2014) Indolosesquiterpene alkaloids from the cameroonian medicinal plant polyalthia oliveri (Annonaceae). Phytochemistry 105:52–59
126. Eagleson M (1994) Concise encyclopedia chemistry. Walter de Gruyter ISBN 3110114518
127. Cordell GA, Quinn-Beattie ML, Farnsworth NR (2001) The potential of alkaloids in drug discovery. Phyther Res An Int J Devoted to Pharmacol Toxicol Eval Nat Prod Deriv 15: 183–205
128. Chaichana N, Dheeranupattana S (2012) Effects of methyl jasmonate and salicylic acid on alkaloid production from in vitro culture of Stemona Sp. Int J Biosci Biochem Bioinforma 2:146
129. Fabricant DS, Farnsworth NR (2001) The value of plants used in traditional medicine for drug discovery. Environ Health Perspect 109:69–75
130. Lahlou M (2013) The success of natural products in drug discovery
131. World Health Organization (2014) The selection and use of essential medicines: report of the WHO expert committee, 2013 (Including the 18th WHO model list of essential medicines and the 4th WHO model list of essential medicines for children); World Health Organization 2014, Vol 985, ISBN 9241209852

132. Son J-K, Chang HW, Jahng Y (2015) Progress in studies on rutaecarpine. II.— synthesis and structure-biological activity relationships. Molecules 20:10800–10821
133. Sheu J-R, Hung W-C, Lee Y-M, Yen M-H (1996) Mechanism of inhibition of platelet aggregation by rutaecarpine, an alkaloid isolated from evodia rutaecarpa. Eur J Pharmacol 318:469–475
134. Pichersky E, Raguso RA (2018) Why do plants produce so many terpenoid compounds? New Phytol 220:692–702
135. Koziol A, Stryjewska A, Librowski T, Salat K, Gawel M, Moniczewski A, Lochynski S (2014) An overview of the pharmacological properties and potential applications of natural monoterpenes. Mini-Rev Med Chem 14:1156–1168
136. Specter M (2009) A life of its own. The New Yorker 28
137. Bergman ME, Davis B, Phillips MA (2019) Medically useful plant terpenoids: biosynthesis, occurrence, and mechanism of action. Molecules 24:3961
138. Santos PL, Matos JPSCF, Picot L, Almeida JRGS, Quintans JSS, Quintans-Júnior LJ (2019) Citronellol, a monoterpene alcohol with promising pharmacological activities-a systematic review. Food Chem Toxicol 123:459–469
139. Varia RD, Patel JH, Modi FD, Vihol PD, Bhavsar SK (2020) In vitro and in vivo antibacterial and anti-inflammatory properties of linalool. Int J Curr Microbiol App Sci 9:1481–1489
140. Weston-Green K, Clunas H, Jimenez Naranjo C (2021) A review of the potential use of pinene and linalool as terpene-based medicines for brain health: discovering novel therapeutics in the flavours and fragrances of cannabis. Front Psychiatry 12:583211
141. Guimarães AG, Quintans JSS, Quintans-Júnior LJ (2013) Monoterpenes with analgesic activity – a systematic review. Phyther Res 27:1–15
142. Salehi B, Upadhyay S, Erdogan Orhan I, Kumar Jugran A, Jayaweera SLD, Dias DA, Sharopov F, Taheri Y, Martins N, Baghalpour N (2019) Therapeutic potential of α-and β-pinene: a miracle gift of nature. Biomol Ther 9:738
143. Allenspach M, Steuer C (2021) α-Pinene: a never-ending story. Phytochemistry 190:112857
144. Santos ES, Abrantes Coelho GL, Saraiva Fontes Loula YK, Saraiva Landim BL, Fernandes Lima CN, Tavares de Sousa Machado S, Pereira Lopes MJ, Soares Gomes AD, Martins da Costa JG, Alencar de Menezes IR (2022) Hypoglycemic, hypolipidemic, and anti-inflammatory effects of beta-pinene in diabetic rats. Evid-Based Complement Altern Med 2022
145. Laurella LC, Mirakian NT, Garcia MN, Grasso DH, Sülsen VP, Papademetrio DL (2022) Sesquiterpene lactones as promising candidates for cancer therapy: focus on pancreatic cancer. Molecules 27:3492
146. Paço A, Brás T, Santos JO, Sampaio P, Gomes AC, Duarte MF (2022) Anti-inflammatory and immunoregulatory action of sesquiterpene lactones. Molecules 27:1142
147. Kriplani P, Guarve K (2022) Recent patents on anticancer potential of sesquiterpene lactones. Stud Nat Prod Chem 73:71–97
148. Saha P, Rahman FI, Hussain F, Rahman SM, Rahman MM (2022) Antimicrobial diterpenes: recent development from natural sources. Front Pharmacol 12:4141
149. Chan EWC, Wong SK, Chan HT (2021) An overview of the chemistry and anticancer properties of rosemary extract and its diterpenes. J Herbmed Pharmacol 11:10–19
150. Eksi G, Kurbanoglu S, Erdem SA (2020) Analysis of diterpenes and diterpenoids. In: Recent advances in natural products analysis. Elsevier, Amsterdam, pp 313–345
151. Petiwala SM, Johnson JJ (2015) Diterpenes from rosemary (rosmarinus officinalis): defining their potential for anti-cancer activity. Cancer Lett 367:93–102
152. Smyrniotopoulos V, Vagias C, Bruyère C, Lamoral-Theys D, Kiss R, Roussis V (2010) Structure and in vitro antitumor activity evaluation of brominated diterpenes from the red alga sphaerococcus coronopifolius. Bioorg Med Chem 18:1321–1330
153. Tirapelli CR, Ambrosio SR, de Oliveira AM, Tostes RC (2010) Hypotensive action of naturally occurring diterpenes: a therapeutic promise for the treatment of hypertension. Fitoterapia 81:690–702

154. Yang J-C, Lu M-C, Lee C-L, Chen G-Y, Lin Y-Y, Chang F-R, Wu Y-C (2011) Selective targeting of breast cancer cells through ROS-mediated mechanisms potentiates the lethality of paclitaxel by a novel diterpene, Gelomulide K. Free Radic Biol Med 51:641–657. https://doi.org/10.1016/j.freeradbiomed.2011.05.012
155. Islam MT, Ali ES, Uddin SJ, Islam MA, Shaw S, Khan IN, Saravi SSS, Ahmad S, Rehman S, Gupta VK et al (2018) Andrographolide, a diterpene lactone from andrographis paniculata and its therapeutic promises in cancer. Cancer Lett 420:129–145. https://doi.org/10.1016/j.canlet.2018.01.074
156. Castellano JM, Guinda A, Delgado T, Rada M, Cayuela JA (1791–1799) Biochemical basis of the antidiabetic activity of oleanolic acid and related pentacyclic triterpenes. Diabetes 2013:62
157. Sohag AAM, Hossain T, Rahaman A, Rahman P, Hasan MS, Das RC, Khan MK, Sikder MH, Alam M, Uddin J (2022) Molecular pharmacology and therapeutic advances of the pentacyclic triterpene lupeol. Phytomedicine 154012
158. Sun B, Wu L, Wu Y, Zhang C, Qin L, Hayashi M, Kudo M, Gao M, Liu T (2020) Therapeutic potential of centella asiatica and its triterpenes: a review. Front Pharmacol 11. https://doi.org/10.3389/fphar.2020.568032
159. Namitha KK, Negi PS (2010) Chemistry and biotechnology of carotenoids. Crit Rev Food Sci Nutr 50:728–760
160. Jantan I, Bukhari SNA, Mohamed MAS, Wai LK, Mesaik MA (2015) The evolving role of natural products from the tropical rainforests as a replenishable source of new drug leads. Drug Discov Dev Mol Med:3–38
161. Rao AV, Rao LG (2007) Carotenoids and human health. Pharmacol Res 55:207–216
162. Zerres S, Stahl W (2020) Carotenoids in human skin. Biochim Biophys Acta (BBA)-Molecular Cell Biol Lipids 1865:158588
163. Ono M, Takeshima M, Nakano S (2015) Chapter six – mechanism of the anticancer effect of lycopene (Tetraterpenoids). In: Bathaie SZ, Tamanoi FBT-TE (eds) Mechanism of the anti-cancer effect of phytochemicals. Academic press, vol 37, pp 139–166, ISBN 1874-6047
164. Paul CW (2011) 15 pressure-sensitive adhesives (PSAs). In: Handbook of adheshion technology. Springer, Berlin, p 341
165. de Barros NR, Miranda MCR, Borges FA, de Mendonça RJ, Cilli EM, Herculano RD (2016) Oxytocin sustained release using natural rubber latex membranes. Int J Pept Res Ther 22:435–444
166. Mendonça RJ, Maurício VB, de Bortolli Teixeira L, Lachat JJ, Coutinho-Netto J (2010) Increased vascular permeability, angiogenesis and wound healing induced by the serum of natural latex of the rubber tree hevea brasiliensis. Phyther Res An Int J Devoted to Pharmacol Toxicol Eval Nat Prod Deriv 24:764–768
167. Chandrasekara A, Januka T, Kumari D, de Camargo AC, Shahidi F (2020) Phenolic antioxidants of bael fruit herbal tea and effects on postprandial glycemia and plasma antioxidant status in healthy adults. J Food Bioact 11
168. Padmanabhan P, Correa-Betanzo J, Paliyath G (2016) Berries and related fruits. In: Caballero B, Finglas PM, Toldrá FBT-E. of F. and H. (eds). Academic Press: Oxford, pp 364–371 ISBN 978-0-12-384953-3
169. Iglesias-Carres L, Mas-Capdevila A, Bravo FI, Aragones G, Arola-Arnal A, Muguerza B (2019) A comparative study on the bioavailability of phenolic compounds from organic and nonorganic red grapes. Food Chem 299:125092
170. Ghasemzadeh A, Jaafar HZE, Rahmat A (2010) Antioxidant activities, total phenolics and flavonoids content in two varieties of malaysia young ginger (zingiber officinale roscoe). Molecules 15:4324–4333
171. Tungmunnithum D, Thongboonyou A, Pholboon A, Yangsabai A (2018) Flavonoids and other phenolic compounds from medicinal plants for pharmaceutical and medical aspects: an overview. Medicine 5:93
172. Oboh G, Agunloye OM, Akinyemi AJ, Ademiluyi AO, Adefegha SA (2013) Comparative study on the inhibitory effect of caffeic and chlorogenic acids on key enzymes linked to

alzheimer's disease and some pro-oxidant induced oxidative stress in rats' brain-in vitro. Neurochem Res 38:413–419
173. Bontpart T, Marlin T, Vialet S, Guiraud J-L, Pinasseau L, Meudec E, Sommerer N, Cheynier V, Terrier N (2016) Two shikimate dehydrogenases, VvSDH3 and VvSDH4, are involved in gallic acid biosynthesis in grapevine. J Exp Bot 67:3537–3550
174. Heleno SA, Martins A, Queiroz MJRP, Ferreira ICFR (2015) Bioactivity of phenolic acids: metabolites versus parent compounds: a review. Food Chem 173:501–513
175. Martínez-Cruz O, Paredes-López O (2014) Phytochemical profile and nutraceutical potential of chia seeds (Salvia Hispanica L.) by ultra high performance liquid chromatography. J Chromatogr A 1346:43–48
176. Kerienė I, Mankevičienė A, Bliznikas S, Jablonskytė-Raščė D, Maikštėnienė S, Česnulevičienė R (2015) Biologically active phenolic compounds in buckwheat, oats and winter spelt wheat. Zemdirbyste-Agriculture 102:289–296
177. Karamać M, Kosińska A, Estrella I, Hernández T, Duenas M (2012) Antioxidant activity of phenolic compounds identified in sunflower seeds. Eur Food Res Technol 235:221–230
178. Adefegha SA, Oboh G, Ejakpovi II, Oyeleye SI (2015) Antioxidant and antidiabetic effects of gallic and protocatechuic acids: a structure–function perspective. Comp Clin Pathol 24:1579–1585
179. Yi W, Fischer J, Akoh CC (2005) Study of anticancer activities of muscadine grape phenolics in vitro. J Agric Food Chem 53:8804–8812
180. Mazzone G (2019) On the inhibition of hydroxyl radical formation by hydroxycinnamic acids: the case of caffeic acid as a promising chelating ligand of a ferrous ion. J Phys Chem A 123: 9560–9566. https://doi.org/10.1021/acs.jpca.9b08384
181. Kalinowska M, Gołębiewska E, Mazur L, Lewandowska H, Pruszyński M, Świderski G, Wyrwas M, Pawluczuk N, Lewandowski W (2021) Crystal structure, spectroscopic characterization, antioxidant and cytotoxic activity of new Mg(II) and Mn(II)/Na(I) complexes of isoferulic acid. Mater (Basel, Switzerland) 14. https://doi.org/10.3390/ma14123236
182. Guzman JD (2014) Natural cinnamic acids, synthetic derivatives and hybrids with antimicrobial activity. Molecules 19:19292–19349
183. Vogt T (2010) Phenylpropanoid biosynthesis. Mol Plant 3:2–20
184. Contardi M, Lenzuni M, Fiorentini F, Summa M, Bertorelli R, Suarato G, Athanassiou A (2021) Hydroxycinnamic acids and derivatives formulations for skin damages and disorders: a review. Pharmaceutics 13:999
185. Tresserra-Rimbau A, Rimm EB, Medina-Remón A, Martínez-González MA, De la Torre R, Corella D, Salas-Salvadó J, Gómez-Gracia E, Lapetra J, Arós F (2014) Inverse association between habitual polyphenol intake and incidence of cardiovascular events in the PREDIMED study. Nutr Metab Cardiovasc Dis 24:639–647
186. Adriouch S, Lampuré A, Nechba A, Baudry J, Assmann K, Kesse-Guyot E, Hercberg S, Scalbert A, Touvier M, Fezeu LK (2018) Prospective association between total and specific dietary polyphenol intakes and cardiovascular disease risk in the nutrinet-santé french cohort. Nutrients 10:1587
187. Grosso G, Stepaniak U, Micek A, Kozela M, Stefler D, Bobak M, Pajak A (2018) Dietary polyphenol intake and risk of hypertension in the polish arm of the HAPIEE study. Eur J Nutr 57:1535–1544
188. Godos J, Sinatra D, Blanco I, Mulè S, La Verde M, Marranzano M (2017) Association between dietary phenolic acids and hypertension in a mediterranean cohort. Nutrients 9:1069
189. Godos J, Castellano S, Ray S, Grosso G, Galvano F (2018) Dietary polyphenol intake and depression: results from the mediterranean healthy eating, lifestyle and aging (meal) study. Molecules 23:999
190. Tresserra-Rimbau A, Guasch-Ferré M, Salas-Salvadó J, Toledo E, Corella D, Castañer O, Guo X, Gómez-Gracia E, Lapetra J, Arós F et al (2016) Intake of total polyphenols and some classes of polyphenols is inversely associated with diabetes in elderly people at high cardiovascular disease risk1, 2, 3. J Nutr 146:767–777. https://doi.org/10.3945/jn.115.223610

191. Salehi B, Quispe C, Sharifi-Rad J, Cruz-Martins N, Nigam M, Mishra AP, Konovalov DA, Orobinskaya V, Abu-Reidah IM, Zam W (2021) Phytosterols: from preclinical evidence to potential clinical applications. Front Pharmacol 11:599959
192. Fernandes P, Cabral JMS (2007) Phytosterols: applications and recovery methods. Bioresour Technol 98:2335–2350. https://doi.org/10.1016/j.biortech.2006.10.006
193. Wang M, Huang W, Hu Y, Zhang L, Shao Y, Wang M, Zhang F, Zhao Z, Mei X, Li T et al (2018) Phytosterol profiles of common foods and estimated natural intake of different structures and forms in China. J Agric Food Chem 66:2669–2676. https://doi.org/10.1021/acs.jafc.7b05009
194. Shahzad N, Khan W, Shadab MD, Ali A, Saluja SS, Sharma S, Al-Allaf FA, Abduljaleel Z, Ibrahim IAA, Abdel-Wahab AF et al (2017) Phytosterols as a natural anticancer agent: current status and future perspective. Biomed Pharmacother 88:786–794. https://doi.org/10.1016/j.biopha.2017.01.068
195. Segura R, Javierre C, Lizarraga MA, Ros E (2006) Other relevant components of nuts: phytosterols, folate and minerals. Br J Nutr 96:S36–S44. https://doi.org/10.1017/BJN20061862
196. Uttu AJ, Sallau MS, Ibrahim H, Iyun ORA (2023) Isolation, characterization, and docking studies of campesterol and β-sitosterol from strychnos innocua (delile) root bark. J Taibah Univ Med Sci 18:566–578. https://doi.org/10.1016/j.jtumed.2022.12.003
197. Khan Z, Nath N, Rauf A, Emran TB, Mitra S, Islam F, Chandran D, Barua J, Khandaker MU, Idris AM et al (2022) Multifunctional roles and pharmacological potential of β-sitosterol: emerging evidence toward clinical applications. Chem Biol Interact 365:110117. https://doi.org/10.1016/j.cbi.2022.110117
198. von Holtz RL, Fink CS, Awad AB (1998) B-sitosterol activates the sphingomyelin cycle and induces apoptosis in LNCaP human prostate cancer cells. Nutr Cancer 32:8–12. https://doi.org/10.1080/01635589809514709
199. Gabay O, Sanchez C, Salvat C, Chevy F, Breton M, Nourissat G, Wolf C, Jacques C, Berenbaum F (2010) Stigmasterol: a phytosterol with potential anti-osteoarthritic properties. Osteoarthr Cartil 18:106–116. https://doi.org/10.1016/j.joca.2009.08.019
200. Hossain MS, Rahman MS, Imon AHMR, Zaman S, Siddiky ASMBA, Mondal M, Sarwar A, Huq TB, Adhikary BC, Begum T (2017) Ethnopharmacological investigations of methanolic extract of pouzolzia Zeylanica (L.) Benn. Clin Phytoscience 2:1–10
201. The Plant List Home – The Plant List (2013) On the Internet
202. Das R, Mitra S, Tareq AM, Emran TB, Hossain MJ, Alqahtani AM, Alghazwani Y, Dhama K, Simal-Gandara J (2022) Medicinal plants used against hepatic disorders in Bangladesh: a comprehensive review. J Ethnopharmacol 282:114588. https://doi.org/10.1016/j.jep.2021.114588
203. Aguirre L, Portillo MP, Hijona E, Bujanda L (2014) Effects of resveratrol and other polyphenols in hepatic steatosis. World J Gastroenterol: WJG 20:7366
204. Parthasarathy M, Evan Prince S (2021) The potential effect of phytochemicals and herbal plant remedies for treating drug-induced hepatotoxicity: a review. Mol Biol Rep 48:4767–4788
205. Jha V, Garcia-Garcia G, Iseki K, Li Z, Naicker S, Plattner B, Saran R, Wang AY-M, Yang C-W (2013) Chronic kidney disease: global dimension and perspectives. Lancet 382:260–272
206. Kunkel SD, Suneja M, Ebert SM, Bongers KS, Fox DK, Malmberg SE, Alipour F, Shields RK, Adams CM (2011) MRNA expression signatures of human skeletal muscle atrophy identify a natural compound that increases muscle mass. Cell Metab 13:627–638
207. Bao H, Peng A (2016) The green tea polyphenol (—)-epigallocatechin-3-gallate and its beneficial roles in chronic kidney disease. J Transl Intern Med 4:99–103
208. Mohan V, Sandeep S, Deepa R, Shah B, Varghese C (2007) Epidemiology of type 2 diabetes: indian scenario. Indian J Med Res 125:217–230
209. Vasanthi HR, ShriShriMal N, Das DK (2012) Phytochemicals from plants to combat cardiovascular disease. Curr Med Chem 19:2242–2251

210. Anandh Babu PV, Liu D (2008) Green tea catechins and cardiovascular health: an update. Curr Med Chem 15:1840–1850
211. Morris DL, Kritchevsky SB, Davis CE (1994) Serum carotenoids and coronary heart disease: the lipid research clinics coronary primary prevention trial and follow-up study. JAMA 272: 1439–1441
212. Knekt P, Reunanen A, Jävinen R, Seppänen R, Heliövaara M, Aromaa A (1994) Antioxidant vitamin intake and coronary mortality in a longitudinal population study. Am J Epidemiol 139: 1180–1189
213. Morrissey KM, Yuraszeck TM, Li C, Zhang Y, Kasichayanula S (2016) Immunotherapy and novel combinations in oncology: current landscape, challenges, and opportunities. Clin Transl Sci 9:89
214. Lestari MLAD, Indrayanto G (2014) Curcumin. Profiles drug Subst excipients Relat Methodol 39:113–204
215. Zulueta A, Caretti A, Signorelli P, Ghidoni R (2015) Resveratrol: a potential challenger against gastric cancer. World J Gastroenterol: WJG 21:10636
216. Jendželovská Z, Jendželovský R, Hiľovská L, Kovaľ J, Mikeš J, Fedoročko P (2014) Single pre-treatment with hypericin, a st. john's wort secondary metabolite, attenuates cisplatin-and mitoxantrone-induced cell death in A2780, A2780cis and HL-60 Cells. Toxicol in Vitro 28: 1259–1273
217. Zhang Y, Zhu X, Huang T, Chen L, Liu Y, Li Q, Song J, Ma S, Zhang K, Yang B (2016) β-carotene synergistically enhances the anti-tumor effect of 5-fluorouracil on esophageal squamous cell carcinoma in vivo and in vitro. Toxicol Lett 261:49–58
218. Mishra AP, Saklani S, Parcha V, Nigam M, Coutinho HDM (2021) Antibacterial activity and phytochemical characterisation of saussurea gossypiphora D. Don. Arch Microbiol 203:5055–5065
219. Soltani S, Shakeri A, Iranshahi M, Boozari M (2021) A review of the phytochemistry and antimicrobial properties of Origanum Vulgare L. and subspecies. Iran J Pharm Res IJPR 20:268
220. Song M, Liu Y, Li T, Liu X, Hao Z, Ding S, Panichayupakaranant P, Zhu K, Shen J (2021) Plant natural flavonoids against multidrug resistant pathogens. Adv Sci 8:2100749
221. Xiao S, Cui P, Shi W, Zhang Y (2020) Identification of essential oils with activity against stationary phase staphylococcus aureus. BMC Complement Med Ther 20:1–10
222. Ghasemian M, Owlia S, Owlia MB (2016) Review of anti-inflammatory herbal medicines. Adv Pharmacol Sci 2016:9130979. https://doi.org/10.1155/2016/9130979
223. Krishnaswamy KM (2008) Traditional indian spices and their health significance. Asia Pac J Clin Nutr 17:265–268
224. Pongprayoon P, Baeckström P, Jacobsson U, Lindström M, Bohlin L (1991) Compounds inhibiting prostaglandin synthesis isolated from ipomoea pes-caprae. Planta Med 57:515–518. https://doi.org/10.1055/s-2006-960196
225. Yang R, Yuan B-C, Ma Y-S, Zhou S, Liu Y (2017) The anti-inflammatory activity of licorice, a widely used chinese herb. Pharm Biol 55:5–18. https://doi.org/10.1080/13880209.2016.1225775
226. Piotrowska H, Kucinska M, Murias M (2012) Biological activity of piceatannol: leaving the shadow of resveratrol. Mutat Res Mutat Res 750:60–82. https://doi.org/10.1016/j.mrrev.2011.11.001
227. Friedrich H (1976) Phenylpropanoid constituents of essential oils. Lloydia 39:1–7
228. Lee SH, Lee SY, Son DJ, Lee H, Yoo HS, Song S, Oh KW, Han DC, Kwon BM, Hong JT (2005) Inhibitory effect of 2′-hydroxycinnamaldehyde on nitric oxide production through inhibition of NF-KB activation in RAW 264.7 cells. Biochem Pharmacol 69:791–799. https://doi.org/10.1016/j.bcp.2004.11.013
229. Li W, Li Y, Zhao Y, Ren L (2020) The protective effects of aloperine against Ox-LDL-induced endothelial dysfunction and inflammation in HUVECs. Artif Cells, Nanomedicine, Biotechnol 48:107–115. https://doi.org/10.1080/21691401.2019.1699816

230. Wei T, Xiaojun X, Peilong C (2020) Magnoflorine improves sensitivity to doxorubicin (DOX) of breast cancer cells via inducing apoptosis and autophagy through AKT/MTOR and P38 signaling pathways. Biomed Pharmacother 121:109139. https://doi.org/10.1016/j.biopha.2019.109139
231. Wenjin C, Jianwei W (2017) Protective effect of gentianine, a compound from Du Huo Ji Sheng Tang, against Freund's complete adjuvant-induced arthritis in rats. Inflammation 40: 1401–1408. https://doi.org/10.1007/s10753-017-0583-8
232. Friedman D, French JA, Maccarrone M (2019) Safety, efficacy, and mechanisms of action of cannabinoids in neurological disorders. Lancet Neurol 18:504–512. https://doi.org/10.1016/S1474-4422(19)30032-8
233. Cristino L, Bisogno T, Di Marzo V (2020) Cannabinoids and the expanded endocannabinoid system in neurological disorders. Nat Rev Neurol 16:9–29. https://doi.org/10.1038/s41582-019-0284-z
234. Wawrosch C, Zotchev SB (2021) Production of bioactive plant secondary metabolites through in vitro technologies—status and outlook. Appl Microbiol Biotechnol 105:6649–6668
235. Wang Y, Tang K (2011) A new endophytic taxol-and baccatin III-producing fungus isolated from taxus chinensis Var. Mairei. Afr J Biotechnol 10:16379–16386
236. Stahlhut R, Park G, Petersen R, Ma W, Hylands P (1999) The occurrence of the anti-cancer diterpene taxol in podocarpus gracilior pilger (podocarpaceae). Biochem Syst Ecol 27:613–622
237. Service RF (2000) Hazel trees offer new source of cancer drug. Science 288:27–28
238. Anarat-Cappillino G, Sattely ES (2014) The chemical logic of plant natural product biosynthesis. Curr Opin Plant Biol 19:51–58. https://doi.org/10.1016/j.pbi.2014.03.007
239. Jacobowitz JR, Weng J-K (2020) Exploring uncharted territories of plant specialized metabolism in the postgenomic era. Annu Rev Plant Biol 71:631–658
240. Nett RS, Lau W, Sattely ES (2020) Discovery and engineering of colchicine alkaloid biosynthesis. Nature 584:148–153
241. Luo H, Vong CT, Chen H, Gao Y, Lyu P, Qiu L, Zhao M, Liu Q, Cheng Z, Zou J (2019) Naturally occurring anti-cancer compounds: shining from chinese herbal medicine. Chin Med 14:48. https://doi.org/10.1186/s13020-019-0270-9
242. Liu D, You M, Xu Y, Li F, Zhang D, Li X, Hou Y (2016) Inhibition of Curcumin on myeloid-derived suppressor cells is requisite for controlling lung cancer. Int Immunopharmacol 39: 265–272
243. Liu Y, Wang X, Zeng S, Zhang X, Zhao J, Zhang X, Chen X, Yang W, Yang Y, Dong Z (2018) The natural polyphenol curcumin induces apoptosis by suppressing STAT3 signaling in esophageal squamous cell carcinoma. J Exp Clin Cancer Res 37:1–12
244. Cai Y-Y, Lin W-P, Li A-P, Xu J-Y (2013) Combined effects of curcumin and triptolide on an ovarian cancer cell line. Asian Pac J Cancer Prev 14:4267–4271
245. Jose A, Labala S, Venuganti VVK (2017) Co-delivery of curcumin and STAT3 SiRNA using deformable cationic liposomes to treat skin cancer. J Drug Target 25:330–341
246. Ravindranathan P, Pasham D, Balaji U, Cardenas J, Gu J, Toden S, Goel A (2018) A combination of curcumin and oligomeric proanthocyanidins offer superior anti-tumorigenic properties in colorectal cancer. Sci Rep 8:13869
247. Rahman MM, Rahaman MS, Islam MR, Rahman F, Mithi FM, Alqahtani T, Almikhlafi MA, Alghamdi SQ, Alruwaili AS, Hossain MS et al (2022) Role of phenolic compounds in human disease: current knowledge and future prospects. Molecules 27(1):233
248. Chalmers J, MacMahon S, Mancia G, Whitworth J, Beilin L, Hansson L, Neal B, Rodgers A, Ni Mhurchu C, Clark T (1999) World health organization-international society of hypertension guidelines for the management of hypertension. Guidelines Sub-Committee of the World Health Organization. Clin Exp Hypertens 1999(21):1009–1060. https://doi.org/10.3109/10641969909061028
249. Al Shukor N, Van Camp J, Gonzales GB, Staljanssens D, Struijs K, Zotti MJ, Raes K, Smagghe G (2013) Angiotensin-converting enzyme inhibitory effects by plant phenolic

compounds: a study of structure activity relationships. J Agric Food Chem 61:11832–11839. https://doi.org/10.1021/jf404641v
250. Hügel HM, Jackson N, May B, Zhang AL, Xue CC (2016) Polyphenol protection and treatment of hypertension. Phytomedicine 23:220–231. https://doi.org/10.1016/j.phymed.2015.12.012
251. Han X, Shen T, Lou H (2007) Dietary polyphenols and their biological significance. Int J Mol Sci 8:950–988
252. Rimbach G, Melchin M, Moehring J, Wagner AE (2009) Polyphenols from cocoa and vascular health—a critical review. Int J Mol Sci 10:4290–4309
253. Bhandari MR, Jong-Anurakkun N, Hong G, Kawabata J (2008) α-glucosidase and α-amylase inhibitory activities of nepalese medicinal herb pakhanbhed (Bergenia Ciliata, Haw.). Food Chem 106:247–252
254. Emerit J, Edeas M, Bricaire F (2004) Neurodegenerative diseases and oxidative stress. Biomed Pharmacother 58:39–46
255. Harman D (2006) Free radical theory of aging: an update. Ann N Y Acad Sci 1067:10–21. https://doi.org/10.1196/annals.1354.003
256. Rizvi SI, Maurya PK (2007) Alterations in antioxidant enzymes during aging in humans. Mol Biotechnol 37:58–61. https://doi.org/10.1007/s12033-007-0048-7
257. Rizvi SI, Maurya PK (2007) Markers of oxidative stress in erythrocytes during aging in humans. Ann N Y Acad Sci 1100:373–382. https://doi.org/10.1196/annals.1395.041
258. Cao G, Booth SL, Sadowski JA, Prior RL (1998) Increases in human plasma antioxidant capacity after consumption of controlled diets high in fruit and vegetables. Am J Clin Nutr 68:1081–1087. https://doi.org/10.1093/ajcn/68.5.1081
259. Shukitt-Hale B, Lau FC, Joseph JA (2008) Berry fruit supplementation and the aging brain. J Agric Food Chem 56:636–641. https://doi.org/10.1021/jf072505f
260. Maurya PK, Rizvi SI (2009) Protective role of tea catechins on erythrocytes subjected to oxidative stress during human aging. Nat Prod Res 23:1072–1079. https://doi.org/10.1080/14786410802267643

Cola accuminata: Phytochemical Constituents, Nutritional Characteristics, Scientific Validated Pharmacological Properties, Ethnomedicinal Uses, Safety Considerations, and Commercial Values

5

Daniel Etim Jacob, Imaobong Ufot Nelson, and Sylvester Chibueze Izah

Contents

D. E. Jacob (✉) · I. U. Nelson
Department of Forestry and Wildlife, University of Uyo, Uyo, Nigeria
e-mail: danieljacob@uniuyo.edu.ng; immanelcin1@gmail.com

S. C. Izah (✉)
Department of Microbiology, Faculty of Science, Bayelsa Medical University, Yenagoa, Nigeria
e-mail: Sylvester.izah@bmu.edu.ng

S. C. Izah et al. (eds.), *Herbal Medicine Phytochemistry*, Reference Series in Phytochemistry, https://doi.org/10.1007/978-3-031-43199-9_59

Abstract

The tropical plant *Cola accuminata*, more often known as Cola nut, has a long history of traditional and ethno-medical applications. Our discussions have covered many topics related to *Cola accuminata*, such as its phytochemical components, nutritional qualities, scientifically proven pharmacological effects, ethnomedicinal usage, safety concerns, and commercial values. Caffeine, theobromine, tannins, flavonoids, and phenolic compounds are only some of the many phytochemicals found in *Cola accuminata*. Its antioxidant, anti-inflammatory, anti-microbial, anti-glucose, neuroprotective, and cancer-fighting activities can be attributed to these components. Scientific studies proving the plant's usefulness in a variety of in vitro and in vivo models give data supporting these pharmacological potentials. *Cola accuminata* has a wide variety of historical and contemporary applications in traditional and folk medicine. It has been used in traditional medicine for a variety of purposes, including as a stimulant, to improve memory and focus, as an aphrodisiac, to treat respiratory diseases, and to control gastrointestinal issues. In addition to its diuretic properties, *Cola accuminata* has been used in wound healing and ulcer care. Nevertheless, when working with *Cola accuminata*, it is essential to prioritize safety. *Cola accuminata* contains caffeine, which in large doses can cause jitteriness, anxiety, and a rapid heartbeat in those who consume it. Care should also be required during pregnancy, lactation, and with certain medical conditions to avoid adverse reactions, drug interactions, gastrointestinal distress, and other complications. In addition, *Cola accuminata* has marketable qualities. Beverages, flavours, and conventional herbal medicines all benefit from their use. Moreover, the pharmaceutical sector is also interested in its features for the creation of new medicines and therapeutic agents.

Keywords

Cola accuminata · Bioactive compounds · Therapeutic potentials · Ethnomedical utilizations · Safety respects · Medicinal plants · Herbal medicine

1 Introduction

The cola nut (*Cola accuminata*), which is the produce of the Cola tree, is considered to be a fruit of paramount significance in West Africa, where it is used for a variety of uses by numerous ethnic backgrounds [1–4]. The fruit tree is not just native to West Africa, where it is cultivated, but it also has its origins in Africa as a whole. According to Unya [5], cola nut (sometimes spelt as Kola nut) belongs to the plant family Sterculiaceae, which has approximately 125 species of trees native to the tropical rainforests of Africa. The most common species in Nigeria are *Cola nitida* and *Cola accuminata* and it takes up an uncommon position among West Africans where it is extensively eaten [5]. The nuts of *Cola nitida* can be identified with its two cotyledons, while *Cola accuminata* can be differentiated with its two to seven cotyledons [6]. Moreover, the colour of *Cola nitida* varies between pink, white, and red, whereas *Cola accuminata* has a colour variation of white and pink [6].

The tree holds special significance in the people's social lives as well as their religious practices throughout the tropical regions of West Africa. Cola nut is highly valued across the entirety of West Africa by people of all socioeconomic backgrounds and religious persuasions, including but not limited to Muslims, Christians, and animists [7]. It is usually offered as a revered sacrifice and as a sign of reverence for the recipient among major tribes in Nigeria. It is also an essential component of getting people together in the community, in addition to playing an important role in a wide variety of rituals of advancement and events, including those that serve to bind legal agreements and treaties [8].

In the report of Sprague and Adega [9, 10], who comment on the vitality and significance of cola nut among the West African people. Portuguese explorer who travelled to the region in the year 1587 reported that a great deal of the inhabitant of the area he came across in his expeditions chewed on the cola fruit to quench their thirst as well as enhance the culinary delights of water [9]. Similar medicinal properties of the cola fruits observed by the explorer were also noted in other publications written by voyageurs at the same time. Other publications have also captured African practices, such as the use of the nut to fortify the digestive tract as well as battle cirrhosis of the liver [11, 12].

In Nigeria, the nuts are not only cultivated in a significant amount, but they are also considered to be an agricultural product that is of fundamental value in the daily routine of those who reside there due to the numerous functions that they play. As a result, the country located in the West African region is the highest production of cola nuts. According to Danquash [13], a significant part of the cola nuts produced in Nigeria are eaten within the borders of the country, while the remaining 10% is shipped abroad. According to Asogwa et al. [14], there is a vast amount of soils with high, moderate, and minimal reproductive rates in Nigeria that have the potential to be capitalized on economically for the production of cola nuts as part of an efficient land use management policy. These kinds of fertile soils have long since been discovered across various regions of the country.

Generally plants have shown that it is a new candidate for the production of drugs. From time immemorial plants have been used to treat and manage disease conditions as well as the preservation [15] and control of insects [16–21]. Therefore this chapter focuses on phytochemical constituents, nutritional characteristics, scientific validated pharmacological properties, ethnomedicinal uses, safety considerations, and commercial values of *Cola accuminata*.

2 History and Cultural Significance of *Cola accuminata*

Various Nigerian ethnic groups, such as the Igbo, Hausa, and Yoruba ethnic groups, make extensive use of *Cola accuminata* in the practice of traditional medical treatments as well as in cultural customs [22]. The plant has been used for medicinal purposes for a very long time, and at one point in history, it had been employed as an instrument of currency. A number of research studies, such as one that was conducted by Odugbemi [23], have provided evidence that it has been utilized in traditional medicine to treat conditions such as headaches and migraines a high fever and gastrointestinal problems.

According to Kammamposal and Laar [24], in the Nigerian culture, cola nuts are frequently given as a token of reverence and openness to visitors, and they are also used within traditional celebrations such as marriage ceremonies and naming rituals throughout a variety of cultural backgrounds. Accordingly, Kanu [25] asserts that the kola nut ceremony is a crucial communal rite in Nigerian culture. The celebration entails those participating in the breaking of kola nuts to share them with one another. The ceremony frequently goes hand in hand with dancing, musical entertainment, and the telling of stories.

The utilization of *Cola accuminata* in Nigerian traditional beliefs is further evidence of the historical significance of this plant in Nigerian society. The nuts are frequently presented as sacrifices to the gods because it is widely held that they possess mystical and defensive qualities. They are additionally employed in clairvoyance celebrations, during which they are tossed onto a special enchantment surface in order to decipher correspondence from the celestial beings [24]. This practice dates back to ancient Africa.

In Igbo land, *Cola accuminata* plays a significant role in their cultural and religious festivities. Cola nuts are broken open and shared during a significant communal rite in Igbo culture known as the kola nut ceremony [26]. Songs, dancing, and oral narration are common parts of the event, which is valued for its ability to bring people together and promote an atmosphere of neighbourhood [27]. Nwankwo et al. [28] noted that the kola nut is also an embodiment of tranquillity and has been utilized to mediate conflicts among various social groups. The fruits are also reported to possess mystical and defensive attributes, and they are frequently employed to make a tribute to the deities in customary faith. Tossed onto a divinity podium, they are employed to decipher information from the celestial beings in religious rituals [27].

The Yoruba have traditionally associated *Cola accuminata* with warmth and close relationships. It is a common gesture of hospitality and reverence. The giving and receiving of Cola nuts are a vital component of social events and customary festivities [29]. The close connection between *Cola accuminata* and Yoruba religious and spiritual practices also demonstrates the plant's societal importance. During practices and rites led by traditional priests and priestesses, the nut is often used as an offering to the gods and ancestral deities. It is commonly used as a channel for interacting with the ethereal world due to its revered metaphysical status. In addition, the Cola nut is used extensively in their indigenous faith for divination purposes. Ifa divination, a multifaceted system that aids in decision-making, enlightenment, and reuniting with past generations and the divine, counts this as one of its key elements. Thus, the nuts are a symbol of the sanctity and significance of divination rituals [30].

Consequently, the historical and cultural significance of *Cola accuminata* is attested by the fact that it is still used and recognized in local communities today. The nut represents many different things, including national pride, friendliness, religious faith, and traditional medicine.

3 Nutritional Properties of *Cola accuminata*

Cola accuminata fruit, also referred to as cola nut, can be used in a number of different ways to improve one's diet because it contains a rich variety of vitamins and minerals. Approximately 2% of catechin, caffeine, theobromine, and kolatin can be found in the fruits [31]. The nuts can be baked, crushed, or gnawed at as well as be used as an additive to beverages like tea or dairy products, as well as breakfast cereal

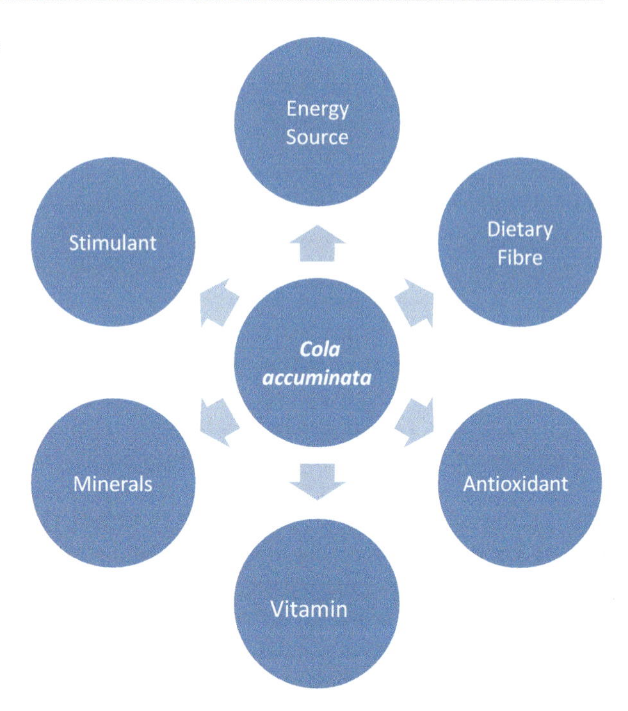

Fig. 1 Nutritional properties of *Cola accuminata*

like porridge [32]. The flavour of the entire nuts is unpleasant whenever they are gnawed at whereas they leave an excellent aftertaste that brings out the flavours and deliciousness of additional food items in the dish [33]. Figure 1 shows a list of a few of the dietary uses of the *Cola accuminata* nuts.

3.1 Stimulant

Cola accuminata is famous for its stimulation qualities, which are mainly attributable to the existence of caffeine and theobromine in the nut. It is commonly eaten for the purpose of enhancing heightened awareness, bettering a person's energy demand, and fighting feelings of fatigue [34]. The caffeine content in *Cola accuminata* varies between 1% and 3%, dependent on the diversity and processing techniques deployed [35]. Theobromine is another stimulant compound that belongs to the xanthine family. It has a similar stimulatory effect as caffeine, but it is less potent. Theobromine levels in *Cola accuminata* varies between 0.5% and 1.5% [36].

3.2 Energy Source

Cola accuminata is an excellent source of carbohydrates, which can be utilized by the body to produce energy [37]. The carbohydrates found in *Cola accuminata* deliver a rapid boost to one's energy levels, which can be helpful during times of extended physical activity or instances of strenuous working out. According to Beck [38], the majority of the carbohydrates in *Cola accuminata* are found primarily in a range of sugars. These sugars include glucose, fructose, and sucrose. *Cola accuminata* also has fatty acids, which are an additional significant source of energy and are present in the plant [39]. In addition to being an excellent source of fatty acids, it is also necessary for the body to be able to absorb fat-soluble vitamins. A wide variety of essential amino acids, which are known as the fundamental constituents of proteins, are also found in *Cola accuminata*. When the body requires more energy, it is capable of converting amino acids into a usable form. According to Dah-Nouvlessounonh et al. [37], the high content of amino acids in *Cola accuminata* adds to the nutritional value of the plant when used as an important form of energy.

3.3 Dietary Fibre

The nut is an excellent source of dietary fibre, which is the portion of plant-based foods that cannot be digested. It is important for maintaining gastrointestinal tract functioning and is also an important factor in overall well-being. Consuming foods rich in dietary fibre can assist in the regulation of stool production, the avoidance of diarrhoea, and the promotion of a digestive system that is in good condition [40]. According to Okwu and Morah [41], it may also play a role in an impression of feeling satiated and contribute to staying on top of one's weight.

The nut has also been reported to contain soluble fibre, which is a type of fibre that disintegrates when immersed in water and produces a substance with the consistency of gel when it reaches the gastrointestinal tract [42]. According to Fernandez-Murga et al.'s [43] research, soluble fibre not only assists in controlling the amount of sugar in the blood but also minimizes blood cholesterol levels and fosters an optimal gut microbiome. Insoluble fibre is another fibre constituent in *Cola accuminata* which assists in adding mass to the diet and delays bloating by keeping bowel movements regular. It favours consistency and accounts for the general well-being of the gastrointestinal tract [43].

3.4 Minerals

The cola nut is an excellent source of a variety of minerals, including potassium, calcium, magnesium, and phosphorus, among others. Minerals such as these are required for the human body to carry out a wide variety of biological functions. Calcium in *Cola accuminata* is essential to keep healthy teeth and bones, as well as for adequate functioning of the muscles and nerves signalling [44]. The nuts of *Cola accuminata* are a good source of magnesium, which is a vital mineral that plays a role in countless biological processes in human beings, as well as the generation of vitality, the functioning of muscles, and ensuring the restoration of a healthy cardiovascular activity [45].

Cola accuminata is also an excellent source of the mineral potassium, which is essential to the regulation of fluid balance, the functioning of nerves, and the twitching of muscles [46]. Adequate potassium consumption has been linked with reduced blood pressure [45, 47]. According to Adeoye et al. [45], the mineral phosphorus can be found in *Cola accuminata*. This mineral plays an important role in maintaining healthy bones, creating energy, and synthesizing deoxyribonucleic acid (DNA). Zinc which is found in the nuts of *Cola accuminata* is very essential for the body's immune system functioning, the healing of wounds, and the production of DNA. According to Adeoye et al. [45], zinc is also important for preserving the health of the skin, hair, and nails. In addition, iron in the nuts is necessary for the production of red blood cells as well as the distribution of oxygen all over the body. Accordingly, Georgieff [48] affirmed that iron is essential for the generation of energy as well as for brain development.

3.5 Antioxidant

These substances are found in cola nuts, and these assist in protecting the body from the oxidative stress that is triggered by free radicals [49, 50]. Some examples of antioxidants found in *Cola accuminata* include phenolic chemical substances. Antioxidants have been linked to a decreased likelihood of developing long-term illnesses in addition to a general improvement in wellness and overall health [51, 52]. According to Adedapo et al. [53], the phenolic compounds found in *Cola accuminata*, which include flavonoids and tannins, are responsible for the plant's powerful antioxidant activity. The nuts of *Cola accuminata* is rich in catechins, which are a type of flavonoid that have powerful antioxidant properties. According to Adeyemi et al. [54], catechins have the ability to eliminate free radicals while safeguarding cells from damage caused by oxidative stress.

Cola accuminata is also an organic source of vitamin C, which is a potent antioxidant that assists in neutralizing free radicals as well as protecting the body cells from it [55]. It contains a sizeable amount of vitamin E, which is an antioxidant that is able to dissolve in fat. The substance helps in protecting the cell membranes of the body from damage caused by oxidative stress [56]. Moreover, *Cola accuminata* is loaded with proanthocyanidins, which are powerful antioxidants recognized to have a variety of medical advantages, such as acting as a shield for the cardiovascular system and soothing effects on inflammation [54].

3.6 Vitamins

These nutrients can be obtained naturally from the nuts. Thus, *Cola accuminata* is a natural source of vitamins, including vitamin C and a number of B vitamins. According to Adedapo et al. [53], *Cola accuminata* contains vitamin C, which is also referred to as ascorbic acid. Vitamin C works as an antioxidant and plays a role in the formation of collagen as well as the primary objective of the immune system and the uptake of iron. The nuts of *Cola accuminata* is also an organic source of B vitamins, including thiamine (B1), riboflavin (B2), niacin (B3), pantothenic acid (B5), pyridoxine (B6), and folate (B9). According to Oboh et al. [55] and Adedapo et al. [53], these vitamins are essential for the formation of red blood cells, the proper functioning of the body's nervous system, and the breakdown of calories. The nuts also contain Vitamin E which is a fat-soluble vitamin with antioxidant abilities that aid in safeguarding the cells against damage from oxidative stress [56]. According to Adaramogye et al. [56], *Cola accuminata* is a good source of vitamin K, which plays a role in the clotting of blood and maintaining bones in good condition. When included in a diet that is otherwise well-balanced, the vitamins that are found in *Cola accuminata* add to the total nutritional worth of the species and can give a variety of nutritional advantages to those who eat it. It is also essential to be aware that the number of vitamins of *Cola accuminata* may differ from one variety of the plant to another, due to its environmental factors as well as handling methods used.

4 Phytochemical Constituents of *Cola accuminata*

4.1 Proximate Composition of *Cola accuminata*

The proximate composition of *Cola accuminata* as reported by several authors varies in percentage composition among the constituent variables assessed as a result of its environmental factors and methods of samples processing. Table 1 gives a summary of the breakdown of the constituent elements as reported by some scholars.

4.1.1 Moisture Content

The moisture content of *Cola accuminata* is reported to vary between 5.80% and 54.40% of its dry matter content [–, 6, 35, 37, 39, 57, 58, 59]. This deviation of moisture content in the various investigations would definitely be associated with the duration the samples were dried before they were used for evaluation. This is in accordance with the report of a study conducted by Lowor et al. [60], who asserted in

Table 1 Proximate composition of *Cola accuminata*

Parameters	% Dry matter	References
Moisture content	3.39–54.4	[6, 35, 37, 39, 57, 58, 59]
Dry matter	45.6–96.62	[6, 35, 37, 39, 57, 58, 59]
Total ash	1.46–4.96	[6, 37, 39, 57]
Crude fat	1.20–15.72	[6, 37, 39, 57, 59]
Crude protein	8.80–11.95	[6, 37, 39, 57]
Total sugars	5.57–6.27	[37, 64]
Reducing sugars	15.22–71.16	[37, 64, 65]
Energy value (Kcal/g)	88.98–98.98	[37, 64]
Crude fibre	7.35–15.80	[6, 37, 39, 59]
Carbohydrate	33.84–67.22	[6, 37, 39, 59]

their report that the moisture content of a test sample varies according to their drying time. Thus, the relatively higher moisture content in *Cola accuminata* nuts in some samples would serve as an indicator that the product was going bad as any food sample that contains an excessive amount of moisture has the potential to support the growth of microbes. This is responsible for the vast majority of the biological, chemical, and physiological activities that take place in a plant [61]. Nevertheless, a low amount of moisture content in the nuts, will make it possible for them to be stored for an extended period of time without them becoming rancid.

4.1.2 Protein Content

Protein content in *Cola accuminata* ranges from 8.80% to 10.64% [6, 37, 39, 57]. The disparity between the protein content of *Cola accuminata* could be attributed to soil and environmental circumstances arising from the fact that the lack of water in the soil would contribute to inadequate amount of nitrogen availability in the plant's tissues [22]. In addition, physiological features of a plant can be innately affected by its genetic composition [62] as it imposes a higher degree of resistance to the nitrogen supply and water deficiency in a plant. Nevertheless, the relatively high protein content of *Cola accuminata* nuts could be used to supplement the amount of protein needed in the body for growth and development [63]. In addition, [6] opined that the nuts of *Cola accuminata* would serve as a suitable alternative for the production of curd or as an additive material to other vital minerals for the formation of a gel in food products especially protein gel which serves as a structural matrix for holding water, flavour, sugars, and food ingredients.

4.1.3 Fat Content

Cola accuminata has a crude fat content that is anywhere from 1.20% to 15.72%. This data indicates that the crude fat content of *Cola accuminata* is highly variable with different studies having reported widely varying amounts. According to Adeniyi et al. [39], analysis of the nutritional profile of *Cola accuminata*, the nut had a crude fat content of 2.87%. The fat content of *Cola accuminata* was calculated to be 1.20% by [57], while the fat content of *Cola accuminata* was reported to be higher at 10.55% by [6].

Similarly, a study by Dah-Nouvlessounon et al. [37] found that *Cola accuminata* has a crude fat content of 15.72% and *Cola accuminata*'s fat content was 5.33%, which was reported according to an analysis by Adeyeye and Ayejuyo [59]. These studies shed light on the fact that the crude fat content in *Cola accuminata* varies, depending on a number of variables such as the variety, the growing conditions, and the processing methods used. These numbers are averages, and there may be some variation in fat content between different samples.

4.1.4 Ash Content

Cola accuminata's total ash content can be anywhere from 1.46% to 4.96%. The presence of mineral components in *Cola accuminata* has been suggested by the fact that various studies have reported varying values for the plant's ash content. In their analysis of *Cola accuminata*'s nutritional profile, Adeniji et al. [39] found that the plant had a total ash content of 2.35%, while according to research by Mustapha [57], *Cola accuminata* has an ash content of 1.46%. Nevertheless, *Cola accuminata*'s total ash content was reported to be 4.96% in a study conducted by Dah-Nouvlessounon et al. [37]. Thus, the mineral composition of *Cola accuminata* has been elucidated by these studies. After plant matter has been completely burned, inorganic mineral residue is left, which is represented by the ash content. It is an indicator that *Cola accuminata* contains beneficial minerals like calcium, potassium, magnesium, and phosphorus and is, therefore, crucial to the body's normal functioning and preservation of health.

4.1.5 Total and Reducing Sugar Content

Both Kingsley et al. [64] and Dah-Nouvlessounon et al. [37] report that the total sugar content of *Cola accuminata* is relatively high compared to their nuts, ranging from 5.57% to 6.27%, while its reducing sugar content varies between 15.22% and 71.16% [37, 64, 65]. These variations in total and reducing sugar content of the nut is attributed to factors such as the cultivar, growing conditions, and maturity of the plant.

4.1.6 Carbohydrate and Energy Value

Carbohydrates, found in nuts like *Cola accuminata*, are an important macronutrient because they provide the body with energy. Carbohydrates, mainly in the forms of dietary fibre and sugars, are present in nuts of *Cola accuminata* in varying amounts of 33.84–67.22 as reported by the studies of Adeniji et al. [39], Mustapha [57], Abulude [6], and Adeyeye and Ayejuyo [59]. In addition, dietary fibre, a complex carbohydrate, is important for digestive function and bowel regularity. It helps with digestion and stool bulk, and it may be useful in warding off constipation. Sugars from carbohydrates are quickly absorbed by the body and used for fuel. They are quickly metabolized and absorbed, supplying energy for workouts and mental processes. Thus, *Cola accuminata* nuts may be a better choice for managing blood sugar levels than other snack options because they contain some natural sugars but are generally low in sugar compared to other snack options [66].

4.2 Antioxidants Properties

4.2.1 Free Radical Scavenging Activity

The high concentration of phenolic compounds and flavonoids in *Cola accuminata* is responsible for its potent free radical scavenging activity. These antioxidants eliminate dangerous free radicals, protecting cells and tissues from oxidative damage. Antioxidant phenolic chemicals present in *Cola accuminata* have been widely acknowledged for their value. Catechins, flavonols, and phenolic acids are just a few examples of these molecules that can boost the plant's antioxidant capacity. The antioxidant and anti-oxidative properties of phenolic compounds such as flavonoids have long been recognized. They are essential in lowering oxidative stress due to their powerful antioxidant capabilities. *Cola accuminata* has a high antioxidant capacity thanks to the presence of flavonoids, which help to scavenge free radicals and lessen oxidative damage. The antioxidant and free radical scavenging properties of tannins are well recognized, and *Cola accuminata* has a lot of them. It has been established that tannins have potent antioxidant properties, protecting cells from oxidative stress-induced damage. This capacity of *Cola* species extract to scavenge free radicals and protect against oxidative stress was reported in a study by Fabunmi and Arotupin [67].

4.2.2 Hydrogen Peroxide Scavenging

Hydrogen peroxide (H_2O_2) is a reactive oxygen species that can damage cells. The scavenging ability of *Cola accuminata* against hydrogen peroxide was reported by Adedapo et al. [53]. According to the results of the study, the oxidative damage caused by H_2O_2 can be reduced by using a methanol extract of *Cola accuminata*'s leaves and stems. The extremely reactive chemicals known as free radicals have been linked to oxidative damage in cells and the progression of illness. *Cola accuminata*'s ability to scavenge free radicals aids in the neutralization of these potentially damaging radicals and protects cells from oxidative stress [68].

4.2.3 Metal Chelating Capacity

Cola accuminata has been shown to have metal chelating capabilities, making it useful for reducing metal-induced oxidative stress. Iron and copper, for example, can be catalysts in reactions that generate free radicals. By binding and sequestering certain metals, *Cola accuminata*'s chelating activity reduces their involvement in oxidative processes. Adedapo et al. [53] found that a methanol extract of *Cola accuminata* leaves and stems had a high metal chelating ability.

4.2.4 Lipid Peroxidation Inhibition

Blocking the chain reaction of lipid peroxidation results in cellular damage when reactive oxygen species are present. The research by Adedapo et al. [53] shows that *Cola accuminata* has anti-oxidant properties that protect against lipid peroxidation. Strong reduction of lipid peroxidation was observed in the methanol extract of *Cola accuminata* leaves and stems, suggesting its potential in protecting cells and tissues from oxidative damage. To prevent cellular damage caused by lipid peroxidation, which takes place when free radicals oxidize lipids in cell membranes, antioxidants are used. Inhibition of lipid peroxidation by *Cola accuminata* protects cell membranes against oxidative stress, as reported by Oboh et al. [68].

4.2.5 DNA Protection

Genomic material is protected from oxidative damage thanks to the DNA-protective qualities included in *Cola accuminata*. The DNA-protective activities of the aqueous extract of *Cola accuminata* seeds against oxidative stress-induced damage were studied by Adeyemi et al. [54]. This outcome suggests that *Cola accuminata* had the potential to aid in DNA maintenance and protect the body against mutations caused by oxidative stress. Consequently, these results emphasize *Cola accuminata*'s antioxidant capability, which makes it a significant natural resource for warding off the health risks associated with oxidative stress. *Cola accuminata* is able to shield cells and tissues from oxidative damage because of the wide variety of antioxidant qualities it possesses.

5 Traditional and Ethno-Medicinal Uses of *Cola accuminata*

The following traditional and ethno-medicinal uses of *Cola accuminata* have been documented in various studies and supported in its historical and cultural significance. However, it is important to note that while these traditional uses provide valuable insights, further scientific research is necessary to validate the efficacy, safety, and appropriate dosage of *Cola accuminata* for each specific use. Some of the traditional and ethno-medicinal uses of *Cola accuminata* are indicated in Fig. 2.

5.1 Digestive Aid

Cola accuminata has traditionally been used as a digestive aid. It has been hypothesized that it stimulates digestion by increasing gastric motility and secretion of

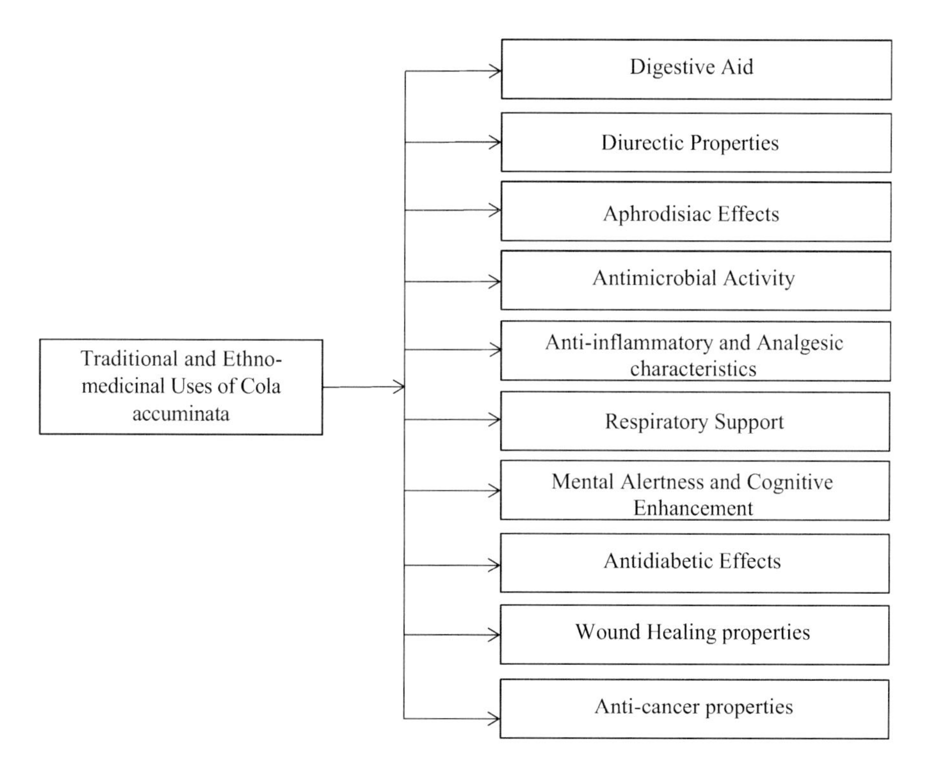

Fig. 2 Traditional and ethno-medicinal uses of *Cola accuminata*

digestive enzymes. Symptoms like bloating and flatulence can be reduced, and digestion can be enhanced [69–71]. *Cola accuminata* has been used traditionally for the treatment of gastric ulcers, and recent research suggests that this use may be justified because of the plant's anti-ulcer efficacy [72].

5.2 Diuretic Properties

Cola accuminata is commonly used as a diuretic because it stimulates the body to produce more pee. This can help the body get rid of accumulated fluids and waste products. *Cola accuminata* has long been used as a diuretic in the treatment of oedema and hypertension [73]. Udegbunam et al. [74] found that taking extracts from the plant *Cola accuminata* led to an increase in urine volume and an increase in the excretion of salt and potassium.

5.3 Aphrodisiac Effects

Cola accuminata has long been known for its aphrodisiacal properties. It may improve virility, libido, and performance in the bedroom. Traditional aphrodisiac use of *Cola accuminata* may be attributable to the plant's stimulating qualities, such as its caffeine content [75]. Nwonuma et al. [76] demonstrate that *Cola accuminata* may have pro-fertility benefits by increasing sperm count and motility.

5.4 Antimicrobial Activity

Significant antibacterial activity against a wide range of infections has been demonstrated by *Cola accuminata*. *Cola accuminata*'s antibacterial characteristics explain why it has been used historically to treat infections and wounds. Omwirhiren et al. [77], Wink [78], and Kamatenesi-Mugisha et al. [79] claim that it inhibits the growth of bacteria, fungi, and viruses, even certain strains that have developed resistance to antibiotics. Authors have variously reported that the bioactive substances such tannins, alkaloids, flavonoids, etc., are responsible for the antimicrobial activity of plants [80–101].

5.5 Anti-inflammatory and Analgesic Characteristics

Cola accuminata has traditionally been used for relief from pain and management of inflammatory diseases due to its anti-inflammatory and painkiller characteristics. Arthritis, rheumatism, and other inflammatory disorders may benefit from its anti-inflammatory and pain-relieving properties [56]. Flavonoids, phenols, and other bioactive chemicals are responsible for its anti-inflammatory and analgesic actions [68].

5.6 Respiratory Support

In the treatment of respiratory disorders like asthma, bronchitis, and cough, *Cola accuminata* has long been traditionally used to aid the body's natural defences. It is reported to have bronchodilatory and expectorant qualities that make it easier to breathe by breaking up mucus and opening airways [102]. *Cola accuminata* has shown promise as an anti-asthmatic agent, with preliminary research indicating it may relax bronchial smooth muscle and lessen inflammation [103].

5.7 Mental Alertness and Cognitive Enhancement

Cola accuminata is well-known for its stimulating features, such as its high caffeine level, which helps keep the mind alert and improves cognitive performance. It has a long history of use to boost memory, focus, and other mental abilities. Reduced weariness, increased alertness, and enhanced cognitive function have all been linked to the stimulating impacts of *Cola accuminata* on the central nervous system [104, 105]. Multiple studies have shown that *Cola accuminata* improves cognitive function, therefore it may be useful in treating dementia and Alzheimer's disease [106].

5.8 Antidiabetic Effects

Cola accuminata has shown remarkable antidiabetic capabilities, which brings us to our seventh point. Studies back up the folk applications of *Cola accuminata* for diabetic management. Blood glucose levels are lowered and glucose tolerance is enhanced, indicating hypoglycaemic activity [107, 108]. *Cola accuminata*'s antidiabetic effects may be due to the presence of bioactive substances such alkaloids, flavonoids, and tannins [68].

5.9 Wound Healing Properties

Cola accuminata has long been used to aid in the recovery from wounds. It has been hypothesized to have tissue-regenerating and wound-protecting capabilities. Extracts of the plant *Cola accuminata* have been found to have healing and regenerative properties in studies [109]. Antimicrobial, anti-inflammatory, and antioxidant characteristics may account for its wound-healing actions [110].

5.10 Anti-cancer Properties

Cola accuminata has gained interest for its possible anti-cancer qualities. *Cola accuminata* extracts have shown anticancer efficacy in preliminary experiments against a number of cancer cell lines [106]. These cancers include breast, colon, prostate, and lung. *Cola accuminata* may have anti-cancer properties due to the presence of bioactive substances such as flavonoids and phenolic acids, which can suppress tumour growth and induce death [65].

6 Modern Scientific Research on Pharmacological Potentials of *Cola accuminata*

The pharmacological effects of *Cola accuminata*, often known as Cola nut, have been the focus of several recent scientific studies. These investigations have been conducted to determine whether or not *Cola accuminata* contains any bioactive chemicals for medicinal use. In Fig. 3, we highlight some of the most exciting recent developments in the study of *Cola accuminata*'s therapeutic potential.

6.1 Antimicrobial Activity

Adebayo et al. [111] tested *Cola accuminata* extracts for antibacterial activity against bacteria and fungus. The extracts inhibited drug-sensitive and drug-resistant

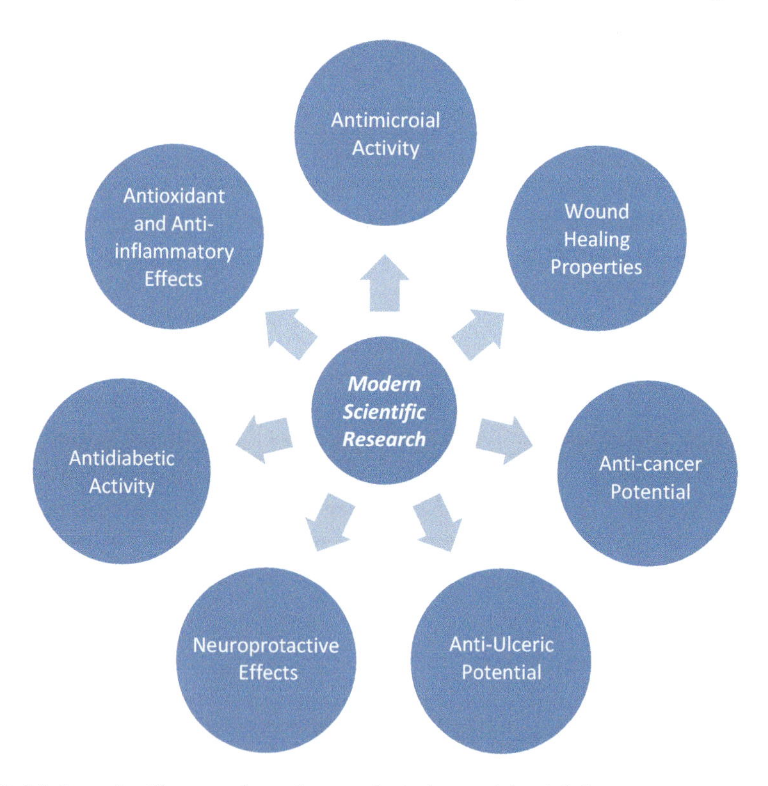

Fig. 3 Modern scientific research on pharmacological potentials of *Cola accuminata*

bacteria in the study. Bioactive substances such as tannins, alkaloids, and flavonoids had broad-spectrum antibacterial properties, according to the study. Also, in a study by Aiyegoro et al. [112], the authors tested *Cola accuminata*'s antibacterial efficacy against a variety of microorganisms. The extracts inhibited Gram-positive, Gram-negative, and drug-resistant microorganisms indicating that tannins, alkaloids, and flavonoids in the extracts had powerful antibacterial effects, according to the study. These studies therefore show that *Cola accuminata* possesses antibacterial capability. Tannins, alkaloids, and flavonoids in the plant provide broad-spectrum antibacterial activity. These chemicals inhibit drug-sensitive and drug-resistant bacteria and fungi, suggesting they could be used to generate new antimicrobials. These findings show promise, but more research is needed to identify and isolate *Cola accuminata*'s antibacterial bioactive components.

6.2 Antioxidant and Anti-inflammatory Effects

Amadi and Nwachukwu [113] and Akinmoladun et al. [114] examined *Cola accuminata* extract antioxidant properties. Phenolic chemicals and flavonoids in the extracts gave them antioxidant properties. These antioxidants neutralize free radicals, reduce oxidative stress, and protect cells from oxidative stress. These effects were linked to *Cola accuminata*'s antioxidant phenolic chemicals and flavonoids. Accordingly, Oboh et al. [115] reported *Cola accuminata* extracts reduced inflammation in animal studies. The extracts greatly suppressed pro-inflammatory mediators and enzymes, showing their anti-inflammatory potential. The extracts modify inflammatory pathways and inhibit pro-inflammatory molecule production, according to the study. Similarly, Olurishe et al. [116] tested *Cola accuminata* extracts for anti-inflammatory properties in a cell model. The extracts inhibited pro-inflammatory mediators dose-dependently, suggesting anti-inflammatory properties. Bioactive chemicals in the extracts disrupt inflammation signalling pathways, according to the study. Therefore, the phenolic chemicals and flavonoids in the plant neutralize free radicals and reduce oxidative stress. The extracts also suppress pro-inflammatory mediators and enzymes suggesting that *Cola accuminata* may alleviate oxidative damage and inflammation.

6.3 Antidiabetic Activity

Victoria et al. [42] and Osadee [117] tested *Cola accuminata* extracts for anti-diabetic effects in animals. The extracts have strong hypoglycaemic effects, reducing blood glucose. Tannins and flavonoids, which increase the generation of insulin and inhibit carbohydrate-digesting enzymes, may explain *Cola accuminata*'s anti-diabetic effects. Also, Ezejiofor et al. [118] examined *Cola accuminata* extracts' anti-diabetic properties in animal models. The extracts dramatically lowered blood glucose and enhanced insulin sensitivity. Bioactive chemicals in *Cola accuminata* improve glucose absorption and metabolism, according to the study. In a similar study, Boudiba et al. [119] examined *Cola accuminata* extracts' anti-diabetic effects in cellular models, and the results showed that the extracts significantly inhibited the metabolism of carbohydrates and glucose absorption enzymes. These findings imply that *Cola accuminata*'s antidiabetic action may be due to its capacity to inhibit carbohydrate-digesting enzymes, lowering blood glucose release. Thus, the *Cola accuminata* possesses anti-diabetic properties. The tannins and flavonoids in the plant increase insulin synthesis, inhibit carbohydrate-digesting enzymes, and increase glucose absorption, making it hypoglycaemic. These findings suggest *Cola accuminata* could be used to produce antidiabetic medicines.

6.4 Neuroprotective Effects

Imam-Fulani et al. [120] examined *Cola accuminata* extract's neuroprotective properties in animals. The extract improved memory and learning. *Cola accuminata*'s antioxidants minimize oxidative stress and preserve neurons, according to the study. Similarly, Oboh et al. [68] investigated *Cola accuminata* bioactive components' neuroprotective properties. The study examined how these components affect memory and cognition. *Cola accuminata* extract bioactive components protected against loss of memory and dementia. Bioactive chemicals alter neurotransmitter systems and increase neural signalling, according to the researchers. Furthermore, Oboh et al. [121] and Ishola et al. [122] examined how *Cola accuminata* extract protects against brain damage. In animal experiments, the extract reduced oxidative stress and protected neurons. The antioxidant qualities of *Cola accuminata* may help treat and prevent neurological illnesses, according to researchers. Thus, these research demonstrate *Cola accuminata*'s neuroprotection. The plant's antioxidant qualities may protect neurons from oxidative stress and death. *Cola accuminata* improves memory, learning, and cognitive function. These findings show that *Cola accuminata* may provide neuroprotective chemicals for treating and preventing neurodegenerative disorders. However, there is still need to identify and isolate these bioactive chemicals and understand their neuroprotective processes in further researches.

6.5 Anti-cancer Potential

Nugraha et al. [123] examined *Cola accuminata* extract's cytotoxicity on breast cancer cell lines. The extract suppressed breast cancer cell multiplication and induced apoptosis. Bioactive chemicals in *Cola accuminata* may fight cancer, according to the study. *Cola accuminata* was also tested in lung cancer cell lines by Alsaeedi et al. [124]. The extract inhibited lung cancer cell growth and induced apoptosis. *Cola accuminata*'s bioactive ingredients selectively kill cancer cells, according to the study. *Cola accuminata* extract was tested on colon cancer cell lines by Percival et al. [125]. The extract killed colon cancer cells and inhibited cell multiplication. The extract's capacity to alter cell growth and survival signalling pathways may explain its anti-cancer action. In Nugraha et al. [123], the authors examined *Cola accuminata*'s anti-cancer effects in prostate cancer cell lines. The extract inhibited prostate cancer cell multiplication and induced apoptosis. The bioactive chemicals in *Cola accuminata* may suppress prostate cancer cell proliferation by interfering with important biological processes.

6.6 Anti-Ulceric Properties

Cola accuminata extract was tested in rats with aspirin-induced stomach ulcers by Holcombe et al. [126]. The extract reduced ulcer size and improved stomach mucosa histology. The study authors hypothesized that the extract's gastroprotective effects came from increased mucus production, antioxidant defences, and oxidative stress reduction. Nkanu [127] tested *Cola accuminata*'s anti-ulcer effects in different rat ulcer models. The extract reduced gastrointestinal lesions, stomach acid output, and pepsin activity, preventing ulcers. Antioxidant enzymes and gastric mucus increased, indicating better stomach lining protection. Suntar et al. [128] evaluated *Cola accuminata* extract's effects on rats' ethanol-induced stomach ulcers. The extract reduced ulcer index, stomach acid output, and lipid peroxidation to treat ulcers. The extract increased gastric mucin and antioxidant enzyme activity, strengthening the mucosal barrier. These findings imply *Cola accuminata* may treat ulcers. The extract may reduce gastric acid output, protect gastrointestinal mucosa, boost mucus formation, and act as an antioxidant. These ways help the drug preserve the gastric mucosa, heal ulcers, and prevent them. These findings support the traditional usage of *Cola accuminata* to treat ulcers.

6.7 Wound Healing Properties

Ishola et al. [122] and Edwards et al. [129] assert that *Cola accuminata* extract helped to heal and improved wound contraction, size, and re-epithelialization. The extract accelerated wound-healing mechanisms like collagen deposition, angiogenesis, and antioxidant activity, according to the researchers. Agbaje and Charles [130] tested *Cola accuminata*'s wound-healing ability in rats using excision and incision wound models. The extract improved wound closure, tensile strength, and collagen formation. Researchers believed the extract's antioxidant and anti-inflammatory properties helped mend wounds. Diallo et al. [131] tested *Cola accuminata* extract for burn wound treatment in rats. The extract improved wound healing by shrinking wounds, re-epithelializing, and depositing collagen. The extract stimulated fibroblast proliferation and migration, improving wound healing. Scientists believed the extract's angiogenesis and anti-inflammatory effects helped mend wounds. These trials confirm *Cola accuminata*'s wound dressing efficacy. This extract promotes wound healing, re-epithelialization, collagen production, antioxidants, and anti-inflammation. Its wound-healing capabilities make it a viable natural therapy.

7 Some Herbal Formulations Involving *Cola accuminata* as Active Ingredients

Traditional and folk medicine uses herbal preparations using *Cola accuminata* as an active ingredient. These compositions mix *Cola accuminata* with other plant-based substances for synergistic effects and improved medicinal results. Some example of *Cola accuminata* herbal formulations in Nigeria are shown below.

7.1 Agbo Formulation

Nigerian herbalists use agbo to cure a variety of diseases. *Cola accuminata* is mixed with medicinal herbs. Akande et al. [132] tested an agbo formulation with *Cola accuminata, Morinda lucida*, and *Alstonia boonei* for antibacterial activity. This herbal formulation inhibited Gram-positive and Gram-negative bacteria, confirming its traditional use.

7.2 Ewuro-Ajekobale Formulation

Yoruba folk medicine practitioners in Nigeria treats diabetes using Ewuro-Ajekobale formulation. This formulation contains *Cola accuminata*, bitter leaf (*Vernonia amygdalina*), and wild tomato (*Solanum torvum*). Musabayane et al. [133] examined the hypoglycaemic effects of the formulation in diabetes animal models and this herbal composition showed considerable blood glucose decreases, indicating its diabetes management potential.

7.3 Ciklavit Herbal Preparation

Traditional Yoruba medicine practitioners treats menstruation issues with this herbal preparation. It has *Cola accuminata, Citrus aurantifolia*, and other therapeutic herbs. Ciklavit was tested on experimentally induced menstruation problems in female rats by Cyril-Olutayo et al. [134], and its anti-inflammatory and analgesic properties supported its use in menstrual problems.

7.4 Agbo-Iba

Yoruba traditional medicine uses a herbal concoction called Agbo-Iba to treat fever and malaria. *Cola accuminata* is combined with *Azadirachta indica* and *Mangifera indica*, two more therapeutic plants. Nwabuisi [135] examined the efficacy of the Agbo-Iba formulation against plasmodium. The findings provided strong evidence for its traditional use as an antimalarial treatment, showing considerable reduction of *Plasmodium falciparum* development.

7.5 Orogbo Formulation

This herbal mixture is used in traditional medicine to treat gastrointestinal problems like diarrhoea and dysentery. It includes *Cola accuminata* together with *Garcinia kola* and *Vernonia amygdalina*, both of which are plants. Odukanmi et al. [136] evaluated the effectiveness of the *Cola accuminata* in preventing diarrhoea in animal models. Supporting its traditional use, the extract significantly reduced the frequency and severity of diarrhoea.

7.6 Opa Eyin Formulation

Traditional healers rely on this herbal concoction to treat skin diseases and wounds. Included are *Newbouldia laevis* and *Cola accuminata*, both of which have medicinal uses. The Opa Eyin formulation's antibacterial efficacy against several bacteria and fungi was studied by Bolawaa et al. [137]. Significant inhibitory effects were seen against the investigated microorganisms, suggesting that the formulation may be useful in the treatment of skin infections.

7.7 Agbo-Jedi Formulation

The formulation is a traditional herbal concoction with aphrodisiac effects, widely used in traditional medicine. *Cola accuminata* is combined with *Mondia whitei* and *Citrullus colocynthis*, two other therapeutic plants. Akinboro et al. [138] and Adeyemi et al. [139] examined the aphrodisiacal efficacy of the Agbo-Jedi formulation. The traditional use of the mixture as an aphrodisiac was supported by the data showing a considerable increase in sexual behaviour and reproductive indices.

7.8 Agbo-Atosi Formulation

This is a herbal concoction used in Nigerian folk medicine to treat hypertension. *Cola accuminata* is combined with other therapeutic herbs like aloe barbadensis and onion. Eiya and Igharo [140] and Akande et al. [132] looked into the antihypertensive effects of the Agbo-Atosi formulation and reported that the formulation had promising antihypertensive effects, with blood pressure levels significantly reduced.

7.9 Oroki Formulation

This herbal formulation with hepatoprotective qualities is widely utilized in traditional medicine in Nigeria. It includes therapeutic herbs like *Cola accuminata* and Garcinia kola. Emeleku [141] evaluated the Oroki formulation's hepatoprotective efficacy in animal models. The formulation showed promising results in protecting against hepatotoxicity, indicating its potential in managing liver health.

7.10 Osomo Formula

Osomo is a local herbal aphrodisiac. It contains *Cola accuminata* and *Allium sativum/cepa*. Adeyemi and Owoseni [142] examined the polyphenolic content and biochemical evaluation of the formulation and reported the formulation showed significant composition of the parameters assessed, thereby supporting the traditional use of this formulation as an aphrodisiac.

7.11 Epa-Ijebu Formulation

This is a traditional herbal concoction with aphrodisiac characteristics that is utilized in most traditional medicine. *Cola accuminata* is combined with *Mondia whitei* and *Carpolobia lutea*, two other therapeutic herbs. The aphrodisiac effects of the Epa-Ijebu combination were tested on animals in a study by Enitan [143]. The combination was shown to significantly improve sexual performance and libido, lending credence to its historical role as a natural aphrodisiac [143]. In summary, *Cola accuminata* has a wide variety of uses in traditional herbal formulations, as evidenced by these examples. It is commonly used for its aphrodisiac, gastrointestinal health, anti-malarial, and anti-inflammatory properties.

8 Safety Considerations and Side Effects of *Cola accuminata*

Some of the safety considerations and side effects of consuming *Cola accuminata* to be discussed under this subhead are shown in Fig. 4.

8.1 Caffeine Content

The plant *Cola accuminata*, among others, has the naturally occurring chemical caffeine. It has been scientifically proven to stimulate the central nervous system. Most people can manage modest amounts of caffeine with no problems, but too much can have negative effects. *Cola accuminata*'s caffeine level is a major cause for concern because of its link to irritability and nervousness. Caffeine in excess can overstimulate the nervous system, causing irritability, nervousness, and even anxiety. Caffeine use has been linked to both difficulty falling asleep and maintaining a restful slumber [144]. One of the negative effects of consuming too much caffeine is a rapid heartbeat. Caffeine causes a transient increase in heart rate due to its stimulatory action on the cardiovascular system. This effect may be more

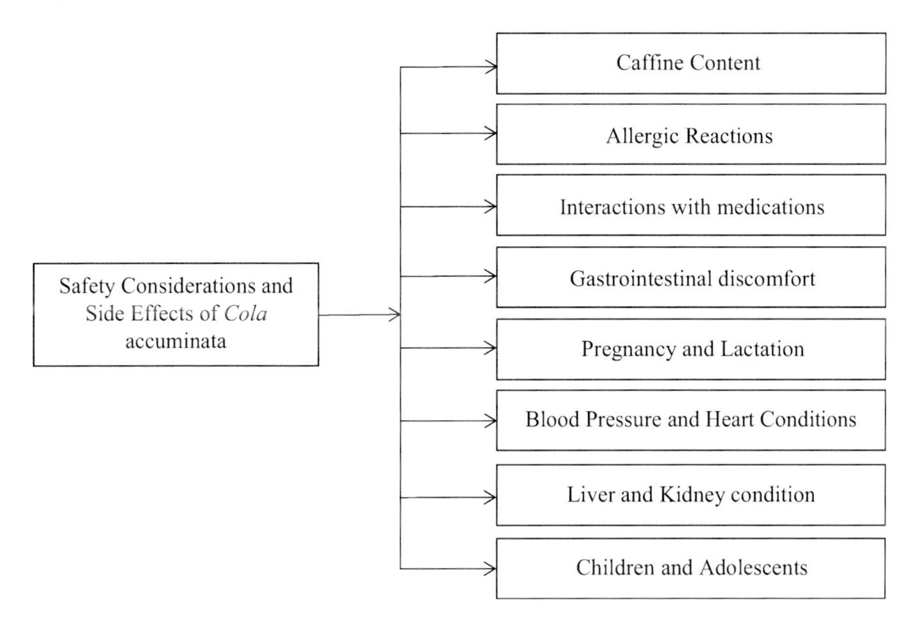

Fig. 4 Safety considerations and side effects of *Cola accuminata*

pronounced in caffeine-sensitive people or people with pre-existing heart issues and may cause palpitations or abnormal heart rhythms [144].

Another common negative reaction to caffeine overdose is gastrointestinal distress. Caffeine increases stomach acid production, which can lead to indigestion, heartburn, and other gastrointestinal distress. It may also have minor diuretic effects, resulting to increased urine production and frequency in certain people [145, 146]. Thus, caffeine tolerance varies greatly from person to person. People with pre-existing problems like heart disease, high blood pressure, or gastrointestinal issues may be particularly vulnerable to the negative effects of caffeine. Nehlig [144] stressed the importance of moderation in caffeine consumption, including *Cola accuminata*, in order to reduce the risk of adverse consequences.

Nevertheless, *Cola accuminata* can be consumed safely provided that consumers are aware of their unique caffeine sensitivity and take into account their total caffeine intake from sources like coffee, tea, and other caffeinated beverages or meals. People who are more susceptible to caffeine's effects or who have health issues that can be made worse by caffeine should cut back on their intake [144].

8.2 Allergic Reactions

Although it is uncommon for people to have an adverse reaction to *Cola accuminata*, it is nonetheless vital to be aware of the possibility. An allergic reaction occurs when the immune system overreacts to a normally harmless chemical. Symptoms of an allergy can present in a number of ways, including on the skin, in the lungs, or throughout the body. Some people have severe allergic reactions to *Cola accuminata*, including hives, swelling of the face and throat, and trouble breathing, which can occur after eating or touching the plant. These reactions to *Cola accuminata*'s allergenic components might manifest themselves quickly or build up over time [147]. Hence, there is a need to stop using *Cola accuminata* and get medical help right away if an allergic reaction is imminent. As such, products containing *Cola accuminata* should be used with caution by people with allergies or a history of adverse reactions. In these circumstances, it is best to talk to a doctor or allergist about getting tested for allergies and getting advice on how to consume *Cola accuminata* safely. *Cola accuminata* allergies appear to be uncommon because people rarely report experiencing negative reactions to the plant. However, when drinking or coming into contact with *Cola accuminata*, it is crucial to remain cautious and respond appropriately to any indicators of allergic reactions.

8.3 Interactions with Medications

Cola accuminata's bioactive ingredients, such as caffeine, may interfere with the effectiveness of several pharmaceuticals. These interactions might lessen or heighten the likelihood of desired results, depending on the drugs involved. If you are currently taking any drugs, you should discuss using *Cola accuminata* with your doctor because of the possibility of drug interactions. One of *Cola accuminata*'s most significant components, caffeine, can interfere with the effectiveness of drugs used to treat heart disease, mental illness, and breathing problems. Because of the potential impact on the safety of patients and treatment outcomes, Baur et al. [148] emphasized the need of taking into account such interactions.

Beta-blockers, used to treat hypertension and other cardiovascular disorders, may have an adverse reaction to caffeine. The blood pressure-lowering benefits of beta-blockers may be diminished by caffeine consumption [149]. Caffeine can also have an influence on the metabolism of, or increase the risk of side effects associated with, certain psychiatric drugs, including selective serotonin reuptake inhibitors (SSRIs) and lithium [150]. In addition, caffeine can interact with bronchodilators like theophylline, which are used to treat respiratory disorders. Theophylline's negative effects can be amplified by caffeine, leading to a faster heart rate and tremors [151]. Given the potential for negative interactions with other medications, it is best to check with a doctor before using *Cola accuminata*.

8.4 Gastrointestinal Discomfort

Intestinal distress has been reported in a small number of people who have used *Cola accuminata* or its extracts. Caffeine and other components of the plant may contribute to uncomfortable side effects such gastrointestinal distress, heartburn, and diarrhoea. One of the primary ingredients of *Cola accuminata* is caffeine, which has been shown to stimulate the digestive system, leading to more gastric acid output and possibly stomach discomfort [145]. In addition, some people may experience diarrhoea as a result of the laxative effects of caffeine [152]. Some people may be more susceptible to caffeine's gastrointestinal side effects than others.

Cola accuminata has a number of chemicals that can affect gastrointestinal function, including caffeine, tannins, and polyphenols. Particular tannins have been linked to astringent characteristics and the potential to cause gastrointestinal discomfort in some people [153]. *Cola accuminata* preparations include several chemicals, which may cause gastrointestinal distress in those who are sensitive to them. It is important to note that the amount of *Cola accuminata* consumed, individual tolerance, and pre-existing gastrointestinal problems may all have a role in the occurrence and severity of gastrointestinal discomfort. As a result, you should pay attention to how your body reacts and make dietary adjustments accordingly.

8.5 Pregnancy and Lactation

Caffeine, a popular stimulant found in *Cola accuminata*, can enter the foetal circulation after crossing the placenta [154]. A higher likelihood of miscarriage, premature birth, low birth weight, and issues with growth have all been linked to maternal caffeine consumption during pregnancy [155]. Infants of moms who drink too much coffee may develop behavioural and sleep problems and experience abdominal distress as a result [144]. Given the current state of knowledge, expectant and nursing mothers should proceed with caution while consuming *Cola accuminata* or items containing it. It is best to talk to a doctor about your specific situation and any hazards involved so you can make an educated choice.

The dearth of particular studies on this plant in pregnant and breastfeeding women is largely to blame for the scarcity of data on the safety of *Cola accuminata* at these stages of a woman's life. In order to protect the health of both mother and child, healthcare providers normally recommend that pregnant and nursing women limit their caffeine intake from all sources, including *Cola accuminata*.

8.6 Blood Pressure and Heart Conditions

Cola accuminata contains caffeine, which has been shown to temporarily increase heart rate and blood pressure. People who have hypertension or cardiovascular disease should use caution when consuming *Cola accuminata* or products containing it, and should keep an eye on their blood pressure and heart rate. In these situations, you should talk to a doctor before using *Cola accuminata*. The stimulant effects of caffeine on the central nervous system are accompanied by a transient elevation in blood pressure and heart rate [156]. This reaction is temporary and is felt by the vast majority of people. However, those who already have hypertension or heart problems may be more susceptible to caffeine's negative effects.

Caffeine can induce a transient increase in blood pressure, thus those with hypertension should be careful about how much they drink [157]. Caffeine, particularly *Cola accuminata* products, should be limited by people with hypertension in order to maintain healthy blood pressure levels. Maintaining a blood pressure reading within a healthy range requires regular monitoring. Caffeine should also be consumed with caution by those who have heart disorders such as arrhythmias or coronary artery disease. Some cardiac arrhythmias may be triggered or made worse by caffeine, and caffeine may also increase the heart's workload [156]. Before taking *Cola accuminata* or products containing it, those with heart issues should talk to a doctor about the possible effects on their condition.

8.7 Liver and Kidney Condition

Cola accuminata or goods containing it should be consumed with caution by people with liver or kidney disorders. *Cola accuminata*'s presence of caffeine and other chemicals may have an impact on several body systems. In these situations, you should talk to a doctor before using *Cola accuminata*. The liver and kidneys are essential organs for detoxification and waste removal. Caffeine, a stimulant, can affect several systems because of its metabolism and excretion. Caffeine and other substances may not be processed properly by people with liver or kidney disorders, which could have negative consequences. Changes in indicators of liver and kidney function were observed in rats given *Cola accuminata* extract in a study by Olatunji et al. [158]. The study authors cautioned against using *Cola accuminata* for anyone who already has liver or kidney disease.

Some people may have a diminished ability to metabolize caffeine due to liver diseases such as liver disease or poor liver function. Extended contact with caffeine, or its build-up, has been linked to hepatic stress [150]. Caffeine and its metabolites may also have an effect on renal function in those with kidney disorders such as chronic kidney disease or impaired kidney function [150]. Individuals with pre-existing liver or kidney issues should not consume *Cola accuminata* or products containing it without first consulting a healthcare expert. A medical specialist is most equipped to analyse an individual's health situation, weigh the pros and downsides, and recommend a safe dosage or alternate treatment.

8.8 Children and Adolescents

Due to its caffeine level and potential effects on developing bodies, children and adolescents should consume *Cola accuminata* moderately. Caffeine can impact sleep, growth, and health in this population. *Cola accuminata* should not be used in children and teenagers without medical advice. Caffeine, a stimulant, affects children and adolescents more than adults. Children and teenagers may be particularly sensitive to caffeine due to their developing bodies. Caffeine can alter sleep patterns and make it hard to fall and remain asleep [159]. Sleep disturbances can harm children and teenagers' growth, cognition, and well-being. In addition to sleep disruptions, high coffee intake in this population may impact growth and development. Caffeine can lower adolescent bone mineral density, which can lead to osteoporosis later in life [159]. Caffeine can also impair nutritional absorption, which could impede growth and development.

Caffeine may potentially harm kids' cardiovascular health. In people with cardiovascular problems, excessive caffeine can raise heart rate and blood pressure [159]. Some people may be more sensitive to caffeine's stimulating effects. Given these concerns, *Cola accuminata* should be limited in children and adolescents and used under medical supervision. A doctor can advise on the right amount of caffeine for this population and weigh the dangers and benefits.

9 Commercial Properties of *Cola accuminata*

Cola accuminata has the potential to be economically valuable and used in a wide range of businesses thanks to its commercial features. This includes its uses in the nourishment and beverage sector, the pharmaceutical sector, and the cosmetics sector to mention but a few (Fig. 5). Let us go into greater depth about these investment properties for businesses, using appropriate sources.

9.1 Food and Beverage Industry

Cola accuminata has been used for decades in the food and beverage industry as a flavouring agent in the creation of carbonated beverages such as cola drinks. The addition of caffeine and other bioactive components gives these drinks their own unique flavour. Candies, chocolates, and desserts all benefit from the addition of *Cola accuminata* extracts and infusions [160]. *Cola accuminata*'s natural components make it a valuable commodity in the food and drink sector.

9.2 Pharmaceutical Industry

The pharmaceutical sector is interested in *Cola accuminata* because of its possible therapeutic characteristics. The pharmacological effects of the bioactive chemicals found in *Cola accuminata* have been extensively researched. As previously mentioned, these effects can be broken down into four categories: antioxidant, anti-inflammatory, antimicrobial, and anti-cancer [31, 68, 122, 161]. Because of these

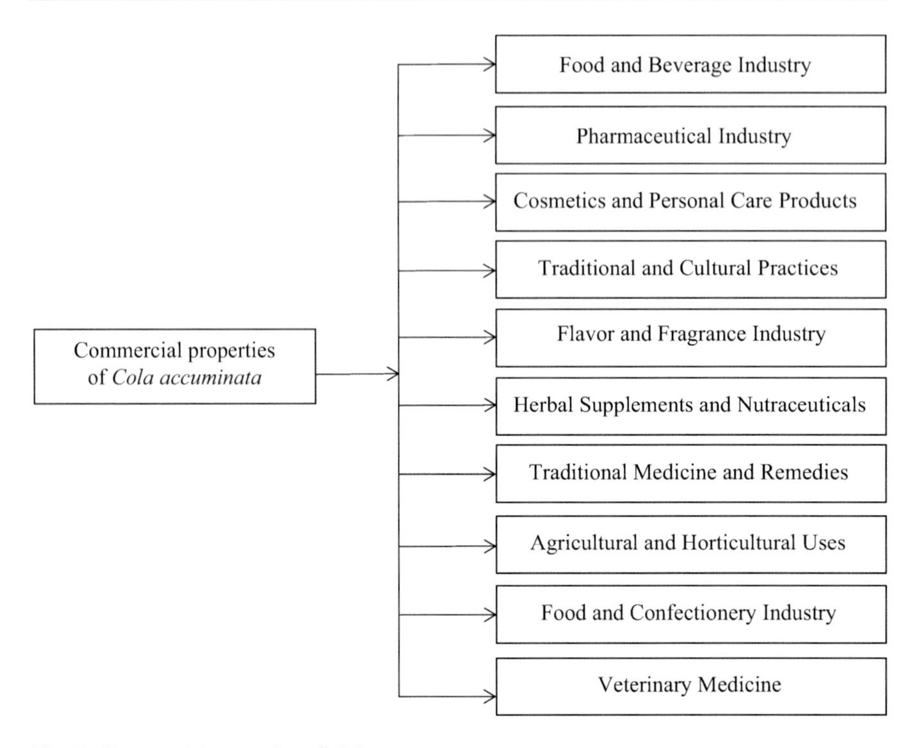

Fig. 5 Commercial properties of *Cola accuminata*

qualities, *Cola accuminata* is a promising ingredient in the creation of natural pharmaceuticals and functional meals.

9.3 Cosmetics and Personal Care Products

Extracts from the *Cola accuminata* plant are used in the cosmetics business. *Cola accuminata*'s antioxidant properties make it a promising addition to cosmetics. The skin benefits from antioxidants because they shield it from oxidative stress, which is linked to aging and environmental damage. To give antioxidant advantages and support healthy skin, *Cola accuminata* extracts can be added into formulations including creams, lotions, etc. The trend towards organic and all-natural cosmetics and toiletries necessitates the usage of natural substances like *Cola accuminata*. Extracts from the *Cola accuminata* plant are used in the production of a wide variety of cosmetics and toiletries. Because of its high antioxidant content, this plant is a welcome addition to many topical skincare formulations. Extracts from the *Cola accuminata* plant have been shown to reduce the effects of oxidative stress on the skin and to boost its appearance in youth [162, 163].

9.4 Traditional and Cultural Practices

Cola accuminata has cultural and traditional value in several locations in addition to the economic purposes described above. *Cola accuminata* has spiritual and ceremonial significance in various societies. *Cola accuminata*, from which the cola nut is harvested, is utilized in religious ceremonies and as a sign of hospitality and brotherhood [164, 165]. *Cola accuminata*'s social and symbolic worth is enhanced by these cultural practices.

9.5 Flavour and Fragrance Industry

Extracts and oils from the *Cola accuminata* plant are used to give products a cola-like aroma and flavour in the flavour and fragrance business. Perfumes, colognes, and other scented goods make use of *Cola accuminata* for its distinctive aroma. *Cola accuminata*'s singular aroma profile enhances the complexity and depth of perfumes in many forms [14, 104].

9.6 Herbal Supplements and Nutraceuticals

Nutraceuticals and herbal supplements made with *Cola accuminata* are becoming increasingly popular. Caffeine and other bioactive chemicals in it suggest it could be used in energizing supplements and medicines. Stimulant effects, cognitive enhancement, and general health benefits are all possible selling points for these items [106]. The incorporation of *Cola accuminata* in such products meets the rising demand for organic and plant-based options in the supplement market.

9.7 Traditional Medicine and Remedies

Cola accuminata plays a significant role in some regions' traditional medicine systems. *Cola accuminata* is used by traditional healers and practitioners to treat a wide range of problems, including gastrointestinal issues, weariness, and lack of concentration [42]. It has aphrodisiac powers, at least in the eyes of some societies. *Cola accuminata* has commercial value since it is used in traditional medicine, which is still common in some cultures.

9.8 Agricultural and Horticultural Uses

The *Cola accuminata* tree has numerous practical applications in agriculture and horticulture. Insects and other pests can be kept away from crops thanks to substances found in the plant's seeds and husks [166, 167]. Because of its historic value and visually appealing foliage, *Cola accuminata* is also occasionally cultivated as an ornamental tree.

9.9 Food and Confectionery Industry

Cola accuminata is used to improve the flavour and scent of many foods and sweets in the food and confectionary industries. To give baked products, candies, chocolates, ice creams, and other sweets with the familiar cola flavour, cola flavouring is often added. Energy bars and other functional food products can benefit from *Cola accuminata*'s naturally occurring caffeine [168].

9.10 Veterinary Medicine

Potential uses of *Cola accuminata* extracts in veterinary medicine have been investigated. Their potential to fight against various animal diseases and antimicrobials are being researched. *Cola accuminata*'s bioactive components have therapeutic potential for animal wellness and good health [158].

10 Conclusion

Cola accuminata, or the Cola nut, is a plant that has been used traditionally and holds great cultural significance in many parts of the world. *Cola accuminata*'s phytochemical components, nutritional profile, scientifically confirmed pharmacological qualities, ethnomedicinal applications, safety concerns, and commercial value have all been discussed at length. These results demonstrate the many uses and benefits of this plant. *Cola accuminata* has been shown to contain a number of bioactive substances, such as alkaloids, tannins, flavonoids, phenolic compounds, and caffeine. The plant's antioxidant, anti-inflammatory, antibacterial, anti-cancer, and neuroprotective benefits can be attributed in part to these components. *Cola accuminata* may have medicinal uses in the treatment and prevention of disease due to the presence of certain phytochemicals.

Carbohydrates, proteins, lipids, vitamins, and minerals are just some of the many nutrients that *Cola accuminata* is a good source of. The energizing effects of the caffeine in it might help you stay awake and focused. Caffeine is a stimulant, but consuming too much of it can cause negative side effects like irritability, nervousness, and a faster heart rate. So, it is best to take it easy sometimes.

The use of *Cola accuminata* in traditional and folk medicine dates back many centuries. It has been used for its stimulating effects, and it may also help with things like memory and focus, digestion, and wound healing, among other things. Several of these traditional applications have been validated by scientific studies, demonstrating the plant's usefulness in a variety of contexts. However, there are a number of factors that must be taken into account while determining the safety of *Cola accuminata*. *Cola accuminata*, like other caffeine-containing drinks, can cause a variety of unwanted side effects if consumed in large quantities. *Cola accuminata* should be used with caution and in consultation with medical specialists by people with certain health issues, including cardiovascular illnesses, psychological conditions, liver or renal conditions, and pregnancy or lactation.

Cola accuminata has substantial commercial value. It is a staple in the beverage industry, where it is used to impart the signature cola flavour and natural caffeine content into classic cola drinks. The plant is used in a wide variety of other contexts, such as aromatherapy, veterinary medicine, the cosmetics and personal care industries, and the food and confectionary industries. Its distinctive aroma, flavour, and bioactive qualities all add to its commercial value and popularity as a formulating component. Nevertheless, more study is needed to determine its full potential and how best to apply it to enhance human health and well-being.

References

1. Jacob DE, Ottong JL, Nelson IU (2018) Determinants of households' dependence on Ikot Ondo Community Forest, Nigeria. J Biol Series 1(3):62–71
2. Daniel KS, Udeagha AG, Umazi Y, Jacob DE (2016) Socio-cultural importance of sacred forests conservation in South Southern Nigeria. Afr J Sustain Dev 6(2):151–162
3. Daniel KS, Jacob DE, Udeagha AU (2015) Tree species composition in selected sacred forests in Nigeria. Intl J Mol Ecol Conserv 5(7):1–10
4. Udeagha AU, Udofia SI, Jacob DE (2013) Cultural and socio-economic perspectives of the conservation of Asanting Ibiono Sacred Forests in Akwa Ibom State, Nigeria. Int J Biodivs Conserv 5(11):696–703
5. Unya IU (2021) The historical significance and role of the Kola nut among the Igbo of Southeastern Nigeria. J Relig Hum Relat 13(1):289–312
6. Abulude FO (2004) Composition and properties of *Cola nitida* and *Cola acuminata* flour in Nigeria. Global J Pure Appl Sci 10(1):11–16
7. Thomas DE (2015) African traditional religion in the modern world. McFarland. ISBN 078649607X, 9780786496075
8. Lovejoy PE (2014) Kola nuts: the 'coffee' of the Central Sudan. In: Goodman J, Goodman J, Sherratt A, Sherratt A, Lovejoy PE, Lovejoy PE (Eds) Consuming habits: global and historical perspectives on how cultures define drugs, Routledge pp. 114–136
9. Sprague K (2018) The kola nut: west African commodity in the Atlantic world. African Study Centre, Los Angeles
10. Adega AP (2016) Kola nut as a symbol of welcome among the Igede of Benue State. AFRREV IJAH 5(4):24–35
11. Wilson T, Temple NJ (eds) (2016) Beverage impacts on health and nutrition. Humana Press, New York
12. Mars B (2001) Addiction-free naturally: liberating yourself from sugar, caffeine, food addictions, tobacco, alcohol, and prescription drugs. Inner Traditions/Bear & Co.
13. Danquah FK (2003) Sustaining a West African cocoa economy: agricultural science and the swollen shoot contagion in Ghana, 1936–1965. Afr Econ Hist 31:43–74
14. Asogwa EU, Anikwe JC, Mokwunye FC (2021) Kola production and utilization for economic development. Afr Sci 7(4):217–222
15. Izah SC, Kigigha LT, Aseibai ER, Okowa IP, Orutugu LA (2016) Advances in preservatives and condiments used in zobo (a food-drink) production. Biotechnol Res 2(3):104–119

16. Izah SC, Youkparigha FO (2019) Larvicidal activity of fresh aqueous and ethanolic extracts of *Cymbopogon citratus* (DC) Stapf on malaria vector, *Anopheles gambiae*. BAOJ Biotech 5:040
17. Youkparigha FO, Izah SC (2019) Larvicidal efficacy of aqueous extracts of *Zingiber officinale* Roscoe (ginger) against malaria vector, *Anopheles gambiae* (Diptera: Culicidae). Int J Environ Agric Sci 3:020
18. Izah SC, Chandel SS, Epidi JO, Venkatachalam T, Devaliya R (2019) Biocontrol of *Anopheles gambiae* larvae using fresh ripe and unripe fruit extracts of *Capsicum frutescens* var. baccatum. Int J Green Pharm 13(4):338–342
19. Seiyaboh EI, Seiyaboh Z, Izah SC (2020) Environmental control of mosquitoes: a case study of the effect of *Mangifera indica* root-bark extracts (family Anacardiaceae) on the larvae of *Anopheles gambiae*. Ann Ecol Environ Sci 4(1):33–38
20. Seiyaboh EI, Odubo TC, Izah SC (2020) Larvicidal activity of *Tetrapleura tetraptera* (Schum and Thonn) Taubert (Mimosaceae) extracts against *Anopheles gambiae*. Int J Adv Res Microbiol Immunol 2(1):20–25
21. Bassey SE, Izah SC (2017) Nigerian plants with insecticidal potentials against various stages of mosquito development. ASIO J Med Health Sci Res 2(1):07–18
22. Durand DN, Hubert AS, Nafan D, Adolphe A, Farid BM, Alphonse S, Lamine BM (2015) Indigenous knowledge and socioeconomic values of three kola species (*Cola nitida*, *Cola acuminata* and *Garcinia kola*) used in southern Benin. Eur Sci J 11(36):206–227
23. Odugbemi T (ed) (2008) A textbook of medicinal plants from Nigeria. University of Lagos Press, Nigeria
24. Kammampoal B, Laar S (2019) The kola nut: its symbolic significance in Chinua Achebe's things fall apart. Int J Stud English Lang Liter 7(8):26–40
25. Kanu IA (2020) The Igbo-African kola nut as a symbolic manifestation of 'Igwebuike' philosophy. AMAMIHE J Appl Philos 18(7):31–46
26. Odo ON, Nwokeocha I, Ezegwu D (2023) 'He who brings kola brings life': communicating the significance of Kolanut among Igbo people of Nigeria, India. Ind J Arts Liter 4(1):24–29
27. Okeke CU, Ezeabara CA, Chimezie H, Udechukwu CD, Aziagba BO (2015) Comparative phytochemical and proximate compositions of *Cola acuminata* (P. Beauv.) Schott and *Cola nitida* (Vent) Schott and Endl. Plant 3(3):26–29
28. Nwankwo EA (2018) Women and heritage preservation in Southeast Nigeria: exploring new approaches. J Herit Manag 3(2):173–191
29. Babalola FD, Agbeja BO (2010) Marketing and distribution of Garcinia kola (Bitter kola) in Southwest Nigeria: opportunity for development of a biological product. Egypt J Biol 12:12–17
30. Adedokun MO, Soaga JAO, Olawumi AT, Oyebanji OO, Oluwalana SA (2013) Socio-economic contribution, marketing and utilization of edible kolanut (*Cola acuminata* and *Cola nitida*) to rural women livelihood in Abeokuta, Nigeria. Int J Mol Ecol Conserv 2(1): 32–38
31. Lowe HI, Watson CT, Badal S, Peart P, Toyang NJ, Bryant J (2014) Promising efficacy of the *Cola acuminata* plant: a mini review. Adv Biol Chem 4(4):240
32. Cabot S (2005) The ultimate detox book. SCB International
33. Donovan T (2013) Fizz: how soda shook up the world. Chicago Review Press
34. Klosterman L (2007) The facts about caffeine. Marshall Cavendish
35. Ajai AI, Ochigbo SS, Jacob MM (2012) Proximate and mineral compositions of different species of kola nuts. Eur J Appl Eng Sci Res 1(3):44–47
36. Burdock GA, Carabin IG, Crincoli CM (2009) Safety assessment of kola nut extract as a food ingredient. Food Chem Toxicol 47(8):1725–1732
37. Dah-Nouvlessounon D, Adjanohoun A, Sina H, Noumavo PA, Diarrasouba N, Parkouda C, Madodé YE, Dicko MH, Baba-Moussa L (2015) Nutritional and anti-nutrient composition of three kola nuts (*Cola nitida*, *Cola acuminata* and *Garcinia kola*) produced in Benin. Food Nutr Sci 6(15):1395

38. Beck HT (2012) Caffeine, alcohol, and sweeteners. In: Beck HT (ed) The cultural history of plants. Routledge, pp 176–193
39. Adeniyi A, Olufunmilayo AD, Akinnuoye AG (2017) Comparative study of the proximate and fatty acid profiles of *Cola nitida, Cola acuminata* and *Garcinia kola*. Am J Food Sci Nutr 4(6): 80–84
40. So D, Yao CK, Ardalan ZS, Thwaites PA, Kalantar-Zadeh K, Gibson PR, Muir JG (2022) Supplementing dietary fibers with a low FODMAP diet in irritable bowel syndrome: a randomized controlled crossover trial. Clin Gastroenterol Hepatol 1:110–116
41. Okwu DE, Morah FNI (2007) Isolation and characterization of flavanone glycoside 4, 5, 7 -trihydroxy flavanone rhamnoglucose from Garcinia kola seed. J Appl Sci 7(2):306–309
42. Victoria AO, Rufina ASY, Oluwaseun OR, Olusola OB, Anderson EL, Sina OJ, Fakunle JB (2016) Phytochemical screening and proximate analysis of young *Cola acuminata* leaves. Unique Res J Med Med Sci 4(5):029–034
43. Fernández-Murga ML, Olivares M, Sanz Y (2020) Bifidobacterium pseudocatenulatum CECT 7765 reverses the adverse effects of diet-induced obesity through the gut-bone axis. Bone 141: 115580
44. Alabi WO, Adeoye AO, Akinlolu AR, Rebecca OO (2020) Proximate and antimicrobial properties of *Garcinia kola*; a significant evidence of a functional food snack. Int J Sci Eng Res 11(5):33–38
45. Adeoye RI, Joel EB, Igunnu A, Arise RO, Malomo SO (2022) A review of some common African spices with antihypertensive potential. J Food Biochem 46(1):e14003
46. Santos RM, Santos RM, Lima DR (2009) An unashamed defense of coffee. Xlibris Corporation
47. Ogbonna AI, Adepoju SO, Ogbonna CIC, Yakubu T, Itelima JU, Dajin VY (2017) Root tuber of *Tacca leontopetaloides* L.(kunze) for food and nutritional security. Microbiol Curr Res 1(1): 7–13
48. Georgieff MK (2011) Long-term brain and behavioral consequences of early iron deficiency. Nutr Rev 69(Suppl_1):S43–S48
49. Alfa HH, Arroo RR (2019) Over 3 decades of research on dietary flavonoid antioxidants and cancer prevention: what have we achieved? Phytochem Rev 18(4):989–1004
50. Joshua PE, Ukegbu CY, Eze CS, Umeh BO, Oparandu LU, Okafor JO, Ogara A (2017) Comparative studies on the possible antioxidant properties of ethanolic seed extracts of *Cola nitida* (kola nut) and *Garcinia kola* (bitter kola) on hydrogen peroxide-induced oxidative stress in rats. J Med Plants Res 12(22):367–372
51. Wilhelmi de Toledo F, Grundler F, Goutzourelas N, Tekos F, Vassi E, Mesnage R, Kouretas D (2020) Influence of long-term fasting on blood redox status in humans. Antioxidants 9(6):496
52. Grundler F, Mesnage R, Goutzourelas N, Tekos F, Makri S, Brack M, de Toledo FW (2020) Interplay between oxidative damage, the redox status, and metabolic biomarkers during long-term fasting. Food Chem Toxicol 145:111701
53. Adedapo AA, Jimoh FO, Koduru S, Afolayan AJ, Masika PJ (2008) Antibacterial and antioxidant properties of the methanol extracts of the leaves and stems of *Calpurnia aurea*. BMC Complement Altern Med 8:1–8
54. Adeyemi OO, Akindele AJ, Yemitan OK, Aigbe FR, Fagbo FI (2010) Anticonvulsant, anxiolytic and sedative activities of the aqueous root extract of Securidaca longepedunculata Fresen. J Ethnopharmacol 130(2):191–195
55. Oboh G, Agunloye OM, Akinyemi AJ, Ademiluyi AO, Adefegha SA (2013) Comparative study on the inhibitory effect of caffeic and chlorogenic acids on key enzymes linked to Alzheimer's disease and some pro-oxidant induced oxidative stress in rats' brain-in vitro. Neurochem Res 38:413–419
56. Adaramoye OA, Achem J, Akintayo OO, Fafunso MA (2007) Hypolipidemic effect of *Telfairia occidentalis* (fluted pumpkin) in rats fed a cholesterol-rich diet. J Med Food 10(2): 330–336

57. Mustapha AO, Olaofe O, Ibrahim HO (2009) Proximate and mineral analysis of kolanuts (*Cola nitida* and *Cola acuminate*). www.globalacademicsgroups.com
58. Dewole EA, Dewumi DFA, Alabi JYT, Adegoke A (2013) Proximate and phytochemical of *Cola nitida* and *Cola acuminata*. Pak J Biol Sci 16:1593–1596
59. Adeyeye EI, Ayejuyo OO (1994) Chemical composition of *Cola acuminata* and *Garcinia kola* seeds grown in Nigeria. Int J Food Sci Nutr 45(4):223–230
60. Lowor ST, Aculey PC, Assuah MK (2010) Analysis of some quality indicators in cured *Cola nitida* (Vent). Agric Biol J N Am 1(6):1206–1214
61. Ruberto G, Baratta MT (2000) Antioxidant activity of selected essential oil components in two lipid model systems. Food Chem 69(2):167–174
62. Vengsarkar AM, Lemaire PJ, Judkins JB, Bhatia V, Erdogan T, Sipe JE (1996) Long-period fiber gratings as band-rejection filters. J Lightwave Technol 14(1):58–65
63. Voet A, Berenger F, Zhang KY (2013) Electrostatic similarities between protein and small molecule ligands facilitate the design of protein-protein interaction inhibitors. PLoS One 8(10):e75762
64. Kingsley N, Martha OE, Ogheneovo OR (2019) Nitida (Vent.) Schott & Endl. Am J Environ Eng Sci 6(1):1–10
65. Chukwuma IF, Apeh VO, Nwora FN, Nkwocha CC, Mba SE, Ossai EC (2023) Phytochemical profiling and antioxidative potential of phenolic-rich extract of *Cola acuminata* nuts. Biointerface Res Appl Chem 13(1):29–39
66. Akhter S, Islam MT, Hossain MT (2015) Proximate analysis, phytochemical screening and antioxidant activity of *Tagetes erecta* leaves. World J Pharm Res 4:1856–1866
67. Fabunmi T, Arotupin D (2015) Antioxidant properties of fermented kolanut husk and testa of three species of kolanut: *Cola acuminata*, *Cola nitida* and *Cola verticillata*. Br Biotechnol J 8(2):1–13
68. Oboh G, Ademosun AO, Ogunsuyi OB, Oyedola ET, Olasehinde TA, Oyeleye SI (2018) In vitro anticholinesterase, antimonoamine oxidase and antioxidant properties of alkaloid extracts from kola nuts (*Cola acuminata* and *Cola nitida*). J Complement Integr Med 16(1):20160155
69. Adewole KE, Gyebi GA, Ibrahim IM (2021) Amyloid β fibrils disruption by kolaviron: molecular docking and extended molecular dynamics simulation studies. Comput Biol Chem 94:107557
70. Nwobodo DC, Okezie UM, Okoye FB, Esimone CO (2022) UV-mediated enhancement of antibacterial secondary metabolites in endophytic *Lasiodiplodia theobromae*. Not Sci Biol 14(4):11284–11284
71. Ibrahim M, Oyebanji E, Fowora M, Aiyeolemi A, Orabuchi C, Akinnawo B, Adekunle AA (2021) Extracts of endophytic fungi from leaves of selected Nigerian ethnomedicinal plants exhibited antioxidant activity. BMC Complement Med Ther 21:98
72. Okwunodulu IN, Ukeje SC (2018) Influence of sprouting on proximate and sensory properties of Gworo (*Cola nitida*) and Ojigbo (*Cola acuminata*) kola nuts. Sustain Food Prod 2:29–36
73. Owoade AO, Alausa AO, Adetutu A, Olorunnisola OS, Owoade AW (2021) Phytochemical characterization and antioxidant bioactivity of *Andrographis paniculata* (Nees). Pan Afr J Life Sci 5(2):246–256
74. Udegbunam SO, Nnaji TO, Udegbunam RI, Okafor JC, Agbo I (2013) Evaluation of herbal ointment formulation of *Milicia excelsa* (Welw) CC berg for wound healing. Afr J Biotechnol 12(21):3351–3359
75. Egbuniwe IC, Uchendu CN, Obidike IR, Michaelis M (2021) Dietary ascorbic acid and betaine improve stress responses, testosterone levels and some sexual traits in male Japanese quails during the dry season. Exper Res 2:e23
76. Nwonuma CO, Adelani-Akande TA, Osemwegie OO, Olaniran AF, Adeyemo TA (2019) Comparative study of the in-vitro phytochemicals and antimicrobial potential of six medicinal plants. F1000 Res 8:81
77. Omwirhiren EM, James SA, Asefon OA (2016) The phytochemical properties and antimicrobial potentials of aqueous and methanolic seed extract of *Cola nitida* (Vent.) and *Cola*

acuminate (Beauvoir) grown in South West, Nigeria. Saudi J Med Pharmaceut Sci 2(12): 354–363
78. Wink M (2015) Modes of action of herbal medicines and plant secondary metabolites. Medicines 2(3):251–286
79. Kamatenesi-Mugisha M, Oryem-Origa H, Odyek O, Makawiti DW (2008) Medicinal plants used in the treatment of fungal and bacterial infections in and around Queen Elizabeth Biosphere Reserve, western Uganda. Afr J Ecol 46:90–97
80. Mahapatra GP, Raman S, Nayak S, Gouda S, Das G, Patra JK (2020) Metagenomics approaches in discovery and development of new bioactive compounds from marine actinomycetes. Curr Microbiol 77:645–656
81. Izah SC, Odubo TC (2023) Effects of *Citrus aurantifolia* fruit juice on selected pathogens of public health importance. ES Food Agroforesty 11:829. https://doi.org/10.30919/esfaf829
82. Enaregha EB, Izah SC, Okiriya Q (2021) Antibacterial properties of Tetrapleura tetraptera pod against some pathogens. Res Rev Insights 5:1–5
83. Epidi JO, Izah SC, Ohimain EI, Epidi TT (2016) Phytochemical, antibacterial and synergistic potency of tissues of *Vitex grandifolia*. Biotechnol Res 2(2):69–76
84. Epidi JO, Izah SC, Ohimain EI (2016) Antibacterial and synergistic efficacy of extracts of *Alstonia boonei* tissues. Br J Appl Res 1(1):0021–0026
85. Izah SC, Chandel SS, Etim NG, Epidi JO, Venkatachalam T, Devaliya R (2019) Potency of unripe and ripe express extracts of long pepper (*Capsicum frutescens* var. baccatum) against some common pathogens. Int J Pharmaceut Phytopharmacol Res 9(2):56–70
86. Izah SC, Etim NG, Ilerhunmwuwa IA, Silas G (2019) Evaluation of crude and ethanolic extracts of *Capsicum frutescens* var. minima fruit against some common bacterial pathogens. Int J Complement Altern Med 12(3):105–108
87. Izah SC, Etim NG, Ilerhunmwuwa IA, Ibibo TD, Udumo JJ (2019) Activities of express extracts of Costus afer Ker–Gawl. [family COSTACEAE] against selected bacterial isolates. Int J Pharmaceut Phytopharmacol Res 9(4):39–44
88. Izah SC, Uhunmwangho EJ, Dunga KE, Kigigha LT (2018) Synergy of methanolic leave and stem-back extract of *Anacardium occidentale* L. (cashew) against some enteric and superficial bacteria pathogens. MOJ Toxicol 4(3):209–211
89. Izah SC, Uhunmwangho EJ, Dunga KE (2018) Studies on the synergistic effectiveness of methanolic extract of leaves and roots of *Carica papaya* L. (papaya) against some bacteria pathogens. Int J Complement Altern Med 11(6):375–378
90. Izah SC, Uhunmwangho EJ, Etim NG (2018) Antibacterial and synergistic potency of methanolic leaf extracts of *Vernonia amygdalina* L. and *Ocimum gratissimum* L. J Basic Pharmacol Toxicol 2(1):8–12
91. Izah SC, Zige DV, Alagoa KJ, Uhunmwangho EJ, Iyamu AO (2018) Antibacterial efficacy of aqueous extract of *Myristica fragrans* (common nutmeg). EC Pharmacol Toxicol 6(4): 291–295
92. Izah SC, Uhunmwangho EJ, Eledo BO (2018) Medicinal potentials of *Buchholzia coriacea* (wonderful kola). Med Plant Res 8(5):27–43
93. Izah SC (2019) Activities of crude, acetone and ethanolic extracts of *Capsicum frutescens* var minima fruit against larvae of *Anopheles gambiae*. J Environ Treat Tech 7(2):196–200
94. Izah SC (2018) Some determinant factors of antimicrobial susceptibility pattern of plant extracts. Res Rev Insight 2(3):1–4
95. Izah SC, Aseibai ER (2018) Antibacterial and synergistic activities of methanolic leaves extract of lemon grass (*Cymbopogon citratus*) and rhizomes of ginger (*Zingiber officinale*) against *Escherichia coli*, *Staphylococcus aureus* and *Bacillus subtilis*. Acta Sci Microbiol 1(6): 26–30
96. Kigigha LT, Izah SC, Uhunmwangho EJ (2018) Assessment of hot water and ethanolic leaf extracts of *Cymbopogon citratus* Stapf (lemon grass) against selected bacteria pathogens. Ann Microbiol Infect Dis 1(3):1–5

97. Kigigha LT, Selekere RE, Izah SC (2018) Antibacterial and synergistic efficacy of acetone extracts of *Garcinia kola* (Bitter kola) and *Buchholzia coriacea* (Wonderful kola). J Basic Pharmacol Toxicol 2(1):13–17
98. Kigigha LT, Biye SE, Izah SC (2016) Phytochemical and antibacterial activities of *Musanga cecropioides* tissues against *Escherichia coli, Pseudomonas aeruginosa Staphylococcus aureus*, Proteus and Bacillus species. Int J Appl Res Technol 5(1):100–107
99. Kigigha LT, Izah SC, Okitah LB (2016) Antibacterial activity of palm wine against Pseudomonas, Bacillus, Staphylococcus, Escherichia, and Proteus spp. Point J Bot Microbiol Res 2(1):046–052
100. Kigigha LT, Apreala A, Izah SC (2016) Effect of cooking on the climbing pepper (*Piper nigrum*) on antibacterial activity. J Environm Treat Tech 4(1):6–9
101. Kigigha LT, Izah SC, Ehizibue M (2015) Activities of *Aframomum melegueta* seed against *Escherichia coli*, *Staphylococcus aureus* and Bacillus species. Point J Bot Microbiol Res 1(2): 23–29
102. Erukainure OL, Ijomone OM, Oyebode OA, Chukwuma CI, Aschner M, Islam MS (2019) Hyperglycemia-induced oxidative brain injury: therapeutic effects of *Cola nitida* infusion against redox imbalance, cerebellar neuronal insults, and upregulated Nrf2 expression in type 2 diabetic rats. Food Chem Toxicol 127:206–217
103. Kenneth EN, Bola AD, Kingsley CI, Mahady G (2014) Phytochemical and antimicrobial properties of crude n-hexane and methanol extracts of *Cola acuminata* nuts. Br J Pharmaceut Res 4(8):920
104. Adelusi AA, Ogunwolu QA, Ugwu CA, Alli MA, Adesanya KA, Agboola-Adedoja MO, Akinpelu AO (2020) Kolanut consumption, its benefits and side effects. World J Adv Res Rev 8(3):356–362
105. Kanoma AI, Muhammad I, Abdullahi S, Shehu K, Maishanu HM, Isah AD (2014) Qualitative and quantitative phytochemical screening of cola nuts (*Cola nitida* and *Cola acuminata*). J Biol Agric Health 45(5):89–97
106. Enogieru AB, Momodu OI (2021) African medicinal plants useful for cognition and memory: therapeutic implications for Alzheimer's disease. Bot Rev 87:107–134
107. Mbembo BM, Inkoto CL, Amogu JJO, Ashande CM, Kutshi NN, Nagahuedi JMS, Mpiana PT, Ngbolua KN (2021) Mini-review on the phytochemistry, pharmacology and toxicology of *Cola nitida* (Vent.) Schott & Endl. (Malvaceae): a medically interesting bio-resource of multiple purposes in Africa. Discov Phytomed 8(4):160–166
108. Nwafor PA, Genesis EU, Dare ST, Adeyemi OI, Odediran SA, Adebajo AC (2020) Evaluation of aphrodisiac activities of four Nigerian ethnomedicinal plants. Ann Complement Altern Med 2(1):1009
109. Oboh G, Oyeleye SI, Ademiluyi AO (2017) The food and medicinal values of indigenous leafy vegetables. Afr Veg Forum 1238:137–156
110. Aina DA, Owolo O, Lateef A, Aina FO, Hakeem AS, Adeoye-Isijola M, Okon V, Asafa TB, Elegbede JA, Olukanni OD, Adediji I (2019) Biomedical applications of *Chasmanthera dependens* stem extract mediated silver nanoparticles as antimicrobial, antioxidant, anticoagulant, thrombolytic, and larvicidal agents. Karbala Int J Mod Sci 5(2):71–80
111. Adebayo EA, Oladipo IC, Badmus JA, Lateef A (2021) Beneficial microbes as novel microbial cell factories in nanobiotechnology: peotentials in nanomedicine. In: Lateef A, Gueguim-Kana EB, Dasgupta N, Ranjan S (eds) Microbial nanobiotechnology. Materials horizons: from nature to nanomaterials. Springer, Singapore, pp 315–342
112. Aiyegoro OA, Akinpelu DA, Okoh AI (2007) In vitro antibacterial potentials of the stem bark of red water tree (*Erythrophleum suaveolens*). J Biol Sci 7(7):1233–1238
113. Amadi CN, Nwachukwu WI (2020) The effects of oral administration of *Cola nitida* on the pharmacokinetic profile of metoclopramide in rabbits. BMC Pharmacol Toxicol 21:1–6
114. Akinmoladun AC, Ibukun EO, Dan-ologe IA (2007) Phytochemical constituents and antioxidant properties of extracts from the leaves of *Chromolaena odorata*. Sci Res Essay 2:191–194

115. Oboh G, Adedayo BC, Adetola MB, Oyeleye IS, Ogunsuyi OB (2022) Characterization and neuroprotective properties of alkaloid extract of *Vernonia amygdalina* Delile in experimental models of Alzheimer's disease. Drug Chem Toxicol 45(2):731–740
116. Olurishe C, Kwanashie H, Zezi A, Danjuma N, Mohammed B (2016) Chronic administration of ethanol leaf extract of *Moringa oleifera* Lam. (Moringaceae) may compromise glycaemic efficacy of Sitagliptin with no significant effect in retinopathy in a diabetic rat model. J Ethnopharmacol 194:895–903
117. Osadebe PO (2021) An investigation into the seasonal variation of the phytochemical and anti-diabetic properties of the eastern Nigerian specie of African mistletoe (*Loranthus Micranthus*) sourced from Kola Acuminata. Diabetes Complicat 5(2):1–5
118. Ezejiofor AN, Udowelle NA, Orisakwe OE (2017) Nephroprotective and antioxidant effect of aqueous leaf extract of Costus afer Ker gawl on cyclosporin-a (Csa) induced nephrotoxicity. Clin Phytosci 2(1):1–7
119. Boudiba S, Küçükaudın S, Tamfu AN, Blaise K, Munvera AM, Ceylan Ö (2023) HPLC-DAD phenolic composition, antioxidant, anticholinesterase, antidiabetic and anti-quorum sensing properties of bitter Kola (*Garcinia kola*) and Kolanut (*Cola acuminata*). Pharm Res 15: 2,373–2,383
120. Imam-Fulani AO, Sanusi KO, Owoyele BV (2018) Effects of acetone extract of *Cola nitida* on brain sodium-potassium adenosine triphosphatase activity and spatial memory in healthy and streptozotocin-induced diabetic female Wistar rats. J Basic Clin Physiol Pharmacol 29(4):411–416
121. Oboh G, Akinyemi AJ, Omojokun OS, Oyeleye IS (2014) Anticholinesterase and anti-oxidative properties of aqueous extract of *Cola acuminata* seed in vitro. Int J Alzheimers Dis 2014:498629. https://doi.org/10.1155/2014/498629
122. Ishola IO, Ikuomola BO, Adeyemi OO (2018) Protective role of *Spondias mombin* leaf and *Cola acuminata* seed extracts against scopolamine-induced cognitive dysfunction. Alexandria J Med 54(1):27–39
123. Nugraha I, Annisa AN, Wibowo AT, Kusuma AM (2018) Chemopreventive activity of Kola (*Cola accuminata*) seed ethanol extract in mice induced by cyclophosphamide. IOP Conf Ser Mater Sci Eng 288(1):012008
124. Alsaeedi H, Qahwaji R, Qadah T (2021) Induction of apoptosis by Kola nut extract as a recent and promising treatment strategy for leukemia. Rev Bionatura 6(2):1725–1732
125. Suntar I, Khan H, Patel S, Celano R, Rastrelli L (2018) An overview on *Citrus aurantium* L.: its functions as food ingredient and therapeutic agent. Oxid Med Cell Longev 2018:7864269. https://doi.org/10.1155/2018/7864269
126. Percival SS, Bukowski JF, Milner J (2008) Bioactive food components that enhance γδ T cell function may play a role in cancer prevention. J Nutr 138(1):1–4
127. Holcombe C, Kaluba J, Lucas SB (1991) Non-ulcer dyspepsia in Nigeria: a case-control study. Trans R Soc Trop Med Hyg 85(4):553–555
128. Nkanu EE (2020) Dihydroquercetin (taxifolin) attenuates dexamethasone activity on prosta-glandin E-2 but potentiates thromboxane-A2 action in gastric acid secretion in Wistar rats. GSC Biol Pharmaceut Sci 13(3):181–188
129. Edwards SE, da Costa RI, Williamson EM, Heinrich M (2015) *Cola nitida* (Vent.) Schott & Endl., *C. acuminata* (P.Beauv.) Schott & Endl. In: Phytopharmacy: an evidence-based guide to herbal medical products. John Wiley & Sons, Ltd., pp 111–113
130. Agbaje EO, Charles OO (2022) *Spondias mombin* Linn. (Anacardiaceous) essential oil ointment enhances healing of excision wounds in rats. J Phytopharmacol 11(4):224–232
131. Diallo D, Sogn C, Samaké FB, Paulsen BS, Michaelsen TE, Keita A (2002) Wound healing plants in Mali, the Bamako region. An ethnobotanical survey and complement fixation of water extracts from selected plants. Pharm Biol 40(2):117–128
132. Akande IS, Adewoyin OA, Njoku UF, Awosika SO (2012) Biochemical evaluation of some locally prepared herbal remedies (Agbo) currently on high demand in Lagos metropolis, Nigeria. J Drug Metab Toxicol 3:118

133. Musabayane CT, Bwititi PT, Ojewole JAO (2006) Effects of oral administration of some herbal extracts on food consumption and blood glucose levels in normal and streptozotocin-treated diabetic rats. Methods Find Exp Clin Pharmacol 28(4):223–228
134. Cyril-Olutayo MC, Akinola NO, Agbedahunsi JM (2021) T2MC-A poly-herbal that inhibits polymerization of intracellular sickle hemoglobin and regulates the expression of erythrocyte Ca2+ activated K+ channel. J Med Herbs 12(2):1–11
135. Nwabuisi C (2002) Prophylactic effect of multi-herbal extract 'Agbo-Iba'on Malaria induced in mice. East Afr Med J 79(7):343–346
136. Odukanmi OA, Salami AT, Morakinyo OL, Yelotan OE, Olaleye SB (2018) Kolaviron modulates intestinal motility and secretion in experimentally altered gut functions of Wistar rats. J Complement Med Res 9(2):55–55
137. Bolawa EO, Adeogun BM, Alani RA (2021) Genotoxic and histopathological alterations in rats exposed to herbal liquors. GSC Biol Pharmaceut Sci 17(3):198–216
138. Akinboro A, Ibrahim A, Muhammed J, Oloyede H, Alimi R (2022) Safety evaluation of an anti-haemorrhoid (pile) herbal recipe (locally called 'agbo jedi-jedi') of southwestern Nigeria using animal genetic assays. Toxicol Adv 5(1):2. https://doi.org/10.53388/TA202305002
139. Adeyemi OO, Alabi AS, Adeyemi OA, Talabi OT, Abidakun OM, Joel IY, Stonehouse NJ (2021) Acute gastroenteritis and the usage pattern of antibiotics and traditional herbal medications for its management in a Nigerian community. PLoS One 16(10):e0257837
140. Eiya BO, Igharo OG (2022) Selected biochemical parameters and oxidative stress status of rats administered antimalaria herbal extract–'agbo'. Afr Sci 21(3):301–307
141. Emaleku AS (2018) Depletion of hepatic antioxidant enzymes in experimental albino rats due to polyherbal medicines administration. J Stem Cell Res Ther 8(3):1–6
142. Adeyemi OS, Owoseni MC (2015) Polyphenolic content and biochemical evaluation of fijk, alomo, osomo and oroki herbal mixtures in vitro. Beni-Suef Univ J Basic Appl Sci 4(3):200–206
143. Enitan SS, Ogechukwu UE, Gotep J, Effiong EJ, Ileoma EO, Mensah-Agyei GO, Adetola AO (2022) Assessment of microbiological quality and efficacy of Gbogbonise Epa Ijebu herbal remedy on some uropathogens. Sch Int J Tradit Complement Med 5(1):7–18
144. Nehlig A (2016) Effects of coffee/caffeine on brain health and disease: what should I tell my patients? Pract Neurol 16(2):89–95
145. De Sanctis V, Soliman N, Soliman AT, Elsedfy H, Di Maio S, El Kholy M, Fiscina B (2017) Caffeinated energy drink consumption among adolescents and potential health consequences associated with their use: a significant public health hazard. Acta Bio Medica Atenei Parmensis 88(2):222–231
146. Rath M (2012) Energy drinks: what is all the hype? The dangers of energy drink consumption. J Am Acad Nurse Pract 24(2):70–76
147. Husain Z, Schwartz RA (2012) Peanut allergy: an increasingly common life-threatening disorder. J Am Acad Dermatol 66(1):136–143
148. Baur DM, Lange D, Elmenhorst EM, Elmenhorst D, Bauer A, Aeschbach D, Landolt HP (2021) Coffee effectively attenuates impaired attention in ADORA2A C/C-allele carriers during chronic sleep restriction. Prog Neuro-Psychopharmacol Biol Psychiatry 109:110232
149. Sirtori CR, Arnoldi A, Cicero AF (2015) Nutraceuticals for blood pressure control. Ann Med 47(6):447–456
150. Cappelletti S, Daria P, Sani G, Aromatario M (2015) Caffeine: cognitive and physical performance enhancer or psychoactive drug? Curr Neuropharmacol 13(1):71–88
151. Pesta DH, Angadi SS, Burtscher M, Roberts CK (2013) The effects of caffeine, nicotine, ethanol, and tetrahydrocannabinol on exercise performance. Nutr Metab 10:1–15
152. Sweetser S (2012) Evaluating the patient with diarrhea: a case-based approach. Mayo Clin Proc 87(6):596–602
153. Scalbert A (1991) Antimicrobial properties of tannins. Phytochemistry 30(12):3875–3883

154. Heckman MA, Weil J, Gonzalez de Mejia E (2010) Caffeine (1, 3, 7-trimethylxanthine) in foods: a comprehensive review on consumption, functionality, safety, and regulatory matters. J Food Sci 75(3):77–87
155. Kumar VH, Lipshultz SE (2019) Caffeine and clinical outcomes in premature neonates. Children 6(11):118
156. Glade MJ (2010) Caffeine – not just a stimulant. Nutrition 26(10):932–938
157. Jee SH, He J, Whelton PK, Suh I, Klag MJ (1999) The effect of chronic coffee drinking on blood pressure: a meta-analysis of controlled clinical trials. Hypertension 33(2):647–652
158. Olatunji GA, Abiodun IO, Saliu OD, Obisesan N (2017) Comparative study of the chemical constituents of Bitter kola (*Garcinia kola*) and cola nut (*Cola acuminata*) seeds extracts. Zimbabwe J Sci Technol 12(1):110–116
159. Temple JL, Bernard C, Lipshultz SE, Czachor JD, Westphal JA, Mestre MA (2017) The safety of ingested caffeine: a comprehensive review. Front Psych 8:80
160. Eteng MU, Eyong EU, Akpanyung EO, Agiang MA, Aremu CY (1997) Recent advances in caffeine and theobromine toxicities: a review. Plant Foods Hum Nutr 51:231–243
161. Oluwaseun AA, Ganiyu O (2008) Antioxidant properties of methanolic extracts of mistletoes (Viscum album) from cocoa and cashew trees in Nigeria. Afr J Biotechnol 7(17):3138–3142
162. Bamigbola EA, Attama AA, Ogeh PC (2018) Evaluation of physico-mechanical and mucoadhesive properties of biopolymer films from *Cola acuminata* gum. Niger J Pharm Res 14(1):1–13
163. Bamigbola EA, Momoh MA, Ikebudu O (2018) Isolation and characterization of *Cola acuminata* gum as a potential pharmaceutical excipient. J Pharmaceut Res Int 24(1):1–14
164. Adedayo LD, Bamidele O, Onasanwo SA (2022) *Cola nitida* and pain relief. In: Rajendram R, Patel VB, Martin CR (eds) Treatments, mechanisms, and adverse reactions of anesthetics and analgesics. Academic Press, pp 375–384
165. Moneim A, Sulieman E (2019) Garcinia Kola (*Bitter kola*): chemical composition. In: Mariod A (ed) Wild fruits: composition, nutritional value and products. Springer, Cham, pp 285–299
166. Duke JA (2018) CRC handbook of nuts. CRC Press
167. Mele PV, Cuc NTT, Seguni Z, Camara K, Offenberg J (2009) Multiple sources of local knowledge: a global review of ways to reduce nuisance from the beneficial weaver ant Oecophylla. Int J Agric Resour Gov Ecol 8(5–6):484–504
168. Adedokun MO, Ojo TM, Idowu SD, Olawumi AT, Oluwalana SA, Ibasanmi T (2007) Socio-economic importance and utilization of *Garcinia kola* (Heckel) in Ore, Ondo State Nigeria. Int J Sci Eng Res 8(11):37–41

Citrus aurantium: Phytochemistry, Therapeutic Potential, Safety Considerations, and Research Needs

6

Olalekan Bukunmi Ogunro, Glory Richard, Sylvester Chibueze Izah, Kurotimipa Frank Ovuru, Oladimeji Taiwo Babatunde, and Moyuri Das

Contents

O. B. Ogunro (✉)
Reproductive and Endocrinology, Toxicology, and Bioinformatics Research Laboratory, Department of Biological Sciences, KolaDaisi University, Ibadan, Oyo State, Nigeria
e-mail: olalekan.ogunro@koladaisiuniversity.edu.ng

G. Richard
Department of Community Medicine, Faculty of Clinical Sciences, Niger Delta University, Wilberforce Island, Bayelsa State, Nigeria

S. C. Izah
Department of Microbiology, Faculty of Science, Bayelsa Medical University, Yenagoa, Bayelsa State, Nigeria
e-mail: sylvester.izah@bmu.edu.ng

K. F. Ovuru
Department of Disease Control and Immunization, Bayelsa State Primary Health Care Board, Yenagoa, Bayelsa State, Nigeria

O. T. Babatunde
Department of Chemical Sciences, Crown-Hill University, Eiyenkorin, Kwara State, Nigeria
e-mail:

M. Das
Department of Zoology, Cooch Behar Panchanan Barma University, Cooch Behar, West Bengal, India

S. C. Izah et al. (eds.), *Herbal Medicine Phytochemistry*, Reference Series in Phytochemistry, https://doi.org/10.1007/978-3-031-43199-9_69

Abstract

Citrus aurantium, commonly referred to as sour orange or bitter orange, holds significant importance both in biological and economic terms. Throughout history, humans have turned to nature in their pursuit of health and wellness. Among the plants that have historically played a role in enhancing fitness, *Citrus aurantium* stands out. A diverse array of phytochemical constituents present in *Citrus aurantium* have been closely tied to its various biological activities, encompassing areas such as gastrointestinal disorders, insomnia, headaches, cancer treatment, antiseptic properties, antioxidant effects, and antispasmodic effects. Beyond its pharmacological relevance, *Citrus aurantium* also boasts numerous non-pharmacological applications that make it particularly intriguing. It serves as a food preservative agent, contributes to aromatherapy practices, acts as a pesticide, provides raw materials for the pulp and paper industry, lends its aromatic qualities to the food processing and cosmetics sectors, and offers potential as an anti-aging agent. Despite its impressive array of properties, the utilization of *Citrus aurantium* and its derivatives has been linked to certain unwanted side effects. While some studies have largely cleared the plant of safety concerns, others have cast doubt on its safety. This chapter delves into the complex aspects surrounding the safety of *Citrus aurantium*'s phytochemistry and its derived products. Ultimately, the chapter advocates for the ongoing use of this plant, but with a careful awareness of its limitations to prevent any undesirable effects.

Keywords

Citrus aurantium · Bitter orange · Phytochemicals · Non-pharmacological uses · Economic relevance

Abbreviations

ALT	Alanine transminase
AMPKα	Activated protein kinase alpha
AMPKα	AMP-activated protein kinase alpha
AST	Aspartate aminotransferase
CAVAPs	Crude polysaccharides of *C. aurantium* L. var. Amara Engls
DPPH	2,2-Diphenyl-1-picrylhydrazyl
ERK	Extracellular signal- regulated kinase
EtOAc	Ethyl acetate
FAS	Fatty acid synthase
FDA	Food and Drug Administration
FRAP	Ferric reducing antioxidant power
GGT	Gamma-glutamyl transferase
GSK3β	Glycogen Synthase Kinase 3 Beta
IL-1β	Interleukin-1β
IL-6	Interleukin 6
iNOS	Inducible nitric oxide synthase
JNK	c-Jun N-terminal kinase
MAPK	Mitogen-activated protein kinase
MTD	Maximum tolerated dose
NAFLD	Non-alcohol fatty liver disease
NASH	Non-alcoholic steatohepatitis
NF-kB	Nuclear factor kappa B
NOAEL	No-observed-adverse-effect-level
NOEL	No-observed-effect-level
Nrf2	Nuclear factor erythroid 2-related factor 2
ORAC	Oxygen radical absorbance capacity
PGC-1α	PPARγ co-activator 1α
PTFC	Pure total flavonoids from citrus
SCD1	Stearoyl-CoA desaturase 1
TAC	Total antioxidant capacity
TBARS	Thiobarbituric acid reactive substances
TNF- α	Tumor necrosis factor α
UCP1	Uncoupling protein-1
WADA	World Anti-Doping Agency

1 Introduction

Citrus fruits stand apart due to their unique structural composition, which includes seeds, pulp, oil cells, and a white membrane layer [1]. These fruits are remarkable sources of essential nutrients such as vitamin C, bioactive flavonoids, and folic acid. These components play diverse roles in reducing inflammation and enhancing the body's immune system. Native to Australia, Malaysia, New Caledonia, and a few other countries, *Citrus aurantium*, commonly known as sour orange, has a significant geographical presence [2]. Over time, it was cultivated in regions including Spain, France, North and South Africa, and various other parts of the tropical and temperate world, owing to its robustness compared to other Citrus species [2]. Among the cultivars of *Citrus aurantium*, '*Citrus aurantium* Changshan-huyou' (CACH) stands out. This cultivar is the result of a hybridization process involving *Citrus grandis* and *C. sinensis* [1].

In recent years, there has been a notable increase in citrus production, leading to a greater focus within the fruit processing industries on creating various products such as juices, essential oils, flavorings, antioxidants, and acidifying ingredients for food applications [2]. However, these advancements generate a substantial amount of solid waste from citrus, including peels, seeds, and membrane residue, during the processing stage. This solid waste from citrus is not without value, as it can be recycled and put to various uses, primarily serving as a source of beneficial phytochemicals, vitamins, and minerals that have potential applications in treating various health conditions [2].

Furthermore, the fruits of *Citrus aurantium* (*C. aurantium*) offer more than just essential oils; they also contain compounds of the flavonoid type, which exhibit a range of biological effects. These compounds contribute to the fruit's health-promoting qualities, including antioxidative, antibacterial, antiallergic, anticancer, and antidiabetic properties [1–3]. *C. aurantium* has been utilized in traditional Chinese medicine to address various health issues such as anxiety, cancer (particularly lung and prostate), digestive problems, and obesity. Additionally, its uses extend to serving as a stimulant, appetite suppressant, and remedy for motion sickness, indigestion, and constipation [2, 4]. Moreover, it has been explored for treating cancer and cardiovascular diseases [2]. Notably, the aroma of its essential oils has been harnessed to induce sleep in individuals prone to insomnia [5–7], and these oils also exhibit potent pesticidal properties [3, 8–10].

Beyond its compelling biological significance, *C. aurantium* possesses intriguing non-pharmacological attributes. Its essential oils, for instance, display antimicrobial and antioxidant characteristics, rendering them valuable as natural food preservatives [11]. Furthermore, due to its exceptional qualities, *C. aurantium* serves as a suitable raw material for producing pulp and paper [12]. Industries such as food processing and cosmetics also harness its aromatic nature [13–15]. Moreover, *C. aurantium* has found relevance in skin treatments, particularly as an anti-aging agent [16].

The prominence of *Citrus aurantium* has increased, notably as a replacement for weight-loss products previously utilizing *Ephedra sinica* Stapf. in Farw., as it contains p-synephrine, an alkaloid of the phenylethanolamine type. This compound has hunger-suppressing effects and shares a chemical relationship with adrenergic medications [3]. The protoalkaloid p-synephrine has garnered attention for its use in weight loss and athletic supplements, as well as potential risks associated with its consumption [17–19]. Although certain studies have largely absolved *Citrus aurantium* and p-synephrine of reported risks, safety concerns remain [18, 20–22]. Recent scientific research has extensively explored the potential effects of various parts of *Citrus aurantium*, including its blossoms, fruits, and essential oils [3].

This chapter provides a comprehensive discussion of the phytochemistry, pharmacological and non-pharmacological potentials, safety considerations, and research gaps related to *Citrus aurantium*. The aim is to evaluate the current scientific endeavors focused on toxicological assessments of the plant, while also setting the groundwork for future research initiatives.

2 Overview of the Non-pharmacological Uses of *Citrus aurantium*

2.1 The Role of *Citrus aurantium* in Food Preservation

The interest in natural preservatives has increased tremendously due to the undesirable tendencies of synthetic preservatives [23]. Owing to its antimicrobial and antioxidant properties, the essential oil of *Citrus aurantium* has become an attractive alternative to the use of synthetic preservatives [11]. This section explores the antimicrobial and antioxidant properties of *Citrus aurantium* in a view to appraising its food preservative potentials.

2.1.1 Antibacterial Properties of *Citrus aurantium* and Its Food Preservative Potential

Citrus aurantium essential oil is a terpenoid-rich mixture with antibacterial effect, claim Anwar et al. [11]. The plant's phytochemicals' close interaction with the target cell and subsequent solubilization of the cell membrane give rise to the plant's antibacterial characteristics. In a study conducted by Ben Hsouna et al. [24], it was discovered that *Citrus aurantium* contains the following monoterpenes: 27.5% limonene, 14% terpineol, 4.3% E-ocimene, and 2.4% delta-3-carene. Sesquiterpenes are said to make up 30.2% of the essential oil, with (E)-nerolidol having the highest concentration (17.5%). Several bacteria and fungi, including *Candida albicans*, *Staphylococcus epidermis*, and *Escherichia coli*, have been demonstrated to be suppressed by terpineol [11]. The essential oils of *Citrus aurantium* were investigated for their antibacterial activity against a panel of 13 bacterial strains using agar diffusion and broth microdilution procedures. According to the findings [24], the plant's essential oils exhibit a moderate-to-strong antibacterial activity. The antibacterial efficacy of *Citrus aurantium* essential oil against the two foodborne pathogens *Escherichia coli* and *Salmonella enterica* in apple juices was studied by Friedman et al. [25] in 2004. The capacity of the essential oil to preserve food was demonstrated by their discovery that it was effective against the two bacteria developing in the apple juice. Additionally, *Pseudomonas fluorescens* and *Aeromonas hydrophila*, two bacteria that can ruin fish, were discovered to be resistant to the antibacterial effects of *Citrus aurantium* essential oils. The antibacterial activity was determined by measuring the MIC using the disc diffusion method at two temperatures, including the bacteria's optimal temperature (30 °C or 37 °C) and the refrigerated temperature (4 °C). As a food preservative, the essential oil was demonstrated to be effective at both temperatures against the intended bacteria [26].

2.1.2 Antifungal Properties of *Citrus aurantium* and Its Food Preservative Potential

Many fungi, including *Aspergillus niger, Aspergillus flavus, Aspergillus nidulans, Aspergillus fumigants, Fusarium graminearum, Fusarium oxysporum, Fusarium culmorum,* and *Alternaria alternate,* are known to be responsible for food spoilage [24]. Therefore, an agent that can inhibit the growth of these fungi will do well as a food preservative. Fortunately, the essential oils of *Citrus aurantium* have shown profound antifungal activity by inhibiting the growth zone of the above-mentioned fungi. The antifungal capacity of *Citrus aurantium* has been attributed to the limonene and (E)-nerolidol content of the essential oils [27]. Therefore, the essential oils of *Citrus aurantium,* particularly the limonene and (E)-nerolidol isolates, would serve well as a food preservative agent.

2.1.3 Antioxidant Properties of *Citrus aurantium* and Its Food Preservative Potential

Antioxidants are known to limit food spoilage by preventing the autoxidation of pigments, flavors, lipids, and vitamins [28]. The essential oils of *Citrus aurantium* have shown profound antioxidant properties that can be harnessed as a food preservation [24, 29, 30]. Furthermore, the peel and fruit juice of *Citrus aurantium* have been demonstrated to possess antioxidant properties [31, 32]. The antioxidant properties of *Citrus aurantium* were confirmed via the 2, 2-diphenyl-1-picrylhydrazyl (DPPH) assay, β-carotene bleaching test, total antioxidant capacity (TAC) assay, and thiobarbituric acid reactive substance (TBARS) assay [24, 29–32]. Indeed, the collective antioxidant properties of *Citrus aurantium* suggest its use as a food preservative agent in the food industry.

2.2 The Role of *Citrus aurantium* in Aromatherapy

The non-pharmacological therapeutic use of essential oils derived from aromatic plants is known as aromatherapy. Aromatherapy has been used to treat a variety of ailments, including body aches, nausea, vomiting, anxiety, melancholy, stress, and insomnia [33]. Due to their high volatility, citrus essential oils can alter the body's hemodynamic parameters or blood flow via controlling circulation via the autonomic nervous system [33]. In particular, the inhalation of *Citrus aurantium* has been observed to possess sedative, hypnotic, and anti-anxiety properties [34, 35]. Furthermore, *Citrus aurantium* essential oils have been used to improve sleep quality in type 2 diabetes mellitus patients, pregnant women, and postmenopausal women [5–7]. Given its ability to improve sleep quality in the above-stated conditions, its effects on several other conditions that distort sleep quality in humans can be assessed.

2.3 The Role of *Citrus aurantium* in Pulp and Paper Production

One major drawback in pulp and paper production is the insufficiency of raw materials [12]. Two main attributes distinguish the *Citrus aurantium* tree as a suitable raw material for pulp and paper production viz its fast growth and chemical composition. In particular, the high cellulose and holocellulose content and low lignin content, which is essential for paper production, stands *Citrus aurantium* out in paper production [12]. Therefore, the *Citrus aurantium* tree stands as a good raw material for pulp and paper production.

2.4 The Role of *Citrus aurantium* in Food Processing and Cosmetics Industries

Citrus aurantium has mostly been investigated as a flavoring additive by the food processing sectors. In addition to alcoholic and non-alcoholic beverages, marmalades, ice cream, sweets and candies, soft drinks, gelatins, and cakes, it has also been used to impart aroma to culinary products [14, 15]. As well as being used to make infusions, *Citrus aurantium* blossoms are also utilized to make perfumes and cosmetics thanks to the essential oil they yield [13]. Additionally, *Citrus aurantium* fruit peel is used to make marmalade, and the fruit's juice is added to salads to provide a tart flavor [36]. Finally, *Citrus aurantium* is a component of some South American and Asian recipes, including yoghurt-based side dishes and hot soups with salt and pepper [37].

2.5 The Role of *Citrus aurantium* in Pesticides Production

Various plant parts have demonstrated pesticidal potentials particularly for the control of mosquito larva [38–42]. Furthermore, essential oils and an isolated part of *Citrus aurantium* have been found to have highly potent pesticidal effects. *Citrus aurantium* and limonene, an isolated component, were found to have fumigant activity against *Bemisia tabaci* (silverleaf whiteflies) in an in vitro experiment and to cause insect mortality between 41.00 and 47.67% within 24 h of exposure at 2.5 and 20.0 l/L air concentration [9, 43, 44]. Additionally, it was found that after 24 h of exposure, *Citrus aurantium* essential oil had a larvicidal effect with LC50 and LC90 values of 22.64 and 83.77 mg/L, respectively [8, 45]. The active component was found to be limonene, and *Citrus aurantium* fresh peeled ripe fruit had potent larvicidal effect against the mosquito vector *Anopheles stephensi* at an LC50 value of 31.20 ppm [3]. *Citrus aurantium* essential oils have also been effective in controlling rice weevils and saw-toothed grain insects [10].

3 Phytochemistry and Bioactive Compounds of *Citrus aurantium*

The well-known pharmacological or therapeutic potentials of plants such as antimicrobial, anti-cancer, antiproliferative, hypolipidemic, and cardio-protective effects are associated with the secondary metabolites, or phytochemicals they possess [42, 46–52]. There are a ton of different chemicals in the *C. aurantium* plant. The

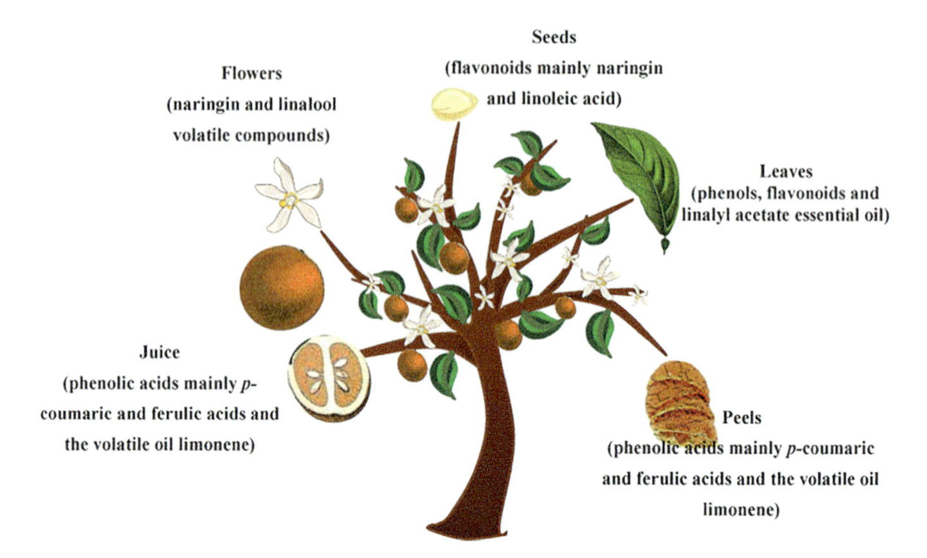

Fig. 1 *Citrus aurantium* with their bioactive compounds. (Source: Maksoud et al. [53] (CC BY 4.0 https://creativecommons.org/licenses/by/4.0/))

primary bioactive substances found in *C. aurantium* are flavonoids, which are further divided into flavanones, flavones, and flavonols. There are also alkaloids such as p-synephrine and other limonoids like limonin and nomilin [53]. Figure 1 depiction of *Citrus aurantium*'s juice, blossoms, seeds, leaves, and peels is followed by a detailed analysis of each part's bioactive chemical concentration. It is important to remember that the location, growing season, and harvesting time all significantly impact the chemical composition or quantity of biomolecules [53].

One of the several chemical components of *C. aurantium*, flavonoids from the phenolic family, have been recognized as significant due to their physiological and pharmacological relevance as well as their health benefits. Flavones, flavanones, flavonols, and anthocyanins are the derivatives of flavonoids commonly found in *C. aurantium* (Fig. 2) [3].

Flavanones are the main flavonoids present in *C. aurantium* and the two most common free flavanones found so far are hesperetin (4-methoxy-3,5,7-tri-hydroxyflavanone) and naringenin (4,5,7-trihydroxyflavanone) [3, 54]. The structural similarities between hesperetin and naringenin include the presence of aglycones or glycosides and two hydroxyl groups at positions C-5 and C-7, respectively [3, 55]. Hesperidin and neohesperdin, which are conjugates with rutinose and neohesperdose, respectively, are the two most common glycosides of hesperetin [3]. Naringenin (naringenin-7-neohesperidoside) and narirutin (naringenin-7-rutinoside) are the two most widely used naringenin glycosyl derivatives [3, 55].

The flavones in their aglycon or/and glycosidic form are the second most prevalent type of flavonoids found in *C. aurantium*. Apigenin, luteolin, and diosmetin are the three free flavones that are most frequently detected in *C. aurantium* [3]. O-glycosides and C-glycosides are the two primary forms of flavone glycosides [54].

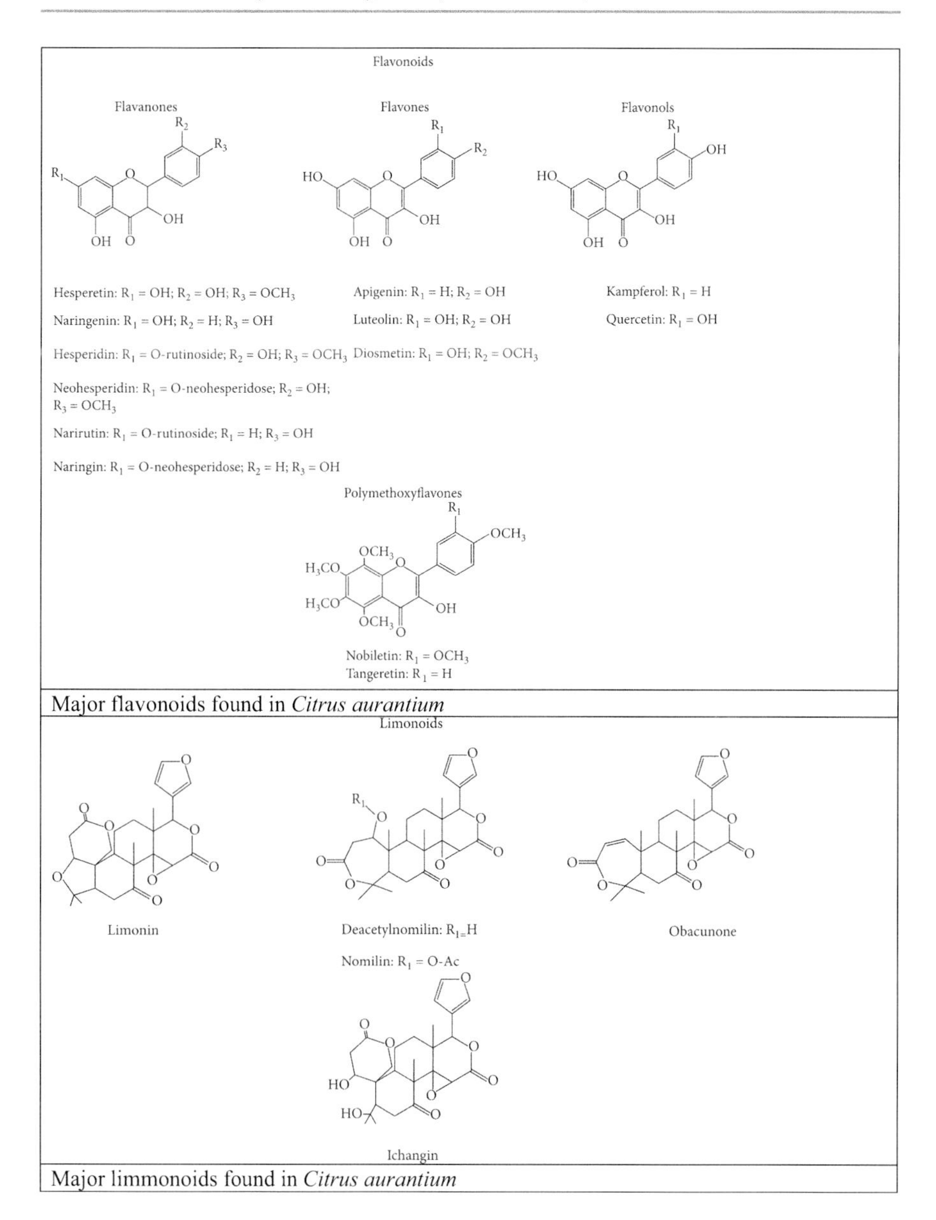

Fig. 2 Structure of major chemical composition of *Citrus aurantium*. (Source: Suntar et al. [3] (CC BY 4.0 https://creativecommons.org/licenses/by/4.0/))

Examples of methoxylated flavones found in *C. aurantium* include nobiletin and tangeretin, meaning that all or almost all their hydroxyls have been capped by methylation [3]. *C. aurantium* may also contain very small quantities of flavonols like quercetin and kaempferol, especially in glycosidic form [3].

The second class of secondary metabolites identified in *C. aurantium* are limonoids (Fig. 2). They are oxygenated triterpenoids because of the exceptionally large quantities of oxygen atoms (7–11) present in their structures [3]. There are glucosidic (bitter taste) and aglyconic (water-insoluble but flavorless) types of limonoids. The most significant limonoid is limonin, which has been recognized as a citrus component [3].

P-synephrine is the chemical that is most common among the phenylethylamine alkaloids found in *C. aurantium* [3]. This substance has a hydroxyl group linked to the para position of the benzene ring and structural similarities with ephedrine [3] (Suntar et al. 2018). The peel of unripe fruits is the part of the plant that has the largest amount of p-synephrine [3].

The chemical constitution of *C. aurantium* is a complex one even though the peel is made up of unique and familiar essential oil. The essential oil of citrus fruit especially *C. aurantium* affords them the strong flavor and smell that have been linked to several medicinal activities. Apart from the essential oil found in the peel of *C. aurantium*, there are also synephrine, flavones, alkaloid, *N*-methyltyramine, and carotenoids, and octopamine. The principal bioactive components of *C. aurantium* are synephrine and *N*-methyltyramine with 0.24–1.45% (g/g) and 0.19–0.83% (g/g) respectively [35].

Other phytoconstituents of *C. aurantium* include neohesperidin, noradrenaline, tyramine quinoline, rhoifoli, 5-odesmethyl nobiletin, narcotin, tryptamine, lonicerin, nobiletin, and naringin.

The essential oil constituent of *C. aurantium* peels contains majorly limonene (about 90%) and then vitamin C, coumarins, triterpenes, pectin, carotene, and flavonoids. The flavonoids constituents are notable for their efficacy against fungi, inflammation, and bacterial infections. There is however variation in the essential oil constitution in the leaves, peel, and flowers of *C. aurantium*. For instance, the essential oil of *C. aurantium* leaves is dominated by linalyl acetate which is about 50% composition while linalool (constituting 35% of the essential oil) is the major makeup of the flowers.

Unripe fruits of *C. aurantium* contains cirantin (which is generally believed to be a contraceptive) including: acetic-acid, decylpelargonate, alphahumulene, gamma-elemene, decanal, alphapinene, alpha-terpineol, geranyl acetate, alpha-terpinyl-acetate, alphaylangene, 5-hydroxyauranetin, ascorbic-acid, isolimonic acid, palmitic acid, aurapten, isoscutellarein, beta-elemene, valencene, betaocimene, beta-pinene, cadinene, benzoic acid, decylaldehyde, geranic acid, caryophyllene, delta-cadinene, dodecen-2-al-(1), dipentene, 4-terpineol, hesperidin, caryophyllene, isotetramethylether, neryl acetate, caprinaldehyde, sabinene, llinalylacetate, nonanol, stachydrine, limonin, nootkatone, tannic acid, octyl acetate, terpinen-4-ol, naringenin, p-cymol, cryptoxanthin, furfurol, tetra-o-methyl-scutellarein, acetaldehyde, pectin, farnesol, linalyl acetate, transhexen-2-al-1, pentanol, umbelliferone, phenol, nerolidol, pyrrol, delta-3-carene, lauric aldehyde, citral, rhoifolin, naringin, aurantiamene, d-citronellic acid, alpha-phellandrene, hexanol, d-limonene, geraniol, citronellic acid, gamma-terpinene, isosinensetin, citronellol, (+)-auraptenal, dodecanal, alphapinene, butanol, dl-linalool, nomilin, methanol, cinnamic acid, l-stachydrine, sinensetin, nonylaldehyde, linalool, tangeretin, octanol, malic acid, terpenyl acetate, citronellal, nerol, pcymene, terpinolene, linalool, cis-ocimene, dlterpineol, thymol, pelargonic acid, cis-ocimene, geranyl oxide, trans-ocimene, phellandrene, carvone, nobiletin, undecanal, phenylacetic acid, duodecylaldehyde, mannose, violaxanthin, limonene, pyrrole, delta-3-carene, neral, d-nerolidol, l-linalool, camphene, gamma-terpinene, indole, formic acid, beta-copaene, alpha-ionone, ethanol [35] (Table 1).

As previously discussed, the phytochemical constituents of *C. aurantium* are majorly the glycosides, flavonoids, carbohydrates, alkaloids, tannins, saponins, terpenoids, steroids, and phenolic compounds [75]. Phytochemical analysis of the leaf extract of *C. aurantium* revealed tannins, phytosterols, terpenoids, proteins, saponins, flavonoids, essential oils, and carbohydrates [58]. Several authors have also reported their findings on the phytochemical constituents of various parts of *C. aurantium*. Variations in percentage of bioactive compounds and chemical composition in their reports have been linked to the period of harvest, type of solvent, extraction method, geographical area, or growing season. For instance, the phytochemical screening by Ishaq et al. [75] was in tandem with that of Rauf et al. [76] where phenols, flavonoids, steroids, and tannins were reported but contravenes the study by Gunwantrao et al. [77] where no terpenoids and flavonoids were reported in ethanol peel extract of *C. aurantium*.

In a recent study by Babajide et al. [78] whereby different solvents (n-hexane, ethanol, and aqueous) were used for extraction, the leaf extract of *C. aurantium* indicated the presence of terpenoid, flavonoid, saponin, tannin, and glycoside but for saponin that was not present in the n-hexane extract. This report was also in agreement with the result from the works of Khudhair et al. [79] where saponins, flavonoids, and tannins were reported in the aqueous leaf extract of *C. auratium* and that of Rao et al. [80] reported the absence of absence of phlobatannins and saponins but the confirmation of flavonoids, tannins, and terpenoids in the ethanol and aqueous extracts of *C. aurantium*. Also, confirmation of flavonoids in this study agrees with the report of He et al. [81] and that of Khudhair et al. [79] where the seeds and peel extracts of *C. auratium* revealed abundant cardiac glycosides, alkaloids, steroids, tannins, saponins, and flavonoids.

Because of the versatility of *C. aurantium*, the phytochemical constituents have been linked to its biological and physiological activities against different ailments. For instance, phenolic compounds in *C. aurantium* have been reported from several studies to have inhibitory potential [82]. In the same way, the flavonoid constituent of *C. aurantium* has been linked to antioxidant, antiproliferative, anti-inflammatory, antimicrobial, anti-diabetic, anticarcinogenic, cardio-protective, and anticholesterolemic activities in promoting the general well-being in humans [35, 83]. The main components in *C. aurantium* seeds have been reported to be flavonoid (56%) at maturity and phenolic acids constituent is 22% [68]. The efficacy of *C. aurantium* as an antioxidant agent is associated with their natural capacity in quenching free radicals and disrupting the radical chains [83].

The extraction of bioactive agents from *C. aurantium* is also very important. As previously said, various factors could be responsible for different chemical components and percentage of the active agents obtained. By and large, extraction of

Table 1 Bioactive compounds and extraction methods of *C. aurantium* plant

S/N	Plant part	Compounds	Extraction means	Class of compound	Solvent	Reference
1	Leaves	α-terpineol, limonene, linalyl acetate, and linalool,	–	Essential oil	–	[56, 57]
		O-cymene, eucalyptol, 4-terpineol, β-pinene, α-pinene, D-limonene, α-terpineol, sabinene, β-linalool, and β-myrcene	Hydro-distillation	Essential oil	–	[58]
		Total phenols, flavonoids, flavonols, proantho-cyanidins, polymerized phenols, hydrolyzable tannins, and soluble phenols	Solvent extraction	Phenolic compounds	Methanol: water (80: 20)	[59]
2	Peels	Sesquiterpenes, monoterpenes	–	Essential oil		[60, 61]
		n-nonadecanoic, abinene, cyclohexene, α-pinene, n-decanal, octyl aldehyde, myrcene, dimethoxybenzylidene, α-terpineol, δ-guaiene, limonene, n-octanol, β-fenchyl alcohol, hexasiloxane, β-pinene, γ-gurjunene, cyclotrisiloxane, octyl aldehyde, and thyocynic acid	Microwave steam distillation	Essential oil	Aqueous	[62]
		Synephrine, N-methylthyramine, octopamine, and carotenoids	–	Flavones and alkaloids	–	[60, 61]
		Geranyl acetate, β-pinene, limonene, linalyl acetate, myrcene, geranial, sabinene, linalool, β-caryophyllene, γ-terpinene, and neral	Peel rinds squeezing	Monoterpene hydrocarbons and oxygenated monoterpenes	n-hexane	
		p-coumaric and ferulic acids	Hydro-distillation	Phenolic compounds	Diethyl ether	[31]

		Galacturonic acid	Microwave-assisted extraction	Pectin	Citric acid aqueous solution	[63, 64]
		Limonene, α-terpinene, octanal, camphor, α-terpinyl acetate, and 1, 8-cineole	Hydro-distillation	–	–	[65]
3	Seeds	Hesperidin, naringin, neohesperidin, and narirutin	Soxhlet extraction	Flavonoids	Acetone and petroleum ether	[66, 67]
		Hesperidin, rutin, naringin, naphtorecinol, quercetin, resorcinol, epigallocatechin, kaempherol, apigenin, neohesperidin, and catechin	Solvent extraction	Flavonoids	Methanol	[68]
		p-coumaric acid, syringic acid, rosmarinic acid, vanillic acid, gallic acid, and trans-2-hydroxycinnamic acid	Solvent extraction	Phenolic acids	Methanol	[68]
		Vanillic acid – Hydroxybenzoic acids; and caffeic acid, *p*-coumaric acid, trans-ferulic acid - Hydroxycinnamic acids	Solvent extraction	Phenolic acid	Methanol	[69]
		Free stigmasterol, esterified β-sitosterol, free β-sitosterol, and campesterol	Soxhlet extraction	Phytosterols	n-hexane	[66]
		Arachidic acid, stearic acid, linoleic acid, cerotic acid, oleic acid, and palmitic acid, and palmitoleic acid	Soxhlet extraction	Fatty acid	n-hexane	[66]
4	Flowers	Rutin, pyrogallol, caffeic acid, gallic acid, quercetin, syringic acid, and naringin	Solvent extraction	Phenolic and flavonoids	Ethanol, methanol, and aqueous	[70]

(continued)

Table 1 (continued)

S/N	Plant part	Compounds	Extraction means	Class of compound	Solvent	Reference
		Hotrienol, linalool, nerol, and terpineol	Soxhlet extraction	Oxygenated monoterpenes	Ethanol	[71]
		Tetrapentacontane and dotriacontan	Soxhlet extraction	Aliphatic hydrocarbons	Ethanol	[71]
		β-pinene, β-myrcene limonene, and β-ocimene	Soxhlet extraction	Monoterpene hydrocarbons	Ethanol	[71]
		Neryl acetate and linalyl acetate	Soxhlet extraction	Esters	Ethanol	[71]
		Hexadecanoic acid, linalool, D-glucuronic acid, daphnetin, octadecenoic acid, D-limonene, phthalic acid, and pyrrolidinone	Solvent extraction	–	80% ethanol	[72]
		Farnesol, linalool, geranyl acetate, linalool acetate, and nerolidol	Microwave-assisted hydro-distillation, Microwave-assisted hydro-distillation solvent microwave extraction and solvent-free microwave extraction	–	Aqueous	[73]
5	Fruit	Synephrine, neohesperidin, and naringin	Ultrasound-assisted aqueous two-phase extraction	Flavonoid	Ethanol	[74]
6	Juice	Limonene, α-thujene, and α-phellandrene	Solvent extraction	Monoterpene hydrocarbons	Ether-pentane (1:1)	[31]
		Caryophyllene oxide	Solvent extraction	Oxygenated sesquiterpenes	Ether-pentane (1:1)	[31]
		Ferulic acids and *p*-coumaric	Solvent extraction	Phenolic acids	Ether-pentane (1:1)	[31]
		Quercetin and rutin	Solvent extraction	Flavonoid	Ether-pentane (1:1)	[31]

bioactive compounds from different parts of *C. aurantium* or their wastes has also been of benefits to the agro-industrial activity while the metabolites serve beneficial roles in various sectors like pharmaceutical, food, cosmetic, beverage, and other industries [53]. Since no single extraction method is regarded as the principal reference point for extraction of active compounds, the following are the established and unconventional extraction means on *C. aurantium* plant parts: Soxhlet extraction, microwave-assisted extraction, supercritical fluid extraction, solvent extraction, hydro-distillation, and ultrasound-assisted extraction [53].

Specifically, bioactive molecules in *C. aurantium* parts like leaves, juice, flowers, seeds, and peels are presented in Table 1.

4 Scientific Evidence-Based Therapeutic Potentials of *Citrus aurantium*

C. aurantium like other species of the citrus family has been attributed to have therapeutic properties in deterring several life-threatening diseases [84] which in recent times has been in high demand due to the shift in the use of natural products for therapeutic purposes as opposed to conventional orthodox medicines. From the early days of human civilization, citrus fruits have been used not only as food but also as medicine, according to Gao et al. [1], there are documented use of Citrus in ancient traditional Chinese medicine hence the need to extensively study and utilize its potential in modern medicine, its use has been reported in Japan and Korea [53] and in Africa [85]. Several species in the Citrus genus like *C. aurantifolia, C. limon, C. reticulata,* and *C. paradisi* are favored for use as medicinal constituents are of their chemical constituents, pleasant aromas, and availability. *C. aurantium* unlike other species of Citrus has just been recognized for its therapeutic properties, the chemical constituents of *C. aurantium* are known to have several benefits to human health, and more benefits have been discovered in recent times.

The ethnomedical uses of *C. aurantium* and CACH are of interest to the scientific community. It promotes the regular, in-depth exploration of the various pharmacological routes that control CACH. Currently known extracts and isolated compounds exhibit a range of pharmacological actions, including anti-inflammatory, antioxidant, anticancer, hypolipidemic, and organ protection. The distinct pharmacological action is explained in this section and then summarized in the following list (Fig. 3).

4.1 Antibacterial Activity

Extracts and essential oils from citrus fruits are known to be natural antimicrobials particularly bacteria used in the food industry as well as in cosmetic industry [86]. Several authors have reported the antibacterial properties of *C. aurantium* extracts against certain species of bacteria which are mostly biostatic for the first 2 days and then biocidal after 7–8 days. Aladekoyi et al. [86] reported a 0.25 mm maximum zone of inhibition against *Pseudomonas aeruginosa* but showed no zone of inhibition against

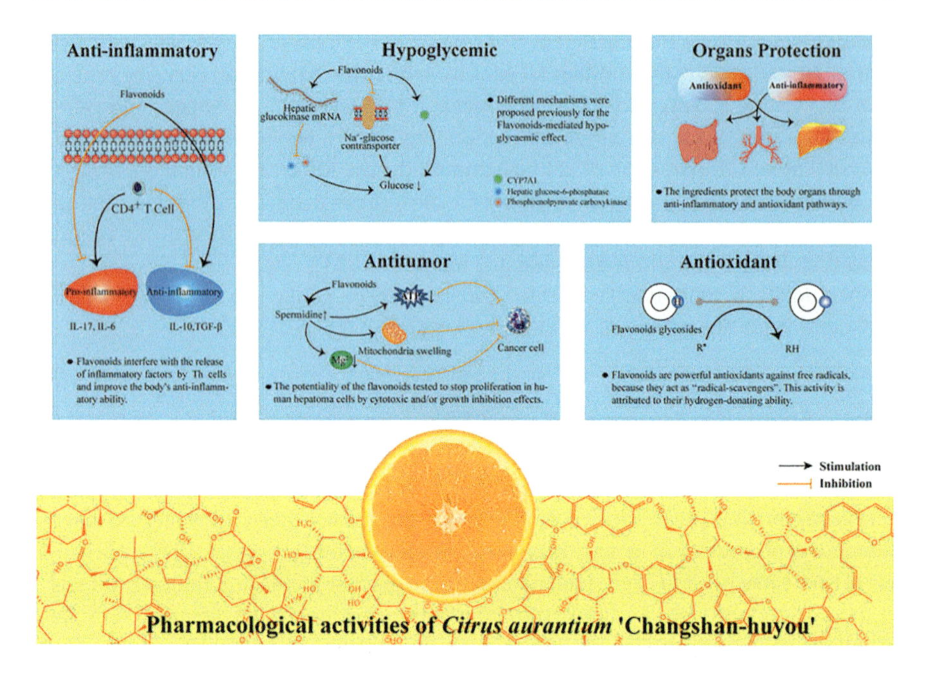

Fig. 3 Distinct therapeutic potentials of *Citrus aurantium* and CACH. (Source: Gao et al. [1] (CC BY 4.0 https://creativecommons.org/licenses/by/4.0/))

Escherichia coli and *Klebsiella* from an antimicrobial susceptibility assay done using oil extracted from the seeds of *C. aurantium*. Essential oils extracted from *C. aurantium* showed a high inhibition zone from 8 to 18 mm when used as an antibiofilm inhibitor against *Bacillus subtilis* and *Stenotrophomonas maltaphilia* [87]. In a study done by Benzaid et al. [88], it was observed that essential oil from the flowers of *C. aurantium* showed a reduction in the growth of *Streptococcus mutans* and the expression of its virulent genes. According to Okla et al. [89], when essential oils from the bark, leaves, and branches of *C. aurantium* were tested against *Erwinia amylovora, Dikeya solani,* and *Agrobacterium tumefaciens*, the notable diameters of inhibition zone came with the increase in the measure of the oil used from the initial 10–25 μl. Karabiyikli et al. [90] studied the effect of *C. aurantium* juice on *Listeria monocytogenes* and *Salmonella typhimurium* and it showed these microbes were able to survive for 2 days in pH neutralized juice and after a seven-day incubation period no growth was found suggesting that the low pH, length of incubation and temperature are enabling factors for the antimicrobial property of *C. aurantium*. Gopal [91] reported significant inhibition zones against Gram-positive *Bacillus subtilis* and *Staphylococcus* and Gram-negative *Klebsiella pneumonia* and *Escherichia coli* using extracts from *C. aurantium* leaves. A study done in Algeria by Haraoui et al. [92] on the antibacterial profile of Citrus plants showed that *C. aurantium* showed a higher antimicrobial activity comparable to *C. limon*; from their observations *C. aurantium* contains certain phytochemicals and the sensitivity and the tolerance level of some bacteria made them more prone to being denatured by these chemicals.

Organisms tested against *C. aurantium* juice that showed high antibacterial activity in their study include Gram positive: *Staphylococcus aureus, Bacillus subtilis, Staphylococcus epidermidis, Micrococcus luteus*, and *Enterococcus faecalis* and Gram negative: *Klebsiella, Pseudomonas aeruginosa,* and *Escherichia coli.* According to Teneva et al. [93] from their study on the antimicrobial activity of essential oils, the antimicrobial activity of the oils is based on their chemical composition; bioactive alcohols like linalool and carveol enhance the antimicrobial activities. From their investigations, Gram-negative bacteria were more resistant to the essential oils because of their outer cell membrane which obstructs the dispersion of the oils to the cytoplasm of the cell through the cell membrane showing zones of inhibition of 9–10 mm while the Gram-positive bacteria were more sensitive to the essential oils with a measured zone of inhibition of 12.5 mm.

4.2 Antifungal Properties

Citrus oils have been harnessed for use in dermatological formulations for topical applications on the skin and scalp to prevent and cure diseases due to their perceived antifungal properties. Several research have shown the efficacy of *C. aurantifolia* but not much has been done for *C. aurantium* over time; citrus oil was studied by Clavaud et al. [94] against the fungal species Malassezia furfur in vitro, it was observed that the oil had a biostatic effect on the organism with zone of exhibition of 2.6 mm, since most *Citrus* spp. has similar phytochemicals though in varying proportions it is assumed that the oil from *C. aurantium* may also have antifungal properties. Kačániová et al. [87] however reported using *C. aurantium* oil in an antibiofilm study resulting in high zones of inhibition against *Penicilium crustosum, P. expansum,* and *P. citrinum.* Ellouze et al. [95] observed a moderate effect of *C. aurantium* leaves essential oil on *Saccharomyces cerevisiae* and *Mucor ramannianus* with the 9.2 and 5 mg/ml minimum inhibition concentration respectively when compared to its antibacterial properties.

4.3 Antioxidant Properties

Antioxidants are stable molecules that donate electrons to free radicals and neutralize them and reducing their capacity to damage cells in the body [96]. According to Maksoud et al. [53], *C. aurantium* is a great source of antioxidants, it has a high level of flavonoid among other phytochemicals [1] whose hydrogen donating ability makes it a strong antioxidant against free radicals. The antioxidant properties of *C. aurantium* may vary according to the plant part used and the antioxidant levels may be dependent on the extraction method. From research by Shi et al. [97], they observed that the dry and immature fruits of *C. aurantium* contain a high flavonoid content of 76.22% and 4 types of flavonoids were observed namely: hesperidin, naringin, narirutin, and neohesperidin. Bendaha et al. [98] worked on the antioxidant activity of essential oil from the peels of *C. aurantium* fruit grown in Eastern

Morocco using 2,2-Diphenyl-1-picrylhydrazyl (DPPH) assay and observed that the essential oil did not show any substantial antioxidant activities as the DPPH capacity ranged from 7–15% displaying a weak DPPH scavenging capability. The essential oil was found to consist of myrcene and more than 90% limonene; myrcene and limonene are not considered to play a major in determining the scavenging activity. Teneva et al. [93] studied the antioxidant activity of *C. aurantium* peel oils using DPPH assay and results show that limonene, β-myrcene, and α-pinene were the main components resulting in 85.22%, 4.3%, and 1.29% respectively. When compared to other antioxidants *C. aurantium* L essential oil showed a higher antioxidant activity at 88.1% in the DPPH assay and this can be explained by the different composition of *C. aurantium* species.

4.4 Anticancer/Cytotoxic Activity

Treatment for cancers through conventional medications can be effective but with many side effects which may lead to damage to other body cells as well as the targeted cancer cells. To this end, new research aimed at using diet for cancer treatment has been on the rise in recent times; by regulating the metabolic niche of the cancer cells and using antitumor (specific active food molecules) that can target the metabolic pathways of the cancer cells [1]. This is aimed at disrupting the increase and spread of the cancer cells after identifying the pathogenesis of the tumor cells. Bian et al. [99] have reported the use of the Chinese herbal drug Weifuchen tablet which contains *Fructus aurantii*; the dried unripe fruits of *C. aurantium* in combination with Red Gingeng (Renshen) and Isoden amethystoides for relief from precancerous lesions in people with gastric cancer through the regulation of intestinal microbial balance. This implies that *C. aurantium* is a good herbal alternative for cancer treatment.

Yao et al. [100] experimented on the use of the active ingredients contained in *C. aurantium* in the treatment of lung cancer, the study which was validated by both molecular and cellular experiments proved that nobiletin in *C. aurantium* inhibits the development of non-small cell lung cancer (NSCLC) as a key ingredient through targets and other related pathways. Though further clinical studies are needed, it was established that nobiletin inhibited the growth of NSCLC and the expression of certain cancer cells was affected.

According to Gao et al. [1], neohesperidin a type of flavonoid has a negative growth impact on human hepatoma cells and *Citrus aurantium* "Changshan-huyou" (CACH) also has anti-tumor activity by intervening with the cell cycle of the cancer cells and impeding their growth.

Shen et al. [101] evaluated the extracts of crude polysaccharides of *C. aurantium* L. var. Amara Engls (CAVAPs) and showed that they enhance the immune system. When RAW264.7 cells were exposed to CAVAPs, the CAVAPS stimulated the production of tumor necrosis factor-α (TNF-α) and interleukin-1β. Further evaluation showed that CAVAPs may possibly activate macrophage signaling pathways through mitogen-activated protein kinase (MAPK) and nuclear factor kappa B (NF-kB).

Research carried out by Han et al. [102] on the apoptotic activities of the peels of *C. aurantium* (CME) on U937 human leukemia cells showed that CME inhibited the growth of U937 cells. CME-induced apoptosis was dose-dependent and triggered by curbing Bcl-2 and IAPs related to the activation of Akt through an intrinsic pathway. Akt is an upstream signal that controls Bcl-2 and IAPs and regulates cell multiplication and programmed death and the study confirmed that CME suppressed Akt phosphorylation, and this is associated with apoptosis.

4.5 Effects on Cardiovascular Activity

Most research on *C. aurantium* are said to be beneficial but some research have looked at the possible toxicity of it. Like other Citrus species, *C. aurantium* contains the alkaloid p-synephrine; known for its anti-obesity properties and treatment of digestive problems because of its fat-decreasing properties in traditional medicines, however according to Arbo et al. [103] it has been flagged as a toxic alkaloid potential risk to the human cardiovascular system by the National Collegiate Athletic Association (NCAA) hence the need for in-depth research. Arbo et al. [103] observed that *C. aurantium* extracts and p-synephrine showed no significant effect on the cardiovascular systems of mice and suggested that the toxicity may arise from the combination of p-synephrine with other stimulants such as caffeine, salicin, and amphetamines. This is in tandem with the observations of other authors including [104] who reported that the association of high doses of p-synephrine together with caffeine and ephedrine caused death due to cardiac complications such as hypertension, strokes, arrhythmias, and heart attack. Hansen et al. [45] also observed that the use of p-synephrine either in its pure state or in *C. aurantium* extract on test animals showed no significant increase in blood pressure and heart rate even at doses much higher than that used in humans. Test animals treated with 95% p-synephrine showed lower increases in blood pressure and heart rate when compared with animals treated with *C. aurantium* extract which suggests that other components of the extracts can be attributed for these changes, there were also significant increases in the cardiovascular parameters when caffeine given together with p-synephrine implicating the addition of the caffeine for the increased heart rate and blood pressure. The reports of the negative effects of p-synephrine have not been confirmed in other studies.

Stohs [105] and Penzak [106] in their independent studies both observed that on its own p-synephrine has low toxicity but when in combination with other components like caffeine, salicin and ephedrine in products can induce cardiovascular effects. Kang et al. [107] (2016) studied the effects of neroli oil on the cardiovascular system as a vasodilator in mice (neroli oil extracted from the flowers of *C. aurantium* var. amara which is rich in limonene and linalool is used to manage anxiety and stress and also to treat high blood pressure and other cardiovascular symptoms) was studied due to the claims that it causes hypotension and bradycardia in adolescents and observed that in the aorta rings of the mice that are already precontracted with prostaglandin f2α, vasodilation was induced, but for rings precontracted with nitric oxide synthase inhibitor the relaxation effects decreased. This induced muscle relaxation was partially reversed by 1H-[1, 2, 4] oxadiazolo [4,3-a] quinoxalin-1-1, a soluble guanylyl cyclase inhibitor. Also, neroli oil suppressed the effect of extracellular Ca^{2+}-dependent contraction in a dose-dependent manner. Ultimately the research indicated that the relaxation effects of neroli affect both the vascular smooth muscle and the vascular endothelium, but no negative effect was observed.

4.6 Anti-obesity Activity

C. aurantium is considered a sport enhancer and swift weight loss ingredient and as an alternative therapeutic dietary supplement as it contains compounds such as flavonoids and polyphenols both of which can reduce the accumulation of lipids [108]. Stohs et al. [109] in their review of human clinical trials involving the use of *C. aurantium* and its alkaloid p-synephrine observed that from over 20 studies carried out with over 50% of the 360 test subjects were obese. P-synephrine was given in combination with caffeine, guarana, green tea, gingko, ginseng, and yerba mate making it difficult to ascertain if weight loss was as a result of p-synephrine or the other components; be that as it may, it was deduced that *C. aurantium* is the most likely causative agent for weight loss because of its effects on metabolic processes including increase in metabolic rate, appetite suppression, and lipolysis. More so 50% of the subjects were caffeine consumers before the commencement of the studies but were still obese. *C. aurantium* was found to show increase in the metabolic rate and energy expenditure of the subjects and when consumed for up to 12 weeks may result in eventual weight loss, hence it can be consumed as dietary supplement at defined doses.

According to research by Li et al. [110], blossoms of *C. aurantium L* var. amara Engl have been reported to inhibit lipid deposits, to better understand the constituents from the blossom extracts responsible for the bioactivity and the mechanism for the lipid breakdown were studied. Using chloroform to harvest the extracts, 14 compounds were identified and nobiletin, trigonelline hydrochloride, and 7-demethysuberosin were more. The chloroform extracts significantly decreased lipid accumulation and improved the plasma biochemical profile in high-fat diet-fed mice by enhancing the gut microbiota. At higher doses, the chloroform extracts increased the abundance of Lachnospiraceae while at lower doses increased the abundance of Erysipelotrichaceae and reduced the ratio of Firmicutes to Bacteroidetes.

Park et al. [108] investigated the anti-obesity mechanism of *C. aurantium* by administering it in high-fat diet-induced obese mice for 8 weeks. Eighty percent ethanol extracts of immature dried *C. aurantium* fruits were analyzed and found to contain naringin 20.1% and neohesperidin 14.4% and other phytochemicals in lesser quantities. The effect of *C. aurantium* on promoting adipogenesis and thermogenesis is dependent on activated protein kinase alpha (AMPKα) activation (the energy sensor and regulator of metabolic homeostasis). From their observations, there were elevated phosphorylation levels of AMPKα and ACC in the 3T3-L1 adipocytes and culture brown adipocytes due to the *C. aurantium*. The anti-obese effects of *C. aurantium* were dependent on the activation of the AMPKα pathway due to the elimination of the anti-adipogenic and pro-thermogenic effects of the *C. aurantium* impeding the AMPKα pathway through the introduction of cell culture pre-treatment. The result was a significant decrease of body weight, serum cholesterol, and adipose tissue weight in the mice suggesting that *C. aurantium* has the potential of being used as an anti-obese agent as it inhibits white adipogenesis and induces brown adipocyte thermogenesis through AMPKα activation.

Verpeut et al. [111] worked on the anti-obesity effects of *C. aurantium* (6% synephrine) and *Rhodiola rosea* L. (3% rosavins and 1% salidroside) in combination on diet-induced obese rats. From their investigations, *C. aurantium* or *R. rosea* alone did not suppress appetite in normal weight rats, but *C. auratium* and *R. rosea* in combination when administered registered a 10.5% appetite suppression by elevating the hypothalamic norepinephrine and the frontal cortex dopamine. It can however be deduced that the combination of *C. aurantium* and *R. rosea* are effective in altering appetite suppression and obesity because of the interaction between the two extracts and their bioactive components.

4.7 Hypoglycemic Activity

According to Gao et al. [1], the use of *C. aurantium* in treating and preventing diabetes and hyperlipidemia is receiving widespread because its mechanism has not been fully understood yet. The assumption is that the hypoglycemic effects of *C. aurantium* are achieved by improving lipid metabolism, regulation of the intestinal flora, and the increased consumption of glucose. Jia et al. [112] researched on the use of neohesperidin isolated from *C. aurantium* in in vivo treatment of hypoglycemia and hypolipidemia in KK-ay diabetic mice. The results showed a significant reduction in fasting glucose, serum glucose, total cholesterol, triglycerides, leptin level, and glycosylated serum protein in the treatment mice when compared to the control at the end of the experiment. At the end of the six-week period Neohesperidin treatment showed a high potential as a hypoglycemic agent in vivo.

From the investigations of Zhang et al. [113], naringin and neohesperidin isolated from *C. aurantium* var. *changshanensis* showed significant effects on glucose consumption by Hep G2 cells through the process of increased phosphorylation of AMPK expression in the cells. Neohesperidin and naringin showed enhancement of glucose consumption in a manner like metformin. Campbell et al. [114] investigated the lipid-lowering effects of an infusion of *Rawolfia vomitoria* leaves and *C. aurantium* fruits, a Nigerian anti-diabetic infusion consumed locally and attributed to decrease blood glucose level when consumed with the patient expected to stick to a low carbohydrate and fat diet with strict no alcohol consumption. The study was conducted on 6–11-week-old lean mice including inbred and outbred mice. There was no significance in their appearance but there was a transitory effect on the motor activity of the mice on the first day of administration. After 6 weeks, there was an observed significant weight loss when the dosage corresponding to ten-fold human daily dosage was administered to genetic diabetic mice which had been maintained on a carbohydrate-deficient diet. The control group did not have a significantly different diet from the treatment group; however, the treatment group had a significantly higher serum triglyceride level proposing that the lipid might be from internal stores. The treatment mice also showed a reduced stearoyl CoA desaturase activity.

4.8 Anti-inflammatory Activity

C. aurantium like other citrus fruits is highly rich in flavonoids and other bioactive natural products. Examples of citrus flavonoids include narirutin, neohesperidine, and naringin and they display strong biological activities like anti-inflammation and are likely therapeutic choices for treatment of metabolic dysregulation [115]. Wang et al. [116], from their research on the fruits of *C. paradisi*; a hybrid of *C. aurantium* as an anti-inflammatory for the inhibition of asthma allergens observed that the dry immature of *C. aurantium* contains narirutin, neohesperidin, and naringin which are the primary active components, and they exhibit powerful inflammatory activities and inhibit airway remodeling in asthma allergens by regulating the Smad2/3 and MAPKs signaling pathways. The overexpression of inflammatory factors is related to the symptoms of diseases like chronic gastritis and peptic ulcer and *C. aurantium* can be used in treating them [1].

Jiang et al. [115] worked on the anti-inflammatory effects of total flavonoids extracted from *C. aurantium* hybrid QZK (TFCH) on non-alcoholic fatty liver disease in vivo and it was observed that TFCH suppresses both systemic and intrahepatic inflammation and attenuates hepatic lesions in obesity-induced rats. The effects were marked by the suppression of NF-Kb and MAPKs. From this result, it can be deduced that the extracted flavonoid component in TFCH is the main activator of anti-non-alcoholic fatty liver disease property of the extract and is a new source for the development of therapeutic medicine for the treatment of anti-non-alcoholic fatty liver disease.

Pallavi et al. [84], from their investigations on the anti-inflammatory and anti-nociceptive properties of the peels of Citrus fruits common in India, observed that all citrus fruits contain naturally high levels of phytochemicals like flavonoids, steroids, terpenoids, glycosides, carotenoids, phenolic compounds, and alkaloids; and showed that *C. medica* fruit peel extract exhibited significant anti-inflammatory and analgesic property in a dose-dependent manner which is similar to the standard indomethacin drug, this was followed by *C. reticulate*, *C. aurantium*, *C. aurantifolia,* and lastly *C. grandis* and generally the anti-inflammatory property was more evident in the later phases of time interval.

CACH can be used to treat both chronic gastritis and peptic ulcers, which are predominantly brought on by the overexpression of inflammatory factors. Jiang et al. [115] on the in vivo anti-inflammatory properties of TFCH showed that TFCH at doses of 50 to 100 mg kg^{-1} significantly lowered TNF-α, IL-6, IL-1, and other inflammatory

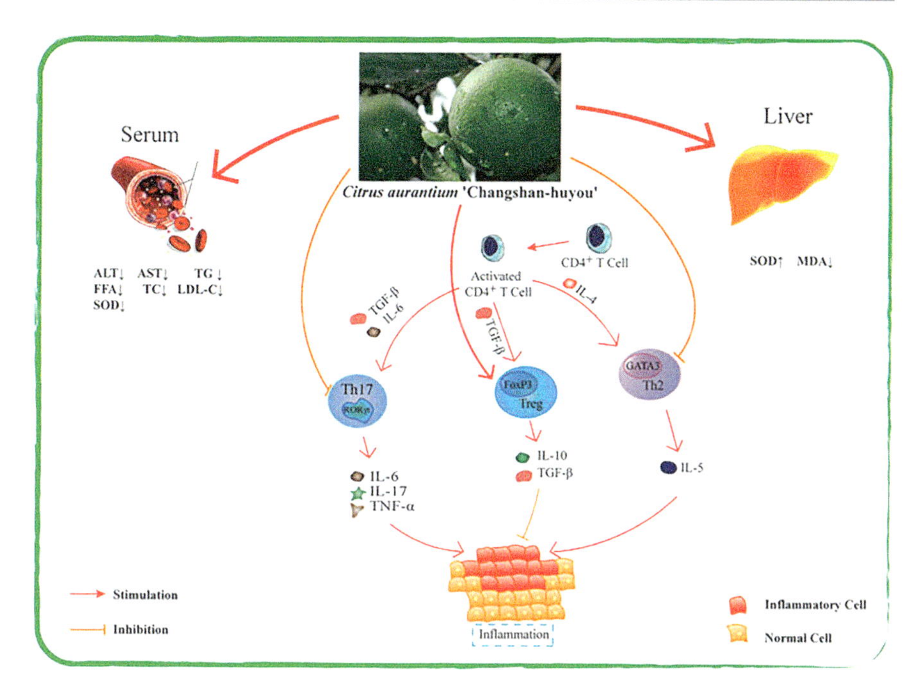

Fig. 4 Anti-inflammatory mechanism of CACH. (Source: Gao et al. [1] (CC BY 4.0 https://creativecommons.org/licenses/by/4.0/))

factors. According to the study, NF-B/MAPK is the main mechanism via which TFCH manifests its anti-inflammatory characteristics [1]. Figure 4 illustrates the most recent research on the anti-inflammatory mechanism of CACH based on the findings [1].

4.9 Hepato-Protective Activity

Jiang et al. [115] stated the protective properties of flavonoids found in *C. aurantium* extracts using the disease model of non-alcohol fatty liver disease (NAFLD) as discussed earlier. According to Yu et al. [117] the pathological features of NAFLD are simple fatty liver, hepatic sclerosis, and steatohepatitis and in some cases liver cancer, the microenvironment of the liver has been implicated for the development and rapid deterioration of NAFLD when it is inflamed and the oxidative stress. To prevent and treat liver diseases using *C. aurantium* is achieved through antioxidant, anti-inflammatory, and regulation of the intestinal microflora. Shi et al. [97] examined and evaluated the effects of TFCH on non-alcoholic steatohepatitis (NASH), a liver disorder involving hepatocellular injury over the course of NAFLD, using dried immature fruits of the *C. aurantium* cultivar known as TFCH. Using the antioxidant response element route of nuclear factor erythroid 2-related factor 2 (Nrf2), the NASH model was assessed in vivo and in vitro for 26 weeks and 8 weeks, respectively. The CCK-8 experiment demonstrated the anti-apoptotic effects of TFCH on

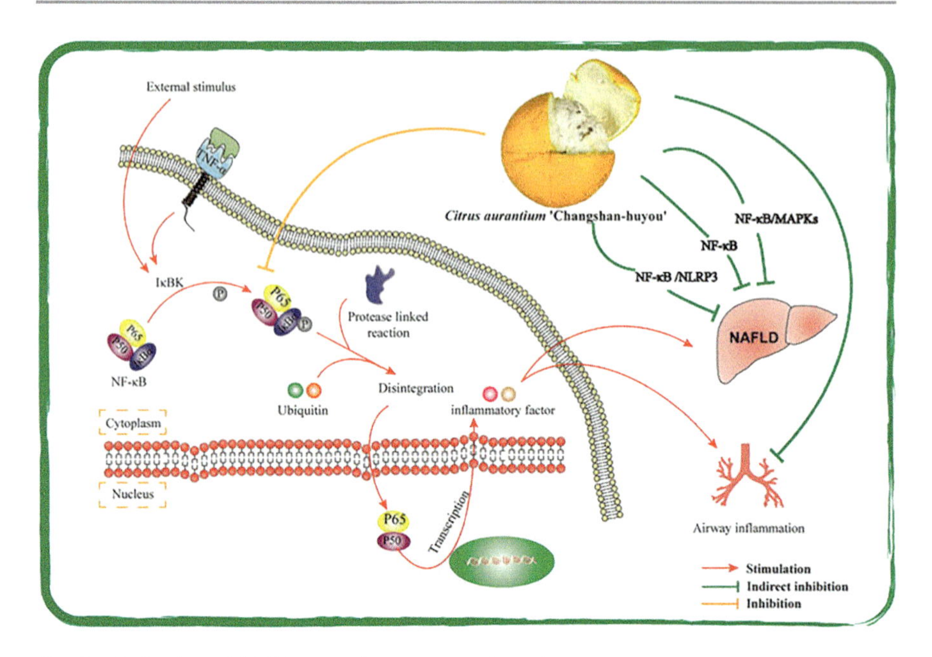

Fig. 5 Mechanism of CACH protection on organs through NF-κB pathway. (Source: Gao et al. [1] (CC BY 4.0 https://creativecommons.org/licenses/by/4.0/))

LX-2 cells grown with FFA. In vivo and in vitro measurements of lipid parameters and oxidative stress markers revealed that TFCH lowers oxidative damage in the NASH model and has a dose-dependent effect on antioxidant capacity. Analysis of Nrf2 and target gene protein expression in mouse liver and human LX-2 cells revealed that TFCH up-regulates these genes' protein expression and significantly alters the Nrf2-ARE signaling pathway. This suggests that TFCH has hepatoprotective and anti-oxidative effects on NASH and that TFCH therapy may eventually offer a unique therapeutic alternative for the treatment of NASH.

For instance, extensive research has been done on NF-B, a crucial protein in the pathways regulating liver inflammation in NAFLD. Jiang et al. [115] (2019) used male SD rats to develop an HFD-induced NAFLD model and isolated and produced TFCH. By limiting the phosphorylation of IB to stop the breakdown of NF-B and so preventing the creation and release of inflammatory components, TFCH reduces the inflammatory milieu in the liver. The results were also corroborated by the research's most successful drug group (polyene phosphatidylcholine capsule group, 196.3 mg/kg) (Fig. 5) [1].

4.10 Intestinal Adjustment

Drug usage should be regulated through treatments in the intestinal milieu since unfavorable drug reactions frequently occur in the intestines [1]. According to Chen et al. [118], consideration must be given to NSAID side effects that cause small intestine injury to take NSAIDs safely. Pure total flavonoids (PTFC) from the *C. aurantium* var. have been used as a form of treatment for digestive disorders, while the benefits and mechanisms of action are yet unknown. To show the protective advantages of PTFC, the study investigated the potential role of autophagy in the process of diclofenac and MSAIDs caused small intestinal injury. In rats with NSAID-induced small intestinal injury and IEC-6 cells treated with diclofenac, the expression levels of the tight junction (TJ) proteins ZO-1, claudin-1, and occludin light chain 3 (LC3)-II, as well as autophagy-related 5 (Atg5), were shown to be lower than in the control group. In the intestinal disease-induced rat group, p-PI3K and p-Akt expression levels increased in comparison to the control group but decreased in the PTFC group. Pretreatment with PTFC preserved the intestinal mucosal barriers and decreased NSAID-induced small intestine damage by increasing autophagy via the PI3K/Akt signaling pathway.

He et al. [119] analyzed the gut microbiota composition in NAFLD mice by pure total flavonoids from citrus (PTFC) treated high-fat diet using PTFC and the outcome of this was that PTFC adjusted the gut microbial structure, the abundance of the intestinal flora and improved the proliferation of Bacteriodeae, Akkermansia, and Christensenellaceae indicating that the advancement of NAFLD is linked to the gut microbiota and PTFC can modulate this target to therapeutic role.

4.11 The Role of *Citrus aurantium* in Skin Treatment

Linalool, linalyl acetate, and terpineol have been shown to have anti-aging properties when present in isolated form [120]. The antioxidant properties of many skin care products are linked to their anti-aging benefits, and these ingredients have been shown to have alluring antioxidant properties. Additionally, exposure to sunshine makes aging-related enzymes like collagenase and elastase work harder. These enzymes particularly degrade the collagen and elastin in the skin, resulting in wrinkles and skin sagging that are associated with getting older [121, 122]. However, *Citrus aurantium* essential oils have been demonstrated to have anti-aging properties since they inhibit the activity of the enzymes collagenase and elastase and are rich in linalool, linalyl acetate, and terpineol [16].

5 Safety Considerations from the Use of *Citrus aurantium* and Its Products

5.1 *Citrus aurantium* and Cardiovascular Toxicity

As asserted earlier, *Citrus aurantium* came to the limelight following the drawback encountered with the use of *Ephedra sinica* as a weight loss and athletic performance supplement, and its subsequent ban by the Food and Drug Administration (FDA) in April 2004 [123–127]. The suggestion of *Citrus aurantium* as a replacement for

Ephedra sinica is due to the structural similarities its main phytochemical, p-synephrine, shares with the Ephedra isolate, ephedrine. Ephedrine was observed to be associated with myocardial infarction, hypertension, and stroke [124, 128]. p-synephrine is a protoalkaloid extracted from the immature fruit or peel of *Citrus aurantium* with widespread use as a weight loss and athletic performance supplement [18]. It is an isomer of synephrine with a hydroxyl group in the para position, while other isomers have a hydroxyl group in the meta and ortho positions even though it is widely accepted that p-synephrine is safer than ephedrine, its toxicity index and range of safety have been controversial [126, 129].

5.2 The Controversies

In 2008, a reported case of a young man who presented with a myocardial infarction after the use of a synephrine-containing substance raised a huge medical concern [19]. The authors of the report hypothesized that myocardial infarction might have been due to coronary spasm followed by thrombosis which developed due to synephrine tablets abuse [19]. Aside the toxicity adduced to synephrine, the authors called the attention of readers to another study that asserted that *Citrus aurantium* contains a number of flavonoids, including 6′,7′-dihydroxybergamottin and furocoumarin, which selectively block the intestinal cytochrome P450 enzyme, CYP3A4, in bioavailability study [124, 130]. Consequently, increasing the serum level of the drugs metabolized by CYP3A4, and resulting in undesirable pharmacological interactions [124]. Indeed, several articles have further discussed the cardiovascular toxicity of *Citrus aurantium* and its product [106, 130–144]. More intriguingly, in 2010, *Citrus aurantium* was listed in the Customer's report as being possibly unsafe and a dietary supplement "to avoid" [145]. However, in many cases of reported adverse effects, the individuals were either exercising or had recently exercised [126, 135, 139, 142]. On the other hand, Stohs et al. [18] dissented from many toxicity reports for a variety of reasons, such as the use of complex mixtures of ingredients in products containing bitter orange extract, the disseminating of false information by governmental organizations, the misunderstandings surrounding the isomeric forms of synephrine and their various pharmacological properties, and the misrepresentations of p-synephrine, particularly its mistaken identity with other substances.

5.3 The Hansen et al. and Dr. Stohs' Debate on *Citrus aurantium* Safety: A Major Controversy

In view of the avalanche of articles attributing major hazardous potentials to *Citrus aurantium* and its product, Hansen et al. [17] intended to investigate the cardiovascular toxicity potentials of *Citrus aurantium* in exercised rats. Sprague-Dawley rats were administered a daily dose of 10 or 50 mg of synephrine/kg body weight (in both extract and pure form) for a total of 28 days, either alone or in combination with

25 mg/kg body weight of caffeine. Additionally, the rats were made to run for 30 min a day, 3 days a week, on a treadmill. Additionally, a group of rats was made to run on a treadmill for the same period while receiving a caffeine dose of 25 mg/kg body weight [17]. The body temperatures were carefully monitored, QT intervals, blood pressures, and heart rates of the rats. It was demonstrated that both dosages of synephrine considerably increased systolic and diastolic blood pressure for up to 8 h following treatment, even though the effects of caffeine seemed to have the largest impact on heart rate and body temperature. The scientists concluded that exercise, coffee, and synephrine all significantly affect blood pressure and heart rate. However, in a Letter to the Editor of the Journal of Cardiovascular Toxicology, Stohs [22] questioned the procedures and outcomes of the trial carried out by Hansen et al. [17]. The authors, according to Stohs [22], did not sufficiently discuss the clinical importance and relevance of *Citrus aurantium* and p-synephrine for humans, did not review the most recent scientific research on these two substances, and did not accurately relate their doses in rats to typically used doses in humans. Regarding the dose, Stohs [22] argued that the 50 mg/kg body weight of synephrine administered to the rats – equivalent to 8.1 mg/kg body weight in a man and roughly 649 mg for a typical 80 kg person – was simply too high for an adverse effect to go undetected [17, 22]. Stohs [22] noted that the authors' dose was 16 times the typical human dose as a result. Even yet, Stohs [22] noted that the trial's increases in systolic and diastolic blood pressure, which were 9.7% and 9.0%, respectively, were insufficient to cause concern. Hansen et al. [17] argued in defense of their methodology by saying that they used 10 mg/kg body weight of synephrine rather than 50 mg/kg body weight. According to the authors, the maximum dose of synephrine that can be given to a person is 1.62 mg/kg body weight. The authors reported that at this dose, systolic and diastolic blood pressure in the rats increased statistically significantly and remained for up to 8 h, indicating a pathogenic situation. Furthermore, Stohs [22] identified certain works that appear to defend synephrine, although Hansen et al. [17] claimed that they were not aware of such. *Citrus aurantium* caused some degree of cardiovascular damage in strenuous-exercising rats, according to the scientists' findings [17].

5.4 Other Toxicity Considerations

In an experiment by Deshmukh et al. [21] to assess the acute and 14-day oral toxicity of *Citrus aurantium* in rats, the LD50 of the plant extract standardized to 50% p-synephrine exceeded 5000 mg/kg body weight in female rats. Additionally, it was reported that no toxicity occurred when the plant extract was given orally to female rats for 4 days straight at a dose of 2000 mg/kg body weight [21]. A 14-day repeated dose of 50% p-synephrine-containing *Citrus aurantium* extract at the doses of 250, 500, 1000, and 2000 mg/kg/day had no appreciable impact on the body weights, relative and absolute organ weights, clinical chemistry and hematological parameters, gross pathological findings, or food intake of the experimental animals [21]. Two male rats with gastrointestinal impaction had perished after taking 2000 mg/kg, it was determined during necropsy. Additionally, during the second week of treatment with 1000 mg/kg and 2000 mg/kg/day for 15 and 45 min, respectively, rats were reported to exhibit brief indications of repeated head-burrowing in the bedding material (hypoactivity) [21]. Despite the short-term and transient clinical symptoms that were noticed at the dose, it was decided that the maximum tolerated dose (MTD) of *Citrus aurantium* extract standardized to 50% p-synephrine was 1000 mg/kg/day [21]. 500 mg/kg body weight was the no-observed-effect-level (NOEL) for *Citrus aurantium* extract in the study. The researchers contended that *Citrus aurantium* extract had a sufficiently large margin of safety to satisfy any serious concerns. *Citrus aurantium* was tested for safety in rats over a sub-chronic 90-day period by Deshmukh et al. [20], who discovered that doses of 100, 300, and 1000 mg/kg/day of the 50% p-synephrine-containing extract exhibited no unfavorable clinical effects. Furthermore, the extract's NOEL was 300 mg/kg, and its no-observed-adverse-effect-level (NOAEL) was 1000 mg/kg, according to the authors' assertions [20]. The authors highlighted that *Citrus aurantium* extract is quite safe as a result.

Despite having useful characteristics, *Citrus aurantium* essential oil has been demonstrated to contain a small amount of toxicity [146]. *Citrus aurantium* essential oil was neither an irritant nor a sensitizer to 25 volunteers when tested at 10%, but it did produce sensitivity in 1.5% of all dermatitis patients when tested at 2% [147–149]. *Citrus aurantium* essential oil has also been linked to an increase in light sensitivity, especially in those with fair skin. When an essential oil is eaten, photo-sensitivity occurs more frequently than when it is administered topically to the skin and exposed to bright light [35]. As a result, it is advisable to avoid staying out in the sun for too long after applying the oil.

6 Research Needs for *Citrus aurantium* as a Therapeutics

Different authors have suggested that in order to fully understand the therapeutic potentials of plants it will be important to identify, isolate, and purify the bioactive compounds that make the plant confer such therapeutic potentials [48, 49, 51, 52, 150–156].

To gain a comprehensive understanding of the therapeutic advantages of *Citrus aurantium*, it is imperative to bridge the information gap among the available studies. Various aspects of toxicity, including acute, sub-acute, and chronic toxicities, alongside the pharmacological safety profiles of the plant, necessitate thorough investigation. Before embarking on clinical trials, it is essential to research the short-term and long-term toxic effects of the plant's compounds on animals. To facilitate this endeavor, a comprehensive grasp of the precise chemical composition of these molecules is crucial. This understanding will facilitate a more profound insight into the metabolic processes that give rise to these beneficial compounds. Moreover, an enhanced comprehension of metabolic engineering can significantly improve the synthesis and accumulation of these compounds. *Citrus aurantium* holds significant importance in sustainable biodiversity preservation and utilization for both current society and future generations. The bitter orange plant not only offers novel opportunities to develop innovative medications to address a spectrum of debilitating ailments [157], but it also remains an extensively untapped source for isolating and characterizing specific chemical components.

Studies investigating the bioactivity potential of *Citrus aurantium* have revealed a diverse array of biological effects attributed to its flowers, fruits, essential oils, and phytoconstituents. These effects encompass antimicrobial, antioxidant, cytotoxic, anxiolytic, antidiabetic, anti-obesity, and anti-inflammatory properties [3]. Nevertheless, further comprehensive and prolonged research is imperative to ensure the plant's safety, particularly with regard to understanding interactions between the plant's components and medications, as well as determining appropriate dosage levels. P-synephrine, characterized as a relatively inefficient adrenergic agonist, is not expected to have notable effects on the cardiovascular system. It's worth noting that despite the absence of concrete evidence pointing to potential toxicity or cardiovascular risks, the NCAA has prohibited its use. In contrast, the World Anti-Doping Agency (WADA), an organization collaborating with Olympic athletes, permits its usage [3].

Multiple scientific studies have underscored the distinct advantages that *Citrus aurantium* holds over other citrus species. Comparative analyses have consistently highlighted bigarade leaves, juice, or flowers as possessing the highest levels of total phenolic content (TPC), total flavonoid content (TFC), and antiradical activity. A significant category of bioactive compounds, including phenols, flavonoids, alkaloids, vitamin C, and essential oils, furnishes compelling evidence. These secondary metabolites confer a range of beneficial biological and therapeutic properties, such as antioxidants, antibacterial agents, anti-cancer agents, anti-diabetic agents, anti-obesity agents, and anxiolytics. Moreover, their utility extends to the cosmetic and functional food sectors, where these compounds find diverse applications. While most of these bioactive components are distributed throughout the entirety of the *Citrus aurantium* plant, distinct identity markers for phytochemical content and proportions have been identified in various plant sections. Remarkably, certain byproducts resulting from the processing of *Citrus aurantium*, such as the peels left behind after juice extraction, emerge as significant reservoirs of valuable biomolecules with promising potential applications. This underscores the importance of repurposing residual biomass, traditionally viewed as waste, within the context of the evolving bioeconomy. Literature also outlines the extraction techniques employed to obtain the bioactive constituents from *Citrus aurantium*. However, to comprehensively grasp their mechanisms of action and ascertain safety via interactions with biological models, advanced clinical research mandates the incorporation of further analytical methods to fractionate and purify these active components [53].

The identification criteria for *Citrus aurantium* present ample opportunities for enhancement. Variations in origins and cultivation practices can lead to significant variations in components due to the distinctive nature of Chinese herbals and diverse cultivation methods. The concurrent presence of multiple chemical elements and compounds often results in varying outcomes in pharmacological activity tests. Because of these conflicting factors, certain extracts utilized in research fail to provide comprehensive details about their extraction procedures or processing conditions. Moreover, some studies have not successfully established accurate identification of *Citrus aurantium* or precise descriptions of extraction sites. These issues contribute to a decline in the overall quality of studies, thereby diminishing their utility as credible references.

Citrus aurantium has yielded over a hundred types of chemicals, many of which exhibit bioactivity. While investigations verifying pharmacological activity have mainly concentrated on specific constituents, such as flavonoids from *Citrus aurantium*'s peel, limited attention has been given to other bioactive components like limonins, organic acids, and additional phenols. Consequently, more reliable, and substantiated data are essential to support the concept of *Citrus aurantium's* "multi-target, comprehensive intervention." Further clarification and comprehensive exploration of the bioactive constituents and pharmacological processes of *Citrus aurantium* are imperative. Notably, researchers frequently prioritize outcomes and overlook the comprehensive mechanisms through which the active components in *Citrus aurantium* contribute to disease prevention and treatment.

The future of medical research on *Citrus aurantium* will revolve around developing activity screening models rooted in pharmacological actions, unearthing novel bioactive ingredients, and dissecting its pharmacological mechanisms. Key areas of focus include establishing robust sources and diversity identification criteria for *Citrus aurantium*, incorporating innovative activity screening models into pharmacological evaluations, and leveraging advanced technologies such as digital light processing and 3D printing to align research with clinical application criteria. Furthermore, despite traditional uses and preliminary evidence from animal models suggesting *Citrus aurantium*'s potential in treating lung and bronchial issues, the current findings lack solid scientific substantiation. The importance of accumulating toxicological and pharmacokinetic data cannot be overstated, as it lays the groundwork for clinical research and the development of *Citrus aurantium*-related products [1].

Citrus peel, leaves, and blossoms harbor essential oils of remarkable value for various reasons. Citrus fruits represent one of the most cultivated fruit crops globally, resulting in the ready availability of these oils as byproducts of citrus processing. The prized attributes of citrus essential oils, which encompass therapeutic benefits related to scent, along with fungicidal, virucidal, and bactericidal activities, render them valuable constituents in cosmetics, fragrances, and potentially eco-friendly substitutes for chemical-based antimicrobials in diverse applications. These characteristics have the potential to drive "green consumerism," promoting the utilization and cultivation of plant products. However, achieving standardization in commercial citrus essential oil composition and conducting further studies to unravel the mechanisms of action of active chemicals remain vital endeavors [158]. There remains ample room for refining their utilization and application.

7 Conclusion

The *Citrus aurantium*, often called bitter or sour orange, is necessary for various utilizations, both in the pharmacological and non-pharmacological world. People have, throughout history, looked to nature as a source of health and well-being for themselves. *Citrus aurantium* is a plant that stands out among traditionally beneficial in increasing physical fitness. Numerous phytochemical components that may be found in *Citrus aurantium* have been shown to have close relationships to the fruit's wide variety of beneficial biological effects, such as the treatment of cancer, gastrointestinal issues, inability to sleep, headaches, and spasticity. *Citrus aurantium* is a very intriguing compound because, in addition to its significant role in the field of pharmacology, it also has a wide variety of applications outside of the pharmaceutical industry. It is used as a pesticide, a source of raw materials for the pulp and paper business, and aromatic properties for the food preservation and processing and cosmetics industries, and it has the potential to work as an anti-aging agent. *Citrus aurantium* and the products derived from it have an incredible number of benefits. Yet, the usage of *Citrus aurantium* has also been connected to a variety of adverse side effects. Even though specific investigations have mostly exonerated the plant of any safety infractions, other research has questioned the matter.

References

1. Gao L, Zhang H, Yuan CH, Zeng LH, Xiang Z, Song JF, Wang HG, Jiang JP (2022) *Citrus aurantium* 'Changshan-huyou' – an ethnopharmacological and phytochemical review. Front Pharmacol 13:1–16. (CC BY 4.0)
2. Benayad O, Bouhrim M, Tiji S et al (2021) Phytochemical profile, α-glucosidase, and α-amylase inhibition potential and toxicity evaluation of extracts from *Citrus aurantium* (L) peel, a valuable by-product from Northeastern Morocco. Biomol Ther. https://doi.org/10.3390/biom11111555
3. Suntar I, Khan H, Patel S, Celano R, Rastrelli L (2018) An overview on *Citrus aurantium* L.: its functions as food ingredient and therapeutic agent. Oxidative Med Cell Longev. https://doi.org/10.1155/2018/7864269. (CC BY 4.0)
4. Stohs S, Shara M (2007) A review of the safety and efficacy of *Citrus aurantium* in weight management. In: Bagchi D, Preuss HG (eds) Obesity: epidemiology, pathophysiology, and prevention. CRC Press, Boca Raton, pp 371–382
5. Abbaspoor Z, Siahposh A, Javadifar N, Faal SS, Mohaghegh Z, Sharifipour F (2022) The effect of *Citrus aurantium* aroma on the sleep quality in postmenopausal women: a randomized controlled trial. IJCBNM 10:86–95
6. Abdollahi F, Mohaddes Ardebili F, Najafi Ghezelje T, Hosseini F (2017) The effect of aaromatherapy with bitter orange extract on sleep quality in patient with type 2 diabetic. Complement Med J 7:1851–1861
7. Mohammadi F, Moradi M, Niazi A, Jamali J (2022) The impact of aromatherapy with *Citrus aurantium* essential oil on sleep quality in pregnant women with sleep disorders: a randomized controlled clinical trial. Int J Community Based Nurs Midwifery 10:160–171
8. da Camara CAG, Akhtar Y, Isman MB, Seffrin RC, Born FS (2015) Repellent activity of essential oils from two species of *Citrus* against *Tetranychus urticae* in the laboratory and greenhouse. Crop Prot 74:110–115
9. El-Akhal F, El Ouali LA, Guemmouh R (2015) Larvicidal activity of essential oils of *Citrus sinensis* and *Citrus aurantium* (Rutaceae) cultivated in Morocco against the malaria vector *Anopheles labranchiae* (Diptera: Culicidae). Asian Pac J Trop Dis 5:458–462
10. Barragan Ferrer D, Rimantas Venskutonis P, Talou T, Zebib B, Manuel Barragan Ferrer J, Merah O (2018) Identification and in vitro activity of bioactive compounds extracted from *Tussilago farfara* (L.) plant grown in Lithuania and France. Free Radic Antioxid 8(1):40–47.

11. Anwar S, Ahmed N, Speciale A, Cimino F, Saija A (2015) Bitter orange (*Citrus aurantium* L.) oils. In: Essential oils in food preservation, flavor and safety. https://doi.org/10.1016/B978-0-12-416641-7.00029-8
12. Tutuş A, Çiçekler M, Küçükbey N (2016) Pulp and paper production from bitter orange (*Citrus aurantium* L.) woods with soda-AQ method. Kastamonu Üniversitesi Orman Fakültesi Dergisi. https://doi.org/10.17475/kujff.29775
13. Ellouze I, Debbabi H, Belkacem A, Rekik B (2011) Variation in physicochemical and sensory quality of sour orange (*Citrus aurantium* L.) marmalade from the Cap Bon region in North-East Tunisia. Fruits 66:315–325
14. Karoui IJ, Wannes WA, Marzouk B (2010) Refined corn oil aromatization by *Citrus aurantium* peel essential oil. Ind Crop Prod 32:202–207
15. Nguyen H, Campi EM, Roy Jackson W, Patti AF (2009) Effect of oxidative deterioration on flavour and aroma components of lemon oil. Food Chem 112:388–393
16. Oulebsir C, Mefti-Korteby H, Djazouli ZE, Zebib B, Merah O (2022) Essential oil of *Citrus aurantium* L. leaves: composition, antioxidant activity, elastase and collagenase inhibition. Agronomy. https://doi.org/10.3390/agronomy12061466
17. Hansen DK, George NI, White GE, Abdel-Rahman A, Pellicore LS, Fabricant D (2013) Cardiovascular toxicity of *Citrus aurantium* in exercised rats. Cardiovasc Toxicol 13:208–219
18. Stohs SJ, Preuss HG, Shara M (2011) The safety of *Citrus aurantium* (bitter orange) and its primary protoalkaloid p-synephrine. Phytother Res 25:1421–1428
19. Louw VJ, Louw H (2008) Citrus aurantium – beware of the bitter orange. S Afr Med J = Suid-Afrikaanse tydskrif vir geneeskunde 98:496
20. Deshmukh NS, Stohs SJ, Magar CC, Kale A, Sowmya B (2017) Bitter orange (*Citrus aurantium* L.) extract subchronic 90-day safety study in rats. Toxicol Rep 4:598–613
21. Deshmukh NS, Stohs SJ, Magar CC, Kadam SB (2017) *Citrus aurantium* (bitter orange) extract: safety assessment by acute and 14-day oral toxicity studies in rats and the Ames Test for mutagenicity. Regul Toxicol Pharmacol 90:318–327
22. Stohs SJ (2013) Questionable conclusions in the article "cardiovascular toxicity of citrus aurantium in exercised rats". Cardiovasc Toxicol 13:180–181
23. Teshome E, Forsido SF, Rupasinghe HPV, Olika Keyata E (2022) Potentials of natural preservatives to enhance food safety and shelf life: a review. Sci World J. https://doi.org/10.1155/2022/9901018
24. Ben HA, Hamdi N, Ben HN, Abdelkafi S (2013) Characterization of essential oil from *Citrus aurantium* L. flowers: antimicrobial and antioxidant activities. J Oleo Sci 62:763–772
25. Friedman M, Henika PR, Levin CE, Mandrell RE (2004) Antibacterial activities of plant essential oils and their components against *Escherichia coli* O157:H7 and *Salmonella enterica* in apple juice. J Agric Food Chem 52:6042–6048
26. Iturriaga L, Olabarrieta I, de Marañón IM (2012) Antimicrobial assays of natural extracts and their inhibitory effect against *Listeria innocua* and fish spoilage bacteria, after incorporation into biopolymer edible films. Int J Food Microbiol 158:58–64
27. Kim J, Marshall MR, Wei C, i. (1995) Antibacterial activity of some essential oil components against five foodborne pathogens. J Agric Food Chem 43:2839–2845
28. Bensid A, El Abed N, Houicher A, Regenstein JM, Özogul F (2022) Antioxidant and antimicrobial preservatives: properties, mechanism of action and applications in food–a review. Crit Rev Food Sci Nutr 62:2985–3001
29. Aazza S, Lyoussi B, Miguel MG (2011) Antioxidant and antiacetylcholinesterase activities of some commercial essential oils and their major compounds. Molecules 16:7672–7690
30. Sarrou E, Chatzopoulou P, Dimassi-Theriou K, Therios I (2013) Volatile constituents and antioxidant activity of peel, flowers and leaf oils of *Citrus aurantium* L. growing in Greece. Molecules 18:10639–10647
31. Jabri Karoui I, Marzouk B (2013) Characterization of bioactive compounds in Tunisian bitter orange (*Citrus aurantium* L.) peel and juice and determination of their antioxidant activities. Biomed Res Int. https://doi.org/10.1155/2013/345415

32. Tounsi MS, Wannes WA, Ouerghemmi I, Jegham S, Ben NY, Hamdaoui G, Zemni H, Marzouk B (2011) Juice components and antioxidant capacity of four Tunisian Citrus varieties. J Sci Food Agric 91:142–151
33. Agarwal P, Sebghatollahi Z, Kamal M, Dhyani A, Shrivastava A, Singh KK, Sinha M, Mahato N, Mishra AK, Baek KH (2022) Citrus essential oils in aromatherapy: therapeutic effects and mechanisms. Antioxidants. https://doi.org/10.3390/antiox11122374
34. Shirzadegan R, Gholami M, Hasanvand S, Beiranvand A (2020) The effects of *Citrus aurantium* aroma on anxiety and fatigue in patients with acute myocardial infarction: a two-center, randomized, controlled trial. J Herb Med 21:100326
35. Suryawanshi JAS (2011) An overview of *Citrus aurantium* used in treatment of various diseases. African J Plant Sci 5:390–395
36. Ersus S, Cam M (2007) Determination of organic acids, total phenolic content, and antioxidant capacity of sour *Citrus aurantium* fruits. Chem Nat Compd 43:607–609
37. Zhang C, Bucheli P, Liang X, Lu Y (2007) Citrus flavonoids as functional ingredients and their role in traditional Chinese medicine. Food 1:287–296. Global Science Books
38. Seiyaboh EI, Seiyaboh Z, Izah SC (2020) Environmental control of mosquitoes: a case study of the effect of *Mangifera indica* root-bark extracts (family Anacardiaceae) on the larvae of *Anopheles gambiae*. Ann Ecol Environ Sci 4:33–38
39. Seiyaboh EI, Odubo TC, Izah SC (2020) Larvicidal activity of *Tetrapleura tetraptera* (Schum and Thonn) Taubert (Mimosaceae) extracts against *Anopheles gambiae*. Int J Adv Res Microbiol Immunol 2:20–25
40. Youkparigha FO, Izah SC (2019) Larvicidal efficacy of aqueous extracts of *Zingiber officinale* Roscoe (ginger) against malaria vector, *Anopheles gambiae* (Diptera: Culicidae). Int J Environ Agric Sci 3:1–6
41. Izah SC (2019) Activities of crude, acetone and ethanolic extracts of *Capsicum frutescens* var. Minima fruit against larva of *Anopheles gambiae*. J Environ Treat Techn 7:196–200
42. Ogunro OB, Salawu AO, Alotaibi SS, Albogami SM, Batiha G-S, Waard MD (2022) Quercetin-3-O-β-D-glucopyranoside-rich fraction from *Spondias mombin* leaves halted responses from oxidative stress, neuroinflammation, apoptosis, and lipid peroxidation in the brain of Dichlorvos-treated Wistar rats. Toxics 10:1–22
43. Siskos EP, Konstantopoulou MA, Mazomenos BE, Jervis M (2007) Insecticidal activity of *Citrus aurantium* fruit, leaf, and shoot extracts against adult olive fruit flies (Diptera: Tephritidae). J Econ Entomol 100:1215–1220
44. Yao WR, Wang HY, Wang ST, Sun SL, Zhou J, Luan YY (2011) Assessment of the antibacterial activity and the antidiarrheal function of flavonoids from bayberry fruit. J Agric Food Chem 59:5312–5317
45. Hansen DK, George NI, White GE, Pellicore LS, Abdel-Rahman A, Fabricant D (2012) Physiological effects following administration of *Citrus aurantium* for 28 days in rats. Toxicol Appl Pharmacol 261:236–247
46. Enaregha EB, Chibueze Izah S, Okiriya Q (2021) Antibacterial properties of *Tetrapleura tetraptera* pod against some pathogens. Res Rev Insights 5:1–4
47. Kigigha LT, Uhunmwangho EJ, Izah SC (2018) Assessment of hot water and ethanolic leaf extracts of *Cymbopogon citrates* Stapf (lemon grass) against selected bacteria pathogens. Ann Microbiol Infect Dis 1:4–8
48. Kigigha LT, Selekere RE, Izah SC (2018) Antibacterial and synergistic efficacy of acetone extracts of *Garcinia kola* (Bitter kola) and *Buchholzia coriacea* (Wonderful kola). Original research article. J Basic Pharmacol Toxicol 2:13–17
49. Chibueze Izah S, Aseibai ER (2018) Antibacterial and synergistic activities of methanolic leaves extract of lemon grass (*Cymbopogon citratus*) and rhizomes of ginger (*Zingiber officinale*) against *Escherichia coli*, *Staphylococcus aureus* and *Bacillus subtilis*. Acta Sci Microbiol 1:26–30
50. Chibueze Izah S (2018) Some determinant factors of antimicrobial susceptibility pattern of plant extracts. Res Rev Insights 2:1–4

51. Yakubu MT, Ogunro OB, Ademola AR, Awakan OJ, Oyewo EB, Muhammad NO, Ajiboye TO (2017) Fadogia Agrestis (Schweinf. Ex Hiern) aqueous stem extract: chemical profile and its effects on acetaminophen-induced oxidative stress in male rats. Nigerian J of Biochem and Mol Bio 32(2):120–133
52. Yakubu MT, Ogunro BO, Ademola RA, Awakan JO, Oyewo EB, Muhammad NO, Ajiboye TO (2017) Fadogia agrestis (Schweinf. Ex Hiern) aqueous stem extract: chemical profile and its effects on acetaminophen-induced oxidative stress in male rats. Niger J Biochem Mol Biol 32:120–133
53. Maksoud S, Abdel-Massih RM, Rajha HN, Louka N, Chemat F, Barba FJ, Debs E (2021) *Citrus aurantium* L. Active constituents, biological effects and extraction methods. An updated review. Molecules 26:1–18. (CC BY 4.0)
54. Lee SH, Yumnam S, Hong GE et al (2015) Flavonoids of Korean *Citrus aurantium* L. induce apoptosis via intrinsic pathway in human hepatoblastoma HepG2 cells. Phytother Res 29: 1940–1949
55. Jayaprakasha GK, Mandadi KK, Poulose SM, Jadegoud Y, Nagana Gowda GA, Patil BS (2008) Novel triterpenoid from *Citrus aurantium* L. possesses chemopreventive properties against human colon cancer cells. Bioorg Med Chem 16:5939–5951
56. De Moraes Pultrini A, Almeida Galindo L, Costa M (2002) Anxiolytic and sedative effects of extracts and essential oil from *Citrus aurantium* L. Biol Pharm Bull 25:1629–1633
57. Gholivand MB, Piryaei M, Abolghasemi MM (2013) Analysis of volatile oil composition of *Citrus aurantium* L. by microwave-assisted extraction coupled to headspace solid-phase microextraction with nanoporous based fibers. J Sep Sci 36:872–877
58. Periyanayagam K, Dhanalakshmi S, Karthikeyan V, Jagadeesan M (2013) Phytochemical studies and GC/MS analysis on the isolated essential oil from the leaves of *Citrus aurantium* Linn. J Nat Prod Plant Resour 3:19–23
59. Khettal B, Kadri N, Tighilet K, Adjebli A, Dahmoune F, Maiza-Benabdeslam F (2017) Phenolic compounds from Citrus leaves: antioxidant activity and enzymatic browning inhibition. J Complement Integr Med. https://doi.org/10.1515/jcim-2016-0030
60. Maurya AK, Mohanty S, Pal A, Chanotiya CS, Bawankule DU (2018) The essential oil from *Citrus limetta* Risso peels alleviates skin inflammation: in-vitro and in-vivo study. J Ethnopharmacol 212:86–94
61. Singh B, Singh JP, Kaur A, Yadav MP (2021) Insights into the chemical composition and bioactivities of citrus peel essential oils. Food Res Int 143:110231
62. Kusuma HS, Putra AFP, Mahfud M (2016) Comparison of two isolation methods for essential oils from orange peel (*Citrus auranticum* L) as a growth promoter for fish: microwave steam distillation and conventional steam distillation. J Aquac Res Dev 7:1–5
63. Hosseini SS, Khodaiyan F, Kazemi M, Najari Z (2019) Optimization and characterization of pectin extracted from sour orange peel by ultrasound assisted method. Int J Biol Macromol 125:621–629
64. Hosseini SS, Khodaiyan F, Yarmand MS (2016) Optimization of microwave assisted extraction of pectin from sour orange peel and its physicochemical properties. Carbohydr Polym 140:59–65
65. Burnett CL, Fiume MM, Bergfeld WF et al (2019) Safety assessment of citrus-derived peel oils as used in cosmetics. Int J Toxicol 38:33S–59S
66. Hamedi A, Zarshenas MM, Jamshidzadeh A, Ahmadi S, Heidari R, Pasdran A (2019) *Citrus aurantium* (bitter orange) seed oil: pharmacognostic, anti-inflam- matory, and anti-nociceptive properties. Herb Prod 5:153–164
67. Abou Baker DH, Ibrahim BMM, Hassan NS, Yousuf AF, El Gengaihi S (2020) Exploiting *Citrus aurantium* seeds and their secondary metabolites in the management of Alzheimer disease. Toxicol Rep 7:723–729
68. Moulehi I, Bourgou S, Ourghemmi I, Tounsi MS (2012) Variety and ripening impact on phenolic composition and antioxidant activity of mandarin (*Citrus reticulate Blanco*) and bitter orange (*Citrus aurantium* L.) seeds extracts. Ind Crop Prod 39:74–80

69. Falcinelli B, Famiani F, Paoletti A, D'egidio S, Stagnari F, Galieni A, Benincasa P (2020) Phenolic compounds and antioxidant activity of sprouts from seeds of Citrus species. Agriculture (Switzerland) 10:1–9
70. Karimi E, Oskoueian E, Hendra R, Oskoueian A, Jaafar HZE (2012) Phenolic compounds characterization and biological activities of *Citrus aurantium* bloom. Molecules 17:1203–1218
71. Değirmenci H, Erkurt H (2020) Relationship between volatile components, antimicrobial and antioxidant properties of the essential oil, hydrosol and extracts of *Citrus aurantium* L. flowers. J Infect Public Health 13:58–67
72. Pasandideh S, Arasteh A (2021) Evaluation of antioxidant and inhibitory properties of *Citrus aurantium* L. on the acetylcholinesterase activity and the production of amyloid nano–bio fibrils. Int J Biol Macromol 182:366–372
73. Mohagheghniapour A, Saharkhiz MJ, Golmakani MT, Niakousari M (2018) Variations in chemical compositions of essential oil from sour orange (*Citrus aurantium* L.) blossoms by different isolation methods. Sustain Chem Pharm 10:118–124
74. Yan Y, Zhou H, Wu C, Feng X, Han C, Chen H, Liu Y, Li Y (2021) Ultrasound-assisted aqueous two-phase extraction of synephrine, naringin, and neohesperidin from *Citrus aurantium* L. fruitlets. Prep Biochem Biotechnol 51:780–791
75. Ishaq A, Sani D, Abdullahi S, Jatau I (2022) In vitro anticoccidial activity of ethanolic leaf extract of *Citrus aurantium* L. against *Eimeria tenella* oocysts. Sokoto J Vet Sci 20:37–43
76. Rauf A, Uddin G, Ali J (2014) Phytochemical analysis and radical scavenging profile of juices of *Citrus sinensis*, *Citrus anrantifolia*, and *Citrus limonum*. Org Med Chem Lett 4:5
77. Gunwantrao BB, Bhausaheb SK, Ramrao BS, Subhash KS (2016) Antimicrobial activity and phytochemical analysis of orange (*Citrus aurantium* L.) and pineapple (*Ananas comosus* (L.) Merr.) peel extract. Ann Phytomed Int J 5:156–160
78. Babajide AB, Adebolu TT, Oladunmoye MK, Oladejo BO (2023) Evaluation of antibacterial activity of *Citrus aurantium* L. leaf extracts on bacteria isolated from blood of hepatitis B positive individuals in Ondo State, Nigeria. Microbes Infect Dis 4:304–311
79. Khudhair AMAAA, Minnat T, Jalyl O (2017) Phytochemical analysis and inhibitory effect of *Citrus aurantium* L. (bitter orange) leaves on some bacterial isolates in vitro. Diyala J Pure Sci 13:115–126
80. Rao NB, Kumari OS, Gajula RG (2016) Anti microbial, anti-oxidant & phytochemical analysis of *Citrus aurantium* (orange) leaf extract. IJRDO-J Biol Sci 2:15–23
81. He XG, Lian LZ, Lin LZ, Bernart MW (1997) High-performance liquid chromatography-electrospray mass spectrometry in phytochemical analysis of sour orange (*Citrus aurantium* L.). J Chromatogr A 791:127–134
82. Arlette NT, Anangmo N, Nadia C, Gertrude MT, Stephanie MT, Pone W (2019) The in vitro anticoccidial activity of aqueous and ethanolic extracts of *Ageratum conyzoide* sand *Vernonia amygdalina* (Asteraceae). World J Pharm Pharm Sci 8:38–49
83. Mannucci C, Calapai F, Cardia L, Inferrera G, D'Arena G, Di Pietro M, Navarra M, Gangemi S, Ventura Spagnolo E, Calapai G (2018) Clinical pharmacology of *Citrus aurantium* and *Citrus sinensis* for the treatment of anxiety. Evid Based Complement Alternat Med. https://doi.org/10.1155/2018/3624094
84. Malleshappa P, Kumaran RC, Venkatarangaiah K, Parveen S (2018) Peels of Citrus fruits: a potential source of anti-inflammatory and anti-nociceptive agents. Pharm J 10:S172–S178
85. Jain S, Arora P, Popli H (2020) A comprehensive review on *Citrus aurantifolia* essential oil: its phytochemistry and pharmacological aspects. Braz J Nat Sci 3:354
86. Aladekoyi G, Omosulis V, OrungbemiO O (2016) Evaluation of antimicrobial activity of oil extracted from three different citrus seeds (*Citrus limon*, *Citrus aurantifolia* and *Citrus aurantium*). Int J Sci Res Eng Stud 3:16–20
87. Kačániová M, Terentjeva M, Galovičová L et al (2020) Biological activity and antibiofilm molecular profile of *Citrus aurantium* essential oil and its application in a food model. Molecules 25:1–21

88. Benzaid C, Belmadani A, Tichati L, Djeribi R, Rouabhia M (2021) Effect of *Citrus aurantium* L. Essential oil on streptococcus mutans growth, biofilm formation and virulent genes expression. Antibiotics 10:1–13
89. Okla MK, Alamri SA, Salem MZM, Ali HM, Behiry SI, Nasser RA, Alaraidh IA, Al-Ghtani SM, Soufan W (2019) Yield, phytochemical constituents, and antibacterial activity of essential oils from the leaves/twigs, branches, branch wood, and branch bark of sour orange (*Citrus aurantium* L.). Processes. https://doi.org/10.3390/pr7060363
90. Karabiyikli Ş, Değirmenci H, Karapinar M (2014) Inhibitory effect of sour orange (*Citrus aurantium*) juice on salmonella typhimurium and listeria monocytogenes. LWT 55:421–425
91. Gopal PV (2012) Evaluation of anti-microbial activity of *Citrus aurantium* against some gram positive and negative bacterial strains. Pharmacia 1:107–109
92. Haraoui N, Allem R, Chaouche TM, Belouazni A (2020) In-vitro antioxidant and antimicrobial activities of some varieties citrus grown in Algeria. Adv Tradit Med 20:23–34
93. Teneva D, Denkova-Kostova R, Goranov B, Hristova-Ivanova Y, Slavchev A, Denkova Z, Kostov G (2019) Chemical composition, antioxidant activity and antimicrobial activity of essential oil from *Citrus aurantium* L zest against some pathogenic microorganisms. Z Naturforsch C J Biosci 74:105–111
94. Clavaud C, Jourdain R, Bar-Hen A et al (2013) Dandruff is associated with disequilibrium in the proportion of the major bacterial and fungal populations colonizing the scalp. PLoS One. https://doi.org/10.1371/journal.pone.0058203
95. Ellouze I, Abderrabba M, Sabaou N, Mathieu F, Lebrihi A, Bouajila J (2012) Season's variation impact on *Citrus aurantium* leaves essential oil: chemical composition and biological activities. J Food Sci. https://doi.org/10.1111/j.1750-3841.2012.02846.x
96. Lobo V, Patil A, Phatak A, Chandra N (2010) Free radicals, antioxidants and functional foods: impact on human health. Pharmacogn Rev 4:118–126
97. Shi Z, Li T, Liu Y, Cai T, Yao W, Jiang J, He Y, Shan L (2020) Hepatoprotective and anti-oxidative effects of total flavonoids from Qu Zhi Qiao (fruit of Citrus Paradisi cv. Changshanhuyou) on nonalcoholic steatohepatitis in vivo and in vitro through Nrf2-ARE signaling pathway. Front Pharmacol 11:1–13
98. Bendaha H, Bouchal B, El Mounsi I, Salhi A, Berrabeh M, El Bellaoui M, Mimouni M (2016) Chemical composition, antioxidant, antibacterial and antifungal activities of peel essential oils of *Citrus aurantium* grown in Eastern Morocco. Pharm Lett 8:239–245
99. Bian Y, Chen X, Cao H et al (2021) A correlational study of Weifuchun and its clinical effect on intestinal flora in precancerous lesions of gastric cancer. Chin Med (United Kingdom) 16:1–13
100. Yao L, Zhang X, Huang C, Cai Y, Wan C (2023) The effect of *Citrus aurantium* on non-small-cell lung cancer: a research based on network and experimental pharmacology. Biomed Res Int. https://doi.org/10.1155/2023/6407588
101. Shen C-Y, Yang L, Jiang J-G, Zheng C-Y, Zhu W (2017) Immune enhancement effects and extraction optimization of polysaccharides from *Citrus aurantium* L. var. amara Engl. Food Funct 8:796–807
102. Han MH, Lee WS, Lu JN, Kim G, Jung JM, Ryu CH, Kim GIY, Hwang HJ, Kwon TK, Choi YH (2012) *Citrus aurantium* L. exhibits apoptotic effects on U937 human leukemia cells partly through inhibition of Akt. Int J Oncol 40:2090–2096
103. Arbo MD, Schmitt GC, Limberger MF, Charão MF, Moro ÂM, Ribeiro GL, Dallegrave E, Garcia SC, Leal MB, Limberger RP (2009) Subchronic toxicity of *Citrus aurantium* L. (Rutaceae) extract and p-synephrine in mice. Regul Toxicol Pharmacol 54:114–117
104. Schmitt GC, Arbo MD, Lorensi AL, Maciel ÉS, Krahn CL, Mariotti KC, Dallegrave E, Leal MB, Limberger RP (2012) Toxicological effects of a mixture used in weight loss products: P-synephrine associated with ephedrine, salicin, and caffeine. Int J Toxicol 31:184–191
105. Stohs SJ (2017) Safety, efficacy, and mechanistic studies regarding *Citrus aurantium* (bitter orange) extract and p-synephrine. Phytother Res 31:1463–1474

106. Penzak SR, Jann MW, Cold JA, Hori YY, Desai HD, Gurley BJ (2001) Seville (sour) orange juice: synephrine content and cardiovascular effects in normotensive adults. J Clin Pharmacol 41:1059–1063
107. Kang P, Ryu KH, Lee JM, Kim HK, Seol GH (2016) Endothelium- and smooth muscle-dependent vasodilator effects of *Citrus aurantium* L. var. amara: focus on Ca2+ modulation. Biomed Pharmacother 82:467–471
108. Park J, Kim HL, Jung Y, Ahn KS, Kwak HJ, Um JY (2019) Bitter orange (*Citrus aurantium* linné) improves obesity by regulating adipogenesis and thermogenesis through AMPK activation. Nutrients 11:1–16
109. Stohs SJ, Preuss HG, Shara M (2012) A review of the human clinical studies involving *Citrus aurantium* (bitter orange) extract and its primary protoalkaloid p-synephrine. Int J Med Sci 9: 527–538
110. Li XY, Hao YF, Hao ZX, Jiang JG, Liu Q, Shen Q, Liu L, Yi YK, Shen CY (2021) Inhibitory effect of chloroform extracts from *Citrus aurantium* L. var. amara Engl. on fat accumulation. Phytomedicine 90:153634
111. Verpeut JL, Walters AL, Bello NT (2013) *Citrus aurantium* and *Rhodiola rosea* in combination reduce visceral white adipose tissue and increase hypothalamic norepinephrine in a rat model of diet-induced obesity. Nutr Res 33:503–512
112. Jia S, Hu Y, Zhang W, Zhao X, Chen Y, Sun C, Li X, Chen K (2015) Hypoglycemic and hypolipidemic effects of neohesperidin derived from *Citrus aurantium* L. in diabetic KK-Ay mice. Food Funct 6:878–886
113. Zhang J, Sun C, Yan Y, Chen Q, Luo F, Zhu X, Li X, Chen K (2012) Purification of naringin and neohesperidin from Huyou (*Citrus changshanensis*) fruit and their effects on glucose consumption in human HepG2 cells. Food Chem 135:1471–1478
114. Campbell JIA, Mortensen A, Mølgaard P (2006) Tissue lipid lowering-effect of a traditional Nigerian anti-diabetic infusion of *Rauwolfia vomitoria* foilage and *Citrus aurantium* fruit. J Ethnopharmacol 104:379–386
115. Jiang J, Yan L, Shi Z, Wang L, Shan L, Efferth T (2019) Hepatoprotective and anti-inflammatory effects of total flavonoids of Qu Zhi Ke (peel of Citrus changshan-huyou) on non-alcoholic fatty liver disease in rats via modulation of NF-κB and MAPKs. Phytomedicine. https://doi.org/10.1016/j.phymed.2019.153082
116. Wang J, Li T, Cai H, Jin L, Li R, Shan L, Cai W, Jiang J (2021) Protective effects of total flavonoids from Qu Zhi Qiao (fruit of *Citrus paradisi* cv. *Changshanhuyou*) on OVA-induced allergic airway inflammation and remodeling through MAPKs and Smad2/3 signaling pathway. Biomed Pharmacother. https://doi.org/10.1016/j.biopha.2021.111421
117. Yu L, Hong W, Lu S, Li Y, Guan Y, Weng X, Feng Z (2022) The NLRP3 inflammasome in non-alcoholic fatty liver disease and steatohepatitis: therapeutic targets and treatment. Front Pharmacol 13:1–19
118. Chen S, Jiang J, Chao G, Hong X, Cao H, Zhang S (2021) Pure Total flavonoids from citrus protect against nonsteroidal anti-inflammatory drug-induced small intestine injury by promoting autophagy in vivo and in vitro. Front Pharmacol 12:1–12
119. He B, Jiang J, Shi Z et al (2021) Pure total flavonoids from citrus attenuate non-alcoholic steatohepatitis via regulating the gut microbiota and bile acid metabolism in mice. Biomed Pharmacother 135:111183
120. Merah O, Sayed-ahmad B, Talou T, Saad Z, Cerny M, Grivot S, Evon P, Hijazi A (2020) Biochemical composition of cumin seeds, and biorefining study. Biomol Ther 10:1–18
121. Garg C, Khurana P, Garg M (2017) Molecular mechanisms of skin photoaging and plant inhibitors. Int J Green Pharm 11:S217–S232
122. Zhang S, Duan E (2018) Fighting against skin aging: the way from bench to bedside. Cell Transplant 27:729–738
123. Opinion S (2013) Scientific Opinion on safety evaluation of Ephedra species for use in food. EFSA J 11:1–79

124. Fugh-berman A, Myeres A (2004) *Citrus aurantium*, an ingredient of dietary supplements marketed for weight loss: current status of clinical and basic research. Exp Biol Med (Maywood) 229:698–704
125. Palamar J (2011) How ephedrine escaped regulation in the United States: a historical review of misuse and associated policy. Health Policy 99:1–9
126. Thomas JE, Munir JA, McIntyre PZ, Ferguson MA (2009) STEMI in a 24-year-old man after use of a synephrine-containing dietary supplement: a case report and review of the literature. Tex Heart Inst J 36:586–590
127. Shekelle PG, Hardy ML, Morton SC, Maglione M, Mojica WA, Suttorp MJ, Rhodes SL, Jungvig L, Gagné J (2003) Efficacy and safety of ephedra and ephedrine for weight loss and athletic performance: a meta-analysis. JAMA 289:1537–1545
128. Haller CA, Benowitz NL (2000) Adverse cardiovascular and central nervous system events associated with dietary supplements containing ephedra alkaloids. N Engl J Med 343:1833–1838
129. Ruiz-Moreno C, Del CJ, Giráldez-Costas V, González-García J, Gutiérrez-Hellín J (2021) Effects of p-synephrine during exercise: a brief narrative review. Nutrients 13:1–9
130. Malhotra S, Bailey DG, Paine MF, Watkins PB (2001) Seville orange juice-felodipine interaction: comparison with dilute grapefruit juice and involvement of furocoumarins. Clin Pharmacol Ther 69:14–23
131. Bent S, Padula A, Neuhaus J (2004) Safety and efficacy of citrus aurantium for weight loss. Am J Cardiol 94:1359–1361
132. Bouchard NC, Howland MA, Greller HA, Hoffman RS, Nelson LS (2005) Ischemic stroke associated with use of an ephedra-free dietary supplement containing synephrine. Mayo Clin Proc 80:541–545
133. Firenzuoli F, Gori L, Galapai C (2005) Adverse reaction to an adrenergic herbal extract (*Citrus aurantium*). Phytomedicine 12:247–248
134. Gange CA, Madias C, Felix-Getzik EM, Weintraub AR, Estes NAM (2006) Variant angina associated with bitter orange in a dietary supplement. Mayo Clin Proc 81:545–548
135. Gray S, Woolf AD (2005) Citrus aurantium used for weight loss by an adolescent with anorexia nervosa. J Adolesc Health 37:414–415
136. Haaz S, Fontaine KR, Cutter G, Limdi N, Perumean-Chaney S, Allison DB (2006) Citrus aurantium and synephrine alkaloids in the treatment of overweight and obesity: an update. Obes Rev 7:79–88
137. Holmes RO, Tavee J (2008) Vasospasm and stroke attributable to ephedra-free xenadrine: case report. Mil Med 173:708–710
138. Inchiosa MA (2011) Experience (mostly negative) with the use of sympathomimetic agents for weight loss. J Obes. https://doi.org/10.1155/2011/764584
139. Nasir JM, Durning SJ, Ferguson M, Barold HS, Haigney MC (2004) Exercise-induced syncope associated with QT prolongation and ephedra-free Xenadrine. Mayo Clin Proc 79: 1059–1062
140. Nykamp DL, Fackih MN, Compton AL (2004) Possible association of acute lateral-wall myocardial infarction and bitter orange supplement. Ann Pharmacother 38:812–816
141. Rossato LG, Costa VM, Limberger RP, Bastos M de L, Remião F (2011) Synephrine: from trace concentrations to massive consumption in weight-loss. Food Chem Toxicol 49:8–16
142. Stephensen TA, Sarlay R (2009) Ventricular fibrillation associated with use of synephrine containing dietary supplement. Mil Med 174:1313–1319
143. Stohs SJ (2010) Assessment of the adverse event reports associated with *Citrus aurantium* (bitter orange) from April 2004 to October 2009. J Funct Foods 2:235–238
144. Sultan S, Spector J, Mitchell RM (2006) Ischemic colitis associated with use of a bitter orange – containing dietary weight-loss supplement. Mayo Clin Proc 81:1630–1631
145. Anon (2010) Dangerous supplements: what you don't know about these 12 ingredients could hurt you. Consum Rep 75:16–20
146. Dosoky NS, Setzer WN (2018) Biological activities and safety of *Citrus* spp. Essential oils. Int J Mol Sci 19:1–25

147. Opdyke DLJ (1973) Monographs on fragrance raw materials. Food Cosmet Toxicol 11:493–494
148. Rudzki E, Grzywa Z, Bruo WS (1976) Sensitivity to 35 essential oils. Contact Dermatitisitis 2: 196–200
149. Rudazki E, Grzywa Z, Bruo WS (1976) Sensitivity to 35 essential oils. Contact Dermatitis 2: 196–200
150. Ogunro OB, Oyeyinka BO, Gyebi GA, Batiha GE-S (2023) Nutritional benefits, ethnomedicinal uses, phytochemistry, pharmacological properties and toxicity of *Spondias mombin* Linn: a comprehensive review. J Pharm Pharmacol. https://doi.org/10.1093/JPP/RGAC086
151. Izah SC, Etim NG, Ilerhunmwuwa IA, Ibibo TD, Udumo JJ (2019) Activities of express extracts of costus afer Ker–Gawl. [Family COSTACEAE] against selected bacterial isolates. Int J Pharm Phytopharmacol Res (eIJPPR) 9:39–44
152. Izah SC, Chandel SS, Epidi JO, Devaliya R (2019) Biocontrol of *Anopheles gambiae* larvae using fresh ripe and unripe fruit extracts of *Capsicum frutescens* var. baccatum. Int J Green Pharm 13:5–7
153. Chibueze Izah S, Singh Chandel S, Etim NG, Epidi O, Venkatachalam T, Devaliya R (2019) Potency of unripe and ripe express extracts of long pepper (*Capsicum frutescens* var. baccatum) against some common pathogens. Int J Pharm Phytopharmacol Res 9:56–70
154. Izah SC, Uhunmwangho EJ, Etim NG (2018) Antibacterial and synergistic potency of methanolic leaf extracts of *Vernonia amygdalina* L. and *Ocimum gratissimum* L. J Basic Pharmacol Toxicol 2:8–12
155. Chibueze Izah S (2018) Studies on the synergistic effectiveness of methanolic extract of leaves and roots of *Carica papaya* L. (papaya) against some bacteria pathogens. Int J Complement Altern Med 11:375–378
156. Chibueze Izah S (2018) Synergy of methanolic leave and stem-back extract of *Anacardium occidentale* L. (cashew) against some enteric and superficial bacteria pathogens. MOJ Toxicol 4:209–212
157. Karthikeyan V, Karthikeyan J (2014) *Citrus aurantium* (bitter orange): a review of its traditional uses, phytochemistry and pharmacology. Int J Drug Discov Herb Res 4:766–772
158. Palazzolo E, Armando Laudicina V, Antonietta Germanà M (2013) Current and potential use of citrus essential oils. Curr Org Chem 17:3042–3049

Medicinal Spice, *Aframomum melegueta*: An Overview of the Phytochemical Constituents, Nutritional Characteristics, and Ethnomedicinal Values for Sustainability

7

Matthew Chidozie Ogwu, Afamefuna Dunkwu-Okafor, Ichehoke Austine Omakor, and Sylvester Chibueze Izah

Contents

M. C. Ogwu (✉)
Goodnight, Family Department of Sustainable Development, Appalachian State University, Boone, NC, USA
e-mail: ogwumc@appstate.edu

A. Dunkwu-Okafor
Department of Microbiology, Faculty of Life Sciences, University of Benin, Benin City, Edo State, Nigeria
e-mail: afamefuna.dunkwu-okafor@uniben.edu

I. A. Omakor
Department of Plant Science and Biotechnology, Faculty of Science, Dennis Osadebay University, Anwai, Asaba, Nigeria
e-mail: Omakorichehoke@dou.edu.ng

S. C. Izah
Department of Microbiology, Faculty of Science, Bayelsa Medical University, Yenagoa, Bayelsa State, Nigeria

S. C. Izah et al. (eds.), *Herbal Medicine Phytochemistry*, Reference Series in Phytochemistry, https://doi.org/10.1007/978-3-031-43199-9_72

Abstract

This chapter explores the botanical characteristics, traditional uses, culinary applications, and medicinal properties of the medicinal spice, *Afromomum melegueta* [Roscoe.] K. Schum., Zingiberaceae (Guinea pepper or Grain of Paradise). The herbaceous plant is part of the cultural identity and heritage of West Africa where it is considered to have originated and has been used and cultivated for centuries and plays vital roles in traditional medicine, rituals, and ceremonies of various Indigenous cultures in the region and other parts of the world. *A. melegueta* is also cultivated in parts of Central and South America, the Caribbean, Southeast Asia, and some regions of Oceania. It was one of the valuable commodities exchanged along with other spices, such as black pepper and cardamom, through the trans-Saharan trade routes. Some of the culinary uses and applications of *A. melegueta* include spice, flavor enhancer, distilled spirit, baking, seasoning blend, craft brewing, and condiments. The herbal medicinal value, distinctive aroma, and fragrance of the plant are attributed to phytochemical constituents such as essential oils (like myristicin, limonene, beta-caryophyllene, and pinene), fatty acids, terpenoids, tannins, glycosides, steroids, carotenoids, phenols, alkaloids, flavonoids, vitamins, and minerals. Some ethnomedicinal uses of the plant seed (oil and extract) include digestive aid, anti-inflammatory, antioxidant properties, weight reduction and management, pain relief, respiratory health, antimicrobial, stomach upset, aphrodisiac, and fever reduction. However, further pharmacognostic and clinical research is required to verify the ethnomedical benefits of the plants. Business opportunities abound in the cultivation, processing and packaging, export and import, culinary ventures, health and wellness products, and fair trade initiative aspects of the plant. As global interest in spice continues to grow, it is essential to prioritize sustainable cultivation practices, fair trade, and phytomedical exploration of vital medicinal spices like *A. melegueta* to ensure its long-term availability, conservation, sustainable utilization, and benefit to local communities.

Keywords

Guinea pepper · Medicinal spice · Herbal products · Herbal medicine · Grain of paradise · Phytochemicals

1 Introduction

Aframomum melegueta [Roscoe.] K. Schum, commonly known as Guinea pepper or Grains of Paradise, is a spice and medicinal plant native to West Africa. It belongs to the Zingiberaceae family, which also includes ginger and turmeric. Guinea pepper has a long history of culinary and medicinal use and plays a significant role in the culture and cuisine of the regions where it grows. *A. melegueta* is believed to have originated in the rainforests of West Africa, specifically in the region that includes present-day Nigeria, Ghana, and Ivory Coast [50, 78]. This perennial plant thrives in the hot and humid tropical climate of the African rainforests, where it can be found growing both in the wild and under cultivation. The name "Grains of Paradise" is often used to refer to the seeds of *A. melegueta*. The term "paradise" in this context alludes to the spice's exotic and prized nature, which was highly valued in medieval Europe [87, 95]. *A. melegueta* is a robust herbaceous plant characterized by its large leaves, tall stalks, and unique flowering structures. The plant's leaves are large, oblong, and typically dark green. They grow alternately along the stems. The stems of Guinea pepper can reach a height of up to 2 meters (6.6 feet). They are sturdy and bear the plant's leaves and flowers. Guinea pepper produces beautiful, red-veined flowers that grow in spike-like clusters. The flowers are tubular and have a distinctive appearance [8, 9]. The fruit of *A. melegueta* is a capsule containing numerous seeds. These seeds, known as Guinea grains or Grains of Paradise, are the primary part of the plant used for culinary and medicinal purposes. Although the seeds are the most famous part of the plant, the roots and rhizomes also contain aromatic compounds and have been used in traditional medicine [6, 29, 30].

In recent years, *A. melegueta* has gained recognition in the global culinary scene. Chefs and food enthusiasts have rediscovered its unique flavor and have started to incorporate it into their dishes [5]. This revival has led to increased demand for Grains of Paradise in international markets. The spice trade, which historically included Grains of Paradise among its prized commodities, has seen a resurgence in interest. Small-scale farmers in West Africa are cultivating Guinea pepper to meet the growing demand from gourmet chefs and artisanal food producers worldwide. Grains of Paradise has a rich history in culinary traditions, both in Africa and beyond. They are known for their complex flavor profile, which includes notes of citrus, pepper, and warmth. Some of the culinary uses and applications of *A. melegueta* include the following:

Spice: Grains of Paradise is primarily used as a spice to season a wide range of dishes. They are often used in small quantities due to their potent flavor. In West African cuisine, they are a key ingredient in spice blends like pepper soup spice and suya spice [69].

Flavor Enhancer: Guinea pepper can enhance the flavor of meats, stews, soups, and sauces. It is particularly well-suited for game meats and hearty, savory dishes [71].

Distilled Spirits: Grains of Paradise is a botanical ingredient in some gin recipes, contributing to the complexity of the spirit's flavor [67–69].

Baking: In European cuisines, Grains of Paradise was historically used in baking, especially in gingerbread and spiced bread recipes [82].

Seasoning Blends: They are sometimes added to spice blends, such as garam masala, to provide a unique and exotic flavor [69, 71].

Craft Brewing: In recent years, craft brewers have experimented with adding Grains of Paradise to beer recipes, creating unique and flavorful brews [90].

Condiments: Guinea pepper can be used to make flavored oils and vinegar, adding a spicy kick to dressings and marinades [90].

In African cuisines, Guinea pepper is more than a spice; it is part of the cultural identity and heritage of the region. However, in addition to its culinary applications, *A. melegueta* has a rich history of traditional uses in various cultures. It has been valued for its medicinal properties and has played a role in religious and cultural practices. In traditional African medicine, Guinea pepper has been used to treat a wide range of ailments. It is believed to have digestive, anti-inflammatory, antimicrobial, and analgesic properties [11, 20, 51]. Grains of Paradise has a reputation as an aphrodisiac in some cultures. They have been used to enhance libido and sexual performance [7]. In some African cultures, Guinea pepper has been used as a protective charm or talisman to ward off evil spirits and negative energy. The seeds have been used in religious rituals and ceremonies in West African traditional religions. Guinea pepper has been associated with prosperity, abundance, and good fortune in some African cultures. While traditional medicine has long recognized the medicinal properties of *A. melegueta*, modern research has begun to shed light on the potential health benefits of this spice [14]. Some areas of interest include the following:

Anti-inflammatory: Grains of Paradise contains compounds with anti-inflammatory properties, which may have applications in reducing inflammation and related conditions.

Antioxidant: The spice is a source of antioxidants, which can help protect cells from oxidative stress and damage.

Digestive Health: In traditional medicine, Guinea pepper has been used to aid digestion. Some research suggests that it may have gastroprotective effects.

Metabolic Health: Preliminary studies have explored the potential impact of Guinea pepper on metabolic health, including its role in regulating blood sugar levels and promoting weight management.

Analgesic: Some traditional uses of Guinea pepper include pain relief. Research into its analgesic properties is ongoing.

This chapter aims to discuss the phytomedicinal value of *A. melegueta*. It presents the origin, distribution, and botany of *A. melegueta* wherein it is shown that the widespread distribution of the plant has led to its incorporation into a variety of culinary and medicinal traditions. This includes culinary and traditional medicine uses. The next sections address the origin, distribution, botanical, nutritional, and phytochemical properties as well as the food uses of *A. melegueta*. Other parts of the chapter are on the ethnobotanical, pharmacognosy, and clinical works, herbal products, toxicity, and commercial values of *A. melegueta*. It is believed to have various medicinal properties, including digestive, anti-inflammatory, and antimicrobial effects [20, 66]. *A. melegueta* is a crop with great potential for sustainable cultivation. It thrives in the agroforestry systems of West Africa, where it can be grown alongside other crops, such as cocoa and oil palms. This intercropping approach contributes to biodiversity and provides additional income to farmers. Sustainability practices, including organic cultivation and fair-trade principles, are being promoted to ensure that the production of Guinea pepper benefits both farmers and the environment.

2 Origin and Distribution of *Aframomum melegueta*

Aframomum melegueta, commonly known as Guinea pepper or Grains of Paradise, is native to West Africa. It is believed to have originated in the tropical rainforests of this region. Specifically, the plant's natural habitat spans several West African countries, including Nigeria, Ghana, Ivory Coast, Liberia, Sierra Leone, and Togo [36]. However, the exact place of origin within this region is difficult to pinpoint precisely, as Guinea pepper has been used and cultivated in West Africa for centuries, making it an integral part of the region's culinary and cultural heritage. Cultivation efforts can be found in parts of Central and South America, the Caribbean, Southeast Asia, and some regions of Oceania [12]. In these areas, it is grown to meet the global demand for this unique spice. Guinea pepper's name (pride of paradise) alludes to the historical trade routes that brought this spice to Europe during the medieval period. It was one of the valuable commodities exchanged along with other spices, such as black pepper and cardamom through trans-Saharan trade routes [33, 34]. These trade networks connected West Africa to Europe, the Middle East, and Asia, introducing the world to the unique flavors of Guinea pepper.

In its native West Africa, Guinea pepper is not only a spice but also a symbol of cultural identity and heritage. It has played a significant role in traditional medicine, rituals, and ceremonies of various Indigenous cultures in the region. Today, Guinea pepper remains an essential part of West African cuisines and is also gaining recognition in the global culinary scene for its distinct and complex flavor profile. While Guinea pepper is now cultivated in other parts of the world due to its increasing popularity, its true origin and cultural significance are deeply rooted in the vibrant and diverse landscapes of West Africa. Guinea pepper is primarily native to West Africa [17, 93]. This region is where the plant has its natural habitat and has been traditionally grown and used for centuries. However, due to its popularity in culinary and medicinal applications, it has garnered interest worldwide, leading to its cultivation in various regions beyond its native habitat. Today, *A. melegueta* is cultivated in other tropical regions around the world. These regions typically share a similar climate to its native West Africa, characterized by a hot and humid tropical climate. Countries in Central and South America, the Caribbean, Southeast Asia, and parts of Oceania have seen cultivation efforts to meet the growing global demand for Guinea pepper. Guinea pepper has gained recognition as a unique and flavorful spice in international cuisines. As a result, it is traded and used by chefs, culinary enthusiasts, and the food industry in many parts of the world. While it may not be grown commercially in all these regions, it is imported and incorporated into dishes and spice blends. In some countries with suitable climates, *A. melegueta* is cultivated as a horticultural or ornamental plant. These efforts may not necessarily be for commercial spice production but for the plant's esthetic and aromatic qualities. In regions with nontropical climates, such as parts of Europe and North America,

Guinea pepper is occasionally grown in controlled environments like greenhouses. This allows for cultivation in climates that would otherwise be unsuitable. The spice is valued for its unique flavor profile, reminiscent of citrus and pepper, and has made it a sought-after ingredient in global cuisine.

3 Botanical Characteristics of *Aframomum melegueta*

The ecology of *A. melegueta* is closely tied to its native habitat in West Africa. This perennial plant thrives in tropical regions with well-drained, fertile soils and adequate rainfall. It typically grows in the understory of rainforests, benefiting from the partial shade provided by taller trees. Physiologically, *A. melegueta* is characterized by its rhizomatous growth, with shoots emerging from underground rhizomes. It produces lance-shaped leaves and distinctive red or orange fruits that contain aromatic seeds. The plant's physiology enables it to adapt to its native rainforest environment, where it plays a role in the ecosystem and serves as a valuable spice in human cultures.

According to Amponsah et al. [9], botanically, *A. melegueta* is a herbaceous perennial plant. Herbaceous plants lack woody stems and persist for several growing seasons. The plant typically reaches a height of 1–2 meters (3.3–6.6 feet) when fully mature. The leaves of Guinea pepper are large, oblong, and dark green. They are arranged alternately along the stems. The leaves contribute to the plant's lush appearance. The stems of *A. melegueta* are sturdy and upright. They support the plant's leaves, flowers, and fruits. The stems are green and become more lignified (woody) as the plant matures. Guinea pepper produces striking, tubular flowers. The flowers are often described as being red-veined and have a distinctive appearance. They grow in clusters on spike-like inflorescences. The flowering period can vary depending on growing conditions. The fruit of *A. melegueta* is a capsule that contains numerous small seeds. These seeds are the part of the plant most commonly used as a spice and are known as Grains of Paradise. The capsules are typically brown or reddish-brown when mature. While the seeds are the most famous part of the plant, the roots and rhizomes of Guinea pepper also contain aromatic compounds. In traditional medicine and some culinary applications, the roots and rhizomes are used along with the seeds. Guinea pepper has a clumping growth habit, meaning that it forms clusters of stems arising from the same underground rhizome. This growth habit helps the plant spread in its natural habitat. The seeds of *A. melegueta* are known for their aromatic and complex fragrance, which includes notes of citrus, pepper, and warmth [3]. This unique fragrance is a key factor in their culinary and medicinal uses. The plant is native to the tropical rainforests of West Africa, where it thrives in the hot and humid climate. It is often found growing in the understory of the rainforest, receiving filtered sunlight with other tropical crops [57, 64, 65]. These morphological characteristics, including the unique appearance of the flowers and the aromatic seeds, make *A. melegueta* easily recognizable. The seeds, in particular, are highly prized for their flavor and fragrance, which have contributed to their historical and contemporary culinary and medicinal uses.

In addition to its ecology within its native range, Guinea pepper is cultivated in tropical regions worldwide. Cultivation practices aim to replicate the plant's native growing conditions to ensure healthy growth and the production of high-quality seeds. The plant has a distinctive appearance, and aromatic seeds have made it a sought-after spice in culinary traditions worldwide. It is known for its complex flavor profile, which combines elements of citrus, pepper, and warmth, making it a valuable addition to a wide range of dishes and spice blends.

4 Nutritional Value and Uses of *Aframomum melegueta*

Aframomum melegueta is not primarily consumed for its nutritional content but for its unique flavor and aromatic properties [40]. Nonetheless, *A. melegueta* contains many vital nutritional components but is not a significant source of macronutrients like carbohydrates, proteins, or fats. However, it does contain various bioactive compounds and phytochemicals that contribute to its flavor and potential health benefits. Some of the nutritional components and bioactive compounds found in Guinea pepper are as follows:

Essential Oils: Guinea pepper seeds contain essential oils, which are responsible for their distinctive aroma and flavor. These oils include compounds like myristicin, limonene, and pinene, which contribute to its complex fragrance [19].

Phenolic Compounds: Phenolic compounds are antioxidants found in Guinea pepper. These compounds have potential health benefits due to their ability to neutralize harmful free radicals in the body [1, 24].

Alkaloids: Some alkaloids, such as guineensine and aframonine, are present in Guinea pepper. These compounds may have bioactive properties and could contribute to their medicinal uses [18].

Flavonoids: Flavonoids are a group of polyphenolic compounds known for their antioxidant properties. While Guinea pepper contains some flavonoids, they are not a primary source [16].

Vitamins and Minerals: Guinea pepper is not a significant source of vitamins and minerals [53]. However, it may contain trace amounts of vitamins and minerals depending on the soil and growing conditions.

The primary use of Guinea pepper is as a spice. Its seeds are used to flavor a wide range of dishes, including soups, stews, sauces, and meat-based dishes. It adds a complex, citrusy, and peppery flavor with a hint of warmth. Grains of Paradise is often included in spice blends, such as garam masala, curry blends, and various regional spice mixtures. It enhances the overall flavor profile of these blends. In historical European cuisines, Guinea pepper was used in baking, particularly in gingerbread and spiced bread recipes. It added depth and complexity to baked goods. Some gin recipes include Guinea pepper as a botanical ingredient. It contributes to the flavor complexity of the spirit. Guinea pepper can be used to make flavored oils, vinegar, and condiments. These are used to add a spicy kick to dressings, marinades, and dipping sauces.

In West African traditional medicine, Guinea pepper has been used to treat various ailments. It is believed to have digestive, anti-inflammatory, and analgesic properties. Some traditional uses of Guinea pepper include its role in promoting digestive health. It has been used to alleviate digestive discomfort and stimulate appetite. Research into the anti-inflammatory properties of Grains of Paradise is ongoing, and it may have potential applications in reducing inflammation. The phenolic compounds in Guinea pepper give it antioxidant properties, which can help protect cells from oxidative stress and damage. In traditional medicine, Guinea pepper has been used for pain relief. While more research is needed, some studies suggest it may have analgesic effects. In some cultures, Guinea pepper has been considered an aphrodisiac, believed to enhance libido and sexual performance. In certain African cultures, Guinea pepper has been used as a protective charm or talisman to ward off evil spirits and negative energy. The seeds of Guinea pepper have been used in religious rituals and ceremonies in West African traditional religions. The spice is associated with prosperity, abundance, and good fortune in some African cultures and has cultural significance in various rituals and celebrations [31, 58–60]. In recent years, Guinea pepper has experienced a culinary revival. Chefs and food enthusiasts have rediscovered its unique flavor and have started to incorporate it into their dishes. This has led to an increased demand for Grains of Paradise in international markets. The global spice trade has also seen a resurgence of interest in Guinea pepper. Small-scale farmers in West Africa are cultivating Guinea pepper to meet the growing demand from gourmet chefs and artisanal food producers worldwide.

5 Global *Aframomum melegueta* Business, Market, and Supply Chain

Aframomum melegueta, commonly known as Guinea pepper or Grains of Paradise, has gained recognition in the global market due to its unique flavor profile and potential health benefits. To understand the global business, market, and supply chain of *A. melegueta*, there is a need to explore various aspects, including production regions, market trends, supply chain dynamics, and business opportunities.

5.1 Production Regions

West Africa is the primary production region of *A. melegueta*: Guinea pepper is primarily cultivated and harvested in West African countries, including Nigeria, Ghana, Ivory Coast, Liberia, Sierra Leone, and Togo. These countries have the ideal tropical rainforest climate for its growth. However, expansion to other tropical regions is ongoing due to increasing global demand. Hence, cultivation efforts have expanded to other tropical regions worldwide. Central and South American countries, the Caribbean, Southeast Asia, and parts of Oceania have seen cultivation initiatives to meet the growing demand for this unique spice.

5.2 Market Trends

Guinea pepper has experienced a culinary revival in recent years, with chefs and food enthusiasts appreciating its complex flavor. It is increasingly used in gourmet cuisine, craft beverages, and artisanal food products [54–56, 76]. The spice's potential health benefits, including anti-inflammatory and antioxidant properties, align with consumer trends toward healthier eating. This has driven interest in incorporating Guinea pepper into wellness-focused products. Some gin producers include Guinea pepper as a botanical ingredient in their recipes. The craft beverage industry's growth has contributed to increased demand. Guinea pepper is integral to various ethnic and regional cuisines, particularly in West African and North African dishes. The growing popularity of these cuisines has expanded its market. Guinea pepper is a key component of spice blends such as garam masala, curry blends, and regional seasoning mixtures. These blends are popular for their convenience and flavor enhancement.

5.3 Supply Chain Dynamics of *A. melegueta*

Cultivation: The supply chain begins with the cultivation of *A. melegueta*. Small-scale farmers in West Africa often grow it alongside other crops like cocoa and oil palms.

Harvesting: Guinea pepper is harvested when the capsules containing the seeds mature and turn brown or reddish-brown. Harvesting methods can vary but typically involve manually plucking the capsules.

Processing: After harvesting, the seeds are separated from the capsules. Depending on the scale of production, this can be done manually or with the help of machinery.

Packaging: Processed seeds are typically packaged in airtight containers to preserve their flavor and aroma. Packaging may vary based on market destinations, including bulk packaging for wholesale and retail-ready packaging for consumer markets.

Distribution: Distribution channels include local markets, international spice traders, and specialty food distributors. Guinea pepper is often imported by countries worldwide.

Retail and Consumer: Grains of Paradise is available in various forms, including whole seeds and ground spice. Consumers can purchase them in stores or online for use in culinary applications and home cooking.

5.4 Business Opportunities with *A. melegueta*

Business opportunities exist in the following areas concerning *A. melegueta:*

Cultivation: Entrepreneurs and agricultural investors can explore opportunities in cultivating Guinea pepper, especially in regions with suitable climates.

Processing and Packaging: Establishing processing and packaging facilities for Guinea pepper seeds can create value-added products for export and local markets.

Export and Import: International traders can participate in the export-import business of Guinea pepper, connecting producers in West Africa with global markets.

Culinary Ventures: Chefs, culinary entrepreneurs, and food manufacturers can incorporate Guinea pepper into their products to meet the demand for unique and exotic flavors.

Health and Wellness Products: Developing health and wellness products that incorporate Guinea pepper's potential health benefits can tap into consumer interest in functional foods.

Sustainable and Fair-Trade Initiatives: Supporting sustainable and fair-trade practices in the supply chain can create business opportunities focused on ethical and environmentally responsible sourcing.

6 Ethnomedicinal Value of *Aframomum melegueta*

Aframomum melegueta contains a variety of primary and secondary phytochemicals, which contribute to its flavor, aroma, and potential health benefits. Some of the primary and secondary phytochemicals found in *Aframomum melegueta* are the following:

Essential Oils: Guinea pepper seeds are rich in essential oils, which are responsible for their distinct aroma and flavor [49, 80]. Some of the primary components of these essential oils include myristicin, limonene, pinene, and beta-caryophyllene [23, 43, 85]. Some of the properties of these essential oils include the following:

- Myristicin: known for its aromatic and spicy flavor
- Limonene: imparts citrusy notes to the spice
- Pinene: adds pine-like and woody aromas
- Beta-caryophyllene: contributes to the spice's peppery and earthy character

Fatty Acids: While not a primary source, Guinea pepper seeds contain small amounts of fatty acids, including oleic acid and linoleic acid [27, 47].

Phenolic Compounds: Phenolic compounds are antioxidants found in Guinea pepper [3, 45, 75]. These secondary metabolites have potential health benefits due to their ability to neutralize harmful free radicals. Examples of phenolic compounds in Guinea pepper include vanillic acid, caffeic acid, syringic acid, quercetin, and kaempferol. Some of the properties of these compounds include the following:

- Vanillic acid: known for its sweet and vanilla-like flavor
- Caffeic acid: has antioxidant properties and contributes to the spice's flavor
- Syringic acid: an antioxidant with potential health benefits

- Quercetin: a flavonoid with anti-inflammatory properties
- Kaempferol: another flavonoid with antioxidant and anti-inflammatory effects

Alkaloids: Some alkaloids are present in Guinea pepper seeds, and while they are not primary sources, they may have bioactive properties [28]. Examples of alkaloids found in *Aframomum melegueta* include guineensine, malic acid, and aframonine [38, 77, 81].

Flavonoids: Flavonoids are a group of polyphenolic compounds with antioxidant properties [73]. While Guinea pepper contains some flavonoids, they are not the primary source of these compounds.

Terpenoids: Terpenoids are secondary metabolites found in Guinea pepper, contributing to its aromatic properties [6]. Some terpenoids include limonene and pinene.

Sesquiterpene Lactones: While not as common as other compounds, some sesquiterpene lactones have been identified in Guinea pepper [26, 41, 91]. These compounds are known for their diverse biological activities.

Tannins: Tannins are polyphenolic compounds that may be present in Guinea pepper, albeit in relatively small quantities [10, 39]. They can contribute to the astringency of the spice.

Glycosides: Some glycosides may be found in Guinea pepper seeds, though their presence and types can vary ([46]; 2017).

Steroids: Steroids are compounds that may be present in small quantities in Guinea pepper, contributing to its chemical diversity [79].

Carotenoids: While not as prominent as in some other spices, Guinea pepper may contain trace amounts of carotenoids, which are responsible for the coloration of some fruits and vegetables [61, 62, 86].

It is important to note that the exact composition of phytochemicals in Guinea pepper can vary based on factors such as the plant's growing conditions, location, and maturation. These phytochemicals collectively contribute to the unique flavor, aroma, and potential health benefits associated with *A. melegueta*. Additionally, ongoing research continues to uncover the diverse chemical constituents present in this spice, further deepening our understanding of its properties and potential uses. While further research is needed to fully understand the bioactive properties of Guinea pepper, it continues to be a popular and intriguing spice in the culinary world with potential applications in functional foods and traditional medicine.

Some of the ethnomedical uses of the bioactive compound in *A. melegueta* include the following:

- **Digestive Aid**: In traditional African medicine, Guinea pepper has been used to aid digestion. It is believed to stimulate the digestive system, relieve indigestion, and alleviate gastrointestinal discomfort [22, 29].
- **Anti-inflammatory**: Guinea pepper has been traditionally employed as an anti-inflammatory agent. It is used to reduce inflammation and soothe inflammatory conditions such as arthritis and joint pain [52].
- **Antioxidant Properties**: The phenolic compounds found in Guinea pepper are antioxidants. Antioxidants help protect cells from oxidative stress and damage

caused by free radicals. Traditional uses may include harnessing these antioxidant properties for overall health [63, 74].

- **Pain Relief**: In some traditional healing practices, Guinea pepper has been used for relief from pain and rheumatism [13, 92]. It is believed to have analgesic properties and may be used topically or consumed to alleviate pain. Aqueous seed extract of grain of paradise has been shown to possess peripheral analgesic activity through actinociceptive activity of the intraperitoneal dose [88, 92].
- **Respiratory Health**: Guinea pepper has been used in traditional remedies for respiratory issues such as coughs and congestion [21]. It may be used as an ingredient in herbal formulations to help ease respiratory discomfort. The study by El Dine et al. [21] showed that the methanolic extract of the plant has antiadhesive concentration-dependent effects on microorganisms in the lung carcinoma line.
- **Antimicrobial and Antihelminthic**: Some cultures have used Guinea pepper as an antimicrobial and antihelminthic agent [37]. It may be employed to help combat infections, although further research is needed to confirm its efficacy [44, 70].
- **Aphrodisiac**: In certain African cultures, Guinea pepper has been considered an aphrodisiac. It is believed to enhance libido and sexual performance, and it is sometimes included in preparations meant to boost sexual vitality [4, 15, 35].
- **Ritual and Spiritual Uses**: Guinea pepper has cultural and spiritual significance in some African traditions. It is used in rituals, ceremonies, and protective charms to ward off evil spirits and promote positive energy [25, 32, 83].
- **Weight Reduction and Management**: The plant seeds and seed extracts are consumed to reduce and manage weight in many Indigenous communities across the world including diabetic patients [48, 84, 89]. The work of Yoneshiro et al. [96] suggest the plant extract adopts adoptive thermogenesis to reduce body fat in humans.
- **Stomach Upset**: Guinea pepper is sometimes used to alleviate stomach upset, including nausea and vomiting, in traditional medicine [42]. López-Ríos et al. [42] reported that the plant seed is taken by postmenopausal women as a remedy for hot flashes, anxiety, and depressive symptoms.
- **Cardiovascular Protection**: The seed and seed extract of the plant is taken to protect and increase cardiovascular activities. It is also used to treat hypertension in Indigenous communities in the Global South [2].
- **Fever Reduction**: In some African cultures, Guinea pepper has been used to help reduce fever and alleviate symptoms of illnesses associated with fever.
- **Nephroprotection:** The seed and seed extract of the plant is taken to reduce kidney injury and protect the organ from adverse damage [2].

Table 1 presents some of the ethnomedicinal properties of *A. melegueta* as well as the parts used and bioactive compounds responsible for the effects.

Table 1 Ethnomedicinal properties of *A. melegueta* as well as the parts used and bioactive compounds responsible for the effects

Ethnomedical uses	Part used	How it is used	Bioactive compound responsible
Digestive aids	Seeds	Crushed or powdered seeds are often added to food or herbal formulations to aid digestion and relieve indigestion	Myristicin, limonene, pinene, and phenolic compounds
Anti-inflammatory	Seeds, oil	Seeds may be ingested, or oil extracts are applied topically to reduce inflammation and alleviate joint pain	Phenolic compounds, and terpenoids
Antioxidant properties	Seeds, extracts	Guinea pepper seeds or extracts are consumed to provide antioxidant support for overall health	Phenolic compounds, flavonoids, and terpenoids
Pain relief	Seeds, oil	Seeds may be crushed and applied topically as poultices or mixed with carrier oils for massage to alleviate pain	Phenolic compounds, terpenoids, and alkaloids
Respiratory health	Seeds, preparations	Seeds are sometimes included in herbal preparations for respiratory issues, promoting clear airways	Terpenoids and essential oils
Antimicrobial agent	Seeds, extracts	Extracts or powdered seeds are used in traditional remedies to combat infections, both topically and internally	Terpenoids, phenolic compounds, and alkaloids
Aphrodisiac	Seeds, preparations	Ground seeds may be included in aphrodisiac preparations believed to enhance libido and sexual vitality	Alkaloids and terpenoids
Ritual and spiritual uses	Seeds, whole	Whole seeds are used in rituals, ceremonies, and protective charms to ward off negative energy and evil spirits	Cultural and spiritual significance
Stomach upset	Seeds	Seeds are occasionally used to soothe stomach upset, including nausea and vomiting	Terpenoids and phenolic compounds
Fever management and reduction	Seeds	Seeds are sometimes used to reduce fever and alleviate symptoms associated with fever	Terpenoids and alkaloid

7 Pharmacognosy and Clinical Research on *Afromomum melegueta*

7.1 Pharmacognosy of *Aframomum melegueta*

Pharmacognosy is the study of natural products from plants and other natural sources to understand their properties, composition, and potential pharmacological uses. *Aframomum melegueta* has been of interest to pharmacognosists and researchers due to its unique bioactive compounds [73]. Some aspects of the pharmacognosy of *Aframomum melegueta*:

Morphological Identification: Pharmacognosists study the morphological characteristics of the plant, including its leaves, flowers, and seeds, to ensure proper identification. Guinea pepper is known for its distinctive seeds enclosed in capsules.

Chemical Composition: Researchers analyze the chemical composition of Guinea pepper, including its essential oil content and specific bioactive compounds. Gas chromatography-mass spectrometry (GC-MS) is often used to identify the chemical constituents.

Quality Control: Pharmacognosists are involved in establishing quality control standards for Guinea pepper products. This helps ensure that commercial products meet certain quality criteria.

Extraction and Isolation: Techniques for extracting and isolating bioactive compounds from Guinea pepper are explored, allowing for further studies on their properties and potential applications.

Phytochemical Screening: Researchers conduct phytochemical screening to identify the presence of various compounds, such as alkaloids, flavonoids, phenolic compounds, and terpenoids, which contribute to the spice's properties.

7.2 Clinical Research on *Aframomum melegueta*

Clinical research involves studying the effects of Guinea pepper on human health and well-being through controlled experiments and trials [72, 94]. While Guinea pepper has a history of traditional uses, clinical research is ongoing to validate these uses and explore potential therapeutic applications. Some areas of clinical research on *Aframomum melegueta*:

- **Anti-inflammatory Properties**: Studies aim to investigate the anti-inflammatory effects of Guinea pepper and its potential in managing conditions characterized by inflammation, such as arthritis.
- **Antioxidant Effects**: Clinical trials may assess the antioxidant properties of Guinea pepper and its ability to protect cells from oxidative damage. This research is relevant to overall health and may have implications for chronic diseases.
- **Digestive Health**: Guinea pepper's traditional use as a digestive aid is a subject of interest. Clinical studies may explore its effects on gastrointestinal health and its role in improving digestion.
- **Pain Management**: Research may investigate the analgesic properties of Guinea pepper and its potential for pain relief, especially in conditions associated with pain and discomfort.
- **Respiratory Health**: Clinical trials may explore the use of Guinea pepper or its extracts in respiratory health, particularly in managing conditions like coughs and congestion.

- **Antimicrobial Activity**: Guinea pepper's potential antimicrobial properties may be examined in clinical studies to assess its effectiveness against infections.
- **Aphrodisiac Effects**: Research may delve into the traditional use of Guinea pepper as an aphrodisiac, exploring its impact on libido and sexual health.
- **Safety and Dosage**: Clinical studies also aim to establish safe dosages and assess any potential side effects or contraindications associated with the use of Guinea pepper.

8 Some Herbal Products and Formulations of *Afromomum melegueta*

Aframomum melegueta has unique flavor profile that has led to its incorporation into various herbal products and formulations. Some examples of herbal products and formulations that may include *A. melegueta* are the following:

Spice Blends: Guinea pepper is a common ingredient in spice blends and seasoning mixtures. It adds depth and complexity to spice mixes used in various culinary applications. For example, it is a component of traditional spice blends like garam masala and curry powder.

Digestive Bitters: Some herbal digestive bitters and digestive aids may include *A. melegueta* among their ingredients. These formulations are designed to support healthy digestion and alleviate digestive discomfort.

Herbal Teas: Guinea pepper may be included in herbal tea blends, often in combination with other herbs and spices known for their digestive or warming properties. These teas are consumed for their flavor and potential digestive benefits.

Aphrodisiac Formulations: Due to its traditional reputation as an aphrodisiac, Guinea pepper may be incorporated into herbal formulations designed to enhance libido and sexual vitality. These products may be marketed as dietary supplements or natural aphrodisiacs.

Aromatherapy Blends: The essential oil derived from *A. melegueta* may be used in aromatherapy blends for its unique aroma. It can be diffused or added to massage oils for its aromatic properties.

Traditional Medicine: In some traditional healing systems, *A. melegueta* is used in herbal formulations to address various health concerns, such as digestive discomfort, joint pain, or respiratory issues.

Flavoring for Alcoholic Beverages: Guinea pepper is sometimes used to flavor alcoholic beverages, particularly craft beers and gins. It contributes to the complex and aromatic profiles of these drinks.

Nutraceuticals: *A. melegueta* extracts or powders may be used as ingredients in nutraceutical products, such as dietary supplements or functional foods. These products may claim various health benefits attributed to the spice's bioactive compounds.

Culinary Oils and Vinegar: Some specialty oils and vinegar may be infused with Guinea pepper to impart their unique flavor and aroma to dressings, marinades, or cooking oils.

Herbal Liqueurs: In some regions, herbal liqueurs or aperitifs are made using Guinea pepper, among other botanicals. These beverages may have a spicy and aromatic character.

9 Toxic Compounds in *Afromomum melegueta*

Aframomum melegueta is generally considered safe for culinary use when consumed in moderation. However, like many spices and plants, it contains certain compounds that, in excessive amounts, can potentially have adverse effects on health [80]. Some of the compounds in Guinea pepper that may have toxic effects if consumed in large quantities are as follows:

Essential Oils: Guinea pepper is rich in essential oils, which contribute to its aroma and flavor. While these oils are generally safe for consumption, excessive consumption of essential oils, especially when not properly diluted, can lead to gastrointestinal discomfort, skin irritation, or allergic reactions in some individuals.

Alkaloids: Some alkaloids have been identified in Guinea pepper. In moderate amounts, alkaloids may have physiological effects, but excessive consumption can lead to toxicity. Common alkaloids found in plants include caffeine and nicotine.

Phenolic Compounds: While phenolic compounds are known for their antioxidant properties, excessive intake of phenolic-rich foods or supplements may lead to gastrointestinal irritation or other adverse effects in sensitive individuals.

Tannins: Guinea peppers may contain tannins, which are polyphenolic compounds. In large amounts, tannins can have an astringent effect and may interfere with the absorption of certain nutrients.

Spice Allergy: Some individuals may be allergic to specific spices, including Guinea pepper. Allergic reactions can range from mild skin irritation to severe respiratory symptoms. Individuals with spice allergies should avoid consuming Guinea pepper.

It is important to emphasize that the potentially toxic effects of these compounds are typically associated with excessive consumption or misuse. When used in culinary applications as a spice, Guinea pepper is considered safe for most people. However, as with any food or spice, moderation is key, and individual tolerance may vary. Additionally, it is essential to be cautious when using essential oils derived from Guinea pepper or any other plant. Essential oils are highly concentrated and should be used sparingly and appropriately diluted when used for aromatherapy, massage, or other applications.

10 Conclusion

Grains of Paradise is a spice with a rich history, diverse culinary uses, and potential medicinal properties. It is native to the rainforests of West Africa and has made its mark in cuisines around the world, adding complexity and depth to a wide range of dishes. Its unique flavor, reminiscent of citrus and pepper, has earned it a special place in the hearts of chefs and food enthusiasts. Beyond the culinary realm, Guinea pepper has deep cultural and traditional significance, serving as a medicinal herb, an aphrodisiac, and a protective charm in various cultures. Modern research is beginning to uncover its potential health benefits, including anti-inflammatory and antioxidant properties. As the global spice trade experiences a resurgence of interest in this unique spice, it is crucial to prioritize sustainable cultivation practices and fair-trade principles to ensure that the production of *A. melegueta* benefits local communities and ecosystems. In a world where culinary diversity is celebrated and the search for new flavors never ends, *A. melegueta* offers a tantalizing journey into the heart of West African cuisine and culture, reminding us of the rich tapestry of flavors and traditions that make our global culinary heritage so vibrant and exciting.

In conclusion, Guinea pepper is a spice with a rich history, diverse culinary uses, and potential medicinal properties. While not a significant source of macronutrients, it contains various bioactive compounds and phytochemicals that contribute to its flavor and potential health benefits. Its unique flavor profile, reminiscent of citrus and pepper, has earned it a special place in the world of culinary arts and traditional medicine. As global interest in this spice continues to grow, it is essential to prioritize sustainable cultivation practices and fair trade to ensure its long-term availability and benefit to local communities. *A. melegueta* has a growing presence in the global culinary and spice market. Its unique flavor, culinary versatility, and potential health benefits have contributed to its popularity. Opportunities abound for those interested in various aspects of the Guinea pepper supply chain, from cultivation and processing to distribution and culinary ventures. Sustainability and fair trade principles are also essential considerations for businesses in this industry.

References

1. Abdel-Naim AB, Alghamdi AA, Algandaby MM, Al-Abbasi FA, Al-Abd AM, Abdallah HM, El-Halawany AM, Hattori M (2017) Phenolics isolated from Aframomum meleguta enhance proliferation and ossification markers in bone cells. Molecules (Basel, Switzerland) 22(9):1467. https://doi.org/10.3390/molecules22091467
2. Abdou RM, El-Maadawy WH, Hassan M, El-Dine RS, Aboushousha T, El-Tanbouly ND, El-Sayed AM (2021) Nephroprotective activity of Aframomum melegueta seeds extract against diclofenac-induced acute kidney injury: a mechanistic study. J Ethnopharmacol 273:113939. https://doi.org/10.1016/j.jep.2021.113939
3. Adefegha SA, Oboh G (2012) Acetylcholinesterase (AChE) inhibitory activity, antioxidant properties and phenolic composition of two Aframomum species. J Basic Clin Physiol Pharmacol 23(4):153–161

4. Adefegha SA, Oboh G, Okeke BM, Oyeleye SI (2017) Comparative effects of alkaloid extracts from Aframomum melegueta (alligator pepper) and Aframomum danielli (Bastered Melegueta) on enzymes relevant to erectile dysfunction. J Diet Suppl 14(5):542–552. https://doi.org/10.1080/19390211.2016.1272661
5. Ajaiyeoba EO, Ekundayo O (1999) Essential oil constituents of Aframomum melegueta (Roscoe) K. Schum. Seeds (alligator pepper) from Nigeria. Flavour Fragr J 14(2):109–111
6. Akpanabiatu MI, Ekpo ND, Ufot UF, Udoh NM, Akpan EJ, Etuk EU (2013) Acute toxicity, biochemical and haematological study of Aframomum melegueta seed oil in male Wistar albino rats. J Ethnopharmacol 150(2):590–594. https://doi.org/10.1016/j.jep.2013.09.006
7. Allas S, Ngoka V, Hartman NG, Owassa S, Ibea M (1995) Aframomum seeds for improving penile activity. http://www.freepatents online.com/5879682.html. 16 July 2010
8. Amadi SW, Zhang Y, Wu G (2016) Research progress in phytochemistry and biology of Aframomum species. Pharm Biol 54(11):2761–2770. https://doi.org/10.3109/13880209.2016.1173068
9. Amponsah J, Adamtey N, Elegba W, Danso KE (2013) In situ morphometric characterization of Aframomum melegueta accessions in Ghana. AoB Plants 5:plt027. https://doi.org/10.1093/aobpla/plt027
10. Anwar WS, Abdel-Maksoud FM, Sayed AM, Abdel-Rahman IAM, Makboul MA, Zaher AM (2023) Potent hepatoprotective activity of common rattan (Calamus rotang L.) leaf extract and its molecular mechanism. BMC Complement Med Ther 23(1):24. https://doi.org/10.1186/s12906-023-03853-9
11. Aranganathan S, Nalini N (2013) Antiproliferative efficacy of hesperetin (citrus flavanoid) in 1, 2-dimethylhydrazine-induced colon cancer. Phytother Res 27(7):999–1005
12. Ataba E, Katawa G, Ritter M, Ameyapoh AH, Anani K, Amessoudji OM, Tchadié PE, Tchacondo T, Batawila K, Ameyapoh Y, Hoerauf A, Layland LE, Karou SD (2020) Ethnobotanical survey, anthelmintic effects and cytotoxicity of plants used for treatment of helminthiasis in the central and Kara regions of Togo. BMC Complement Med Ther 20(1):212. https://doi.org/10.1186/s12906-020-03008-0
13. Biobaku KT, Azeez OM, Amid SA, Asogwa TN, Abdullahi AA, Raji OL, Abdulhamid JA (2021) Thirty days oral Aframomum melegueta extract elicited analgesic effect but influenced cytochrome p4501BI, cardiac troponin T, testicular alfa-fetoprotein and other biomarkers in rats. J Ethnopharmacol 267:113493. https://doi.org/10.1016/j.jep.2020.113493
14. Bravo L (1998) Polyphenols: chemistry, dietary sources, metabolism, and nutritional significance. Nutr Rev 56(11):317–333
15. Chauhan NS, Sharma V, Dixit VK, Thakur M (2014) A review on plants used for improvement of sexual performance and virility. Biomed Res Int 2014:868062. https://doi.org/10.1155/2014/868062
16. Chung WY, Jung YJ, Surh YJ, Lee SS, Park KK (2001) Antioxidative and antitumor promoting effects of [6]-paradol and its homologs. Mutat Res 496(1–2):199–206
17. Clifford AJ, Ebeler SE, Ebeler JD, Bills ND, Hinrichs SH, Teissedre PL, Waterhouse AL (1996) Delayed tumor onset in transgenic mice fed an amino acid–based diet supplemented with red wine solids. Am J Clin Nutr 64(5):748–756. [16]
18. Dibwe DF, Awale S, Kadota S, Tezuka Y (2012) Damnacanthal from the Congolese medicinal plant Garcinia huillensis has potent preferential cytotoxicity against human pancreatic cancer PANC-1 cells. Phytother Res 26(12):1920–1926
19. Dibwe DF, Awale S, Morita H, Tezuka Y (2015) Anti-austeritic constituents of the Congolese medicinal plant Aframomum melegueta. Nat Prod Commun 10(6):997–999
20. Doherty VF, Olaniran O, Kanife UC (2010) Antimicrobial activities of Aframomum melegueta (alligator pepper). Int J Biol 2(2):126–131
21. El Dine RS, Elfaky MA, Asfour H, El Halawany AM (2021) Anti-adhesive activity of *Aframomum melegueta* major phenolics on lower respiratory tract pathogens. Nat Prod Res 35(4):539–547. https://doi.org/10.1080/14786419.2019.1585843

22. El-Halawany AM, El Dine RS, El Sayed NS, Hattori M (2014) Protective effect of Aframomum melegueta phenolics against CCl_4-induced rat hepatocytes damage; role of apoptosis and pro-inflammatory cytokines inhibition. Sci Rep 4:5880. https://doi.org/10.1038/srep05880
23. Essien EE, Thomas PS, Oriakhi K, Choudhary MI (2017) Characterization and antioxidant activity of volatile constituents from different parts of Aframomum danielli (Hook) K. Schum. Medicines (Basel, Switzerland) 4(2):29. https://doi.org/10.3390/medicines4020029
24. Gastaldelli A (2011) Role of beta-cell dysfunction, ectopic fat accumulation and insulin resistance in the pathogenesis of type 2 diabetes mellitus. Diabetes Res Clin Pract 93(Suppl 1):S60–S65
25. Gruca M, van Andel TR, Balslev H (2014) Ritual uses of palms in traditional medicine in sub-Saharan Africa: a review. J Ethnobiol Ethnomed 10:60. https://doi.org/10.1186/1746-4269-10-60
26. Habtamu A, Melaku Y (2018) Antibacterial and antioxidant compounds from the flower extracts of *Vernonia amygdalina*. Adv Pharmacol Sci 2018:4083736. https://doi.org/10.1155/2018/4083736
27. Hattori H, Mori T, Shibata T, Kita M, Mitsunaga T (2021) 6-paradol acts as a potential anti-obesity vanilloid from grains of paradise. Mol Nutr Food Res 65(16):e2100185. https://doi.org/10.1002/mnfr.202100185
28. Ibekwe H (2019) Effect of methanolic crude extract of Aframomum melegueta (A.m) seeds on selected lactogenic hormones of albino rats. Int J Biochem Mol Biol 10(2):9–16
29. Ilic N, Schmidt BM, Poulev A, Raskin I (2010) Toxicological evaluation of grains of paradise (Aframomum melegueta) [Roscoe] K. Schum. J Ethnopharmacol 127(2):352–356. https://doi.org/10.1016/j.jep.2009.10.031
30. Ilic NM, Dey M, Poulev AA, Logendra S, Kuhn PE, Raskin I (2014) Anti-inflammatory activity of grains of paradise (Aframomum melegueta Schum) extract. J Agric Food Chem 62(43):10452–10457. https://doi.org/10.1021/jf5026086
31. Imarhiagbe O, Ogwu MC (2022) Sacred groves in the global south: a panacea for sustainable biodiversity conservation. In: Izah SC (ed) Biodiversity in Africa: potentials, threats and conservation, Sustainable development and biodiversity, vol 29. Springer, Singapore, pp 525–546. https://doi.org/10.1007/978-981-19-3326-4_20
32. Inegbenebor U, Ebomoyi MI, Onyia KA, Amadi K, Aigbiremolen AE (2009) Effect of alligator pepper (Zingiberaceae aframomum melegueta) on first trimester pregnancy in Sprague Dawley rats. Niger J Physiol Sci 24(2):161–164. https://doi.org/10.4314/njps.v24i2.52901
33. Jimenez-Fernandez R, Rodriguez Vázquez R, Marín-Morales D, Herraiz-Soria E, Losa-Iglesias ME, Becerro-de-Bengoa-Vallejo R, Corral-Liria I (2023) Exploring knowledge about fang traditional medicine: an informal health seeking behaviour for medical or cultural afflictions in Equatorial Guinea. Healthcare (Basel, Switzerland) 11(6):808. https://doi.org/10.3390/healthcare11060808
34. Kafoutchoni KM, Idohou R, Egeru A, Salako KV, Agbangla C, Adomou AC, Assogbadjo AE (2018) Species richness, cultural importance, and prioritization of wild spices for conservation in the Sudano-Guinean zone of Benin (West Africa). J Ethnobiol Ethnomed 14(1):67. https://doi.org/10.1186/s13002-018-0267-y
35. Kamtchouing P, Mbongue GY, Dimo T, Watcho P, Jatsa HB, Sokeng SD (2002) Effects of Aframomum melegueta and Piper guineense on sexual behaviour of male rats. Behav Pharmacol 13(3):243–247. https://doi.org/10.1097/00008877-200205000-00008
36. Kamte SLN, Ranjbarian F, Campagnaro GD, Nya PCB, Mbuntcha H, Woguem V, Womeni HM, Ta LA, Giordani C, Barboni L, Benelli G, Cappellacci L, Hofer A, Petrelli R, Maggi F (2017) Trypanosoma brucei inhibition by essential oils from medicinal and aromatic plants traditionally used in Cameroon (Azadirachta indica, Aframomum melegueta, Aframomum daniellii, Clausena anisata, Dichrostachys cinerea and Echinops giganteus). Int J Environ Res Public Health 14(7):737. https://doi.org/10.3390/ijerph14070737
37. Katawa G, Ataba E, Ritter M, Amessoudji OM, Awesso ER, Tchadié PE, Bara FD, Douti FV, Arndts K, Tchacondo T, Batawila K, Ameyapoh Y, Hoerauf A, Karou SD, Layland LE (2022) Anti-Th17 and anti-Th2 responses effects of hydro-ethanolic extracts of Aframomum

melegueta, Khaya senegalensis and Xylopia aethiopica in hyperreactive onchocerciasis individuals' peripheral blood mononuclear cells. PLoS Negl Trop Dis 16(4):e0010341. https://doi.org/10.1371/journal.pntd.0010341

38. Keihanian F, Moohebati M, Saeidinia A, Mohajeri SA (2023) Iranian traditional medicinal plants for management of chronic heart failure: a review. Medicine 102(19):e33636. https://doi.org/10.1097/MD.0000000000033636
39. Kouitcheu Mabeku LB, Nanfack Nana B, Eyoum Bille B, Tchuenteu Tchuenguem R, Nguepi E (2017) Anti-helicobacter pylori and antiulcerogenic activity of Aframomum pruinosum seeds on indomethacin-induced gastric ulcer in rats. Pharm Biol 55(1):929–936. https://doi.org/10.1080/13880209.2017.1285326
40. Kuete V, Krusche B, Youns M, Voukeng I, Fankam AG, Tankeo S, Efferth T (2011) Cytotoxicity of some Cameroonian spices and selected medicinal plant extracts. J Ethnopharmacol 134(3):803–812
41. Lans C (2007) Comparison of plants used for skin and stomach problems in Trinidad and Tobago with Asian ethnomedicine. J Ethnobiol Ethnomed 3:3. https://doi.org/10.1186/1746-4269-3-3
42. López-Ríos L, Barber MA, Wiebe J, Machín RP, Vega-Morales T, Chirino R (2021) Influence of a new botanical combination on quality of life in menopausal Spanish women: results of a randomized, placebo-controlled pilot study. PLoS One 16(7):e0255015. https://doi.org/10.1371/journal.pone.0255015
43. Luna EC, Luna IS, Scotti L, Monteiro AFM, Scotti MT, de Moura RO, de Araújo RSA, Monteiro KLC, de Aquino TM, Ribeiro FF, Mendonça FJB (2019) Active essential oils and their components in use against neglected diseases and arboviruses. Oxidative Med Cell Longev 2019:6587150. https://doi.org/10.1155/2019/6587150
44. Mahoney O, Melo C, Lockhart A, Cornejal N, Alsaidi S, Wu Q, Simon J, Juliani R, Zydowsky TM, Priano C, Koroch A, Fernández Romero JA (2022) Antiviral activity of aframomum melegueta against severe acute respiratory syndrome coronaviruses type 1 and 2. S Afr J Bot 146:735–739. https://doi.org/10.1016/j.sajb.2021.12.010
45. Mickymaray S, Al Aboody MS (2019) In vitro antioxidant and bactericidal efficacy of 15 common spices: novel therapeutics for urinary tract infections? Medicina (Kaunas) 55(6):289. https://doi.org/10.3390/medicina55060289
46. Mohammed A, Koorbanally NA, Islam MS (2015) Ethyl acetate fraction of Aframomum melegueta fruit ameliorates pancreatic β-cell dysfunction and major diabetes-related parameters in a type 2 diabetes model of rats. J Ethnopharmacol 175:518–527. https://doi.org/10.1016/j.jep.2015.10.011
47. Mohammed A, Koorbanally NA, Islam MS (2016) Phytochemistry, antioxidative activity and inhibition of key enzymes linked to type 2 diabetes by various parts of aframomum melegueta in vitro. Acta Pol Pharm 73(2):403–417
48. Mohammed A, Gbonjubola VA, Koorbanally NA, Islam MS (2017) Inhibition of key enzymes linked to type 2 diabetes by compounds isolated from Aframomum melegueta fruit. Pharm Biol 55(1):1010–1016. https://doi.org/10.1080/13880209.2017.1286358
49. Morais MC, Souza JV, da Silva Maia Bezerra Filho C, Dolabella SS, Sousa DP (2020) Trypanocidal essential oils: a review. Molecules (Basel, Switzerland) 25(19):4568. https://doi.org/10.3390/molecules25194568
50. Moret ES (2013) Trans-Atlantic diaspora ethnobotany: legacies of west African and Iberian Mediterranean migration in Central Cuba. In: African ethnobotany in the Americas Springer, pp 217–245. https://doi.org/10.1007/978-1-4614-0836-9_9. ISBN 978-1-4614-0835-2
51. Ngwoke KG, Chevallier O, Wirkom VK, Stevenson P, Elliott CT, Situ C (2014) In vitro bactericidal activity of diterpenoids isolated from Aframomum melegueta K. Schum against strains of Escherichia coli, listeria monocytogenes and Staphylococcus aureus. J Ethnopharmacol 151(3):1147–1154
52. Nwakiban APA, Fumagalli M, Piazza S, Magnavacca A, Martinelli G, Beretta G, Magni P, Tchamgoue AD, Agbor GA, Kuiaté JR, Dell'Agli M, Sangiovanni E (2020) Dietary

Cameroonian plants exhibit anti-inflammatory activity in human gastric epithelial cells. Nutrients 12(12):3787. https://doi.org/10.3390/nu12123787
53. Odetunde SK, Adekola IT, Avungbeto MO, Lawal AK (2015) Antimicrobial effect and phytochemical analysis of Aframomum melegueta on some selected bacteria and fungi. Eur J Biotechnol Biosci 3(4):15–19
54. Ogwu MC (2019) Lifelong consumption of plant-based GM foods: is it safe? In: Papadopoulou P, Misseyanni A, Marouli C (eds) Environmental exposures and human health challenges. IGI Global, Pennsylvania, pp 158–176. https://doi.org/10.4018/978-1-5225-7635-8.ch008
55. Ogwu MC (2019) Towards sustainable development in Africa: the challenge of urbanization and climate change adaptation. In: Cobbinah PB, Addaney M (eds) The geography of climate change adaptation in urban Africa. Springer Nature, Cham, pp 29–55. https://doi.org/10.1007/978-3-030-04873-0_2
56. Ogwu MC (2019) Understanding the composition of food waste: an "-omics" approach to food waste management. In: Gunjal AP, Waghmode MS, Patil NN, Bhatt P (eds) Global initiatives for waste reduction and cutting food loss. IGI Global, Pennsylvania, pp 212–236. https://doi.org/10.4018/978-1-5225-7706-5.ch011
57. Ogwu MC (2020) Value of *Amaranthus* [L.] species in Nigeria. In: Waisundara V (ed) Nutritional value of Amaranth. IntechOpen, London, pp 1–21. https://doi.org/10.5772/intechopen.86990
58. Ogwu MC (2023) Local food crops in Africa: sustainable utilization, threats, and traditional storage strategies. In: Izah SC, Ogwu MC (eds) Sustainable utilization and conservation of Africa's biological resources and environment, Sustainable development and biodiversity, vol 888. Springer, Singapore, pp 353–376. https://doi.org/10.1007/978-981-19-6974-4_13
59. Ogwu MC, Osawaru ME (2022) Traditional methods of plant conservation for sustainable utilization and development. In: Izah SC (ed) Biodiversity in Africa: potentials, threats and conservation, Sustainable development and biodiversity, vol 29. Springer, Singapore, pp 451–472. https://doi.org/10.1007/978-981-19-3326-4_17
60. Ogwu MC, Osawaru ME (2023) Disease outbreaks in ex-situ plant conservation and potential management strategies. In: Izah SC, Ogwu MC (eds) Sustainable utilization and conservation of Africa's biological resources and environment, Sustainable development and biodiversity, vol 888. Springer, Singapore, pp 497–518. https://doi.org/10.1007/978-981-19-6974-4_18
61. Ogwu MC, Osawaru ME, Aiwansoba RO, Iroh RN (2016) Ethnobotany and collection of West African Okra [*Abelmoschus caillei* (A. Chev.) Stevels] germplasm in some communities in Edo and Delta states, southern Nigeria. Borneo J Resource Sci Technol 6(1):25–36. https://doi.org/10.33736/bjrst.212.2016
62. Ogwu MC, Osawaru ME, Aiwansoba RO, Iroh RN (2016) Status and prospects of vegetables in Africa. In: Borokini IT, Babalola FD (eds) Conference proceedings of the joint biodiversity conservation conference of Nigeria Tropical Biology Association and Nigeria chapter of Society for Conservation Biology on MDGs to SDGs: toward sustainable biodiversity conservation in Nigeria. University of Ilorin, Ilorin, pp 47–57pp
63. Ogwu MC, Osawaru ME, Obahiagbon GE (2017) Ethnobotanical survey of medicinal plants used for traditional reproductive care by Usen people of Edo state, Nigeria. Malaya J Biosci 4(1):17–29
64. Ogwu MC, Osawaru ME, Owie MO (2023) Effects of storage at room temperature on the food components of three cocoyam species (*Colocasia esculenta, Xanthosoma atrovirens, and X. Sagittifolium*). Food Stud 13(2):59–83. https://doi.org/10.18848/2160-1933/CGP/v13i02/59-83
65. Ogwu MC, Osawaru ME, Amodu E, Osamo F (2023) Comparative morphology, anatomy, and chemotaxonomy of two *Cissus* Linn. species. Braz J Bot. https://doi.org/10.1007/s40415-023-00881-0
66. Okoli CO, Akah PA, Nwafor SV, Ihemelandu UU, Amadife C (2007) Anti-inflammatory activity of seed extracts of Aframomum melegueta. J Herbs Spices Med Plants 13(1):11–21

67. Okwu DE (2004) Phytochemicals and vitamin content of indigenous spices of south eastern Nigeria. J Sustain Agric Environ 6(2):30–34
68. Okwu DE (2005) Phytochemicals, vitamins and mineral contents of two Nigerian medicinal plants. Int J Mol Med Adv Sci 1(4):375–381
69. Okwu DE, Okpara M (2005) Phytochemicals, vitamins and mineral contents of two Nigerian medicinal plants. Int J Mol Med Adv Sci 1:375–381
70. Olajuyigbe OO, Adedayo O, Coopoosamy RM (2020) Antibacterial activity of defatted and nondefatted Methanolic extracts of *Aframomum melegueta* K. Schum. Against multidrug-resistant bacteria of clinical importance. ScientificWorldJOURNAL 2020:4808432. https://doi.org/10.1155/2020/4808432
71. Olowokudejo JD, Kadiri AB, Travih VA (2005) An ethnobotanical survey of herbal markets and medicinal plants in Lagos state of Nigeria. Ethnobot Leaflets 12:851–865
72. Omoboyowa DA, Aja AO, Eluu F, Ngobidi KC (2017) Effects of methanol seed extract of Aframomum melegueta (alligator pepper) on Wistar rats with 2,4 dinitro phenyl hydrazine-induced hemolytic anemia. Recent Adv Biol Med 3:11–17
73. Omotuyi IO, Nash O, Ajiboye BO, Olumekun VO, Oyinloye BE, Osuntokun OT, Olonisakin A, Ajayi AO, Olusanya O, Akomolafe FS, Adelakun N (2021) Aframomum melegueta secondary metabolites exhibit polypharmacology against SARS-CoV-2 drug targets: in vitro validation of furin inhibition. Phytother Res PTR 35(2):908–919. https://doi.org/10.1002/ptr.6843
74. Onoja SO, Omeh YN, Ezeja MI, Chukwu MN (2014) Evaluation of the in vitro and in vivo antioxidant potentials of Aframomum melegueta Methanolic seed extract. J Trop Med 2014: 159343. https://doi.org/10.1155/2014/159343
75. Osawaru ME, Ogwu MC (2014) Ethnobotany and germplasm collection of two genera of cocoyam (*Colocasia* [Schott] and *Xanthosoma* [Schott], Araceae) in Edo state Nigeria. Sci Technol Arts Res J 3(3):23–28. https://doi.org/10.4314/star.v3i3.4
76. Osawaru ME, Ogwu MC (2020) Survey of plant and plant products in local markets within Benin City and environs. In: Filho LW, Ogugu N, Ayal D, Adelake L, da Silva I (eds) African handbook of climate change adaptation. Springer Nature, Cham, pp 1–24. https://doi.org/10.1007/978-3-030-42091-8_159-1
77. Osawaru ME, Ogwu MC, Omoigui ID, Aiwansoba RO, Kevin A (2016) Ethnobotanical survey of vegetables eaten by Akwa Ibom people residing in Benin City, Nigeria. Univ Benin J Sci Technol 4(1):70–93
78. Osuntokun OT (2020) Aframomum Melegueta (grains of paradise). Ann Microbiol Infect Diseases 3(1):1–6
79. Osuntokun OT, Oluduro AO, Idowu TO, Omotuyi AO (2017) Assessment of Nephro toxicity, anti-inflammatory and antioxidant properties of epigallocatechin, Epicatechin and Stigmasterol phytosterol (synergy) derived from ethyl acetate stem bark extract of Spondias mombin on Wister rats using molecular method of analysis. J Mol Microbiol 1(103):1–11
80. Owokotomo IA, Ekundayo O, Abayomi TG, Chukwuka AV (2015) In-vitro anti-cholinesterase activity of essential oil from four tropical medicinal plants. Toxicol Rep 2:850–857. https://doi.org/10.1016/j.toxrep.2015.05.003
81. Palanisamy CP, Cui B, Zhang H, Panagal M, Paramasivam S, Chinnaiyan U, Jeyaraman S, Murugesan K, Rostagno M, Sekar V, Natarajan SP (2021) Anti-ovarian cancer potential of phytocompound and extract from south African medicinal plants and their role in the development of chemotherapeutic agents. Am J Cancer Res 11(5):1828–1844
82. Queneherve P, Serge ML, Frederic S, Virginie B (2010) Xenic culturing of plant-parasitic nematodes: artificial substrates better than soilbased culture systems. Nematropica 40:269–274
83. Quiroz D, van Andel T (2018) The cultural importance of plants in Western African religions. Econ Bot 72(3):251–262. https://doi.org/10.1007/s12231-018-9410-x
84. Rafeeq M, Murad HAS, Abdallah HM, El-Halawany AM (2021) Protective effect of 6-paradol in acetic acid-induced ulcerative colitis in rats. BMC Complement Med Ther 21(1):28. https://doi.org/10.1186/s12906-021-03203-7

85. Riera CE, Menozzi-Smarrito C, Affolter M, Michlig S, Munari C, Robert F, Vogel H, Simon SA, le Coutre J (2009) Compounds from Sichuan and Melegueta peppers activate, covalently and non-covalently, TRPA1 and TRPV1 channels. Br J Pharmacol 157(8):1398–1409. https://doi.org/10.1111/j.1476-5381.2009.00307.x
86. Sathasivam R, Ki JS (2018) A review of the biological activities of microalgal carotenoids and their potential use in healthcare and cosmetic industries. Mar Drugs 16(1):26. https://doi.org/10.3390/md16010026
87. Sekercioglu CH, Daily GC, Ehrlich PR (2004) Ecosystem consequences of bird declines. Proc Nat Acad Sci USA 101:18042–18047
88. Sudeep HV, Aman K, Jestin TV, Shyamprasad K (2022) *Aframomum melegueta* seed extract with standardized content of 6-paradol reduces visceral fat and enhances energy expenditure in overweight adults – a randomized double-blind, placebo-controlled clinical study. Drug Des Devel Ther 16:3777–3791. https://doi.org/10.2147/DDDT.S367350
89. Sugita J, Yoneshiro T, Sugishima Y, Ikemoto T, Uchiwa H, Suzuki I, Saito M (2014) Daily ingestion of grains of paradise (Aframomum melegueta) extract increases whole-body energy expenditure and decreases visceral fat in humans. J Nutr Sci Vitaminol 60(1):22–27. https://doi.org/10.3177/jnsv.60.22
90. Tirado-Kulieva VA, Hernández-Martínez E, Minchán-Velayarce HH, Pasapera-Campos SE, Luque-Vilca OM (2023) A comprehensive review of the benefits of drinking craft beer: role of phenolic content in health and possible potential of the alcoholic fraction. Curr Res Food Sci 6:100477. https://doi.org/10.1016/j.crfs.2023.100477
91. Titanji VP, Zofou D, Ngemenya MN (2008) The antimalarial potential of medicinal plants used for the treatment of malaria in Cameroonian folk medicine. Afr J Tradit Complement Altern Med 5(3):302–321
92. Umukoro S, Ashorobi RB (2007) Further studies on the antinociceptive action of aqueous seed extract of Aframomum melegueta. J Ethnopharmacol 109(3):501–504. https://doi.org/10.1016/j.jep.2006.08.025
93. Urquiaga I, Leighton F (2000) Plant polyphenol antioxidants and oxidative stress. Biol Res 33(2):55–64
94. Venugopal R, Liu RH (2012) Phytochemicals in diets for breast cancer prevention: the importance of resveratrol and ursolic acid. Food Sci Human Wellness 1(1):1–13
95. Villarreal JC, Cargill DC, Hagborg A, Söderstrom L, Renzaglia KS (2010) A synthesis of hornwort diversity: patterns, causes and future work. Phytotaxa 9(1):150–166
96. Yoneshiro T, Matsushita M, Sugita J, Aita S, Kameya T, Sugie H, Saito M (2021) Prolonged treatment with grains of paradise (Aframomum melegueta) extract recruits adaptive thermogenesis and reduces body fat in humans with low Brown fat activity. J Nutr Sci Vitaminol 67(2):99–104. https://doi.org/10.3177/jnsv.67.99

Turkey Berry (*Solanum torvum* Sw. [Solanaceae]): An Overview of the Phytochemical Constituents, Nutritional Characteristics, and Ethnomedicinal Values for Sustainability

8

Matthew Chidozie Ogwu, Afamefuna Dunkwu-Okafor, Ichehoke Austine Omakor, and Sylvester Chibueze Izah

Contents

M. C. Ogwu (✉)
Goodnight, Family Department of Sustainable Development, Appalachian State University, Boone, NC, USA
e-mail: ogwumc@appstate.edu

A. Dunkwu-Okafor
Department of Microbiology, Faculty of Life Sciences, University of Benin, Benin City, Edo State, Nigeria
e-mail: afamefuna.dunkwu-okafor@uniben.edu

I. A. Omakor
Department of Plant Science and Biotechnology, Faculty of Science, Dennis Osadebay University, Anwai, Asaba, Nigeria
e-mail: Omakorichehoke@dou.edu.ng

S. C. Izah
Department of Microbiology, Faculty of Science, Bayelsa Medical University, Yenagoa, Bayelsa State, Nigeria

S. C. Izah et al. (eds.), *Herbal Medicine Phytochemistry*, Reference Series in Phytochemistry, https://doi.org/10.1007/978-3-031-43199-9_73

Abstract

This chapter aims to discuss the phytomedicinal value of turkey berry (*Solanum torvum* Sw.). The shrub or small tree, turkey berry, is native to tropical and subtropical regions of Africa, Asia, and the Americas and has a long history of use in various culinary and traditional medicinal practices across different cultures. However, it is highly adaptable to different ecological niches, which allows it to grow in a variety of habitats, from coastal areas to highlands. The eggplant or tomato-like fruits of *S. torvum* have a unique aroma, sweet taste, and high phenolic contents that are essential to the dietary and medicinal properties of the plant. The plant is rich in vitamins, minerals, dietary fiber, protein, and secondary phytochemicals like flavonoids, alkaloids, tannins, saponins, essential oils, glycosides, and steroids. It is used to prepare curry and gravy dishes, stir-fries, pickles and chutneys, soups and stews, sauces and sambals, fried snacks, stuffed dishes, rice-based dishes, and other traditional dishes. The ethnobotanical values of *S. torvum* include rituals and symbolism, culinary traditions, economic uses, and traditional medicine. The plant is used to aid digestion as well as an anti-inflammatory, antioxidant, antimicrobial, immune modulation, cardiovascular health maintenance, wound healing, antidiabetic, weight management, and fever reduction. Clinical research on turkey berry is ongoing, and there is still much to learn and explore regarding the potential health benefits and therapeutic applications of the plant. Some herbal products containing turkey berry include herbal teas, herbal powders, herbal tinctures and extracts, capsules, herbal blends, and infusions. Turkey berry is generally considered safe for consumption when used in moderation as a food or traditional herbal remedy. However, like many botanicals, excessive or improper use can potentially lead to adverse effects.

Keywords

Ethnobotany · Clinical research · Herbal products · Phytochemicals · Toxicity · Antimicrobials

1 Introduction

Turkey berry (*Solanum torvum* Sw.), also known as wild eggplant, devil's fig, or pea eggplant, Sundakkai in Tamil Nadu, India, Makheua Phuang in Thai, and Izote in Mexico, is a plant that belongs to the Solanaceae family, which includes other well-known members such as tomatoes, potatoes, and bell peppers [5, 42]. This plant is native to tropical and subtropical regions of Africa, Asia, and the Americas. Turkey berry has a long history of use in various culinary and traditional medicinal practices across different cultures [3, 25, 68].

Understanding the botanical characteristics of turkey berry is essential for recognizing, cultivating, utilizing, and conserving the plant. Turkey berry is a perennial plant that can grow as a shrub or small tree, typically reaching heights of 1 to 3 m (3 to 10 ft). It features woody stems and branches with numerous leaves and fruit clusters [25]. The leaves of turkey berry are simple, alternate, and dark green. They are generally elliptical or ovate, with a slightly serrated margin [66]. The flowers of *S. torvum* grow in clusters on the plant and are typically small, star-shaped, and white to pale lavender. The fruit of turkey berry is a small, round, or oval berry, about 1–2 cm (0.4–0.8 in.) in diameter [110]. When mature, the fruits turn from green to yellow or orange and contain numerous seeds. The fruit resembles miniature eggplants or tomatoes. They have a unique aroma, sweet taste, and high phenolic contents that are essential to the dietary and medicinal properties of the plant [126]. It is worth noting that different cultivars of turkey berry exist with variations in fruit size and color. Some have green or white fruits, while others have red or purple ones. The plant is well adapted to tropical and subtropical climates and can thrive in a variety of soil types, including sandy, loamy, and clay soils [116]. It is often found in disturbed habitats, such as roadsides and fallow fields, and it often grows as a weed in agricultural fields [64, 130].

Turkey berry is a widely distributed plant, with native populations found in various regions around the world. In Africa, turkey berry is native to several African countries, including Nigeria, Ghana, Cameroon, Madagascar, and other tropical and subtropical regions across the continent [57, 132]. In Asia, *S. torvum* is prevalent in countries such as India, Sri Lanka, Malaysia, Thailand, and Indonesia. Turkey berry has also been reported in the Americas, particularly in tropical regions of the Caribbean, Central America, and South America [8, 42]. Due to its adaptability and hardiness, turkey berry has become naturalized in many other parts of the world, including parts of the United States, Australia, and various Pacific islands [46].

Turkey berry has a rich history of ethnobotanical uses in traditional medicine and culinary traditions across different cultures. Its diverse applications include:

Culinary Uses: Jena et al. [63] reported that turkey berry is commonly used in cooking various culinary dishes, particularly in Asian (South Indian) and African cuisines. It is added to soups, stews, curries, and stir-fries. The bitter and slightly tangy taste of turkey berry adds a unique flavor dimension to dishes. It is often used to balance the flavors of other ingredients.

Traditional Medicine: This includes

- Digestive Aid: In traditional medicine, turkey berry is used to aid digestion and alleviate gastrointestinal discomfort [19].
- Anti-inflammatory and Antioxidant: Some cultures use turkey berry as a natural remedy for inflammatory conditions and the plant's antioxidant properties have led to its use as a remedy for various health issues [25, 86].
- Fever Reduction: In some regions, turkey berry is employed to help reduce fever and symptoms of illnesses [5].

Pest Control: The leaves and fruits of turkey berry contain compounds that can act as natural insect repellents [33, 66]. For this role, they are sometimes placed in storage containers to protect grains and pulses from pests [94–96, 100].

Livestock Feed: In some areas, turkey berry leaves and fruits are used as supplemental fodder for livestock [12, 38, 90, 104].

This chapter aims to discuss the phytomedicinal value of turkey berry. It will present the origin, distribution, and botany of *S. torvum* wherein it is shown that the widespread distribution of the plant has led to its incorporation into a variety of culinary and medicinal traditions. This includes culinary and traditional medicine uses. The next sections address the nutritional and phytochemical properties as well as the food uses of turkey berry. Turkey berry is a versatile ingredient in many cuisines. It is used in curries, soups, stews, and stir-fries. Its slightly bitter and tangy flavor adds depth to dishes. Turkey berry has a history of use in traditional medicine across its distribution areas. Other parts of the chapter are on the ethnobotanical, pharmacognosy, and clinical works, herbal products, toxicity, and commercial values of turkey berry. It is believed to have various medicinal properties, including digestive, anti-inflammatory, and antimicrobial effects. The adaptability of turkey berry to various regions and environmental conditions has allowed them to become an integral part of diverse culinary traditions and traditional healing practices. The plant's versatility, from its multiple uses in regional cuisines to its potential health benefits, underscores its significance in global biodiversity and human culture.

2 Origin and Distribution of Turkey Berry (*Solanum torvum*)

The exact origin of *S. torvum* is unknown but researchers believe that the plant likely originated in tropical and subtropical regions of Africa, Asia, and South America where it can be found in the wild [48, 132]. In addition to being found in the wild in these regions, it has cultural and agricultural significance through a long history of cultivation and use in traditional cuisines and medicinal practices. Turkey berry is well-adapted to tropical and subtropical climates, where it thrives in warm and humid conditions. It is known for its ability to grow in a variety of soil types, including sandy, loamy, and clay soils, as long as there is good drainage. The plant can be found in disturbed habitats such as roadsides, fallow fields, and areas with disrupted vegetation. Turkey berry can be cultivated in gardens and farms, but it also grows as a wild plant in many regions. Turkey berry is commonly found in tropical and subtropical regions of Africa, Asia, the Americas, and other parts of the world. Its adaptability to different ecological niches allows it to grow in a variety of habitats, from coastal areas to highlands. Genetic studies have indicated high genetic diversity in *S. torvum* populations in Africa and Asia, reinforcing the idea of its long-term presence in these areas [134, 135]. Also, there have been archaeological findings of *S. torvum* seeds in ancient sites in Africa, Asia, and South America providing further evidence of its historical cultivation and consumption. Within these regions, turkey berry goes by a multitude of local names in different regions, reflecting its cultural significance and versatility in traditional dishes. Some of these names include Igbo or Oha (Igbo language) in Nigeria where it is a key ingredient in soups like Oha soup, Sundakkai and Bhatkatiya in Tamil and Hindi (India) where it is a popular addition to South Indian cuisine [10, 124]. It is called Makheua Phuang in Thailand and is used in various dishes, including curries and stir-fries. It is commonly called wild eggplant or bitter ball in the Caribbean and is used in traditional recipes. In Brazil, turkey berry is called Berinjela-do-mato and is used in many Brazilian local cuisines [79].

3 Distribution and Naturalized Regions of Turkey Berry (*Solanum torvum*)

Turkey berry has an extensive distribution that spans multiple continents and regions because it thrives in a variety of climates and habitats, making it a resilient and adaptable plant. Due to its adaptability and ability to grow in a wide range of conditions, turkey berry has become naturalized in various other regions around the world. Some examples include the United States (found in states like Florida and Hawaii, where it grows as a weed), Australia, and the Pacific Islands [47, 50]. Turkey berry exhibits adaptability to diverse habitats and climatic conditions, which contributes to its widespread distribution. It can thrive in:

(a) **Africa:**

West Africa: Turkey berry is native to several West African countries, including Nigeria, Ghana, Cameroon, and Ivory Coast. It is a staple in many West African cuisines, where it is used in dishes such as soups and stews [49, 99].

East Africa: The plant is also found in East African countries like Kenya, Tanzania, and Uganda, where it is used both in cooking and traditional medicine.

Southern Africa: Turkey berry is present in Southern African countries like Zambia and Zimbabwe, where it is known by various local names.

(b) **Asia:**

India: Turkey berry is widely cultivated and consumed in India, where it is a common ingredient in various regional cuisines.

Southeast Asia: In countries like Thailand, Malaysia, and Indonesia, turkey berry is used in curries, stir-fries, and traditional dishes.

Sri Lanka: Known as Waran, it is a popular vegetable in Sri Lankan cuisine.

Philippines: In the Philippines, it is called tabatsoy and is used in local dishes.

China: Turkey berry is found in parts of southern China.

(c) **Americas:**

Caribbean: Turkey berry is native to the Caribbean region, where it is used in local dishes and traditional remedies. It is known as a wild eggplant in some Caribbean countries.

Central America: In countries like Costa Rica and Nicaragua, it is called berenjena cimarrona and is used in traditional dishes.

South America: Turkey berry is present in South American countries such as Brazil, Colombia, and Venezuela, where it is known as berenjena cimarrona or jilo.

(d) **Tropical and Subtropical Regions**: Turkey berry prefers warm and humid tropical and subtropical climates, making it well-suited for regions with consistent temperatures and rainfall [34, 125].

The distribution and naturalization of Turkey berry in these regions is connected to

- **Various Soil Types**: The plant can grow in a variety of soil types, including sandy, loamy, and clay soils, as long as there is good drainage.
- **Disturbed Habitats**: It often thrives in disturbed habitats, such as roadsides, fallow fields, and areas with disrupted vegetation.

– **Cultivated Settings**: While it can grow as a wild plant, turkey berry is also cultivated in gardens and farms in many regions for its culinary and medicinal uses.

4 Botany and Cultivation of Turkey Berry (*Solanum torvum*)

Solanum torvum is a perennial plant that can grow as either a shrub or a small tree. Its growth habit depends on environmental conditions and cultivation practices [7, 121]. Depending on the growth conditions and regional variations, turkey berry plants can appear as compact shrubs with dense foliage or as small trees with a more open canopy. The plant's overall appearance can be bushy or somewhat spindly, with multiple branches and leaves (Fig. 1). The plant has a fibrous root system. The size of turkey berry plants can vary widely. They typically range from 1 to 3 m (3 to 10 ft) in height, but some individuals can grow taller under favorable conditions [120]. The plant features woody stems and branches. These stems are often green when young, turning brown or gray as they mature. Turkey berry has simple, alternate leaves that are dark green and slightly glossy. The leaves are typically elliptical or ovate with a slightly serrated margin [70]. The leaves are typically dark green, but leaf color can vary depending on factors like age and environmental conditions. The leaf blades are usually elliptical or ovate, with a smooth or slightly serrated margin. The leaves are generally smooth and slightly glossy [9, 51]. The leaves of turkey berry are simple, meaning they consist of a single leaf blade. They are arranged alternately along the stems. The flowers of *S. torvum* are small, star-shaped, and white to pale lavender. They are typically arranged in clusters or inflorescences [52]. Turkey berry produces small, star-shaped flowers that are typically white to pale lavender. The flowers are arranged in clusters or inflorescences at the tips of branches. These flowers are hermaphroditic, meaning they have both male and female reproductive structures within the same flower. The fruit of turkey berry is a small, round, or oval berry, typically measuring

Fig. 1 Leaves of turkey berry (*S. torvum*)

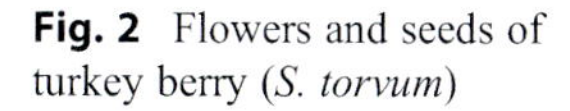

Fig. 2 Flowers and seeds of turkey berry (*S. torvum*)

about 1 to 2 cm (0.4 to 0.8 in.) in diameter [18, 80, 84]. When ripe, the fruit changes from green to yellow or orange and contains numerous small seeds (Fig. 2). The fruit bears a resemblance to miniature eggplants or tomatoes. The fruit of *S. torvum* is a small berry that starts green and matures to yellow or orange. When the fruit is ripe, it is typically 1 to 2 cm (0.4 to 0.8 in.) in diameter. Each fruit contains numerous tiny seeds [61, 74]. The fruit has a slightly bitter and tangy taste, which contributes to its culinary and culinary appeal. It is worth noting that *S. torvum* variants and cultivars exist with differences in fruit size, color, and taste, depending on the region and local preferences. These features are linked to factors such as soil type, climate, and regional genetic variations [59, 133]. These characteristics contribute to the plant's adaptability and versatility in different environments and culinary traditions.

Cultivating turkey berry can be a rewarding endeavor due to the increasing culinary and medical application of the crop. It is also not challenging because the crop is highly versatile [76, 105]. Turkey berry is well-suited to tropical and subtropical climates, and it can thrive in a variety of soil types. It thrives in warm and humid tropical and subtropical climates but prefers temperatures between 25 °C and 35 °C (77 °F to 95 °F) and cannot tolerate frost. The planting location must also receive plenty of sunlight, as turkey berry requires full sun for optimal growth [118]. Turkey berry can grow in various soil types, but it prefers well-drained soil with a pH level between 6.0 and 7.5 [122]. Soil fertility may also be improved by adding organic matter, such as compost or well-rotted manure, to enhance nutrient availability [106]. Turkey berry can be grown from seeds or propagated from stem cuttings. If planting from seeds, soak the seeds in water for 24 h before planting to improve germination rates. Sow the seeds directly into the prepared soil, spacing them about 1 to 2 ft apart in rows or a well-prepared garden bed [117].

Water the seeds or cuttings thoroughly after planting. Turkey berry requires consistent moisture for optimal growth. Keep the soil evenly moist but not waterlogged. Water deeply when the soil surface begins to dry and ensure good drainage to prevent waterlogging, which can lead to root rot. Applying a layer of mulch around the base of the plants helps retain moisture, suppress weeds, and regulate soil temperature [129]. Organic mulch like straw or compost is ideal. Turkey berry benefits from regular fertilization with balanced, slow-release fertilizers [123, 127]. Farmers can use a general-purpose fertilizer or one specifically formulated for vegetables. Follow the recommended dosage on the fertilizer label and avoid over-fertilizing, as excessive nitrogen can result in excessive vegetative growth at the expense of fruit production. Pruning can help manage the growth of turkey berry plants and encourage better fruit production. Pinch back the growing tips of young plants to encourage bushier growth [30, 31]. Provide support or trellises for the plants, especially if they are grown as tall shrubs or small trees, to prevent sprawling and improve air circulation. Turkey berry is relatively resistant to pests and diseases, but occasional issues may arise. Monitor your plants regularly and take appropriate measures if pests or diseases are detected [32, 62]. Common pests of turkey berry include aphids, whiteflies, and fruit flies. Use organic pest control methods or insecticidal soap if necessary. Inspect the plants regularly for signs of diseases like powdery mildew or leaf spot, and treat promptly with appropriate fungicides if needed [20, 58]. Turkey berry typically begins to produce fruits within a few months after planting [119, 128]. Harvest the fruits when they are firm, green, and about 1 to 2 cm (0.4 to 0.8 in.) in diameter. Ripe fruits turn yellow or orange but should be harvested before they become too ripe to maintain optimal flavor. Harvesting regularly encourages continued fruit production. It is a versatile and flavorful addition to many culinary dishes, and its adaptability to various soil types and climates makes it an attractive option for home gardeners and farmers in tropical and subtropical regions.

The diversity of *S. torvum* encompasses a range of aspects, including its botanical characteristics, cultivars, regional variations, and uses in different cultures [1]. This diversity highlights the plant's adaptability and versatility in various environments and its rich history of culinary and medicinal applications.

Botanical Diversity of *Solanum torvum*:

- Varieties: *Solanum torvum* variants and cultivars have differences in fruit size, color, and taste. These variations are found in different regions and have led to local preferences for specific types [6, 29].
- Plant Morphology: The size and shape of turkey berry plants can vary. They can grow as small shrubs or small trees, depending on environmental conditions and local cultivation practices.

Regional Diversity of *Solanum torvum*:

- Culinary Uses: Turkey berry is used differently in various culinary traditions around the world. It is incorporated into diverse dishes, including soups, stews, curries, and stir-fries, each with its unique flavors and cooking methods.
- Local Names: Turkey berry is known by numerous local names in different regions, reflecting its cultural significance. For example, it is called Sundakkai in Tamil Nadu, India, Makheua Phuang in Thailand, and njama njama in Cameroon. These names emphasize its role in local cuisines.
- Traditional Medicine: The uses of turkey berry in traditional medicine also vary by region. Different cultures have developed remedies based on their

understanding of the plant's properties, resulting in diverse applications for ailments ranging from digestive issues to inflammation [87].

Culinary Diversity of *Solanum torvum*:

- Preparation Methods: The culinary diversity of turkey berry is evident in various ways it is prepared. It can be fried, roasted, boiled, or used fresh, depending on the dish and cultural practices.
- Complementary Ingredients: The choice of ingredients that accompany turkey berry varies by region. Some cultures pair it with spices, coconut, or other vegetables to create unique flavor profiles.
- Dish Types: Turkey berry is used in a wide array of dishes, such as Indian sambar, Thai green curry, Filipino pinakbet, and Ghanaian kontomire stew. The plant's adaptability to different cuisines underscores its versatility.

Variability in Traditional Medicine:

- Remedial Practices: Traditional medicinal uses of turkey berry are diverse and can include treatments for digestive problems, fever, inflammation, and more. Local herbalists and healers may have distinct methods of preparation and application.
- Preparation Techniques: The methods for preparing turkey berry-based remedies can vary, such as infusions, decoctions, poultices, or direct application of crushed leaves or fruits.

Environmental Adaptability:

- Habitat Range: Turkey berry exhibits adaptability to various ecological niches. It can thrive in tropical and subtropical regions, ranging from coastal areas to highlands, and in different soil types [22].
- Cultivation Practices: Depending on local agricultural practices, turkey berry may be cultivated in gardens, farms, or allowed to grow naturally in fields and disturbed habitats. Its adaptability makes it suitable for both subsistence and commercial cultivation.

Nutritional Diversity of *Solanum torvum*:

- Nutrient Content: The nutritional composition of turkey berry can vary based on factors like soil quality, cultivation methods, and maturity at harvest. Variations in vitamins, minerals, and phytochemicals are observed [14].
- Dietary Significance: Turkey berry provides essential nutrients and dietary fiber, contributing to the nutritional diversity of diets in regions where it is consumed regularly.

Medicinal Versatility of *Solanum torvum*:

- Health Conditions: The potential health benefits of turkey berry are diverse, including antioxidant, anti-inflammatory, and digestive properties. Different cultures may emphasize specific health aspects based on their traditional knowledge.

- Application Methods: Traditional healers and practitioners may employ various parts of the plant, including leaves, fruits, or roots, for medicinal purposes, leading to a diverse range of applications.

Variability in Preparation and Presentation:

- Culinary Creativity: The culinary diversity of turkey berry is marked by the creativity in its preparation. It can be a prominent ingredient or a complementary one, lending unique flavors and textures to dishes.
- Regional Specialties: Specific dishes featuring turkey berry become regional specialties, celebrated for their unique taste and cultural significance.

5 Nutritional Properties and Food Uses of Turkey Berry (*Solanum torvum*)

Turkey berry is not only valued for its culinary and medicinal uses but also for its nutritional content. It is important to note that turkey berry has a slightly bitter and tangy taste, which can be an acquired taste for some. Its culinary use varies by region, and it is prized for its ability to add depth of flavor to dishes [2, 4, 91–93]. Turkey berry is not only valued for its culinary and medicinal uses but also for its nutritional content [11]. While the nutrient composition can vary depending on factors such as maturity and cultivation methods, these are some key nutritional properties:

1. **Vitamins:** Turkey berry is a source of various vitamins, including vitamin C, vitamin A, and vitamin B-complex vitamins like niacin and riboflavin.
2. **Minerals:** It contains essential minerals such as potassium, calcium, and iron, which are important for maintaining overall health.
3. **Dietary Fiber:** Turkey berry is a good source of dietary fiber, which aids in digestion and supports gastrointestinal health [60].
4. **Protein:** While not a primary source of protein, turkey berry contains a moderate amount, making it a valuable addition to vegetarian diets.
5. **Secondary Phytochemicals:** The plant is rich in phytochemicals, including flavonoids, alkaloids, and phenolic compounds, which contribute to its potential health benefits [13].

Solanum torvum is widely used in culinary traditions across different cultures. It is appreciated for its unique flavor and versatility in a variety of dishes. Some common food uses of turkey berry include:

1. **Curry and Gravy Dishes:** Turkey berry is a popular ingredient in many curry and gravy dishes, particularly in South Asian and Southeast Asian cuisines. It adds a slightly bitter and tangy flavor to the dishes, complementing the richness of the sauce.

2. **Stir-Fries:** In stir-fry dishes, turkey berry is often used alongside other vegetables and protein sources. Its firm texture holds up well to stir-frying, and it adds a distinct taste to the overall dish.
3. **Pickles and Chutneys:** In some regions, turkey berry is used to prepare pickles and chutneys. The berries are typically marinated with spices, oil, and other seasonings. These pickles and chutneys can be served as condiments or side dishes.
4. **Soups and Stews:** Turkey berry can be added to soups and stews for extra flavor and nutrition. It is often used in traditional dishes like sambar (a South Indian lentil stew) and sinigang (a Filipino sour soup).
5. **Sauces and Sambals:** Turkey berry is sometimes included in sauces and sambals (spicy condiments) to enhance their flavor profiles. It can be used in combination with other ingredients like chilies, garlic, and herbs.
6. **Fried Snacks:** In some cultures, turkey berry is coated in batter and deep-fried to make crispy snacks. The fried turkey berries are enjoyed for their crunchy texture and savory taste.
7. **Stuffed Dishes:** Turkey berry can be stuffed with various fillings, such as spices, ground meat, or lentils, and then cooked. Stuffed turkey berries are common in Indian cuisine and other regional cuisines.
8. **Rice and Rice Dishes:** In certain dishes, turkey berry is added to rice during cooking to infuse its flavor into the grains. It is also used in rice-based dishes like biryani and pilaf.
9. **Traditional Dishes:** Turkey berry is a key ingredient in traditional dishes like njama njama in Ghana, Terong Belanda in Indonesia, and Koottu in South India. These dishes are celebrated for their unique combination of flavors.
10. **Medicinal Preparations:** While not a food use, turkey berry is sometimes used in medicinal preparations, including herbal teas and infusions.

6 Phytochemical Constituents and Potential Health Benefits of Turkey Berry (*Solanum torvum*)

Turkey berry (*S. torvum*) contains a variety of phytochemical constituents, which are natural compounds found in plants. These phytochemicals contribute to the plant's flavor, aroma, and potential health benefits. Some of the notable phytochemical constituents found in turkey berry include:

1. **Alkaloids:** Alkaloids are naturally occurring organic compounds that often have physiological effects on humans and animals [97, 103]. Turkey berry contains various alkaloids, including solanidine and solasonine. Alkaloids may have diverse effects, ranging from antimicrobial to anti-inflammatory properties [15, 101].
2. **Flavonoids:** Flavonoids are a group of polyphenolic compounds found in many fruits and vegetables, including turkey berry. These compounds are known for

their antioxidant properties and potential health benefits, such as reducing oxidative stress and inflammation.

3. **Glycosides:** Glycosides are compounds composed of a sugar molecule (glycone) and a non-sugar moiety (aglycone). Turkey berry contains various glycosides, which can have different biological activities. Some glycosides may play a role in the plant's medicinal properties [23].
4. **Tannins:** Tannins are polyphenolic compounds found in turkey berry and other plants. They can contribute to the astringent taste of certain fruits and are often used in traditional medicine. Tannins have antioxidant properties and may have applications in wound healing and digestive health [35].
5. **Phenolic Compounds:** Phenolic compounds, including phenolic acids and phenolic glycosides, are present in turkey berry [36, 102]. These compounds are known for their antioxidant and anti-inflammatory activities. They may contribute to the potential health benefits of the plant.
6. **Steroids and Triterpenoids:** Turkey berry contains steroidal compounds and triterpenoids. These compounds are part of the plant's chemical makeup. Some steroidal compounds may have anti-inflammatory properties [28, 75].
7. **Saponins:** Saponins are glycosides with a characteristic foaming property. They are found in various plant species, including turkey berry. Saponins may have immune-modulating and cholesterol-lowering effects [21, 78].
8. **Essential Oils:** Some varieties of turkey berry may contain essential oils, which are volatile aromatic compounds responsible for the plant's fragrance. These oils can vary in composition and may contribute to the plant's flavor.

It is important to note that the phytochemical composition of turkey berry can vary depending on factors such as plant variety, growing conditions, and maturity at harvest [37]. While these phytochemicals contribute to the plant's potential health benefits, further research is needed to better understand their specific effects and mechanisms of action. Additionally, the presence of these compounds in turkey berry has made it a subject of interest in both traditional and modern herbal medicines, as well as in culinary applications. The consumption of turkey berry has been associated with several potential health benefits, although further research is needed to fully understand its mechanisms and efficacy. Some of these benefits include:

1. **Antioxidant Properties:** The presence of phytochemicals in turkey berry gives it the typical antioxidant properties, which can help combat oxidative stress and reduce the risk of chronic diseases.
2. **Anti-inflammatory Effects:** Turkey berry may possess anti-inflammatory properties, which could be beneficial for individuals dealing with inflammatory conditions.
3. **Digestive Health:** Traditionally, turkey berry has been used to support digestive health and alleviate gastrointestinal discomfort [53].
4. **Antimicrobial Activity:** Some studies suggest that turkey berry may have antimicrobial properties, which can help protect against various infections [39].

5. **Antidiabetic Potential:** Preliminary research indicates that turkey berry may play a role in blood sugar regulation and could be explored as an adjunct in diabetes management.
6. **Weight Management:** The dietary fiber in turkey berry can contribute to feelings of fullness, potentially aiding in weight management.
7. **Nutrient Boost:** Incorporating turkey berry into the diet can provide essential vitamins and minerals that support overall health [77].

7 Ethnobotanical Value of Turkey Berry (*Solanum torvum*)

The ethnobotany of turkey berry (*S. torvum*) reveals its rich history of traditional uses across various cultures and regions. It underscores the multifaceted role of the plant in various cultures and regions. From culinary traditions to traditional medicine and cultural symbolism, this plant has a long history of diverse uses. Its adaptability and versatility make it an integral part of many culinary and healing traditions, contributing to its enduring importance in the ethnobotanical landscape.

This versatile plant has been valued for centuries for its culinary, medicinal, and cultural significance.

Rituals and Symbolism: In certain cultures, turkey berry holds symbolic significance and is used in rituals, ceremonies, and offerings. It may be included in traditional healing rituals or as a protective charm.

Culinary Traditions: Turkey berry is a staple ingredient in many traditional dishes and contributes to the cultural identity of the cuisine in which it is used [54].

Economic Uses: Turkey berry is cultivated and harvested for both domestic consumption and commercial purposes, contributing to the livelihoods of farmers and local economies in regions where it is grown [16, 56, 95].

Traditional Medicine: Turkey berry has a long history of medicinal uses in traditional herbal medicine systems across different cultures and regions. While scientific research is ongoing to validate some of these traditional uses, some of the common medicinal uses attributed to turkey berry include:

- **Digestive Health**: This is one of the most widely recognized traditional uses of turkey berry is its role in promoting digestive health. It is believed to have mild laxative properties and is used to alleviate constipation and improve bowel regularity. Turkey berry is also used to relieve indigestion, bloating, and gas [55].
- **Anti-inflammatory Properties**: In traditional medicine, turkey berry is used to reduce inflammation and alleviate pain associated with conditions like arthritis and joint pain. Poultices made from crushed leaves or heated leaves are sometimes applied topically to reduce inflammation.
- **Antipyretic (Fever-Reducing) Effects**: In some cultures, turkey berry is used to lower fever and manage symptoms of fever-related illnesses. It is believed to have antipyretic properties that help reduce body temperature during fever [17, 131].

- **Diuretic and Kidney Health**: Turkey berry has diuretic properties and is used in some regions to promote urination and support kidney health. It is believed to help in the elimination of toxins and excess fluids from the body [88].
- **Wound Healing**: In certain traditional medicine systems, turkey berry is used topically as a poultice or ointment to promote wound healing. It is believed to have antimicrobial properties that may help prevent infection.
- **Respiratory Health**: Some traditional remedies include turkey berry for managing respiratory conditions such as coughs and colds. It is believed to have properties that can help alleviate symptoms of respiratory ailments [45].
- **Antioxidant Effects**: Turkey berry contains antioxidants, including flavonoids and polyphenols, which may help combat oxidative stress and reduce the risk of chronic diseases. These antioxidants are thought to play a role in overall health and wellness.
- **Potential Antidiabetic Effects**: Preliminary studies have suggested that turkey berry may have potential antidiabetic effects. Some research indicates that it may help lower blood sugar levels, making it of interest in the management of diabetes.

It is important to note that while turkey berry has a history of traditional medicinal use, scientific research is ongoing to better understand its active compounds and their potential therapeutic benefits [40]. As with any herbal remedy, it is advisable to consult with a healthcare professional before using turkey berry or its extracts for medicinal purposes, especially if you have underlying health conditions or are taking other medications. Additionally, the quality and safety of herbal preparations can vary, so it is essential to use them with caution and under appropriate guidance.

8 Pharmacognosy and Clinical Research on Turkey Berry (*Solanum torvum*)

One way by which modern scientific research is exploring the potential health benefits of turkey berry and its bioactive compounds for better understanding and to lead to new applications and products is pharmacognosy and clinical research trials. Pharmacognosy is the study of natural products obtained from plants and their applications in medicine. *S. torvum* has been a subject of interest in pharmacognosy due to its traditional medicinal uses and the presence of various phytochemical constituents. Pharmacognostic works on turkey berry typically involve one of the following as outlined in Nurit-Silva et al. [89]:

1. **Morphological Identification:** Pharmacognosy begins with the accurate identification and authentication of the plant material. In the case of turkey berry, this involves recognizing its distinct morphological characteristics, including the growth habit, leaves, flowers, and fruits.
2. **Phytochemical Analysis:** Pharmacognostic studies of turkey berry involve the analysis of its phytochemical constituents. This includes identifying and

quantifying alkaloids, flavonoids, glycosides, tannins, phenolic compounds, steroids, triterpenoids, saponins, and other secondary metabolites present in the plant. Various analytical techniques, such as chromatography and spectroscopy, may be employed to determine the composition of phytochemicals.

3. **Quality Control:** Pharmacognosy plays a vital role in the quality control and standardization of herbal medicines and dietary supplements containing turkey berry extracts. Establishing quality parameters ensures the consistency and safety of herbal products.
4. **Medicinal Uses:** Pharmacognostic studies provide insights into the traditional medicinal uses of turkey berry. This includes documenting its historical applications in treating digestive disorders, inflammatory conditions, fever, and more. The identification of specific bioactive compounds helps understand the plant's pharmacological properties.
5. **Pharmacological Studies:** Pharmacognosy contributes to the exploration of turkey berry's pharmacological activities. Researchers investigate its potential as an anti-inflammatory, antioxidant, analgesic, diuretic, and more. Animal and in vitro studies are conducted to evaluate the plant's bioactivity.
6. **Formulation Development:** Turkey berry may be incorporated into various herbal formulations, traditional medicines, or dietary supplements. Pharmacognostic data assist in selecting the appropriate parts of the plant, extracting methods, and dosage forms for formulation.
7. **Toxicological Studies:** Assessing the safety of turkey berry is an essential aspect of pharmacognosy. Toxicological studies help determine the plant's potential adverse effects and safe usage levels.
8. **Standardization and Regulation:** Pharmacognostic data contribute to the development of monographs and standards for turkey berry in pharmacopeias and regulatory frameworks.
   - Standardization ensures the consistency and quality of herbal products containing turkey berry.
9. **Future Research:** Ongoing pharmacognostic research may uncover new phytochemical constituents and potential medicinal uses of turkey berry. Modern pharmacological and clinical studies can validate its traditional applications and explore new therapeutic avenues.

Clinical research on *S. torvum* is ongoing, and while there is still much to learn, several studies have explored its potential health benefits and therapeutic applications. Some areas of clinical research related to turkey berry include:

1. **Antioxidant and Anti-inflammatory Properties**: Some studies have investigated the antioxidant and anti-inflammatory effects of turkey berry. Antioxidants can help protect cells from oxidative stress, while anti-inflammatory properties may have implications for conditions like arthritis [65].
2. **Antidiabetic Effects**: Research has examined the potential antidiabetic properties of turkey berry. It has been suggested that it may help lower blood sugar levels, making it of interest in diabetes management [107].

3. **Gastrointestinal Health**: Turkey berry's traditional use as a digestive aid has prompted studies to assess its effects on gastrointestinal health. Research has explored its potential role in alleviating indigestion, promoting regular bowel movements, and reducing symptoms of gastrointestinal disorders [67].
4. **Antimicrobial Activity**: Some studies have investigated the antimicrobial properties of turkey berry. It has been tested against various bacteria and fungi, suggesting potential applications in controlling microbial infections [24, 26].
5. **Hepatoprotective Effects**: Turkey berry has been studied for its hepatoprotective effects, indicating its potential to protect the liver from damage caused by toxins or certain diseases [69].
6. **Wound Healing**: Research has explored the wound-healing properties of turkey berry. It has been tested in both animal and clinical studies to assess its potential to promote the healing of wounds and injuries [41].
7. **Immune Modulation**: Some studies have looked at the immunomodulatory effects of turkey berry [27, 111]. This research aims to understand its impact on the immune system and its potential applications in immune-related conditions [43].
8. **Anticancer Properties**: Preliminary studies have investigated turkey berry's potential anticancer properties, particularly its effects on cancer cell proliferation and apoptosis (cell death). However, more research is needed in this area [108].
9. **Cardiovascular Health**: Research has explored the effects of turkey berry on cardiovascular health markers, such as blood pressure and lipid profiles. These studies aim to determine its potential role in heart health [71].
10. **Toxicological Studies**: Clinical research includes safety and toxicity assessments to evaluate the potential adverse effects of turkey berry and to establish safe usage levels [44, 109].

These studies provide valuable insights into the potential health benefits of turkey berry, more research is needed to confirm and fully understand its therapeutic applications [112]. Additionally, the results of clinical studies can vary, and individual responses to herbal remedies may differ. As a result, it is advisable to consult with a healthcare professional before using turkey berry or its extracts for medicinal purposes, especially if you have underlying health conditions or are taking other medications.

9 Herbal Products of Turkey Berry (*Solanum torvum*)

Turkey berry is utilized in various herbal products and traditional preparations across different cultures. Its diverse phytochemical composition and potential health benefits have made it a valuable ingredient in herbal remedies and dietary supplements. Some herbal products and traditional preparations that incorporate turkey berry:

- **Herbal Teas:** Turkey berry leaves and fruits can be dried and used to prepare herbal teas. These teas are often consumed for their potential digestive and antioxidant properties [113].

- **Herbal Powders:** Dried and powdered turkey berry can be used as a dietary supplement. It is available in capsule or powder form and is sometimes marketed for its potential health benefits [113].
- **Tinctures and Extracts:** Turkey berry extracts, in the form of tinctures or liquid extracts, may be used as herbal remedies. These concentrated forms allow for easy dosage and administration.
- **Traditional Remedies:** In traditional medicine systems, turkey berry is used in various herbal remedies for digestive disorders, fever, inflammation, and other ailments. For example, in some regions of Africa, it is used in concoctions and infusions to treat stomachaches and indigestion.
- **Culinary Herbs and Spices:** Turkey berry is a popular culinary herb in some regions. It is used to add flavor to various dishes, particularly in South Indian and Southeast Asian cuisines. It is often used in curries, stews, and stir-fries to impart a slightly bitter and tangy taste.
- **Herbal Ointments and Topical Preparations:** Turkey berry leaves may be used in the preparation of herbal ointments or poultices for topical applications. They are believed to have anti-inflammatory and wound-healing properties.
- **Herbal Capsules and Tablets:** Some herbal supplement manufacturers offer turkey berry capsules and tablets as part of their product line. These products are often marketed for their potential health benefits.
- **Herbal Blends:** Turkey berry may be combined with other herbs and botanicals to create specialized herbal blends targeting specific health concerns, such as digestive health or immune support.
- **Herbal Infusions:** Turkey berry leaves and fruits can be steeped in hot water to make herbal infusions. These infusions are often consumed for their potential medicinal properties and pleasant flavor.
- **Traditional Tonics:** In some cultures, turkey berry is used as an ingredient in traditional health tonics or elixirs believed to boost overall wellness [73, 80].

Turkey berry is a part of various herbal products and traditional preparations, but scientific research is ongoing to better understand their specific health benefits and mechanisms of action. Also, as with any herbal product, it is advisable to consult with a healthcare professional before using turkey berry or its extracts for medicinal purposes, especially if you have underlying health conditions or are taking other medications. Additionally, the quality and safety of herbal products can vary, so it is essential to choose reputable sources and products.

10 Toxicity of Turkey Berry (*Solanum torvum*)

Turkey berry is generally considered safe for consumption when used in moderation as a food or traditional herbal remedy [72]. However, like many botanicals, excessive or improper use can potentially lead to adverse effects. Some considerations regarding the toxicity and safety of turkey berry include:

Solanine Content: Turkey berry, like other plants in the Solanaceae family (nightshades), contains certain alkaloids, including solanine and solasonine. Solanine is known for its potential toxicity when consumed in large quantities. The solanine content in turkey berry can vary depending on factors such as the plant's age and maturity. To minimize the risk of solanine toxicity, it is advisable to cook turkey berry before consumption, as cooking can help reduce the alkaloid content. Overripe or unripe fruits may have higher solanine levels [82, 114].

Allergic Reactions: Some individuals may be allergic to certain compounds present in turkey berry. Allergic reactions, though rare, can include symptoms like skin rashes, itching, and gastrointestinal discomfort. If you have known allergies to plants in the Solanaceae family, exercise caution when trying turkey berry for the first time.

Gastrointestinal Effects: When consumed in large amounts, turkey berry may cause gastrointestinal upset, including nausea, vomiting, and diarrhea. This is more likely to occur with excessive intake [115].

Interaction with Medications: Turkey berry may interact with certain medications or medical conditions. For example, it may affect blood sugar levels, which could be problematic for individuals with diabetes. If you are taking medications or have underlying health conditions, it is important to consult with a healthcare professional before incorporating turkey berry into your diet or as a supplement.

Raw Consumption: As a precaution, it is recommended to avoid consuming raw turkey berry, especially if you are unsure about its ripeness. Cooking the berries can help neutralize potentially harmful compounds and improve their palatability.

Safe Dosage: There is no established standard dosage for turkey berry as an herbal remedy. If you plan to use it for medicinal purposes, start with small amounts and monitor your body's response. Avoid excessive or prolonged use.

Consultation with Healthcare Professionals: Before using turkey berry or any herbal remedy for medicinal purposes, consult with a qualified healthcare professional, especially if you have pre-existing health conditions, are pregnant or nursing, or are taking medications.

Although turkey berry is generally considered safe for culinary use and traditional herbal remedies when consumed in moderation, it is essential to exercise caution, especially when using it for therapeutic purposes. If any adverse effects or concerns about its safety arise it is recommended to seek medical advice promptly. Individuals with known allergies, sensitivities, or specific medical conditions should be particularly vigilant and consult healthcare professionals.

11 Commercial Value of Turkey Berry (*Solanum torvum*)

Solanum torvum holds commercial value in various ways, particularly in regions where it is cultivated and used for culinary and medicinal purposes. Some aspects of its commercial value include:

Culinary Use: Turkey berry is an integral part of the cuisine in many regions, particularly in South Asia, Southeast Asia, and Africa. It is used in a wide range of dishes, adding a unique flavor and texture to curries, stews, and stir-fries. In regions where it is in high demand, the sale of fresh turkey berries in local markets can contribute to the income of small-scale farmers.

Export and Trade: In countries where turkey berry is grown commercially, it may be exported to international markets to cater to the culinary preferences of immigrant populations and consumers interested in global cuisines. Turkey berries are sometimes dried and packaged for export.

Traditional Medicine: Turkey berry is used in traditional medicine systems in various cultures for its potential health benefits [84]. The sale of turkey berry-based herbal remedies and dietary supplements can contribute to the herbal products industry.

Herbal Products: Turkey berry is incorporated into herbal products, including capsules, tablets, herbal teas, and tinctures, which are sold in health food stores and online markets. These products are marketed for their potential digestive, antioxidant, and anti-inflammatory properties.

Agroforestry and Farming: Turkey berry cultivation can be part of agroforestry systems, contributing to diversified farming practices. It can be grown alongside other crops, providing additional income and resources to farmers [81].

Traditional Tonics and Elixirs: In some cultures, turkey berry is used as an ingredient in traditional health tonics and elixirs. These tonics are sold in local markets and may contribute to the livelihoods of traditional healers [83].

Pharmaceutical and Cosmetics Industry: Some phytochemicals found in turkey berry have potential applications in the pharmaceutical and cosmetics industries. Extracts or compounds from turkey berry may be used in the development of new products.

Sustainable Agriculture: Turkey berry cultivation can be part of sustainable agriculture practices, as it is a hardy plant that can thrive in various conditions. Its cultivation may contribute to soil fertility and crop rotation strategies [85].

Food Processing Industry: Processed turkey berry products, such as pickles and chutneys, are available in some markets and cater to consumer preferences for convenience.

Research and Development: Ongoing research on the phytochemicals and potential health benefits of turkey berry may lead to the development of new commercial products and applications.

The commercial value of turkey berry can vary by region and market demand. Additionally, the quality and safety of turkey berry products are essential considerations, especially in the herbal products and dietary supplement industries. Regulations and quality control measures play a role in ensuring the integrity of turkey berry-based products.

12 Conclusion

Solanum torvum is a versatile plant with a rich history of use in culinary and traditional medicinal practices. Its botanical characteristics, wide distribution, ethnobotanical uses, nutritional properties, potential health benefits, and culinary applications make it a valuable and intriguing plant. As interest in natural and traditional remedies continues to grow, turkey berry remains an important resource with the potential for further scientific exploration and utilization in modern healthcare and cuisine. Turkey berry (*S. torvum*) is a remarkable plant with a rich history of cultivation, culinary use, and traditional medicine across diverse cultures and regions. Its origin in tropical and subtropical regions of Africa and Asia has given rise to its widespread distribution and adaptability to different climates and habitats. Its versatility in culinary applications, potential health benefits, and cultural significance make it a valuable and intriguing plant that continues to thrive and contribute to various aspects of human life around the world. Understanding its origin and distribution provides insight into the global significance of this unique plant. The botany of turkey berry (*S. torvum*) encompasses its growth habit, leaves, flowers, fruits, and adaptability to various environments. Its simple, yet distinctive, botanical features make it easily recognizable, and its versatility in terms of growth conditions and culinary uses has contributed to its significance in different cultures around the world. Pharmacognosy plays a crucial role in the comprehensive study of turkey berry, encompassing its morphological identification, phytochemical analysis, medicinal uses, pharmacological properties, and quality control. This multidisciplinary approach contributes to the understanding and utilization of turkey berry in traditional and modern medicine.

References

1. Abbas I, Nakamura K, Hasym A (1985) Survivorship and fertility schedules of a sumatran epilachnine “species” feeding on *Solanum torvum* under laboratory conditions (Coleoptera: Coccinellidae). Appl Entomol Zool 20(1):50–55. https://doi.org/10.1303/aez.20.50
2. Abhilash PC, Jamil S, Singh V, Singh A, Singh N, Srivastava SC (2008) Occurrence and distribution of hexachlorocyclohexane isomers in vegetation samples from a contaminated area. Chemosphere 72(1):79–86
3. Abraham JD, Sekyere EK, Gyamerah I (2022) Effect of boiling on the nutrient composition of *Solanum Torvum*. Int J Food Sci 2022:7539151. https://doi.org/10.1155/2022/7539151
4. Adedeji O, Ajuwon OY, Babawale OO (2007) Foliar epidermal studies, organographic distribution and taxonomic importance of trichomes in the family Solanaceae. Int J Bot 3(3): 276–282
5. Agbemafle I, Francis S, Jensen H, Reddy M (2020) Drivers of perceptions about Turkey Berry and Palm Weevil larvae among Ghanaian women of reproductive age: a mixed methods approach. Curr Dev Nutr 4(Suppl 2):1284. https://doi.org/10.1093/cdn/nzaa059_001
6. Akinyemi DS, Oke SO (2013) Soil seedbank dynamics and regeneration in three different physiognomies in Shasha Forest Reserve in southwestern Nigeria. Res J Bot 15(2):76–85. https://doi.org/10.3923/rjb.2007.76.85
7. Alfarabi M, Widyadhari G (2018) Toxicity test and phytochemical identification of Rimbang (*Solanum torvum* Swartz) extract. AL KAUNYAH: Journal of Biology 11(2):109–115
8. Aljabri M, Alharbi K, Alonazi M (2023) *In vitro* and *in silico* analysis of *Solanum torvum* fruit and methyl caffeate interaction with cholinesterases. Saudi J Biol Sci 30(10):103815. https://doi.org/10.1016/j.sjbs.2023.103815
9. Almaguel L, Machado LR, Caceres I (1984) New food-plants of the mite *Polyphagotarsonemus latus*. (Nuevas plantas hospedantes del ácaro *Polyphagotarsonemus latus*.). Ciencia Técnica Agric Protección Plantas 7(1):99–108

10. Ameh SJ, Tarfa FD, Ebeshi BU (2012) Traditional herbal management of sickle cell anemia: lessons from Nigeria. Anemia 2012:607436. https://doi.org/10.1155/2012/607436
11. Anon (1998) Florida exotic pest plant council's 1997 List of Florida's most invasive species
12. Arriaga ER, Hernández EM (1998) Resources foraged by *Euglossa atroveneta* (Apidae: Euglossinae) at Unión Juárez, Chiapas, Mexico. A palynological study of larval feeding. Apidologie 29(4):347–359. https://doi.org/10.1051/apido:19980405
13. Arthan D, Kittakoop P, Esen A, Svasti J (2006) Furostanol glycoside 26-O-ß-glucosidase from the leaves of *Solanum torvum*. Phytochemistry 67(1):27–33
14. Barbosa QPS, da Camara CAG, Silva TMS, Ramos CS (2012) Chemical constituents of essential oils from *Solanum torvum* leaves, stems, fruits, and roots. Chem Nat Compd 48(4): 698–699. https://doi.org/10.1007/s10600-012-0355-5
15. Bari MA, Islam W, Khan AR, Mandal A (2010) Antibacterial and antifungal activity of *Solanum torvum* (Solanaceae). Int J Agric Biol 12(3):386–390
16. Blankespoor GW (1991) Slash-and-burn shifting agriculture and bird communities in Liberia. West Afr Biol Conserv 57(1):41–71. https://doi.org/10.1016/0006-3207(91)90107-K
17. Bora U, Sahu A, Saikia AP, Ryakala VK, Goswami P (2007) Medicinal plants used by the people of Northeast India for curing malaria. Phytother Res 21(8):800–804. https://doi.org/10.1002/ptr.2178
18. Breedlove DE (1998) Floristic list for Mexico. IV. Flora of Chiapas. http://www.ibiologia.unam.mx/publicaciones/lfl4
19. Buathong R, Duangsrisai S (2023) Plant ingredients in Thai food: a well-rounded diet for natural bioactive associated with medicinal properties. PeerJ 11:e14568. https://doi.org/10.7717/peerj.14568
20. Cantelo WW, Smith JS Jr, Baumhover AH, Stanley JM, Henneberry TJ (1972) Suppression of an isolated population of the tobacco hornworm with blacklight traps unbaked or baited with virgin female moths. Environ Entomol 1(2):253–258
21. Carabot Cuervo A, Blunden G, Patel AV (1991) Chlorogenone and neochlorogenone from the unripe fruits of *Solanum torvum*. Phytochemistry 30(4):1339–1341. https://doi.org/10.1016/S0031-9422(00)95233-6
22. Chadhokar PA (1976) Control of devil's fig (*Solanum torvum* Sw.) in tropical pastures. PANS 22(1):75–78
23. Challa S, Buenafe OEM, Queiroz EF, Maljevic S, Marcourt L, Bock M, Kloeti W, Dayrit FM, Harvey AL, Lerche H, Esguerra CV, de Witte PAM, Wolfender J-L, Crawford AD (2014) Zebrafish bioassay-guided microfractionation identifies anticonvulsant steroid glycosides from the Philippine medicinal plant *Solanum torvum*. ACS Chem Neurosci 5(10):993–1004. https://doi.org/10.1021/cn5001342
24. Chanchaichaovivat A, Ruenwongsa P, Panijipan B (2007) Screening and identification of yeast strains from fruits and vegetables: potential for biological control of postharvest chilli anthracnose (*Colletotrichum capsici*). Biol Control 42(3):326–335
25. Chidambaram K, Alqahtani T, Alghazwani Y, Aldahish A, Annadurai S, Venkatesan K, Dhandapani K, Thilagam E, Venkatesan K, Paulsamy P, Vasudevan R, Kandasamy G (2022) Medicinal plants of *Solanum* species: the promising sources of phyto-insecticidal compounds. J Trop Med 2022:4952221. https://doi.org/10.1155/2022/4952221
26. Clain C, da Silva D, Fock I, Vaniet S, Carmeille A, Gousset C, Sihachakr D, Luisetti J, Kodja H, Besse P (2004) RAPD genetic homogeneity and high levels of bacterial wilt tolerance in *Solanum torvum* Sw. (Solanaceae) accessions from Reunion Island. Plant Sci 166(6):1533–1540. https://doi.org/10.1016/j.plantsci.2004.02.006
27. Coe FG, Anderson GJ (2005) Snakebite ethnopharmacopoeia of eastern Nicaragua. J Ethnopharmacol 96(1/2):303–323. https://doi.org/10.1016/j.jep.2004.09.026
28. Colmenares AP, Rojas LB, Mitaine-Offer AC, Pouységu L, Quideau S, Miyamoto T, Tanaka C, Paululat T, Usubillaga A, Lacaille-Dubois MA (2013) Steroidal saponins from the fruits of *Solanum torvum*. Phytochemistry 86:137–143. https://doi.org/10.1016/j.phytochem.2012.10.010

29. Comeau PL (1993) The vegetation surrounding Mud Volcanoes in Trinidad. Living World J Trinidad Tobago Field Nat Club 17–27. http://ttfnc.org/livingworld/index.php/lwj/article/view/366
30. Corlett RT (1988) The naturalized flora of Singapore. J Biogeogr 15(4):657–663. https://doi.org/10.2307/2845443
31. Corlett RT (2005) Interactions between birds, fruit bats and exotic plants in urban Hong Kong. South China Urban Ecosyst 8(3):275–283. https://doi.org/10.1007/s11252-005-3260-x
32. Deb D, Sarkar A, Barma BD, Datta BK, Majumdar K (2013) Wild edible plants and their utilization in traditional recipes of Tripura, Northeast India. Adv Biol Res 7(5):203–211. https://doi.org/10.5829/idosi.abr.2013.7.5.11895
33. Dong CZ, Wu QZ, Xu HH, Xie CL, Rui W (2011) Insecticidal activity of the extracts from 40 species of plants in Hainan Island against *Musca domestica* Linnaeus. Acta Agric Univ Jiangxiensis 33(3):476–481
34. Dowsett-Lemaire F (1988) The forest vegetation of Mt Mulanje (Malawi): a floristic and chorological study along an altitudinal gradient (650–1950 m). Bull Jardin Bot Natl Belg 58(1–2):77–107. https://doi.org/10.2307/3668402
35. Fu L, Xu B-T, Xu X-R, Qin X-S, Gan R-Y, Li H-B (2010) Antioxidant capacities and total phenolic contents of 56 wild fruits from South China. Molecules 15(12):8602–8617. https://doi.org/10.3390/molecules15128602
36. Gandhi GR, Ignacimuthu S, Paulraj MG (2011) *Solanum torvum* Swartz. fruit containing phenolic compounds shows antidiabetic and antioxidant effects in streptozotocin induced diabetic rats. Food Chem Toxicol 49(11):2725–2733. https://doi.org/10.1016/j.fct.2011.08.005
37. Garibaldi A, Minuto A, Gullino ML (2005) Verticillium wilt incited by *Verticillium dahliae* in eggplant grafted on *Solanum torvum* in Italy. Plant Dis 89(7):777
38. Gautier-Hion A, Emmons LH, Dubost G (1980) A comparison of the diets of three major groups of primary consumers of Gabon (primates, squirrels and ruminants). Oecologia 45(2): 182–189. https://doi.org/10.1007/BF00346458
39. Ge H, Ma L (2007) Baseline survey, assessment and protection for eco-environment in a proposed gold mining area in Lannigou, Guizhou Provine, China. Int J Min Reclam Environ 21(3):173–184
40. Götz M, Winter S (2016) Diversity of *Bemisia tabaci* in Thailand and Vietnam and indications of species replacement. J Asia Pac Entomol 19(2):537–543. https://doi.org/10.1016/j.aspen.2016.04.017
41. Gousset C, Collonnier C, Mulya K, Mariska I, Rotino GL, Besse P, Servaes A, Sihachakr D (2005) *Solanum torvum*, as a useful source of resistance against bacterial and fungal diseases for improvement of eggplant (*S. melongena* L.). Plant Sci 168(2):319–327. https://doi.org/10.1016/j.plantsci.2004.07.034
42. Govender N, Zulkifli NS, Badrul Hisham NF, Ab Ghani NS, Mohamed-Hussein ZA (2022) Pea eggplant (*Solanum torvum* Swartz) is a source of plant food polyphenols with SARS-CoV inhibiting potential. PeerJ 10:e14168. https://doi.org/10.7717/peerj.14168
43. Govindaraju K, Tamilselvan S, Kiruthiga V, Singaravelu G (2010) Biogenic silver nanoparticles by *Solanum torvum* and their promising antimicrobial activity. J Biopest 3(1):394–399
44. Grice AC, Lawes RA, Abbott BN, Nicholas DM, Whiteman LV (2004) How abundant and widespread are riparian weeds in the dry tropics of north-east Queensland? In: Sindel BM, Johnson SB (eds) Weed management: balancing people, planet, profit. 14th Australian Weeds Conference, Wagga Wagga, New South Wales, Australia, 6–9 September 2004: papers and proceedings. Weed Society of New South Wales, Sydney, pp 173–175
45. Ha C, Coombs S, Revill PA, Harding RM, Vu M, Dale JL (2008) Design and application of two novel degenerate primer pairs for the detection and complete genomic characterization of potyviruses. Arch Virol 153(1):25–36. https://doi.org/10.1007/s00705-007-1053-7
46. Hancock IR, Henderson CP (1988) Flora of the Solomon Islands. Research bulletin no 7. Dodo Creek Research Station, Honiara

47. Hapairai LK, Joseph H, Sang MAC, Melrose W, Ritchie SA, Burkot TR, Sinkins SP, Bossin HC (2013) Field evaluation of selected traps and lures for monitoring the filarial and arbovirus vector, *Aedes polynesiensis* (Diptera: Culicidae), in French Polynesia. J Med Entomol 50(4): 731–739. https://doi.org/10.1603/ME12270
48. Hayta S, Polat R, Selvi S (2014) Traditional uses of medicinal plants in Elazig (Turkey). J Ethnopharmacol 154:613–623. https://doi.org/10.1016/j.jep.2014.04.026
49. Heine H (1963) Solanaceae. In: Hutchison J, Dalziel JM, Hepper FN (eds) Flora of West Tropical Africa, vol 2. Crown Agents, London, pp 325–335
50. Henty EE, Pritchard GH (1975) Weeds of New Guinea and their control. In: Weeds of New Guinea and their control. Department of Forests, Division of Botany, Lae. 180 pp
51. Hnatiuk RJ (1990) Census of Australian vascular plants. Australian Flora and Fauna series number 11. Australian Government Publishing Service, Canberra. xvi + 650 pp
52. Holm LG, Pancho JV, Herberger JP, Plucknett DL (1991) A geographic atlas of world weeds. Krieger Publishing Co, Malabar. 391 pp
53. Hopkins MS, Graham AW (1983) The species composition of soil seed banks beneath lowland tropical rainforests in North Queensland, Australia. Biotropica 15(2):90–99. https://doi.org/10.2307/2387950
54. Hossain M (1973) Observations on stylar heteromorphism in *Solanum torvum* Sw. (Solanaceae). Bot J Linn Soc 66(4):291–301. https://doi.org/10.1111/j.1095-8339.1973.tb02176.x
55. Hsu Y-M, Weng J-R, Huang T-J, Lai C-H, Su C-H, Chou C-H (2010) *Solanum torvum* inhibits *Helicobacter pylori* growth and mediates apoptosis in human gastric epithelial cells. Oncol Rep 23:1401–1405. https://doi.org/10.3892/or_00000777
56. Imarhiagbe O, Ogwu MC (2022) Sacred groves in the Global South: a panacea for sustainable biodiversity conservation. In: Izah SC (ed) Biodiversity in Africa: potentials, threats and conservation. Sustainable development and biodiversity, vol 29. Springer, Singapore, pp 525–546. https://doi.org/10.1007/978-981-19-3326-4_20
57. Irakoze ML, Wafula EN, Owaga E (2021) Potential role of African fermented indigenous vegetables in maternal and child nutrition in Sub-Saharan Africa. Int J Food Sci 2021: 3400329. https://doi.org/10.1155/2021/3400329
58. Isahaque NMM, Chaudhuri RP (1983) A new alternate host plant of brinjal shoot and fruit borer *Leucinodes orbonalis* Guen. Assam J Res Assam Agric Univ 4(1):83–85
59. Isahaque NMM, Chaudhuri RP (1985) A new alternate host plant of brinjal shoot and fruit borer *Leucinodes orbonalis* Guen. Assam J Res Assam Agric Univ 4(1):83–85
60. Israf DA, Lajis NH, Somchit MN, Sulaiman MR (2004) Enhancement of ovalbumin-specific IgA responses via oral boosting with antigen co-administered with an aqueous *Solanum torvum* extract. Life Sci 75(4):397–406. https://doi.org/10.1016/j.lfs.2003.10.038
61. Ivens GW, Moody K, Egunjobi JK (1978) West African weeds. Oxford University Press, Ibadan, pp 178–179
62. Janzen DH (1966) Notes on the behavior of the carpenter bee *Xylocopa fimbriata* in Mexico (Hymenoptera: Apoidea). J Kansas Entomol Soc 39(4):633–641. https://www.jstor.org/stable/25083568
63. Jena AK, Deuri R, Sharma P, Singh SP (2018) Underutilized vegetable crops and their importance. J Pharmacogn Phytochem 7(5):402–407
64. Kellman MC (1973) Dry season weed communities in the upper Belize valley. J Appl Ecol 10(3):683–694. https://doi.org/10.2307/2401862
65. Kelly DL, Dickinson TA (1985) Local names for vascular plants in the John Crow Mountains. Jam Econ Bot 39(3):346–362. https://doi.org/10.1007/BF02858806
66. Khunbutsri D, Naimon N, Satchasataporn K, Inthong N, Kaewmongkol S, Sutjarit S, Setthawongsin C, Meekhanon N (2022) Antibacterial activity of *Solanum torvum* leaf extract and its synergistic effect with oxacillin against methicillin-resistant staphyloccoci isolated from dogs. Antibiotics 11(3):302. https://doi.org/10.3390/antibiotics11030302

67. Kimpouni V (2008) First data on the floristic diversity of Aubeville forest (Congo – Brazzaville). (Premières données sur la diversité floristique de la forêt d'Aubeville (Congo – Brazzaville).). Syst Geogr Plants 78(1):47–62
68. Koffuor GA, Amoateng P, Andey TA (2011) Immunomodulatory and erythropoietic effects of aqueous extract of the fruits of *Solanum torvum* Swartz (Solanaceae). Pharm Res 3(2): 130–134. https://doi.org/10.4103/0974-8490.81961
69. Lee CL, Hwang TL, He WJ, Tsai YH, Yen CT, Yen HF, Chen CJ, Chang WY, Wu YC (2013) Anti-neutrophilic inflammatory steroidal glycosides from *Solanum torvum*. Phytochemistry 95:315–321. https://doi.org/10.1016/j.phytochem.2013.06.015
70. Lely N (2016) Antimicrobial activity test of Turkey berry leaf extract (*Solanum torvum* Swartz) against *Staphylococcus aureus*, *Escherichia coli* and fungi *Candida albicans*. Jurnal ilmiah bakti farmasi 1(2):55–58
71. Li JS, Zhang L, Huang C, Guo FJ, Li YM (2014) Five new cyotoxic steroidal glycosides from the fruits of *Solanum torvum*. Fitoterapia 93:209–215. https://doi.org/10.1016/j.fitote.2014.01.009
72. Liquido NJ, Harris EJ, Dekker LA (1994) Ecology of *Bactrocera latifrons* (Diptera: Tephritidae) populations: host plants, natural enemies, distribution, and abundance. Ann Entomol Soc Am 87(1):71–84. https://doi.org/10.1093/aesa/87.1.71
73. Lobova TA, Mori SA (2004) Epizoochorous dispersal by bats in French Guiana. J Trop Ecol 20(5):581–582. https://doi.org/10.1017/S0266467404001634
74. Lorenzi H (1982) Weeds of Brazil, terrestrial and aquatic, parasitic, poisonous and medicinal (Plantas daninhas de Brasil, terrestres, aquaticas, parasitas, toxicas e medicinais). H. Lorenzi, Nova Odessa. 425 pp.
75. Lu YY, Luo JG, Huang XF, Kong LY (2009) Four new steroidal glycosides from *Solanum torvum* and their cytotoxic activities. Steroids 74(1):95–101. https://doi.org/10.1016/j.steroids.2008.09.011
76. MacKee HS (1985) Les Plantes Introduites et Cultivees en Nouvelle-Caledonie. Volume hors series, Flore de la Nouvelle-Caledonie et Dependances. Museum Nationelle d'Histoire Naturelle, Paris
77. Mahapatra AK, Mishra S, Basak UC, Panda PC (2012) Nutrient analysis of some selected wild edible fruits of deciduous forests of India: an explorative study towards non conventional bio-nutrition. Adv J Food Sci Technol 4(1):15–21
78. Mahmood U, Agrawal PK, Thakur RS (1985) Torvonin-A, a spirostane saponin from *Solanum torvum* leaves. Phytochemistry 24(10):2456–2457. https://doi.org/10.1016/S0031-9422(00)83069-1
79. Maro LAC, Pio R, Guedes MNS, Abreu CMP, Moura PHA (2014) Environmental and genetic variation in the post-harvest quality of raspberries in subtropical areas in Brazil. Acta Sci 36(3):323–328
80. Matsuzoe N, Aida H, Hanada K, Ali M, Okubo H, Fujieda K (1996) Fruit quality of tomato plants grafted on Solanum rootstocks. J Jpn Soc Hortic Sci 65(1):73–80. https://doi.org/10.2503/jjshs.65.73
81. Mavoungou JF, Picard N, Kohagne LT, M'Batchi B, Gilles J, Duvallet G (2013) Spatio-temporal variation of biting flies, Stomoxys spp. (Diptera: Muscidae), along a man-made disturbance gradient, from primary forest to the city of Makokou (North-East, Gabon). Med Vet Entomol 27(3):339–345. https://doi.org/10.1111/j.1365-2915.2012.01064.x
82. Muhammad A, Sheeba F (2011) Pharmacognostical studies and evaluation of total phenolic and flavonoid contents of traditionally utilized fruits of *Solanum torvum* Sw. Indian J Nat Prod Resour 2(2):218–224
83. Myster RW, Walker LR (1997) Plant successional pathways on Puerto Rican landslides. J Trop Ecol 13(2):165–173. https://doi.org/10.1017/S0266467400010397
84. Narikawa T, Sakata Y, Komochi S, Melor R, Heng CK, Jumali S (1988) Collection of Solanaceous plants in Malaysia and screening for disease resistance. Jpn Agric Res Q 22(2): 101–106

85. Nayak SK, Satapathy KB (2015) Diversity, uses and origin of invasive alien plants in Dhenkanal district of Odisha, India. Int Res J Biol Sci 4(2):21–27. http://www.isca.in/IJBS/Archive/v4/i2/4.ISCA-IRJBS-2014-223.pdf
86. Ndebia EJ, Kamgang R, Nkeh-ChungagAnye BN (2006) Analgesic and anti-inflammatory properties of aqueous extract from leaves of *Solanum torvum* (Solanaceae). Afr J Tradit Complement Alternat Med 4(2):240–244. https://doi.org/10.4314/ajtcam.v4i2.31214
87. Ndebia EJ, Kamgang R, Nkeh-ChungagAnye BN (2007) Analgesic and anti-inflammatory properties of aqueous extract from leaves of *Solanum torvum* (Solanaceae). Afr J Tradit Complement Altern Med 4(2):240–244
88. Nguelefack TB, Feumebo CB, Ateufack G, Watcho P, Tatsimo S, Atsamo AD, Tane P, Kamanyi A (2008) Anti-ulcerogenic properties of the aqueous and methanol extracts from the leaves of *Solanum torvum* Swartz (Solanaceae) in rats. J Ethnopharmacol 119(1):135–140. https://doi.org/10.1016/j.jep.2008.06.008
89. Nurit-Silva K, Costa-Silva R, Coelho VPM, de Agra MF (2011) A pharmacobotanical study of vegetative organs of *Solanum torvum*. Rev Bras 21(4):568–574. https://doi.org/10.1590/S0102-695X2011005000101
90. Nwizugbo KC, Ogwu MC, Eriyamremu GE, Ahana CM (2023) Alterations in energy metabolism, total protein, uric and nucleic acids in African sharptooth catfish (*Clarias gariepinus* Burchell.) exposed to crude oil and fractions. Chemosphere 316:137778. https://doi.org/10.1016/j.chemosphere.2023.137778
91. Ogwu MC (2019) Lifelong consumption of plant-based GM foods: is it safe? In: Papadopoulou P, Misseyanni A, Marouli C (eds) Environmental exposures and human health challenges. IGI Global, Pennsylvania, pp 158–176. https://doi.org/10.4018/978-1-5225-7635-8.ch008
92. Ogwu MC (2019) Towards sustainable development in Africa: the challenge of urbanization and climate change adaptation. In: Cobbinah PB, Addaney M (eds) The geography of climate change adaptation in urban Africa. Springer Nature, Cham, pp 29–55. https://doi.org/10.1007/978-3-030-04873-0_2
93. Ogwu MC (2019) Understanding the composition of food waste: an "-Omics" approach to food waste management. In: Gunjal AP, Waghmode MS, Patil NN, Bhatt P (eds) Global initiatives for waste reduction and cutting food loss. IGI Global, Hershey, pp 212–236. https://doi.org/10.4018/978-1-5225-7706-5.ch011
94. Ogwu MC (2023) Local food crops in Africa: sustainable utilization, threats, and traditional storage strategies. In: Izah SC, Ogwu MC (eds) Sustainable utilization and conservation of Africa's biological resources and environment. Sustainable development and biodiversity, vol 888. Springer, Singapore, pp 353–374. https://doi.org/10.1007/978-981-19-6974-4_13
95. Ogwu MC, Osawaru ME (2022) Traditional methods of plant conservation for sustainable utilization and development. In: Izah SC (ed) Biodiversity in Africa: potentials, threats and conservation. Sustainable development and biodiversity, vol 29. Springer, Singapore, pp 451–472. https://doi.org/10.1007/978-981-19-3326-4_17
96. Ogwu MC, Osawaru ME (2023) Disease outbreaks in ex-situ plant conservation and potential management strategies. In: Izah SC, Ogwu MC (eds) Sustainable utilization and conservation of Africa's biological resources and environment. Sustainable development and biodiversity, vol 888. Springer, Singapore, pp 497–518. https://doi.org/10.1007/978-981-19-6974-4_18
97. Ogwu MC, Osawaru ME, Aiwansoba RO, Iroh RN (2016) Ethnobotany and collection of West African Okra [*Abelmoschus caillei* (A. Chev.) Stevels] germplasm in some communities in Edo and Delta States, Southern Nigeria. Borneo J Resour Sci Technol 6(1):25–36. https://doi.org/10.33736/bjrst.212.2016
98. Ogwu MC, Osawaru ME, Aiwansoba RO, Iroh RN (2016) Status and prospects of vegetables in Africa. In: Borokini IT, Babalola FD (eds) Conference proceedings of the joint biodiversity conservation conference of Nigeria Tropical Biology Association and Nigeria chapter of Society for Conservation Biology on MDGs to SDGs: toward sustainable biodiversity conservation in Nigeria. University of Ilorin, Nigeria, pp 47–57

99. Ogwu MC, Osawaru ME, Obahiagbon GE (2017) Ethnobotanical survey of medicinal plants used for traditional reproductive care by Usen people of Edo State, Nigeria. Malaya J Biosci 4(1):17–29
100. Ogwu MC, Osawaru ME, Owie MO (2023) Effects of storage at room temperature on the food components of three Cocoyam species (*Colocasia esculenta, Xanthosoma atrovirens, and X. sagittifolium*). Food Stud Interdiscip J 13(2):59–83. https://doi.org/10.18848/2160-1933/CGP/v13i02/59-83
101. Ogwu MC, Osawaru ME, Amodu E, Osamo F (2023) Comparative morphology, anatomy, and chemotaxonomy of two *Cissus* Linn. species. Braz J Bot 46:397. https://doi.org/10.1007/s40415-023-00881-0
102. Osawaru ME, Ogwu MC (2014) Ethnobotany and germplasm collection of two genera of cocoyam (*Colocasia* [Schott] and *Xanthosoma* [Schott], Araceae) in Edo State Nigeria. Sci Technol Arts Res J 3(3):23–28. https://doi.org/10.4314/star.v3i3.4
103. Osawaru ME, Ogwu MC, Omoigui ID, Aiwansoba RO, Kevin A (2016) Ethnobotanical survey of vegetables eaten by Akwa Ibom people residing in Benin City, Nigeria. Univ Benin J Sci Technol 4(1):70–93
104. Ovuru KF, Izah SC, Ogidi OI, Imarhiagbe I, Ogwu MC (2023) Slaughterhouse facilities in developing nations: sanitation and hygiene practices, microbial contamination and sustainable management system. Food Sci Biotechnol. https://doi.org/10.1007/s10068-023-01406-x
105. Parham JW (1958) The weeds of Fiji. Bulletin Fiji Department of Agriculture, 35. Government Press, Suava
106. Peregrine WTH, Kbin A (1982) Grafting – a simple technique for overcoming bacterial wilt in tomato. Trop Pest Manage 28(1):71–76
107. Pi JS (1977) Contribution to the study of alimentation of lowland gorillas in the natural state, in Río Muni, Republic of Equatorial Guinea (West Africa). Primates 18(1):183–204
108. Rahman MA, Sultana R, Emran T, Islam MS, Rahman MA, Chakma JS, Harun-ur-Rashid, Hasan CMM (2013) Effects of organic extracts of six Bangladeshi plants on in vitro thrombolysis and cytotoxicity. BMC Complement Altern Med 13:25. http://www.biomedcentral.com/1472-6882/13/25
109. Raimondo FM, Orlando A (1978) First finding in Italy of *Solanum torvum* Sw. Inform Bot Ital 10(1):43–45
110. Ramamurthy CH, Subastri A, Suyavaran A, Subbaiah KC, Valluru L, Thirunavukkarasu C (2016) *Solanum torvum* Swartz. fruit attenuates cadmium-induced liver and kidney damage through modulation of oxidative stress and glycosylation. Environ Sci Pollut Res Int 23(8): 7919–7929. https://doi.org/10.1007/s11356-016-6044-3
111. Rammohan M, Reddy CS (2011) Anti inflammatory activity of the seed and fruit wall extracts of *Solanum torvum*. Hygeia J Drugs Med 2(2):54–58
112. Sabatino L, Palazzolo E, D'Anna F (2013) Grafting suitability of Sicilian eggplant ecotypes onto *Solanum torvum*: fruit composition, production and phenology. J Food Agric Environ 11: 1195–1200
113. Sajeev KK, Sasidharan N (1997) Ethnobotanical observations on the tribals of Chinnar Wildlife Sanctuary. Anc Sci Life 16(4):284–292. https://www.ncbi.nlm.nih.gov/pmc/articles/PMC3331175/
114. Sakthivel P, Karuppuchamy P, Kalyanasundaram M, Srinivasan T (2012) Host plants of invasive papaya mealybug, *Paracoccus marginatus* (Williams and Granara de Willink) in Tamil Nadu. Madras Agric J 99(7/9):615–619
115. Sánchez JA, Montejo L, Gamboa A, Albert-Puentes D, Hernández F (2015) Germination and dormancy of shrubs and climbing plants of the evergreen forest of Sierra del Rosario, Cuba (Germinación y dormancia de arbustos y trepadoras del bosque siempreverde de la Sierra del Rosario, Cuba). Pastos Forrajes 38(1):11–28
116. Sato K, Uehara T, Holbein J, Sasaki-Sekimoto Y, Gan P, Bino T, Yamaguchi K, Ichihashi Y, Maki N, Shigenobu S, Ohta H, Franke RB, Siddique S, Grundler FMW, Suzuki T, Kadota Y, Shirasu K (2021) Transcriptomic analysis of resistant and susceptible responses in a new

model root-knot nematode infection system using *Solanum torvum* and *Meloidogyne arenaria*. Front Plant Sci 12:680151. https://doi.org/10.3389/fpls.2021.680151
117. Shetty KD, Reddy DDR (1985) Resistance in Solanum species to root-knot nematode *Meloidogyne incognita*. Indian J Nematol 15:230
118. Singh PK, Gopalakrishnan TR (1997) Grafting for wilt resistance and productivity in brinjal (*Solanum melongena* L.). Hortic J 10(2):57–64. 5 ref
119. Singh AK, Kamal (1985) Fungi of Gorakhpur XXXVI. Indian J Mycol Plant Pathol 15(2): 121–124
120. Sirait N (2009) Eggplant Cepoka (*Solanum torvum*) Herbs that are efficacious as medicine. Rep Res Dev Ind Plants 15(1):11–13
121. Smith SW, Giesbrecht E, Thompson M, Nelson LS, Hoffman RS (2008) Solanaceous steroidal glycoalkaloids and poisoning by *Solanum torvum*, the normally edible susumber berry. Toxicon 52(6):667–676. https://doi.org/10.1016/j.toxicon.2008.07.016
122. Space JC, Flynn T (2002) Report to the Government of The Cook Islands on invasive plant species of environmental concern. USAL USDA Forest Service, Honolulu. 146 pp
123. Symon DE (1981) Solanum in Australia. J Adelaide Bot Gard 4:115–116
124. Takoukam Kamla A, Gomes DGE, Beck CA, Keith-Diagne LW, Hunter ME, Francis-Floyd R, Bonde RK (2021) Diet composition of the African manatee: spatial and temporal variation within the Sanaga River Watershed, Cameroon. Ecol Evol 11(22):15833–15845. https://doi.org/10.1002/ece3.8254
125. Tsouh Fokou PV, Nyarko AK, Appiah-Opong R, Tchokouaha Yamthe LR, Ofosuhene M, Boyom FF (2015) Update on medicinal plants with potency on *Mycobacterium ulcerans*. Biomed Res Int 2015:917086. https://doi.org/10.1155/2015/917086
126. Vahapoglu B, Erskine E, Gultekin Subasi B, Capanoglu E (2021) Recent studies on berry bioactives and their health-promoting roles. Molecules 27(1):108. https://doi.org/10.3390/molecules27010108
127. Wang ZR (1990) Farmland weeds in China. Agricultural Publishing House, Beijing
128. Waterhouse DF (1997) The major invertebrate pests and weeds of agriculture and plantation forestry in the southern and western Pacific. ACIAR monograph no 44. Australian Centre for International Agricultural Research, Canberra. 93 pp
129. Westbrooks RG, Eplee RE (1988) Federal noxious weeds in Florida. In: Proceedings of the 42nd annual meeting of the Southern Weed Science Society, pp 316–321
130. Whistler WA (1983) Weed handbook of Western Polynesia. Schriftenreihe der Deutschen Gesellschaft fnr Technische Zusammenarbeit, 157 pp
131. Wiart C, Mogana S, Khalifah S, Mahan M, Ismail S, Buckle M, Narayana AK, Sulaiman M (2004) Antimicrobial screening of plants used for traditional medicine in the state of Perak, Peninsular Malaysia. Fitoterapia 75(1):68–73
132. Yang X, Cheng YF, Deng C, Ma Y, Wang ZW, Chen XH, Xue LB (2014) Comparative transcriptome analysis of eggplant (*Solanum melongena* L.) and Turkey berry (*Solanum torvum* Sw.): phylogenomics and disease resistance analysis. BMC Genomics 15(1):412. https://doi.org/10.1186/1471-2164-15-412
133. Yang X, Deng C, Zhang Y, Cheng Y, Huo Q, Xue L (2015) The WRKY transcription factor genes in eggplant (*Solanum melongena* L.) and Turkey Berry (*Solanum torvum* Sw.). Int J Mol Sci 16(4):7608–7626. https://doi.org/10.3390/ijms16047608
134. Zhang M, Zhang H, Tan J, Huang S, Chen X, Jiang D, Xiao X (2021) Transcriptome analysis of eggplant root in response to root-knot nematode infection. Pathogens 10(4):470. https://doi.org/10.3390/pathogens10040470
135. Zhang H, Chen H, Tan J, Huang S, Chen X, Dong H, Zhang R, Wang Y, Wang B, Xiao X, Hong Z, Zhang J, Hu J, Zhang M (2023) The chromosome-scale reference genome and transcriptome analysis of *Solanum torvum* provides insights into resistance to root-knot nematodes. Front Plant Sci 14:1210513. https://doi.org/10.3389/fpls.2023.1210513

Garcinia kola Heckel. (Clusiaceae): An Overview of the Cultural, Medicinal, and Dietary Significance for Sustainability

9

Matthew Chidozie Ogwu, Happiness Isioma Ogwu, Moses Edwin Osawaru, and Sylvester Chibueze Izah

Contents

M. C. Ogwu (✉)
Goodnight, Family Department of Sustainable Development, Appalachian State University, Boone, NC, USA
e-mail: ogwumc@appstate.edu

H. I. Ogwu
Department of Microbiology, Faculty of Life Sciences, University of Benin, Benin City, Edo State, Nigeria
e-mail: ogwuhi@appstate.edu

M. E. Osawaru
Department of Plant Biology and Biotechnology, Faculty of Life Sciences, University of Benin, Benin City, Edo State, Nigeria
e-mail: moses.osawaru@uniben.edu

S. C. Izah
Department of Microbiology, Faculty of Science, Bayelsa Medical University, Yenagoa, Bayelsa State, Nigeria
e-mail: sylvester.izah@bmu.edu.ng

S. C. Izah et al. (eds.), *Herbal Medicine Phytochemistry*, Reference Series in Phytochemistry, https://doi.org/10.1007/978-3-031-43199-9_74

Abstract

This chapter aims to discuss the cultural, medicinal, and dietary significance of *Garcinia kola* Heckel (Clusiaceae) from a sustainability standpoint. The small- to medium-sized evergreen tree is native to the dense humid tropical rainforest ecosystems of West and Central Africa and grows alongside other traditionally valued forest crops. Also known as bitter kola or kola, the plant is deeply rooted in the cultures and traditions of West and Central Africa and has played a significant role in the lives of indigenous communities for centuries by offering a rich tapestry of uses and cultural significance. It bears bitter-tasting kola seeds. These seeds, with their intense bitterness, have become iconic and emblematic of the region's cultural heritage. Some food uses of bitter kola include snacks, flavoring agents, preparation of traditional soup, beverage infusion, medicinal concoctions added to food, and condiments and seasoning. Some of the cultural significance of *G. kola* include as a symbol of hospitality, traditional rituals and ceremonies, traditional medicine, food, and flavoring (as seasoning), stimulant and energy boost, social and ceremonial uses (like weddings and religious gatherings), offering and prayers, cosmetic and skin care, trade and economic value, and cultural conservation. Bitter kola is renowned for its various medicinal properties, and it has been used in traditional African medicine for centuries. It contains garcinol which has shown promise in research for its anticancer effects as well as kolaviron which has therapeutic effects. Some medicinal uses and potential health benefits of bitter kola include antimicrobial properties, antioxidant activity, anti-inflammatory effects, hepatoprotective properties, antidiabetic, antimalarial activity, cardiovascular health, analgesic and anti-inflammatory effects, anticancer potential, digestive aid and gut health improvement, aphrodisiac, and weight loss and management. The use of *Garcinia kola* for health management is generally considered safe when consumed and used in moderation. However, like many natural products, bitter kola should be used with caution, and excessive consumption can lead to potential adverse effects due to the bitter taste and sensitivity, caffeine content, potential allergic reactions, gastrointestinal distress, interactions with medications, and liver distress from overconsumption. Some herbal formulations prepared based on traditional knowledge that contain bitter kola include bitter kola extracts or tinctures, capsules, supplements, traditional herbal remedies, chewing sticks (snacks), traditional drinks and tonics (bitters), and honey infusions. This chapter suggests the need to incorporate sustainable practices and further research into the phytomedicinal and traditional benefits of the plant.

Keywords

Bitter kola · Kola · Nuts · Sustainability · Ethnomedicine · Cultural value · Molecular docking · Herbal formulations · Herbal medicine

1 Introduction

Garcinia kola Heckel, commonly known as bitter kola, is a tropical flowering plant from the plant family, Clusiaceae. It is native to West and Central Africa and is particularly found in countries like Nigeria, Cameroon, Ghana, and Cote d'Ivoire. Bitter kola is a small, evergreen tree that produces fruit pods containing seeds known as kola nuts or simply kola. The intensely bitter-tasting kola may not be to everyone's liking and consumption is often limited to small quantities or specific preparations. These seeds are revered for their cultural, medicinal, and dietary significance in the regions where they grow and beyond [81]. Other diverse uses are linked to the phytochemical constituents of the fruits and tree bark which include several bioactive compounds [67, 69, 70]. They may be incorporated into some food products because of their medicinal and cultural value but not used as a staple food source in most diets within the native range of the crop [47]. Instead, bitter kola is valued for its role in traditional practices, customary events, festivities, and family celebrations as well as health benefits. Therefore, the tree is celebrated for its cultural, medicinal, and dietary significance in the regions where it grows. The seeds of *G. kola* are the most celebrated part of the plant and are commonly referred to as kola nuts. These seeds are known for their cultural significance, traditional uses, and potential medicinal properties [48]. Bitter kola seeds are used as a symbol of hospitality, a traditional remedy for various ailments, and an ingredient in traditional rituals and ceremonies [56]. Bitter kola holds a special place in the cultural and social traditions of many West and Central African communities. It is often used as a symbol of hospitality and is offered to guests as a sign of respect and welcome. In some cultures, presenting a kola nut during gatherings or ceremonies is a customary gesture.

The phytochemical compounds found in *G. kola* are not only beneficial for human health but also contribute to the defense mechanisms, ecophysiology, and adaptive capabilities of the plants, helping them ward off pests, diseases, and environmental stressors [95]. Bitter kola is rich in phytochemical constituents like flavonoids, polyphenols, alkaloids, saponins, tannins, terpenes, and many others, and these compounds have actual and potential health benefits against chronic diseases such as cancer, heart disease, and diabetes [71, 81]. Bitter kola is typically consumed in small quantities due to its intense bitterness. Its primary use is for its cultural and traditional significance, as well as its potential role in traditional medicine for various ailments [64, 80]. As such, bitter kola is not considered a significant dietary source of essential nutrients, but it may offer certain health benefits in the context of traditional medicine and folklore [65, 82]. However, scientific research on its specific nutritional content and health effects is limited, and further studies are needed to better understand its potential contributions to human health. Additionally, the ethnobotany of the bitter kola tree is a testament to the intricate relationship between people and plants in West and Central Africa [97–99, 114, 118]. The plant has deep roots in the traditions and folklore of West and Central Africa, which include:

Symbol of Hospitality: Bitter kola is considered a symbol of hospitality in many African cultures. It is often presented to guests as a gesture of respect and welcome. In some regions, offering a kola nut during social gatherings or ceremonies is a customary practice.

Traditional Medicine: Bitter kola has a rich history in traditional African medicine. It has been used for centuries to treat various ailments and is renowned for its medicinal properties [63, 106]. The seeds are believed to possess numerous health benefits, and they are used in various remedies.

Rituals and Ceremonies: Bitter kola is incorporated into various traditional rituals and ceremonies. The presence of kola is often associated with spiritual and cultural practices, including naming ceremonies, weddings, and religious rites.

Economic Value: The kola nut trade has economic significance in some regions, with bitter kola being an important commodity for trade and exchange [89–91]. Beyond its botanical characteristics, *G. kola* is culturally, medicinally, and economically significant. The seeds of the plant, known as kola nuts, have been used traditionally for various purposes, including:

Cultural and Social Traditions: Kola nuts are often presented as a symbol of hospitality and respect in many African cultures [62]. They are commonly offered to guests during social gatherings and ceremonies.

Traditional Medicine: Bitter kola has a rich history in traditional African medicine. It is believed to have various medicinal properties and has been used to treat ailments such as digestive issues, infections, and coughs [25]. Some people also associate it with aphrodisiac properties and improved energy levels [12].

Clearly, from its role as a symbol of hospitality to its use in traditional medicine, bitter kola weaves itself into the fabric of daily life and cultural practices. Its bitter taste is not just a flavor but a reflection of its deep-rooted significance. The use of bitter kola extends far beyond its nutritional value; it embodies a cultural identity, a source of livelihood, and a bridge between the physical and spiritual realms. As communities continue to cherish and celebrate bitter kola, they also recognize the importance of conserving this invaluable botanical treasure for future generations. In addition to its medicinal and cultural significance, bitter kola is consumed as a dietary supplement in some regions. The kola nuts are chewed or ground into a powder to make a paste that can be mixed with water or other beverages. Bitter kola is known for its intensely bitter taste, which is attributed to the presence of compounds like caffeine and theobromine [105].

G. kola, like many other rainforest plants, plays a significant role in its native ecosystem. It contributes to the ecological balance and provides essential services to the environment and wildlife [83]. The kola nut seeds are consumed by various animals in the rainforest, including rodents, primates, and birds. These animals aid in the dispersal of the seeds, helping the plant propagate and colonize new areas. Bitter kola contributes to the biodiversity of the rainforest by providing habitat and sustenance for a variety of wildlife species. It is part of the intricate web of life that characterizes these lush ecosystems. Some animals in the rainforest may also benefit from the medicinal properties of bitter kola. For instance, primates that consume the seeds may experience some of the plant's therapeutic effects [81]. While *Garcinia kola* thrives in its natural habitat, it faces several conservation challenges, primarily due to deforestation and habitat destruction in African rainforests. As human populations and economic activities expand, the rainforests are increasingly threatened by logging, agriculture, and urbanization. This poses a risk to the biodiversity and survival of plant species like bitter kola. Efforts are being made by conservation organizations and governments in African countries to address these challenges [92, 96, 115, 117]. These initiatives include promoting sustainable forestry practices, establishing protected areas, and raising awareness about the importance of preserving the rainforest ecosystem.

This chapter aims to discuss the cultural, medicinal, and dietary significance of *G. kola* from a sustainability standpoint. The chapter presents the origin, distribution, and botany of *G. kola* as well as the diverse food and traditional medicine uses. Other parts of the chapter focus on the ethnobotanical, pharmacognosy, and clinical works, herbal products, toxicity, and commercial values of *G. kola*. It is believed that bitter kola has various medicinal properties, including anti-inflammatory, antioxidant, anti-inflammatory, and antimicrobial effects [53]. *G. kola* is a crop with great potential for sustainable cultivation. The small- to medium-sized evergreen tree is native to the dense humid rainforest ecosystems of West and Central Africa and grows alongside other traditionally valued forest crops. This chapter suggests the need to incorporate sustainable practices, including organic cultivation, agroforestry, and further research into the phytomedicinal and traditional benefits of the plant such as the use of molecular docking techniques with compounds from *G. kola* using the 3D structures of these compounds and target proteins, as well as specialized software and expertise in computational chemistry.

2 Origin and Distribution of *Garcinia kola*: The Bitter Kola Tree

The center of origin of *G. kola* is believed to be in the rainforests of West and Central Africa where the plant has enjoyed a long history of cultivation and widespread distribution throughout the region (Fig. 1). However, it is generally accepted that bitter kola's natural habitat is within the lush, tropical rainforests of Africa, which includes several countries like Nigeria, Cameroon, Ghana, Cote d'Ivoire (Ivory Coast), Sierra Leone, Liberia, Equatorial Guinea, Gabon, Congo, and the Democratic Republic of Congo in lush, humid environments [107]. These areas are also known for their rich biodiversity and diverse flora within which *G. kola* has naturally thrived for centuries [83]. In these rainforests, bitter kola has deep cultural, medicinal, and ecological significance. It has been used in traditional medicine, is a symbol of hospitality, and plays a role in the local ecosystems as a source of sustenance for various wildlife species. Wild stands of *G. kola* are also supported by present and ongoing cultivation efforts.

In Nigeria, the plant is extensively grown in various southern states such as Ogun, Ondo, Edo, and Cross River. Nigerian farmers have played a significant role in the production of bitter kola, making it a substantial part of the country's agricultural economy [61]. In Cameroon, *G. kola* is also prevalent in the southern

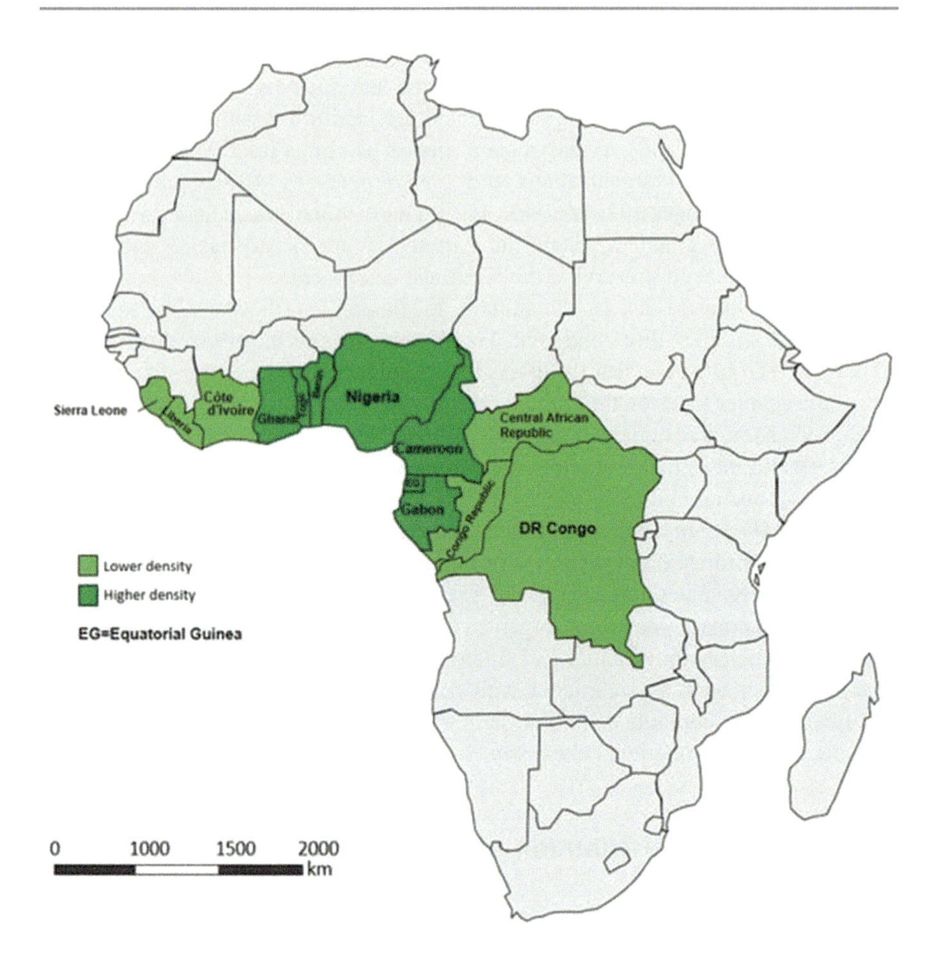

Fig. 1 Distribution of *Garcinia kola* in Africa. (Source: Maňourová et al. [81] (CC BY 4.0))

and central regions of the country. The plant is found in the dense rainforests that characterize these areas. Bitter kola is cultivated and harvested in regions with suitable rainforest conditions in Ghana and has a long history of traditional use especially in traditional ceremonies. In Cote d'Ivoire, kola can be found in the rainforests of the southwestern part of the country. Nigeria, Cameroon, Ghana, and Cote d'Ivoire are among the primary producers of bitter kola but the plant is also present in other West and Central African nations with suitable rainforest habitats including Sierra Leone, Liberia, Equatorial Guinea, Gabon, Congo, and the Democratic Republic of Congo. It thrives in the humid, equatorial climate and is well adapted to the dense, evergreen rainforests where it receives abundant rainfall and high levels of humidity.

The life span of *G. kola* can vary depending on a range of factors, including environmental conditions, disease, and other stressors [108]. Generally, bitter kola is a long-lived tree with a potential for longevity under favorable circumstances. In the wild, *G. kola* life span can extend from several decades to potentially over a century. However, in cultivated settings or agroforestry systems, the life span may be influenced by factors such as management practices and environmental conditions. The longevity of *G. kola* is closely tied to the environmental conditions in which they grow. Adequate rainfall, suitable soil conditions, and a lack of significant stressors can contribute to longer life spans. Conversely, factors like drought, disease, or damage from pests may shorten a tree's life. The susceptibility of bitter kola trees to diseases and pests can influence their lifespan. Healthy trees that are well-maintained and protected from threats are more likely to live longer. Human activities, such as logging and habitat destruction, can significantly impact the life span of bitter kola trees in the wild. Deforestation and logging can lead to prematurely removing these trees from their natural habitat. The presence of young trees and successful regeneration through seed dispersal, as well as the presence of both male and female trees for pollination (as bitter kola is dioecious), can contribute to the perpetuation of the species. It is worth noting that *G. kola* has no specific fixed life span, individual trees can live for several decades or more under favorable conditions. Bitter kola can thrive in its natural rainforest habitat due to the long reproductive cycle and adaptive capabilities of the plant. However, like many species in tropical rainforests, the ongoing challenges of habitat loss and environmental changes due to human activities can pose threats to its continued existence in the wild. Conservation efforts are essential to protect and sustain populations of *G. kola* and other valuable rainforest species.

3 Botanical Overview of *Garcinia kola*

G. kola belongs to the plant family Clusiaceae (formerly known as Guttiferae) within the broader classification of angiosperms, eudicots, and Malpighiales. The *Garcinia* genus comprises over 200 species of flowering plants. The genus is known for containing numerous species, many of which are tropical evergreen trees and shrubs. *Garcinia kola* shares its genus with other *Garcinia* species, some of which have also been of cultural and economic importance. Some relatives of *G. kola* are *Garcinia mangostana* (mangosteen), *Garcinia cambogia* which has small, pumpkin-shaped fruit and is native to Southeast Asia and India, *Garcinia gummi-gutta*, which is widely used in Southeast Asia in traditional cooking due to the strong sour flavor to dishes, *Garcinia indica* (kokum or kokam), which is native to the Western Ghats region of India and valued for its fruit that is used in culinary applications and traditional medicine, *Garcinia atroviridis* (asam gelugur), a tropical fruit tree native to Southeast Asia with sour fruits used in traditional cooking, especially in Malaysian and Thai cuisine and *Garcinia livingstonei* (imbe) that is native to Southern Africa and has edible fruits with a sweet and tangy flavor. *G. kola* is a small- to medium-sized evergreen tree with a straight trunk and dense, glossy leaves [45]. The tree produces fruit pods containing seeds, which are commonly referred to as kola nuts. These seeds are the most renowned part of the plant and are known for their intensely bitter taste.

3.1 Morphological Characteristics of *Garcinia kola*

The plant is a small- to medium-sized evergreen tree that typically reaches heights ranging from 7 to 15 m, although some specimens can grow up to 30 m under favorable conditions. The leaves of *G. kola* are simple, glossy, and dark green [45]. Leaves are arranged oppositely on the branches and are elliptical or oblong. The leaves have a prominent midrib and lateral veins. Bitter kola trees produce small, greenish-yellow to pale-green flowers [24]. The flowers are unisexual, meaning they occur as either male or female on the same tree. They are typically solitary or arranged in clusters at the leaf axils. The fruit of *G. kola* is a berry-like capsule. It is approximately 5–7 cm in diameter and contains several seeds. When ripe, the fruit turns brown or reddish-brown. Inside the fruit are the seeds, which are commonly referred to as kola nuts. These seeds are the most celebrated part of the plant and are known for their intensely bitter taste. Seeds of *G. kola* are brown, oblong, and roughly 2–4 cm in length. They have a hard outer shell and are divided into segments, each containing a cotyledon. The bitterness of the seeds is due to the presence of compounds such as caffeine and theobromine [56, 102].

3.2 Reproductive Features of *Garcinia kola*

Garcinia kola is a dioecious plant, meaning it has separate male and female flowers on different trees. To produce fruit, both male and female trees need to be present in close proximity for pollination to occur. The reproductive features of *G. kola* are essential for understanding how this plant reproduces and produces seeds for propagation. Kola is a dioecious plant, which means it has separate male and female trees. In dioecious species, individual trees are either male or female, and both types are required for successful reproduction. This reproductive strategy is common in many tree species and ensures genetic diversity in the population. The flowers of *G. kola* are unisexual, meaning that individual flowers are either male or female. This characteristic is typical in dioecious plants. The male flowers produce pollen, which contains the male reproductive cells (sperm). These flowers typically have stamens with anthers that release pollen. Female flowers have structures called carpels that contain the ovules, which eventually develop into seeds if fertilized. The female flowers receive pollen from the male flowers to initiate fertilization. The flowers are typically pollinated by insects, particularly bees and butterflies. These insects transfer pollen from the male flowers to the female flowers while foraging for nectar. After successful pollination, the female flowers develop into fruit. The fruit of *G. kola* is a berry-like capsule. Inside the fruit, the fertilized ovules develop into seeds. These seeds are commonly referred to as kola nuts. The number of seeds produced can vary depending on various factors, including the health of the tree and the effectiveness of pollination. Bitter kola seeds are primarily dispersed by animals. Various wildlife species, including rodents, primates, and birds, consume the seeds and then disperse them in different locations [88]. This process aids in seed dispersal and allows new plants to be established in different areas. Once a bitter kola seed is dispersed and finds a suitable environment, it germinates. Germination is the process by which a seed sprouts and begins to grow into a new plant. As the seedling grows, it develops into a juvenile tree, eventually reaching maturity and producing flowers and fruits. This reproductive cycle continues as long as there are both male and female trees in proximity to facilitate pollination. This reproductive strategy enhances genetic diversity within the species and contributes to the sustainability of bitter kola populations in their natural habitat.

4 Phytochemical Constituents of *Garcinia kola*

Bitter kola is rich in biologically active compounds and some of them are responsible for the color, flavor, and disease resistance of the plants. These chemical compounds are not necessarily considered essential nutrients like vitamins and minerals, but they play a crucial role in human health [6, 34, 35]. Consuming a diet rich in phytochemicals has been associated with numerous health benefits, including antioxidant and anti-inflammatory effects, as well as potential protection against chronic diseases such as cancer, heart disease, and diabetes [43, 71]. The phytochemicals in *G. kola* include flavonoids, polyphenols, alkaloids, saponins, tannins, terpenes, and many others [46].

Some of the alkaloids found in *G. kola* are caffeine and theobromine and they are known to possess pharmacological effects on humans [58, 59]. For instance, caffeine has a stimulating effect on the central nervous system and can increase alertness and energy levels when consumed while theobromine has mild stimulant properties and can dilate blood vessels. Although the levels of caffeine and theobromine in bitter kola may not be as high as in coffee or chocolate, their presence contributes to the plant's stimulant properties [87]. Tannins found in *G. kola are* polyphenolic compounds that are known for their astringent taste [84]. They can bind to proteins and minerals, and they have antioxidant properties. Polyphenols are a group of naturally occurring compounds known for their antioxidant properties. The plant also contains ellagic acids, which are a kind of polyphenol with antioxidant properties [85]. Ellagic acid has been studied for its potential health benefits, including its role in protecting cells from oxidative damage. Examples of flavonoids found in *G. kola* are quercetin and kaempferol [18]. These compounds have antioxidant, anti-inflammatory, and potential anticancer properties [109]. Quercetin has been studied for its potential role in reducing the risk of chronic diseases while kaempferol has been associated with various health benefits, including cardiovascular health [55]. Saponins are natural compounds also found in kola with diverse biological activities and are often responsible for the foaming properties of certain plant extracts. Garcinol is a type of polyisoprenylated benzophenone found in *Garcinia* species, including bitter kola [27, 51, 52]. It has been studied for its potential anti-inflammatory and antioxidant properties. Garcinol has also shown promise in research for its anticancer effects. Kolaviron is a complex bioactive compound found in bitter kola [86]. It is a mixture of flavonoids and other bioactive compounds. Kolaviron has been extensively studied for its potential therapeutic effects, including its antioxidant, anti-inflammatory, and hepatoprotective properties [28]. It is considered one of the key

Table 1 Health benefits of phytochemicals found in *Garcinia kola*

Phytochemicals in *G. kola*	Health benefits and uses
Alkaloids	
Caffeine	Natural stimulant, increases alertness and energy. Potential to improve mental focus and concentration. May alleviate fatigue and drowsiness. Used in some traditional remedies for its stimulating effects.
Theobromine	Potential vasodilatory properties may support cardiovascular health. Mild stimulants can have a relaxing effect on smooth muscles.
Tannins	
Ellagic	Potential role in cancer prevention and treatment. May have anti-inflammatory effects. Strong antioxidant, helps protect cells from oxidative damage.
Flavonoids	
Quercetin	Potential for reducing the risk of chronic diseases. May support heart health by improving blood vessel function. Anti-inflammatory properties may reduce inflammation. Potent antioxidant, protects cells from free radical damage.
Kaempferol	Potential anticancer properties. Cardiovascular benefits, support heart health. Antioxidant helps neutralize harmful free radicals. Anti-inflammatory effects may reduce inflammation.
Saponins	
Garcinol	Investigated for its potential in cancer prevention and treatment. Antioxidant effects, help combat oxidative stress. Potential anti-inflammatory properties.
Kolaviron	Strong antioxidant, that protects cells and tissues from oxidative damage. May have antimicrobial and antiviral activities. Hepatoprotective properties support liver health. Anti-inflammatory effects, may alleviate inflammation.
Polyphenols	
Gallic acid	Potent antioxidant, that protects cells from oxidative stress. Potential for cancer prevention and treatment. May support heart health by improving blood vessel function. Anti-inflammatory effects, reduce inflammation.

compounds responsible for some of the medicinal properties attributed to bitter kola. Bitter kola contains gallic acid, a polyphenolic compound with antioxidant properties. Gallic acid has been studied for its potential health benefits, including its role in protecting cells from reactive oxygen species. Some of the health benefits of secondary phytochemicals found in bitter kola are presented in Table 1.

5 Nutritional Properties of *Garcinia kola*

Bitter kola is used as a food and flavoring ingredient in various cultural contexts and is primarily known for its intense bitterness, which is not always suitable for all tastes [17, 93, 103, 104]. Additionally, it is consumed in moderation due to its bitter taste and potential effects on digestion. Common food uses of *Garcinia kola* are presented in Table 2.

Table 2 Common food uses of bitter kola

Food and related uses	Description
Snack (chewing snack)	Bitter kola seeds are often chewed as a snack. They have a distinctly bitter taste and are enjoyed for their flavor and potential health benefits.
Flavoring agent	Ground or powdered bitter kola is sometimes used as a natural flavoring agent in traditional African cuisine, adding a unique bitterness to dishes.
Traditional soup	Bitter kola may be added to traditional soups and stews, especially in West African cuisine, to enhance the flavor and impart its distinctive bitterness.
Beverage infusion	Bitter kola can be infused in water or other liquids to create a bitter, aromatic beverage. This infusion is sometimes used for its potential medicinal properties.
Medicinal concoction	In some cultures, bitter kola is used to prepare medicinal concoctions or teas to address various health issues, such as digestive discomfort or coughs.
Condiments and seasoning	Ground or crushed bitter kola can be used as a condiment or spice to season dishes, similar to how other spices like pepper or chili are used.

Garcinia kola is consumed for its cultural, traditional, and potential medicinal properties rather than its nutritional content. However, bitter kola is not typically consumed as a primary source of nutrition, it does contain some nutrients and bioactive compounds. Some nutritional properties of *G. kola* include:

Bitterness: The most prominent characteristic of bitter kola is its intense bitterness, which is attributed to the presence of compounds like caffeine and theobromine. This bitterness is not related to its nutritional value but rather to its flavor profile [110].

Phytochemicals: Bitter kola contains bioactive phytochemicals many of which have known health benefits like stimulatory effects on the central nervous system and antioxidant properties. Bitter kola contains antioxidants, which can help protect cells from oxidative damage. Antioxidants play a role in overall health and may contribute to the plant's potential health benefits [111–113]. Kolaviron which is also present in bitter kola has therapeutic effects. These compounds are more related to their traditional and medicinal uses rather than their nutritional content [119].

Vitamins and Minerals: Bitter kola contains trace amounts of certain vitamins and minerals, although not in significant quantities to meet daily dietary requirements. These may include small amounts of vitamin C, vitamin A, and calcium [6, 34, 35].

Fiber: Bitter kola may contain dietary fiber, although the fiber content is relatively low. Fiber is important for digestive health and can help regulate bowel movements.

6 Traditional Uses of *Garcinia kola*

The ethnobotany of *Garcinia kola* is deeply rooted in the cultures and traditions of West and Central Africa. This remarkable plant has played a significant role in the lives of indigenous communities for centuries, offering a rich tapestry of uses and cultural significance. The small- to medium-sized evergreen tree that is native to the

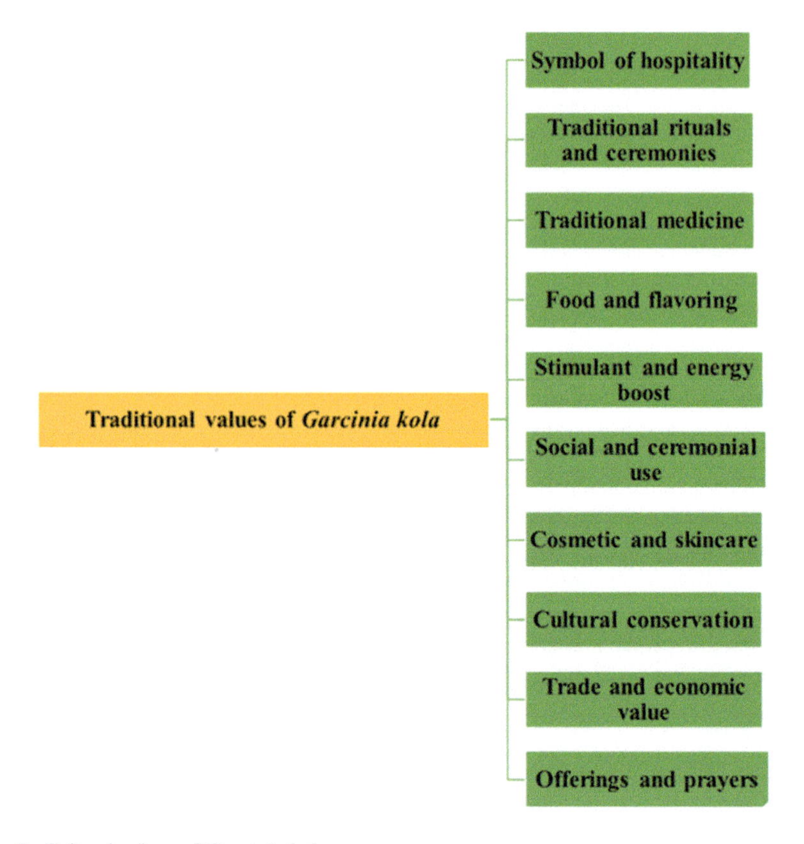

Fig. 2 Cultural values of *Garcinia kola*

dense humid rainforests of West and Central Africa bears bitter-tasting kola seeds. These seeds, with their intense bitterness, have become iconic and emblematic of the region's cultural heritage. Some of the cultural significance of *G. kola* include (Fig. 2):

- **Symbol of Hospitality**: Bitter kola is synonymous with hospitality in many West African cultures. It is customary to offer kola nuts to guests as a gesture of warm welcome and respect. The act of sharing bitter kola during social gatherings and ceremonies signifies unity and goodwill.
- **Traditional Rituals and Ceremonies**: Bitter kola plays a pivotal role in various traditional rituals and ceremonies across West and Central Africa. Its presence can be seen in rites of passage, ancestral worship, and spiritual offerings. The nuts are often included in sacrificial offerings to deities, ancestors, or spirits, depending on the cultural context.
- **Traditional Medicine**: One of the most profound ethnobotanical aspects of bitter kola is its esteemed place in traditional African medicine [22]. The plant has been harnessed for centuries to address a wide array of health issues. It is traditionally

believed to have therapeutic properties for ailments such as coughs, colds, malaria, and stomach aches [38, 44]. Others are:

- **Digestive Aid**: The intense bitterness of bitter kola is thought to be beneficial for digestion [25]. It is often consumed after meals to alleviate digestive discomfort and promote healthy digestion.
- **Antimicrobial Properties**: In traditional practices, bitter kola has been used for its potential antimicrobial properties [42, 54]. It is regarded as a natural remedy to combat certain infections.
- **Anti-inflammatory Effects**: Some traditional uses involve the application of bitter kola for its anti-inflammatory effects [1]. It is used externally to reduce inflammation and promote healing.

- **Food and Flavoring**
  - **Chewing Snack**: Bitter kola seeds are occasionally chewed as a snack. The unique and intense bitterness of the seeds is appreciated by those who have acquired a taste for them. It is a common sight to see people chewing bitter kola, especially during social gatherings and leisure time.
  - **Seasoning**: Bitter kola can be ground or crushed and used as a seasoning or condiment in traditional African dishes. It imparts a distinctive bitterness to the food, enhancing its flavor. Bitter kola-infused dishes are celebrated for their unique taste.
- **Stimulant and Energy Boost**
  - **Caffeine Content**: Bitter kola contains caffeine, a natural stimulant. This caffeine content contributes to its traditional use as a source of increased alertness and energy. It is often consumed to combat fatigue and drowsiness, especially during long journeys or night vigils.
- **Social and Ceremonial Use**
  - **Weddings and Celebrations**: Bitter kola occupies a prominent place in weddings and celebratory ceremonies. It is shared among participants as part of the cultural festivities, symbolizing joy and unity.
  - **Religious Gatherings**: In both cultural and religious gatherings, bitter kola is distributed among attendees as a symbol of unity and shared values. It is believed to foster harmony and goodwill among the participants.
  - Traditional ritual: Bitter kola is used in various traditional rituals and ceremonies in Africa, often associated with spiritual and cultural practices.
- **Offerings and Prayers**: Bitter kola is considered a sacred plant of spiritual significance by some communities. It is used in offerings and prayers to seek protection, blessings, or favor from the spiritual realm [60]. The intense bitterness of the seeds is believed to make them particularly potent in spiritual contexts.
- **Cosmetics and Skincare**: Bitter kola has been used topically for skin care and cosmetic purposes. It is believed to have properties that benefit the skin, such as reducing blemishes and promoting a healthy complexion.
- **Trade and Economic Value**: Bitter kola holds commercial economic significance beyond its cultural and traditional roles. It is traded and sold in local markets, serving as a source of income for communities engaged in its cultivation and trade [116]. This economic value has contributed to its continued cultivation and importance.

- **Cultural Conservation**: The cultural and traditional importance of bitter kola has sparked efforts to conserve and protect this species and its natural habitat. Communities recognize the need to preserve the tree that holds such a central place in their heritage [94, 100, 101].

7 Modern Scientific Research on Ethnomedical Uses and Pharmacological Potentials of *Garcinia kola*

Garcinia kola has also been studied for its potential pharmacological properties. These properties are largely attributed to the various phytochemical constituents found in the plant. Pharmacological properties of *Garcinia kola* include (Fig. 3):

Fig. 3 Health benefits of *Garcinia kola*

Antimicrobial Properties: Bitter kola has been investigated for its antibacterial properties [68]. Some studies suggest that extracts from bitter kola may have inhibitory effects against certain pathogenic bacteria, potentially making it useful in combating bacterial infections [23, 42]. Research has shown that bitter kola contains compounds with antimicrobial properties, which may help combat certain bacteria and viruses [57]. It has been used traditionally to treat infections, coughs, and colds. Research has also indicated that bitter kola extracts may exhibit antifungal activity [66]. This property could have applications in addressing fungal infections.

Antioxidant Activity: Bitter kola contains various phytochemicals, including flavonoids, polyphenols, and tannins, that possess antioxidant properties [68]. Antioxidants help neutralize harmful free radicals, reducing oxidative stress and potentially contributing to overall health. The presence of compounds like ellagic acid, quercetin, and gallic acid in bitter kola contributes to its antioxidant effects [72]. Bitter kola can help protect the body's cells from oxidative damage, potentially contributing to overall health and well-being [4].

Anti-Inflammatory Effects: Some studies have explored the anti-inflammatory potential of bitter kola. Inflammation is a key factor in various chronic diseases, and the anti-inflammatory properties of bitter kola compounds like garcinol and kolaviron may have therapeutic implications [2, 3, 5, 73]. Some compounds found in bitter kola have anti-inflammatory properties, which may make it beneficial for alleviating pain and inflammation in conditions like arthritis.

Hepatoprotective Properties: Kolaviron, a bioactive compound found in bitter kola, has been studied for its hepatoprotective (liver-protective) effects [2, 3]. It may help protect the liver from damage caused by toxins and oxidative stress ([8, 26]; Adaramoye and Adeyemi 2012). This hepatoprotective property is particularly significant in regions where liver diseases are prevalent.

Antidiabetic Potential: Some research has suggested that bitter kola may have antidiabetic properties. Extracts from bitter kola have been studied for their potential to lower blood glucose levels and improve insulin sensitivity [7, 19, 63]. These effects could be beneficial for individuals with diabetes or those at risk of developing the condition.

Antimalarial Activity: Traditional medicine practices have long used bitter kola for its potential antimalarial properties [38]. While more research is needed, some studies have explored the plant's efficacy against the malaria parasite. Bitter kola's antimalarial potential may be attributed to its phytochemical constituents.

Cardiovascular Health: The antioxidant and anti-inflammatory properties of certain bitter kola compounds may have positive effects on cardiovascular health [20, 22, 49]. These properties can help reduce the risk of cardiovascular diseases by protecting blood vessels and reducing inflammation.

Analgesic and Anti-Inflammatory Effects: Some traditional uses of bitter kola involve its application as an analgesic (pain reliever) and anti-inflammatory agent, particularly for addressing musculoskeletal pain and inflammation [79]. These effects may be attributed to its anti-inflammatory compounds, which can provide relief from various painful conditions.

Anticancer Potential: While research is in its early stages, some studies have explored the potential anticancer properties of bitter kola compounds, including garcinol and kolaviron [9, 10, 21]. These compounds have demonstrated anticancer effects in laboratory studies, but further research is needed to determine their clinical applications [29, 74].

Digestive Aid and Gut Health Improvement: Bitter kola is often consumed to relieve digestive discomfort, such as bloating, indigestion, and constipation [75]. It is believed to stimulate digestion and promote the flow of gastric juices.

Aphrodisiac: Bitter kola is sometimes regarded as an aphrodisiac and is believed to enhance sexual performance and libido in some cultures [14–16]. However, scientific evidence supporting these claims is limited.

Weight Loss: Some individuals have explored the potential of bitter kola for weight loss due to its reputation as an appetite suppressant [72, 75, 76]. However, more research is needed to confirm its effectiveness for this purpose.

Although bitter kola shows promise in various pharmacological areas, much of the research is preliminary, and further studies are required to fully understand its mechanisms of action and therapeutic potential [82]. Additionally, the traditional uses of bitter kola should not replace conventional medical treatments, and individuals with specific medical conditions should consult healthcare professionals for appropriate guidance and treatment.

8 Toxicity of *Garcinia kola*: Safety Considerations and Side Effects

The use of *Garcinia kola* for health management is generally considered safe when consumed and used in moderation. It has a long history of traditional use in West and Central Africa for various purposes, including as a cultural symbol, a traditional remedy, and a dietary component. However, like many natural products, bitter kola should be used with caution, and excessive consumption can lead to potential adverse effects. Some potential toxic considerations associated with bitter kola include:

- **Bitter Taste and Sensitivity**: The most notable characteristic of bitter kola is its intensely bitter taste. Some individuals may find the taste unpleasant or intolerable, which can limit their consumption. Additionally, individuals with heightened sensitivity to bitter flavors may experience discomfort when consuming bitter kola.
- **Caffeine Content**: Bitter kola contains caffeine, a natural stimulant. While the caffeine content in bitter kola is not as high as in coffee or certain other sources, excessive consumption of caffeine can lead to side effects such as nervousness, restlessness, insomnia, and increased heart rate. People who are sensitive to caffeine should consume bitter kola in moderation.
- **Potential Allergic Reactions**: As with any food or natural product, some individuals may be allergic to specific components of bitter kola. Allergic reactions

can range from mild symptoms like itching or hives to more severe reactions that may require medical attention. If you have known food allergies or experience allergic symptoms after consuming bitter kola, you should avoid it.

- **Gastrointestinal Distress**: Excessive consumption of bitter kola seeds can sometimes lead to gastrointestinal discomfort, including stomachaches, indigestion, or diarrhea. The bitter taste of the seeds may also trigger nausea in some individuals.
- **Interactions with Medications**: Bitter kola contains bioactive compounds that could potentially interact with certain medications. If you are taking prescription medications or have underlying health conditions, it is advisable to consult with a healthcare professional before adding bitter kola to your diet to avoid potential interactions.
- **Liver Health**: While bitter kola has been studied for its potential hepatoprotective (liver-protective) properties, consuming extremely large quantities of bitter kola or its extracts may have unintended effects on liver function [11, 13]. Moderation in consumption is key.
- **Pregnancy and Lactation**: Pregnant and breastfeeding women should be cautious when consuming bitter kola. Limited information is available on the safety of bitter kola during pregnancy and lactation, so it is advisable to consult a healthcare provider before using it during these periods.
- **Children and Seniors**: Due to differences in sensitivity and metabolism, children and seniors may be more vulnerable to the effects of bitter kola. Care should be taken to limit consumption and ensure that it does not adversely affect their health.
- **Individual Variability (Quality and Source)**: The quality and safety of bitter kola products can vary. It's important to obtain bitter kola from reputable sources to ensure it has been handled and processed appropriately. Individual tolerance and sensitivity to bitter kola can vary widely. What is well-tolerated by one person may cause discomfort in another. Start with small quantities to assess your response.

As with any dietary supplement or natural product, it is important to use bitter kola in moderation and be aware of your tolerance. However, individual tolerance and sensitivity can vary, so it's wise to exercise caution and practice moderation. If you have specific health concerns, allergies, or are taking medications, it is advisable to consult with a healthcare professional before incorporating bitter kola into your diet or using it for medicinal purposes. If you have specific health concerns or are taking medications, consult with a healthcare professional before incorporating bitter kola into your diet. Additionally, if you experience any adverse reactions after consuming bitter kola, discontinue use and seek medical advice.

9 Recommendation for Further Research: Molecular Docking Studies with Compound from *Garcinia kola*

Molecular docking is a computational technique used in drug discovery and bioinformatics to predict the binding interactions between small molecules (ligands) and target proteins (receptors) [36]. Conducting specific molecular docking studies with compounds from *G. kola* using the 3D structures of these compounds and target proteins, as well as specialized software and expertise in computational chemistry. General steps involved in molecular docking as suggested in Adewole et al. [18] include:

1. **Data Preparation**: Obtain the 3D structures of the compounds from Garcinia kola (e.g., using chemical structure databases or software tools). Acquire the 3D structure of the target protein or receptor of interest (e.g., from protein structure databases or experimental methods like X-ray crystallography).
2. **Ligand Preparation**: Prepare the ligands (compounds from Garcinia kola) by assigning partial charges and optimizing their 3D conformations. This step ensures that the ligands are in a suitable format for docking.
3. **Receptor Preparation**: Prepare the receptor (target protein) by assigning partial charges, adding hydrogen atoms, and optimizing its 3D conformation. This step ensures that the receptor is in a suitable state for docking.
4. **Grid Generation**: Create a grid or docking box around the active site of the target protein. The grid defines the region where ligands will be docked and scored for binding affinity.
5. **Docking**: Use molecular docking software (e.g., AutoDock, AutoDock Vina, or other specialized tools) to dock the prepared ligands into the binding site of the target protein. The software explores various conformations and orientations of the ligands within the binding site and evaluates their interactions with the receptor based on scoring functions that account for van der Waals forces, electrostatic interactions, and other factors. Docking simulations may include energy minimization and molecular dynamics simulations to refine the ligand-receptor complexes.
6. **Scoring and Analysis**: After docking, the software calculates binding scores and ranks the ligands based on their predicted binding affinities. Lower scores typically indicate stronger binding interactions. Researchers analyze the docking results to identify potential ligands from Garcinia kola that have favorable binding interactions with the target protein. Visualization tools are often used to examine the ligand-receptor complexes and understand the nature of the interactions, such as hydrogen bonds or hydrophobic contacts.
7. **Validation and Interpretation**: Molecular docking results should be validated through experimental studies, such as binding assays or structural biology techniques, to confirm the predicted binding interactions. Interpret the docking results in the context of the specific biological or pharmacological question being investigated.

It is important to emphasize that molecular docking is a computational tool for predicting binding interactions, and its accuracy depends on the quality of the input data and the reliability of the scoring functions used. Experimental validation is crucial to confirm the predictions made through docking studies. Additionally, conducting molecular docking studies with compounds from *G. kola* would require access to the 3D structures of these compounds, which may not be readily available in public databases. Therefore, specialized research efforts would be needed to obtain or generate these structures for docking studies.

10 Some Herbal Formulations Involving *Garcinia kola* and their Active Ingredients

Garcinia kola is often consumed as a whole seed or used in traditional remedies. However, it is less common to find standardized herbal formulations involving bitter kola, especially within commercial herbal products in Western markets [116]. However, some traditional and contemporary formulations do exist, often leveraging the bioactive compounds found in bitter kola for specific health benefits. Some examples include:

1. **Bitter Kola Extracts or Tinctures**: Some herbal product manufacturers offer bitter kola extracts or tinctures. These liquid formulations typically involve the extraction of bioactive compounds from bitter kola seeds using solvents. They may be marketed for their potential health benefits, including their antioxidant properties and use in traditional medicine.
2. **Bitter Kola Capsules or Supplements**: In some regions, you can find bitter kola capsules or dietary supplements. These products are designed to provide a standardized dose of bitter kola's bioactive compounds, often with a focus on its potential health benefits.
3. **Traditional Herbal Remedies**: In many traditional healing practices across West and Central Africa, bitter kola is used as an ingredient in various herbal remedies. These formulations may include other locally available medicinal plants and are often used to address specific health conditions. For example, bitter kola may be combined with other herbs to create a remedy for coughs or digestive issues.
4. **Bitter Kola Chewing Sticks**: Bitter kola is sometimes incorporated into chewing sticks or "oral health sticks." These sticks are used for oral hygiene and teeth cleaning. The natural bitterness of bitter kola is believed to help with oral health.
5. **Bitter Kola in Traditional Drinks and Tonics**: In some traditional cultures, bitter kola is added to herbal drinks, tonics, or concoctions. These preparations are believed to have various health benefits, such as energy enhancement, digestive support, and immune system boosting.
6. **Bitter Kola and Honey Infusions**: Bitter kola can be infused in honey to create a sweet and bitter combination. This infusion is sometimes used for its potential health benefits, including its supposed immune-boosting and energy-enhancing properties.

It is important to note that while these formulations may be based on traditional knowledge and practices, scientific research on the specific health effects of bitter kola and its active compounds is ongoing. If you're interested in using herbal formulations involving Garcinia kola, it's advisable to consult with a healthcare professional or herbalist who is knowledgeable about traditional remedies. They can guide the appropriate use and dosage of such products, as well as any potential interactions or contraindications with other medications or health conditions.

11 Commercial Value of *Garcinia kola*

Garcinia kola holds commercial value primarily due to its cultural significance and traditional uses, as well as its potential medicinal properties. Although it may not be a major cash crop like some other agricultural products, bitter kola plays a vital role in the economies of certain regions in West and Central Africa and some aspects of the commercial value of *G. kola* include:

Cultural Significance: Bitter kola is deeply embedded in the cultures and traditions of West and Central Africa. It is often used as a symbol of hospitality and goodwill. The exchange of bitter kola during social gatherings, ceremonies, and rituals is a common practice, contributing to its cultural value.

Traditional Medicine: Bitter kola has a long history of use in traditional African medicine. Many indigenous communities believe in its therapeutic properties for addressing various health issues, including coughs, colds, malaria, and digestive problems. This traditional use adds to its commercial value as a natural remedy.

Trade and Local Markets: Bitter kola is traded and sold in local markets across West and Central Africa. It serves as a source of income for local communities engaged in its cultivation and trade. The sale of bitter kola in local markets contributes to the livelihoods of many people.

Export Market: Bitter kola has gained some recognition in international markets due to its potential medicinal properties and as a unique botanical product. In recent years, there has been a growing interest in the export of bitter kola to other countries, including Europe and the United States, where it is sometimes used in traditional medicine practices or as a dietary supplement.

Pharmaceutical and Nutraceutical Industries: Some pharmaceutical and nutraceutical companies have shown interest in the potential medicinal properties of bitter kola's bioactive compounds. They may use bitter kola extracts or compounds in the development of herbal supplements or traditional medicine products, adding to its commercial value.

Food and Beverage Industry: Bitter kola is occasionally used as a flavoring or seasoning in traditional African dishes, contributing to its value in the culinary industry. It imparts a distinctive bitterness to food, enhancing its flavor.

Oral Health Products: Bitter kola is sometimes incorporated into oral hygiene products, such as toothpaste and chewing sticks. These products may be marketed for their potential oral health benefits.

Cosmetics and Skincare: In some regions, bitter kola is used topically for skincare and cosmetic purposes. It is believed to have properties that benefit the skin, such as reducing blemishes and promoting a healthy complexion.

Research and Development: Scientific research on bitter kola's potential medicinal properties and bioactive compounds has spurred interest in its commercial applications. This includes the development of herbal formulations and products.

The commercial value of bitter kola is not as prominent as that of major cash crops, its importance lies in its cultural heritage, traditional uses, and potential in various industries. As interest in natural remedies and traditional medicine grows, bitter kola continues to find a place in both local and international markets, contributing to the economies of the regions where it is cultivated and traded.

12 Future Prospects of Therapeutic Roles of Garcinia kola

The therapeutic roles of *Garcinia kola* have garnered significant interest due to its traditional uses and the presence of bioactive compounds with potential health benefits. Yet much research is still needed to fully understand its mechanisms of action and therapeutic potential because bitter kola holds promising prospects in various areas of healthcare and wellness. Some potential future therapeutic roles of bitter kola include:

- **Antioxidant and Anti-Inflammatory Properties**: Bitter kola is rich in bioactive compounds like kolaviron, quercetin, and ellagic acid, which possess antioxidant and anti-inflammatory properties [30, 31]. These properties make bitter kola a potential candidate for managing oxidative stress-related diseases and chronic inflammatory conditions [39].
- **Antimicrobial and Antiviral Activity**: Preliminary studies have suggested that extracts from bitter kola may exhibit antibacterial, antifungal, and antiviral properties [50]. These findings raise the possibility of using bitter kola as a natural remedy for infectious diseases.
- **Antidiabetic Potential**: Research has indicated that bitter kola extracts may help lower blood glucose levels and improve insulin sensitivity. With the increasing prevalence of diabetes worldwide, bitter kola may have a role in managing this chronic condition.
- **Hepatoprotective Effects**: Kolaviron, a bioactive compound in bitter kola, has been studied for its hepatoprotective (liver-protective) properties [32, 40]. As liver diseases remain a global health concern, bitter kola could have applications in supporting liver health.
- **Cardiovascular Health**: Bitter kola's antioxidant and anti-inflammatory properties may have positive effects on cardiovascular health. It could potentially contribute to reducing the risk of heart disease by protecting blood vessels and reducing inflammation [41, 51].
- **Antimalarial Activity**: Traditional medicine practices have long used bitter kola for its potential antimalarial properties [33]. Further research may lead to the development of natural antimalarial remedies.
- **Analgesic Uses**: Bitter kola has been traditionally used as an analgesic and anti-inflammatory agent for addressing musculoskeletal pain and inflammation. Its potential role in pain management warrants further exploration.
- **Cancer Research**: Some studies have investigated the anticancer properties of bitter kola compounds, such as garcinol and kolaviron [37, 51]. While research is in its early stages, these compounds have shown promise in laboratory studies as potential agents for cancer prevention and treatment [74].

- **Nutraceuticals and Dietary Supplements**: Bitter kola extracts and compounds may find applications in the development of nutraceuticals and dietary supplements aimed at promoting overall health and well-being [82].
- **Cosmeceuticals and Skincare Products**: Bitter kola's potential benefits for skin health and its antioxidant properties may lead to its incorporation into cosmeceuticals and skincare products.
- **Functional Foods and Beverages**: Bitter kola may be used as an ingredient in functional foods and beverages, offering consumers a natural source of bioactive compounds with potential health benefits.
- **Continued Research and Clinical Trials**: Future research endeavors, including well-designed clinical trials, will help validate the therapeutic roles of bitter kola and establish its safety and efficacy in specific medical conditions [77, 78].

Although bitter kola shows promise in various therapeutic areas, much of the research is still in the early stages, and further studies are required to establish its clinical applications. As research progresses, bitter kola may find its place in modern medicine and complementary healthcare, offering potential alternatives or adjuncts to existing treatments.

13 Conclusion

The evergreen foliage, dioecious reproductive system, and distinctive kola nuts make *G. kola* a unique and valuable plant species in the tropical rainforest ecosystem. Beyond its ecological role, the plant's cultural and medicinal importance has solidified its place in the hearts and traditions of the people of the region. Bitter kola is a culturally significant plant in West and Central Africa with a history of traditional medicinal use. It offers potential health benefits, but it is important to approach its consumption with caution, especially if you have underlying medical conditions or are taking medications. Always consult with a healthcare professional before incorporating bitter kola or any herbal remedy into your diet or wellness routine. Kola is a culturally, medicinally, and ecologically significant plant that has been integral to the local traditions and ecosystems in West and Central Africa and beyond for centuries. However, the plant is facing numerous sustainability and conservation challenges due to habitat destruction, efforts are underway to protect this valuable plant and the rich biodiversity of African rainforests. The cultural and historical importance of *G. kola*, combined with its potential health benefits, make it a captivating subject of study and a testament to the intricate relationship between people, plants, and the environment in which they thrive. Preserving the natural habitats of plants like bitter kola is not only essential for the well-being of local communities but also for the conservation of the Earth's biodiversity and the continued exploration of the secrets hidden within the rainforests of Africa.

References

1. Abarikwu SO (2014) Anti-inflammatory effects of kolaviron modulate the expressions of inflammatory marker genes, inhibit transcription factors ERK1/2, p-JNK, NF-jB, and activate Akt expressions in the 93RS2 Sertoli cell lines. Mol Cell Biochem 401:197–208
2. Abarikwu SO, Farombi EO, Kashyap MP, Pant AB (2011) Kolaviron protects apoptotic cell death in PC12 cells exposed to atrazine. Free Radic Res 45:1061–1073
3. Abarikwu SO, Farombi EO, Pant AB (2011) Biflavanone kolaviron protects human dopaminergic SH-SY5Y cells against atrazine induced toxic insult. Toxicol In Vitro 25:848–858. https://doi.org/10.1016/j.tiv.2011.02.005
4. Abarikwu SO, Njoku R-CC, John IG et al (2021) Antioxidant and anti-inflammatory protective effects of rutin and kolaviron against busulfan-induced testicular injuries in rats. Syst Biol Reprod Med. https://doi.org/10.1080/19396368.2021.1989727
5. Abdallah HM, Almowallad FM, Esmat A et al (2015) Anti-inflammatory activity of flavonoids from Chrozophora tinctoria. Phytochem Lett 13:74–80. https://doi.org/10.1016/j.phytol.2015.05.008
6. Adaramoye OA (2009) Comparative effects of vitamin E and kolaviron (a biflavonoid from Garcinia kola) on carbon tetrachloride-induced renal oxidative damage in mice. Pak J Biol Sci 12:1146–1151. https://doi.org/10.3923/pjbs.2009.1146.1151
7. Adaramoye OA (2012) Antidiabetic effect of kolaviron, a biflavonoid complex isolated from Garcinia kola seeds, in Wistar rats. Afr Health Sci 12:498–506
8. Adaramoye OA, Adeyemi EO (2006) Hypoglycaemic and hypolipidaemic effects of fractions from kolaviron, a biflavonoid complex from Garcinia Kola in streptozotocininduced diabetes mellitus rats. J Pharm Pharmacol 58:121–128. https://doi.org/10.1211/jpp.58.1.0015
9. Adaramoye OA, Arisekola M (2012) Kolaviron, a biflavonoid complex from Garcinia kola seeds, ameliorates ethanol-induced reproductive toxicity in male wistar rats. Niger J Physiol Sci 28:9–15
10. Adaramoye OA, Nwaneri VO, Anyanwo KC et al (2005) Possible anti-atherogenic effect of kolaviron (a Garcinia kola seed extract) in hypercholesterolaemic rats. Clin Exp Pharmacol Physiol 32:40–46
11. Adaramoye OA, Farombi EO, Nssien M et al (2008) Hepato-protective activity of purified fractions from Garcinia kola seeds in mice intoxicated with carbon tetrachloride. J Med Food 11:544–550. https://doi.org/10.1089/jmf.2007.0539
12. Adaramoye OA, Akanni OO, Farombi EO (2013) Nevirapine induces testicular toxicity in wistar rats: reversal effect of kolaviron (biflavonoid from Garcinia kola seeds). J Basic Clin Physiol Pharmacol 24:313–320. https://doi.org/10.1515/jbcpp-2012-0078
13. Adaramoye OA, Kehinde AO, Adefisan A et al (2016) Ameliorative effects of kolaviron, a biflavonoid fraction from Garcinia kola seed, on hepato-renal toxicity of anti-tuberculosis drugs in wistar rats. Tokai J Exp Clin Med 41:14–21
14. Adedara IA, Farombi EO (2012) Chemoprotection of ethylene glycol monoethyl ether-induced reproductive toxicity in male rats by kolaviron, isolated biflavonoid from Garcinia kola seed. Hum Exp Toxicol 31:506–517. https://doi.org/10.1177/0960327111424301
15. Adedara IA, Farombi EO (2013) Chemoprotective effects of kolaviron on ethylene glycol monoethyl ether-induced pituitary-thyroid axis toxicity in male rats. Andrologia 45:111–119. https://doi.org/10.1111/j.1439-0272.2012.01321.x
16. Adedara IA, Farombi EO (2014) Kolaviron protects against ethylene glycol monoethyl ether-induced toxicity in boar spermatozoa. Andrologia 46:399–407. https://doi.org/10.1111/and.12095
17. Adesuyi AO, Elumm IK, Adaramola FB, Nwokocha AGM (2012) Nutritional and phytochemical screening of Garcinia kola. Adv J Food Sci Technol 4:9–14
18. Adewole KE, Gyebi GA, Ibrahim IM (2021) Amyloid b fibrils disruption by kolaviron: molecular docking and extended molecular dynamics simulation studies. Comput Biol Chem. https://doi.org/10.1016/j.compbiolchem.2021.107557

19. Adoga JO, Channa ML, Nadar A (2021) Kolaviron attenuates cardiovascular injury in fructose-streptozotocin induced type-2 diabetic male rats by reducing oxidative stress, inflammation, and improving cardiovascular risk markers. Biomed Pharmacother. https://doi.org/10.1016/j.biopha.2021.112323
20. Agboola OS, Oyagbemi AA, Omobowale TO et al (2016) Modulatory role of Kolaviron (KV), a biflavonoid from Garcinia kola, in sodium arsenite-induced hepatotoxicity and haematotoxicity in rats. Toxicol Int 23:54–62. https://doi.org/10.22506/ti/2016/v23/i1/146671
21. Aggarwal V, Tuli HS, Kaur J et al (2020) Garcinol exhibits anti-neoplastic effects by targeting diverse oncogenic factors in tumor cells. Biomedicine 8:103. https://doi.org/10.3390/biomedicines8050103
22. Ajani EO, Shallie PD, Adegbesan BO et al (2008) Protective effect of Garcinia Kola (Kolaviron) extract on predisposition of rats to cardiovascular diseases following separate administration of amodiaquine and artesunate. Afr J Tradit Complement Altern Med 5:180–186
23. Ajayi TO, Moody JO, Fukushi Y et al (2014) Antimicrobial activity of Garcinia kola (Heckel) seed extracts and isolated constituents against caries-causing microorganisms. Afr J Biomed Res 17:165–171
24. Akihisa T, Yasukawa K, Oinuma H et al (1996) Triterpene alcohols from the flowers of compositae and their anti-inflammatory effects. Phytochemistry 43:1255–1260. https://doi.org/10.1016/S0031-9422(96)00343-3
25. Akinrinde AS, Olowu E, Oyagbemi AA, Omobowale OT (2015) Gastrointestinal protective efficacy of Kolaviron (a bi-flavonoid from Garcinia kola) following a single administration of sodium arsenite in rats: biochemical and histopathological studies. Pharm Res 7:268–276. https://doi.org/10.4103/0974-8490.157978
26. Akinrinde AS, Omobowale O, Oyagbemi A et al (2016) Protective effects of kolaviron and gallic acid against cobaltchloride-induced cardiorenal dysfunction via suppression of oxidative stress and activation of the ERK signaling pathway. Can J Physiol Pharmacol 94:1276–1284. https://doi.org/10.1139/cjpp-2016-0197
27. Akoro SM, Aiyelaagbe OO, Onocha PA, Gloer JB (2020) Gakolanone: a new benzophenone derivative from Garcinia kola Heckel stem-bark. Nat Prod Res 34:241–250. https://doi.org/10.1080/14786419.2018.1528583
28. Alabi QK, Akomolafe RO, Olukiran OS et al (2018) Kolaviron attenuates diclofenac-induced nephrotoxicity in male wistar rats. Appl Physiol Nutr Metab 43:956–968. https://doi.org/10.1139/apnm-2017-0788
29. Altmann K-H, Gertsch J (2007) Anticancer drugs from nature: natural products as a unique source of new microtubule stabilizing agents. Nat Prod Rep 24:327–357. https://doi.org/10.1039/b515619j
30. Awogbindin IO, Olaleye DO, Farombi EO (2015) Kolaviron improves morbidity and suppresses mortality by mitigating oxido-inflammation in BALB/c mice infected with influenza virus. Viral Immunol 28:367–377. https://doi.org/10.1089/vim.2015.0013
31. Awogbindin IO, Maduako IC, Adedara IA et al (2021) Kolaviron ameliorates hepatic and renal dysfunction associated with multiwalled carbon nanotubes in rats. Environ Toxicol 36:67–76. https://doi.org/10.1002/tox.23011
32. Ayepola OR, Brooks NL, Oguntibeju OO (2014) Kolaviron improved resistance to oxidative stress and inflammation in the blood (erythrocyte, serum, and plasma) of streptozotocin-induced diabetic rats. Sci World J. https://doi.org/10.1155/2014/921080
33. Azebaze AGB, Teinkela JEM, Nguemfo EL et al (2015) Antiplasmodial activity of some phenolic compounds from cameroonians allanblackia. Afr Health Sci 15:835–840. https://doi.org/10.4314/ahs.v15i3.18
34. Azzi A, Zingg J-M (2005) Vitamin E: textbooks require updating. Biochem Mol Biol Educ 33:184–187. https://doi.org/10.1002/bmb.2005.494033032451
35. Baggett S, Protiva P, Mazzola EP et al (2005) Bioactive benzophenones from Garcinia xanthochymus fruits. J Nat Prod 68:354–360. https://doi.org/10.1021/np0497595

36. Bajorath J (2021) Evolution of assay interference concepts in drug discovery. Expert Opin Drug Discov 16:719–721. https://doi.org/10.1080/17460441.2021.1902983
37. Bartolini D, De Franco F, Torquato P et al (2020) Garcinoic acid is a natural and selective agonist of pregnane X receptor. J Med Chem 63:3701–3712. https://doi.org/10.1021/acs.jmedchem.0c00012
38. Bickii J, Tchouya GRF, Tchouankeu JC, Tsamo E (2006) The antiplasmodial agents of the stem bark of *Entandrophragma angolense* (Meliaceae). Afr J Tradit Complement Altern Med 4:135–139
39. Birringer M, Lington D, Vertuani S et al (2010) Proapoptotic effects of long-chain vitamin E metabolites in HepG2 cells are mediated by oxidative stress. Free Radic Biol Med 49: 1315–1322. https://doi.org/10.1016/j.freeradbiomed.2010.07.024
40. Braide VB (1991) Antihepatotoxic biochemical effects of kolaviron, a biflavonoid of Garcinia kola seeds. Phytother Res 5:35–37. https://doi.org/10.1002/ptr.2650050110
41. Brigelius-Flohé R (2007) Adverse effects of vitamin E by induction of drug metabolism. Genes Nutr 2:249–256. https://doi.org/10.1007/s12263-007-0055-0
42. Chatterjee A, Bagchi D, Yasmin T, Stohs SJ (2005) Antimicrobial effects of antioxidants with and without clarithromycin on *Helicobacter pylori*. Mol Cell Biochem 270:125–130. https://doi.org/10.1007/s11010-005-5277-0
43. Chetia Phukan B, Dutta A, Deb S et al (2022) Garcinol blocks motor behavioural deficits by providing dopaminergic neuroprotection in MPTP mouse model of Parkinson's disease: involvement of anti-inflammatory response. Exp Brain Res 240:113–122
44. Dozie-Nwakile OC, Dozie NC, Kingsley UI et al (2021) Effects of kolaviron on pneumonia-like infection induced in albino wistar rats. Anti-Inflamm Anti-Allergy Agents Med Chem 20: 219–227
45. Eleazu CO, Eleazu KC, Awa E, Chukwuma SC (2012) Comparative study of the phytochemical composition of the leaves of five Nigerian medicinal plants. J Biotechnol Pharm Res 3: 42–46
46. Eleyinmi AF, Bressler DC, Amoo IA et al (2006) Chemical composition of bitter cola (Garcinia kola) seed and hulls. Pol J Food Nutr Sci 15:395
47. Emmanuel O, Uche ME, Dike ED et al (2022) A review on garcinia kola heckel: traditional uses, phytochemistry, pharmacological activities, and toxicology. Biomarkers 27:101–117
48. Farombi EO, Tahnteng JG, Agboola AO et al (2000) Chemoprevention of 2-acetylamino-fluorene-induced hepatotoxicity and lipid peroxidation in rats by kolaviron: a Garcinia kola seed extract. Food Chem Toxicol 38:535–541
49. Farombi EO, Awogbindin IO, Farombi TH et al (2019) Neuroprotective role of kolaviron in striatal redo-inflammation associated with rotenone model of Parkinson's disease. Neurotoxicology 73:132–141. https://doi.org/10.1016/j.neuro.2019.03.005
50. Hioki Y, Onwona-Agyeman S, Kakumu Y et al (2020) Garcinoic acids and a benzophenone derivative from the seeds of Garcinia kola and their antibacterial activities against oral bacterial pathogenic organisms. J Nat Prod 83:2087–2092. https://doi.org/10.1021/acs.jnatprod.9b01045
51. Hu H, Zhao H, Wu Z, Rao M (2020) Effects of garcinoic acid on cardiac function and its mechanism in post-myocardial infarction rats. Med J Wuhan Univ 41:188–192. https://doi.org/10.14188/j.1671-8852.2019.0924
52. Huang W-C, Kuo K-T, Adebayo BO et al (2018) Garcinol inhibits cancer stem cell-like phenotype via suppression of the Wnt/b-catenin/STAT3 axis signalling pathway in human non-small cell lung carcinomas. J Nutr Biochem 54:140–150. https://doi.org/10.1016/j.jnutbio.2017.12.008
53. Hussain RA, Owegby AG, Waterman PG (1982) Kolanone, a novel polyisoprenylated benzophenone with antimicrobial properties from the fruit of Garcinia kola. Planta Med 44:78–81. https://doi.org/10.1055/s-2007-971406
54. Hyun JJ, Woo SS, Yeo S-H et al (2006) Antifungal effect of amentoflavone derived from *Selaginella tamariscina*. Arch Pharm Res 29:746–751. https://doi.org/10.1007/bf02974074

55. Ibironke GF, Fasanmade AA (2015) Analgesic and central nervous system depressant activities of kolaviron (a Garcinia kola biflavonoid complex). Afr J Biomed Res 18:217–223
56. Igado OO, Olopade JO, Adesida A et al (2012) Morphological and biochemical investigation into the possible neuroprotective effects of kolaviron (Garcinia kola bioflavonoid) on the brains of rats exposed to vanadium. Drug Chem Toxicol 35:371–380. https://doi.org/10.3109/01480545.2011.630005
57. Iinuma M, Tosa H, Tanaka T et al (1996) Antibacterial activity of some Garcinia benzophenone derivatives against methicillin-resistant *Staphylococcus aureus*. Biol Pharm Bull 19: 311–314. https://doi.org/10.1248/bpb.19.311
58. Ijomone OM, Nwoha PU, Olaibi OK et al (2012) Neuroprotective effects of kolaviron, a biflavonoid complex of Garcinia kola, on rats hippocampus against methamphetamine-induced neurotoxicity. Maced J Med Sci 5:10–16. https://doi.org/10.3889/MJMS.1857-5773.2011.0203
59. Ikpesu TO, Tongo I, Ariyo A (2014) Restorative prospective of powdered seeds extract of Garcinia kola in *Chrysichthys furcatus* induced with glyphosate formulation. China J Biol. https://doi.org/10.1155/2014/854157
60. Imarhiagbe O, Ogwu MC (2022) Sacred groves in the global south: a panacea for sustainable biodiversity conservation. In: Izah SC (ed) Biodiversity in Africa: potentials, threats and conservation. Sustainable development and biodiversity, vol 29. Springer, Singapore, pp 525–546. https://doi.org/10.1007/978-981-19-3326-4_20
61. Ishola IO, Adamson FM, Adeyemi OO (2017) Ameliorative effect of kolaviron, a biflavonoid complex from Garcinia kola seeds against scopolamine-induced memory impairment in rats: role of antioxidant defense system. Metab Brain Dis 32:235–245. https://doi.org/10.1007/s11011-016-9902-2
62. Iwu M, Igboko O (1982) Flavonoids of Garcinia kola seeds. J Nat Prod 45:650–651. https://doi.org/10.1021/np50023a026
63. Iwu MM, Igboko OA, Okunji CO, Tempesta MS (1990) Antidiabetic and aldose reductase activities of biflavanones of Garcinia kola. J Pharm Pharmacol 42:290–292. https://doi.org/10.1111/j.2042-7158.1990.tb05412.x
64. Iwu M, Duncan C, Okunji CO (1999) Perspectives on new crops and new uses. ASHS Press, Alexandria
65. Iwu MM, Diop AD, Meserole L, Okunji CO (2002) Garcinia kola: a new look at an old adaptogenic agent. In: Iwu MM, Wootton J (eds) Advances in phytomedicine, 1st edn. Elsevier, Amsterdam, pp 191–199
66. Jackson DN, Yang L, Wu S et al (2015) Garcinia xanthochymus benzophenones promote hyphal apoptosis and potentiate activity of fluconazole against *Candida albicans* biofilms. Antimicrob Agents Chemother 59:6032–6038. https://doi.org/10.1128/AAC.00820-15
67. Jamila N, Khairuddean M, Khan SN, Khan N (2014) Complete NMR assignments of bioactive rotameric (3?8) biflavonoids from the bark of Garcinia hombroniana. Magn Reson Chem 52: 345–352. https://doi.org/10.1002/mrc.4071
68. Jouda J-B, Tamokou J-D, Mbazoa CD et al (2016) Antibacterial and cytotoxic cytochalasins from the endophytic fungus Phomopsis sp. harbored in Garcinia kola (Heckel) nut. BMC Complement Altern Med. https://doi.org/10.1186/s12906-016-1454-9
69. Kabangu K, Galeffi C, Aonzo E (1987) A new biflavanone from the bark of Garcinia kola. Planta Med 53:275–277. https://doi.org/10.1055/s-2006-962704
70. Kalgutkar AS, Crews BC, Rowlinson SW et al (2000) Biochemically based design of cyclooxygenase-2 (COX-2) inhibitors: facile conversion of nonsteroidal antiinflammatory drugs to potent and highly selective COX-2 inhibitors. Proc Natl Acad Sci U S A 97: 925–930. https://doi.org/10.1073/pnas.97.2.925
71. Kalu WO, Okafor PN, Ijeh II, Eleazu C (2016) Effect of kolaviron, a biflavanoid complex from Garcinia kola on some biochemical parameters in experimentally induced benign prostatic hyperplasic rats. Biomed Pharmacother 83:1436–1443. https://doi.org/10.1016/j.biopha.2016.08.064

72. Kilpatrick IC, Traut M, Heal DJ (2001) Monoamine oxidase inhibition is unlikely to be relevant to the risks associated with phentermine and fenfluramine: a comparison with their abilities to evoke monoamine release. Int J Obes 25:1454–1458. https://doi.org/10.1038/sj.ijo.0801732
73. Kim HK, Son KH, Chang HW et al (1998) Amentoflavone, a plant biflavone: a new potential anti-inflammatory agent. Arch Pharm Res 21:406–410. https://doi.org/10.1007/BF02974634
74. Kopytko P, Piotrowska K, Janisiak J, Tarnowski M (2021) Garcinol – a natural histone acetyltransferase inhibitor and new anti-cancer epigenetic drug. Int J Mol Sci 22:1–11. https://doi.org/10.3390/ijms22062828
75. Lee P-S, Teng C-Y, Kalyanam N et al (2019) Garcinol reduces obesity in high-fat-diet-fed mice by modulating gut microbiota composition. Mol Nutr Food Res 63:1800390. https://doi.org/10.1002/mnfr.201800390
76. Lee P-S, Nagabhushanam K, Ho C-T, Pan M-H (2021) Inhibitory effect of garcinol on obesity-exacerbated, colitis-mediated colon carcinogenesis. Mol Nutr Food Res 65:2100410. https://doi.org/10.1002/mnfr.202100410
77. Li F, Shanmugam MK, Chen L et al (2013) Garcinol, a polyisoprenylated benzophenone modulates multiple proinfl ammatory signaling cascades leading to the suppression of growth and survival of head and neck carcinoma. Cancer Prev Res (Phila Pa) 6:843–854. https://doi.org/10.1158/1940-6207.CAPR-13-0070
78. Li X, Ai H, Sun D et al (2016) Anti-tumoral activity of native compound morelloflavone in glioma. Oncol Lett 12:3373–3377. https://doi.org/10.3892/ol.2016.5094
79. Luzzi R, Guimarães CL, Verdi LG et al (1997) Isolation of biflavonoids with analgesic activity from *Rheedia gardneriana* leaves. Phytomedicine 4:141–144. https://doi.org/10.1016/S0944-7113(97)80060-8
80. Madubunyi II (1995) Antimicrobial activities of the constituents of Garcinia kola seeds. Int J Pharmacogn 33:232–237
81. Maňourová A, Leuner O, Tchoundjeu Z, Van Damme P, Verner V, Přibyl O, Lojka B (2019) Medicinal potential, utilization and domestication status of bitter Kola (*Garcinia kola* Heckel) in West and Central Africa. Forests 10:124. https://doi.org/10.3390/f10020124. (CC BY 4.0)
82. Marinelli R, Torquato P, Bartolini D et al (2020) Garcinoic acid prevents b-amyloid (Ab) deposition in the mouse brain. J Biol Chem 295:11866–11876. https://doi.org/10.1074/jbc.RA120.013303
83. Mbwambo ZH, Kapingu MC, Moshi MJ et al (2006) Antiparasitic activity of some xanthones and biflavonoids from the root bark of *Garcinia livingstonei*. J Nat Prod 69:369–372. https://doi.org/10.1021/np050406v
84. Miladiyah I, Jumina J, Haryana SM, Mustofa M (2018) Biological activity, quantitative structure-activity relationship analysis, and molecular docking of xanthone derivatives as anticancer drugs. Drug Des Devel Ther 12:149–158. https://doi.org/10.2147/DDDT.S149973
85. Niwa M, Terashima K, Aqil M (1993) Garcinol, a novel arylbenzofuran derivative from Garcinia kola. Heterocycles 36:671–673. https://doi.org/10.3987/COM-92-6291
86. Nokhala A, Siddiqui MJ, Ahmed QU et al (2020) Investigation of a-glucosidase inhibitory metabolites from *Tetracera scandens* leaves by GC–MS metabolite profiling and docking studies. Biomol Ther. https://doi.org/10.3390/biom10020287
87. Nworu CS, Akah PA, Esimone CO et al (2008) Immunomodulatory activities of kolaviron, a mixture of three related biflavonoids of Garcinia kola Heckel. Immunopharmacol Immunotoxicol 30:317–332. https://doi.org/10.1080/08923970801925430
88. Offor U, Ajayi SA, Jegede IA et al (2017) Renal histoarchitectural changes in nevirapine therapy: possible role of kolaviron and vitamin C in an experimental animal model. Afr Health Sci 17:164–174. https://doi.org/10.4314/ahs.v17i1.21
89. Ogwu MC (2019) Lifelong consumption of plant-based GM foods: is it safe? In: Papadopoulou P, Misseyanni A, Marouli C (eds) Environmental exposures and human health challenges. IGI Global, Philadelphia, pp 158–176. https://doi.org/10.4018/978-1-5225-7635-8.ch008

90. Ogwu MC (2019) Towards sustainable development in Africa: the challenge of urbanization and climate change adaptation. In: Cobbinah PB, Addaney M (eds) The geography of climate change adaptation in urban Africa. Springer Nature, pp 29–55. https://doi.org/10.1007/978-3-030-04873-0_2
91. Ogwu MC (2019) Understanding the composition of food waste: an "-omics" approach to food waste management. In: Gunjal AP, Waghmode MS, Patil NN, Bhatt P (eds) Global initiatives for waste reduction and cutting food loss. IGI Global, Philadelphia, pp 212–236. https://doi.org/10.4018/978-1-5225-7706-5.ch011
92. Ogwu MC (2020) Value of *Amaranthus* [L.] species in Nigeria. In: Waisundara V (ed) Nutritional value of Amaranth. IntechOpen, pp 1–21. https://doi.org/10.5772/intechopen.86990
93. Ogwu MC (2023) Local food crops in Africa: sustainable utilization, threats, and traditional storage strategies. In: Izah SC, Ogwu MC (eds) Sustainable utilization and conservation of Africa's biological resources and environment. Sustainable development and biodiversity, vol 888. Springer, Singapore, pp 353–376. https://doi.org/10.1007/978-981-19-6974-4_13
94. Ogwu MC, Osawaru ME (2022) Traditional methods of plant conservation for sustainable utilization and development. In: Izah SC (ed) Biodiversity in Africa: potentials, threats and conservation. Sustainable development and biodiversity, vol 29. Springer, Singapore, pp 451–472. https://doi.org/10.1007/978-981-19-3326-4_17
95. Ogwu MC, Osawaru ME (2023) Disease outbreaks in ex-situ plant conservation and potential management strategies. In: Izah SC, Ogwu MC (eds) Sustainable utilization and conservation of Africa's biological resources and environment. Sustainable development and biodiversity, vol 888. Springer, Singapore, pp 497–518. https://doi.org/10.1007/978-981-19-6974-4_18
96. Ogwu MC, Osawaru ME, Ahana CM (2014) Challenges in conserving and utilizing plant genetic resources (PGR). Int J Genet Mol Biol 6(2):16–22. https://doi.org/10.5897/IJGMB2013.0083
97. Ogwu MC, Osawaru ME, Aiwansoba RO, Iroh RN (2016) Ethnobotany and collection of west African Okra [*Abelmoschus caillei* (A. Chev.) Stevels] germplasm in some communities in Edo and Delta States, Southern Nigeria. Borneo J Resour Sci Technol 6(1):25–36. https://doi.org/10.33736/bjrst.212.2016
98. Ogwu MC, Osawaru ME, Aiwansoba RO, Iroh RN (2016) Status and prospects of vegetables in Africa. In: Borokini IT, Babalola FD (eds) Conference proceedings of the joint biodiversity conservation conference of Nigeria Tropical Biology Association and Nigeria chapter of Society for Conservation Biology on MDGs to SDGs: toward sustainable biodiversity conservation in Nigeria. University of Ilorin, Nigeria. 47–57pp
99. Ogwu MC, Osawaru ME, Obahiagbon GE (2017) Ethnobotanical survey of medicinal plants used for traditional reproductive care by Usen people of Edo State, Nigeria. Malaya J Biosci 4(1):17–29
100. Ogwu MC, Osawaru ME, Owie MO (2023) Effects of storage at room temperature on the food components of three cocoyam species (*Colocasia esculenta, Xanthosoma atrovirens, and X. Sagittifolium*). Food Stud Interdiscip J 13(2):59–83. https://doi.org/10.18848/2160-1933/CGP/v13i02/59-83
101. Ogwu MC, Osawaru ME, Amodu E, Osamo F (2023) Comparative morphology, anatomy, and chemotaxonomy of two *Cissus* Linn. species. Braz J Bot. https://doi.org/10.1007/s40415-023-00881-0
102. Ojo OA, Ojo AB, Maimako RF et al (2021) Exploring the potentials of some compounds from Garcinia kola seeds towards identification of novel PDE-5 inhibitors in erectile dysfunction therapy. Andrologia. https://doi.org/10.1111/and.14092
103. Okoko T (2009) In vitro antioxidant and free radical scavenging activities of Garcinia kola seeds. Food Chem Toxicol 47:2620–2623. https://doi.org/10.1016/j.fct.2009.07.023
104. Okoko T (2018) Kolaviron and selenium reduce hydrogen peroxide-induced alterations of the inflammatory response. J Genet Eng Biotechnol 16:485–490. https://doi.org/10.1016/j.jgeb.2018.02.004

105. Okoko T, Ndoni SA (2021) Protective effect of kolaviron on bromate-induced toxicity on raw u937 cells and macrophages. Malays J Biochem Mol Biol 24:169–174
106. Okoye TC, Uzor PF, Onyeto CA, Okerere EK (2014) Safe African medicinal plants for clinical studies. In: Kuete V (ed) Toxicological survey of African medicinal plants. Elsevier, London, pp 535–555
107. Okwu DE (2005) Phytochemicals, vitamins and mineral contents of two Nigerian medicinal plants. Int J Mol Med Adv Sci 1:375–381
108. Ola OS, Adewole KE (2021) Anticlastogenic and hepatoprotective effects of Kolaviron on sodium valproate-induced oxidative toxicity in Wistar rats. Egypt J Basic Appl Sci 8:167–179. https://doi.org/10.1080/2314808X.2021.1928974
109. Olaleye SB, Onasanwo SA, Ige AO et al (2010) Anti-inflammatory activities of a kolaviron-inhibition of nitric oxide, prostaglandin E2 and tumor necrosis factor-alpha production in activated macrophage-like cell line. Afr J Med Med Sci 39(Suppl):41–46
110. Olatoye FJ, Akindele AJ, Onwe S (2021) Ameliorative effect of Kolaviron, an extract of Garcinia kola seeds, on induced hypertension. J Complement Integr Med. https://doi.org/10.1515/jcim-2020-0354
111. Omotoso GO, Ukwubile II, Arietarhire L et al (2018) Kolaviron protects the brain in cuprizone-induced model of experimental multiple sclerosis via enhancement of intrinsic antioxidant mechanisms: possible therapeutic applications? Pathophysiology 25:299–306. https://doi.org/10.1016/j.pathophys.2018.04.004
112. Omotoso GO, Olajide OJ, Gbadamosi IT et al (2019) Cuprizone toxicity and Garcinia kola biflavonoid complex activity on hippocampal morphology and neurobehaviour. Heliyon. https://doi.org/10.1016/j.heliyon.2019.e02102
113. Onyekwelu JC, Oyewale O, Stimm B, Mosandl R (2015) Antioxidant, nutritional and anti-nutritional composition of Garcinia kola and *Chrysophyllum albidum* from rainforest ecosystem of Ondo State, Nigeria. J For Res 26:417–424. https://doi.org/10.1007/s11676-015-0068-2
114. Osawaru ME, Ogwu MC (2014) Ethnobotany and germplasm collection of two genera of cocoyam (*Colocasia* [Schott] and *Xanthosoma* [Schott], Araceae) in Edo State Nigeria. Sci Technol Arts Res J 3(3):23–28. https://doi.org/10.4314/star.v3i3.4
115. Osawaru ME, Ogwu MC (2014) Conservation and utilization of plant genetic resources. In: Omokhafe K, Odewale J (eds) Proceedings of 38th annual conference of the Genetics Society of Nigeria. Empress Prints Nigeria Limited, pp 105–119
116. Osawaru ME, Ogwu MC (2020) Survey of plant and plant products in local markets within Benin City and environs. In: Filho LW, Ogugu N, Ayal D, Adelake L, da Silva I (eds) African handbook of climate change adaptation. Springer Nature, Switzerland, pp 1–24. https://doi.org/10.1007/978-3-030-42091-8_159-1
117. Osawaru ME, Ogwu MC, Ahana CM (2013) Current status of plant diversity and conservation in Nigeria. Nigerian J Life Sci 3(1):168–178
118. Osawaru ME, Ogwu MC, Omologbe J (2014) Characterization of three Okra [*Abelmoschus* (L.)] accessions using morphology and SDS-PAGE for the basis of conservation. Egypt Acad J Biol Sci 5(1):55–65
119. Tauchen J, Frankova A, Manourova A, Valterova I, Lojka B, Leuner O (2023) *Garcinia kola*: a critical review on chemistry and pharmacology of an important West African medicinal plant. Phytochem Rev 1–47. https://doi.org/10.1007/s11101-023-09869-w. Advance online publication

Vernonia amygdalina Delile (Asteraceae): An Overview of the Phytochemical Constituents, Nutritional Characteristics, and Ethnomedicinal Values for Sustainability

10

Matthew Chidozie Ogwu and Beckley Ikhajiagbe

Contents

Abstract

This chapter focuses on the ethnomedicinal values of bitter leaf (*Vernonia amygdalina* Delile (Asteraceae)) as well as the nutritional and phytochemical properties of the plant. The chapter discusses ongoing pharmacognostic and clinical research works on the plant as well as herbal products containing bitter leaf or its active ingredients, toxicity and safety considerations, and commercial

M. C. Ogwu (✉)
Goodnight, Family Department of Sustainable Development, Appalachian State University, Boone, NC, USA
e-mail: ogwumc@appstate.edu

B. Ikhajiagbe
Department of Plant Biology and Biotechnology, Faculty of Life Sciences, University of Benin, Benin City, Nigeria
e-mail: Beckley.ikhajiagbe@uniben.edu

S. C. Izah et al. (eds.), *Herbal Medicine Phytochemistry*, Reference Series in Phytochemistry, https://doi.org/10.1007/978-3-031-43199-9_75

values. *V. amygdalina* is a tropical perennial plant native to West Africa. The plant is widely valued for its culinary, medicinal, and nutritional uses and has a distinctive bitter taste. Its bitter taste, once moderated, adds a unique flavor profile to traditional dishes and soups. Bitter leaf is often exchanged as a symbol of hospitality and goodwill during traditional events in West Africa. In traditional African medicine, bitter leaf has a long history of use for its health benefits. It is used as a remedy for various ailments including malaria, fever, gastrointestinal issues, pain relief, immune system boosting, antimicrobial and anti-parasitic properties, nervous system regulation, cardiovascular health maintenance, digestive aid, fever reduction, diabetes management, stomachache, and others. These health benefits are linked to the phytochemical constituents of the plant. Prominent phytochemicals in bitter leaf include vernodalin, vernolepin, vernonioside, vernomygdin, and others. Further research to ensure the conservation of the plant and the sustainability of practices associated with it is paramount.

Keywords

Bitter leaf · Ethnomedicine · Phytochemistry · Bioactive compounds · Antioxidants · Antimicrobial · Culinary uses · Traditional medicine

1 Introduction

Vernonia amygdalina Delile (Asteraceae), also known as bitter leaf or African bitter leaf, is a tropical plant native to West Africa. Bitter leaf is cultivated in many West African countries, including Nigeria, Ghana, Cameroon, and others. It is often grown in home gardens and small farms. The plant is widely valued for its culinary, medicinal, and nutritional uses [5, 6]. Bitter leaf is characterized by its distinctive bitter taste and is a staple in many West African cuisines and traditional medicine practices. Different cultures prepare it in various ways, incorporating it into a wide range of dishes and soups [76, 79, 80]. However, bitter leaf is the key ingredient in the popular native soup in Southern Eastern Nigeria called ofe onugbu (bitter leaf Soup). The soup is typically made with a mixture of vegetables, meat or fish, and seasonings. The bitterness of the leaves adds a unique flavor to the soup. Bitter leaf can also be used in various vegetable dishes, stir-fries, and side dishes [9, 10, 64, 65]. It is often cooked with other vegetables and spices to balance its bitterness. To make the leaves less bitter, they are typically washed and sometimes soaked in water before use in culinary preparations. This process helps reduce the intensity of the bitter taste.

In traditional African medicine, bitter leaf has a long history of use for its potential health benefits. It is believed to have medicinal properties and is used in various remedies to address ailments such as malaria, fever, gastrointestinal issues, and more [18]. Bitter leaf contains antioxidants, which can help protect cells from oxidative damage [11]. These properties are attributed to the presence of bioactive compounds like flavonoids, polyphenols, and others [16, 17]. Bitter leaf is consumed to alleviate digestive discomfort and promote healthy digestion. It is believed to have mild laxative properties. Also, research has indicated that extracts from bitter leaf may possess antimicrobial properties, which could be valuable in combating certain infections. Bitter leaf has been studied for its anti-inflammatory effects, which may have implications for managing inflammatory conditions. Further, bitter leaf is a source of essential vitamins and minerals, including vitamins A, C, and K, as well as minerals (like calcium, magnesium, and iron) and dietary fiber [23]. Bitter leaf contains dietary fiber, which is important for digestive health and can help regulate blood sugar levels. While not a primary source of protein, bitter leaf contains some protein content [19, 22]. Bitter leaf is a versatile plant with cultural and culinary significance, and it continues to be studied for its potential health benefits. Its bitter taste, once moderated, adds a unique flavor profile to traditional dishes and soups. Additionally, as interest in natural remedies and traditional medicine grows, bitter leaf may find broader applications in wellness and healthcare [12]. Bitter leaf has cultural significance in social gatherings, ceremonies, and rituals across West Africa. It is often exchanged as a symbol of hospitality and goodwill during these events.

The diversity of *Vernonia amygdalina* reflects its adaptability to a range of ecological conditions and its cultural significance in various West African communities. This diversity is a valuable resource for both culinary traditions and potential medicinal applications, as different plant characteristics may offer unique flavor profiles and bioactive properties. Scientists and researchers continue to explore and document this diversity to better understand and utilize the potential of bitter leaf in agriculture, nutrition, and healthcare [63, 77]. *V. amygdalina* exhibits considerable diversity in terms of its morphology, habitat, and genetic makeup across its native range in West Africa. This diversity is shaped by various environmental factors, such as climate, soil conditions, and altitude, as well as genetic variations within the species. The leaves of bitter leaf plants display variations in the size and shape of their leaves. Leaves are typically broad and serrated, but variations exist. The plants can vary in growth habit, with some individuals growing as small shrubs and others as taller, more tree-like plants. The flowers of *V. amygdalina* can vary in color, size, and petal arrangement, although they are generally small and clustered. It is highly adaptable to a range of ecological niches and can be found in both forested and non-forested areas. It grows in lowland rainforests, savannas, and cultivated fields and can be found at varying altitudes, from lowland areas to higher elevations in some regions. Genetic diversity within the species contributes to variations in traits such as leaf chemistry, disease resistance, and adaptability to different environmental conditions [52]. Genetic studies have shown that different populations of bitter leaf may exhibit variations in their genetic makeup, which can influence the plant's characteristics and adaptability to local conditions. Traditional medicinal practices involving bitter leaf can vary from one region to another. Different communities may use it to address specific health concerns, resulting in variations in the formulations and remedies. Bitterness levels in bitter leaf can vary among individual plants and populations. Some plants may have more intensely bitter leaves, while others may be less bitter. Techniques to reduce bitterness are also variable.

This chapter aims to discuss the ethnomedicinal value of bitter leaf. The chapter presents the origin, distribution, and botany of *V. amygdalina*. The next sections address the nutritional and phytochemical properties as well as the food uses of bitter leaf. *V. amygdalina* has a long history of use in traditional medicine across West Africa. Other parts of the chapter are on the ethnobotanical, pharmacognosy, and clinical works, herbal products, toxicity, and commercial values of *V. amygdalina*. It is believed to have various medicinal properties, including digestive, anti-inflammatory, and antimicrobial effects. The environmental remediation and high adaptability of bitter leaf contribute to their growing widespread distribution whereas the efficacy associated with herbal remedies prepared from the plant makes it an integral part of traditional healing practices. However, further research to ensure the conservation of the plant and the sustainability of practices associated with it is paramount.

2 Origin and Distribution of *Vernonia amygdalina*

Vernonia amygdalina is native to tropical and subtropical regions of West Africa, but the exact origin within this region is not definitively known. It is largely hypothesized to have originated in the forested areas of West Africa, which encompass several countries along the western coast of the continent and a range of ecological zones, including lowland rainforests, savannas, and cultivated fields, where *V. amygdalina* has adapted and thrives [28]. The plant has a long history of traditional use in various West African cultures, and its use predates recorded history. It has been cultivated and harvested for its leaves, which are valued for their distinctive bitter taste and potential health benefits. Today, *V. amygdalina* is widely cultivated and grown in home gardens, small farms, and agricultural fields across West Africa. It is a significant part of the culinary and medicinal traditions in the region and continues to be an important cultural and economic resource. The plant's adaptability to various ecological conditions and its cultural significance have contributed to its spread and cultivation in different West African countries. It is used in a variety of traditional dishes and is an integral part of many cuisines in the region. Additionally, its medicinal properties have led to its continued use in traditional African medicine for various health concerns [69].

The distribution of bitter leaf is closely tied to the indigenous peoples and culture of West Africa, who have utilized it for generations as a valuable resource for both sustenance and traditional medicine [34]. Bitter leaf has played a vital role in the culinary and medicinal traditions of these communities, and its history is deeply embedded in their cultures. While *V. amygdalina* is native to West Africa, it has been cultivated and distributed beyond its native range due to its cultural and economic importance. Bitter leaf is cultivated in many West African countries, including Nigeria, Ghana, Cameroon, Ivory Coast, and others. It is often grown in home gardens, small farms, and agricultural fields. The plant is adapted to a range of ecological conditions, from lowland rainforests to savannas, and it is known for its hardiness and ability to thrive in diverse environments. *V. amygdalina* is distributed across multiple countries in West and Central Africa, with Nigeria being one of the primary producers and consumers of bitter leaf. In addition to its widespread distribution within West Africa, bitter leaf has garnered some recognition in international markets, particularly among African diaspora communities and those interested in African cuisine and traditional medicine [16, 17]. The distribution of *V. amygdalina* across different regions and ecological zones has led to variations in the plant's characteristics. Different populations of bitter leaf may exhibit variations in leaf size, bitterness levels, and adaptability to local conditions. These variations have contributed to the plant's adaptability and resilience, allowing it to flourish in a range of settings. As *V. amygdalina* is of significant cultural and economic importance, efforts have been made to promote its sustainable cultivation and harvesting practices. Organizations and initiatives have emerged to support small-scale farmers and promote sustainable agricultural practices for bitter leaf production.

African diaspora communities around the world, including those in Europe, North America, and other continents, have brought their culinary traditions with them. As a result, *V. amygdalina* is grown and used in some of these communities to recreate traditional dishes. In recent years, there has been an increasing global awareness of African cuisine and traditional ingredients. This has led to the availability of *V. amygdalina* in international markets, particularly in areas with a significant African diaspora. The plant's potential medicinal properties have attracted the interest of herbalists and practitioners of traditional medicine beyond Africa. As a result, bitter leaf extracts and preparations have been explored and used in various parts of the world for their potential health benefits. Scientific research on *V. amygdalina* has gained recognition globally. Researchers and scientists from different parts of the world have studied the plant's bioactive compounds, potential health benefits, and medicinal applications. While the distribution of *V. amygdalina* has expanded beyond its native range, its cultural and culinary significance remains strongest in West and Central Africa. In these regions, the plant continues to be an essential ingredient in traditional dishes, a symbol of hospitality and goodwill, and a valuable resource in traditional medicine. The global distribution of *V. amygdalina* reflects its versatility and adaptability as well as its potential to contribute to diverse culinary traditions and potential health benefits worldwide.

3 Botanical Characteristics of *Vernonia amygdalina*

Vernonia amygdalina belongs to the genus *Vernonia*, which is part of the large Asteraceae family. The genus *Vernonia* includes over 1000 species distributed across various continents, but *Vernonia amygdalina* is one of the most well-known and widely used species within this genus. Its vernacular names, such as bitter leaf, aptly describe its defining characteristic: an intensely bitter taste. The plant has a rich history deeply intertwined with the cultures of West Africa, where it is native. Its leaves, known for their bitterness, are used in a wide range of culinary dishes, traditional remedies, and cultural practices.

V. amygdalina is a perennial plant that can vary in size from a small shrub to a taller, more tree-like plant. Its growth habit can be erect or spreading. The stem of *V. amygdalina* is typically green, herbaceous, and somewhat woody at the base, especially in older plants. The leaves are the most distinctive feature of the plant [31]. They are broad, elliptical to ovate, and typically measure about 5–12 centimeters in length. The leaves have serrated or toothed margins. The upper surface of the leaves is usually dark green, while the lower surface may be lighter in color, often with a slightly pubescent (hairy) texture [32]. The leaves are arranged alternately along the stems of the plant [31]. The plant produces small, tubular flowers that are typically lavender to purple. These flowers are clustered in inflorescences, forming composite flower heads. The inflorescences vary in size and shape and can be found at the ends of branches. The plant typically blooms during its flowering season, which may vary depending on local climatic conditions [35]. The fruits of *V. amygdalina* are small achenes, often with a tuft of hairs attached to aid in wind dispersal. The defining characteristic of bitter leaf is its intense bitterness. This bitterness is attributed to the presence of certain compounds in the leaves. The roots and stem of the plant are less commonly used than the leaves. The stem is typically green and shrubby. Bitter leaf exhibits a variable growth habit. It can range from a small shrub with a bushy appearance to a taller, more tree-like plant. It can grow as a single stem or have multiple stems, depending on the conditions [47]. *V. amygdalina* possesses reproductive features typical of flowering plants. These features enable the plant to reproduce and propagate itself. It is primarily pollinated by insects, particularly bees and butterflies. These pollinators visit the flowers in search of nectar and, in the process, transfer pollen from one flower to another, facilitating fertilization. These reproductive features ensure the continuation of *V. amygdalina*'s life cycle and its ability to produce new generations of plants. While the plant is primarily valued for its culinary and medicinal uses, understanding its reproductive biology is essential for its conservation and sustainable cultivation [62] (Fig. 1).

Some botanical relatives of *V. amygdalina* include:

Vernonia calvoana (African Ironweed): *V. calvoana* is a closely related species to *V. amygdalina and* is native to Central and West Africa. Both plant species share some botanical characteristics, and like bitter leaf, *V. calvoana* has been used in traditional medicine practices.

Vernonia cinerea (Little Ironweed): *V. cinerea* is a species found in various parts of the world, including Asia, Africa, and Australia. While it is not native to Africa, it is part of the *Vernonia* genus and shares a common genus with *V. amygdalina*.

Vernonia guineensis (Guinea Ironweed): *V. guineensis* is another species within the *Vernonia* genus that is found in West and Central Africa. It is closely related to *V. amygdalina* and is used in similar ways in traditional medicine and sometimes in culinary preparations.

Vernonia galamensis (African Ironweed): *V. galamensis* is native to West Africa and is cultivated for its seeds, which are used for oil extraction.

Vernonia anthelmintica (Bitter Weed or Wormseed): While not native to Africa, *V. anthelmintica* is another species within the *Vernonia* genus that is known for its seeds, which have been used traditionally for their anthelmintic (anti-parasitic) properties.

Fig. 1 *Vernonia amygdalina* plant

Vernonia diversifolia (Variable-leaved Ironweed): *V. diversifolia* is a species found in various parts of the world, including Africa with diverse characteristics and traditional medicinal uses.

Vernonia adoensis (Ethiopian Ironweed): *V. adoensis* is native to East Africa, including Ethiopia, and has potential ethnomedicinal uses.

4 Environmental Value of *Vernonia Amygdalina*

Vernonia amygdalina possesses several environmental values and benefits, contributing to the ecosystems where it grows [3, 9]. While it is primarily known for its culinary and medicinal uses, it also plays a role in the environment in various ways. According to Ikhajiagbe et al. [33] some of these roles include:

Soil Stabilization: Like many other plants, *V. amygdalina* helps stabilize soil through its root system. The roots of the plant bind the soil particles, reducing erosion caused by rainfall and runoff.

Biodiversity Support: Bitter leaf, as a native plant in West and Central Africa, provides habitat and forage for local wildlife. Insects, including pollinators like bees and butterflies, visit its flowers for nectar, contributing to local biodiversity.

Carbon Sequestration: *V. amygdalina* absorbs carbon dioxide from the atmosphere through photosynthesis. This process helps mitigate climate change by sequestering carbon and converting it into organic matter.

Nitrogen Fixation: Some plants, including certain species within the Asteraceae family (to which *Vernonia amygdalina* belongs), have associations with nitrogen-fixing bacteria in their root nodules [19]. These bacteria convert atmospheric nitrogen into a form that plants can use as a nutrient. While not all members of the family have this ability, some may contribute to nitrogen enrichment in the soil.

Erosion Control: In its native region (West and Central Africa), flood and erosion are a growing challenge but the growth of *Vernonia amygdalina* can help prevent soil loss by stabilizing slopes and riverbanks.

Medicinal Plant Conservation: *V. amygdalina*'s value as a medicinal plant has led to conservation efforts in some regions to ensure its continued existence [61]. Preserving its natural habitat helps protect the plant and other species that depend on it.

Traditional Agroforestry Systems: In some traditional agroforestry systems, bitter leaf may be grown alongside other crops and trees, contributing to diversified and sustainable agricultural practices that are beneficial for both the environment and local communities.

The presence of *V. amygdalina* in local ecosystems contributes to maintaining ecological balance and supporting biodiversity. Recognizing the environmental values of this plant can contribute to its conservation and sustainable use in its native habitat and land management. The plant also exhibits some weedy characteristics through their

Rapid Growth: *V. amygdalina* is a fast-growing plant. Under favorable conditions, it can establish itself quickly and spread in an area.

Seed Production: The plant produces seeds in abundance. These seeds have adaptations, such as tufts of hairs that aid in wind dispersal. This can facilitate the plant's spread in the wild.

Adaptability: Bitter leaf is adaptable to a range of ecological conditions, from lowland rainforests to savannas. This adaptability allows it to thrive in diverse environments, including disturbed or cultivated areas.

Vegetative Propagation: In addition to sexual reproduction through seeds, *V. amygdalina* can propagate vegetatively. This means that new plants can sprout from stem cuttings or even root fragments under certain conditions.

Resilience: The plant is hardy and can tolerate some degree of environmental stress, including drought conditions.

Naturalized Populations: In areas where it is cultivated, *V. amygdalina* may escape cultivation and become naturalized. This can lead to the plant growing in unintended locations.

Competitive Ability: In some situations, *V. amygdalina* may outcompete other plant species for resources such as light, water, and nutrients.

In some cases, *V. amygdalina* may be intentionally cultivated as a crop, and its rapid growth and adaptability can be advantageous. In other cases, particularly in natural ecosystems, it may be considered invasive if it displaces native vegetation. Efforts to manage the spread of *V. amygdalina* in natural ecosystems should be guided by principles of conservation and biodiversity preservation. In cultivated areas, appropriate agricultural practices can help control its growth and prevent it from becoming a weed.

4.1 Bioremediation Value of *Vernonia amygdalina*

Vernonia amygdalina has demonstrated potential bioremediation value in certain environmental contexts. Bioremediation is the use of living organisms, including plants, to mitigate or clean up environmental contaminants, such as heavy metals and organic pollutants. Some bioremediation roles attributed to *V. amygdalina* in Shittu and Ikhajiagbe [81] include:

Heavy Metal Accumulation: Bitter leaf can accumulate heavy metals from the soil. Some studies have shown that it can accumulate metals such as lead, cadmium, and nickel from contaminated soil. This characteristic makes it a candidate for phytoremediation, a specific form of bioremediation that uses plants to remove, stabilize, or detoxify contaminants in soil and water.

Hyperaccumulation Potential: Some plants, including certain species within the Vernonia genus, are known as hyperaccumulators. These plants can accumulate exceptionally high concentrations of heavy metals in their tissues without displaying toxicity symptoms. *V. amygdalina* has shown the potential for hyperaccumulation of heavy metals, particularly in its leaves. This ability is advantageous in removing and concentrating contaminants for subsequent remediation.

Soil Improvement: Bitter leaf can improve soil quality through its root system and organic matter deposition. As it grows and matures, the plant's roots help enhance soil structure and nutrient content. Improved soil conditions can indirectly support bioremediation efforts by promoting the growth of other remediation plants and microorganisms.

Sustainable Remediation: The use of *V. amygdalina* for bioremediation aligns with sustainable and environmentally friendly remediation practices. It reduces the need for harsh chemicals and mechanical interventions, minimizing potential environmental harm.

Although *V. amygdalina* has shown promise in certain phytoremediation studies, its use in practical remediation applications depends on several factors, including the specific contaminants involved, site conditions, and regulatory considerations. Bioremediation is a complex process that requires careful planning and monitoring. Additionally, the effectiveness of bioremediation using *V. amygdalina* may vary depending on the extent of contamination and the suitability of the plant for the specific site. Therefore, it is essential to conduct site-specific assessments and consider the broader ecological implications when using plants for bioremediation purposes.

5 Cultural Significance of *Vernonia amygdalina*

Vernonia amygdalina holds significant historical and cultural importance in West and Central Africa. Its multifaceted significance is deeply woven into the cultural, culinary, and traditional medicinal practices of the region. Some cultural significance of bitter leaf include (Fig. 2):

- **Traditional Medicine**: Bitter leaf has a long history of use in traditional African medicine systems. Indigenous communities have employed it as a remedy for various health conditions, including malaria, fever, gastrointestinal issues, and more [18]. The plant's phytochemical composition, including bitter compounds and bioactive constituents, is believed to contribute to its potential medicinal properties.
- **Culinary Traditions**: Bitter leaf is a fundamental ingredient in West and Central African cuisine. It imparts a unique and cherished flavor to numerous dishes and soups. One of the most famous dishes featuring bitter leaf is bitter leaf soup (ofe onugbu), a delicacy enjoyed across the region. The plant is also used in vegetable dishes, stews, and side dishes.
- **Cultural Symbolism**: Bitter leaf has symbolic significance in various cultural practices and rituals. It is exchanged as a symbol of hospitality, goodwill, and friendship during social gatherings, ceremonies, and celebrations. Its presence in

Fig. 2 Cultural values of *Vernonia amygdalina*

traditional offerings and exchanges strengthens social bonds and fosters community ties.

- **Medicinal Knowledge Transfer**: The use of *V. amygdalina* in traditional medicine has been passed down through generations within African communities. Elders and traditional healers play a crucial role in transferring knowledge about the plant's medicinal properties and applications to the younger generation.
- **Cultural Identity**: Bitter leaf contributes to the cultural identity of West and Central African communities. Its inclusion in traditional dishes and remedies reflects the region's culinary heritage and medical practices. It distinguishes West and Central African cuisine from other global culinary traditions, adding a distinctive flavor profile.
- **Ethnobotanical Heritage**: *V. amygdalina* is an integral part of the ethnobotanical heritage of West and Central Africa. It exemplifies the region's rich botanical knowledge and the relationship between people and plants. Ethnobotanical studies have further illuminated the plant's uses and cultural significance within indigenous communities.
- **Cultural Conservation**: The cultural and historical importance of *V. amygdalina* has spurred efforts to conserve and protect the plant's natural habitat. Conservation initiatives aim to safeguard this valuable botanical resource.

1. **Phytochemical Constituents of *Vernonia amygdalina***

Vernonia amygdalina contains a diverse array of phytochemical constituents, which contribute to its medicinal and nutritional properties. These phytochemicals are responsible for the plant's characteristic bitterness and its potential health benefits.

Sesquiterpene Lactones: These are bitter compounds that are primarily responsible for the plant's intense bitterness. Sesquiterpene lactones, such as vernodalin, vernolepin, and vernomygdin, are characteristic phytochemicals of bitter leaf [24, 43].

Alkaloids: Bitter leaf contains alkaloids, including vernodalin alkaloid, which may contribute to its medicinal properties [18, 43].

Flavonoids: Flavonoids are a group of polyphenolic compounds found in bitter leaf. They have antioxidant properties and may play a role in protecting cells from oxidative damage [11, 29, 30].

Steroids and Triterpenoids: Some steroids and triterpenoids, such as vernonioside A1 and vernonioside B1, have been isolated from *V. amygdalina*. These compounds may have potential pharmacological activities [20, 43].

Phenolic Compounds: Phenolic compounds, including caffeic acid, chlorogenic acid, and quercetin, are present in bitter leaf. These compounds have antioxidant and anti-inflammatory properties [11].

Saponins: Bitter leaf contains saponins, which are glycosides with surfactant properties. Saponins have been studied for their potential health benefits, including cholesterol-lowering effects [24].

Glycosides: Glycosides are compounds that contain a sugar molecule bonded to another molecule. Bitter leaf contains various glycosides, and some may have biological activities.

Tannins: Tannins are a group of polyphenolic compounds found in bitter leaf. They have astringent properties and may have antimicrobial effects [25, 26].

Minerals and Vitamins: Bitter leaf is a source of essential minerals such as calcium, magnesium, and iron. It also contains vitamins like vitamin A, vitamin C, and vitamin K.

Essential Oils: Bitter leaf may contain essential oils that contribute to its aroma and potential therapeutic properties [41].

The phytochemical composition of *V. amygdalina* can vary depending on factors such as the plant's age, environmental conditions, and geographic location. The combination of these phytochemical constituents gives bitter leaf its characteristic bitterness and makes it a valuable resource in traditional medicine and culinary practices in West and Central Africa. Additionally, ongoing research is uncovering the potential health benefits and pharmacological activities of these phytochemicals. Some of the essential phytochemical constituents of bitter leaf, parts found, and their usefulness are presented in Table 1.

V. amygdalina is a leafy green vegetable that is valued for its potential medicinal properties and nutritional content. While bitter leaf is primarily recognized for its bitterness and medicinal uses, it also contains various essential nutrients that contribute to its nutritional value. The approximate concentrations of mineral nutrients in bitter leaf are presented in Table 2. Bitter leaf is a source of several vitamins, including vitamin A, vitamin C, and vitamin K. These vitamins are essential for various bodily functions, including immune support, vision, and blood clotting. Bitter leaf contains important minerals such as calcium, magnesium, and iron. Calcium is crucial for bone health, magnesium is involved in various enzymatic processes, and iron is essential for red blood cell formation and oxygen transport. Bitter leaf is a good source of dietary fiber, which is important for digestive health. Fiber helps regulate bowel movements, prevent constipation, and may assist in managing weight loss [12]. Bitter leaf contains various antioxidants, including flavonoids and phenolic compounds. These antioxidants help protect cells from oxidative stress and may have anti-inflammatory properties [75]. Bitter leaf is rich in phytochemicals, such as sesquiterpene lactones and alkaloids, which are responsible for its characteristic bitterness. Some of these phytochemicals have been studied for their potential health benefits. Bitter leaf is a low-calorie vegetable, making it a suitable choice for those looking to manage their calorie intake [48]. While not a high-protein source, bitter leaf does contain some protein, which is essential for tissue repair and overall growth and development. Bitter leaf contains carbohydrates, primarily in the form of dietary fiber and small amounts of sugars. Carbohydrates provide energy for the body. Bitter leaf is naturally low in fat, making it a heart-healthy choice when included in a balanced diet. It is important to note that the nutritional content of bitter leaf can vary based on factors such as plant variety, growing conditions, and how it is prepared and cooked. Bitter leaf is often

Table 1 Some phytochemical constituents of bitter leaf, parts found, and usefulness

Phytochemical constituent	Types	Medicinal uses and potential health benefits	Part of the plant found
Sesquiterpene lactones	Vernodalin, Vernolepin, and Vernomygdin	Bitterness, anti-inflammatory, antioxidant, pain relief, and may support the immune system	Leaves
Alkaloids	Vernodalin, Vernomenin, Vernolepin, Epivernodaline, cyanogenic alkaloids	Pain relief, anti-parasitic properties, and regulations of the nervous system	Various parts
Flavonoids	Kaempferol, Quercetin, Luteolin, Rutin, Myricetin, Catechins	Antioxidant, anti-inflammatory, cardiovascular health, and support of the immune system	Leaves and stems
Steroids and triterpenoids	β-Sitosterol, Campesterol, Stigmasterol, Lupeol, β-Amyrin, Ursolic acid, Oleanolic acid, β-Amyrin, Friedelin, Betulinic acid	Anti-inflammatory, immunomodulatory, anti-cancer effects, and support skin health	Leaves
Phenolic compounds	Caffeic acid, Chlorogenic acid, Epicatechin	Antioxidant, anti-inflammatory, antioxidant, anti-inflammatory, and cardiovascular health as well as aid digestion	Leaves
Saponins	Vernoniosides, bitter saponins, steroidal saponins	Cholesterol reduction, antioxidant, and immune support	Leaves
Glycosides	Cardiac glycosides; Iridoid glycosides and Anthraquinone glycosides	Cardiac health, diuretic properties, and antioxidant effects	Leaves
Tannins	Gallotannins, proanthocyanidins, and ellagitannins	Astringent, antimicrobial effects, astringent, wound healing, antioxidant, and antimicrobial properties	Leaves
Minerals and vitamins	Calcium, potassium, sodium, phosphorus, iron, etc.	Nutritional and health value. Calcium for bone health, iron for anemia prevention, and vitamins for overall health	Leaves
Essential oils	Terpenes, eugenol	Aroma, and potential therapeutic properties. Aroma therapy has potential for skin care as well as calming effects	Leaves

used in traditional dishes in West and Central Africa, where it is typically cooked or added to soups and stews.

Note that bitter leaf's bitterness may require certain culinary techniques to make it more palatable, its potential nutritional and medicinal properties make it a valuable addition to traditional diets in many regions. However, individuals with certain medical conditions or sensitivities may need to consume bitter leaf in moderation,

Table 2 Approximate concentrations of mineral nutrients in bitter leaf

Mineral nutrient	Approximate concentration (per 100 g of fresh leaves) in mg
Calcium	309
Magnesium	109
Potassium	470
Sodium	11
Phosphorus	66
Iron	1.9
Zinc	1.6
Copper	0.14
Manganese	1.3

and it is advisable to consult with a healthcare professional or nutritionist for personalized dietary recommendations.

6 Ethnomedicinal Uses of *Vernonia amygdalina*

Vernonia amygdalina is a versatile and culturally significant plant with a rich history of ethnomedicinal uses in various regions of West and Central Africa. This plant is celebrated not only for its culinary contributions but also for its potential therapeutic properties. Bitter leaf has been a staple in the diets of various West and Central African communities for centuries. Its bitter taste, which gives the plant its name, is a defining characteristic that has shaped its culinary and medicinal roles. In many African cultures, bitter leaf is not merely an edible green but also a symbol of resilience and adaptability. The plant thrives in diverse ecological conditions, and this resilience is often associated with qualities that are admired and respected by people. Additionally, the bitterness of the leaves is seen as a metaphor for life's challenges, suggesting that strength and healing can be found even in adversity. The ethnomedicinal uses of bitter leaf are deeply intertwined with cultural practices and belief systems. Traditional healers, often referred to as herbalists or traditional medicine practitioners, play a vital role in preserving and passing down the knowledge of using bitter leaf and other medicinal plants. These healers are highly respected in their communities and are considered custodians of invaluable wisdom. Some of the prominent ethnomedicinal uses of *V. amygdalina* include (Fig. 3):

- **Malaria Treatment and Prevention**: Perhaps one of the most well-known traditional uses of bitter leaf is its role in the management of malaria. Various parts of the plant, including the leaves and stem, are used to prepare decoctions and infusions believed to have anti-malarial properties [2, 18]. The bitter leaf's role in malaria treatment aligns with its historical significance, as malaria is a prevalent and debilitating disease in many parts of Africa.
- **Fever Reduction**: Bitter leaf is often employed as a natural remedy to reduce fever. The leaves are typically brewed into a medicinal tea or used in poultices for

Fig. 3 Ethnomedicinal values of *Vernonia amygdalina*

external application. The plant's ability to lower body temperature is valued in regions where fever is a common symptom of various illnesses.

- **Digestive Health**: Bitter leaf is recognized for its digestive benefits. It is believed to aid in digestion, alleviate indigestion, and provide relief from gastrointestinal discomfort. It is sometimes consumed as a tea or added to soups and stews to promote digestive well-being.
- **Diabetes Management**: Traditional medicine practitioners in some regions use bitter leaf as an adjunct treatment for diabetes [12]. While modern scientific research on this aspect is ongoing, bitter leaf is believed to have properties that help regulate blood sugar levels.
- **Skin Care**: Bitter leaf's anti-inflammatory and antimicrobial properties make it a popular choice for treating various skin conditions. It is used topically to alleviate itching, soothe rashes, and support wound healing.

- **Cough and Respiratory Health**: Bitter leaf is often included in herbal preparations to alleviate coughs, colds, and respiratory infections. Its potential anti-inflammatory and immune-boosting properties make it a valuable component in traditional cough remedies.
- **Pain Relief**: Bitter leaf is considered an analgesic, and it is used to alleviate various types of pain, including headaches and body aches [54]. Some traditional preparations may be applied externally to painful areas.
- **Stomach Ulcer Soothing**: In some traditional systems, bitter leaf is used to soothe and potentially promote healing in cases of stomach ulcers [45]. Its reported anti-inflammatory properties may contribute to this application [66].
- **Hepatitis Support**: Traditional healers in certain regions use bitter leaf as a supportive therapy for hepatitis. It is believed to promote liver health and assist in recovery.
- **Immune System Support**: Bitter leaf is seen as a natural immune booster. Its regular consumption is thought to strengthen the immune system and enhance overall health.
- **Weight Management**: In some cultures, bitter leaf is incorporated into weight management strategies. It is believed to play a role in appetite regulation and metabolism [12].
- **Anemia Management**: Bitter leaf contains iron and other nutrients, making it a potential remedy for anemia. It may be used to support individuals with iron-deficiency anemia.
- **Hypertension**: Some traditional practitioners use bitter leaf to help manage hypertension. While research is ongoing in this area, it is believed that the plant's compounds may contribute to blood pressure regulation.

It is important to note that the ethnomedicinal uses of bitter leaf are often passed down through generations and are deeply rooted in cultural traditions. These traditional remedies are valued for their accessibility, affordability, and perceived effectiveness, particularly in regions where access to modern healthcare may be limited. Modern scientific research are ongoing into the ethnomedicinal uses of *V. amygdalina* and are well-documented in traditional systems. Modern scientific research has begun to explore the plant's potential therapeutic properties. Studies have investigated the bioactive compounds present in bitter leaf and their mechanisms of action. Some of the key findings and ongoing research areas include:

Antioxidant Properties: Bitter leaf contains various phytochemicals, such as flavonoids and phenolic compounds, which exhibit antioxidant activity. These antioxidants are believed to protect cells from oxidative stress, which is linked to various chronic diseases.

Anti-inflammatory Effects: Bitter leaf's anti-inflammatory properties have been studied, and it is considered a potential natural remedy for inflammatory conditions.

Antimicrobial Activity: Research has indicated that bitter leaf extracts may have antimicrobial properties, which can be useful in combating infections.

Hepatoprotective Potential: Studies suggest that bitter leaf may have hepatoprotective effects, potentially supporting liver health.

Immunomodulatory Effects: Bitter leaf's impact on the immune system is a subject of interest, and research is ongoing to better understand its immunomodulatory properties.

Antidiabetic Potential: Bitter leaf's role in blood sugar regulation has garnered attention, and investigations into its potential as an antidiabetic agent continue [12, 49].

It is worth emphasizing that while traditional knowledge is valuable, scientific validation is critical to fully understand the safety and efficacy of bitter leaf in modern healthcare. Researchers are working to bridge the gap between traditional and evidence-based medicine by conducting rigorous studies to assess the plant's potential benefits. Despite the promise of *V. amygdalina* as a source of natural remedies, several challenges and considerations exist:

Standardization: Traditional preparations of bitter leaf vary in terms of plant part used, preparation methods, and dosages. Standardization is crucial for ensuring consistent and safe therapeutic outcomes.

- **Safety Concerns**: While bitter leaf is generally considered safe for most individuals, there can be adverse effects if consumed in excessive quantities. Specific populations, such as pregnant women and individuals with certain medical conditions, should exercise caution.
- **Integration with Modern Medicine**: Integrating traditional remedies like bitter leaf into modern healthcare systems requires careful evaluation, regulation, and collaboration between traditional healers and healthcare professionals.
- **Preservation of Traditional Knowledge**: Preserving and passing down traditional knowledge of medicinal plants, like bitter leaf, is vital. Efforts to document and protect this knowledge are ongoing.
- **Sustainability**: The popularity of bitter leaf and other medicinal plants can lead to overharvesting and habitat degradation. Sustainable harvesting practices must be promoted.

The future of *Vernonia amygdalina* in modern healthcare lies in multidisciplinary research that combines traditional wisdom with scientific rigor. Collaborations between ethnobotanists, pharmacologists, healthcare practitioners, and local communities are essential to harness the plant's potential for the benefit of public health.

7 Pharmacognostic and Clinical Research Involving *Vernonia amygdalina*

Clinical research on *Vernonia amygdalina* will culminate in an understanding of the potential therapeutic properties and safety for human use. While traditional and ethnobotanical knowledge has long recognized the plant's medicinal value, modern clinical studies aim to provide scientific evidence to support its traditional uses and explore new applications.

- **Antimalarial Properties**: Some clinical studies have investigated the antimalarial properties of *Vernonia amygdalina*. These studies aim to assess its efficacy in the treatment and prevention of malaria, which is a prevalent and serious disease in many regions where the plant is used traditionally [2].
- **Diabetes Management**: Research has explored the effects of bitter leaf extracts or preparations on blood sugar regulation [12]. Clinical trials may investigate the potential of bitter leaf to help manage diabetes and its complications.
- **Liver Health**: Clinical studies may assess the impact of bitter leaf on liver health, including its hepatoprotective properties. Researchers aim to understand its role in supporting liver function and mitigating liver-related conditions.
- **Antioxidant and Anti-inflammatory Effects**: Investigations into the antioxidant and anti-inflammatory properties of bitter leaf are ongoing. Clinical trials may explore its potential in reducing oxidative stress and inflammation, which are associated with various chronic diseases.
- **Immune Modulation**: Some research focuses on the immunomodulatory effects of *Vernonia amygdalina* [50, 71]. Clinical studies aim to understand how it may enhance immune responses and strengthen the body's defense mechanisms.
- **Antimicrobial Activity**: Clinical research may assess the efficacy of bitter leaf extracts against specific pathogens, including bacteria, viruses, and parasites [28]. This research could have implications for infectious disease management.
- **Cancer Research**: Studies have explored the potential anticancer properties of bitter leaf compounds [38]. Clinical trials may investigate its effects on specific types of cancer and its mechanisms of action.
- **Blood Pressure Regulation**: Clinical investigations may assess the impact of bitter leaf on blood pressure regulation, potentially leading to therapeutic applications for individuals with hypertension [1, 46].
- **Safety and Tolerance**: Clinical studies also focus on evaluating the safety and tolerability of bitter leaf preparations in human subjects [73]. This includes assessing potential side effects and interactions with medications.
- **Dosage and Formulation**: Research may determine the optimal dosage and formulation of bitter leaf extracts or supplements for therapeutic use [73]. This information is important for developing standardized products.
- **Nutritional Benefits**: Clinical trials can examine the nutritional properties of bitter leaf, including its nutrient content and potential as a dietary supplement [55].
- **Traditional Medicine Validation**: Clinical research plays a role in validating the traditional uses of bitter leaf in various traditional medicine systems [7]. This can help bridge the gap between traditional and evidence-based medicine.

Note that the findings from this research work may vary depending on factors such as the specific preparation of bitter leaf, dosage, and patient populations. Collaborations between researchers, healthcare professionals, traditional healers, and local communities are crucial for conducting meaningful and culturally sensitive clinical studies on *V. amygdalina*. Additionally, regulatory considerations and ethical guidelines play a role in the design and implementation of clinical trials involving medicinal plants.

8 Toxicity and Safety Considerations of *Vernonia amygdalina*

Vernonia amygdalina is generally considered safe for consumption when used in culinary and traditional medicinal practices. It has a long history of use as a food ingredient and in traditional medicine across West and Central Africa [27, 60]. However, like many plants, bitter leaf may have certain toxic or adverse effects under certain circumstances or when consumed in excessive amounts. According to Eze et al. [27] some of the toxic considerations of bitter leaf include:

- **Bitterness and Palatability**: Bitter leaf gets its name from its bitter taste, which is a characteristic feature. While the bitterness is generally tolerated in culinary preparations, some individuals may find it unpalatable in large quantities. Bitterness can also vary among different plant varieties.
- **Cyanogenic Glycosides**: Like many plants, bitter leaf may contain small amounts of cyanogenic glycosides, which can release toxic cyanide when metabolized [16, 17]. However, the levels of these compounds are usually very low and not considered a significant health risk when consuming bitter leaf as part of a balanced diet. Proper cooking and preparation methods can further reduce the potential for cyanide release.
- **Pregnancy and Lactation**: Pregnant and lactating women are often advised to consume bitter leaf in moderation. Excessive consumption of certain phytochemicals found in bitter leaf, such as sesquiterpene lactones, during pregnancy may have adverse effects [39, 40]. It is advisable for pregnant and nursing women to consult with healthcare professionals before incorporating bitter leaf into their diets.
- **Allergies**: Some individuals may be allergic to specific compounds found in bitter leaf, leading to allergic reactions. Symptoms can include skin rashes, itching, or gastrointestinal discomfort. If allergic reactions occur, discontinuing the consumption of bitter leaf is recommended.
- **Digestive Discomfort**: In some cases, excessive consumption of bitter leaf may lead to digestive discomfort, including diarrhea or stomach upset [42, 45]. It is advisable to consume bitter leaf in moderation and observe how your body responds.
- **Interactions with Medications**: Bitter leaf may interact with certain medications due to its phytochemical content. It is important to inform healthcare providers about the use of bitter leaf, especially when taking prescription medications, to avoid potential drug interactions.
- **Individual Sensitivity**: People may vary in their sensitivity to the bitterness of bitter leaf and its potential effects on digestion. Some individuals may tolerate it well, while others may find it less agreeable.

It is important to note that the vast majority of people in regions where bitter leaf is traditionally consumed incorporate it into their diets without experiencing adverse effects. Bitter leaf is valued for its potential health benefits and nutritional value, and it is considered a safe and nutritious food when used appropriately. To mitigate potential risks and ensure safe consumption:

Cook Bitter Leaf Properly: Cooking, boiling, or blanching bitter leaf can help reduce bitterness and minimize the potential release of cyanide compounds [17]. Traditional culinary practices often involve cooking bitter leaf in soups, stews, or as a side dish.

Moderation: As with any food or herbal remedy, consuming bitter leaf in moderation is advisable. Excessive consumption of any substance can lead to adverse effects.

Consult Healthcare Professionals: It is recommended to consult certified and practicing healthcare professionals before using bitter leaf to treat or manage illness if you have underlying medical conditions, are pregnant or nursing, or are taking medications.

Although *V. amygdalina* is generally safe for most people when used in culinary and traditional medicinal practices, it is essential to be mindful of individual sensitivities and potential interactions with medications. As with any dietary ingredient, balance and moderation are key to safe and enjoyable consumption.

9 Some Herbal Formulations Involving *Vernonia amygdalina* and Their Active Ingredients

Vernonia amygdalina is a versatile plant that is often incorporated into various herbal formulations in traditional medicine across West and Central Africa. These formulations may combine bitter leaf with other herbs, roots, or natural ingredients to address specific health concerns.

Malaria Remedies: Bitter leaf is frequently used in herbal formulations to combat malaria [13]. It is often combined with other antimalarial herbs, such as *Artemisia annua* (sweet wormwood) or *Morinda lucida* (brimstone tree). These formulations may be consumed as decoctions or infusions to reduce fever and alleviate malaria symptoms [2, 15].

Digestive Tonic: Bitter leaf is sometimes included in digestive tonics or bitters. These formulations often contain a blend of bitter herbs and spices to promote digestion, alleviate indigestion, and stimulate appetite [8, 37]. Bitter leaf's natural bitterness contributes to its role in these remedies.

Cough and Cold Remedies: Bitter leaf is used in herbal preparations aimed at relieving coughs, colds, and respiratory infections [70]. It may be combined with other respiratory herbs like ginger, garlic, and honey to create soothing and immune-boosting remedies.

Anti-Inflammatory Mixtures: Bitter leaf's potential anti-inflammatory properties make it a valuable component of herbal formulations aimed at reducing inflammation [13, 53, 83]. These mixtures may include other anti-inflammatory herbs like turmeric, ginger, and neem.

Liver Support Formulas: Bitter leaf is used in herbal formulations to support liver health and promote detoxification [13, 36]. These remedies may include herbs, such as dandelion root, milk thistle, and burdock root.

Wound Healing Salves: Bitter leaf extracts or poultices are sometimes applied topically to wounds, cuts, and skin irritations to promote healing and prevent infection [44]. The plant's antimicrobial properties are believed to contribute to its effectiveness in this regard.

Diabetes Management: In some traditional systems, herbal formulations are prepared to help manage diabetes [14, 74]. Bitter leaf may be a key ingredient, as it is believed to have blood sugar-regulating properties. Other diabetes-supportive herbs such as fenugreek or cinnamon may be included.

Immune-Boosting Tonics: Bitter leaf is often incorporated into immune-boosting tonics to enhance overall health and strengthen the immune system [50, 72]. These formulations may contain a combination of immune-enhancing herbs like garlic, ginger, and echinacea.

Anti-hypertensive Mixtures: Some herbal preparations aim to help manage hypertension. Bitter leaf's potential role in blood pressure regulation is considered when formulating these remedies, and it may be combined with other blood pressure-supportive herbs like hawthorn or olive leaf [51].

Anti-parasitic Remedies: Bitter leaf is used in herbal formulations designed to combat intestinal parasites [21]. These remedies may include other anti-parasitic herbs like wormwood, black walnut, or cloves.

Note that specific herbal formulations and recipes can vary widely across different cultures and regions. Traditional medicine practitioners, often referred to as herbalists or traditional healers in West and Central Africa, possess unique knowledge and expertise in crafting these remedies based on local traditions and available resources. While traditional herbal formulations can offer potential health benefits, it is essential to exercise caution and consult with healthcare professionals, especially when using herbal remedies alongside prescription medications or for the treatment of serious medical conditions. Additionally, the safety and efficacy of these formulations should be considered in light of scientific research and evidence-based practices.

10 Commercial Value of *Vernonia amygdalina*

Vernonia amygdalina holds commercial value in several ways, primarily driven by its culinary and potential medicinal uses. Its commercial significance varies across different regions and markets [59, 78].

Culinary Use: Bitter leaf is a popular and essential ingredient in the cuisines of many West and Central African countries. It is used in a wide range of traditional dishes, soups, stews, and sauces. Due to its unique bitter flavor, it adds depth and complexity to the taste of these dishes, making them culturally significant. In some regions, processed or packaged bitter leaf products are available for convenience.

Traditional Medicine: Bitter leaf has a long history of use in traditional medicine across Africa [67, 68]. As interest in herbal and natural remedies grows globally, there is potential commercial value in the development and marketing of herbal products containing bitter leaf extracts or formulations. These products may be promoted for their potential health benefits, including immune support, anti-inflammatory properties, and digestive health.

Pharmaceutical and Nutraceutical Industries: The phytochemical compounds found in bitter leaf, such as flavonoids, phenolic compounds, and sesquiterpene lactones, have attracted the attention of researchers and the pharmaceutical industry. There is ongoing exploration of the potential use of these compounds in drug development and nutraceutical products.

Export Market: Bitter leaf has gained recognition in international markets, especially among African diaspora communities. It is exported to countries with large African populations or those interested in African cuisine. In some regions, dried or frozen bitter leaf is exported to meet the demand of African expatriates and consumers interested in African culinary traditions.

Herbal and Dietary Supplements: Bitter leaf is sometimes used as an ingredient in herbal supplements and dietary products. These supplements are marketed for various purposes, such as immune support, detoxification, and digestive health.

Cultivation and Farming: As the demand for bitter leaf increases, there may be opportunities for commercial cultivation and farming of *Vernonia amygdalina*. Controlled cultivation can ensure a consistent supply of high-quality bitter leaf for domestic and international markets.

Traditional Food Processing: Traditional food processors who specialize in preparing and packaging bitter leaf for sale may contribute to its commercial value. These businesses may operate in local markets and play a role in preserving culinary traditions.

Research and Development: Research institutions and organizations may invest in the research and development of bitter leaf-based products and formulations. This can lead to innovations in the food, pharmaceutical, and nutraceutical industries.

Herbal Tea and Beverage Industry: Bitter leaf can be used as an ingredient in herbal teas and beverages. Its unique flavor and potential health benefits make it a candidate for inclusion in herbal tea blends targeting health-conscious consumers.

Export Regulations: Regulatory bodies in some countries may establish standards and regulations for the export and import of bitter leaf and related products. Compliance with these regulations is essential for market access and commercial success.

The commercial value of *Vernonia amygdalina* makes it evident that it is essential to consider sustainable cultivation, harvesting practices, quality control, and adherence to safety standards to ensure the long-term viability of the bitter leaf industry. Additionally, collaborations between researchers, traditional healers, farmers, and entrepreneurs can contribute to the responsible and profitable commercialization of this valuable plant.

11 Future Prospects of Therapeutic Roles of *Vernonia amygdalina*

The prospects of *V. amygdalina* in therapeutic roles are promising and multifaceted. As interest in natural and plant-based remedies continues to grow, coupled with ongoing scientific research, *Vernonia amygdalina* holds great potential for contributing to various aspects of modern healthcare. Research has shown that *Vernonia amygdalina* contains a variety of phytochemicals, including flavonoids and phenolic compounds, which exhibit antioxidant and anti-inflammatory properties. These properties are important in combating oxidative stress and chronic inflammation, which are underlying factors in many chronic diseases [41]. The potential immunomodulatory effects of bitter leaf suggest its use in enhancing immune function. As immune support becomes increasingly important, bitter leaf formulations may find a place in promoting overall health and resilience. Bitter leaf has been traditionally used for its antimicrobial properties [4]. Research is ongoing to understand its effectiveness against various microorganisms, including bacteria, viruses, and parasites. This could have implications for infectious disease management. Preliminary studies have explored the antidiabetic potential of *Vernonia amygdalina*. As diabetes remains a global health concern, there is interest in developing bitter leaf-based products or therapies for blood sugar regulation. Bitter leaf is used traditionally to support liver health and detoxification. Further research into its hepatoprotective properties may lead to the development of liver support products or therapies. Some studies suggest that bitter leaf may play a role in blood pressure regulation. This could be valuable for individuals with hypertension or those at risk of high blood pressure. Bitter leaf has attracted attention for its potential anticancer properties. Investigations into the mechanisms of action of its bioactive compounds may lead to the development of novel cancer therapies or preventive strategies. Bitter leaf may be incorporated into nutraceuticals and dietary supplements targeting specific health concerns. These products could offer convenient and standardized ways to access its therapeutic benefits. The isolation and characterization of bioactive compounds from bitter leaf may pave the way for the development of pharmaceutical drugs derived from its natural compounds [82]. This could result in innovative treatments for various health conditions. Beyond its medicinal properties, *V. amygdalina* continues to be a culinary staple. Innovative chefs and food scientists may explore new ways to incorporate bitter leaf into culinary creations, promoting both its unique flavor and potential health benefits [56–58]. As the demand for bitter leaf increases, there will be a growing need for sustainable cultivation practices to ensure a consistent supply. Conservation efforts to protect wild populations of *Vernonia amygdalina* may also become essential. Collaborations between traditional healers and modern researchers can help preserve and document the rich traditional knowledge associated with bitter leaf. This knowledge transfer can inform future therapeutic developments. It is important to emphasize that the therapeutic roles of *Vernonia amygdalina* are an evolving field of study, and further research is needed to validate its efficacy and safety for specific health conditions. Collaboration between scientists, healthcare professionals, traditional healers, and local communities will be crucial in harnessing the full potential of this remarkable plant. Additionally, regulatory considerations, quality control, and standardized formulations will play a role in its future therapeutic applications.

12 Conclusion

The use of bitter leaf in ethnomedicine expresses the deep connection between nature, culture, and healthcare. The growing ethnomedicinal use reflects the wisdom passed down through generations in West and Central Africa. While its bitterness defines its taste, it is the plant's potential therapeutic properties that make it truly remarkable. From malaria treatment to immune system support and beyond, bitter leaf continues to play a vital role in the lives of many people. As modern science sheds light on its bioactive compounds and mechanisms of action, there is optimism about its potential contributions to modern healthcare. However, the complex intersection of traditional knowledge, scientific inquiry, and cultural heritage makes it essential to respect and value the lessons of the past while embracing the possibilities of the future. Therefore, the story of *V. amygdalina* is a testament to the enduring power of nature to heal, nurture, and inspire. *V. amygdalina* is a remarkable plant with a rich history deeply rooted in West Africa. Its origin within the region can be traced back to the tropical and subtropical zones where it has thrived for centuries. Bitter leaf's significance transcends its botanical characteristics, encompassing cultural, culinary, and medicinal dimensions that have been integral to the lives of West African communities. From the renowned bitter leaf soup (ofe onugbu) to its use in traditional medicine, bitter leaf has left an indelible mark on the cultural landscape of the region. Its cultivation and distribution have expanded beyond its native range, making it a valuable resource not only within West Africa but also in international markets. As awareness of the actual and potential health benefits grows, *Vernonia amygdalina* continues to be a subject of scientific inquiry, further unraveling its diverse uses and potential applications in modern medicine and nutrition.

References

1. Abdulmalik O, Oladapo OO, Bolaji MO (2016) Effect of aqueous extract of Vernonia amygdalina on atherosclerosis in rabbits. ARYA Atheroscler 12(1):35–40
2. Abosi AO, Raseroka BH (2003) In vivo antimalarial activity of Vernonia amygdalina. Br J Biomed Sci 60(2):89–91. https://doi.org/10.1080/09674845.2003.11783680
3. Aboyeji CM (2019) Impact of green manures of Vernonia amygdalina and Chromolaena odorata on growth, yield, mineral and proximate composition of radish (*Raphanus sativus* L.). Sci Rep 9(1):17659. https://doi.org/10.1038/s41598-019-54071-8
4. Achuba FI (2018) Role of bitter leaf (*Vernonia amygdalina*) extract in prevention of renal toxicity induced by crude petroleum contaminated diets in rats. Int J Vet Sci Med 6(2):172–177. https://doi.org/10.1016/j.ijvsm.2018.07.002
5. Adaramoye OA, Akintayo O, Achem J, Fafunso MA (2008) Lipid-lowering effects of methanolic extract of *Vernonia amygdalina* leaves in rats fed on high cholesterol diet. Vasc Health Risk Manag 4:235–241

6. Adaramoye O, Ogungbenro B, Anyaegbu O, Fafunso M (2008) Protective effects of extracts of *Vernonia amygdalina*, *Hibiscus sabdariffa* and vitamin against radiationinduced liver damage in rats. J Radiat Res 49:123–131
7. Adedapo AA, Aremu OJ, Oyagbemi AA (2014) Anti-oxidant, anti-inflammatory and anti-nociceptive properties of the acetone leaf extract of *Vernonia amygdalina* in some laboratory animals. Adv Pharm Bull 4(Suppl 2):591–598. https://doi.org/10.5681/apb.2014.087
8. Adefisayo MA, Akomolafe RO, Akinsomisoye OS, Alabi QK, Ogundipe L, Omole JG, Olamilosoye KP (2018) Protective effects of methanol extract of *Vernonia amygdalina (del.)* leaf on aspirin-induced gastric ulceration and oxidative mucosal damage in a rat model of gastric injury. Dose Response 16(3):1559325818785087. https://doi.org/10.1177/1559325818785087
9. Adugna W, Ayele S, Addis M (2023) Different proportions of dried Vernonia amygdalina leaves and wheat bran mixture supplementation on feed intake, digestibility and body weight change of Arsi-Bale sheep fed with natural pasture hay as basal diet. Veterinary medicine and science 9 (2):957–966. https://doi.org/10.1002/vms3.1068
10. Alawa CBI, Adamu AM, Gefu JO, Ajanusi OJ, Abdu PA, Chiezey NP, Bowman DD (2003) In vitro screening of two Nigerian medicinal plants (*Vernonia amygdalina* and *Annona senegalensis*) for anthelmintic activity. Vet Parasitol 113:73–81
11. Andrzejewska J, Sadowska K, Klóska L, Rogowski L (2015) The effect of plant age and harvest time on the content of chosen components and antioxidative potential of black chokeberry fruit. Acta Scientiarum Hortorum Cultus 14:105–114
12. Asante DB, Effah-Yeboah E, Barnes P, Abban HA, Ameyaw EO, Boampong JN, Dadzie JB (2016) Antidiabetic effect of young and old ethanolic leaf extracts of *Vernonia amygdalina*: A comparative study. J Diabetes Res 2016:1–13
13. Asante DB, Henneh IT, Acheampong DO, Kyei F, Adokoh CK, Ofori EG, Domey NK, Adakudugu E, Tangella LP, Ameyaw EO (2019) Anti-inflammatory, anti-nociceptive and antipyretic activity of young and old leaves of *Vernonia amygdalina*. Biomed Pharmacother 111:1187–1203. https://doi.org/10.1016/j.biopha.2018.12.147
14. Atangwho IJ, Yin KB, Umar MI, Ahmad M, Asmawi MZ (2014) *Vernonia amygdalina* simultaneously suppresses gluconeogenesis and potentiates glucose oxidation via the pentose phosphate pathway in streptozotocin-induced diabetic rats. BMC Complement Altern Med 14: 426. https://doi.org/10.1186/1472-6882-14-426
15. Bihonegn T, Giday M, Yimer G, Animut A, Sisay M (2019) Antimalarial activity of hydromethanolic extract and its solvent fractions of *Vernonia amygdalina* leaves in mice infected with *plasmodium berghei*. SAGE Open Med 7:2050312119849766. https://doi.org/10.1177/2050312119849766
16. Butterly BR, Buzzell RI (1977) The relationship between chlorophyll content and rate of photosynthesis in soybeans. Can J Plant Sci 57:1–5
17. Chalker-Scott L (1999) Environmental significance of anthocyanins in plant stress responses. Photochem Photobiol 70:1–9
18. Challand S, Willcox M (2009) A clinical trial of the traditional medicine *Vernonia amygdalina* in the treatment of uncomplicated malaria. J Altern Complement Med 15:1231–1237
19. Chenard CH, Kopsell DA, Kopsell DE (2005) Nitrogen concentration affects nutrient and carotenoid accumulation in parsley. J Plant Nutr 28:285–297
20. Christensen LP, Brandt K (2007) In: Crozier A, Clifford MN, Ashihara H (eds) Acetylenes and psoralens in "plant secondary metabolites: occurrence, structure and role in the human diet". Blackwell Publishing Ltd., Oxford, UK, pp 137–173
21. Dégbé M, Debierre-Grockiego F, Tété-Bénissan A, Débare H, Aklikokou K, Dimier-Poisson I, Gbeassor M (2018) Extracts of Tectona grandis and *Vernonia amygdalina* have anti-toxoplasma and pro-inflammatory properties in vitro. In: Les extraits de Tectona grandis et de *Vernonia amygdalina* ont des propriétés anti-Toxoplasma et pro-inflammatoires in vitro. *Parasite*, vol 25, (Paris, France), p 11. https://doi.org/10.1051/parasite/2018014
22. Delgado R, Gonzalez MR, Martin P (2006) Interaction effects of nitrogen and potassium fertilization on anthocyanin composition and chromatic features of tempranillo grapes. J Int Sci Vigne Vin 40:141–150
23. Demmig-Adams B, Adams WW (2000) Harvesting sunlight safely. Nature 403:371–374

24. Erasto P, Grierson DS, Afolayan AJ (2006) Bioactive sesquiterpene lactones from the leaves of Vernonia amygdalina. J Ethnopharmacol 106:117–120
25. Erasto P, Grierson DS, Afolayan AJ (2007) Antioxidant constituents in *Vernonia amygdalina* leaves. Pharm Biol 45:195–199
26. Erasto P, Grierson DS, Afolayan AJ (2007) Evaluation of antioxidant activity and the fatty acid profile of the leaves of *Vernonia amygdalina* growing in South Africa. Food Chem 104: 636–642
27. Eze JC, Okafor F, Nwankwo NE, Okeke ES, Onwudiwe NN (2020) Schistosomiasis prevention option: toxicological evaluation of *Vernonia amygdalina* on the tissues of *Bulinus truncatus* at different pH conditions. Heliyon 6(8):e04796. https://doi.org/10.1016/j.heliyon.2020.e04796
28. Farombi EO, Owoeye O (2011) Antioxidative and chemopreventive properties of *Vernonia amygdalina* and *Garcinia biflavonoid*. Int J Environ Res Public Health 8(6):2533–2555. https://doi.org/10.3390/ijerph8062533
29. Fleischer WE (1935) The relation between chlorophyll content and rate of photosynthesis. J Gen Physiol 18:573–597
30. Gross M (2004) Flavonoids and cardiovascular disease. Pharm Biol 42:21–35
31. Hughes NM, Morley CB, Smith WK (2007) Coordination of anthocyanin decline and photosynthetic maturation in juvenile leaves of three deciduous tree species. New Phytol 175: 675–685
32. Igile GO, Oleszek W, Jurzysta M, Burda S, Fafunso M, Fasanmade AA (1994) Flavonoids from *Vernonia amygdalina* and their antioxidant activities. J Agric Food Chem 42:2445–2448
33. Ikhajiagbe B, Saheed M, Ogunro J (2020) Studies of foliar bioconcentration of metals by Vernonia amygdalina in a model heavy metal-polluted soil. Stud Univ Vasile Goldis Arad Ser Stiint Vietii 30:64–68
34. Imarhiagbe O, Ogwu MC (2022) Sacred groves in the global south: A panacea for sustainable biodiversity conservation. In: Izah SC (ed) Biodiversity in Africa: potentials, *Threats and conservation.* Sustainable development and biodiversity, vol 29. Springer, Singapore, pp 525–546. https://doi.org/10.1007/978-981-19-3326-4_20
35. Iroanaya O, Okpuzor J, Mbagwu H (2010) Anti-nociceptive and anti phlogistic actions of a polyherbal decoction. Int J Pharmacol 6:31–36
36. Iwalokun BA, Efedede BU, Alabi-Sofunde JA, Oduala T, Magbagbeola OA, Akinwande AI (2006) Hepatoprotective and antioxidant activities of *Vernonia amygdalina* on acetaminophen-induced hepatic damage in mice. J Med Food 9(4):524–530. https://doi.org/10.1089/jmf.2006.9.524
37. Iwo MI, Sjahlim SL, Rahmawati SF (2017) Effect of *Vernonia amygdalina* Del. leaf ethanolic extract on intoxicated male wistar rats liver. Sci Pharm 85(2):16. https://doi.org/10.3390/scipharm85020016
38. Izevbigie EB, Bryant JL, Walker A (2004) A novel natural inhibitor of extracellular signalregulated kinases and human breast cancer cell growth. Exp Biol Med 229:163–169
39. Jisaka M, Kawanaka M, Sugiyama H, Takegawa K, Huffman MA, Ohigashi H, Koshimizu K (1992) Antischistosomal activities of sesquiterpene lactones and steroid glucosides from *Vernonia amygdalina*, possibly used by wild chimpanzees against parasiterelated diseases. Biosci Biotechnol Biochem 56:845–846
40. Jisaka M, Ohigashi H, Takegawa K, Hirota M, Irie R, Huffman MA, Koshimizu K (1993) Steroid glucosides from *Vernonia amygdalina*, a possible chimpanzee medicinal plant. Phytochemistry 34:409–413
41. Johnson W, Tchounwou PB, Yedjou CG (2017) Therapeutic mechanisms of *Vernonia amygdalina* Delile in the treatment of prostate cancer. Molecules (Basel, Switzerland) 22(10): 1594. https://doi.org/10.3390/molecules22101594
42. Kopsell DA, Kopsell DE, Curran-Celentano J (2007) Carotenoid pigments in kale are influenced by nitrogen concentration and form. J Sci Food Agric 87:900–907
43. Kupchan SM, Hemingway RJ, Karim A, Werner D (1969) Vernodalin and vernomygdin, two new cytotoxic sesquiterpene lactones from *Vernonia amygdalina* Del. J Org Chem 34: 3908–3911

44. Lambebo MK, Kifle ZD, Gurji TB, Yesuf JS (2021) Evaluation of wound healing activity of Methanolic crude extract and solvent fractions of the leaves of *Vernonia auriculifera* Hiern (Asteraceae) in mice. J Exp Pharmacol 13:677–692. https://doi.org/10.2147/JEP.S308303
45. Leonidas M, Faye D, Justin KN, Viateur U, Angélique N (2013) Evaluation of the effectiveness of two medicinal plants *Vernonia amygdalina* and *Leonotis nepetaefolia* on the gastrointestinal parasites of goats in Rwanda: case study of Huye and Gisagara districts. J Vet Med Anim Health 5:229–236
46. Liwa AC, Smart LR, Frumkin A, Epstein HA, Fitzgerald DW, Peck RN (2014) Traditional herbal medicine use among hypertensive patients in sub-Saharan Africa: a systematic review. Curr Hypertens Rep 16(6):437. https://doi.org/10.1007/s11906-014-0437-9
47. Longstreth DJ, Nobel PS, Deb G (1980) Nutrient influences on leaf photosynthesis. Plant Physiol 65:541–543
48. Marschner P (2012) Marschner's mineral nutrition of higher plants". 651pp, 3rd edn. Academic Press: Elsevier Ltd
49. Michael UA, David BU, Theophine CO, Philip FU, Ogochukwu AM, Benson VA (2010) Antidiabetic effect of combined aqueous leaf extract of Vernonia amygdlina and metformin in rats. Journal of Basic and Clinical Pharmacy 1:197–202
50. Momoh MA, Muhamed U, Agboke AA, Akpabio EI, Osonwa UE (2012) Immunological effect of aqueous extract of *Vernonia amygdalina* and a known immune booster called immunace(®) and their admixtures on HIV/AIDS clients: a comparative study. Asian Pac J Trop Biomed 2(3): 181–184. https://doi.org/10.1016/S2221-1691(12)60038-0
51. Mushagalusa Kasali F, Ahadi Irenge C, Murhula Hamuli P, Birindwa Mulashe P, Murhula Katabana D, Mangambu Mokoso JD, Mpiana PT, Ntokamunda Kadima J (2021) Ethnopharmacological survey on treatment of hypertension by traditional healers in Bukavu City, DR Congo. Evid Based Complement Alternat Med 2021:6684855. https://doi.org/10.1155/2021/6684855
52. Nantongo JS, Odoi JB, Agaba H, Gwali S (2023) Genetic diversity and population structure of *Vernonia amygdalina* Del. In Uganda based on genome wide markers. PLoS One 18(7): e0283563. https://doi.org/10.1371/journal.pone.0283563
53. Nguyen TXT, Dang DL, Ngo VQ, Trinh TC, Trinh QN, Do TD, Thanh TTT (2021) Anti-inflammatory activity of a new compound from *Vernonia amygdalina*. Nat Prod Res 35(23): 5160–5165. https://doi.org/10.1080/14786419.2020.1788556
54. Njan AA, Adzu B, Agaba AG, Byarugaba D, Díaz-Llera S, Bangsberg DR (2008) The analgesic and antiplasmodial activities and toxicology of *Vernonia amygdalina*. J Med Food 11:574–581
55. Oboh G (2006) Nutritive value and haemolytic properties (in vitro) of the leaves of *Vernonia amygdalina* on human erythrocyte. Nutr Health 18(2):151–160. https://doi.org/10.1177/026010600601800207
56. Ogwu MC (2019) Lifelong consumption of plant-based GM foods: is it safe? In: Papadopoulou P, Misseyanni A, Marouli C (eds) Environmental exposures and human health challenges. IGI Global, Pennsylvania, pp 158–176. https://doi.org/10.4018/978-1-5225-7635-8.ch008
57. Ogwu MC (2019) Towards sustainable development in Africa: the challenge of urbanization and climate change adaptation. In: Cobbinah PB, Addaney M (eds) The geography of climate change adaptation in urban Africa. Springer Nature, Switzerland, pp 29–55. https://doi.org/10.1007/978-3-030-04873-0_2
58. Ogwu MC (2019) Understanding the composition of food waste: an "-omics" approach to food waste management. In: Gunjal AP, Waghmode MS, Patil NN, Bhatt P (eds) Global initiatives for waste reduction and cutting food loss. IGI Global, Pennsylvania, pp 212–236. https://doi.org/10.4018/978-1-5225-7706-5.ch011
59. Ogwu MC (2020) Value of *Amaranthus* [L.] species in Nigeria. In: Waisundara V (ed) Nutritional value of Amaranth. IntechOpen, UK, pp 1–21. https://doi.org/10.5772/intechopen.86990
60. Ogwu MC (2023) Local food crops in africa: sustainable utilization, threats, and traditional storage strategies. In: Izah SC, Ogwu MC (eds) *Sustainable utilization and conservation of*

Africa's biological resources and environment. Sustainable development and biodiversity, vol 888. Springer, Singapore, pp 353–376. https://doi.org/10.1007/978-981-19-6974-4_13

61. Ogwu MC, Osawaru ME (2022) Traditional methods of plant conservation for sustainable utilization and development. In: Izah SC (ed) Biodiversity in Africa: potentials, *threats and conservation*, Sustainable development and biodiversity, vol 29. Springer, Singapore, pp 451–472. https://doi.org/10.1007/978-981-19-3326-4_17
62. Ogwu MC, Osawaru ME (2023) Disease outbreaks in ex-situ plant conservation and potential management strategies. In: Izah SC, Ogwu MC (eds) Sustainable utilization and conservation of Africa's biological resources and environment, Sustainable development and biodiversity, vol 888. Springer, Singapore, pp 497–518. https://doi.org/10.1007/978-981-19-6974-4_18
63. Ogwu MC, Osawaru ME, Ahana CM (2014) Challenges in conserving and utilizing plant genetic resources (PGR). Int J Genet Mol Biol 6(2):16–22. https://doi.org/10.5897/IJGMB2013.0083
64. Ogwu MC, Osawaru ME, Aiwansoba RO, Iroh RN (2016) Ethnobotany and collection of west African okra [*Abelmoschus caillei* (A. Chev.) Stevels] germplasm in some communities in Edo and Delta states, southern Nigeria. Borneo J Resour Sci Technol 6(1):25–36. https://doi.org/10.33736/bjrst.212.2016
65. Ogwu MC, Osawaru ME, Aiwansoba RO, Iroh RN (2016) Status and prospects of vegetables in Africa. In: Borokini IT, Babalola FD (eds) Conference proceedings of the joint biodiversity conservation conference of Nigeria Tropical Biology Association and Nigeria chapter of Society for Conservation Biology on MDGs to SDGs: toward sustainable biodiversity conservation in Nigeria. University of Ilorin, Nigeria, pp 47–57
66. Ogwu MC, Osawaru ME, Obahiagbon GE (2017) Ethnobotanical survey of medicinal plants used for traditional reproductive care by Usen people of Edo state, Nigeria, Malaya. J Biosci 4(1):17–29
67. Ogwu MC, Osawaru ME, Owie MO (2023) Effects of storage at room temperature on the food components of three cocoyam species (*Colocasia esculenta, Xanthosoma atrovirens, and X. Sagittifolium*). Borneo J Resour Sci Technol 13(2):59–83. https://doi.org/10.18848/2160-1933/CGP/v13i02/59-83
68. Ogwu MC, Osawaru ME, Amodu E, Osamo F (2023) Comparative morphology, anatomy, and chemotaxonomy of two *Cissus* Linn. species. Rev Bras Bot. https://doi.org/10.1007/s40415-023-00881-0
69. Okwuzu JO, Odeiga P, AdetoroOtubanjo O, Ezechi OC (2017) Cytotoxicity testing of aqueous extract of bitter leaf (*Vernonia amygdalina Del*) and sniper 1000EC (2,3 dichlorovinyl dimethyl phosphate) using the Alium cepa test. Afr Health Sci 17(1):147–153. https://doi.org/10.4314/ahs.v17i1.19
70. Oladele JO, Oyeleke OM, Oladele OT, Oladiji AT (2021) Covid-19 treatment: investigation on the phytochemical constituents of *Vernonia amygdalina* as potential Coronavirus-2 inhibitors. Comput Toxicol (Amsterdam, Netherlands) 18:100161. https://doi.org/10.1016/j.comtox.2021.100161
71. Omoregie ES, Pal A (2016) Antiplasmodial, antioxidant and immunomodulatory activities of ethanol extract of *Vernonia amygdalina* del. leaf in Swiss mice. Avicenna J Phytomedicine 6(2):236–247
72. Onah IA, Onuigbo EB, Odimegwu DC (2019) Adjuvant effect of *Vernonia amygdalina* leaf extract on host immune response to hepatitis B virus subunit vaccine. Pharmazie 74(3):179–185. https://doi.org/10.1691/ph.2019.8920
73. Onasanwo SA, Oyebanjo OT, Ajayi AM, Olubori MA (2017) Anti-nociceptive and anti-inflammatory potentials of *Vernoniaamygdalina* leaf extract via reductions of leucocyte migration and lipid peroxidation. J Intercult thnopharmacol 6(2):192–198. https://doi.org/10.5455/jice.20170330010610
74. Ong KW, Hsu A, Song L, Huang D, Tan BK (2011) Polyphenols-rich *Vernonia amygdalina* shows anti-diabetic effects in streptozotocin-induced diabetic rats. J Ethnopharmacol 133(2):598–607. https://doi.org/10.1016/j.jep.2010.10.046

75. Oriakhi K, Oikeh EI, Ezeugwu N, Anoliefo O, Aguebor O, Omoregie ES (2013) Comparative antioxidant activities of extracts of *Vernonia amygdalina* and Ocimum gratissimum leaves. J Agric Sci 6:13–20
76. Osawaru ME, Ogwu MC (2014) Ethnobotany and germplasm collection of two genera of cocoyam (*Colocasia* [Schott] and *Xanthosoma* [Schott], Araceae) in Edo state Nigeria. Sci Technol Arts Res J 3(3):23–28. https://doi.org/10.4314/star.v3i3.4
77. Osawaru ME, Ogwu MC (2014) Conservation and utilization of plant genetic resources. In: Omokhafe K, Odewale J (eds) Proceedings of 38th annual conference of the genetics society of Nigeria. Empress Prints Nigeria Limited, pp 105–119
78. Osawaru ME, Ogwu MC (2020) Survey of plant and plant products in local markets within Benin City and environs. In: Filho LW, Ogugu N, Ayal D, Adelake L, da Silva I (eds) African handbook of climate change adaptation. Springer Nature, Cham. pp 1–24. https://doi.org/10.1007/978-3-030-42091-8_159-1
79. Osawaru ME, Ogwu MC, Ahana CM (2013) Current status of plant diversity and conservation in Nigeria. Nigerian J Life Sci 3(1):168–178
80. Osawaru ME, Ogwu MC, Omologbe J (2014) Characterization of three okra [*Abelmoschus* (L.)] accessions using morphology and SDS-PAGE for the basis of conservation. *Egyptian academic*. J Biol Sci 5(1):55–65
81. Shittu H, Ikhajiagbe B (2013) Phytoaccumulation of heavy metals in an oil-spiked soil by *Vernonia amygdalina* after exposure to sodium azide solutions. Niger J Life Sci (Official J Faculty Life Sci, University of Benin, Nigeria) 3:169–179. https://doi.org/10.52417/njls.v3i2.145
82. Syahputra RA, Harahap U, Dalimunthe A, Pandapotan M, Satria D (2021) Protective effect of *Vernonia amygdalina* Delile against doxorubicin-induced cardiotoxicity. Heliyon 7(7):e07434. https://doi.org/10.1016/j.heliyon.2021.e07434
83. Yousseu Nana W, Billong Mimb JR, Atsamo AD, Tsafack EG, Djuichou Nguemnang SF, Fagni Njoya ZL, Matah Marthe VM, Madjo Kouam YK, Mbiantcha M, Ateufack G (2023) *In vitro* and *in vivo* anti-inflammatory properties of the Hydroethanolic extract of the roots of *Vernonia guineensis* (Asteraceae). Int J Inflamm 2023:7915367. https://doi.org/10.1155/2023/7915367

Citrus aurantifolia: Phytochemical Constituents, Food Preservative Potentials, and Pharmacological Values

11

Sylvester Chibueze Izah, Glory Richard, and
Tamaraukepreye Catherine Odubo

Contents

S. C. Izah (✉) · T. C. Odubo
Department of Microbiology, Faculty of Science, Bayelsa Medical University, Yenagoa, Bayelsa State, Nigeria
e-mail: sylvester.izah@bmu.edu.ng

G. Richard
Department of Clinical Sciences, Faculty of Clinical Sciences, Niger Delta University, Wilberforce Island, Bayelsa State, Nigeria

S. C. Izah et al. (eds.), *Herbal Medicine Phytochemistry*, Reference Series in Phytochemistry, https://doi.org/10.1007/978-3-031-43199-9_76

Abstract

This chapter focuses on the phytochemical constituents, food preservative potentials, pharmacological values, safety considerations, and research needs of *Citrus aurantifolia*. The phytochemical components of *Citrus aurantifolia* offer diverse health benefits, including strengthening the immune system, supporting collagen synthesis, enhancing iron absorption, reducing inflammation (vitamin C), providing anti-inflammatory and antioxidant effects, and antiviral activities (Flavonoids). Also, the other phytochemicals, including limonoids, essential oils, alkaloids, coumarins, terpenes, and phenolics, have specific health effects. *Citrus aurantifolia* is an antioxidant and anti-inflammatory compound that may alleviate conditions like arthritis. The essential oils from the fruit can inhibit pathogenic microbes, while limonoids and coumarins possess anticancer potentials. *Citrus aurantifolia* extracts can also be used to reduce stress and manage diabetes. They also function as an analgesic, anti-cholinesterase agent, immune modulator, antidiarrheal, antiulcer, anthelmintic, and, surprisingly, a contraceptive method for some women. *Citrus aurantifolia* offers natural antimicrobial properties effective against foodborne pathogens, preserving freshness and flavor while reducing reliance on synthetic preservatives. On safety considerations, *Citrus aurantifolia* can trigger allergies, irritate the stomach, erode dental health, heighten photosensitivity, and interact with medications. Overconsumption may lead to dehydration and peel pesticide residues may transfer to the juice. Proper handling and storage are crucial. Potential research needs include comprehensive phytochemical profiling and isolation, investigation of antioxidant, antibacterial, and antifungal properties, research on bioavailability and pharmacokinetics, clinical trials for health applications, safety and toxicology studies, synergy with other compounds, and sustainable cultivation and processing methods.

Keywords

Disease management · Medicinal plants · Phytochemistry · Human health · Safety considerations · Plant-based therapeutics

1 Introduction

Ethnobotany plays a pivotal role in unraveling the traditional applications of plants for medicinal and therapeutic purposes. This field, which investigates the dynamic interplay between humans and plants, enables us to tap into the treasure trove of ancestral knowledge. Herbal medicine, a specialized branch of ethnobotany, takes full advantage of this wisdom, leveraging it to harness the therapeutic potential of various plant species [33]. By uniting the insights of traditional practices with the rigor of scientific exploration, ethnobotany safeguards the invaluable knowledge of indigenous cultures.

Indigenous communities often reside in remote environments and profoundly connect with their natural surroundings. Through generations living in harmony with the land, they have cultivated a deep understanding of the indigenous flora and the myriad ways these plants can be used. Ethnobotanical studies serve as a means of preserving this collective wisdom [22].

Herbal medicine, rooted in ethnobotanical knowledge, involves applying compounds sourced from plants for therapeutic purposes [58]. *Citrus aurantifolia* emerges as a prime exemplar of the vast potential inherent in herbal therapy. The fruit, replete with various health-enhancing phytochemical constituents, takes center stage in numerous scientific investigations.

Researchers are particularly drawn to the juice extracts of *Citrus aurantifolia* (Izah and Odubo). The juice is rich in phytochemicals that provide numerous health benefits and has become the central focus of extensive research [47]. In essence, this citrus fruit embodies the marriage of traditional herbal wisdom with contemporary scientific exploration, offering a compelling case for the potential of herbal medicine to enhance human well-being.

Citrus aurantifolia, commonly known as lime, is studied for its delightful flavor and rich composition of bioactive compounds. The antimicrobial properties of *Citrus aurantifolia*, primarily attributed to essential oils like citral and limonene, position it as a viable natural option for food preservation. These compounds exhibit efficacy in inhibiting the growth of harmful bacteria and fungi, extending the shelf life of food products without the need for synthetic preservatives [47]. Incorporating *Citrus aurantifolia* into food preservation methods aligns with the growing consumer demand for unprocessed and safer food choices.

Beyond its role in food preservation and culinary applications, *Citrus aurantifolia* offers significant pharmacological properties. It possesses Vitamin C, with its anti-inflammatory and antioxidant attributes, is valuable in addressing chronic diseases and combating oxidative stress. *Citrus aurantifolia* contains limonoids, flavonoids, and essential oils such as citral and limonene, each with a spectrum of health advantages, including anti-inflammatory, antibacterial, and even anticancer effects [39]. Limonoids, mainly, show promise in cancer prevention and treatment, making *Citrus aurantifolia* a valuable addition to a well-rounded diet. Furthermore, essential oils of *Citrus aurantifolia* have even been used in aromatherapy to alleviate stress and anxiety.

This chapter looks into the capacity of natural products to enhance human well-being, exemplifying the harmonious blend of traditional wisdom and modern science in harnessing plant-based remedies for the betterment of health.

2 Phytochemical Constituents of *Citrus aurantifolia*

Citrus aurantifolia fruit has a relatively high citric acid content in comparison to *Citrus limon* (lemon) [59, 71]. *Citrus aurantifolia* has remarkable phytochemical constituents. These natural compounds are found in various plant parts, including the fruit, leaves, and essential oils [47].

Secondary metabolites are chemical compounds produced by plants, and they play a crucial role in defending against microbial and insect attacks [46, 55–57]. These secondary metabolites encompass various categories, including flavonoids, terpenoids, phenolics, limonoids, and alkaloids, which can be further categorized into volatile and non-volatile components. In the case of *Citrus aurantifolia*, the various parts, such as stems, leaves, flowers, seeds, and fruit, serve as sources for bioactive components [45, 47]. In a review study by Indriyani et al. [45], it was revealed that flavonoids and alkaloids are found in all parts of the plant (leaf, fruit, rind, seed, stem, root, and bark), while essential oil is found in the leaf and peel; terpenoids in the leaf, peel, seed, root, and bark; limonoids in the seed; and phenolics in the leaf, fruit, rind, seed, stem, and bark. The authors further reported that the peels of citrus fruits that are not commonly used contain secondary metabolites with significant antioxidant properties compared to other plant parts.

2.1 Alkaloids

Alkaloids constitute a broad category of naturally occurring organic compounds primarily found within plant materials [13, 24, 82]. They often trigger pharmacological responses in both humans and various other organisms. The vastness of this compound class is evident from the isolation of more than 10,000 distinct alkaloids in their natural contexts, underscoring their immense diversity. Alkaloids are present in a wide array of plant species, and it is believed that plants have evolved this diversity of alkaloids as a defense mechanism against herbivores and various threats.

These alkaloids can manifest pharmacological effects when administered internally or applied externally for medicinal purposes. Among the alkaloids commonly associated with plants from the *Citrus* genus, synephrine, tyramine, and octopamine stand out as the most prevalent [98]. Alkaloids have been reported in different parts of the *Cirtus auratifolia*, including peels [40], leaves [1], and fruit juice [47]. Due to their remarkable biological properties, citrus alkaloids and glycosides exhibit significant potential as pharmacological supplements.

2.2 Phenolic Compounds

Research indicates that the presence of phenolic compounds plays a significant role in how plants respond to various pathogens, including fungi, viruses, and bacteria, thereby enhancing their ability to fend off infections [90]. In the case of *Citrus aurantifolia* fruit, the level of phenolic compounds can be influenced by the fruit's degree of maturity [45]. It's noteworthy that *Citrus aurantifolia* fruit exhibits the highest phenolic content. This content varies as the fruit matures, and the extract solvent could have an impact (positive or negative) on the phenolic compounds [45]. Singh et al. [90] reported the presence of four phenolic acids: gallic acid (26.85 g/g), tannic acid (14.27 g/g), and traces of ferulic and coumaric acids in the pulp of *Citrus aurantifolia*. The authors reported that tannic acid has the highest concentration at 26.85 g/g, while gallic acid is the predominant phenolic acid in the pulp.

2.3 Limonoids

Limonoids, a group of chemical constituents, are prevalent in citrus fruits. Citrus fruits serve as rich sources of limonoids, with relatively high concentrations of these compounds. Two distinct forms of limonoids can be identified in citrus fruits: glycosides and aglycones. Aglycones can break down into acidic mono-lactones, neutral dilactones, and neutral dicarboxylic acids [45]. Limonoid aglycones are known for their bitter flavor and insolubility in water, contributing to their nomenclature [45, 66]. Oranges, in particular, are associated with a specific chemical responsible for their bitter taste. Both limonin and nomilin are glycosides, highlighting their significance [6, 45].

Limonoids belong to the subgroup of oxygenated tetracyclic triterpenoids within the broader category of triterpenoids. These compounds possess diverse biological effects, such as anticarcinogenic, antioxidative, antibacterial, larvicidal, antimalarial, and antiviral properties, as well as hypoallergenic, anti-inflammatory, anti-proliferative, antimutagenic, anticarcinogenic, antitumor, and anti-obesity and hyperglycemic effects [45]. In a study by Castillo-Herrera et al. [19], methanol and acetone were used to extract limonoids from *Citrus aurantifolia* seeds, and findings showed that methanol yielded a limonoid concentration of 1.65 mg/g, while acetone resulted in a lower concentration of 0.779 mg/g.

2.4 Terpenoids

Terpenoids are a class of chemical compounds known for their low molecular weight and volatility. They are commonly present in essential oils and play significant roles, exhibiting antibacterial properties [45]. Terpenoids exert their antibacterial effects by interacting with the cell membrane, characterized by its lipophilic nature, resulting in membrane rupture. Their hydrophobic nature makes the cytoplasmic membrane a primary target for these compounds. Authors have reported that *Citrus aurantifolia* parts contain α-pinene and d-limonene [10, 15, 18]. These two compounds are characteristic components of monoterpenes, a subgroup of terpenoids, and are known for their versatile biological functions, such as antibacterial, antiseptic, and anticancer properties [45]. *Citrus aurantifolia* is rich in various types of terpenoids, including monoterpenes, alcoholic terpenes, aldehyde terpenes, ketone terpenes, and ester terpenes [45].

2.5 Flavonoids

Flavonoids, a typical class of secondary metabolites in many plant species, are prevalent among various plants, including *Citrus aurantifolia*. Specifically, *Citrus aurantifolia* is known for its rich production of flavonoids, such as 2-phenylbenzyl-pyrone. Citrus flavonoids exhibit a wide range of biological activities, acting as antioxidants, altering enzyme functions, reducing cell proliferation, and serving as antibiotics, hypoallergenic, antiulcer, antidiarrheal, and anti-inflammatory agents [45, 67]. Flavonoids such as apigenin, rutin, kaempferol, quercetin, and nobiletin have been reported in *Citrus aurantifolia* extract [45, 67]. Additionally, flavonoids like eriocitrin and hesperidin (which possess antioxidant properties) have been reported in *Citrus aurantifolia* extract [60]. Different solvents have been used to extract flavonoids from other parts of *Citrus aurantifolia* including ethanol on the leaves [42] and methanol [67, 94].

2.6 Essential Oils

Citrus aurantifolia oil, exclusively sourced from *Citrus aurantifolia* plants, constitutes a vital essential oil. The peel of *Citrus aurantifolia* yields an essential oil composed mainly of terpenes (75%), oxygenated compounds (12%), and sesquiterpenes (3%), with steam distillation being the primary extraction method [45]. The volatile constituents of *Citrus aurantifolia* include, monoterpenes such as d-limonene and terpinene, sesquiterpenes, hydrocarbons, and their oxygenated derivatives like geranial, nonanal, neryl, and linalool, as well as aldehydes, ketones, acids, alcohols, and esters [45]. Terpenes, a subgroup within this category, play a pivotal role in the essential oil's composition [45].

Furthermore, non-volatile components responsible for various qualities include fatty acids, long-chain hydrocarbons, sterols, waxes, and limonoids. Findings indicate that the primary constituents of *Citrus aurantifolia* essential oil are d-limonene (35.98%), a-pinene (9.02%), a-terpineol (8.12%), and citral (7.49%) [45, 83, 93]. Furthermore, linalyl acetate, geraniol, citral, a-pinene, a-terpineol, felandrene, sesquiphellandrene, citronellol, neryl acetate, fencone, geranyl acetate, and farnesene could also be present in the essential oil [45]. The two primary components of *Citrus aurantifolia* essential oil are citral (which gives the distinctive citrus flavor and aroma) and limonene (which can potentially inhibit the growth of harmful bacteria and fungi, owing to its potent antibacterial properties).

2.7 Vitamin C (Ascorbic Acid)

Vitamin C, also known as ascorbic acid, is crucial in supporting the body's ability to defend against illnesses and diseases [9]. Vitamin C is pivotal in helping various physiological processes, including bolstering the immune system, stimulating collagen production to maintain healthy skin and connective tissues, facilitating the absorption of dietary iron, and acting as an effective scavenger of free radicals to mitigate inflammation and oxidative stress.

Citrus aurantifolia is esteemed for its rich vitamin C content, also referred to as ascorbic acid [88]. The quality of acid *Citrus aurantifolia* fruit is primarily determined by its vitamin C content, which varies depending on the side of the tree from which the fruit is harvested [88].

In the high and mid hills zone of Nepal, fruits from the south side had the highest levels of ascorbic acid (79.6 mg and 69.9 mg), while fruits from the central part had the lowest levels (62.8 mg and 55.1 mg) [88]. Conversely, in the Terai region, fruits from the north side had the highest concentrations (58.7 mg), and fruits from the center had the lowest concentrations (41.8 mg) [88]. Al-Aamri et al. [9] reported variations in ascorbic acid content in *Citrus aurantifolia* can be attributed to factors like the nature of the fruit, geographical origin, or the extraction solvent used. The study further showed that the amount of ascorbic acid in fruit juice was considerably lower than that in the extract. The ascorbic acid content in the juice was higher in citrus species collected from Lamjung at a higher altitude (760 m) than in those from Nawalparasi at a lower height (300–400 m) [9]. This variation may be influenced by fruit position on the tree, environmental conditions, ripeness, citrus species, variety, temperature, and other factors, as ascorbic acid levels in citrus fruits are not stable and fluctuate over time [9] (Table 1).

3 Food Preservative Potentials of *Citrus aurantifolia*

Citrus aurantifolia is a desirable option as a food preservative due to its natural antibacterial and antioxidant properties. It excels in inhibiting the growth of bacteria, guarding against foodborne illnesses, and maintaining the flavor and freshness of food products. *Citrus aurantifolia* has the ability to reduce reliance on artificial preservatives aligns with the growing consumer demand for healthier and more natural food options, driven by increasing awareness of the risks associated with synthetic preservatives. *Citrus aurantifolia* is a compelling alternative to chemical preservatives with significant potential in ensuring the safety, quality, and longevity of a diverse range of food products as we explore eco-friendly and clean-label food

Table 1 Key phytochemicals found in *Citrus aurantifolia*

Phytochemical component	Description and health benefits
Vitamin C (ascorbic acid)	Strengthens the immune system Promotes collagen synthesis for healthy skin and connective tissues Enhances dietary iron absorption It acts as a free radical scavenger, reducing inflammation and oxidative stress
Flavonoids	It can impact the characteristic color of the fruit. Offers anti-inflammatory and antioxidant effects. Exhibits antiviral activities May reduce the risk of chronic diseases like cancer and heart disease
Limonoids	Abundant in the peel and seeds Under investigation for potential anticancer effects It could impede the growth of specific cancer cells
Essential oils	It is derived from the peel and leaves and is highly fragrant Primary components include citral and limonene Citral contributes to the citrus flavor and is used in aromatherapy for stress reduction Limonene has antibacterial properties and can inhibit harmful bacteria and fungi
Alkaloids	Alkaloids can have various effects on the body depending on type and concentration
Coumarins	Coumarins are associated with anti-inflammatory and anticoagulant properties
Terpenes	Terpenes are responsible for the distinctive aroma and may have potential anticancer and anti-inflammatory actions
Phenolic	Phenolic compounds in plants offer potent antioxidant properties, reducing inflammation and lowering the risk of chronic diseases, such as heart disease

Table 2 Food preservative potentials of *Citrus aurantifolia*

Aspect	Food preservative potentials of *Citrus aurantifolia*
Natural antimicrobial properties	*Citrus aurantifolia* has inherent antimicrobial properties. This makes it valuable for food preservation and extending the shelf life of various food products
Inhibition of foodborne pathogens	*Citrus aurantifolia* has antimicrobial properties that are effective against common foodborne pathogens, including *Staphylococcus aureus*, *Escherichia coli*, *Pseudomonas aeruginosa*, *Enterobacter aerogenes*, *Proteus*, and *Salmonella* species
Preservation of freshness and flavor	*Citrus aurantifolia* is rich in essential oils and phytochemicals, which preserve the sensory quality of food by inhibiting microbial growth and oxidative processes
Reduced dependence on synthetic preservatives	*Citrus aurantifolia* offers a natural alternative to synthetic preservatives, aligning with the consumer trend towards minimally processed foods with clear labels
Potential applications in different food products	*Citrus aurantifolia* is versatile and adaptable, suitable for a wide range of food products

preservation solutions. Table 2 summarizes the food preservative potentials of *Citrus aurantifolia*.

3.1 Natural Antimicrobial Properties

Citrus aurantifolia possesses inherent antimicrobial properties, making it a valuable component in food preservation [47]. Extensive research has focused on the antibacterial capabilities of essential oils derived from this *Citrus aurantifolia* species, such as citral and limonene [45]. These essential oils exhibit potent antimicrobial effects by disrupting various organisms' cellular membranes and metabolic processes. This attribute renders *Citrus aurantifolia* an ideal choice for extending the shelf life of diverse food products, thereby lowering the risk of contamination and spoilage.

A study by Olaniran et al. [75] yielded significant findings regarding using *Citrus aurantifolia* juice as a preservative. The study demonstrated that when *Citrus aurantifolia* juice was applied at a concentration of 4%, it exhibited the highest bacteriostatic and fungistatic efficacy. The authors concluded that *Citrus aurantifolia* juice was an effective preservative and demonstrated antimicrobial properties. This suggests that *Citrus aurantifolia* juice has the potential to be used as an alternative to chemical preservatives, offering a natural and effective means of preserving food products.

3.2 Inhibition of Foodborne Pathogens

Foodborne pathogens, bacteria, parasites, and viruses that can induce illness when ingested via contaminated food substantially threaten food safety and public health. Fortunately, several species of plants and plant-derived compounds harbor essential antimicrobial properties that can be harnessed to inhibit the proliferation of these diseases. *Citrus aurantifolia*'s antimicrobial properties make it effective in inhibiting the growth of common foodborne pathogens, including *Staphylococcus aureus, Escherichia coli, Pseudomonas aeruginosa, Enterobacter aerogenes, Proteus,* and *Salmonella* species [47]. Therefore, incorporating *Citrus aurantifolia* extracts or essential oils into food products can help reduce the risk of foodborne illnesses, ensuring food safety for consumers.

3.3 Preservation of Freshness and Flavor

Citrus aurantifolia exhibits the potential to serve as a natural preservative, capable of preserving the flavor and freshness of food. This citrus fruit is rich in essential oils and phytochemicals, both of which contribute to protecting the sensory quality of food products by inhibiting microbial growth and oxidative processes. This quality is particularly advantageous in maintaining the aroma, taste, and texture of a diverse range of culinary items, from sauces and dressings to baked goods and dairy products [49].

3.4 Reduced Dependence on Synthetic Preservatives

The widespread use of artificial chemical preservatives in the food industry has raised concerns about potential adverse health effects. In response, *Citrus aurantifolia* offers a natural alternative to synthetic preservatives, aligning with the increasing consumer preference for minimally processed food products with clear labels. By reducing our dependence on artificial preservatives, *Citrus aurantifolia* contributes to creating healthier and more natural food choices, reflecting the broader trend of favoring unprocessed options.

3.5 Potential Applications in Different Food Products

Citrus aurantifolia holds promise as a versatile food preservative suitable for a wide range of food products. Its adaptability allows for diverse applications, including beverages such as zobo [49], condiments, sauces, baked goods, dairy items, and even the preservation of fruits and vegetables in their fresh state. This adaptability makes it a valuable asset, benefiting businesses seeking natural and safe food preservation methods and consumers seeking such options.

4 Pharmacological Values of *Citrus aurantifolia*

Plants have played an indispensable role in managing and treating an array of conditions, encompassing peptic/gastric ulcers, malaria, anemia, diarrhea, diabetes, obesity, cancer, inflammation, hypertension, rheumatism, microbial infections, and a plethora of other ailments. One such plant that stands out is *Citrus aurantifolia*.

Kawaii et al. [61] have reported that all parts of *Citrus aurantifolia*, including the stem, leaves, roots, flowers, and fruit, can be employed to treat microbial infections. Different reviews on the traditional uses of *Citrus aurantifolia* highlight its antibacterial, antidiabetic, antifungal, antihypertensive, anti-inflammatory, antilipidemia, antioxidant, anti-parasitic, and antiplatelet activities [71]. Additionally, it has been documented for its efficacy in treating cardiovascular, hepatic, osteoporosis, and urolithiasis diseases, as well as its potential as a fertility promoter [36, 74, 86, 96, 100].

The juice and essential oil of *Citrus aurantifolia* have found widespread application in the pharmaceutical, food, and cosmetic industries due to their medicinal properties and fragrant characteristics. Aibinu et al. [6] have reported that when the juice is combined with sugar and palm oil or honey, it forms a potent cough-relieving mixture, further underscoring the versatile utility of this plant in healthcare and wellbeing. This section focuses on the pharmacological values of *Citrus aurantifolia* and its potential contributions to human health (Table 3).

Table 3 Pharmacological properties and benefits of *Citrus aurantifolia*

Pharmacological property	Properties and uses of *Citrus aurantifolia*
Antioxidant	*Citrus aurantifolia* is a rich source of vitamin C, offering antioxidant protection against oxidative stress and free radical damage, reducing the risk of chronic diseases
Anti-inflammatory	Phytochemicals, including flavonoids, in *Citrus aurantifolia* exhibit anti-inflammatory effects, potentially alleviating symptoms in conditions like arthritis and inflammatory bowel disease
Antimicrobial	Essential oils in *Citrus aurantifolia*, such as citral and limonene, have strong antimicrobial properties and are effective against many pathogenic bacteria and fungi
Anticancer	The limonoids and coumarins in *Citrus aurantifolia* have demonstrated anticancer potential by inhibiting the growth of various cancer cells
Stress and anxiety	Essential oils from *Citrus aurantifolia* are used in aromatherapy to reduce stress and anxiety
Antidiabetic	*Citrus aurantifolia* extracts can be used in managing diabetes by modulating glycogen metabolism
Anti-plasmodial	*Citrus aurantifolia* leaf extracts have anti-plasmodial activity, potentially contributing to the treatment of malaria
Analgesic	*Citrus aurantifolia* leaves have been found to alleviate pain without adverse effects, offering a natural analgesic option
Anti-cholinesterase	*Citrus aurantifolia* methanol extracts exhibit anti-cholinesterase activity by inhibiting acetylcholinesterase and butyrylcholinesterase (BChe), potentially impacting cognitive health
Anti-modulatory	*Citrus aurantifolia* juice has been shown to modulate the immune response
Anti-diarrhea	*Citrus aurantifolia* has been traditionally used to alleviate diarrhea, with bioactive compounds potentially reducing symptoms of indigestion.
Antiulcer	Specific components of *C. aurantifolia*, such as citral and 4-hexen-3-one, have demonstrated inhibitory effects on *Helicobacter pylori* growth, which could be valuable in addressing ulcer-related conditions
Anthelmintic	*Citrus aurantifolia* peel constituents have shown the ability to inhibit the reproduction of parasites such as certain helminths
Antifertility	Some women have used *Citrus aurantifolia* juice as a contraceptive method due to its acidity and phytochemical constituents that can destroy human sperm cells

4.1 Antioxidant Properties

One of the critical pharmacological values of *Citrus aurantifolia* lies in its antioxidant properties. The fruit is an abundant source of vitamin C, a well-known antioxidant that helps combat oxidative stress and free radical damage. These harmful molecules are implicated in various chronic diseases, including cardiovascular disease, cancer, and aging. Regular consumption of *Citrus aurantifolia* can help protect cells and tissues from oxidative damage, reducing the risk of these diseases.

Numerous studies have highlighted using essential oils derived from *Citrus aurantifolia* leaves as natural antioxidants [71], particularly in the pharmaceutical and food industries, owing to their relatively safe toxicological profiles. In a study by Al-Aamri et al. [9], it was found that 31 essential oils sourced from citrus fruits exhibited superior antioxidant properties compared to Trolox. Authors have documented that the presence of D-limonene in both the leaves and rind of *Citrus aurantifolia* accounts for its antioxidant properties and the ability to combat cell damage induced by free radicals [39, 65, 81]. Furthermore, the flavonoids found in *Citrus aurantifolia* have demonstrated antioxidant and anticancer attributes by inhibiting cell division in cancerous cells, as noted in the research conducted by Abubakar et al. [3].

4.2 Anti-inflammatory Effects

Inflammation is a natural defense mechanism in the human body designed to eliminate or restrict the spread of harmful agents. While it is a typical response to tissue injury, it can become uncontrolled in chronic autoimmune conditions like rheumatoid arthritis, Crohn's disease, and allergic reactions such as asthma and anaphylactic shock [8, 84]. The phytochemical constituents of *Citrus aurantifolia*, such as flavonoids, have demonstrated anti-inflammatory properties [71]. Inflammation is a central component of many chronic conditions, including arthritis and inflammatory bowel disease. The natural compounds in *Citrus aurantifolia* can help mitigate inflammation, relieving individuals suffering from these ailments. *Citrus aurantifolia* leaves and rind contain D-limonene as a significant constituent, which has been reported to reduce inflammatory markers associated with conditions characterized by chronic inflammation, such as osteoarthritis.

4.3 Antimicrobial Activity

Plant has emerged as a source of new antimicrobial probability due to their ability to inhibit gram-positive and harmful microbes [50–54]. The essential oils found in *Citrus aurantifolia*, particularly citral and limonene, exhibit potent antimicrobial properties. These compounds have traditionally been used to inhibit the growth of pathogenic bacteria and fungi. As a result, *Citrus aurantifolia* has applications in the field of pharmacology as a potential source of natural antimicrobial agents. This can be particularly valuable when antibiotic resistance is a growing concern.

Numerous reports have highlighted the potent antimicrobial properties of *Citrus aurantifolia* juice against a broad spectrum of pathogenic microorganisms (bacterial and fungal) [7, 31, 32, 63, 77, 87, 95, 97], including *V. cholerae*, *S. aureus*, *K. pneumonia*, *S. pyogenes*, *Actinobacillus* sp., *Citrobacter* sp., *Shigella* sp., as documented by Jayana et al. [59] and Adebayo-Tayo et al. [4]. The effectiveness of this antimicrobial activity can be attributed mainly to the complex chemical composition of the juice, which encompasses various chemical classes such as terpenes, aldehydes, alcohols, esters, phenols, ethers, and ketones, as highlighted by Al-Aamri et al. [9]. Moreover, extracts derived from *Citrus aurantifolia* fruit have been found to exhibit antimicrobial properties against microorganisms like *Candida albicans*, *Aspergillus niger*, and *Aspergillus fumigatus*, as reported by Aibinu et al. [6] and Abubakar et al. [2].

In addition to the fruit extracts, the ethanolic extracts obtained from *Citrus aurantifolia* leaves have demonstrated bactericidal effects against a range of bacterial species, including *Bacillus subtilis*, *Escherichia coli*, *Salmonella*, and *Streptococcus feacalis*, as observed by Oboh and Abulu [73].

These antimicrobial properties are attributed to bioactive phytochemicals in different parts of *Citrus aurantifolia*. Flavonoids, in particular, have been recognized for their influence on arachidonic acid metabolism, enabling them to combat a wide array of bacteria by forming complexes with extracellular and soluble proteins within bacterial cell walls, as explained by Adebayo-Tayo et al. [3].

4.4 Anticancer Potential

Cancer is a primary global public health concern, the second leading cause of death. In the United States in 2015, there were approximately 1.6 million new cancer cases, resulting in roughly 500 thousand cancer-related deaths [89]. Perhaps one of the most exciting pharmacological values of *Citrus aurantifolia* is its potential for cancer prevention and treatment [71]. The limonoids in this citrus fruit have shown anticancer properties by inhibiting the growth of cancer cells. These compounds may help reduce the risk of cancer development and support treating various cancer types. An intriguing study by Patil et al. [78, 79, 80] revealed the anticancer potential of *Citrus aurantifolia* fruit extract. They found that at a concentration of 100 μg/ml, this extract inhibited the growth of human colon SW-480 cancer cells by an impressive 78% after 48 h of exposure. This remarkable effect can be attributed to the fruit extract's substantial amounts of limonene and dihydrocarvone. Additionally, when a 25 μM concentration of *Citrus aurantifolia* peel extract was used, it displayed the capability to inhibit the growth of colon SW-480 cancer cells in humans by 67% after 72 h of exposure. This inhibition is attributed to three key coumarins: 5-geranyloxy-7-methoxycoumarin, *Citrus aurantifolia* tin, and isopimpinellin found in the peel extract.

Patil et al. [78, 79] reported that 100 μg/ml of *Citrus aurantifolia* juice extract effectively halted approximately 73–89% of pancreatic Panc-28 cancer cell growth in humans after 96 h of exposure. This remarkable effect can be attributed to limonoid substances, such as limonexic acid, isolimonexic acid, and limonin in the juice extract. Meanwhile, *Citrus aurantifolia* seeds extract exhibited the ability to stop the growth of pancreatic Panc-28 cancer cells in humans with an inhibitory concentration of 50% (IC50) ranging from 18 to 42 μM after 72 h of exposure. This inhibition was achieved through compounds like limonin, limonexic acid, isolimonexic acid, β-sitosterol glucoside, and limonin glucoside.

In another study, Gharagozloo et al. [35] reported that *Citrus aurantifolia* fruit juice extract from Iran, at concentrations ranging from 125 to 500 μg/ml, effectively inhibited the growth of breast MDA-MB-453 cancer cells after 24 h of exposure. Furthermore, Adina et al. [4] found that a concentration of 6 and 15 μg/ml of *Citrus aurantifolia* peel extract from Indonesia was capable of inhibiting the growth of breast MCF-7 cancer cells, targeting different phases of the cell cycle, specifically G1 and G2/M, respectively, after 48 h of exposure.

Additionally, Castillo-Herrera et al. [19] reported that limonin extract derived from *Citrus aurantifolia* seeds in Mexico displayed remarkable inhibitory effects on the growth of L5178Y lymphoma cells, with an IC50 ranging from 8.5 to 9.0 μg/ml. These findings underscore the promising potential of *Citrus aurantifolia* extracts in the fight against various types of cancer.

4.5 Stress and Anxiety Relief

Anxiety, a common mental health condition often associated with comorbidity, poses a significant psychological, social, and economic burden and an elevated risk of physical illness [62, 76]. The pharmacological treatment of anxiety and mood disorders includes drugs like beta blockers, benzodiazepines, monoamine oxidase inhibitors, tricyclic antidepressants, and selective serotonin reuptake inhibitors. Still, they have notable limitations, such as comorbid psychiatric disorders, acute poisoning, and addiction [91]. The essential oils from *Citrus aurantifolia* are often used in aromatherapy for their calming and stress-reducing effects. These essential oils can be explored in pharmacology as a natural alternative to pharmaceutical anxiolytics. The aroma of *Citrus aurantifolia* can help alleviate stress and anxiety, promoting mental well-being. Furthermore, Guzmán et al. [37] reported that in China, Brazil, and Mexico, the fruit peel and juice of *Citrus aurantifolia* have been employed as tranquilizers and anxiolytic medications.

4.6 Antidiabetic Properties

Diabetes mellitus, a metabolic disorder characterized by elevated blood sugar levels, stems from issues with insulin secretion or action, affecting the metabolism of carbohydrates, fats, and proteins [23]. Type 1 diabetes results from the autoimmune destruction of pancreatic β-cells, necessitating external insulin supplementation for patients to maintain their well-being [20]. Conversely, Type 2 diabetes is associated with dysfunctional insulin levels that impair glucose uptake.

Type 2 diabetes, which can manifest as a chronic or mild condition, can eventually lead to vascular complications, dysfunction, and an increased risk of coronary artery and peripheral vascular diseases. This ailment is a pressing global public health concern, as the World Health Organization reports that approximately 422 million individuals have diabetes, with around 90% of these cases being Type 2 diabetes [44].

Ethnobotanical research has revealed a notable preference for extracts from the Rutaceae family in diabetes management. For instance, the essential oil of *Citrus aurantifolia* has shown potential in inducing apoptosis in pancreatic cell lines and reducing hyperglycemia in alloxan-induced rats. Furthermore, it may possess anti-diabetic properties through its ability to modulate glycogen metabolism [44]. Narang and Jiraungkoorskul [71] have reported the antidiabetic potential of *Citrus aurantifolia*.

4.7 Antidepressant

In the current era, mental and neurological disorders stand out as one of the foremost global health concerns. Among these, anxiety and mood disorders are particularly prevalent and carry a heavy psychological and socio-economic burden, increasing the risk of physical ailments. Conventional medications for anxiety and mood disorders include beta-blockers, benzodiazepines, monoamine oxidase inhibitors, tricyclic antidepressants, and selective serotonin reuptake inhibitors. However, these medications have limitations, such as the potential for co-existing psychiatric disorders, acute toxicity, drug interactions, and the development of physical dependence, leading to intolerable side effects [91]. As a result, there has been a quest for alternative and innovative treatments, and promising results have emerged from using plants. For instance, Chinese, Brazilians, and Mexicans have employed the fruit peel and juice extract of *Citrus aurantifolia* as tranquilizers and anxiolytic therapies. At the same time, the leaves can be used as tranquilizers. Additionally, the flowers of *Citrus aurantifolia* are used in Mexico to alleviate anxiety and nervousness. In some parts of Latin America and Africa, an infusion of dried *Citrus aurantifolia* flowers is a nerve-soothing sedative [91].

4.8 Anti-plasmodial

Turning to anti-plasmodial treatments, malaria, primarily transmitted through the bite of female Anopheles mosquitoes carrying *Plasmodium falciparum*, poses a significant global health threat. The parasite is introduced into human red blood cells during the mosquito's bite, leading to a malarial infection once it multiplies in the bloodstream. Malaria claims the lives of a staggering live in many region of the word with Africa being the most endemic region of the diseases [17, 26, 28, 48, 72, 101]. Furthermore, malaria kills 235,000 to 639,000 individuals worldwide annually [43, 70, 99]. Furthermore, resistance to conventional antimalarial drugs has been documented, a significant breakthrough emerged from Chinese researchers who discovered the efficacy of Artemisinin (Qinghao), a traditional Chinese medicine, in combatting drug-resistant forms of malaria infection [30]. In Africa, *Citrus aurantifolia* has been employed to treat various illnesses, with the fruit juice being recognized for its ability to alleviate swelling resulting from mosquito bites. Furthermore, Ettebong et al. [30] have reported that *Citrus aurantifolia* leaf extract exhibits substantial anti-plasmodial activity in vivo.

4.9 Analgesic

The management of analgesic conditions often necessitates synthetic molecules, which, unfortunately, can lead to various adverse effects on the body. These effects may vary in severity and impact critical systems like the kidneys and digestive tract. Plant-based treatments have gained attention as an alternative approach to addressing analgesic ailments [21]. In a recent study by Al-Snafi [12], it was found that administering an extract of *Citrus* species to mice effectively alleviated their analgesic condition. Significantly, this natural remedy exhibited no adverse effects and did not result in mortality among the test subjects.

4.10 Anticholinesterase Activity

Alzheimer's disease is a progressive neurological disorder characterized by cognitive deficits and behavioral abnormalities. It has been well-documented that the progression of Alzheimer's disease is linked to the presence of reactive oxygen species, emphasizing the importance of using antioxidants to reduce disease advancement and minimize neuronal degeneration. Additionally, a cholinergic deficit resulting from basal forebrain degeneration calls for strategies to enhance acetylcholine production – an essential neurotransmitter for cognitive functions – to restore cognitive deficits. This approach is known as the cholinergic hypothesis [14, 85, 92].

Notably, *Citrus aurantifolia* methanol extracts in n-hexane fractions have significant anti-cholinesterase effects. This effect is achieved through the inhibition of acetylcholinesterase (ACHe) and butyrylcholinesterase (BChe), as demonstrated in plasma assays, as reported by Loizzo et al. [67].

4.11 Anti-modulatory Activity

Citrus aurantifolia, commonly known as lime, has been found to exhibit immune-modulatory activity, as demonstrated in a trial utilizing mitogen-activated culture mononuclear cells. In this study conducted by Gharagozloo and Ghaderi [34], the juice of *Citrus aurantifolia* was assessed for its impact on the immune response.

The results indicated that the application of 250 and 500 micrograms of *Citrus aurantifolia* juice led to a significant inhibition of the proliferation of phytohemagglutinin-activated mononuclear cells. This suggests that *Citrus aurantifolia* juice can modulate the immune response, potentially by regulating the activation and proliferation of immune cells.

This finding underscores the potential immunomodulatory properties of *Citrus aurantifolia*, which may have implications for its use in various contexts, such as supporting immune health or developing therapies to regulate immune responses. Further research and exploration of these properties could provide valuable insights into the potential applications of *Citrus aurantifolia* juice in immunology and health.

4.12 Anti-diarrhea Activity

Diarrhea is a condition that can pose a significant threat to patients, potentially leading to severe pain and, in extreme cases, even death if it remains untreated or is inadequately managed [25, 52]. While diarrhea can affect individuals of all age groups, it is especially concerning in infants and children, necessitating special attention. Several factors are known to be direct or indirect triggers for diarrhea, including bacterial and viral infections, allergies, malabsorption issues, and food poisoning [25].

Citrus aurantifolia has been traditionally used to support digestive health. Its bioactive compounds may aid in reducing symptoms of indigestion and promoting overall gastrointestinal well-being. This pharmacological value is particularly relevant in digestive disorders and conditions. He et al. [41] conducted a study revealing that pre-treatment with dried fruits of *Citrus aurantifolia* significantly alleviated 2,4,6-trinitrobenzene sulfonic acid-induced diarrhea in rats. This finding suggests a potential anti-diarrhea effect associated with *Citrus aurantifolia*, offering promise for managing this distressing condition.

4.13 Antiulcer

Helicobacter pylori, a prominent pathogen associated with developing various ulcer-related conditions such as gastritis, peptic ulcers, gastric adenocarcinoma, and mucosa-associated lymphoid tissue lymphoma, has become widely prevalent in the human population [11, 27]. This is primarily due to its high level of antibiotic resistance, which poses a significant challenge to the effective treatment and eradication of *Helicobacter pylori*. Consequently, there is a growing need for new alternatives to address this issue.

In regions with a high prevalence of *Helicobacter pylori*, citrus fruit has been reported to reduce the occurrence of gastric cancer [16, 64]. Lee et al. [64] further demonstrated that specific components of *Citrus aurantifolia*, such as citral and 4-hexen-3-one, can inhibit the growth of triple drug resistant *Helicobacter pylori* strains. Notably, the 4-hexen-3-one constituent exhibited a more substantial inhibitory effect on the urease activity of triple drug resistant *Helicobacter pylori* strains as its concentration increased, indicating the potential of these citrus components as a promising approach in addressing *Helicobacter pylori*-related conditions.

4.14 Anthelmintic

Helminthosis is a parasitic infection known for its detrimental impact on animal health, causing growth impairment, reducing survival rates, hindering reproductive performance, and eventually leading to the death of infected animals. This poses a substantial economic burden on humans, especially livestock farmers. Currently, chemotherapy is the most effective means of treating helminthosis, but it has significant cost implications and raises concerns regarding developing resistant helminth parasites.

While the fruit peels of *Citrus aurantifolia* have gained recognition for their medicinal properties in managing various illnesses and infections, they are rarely included in the diet of livestock [29, 91]. Enejoh et al. [29] conducted a study revealing that the constituents found in *Citrus aurantifolia* peel exhibit a noteworthy ability to inhibit the hatching of *Heligmosomoides bakeri* eggs, a trichostrongyloid parasite naturally occurring in wild mice. Additionally, these constituents were effective in killing the helminth larvae in vitro. This suggests a potential role for *Citrus aurantifolia* peel in combating helminth infections and highlights the significance of exploring this natural remedy in livestock health and management.

4.15 Antifertility

The ability to reproduce offspring is a fundamental and vital aspect of life. However, various health conditions can interfere with this function. When individuals face challenges in their ability to conceive and reproduce, it can give rise to multiple concerns and potentially lead to additional health-related issues, such as anxiety and depression. Interestingly, reports have shown that infertility can also be induced intentionally by using various plant extracts by individuals who do not wish to have children. For example, the juice of *Citrus aurantifolia*, commonly known as *Citrus aurantifolia*, has been employed by African women as an alternative contraceptive method, along with douching, to prevent pregnancy and sexually transmitted diseases [74].

Furthermore, in vitro studies conducted by Okon and Etim [74], Hair et al. [38], Masud et al. [69], and Longo et al. [68] have unveiled antifertility findings regarding the juice of *Citrus aurantifolia*. This citrus juice can destroy human immunodeficiency virus (HIV) and human sperm cells. This effect is attributed to the juice's high acidity and phytochemical constituents. These studies shed light on the potential applications and consequences of using *Citrus aurantifolia* juice in the context of antifertility and protection against certain health risks.

In summary, *Citrus aurantifolia* offers a range of pharmacological values due to its rich composition of bioactive compounds. Its antioxidant, anti-inflammatory, antimicrobial, and anticancer properties make it a valuable addition to pharmacological research in preventing and treating chronic diseases. Additionally, its stress-relieving and digestive health benefits contribute to its potential role in enhancing overall well-being. As we continue to explore the health benefits of natural products, *Citrus aurantifolia* stands as a promising candidate in the field of pharmacology for its diverse and potent pharmacological values.

5 Safety Considerations of *Citrus aurantifolia* Juice Extracts

Even though juice extracts from *Citrus aurantifolia* can enhance culinary capabilities and offer potential health benefits, it is essential to be aware of the associated risks. Some of the related risks of *Citrus aurantifolia* are as follows:

allergies, stomach sensitivity, dental health, photosensitivity, medication interactions, hydration, etc. (Table 4). Therefore, achieving a safe and enjoyable experience hinges on consuming these extracts in moderation and making well-informed choices.

6 Research Needs on the Juice Extracts of *Citrus aurantifolia*

Citrus aurantifolia juice extracts hold significant promise as a valuable natural resource with various potential health benefits. To harness research, it is essential to encompass the identification and isolation of bioactive compounds, followed by comprehensive clinical trials and safety assessments. Researching this will advance the understanding of the natural compounds in plants and pave the way for innovative preventive and therapeutic applications in nutrition, medicine, and food science. Table 5 shows the potential research needs on the juice extracts of *Citrus aurantifolia*.

Table 4 Safety considerations of *Citrus aurantifolia* juice extracts

Consideration	Safety considerations of *Citrus aurantifolia* juice extracts
Allergies	*Citrus aurantifolia* can trigger allergic reactions, ranging from mild skin rashes to severe anaphylaxis. Therefore, individuals with citrus allergies should use extreme caution when consuming *Citrus aurantifolia* juice
Stomach sensitivity	*Citrus aurantifolia* juice contains high acidity, which can irritate the stomach and digestive tract, potentially causing discomfort, heartburn, or acid reflux. Therefore, those with digestive disorders should consume *Citrus aurantifolia* juice cautiously
Dental health	The acidity in *Citrus aurantifolia* juice can erode tooth enamel, increasing the risk of dental issues like tooth sensitivity and cavities. Therefore, rinsing with water after consumption is advised, and avoid brushing teeth immediately afterward
Photosensitivity	*Citrus aurantifolia* juice may contain compounds that make the skin more sensitive to sunlight, potentially leading to phytophotodermatitis and skin damage when exposed to the sun. Therefore, caution must be exercised when applying *Citrus aurantifolia* juice to the skin or ingesting it before sun exposure
Medication interactions	*Citrus aurantifolia* juice can interact unfavorably with certain medications, altering their effectiveness or causing side effects
Hydration and electrolytes	*Citrus aurantifolia* juice has a natural diuretic effect due to its potassium content, which can lead to dehydration and electrolyte imbalances when consumed excessively
Pesticide residues	*Citrus aurantifolia* peels may carry pesticide residues that can transfer into the juice from the soil
Safe handling and storage	Proper handling and storage practices are essential for food safety

Table 5 Research needs on the juice extracts of *Citrus aurantifolia*

Research area	Potential research needs on the juice extracts of *Citrus aurantifolia*
Phytochemical profiling and isolation	Comprehensive profiling of phytochemicals in *Citrus aurantifolia* juice extracts Isolation and purification of these compounds to understand their specific health contributions
Antioxidant effects	Research on how *Citrus aurantifolia* reduces oxidative damage and inflammation
Antibacterial and antifungal activities	Studying the extent to which *Citrus aurantifolia* juice extracts inhibit the growth of pathogenic microorganisms – potential applications as natural food preservation and antibiotic alternatives
Bioavailability and pharmacokinetics	Research on how the body absorbs, distributes, metabolizes, and eliminates the bioactive constituents of the juice. Essential for optimal dosage and administration
Clinical trials and health uses	Conducting well-designed clinical trials to assess the effectiveness of *Citrus aurantifolia* juice extracts in preventing and treating conditions such as cancer and chronic inflammatory disorders
Studies on the product safety and toxicology	Comprehensive toxicology studies to establish safe consumption dose levels for *Citrus aurantifolia* juice extracts
Synergistic effects with other components	Study how *Citrus aurantifolia* juice extracts interact with other natural compounds and medications
Research into sustainable cultivation practices and processing technologies	Research into sustainable cultivation methods and processing technologies to preserve the integrity of bioactive components and ensure a consistent supply of high-quality raw materials

7 Conclusion

Investigating *Citrus aurantifolia* and its juice extracts reveals a multifaceted natural resource with significant potential in nutrition, medicine, and food preservation. The phytochemical constituents present in *Citrus aurantifolia* contribute to its distinct properties, and research in several areas is vital to unlock its full potential. Firstly, the phytochemical profiling and isolation of specific compounds, such as vitamin C, flavonoids, and essential oils, form a fundamental research need. Understanding the bioactive compounds in *Citrus aurantifolia* enables us to harness their potential in various applications. Moreover, the antioxidant properties of *Citrus aurantifolia* juice extracts are a noteworthy avenue of research. Exploring these properties in the context of oxidative stress-related diseases has promising implications for preventive and therapeutic measures. These extracts' antimicrobial and antifungal activities are another significant area of study. These natural extracts show potential for use as food preservatives and alternatives to synthetic antibiotics. Bioavailability and pharmacokinetics studies are essential to optimize the health benefits of *Citrus aurantifolia*. Understanding how the body processes these compounds ensures effective administration. To validate the practical health applications of *Citrus aurantifolia* juice extracts, transitioning from in vitro and animal studies to well-designed clinical trials is necessary. This research can help us unlock their potential in preventing and managing various health conditions. Additionally, safety and toxicology studies are crucial to establishing the safe dosage levels of these extracts, assuring consumers and healthcare providers of their safety. Considering potential synergistic effects with other compounds, particularly in combination therapy, offers exciting prospects for enhancing therapeutic outcomes. Finally, sustainable cultivation and efficient processing methods should be researched to ensure a consistent supply of high-quality raw materials while maintaining the integrity of the bioactive compounds.

References

1. Abdallah M, Ahmed I (2018) Comparative study of antibacterial and phytochemical screening of ethanolic extracts of *Citrus aurantifolia* and *Psidium guajava* on some clinical isolates. East Afr Sch J Med Sci 1:70–76
2. Abubakar ZU, Sani TT, Muhammad A (2018) Antibacterial activity of *Citrus aurantifolia* leaves extracts against some enteric bacteria of public health importance. Modern Approac Mater Sci 1(2):33–38
3. Adebayo-Tayo B, Akinsete TO, Odeniyi OA (2016) Phytochemical composition and comparative evaluation of antimicrobial activities of the juice extract of *Citrus aurantifolia* and its silver nanoparticles. Niger J Pharm Res 12(1):59–64
4. Adina AB, Goenadi FA, Handoko FF, Nawangsari DA, Hermawan A, Jenie RI et al (2014) Combination of ethanolic extract of *Citrus aurantifolia* peels with doxorubicin modulate cell cycle and increase apoptosis induction on MCF-7 cells. Iran J Pharm Res 13:919–926
5. Ahmed M, Saeid A (2021) Citrus-research, development and biotechnology. IntechOpen, London. Citrus Fruits: Nutritive Value and Value-Added Products; p. 13
6. Aibinu I, Adenipekun T, Adelowotan T, Ogunsanya T, Odugbemi T (2007) Evaluation of the antimicrobial properties of different parts of *Citrus aurantifolia* (lime fruit) as used locally. Afr J Tradit Complement Alternat Med 4(2):185–190
7. Ajayi-Moses OB, Ogidi CO, Akinyele BJ (2019) Bioactivity of *Citrus* essential oils (CEOs) against microorganisms associated with spoilage of some fruits. Chem Biol Technol Agric 6:22
8. Akdis M, Aab A, Altunbulakli C, Azkur K, Costa RA, Crameri R, Duan S, Eiwegger T, Eljaszewicz A, Ferstl R, Akdis CA (2016) Interleukins (from IL-1 to IL-38), interferons, transforming growth factor β, and TNF-α: receptors, functions, and roles in diseases. J Allergy Clin Immunol 138(4):984–1010
9. Al-Aamri MS, Al-Abousi NM, Al-Jabri SS, Alam T, Khan SA (2018) Chemical composition and in-vitro antioxidant and antimicrobial activity of the essential oil of *Citrus aurantifolia L.* leaves grown in Eastern Oman. J Taibah Univ Med Sci 13(2):108–112
10. Al-Breiki AM, Al-Brashdi HM, Al-Sabahi JN, Khan SA (2018) Comparative GC-MS analysis, in-vitro antioxidant and antimicrobial activities of the essential oils isolated from the peel of Omani lime. Chiang Mai J Sci 45(4):1782–1795
11. Allagoa DO, Eledo BO, Dunga KE, Izah SC (2018) Assessment of some immune system related parameters on helicobacter pylori infected students in a Nigerian tertiary educational institution. Int J Gastroenterol 2(2):24–27

12. Al-Snafi AE (2016) Nutritional value and pharmacological importance of citrus species grown in Iraq. IOSR J Pharm 6(8):76–108
13. Atanasov AG, Waltenberger B, Pferschy-Wenzig EM, Linder T, Wawrosch C, Uhrin P, Temml V, Wang L, Schwaiger S, Heiss EH, Stuppner H (2015) Discovery and resupply of pharmacologically active plant-derived natural products: a review. Biotechnol Adv 33(8): 1582–1614
14. Atta-ur-Rahman, Choudhary MI (2001) Bioactive natural products as a potential source of new pharmacophores. A theory of memory. Pure Appl Chem 73:555–560
15. Azghar A, Dalli M, Azizi SE, Benaissa EM, Lahlou YB, Elouennass M, Maleb A (2023) Chemical composition and antibacterial activity of citrus peels essential oils against multidrug-resistant bacteria: a comparative study. J Herbal Med 42:100799
16. Bae JM, Kim EH (2016) Dietary intakes of citrus fruit and risk of gastric cancer incidence: an adaptive meta-analysis of cohort studies. Epidemiol Health 38:e2016034
17. Bassey SE, Izah SC (2017) Some determinant factors of malaria prevalence in Nigeria. J Mosquito Res 7(7):48–58
18. Benayad O, Bouhrim M, Tiji S, Kharchoufa L, Addi M, Drouet S, Hano C, Lorenzo JM, Bendaha H, Bnouham M, Mimouni M (2021) Phytochemical profile, α-glucosidase, and α-amylase inhibition potential and toxicity evaluation of extracts from *Citrus aurantium* (L) peel, a valuable by-product from northeastern Morocco. Biomol Ther 11(11):1555
19. Castillo-Herrera G, Farias-Alvarez L, Garcia-Fajardo J, Delgado-Saucedo J, Puebla-Perez A, Lugo-Cervantes E (2015) Bioactive extracts of *Citrus aurantifolia* swingle seeds obtained by supercritical CO2 and organic solvents comparing its cytotoxic activity against L5178Y leukemia lymphoblasts. J Supercrit Fluids 101:81–86
20. Clark AL, Urano F (2016) Endoplasmic reticulum stress in beta cells and autoimmune diabetes. Curr Opin Immunol 43:60–66
21. Das K, Tiwari RKS, Shrivastava DK (2010) Techniques for evaluation of medicinal plant products as antimicrobial agent: current methods and future trends. J Med Plants Res 4(2): 104–111
22. de Albuquerque UP, Hanazaki N (2009) Five problems in current ethnobotanical research—and some suggestions for strengthening them. Hum Ecol 37:653–661
23. Deshpande TA, Isshak M, Priefer R (2020) PTP1B inhibitors as potential target for type II diabetes. Curr Res Diabetes Obes J 14:555876
24. Dey P, Kundu A, Kumar A, Gupta M, Lee BM, Bhakta T, Dash S, Kim HS (2020) Analysis of alkaloids (indole alkaloids, isoquinoline alkaloids, tropane alkaloids). In: Recent advances in natural products analysis. Elsevier, pp 505–567
25. Ekawati ER, Darmanto W (2019) Lemon (Citrus limon) juice has antibacterial potential against diarrhea-causing pathogen. IOP Conf Ser Earth Environ Sci 217:1–6
26. Eledo BO, Izah SC (2018) Studies on some haematological parameters among malaria infected patients attending a tertiary Hospital in Nigeria. Open Access Blood Res Transfusion J 2(3):555586
27. Eledo BO, Allagoa DO, Onuoha EC, Okamgba OC, Ihedioha AU, Izah SC, Orutugu LA (2017) Effect of helicobacter pylori on some haemostatic parameters among students of a tertiary institution in Nigeria. Clin Biotechnol Microbiol 1(5):219–224
28. Eledo BO, Allagoa DO, Egwugha CT, Dunga KE, Izah SC (2019) Some haemostatic indicators among malaria infected adolescents attending a Nigerian university teaching hospital. Open Access Blood Res Transfusion J 3(1):555602. https://doi.org/10.19080/OABTJ.2019.03.555602
29. Enejoh O, Ogunyemi I, Bala M, Oruene I, Suleiman M, Ambali S (2015) Ethnomedical importance of *Citrus aurantifolia* (Christm) Swingle. Pharm Innov J 4:1–6
30. Ettebong E, Ubulom P, Etuk A (2019) Antiplasmodial activity of methanol leaf extract of *Citrus aurantifolia* (Christm) swingle. J Herbal Med Pharmacol 8(4):274–280

31. Flores RC, Audicio N, Sanz MK, Ponzi M (2014) Antibacterial activity of lime (Citrus x aurantifolia) essential oil against listeria monocytogenes in tyndallised apple juice. Rev Soc Venez Microbiol 34(1):10–14
32. George-Okafor UO, Anosike EE (2010) The potentials of lime (*Citrus aurantifolia*) for improving traditional corn fermentation for probiotic lactic acid bacterial proliferation. Pak J Nutr 9(12):1207–1213
33. Gewali MB, Awale S (2008) Aspects of traditional medicine in Nepal. Institute of Natural Medicine University of Toyama, pp 140–142
34. Gharagozloo M, Ghaderi A (2001) Immunomodulatory effect of concentrated lime juice extract on activated human mononuclear cells. J Ethnopharmacol 77(1):85–90
35. Gharagozloo M, Doroudchi M, Ghaderi A (2002) Effects of *Citrus aurantifolia* concentrated extract on the spontaneous proliferation of MDA-MB-453 and RPMI-8866 tumor cell lines. Phytomedicine 9:475–477
36. Gokulakrishnan K, Senthamilselvan P, Sivakumari V (2010) Regenerating activity of *Citrus aurantifolia* on paracetamol induced hepatic damage. Asian J Biol Sci 4:176–179
37. Guzmán Gutiérrez SL, Chilpa RR, Jaime HB (2014) Medicinal plants for the treatment of "nervios", anxiety, and depression in Mexican Traditional Medicine. Braz J Pharmacog 24(5): 591–608
38. Hair WM, Gubbay O, Jabbour HN, Lincoln GA (2002) Prolactin receptor expression in human testis and accessory tissues: localization and function. Mol Hum Reprod 8(7):606–611
39. Haokip SW, Sheikh KA, Das S, Devi OB, Singh YD, Wangchu L, Heisnam P (2023) Unraveling physicochemical profiles and bioactivities of citrus peel essential oils: a comprehensive review. Eur Food Res Technol 249(11):2821–2834
40. He D, Shan Y, Wu Y, Liu G, Chen B, Yao S (2011) Simultaneous determination of flavanones, hydroxycinnamic acids and alkaloids in Citrus fruits by HPLC-DAD-ESI/MS. Food Chem 127:880–885
41. He W, Li Y, Liu M, Yu H, Chen Q, Chen Y, Ruan J, Ding Z, Zhang Y, Wang T (2018) *Citrus aurantium* L. and its flavonoids regulate TNBS-induced inflammatory bowel disease through anti-inflammation and suppressing isolated jejunum contraction. Int J Mol Sci 19(10):1–14
42. Herawati D, Ekawati ER, Yusmiati SNH (2020) Identification of Saponins and flavonoids in lime (*Citrus aurantifolia*) Peel extract. Proc. Int. Conf. Ind. Eng. Oper Manag:3661–3666
43. Howard N (2017) Malaria control for afghans in Pakistan and Afghanistan (1990–2005): a mixed-methods assessment considering effectiveness, efficiency, equity, and humanity (Doctoral dissertation, London School of Hygiene & Tropical Medicine)
44. Ibrahim FA, Usman LA, Akolade JO, Idowu OA, Abdulazeez AT, Amuzat AO (2019) Antidiabetic potentials of *Citrus aurantifolia* leaf essential oil. Drug Res 69:201–206
45. Indriyani NN, Anshori JA, Permadi N, Nurjanah S, Julaeha E (2023) Bioactive components and their activities from different parts of *Citrus aurantifolia* (Christm.) Swingle for food development. Foods *12*(10):2036
46. Izah SC (2019) Activities of crude, acetone and Ethanolic extracts of Capsicum frutescens var. minima fruit against larvae of Anopheles gambiae. Journal of environmental treatment. Techniques 7(2):196–200
47. Izah SC, Odubo TC (2023) Effects of *Citrus aurantifolia* fruit juice on selected pathogens of public health importance. ES Food Agrofor 11:829. https://doi.org/10.30919/esfaf829
48. Izah SC, Youkparigha FO (2019) Larvicidal activity of fresh aqueous and ethanolic extracts of Cymbopogon citratus (DC) Stapf on malaria vector, anopheles gambiae. BAOJ Biotech 5:040
49. Izah SC, Kigigha LT, Aseibai ER, Okowa IP, Orutugu LA (2016) Advances in preservatives and condiments used in zobo (a food-drink) production. Biotechnol Res 2(3):104–119
50. Izah SC, Uhunmwangho EJ, Dunga KE (2018) Studies on the synergistic effectiveness of methanolic extract of leaves and roots of *Carica papaya* L. (papaya) against some bacteria pathogens. Int J Complement Alternat Med 11(6):375–378

51. Izah SC, Uhunmwangho EJ, Dunga KE, Kigigha LT (2018) Synergy of methanolic leave and stem-back extract of Anacardium occidentale L. (cashew) against some enteric and superficial bacteria pathogens. MOJ Toxicol 4(3):209–211
52. Izah SC, Uhunmwangho EJ, Eledo BO (2018) Medicinal potentials of Buchholzia coriacea (wonderful kola). Medicinal Plant Res 8(5):27–43
53. Izah SC, Uhunmwangho EJ, Etim NG (2018) Antibacterial and synergistic potency of methanolic leaf extracts of Vernonia amygdalina L. and Ocimum gratissimum L. J Basic Pharmacol Toxicol 2(1):8–12
54. Izah SC, Zige DV, Alagoa KJ, Uhunmwangho EJ, Iyamu AO (2018) Antibacterial efficacy of aqueous extract of Myristica fragrans (Common nutmeg). EC Pharmacol Toxicol 6(4): 291–295
55. Izah SC, Chandel SS, Etim NG, Epidi JO, Venkatachalam T, Devaliya R (2019) Potency of unripe and ripe express extracts of long pepper (Capsicum frutescens var. baccatum) against some common pathogens. Int J Pharm Phytopharmacol Res 9(2):56–70
56. Izah SC, Etim NG, Ilerhunmwuwa IA, Ibibo TD, Udumo JJ (2019) Activities of express extracts of Costus afer Ker–Gawl. [Family COSTACEAE] against selected bacterial isolates. Int J Pharm Phytopharmacol Res 9(4):39–44
57. Izah SC, Etim NG, Ilerhunmwuwa IA, Silas G (2019) Evaluation of crude and ethanolic extracts of Capsicum frutescens var. minima fruit against some common bacterial pathogens. Int J Complement Alternat Med 12(3):105–108
58. Izah SC, Aigberua AO, Richard G (2022) Concentration, source, and health risk of trace metals in some liquid herbal medicine sold in Nigeria. Biol Trace Elem Res 200:3009–3302
59. Jayana BL, Prasai T, Singh A, Yami DK (2010) Study of antimicrobial activity of lime juice against Vibro cholera. Sci World 8(8):44–46
60. Karimi A, Nasab NK (2014) Effect of garlic extract and *Citrus aurantifolia* (lime) juice and on blood glucose level and activities of aminotransferase enzymes in Streptozotocin-induced diabetic rats. World J Pharm Sci 2:821–827
61. Kawaii S, Tomono Y, Katase E (2000) Quantitative study of flavonoids in leaves of citrus plants. J Agric Food Chem 48(9):3865–3871
62. Kessler RC, Ormel J, Petukhova M, McLaughlin KA, Green JG, Russo LJ et al (2011) Development of lifetime comorbidity in the world health organization world mental health surveys. Arch Gen Psychiatry 68(1):90–100
63. Khan SU, Anjum SI, Ansari MJ, Khan MHU, Kamal S, Rahman K, Shoaib M, Man S, Khan AJ, Khan SU, Khan D (2019) Antimicrobial potentials of medicinal plant's extract and their derived silver nanoparticles: a focus on honey bee pathogen. Saudi J Biol Sci 26(7):1815–1834
64. Lee SM, Park SY, Kim MJ, Cho EA, Jun CH, Park CH, Kim HS, Choi SK, Rew JS (2018) Key lime (*Citrus aurantifolia*) inhibits the growth of triple drug resistant Helicobacter pylori. Gut Pathogens 10:16–24
65. Lin LY, Chuang CH, Chen HC, Yang KM (2019) Lime (Citrus aurantifolia (Christm.) Swingle) essential oils: volatile compounds, antioxidant capacity, and hypolipidemic effect. Foods 8(9):398
66. Liu N, Li X, Zhao P, Zhang X, Qiao O, Huang L, Guo L, Gao W (2021) A review of chemical constituents and health-promoting effects of Citrus peels. Food Chem 365:130585
67. Loizzo MR, Tundis R, Bonesi M, Menichini F, De Luca D, Colica C, Menichini F (2012) Evaluation of *Citrus aurantifolia* Peel and leaves extracts for their chemical composition, antioxidant and anti-cholinesterase activities. J Sci Food Agric 92:2960–2967
68. Longo D, Fauci A, Kasper D, Hauser S, Jameson J, Loscalzo J (2011) Alterations in sexual function and reproduction. In: Longo D, Fauci A, Kasper D, Stephen Hauser J, Jameson JL (eds) Harrison's principles of internal medicine, 18th edn. McGrawHill, New York, p 2887
69. Masud S, Mehboob F, Bappi MU (2007) Severe hyperprolactinaemia directly depresses the gonadal activity causing infertility. Esculapio J Services Inst Med Sci 2:25–27
70. Miguel E, Fournet F, Yerbanga S, Moiroux N, Yao F, Vergne T, Cazelles B, Dabiré RK, Simard F, Roche B (2018) Optimizing public health strategies in low-income countries:

epidemiology, ecology and evolution for the control of malaria. In: Ecology and evolution of infectious diseases: pathogen control and public health management in low-income countries, pp 253–268

71. Narang N, Jiraungkoorskul W (2016) Anticancer activity of key lime, *Citrus aurantifolia*. Pharmacogn Rev 10:118–122
72. Ndiok EO, Ohimain EI, Izah SC (2016) Incidence of malaria in type 2 diabetic patients and the effect on the liver: a case study of Bayelsa state. J Mosqu Res 6(15):1–8
73. Oboh PA, Abulu EO (1997) The antimicrobial activities of extracts of *Sidium guajava* and *Citrus aurantifolia*. Niger J Biotechnol 8(1):25–29
74. Okon U, Etim B (2014) *Citrus aurantifolia* impairs fertility facilitators and indices in male albino wistar rats. Int J Reprod Contracep Obstet Gynecol 3:640–645
75. Olaniran AF, Afolabi RO, Abu HE, Owolabi A, Iranloye YM, Okolie CE (2020) Lime potentials as biopreservative as alternative to chemical preservatives in pineapple, orange and watermelon juice blend. Food Res 4(6):1878–1884
76. Olatunji BO, Cisler JM, Tolin DF (2007) Quality of life in the anxiety disorders: a meta-analytic review. Clin Psychol Rev 27:572–581
77. Pathan RK, Gali PR, Pathan P, Gowtham T, Pasupuleti S (2012) In vitro antimicrobial activity of *Citrus aurantifolia* and its phytochemical screening. Asian Pac J Trop Dis 2:S328–S331
78. Patil J, Jayaprakasha G, Chidambara MK, Tichy S, Chetti M, Patil B (2009) Apoptosis-mediated proliferation inhibition of human colon cancer cells by volatile principles of *Citrus aurantifolia*. Food Chem 114:1351–1358
79. Patil JR, Chidambara MKN, Jayaprakasha GK, Chetti MB, Patil BS (2009) Bioactive compounds from Mexican lime (*Citrus aurantifolia*) juice induce apoptosis in human pancreatic cells. J Agric Food Chem 57:10933–10942
80. Patil JR, Jayaprakasha GK, Kim J, Murthy KN, Chetti MB, Nam SY et al (2013) 5-geranyloxy-7-methoxycoumarin inhibits colon cancer (SW480) cells growth by inducing apoptosis. Planta Med 79:219–226
81. Petretto GL, Vacca G, Addis R, Pintore G, Nieddu M, Piras F, Sogos V, Fancello F, Zara S, Rosa A (2023) Waste Citrus Limon leaves as source of essential oil rich in limonene and Citral: chemical characterization, antimicrobial and antioxidant properties, and effects on cancer cell viability. Antioxidants 12(6):1238
82. Porras G, Chassagne F, Lyles JT, Marquez L, Dettweiler M, Salam AM, Samarakoon T, Shabih S, Farrokhi DR, Quave CL (2020) Ethnobotany and the role of plant natural products in antibiotic drug discovery. Chem Rev 121(6):3495–3560
83. Puspita S, Eddy DR, Wahyudi T, Julaeha E (2020) Microencapsulation of lime Peel essential oils (*Citrus aurantifolia*) with complex coacervation methods using gelatin/sodium alginate coating. J Kim Val 6:106–112
84. Riccaboni M, Bianchi I, Petrillo P (2010) Spleen tyrosine kinases: biology, therapeutic targets and drugs. Drug Discov Today 15(13-14):517–530
85. Sastre J, Pallardo FV, Vina J (2000) Mitochondrial oxidative stress plays a key role in aging and apoptosis. IUBMBLife 49:427–435
86. Shalaby N, Howaida A, Hanaa H, Nour B (2011) Protective effect of *Citrus sinensis* and *Citrus aurantifolia* against osteoporosis and their phytochemical constituents. J Medicinal Plants Res 5:579–588
87. Sharifian MR, Shokouhinejad N, Monsef Esfahani HR, Aligholi M, Amjadi M (2011) Antimicrobial effect of *Citrus aurantifolia* extract on Enterococcus faecalis within the dentinal tubules in the presence of smear layer. J Dental Med 24(2):148–155
88. Shrestha RL, Dhakal DD, Gautum DM, Paudyal KP, Shrestha S (2012) Variation of physiochemical components of acid lime (*Citrus aurantifolia* swingle) fruits at different sides of the tree in Nepal. Am J Plant Sci 3(12):Article ID:25781,5 pages. https://doi.org/10.4236/ajps.2012.312206
89. Siegel RL, Miller KD, Jemal A (2015) Cancer statistics, 2015. CA Cancer J Clin 65:5–29

90. Singh A, Maurya S, Singh UP, Singh KP (2014) Chromatographic analysis of phenolic acids in the fruit pulp of some Citrus varieties and their therapeutic importance in human health. Int J Appl Sci Rev 1:150–154
91. Sohi S, Shri R (2018) Neuropharmacological potential of the genus Citrus: a review. J Pharmacogn Phytochem 7:1538–1548
92. Soholm B (1998) Clinical improvement of memory and other cognitive functions by Ginkgo biloba: review of relevant literature. Adv Ther 15:54–65
93. Spadaro F, Costa R, Circosta C, Occhiuto F (2012) Volatile composition and biological activity of key lime *Citrus aurantifolia* essential oil. Nat Prod Commun 7:1523–1526. https://doi.org/10.1177/1934578X1200701128
94. Swandiny GF, Nafisa S, Gangga E (2021) Standardization of 70% ethanol extract and 96% lime leaves as antioxidants with DPPH and FRAP. J Pharmacogn Phytochem 10:47–52
95. Taiwo SS, Oyekanmi BA, Adesiji YO, Opaleye OO, Adeyeba OA (2007) In vitro antimicrobial activity of crude extracts of *Citrus aurantifolia* Linn and Tithonia diversifolia Poaceae on clinical bacterial isolates. Int J Trop Med 2:113–117
96. Tosukhowong P, Yachantha C, Sasivongsbhakdi T, Ratchanon S, Chaisawasdi S, Boonla C (2008) Citraturic, alkalinizing and antioxidative effects of limeade-based regimen in nephrolithiasis patients. Urol Res 36:149–155
97. Tran TKN, Ngo TCQ, Tran TH, Bach LG, Tran TT, Huynh XP (2021) Comparison of volatile compounds and antibacterial activity of *Citrus aurantifolia*, Citrus latifolia, and Citrus hystrix shell essential oils by pilot extraction. In: IOP conference series: materials science and engineering, vol 1092. IOP Publishing, p 012076
98. Wheaton TA, Stewart I (1969) Biosynthesis of synephrine in citrus. Phytochemistry 8:85–92
99. Wijesundere DA, Ramasamy R (2017) Analysis of historical trends and recent elimination of malaria from Sri Lanka and its applicability for malaria control in other countries. Front Public Health 5:212
100. Yamada T, Hayasaka S, Shibata Y, Ojima T, Saegusa T, Gotoh T (2011) Frequency of Citrus fruit intake is associated with the incidence of cardiovascular disease: The Jichi Medical School cohort study. J Epidemiol 21:169–175
101. Youkparigha FO, Izah SC (2019) Larvicidal efficacy of aqueous extracts of *Zingiber officinale* Roscoe (ginger) against malaria vector, *Anopheles gambiae* (Diptera: Culicidae). Int J Environ Agric Sci 3:020

Phytochemical Constituents of *Rosmarinus officinalis Linn.* and Their Associated Role in the Management of Alzheimer's Disease

12

Marcella Tari Joshua

Contents

Abstract

Neurodegenerative disorders, such as Alzheimer's disease, lead to a gradual and irreversible loss of neurons and brain function, resulting in progressive dementia characterized by disoriented plaques and neurofibrillary tangles composed of beta-amyloid and hyperphosphorylated tau proteins, respectively. Currently, there is no permanent treatment for these disorders. However, herbal plants and formulations, along with certain research findings, have reported the potential therapeutic use of rosemary leaves to mitigate these conditions more pleasantly. *Rosmarinus*

M. T. Joshua (✉)
Department of Medical Laboratory Science, Faculty of Health Sciences, Bayelsa Medical University, Yenagoa, Nigeria

S. C. Izah et al. (eds.), *Herbal Medicine Phytochemistry*, Reference Series in Phytochemistry, https://doi.org/10.1007/978-3-031-43199-9_14

officinalis Linn., commonly known as rosemary, holds significant economic importance and is often used as a flavor enhancer in soups. In the field of phytomedicine, this herb is gaining recognition as researchers investigate not only its unique taste and aroma but also the various health benefits of its constituents when extracted. The phytotherapeutic components of rosemary, such as rosmarinic acid and other derivatives found in its essential oils, including caffeic acid, p-coumaric acid, rosmanol, carnosol, carsonic acid, 4-hydroxybenzoic acid, beta-amyrin, oleanic acid, and urosolic acid, offer numerous advantages. These constituents possess anti-memory loss properties and various other health benefits, primarily due to their inhibitory effects on factors contributing to memory problems. This chapter sheds light on the impact of *Rosmarinus officinalis Linn.*, specifically rosemary leaves and its essential constituents, on health benefits. These benefits include its potential as an anti-memory loss agent, possibly attributed to the presence of flavonoids, triterpenoids, diterpenoids, and antioxidants. The review chapter critically examines the protective systematic effect this natural medicinal plant produces, such as being a stimulus as well as a protective agent that promotes healthier brain functionality.

Keywords

Evidence- Rosemary Leaves · *Rosmarinus officinalis Linn.* Rosmarinic Acid · Anti-memory loss · Flavonoids · Amyloid Beta-Protein · Dementia

Abbreviations

AChE	Acetylcholinesterase
AD	Alzheimer's Disease
AGE	Advanced Glycation End Product
AGE-RAGE	Advanced Glycation End Product-Receptors of Advanced Glycation End Product
ALS	Amyotrophic Lateral Sclerosis
BChE	Butyrylcholinesterase
CA	Carsonoic Acid
CJD	Creutzfeldt-Jakob Disease
CORT	Corticosterone
FDA	Food and Drug Administration
FTD	Frontotemporal Dementia
KEAP1	Kelch-Like Associated Protein 1
LBD	Lewy Body Dementia
MS	Multiple Sclerosis
NRF2	Nuclear Factor Erythroid-2-Related Factor
p-CA	P-Coumaric Acid
PSP	Progressive Supranuclear Palsy
RA	Rosmarinic Acid
RAW 264.7	Murine Macrophage Cell Line
UA	Ursolic Acid

1 Introduction

Phytochemistry is the branch of chemistry that focuses on the study of the chemical compounds found in plants, particularly those with medicinal properties. It involves the identification, isolation, characterization, and analysis of the various chemical constituents present in plant species. Phytochemistry plays a crucial role in understanding the therapeutic potential of medicinal plants and in developing herbal medicines.

Herbal medicine, also known as phytotherapy or botanical medicine, is a traditional or alternative healthcare system that utilizes medicinal plants and their extracts to prevent, alleviate, or treat various health conditions [3, 4, 24, 48]. It relies on the knowledge of the phytochemicals present in plants and their pharmacological effects on the human body. Herbal medicine has been practiced for centuries in different cultures and is still widely used today as a complementary or alternative approach to conventional medicine.

Medicinal plants are plants that contain bioactive compounds with therapeutic properties [45–47, 50–52]. These plants have been used for centuries in traditional medicine systems around the world to treat a wide range of ailments and diseases. Medicinal plants serve as the primary source of ingredients for herbal medicines. They are valued for their natural compounds, which can have anti-inflammatory, antioxidant, analgesic, antimicrobial, and other beneficial effects on health [40–44].

Therefore, phytochemistry delves into the chemical composition of plants, herbal medicine employs medicinal plants and their extracts for health purposes, and medicinal plants are the natural sources of bioactive compounds used in herbal remedies and traditional medicine systems. These fields are interconnected and contribute to our understanding of the healing potential of the natural world.

Rosemary, scientifically known as "Rosmarinus officinalis Linn.," is an evergreen shrub known for its aromatic and medicinal properties. It has received approval from the European Union [8] and is commonly found growing around the Mediterranean region. This plant is classified under the Lamiaceae plant family and is characterized by slender, needle-like leaves on erect woody stems. The leaves are green on the upper side and white underneath. Rosemary branches are firm with a cracked bark, and the woody stems are brown and square-shaped. It produces small light blue to white flowers in a distinctive inflorescence, typically blooming from late spring to early summer.

Various parts of the rosemary plant, including its leaves, flowers, and twigs, are processed to yield phytonutraceuticals [75]. Rosemary is cultivated in different regions worldwide and holds multiple uses. It is a staple in Mediterranean cuisine, used to make tea, incorporated into cosmetology products, and has a history in traditional and contemporary medicine for treating issues such as colds, coughs, and rheumatoid problems. Given its diverse applications, exploring the neuron-protective properties of rosemary leaves as a potential treatment for memory loss associated with neurodegenerative disorders is essential.

Numerous epidemiological studies have shown that a diet rich in polyphenols from rosemary leaves (*Rosmarinus officinalis* L.) can reduce the risk of Alzheimer's disease, a prevalent form of dementia in elderly individuals [7, 60, 86]. Among the various phyto-constituents found in *Rosmarinus officinalis*, compounds like caffeic acid, rosmarinic acid, oleanolic acid, urosolic acid, rosmanol, carnosol, carsonic acid, piccrosalvin, beta-amyrin, and rosmarinic acid stand out as promising anti-memory agents. These compounds appear to mitigate the progressive development of dementia by inhibiting alpha-beta aggregation and reducing synaptic toxicity in vitro [64].

According to Farmanfarma et al. [27], the accumulation of certain proteins, particularly in developed countries, has been linked to the development of brain cancer, accounting for approximately 3% of all cancer types, especially within the category of nervous system malignancies. Among these, 48% are classified as primary tumors in the central nervous system, with about 57% falling into the category of glioblastoma, a type of glioma that predominantly affects males between the ages of 75 and 84 and is also potentially linked to Alzheimer's disease. Extracts from *Rosmarinus officinalis* L. have been utilized to counteract these effects.

Reports by Sayorwan et al. [76] and Czajkowski and Nazaruk [19] regarding essential oil extracts from rosemary herb highlight their therapeutic benefits in slowing memory loss and preventing Alzheimer's disease, supported by evidence-based research. Sayorwan et al. [76] demonstrated that inhaling essential oils extracted from the Rosmarinic acid found in rosemary leaves via the olfactory nerve had a significant impact on brainwave activity. This stimulation influenced mood states and cognitive processes, enhancing alertness and mental activity, ultimately leading to reduced sleepiness, increased productivity, and a sense of refreshment.

Furthermore, Czajkowski and Nazaruk [19] reported that Rosmarinic acid from rosemary leaves had protective properties against neurotoxicity induced by the deposition of tau proteins and beta-deposits in the brain. This therapeutic potential of Rosmarinic acid had a gradual effect on the development of dementia, a prevalent form of neurodegenerative disorder characterized by the aggregation of tau proteins and beta-amyloid in the brain.

Regarding gliomas, which are highly toxic brain tumors, the use of rosmarinic acid as a treatment therapy for glioblastoma has been reported but has shown limited success in preventing neurological dysfunction and death [27]. Frontotemporal disorders, resulting from damage to neurons in the frontal and temporal regions of the brain, manifest as emotional difficulties, communication problems, and impaired working and walking abilities.

In a study conducted by Rasoolijazi et al. [71], the anti-memory effects of rosemary leaves were assessed using a behavioral evaluation involving eight rats in a Morris water maze test. The experiment involved dividing the rats into two groups: a positive group induced with neurotoxins (scopolamine) and a control group provided with regular rat chow and water ad libitum. The positive group exhibited signs of depression and reduced mobility due to the induced neurotoxins. Subsequently, the induced rats were orally treated with extracts from *Rosmarinus officinalis* L. Hidden cameras were strategically placed in different sections of the animal housing facility to record the rats' swimming patterns and behavioral activities. The rats underwent learning and spatial memory retention phases.

The rats received varying doses of rosemary leaf extracts dissolved in distilled water, with doses of 50, 100, and 200 mg/kg administered through oral gavage

(1 mL). After treatment, the laboratory mice were induced with a single dose of scopolamine intraperitoneally. The rats were then observed for their ability to locate the platform within 60 s, considering the process of navigating to the correct environment and the reduction of dormancy in each mouse. The results of the experiment highlighted the potential neuroprotective effect of the rosemary leaf extracts on the proper functioning of the hippocampus, particularly emphasizing the effectiveness of rosmarinic acid, a key constituent of rosemary leaves.

Currently, there is a growing need for further research on the therapeutic potential of *Rosmarinus officinalis Linn.* in the treatment of cognitive impairment. Existing studies in this area are limited, and there is a growing interest among researchers to conduct more comprehensive research in the future to gain deeper insights into the various aspects of the rosemary plant [5]. The objective of this chapter is to explore evidence-based research that demonstrates the therapeutic properties of Rosmarinic acid and other phytochemical constituents found in rosemary leaves as potential candidates for mitigating memory loss in individuals.

2 Overview of Neurodegenerative Diseases

Neurodegenerative diseases are a group of conditions characterized by a gradual decline in the function and structure of the nervous system, particularly affecting neurons in the brain and spinal cord. These illnesses can significantly diminish a person's quality of life and often lead to the progressive loss of cognitive and physical abilities. Neurodegenerative diseases typically involve the accumulation of abnormal proteins, oxidative stress, and inflammation within the brain, although their precise causes are not always fully understood. Several well-known examples of neurodegenerative diseases are summarized in Table 1.

1. Alzheimer's disease: This is the most common neurodegenerative disorder, primarily impacting memory and cognitive functions. It is characterized by the buildup of tau protein tangles and beta-amyloid plaques in the brain.
2. Parkinson's disease: Parkinson's disease affects motor coordination and movement and is associated with the loss of dopamine-producing neurons in the brain. Symptoms often include tremors, stiffness, and slow movements.
3. Huntington's disease: This genetic disorder leads to the degeneration of neurons in the basal ganglia and cortex, resulting in cognitive decline, psychiatric symptoms, and motor impairment.
4. Amyotrophic lateral sclerosis (ALS), also known as Lou Gehrig's disease: ALS affects motor neurons in the brain and spinal cord, causing issues with breathing, speaking, swallowing, muscle weakness, and paralysis.
5. Multiple sclerosis (MS): In this autoimmune disease, the immune system targets the myelin sheath surrounding nerve fibers in the central nervous system, leading to various neurological symptoms like fatigue, coordination problems, and sensory disturbances.

Table 1 Common neurodegenerative diseases

Neurodegenerative disease	Description	Key features/symptoms
Alzheimer's disease	Affects memory and cognition, characterized by beta-amyloid plaques and tau protein tangles in the brain.	Memory loss, cognitive decline.
Parkinson's disease	Impairs movement, linked to dopamine neuron loss.	Tremors, rigidity, bradykinesia (slow movement).
Huntington's disease	Genetic disorder, causes basal ganglia and cortex neuron degeneration.	Motor dysfunction, psychiatric symptoms.
Amyotrophic Lateral Sclerosis	ALS or Lou Gehrig's disease, affects motor neurons in brain and spinal cord.	Muscle weakness, paralysis, speaking/swallowing difficulties.
Multiple Sclerosis (MS)	Autoimmune disease attacks myelin sheath, central nervous system.	Fatigue, coordination problems, sensory disturbances.
Frontotemporal Dementia (FTD)	Degeneration of frontal and temporal brain lobes.	Behavior/personality changes, language deficits.
Amyloidosis	Group of rare diseases, amyloid protein deposits in organs, including the nervous system.	Organ dysfunction, neurological symptoms.
Creutzfeldt-Jakob Disease (CJD)	Rare, fatal prion disease affecting the brain.	Rapid cognitive decline, muscle stiffness/twitching.
Progressive Supranuclear Palsy	Rare disorder affecting movement and balance.	Eye movement issues, speech/swallowing problems.
Lewy Body Dementia (LBD)	Characterized by Lewy body protein deposits in the brain, similar to Parkinson's symptoms.	Cognitive decline, visual hallucin

6. Frontotemporal Dementia (FTD): FTD involves the degeneration of the frontal and temporal lobes of the brain, resulting in changes in behavior, personality, and language skills.
7. Amyloidosis: Amyloidosis encompasses a group of rare disorders where abnormal proteins called amyloids accumulate in various organs, including the nervous system, causing neurological symptoms and organ dysfunction.
8. Creutzfeldt-Jakob disease (CJD): CJD is a fatal prion disease that damages the brain, causing muscle twitching, stiffness, and rapid cognitive decline.
9. Progressive supranuclear palsy (PSP): PSP is a rare neurological condition that impairs balance and movement, often leading to difficulties in swallowing, speaking, and eye movements.
10. Lewy body dementia (LBD): LBD is a type of dementia characterized by the presence of abnormal protein deposits called Lewy bodies. It leads to cognitive impairment, visual hallucinations, and movement symptoms resembling Parkinson's disease.

While the causes, symptoms, and progression rates of these neurodegenerative diseases vary, they all share a common feature: the deterioration of the nervous system, resulting in significant disability, often necessitating specialized care and treatment.

The diagnosis of neurodegenerative diseases is characterized by the prevalence of ongoing memory loss, which is most commonly associated with dementia. This condition often manifests as a transition from the normal aging process, affecting many older individuals, but it is increasingly being observed in younger people as well. Dementia is influenced by various factors, including depression, anxiety, physical health, sleep patterns, lifestyle choices, aging, gender, and the cognitive capacity of individuals, particularly in terms of reasoning, as assessed by the function of the Hippocampus [39].

The amygdala, a brain region, plays a vital role in cognition, learning, memory, emotional regulation, and response to stress. However, many aging individuals experience issues related to thinking, changes in behavior, reasoning, and memory, resulting in reduced productivity in their daily lives.

Neurodegenerative diseases like Alzheimer's disease, Parkinson's disease, Huntington's disease, and motor neuron diseases are considered multifactorial age-related conditions that affect the intricate central nervous system of the brain, typically occurring with aging. Alzheimer's disease, in particular, is the most common form of dementia and is closely associated with a decline in cholinergic signaling to the brain. It currently affects approximately 50 million people worldwide, with expectations of this number tripling by 2050.

Several modifiable risk factors are linked to the development of Alzheimer's disease and dementia, including depression, obesity, diabetes, low levels of physical and mental activity, and hypertension. Additionally, factors such as sleep disturbances and anxiety have been reported to have connections to memory loss. Alzheimer's disease is characterized by irreversible neuron loss and impaired brain function, potentially due to neurotoxicity induced by beta-amyloid proteins and etau and beta-amyloid protein plaques buildup.

In summary, neurodegenerative diseases are associated with memory loss and cognitive decline, affecting individuals across various age groups. Alzheimer's disease is a prevalent form of dementia with modifiable risk factors, and its impact on global health is expected to increase significantly in the coming decades.

3 Overview of Alzheimer's Disease

Alzheimer's disease is a progressive and degenerative condition that primarily affects memory, thinking, and behavior. It falls under the umbrella of dementia, which describes significant cognitive decline interfering with daily life. While it's most commonly associated with older individuals, Alzheimer's, also known as early-onset Alzheimer's, can impact people in their 40 and 50 s. The key aspects of Alzheimer's disease are summarized in Table 2.

Table 2 Concise summary of key aspects of Alzheimer's disease

Aspect	Description
Definition	Progressive and degenerative brain disorder
Common symptoms	Memory loss, cognitive decline, behavior changes
Pathology	Accumulation of beta-amyloid plaques and tau tangles in the brain
Risk factors	Age, genetics, family history, cardiovascular health, head injuries, lifestyle
Stages	Early stage, middle stage, late stage
Diagnosis	Comprehensive evaluation, ruling out other causes
Treatment	No cure, medications to manage symptoms, non-drug interventions
Caregiving	Significant burden on caregivers, support needed
Research	Ongoing efforts to understand causes and develop treatments

1. **Symptoms:** Alzheimer's often begins with mild memory problems and difficulty performing cognitive tasks. Over time, these symptoms worsen, leading to confusion, disorientation, language difficulties, impaired judgment, and personality changes. In advanced stages, individuals may struggle to recognize loved ones or perform everyday tasks.
2. **Pathology:** A hallmark of Alzheimer's is the accumulation of abnormal protein deposits in the brain. These include tau tangles within cells and beta-amyloid plaques between nerve cells. These abnormalities disrupt communication among brain cells and lead to their death.
3. **Risk Factors:** While the exact cause of Alzheimer's remains uncertain, several risk factors have been identified. Age is a primary factor, with risk increasing significantly after the age of 65. Family history and genetics also play a role. Other risk factors include cardiovascular health, head injuries, and lifestyle factors like diet and exercise.
4. **Stages:** Alzheimer's typically progresses through three stages: early, middle, and late. Early-stage symptoms include mild memory issues, progressing to greater confusion and memory loss in the middle stage. In the late stage, significant cognitive decline, loss of communication skills, and round-the-clock care are often required.
5. **Diagnosis:** Diagnosis involves a comprehensive evaluation, including medical history, cognitive tests, neurological assessments, and sometimes brain imaging. While there's no definitive test for Alzheimer's, diagnosis is typically reached by ruling out other potential causes of cognitive impairment.
6. **Treatment:** Currently, there's no cure for Alzheimer's disease. However, some medications can help manage symptoms and slow disease progression, particularly in the early stages. Non-pharmacological interventions such as cognitive training and lifestyle changes can also enhance the quality of life.
7. **Caregiving:** Alzheimer's places a significant burden on caregivers, often family members. Providing physical and emotional care for someone with Alzheimer's

can be challenging. Support groups, respite care, and professional assistance can help reduce caregiver stress.

8. **Research:** Ongoing research aims to better understand the causes of Alzheimer's, develop effective treatments, and ultimately find a cure or preventive measures. Studies focus on genetics, inflammation, and lifestyle choices contributing to the disease.

4 Biology, Cultivation, Nutritional and Proximate Composition of Rosemary Leaves

Rosemary leaves, belonging to the Lamiaceae family, are characteristic of the Mediterranean region and are widely recognized for their diverse biological properties, making them valuable in both traditional medicine and culinary applications [36]. While rosemary cultivation is not as abundant in places like Nigeria compared to more developed tropical regions, efforts are underway to encourage farmers to cultivate this versatile herb in larger quantities.

Rosemary thrives in warm climates with moderate humidity levels and exhibits good heat tolerance. It prefers temperatures between 55° and 80° Fahrenheit and requires well-draining soil to flourish [10]. Therefore, successful cultivation should take place in locations with favorable climatic conditions.

Cultivating rosemary involves sexual reproduction and requires loamy-rich soil that is well-moistened and offers excellent drainage to enhance the plant's aromatic intensity [49]. Alternatively, rosemary can be propagated from non-woody sprigs. To achieve this, softwood cuttings are taken in spring, preferably after the plant has flowered, without including the flowers themselves. These cuttings, measuring around 3–5 in. and taken just below a branching point or leaf node, are placed in a compost mix with a significant portion of vermiculite. The container is then kept in a suitable environment, such as a propagator, cold frame, or windowsill, ensuring the soil remains consistently moist but not overly wet. After approximately 4 weeks, when the plant offers firm resistance upon a gentle tug, it is considered mature and ready for harvest [58].

Cultivating rosemary from seeds is typically done in the spring, although it requires patience due to the slow germination process and relatively low success rate. To start, the seeds should be placed in a well-drained compost pot indoors, ideally at least 3 months before the growing season begins. It's recommended to use a mix of loam compost with plenty of vermiculite or perlite. After moistening the seeds, the pot can be placed in a propagator. Seedlings typically emerge within 14–28 days and can then be removed from the propagator. Once they reach a height of around 3 in., they should be transferred to a larger pot to continue maturing. Harvesting can occur when the stems become strong [58].

Rosemary leaves are highly valued as a culinary herb, adding flavor to various dishes, particularly meats like lamb and fish, lemon sauces, stews, and beans. Additionally, they possess numerous medicinal properties, including antioxidant and antibacterial qualities. The key chemical constituents responsible for these properties are Rosmarinic acid, Carnosol, Carnosic acid, and essential oils rich in terpenes and terpenoids, including alpha-pinene, 1,8-Cineole, alpha-terpineol, camphor, borneol, and myrcene [16].

In terms of nutritional content, fresh rosemary leaves (*Rosmarinus officinalis Linn.*) contain essential minerals such as manganese, potassium, magnesium, copper, calcium, iron, along with vitamins A, C, B6, thiamine, riboflavin, and folate. Rosemary is cholesterol-free and provides around 131 calories per gram serving when consumed. In addition to these nutrients, it contains pyridoxine, folate, ursolic acid, carnosol, carnosic acid, rosmanol, 4-hydroxybenzoic acid, and riboflavin [59].

Furthermore, research by Ezza and Malia [25] suggests that rosemary extract may have memory-enhancing properties. This effect is likely attributed to its primary constituent, rosmarinic acid, a potent flavonoid known for its potential to reduce the risk of memory loss. Rosmarinic acid works in a complex manner within the brain by modulating the neurotransmission of acetylcholinesterase, an enzyme that plays a crucial role in the production of the neurotransmitter acetylcholine. An imbalance in acetylcholinesterase levels can lead to cognitive issues. Table 3 summarizes the biology, cultivation, nutritional, and some uses of rosemary leaves.

Table 3 Examples of some common neurodegenerative disorders and their symptoms

Aspect	Description
Biology	*Rosmarinus officinalis* – Rosemary is an aromatic, evergreen herb native to the Mediterranean region. It belongs to the Lamiaceae family.
Cultivation	Rosemary is grown as a perennial shrub. It prefers well-drained soil and full sun exposure. It can be cultivated in gardens, pots, or containers.
Nutritional composition	Rosemary leaves are low in calories. They contain dietary fiber, vitamins, and minerals. Notable vitamins include vitamin A, vitamin C, and vitamin B6. Important minerals include calcium, iron, and magnesium.
Bioactive compounds	Rosemary contains several bioactive compounds, including rosmarinic acid and antioxidants like flavonoids and phenolic acids.
Flavor and aroma	Rosemary has a strong, woody aroma and a slightly bitter, pine-like flavor. It is commonly used as a seasoning in various culinary dishes.
Medicinal and culinary uses	Rosemary is used in traditional medicine for its potential health benefits, including improved digestion and cognitive function. It is a popular herb in cooking, enhancing the flavor of dishes like roasted meats, soups, and stews.

5 Active Chemical Composition of Rosemary (Rosmarinic Acid)

In recent times, there has been a noticeable increase in the recognition of the value of using natural medicinal plant products to address various health concerns. This growing awareness has led to a heightened interest in understanding the chemical compounds present in these plants, particularly those with properties that can mimic the effects of certain diseases, such as Alzheimer's disease. This discussion focuses on the rosemary plant, which contains essential oils known as Rosmarinic acid and is recognized for its potential in mitigating the effects of Alzheimer's disease.

Rosmarinic acid is a group of phytochemicals found in rosemary leaves, and they encompass flavonoids and other phenolic compounds, including terpenes. These compounds can be extracted from rosemary using solvents like ethanol, acetone, and hexane [1]. The extraction process often involves the use of supercritical carbon dioxide, highlighting the bioactive nature of these substances [59, 62].

Research has shown that these isolated compounds play a role in inhibiting structural neurodegenerative changes [15]. Furthermore, the removal of beta and tau proteins has been associated with improved Alzheimer's disease management in elderly individuals [53]. Some specific compounds identified in rosemary include rosmaridiphenol, alpha-amyrin, botulin, luteolin, diterpenes, and beta-amyrin.

It's worth noting that the composition and concentration of these isolates can vary based on factors such as the plant's country of origin, drying and storage methods, weather conditions, cultivation practices, genetic variation, and harvest timing [23].

Examining each of these constituents in detail, it's noteworthy that phenolics and flavonoids have been reported to mimic the action of insulin in the treatment of diabetes mellitus [49]. The phytochemical active constituents, including rosmarinic acid, have unique mechanisms of action that contribute to their therapeutic functions as anti-memory agents. Understanding these individual properties is crucial for appreciating their potential health benefits.

6 Some Phytochemical Constituents of Rosemary Leaf (*Rosmarinus officinalis Linn.*) for the Alleviation of Alzheimer's Disease

Rosemary leaves contain a range of active bioactive chemical constituents, each with its own unique properties and potential health benefits, particularly in alleviating Alzheimer's disease. Some of these constituents, including rosmarinic acid, caffeic acid, p-coumaric acid, carnosol, ursolic acid, 4-Hydroxybenzoic acid, and carnosic acid, belong to various chemical classes, such as monoterpenes, diterpenes, and polyphenolic compounds. The chemical names, molecular formulas, molecular weights, and chemical classifications of these constituents are detailed in Table 4. Additionally, the chemical structures of these phytochemical constituents are illustrated in Table 5.

Table 4 Summary of some identified phytochemical constituents of *Rosmarinus Officinalis Linn*

Phytochemical constituents	Compound name	Molecular formula	Molecular weight (g/mol)	Chemical classification	References
Rosmanol	20-Deoxocarnosol	$C_{20}H_{26}O_5$	346.4	Phenolic diterpenes	[15, 74]
Carnosol	4aR-(epoxymethano)phenanthrene-2-one	$C_{20}H_{26}O_4$	330.4	Ortho-phenolic diterpene	[15]
Carnosic acid	Abieta-8,11,13-triene	$C_{20}H_{28}O_4$	332.4	Benzenediol abietane diterpene	[15]
Rosmarinic acid	3,4,Dihydroxyphenylactic acid	$C_{18}H_{16}O_8$	360.1	Flavonoids	[15, 79]
Caffeic acid	3-(3,4,Dihydroxyphenyl)-2-propenoic acid	$C_9H_8O_4$	180.2	Phenols	[9]
4-Hydroxybenzoic acid	Para-Hydroxybenzoic acid	$C_7H_6O_3$	260.2	Phenols	[34, 68, 83]
P-Coumaric acid	(2E)-3-(4-Hydroxyphenyl)prop-2-enoic acid	$C_9H_8O_3$	164.2	Phenols	[73]
Oleanic acid	4a(2H)-carboxylic acid	$C_{18}H_{34}O_2$	456.7	Pentacyclic triterpenoid	[26, 35, 78]
Ursolic acid	Urs-12-en-28-oic-acid	$C_{30}H_{48}O_3$	456	Pentacyclic triterpenoid	[91]
Beta-Amyrin	(3-beta)-oliean-12-en-3-ol	$C_{33}H_{58}O_5$	498.9	Pentacyclic triterpenoid	[12, 85]

Table 5 Chemical structures of the phytochemical constituents of *Rosmarinus Officinalis Linn.*

Phytochemical constituents	Chemical structures	References
Rosmanol	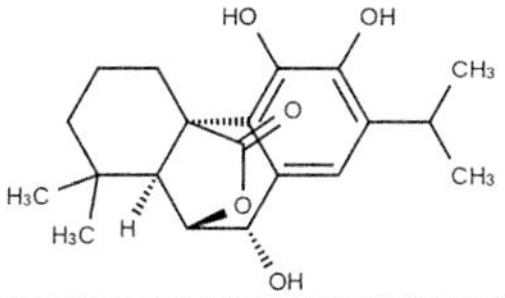 HO, OH, CH_3, O, H_3C, O, CH_3, H_3C, H, OH	[67]
Carnosol	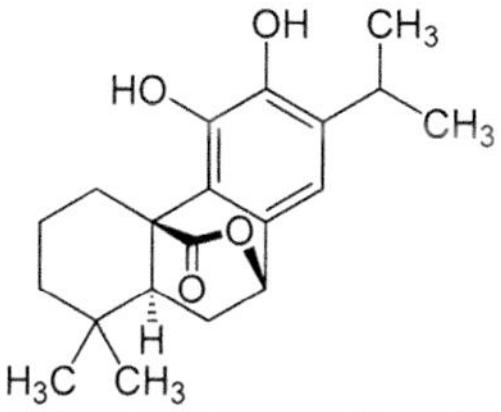 OH, CH_3, HO, CH_3, O, O, H, H_3C, CH_3	[87]
Carnosic acid	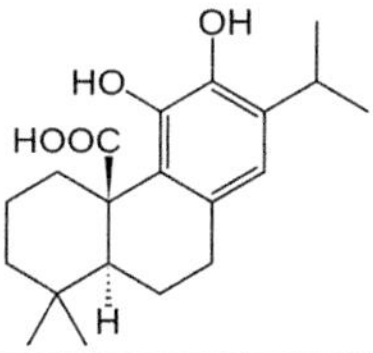 OH, HO, HOOC, H	[87]
Rosmarinic acid	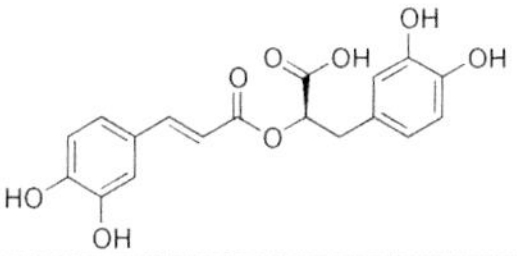 OH, O, OH, OH, O, O, HO, OH	[22]
Caffeic acid	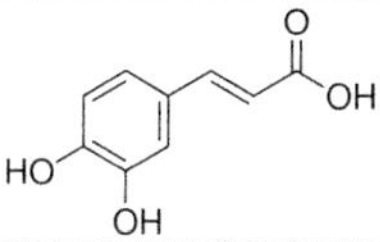 O, OH, HO, OH	[89]
4-Hydroxybenzoic acid	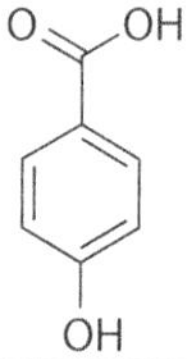 O, OH, OH	[56]
P-Coumaric acid	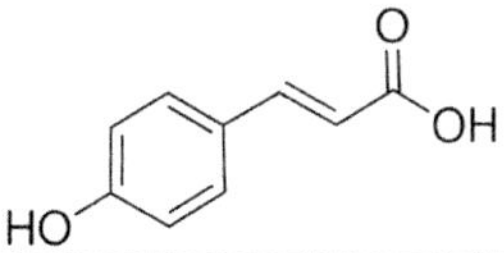 O, OH, HO	[28]
Oleanic acid	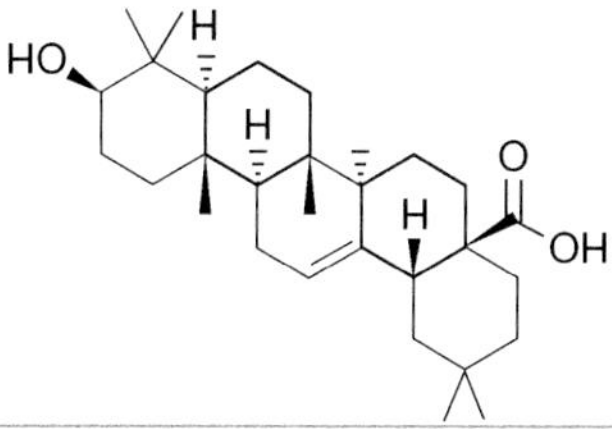 H, HO, H, O, H, OH	[69]

(continued)

Table 5 (continued)

Phytochemical constituents	Chemical structures	References
Ursolic acid	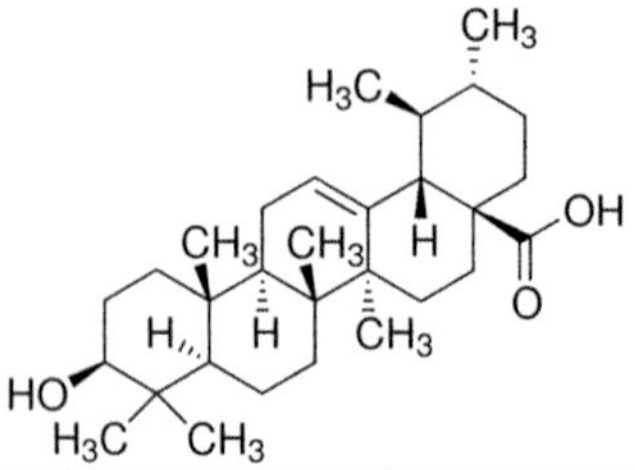	[20]
Beta-Amyrin	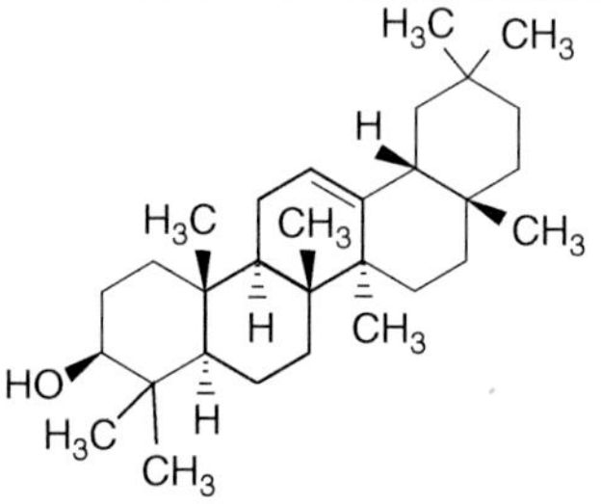	[17]

To obtain these valuable constituents, rosemary leaves sourced from local farms are harvested and carefully extracted. Various extraction methods can be employed, depending on the polarity of the specific compounds being targeted. These methods may include traditional hydro-distillation, maceration, decoction, and solvent extraction using solvents like ethanol, methanol, or n-hexane. Alternatively, more modern techniques, such as "Supercritical Fluid Extraction" (SFE), can be utilized. The choice of extraction method depends on the specific compounds of interest and the desired clinical application as a phytomedicine product [8, 36].

6.1 Caffeic Acid

Caffeic Acid (CA) has been identified as a metabolite derived from phenolic acid and has shown promise in reducing oxidative stress and brain damage by inhibiting 5-Lipoxygenase, according to reports by Cai et al. [14]. Additionally, Andrade et al. [9] conducted research indicating that CA has significant potential as it inhibits the fibrillation of amyloid alpha-beta peptides within neuronal membranes, as observed in an in-vitro analysis using an experimental rat model. It's worth noting that while CA did disaggregate amyloid-beta peptides, its activity exhibited a step-wise directional increase over time due to the presence of lipid membranes in the rat brain. This characteristic makes caffeic acid a potential constituent in addressing memory loss, particularly in diseases like Alzheimer's.

6.2 4-Hydroxybenzoic Acid

Also known as p-hydroxybenzoic acid, 4-Hydroxybenzoic acid is a monohydroxy benzoic crystalline salt that is moderately soluble in chloroform and highly soluble in alcohols and acetone, classified as a "polar organic solvent." It serves as a precursor to aspirin and is naturally found in substances like rosmarinic acid and coconuts [21]. This compound possesses several biological properties, including neuroprotective and antioxidative effects, which contribute to combating reactive oxygen species in the brain [32].

6.3 P-Coumaric Acid

P-coumaric acid, a natural compound with the chemical formula $C_9H_8O_3$, is a hydroxycinnamic acid [82] that forms an esterified system with tartaric acid known as tartaric p-coumaroyl ester [73]. This compound has immunoregulatory properties and is soluble in water and other solvents like diethyl ether and ethanol. P-coumaric acid can cross the blood-brain barrier and has a neuroprotective role, as evidenced by its ability to reduce neurotoxicity induced by alpha-beta amyloid proteins [33] in the hippocampal neuronal region of the brain, mitigating cognitive impairment [72]. It is also found in vegetables like rosemary leaves and is biosynthesized from cinnamic acid. Recent research by Yu et al. [88] demonstrated that p-coumaric acid has a neuroprotective effect against memory impairment induced by chronic corticosterone [28]. This effect was attributed to the reduction of oxidative stressors in the brain, along with interactions with neuronal receptors in the hippocampus [65]. Specifically, p-coumaric acid appeared to interfere with the interaction between corticosterone-induced advanced glycation end products and their receptors, restoring balance and protecting against depression and memory impairment [88].

6.4 Rosmarinic Acid

Rosmarinic Acid is a naturally occurring compound found in rosemary leaves. It has a chemical structure represented as $C_{18}H_{16}O_8$ and belongs to the Lamiaceae family. This compound has distinct characteristics, including a molecular weight of 360.1 g/mol, solubility in various solvents like water and formaldehyde, a melting point of 171–175 °C, and a boiling point of 694.7 °C [79].

Rosmarinic acid is known for its complex actions on inflammatory processes, including inhibiting cytokine release from activated T-cells and preventing T-cell activation. These properties make it relevant in addressing memory loss and allergic inflammatory reactions [63]. As the field of immunology and research techniques advances, the mechanisms of action of rosmarinic acid are becoming increasingly intricate.

In terms of memory loss, studies suggest that rosmarinic acid has anti-inflammatory and neuroprotective properties. It may protect against cognitive damage and aid in recovery [35]. The mechanism behind its action involves suppressing the accumulation of beta-amyloid plaques in the brain, which is associated with Alzheimer's disease. Rosmarinic acid accomplishes this by modulating the neurotransmission of acetylcholinesterase, an enzyme that affects the production of the neurotransmitter acetylcholine. By normalizing acetylcholinesterase and acetylcholine levels in the brain, rosmarinic acid contributes to improved cognitive function [25].

Furthermore, Rosmarinic Acid is considered the most active component of rosemary leaves and one of the most important polyphenols. It acts on receptors in the hippocampus, enhancing memory and combating memory loss caused by the accumulation of beta-amyloid proteins. It works by cleaving the amyloid precursor protein (APP) in the presence of the enzyme alpha-secretase, preventing the formation of amyloid plaques [77].

Rosmarinic acid is classified as an aqueous phenolic compound [34] and is known for its anti-inflammatory properties, primarily due to the presence of the RosA moiety. Given the rising concerns about memory impairment, incorporating rosemary leaves into our diet may enhance brain activity. Rosmarinic acid plays a vital role in addressing memory-related issues and neuroprotection ([83]; Guo et al. [31]) offering potential benefits in treating and preventing memory loss disorders.

6.5 Rosmanol

Rosmanol, with a chemical structure represented as $C_{20}H_{26}O_5$, is a compound found naturally in rosemary leaves. Traditionally, it was extracted using organic solvents like n-hexane, and the bioactive compounds were purified [55]. However, modern extraction methods such as Pasteurized Liquid Extraction (PLE) and Supercritical Fluid Extraction (SFE), using carbon dioxide (CO_2) and co-solvents like ethanol or methanol, have replaced these older methods. SFE offers improved selectivity, better fractionation capabilities, and higher extraction yields [74].

For enhanced absorption and potential anti-memory effects, combining Rosmanol and carnosol is recommended. Rosmanol is commonly found in the Lamiaceae family, such as in rosemary leaves, and is a polyphenol of 3,4-Dihydroxyphyllique, synthesized from carnosol in the human metabolic pathway [18]. Previous studies have identified Rosmanol as a potent anti-inflammatory compound that can inhibit the activation of nuclear factor pathways responsible for cell damage [54].

In terms of its mechanism of action in memory loss, research by Ayodeji et al. [11] on Wistar rats examined its inhibitory effects on iron-induced lipid peroxidation and key enzymes like Acetylcholinesterase (AChE) and Butyrylcholinesterase (BChE), which play crucial roles in brain neurotransmission. The study evaluated the antioxidant activity of Rosmanol as compared to galantamine, a typical cholinesterase inhibitor. The results showed that Rosmanol significantly increased scavenging activity in the brain cells of the rats. Molecular docking evaluations revealed a binding activity with AChE and BChE, indicating its potential affinity for inhibiting cholinergic enzymes and lipid peroxidation in brain neurons [2].

The scavenging activity of Rosmanol, acting as a cholinesterase inhibitor for improved memory function, is attributed to secondary metabolites known as flavonoids. These flavonoids are considered safe and have been shown to have neuroprotective effects. They prevent the action of acetylcholinesterase and enhance memory [13]. In the experiments, Rosmanol exhibited a binding energy of −11.1 kcal/mol, surpassing that of the galantamine inhibitor, which had a binding energy of −10.8 kcal/mol [80].

Rosmanol, with its neuroprotective and cholinesterase-inhibiting properties, holds promise in addressing memory loss and cognitive function, potentially offering therapeutic benefits for individuals experiencing memory-related disorders.

6.6 Carsonic Acid

Carsonic acid, an abietane diterpenoid extracted from rosemary leaves, has been associated with anti-memory effects [38]. It is obtained through an evaporation extraction method using hexane and subsequent dissolution in methanol, followed by complete evaporation. This compound is a carboxytricyclic compound categorized as a member of catechols and a monocarboxylic acid. It features a hydroxyl and carboxylic group at positions 11 and 12, and it is also substituted at position 20. As a member of the carboxytricyclic group, it falls under the category of catechols and is a conjugated acid of carnosate.

In terms of its mechanism of action in addressing memory loss problems, Carsonic Acid (CA) has demonstrated neuroprotective properties (Guo et al. [31]) in in-vivo experiments conducted by Takumi et al. [84] and Lipton et al. [57]. In these studies, mice were subjected to a downregulation of microglia and increased serum levels of proinflammatory cytokines, including tumor necrotic factor alpha. This proinflammatory cytokine activated the KEAP1/NRF2 pathway, leading to an increase in anti-inflammatory factors. The mice were then treated with 10 mg/kg body weight of CA extracts administered via the transnasal route for periods ranging from 14 days to 3 months.

The results showed significant behavioral and histological improvements in the treated mice, suggesting a translational progress. Histological examination of the brain indicated strong evidence that the formation of reactive oxygen species, responsible for oxidative stress and contributing to neuronal and synaptic damage, plays a role in the pathogenesis of memory impairment [57].

6.7 Oleanic Acid

Oleanic acid, also known as oleanolic acid ($C_{30}OH_{48}O_3$) [78], also possesses a chemical modification involving triple bonds, specifically the C-3 hydroxy, C-12,

C-13, and the addition of a fourth one, C-28 carboxylic acids. This leads to the creation of a sequence of new synthetic oleanane triterpenoids [37, 81]. This compound is found naturally as 3, beta-hydroxyolean-12-en-28-oic acid and is biologically active in approximately 1620 plant species, including *Rosmarinus officinalis Linn.* [26, 30], making it widely distributed in the plant kingdom.

Oleanic acid is a phytochemical constituent of saponins, which act as anti-memory agents, particularly in Alzheimer's disease, by blocking the action of the enzyme nitric oxide synthase involved in brain inflammation. Its affordability and availability make it a suitable candidate for further synthetic modification and production through in-vitro techniques such as Heterological Biogenesis or Artificial Biology [70, 90].

6.8 Ursolic Acid

Ursolic acid has been reported to alleviate early subarachnoid hemorrhagic brain injury in experimental rat models [91]. When rats were fed with 25–50 mg per body weight per day (25/50 mg/kg/day) of ursolic acid extract, significant reductions in brain inflammation were observed after a 48-h observation period. This effect was attributed to anti-inflammatory constituents, including intercellular adhesion molecule-1, inducible nitric oxide synthase, and matrix metalloproteinase-9, along with a decrease in apoptosis scores, showcasing the neuroprotective properties of ursolic acid.

6.9 Beta-Amyrin

Beta-amyrin is a natural chemical compound widely distributed in plants, including rosemary leaves. It is classified as a pentacyclic triterpenoid and is considered a component of glycyrrhizin, used as a sweetener and having pharmacological activity. Beta-amyrin originates from the precursor oleanolic acid and is positioned at the 3 beta-position between positions 12 and 13 [12, 85].

Beta-amyrin exhibits neurological effects, including memory regulation [66]. Research has demonstrated its neuroprotective properties, particularly in an Alzheimer's mouse model with synaptic plasticity in the hippocampus. In clinical experiments, mice with Alzheimer's disorder were treated with beta-amyrin extract from rosemary leaves, showing positive effects. This treatment was compared with a positive control group that received minocycline, which has been reported to have both positive and negative outcomes in Alzheimer's rat models [6, 29, 61].

7 Conclusion

Alzheimer's disease, a form of memory loss known as Vascular Dementia, is becoming a growing concern, not only among the elderly but also in younger individuals. Various factors, including physical inactivity, aging, stress, and depression, contribute to this alarming trend. The conventional approach of using drugs that interact with multiple receptors, resulting in low bioavailability, has led to a shift toward exploring phytomedicine. Rising healthcare costs and a shortage of specialized healthcare professionals have prompted scientists to consider phytomedicine as an alternative therapeutic approach for individuals with conditions like Alzheimer's.

Phytomedicine, which involves investigating the therapeutic properties of traditional plants, offers a promising avenue for enhancing drug efficacy and therapy. Scientists are increasingly recognizing its potential, as nature often provides abundant solutions to the challenges faced in pharmacological interactions. Within the spectrum of phytomedicine, the constituents of *Rosmarinus officinalis Linn.*, including caffeic acid, rosmarinic acid, rosmanol, p-coumaric acid, beta-amyrin, among others, have been highlighted for their potential in alleviating memory loss associated with Alzheimer's and dementia.

Numerous research studies have reported the anti-memory effects of rosmarinic acid on Alzheimer's disease and dementia. These studies have elucidated the mechanisms by which these constituents can aid in the treatment of memory impairment, particularly in the hippocampus and other affected regions. In light of this extensive body of scientific evidence, it is imperative to emphasize the need for further exploration of the therapeutic potential of *Rosmarinus officinalis Linn.*, commonly known as rosemary leaf, within the context of Ayurveda medicine.

Furthermore, advancements in analytical methods, including molecular and genetic techniques, enable the precise isolation and quantification of rosmarinic acid and other constituents of *Rosmarinus officinalis Linn.* This knowledge is critical for developing dosage formulations for drug therapy. Given the favorable conditions for cultivating rosemary leaves, it is advisable to encourage widespread cultivation of this plant. Following harvest, regular consumption of rosemary leaves in accordance with recommended dosages by regulatory authorities such as the Food and Drug Administration (FDA) can significantly contribute to maintaining optimal brain function and cognitive abilities.

Moreover, there is a need for further clinical trials to establish detailed guidelines for incorporating rosemary leaf as a form of drug therapy, possibly in the form of nutraceuticals. Rosmarinic acid's inhibition of acetylcholinesterase in the brain positions it as an herb with anti-memory loss properties in adult therapeutic medicine. If these principles are well integrated into the clinical realm of drug therapy with FDA authorization, older individuals can effectively adopt these practices to ensure they receive the right nutritional benefits from this herb in their daily lives. This could potentially have a substantial impact on preserving good memory and cognitive function.

References

1. Aguilar F, Autrup H, Barlow S, Castle J, Crebelli R, Dekant W (2008) Use of rosemary extracts as a food additive. Scientific opinion of the panel or food additives, flavourings, processing aids and materials in contact with food. Eur Food Saf Auth 721:1–29
2. Ahmed HM, Babakir-Mina M (2020) Investigation of rosemary herbal extracts (*Rosmarinus Officinalis*) and their potential effects on immunity. Phototherm Resour 34(8):1829–1837

3. Aigberua AO, Izah SC (2019) pH variation, mineral composition and selected trace metal concentration in some liquid herbal products sold in Nigeria. Int J Res Stud Biosci 7(1):14–21
4. Aigberua AO, Izah SC (2019) Macro nutrient and selected heavy metals in powered herbal medicine sold in Nigeria. Int J Med Plants Nat Prod 5(1):23–29
5. Al Shawabkeh MJ, Al MJA (2018) Effect of rosemary on fasting blood glucose, hemoglobin A1c and vitamin B12 in healthy person and type 2 diabetic patients taking glucomid or/and metformin. Natl J Physiol Pharm 8(1):1–4
6. Amani M, Shokouhi G, Salari AA (2019) Minocycline prevents the development of depression-like behaviour and hippocampal inflammation in a rat -model of Alzheimer's disease. Psychopharmacology 236:1281–1292
7. Anastasiou CA (2017) Mediterranean diet and cognitive health: initial results from the hellenic longitudinal investigation of ageing and diet. PLoS One 12:e0182048
8. Andrade JM, Faustino C, Garcia C, Ladeiras D, Reis CP, Rijo P (2018) *Rosmarinus Officinalis* L.: an update review of its phytochemistry and biological activity. Future Sci Open Access 4(4): FSO283. https://doi.org/10.4155/fsoa-2017-0124
9. Andrade S, Loureiro JA, Pereira MC (2021) Caffeic acid for the prevention and treatment of Alzheimer's disease. The effect of lipid membranes on the inhibition of aggregation and disruption of beta-amyloid fibrils. Int J Biol Macromol 190(1):853–861
10. Arturo A, Irma A, Maria-Rosa M-L, Maria-Aranzazu M (2021) Interactions between nutraceuticals/nutrients and nutrients and therapeutic drugs, Nutraceuticals, 2nd Edition. In: Efficacy, safety and toxicity. pp 1175–1197
11. Ayodeji OF, Goodness IO, Kayode E, Adewole OM, Odunnayo MA, Ahmed AI, Kunle O, Ganiyu O, Oluwafemi OO (2020) Aqueous extracts of bay leaf (*Laurus Nobilis*) and rosemary (*Rosmarinus Officinalis*) inhibit iron-induced lipid peroxidation and key -enzymes implicated in Alzheimer's disease in rat brain in-vivo. Am J Biochem Biotechnol 18(1):9–23
12. Babalola IT, Shode FO (2013) Ubiquitous ursolic acid: a potential pentacyclic triterpene natural product. Pharmacogn Phytochem 2(2):214–222
13. Boudouda HB, Zaghib A, Karioti A (2015) Antibacterial, antioxidant, anticholinesterase potential and flavanol glycosides of *Biscusellsraphanifolia,* (*Brassicaceae*). Pak J Pharm Sci 28(1):153–158
14. Cai H, Huang XJ, Xu ST, Shen H, Zhang PF, Huang Y, Jiang JY, Sun YJ, Jiang B, Wu XM (2016) Discovery of novel hybrids of diary 1-2,4-triazoles and caffeic acid as dual inhibitors of cyclooxygenases for cancer therapy. Eur J Med Chem 108:89–103
15. Carnejo A, Aguilar SF, Caballero L, Machura I, Munoz P, Caballero I (2017) Rosmarinic acid prevents fibrillization and diminishes vibrational modes associated to beta sheet in tau protein linked to Alzeihmer's disease. J Enzyme Inhib Med Chem 32(1):945–953
16. Chavez-Gonzalez ML (2016) *Rosmarinus offinalis* essential oil: a review of its phytochemistry, anti-inflammatory activity and mechanisms of action involved. J Ethnopharmacol 229:29–45
17. Christopher B, Reinhard J (2012) Composition and physiological function of the wax layers coating *Arabidopsis* leaves: beta-amyrin negatively affects the intracuticular water barrier. Plant Physiol 160(2):1120–1129
18. Conde-Hernandez LA, Espinosa-Victoria JR, Trejo A, Guerrero-Beltran J (2016) CO_2 supercritical extraction hydro distillation and steam distillation of essential oil of rosemary (*Rosmarinisofficinalis*). J Food Eng 112(3):712–856
19. Czajkowski P, Nazaruk J (2014) Rola skladnikow naturalnych w zapobieganiu chorobum neurodegeneracyjnym. Geriatria 8:258–263
20. Dawei MU, Gaobiao Z, Jianye L, Bin S, Heqing G (2018) Ursolic acid activates the apoptosis of prostate cancer via ROCK/PTEN mediated mitochondrial translocation of Cofilin-1. Oncol Lett 15(3):3202–3206
21. Dey G, Chakraborty M, Miltra A (2005) Profiling C6-C3 and C6-C1 phenolic metabolites in *Cocos nucifera*. J Plant Physiol 162(4):375–381
22. Edgar181 (2006) Structure of rosmarinic acid. http://www.absource.de.com. Assessed 25 Aug 2006

23. Elhassan IA, Osman NM (2014) New chemotype *Rosmarinus officinalis* L. (rosemary) '*R. officinalis* ct. bornyl acetate. Am J Res Commun 2(4):232–240
24. Enaregha EB, Izah SC, Okiriya Q (2021) Antibacterial properties of Tetrapleura tetraptera pod against some pathogens. Res Rev Insights 5:1–5. https://doi.org/10.15761/RRI.1000165
25. Ezza I, Malia F (2021) The health benefits of rosemary. The savoury herb may boost memory and more. Nutrition 3:337–425
26. Fai YM, Tao CC (2009) A review of presence of oleanolic acid in natural products. NaturaProda Med 2:77–290
27. Farmanfarma KK, Mohammadian M, Shahabinia S, Hassanipour H, Salehinya H (2019) Brain cancer in the world: an epidemiological review. World Cancer Res J 6(5):110–125
28. Fatimah S, Wan M, Md S, Rasdi R, Nur S, Md L, Nur L, Mohd R (2019) Determination and quantification of P-coumaric acid in pineapples (*Ananas comosus*) extracts using gradient mode RP-HPLC. Pharm Res 11(1):78–82
29. Ferretti MT, Allard S, Partridge V, Duccatenzeiler A, Cuello AC (2012) Minocycline corrects early pre-plaque neuroinflammation and inhibits BACE-1 in a transgenic model of Alzheimer's disease-like amyloid pathology. J Neuroinflammation 9:62
30. Fukushima EO, Seki H, Obyama K, Ono E, Umemoto N, Mizutani M, Santo K, Muranaka T (2021) CYP716A sub-family members are multifunctional oxidases in triterpenoid biosynthesis. Plant Cell Physiol 52:2050–2061
31. Guo Q, Shen Z, Yu H, Lu G, Yu Y, Liu X, Zheng P (2016) Carsonic acid protect against acetaminophen-induced hepatoxicity by potentiating NRF2- mediated antioxidant capacity in mice. Korean J Physiol Pharmacol 20:15–23
32. Guzman-Villanueva D, Mendiola M, Nguyen H, Weissig V (2015) Influence of Triphenylphosphonium (TPP) cation hydrophobization with phospholipids on cellular toxicity and mitochondrial selectivity. J Pharm Sci 2:1–9
33. Hardy J, Selkoe DJ (2002) The amyloid hypothesis of Alzheimer's disease program and problems on the road to therapeutics. Science 297:355–356
34. Heleno SA, Martins A, Queiroz MJ, Ferreira IC (2015) Bioactivity of phenolic acids: metabolites versus parent compounds: a review. Food Chem 173:501–513
35. Helmut M (2015) Brain food for Alzheimer's-free ageing: focus on herbal medicines. Adv Exp Med Biol 6:51–110
36. Hernandez MD, Sotomayor JA, Hernandez A, Jordan MJ (2016) Rosemary (*Rosmarinus officinalis* L.) oils, chapter 77. In: Preedy V (ed) Essential oils in food preservation, flavour and safety, 5th edn. Academic, London, pp 677–688
37. Honda T, Honda Y, Favaloro FG Jr, Gribble GW, Suh N, Place AF, Rendi MH, Sporn MB (2002) A novel dicyanotriterpenoid, 2-cyano-3,12-dioxooleana-1-9(11)-dien-28-onitrile, active at picomolar concentrations for inhibition of nitric oxide production. Bioorg Med Chem Lett 12: 1027–1030
38. Huang MT, Ho CT, Wang ZY, Ferraro T, Lou YR, Stauber K, Ma W, Conney AH (1994) Inhibition of skin tumorigenesis by rosemary and its constituents carsonol and urolic acid. Cancer Res 54:701–708
39. Hye JP, Huiyoung K, Ji HL, Eunbi C, Young CL, Minho M, Mia J, Dong HK, Ji WJ (2019) Beta amyrin ameliorates Alzheimer's disease-like aberrant synaptic plasticity in the mouse hippocampus. Biomed Res Ther 28(1):74–82
40. Izah SC, Uhunmwangho EJ, Dunga KE, Kigigha LT (2018) Synergy of methanolic leave and stem-back extract of *Anacardium occidentale* L. (cashew) against some enteric and superficial bacteria pathogens. MOJ Toxicol 4(3):209–211
41. Izah SC, Uhunmwangho EJ, Dunga KE (2018) Studies on the synergistic effectiveness of methanolic extract of leaves and roots of *Carica papaya* L. (papaya) against some bacteria pathogens. Int J Complement Altern Med 11(6):375–378
42. Izah SC, Uhunmwangho EJ, Etim NG (2018) Antibacterial and synergistic potency of methanolic leaf extracts of *Vernonia amygdalina* L. and *Ocimum gratissimum* L. J Basic Pharmacol Toxicol 2(1):8–12

43. Izah SC, Zige DV, Alagoa KJ, Uhunmwangho EJ, Iyamu AO (2018) Antibacterial efficacy of aqueous extract of myristica fragrans (common nutmeg). EC Pharmacol Toxicol 6(4):291–295
44. Izah SC, Uhunmwangho EJ, Eledo BO (2018) Medicinal potentials of *Buchholzia coriacea* (wonderful kola). Med Plant Res 8(5):27–43
45. Izah SC, Chandel SS, Etim NG, Epidi JO, Venkatachalam T, Devaliya R (2019) Potency of unripe and ripe express extracts of long pepper (*Capsicum frutescens var. baccatum*) against some common pathogens. Int J Pharm Phytopharmacol Res 9(2):56–70
46. Izah SC, Etim NG, Ilerhunmwuwa IA, Silas G (2019) Evaluation of crude and ethanolic extracts of *Capsicum frutescens var. minima* fruit against some common bacterial pathogens. Int J Complement Altern Med 12(3):105–108
47. Izah SC, Etim NG, Ilerhunmwuwa IA, Ibibo TD, Udumo JJ (2019) Activities of express extracts of Costus afer Ker–Gawl. [Family COSTACEAE] against selected bacterial isolates. Int J Pharm Phytopharmacol Res 9(4):39–44
48. Izah SC, Aigberua AO, Richard G (2022) Concentration, source, and health risk of trace metals in some liquid herbal medicine sold in Nigeria. Biol Trace Elem Res 200:3009–3302
49. Katarzyna P, Katarzyna J, Karolina J (2020) Properties and use of rosemary (*Rosmarinus officinalis* L.). J Life Sci 66(3):76–82
50. Kigigha LT, Biye SE, Izah SC (2016) Phytochemical and antibacterial activities of Musanga cecropioides tissues against *Escherichia coli*, *Pseudomonas* aeruginosa *Staphylococcus* aureus, *Proteus* and *Bacillus* species. Int J Appl Res Technol 5(1):100–107
51. Kigigha LT, Izah SC, Uhunmwangho EJ (2018) Assessment of hot water and ethanolic leaf extracts of *Cymbopogon citratus Stapf* (Lemon grass) against selected bacteria pathogens. Ann Microbiol Infect Dis 1(3):1–5
52. Kigigha LT, Selekere RE, Izah SC (2018) Antibacterial and synergistic efficacy of acetone extracts of *Garcinia kola* (Bitter kola) and *Buchholzia coriacea* (Wonderful kola). J Basic Pharmacol Toxicol 2(1):13–17
53. Kuna A, Lakshmiprasanna K (2020) Management of memory loss in elderly. Encycl Food Health 1:1–10
54. Lai CS, Lee JH, Ho CT, Lau CB, Wang JM, Wang YJ, Pan MH (2009) Rosmanol potentially inhibits lipolysaccharide-induced INOS and COX-2 expression through downregulating MARK, NF-kb, STAT3 and C/EBP signalling pathways. J Agric Food Chem 57:10990–10998
55. Lang Q, Wai CM (2001) Supercritical fluid extraction in herbal and natural product services: a practical review. Talanta 53(4):771–782
56. Lemini C, Silva G, Luque D, Valverde A, Rubio-Poo C, Chavez-Lara B, Valenzuela F (1997) Estrogenic effects of P-hydroxybenzoic acid in CD1 mice. Environ Res 75(2):130–134
57. Lipton SA, Rezaie T, Nutter A, Lopez KM, Parker J, Kosaka K, Satoh T, Mckercher SR, Masliah E, Nakanishi N (2016) Therapeutic advantage of pro-electrophilic drugs to activate the NRF2/ARE pathway in Alzheimer's disease models. Cell Death Dis 7(12):389
58. Malenie G (2021) How to grow rosemary from cuttings and from seed with top care tips. http://www.homesandgardens.com. Assessed 31 Aug 2022
59. Mena P, Cirlini M, Tassotti M, Herrlinger KA, Dall'Asta C, Del Rio D (2016) Phytochemical profiling of flavonoids, phenolic acids, terpenoids and volatile fraction of a rosemary (*Rosmarinus offinalis* L.) extract. Molecules 21(11):1576
60. Morris MC (2015) MIND diet associated with reduced incidence of Alzheimer's disease. Alzheimers Dement 11:1007–1014
61. Nobel W, Garwood C, Stephenson J, Kinsey AM, Hanger DP, Anderton BH (2009) Minocycline reduces the development of abnormal tau species in mice of Alzheimer's. Sci Rep 23:739–750
62. Nuwak K, Jawerska M, Ogunowski I (2013) Rosemary-rusletabrgata w zwlazksbiolegiezalec-zynasechemik 67(2):11–13
63. Oh HA, Park CS, Ahn HJ, Park CS, Kim HM (2011) Effect of Perilla *frutescens var acuta kudo* and rosmarinic acid on allergic inflammatory reactions. Exp Biol Med (Maywood) 236(11):99–106

64. Ona K, Takamura Y, Yoshike Y, Zhu L, Han F, Mao X, Ikeda T, Taksaki J, Nishijo H, Takashima A, Teplow D, Zagorski M (2012) Phenolic compounds prevent amyloid beta-protein oligomerization and synaptic dysfunction by site-specific binding. J Biol Chem 287: 14631–14643
65. Ou SY, Teng JW, Zhao YY, Zhao J (2012) P-Coumaric acid production from Lignocelluloses. In: Phenolic acids: composition, applications and health benefits. Nova Science Publishers, New York, pp 63–71
66. Park SJ, Young LA, Oh SR, Youngwan L, Guyoung K, Hyun W, Hyung EL, Dae SJ, Ji WJ, Jong HR (2014) Amyrin attenuates scopolamine-induced cognitive impairment in mice. Biol Pharm Bull 37(7):1207–1213
67. Petiwala S, Johnson JJ (2015) Diterpene from rosemary (*Rosmarinus officinalis*) defining their potential for anti-cancer activity. J Cancer 367(2):97–102
68. Pezzini I, Mattoli V, Ciofani G (2017) Mitochondria and neurodegenerative diseases: the promising role of nanotechnology in targeted drug delivery. Expert Opin Drug Deliv 14: 513–523
69. Pollier J, Goossens A (2011) Oleanolic acid. Mol Interest 21(4):816–822
70. Pollier J, Moses T, Goosens A (2011) Combinatorial biosynthesis in plants, a CPI review on it's potential and future exploitation. Nat Prod Rep 28:1897–1916
71. Rasoolijazi H, Mehdizadeh M, Soleimani M, Nikbakhte F, Farsani ME (2015) The effect of rosemary extract on spatial memory, learning and antioxidant enzymes activities in the hippocampus of middle-aged rats. Med J Islam Repub Iran 29:187
72. Sakamula R, Thong-Asa W (2018) Neuroprotective effect of P-coumaric acid in mice with cerebral ischemia reperfusion injuries. Metab Brain Dis 33:765–773
73. Salameh D, Brandam C, Medawar W, Lteif R, Strechaiano P (2008) Highlight on the problems generated by P-coumaric acid analysis in wine fermentations. Food Chem 107:1661–1667
74. Sanchez-Camargo AP, Herrero M (2014) Reference module in chemistry. In: Molecular sciences and chemical engineering. Institute of Food Science Research (CIAL, CSIC), Madrid. http://www.sciencedirect.com. Accessed 31 Dec 2022
75. Sasikumar B (2004) Handbook of herbs and spices. Woodhead publishing series in food science, technology and nutrition, vol 2. Woodhead Publishing, Oxford, pp 243–255
76. Sayorwan W, Ruangrungsi N, Piriyapunporn T, Hongratanaworakit T, Kotchabhakdi N, Siripornpanich V (2013) Effects of inhaled rosemary oil on subjective feelings and activities of the nervous system. Sci Pharm 81(2):531–542
77. Schedin-Weiss S, Winblad B, Tjernberg LO (2014) The role of protein glycosylation in Alzheimer's disease. J FEBS 281(1):46–62
78. Sigma-Aldrich (2018) Oleanolic acid, Merck. http://www.wikipedia.org/oleanolicacid. Accessed 3 Mar 2023
79. Silva EK, Zabot GL, Naithia-Neves G, Nogueira GC, Meireles MA (2018) Process engineering applying supercritical technology for obtaining functional and therapeutic products, chapter 12. In: Advances in biotechnology for food industry. A handbook of food bioengineering, chemical book. Academic, London, pp 327–358
80. Solanki J, Parihar P, Mansuri MJ, Sparihar MS (2015) Flavonoid-based therapies in the early management of neurodegenerative disease. Adv Nutr 6(1):64–72
81. Sporn MB, Liby KT, Yore MM, Fu L, Lopchuk JM, Gribble GW (2011) New synthetic triterpenoids potent agent for prevention and treatment of tissue injured and oxidative stress. Nat Prod Rep 74:537–545
82. Sussrichavalit T, Pracheyasittikul S, Isarankura N, Ayudhya C, Prachayasittikul V (2014) Synthesis of a "clickable" angiopep-conjugated, p-coumaric acid for brain-targeted delivery. J Mater Sci 49:8204–8213
83. Swajgier D, Borowiec K, Pustelniak K (2017) The neuroprotective effects of phenolic acids, molecular mechanisms of action. Nutrients 9:E477

84. Takumi S, Dorit T, Chang-Ki O, Stuart L (2022) Potential therapeutic use of the rosemary diterpene carsonic acid for Alzheimer's disease and long- COVID through NRF2 activation to counteract the NLRP3 inflammation. Antioxidants (Basel) 11(1):124
85. Tsansakul P, Shibuya M, Kushiro T, Ebizuka Y (2006) Dammarenediol-11-synthase, the first dedicated enzyme for ginsenoside biosynthesis in *Panex ginseng*. Fed Eur Biochem Soc Lett 580(22):5143–5149
86. Uddin MS, Mamun AA, Hossain MS, Ashaduzzaman M, Noor MAA, Hossain MS, Uddin MJ, Sarker J, Asaduzzaman M (2016) Neuroprotective effect of *Phyllanthusacidus* L. on learning and memory impairment in scopolamine-induced animal model of dementia and oxidative stress: natural wonder for regulating the development and progression of Alzheimer's disease. Adv Alzheimers Dis 5(2):53–72
87. Wells AT (2012) Structural inorganic chemistry, 5th edn. Oxford University Press, Oxford, UK, p 51
88. Yu X-D, Zhang D, Xiao C-L, Zhou Y, Li X, Wang L, He Z, Reilly J, Xiao Z-Y, Shu X (2022) P-coumaric acid reverses depression-like behaviour and memory deficit via inhibiting AGE-RAGE mediated neuroinflammation. Cell 10:1594
89. Yunliang W, Yutong W, Jinfeng L, Linlin H, Yuzhan Z, Xiaopeng Y, Zheilei Z, Hongying B, Honglei-Jiyu L (2016) Effects of caffeic acid in learning deficits in model of Alzheimer's disease. Int J Mol Med 38:869–875
90. Zhang H, Boghigian BA, Armando J, Pfeifer BA (2011) Methods and options for the heterologous production of complex natural products. Nat Prod Rep 28:125–151
91. Zhang T, Su J, Guo T, Zhu K, Wang X, Li X (2014) Ursolic acid alleviates early brain injury after experimental subarachnoid haemorrhage by suppressing TLR4-mediated inflammatory pathway. Int Immunopharmacol 23(2):585–591

Diversity of Medicinal Plants Used in the Treatment and Management of Viral Diseases Transmitted by Mosquitoes in the Tropics

13

Maduamaka Cyriacus Abajue, Wisdom Ebiye Sawyer, Sylvester Chibueze Izah, and Mathew Chidozie Ogwu

Contents

M. C. Abajue (✉)
Department of Animal and Environmental Biology, Faculty of Science, University of Port Harcourt, Port Harcourt, Nigeria
e-mail: maduamaka.abajue@uniport.edu.ng

W. E. Sawyer
Department of Community Medicine, Faculty of Clinical Sciences, Niger Delta University, Wilberforce Island, Bayelsa State, Nigeria

S. C. Izah
Department of Microbiology, Faculty of Science, Bayelsa Medical University, Yenagoa, Bayelsa State, Nigeria

M. C. Ogwu (✉)
Goodnight, Family Department of Sustainable Development, Appalachian State University, Boone, NC, USA
e-mail: ogwumc@appstate.edu

S. C. Izah et al. (eds.), *Herbal Medicine Phytochemistry*, Reference Series in Phytochemistry, https://doi.org/10.1007/978-3-031-43199-9_60

Abstract

Mosquitoes are insects with slender bodies, three pairs of jointed legs, a pair of wings, and an elongated mouthpart for sucking the blood of vertebrate animals including humans. The saliva of mosquitoes causes itchy rashes when deposited on the skin of humans during blood feeding. Many female mosquito species are capable of ingesting and/or transmitting infectious pathogens such as viruses that cause diseases like chikungunya fever, dengue fever, Japanese encephalitis, West Nile fever, yellow fever, Zika fever, and others to humans. These viral diseases are primarily transmitted through the bites of infected female mosquito (*Aedes*, *Anopheles*, and *Culex*) species in the Culicidae family. The listed viruses are endemic in the tropics and subtropics of Africa and Southeast Asia, where they cause the emergence and re-emergence of mosquito-borne viral diseases. These diseases are responsible for thousands of deaths reported annually within the affected regions. Treatment and management of mosquito-borne viral diseases with orthodox medicines are common in several parts of the world. But due to the toxicity of synthetic drugs, their alternatives (mainly from medicinal plants) are beginning to receive wider attention. Therefore, the age-long practice of using plants and herbs by herbalists in the tropics to treat different ailments with believable reports from patients and herbalists is being empirically evaluated presently. Based on verifiable reports of antiviral properties of some medicinal plants, this chapter intends to discuss medicinal plants diversity in the tropics, highlight recent updates on their use in the treatment and management of mosquito-borne viral diseases, and succinctly espouse some isolated and identified phytochemicals that are potentially active against the listed mosquito-borne viral diseases.

Keywords

Mosquito · Viruses · Diseases · Medicinal plants · Recent updates · Isolated phytochemicals

Abbreviations

CDCP	Centers for Disease Control and Prevention
DENV-1	Dengue virus 1
DENV-2	Dengue virus 2
DENV-3	Dengue virus 3
DENV-4	Dengue virus 4
HIV	Human immunodeficiency virus
IDPs	Internally displaced persons
RdRp	RNA dependable RNA polymerase

RNA	Ribonucleic acid
UV	Ultra violet
WHO	World Health Organization

1 Introduction

Mosquito-borne viral diseases are spread through bites of female mosquitoes that are infected with viruses. Mosquito-borne viral diseases in the tropics include chikungunya, dengue, Japanese encephalitis, West Nile fever, yellow fever, Zika, and others. Mosquitoes that carry these viruses (vectors) belong to a few genera – *Aedes*, *Anopheles*, and *Culex*. These mosquito genera are well distributed in all the countries of Africa, Asia, South America, and parts of Europe. Over seven hundred million people globally suffer from mosquito-borne diseases with more than one million deaths annually [1–3]. Among the diseases, chikungunya is caused by an alphavirus that is primarily transmitted through the bites of infected female *Aedes* mosquitoes (*Aedes aegypti*, *Aedes albopictus*, and *Aedes furcifer*) [4]. The virus is presently endemic in the tropics and subtropics of Africa and Southeast Asia. The transmission cycles in these regions are between non-human primates and *Aedes* mosquito species. For dengue, the disease is rather caused by flavivirus which is transmitted through the bites of infected female *Aedes* mosquitoes (*Aedes aegypti* and *Aedes albopictus*). Dengue virus is a positive-stranded ribonucleic acid in the family Flaviviridae that causes one of the major morbidity and mortality in the tropics and subtropics. The disease is causing approximately ten thousand deaths with about 100 million symptomatic infections every year in more than 125 countries of the world and is being transmitted by *Aedes* mosquitoes [5, 6]. Japanese encephalitis is also another flavivirus disease that is spread and transmitted by *Culex* mosquito species. The virus is the leading cause of encephalitis in many tropical Asian countries with about 68,000 cases of symptomatic infections annually. The disease is endemic in about 24 WHO countries in Southeast Asia and Western Pacific regions with about three billion people at risk of being infected [7].

For West Nile, the disease is caused by a positive-stranded ribonucleic acid flavivirus in the family Flaviviridae. The virus is transmitted through the bites of infected *Culex* mosquito species (*Culex quinquefasciatus*, *Culex stigmatosoma*, *Culex thriambus*, *Culex pipiens*, and *Culex nigripalpus*). These *Culex* species are responsible for infecting birds such as blue jays and crows with West Nile virus and can occasionally infect other mammals such as dogs, horses, and humans. Unfortunately, West Nile virus can as well be transmitted by other species of birds [8–11]. Yellow fever is also another disease caused by flavivirus. The virus is transmitted through the bites of infected female *Aedes* mosquitoes (*Aedes aegypti* and *Aedes albopictus*). The virus is endemic in both North and South America and in the tropics of Africa. For clearer distinction, yellow fever is capable of causing epizootic infections in non-human primates and yellow fever outbreaks in humans [12]. Thus, a section referred to as the sylvatic reservoir system is in the primates; hence, yellow fever transmission encompasses urban (human and urban *Aedes* mosquitoes) and sylvatic (non-human primates and forest-restricted *Aedes* mosquitoes) cycles. The sylvatic cycle is responsible for the major cases reported in South America while the transmission of the virus from monkeys to humans or from human to human through bites of infected mosquitoes is currently reported in Africa alone [13–17]. In the case of Zika, the disease is also caused by a flavivirus that is transmitted through the bites of infected female *Aedes* mosquito species (*Aedes aegypti*, *Aedes africanus*, *Aedes albopictus*, and *Aedes hensilli*). *Aedes albopictus* has been noted to be the likely major vector due to its link to the Zika virus outbreak in Gabon in 2007. Unfortunately, the virus has been reported to be transmitted from mother to foetus, through blood transmission, organ transplantation, laboratory exposure, and sexual transmission [18–20]. Outside the 2007 outbreak of Zika virus in Gabon, other outbreaks have occurred in Micronesia and Brazil [21]. Furthermore, over 6000 cases of symptomatic Zika virus infections have been reported between 2015 and 2017 in the USA [22].

These mosquito-borne viral diseases have caused a reasonable economic loss for both families and government and thus aggravate poverty in developing countries. For instance, a dengue episode causes an ambulatory patient to lose 14.8 working days and $514 while non-fatal hospitalized patients lose 18.9 working days and $1491 in eight burdened countries [23], hence making mosquitoes the most arthropod of medical and veterinary importance. Mosquitoes are insects in the family of Culicidae that are grouped with the common flies. They have slender bodies, three pairs of jointed legs, a pair of wings, and an elongated mouthpart for sucking the blood of vertebrate animals including humans. The saliva of a mosquito when transferred to humans during blood sucking can cause itchy rashes. Many female mosquito species are capable of ingesting and/or transmitting infectious diseases between humans or from animals to humans. Thus, different female mosquito species are transmitters of alphavirus and flavivirus pathogens, (Fig. 1) that cause mosquito-borne viral diseases such as chikungunya fever, dengue fever, Japanese encephalitis, West Nile fever, yellow fever, and Zika fever to humans [24].

Treatment and management of mosquito-borne viral diseases with orthodox medicines are very common. But due to the toxicity of orthodox-based medicine, their alternatives have gained attention. Based on the traditional use of plants and herbs by herbalists in the tropics to treat different ailments with believable reports from patients and herbalists, attention is being shifted to using medicinal plants, as a novel approach to treating viral diseases. Medicinal plants are highly diversified in the tropics and hence, have been used to treat and manage symptomatic viral infections by traditional healers. Medicinal plants in the tropics with complex bioactive phytochemicals that are rich for pharmaceutical purposes have been carefully selected and used in their crude forms by past generations, and the practice was handed over to the present generation based on folklores. There are many plant extracts and isolated compounds that have been reported to be antiviral agents and are less harmful, and less expensive as compared to orthodox medicines [25–28]. Based on verifiable reports of the antiviral properties of medicinal plants against viral diseases in the literature, this chapter shall discuss and tabulate medicinal plant

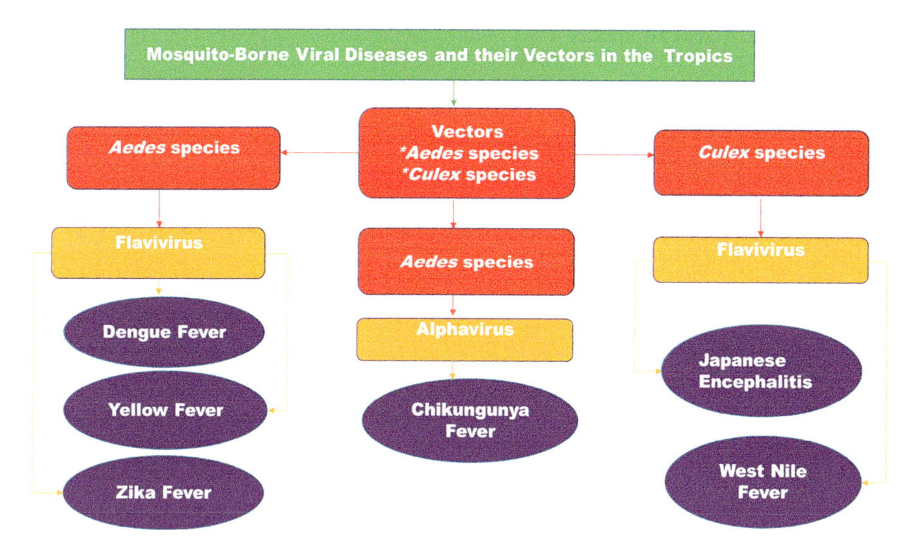

Fig. 1 Overview of viral diseases transmitted by mosquitoes in the tropics

diversity that is used in the treatment of mosquito-borne viral diseases in the tropics. This chapter will also highlight recent updates on the use of medicinal plants in the treatment and management of mosquito-borne viral diseases and as well espouse the isolated and identified phytochemicals that are potentially active against mosquito-borne viral diseases.

2 Overview of Distribution of Mosquito-Borne Viral Diseases in the Tropics

Environmental changes as a significant factor have induced viral diseases such as dengue fever which is an emerging serious public health importance and ranked the most concerned mosquito-borne viral disease with the potential of causing an epidemic. Dengue cases have recorded a multifold increase within the last 50 years with a staggering record of impacts on human and economic costs. One of the major fears about dengue is the fast spread of its vector: *Aedes aegypti* mosquito which has spread to more than 20 countries of Europe and to the Caribbean Islands where it was depicted to also vector chikungunya [24]. Mosquitoes are continually causing the emergence and re-emergence of mosquito-borne viral diseases that are threatening the world's public health systems and status. For instance, dengue fever estimated at 2.5 billion globally [29] affected about 75,454 people, and chikungunya fever affected 18,639 in India [30]. Other mosquito-borne viral diseases such as West Nile, Yellow fever, Zika, and others have also impacted human health and their socioeconomic importance is severely felt.

2.1 Chikungunya Fever

Chikungunya is a viral disease and its vectors are *Aedes aegypti* and *Aedes albopictus* mosquitoes. The virus which these *Aedes* mosquito species carry belongs to the Alphavirus genus in the family Togaviridae [31]. Chikungunya fever is thus a mosquito-borne disease that is now a worldwide public health threat. In the tropics, there are reports of outbreaks due to a high prevalence of *Aedes* mosquito species that transmit the virus to humans, circulation of the virus, and high population density in the urban cities of tropical countries [32]. *Aedes aegypti* is the most prevalent species in urban centres and is the major vector of the virus but as of 2006, *Aedes albopictus* was depicted to contribute to the spread of the virus to Europe and the Americas [33–36]. *Aedes aegypti* and *Aedes albopictus* are the major vectors of chikungunya in Asia and Indian Ocean region but other species of *Aedes*, *Culex annulirostris*, *Mansonia uniformis*, and *Anopheles* species are denoted to vector chikungunya virus [24].

There are three poly groups of the chikungunya virus (West African poly group; Eastern, Central, and Southern Africa poly groups; and the Asian poly group) that are distinctly separated geographically [37]. The virus causes chikungunya fever in man and was first diagnosed in Tanzania in the early 1950s. Since then, there have been subsequent outbreaks of the disease in sub-Saharan African countries and Asia [38]. Chikungunya epidemics have occurred in South-East Asia, India, Islands of the Pacific, and the Indian Ocean respectively in the last two decades leading to more than six million infections [4]. The virus has also been recorded in the Western Hemisphere with over two million cases of infection in about 50 American countries and 114 countries globally [39]. Chikungunya outbreaks were restricted in Africa (Democratic Republic of the Congo, Uganda, Central Africa Republic, Guinea, Angola, South Africa, Malawi, Nigeria, and other sub-Saharan countries) and Asia (Thailand, Cambodia, India, Myanmar, Philippines, Laos, Vietnam, Indonesia, Malaysia, and Sri Lanka) between the 1960s and 1980s [40]. There was an outbreak in a French territory in the Indian Ocean between 2005 and 2006, and about 266,000 inhabitants were affected. In 2007, the outbreak got to countries within the WHO South-East Asia Region and India with about 1.4 million cases. The disease was reported for the first time in Europe in 2007 in Northeast Italy. Similarly, the disease transmission was reported for the first time in 2013 at the WHO Region of the Americas, the Caribbean Island of Saint Martin, and other islands of the region [40].

There was a report of an average loss of more than 106,000 disability-adjusted life yearly, between 2010 and 2019, and in the year 2020, about 170,000 cases of chikungunya virus fever were reported globally, causing a higher rate of infant morbidity [41–43]. From 2004, chikungunya virus epidemics occurred in Kenya with 70% of her population being infected. Since then, it has spread to other countries like the Reunion Island, Thailand, Malaysia, India, Hong Kong, Taiwan, Sri Lanka, and other European and American countries [44–48]. Symptoms associated with acute infection include arthralgia/polyarthralgia, high fever, headache, myalgia, skin rashes, joint swelling, and nausea [49]. Chikungunya virus though causes low mortality with 0.07% death but patients with polyarthralgia can nurse it for months or even years after the acute phase of infection has been resolved and with joint pains that can last for months [41, 50].

2.2 Dengue Fever

Dengue fever is one of the viral infections imposing enormous economic burdens on the people of the tropics and subtropical countries. It is also referred to as backbone fever because it causes severe body pain mainly in the joints. The disease is caused by four flavivirus serotypes (dengue virus 1, dengue virus 2, dengue virus 3, and dengue virus 4) that belong to Flaviviridae family with dengue virus 2 being the most deadly virus. The virus is being harboured by female *Aedes aegypti*, *Aedes albopictus*, and *Aedes polynesiensis* mosquitoes which are known to thrive more in urban and semi-urban areas [51, 52]. Though the infection is asymptomatic and mild, the number of carriers is increasing considerably at the global level. Moreover, the actual number of infected cases is not comprehensively reported, or wrongly diagnosed with other febrile infections. The infected female mosquito bites in the day time and infect humans with the virus. More than 300 million people are reported to be infected every year, and about 22,000 of them are likely not to survive the infection [53].

Global infection of dengue fever is estimated at 2.5 billion with yearly recurrent or new cases at 50 to 100 million [51, 54, 55]. More than 40% of the world's population (amounting to about 2.5 billion people) are currently at risk of dengue infection. Going by WHO estimate, over 100 million people globally are infected annually, and about 500,000 of them especially children must be treated in the hospital while 2.5% of the infected persons cannot survive the disease. Dengue was earlier localized in just 9 countries of the world before 1970 but has now risen to endemic status in more than 100 countries (Africa, the Americas, South-East Asia, Western Pacific, and the Eastern Mediterranean). A report in 2010 showed that about 2.3 million in the Americas, Western Pacific regions, and South-East Asia are the ones mostly affected [24]. The disease is endemic in over 100 countries of Africa, Asia, America, Eastern Mediterranean, and Western Pacific regions. About 20 countries in Africa recorded dengue epidemics between 1960 and 2017 [56].

Due to the international trade of used tyres, timber, bamboo, and other similar stuff that provide breeding space for *Aedes*, another secondary dengue vector (*Aedes albopictus*) seen in Asia has spread to North America and Europe. *Aedes albopictus* is noted to have a wider geographical distribution because of its resilience and ability to survive in urban and rural environments. The eggs are resistant to harsh weather and therefore, can be viable all through the length of the dry season [24]. Female *Aedes aegypti* is the main vector of the dengue virus which is transmitted to humans when the female *Aedes* mosquito sucks blood from human beings. Once a female *Aedes* is infected, it can transmit the virus throughout its life cycle. Interestingly, *A. aegypti* are urban-restricted mosquitoes that breed in most man-made containers such as used tyres, beverage cans, plastic bottles, and the like. *Aedes* usually bite in the early hours of the morning and evening before dusk. The female can bite many people during its feeding phase. Infected individuals show signs of headache, fever, severe muscle and joint pains, and hemorrhagic fever which may lead to dengue shock syndrome which is a resultant effect of severe cases of dengue fever [57, 58].

Dengue fever is a severe flu-like illness that manifests with high fever, severe headache, muscle and joint pains, nausea, vomiting, swollen glands, or rash. Dengue is rarely fatal but severe ones come with possible serious complications, with symptoms that include low temperature, severe abdominal pains, and rapid breathing. Dengue virus has four serotypes (DENV-1, DENV-2, DENV-3, and DENV-4). Recovery from any of the serotypes confers a lasting immunity to the serotype. However, if infected by another serotype, it will still cause a severe dengue fever if not managed on time. Dengue infection presents symptoms that are related to malaria and other febrile conditions; hence, it is difficult to diagnose or predict if a patient is at risk or not at the time of medical examination [59]. Hemorrhagic fever was first reported in the 1950s during the dengue epidemics in Thailand and the Philippines. Presently, it is one of the primary roots of illnesses and mortality in many countries in Asia, India, and Latin America [60]. Untreated dengue cases may lead to complications such as respiratory distress, fluid accumulation, severe bleeding, organ damage, and others [61].

2.3 Japanese Encephalitis

Japanese encephalitis is a zoonotic disease (infectious disease transmitted from animals to humans) that contributes to one of the cases of morbidity and mortality in Southeast Asia and the Western Pacific area [62]. Japanese encephalitis is caused by a virus in the flavivirus (Flaviviridae) family. The virus is carried and transmitted to man by a *Culex* mosquito species and could cause permanent neuropsychiatric disorder in over 40% of cases and 20% death [63]. In some cases, the virus is transmitted by a tick. The disease affects mainly children and elderly people with compromised immunity [64].

Japanese encephalitis prevalence is highest in the tropics (China, Southeast Asia, Philippines, India, Japan, and Korea) where some major outbreaks have been reported [65]. Cases of Japanese encephalitis (mostly children below 5 years) are estimated at 50,000 leading to 10,000 deaths in a year. The disease exists in Asia at the islands of the Western Pacific in the East and extends to the Pakistanis border in the West and from the Republic of Korea in the North to Papua New Guinea in the South. Countries like Japan, the Korea Republic, Taiwan, and China are currently having considerably reduced cases but the disease has extended to India, Napa, and Sri Lanka where the disease is an important public health problem. Japanese encephalitis is mainly transmitted in rural agricultural provinces, especially among those living in flooded rice farms. It can as well occur near urban centres, and in temperate zones of Asia, as the virus transmission is seasonal [24].

Culex tritaeniorhynchus is the primary vector and transmits the virus; however, animals such as pigs and wading birds are potential transmitters of the virus. Humans are minor hosts as they may not be able to develop considerable concentrations of the virus in their bloodstreams that could infect female mosquitoes that are feeding. Japanese encephalitis virus when transmitted by an infected female *Culex* species to humans will usually show mild or no symptoms. Only a small fraction of humans infected develop inflammation of the brain that is associated with high fever, sudden onset of headache, disorientation, coma, tremors, and convulsions. About 25% of severe cases can be fatal while 30% of survivors would have lasting central nervous system damage [24].

2.4 West Nile Virus Disease

West Nile virus is a flavivirus-causing disease in the family Flaviviridae; the virus is transmitted by a *Culex* mosquito. West Nile virus disease presents fever, muscle weakness, encephalitis, and meningitis as symptoms without a known or medically acceptable treatment and no available vaccine yet. West Nile virus extends its distribution to Africa, Asia, Australia, and Europe and has spread to the USA, Mexico, Canada, and the Caribbean after being detected in New York City in 1999 [66]. It is maintained naturally in a *Culex* mosquito-bird-mosquito transmission cycle while man, horses, and other mammals serve as incidental hosts [67]. The majority of the individuals infected with the virus present no symptoms while about 20% of infected people experience some level of clinical symptoms such as West Nile fever but only a few will have severe West Nile neuro-invasive diseases, potentially fatal cases such as encephalitis, meningitis, acute fluid paralysis, fatigue, muscle weakness, and persistent tremors [68–70].

2.5 Yellow Fever

This disease is a known viral haemorrhagic fever that was before now regarded as a lethal disease until its vaccine was effectively developed. The disease is known to have an acute phase with symptoms such as fever, muscle pain, headache, shivers, loss of appetite, nausea, and vomiting. Most patients usually get relieved after 3–4 days but 15% of them enter a toxic phase making the patient to have jaundice and may be bleeding which could appear in the vomit. Factors such as less vaccinated population, urbanization, deforestation, climate change, and population movements have led to an increase in yellow fever cases; hence, about 200,000 infections and 30,000 cases of death are recorded yearly [24].

Yellow fever virus is transmitted by species of *Haemagogus* and *Aedes* mosquitoes to monkeys and humans. In African countries, yellow fever outbreaks have occurred in recent years as seasonal workers, nomadic, and displaced people were affected. North and Central America, the Caribbean, and Asia have had their share of the disease in the last 30 years due to an increase in air travel and sporadic increases in the distribution of *Aedes aegypti* in urban areas [24].

2.6 Zika Virus Disease

Zika virus is another flavivirus disease transmitted by *Aedes* species, which was identified in 1952 and circulated in Brazil in 2015 but was discovered in 1947 in Uganda with a transmission cycle between primates and wild *Aedes aegypti* which later infected humans [71]. The viral infection is now a serious threat to public health because of its suspected effect on the development of foetuses in pregnant women and malformations of congenital organs. It can as well cause meningitis and Guillain-Barre syndrome in severe cases [72–75].

Symptoms associated with Zika virus disease are similar to other febrile illnesses earlier stated, thus making its early diagnosis difficult. Zika virus infection is as well reported to be transmitted sexually; hence its global emerging epidemic threat has been warned by World Health Organization [76–78]. Proposed plans to fight the virus are to develop vaccines and screen molecules to inhibit different phases of the viral life-cycle [79–81].

3 Diversity of Medicinal Plants Reported to Be Used in the Treatment and Management of Mosquito-Borne Diseases in the Tropics

Based on the biology and life history of mosquitoes, it won't be out of place to say that mosquitoes are the most deadly insects globally. The assertion is because of many debilitating diseases transmitted to over one billion people every year [24]. Mosquitoes are vectors of many diseases such as malaria, dengue, yellow fever, lymphatic filariasis, chikungunya, Japanese encephalitis, West Nile fever, and Zika which collectively kill over one million people globally every year. Larvicidal and ovicidal activity of leaf and seed extract of plants against mosquitoes (*Aedes aegypti* and *Anopheles stephensi*) have been reported in some countries (India and Nigeria) [80–85].

Based on the World Health Organization's estimate, about 80% of the global population resort to medicinal plants as the basic source of their medical/healthcares [86]. Plants are known to produce phytochemicals such as alkaloids, flavonoids, essential oils, terpenoids, and polyphenols with different chemical structures. These chemicals are what plants use to defend themselves against microbial infection, herbivorous and insects attack, and environmental stress [87]. Phytochemicals have several organic activities that are antiviral, such as influenza, herpes simplex, hepatitis, and dengue, hence benefiting man, in fighting the above diseases [88–94].

Africa and countries in the tropics have a diverse range of medicinal/herbal plants that are used by traditional medicine herbalists for different ailments. However, there is a paucity of documented data on medicinal plants of the tropics in literature. In this section of the chapter, different reports of traditional medicines against mosquito-borne diseases in Africa and other countries in the tropics will be discussed. A concise review of scientific evaluation and validation of medicinal plants that have therapeutic and prophylactic properties against viral-causing diseases such as chikungunya, dengue, Japanese encephalitis, West Nile fever, yellow fever, and Zika fever. By extension, the records have shown that many plants in the tropics are medicinal and therefore, could be extended to the treatment and management of mosquito-borne viral diseases. This section, therefore, will highlight reports of different medicinal plants used in the treatment and management of mosquito-borne viral diseases that are documented in scientific literature.

3.1 Medicinal Plants for Treating and Managing Chikungunya Virus Fever

There are plant extracts that are reported to have anti-chikungunya virus properties (Table 1) with an interesting report that epigallocatechin gallate, derived from *Camellia sinensis* and *Curcuma longa* plants, can prevent the chikungunya virus from attaching to the host's cells and harringtonine from *Cephalotaxus harringtonia* blocking replication of the chikungunya virus in vitro study [95–97]. A study noted that extraction of phytochemicals for the treatment of diseases is dependent on the polarity of solvents; hence, hexane and chloroform are less polar solvents and therefore, could only extract alkaloids, coumarins, fatty acids, and terpenoids while more polar solvents such as ethyl acetate, ethanol, methanol, and water will extract saponins, tannins, flavones, polyphenols, terpenoids, anthocyanins, polypeptides, and lectins from plants [98].

Plants that have been reported to inhibit chikungunya virus at different levels of concentration include *Picrorhiza kurroa*, *Ocimum tenuiflorum*, *Terminalia chebula*, *Commiphora wightii*, *Cedrus deodara*, and *Zingiber officinale.* Among these plant

Table 1 Isolated phytochemicals reported to be used in the treatment and management of chikungunya

Isolated compound	Plant species	Plant's part	Medicinal activity	Reference
Coumarin (A and B)	*Mammea americana*	Seeds	*Active in Vero cells*	[37]
Curcumin	*Curcuma longa*	Bulb	Anti-chikungunya virus	[96, 99]
Berberine	*Berberis vulgaris*	Roots, stem-bark	Anti-chikungunya virus	[100–102]
Andrographolide	*Andrographis paniculata*	NS	Anti-chikungunya virus	[99]
Silymarin	*Silybum marianum*	Fruits, seeds	Anti-chikungunya virus	[103]
Baicalein	*Scutellaria baicalensis*	Roots	Anti-chikungunya virus	[104]
Silvestrol	*Aglaia foveolata*	NS	Anti-chikungunya virus	

Note: Phenols, saponins, alkaloids, flavonoids, and tannins are bioactive compounds contained in *Oroxylum indicum* plant which are anti-chikungunya virus [105, 106]; *NS* not stated

Table 2 Medicinal plants reported in India for the treatment and management of chikungunya fever

Plant family	Plant species
Acanthaceae	*Andrographis paniculata*
Asteraceae	*Pluchea lanceolata*
Brassicaceae	*Lepidium sativum*
Burseraceae	*Commiphora wightii*
Burseraceae	*Boswellia serrata*
Combretaceae	*Terminalia chebula*
Combretaceae	*Terminalia bellerica*
Cyperaceae	*Cyperus rotundus*
Euphorbiaceae	*Emblica officinalis*
Fabaceae	*Glycyrrhiza glabra*
Lamiaceae	*Ocimum sanctum*
Menispermaceae	*Tinospora cordifolia*
Ranunculaceae	*Nigella sativa*
Rubiaceae	*Rubia cordifolia*
Scrophulariaceae	*Picrorhiza kurroa*
Solanaceae	*Withania somnifera*
Verbenaceae	*Vitex negundo*
Zingiberaceae	*Zingiber officinale*
Zingiberaceae	*Curcuma longa*

Source: [107]

species, *Ocimum tenuiflorum* and *Terminalia chebula* inhibited 100% chikungunya virus at a concentration of 200 μg/ml [32]. Plants such as *Gynura bicolor*, *Hydrocotyle sibthorpioides*, *Ocimum americanum*, and others listed in Table 2 have anti-chikungunya virus; hence, their phytochemicals could be isolated and characterized as lead compounds in developing anti-chikungunya medicine. Based on extractants, water-extracted phytochemicals are weak against the chikungunya virus compared to other extractants [31]. Resistance and cost are some of the factors necessitating alternative means of anti-chikungunya virus. Thus, plant sources become a target due to their availability, cheapness, and diverse phytochemicals.

3.2 Medicinal Plants for Treating and Managing Dengue Virus Fever

Due to resistance, cost, and other medical logistics, the need to search for novel anti-dengue medications that are organically produced is voting for the use of medicinal plants which have been used widely by traditional medicine herbalists to treat febrile ailments. This option is anchored on safety, non-toxicity, and less harmful reports of the use of medicinal plants in the tropics. About 80% of Africans and Asians depend on traditional medicine as the basic healthcare for febrile diseases due to cost and geographical constraints [108]. At the moment, there is no specific treatment for dengue infection that is globally accepted among the medical professions. However, strategies to eliminate the disease are currently going on such as immunization, modulation of the host's immunity, and interfering with the host's ribonucleic acid [109]. Until these scientific innovations and breakthrough are presented, the traditional herbalists in the tropics have started using herbal formulations of plant-based materials to treat dengue infection which has been applauded by modern scientific reports. An example is the use of *Carica papaya*, a plant that has been used widely by traditional medicine herbalists to treat many febrile illnesses including dengue fever [59]. Hence, medicinal plants have been employed by traditional medicine herbalists, especially as one of the therapeutic approaches to treating and managing the disease. The practice of using medicinal plants in India to treat dengue viral infection is copious and has shown to be efficacious and better than using a single isolated compound. For instance, azadirachtin from *Azadirachta indica* was inefficient against dengue infection but the whole extracts strongly inhibited the dengue infection [60] which may be attributed to the synergistic effect of the compounds in the plant.

Medicinal plants have been used in developing countries to treat all known diseases and are as well becoming popular in developed countries [110]. Hence, 80% of the global population attends to their healthcare requirements by using medicinal plants [111]. Medicinal plants (Table 3) such as *Euphorbia hirta* and *Carica papaya* among other 31 species of plants are among the folkloric reports of medicinal plants against dengue infection in many African communities [112, 113]. The diversity of medicinal plants used in treating human-associated febrile infections such as dengue is estimated at 52,000 species [114]. These plants are reported to reverse and prevent dengue infection from reaching a critical stage by possessing different potentials to inactivate, inhibit, and increase blood platelet count, white blood cells, and neutrophils. Some can act as anti-dengue, prevent bleeding, and downregulate dengue infection [115].

Medicinal plants have diverse chemicals that are biologically capable to inhibit the virus replication cycle. As noted earlier, no medically approved drug is in the public domain for targeting the dengue virus. Other drugs that are approved to be positive for other viruses have been reported not to be effective as dengue virus is resistant to them, thus necessitating the use of medicinal plants to combat dengue disease. A study has presented isolated phytochemicals and their molecular structures that are depicted as anti-dengue (Table 4). The way these structured compounds work includes inhibition of the dengue virus at the post-infection stage, virus budding and secretion, binding of the peptide to the virus, and inhibition of the virus at an early stage of infection. Some will inhibit the dengue virus or have immunomodulatory activity, inhibit protease activity and secretion of virus particles, and inhibit entry. Some will block the activity of non-structural proteins (NS1, NS2, and NS3) especially NS3 protease, inhibit virus adsorption and host cell internalization, early stage of the virus cycle, and from binding to receptors. They can as well

Table 3 Medicinal plants reported for the treatment and management of dengue in the tropics

Plant family	Plant species	Reference
Acanthaceae	*Andrographis paniculata*	[115–117]
Acoraceae	*Acorus calamus*	[116]
Amaranthaceae	*Alternanthera philoxeroides*	[116, 117]
Caprifoliaceae	*Lonicera japonica*	[116]
Caricaceae	*Carica papaya*	[115–117]
Chordariaceae	*Cladosiphon okamuranus*	[113, 114]
Clusiaceae	*Garcinia mangostana*	[116]
Cucurbitaceae	*Momordica charantia*	[115]
Elaeagnaceae	*Hippophae rhamnoides*	[116, 117]
Euphorbiaceae	*Euphorbia hirta*	[116]
Euphorbiaceae	*Cladogynos orientalis*	[115]
Euphorbiaceae	*Euphorbia hirta*	[116, 117]
Fabaceae	*Acacia catechu*	[116]
Fabaceae	*Glycyrrhiza glabra*	
Fabaceae	*Leucaena leucocephala*	[115]
Fabaceae	*Mimosa scabrella*	
Fabaceae	*Tephrosia madrensis*	
Fabaceae	*Tephrosia crassifolia*	
Fabaceae	*Tephrosia viridiflora*	
Fabaceae	*Castanospermum australe*	[117]
Fabaceae	*Mimosa scabrella*	
Fagaceae	*Quercus lusitanica*	[116, 117]
Flagellariaceae	*Flagellaria indica*	[115]
Gramineae	*Cymbopogon citratus*	[116]
Halymeniaceae	*Cryptonemia crenulata*	[115]
Labiatae	*Ocimum sanctum*	
Lamiaceae	*Basilicum polystachyon*	[116]
Lamiaceae	*Ocimum tenuiflorum*	
Liliaceae	*Allium sativum*	
Marcgraviaceae	*Schwartzia brasiliensis*	
Meliaceae	*Azadirachta indica*	[116, 117]
Menispermaceae	*Cissampelos pareira*	[117]
Moraceae	*Ficus septica*	[116]
Myrtaceae	*Psidium guajava*	[115, 116]
Nyctaginaceae	*Boerhavia diffusa*	[116]
Orchidaceae	*Gastrodia elata*	[117]
Phyllanthaceae	*Phyllanthus urinaria*	
Phyllophoraceae	*Gymnogongrus torulosus*	[115]
Phyllophoraceae	*Gymnogongrus griffithsiae*	
Piperaceae	*Piper retrofractum*	
Poaceae	*Cymbopogon citratus*	
Rhizophoraceae	*Rhizophora apiculata*	[116, 117]
Rubiaceae	*Oldenlandia uniflora*	[116]

(continued)

Table 3 (continued)

Plant family	Plant species	Reference
Rubiaceae	*Pavetta canescens*	
Rubiaceae	*Tarenna asiatica*	
Rubiaceae	*Uncaria tomentosa*	[115]
Sapindaceae	*Doratoxylon apetalum*	[116]
Sapindaceae	*Nephelium lappaceum*	
Saururaceae	*Houttuynia cordata*	[115]
Schisandraceae	*Schisandra chinensis*	[116]
Solieriaceae	*Meristiella gelidium*	[115]
Verbenaceae	*Lippia alba*	
Verbenaceae	*Lippia citriodora*	
Verbenaceae	*Lippia citriodora*	[117]
Zingiberaceae	*Curcuma longa*	[116]
Zingiberaceae	*Kaempferia parviflora*	
Zingiberaceae	*Boesenbergia rotunda*	[116, 117]
Zosteraceae	*Zostera marina*	[115–117]

decrease viral RNA production, act as virucidal such as catechin, and inhibition of viral particle assemblage [110].

3.3 Medicinal Plants for Treating and Managing Japanese Encephalitis

In a quest to also treat patients infected with the Japanese encephalitis virus with herbal formulations, a study [157] evaluated the antiviral activity of *Isatis indigotica* plant. The author reported low cytotoxicity of the plant but potentially inhibited Japanese encephalitis virus replication in vitro study. The plant's extract reduced the virus yield, block the virus attachment to the host, and showed virucidal activity. *Isatis indigotica* extracts were reported to contain active compounds such as indigotin, indirubin, indican, and isatin. Among these isolated phytochemical compounds of *I. indigotica*, indirubin was appraised to have strong protective potential against the Japanese encephalitis virus and proved less toxic to the host's cells.

Because of the need to develop safe and cheaper antiviral medicines, Roy et al. [158] also evaluated *Trachyspermum ammi* plant against the Japanese encephalitis virus. The authors reported that the essential oil obtained from the plant through distillation was able to inhibit the virus in vitro study, thus showing the anti-Japanese encephalitis virus activity. The essential oil of *Trachyspermum ammi* is reported to contain thymol, α-pipene, p-cymen, and limonene, which are its major curative agents against many diseases caused by fungi, bacteria, and viruses. The essential oil inhibited the Japanese encephalitis virus with low cytotoxicity and without the formation of plagues [159].

Table 4 Isolated phytochemicals reported to be used in the treatment and management of dengue

Isolated compound	Plant species	Plant's part	Medicinal activity	Reference
Castanospermine	*Castanospermum australe*	NS	Against dengue activity	[118]
Quercetin, quercetinin, chlorogenic acid, hyperoside	**Houttuynia cordata*	Leaves	Against dengue activity	[119, 120]
Rutin	*Spondias mombin*		Inhibition, anti-replication of DEV-2	[121]
Quercetin, ellagic acid	*Spondias tuberosa*	NS	Inhibition, anti-replication of DEV-2	
Fucoidan	*Cladosiphon okamuranus*	NS	Inhibition of dengue virus 2	[122]
Galactans	*Cryptonemia crenulata*, *Gymnogongrus torulosus*	NS	Inhibition of DEV-1,3, anti-replication in Vero cell	[123]
Kappa-carrageenan	*Gymnogongrus griffithsiae*, *Meristiella gelidium*	NS	Inhibition of dengue virus 2 and 3	
Trigocherrins A, trigocherriolides B and C	*Trigonostemon cherrieri*	Bark, wood	Dengue RdRp	[124]
Tatanan A	*Acorus calamus*	Roots	Anti-dengue virus	[125]
Andrographolide	*Andrographis paniculata*	Leaves	Inhibition of dengue virus 2	[126–128]
Nimbin, desacetyl nimbin, salanin	*Azadirachta indica*	Leaves	Inhibition of dengue virus 2	[129, 130]
Kaempferol 3-O-rutinoside, epicatechin	*Azadirachta indica*	Leaves	Inhibition of dengue virus 2	[131]
Gedunin, pigmol	*Azadirachta indica*	Leaves	Anti-dengue virus	[132]
Pinostrobin, pinocembrin, alpinetin, cardamonin, panduratin A	*Boesenbergia rotunda*	Rhizomes	Anti-dengue virus, dengue virus 2	[133–135]
Pareirarine, cissamine, magnoflorine	*Cissampelos pareira*	Flowers	Inhibition of dengue virus 2, 3	[93, 133–136]
Rutin, scopoletin	*Cladogynos orientalis*	Leaves, roots/stem	Anti-dengue virus, dengue virus 2	[137, 138]
Galatan	*Cryptonemia crenulata*	Alga	Anti-dengue virus, dengue virus 2, 3, 4	[135]
Curcumin	*Curcuma longa*	Rhizomes	Inhibition of dengue virus	[135, 140]
Pectolinarin, acacetin-7-O-rutinoside	*Distictella elongata*	Leaves, fruits	Anti-dengue virus 2	[93, 139]
Quercetin	*Hippophae rhamnoides*	Leaves	Anti-dengue virus 2	[93, 115, 135, 140]

(continued)

Table 4 (continued)

Isolated compound	Plant species	Plant's part	Medicinal activity	Reference
Linoleum, limonene	*Lippia alba*	NS	Inhibition of dengue virus in Vero cells	[141]
Catechin, gallic acid, naringin, quercetin	*Psidium guajava*	Roots, leaves, bark	Anti-dengue	[142]
Baicalein	*Scutellaria baicalensis*	Root	Inhibition of dengue virus 1, 2, 3, 4	[143, 144]
Fucoidan, glucuronic, fucose	*Cladosiphon okamuranus*	Alga	Impede dengue virus 2; anti-dengue effect	[122, 145]
Azadirachtin	*Azadirachta indica*	Leaves	Anti-dengue virus 2	[130]
Castanospermine	*Castanospermum austral*	Seed	Anti-dengue virus 1,2,3,4	[118]
Chartaceones A-F	*Cryptocarya chartacea*	Bark	Anti-dengue virus 2	[146]
Betulinic acid 3 β-caffeate	*Flacourtia ramontchi*	Stem/bark	Anti-dengue virus	[147]
DL-galatan	*Gymnogongrus torulosus*	Red seaweed	Anti-dengue virus 2	[148]
α mangostin	*Garcinia mangostana*	Fruit	Anti-dengue virus 1,2,3, 4	[149]
Galactomannans	*Mimosa scabrella*	Seeds	Anti-dengue virus 1	[150]
Geranin	*Nephelium lappaceum*	NS	Anti-dengue virus 2	[151]
Catechin	*Psidium guajava*	Bark	Anti-dengue virus 2	[152]
Schisandrin A	*Schisandra chinensis*	NS	Anti-dengue virus 1,2,3, 4	[153]
Trigocherrin A1, A, B	*Trigonostemon cherrieri*	Bark, wood	NS	[124]
Hirsutine	*Uncaria rhynchophylla*	NS	Anti-dengue virus 1,2,3,4	[154]
Zosteric acid	*Zoster marina*	NS	Anti-dengue virus 1,2,3,4	[155]
Diallyl disulphide	*Allium sativum*	NS	Anti-dengue virus 2	[156]

Note: * when the concentration of *Houttuynia cordata* compound is used individually, the inhibition rate would be moderate but when combined with others, it will result in a greater rate of inhibition activity; *NS* not stated, *RdRp* RNA dependable RNA polymerase

3.4 Medicinal Plants for Treating and Managing West Nile Virus Disease

Medicinal plants with their isolated and purified organic products are rich resources of anti-virus which have been shown to target viral life cycles either as inhibitors of virus entry, prevention of replication, gathering, or release of virus [160]. West Nile virus is less lethal but its wild spread and re-emergence is posing a global health

Table 5 Isolated phytochemicals reported to be used in the treatment and management of West Nile virus

Isolated compound	Plant species	Plant's part	Medicinal activity	Reference
Delphinidin	NS	Flowers, fruits	Alteration of WNV virus replication, virucidal	[162]
Epigallocatechin	*Camellia sinensis*	Leaves	Alteration of WNV virus replication, virucidal	
Palmatine	*Coptis chinensis*	NS	Inhibition of NS2B-NS3	

Note: *NS* not stated, *WNV* West Nile virus, *NS2B* non-structural protease 2B, *NS3* non-structural protease 3

threat due to its asymptomatic characteristics; hence, Goh et al. [161] reported few phytochemical compounds that are capable of altering, preventing, and inhibiting West Nile virus (Table 5).

3.5 Medicinal Plants for Treating and Managing Yellow Fever

A few surveys of medicinal plants used by traditional medicinal herbalists used in the treatment and management of yellow fever showed about 95 species of plants in 52 families (Table 6). The bark and leaves of these plants are the most frequently used. The plants are reported not to only be cheap and affordable but prevent, and treat, yellow fever patients with acceptable tolerance and no serious side effects [163]. The majority of the African population are sceptical about receiving vaccines, especially in rural communities; hence, they prefer to treat and manage yellow fever with herbal medicine. For instance, *Cochlospermum tinctorium*, *Eucalyptus globulus*, *Mangifera indica*, and *Musa sapientum* are reported as medicinal plants for treating and managing yellow fever in some parts of Central and Northern parts of Nigeria. *Musa sapientum* is the most reported plant in the treatment of yellow fever. Other plants used in the treatment of yellow fever include *Azadirachta indica*, *Alstonia boonei*, *Anacardium occidentale*, *Carica papaya*, *Citrus aurantifolia*, *Senna occidentalis*, *Zingiber officinale*, and others [163].

3.6 Medicinal Plants for Treating and Managing Zika Virus Disease

A few medicinal plants evaluated against Zika virus are *Berberis vulgaris* which contains berberine compound and *Rheum palmatum* and *Coptis* species, which contain emodin compound. These compounds have been shown to be virucidal, reducing about 80% of virus infectivity in Vero cells. Emodin, for instance, can block the entry of viruses and inhibit the processes associated with viral particle release from cells [166] and also has a modest effect as a pre-treatment medicine by

Table 6 Medicinal plants reported for the treatment and management of yellow fever in the tropics

Family	Species	Reference
Acanthaceae	*Hygrophila auriculata*	[163]
Aizoaceae	*Trianthema pentandra*	
Alliaceae	*Allium cepa*	[164]
	Allium sativum	
Anacardiaceae	*Mangifera indica*	[163, 164]
	Anacardium occidentale	[163]
	Sclerocarya birrea	
	Spondias mombin	
Annonaceae	*Enantia chlorantha*	[163, 164]
	Annona senegalensis	[163]
Apocynaceae	*Strophanthus kombe*	[164]
	Leptadenia hastata	[163]
	Alstonia boonei	[165]
Arecaceae	*Cocos nucifera*	[164]
	Elaeis guineensis	[163]
Aristolochiaceae	*Aristolochia ringens*	
Asteraceae	*Chromolaena odorata*	[164]
Bignoniaceae	*Stereospermum kunthianum*	[163]
	Kigelia africana	
Bixaceae	*Cochlospermum tinctorium*	
Bombacaceae	*Adansonia digitata*	
Boraginaceae	*Cordia africana*	
Bromeliaceae	*Ananas comosus*	[163, 164]
Caesalpiniaceae	*Senna alata*	[164]
	Senna occidentalis	
Caricaceae	*Carica papaya*	[163, 164]
Celastraceae	*Hippocratea indica*	[164]
Celastraceae	*Celastrus indica*	[163]
Clusiaceae	*Garcinia kola*	
Combretaceae	*Guiera senegalensis*	
	Terminalia avicennioides	
	Anogeissus leiocarpus	
	Combretum micranthum	
Compositae	*Vernonia amygdalina*	
Connaraceae	*Byrsocarpus coccineus*	[163, 164]
Cucurbitaceae	*Citrullus lanatus*	[163]
	Momordica charantia	[164]
Ebenaceae	*Diospyros mespiliformis*	[163]
Euphorbiaceae	*Jatropha curcas*	[164]
	Sapium grahamii	[163]
	Euphorbia hirta	
	Euphorbia unispina	
	Croton lobatus	[164]

(continued)

Table 6 (continued)

Family	Species	Reference
Euphorbiaceae	*Ricinus communis*	[163]
Fabaceae	*Mucuna solan*	[164]
	Parkia biglobosa	[163]
	Prosopis africana	
	Cassia occidentalis	
	Detarium senegalense	
	Tamarindus indica	[163]
	Abrus precatorius	[164]
	Cassia tora	
Gentianaceae	*Anthocleista djalonensis*	[164]
Gramineae	*Cymbopogon citratus*	[165]
Lamiaceae	*Ocimum gratissimum*	[164]
Lauraceae	*Cinnamomum verum*	[163]
Lythraceae	*Lawsonia inermis*	[164]
Malvaceae	*Sida acuta*	[165]
Meliaceae	*Khaya senegalensis*	[164]
Meliaceae	*Azadirachta indica*	[163]
	Entandrophragma utile	
Mimosaceae	*Parkia biglobosa*	
	Acacia nilotica	
Moraceae	*Ficus asperifolia*	[164]
	Ficus sycomorus	[163]
	Ficus platyphylla	
	Ficus polita	
Moringaceae	*Moringa oleifera*	[163]
Musaceae	*Musa sapientum*	
Myristicaceae	*Pycnanthus angolensis*	
Myrtaceae	*Eucalyptus globulus*	
	Psidium guajava	
	Syzygium aromaticum	
Nymphaeaceae	*Nymphaea lotus*	
Oleaceae	*Olea europaea*	
Phyllanthaceae	*Securinega virosa*	
Piperaceae	*Piper guineense*	
Plumbaginaceae	*Plumbago zeylanica*	
Poaceae	*Sorghum caudatum*	[164]
	Cymbopogon citratus	[163]
	Saccharum officinarum	
Rhamnaceae	*Ziziphus mauritiana*	
Rubiaceae	*Morinda lucida*	
	Nauclea diderrichii	
Rutaceae	*Citrus aurantifolia*	[163, 164]
	Citrus aurantium	[164]

(continued)

Table 6 (continued)

Family	Species	Reference
Sapotaceae	*Nicotiana tabacum*	
	Solanum torvum	[165]
Tiliaceae	*Corchorus olitorius*	[164]
	Triumfetta cordifolia	
Xanthorrhoeaceae	*Aloe barteri*	[163]
Zingiberaceae	*Aframomum melegueta*	
	Curcuma longa	[164]
	Zingiber officinale	

interfering with virus entry [167]. At low concentrations, berberine and emodin are effective prophylaxis agents as berberine has the potential to spread to organs (liver, kidneys, brain), and muscles, and can remain in those places [168], which is a preferred distinction needed in the treatment and management of Zika virus [167]. Berberine is also reported to have low toxicity as no serious side effects have been reported in humans [169–171]. Thus, berberine and emodin phytochemical compounds are considered for their virucidal effect and impair Zika virus particles in vitro study. In endemic regions, both berberine and emodin are acceptable recommendations for the prevention, treatment, and management of patients infected with the Zika virus [167].

Apart from the general reports that plant products (polyphenols, flavonoids, alkaloids, terpenes, and others) have strong potential to inhibit the replication of the Zika virus, some isolated phytochemical compounds (Table 7) have been reported to work against Zika virus in humans. For instance, curcumin which is a phytochemical compound from *Zingiber officinale*, has been shown to have a strong potential against the Zika virus by possessing inhibiting potential on the viral envelope and can be very useful, especially during outbreaks of Zika infection [162, 172]. Similarly, phytochemical extracts and compounds isolated from seaweed plants are evaluated for anti-Zika virus prospect [173] while another study [174] reported that *Aphloia theiformis* extracts affected the organization of Zika virus particles severely, based on electron microscope examination.

4 Recent Updates on the Use of Medicinal Plants in the Treatment and Management of Mosquito-Borne Viral Diseases

Plants have emerged as credible candidate for replacing synthetic drugs. Hence, their discovery is a major boost to the fields of pharmaceutical chemistry and microbiology, herbal sciences, and humanity at large. Several plants have been reported to be active in the treatment of several diseases, especially those caused by microbes [176–195]. Viruses are seen as microbes in the field of microbiology, and the use of

Table 7 Isolated phytochemicals reported for the treatment and management of Zika virus in the tropics

Isolated compound	Plant species	Plant's part	Medicinal activity	Reference
Berberine	*Berberis vulgaris*	NS	Prevent virus entry; replication	[167]
Emodin	*Rheum palmatum*	NS	Anti-Zika virus, virucidal	
	Polygonum multiflorum	NS		
	Cassia obtusifolia			
Curcumin	*Zingiber officinale*	Root	Inhibit Zika virus	[166]
Lycorine	*Hippeastrum puniceum*	Flower	Anti-Zika virus	[175]
Pretazettine		Root		
Narciclasine		Bulb		
Narciclasine-4-O		NS		
β-D-xylopyranoside		NS		

Note: *NS* not stated

plants for the treatment of diseases caused by viruses has been a mere claim until now.

Basic scientific inquiries on the use of medicinal plants in the treatment and management of mosquito-borne viral diseases have been accentuated within the last few years due to viral infections such as COVID-19 and Ebola virus diseases leading to global pandemics and epidemics in some West African countries, respectively. Hence, many medicinal plants are depicted to have reasonable secondary metabolites against viral diseases. Empirical reports about isolated phytochemical compounds from medicinal plants which have performed optimally in many medical experiments have been reported. Thus, the formation of novel medicine from phytochemical constituents and plant-isolated compounds is beginning to be a major part of treating mosquito-borne viral diseases, especially in developing countries.

Based on the research reports succinctly discoursed above, it is glaring that a greater percentage of the population in the tropics use plant-based medicine as a basic form of their healthcare. Reasons for this practice are but not limited to the cost and safety of plant-based medicines which have been used by the generations past with little or no negative reports, culture and religious beliefs, inaccessibility of orthodox medicine due to higher cost, lack of primary healthcare centres, and trained professionals especially in rural communities [196]. Fortunately, a greater percentage of infected victims survive upsurges of viral infections (epidemics and pandemics) by relying on plant-based medicines which have drawn the attention of researchers to understudy the potentials of these plants that are used to treat and manage mosquito-borne viral diseases. In West Africa alone, about 124 species in 50 families of medicinal plants are reported to have antiviral properties including viral diseases transmitted by mosquito species, hence concurring with a report of higher medicinal plant diversity in the tropics [197–199].

Recent updates on the use of medicinal plants in the control [200–204], treatment, and management of mosquito-borne viral diseases are showing that about 52% which translate to 65 out of the 124 species are the ones that have been scientifically ascertained to have therapeutic potentials against mosquito-borne viral diseases especially dengue and other non-mosquito-borne viral diseases [196]. Progressively, 4 of the 65 plant species have their essential phytochemicals such as flavonoids, quercetin, morin, fisetin, naringenin, and hesperidin from *Citrus aurantifolia* and *Citrus paradise*. Alkaloids such as salidroside 2(4-hydroxyphenyl) from *Cucumis metuliferus* and ethyl β-D-glucopyranoside from *Loranthus micranthus*. Flavonoids (3,5-dicaffeoylquinic acid, acteoside, kaempferol 7-O-glucoside, bastadin-11) and stilbenes (vedelianin, schweinfurthin G, mappain) from *Macaranga barteri* are identified and isolated compounds responsible to potentially having countering activity against viral diseases. Other natural antiviral compounds isolated but not from plants such as mushroom (*Hypoxylon fuscum*) and lichen (*Ramalina farinacea*) include dihydropenicillic acid and sekikaic acid, respectively, with all the structures presented for further investigation and possible replication and production for treatment and management of viral diseases, including mosquito-borne viral diseases [196].

The use of medicinal plants in the treatment of mosquito-borne viral diseases is, therefore, becoming famous due to the bioactive compounds which are a very rich source of pharmaceuticals that contain reasonable antiviral properties. Hence, the World Health Organisation (WHO) and other traditional-based medicinal professionals have proposed that using medicinal plant extracts and their isolated compounds have been very useful in dealing with diseases especially dengue and other related viral diseases [205]. The WHO has gone ahead to consider isolated compounds from medicinal plants to be largely safe, non-toxic, and by comparison, assumed to be less harmful, and cheaper than orthodox medicines against viral diseases [205]. Currently, many researchers are scientifically exploring medicinal plants in an attempt to identify the active compounds that have antiviral properties since many of the secondary metabolites of the medicinal plants have been widely reported to have been used to treat and manage many viral ailments including mosquito-borne diseases such as chikungunya, dengue, Japanese encephalitis, West Nile fever, yellow, Zika, and others [206, 207].

5 Conclusion

Mosquito-borne diseases are contributing to a greater percentage of deaths in the world especially in the tropics, with their emerging and re-emerging infections being an untiring peril to the human population. Mosquito-borne diseases especially viral infections are seriously challenging the survival of the human population, especially children, gravid mothers, and immune-compromised adults. Mosquito-borne viral diseases remain one of the major causes of deaths globally with leading records of outbreaks in the human population and epizootic in non-human vertebrates. Mosquito-borne viral infections have caused severe economic problems in the tropics and subtropical regions of the world but are more endemic in Africa, America, Asia, Eastern Mediterranean Province, and Western Pacific. Regrettably, some of these diseases are prevalent in urban and semi-urban cities and are spread and transmitted by infected female *Aedes*, *Anopheles*, and *Culex* species. Symptomatic cases are generally presented with febrile characteristics, and their treatments vary depending on the orthodox therapeutic agents available. However, orthodox medicines are beginning to fail to meet the required expectations.

Presently, there are reports that medicinal plants and their bioactive compounds are effective against viral infections by either reducing viral loads or inhibiting viral multiplication. The diversity of the reported medicinal plants is very high in the tropics; thus, they have been used in the ancient days till now by herbalists who inherited or learned the art of using such plants to treat different ailments. Hence, a greater percentage of people in the tropics rely mainly on medicinal plants to treat and prevent diseases including mosquito-borne viral infections. Based on folklore and testimonies of patients and herbal medicinal practitioners, medicinal plants in the tropics could be a major source of producing natural antiviral medicines. Many medicinal extracts from various parts of plants such as leaves, flowers, bark, roots, stems, seeds, and whole plant and their isolated phytochemicals have been extensively reported to have reasonable antiviral properties based on; in vitro and in vivo studies. These empirical reports are corroborative of folkloric reports of medicinal plant potentials, which are used in dealing with viral infections and thus form the basis of recent reports of the use of medicinal plants to treat and manage mosquito-borne viral diseases in the tropics.

References

1. Kraemer MUG, Sinka ME, Duda KA, Mylne AQN, Shearer FM, Barker CM, Moore CG, Carvalho RG, Coelho GE, Bortel WV et al (2015) The global distribution of the arbovirus vectors *Aedes aegypti* and *Ae. Albopictus*. elife 4:e08347
2. Sinka ME, Bangs MJ, Manguin S, Rubio-Palis Y, Chareonviriyaphap T, Coetzee M, Mbogo CM, Hemingway J, Patil AP, Temperley WH et al (2012) A global map of dominant malaria vectors. Parasit Vectors 5:69
3. Piovezan-Borges AC, Valente-Neto F, Urbieta GL, Laurence SGW, de Oliveira RF (2022) Global trends in research on the effects of climate change on *Aedes aegypti*: international collaboration has increased, but some critical countries lag behind. Parasit Vectors 15:346
4. Silva LA, Dermody TS (2017) Chikungunya virus: epidemiology, replication, disease mechanisms, and prospective intervention strategies. J Clin Investig 127:73–749
5. Benelli G, Jeffries CL, Walker T (2016) Biological control of mosquito vectors: past, present, and future. Insects 7:52
6. Benelli G, Mehlhorn H (2016) Declining malaria, rising of dengue and Zika virus: insights for mosquito vector control. Parasitol Res 115:1747–1754
7. WHO (2019) Japanese encephalitis fact sheets. https://www.who.int/news-room/fact-sheets/details/japanese-encephalitis. Accessed 6 May 2023
8. McDonald E, Mathis S, Martin SW, Erin Staples J, Fischer M, Lindsey NP (2021) Surveillance for West Nile virus disease – United States, 2009–2018. Am J Transplant 21:1959–1974

9. Ahlers LRH, Goodman AG (2018) The immune responses of the animal hosts of West Nile virus: a comparison of insects, birds, and mammals. Front Cell Infect Microbiol 8:96
10. Mencattelli G, Ndione MHD, Rosà R, Marini G, Diagne CT, Diagne MM, Fall G, Faye O, Diallo M, Faye O et al (2022) Epidemiology of West Nile virus in Africa: an underestimated threat. PLoS Negl Trop Dis 16:e0010075
11. Kain MP, Bolker BM (2019) Predicting West Nile virus transmission in North American bird communities using phylogenetic mixed effects models and eBird citizen science data. Parasit Vectors 12:395
12. Mboussou F, Ndumbi P, Ngom R, Kassamali Z, Ogundiran O, Beek J, Williams GS, Okot C, Hamblion EL, Impouma B et al (2019) Infectious disease outbreaks in the African region: overview of events reported to the World Health Organization in 2018. Epidemiol Infect 147: e307
13. Valentine MJ, Murdock CC, Kelly PJ (2019) Sylvatic cycles of arboviruses in non-human primates. Parasit Vectors 12:463
14. Cavalcante KRLJ, Tauil PL (2017) Risk of re-emergence of urban yellow fever in Brazil. Epidemiol Serv Saude Brasília 26:617–620
15. Couto-Lima D, Madec Y, Bersot MI, Campos SS, Motta MDA, Dos Santos FB, Vazeille M, Vasconcelos P, Lourençode-Oliveira R, Failloux AB (2017) Potential risk of re-emergence of urban transmission of yellow fever virus in Brazil facilitated by competent *Aedes* populations. Sci Rep 7:4848
16. Câmara FP, Gomes ALBB, Carvalho LMF, Castello LGV (2011) Dynamic behavior of sylvatic yellow fever in Brazil (1954–2008). Rev Soc Bras Med Trop 44:297–299
17. WHO (2018) A global strategy to eliminate yellow fever epidemics 2017–2026. WHO, Geneva. https://www.who.int/csr/disease/yellowfev/eye-strategy/en/. Accessed 12 Apr 2023
18. Plourde AR, Bloch EM (2016) A literature review of Zika virus. Emerg Infect Dis 22:1185–1192
19. Grard G, Caron M, Mombo IM, Nkoghe D, Mboui Ondo S, Jiolle D, Fontenille D, Paupy C, Leroy EM (2014) Zika virus in Gabon (Central Africa) – 2007: a new threat from *Aedesalbopictus*. PLoS Negl Trop Dis 8:e2681
20. Brent C, Dunn A, Savage H, Faraji A, Rubin M, Risk I, Garcia W, Cortese M, Novosad S, Krow-Lucal ER et al (2016) Preliminary findings from an investigation of Zika virus infection in a patient with no known risk factors. MMWR Morb Mortal Wkly Rep 65:981–982
21. Gatherer D, Kohl A (2016) Zika virus: a previously slow pandemic spreads rapidly through the Americas. J Gen Virol 97:269–273
22. Lee S, Ezinwa N (2018) Mosquito-borne diseases. Prim Care Clin Off Pract 45:393–407
23. Suaya JA, Shepard DS, Siqueira JB, Martelli CT, Lum LC, Tan LH et al (2009) Costs of dengue cases in eight countries in the Americas and Asia: a prospective study. Am J Trop Med Hyg 80(5):846–855
24. WHO (2014) A global brief on vector-borne diseases. World Health Organisation, Geneva. Accessed 12 Mar 2023
25. Okonkwo EE (2012) Traditional healing systems among Nsukka Igbo. J Tour Herit Stud 1: 69–81
26. Agnaniet H, Mbot EJ, Keita O, Fehrentz JA, Ankli A, Gallud A, Garcia M, Gary-Bobo M, Lebibi J, Cresteil T, Menut C (2016) Antidiabetic potential of two medicinal plants used in Gabonese folk medicine. BMC Complement Altern Med 16:71
27. Wang M, Tao L, Xu H (2016) Herbal medicines as a source of molecules with anterovirus 71 activity. Chin Med 11:2
28. Babar M, Najam-us-Sahar SZ, Ashraf M, Kazi AG (2013) Antiviral drug therapy-exploiting medicinal plants. J Antivir Antiretrovir 5:28–36
29. Kumar A, Valecha N, Jain T, Dash AP (2007) Burden of malaria in India: retrospective and prospective view. Am J Trop Med Hyg 77:69–78
30. Tyagi BK (2016) Advances in vector mosquito control technologies, with particular reference to herbal products. In: Veer V, Gopalakrishnan R (eds) Herbal insecticides, repellents and biomedicines: effectiveness and commercialization. Springer Nature India, pp 1–9

31. Chan SM, Khoo KS, Sekaran SD, Sit NW (2021) Mode-dependent antiviral activity of medicinal plant extracts against the mosquito-borne chikungunya virus. Plan Theory 10:1658
32. Raghavendhar S, Tripati PK, Ray P, Patel AK (2019) Evaluation of medicinal herbs for anti-CHIKV activity. Virology 533(2019):45–49
33. Dhimal M, Gautam I, Joshi HD, O'Hara RB, Ahrens B, Kuch U (2015) Risk factors for the presence of chikungunya and dengue vectors (*Aedes aegypti* and *Aedesalbopictus*), their altitudinal distribution and climatic determinants of their abundance in central Nepal. PLoS Negl Trop Dis 9(3):e0003545
34. Palha N, Guivel-Benhassine F, Briolat V, Lutfalla G, Sourisseau M, Ellett F et al (2013) Real-time whole-body visualization of chikungunya virus infection and host interferon response in zebra fish. PLoS Pathog 9(9):1003619
35. Plante K, Wang E, Partidos CD, Weger J, Gorchakov R, Tsetsarkin K et al (2011) Novel chikungunya vaccine candidate with an IRES-based attenuation and host range alteration mechanism. PLoS Pathog 7(7):e1002142
36. Schwartz O, Albert ML (2010) Biology and pathogenesis of chikungunya virus. Nat Rev Microbiol 8(7):491–500
37. Mohamat SA, Mat NFC, Barkhadle NI, Jusoh TNAM, Shueb RH (2020) Chikungunya and alternative treatment from natural products: a review. Mal J Med Health Sci 16(1):304–311
38. Ross RW (1956) The Newala epidemic: III. The virus: isolation, pathogenic properties and relationship to the epidemic. J Hyg 54:177–191
39. CDCP Centers for Disease Control and Prevention. Chikungunya virus. https://www.cdc.gov/chikungunya/index.html. Accessed 12 Mar 2023
40. Wahid B, Ali A, Rafique S, Idrees M (2017) Global expansion of chikungunya virus: mapping the 64-year history. Int J Infect Dis 58:69–76
41. Puntasecca CJ, King CH, LaBeaud AD (2021) Measuring the global burden of chikungunya and Zika viruses: a systematic review. PLoS Negl Trop Dis 15:e0009055
42. ECDP 2020 European Centre for Disease Prevention and Control. Commun Dis Threats Rep Week 51:13–19. https://www.ecdc.europa.eu/en/publications-data/communicable-disease-threats-report-13-19-december- 2020-week-51. Accessed 12 Mar 2023
43. Ferreira FC, da Silva AS, Recht J, Guaraldo L, Moreira ME, de Siqueira AM, Gerardin P, Brasil P (2021) Vertical transmission of chikungunya virus: a systematic review. PLoS One 16:e0249166
44. Schuffenecker I, Iteman I, Michault A, Murri S, Franguel L, Vaney MC et al (2006) Genome microevolution of chikungunya viruses causing the Indian Ocean outbreak. PLoS Med 3(7):e263
45. Zeller H, Bortel WV, Sudre B (2016) Chikungunya: its history in Africa and Asia and its spread to new regions in 2013–2014. J Infect Dis 214(5):S436–S440
46. Chusri S, Siripaitoon P, Hirunpat S, Silpapojakul K (2011) Case reports of neuro-chikungunya in southern Thailand. Am J Trop Med Hyg 85(2):386–389
47. Azami NAM, Salleh SA, Shah SA, Neoh HM, Othman Z, Zakaria SZS et al (2013) The emergence of chikungunya seropositivity in healthy Malaysian adults residing in outbreak-free locations: Chikungunya seroprevalence results from the Malaysian Cohort. BMC Infect Dis 13(1):1–9
48. Mavalankar D, Shastri P, Raman P (2007) Chikungunya epidemic in India: a major public-health disaster. Lancet Infect Dis 7(5):306–307
49. Tanabe ISB, Tanabe ELL, Santos EC, Martins WV, Araújo IMTC, Cavalcante MCA, Lima ARV, Câmara NOS, Anderson L, Yunusov D et al (2018) Cellular and molecular immune response to chikungunya virus infection. Front Cell Infect Microbiol 8:345
50. Van Aalst M, Nelen CM, Goorhuis A, Stijnis C, Grobusch MP (2017) Long-term sequelae of chikungunya virus disease: a systematic review. Travel Med Infect Dis 15:8–22
51. WHO (2023) Dengue and severe dengue. www.who.int/mediacentre/factsheets/fs117/en/index.html/. Accessed 12 Mar 2023

52. Cao-Lormeau VM (2009) Dengue viruses binding proteins from *Aedes aegypti* and *Aedes polynesiensis* salivary glands. Virol J 6:35
53. Waggoner JJ, Gresh L, Vargas MJ, Ballesteros G, Tellez Y, Soda KJ et al (2016) Viremia and clinical presentation in Nicaraguan patients infected with Zika virus, chikungunya virus, and dengue virus. Clin Infect Dis 63:1584–1590
54. Perera SD, Jayawardena UA, Jayasinghe CD (2018) Potential use of *Euphorbia hirta* for dengue: a systematic review of scientific evidence. J Trop Med 2048537:1–7
55. Raja DB, Mallol R, Ting CY, Kamaludin F, Ahmad R, Ismail S, Jayaraj VJ, Sundram BM (2019) Artificial intelligence model as predictor for dengue outbreaks. Malays J Public Health Med 2019:103–108
56. Jiang X, Kanda T, Nakamoto S, Saito K, Nakamura M, Wu S, Haga Y, Sasaki R, Sakamoto N, Shirasawa H et al (2015) The JAK2 inhibitor AZD1480 inhibits hepatitis A virus replication in Huh7 cells. Biochem Biophys Res Commun 458:908–912
57. Gabler DJ (2006) Dengue/dengue haemorrhagic fever: history and current status. In: Proceedings of the Novartis Foundation Symposium, 277:3. Wiley Online Library, Chichester
58. Guha-Sapir D, Schimmer B (2005) Dengue fever: new paradigms for a changing epidemiology. Emerg Themes Epidemiol 2:1–10
59. Rozera R, Verma S, Kumar R, Haque A, Attri A (2019) Herbal remedies, vaccines and drugs for dengue fever: emerging prevention and treatment strategies. Asian Pac J Trop Med 12(4): 147–152
60. Singh PK, Rawat P (2017) Evolving herbal formulations in management of dengue fever. J Ayurveda Integr Med 8:207–210
61. WHO (2023) Dengue and severe dengue. https://www.who.int/news-room/fact-sheets/detail/dengue-and-severe-dengue. Accessed 12 May 2023
62. Chung CC, Lee SSJ, Chen YS et al (2007) Acute flaccid paralysis as an unusual presenting symptom of Japanese encephalitis: a case report and review of the literature. Infection 35(1): 30–32
63. Kaur R, Vrati S (2003) Development of a recombinant vaccine against Japanese encephalitis. J Neurovirol 9(4):421–431
64. Griffiths MJ, Lemon JV, Rayamajhi A, Poudel P, Shrestha P, Srivastav V et al (2013) The functional, social and economic impact of acute encephalitis syndrome in Nepal-a longitudinal follow-up study. PLoS Negl Trop Dis 7:e2383
65. Srivastava S, Khanna N, Saxena SK, Singh A, Mathur A, Dhole TN (1999) Degradation of Japanese encephalitis virus by neutrophils. Int J Exp Pathol 80:17–24
66. Nash D, Mostashari F, Fine A, Miller J, O'Leary D, Murray K, Huang A, Rosenberg A, Greenberg A, Sherman M et al (2001) The outbreak of West Nile virus infection in the New York City area in 1999. N Engl J Med 344:1807–1814
67. Nasci RS, Savage HM, White DJ, Miller JR, Cropp BC, Godsey MS, Kerst AJ, Bennett P, Gottfried K, Lanciotti RS (2001) West Nile virus in overwintering *Culex* mosquitoes, New York City. Emerg Infect Dis 7:742–744
68. Hayes EB, Sejvar JJ, Zaki SR, Lanciotti RS, Bode AV, Campbell GL (2005) Virology, pathology, and clinical manifestations of West Nile virus disease. Emerg Infect Dis 11: 1174–1179
69. Mostashari F, Bunning ML, Kitsutani PT, Singer DA, Nash D, Cooper MJ, Katz N, Liljebjelke KA, Biggersta BJ, Fine AD et al (2001) Epidemic West Nile encephalitis, New York, 1999: results of a household-based seroepidemiological survey. Lancet 358:261–264
70. WHO (2023) World Health Organization. Fact sheet: West Nile. https://www.who.int/news-room/fact-sheets/detail/west-nile-virus. Accessed 12 Mar 2023
71. Campos GS, Antonio CB, Sardi SI (2015) Zika virus outbreak, Bahia, Brazil. Emerg Infect Dis 21(10):1885–1886
72. Filho M, Celina Maria TM, Ricardo Arraes DAX, Thalia Velho BA, Maria Angela WR et al (2016) Initial description of the presumed congenital Zika syndrome. Am J Public Health 106 (4):598–600

73. Rubin EJ, Greene MJ, Baden LR (2016) Zika virus and microcephaly. N Engl J Med 374: 984–985
74. Brasil P, Patricia CS, Andrea DAF, Heruza EZ, Guilherme AC et al (2016) Guillain-Barré syndrome associated with Zika virus infection. Lancet 387(10026):1482
75. Dos Santos T, Angel R, Maria A, Antonio S, Pilar R et al (2016) Zika virus and the Guillain-Barré syndrome-case series from seven countries. N Engl J Med 375(16):1598–1601
76. Duggal NK, Ritter JM, Pestorious SE, Zaki SR, Davis BS, Chang GJ et al (2017) Frequent Zika sexual transmission and prolonged viral RNA shedding in an immunodeficient mouse model. Cell Rep 18:1751–1760
77. Govero J, Esakky P, Scheaffer SM, Fernandez E, Drury A, Platt DJ, Gorman MJ, Richner JM, Caine EA, Salazer V et al (2016) Zika virus infection damages the testes in mice. Nature 540: 438–442
78. Ma W, Li S, Ma S, Jia L, Zhang F, Zhang Y, Zhang J et al (2016) Zika virus causes testis damage and leads to male infertility in mice. Cell 167(6):1511–1524.e10
79. Dowall SD, Graham VA, Rayner E, Atkinson B, Hall G, Watson RJ, Bosworth A, Bonney LC, Kitchen S, Hewson RA (2016) Susceptible mouse model for Zika virus infection. PLoS Negl Trop Dis 10:e0004658
80. Eyer L, Nencka R, Huvarova I, Palus M, Joao Alves M, Gould EA, De Clercq E, Ruzek D (2016) Nucleoside inhibitors of Zika virus. J Infect Dis 214:707–711
81. Zmurko J, Marques RE, Schols D, Verbeken E, Kaptein SJ, Neyts J (2016) The viral polymerase inhibitor 7-deaza-20-c-methyladenosine is a potent inhibitor of in vitro Zika virus replication and delays disease progression in a robust mouse infection model. PLoS Negl Trop Dis 10:e0004695
82. Marimuthu G, Rajamohan S, Mohan R, Krishnamoorthy Y (2012) Larvicidal and ovicidal properties of leaf and seed extracts of *Delonix elata* (L.) Gamble (family: Fabaceae) against malaria (*Anopheles stephensi* Liston) and dengue (*Aedes aegypti*) (Diptera: Culicidae) vector mosquitoes. Parasitol Res 111(1):65–77
83. Ghosh A, Chowdhury N, Chandra G (2012) Plant extracts as potential mosquito larvicides. Indian J Med Res 13(5):581–598
84. Hatil HE, Kamali L (2009) Medicinal plants in East and Central Africa: challenges and constraints. Ethnobot Leaflets 13:364–369
85. Adebajo AC, Famuyiwa FG, John JD, Idem ES, Adeoye AO (2012) Activities of some Nigerian medicinal plants against *Aedes aegypti*. Chinese Med 3:151–156
86. WHO (1993) World Health Organization guidelines on the conservation of medicinal plants. International Union for the Conservation of Nature, Geneva, pp 1–38
87. Zaynab M, Fatima M, Abbas S, Sharif Y, Umair M, Zafar MH, Bahadar K (2018) Role of secondary metabolites in plant defense against pathogens. Microb Pathog 124:198–202
88. Prinsloo G, Marokane CK, Street RA (2018) Anti-HIV activity of southern African plants: current developments, phytochemistry and future research. J Ethnopharmacol 210:133–155
89. Salehi B, Kumar NVA, Sener B, Sharifi-Rad M, Kılıç M, Mahady GB, Vlaisavljevic S, Iriti M, Kobarfard F, Setzer WN et al (2018) Medicinal plants used in the treatment of human immunodeficiency virus. Int J Mol Sci 19:1459
90. Akram M, Tahir IM, Shah SMA, Mahmood Z, Altaf A, Ahmad K, Munir N, Daniyal M, Nasir S, Mehboob H (2018) Antiviral potential of medicinal plants against HIV, HSV, influenza, hepatitis, and coxsackievirus: a systematic review. Phytother Res 32:811–822
91. Mohan S, Elhassan Taha MM, Makeen HA, Alhazmi HA, Al Bratty M, Sultana S, Ahsan W, Najmi A, Khalid A (2020) Bioactive natural antivirals: an updated review of the available plants and isolated molecules. Molecules 25:4878
92. Geng CA, Chen JJ (2018) The progress of anti-HBV constituents from medicinal plants in China. Nat Prod Bioprospect 8:227–244
93. Saleh MSM, Kamisah Y (2021) Potential medicinal plants for the treatment of dengue fever and severe acute respiratory syndrome-coronavirus. Biomol Ther 11(1):1–25

94. Frederico ÉHFF, Cardoso ALBD, Moreira-Marconi E, de Sá-Caputo DDC, Guimarães CAS, Dionello CDF, Morel DS, Paineiras-Domingos LL, de Souza PL, Brandão-Sobrinho-Neto S et al (2017) Anti-viral effects of medicinal plants in the management of dengue: a systematic review. Afr J Tradit Complement Altern Med 14:33–40
95. Webber C, Sliva K, von Rhein C, Kümmerer BM, Schnierle BS (2015) The green tea catechin, *Epigallocatechingallate* inhibits chikungunya virus infection. Antivir Res 113:1–3
96. Mounce BC, Cesaro T, Carrau L, Vallet T, Vignuzzi M (2017) Curcumin inhibits Zika and chikungunya virus infection by inhibiting cell binding. Antivir Res 42:148–157
97. Kaur P, Thiruchelvan M, Lee RC, Chen H, Chen KC, Ng ML, Chu JJ (2013) Inhibition of chikungunya virus replication by harringtonine, a novel antiviral that suppresses viral protein expression. Antimicrob Agents Chemother 57:155–167
98. Cowan MM (1999) Plant products as antimicrobial agents. Clin Microbiol Rev 12:564–582
99. Von Rhein C, Weidner T, Henß L, Martin J, Weber C, Silva K et al (2016) Curcumin and *Boswellia serrata* gum resin extract inhibit chikungunya and vesicular stomatitis virus infections in vitro. Antiviral Res 125:51–57
100. Varghese FS, Thaa B, Amrun SN, Simarmata D, Rausalu K, Nyman TA et al (2016) The antiviral alkaloid berberine reduces chikungunya virus-induced mitogen-activated protein kinase signaling. J Virol 90:9743–9757
101. Varghese FS, Kaukinen P, Gläsker S, Bespalov M, Hanski L, Wennerberg K et al (2016) Discovery of berberine, abamectin and ivermectin as antivirals against chikungunya and other alphaviruses. Antivir Res 126:117–124
102. Wintachai P, Kaur P, Lee RC, Ramphan S, Kuadkitkan A, Wikan L et al (2015) Activity of andrographolide against chikungunya virus infection. Sci Rep 5:14179
103. Lani R, Hassandarvish P, Chiam CW, Moghaddam E, Chu JJH, Rausalu K et al (2015) Antiviral activity of silymarin against chikungunya virus. Sci Rep 5(11421):1–10
104. Lani R, Hassandarvish P, Shu MH, Phoon WH, Chu JJH, Higgs S et al (2016) Antiviral activity of selected flavonoids against chikungunya virus. Antiviral Res 133:50–61
105. Mohamat SA, Shueb RH, Che Mat NF (2018) Anti-viral activities of Oroxylum indicum extracts on chikungunya virus infection. Indian J Microbiol 58(1):68–75
106. Dinda B, SilSarma I, Dinda M, Rudrapaul P (2015) *Oroxylum indicum* (L.) Kurz, an important Asian traditional medicine: from traditional uses to scientific data for its commercial exploitation. J Ethnopharmacol 161:255–278
107. Arora A, Tomar PC, Kumari P, Kumari A (2020) Medicinal alternative for chikungunya cure: a herbal approach. J Microbiol Biotech Food Sci 9:970–978
108. WHO World Health Organization (2008) Traditional medicine. Fact Sheet. http://www.who.int/mediacentre/factsheets/fs134/en/. Accessed 12 Mar 2023
109. Fusco DN, Chung RT (2014) Review of current dengue treatment and therapeutics in development. J Bioanal Biomed 1(8):1
110. Altamish M, Khan M, Baig MS, Pathak B, Rani V, Akhtar J, Khan AA, Ahmad S, Krishnan A (2022) Therapeutic potential of medicinal plants against dengue infection: a mechanistic viewpoint. ACS Omega 7:24048–24065
111. Hunsperger EA, Yoksan S, Buchy P, Nguyen VC, Sekaran SD, Enria DA, Pelegrino JL, Vazquez S, Artsob H, Drebot M, Gubler DJ, Halstead SB, Guzman MG, Margolis HS, Nathanson CM, Lic NRR, Bessoff KE, Kliks S, Peeling RW (2009) Evaluation of commercially available anti-dengue virus immunoglobulin M tests. Emerg Infect Dis 15(3):436–440
112. Philippine Medicinal Plants (2011) Gatas-gatas. http://www.stuartxchange.org/GatasGatas.html. Accessed 12 Mar 2023
113. The Cure Library (2007) Dengue fever cure using tawa–tawa aka gatas–gatas weed. http://www.curelibrary.com/blog/2007/04/. Accessed 12 Mar 2023
114. Kala CP (2012) Leaf juice of *Carica papaya L.*: a remedy of dengue fever. Med Aromat Plants 1:109
115. Abd-Kadir LS, Yaakob H, Zulkifli MR (2013) Potential anti-dengue medicinal plants: a review. J Nat Med 67(4):677–689

116. Dhiman M, Sharma L, Dadhich A, Dhawan P, Sharma MM (2022) Traditional knowledge to contemporary medication in the treatment of infectious disease dengue: a review. Front Pharmacol 13:750494
117. Riaz T, Rashid A, Akram M, Khalil MT, Anwar H, Laila U, Zainab R, Ugariogu S (2023) The anti-viral potential of medicinal plants and herbal formulations for the treatment of dengue fever. Nat Ayurvedic Med 7(1):000377
118. Whitby K, Pierson TC, Geiss B, Lane K, Engle M, Zhou Y, Doms RW, Diamond MS (2005) Castanospermine, a potent inhibitor of dengue virus infection in vitro and in vivo. J Virol 79: 8698–8706
119. Chiow KH, Phoon MC, Putti T, Tan BKH, Chow VT (2016) Evaluation of antiviral activities of *Houttuynia cordata* Thunb. extract, quercetin, quercetrin and cinanserin on murine coronavirus and dengue virus infection. Asian Pac J Trop Med 9:1–7
120. Xie ML, Phoon MC, Dong SX, Tan BKH, Chow VT (2013) Houttuynia cordata extracts and constituents inhibit the infectivity of dengue virus type 2 in vitro. Int J Integr Biol 14:78–85
121. Silva ARA, Morais SM, Marques MMM, Lima DM, Santos SCC, Almeida RR, Vieira IGP, Guedes MIF (2011) Antiviral activities of extracts and phenolic components of two spondias species against dengue virus. J Venom Anim Toxins Incl Trop Dis 17:406–413
122. Hidari KIPJ, Takahashi N, Arihara M, Nagaoka M, Morita K, Suzuki T (2008) Structure and anti-dengue virus activity of sulfated polysaccharide from a marine alga. Biochem Biophys Res Commun 376:91–95
123. De Sf-Tischer PC, Talarico LB, Noseda MD, Silvia SM, Damonte EB, Duarte MER (2006) Chemical structure and antiviral activity of carrageenans from *Meristiella gelidium* against herpes simplex and dengue virus. Carbohydr Polym 63:459–465
124. Allard PM, Leyssen P, Martin MT, Bourjot M, Dumontet V, Eydoux C, Guillemot JC, Canard B, Poullain C, Guéritte F et al (2012) Antiviral chlorinated daphnane diterpenoid orthoesters from the bark and wood of *Trigonostemon cherrieri*. Phytochemistry 84:160–168
125. Yao X, Ling Y, Guo S, Wu W, He S, Zhang Q, Zou M, Nandakumar KS, Chen X, Liu S (2018) Tatanan A from the *Acorus calamus* root inhibited dengue virus proliferation and infections. Phytomedicine 42:258–267
126. Adiguna SP, Panggabean JA, Atikana A, Untari F, Izzati F, Bayu A, Rosyidah A, Rahmawati SI, Putra MY (2021) Antiviral activities of andrographolide and its derivatives: mechanism of action and delivery system. Pharmaceuticals (Basel) 14(11):1102
127. Neelawala D, Rajapakse S, Kumbukgolla W (2019) Potential of medicinal plants to treat dengue. Pdfs. Semanticscholar Org 5:2455–8931
128. Kaushik S, Dar L, Kaushik S, Yadav JP (2021) Identification and characterization of new potent inhibitors of dengue virus NS5 proteinase from *Andrographis paniculata* supercritical extracts on in animal cell culture and in silico approaches. J Ethnopharmacol 267:113541
129. Dwivedi VD, Tripathi IP, Mishra SK (2016) In silico evaluation of inhibitory potential of triterpenoids from *Azadirachta indica* against therapeutic target of dengue virus, NS2B-NS3 Protease. J Vector Borne Dis 53(2):156–161
130. Parida MM, Upadhyay C, Pandya G, Jana AM (2002) Inhibitory potential of neem (*Azadirachta indica* Juss) leaves on dengue virus type-2 replication. J Ethnopharmacol 79(2):273–278
131. Dwivedi VD, Bharadwaj S, Afroz S, Khan N, Ansari MA, Ya-dava U, Tripathi RC, Tripathi IP, Mishra SK, Kang SG (2021) Anti-dengue infectivity evaluation of bioflavonoid from *Azadirachta indica* by dengue virus serine protease inhibition. J Biomol Struct Dyn 39(4): 1417–1430
132. Rao VB, Yeturu K (2020) Possible anti-viral effects of neem (*Azadirachta indica*) on dengue virus. BioRxiv. https://doi.org/10.1101/2020.04.29.069567
133. Liew YJM, Lee YK, Khalid N, Rahman NA, Tan BC (2020) Enhancing flavonoid production by promiscuous activity of prenyltransferase, BrPT2 from *Boesenbergia rotunda*. Peer J 8: e9094

134. Abduraman MA, Hariono M, Yusof R, Rahman NA, Wahab HA, Tan ML (2018) Development of a NS2B/NS3 protease inhibition assay using AlphaScreen(®) beads for screening of anti-dengue activities. He-liyon 4(12):e01023
135. Lim SYM, Chieng JY, Pan Y (2021) Recent insights on anti-dengue virus (DENV) medicinal plants: review on in vitro, in vivo and in silico discoveries. All Life 14(1):1–33
136. Haider M, Dholakia D, Panwar A, Garg P, Anand V, Gheware A, Singha K et al (2021) Traditional use of *Cissampelos pareira* for hormone disorder and fever provides molecular links of modulation to viral inhibition. BioRxiv. https://doi.org/10.1101/2021.02.17.431579
137. Pong LY, Yew PN, Lee WL, Lim YY, Sharifah SH (2020) Anti-dengue virus serotype 2 activity of tannins from porcupine dates. Chin Med 15:49
138. Balasubramanian A, Pilankatta R, Teramoto T, Sajith AM, Nwulia E, Kulkarni A, Padmanabhan R (2019) Inhibition of dengue virus by curcuminoids. Antivir Res 162:71
139. Simões LR, Maciel GM, Brandão GC, Kroon EG, Castilho RO, Oliveira AB (2011) Antiviral activity of *Distictella elongata* (Vahl) Urb. (Bignoniaceae), a potentially useful source of anti-dengue drugs from the state of Minas Gerais, Brazil. Lett Appl Microbiol 53(6):602–607
140. Ali F, Chorsiya A, Anjum V, Khasimbi S, Ali A (2021) A systematic review on phytochemicals for the treatment of dengue. Phytother Res 35(4):1782–1816
141. Teixeira RR, Pereira WL, Oliveira AFC et al (2014) Natural products as source of potential dengue antivirals. Molecules 19(6):8151
142. Chavda VP, Kumar A, Banerjee R, Das N (2022) Ayurvedic and other herbal remedies for dengue: an update. Clin Complement Med Pharmacol 2:100024
143. Sucipto TH, Churrotin S, Setyawati H, Martak F, Mulyatno KC, Amarullah IH, Kotaki T, Kameoka M, Yotopranoto S, Soegijanto AS (2018) A new copper (ii)- imidazole derivative effectively inhibits replication of denv-2 in vero cell. Afr J Infect Dis 12(1 Suppl):116–119
144. Zandi K, Lim TH, Rahim NA, Shu MH, Teoh BT, Sam SS, Danlami MB, Tan KK, Abubakar S (2013) Extract of *Scutellaria baicalensis* inhibits dengue virus replication. BMC Complement Altern Med 13(1):1–10
145. Elizondo-Gonzalez R, Cruz-Suarez LE, Ricque-Marie D, Mendoza-Gamboa E, Rodriguez-Padilla C, Trejo-Avila LM (2012) In vitro characterization of the antiviral activity of fucoidan from *Cladosiphon okamuranus* against Newcastle disease virus. Virol J 9:307
146. Allard PM, Dau ETH, Eydoux C, Guillemot JC, Dumontet V, Poullain C, Canard B, Gueritte F, Litaudon M (2011) Alkylated flavanones from the bark of *Cryptocarya chartacea* as dengue virus NS5 polymerase inhibitors. J Nat Prod 74(11):2446–2453
147. Bourjot M, Leyssen P, Eydoux C, Guillemot JC, Canard B, Rasoanaivo P, Gueritte F, Litaudon M (2012) Chemical constituents of *Anacolosa pervilleana* and their antiviral activities. Fitoterapia 83(6):1076–1080
148. Rothan HA, Zulqarnain M, Ammar YA, Tan EC, Rahman NA, Yusof R (2014) Screening of antiviral activities in medicinal plants extracts against dengue virus using dengue NS2B-NS3 protease assay. Trop Biomed 31(2):286–296
149. Tarasuk M, Songprakhon P, Chimma P, Sratongno P, Na-Bangchang K, Yenchitsomanus P (2017) Alpha-Mangostin inhibits both dengue virus production and cytokine/chemokine expression. Virus Res 240:180–189
150. Ono L, Wollinger W, Rocco IM, Coimbra TL, Gorin PA, Sierakowski MR (2003) In vitro and in vivo antiviral properties of sulfated galactomannans against yellow fever virus (BeH111 Strain) and dengue 1 virus (Hawaii Strain). Antiviral Res 60(3):201–208
151. Abdul Ahmad SA, Palanisamy UD, Tejo BA, Chew MF, Tham HW, Hassan S, Geraniin S (2017) Extracted from the rind of *Nephelium lappaceum* binds to dengue virus type-2 envelope protein and inhibits early stage of virus replication. Virol J 14(1):229
152. Trujillo-Correa AI, Quintero-Gil DC, Diaz-Castillo F, Quiñones W, Robledo SM, Martinez-Gutierrez M (2019) In vitro and in silico anti-dengue activity of compounds obtained from *Psidium guajava* through bioprospecting. BMC Complement Altern Med 19(1):298

153. Yu JS, Wu YH, Tseng CK, Lin CK, Hsu YC, Chen YH, Lee JC (2017) Schisandrin A inhibits dengue viral replication via up regulating antiviral interferon responses through STAT signaling pathway. Sci Rep 7(1):45171
154. Hishiki T, Kato F, Tajima S, Toume K, Umezaki M, Takasaki T, Miura T (2017) Hirsutine, an indole alkaloid of *Uncaria rhynchophylla*, inhibits late step in dengue virus lifecycle. Front Microbiol 8:1674
155. Rees C, Costin J, Fink R, Mcmichael M, Fontaine K, Isern S, Michael S (2008) In vitro inhibition of dengue virus entry by p-sulfoxy-cinnamic acid and structurally related combinatorial chemistries. Antivir Res 80(2):135–142
156. Hall A, Troupin A, Londono-Renteria B, Colpitts T (2017) Garlic organosulfur compounds reduce inflammation and oxidative stress during dengue virus infection. Viruses 9(7):159
157. Chang S, Chang Y, Lu K, Tsou Y, Lin C (2012) Antiviral activity of *Isatis indigotica* extract and its derived Indirubin against Japanese encephalitis virus. Evid-Based Complement Altern Med 2012:1–7
158. Roy S, Chaurvedi P, Chowdhary A (2015) Evaluation of antiviral activity of essential oil of *Trachyspermum ammi* against Japanese encephalitis virus. Pharmacogn Rev 7:263–267
159. Bairwa R, Sodha RS, Rajawat BS (2012) *Trachyspermum ammi*. Pharmacogn Rev 6: 56–60
160. Lin LT, Hsu WC, Lin CC (2014) Antiviral natural products and herbal medicines. J Tradit Complement Med 4(1):24–35
161. Goh VSL, Mok C, Chu JJH (2020) Antiviral natural products for arbovirus infections. Molecules 25(2796):1–22
162. Vázquez-Calvo Á, Jiménez de Oya N, Martín Acebes MA, Garcia Moruno E, Saiz JC (2017) Antiviral properties of the natural polyphenols delphinidin and epigallocatechin gallate against the flaviviruses West Nile virus, Zika virus, and dengue virus. Front Microbiol 8:1314
163. Abubakar IB, Kankara SS, Malami I, Danjuma JB, Muhammad YZ, Yahaya H, Singh D, Usman UJ, Ukwuani-Kwaja AN, Muhammad A, Ahmed SJ, Folami SO, Falana MB, Nurudeen QO (2022) Traditional medicinal plants used for treating emerging and re-emerging viral diseases in northern Nigeria. Eur J Integr Med 49:1–28
164. Adeyemi AA, Gbolade AA, Moody JO, Ogbole OO, Fasanya MT (2010) Traditional anti-fever phytotherapies in Sagamu and Remo North Districts in Ogun State, Nigeria. J Herbs Spices Med Plants 16(3–4):203–218
165. Ajaiyeoba EO, Oladepo O, Fawole OI, Bolaji OM, Osowole O, Bolaji O, Akinboye DO, Ogundahunsi OAT, Falade CO, Gbotosho GO, Itiola OA, Happi TC, Ebong OO, Ononiwu IM, Osowole OS, Oduola A, Ashidi JS, Oduola AMA (2006) Cultural categorization of febrile illnesses in correlation with herbal remedies used for treatment in Southwestern Nigeria. J Ethnopharmacol 85:179–185
166. Schwarz S, Wang K, Yu W, Sun B, Schwarz W (2011) Emodin inhibits current through sars-associated coronavirus 3a protein. Antivir Res 90:64–69
167. Batista MN, Braga ACS, Campos GRF, Souza MM, de Matos RPA, Lopes TZ, et al. (2019) Natural products isolated from oriental medicinal herbs inactivate Zika virus. Viruses 11(49): 1–10
168. Tan XS, Ma JY, Feng R, Ma C, Chen WJ, Sun YP, Fu J, Huang M, He CY, Shou JW et al (2013) Tissue distribution of berberine and its metabolites after oral administration in rats. PLoS One 8:e77969
169. Rad SZK, Rameshrad M, Hosseinzadeh H (2017) Toxicology effects of *Berberis vulgaris* (barberry) and its active constituent, berberine: a review. Iran J Basic Med Sci 20:516–529
170. Yin J, Xing H, Ye J (2008) Efficacy of berberine in patients with type 2 diabetes mellitus. Metab Clin Exp 57:712–717
171. Ahmed T, Gilani AU, Abdollahi M, Daglia M, Nabavi SF, Nabavi SM (2015) Berberine and neurodegeneration: a review of literature. Pharmacol Rep PR 67:970–979
172. Fernando S, Fernando T (2017) Antivirals for allosteric inhibition of Zika virus using a homology model and experimentally determined structure of envelope protein. BMC Res Notes 10(1):354

173. Datta A, Sukul NC (1987) Antifilarial effect of *Zinger officinale* on *Dirofilaria immitis*. J Helminthol 61:268–670
174. Clain E, Sinigaglia L, Koishi AC, Gorgette O, Gadea G, Viranaicken W, Krejbich-Trotot P, Mavingui P, Desprès P, dos Santos CND et al (2018) Extract from *Aphloia theiformis*, an edible indigenous plant from Reunion Island, impairs Zika virus attachment to the host cell surface. Sci Rep 8:1
175. Barbosa EC, Alves TMA, Kohlhoff M, Soraya Jangola TG, Pires DEV, Figueiredo ACC, Alves EAR, Calzavara-Silva CE, Sobral M, Kroon EG, Rosa LH, Zani CL, de Oliveira JG (2022) Searching for plant-derived antivirals against dengue virus and Zika virus. Virol J 19 (31):1–15
176. Epidi JO, Izah SC, Ohimain EI, Epidi TT (2016) Phytochemical, antibacterial and synergistic potency of tissues of *Vitex grandifolia*. Biotechnol Res 2(2):69–76
177. Epidi JO, Izah SC, Ohimain EI (2016) Antibacterial and synergistic efficacy of extracts of *Alstonia boonei* tissues. Br J Appl Res 1(1):0021–0026
178. Izah SC, Odubo TC (2023) Effects of *Citrus aurantifolia* fruit juice on selected pathogens of public health importance. ES Food Agroforest 11:829. https://doi.org/10.30919/esfaf829
179. Enaregha EB, Izah SC, Okiriya Q (2021) Antibacterial properties of *Tetrapleura tetraptera* pod against some pathogens. Res Rev Insights 5:1–5
180. Izah SC, Chandel SS, Etim NG, Epidi JO, Venkatachalam T, Devaliya R (2019) Potency of unripe and ripe express extracts of long pepper (*Capsicum frutescens* var. baccatum) against some common pathogens. Int J Pharmaceut Phytopharmacol Res 9(2):56–70
181. Izah SC, Etim NG, Ilerhunmwuwa IA, Silas G (2019) Evaluation of crude and ethanolic extracts of *Capsicum frutescens* var. minima fruit against some common bacterial pathogens. Int J Complement Altern Med 12(3):105–108
182. Izah SC, Etim NG, Ilerhunmwuwa IA, Ibibo TD, Udumo JJ (2019) Activities of express extracts of Costus afer Ker–Gawl. [Family COSTACEAE] against selected bacterial isolates. Int J Pharmaceut Phytopharmacol Res 9(4):39–44
183. Izah SC, Uhunmwangho EJ, Dunga KE, Kigigha LT (2018) Synergy of methanolic leave and stem-back extract of *Anacardium occidentale* L. (cashew) against some enteric and superficial bacteria pathogens. MOJ Toxicol 4(3):209–211
184. Izah SC, Uhunmwangho EJ, Dunga KE (2018) Studies on the synergistic effectiveness of methanolic extract of leaves and roots of *Carica papaya* L. (papaya) against some bacteria pathogens. Int J Complement Altern Med 11(6):375–378
185. Izah SC, Uhunmwangho EJ, Etim NG (2018) Antibacterial and synergistic potency of methanolic leaf extracts of *Vernonia amygdalina* L. and *Ocimum gratissimum* L. J Basic Pharmacol Toxicol 2(1):8–12
186. Izah SC, Zige DV, Alagoa KJ, Uhunmwangho EJ, Iyamu AO (2018) Antibacterial efficacy of aqueous extract of *Myristica fragrans* (common nutmeg). EC Pharmacol Toxicol 6(4): 291–295
187. Izah SC, Uhunmwangho EJ, Eledo BO (2018) Medicinal potentials of *Buchholzia coriacea* (wonderful kola). Med Plant Res 8(5):27–43
188. Izah SC (2018) Some determinant factors of antimicrobial susceptibility pattern of plant e extracts. Res Rev Insight 2(3):1–4
189. Izah SC, Aseibai ER (2018) Antibacterial and synergistic activities of methanolic leaves extract of lemon grass (*Cymbopogon citratus*) and rhizomes of ginger (*Zingiber officinale*) against *Escherichia coli*, *Staphylococcus aureus* and *Bacillus subtilis*. Acta Sci Microbiol 1(6): 26–30
190. Kigigha LT, Izah SC, Uhunmwangho EJ (2018) Assessment of hot water and ethanolic leaf extracts of *Cymbopogon citratus* Stapf (Lemon grass) against selected bacteria pathogens. Ann Microbiol Infect Dis 1(3):1–5
191. Kigigha LT, Selekere RE, Izah SC (2018) Antibacterial and synergistic efficacy of acetone extracts of Garcinia kola (Bitter kola) and *Buchholzia coriacea* (Wonderful kola). J Basic Pharmacol Toxicol 2(1):13–17

192. Kigigha LT, Biye SE, Izah SC (2016) Phytochemical and antibacterial activities of *Musanga cecropioides* tissues against *Escherichia coli, Pseudomonas aeruginosa Staphylococcus aureus, Proteus* and *Bacillus* species. Int J Appl Res Technol 5(1):100–107
193. Kigigha LT, Izah SC, Okitah LB (2016) Antibacterial activity of palm wine against *Pseudomonas, Bacillus, Staphylococcus, Escherichia,* and *Proteus* spp. Point J Bot Microbiol Res 2(1):046–052
194. Kigigha LT, Apreala A, Izah SC (2016) Effect of cooking on the climbing pepper (*Piper nigrum*) on antibacterial activity. J Environ Treat Techniq 4(1):6–9
195. Kigigha LT, Izah SC, Ehizibue M (2015) Activities of *Aframomum melegueta* seed against *Escherichia coli, Staphylococcus aureus* and *Bacillus* species. Point J Bot Microbiol Res 1(2): 23–29
196. Popoola TD, Segun PA, Ekuadzi E, Dickson RA, Awotona OR, Nahar L, Sarker SD, Fatokun AA (2022) West African medicinal plants and their constituent compounds as treatments for viral infections, including SARS-CoV-2/COVID-19. DARU J Pharmaceut Sci 30:191–210
197. Cole N (1996) Diversity of medicinal plants in West African habitats. In: The biodiversity of African plants. Springer, pp 704–713
198. Sawadogo WR, Schumacher M, Teiten M-H, Dicato M, Diederich M (2012) Traditional West African pharmacopeia, plants and derived compounds for cancer therapy. Biochem Pharmacol 84(10):1225–1240
199. Ekanem AP, Udoh FV (2009) The diversity of medicinal plants in Nigeria: an overview. In: African natural plant products: new discoveries and challenges in chemistry and quality. ACS Symp Ser 1021:135–147
200. Izah SC, Chandel SS, Epidi JO, Venkatachalam T, Devaliya R (2019) Biocontrol of *Anopheles gambiae* larvae using fresh ripe and unripe fruit extracts of *Capsicum frutescens* var. baccatum. Int J Green Pharm 13(4):338–342
201. Izah SC (2019) Activities of crude, acetone and ethanolic extracts of *Capsicum frutescens* var. minima fruit against larvae of *Anopheles gambiae*. J Environ Treat Techniq 7(2):196–200
202. Seiyaboh EI, Seiyaboh Z, Izah SC (2020) Environmental control of mosquitoes: a case study of the effect of Mangifera Indica root-bark extracts (family Anacardiaceae) on the larvae of *Anopheles gambiae*. Ann Ecol Environ Sci 4(1):33–38
203. Seiyaboh EI, Odubo TC, Izah SC (2020) Larvicidal activity of *Tetrapleura tetraptera* (Schum and Thonn) Taubert (Mimosaceae) extracts against *Anopheles gambiae*. Int J Adv Res Microbiol Immunol 2(1):20–25
204. Youkparigha FO, Izah SC (2019) Larvicidal efficacy of aqueous extracts of *Zingiber officinale* Roscoe (ginger) against malaria vector, *Anopheles gambiae* (Diptera: Culicidae). Int J Environ Agric Sci 3(1):1–6
205. Herrmann EC Jr, Kucera LS (1967) Antiviral substances in plants of the mint family (Labiatae). 3. Peppermint (*Mentha piperita*) and other mint plants. Proc Soc Exp Biol Med 124:874–878
206. Tang LI, Ling AP, Koh RY, Chye SM, Voon KG (2012) Screening of anti-dengue activity in methanolic extracts of medicinal plants. BMC Complement Altern Med 12:3
207. Kaushik S, Kaushik S, Sharma V, Yadav JP (2018) Antiviral and therapeutic uses of medicinal plants and their derivatives against dengue viruses. Pharmacogn Rev 12:177–185

Medicinal Plants in the Tropics Used in the Treatment and Management of Parasitic Diseases Transmitted by Mosquitoes: Administration, Challenges, and Strategic Options for Management

14

Maduamaka Cyriacus Abajue and Michael Ndubuisi Wogu

Contents

Abstract

Mosquitoes are highly diversified insect group, globally distributed, and carry and transmit parasitic infections that are recognized threats to the world public health. Infections such as lymphatic filariasis and malaria are caused by filarial nematodes (*Wuchereria bancrofti*, *Brugia malayi*, and *Brugia timori*) and protozoans (*Plasmodium falciparum*, *P. vivax*, *P. ovale*, *P. malariae*, and *P. knowlesi*), respectively, which are transmitted to humans by mosquitoes. Filarial parasites are transmitted to humans and are spread by infected female mosquitoes (*Anopheles*, *Culex*, *Aedes*, *Mansonia*) while malaria parasites are spread and transmitted by infected female

M. C. Abajue (✉) · M. N. Wogu
Department of Animal and Environmental Biology, Faculty of Science, University of Port Harcourt, Port Harcourt, Nigeria
e-mail: maduamaka.abajue@uniport.edu.ng; michael.wogu@uniport.edu.ng

S. C. Izah et al. (eds.), *Herbal Medicine Phytochemistry*, Reference Series in Phytochemistry, https://doi.org/10.1007/978-3-031-43199-9_61

Anopheles mosquitoes to humans. Over 1.4 billion people in about 73 countries are at risk of lymphatic filarial infection while about 3.4 billion people in about 97 countries are at risk of malaria, contributing to about 90% of malaria deaths in developing countries. Orthodox means of treating the diseases have encountered some setbacks due to drug resistance, cost, accessibility, and other logistics, hence becoming necessary to employ sustainable and affordable alternatives. In this chapter, we shall emphasize entomological approach to managing mosquito-borne diseases which is rooted in controlling mosquito species at different stages of development. For the disease-causing parasites, various extracts of medicinal plants which are abound in the tropics and reported to have anti-filarial and anti-plasmodial properties shall be succinctly discussed. Biodiversity loss occasioned by continuous harvesting of plants and root structures, inefficient scientific production tools, paucity of documented data on medicinal plants, and lack of subsidy from government and other stakeholders shall be highlighted as some of the challenges against the use of medicinal plants for the treatment and management of mosquito-borne parasitic diseases. We shall also espouse strategic options such as creation of platforms that will link the health sectors in the ministry of health with traditional medicine practitioners in a view to establishing a reliable and responsive national agency that will protect plant-based medicines and her members. Also, to champion a common front for acceptance of plant-based medicines, product patency, and advocate for sustainable gathering of medicinal plants in the tropics is highly needed.

Keywords

Mosquito · Parasites · Diseases · Medicinal plants · Administration · Challenges · Strategic options

Abbreviations

ACT	Artemisinin-combined therapy
CDCP	Centers for Disease Control and Prevention
HIV	Human immunodeficiency virus
IDPs	Internally displaced persons
UV	Ultraviolet

1 Introduction

Mosquito-borne parasitic diseases are illnesses that are caused by parasites found in human populations which are transmitted from animal to humans or from human to humans by mosquito species. More than one billion people globally contract mosquito-borne parasitic diseases that cause about one million deaths every year. Mosquito-borne and other vector-borne diseases contribute about 16.66% of illnesses and disabilities suffered by children and pregnant women especially the poorest people in the society living in the developing countries [1]. Mosquito-borne parasitic diseases affect both urban and rural villages but most prevalent in villages that lack adequate living conditions such as housing and sanitation and more devastating for those with compromised immunity.

Mosquito is an insect with jointed legs, a pair of wing, and a modified mouthpart for sucking of blood of vertebrate animals including humans. Many female mosquito species are capable of transmitting infectious diseases between humans or from animals to humans. Thus, female mosquitoes are the most dreadful insects that transmit debilitating diseases to man and animals. They harbour infectious parasites such as protozoa, and nematodes in their midgut, which are transmitted from their salivary glands to their hosts (human beings and other vertebrate animals) when sucking their blood. The transmitted parasites (filariae and plasmodia) as outlined in (Fig. 1) cause filarial and malaria diseases, respectively, to man [1] and are known as mosquito-borne parasitic diseases.

Based on disease control (prevention, treatment, and management), medicinal plants are carefully selected from groups of plants which have been utilized in the tropics for years, for treatment and management of diseases from the ancient time till now. In traditional medicine, herbs and plant products are still being used mainly in Africa, Asia, and Latin America, and their use is currently extending to Australia and the USA. Inference from traditional medicine practitioners shows that the use of medicinal plants against diseases such as mosquito-borne parasitic diseases is economical, simple to use, easily accessible, and cheap while synthetic drugs require a lot of time, resources, and efforts (design and validation) from the onset. Plants are reported to have natural compounds in them which are anti: viruses, bacteria, protozoa, nematodes, fungi, oxidant, inflammatory, and carcinogenic [2–4]. Chemical compounds found in plants are generally known as phytochemicals and are secondary metabolites that protect plants against UV radiation, microbial infection, and insect attack [5], hence their use in the treatment and management of insect-borne parasitic diseases.

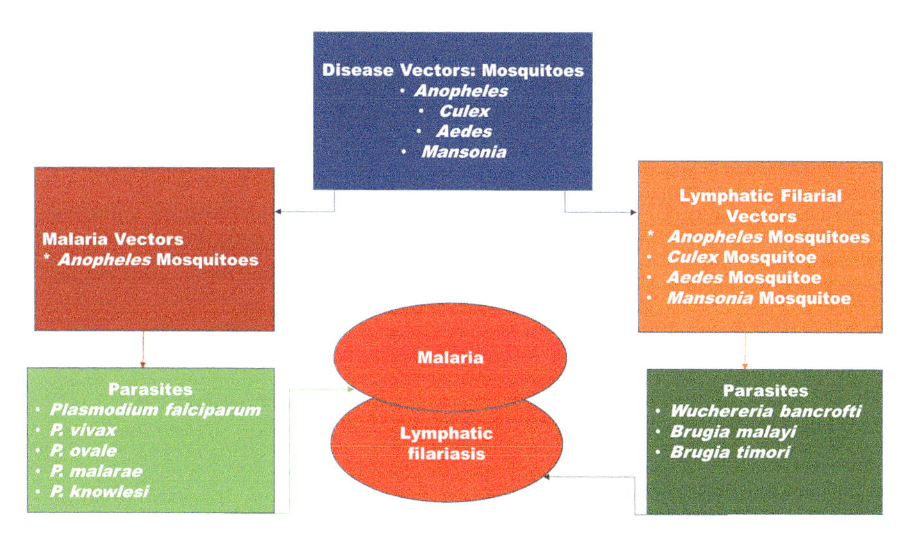

Fig. 1 Overview of mosquito-borne parasitic diseases and their vectors in the tropics

Some approaches employed in dealing with mosquito-borne diseases involve killing or repelling the mosquitoes with synthetic insecticides. Insecticides may be used to lace bed nets or spray the habitable homes. However, the use of insecticides impact negatively on the environment while the mosquitoes develop resistance to the killing powers of the insecticides. Mosquitoes' resistance to insecticides is a major factor driving against mosquito-borne diseases control strategies. Therefore, efforts are on for reasonable alternative remedies to deal with the population of mosquitoes at all stages of their life development. The most voted alternative is to use plant materials to control mosquitoes. Many researches have reported that some plants have the potentials to kill the larval or the adult stages of mosquitoes. For instance, natural products isolated from Chinese herbs and synthetic analogs of *Curcumin* killed the larval stages of *Aedes aegypti* [26]. Similarly, plants such as *Aframomum melegueta*, *Alstonia boonei*, *Croton zambesicus*, and *Newbouldia laevis* are reported in Nigeria as mosquito larvicidal agents [16] which are to be championed in researches related to management of mosquito-borne disease vectors. By extension, the management of mosquito-borne diseases may have been inspired by evidential reports of plasmodial resistance to malaria drugs from the quinines, chloroquinines, and now artemisinin combination therapy, thus requiring that new and affordable alternatives should be sought for. Thus, plant-based medicines stand a high chance of consideration in pursuant of novel mosquito-borne disease management that are rooted to natural resources.

Africa and countries in the tropics have a diverse range of medicinal/herbal plants which are used by traditional medicine practitioners or herbalists. However, there is a paucity of documented data of such medicinal plants from the tropics in open access literatures. In this chapter, we shall look at different reports of traditional medicines against mosquito-borne diseases in Africa and other countries in the tropics. A concise review of scientific evaluation and validation of medicinal plants that have therapeutic and prophylactic properties against parasite-causing diseases such as lymphatic filariasis and malaria shall be presented. Though few reports have shown that many plants in the tropics have various medicinal properties, this section will, therefore, limit discussion to mosquito-borne parasitic diseases such as lymphatic filariasis and malaria. The use of medicinal plants that are abound in the tropics to treat mosquito-borne parasitic diseases shall form the basis of discussion. Elaborate emphasis on phytochemicals and isolated compounds from the plants which have therapeutic and prophylactic potentials against mosquito-borne parasitic diseases (lymphatic filariasis and malaria) shall be highlighted as well. The subsequent section will highlight different preparation methods and administration of medicinal plants in the treatment and management of lymphatic filariasis and malaria, respectively.

2 Overview of Distribution of Mosquito-Borne Parasitic Diseases in the Tropics

Mosquito-borne parasitic diseases are previously envisaged as a peculiar problem in the tropical countries, but today, they are recognized threats to global public health because of mosquito diversity and their geographical spread, hence culminating to different human and animal diseases. Mosquito-borne parasitic diseases are globally spreading due to climate change, pattern of land use, and the continued increase in number of people moving from one country to the other, hence threatening almost half of the global population. Mosquitoes are responsible for majority of the vector-borne parasitic diseases in about 125 countries of the world especially in the tropics [6]. For instance, malaria prevalence alone in the world is estimated at 3.3 billion, and in India as a reference point, it causes about 1.85 million disability adjusted life per year, affected 1.95 million people between 2012 and 2013, while lymphatic filariasis affected 6 million people, thus impacting the socio-economic development of developing countries [7].

2.1 Lymphatic Filariasis

Lymphatic filariasis is commonly known as elephantiasis as a result of a man's skin being thickened and hardened because of excessive swelling due to lymphedema (lymph accumulation). This characteristic lymphedema is commonly obvious in the lower limbs and may affect the breasts, arms, and scrotum, hence making the disease one of the most painful and regrettably disfiguring disease. The disease is caused by round parasitic worms known as filariae (phylum Nematoda). There are three types of filariae (*Wuchereria bancrofti*, *Brugia malayi*, and *Brugia timori*) but *W. bancrofti* is responsible for about 90% of the disease [8]. Filarial parasites are likely to be the only multicellular organisms capable of infecting man with a debilitating disease. Filarial parasites are transmitted and spread by mosquitoes (*Anopheles*, *Culex*, *Aedes*, *Mansonia*) when they take up the microfilariae (the first larval stage of the filariae) in the bloodstream of man during blood feeding. These microfilariae will moult twice to become adult larvae (infective third larval stage) in the gut of mosquitoes which will then be transmitted to man during blood meal by an infected mosquito. The infective adult larvae wriggle into the tissues and enter the lymphatic vessels where they undergo two moults to become reproductive adults (males and females). The reproductive adults exchange their gametes during mating to produce microfilarial larvae which enter the bloodstream of man and are taken up by other uninfected mosquitoes for development [9].

Over 1.4 billion people in about 73 countries are at risk of lymphatic filariasis while more than 120 million of lymphatic filariasis have been reported, with about 40 million of them showing clinical symptoms [10, 11]. Among the total infected people, 40 million live in Africa and India, respectively, while South East Asia, Pacific, New Guinea and America share the remaining [12]. Endemicity of the disease is highest in Africa with about 39 countries having an estimated 390 million people being at risk of infection [13]. Lymphatic filariasis causes a devastating health challenge in addition to its socio-economic burden on infected population. It also causes acute and chronic morbidity among the people living in tropical and sub-tropical countries in Africa, Asia, Western Pacific, and parts of America, with its awful social significance [14]. The disease occurs when filarial parasites (nematodes) are transmitted to humans through a bite of infected mosquito species (*Aedes* spp., *Anopheles* spp., *Culex* spp., and *Mansonia* sp.). The parasites can live in the lymphatic system of an infected human between 6 and 8 years, produce millions of microfilariae that spread in the bloodstream, and significantly disrupt/weaken the immune system.

Lymphatic filariasis mostly presents no immediate identified symptoms but mutely damages the lymphatic system, kidneys, and immune system. Acute cases of lymphatic filariasis cause local inflammation on the skin, lymph nodes, and lymphatic vessels usually associated with tissue swelling known as lymphedema. Lymphatic filariasis occurs mainly in childhood but the pitiable symptom or condition is noted later or manifests in adulthood especially in males. The disease disorder is associated with damage of the lymphatic system, arms, legs, and genitals. These pitiable conditions cause severe pains, loss of optimum productivity, and social exclusion/stigma. Thus, 40 million out of the 120 million infected people with lymphatic filariasis are already defaced, disabled, and socially excluded.

The disease is known to upset over 25 million men mainly with genital disease, and over 15 million are with lymphedema. The South-East Asia region has the highest share of the disease at 65% while the African region has 30%, and other tropical areas share the remaining 5%. Different mosquito groups such as *Culex*, *Anopheles*, and *Aedes* are responsible for transmitting the filarial worms. *Culex* mosquitoes are the main culprit with a widespread distribution across urban and semi-urban areas. *Anopheles* are capable but mainly in rural areas while *Aedes* are responsible in the Pacific Islands and some parts of the Philippines [1].

2.2 Malaria

Malaria is a disease caused by protozoan parasites in the genus *Plasmodium* which accentuates fever, chills, a flu-like illness, and other symptoms. The symptoms are usually manifested in individuals, at least 7 days after exposure to bites of infected mosquitoes. A particular report showed that about 207 million people were down with malaria in 2012 alone, thus leading to unfortunate death of 627,000 of them. However, death rate is said to be decreasing by 42% in the world and 49% in the WHO region of Africa due to increase in prevention and control strategies by government and non-government organization. Nevertheless, about 3.4 billion people in about 97 countries of the world are at risk of malaria infection. Unfortunately, its burden is extremely peaked in sub-Saharan Africa and contributing to about 90% of malaria deaths. For instance, the Democratic Republic of the Congo and Nigeria are the countries with the highest burden of malaria, where four among ten malaria-infected patients die of the disease [1]. Moreover, young children, gravid women, HIV patients, internally displaced persons (IDPs), and non-immune migrants are mostly at risk of malaria infection. The poorest people living in remote villages where access to healthcare facilities is lacking are the most vulnerable to malaria infection.

Hitherto to the report above, about 40% of the world population is at risk of malaria, with half of the estimated 1.2 billion people that are at high risk of transmission found in Africa. They are concentrated in 13 countries with Nigeria, Congo, Ethiopia, Tanzania, and Kenya having over 50% of the cases [15] with Nigeria alone parading 25% of the cases. Malaria occurs all through the year in Southern Nigeria but seasonal in the northern region. *Plasmodium falciparum* (one of the protozoan parasites) is depicted as the major cause of malaria in Nigeria and responsible for malaria death cases. Thus, the estimated malaria cases in Nigeria as at 2006 in every 1000 individuals was 397 leading to 16 deaths and 3.9 fatality [16]. Reports have shown that malaria as a case study is still a global burden with over 3.4 billion people estimated to be at risk. The approximately 207 million cases of malaria recorded in 2012 with 80% of them occurring in Africa is an indication that the disease is though a global burden but more prevalent in the sub-Saharan countries. Hence, about 90% of infected persons in Africa don't survive the disease and 70% of them are children who are less than 5 years [17]. In 2019, about 229 million new cases were reported globally, out of which 409,000 of the patients died with about 94% occurring in the WHO African region. Demographically, the prevalence is more in pregnant women with 16.3% and children 0.6% [18–20]. Countries like Botswana, Namibia, Swaziland, Zimbabwe, and South Africa have approximately 15 million people that are at risk but 10 million of them are at a higher risk with 437 deaths [21].

Mosquito species such as *Anopheles arabiensis* is the main vector of *Plasmodium* responsible for transmission of malaria in the region. Female anopheline mosquito species are the vectors of five *Plasmodium* species (*Plasmodium falciparum*, *P. vivax*, *P. ovale*, *P. malariae*, and *P. knowlesi*) that are causing malaria in humans. *Plasmodium falciparum* and *P. vivax* are the most common but *P. falciparum* is more problematic with the highest rates of complications and mortality especially in sub-Saharan Africa [1]. *Anopheles gambiae* and *A. funestus* are the main vectors of malaria plasmodia species (*Plasmodium falciparum* and *P. vivax*) which are the deadliest parasites in sub-Saharan Africa [22].

3 Diversity of Medicinal Plants Used in the Treatment and Management of Mosquito-Borne Parasitic Diseases in the Tropics

Based on mosquito biology and ecology, the authors of this chapter who are inclined in entomology and parasitology, respectively, are propelled to describe mosquitoes as the most deadly insects existing on earth. The description is in connection with many debilitating diseases they transmit globally to over one billion people every year. Mosquitoes are vectors of many diseases such as malaria, dengue, yellow fever, lymphatic filariasis, chikungunya, Japanese encephalitis, West Nile fever, zika, and others which are collectively killing over one million people globally on a yearly basis. Management of mosquito-borne diseases is geared towards its vector (mosquito species) control and elimination at different stages of the vectors' development. Thus, larvicidal activity of six Indian plants against *Aedes aegypti* and *Anopheles stephensi* has been reported [23]. Others include larvicidal and ovicidal potentials of *Eclipta alba* against *Aedes aegypti* and leaf and seed extract of *Delonix elata* against *Anopheles stephensi* and *Aedes aegypti* [24, 25].

3.1 Lymphatic Filariasis and Medicinal Plants Used for Its Treatment

The diversity of medicinal plants against lymphatic filariasis, in particular, is estimated at 46 plant species, and many of them have the potentials to work against filarial infection in man [27]. Just like other herbal remedies against mosquito-borne diseases, the parts of plants used mostly to treat lymphatic filariasis are leaves, barks, roots, stems, bark and bulb but herbalists claim that active phytochemicals of the plants are concentrated in the roots underground [28, 29]. Extracts of medicinal plants (*Azadirachta indica*, *Polyalthia suaveolens*, *Andrographis paniculata*, *Bauhinia racemosa*, and *Haliclona oculata*) and others (Table 1) have shown to exert bio-efficacy through immunomodulatory elicitation response, either as alone treatment or as adjuvant [30]. The demand for medicinal plant-based formulations is increasing, hence culminating to studies that have ascertained the molecular structures and isolated phytochemicals (Table 2), which are effective against lymphatic filariasis. This has given rise to a scintillating impulse at which isolation of active compounds are top priority to meeting the pharmaceutical requirements to alternate or compensate the synthetic formulations against lymphatic filariasis.

3.2 Malaria and the Medicinal Plants Used for Its Treatment in the Tropics

Many tropical plants are endowed with phytochemical compounds which are bioactive agents that are effective against many ailments such as malaria. The phytochemicals (alkaloids, flavonoids, and terpenoids) are schizonticides. Anti-malaria activity of medicinal plants in the tropics especially the alcoholic-based extracts have been consistent in the reports of traditional medicinal herbalists [56]. It is noted that alkaloids target the apicoplast in the plasmodial parasites and inhibit their protein synthesis, while flavonoids obstruct the invasion of myoinositol and L-glutamine produced in an infected erythrocytes [57]. They can as well accelerate the level of erythrocytic oxidation, prevent the *Plasmodium* from synthesizing proteins, and prevent fatty acid biosynthesis [58]. In vivo screening for anti-malaria of 15 plant species against plasmodial parasites (*Plasmodium berghei*, *P. yoelii*) are reported to be active and successful. However, in in vivo studies, lower doses are reportedly not efficient as they can only suppress *Plasmodium* load at a higher dosage [16]. *Azadirachta indica* which is one of the widely used medicinal plants for different diseases was reported to deal with all the stages of maturation of plasmodium development by releasing oxidative stress on plasmodial gametocytes. Similarly, *Picralima nitida* plant has a reasonable in vivo anti-malaria action against early and old malaria infections while hexane extract of *Cocos nucifera* dealt with

Table 1 Medicinal plants reported to be used in the treatment of lymphatic filariasis in the tropics

Plant family	Plant species	References
Acanthaceae	*Andrographis paniculata*	[30, 31]
Alliaceae	*Tulbaghia alliacea*	[27, 32]
Amaranthaceae	*Achyropsis avicularis*	[27, 32]
Apiaceae	*Alepidea amatymbica*	[27]
	Trachyspermum ammi	[30, 31]
Aquifoliaceae	*Ilex mitis*	[27]
Araceae	*Acorus calamus*	
Araliaceae	*Cussonia paniculata*	
Asteraceae	*Aster bakerianus*	
	Berkheya setifera	
	Dicoma anomala	
	Eriocephalus L. sp.	
	Platycarpha glomerata	
	Senecio speciosus	
	Centratherum anthelminticum	[31]
	Neurolaena lobata	
	Sphaeranthus indicus	
	Aster bakerianus	[32]
	Dicoma anomala	
	Felicia erigeroides	[27]
Balanophoraceae	*Sarcophyte sanguinea*	
Betulaceae	*Alnus nepalensis*	[31]
Bignoniaceae	*Kigelia africana*	[27, 32]
Caesalpiniaceae	*Bauhinia racemosa*	[31]
	Caesalpinia bonducella	
Capparaceae	*Capparis tomentosa*	[27, 32]
Caricaceae	*Carica papaya*	[32]
Chalinidae*	*Haliclona oculata*	[30]
	Haliclona exigua	
Chenopodiaceae	*Chenopodium ambrosioides*	[27]
Combretaceae	*Terminalia chebula*	[31]
Convolvulaceae	*Ipomoea oblongata*	[27]
	Argyreia speciosa	[31]
Dioscoreaceae	*Dioscorea sylvatica*	[27]
Euphorbiaceae	*Croton sylvaticus*	
	Euphorbia clavarioides	
	Ricinus communis	[31]
	Mallotus philippensis	
	Excoecaria agallocha	
	Croton sylvaticus	[32]
	Euphorbia gorgonis	[27]
	Ricinus communis	[27, 30, 32]

(continued)

Table 1 (continued)

Plant family	Plant species	References
Fabaceae	*Elephantorrhiza elephantina*	[27]
	Albizia anthelmintica	[27]
	Acacia auriculiformis	[31]
	Butea monosperma	
	Pongamia pinnata	
	Psoralea corylifolia	
Ganodermataceae	*Ganoderma* sp.	[27]
Gunneraceae	*Gunnera perpensa*	
Hyacinthaceae	*Drimia depressa*	
	Eucomis autumnalis	[27, 32]
	Eucomis comosa	[27
	Ledebouria sp.	
	Eucomis bicolor	
	Urginea delagoensis	
Hypericaceae	*Psorospermum baumii*	[32]
Hypoxidaceae	*Hypoxis latifolia*	[27]
Lamiaceae	*Leucas cephalotes*	[31]
	Leucas aspera	
Malvaceae	*Hibiscus sabdariffa*	[30, 31]
	Hibiscus mutabilis	[31]
	Hermannia geniculata	[27]
Meliaceae	*Azadirachta indica*	[30]
	Xylocarpus granatum	[30, 31]
	Azadirachta indica	[31, 32]
Menispermaceae	*Tinospora crispa*	[31]
Mesembryanthemaceae	*Aptenia cordifolia*	[27]
Moraceae	*Streblus asper*	[31]
	Ficus racemosa	
Moringaceae	*Moringa oleifera*	[31, 32]
Pinaceae	*Cedrus deodara*	[31]
Piperaceae	*Piper betle*	
Plumbaginaceae	*Plumbago indica*	
Polygonaceae	*Rumex obtusifolius*	[27]
Portulacaceae	*Portulaca oleracea*	
Rhizophoraceae	*Cassipourea flanaganii*	
Rosaceae	*Cliffortia linearifolia*	
Rubiaceae	*Pentanisia prunelloides*	
	Morinda citrifolia	[30]
Rutaceae	*Aegle marmelos*	[31]
Sapindaceae	*Cardiospermum halicacabum*	[31, 32]
Saxifragaceae	*Saxifraga stracheyion*	[31]

(continued)

Table 1 (continued)

Plant family	Plant species	References
Solanaceae	*Datura stramonium*	[27, 31]
	Solanum aculeastrum	
	Withania somnifera	
	Solanum khastanum	
Verbenaceae	*Lantana camara*	[31, 32]
	Vitex negundo	[31]
Vitaceae	*Rhoicissus tridentata*	[27]
	Rhoicissus tomentosa	
Zingiberaceae	*Zingiber officinale*	[32]

Note: * = not plant but sponges

P. falciparum and *P. berghei* in both in vivo and vitro studies [59]. Empirically, these medicinal plants have reputable anti-malaria properties in in vitro experiment but not too strong in in vivo study except when the dosages are higher, which therapeutically are not meaningful [16].

To cut down on resources and time wastages on assumed medicinal plants, traditional medicinal herbalists and the old native people are usually consulted for anti-malarial plants. In one such consultations, 16 species of plants used by traditional medicine herbalists and the natives were beneficial as their voted anti-malaria plants, were chemically evaluated and proven to contain different phytochemicals (fagoronine, palmatine, jatrorrhizine, picraline, astonine, gendunin, meldenin, and others) that are anti-plasmodial; hence, they have been isolated and characterized [16]. Table 3 showcases some of the isolated and characterized phytochemicals contained in different medicinal plants, which are reported as effective remedies against malaria-causing parasites.

Phytochemicals contained in the plants include alkaloids, flavonoids, quassinoids, limonoids, terpenes, chalcones, coumarines, and others [76]. The concentration of these compounds is dependent on the solvents used to extract them. For instance, methanol extract yields more chemicals than water extract because it contains lipophilic compounds [77]. Hence, concentration of the chemicals determine how effective they will be against plasmodial parasites. Secondary metabolites of these chemicals are synergistically more effective against plasmodial parasites. The synergy of the compounds contained in a plant confers them with anti-malaria property, pointing that any plant without anti-plasmodial activity lack active metabolic compounds [78]. Medicinal plants used in the treatment of malaria in Uganda, for instance, are effective against malaria parasites and possess low level of toxicity such as leaves of *Artemisia annua*, *A. africana*, and all plant parts of *Carica papaya* and *F. virosa*. Those with high toxicity are that of petroleum ether leaf extract of *V. amygdalina* and dichloromethane leaf extract of *Microglossa pyrifolia* [79, 80].

In Africa, medicinal plants used in the treatment of malaria and other mosquito-borne diseases are not regularly documented and regulated by appropriate

Table 2 Isolated phytochemicals reported to be used in the treatment of lymphatic filariasis

Isolated compound	Plant species	Plant's part	Medicinal activity	References
Azadirachtin	*Azadirachta indica*	All parts	Inhibition of microfilariae	[33]
Andrographolide	*Andrographis paniculata*	NS	Anti-filarial, microfilariae	[34]
Minosamycin, xestospongin C, D, araguspongin C	*Haliclona oculata*	NS	Anti-filarial	
Gedunin, photogedunin	*Xylocarpus granatum*	Fruits	Anti-*B. malayi*, *B. pahangi*, *microfilarial activity*	[35–38]
Gossypetin, hibiscetine, sabdaretine, delphinidin	*Hibiscus sabdariffa*	Seeds	Anti-*B. malayi*, *female sterility*	[39, 40]
Platyphyllenone, alusenone, hirustenone, hirsutanonol	*Alnus nepalensis*	Leaves	Anti-*B. malayi*, anti-*microfilarial*	[41]
Galactolipid 1–3	*Bauhinia racemosa*	Leaves	Anti-*B. malayi*, anti-*microfilarial*	[42]
Cinnamic acid	*Cinnamomum* sp.	NS	Inhibition of microfilarial	[43]
Withanolide	*Withania somnifera*	Roots	Prevent *B. malayi*	[38, 44, 45]
Acaciaside- A, B	*Acacia auriculiformis*	Funicle	Anti-microfilarial, larvicidal	[46]
Coumarins	*Aegle marmelos*	Leaves	Inhibition of *B. malayi*	[47]
Diarylheptanoid	*Alnus nepalensis*	Leaves	Anti-*B. malayi*, anti-filarial	
Ferulic acid	*Hibiscus mutabilis*	Leaves	Anti-microfilarial	[48]
Glycyrrhetinic acid	*Glycyrrhiza glabra*	Roots	Anti-*B. malayi*	[49]
Oleanonic acid	*Lantana camara*	Stem	Anti-*B. malayi*, female sterility	[50]
Plumbagin	*Plumbago indica*	Roots	Worm immobility	[51]
B-sitosterol, +catechin-3-gallate, bergenin	*Saxifraga stracheyion*	Roots	Inhibition and movement of whole worms	[52]
Solamargine	*Solanum khastanum*	Fruits	Anti-filarial and microfilarial	[53]
Asperoside, strebloside	*Streblus asper*	Stem, bark	Anti-*B. malayi*, worm mortality, inhibition, immobility	[54]

(continued)

Table 2 (continued)

Isolated compound	Plant species	Plant's part	Medicinal activity	References
Withaferin A	*Withania somnifera*	NS	Anti-*B. malayi*, larvicidal	[44]
Eranoid, limonenes, β clemene, zingiberol, linalool	*Zingiber officinale*	Rhizome	Microfilarial reduction	[55]

Note: *NS* not stated

professional body or governmental agencies but still, traditional herbs and herbalists are being patronized for primary healthcare in relation to treating malaria symptoms. Providing and making documentations for medicinal plants that are used for prevention and treatment of malaria and other mosquito-borne diseases in the tropics would create a reasonable step to preserving and improving access to the use of medicinal plants. The goal of documenting medicinal plants used in the treatment of malaria disease will expedite prospective studies on the security and usefulness of medicinal plants and can provide a base for classifying particular chemical characters with anti-malaria activity which could be central in the development of standardized phytomedicine [21]. It has become imperative to search for organically alternative approach for the treatment of malaria since the causative agents (*Plasmodium* species) and the vectors (*Anopheles* species) have developed resistance to many treatment drugs and insecticides, respectively. Chloroquine resistance, for instance, is due to misuse while artemisinin combination therapy is a good and available option for treating malaria but quinolones (quinine, chloroquine, and mefloquinine) are known to cause cardiotoxicity. *Plasmodium* has also been reported to develop resistance to artemisinin combination therapy [81, 82].

In Nigeria, *Icacina senegalensis* Juss which undergone methanolic extraction competed favourably against chloroquine by suppressing about 80% of *P. berghei* while chloroquine suppressed about 92% of *P. berghei* after 5 days of mice infection with *P. berghei*. Though *I. senegalensis* for now has not been characterized, its anti-malaria potency was attributed to having alkaloids, flavonoids, and terpenes which are phytochemicals depicted to have anti-plasmodial agents [83]. Again, ethanolic leaf extract of *Pseudocedrela kotschyi* has been evaluated in; in vivo experiment for anti-malaria potential. Hence, the plant extracts were able to suppress and cure mice of *Plasmodium berghei berghei* infection at different dosages as the lethal dose of the extract was ascertained at 5000 mg/kg without acute toxicity. By comparison, the leaf extract at 400 mg/kg suppressed *P. berghei berghei* parasitaemia at 91% against chloroquine at 10 mg/kg which suppressed 94% of the *P. berghei berghei* parasitaemia density after 5 days of mice exposure [56]. In Western Nigeria, some authors [84] reported 37 species of plants in 25 families, which traditional herbalists have used to treat insect-transmitted diseases. Based on plant parts used, leaves constitute the most utilized parts at 43.2%, while bark is 32.4%, a whole plant is 10.8%, fruits and seeds are 5.4%, while root and whole plant are 7%. Both fresh and dry plants are used except occasionally where fresh is preferred [84]. In Southern

Table 3 Isolated phytochemicals reported to be used in the treatment of malaria in the tropics

Isolated compound	Plant species	Plant's part	Medicinal activity	References
Fagaronine	*Fagara zanthoxyloides*	Roots	Anti-*P. falciparum*	[60]
Akuammiline, akuammidine, akuammine, akuammine, akuammicine, picraline, alstonine, and β limonoids	*Picralina nitida*	NS	Anti-*P. falciparum*	[61]
Gendunin	*Khaya grandifoliola, Azadirachta indica*	NS, wood	Less active anti-*P. berghei*	[62, 63]
Meldenin	*Azadirachta indica*	Leaves	Anti-*P. falciparum*	[64]
Fissinolide	*Khaya grandifoliola*	Bark, roots, seeds	Anti-*P. falciparum*	[65]
Ursolic acid	*Spathodea campanulata*	Bark	Anti-*P. berghei*	[66]
Anthraquinones	*Morinda lucida*	NS	Anti-*P. falciparum*	[67, 68]
Sesquiterpene lactone, Tagitinin C	*Tithonia diversifolia*	Leaves	Anti-*P. falciparum*	[69]
Ajoene	*Allium sativum*	Bulb	Anti-*P. berghei*	[70]
Simalikalactone D	*Quassia Amara*	Leaves	NS	[71]
Diterpene	*Microglossa pyrifolia*	NS	Anti-*P. falciparum*	[20]
Nitidine	*Zanthoxylum chalybeum*	NS	Anti-*P. falciparum*	
Pristimarin	*Maytenus senegalensis*	NS	Anti-*P. falciparum*	[72]
Pinocembrin, flavonoid santin, clerodane diterpene	*Dodonaea angustifolia*	Leaves	Anti-*P. berghei*	[73]
Aloinoside	*Aloe macrocarpa*	Leaves	Anti-*P. berghei*	[74]
Aloin	*Aloe debrana*	Leaves	Anti-*P. berghei*	[75]
Otostegindiol	*Otostegia integrifolia*	NS	Anti-*P. berghei*	

Note: *NS* not stated

Nigeria, malaria transmission is experienced all through the year because it is a rain forest region that is characterized with a humid tropical climate. Thus, it supports greater number of plant diversity including medicinal plants that the natives and traditional medicinal herbalists use for different ailments. According to a review report [16], 24 plant species have been tested in in vitro studies and were found to have anti-plasmodial properties and low cytotoxicity values. *Nauclea latifolia*, for instance, was reported to inhibit *Plasmodium falciparum* erythrocytic cycle towards the last phase of schizogony between 32 and 48 h. To break some logistic barriers in the experimentation of medicinal plants against malaria, developing countries like Nigeria occasionally collaborate with foreign laboratories to conduct some in vitro and vivo studies to close the gap of lack of requisite laboratories in their native countries [16].

In Zimbabwe, 28 species of plants were reported to have anti-malaria properties but only 26 were taxonomically identified in 16 families [21]. Detailed analysis show that out of the 26 species, 38.5%, 30.8%, 23.0%, and 7.7% are trees, shrubs, climbers, and herbs, respectively. Twenty of the 26 species have anti-malaria (therapeutic) properties, 4 have preventive (prophylactic) properties, while the remaining 2 species serve the dual purpose. Among these species, only *Momordica foetida* and *Capsicum annuum* can be cultivated while others can only be harvested from the wild, a scenario that could lead to loss of biodiversity. Interesting report directly from the traditional medicine healers stated that some plants such as *Euclea natalensis* (Ebenaceae) may not have therapeutic or prophylactic values but may be added to boost strength and increase appetite in malaria patients [21].

In Ethiopia, 51 species of medicinal plants in 28 families against malaria have been reported with the family Fabaceae, Asteraceae, Euphorbiaceae, Aloaceae, Alliaceae, and Meliaceae having 6, 5, 4, 3, 3, 2, and 2 species, respectively, while other family members have 1 species each. Among these medicinal plants against malaria, *Allium sativum*, *Croton macrostachyus*, *Carica papaya*, and *Lepidium sativum* are the most frequently used plants to treat malaria [85]. Phytochemical compounds such as alkaloids, terpenes, limonoids, chromones, xanthones, flavonoids, and anthraquinones have been reported to be successful in the treatment of malaria. Anthraquinones and naphthalene derivatives are the most reported compound against malaria. These compounds are effectively reported in in vivo and vitro studies at 61% and 39%, respectively, against *Plasmodium falciparum* [85].

In Uganda, about 182 species of medicinal plants in 63 families, which have anti-malaria properties, but only 17 species are frequently used in malaria healthcare. The leaves constitute the highest parts that are used at 54.4%, and followed by roots at 17.4%, bark at 16%, while whole plant and other parts constitute 12.2% [86]. Among the 182 species of medicinal plants, 112 have been evaluated for treatment of malaria, but 108 were ascertained to be effective for malaria treatment.

In Ghana, 422 species of medicinal plants in 25 families that are highly diversified are reportedly used in malaria treatment, with Euphorbiaceae constituting 14.3%, Asteraceae 9.5%, and Poaceae and Fabaceae having 7.1% each as the most used families [87]. Trees are mostly used at 42.9%, herbs 28.6%, shrubs 16.7%, grasses 7.1%, and climbers 2.4%. About 75% of the plant materials are collected from the neighbourhood surroundings, while 25% are harvested from the wild and elsewhere. Based on plant parts used, leaves constitute 80%, fruits 13.3%, stem-bark 4.4%, and flowers 2.2%. Threats to medicinal plants include drought, farming, over harvesting, bush fire, and others [87].

In Kenya, 286 plant species in 75 families are reported to be useful in the treatment and management of malaria with Asteraceae, Fabaceae, Lamiaceae, Euphorbiaceae, Rutaceae, and Rubiaceae families. The percentage use of these listed plant families are 36.5%, 29.7%, 24.3%, 21.6%, and 17.6%, respectively [22]. Plant parts used are shrubs, trees, herbs, roots, bark, root-bark, and stem-bark at 33.2%, 30.1%, 29.7%, 19.4%, 10.8%, 10.5%, and 6.9%, respectively, hence similar to reports from other African countries. No serious report of threat associated with leaves usage but root and root structures (tubers and rhizomes) are potentially driving the plants to extinction not minding reports that plant parts below the ground contain more phytochemicals.

From the time immemorial, medicinal herbs have been used to treat malaria in Africa and other tropical countries. Medicinal plants are abound in the tropics with higher concentration of organic chemicals that have been used for treatment of malaria (Table 4). For instance, quinine and artemisinin are isolates of *Cinchona* and *Artemisia* plants that are natives to sub-Saharan Africa. Thus, the first plant-based anti-malaria drug was extracted from *Cinchona* plant as quinine in the form of alkaloid as reported by Baird et al. [88] and was later isolated and characterized (Saxene et al. 2003). Later in the nineteenth century, another medicinal plant *Artemisia annua* became another source of malaria treatment known as artemisinin which was combined with other therapies and referred to as artemisinin-combined therapy (ACT) [89]. This approach of malaria treatment has been the medically approved treatment in Nigeria since 2005 [90]. However, a lot of factors of production militate against artemisinin combination therapy production and therefore, the drug remains expensive that many people in the developing countries cannot afford it. It is therefore imperative that plants from the tropics especially in sub-Saharan African regions should be screened for anti-malaria properties since greater percentages of malaria death cases occur in the regions.

4 Preparations and Administration of Herbal Medicine Used in the Treatment of Mosquito-Borne Parasitic Diseases in the Tropics

Some of the parts of the medicinal plants used in the treatment of malaria disease are processed into powder and preserved in closed bottles. When in need, the powder is either soaked in cold or hot water to extract the active ingredients. After a while, the water is decanted or sieved and given to a malaria patient to drink. To make concoctions from the powders, the recommendations are usually a full, half, or quarter of either teaspoon or tablespoon or may be a pinch. The concoction is either prescribed to be taken twice or thrice a day as from 3 to 7 days or till the patient is certified healed. Where some fruits are to be used, they are swallowed whole. For instance, the fruit of *Capsicum annuum* (*Capsicum frutenscens*) is given to a patient to swallow without chewing because of its bitterness. In some case, especially for preventive measures, some leaves are eaten as vegetables such as *Momordica balsamina* and *M. foetida* [21].

Table 4 Medicinal plants reported in Africa for the treatment of malaria

Plant family	Plant species	References
Acanthaceae	*Justicia betonica*	[86, 91]
	Justicia anselliana	[86]
	Monechma subsessile	
	Thunbergia alata	
Alliaceae	*Allium sativum*	[16]
	Allium cepa	[86]
	Aloe kedongensis	
	Aloe volkensii	
	Aloe ferox	
	Aloe lateritia	
Aloeaceae	*Aloe dawei*	
Amaranthaceae	*Amaranthus hybridus*	
Anacardiaceae	*Anacardium occidentale*	[16, 91]
	Mangifera indica	[16, 86, 91]
	Rhus natalensis	[86]
	Rhus vulgaris	
Annonaceae	*Uvaria chamae*	[16]
	Xylopia aethiopica	[91]
	Annona senegalensis	[16]
	Enantia chlorantha	[16, 91]
Apiaceae	*Heteromorpha trifoliata*	[86]
	Centella asiatica	
	Carissa edulis	
	Carissa spinarum	
Apocynaceae	*Catharanthus roseus*	
	Funtumia africana	[16, 91]
	Picralima nitida	[16]
	Rauvolfia vomitoria	[91]
	Alstonia boonei	[16, 86, 91]
	Rauvolfia vomitoria	[16]
Araceae	*Culcasia faleifolia*	[86]
Aristolochiaceae	*Aristolochia elegans*	
	Aristolochia tomentosa	
	Cryptolepis sanguinolenta	[16]
Asclepiadaceae	*Gomphocarpus physocarpus*	[86]
	Pergularia daemia	[91]
	Aloe vera	[86]
Asteraceae	*Aspilia africana*	[16]
	Ageratum conyzoides	[86]
	Artemisia annua	
	Artemisia afra	
	Aspilia africana	
	Baccharoides adoensis	

(continued)

Table 4 (continued)

Plant family	Plant species	References
	Bidens grantii	
	Bothriocline longipes	
	Conyza bonariensis	
	Conyza sumatrensis	
	Crassocephalum vitellinum	
	Emilia javanica	
	Guizotia scabra	
	Gynura scandens	
	Melanthera scandens	
	Pluchea ovalis	
	Microglossa pyrifolia	
	Schkuhria pinnata	
	Sigesbeckia orientalis	
	Solanecio mannii	
	Sonchus oleraceus	
	Tagetes minuta	
	Tithonia diversifolia	
	Vernonia adoensis	
	Vernonia amygdalina	
	Vernonia cinerea	
	Vernonia lasiopus	
Bignoniaceae	*Spathodea campanulata*	[16]
	Stereospermum kunthianum	
	Markhamia lutea	[86]
	Spathodea campanulata	
	Newbouldia laevis	[16]
Bombacaceae	*Adansonia digitata*	
	Ceiba pentandra	[91]
Boraginaceae	*Heliotropium indicum*	
	Ananas comosus	
Caesalpiniaceae	*Cassia didymobotrya*	[86]
	Chamaecrista nigricans	
	Erythrophleum pyrifolia	
	Senna spectabilis	
	Senna siamea	[91]
	Senna podocarpa	
	Cassia hirsuta	[86]
Canellaceae	*Warburgia ugandensis*	
Cannaceae	*Canna indica*	[16, 91]
Caricaceae	*Carica papaya*	[16, 86, 91]
Celastraceae	*Hippocratea africana*	[16]
	Maytenus senegalensis	[86]
Chenopodiaceae	*Chenopodium ambrosioides*	
	Chenopodium opulifolium	

(continued)

Table 4 (continued)

Plant family	Plant species	References
Combretaceae	*Anogeissus leiocarpus*	[16]
	Guiera senegalensis	
	Terminalia avicennioides	
	Terminalia latifolia	
	Combretum molle	[86]
Compositae	*Chromolaena odorata*	[16, 91]
	Tithonia diversifolia	
	Ageratum conyzoides	[16]
	Eupatorium odoratum	
	Vernonia amygdalina	[16, 91]
	Vernonia cinerea	[16]
	Acanthospermum hispidum	[91]
Crassulaceae	*Kalanchoe densiflora*	[86]
Cucurbitaceae	*Cucurbita maxima*	
	Momordica foetida	
	Momordica balsamina	[16]
Dracaenaceae	*Dracaena steudneri*	[86]
Ebenaceae	*Euclea latideus*	
	Bridelia micrantha	
	Clutia abyssinica	
	Croton macrostachyus	
	Flueggea virosa	
	Diospyros mespiliformis	[91]
Euphorbiaceae	*Alchornea cordifolia*	[16]
	Bridelia ferruginea	
	Bridelia micrantha	
	Croton zambesicus	
	Phyllanthus amarus	
	Alchornea cordifolia	[86]
	Jatropha curcas	
	Macaranga schweinfurthii	
	Phyllanthus niruri	
	Shirakiopsis elliptica	
	Tetrorchidium didymostemon	
	Cajanus cajan	
	Crotalaria agatiflora	
	Crotalaria ochroleuca	
	Entada abyssinica	
	Entada africana	
	Erythrina abyssinica	
	Erythrina excelsa	
	Bridelia ferruginea	[91]

(continued)

Table 4 (continued)

Plant family	Plant species	References
Fabaceae	*Arachis hypogea*	[86]
	Indigofera arrecta	
	Indigofera congesta	
	Indigofera emerginella	
	Macrotyloma axillare	
	Pseudarthria hookeri	
	Rhynchosia viscosa	
	Senna absus	
	Senna didymobotrya	
	Senna siamea	
	Tamarindus indica	
Flacourtiaceae	*Oncoba spinosa*	
	Trimeria bakeri	
	Homalium letestui	[16]
Guttiferae	*Harungana madagascariensis*	[16, 86, 91]
	Allanblackia floribunda	[16]
Labiatae	*Ocimum gratissimum*	[16]
	Solenostemon monostachyus	[16]
	Hyptis suaveolens	[91]
	Ocimum gratissimum	
	Hyptis pectinata	[86]
Lamiaceae	*Aeollanthus repens*	
	Ajuga remota	
	Clerodendrum myricoides	
	Clerodendrum rotundifolium	
	Hoslundia opposita	
	Leonotis nepetifolia	
	Ocimum basilicum	
	Ocimum gratissimum	
	Ocimum lamiifolium	
	Plectranthus barbatus	
	Plectranthus caninus	
	Plectranthus cf. *forskohlii*	
	Rosmarinus officinalis	
	Tetradenia riparia	
Lauraceae	*Cassytha filiformis*	[16]
	Persea americana	[86]
Leguminosae-Caesalpinioideae	*Abrus precatorius*	[16]
	Afzelia africana	
	Cajanus cajan	
	Cassia occidentalis	
	Cassia siamea	
	Cassia sieberiana	

(continued)

Table 4 (continued)

Plant family	Plant species	References
	Cassia singueana	
	Daniellia ogea	
	Daniellia oliveri	
	Piliostigma thonningii	
Leguminosae-Mimosoideae	*Cylicodiscus gabunensis*	
	Parkia biglobosa	
	Prosopis africana	
	Tetrapleura tetraptera	
Leguminosae-Papilionoideae	*Erythrina senegalensis*	
	Indigofera pulchra	
	Lonchocarpus cyanescens	
	Pericopsis elata	
Loganiaceae	*Anthocleista djalonensis*	
	Anthocleista vogelii	
Loranthaceae	*Tapinanthus sessilifolius*	
	Tapinanthus constrictiflorus	[86]
Malvaceae	*Hibiscus surattensis*	
	Gossypium barbadense	[91]
	Gossypium arboreum	[16]
	Gossypium barbadense	
	Gossypium hirsutum	[16, 91]
	Sida acuta	[16]
Meliaceae	*Azadirachta indica*	[16, 86, 91]
	Khaya grandifoliola	[16]
	Khaya senegalensis	
	Khaya grandifoliola	[91]
	Carapa grandiflora	[86]
	Melia azedarach	
Menispermaceae	*Sphenocentrum jollyanum*	[16, 91]
	Cissampelos mucronata	[86]
	Acacia nilotica	
	Acacia sieberiana	
Mimosaceae	*Acacia hockii*	
	Albizia coriaria	
	Albizia grandibracteata	
	Albizia zygia	
	Newtonia buchananii	
Moraceae	*Ficus thonningii*	[16]
	Milicia excelsa	[16, 86]
	Antiaris toxicaria	[86]
	Ficus natalensis	
	Ficus saussureana	
	Ficus platyphylla	[16]

(continued)

Table 4 (continued)

Plant family	Plant species	References
Moringaceae	*Moringa oleifera*	[16, 86]
Musaceae	*Musa paradisiaca*	[86]
	Musa sapientum	[91]
Myricaceae	*Myrica kandtiana*	[86]
Myristicaceae	*Pycnanthus angolensis*	[86, 91]
Myrsinaceae	*Maesa lanceolata*	[86]
Myrtaceae	*Eucalyptus grandis*	
	Syzygium cordatum	
	Syzygium cumini	
	Syzygium guineense	
	Ormocarpum trachycarpum	
	Psidium guajava	[16, 86, 91]
Ochnaceae	*Lophira alata*	[16]
	Lophira lanceolata	
Palmae	*Elaeis guineensis*	
Papilionaceae	*Butyrospermum paradoxum*	[86]
Passifloraceae	*Passiflora edulis*	
	Mondia whitei	[91]
	Parquetina nigrescens	
	Pittosporum brachcalya	[86]
	Pittosporum mannii	
Poaceae	*Cymbopogon giganteus*	[16]
	Setaria megaphylla	
	Digitaria scalarum	[86]
	Imperata cylindrica	
	Zea mays	
	Cymbopogon citratus	[16, 86, 91]
	Saccharum officinarum	[16]
Polygalaceae	*Securidaca longipedunculata*	[86]
	Maesopsis eminii	
Portulacaceae	*Talinum portulacifolium*	
Rosaceae	*Prunus africana*	
	Rubus steudneri	
	Hallea rubrostipulata	
	Crossopteryx febrifuga	[16]
	Mitragyna inermis	
Rubiaceae	*Nauclea latifolia*	[16, 91]
	Coffea canephora	[86]
	Pentas longiflora	
	Vangueria apiculata	
	Citrus sinensis	
	Morinda lucida	[91]

(continued)

Table 4 (continued)

Plant family	Plant species	References
Rutaceae	*Citrus reticulata*	[86]
	Citrus aurantium	[91]
	Citrus sinensis	[16]
	Fagara zanthoxyloides	
	Citrus aurantifolia	[91]
	Citrus paradisi	
Salicaceae	*Trimeria grandifolia*	[86]
Sapindaceae	*Blighia unijugata*	
	Blighia sapida	[16]
	Lecaniodiscus cupanioides	[91]
Sapotaceae	*Manilkara obovata*	[86]
	Chrysophyllum albidum	[91]
Scrophulariaceae	*Striga hermonthica*	[16]
	Sopubia ramosa	[86]
	Scoparia dulcis	[16]
Simaroubaceae	*Quassia undulata*	
	Harrisonia abyssinica	[86]
	Quassia amara	[16]
Solanaceae	*Solanum erianthum*	
	Datura stramonium	[86]
	Physalis angulata	[91]
	Physalis peruviana	[86]
	Solanum nigrum	
	Capsicum frutescens	[91]
	Solanum nigrum	
Sterculiaceae	*Sterculia setigera*	[16]
Tiliaceae	*Triumfetta rhomboidea*	[86]
Ulmaceae	*Celtis durandii*	[16]
	Celtis africana	[86]
	Trema orientalis	[91]
Umbelliferae	*Steganotaenia araliacea*	[86]
Rutaceae	*Teclea nobilis*	
	Toddalia asiatica	
	Zanthoxylum chalybeum	
	Zanthoxylum leprieurii	
Verbenaceae	*Lippia multiflora*	[16]
	Stachytarpheta cayennensis	[16]
	Stachytarpheta indica	
	Lantana camara	[86]
	Lantana trifolia	
Vitaceae	*Cissus populnea*	[16]
Zingiberaceae	*Zingiber officinale*	[16, 91]
	Curcuma longa	[16, 86, 91]

Decoction which is the act of boiling of plants, and/or plant parts, in water is the commonest way of making the medicine out of the plants available for drinking. Concoction, on the other hand, is the processing of medicinal plants into powder and/or ingestion of fresh extract from plants. For decoction, and infusion (steam bath), preparations depend on plant type, parts, and choice which is mainly related to geographical communities. Decoction goes with water measurement and is drank orally. For concoction, some plants are dried at room temperature, pulverized, and transformed to powder. When in need, 2–3 tablespoon is added to water and boiled to obtain a decoction. Boiling of plants or plant parts has no reference time limit but can be estimated with the initial volume of water reducing to half of its original volume. Some are prepared by squeezing fresh leaves of a chosen plant to obtain 2–3 teaspoon of a juice-like liquid daily. In some cases, different plant leaves can be used to make an infusion with fresh roots that are pulverized and taken orally. Some squeezed juice of medicinal plant can be mixed with cold or warm water for bathing while some are eaten as vegetable for preventive measures [92]. For lymphatic filariasis, desired plants or parts are grinded into powder to increase the surface area of the plant so that a chosen extractant can easily extract reasonable quantity of active ingredient in the plant. Decoction and infusion methods of preparation are mostly employed while orally drinking of the prepared medicine for malaria treatment while topical application of concoction on the affected limbs and bathing with decoction are common methods of treating lymphatic filariasis.

In Ethiopia, crushing, powdering, maceration, decoction, and others are the commonest means of formulating medicinal plants. The medicines are orally administered as a route of delivery into the body. Water and some ingredients (honey, butter, salt, coffee, tea, and milk) are added while making the preparation to reduce the bitter taste of the plants, hence spicing the medicine and reducing toxicity [85].

In Ghana, leaves are more frequently used, mainly for decoction, hence ranking 95.35%, and decoction for steam bath and rubbing or massaging plant tissue on the body constitute 2.3% each [87]. Decoction of medicinal plants constitute 70.5%, infusion 5.4%, oil treatment and steaming 1.3%, and roasting 0.3%. Plant materials may be chewed, pounded, ground, or crushed while dry materials are burnt, smoked, and inhaled. Extracting solvents include methanol, water, ether, ethanol, dichloromethane, chloroform, ethyl acetate, and petroleum ether [22].

In Nigeria, a popular concoction known as Agbo-Iba is being reported to protect mice against *Plasmodium yoelii nigeriensis*. The concoction is formulated with leaves of *Cajanus cajan* (pigeon pea), *Latifolia* sp., *Cymbopogon giganteus*, *Nauclea latifolia*, *Euphorbia lateriflora*, *Mangifera indica*, and barks of *Cassia alata* and *Uvaria chamae* [93]. The experimentation of the Agbo-Iba formulation is decocted for 3 h and administered to mice orally. The result showed protection of the mice against *P. yoelii nigeriensis*. However, a report has it that Agbo-Iba was more of prophylactic instead of therapeutic for malaria treatment [93]. Similarly, a multi-herbal combination mixture containing the leaves of *Carica papaya*, *Azadirachta indica*, *Cymbopogon citratus*, and *Anacardium occidentale* are used as a steam therapy against malaria. The steam from the concoction in a cooking pot is guided for a patient to inhale a hot steam from the pot by covering both the patient and the hot pot.

Administration dosages of medicinal decoctions and concoctions vary, and no stringent precision exist among traditional medicine practitioners for the same ailment but are very certain about their recipes. None conformity and compliance to dosage by traditional medicine practitioners and patients, respectively, may hamper the goal of malaria treatment with medicinal plants. For instance, some patients may stop taking a *Cissampelos mucronata* roots concoction half way because of its bitterness. This concoction is usually prescribed to be taken thrice a day for 7 days [21]. However, dosages and strength of the medicine are peculiarly age dependent. In some cases, the volume of consumption range is between 100 and 500 ml, 250, and 100 ml for adults, older children above 5 years, and children less than 5 years, respectively, or from 1 to 3 tablespoon in reverse order. The chosen quantity could be taken one to three times daily for 7 days or until full recovery is ascertained [94, 78, 92].

In discussing medicinal remedies of plants about malaria, for instance, the terms anti-malaria and anti-plasmodia are mentioned but does not mean the same thing. Anti-plasmodium is related to in vitro study of medicinal plants while anti-malaria is related to in vivo study of medicinal plants. For instance, anti-plasmodial action of plant extracts in in vitro study will be rated 'very good' when the IC_{50} is less than 5 μg/ml, good when IC_{50} is greater than 5 μg/ml but less than 10 μg/ml, and moderate when the IC_{50} is 10ug/ml or less than 20 μg/ml. For anti-malaria, action of plant extracts in in vivo study will be rated 'very good' when the extracts suppress about or more than 50% at 100 mg/kg of a body weight in a day. It will be good when the extracts suppress malaria by more than 50% at 250 mg/kg of a body weight in a day and moderate when more than 50% of malaria is suppressed at 500 mg/kg of a body weight in a day [95]. Basic febrile symptom indicators for malaria, which traditional medicine practitioners depict to commence treatment, include feeling of cold/goose pimples, headache, fever, sweating, loss of appetite, body weakness/feeling sleepy, dizziness, vomiting, and body pains. Medicinal plant-based preparations are orally drunk, steam inhaled, topically applied, and rubbing of essential oils on the skin [21, 22].

5 Challenges in the Use of Medicinal Plants in the Treatment of Mosquito-Borne Parasitic Diseases in the Tropics

The use of medicinal plants to treat different diseases especially infectious ones [96–116] and control of vectors of some parasitic diseases [117–121] has transcended from folkloric phase to a novel practice that involves heuristic exploration and taking cognizance of all the processes (formulations and dosages) to be documented for posterity and reproducibility. However, like every other sector of human endeavour, the use of medicinal plants especially the act of harvesting, formulations, standardized dosages, and others present some challenges in their use. For instance, harvesting whole plant and making use of root structures are not sustainable practices since many of the medicinal plants are sourced directly from the wild and hence, could lead to biodiversity loss. Seasonal variation, environmental geography, and methods of extraction of active compounds are challenging factors that can influence the concentration of phytochemical extracts which are essential in the treatment of mosquito-borne parasitic diseases [122].

The process of extracting active compounds from medicinal plants for further examination and characterization poses some difficulties for researchers. For example, poor aqueous solubility, and rapid breakdown of compounds such as berberine and emodin (characterized plant compounds for treatment of mosquito-borne diseases) have been a challenge in pharmaceutical development of such compounds into readily available medicine [123–125]. Others include having enough quantity of the extracts to justify production cost and having the compounds that can work in both in vitro and vivo because some compounds are active in vitro at a small quantity which could not be enough for isolation, hence limiting their use in in vivo studies. Again, some parasites lack animal models for their empirical evaluation. For instance, filarial parasite such as *Wuchereria bancrofti* cannot survive or be maintained in any known experimental animal for in vitro and vivo studies, thus no screening model to be used to evaluate the potency of medicinal plants on the parasite. In some cases where phytochemical compounds are active in in vivo study, it occurs at higher dosages which are therapeutically not meaningful.

In most cases, the toxicity of most of the plant compounds are not yet evaluated to determine their toxicity level, hence limiting the acceptance of their medicinal potentials. The data necessary for selective index of the compounds are not worked out or not readily available at the disposal of professionals, hence their cytotoxicity importance is still questionable. More still, there is paucity of clinical data about safety and efficacy occasioned by lack of requisite laboratories especially in the developing countries, hence making practitioners not to have a common ground for standard preparations and dosages and contraindications of plant-based medicines.

6 Strategies for the Overcoming the Hurdles in the Use of Herbal Medicine in the Treatment of Parasitic Diseases Transmitted by Mosquito in the Tropics

Motivational incentives are spices that drive the wheels of production; hence, herbal medicine practitioners should be encouraged and promoted by organs of governments and multinationals. They should also be enlightened on the sustainable means of conserving biodiversity while providing healthcare services. It is therefore, imperative to cooperate with herbal medicine practitioners when carrying out scientific researches on medicinal plants. Their inclusion in the researches will reduce wastages in using many acclaimed medicinal plants, hence restricting scientific evaluation on the selected plants that are efficacious against parasitic diseases. For a sincere cooperation and willingness, the herbal medicine practitioners who are likely to hoard medicinal information of different plants should be enticed with some level of patency of the outcome of the researches.

There should be need to have policies on the use of traditional or herbal medicines to regulate traditional medicine practitioners with a goal to having standards and fine-tune researches that are leaned to their traditional claims. The policies shall convey the objectives of propagation, protection, and sustainable use of medicinal plants in the treatment of various ailments in their communities. The health sector should be organized through ministry of health to champion a bill for collaboration between healthcare sectors and traditional medicine practitioners to have a national council of traditional medicine practitioners and other medical practitioners to act as a body that would protect herbal products of her members. The isolated or purified compounds that emanate from scientific evaluations should undergo further investigation to cut down the limitations in the quest to develop novel plant-based medicines that are sustainable, cheap, and affordable. For parasites without reliable animal models, new animal models ought to be sought for so as to understand the host/parasite interactions for effective testing of medicinal plants in both in vitro and vivo studies.

7 Conclusion

Diseases associated with mosquitoes are previously seen as a peculiar problem in the tropical countries, but today, they are accepted threat to the world public health because of mosquito species diversity and their geographical spread. Mosquitoes are spreading globally due to climate change, pattern of land use, and the continued increase in number of people moving from one country to the other, hence threatening almost half of the global population. Mosquito-borne parasitic diseases affect both people in the urban and rural villages but most prevalent in villages without adequate living conditions such as housing and sanitation and more disturbing to those with low immunity. Many female mosquito species are culprits in transmitting infectious parasitic diseases such as lymphatic filariasis and malaria between humans and/or from animals to humans.

The use of medicinal plants to treat and manage mosquito-borne parasitic diseases in the tropics mainly in Africa is linked to cultural acceptability, efficacy, accessibility, and affordability. Greater percentage of people living in the rural communities in the tropical countries treat almost every ailment with traditional medicines that originate from plants which are prescribed by traditional medicine practitioners or herbalists. Medicinal plants are cautiously selected plants which have been used in the tropics for years, for treatment and management of diseases from the ancient time till now. In traditional medicine, herbs and plant products are still being used among the African, Asian, and Latin American people, and their use is currently extending to Australia and the USA.

Information from traditional medicine practitioners show that the use of medicinal plants against diseases such as mosquito-borne diseases is economical, simple to use, easily accessible, and cheap while synthetic drugs require a lot of time, resources, and efforts from the onset. Medicinal plants are reported to have natural compounds in them which are anti: viruses, bacteria, protozoa, nematodes, fungi, oxidant, inflammatory, and carcinogenic. Their use in traditional medicine is beginning to attract consideration among the healthcare professionals in the international community because of its current documentation in print and other forms of media.

Tropical countries are biodiversity hotspot for medicinal plants which traditional medicine practitioners and/or herbalists have used to treat mosquito-borne parasitic diseases with reliable successes. However, biodiversity loss occasioned by continuously harvesting of whole plants and root structure, lack of support and motivation from appropriate quarters, adequate scientific production tools, and paucity of documented data on medicinal plants are among the challenges against the use of medicinal plants used in the treatment and management of mosquito-borne parasitic diseases. Defined policies for collaboration between the organized health sectors through the ministry of health to have a consolidated national council of traditional medicine practitioners and other medicine practitioners ought to act as a body to protect medicinal products, and council members shall form alliance for acceptance, patronage, and sustainability of use of medicinal plants for treatment of mosquito-borne parasitic diseases in the tropics.

References

1. WHO (2014) A global brief on vector-borne diseases. World Health Organisation, Geneva. Accessed 12 Mar 2023
2. Clain E, Sinigaglia L, Koishi AC, Gorgette O, Gadea G, Viranaicken W, Krejbich-Trotot P, Mavingui P, Desprès P, dos Santos CND et al (2018) Extract from *Aphloia theiformis*, an edible indigenous plant from Reunion Island, impairs zika virus attachment to the host cell surface. Sci Rep 8:1–12
3. Dos Santos AE, Kuster RM, Yamamoto KA, Salles TS, Campos R, De Meneses MD, Soares MR, Ferreira D (2014) Quercetin and quercetin 3-*O*-glycosides from Bauhinia longifolia (Bong.) Steud. show anti-Mayaro virus activity. Parasites Vectors 7:130
4. Johari J, Kianmehr A, Mustafa M, Abubakar S, Zandi K (2012) Antiviral activity of baicalein and quercetin against the Japanese encephalitis virus. Int J Mol Sci 13:16785–16795
5. Manach C, Scalbert A, Morand C, Rémésy C, Jiménez L (2004) Polyphenols: food sources and bioavailability. Am J Clin Nutr 79:727–747
6. WHO (2014) Vector borne diseases. World Health Organisation, Geneva. Available via DIALOG. http://www.who.int/mediacentre/factsheets/fs387/en/. Accessed 12 Mar 2023
7. Tyagi BK (2016) Advances in vector mosquito control technologies, with particular reference to herbal products. In: Veer V, Gopalakrishnan R (eds) Herbal insecticides, repellents and biomedicines: effectiveness and commercialization. Springer Nature, New Delhi, pp 1–9
8. WHO (2016) Lymphatic filariasis. Fact Sheet Updated October 2016, pp 1–4
9. Maizels RM, Gomez-Escobar N, Gregory WF, Murray J, Zang X (2001) Immune evasion genes from filarial nematodes. Int J Parasitol 31:889–898
10. WHO (2009) Global programme to eliminate lymphatic filariasis. Wkly Epidemiol Rec 8:437–444
11. Maurya SK, Singh AK, Seth A (2015) Potential medicinal plants for lymphatic filariasis. J Crit Rev 2:1–6
12. Wynd S, Melrose WD, Durrheim DN, Carron J, Gyapong M (2007) Understanding the community impact of lymphatic filariasis: a review of sociocultural literature. Bull World Health Organ 85:421–500

13. Simonsen PE, Malecela MN, Michael E, Mackenzie CD (2008) Lymphatic filariasis research and control in Eastern and Southern Africa. DBL- Centre for Health Research and Development, Denmark, p 185p
14. Shrivastava S, Gidwani B, Gupta A, Kaur CD (2016) Ethnopharmacological approaches to treat lymphatic filariasis. Int J Pharm Anal Res 5(3):1–16
15. WHO (2008) World malaria report 2008. World Health Organization, Geneva, pp 7–15, 99–101
16. Adebayo JO, Krettli AU (2011) Potential antimalarials from Nigerian plants: a review. J Ethnopharmacol 133:289–302
17. WHO (2013) World malaria report. http://apps.who.int/iris/bitstream/10665/97008/1/9789241564694_eng.pdf. Accessed 12 Mar 2023
18. WHO (2020) World malaria report. World Health Organization, Geneva. https://www.who.int/
19. Tsegaye AT, Ayele A, Birhanu S (2021) Prevalence and associated factors of malaria in children under the age of five years in Wogera district, northwest Ethiopia: a cross-sectional study. PLoS One 16:e0257944
20. Gontie GB, Wolde HF, Baraki AG (2020) Prevalence and associated factors of malaria among pregnant women in Sherkole district, Benishangul Gumuz regional state West Ethiopia. BMC Infect Dis 20:573
21. Ngarivhume T, Klooster CEAV, deJong JTVM, Westhuizen JHV (2015) Medicinal plants used by traditional healers for the treatment of malaria in the Chipinge district in Zimbabwe. J Ethnopharmacol 159:224–237
22. Omara T (2020) Antimalarial plants used across Kenyan communities. Evid Based Complement Alternat Med 4538602:1–31
23. Marimuthu G, Rajamohan S, Mohan R, Krishnamoorthy Y (2012) Larvicidal and ovicidal properties of leaf and seed extracts of *Delonix elata* (L.) Gamble (family: Fabaceae) against malaria (*Anopheles stephensi* Liston) and dengue (*Aedes aegypti*) (Diptera: Culicidae) vector mosquitoes. Parasitol Res 111(1):65–77
24. Ghosh A, Chowdhury N, Chandra G (2012) Plant extracts as potential mosquito larvicides. Indian J Med Res 135(5):581–598
25. Hatil HE, Kamali L (2009) Medicinal plants in east and Central Africa: challenges and constraints. Ethnobot Leafl 13:364–369
26. Govindarajan M, Karuppannan P (2011) Mosquito larvicidal and ovicidal properties of *Eclipta alba* (L.) Hassk (Asteraceae) against chikungunya vector, *Aedes aegypti* (Linn.) (Diptera: Culicidae). Asian Pac J Trop Med 4(1):24–28
27. Komoreng L, Thekisoe O, Lehasa S, Tiwani T, Mzizi N, Mokoena N, Khambule N, Ndebele S, Mdletshe N (2017) An ethnobotanical survey of traditional medicinal plants used against lymphatic filariasis in South Africa. S Afr J Bot 111:12–16
28. Shale TL, Stirk WA, Van Staden J (1999) Screening of plants used by southern African traditional healers in the treatment of dysmenorrhea for prostaglandin-synthesis inhibitors and uterine relaxing activity. J Ethnopharmacol 64:9–14
29. Apidi JR, Grierson DS, Afolayan AJ (2008) Ethnobotanical study of plant used for the treatment of diarrhoea in the Eastern Cape, South Africa. Pak J Biol Sci 11:1961–1963
30. Al-Abd NM, Nor ZM, Al-Adhroey AH, Suhaimi A, Sivanandam S (2013) Recent advances on the use of biochemical extracts as filaricidal agents. Evid Based Complement Alternat Med 2013:1–13
31. Behera DR, Bhatnagar S (2018) Filariasis: role of medicinal plant in lymphatic filariasis. Int J Herb Med 6(1):40–46
32. Chimbwali L (2022) Ethnobotanical survey of plants traditionally used to treat lymphatic filariasis in southern, Western and Northwestern provinces of Zambia. Syst Rev Pharm 13(1): 34–37
33. Mishra V, Parveen N, Singha KC, Khan NU (2005) Antifilarial activity of *Azadirachta indica* on cattle filarial parasite *Setaria cervi*. Fitoterapia 76(1):54–61

34. Lakshmi V, Srivastava S, Kumar Mishra S, Misra S, Verma M, Misra-Bhattacharya S (2009) In vitro and in vivo antifilarial potential of marine sponge, *Haliclona exigua* (Kirkpatrick), against human lymphatic filarial parasite *Brugia malayi*. Parasitol Res 105(5):1295–1301
35. Arunadevi R, Sudhakar S, Lipton AP (2010) Assessment of antibacterial activity and detection of small molecules in different parts of *Andrographis paniculata*. J Theor Biol 6:192
36. Singha PK, Roy S, Dey S (2003) Antimicrobial activity of *Andrographis paniculata*. Fitoterapia 74(7–8):692–694
37. Omar W, Ibrahim AJ, Sulaiman O, Hashim Y (1997) Screening of local plants for antifilarial activity against adult worm and microfilaria of *Brugia pahangi*. J Trop For Sci 3:216–219
38. Ndjonka D, Rapado LN, Silber AM, Liebau E, Wrenger C (2013) Natural products as a source for treating neglected parasitic diseases. Int J Mol Sci 14:3395–3439
39. Ali BH, Wabel NA, Blunden G (2005) Phytochemical, pharmacological and toxicological aspects of *Hibiscus sabdariffa* L.: a review. Phytother Res 19(5):369–375
40. Saxena K, Dube V, Kushwaha V et al (2011) Antifilarial efficacy of *Hibiscus sabdariffa* on lymphatic filarial parasite *Brugia malayi*. Med Chem Res 20(9):1594–1602
41. Yadav D, Singh SC, Verma RK et al (2013) Antifilarial diarylheptanoids from *Alnus nepalensis* growing in high altitude areas of Uttarakhand, India. Phytomedicine 20:124–132
42. Sashidhara KV, Singh SP, Misra S, Gupta J, Misra-Bhattacharya S (2012) Galactolipids from *Bauhinia racemosa* as a new class of antifilarial agents against human lymphatic filarial parasite, *Brugia malayi*. Eur J Med Chem 50:230–235
43. Azeez S, Babu RO, Aykkal R, Narayanan R (2012) Virtual screening and *in vitro* assay of potential drug like inhibitors from spices against glutathione-S-transferase of filarial nematodes. J Mol Model 18:151–163
44. Kushwaha S, Roy S, Maity R, Mallick A, Soni VK, Singh PK, Chaurasiya ND, Sangwan RS, Misra-Bhattacharya S, Mandal C (2012) Chemotypical variations in *Withania somnifera* lead to differentially modulated immune response in BALB/c mice. Vaccine 30:1083–1093
45. Kushwaha S, Soni VK, Singh PK, Bano N, Kumar A, Sangwan RS, Misra-Bhattacharya S (2012) *Withania somnifera* chemotypes NMITLI 101R, NMITLI 118R, NMITLI 128R and withaferin A protect *Mastomys coucha* from *Brugia malayi* infection. Parasite Immunol 34: 199–209
46. Ghosh M, Babu SP, Sukul NC, Mahato SB (1993) Antifilarial effect of two triterpenoid saponins isolated from *Acacia auriculiformis*. Indian J Exp Biol 31:604–606
47. Sharma RD, Veerpathran AR, Dakshinamoorthy G, Sahare KN, Goswami K, Reddy MVR (2010) Possible implication of oxidative stress in antifilarial effect of certain traditionally used medicinal plants *in vitro* against *Brugia malayi* microfilariae. Pharm Res 2(6):350–354
48. Saini P, Gayen P, Nayak A, Kumar D, Mukherjee N, Pal BC, etal. 2012 Effect of ferulic acid from *Hibiscus mutabilis* on filarial parasite *Setaria cervi*: molecular and biochemical approaches. Parasitol Int 61: 520–531
49. Kalani K, Kushwaha V, Verma R, Murthy PK, Srivastava SK (2013) Glycyrrhetinic acid and its analogs: a new class of antifilarial agents. Bioorg Med Chem Lett 23:2566–2570
50. Misra N, Sharma M, Raj K, Bhattacharya SM (2007) Chemical constituents and antifilarial activity of *Lantana camara* against human lymphatic filariid *Brugia malayi* and rodent filariid *Acanthocheilonema viteae* maintained in rodent hosts. Parasitology 100(3):439–448
51. Mathew N, Paily Abidha KP, Vanamail P, Kalyansundram M, Balraman K (2002) Macrofilaricidal activity of plant *Plumbago indica/rosea in vitro*. Drug Dev Res 56(1):33–39
52. Singh R, Singhal KC, Khan NU (2000) Exploration of antifilarial potential and possible mechanism of action of the root extracts of *Saxifraga stracheyi* on cattle filarial parasite *Setaria cervi*. Phytother Res 14(1):63–66
53. Ghosh M, Shinhababu SP, Sukul NC, Sahu NP, Mahato SB (1994) Antifilarial effect of solamargine isolated from *Solanum Khastanum*. Int J Pharm 32:1–7
54. Chatterjee RK, Fatma N, Murthy PK, Sinha P, Kulshrestha DK, Dhawan BN (1992) Macrofilaricidal activity of the stem bark of *Streblus asper* and its major active constituents. Drug Dev Res 26:67–78

55. Datta A, Sukul NC (1987) Antifilarial effect of *Zinger officinale* on *dirofilaria immitis*. J Helminthol 61:268–670
56. Akuodor GC, Ajoku GA, Ezeunala MN, Chilaka KC, Asika EC (2015) Antimalarial potential of the ethanolic leaf extract of *Pseudocedrela kotschyi*. J Acute Dis 2015:23–27
57. Chinwuba P, Akah PA, Iiodigwe EE (2015) In vivo antiplasmodial activity of the ethanol stem extract and fractions of *Citrus sinensis* in mice. Merit Res J Med Med Sci 3(4):140–146
58. Freundlich JS, Anderson JW, Sarantakis D et al (2005) Synthesis, biological activity, and x-ray crystal structural analysis of diaryl ether inhibitors of malarial enoyl acyl carrier protein reductase. Part 1: 4′-substituted triclosan derivatives. Bioorg Med Chem Lett 15(23):5247–5252
59. Okokon JE, Ubulom PM, Udokpoh AE (2007) Antiplasmodial activity of *Setaria megaphylla*. Phytother Res 21:366–368
60. Kassim OO, Loyevsky M, Elliott B, Geall A, Amonoo H, Gordeuk VR (2005) Effects of root extracts of *Fagarazan thoxyloides* on the in vitro growth and stage distribution of *Plasmodium falciparum*. Antimicrob Agents Chemother 49:264–268
61. Ansa-Asamoah R, Kapadia GJ, Lloyd HA, Sokoloski EA (1990) Picratidine, a new indole alkaloid from *Picralima nitida* seeds. J Nat Prod 53:975–977
62. Bickii J, Nijifutie N, Foyere JA, Basco LK, Ringwald P (2000) In vitro antimalarial activity of limonoids from *Khaya grandifoliola* C.D.C (Meliaceae). J Ethnopharmacol 69:27–33
63. MacKinnon S, Durst T, Arnason JT, Angerhofer C, Pezutto J, Sanchez-Vindas PE, Poveds LJ, Gbeassor M (1997) Antimalarial activity of tropical Meliaceae extracts and gedunin derivatives. J Nat Prod 60:336–341
64. Joshi SP, Rojatkar SR, Nagasampagi BA (1998) Antimalarial activity of neem (*Azadirachta indica*). J Med Aromat Plant Sci 20:1000–1004
65. Khalid SA, Friedrichsen GM, Kharazmi A, Theander TG, Olsen CE, Christensen SB (1998) Limonoids from *Khaya senegalensis*. Phytochemistry 49:1769–1772
66. Amusan OOG, Adesogan EK, Makinde JM (1996) Antimalarial active principles of *Spathodea campanulata* stem bark. Phytother Res 10:692–693
67. Koumaglo K, Gbeassor M, Nikabu O, de Souza C, Werner W (1992) Effects of three compounds extracted from *Morinda lucida* on *Plasmodium falciparum*. Planta Med 58:533–534
68. Sittie AA, Lemmich E, Olsen CE, Hviid L, Kharazmi A, Nkrumah FK, Christensen SB (1999) Structure–activity studies: in vitro antileishmanial and antimalarial activities of anthraquinones from *Morinda lucida*. Planta Med 65:259–261
69. Goffin E, Ziemons E, De Mol P, do Ceu de Madureira M, Martins AP, Proenca da Cunha A, Philippe G, Tits M, Angenot L, Frederich M (2002) In vitro antiplasmodial activity of *Tithonia diversifolia* and identification of its main active constituent: tagitinin C. Planta Med 68:543–545
70. Perez HA, De La Rosa M, Apitz R (1994) In vivo activity of ajoene against rodent malaria. Antimicrob Agents Chemother 38:337–339
71. Bertani S, Houël E, Stien D, Chevolot L, Jullian V, Garavito G, Bourdy G, Deharo E (2006) Simalikalactone D is responsible for the antimalarial properties of an Amazonian traditional remedy made with *Quassia amara* L. (Simaroubaceae). J Ethnopharmacol 108:155–157
72. Mulaw T, Wubetu M, Dessie B, Demeke G, Molla Y (2019) Evaluation of antimalarial activity of the 80% methanolic stem bark extract of *Combretum molle* against *Plasmodium berghei* in mice. J Evid Based Integr Med. https://doi.org/10.1177/2515690X19890866
73. Melaku Y, Worku T, Tadesse Y, Mekonnen Y, Schmidt J, Arnold N et al (2017) Antiplasmodial compounds from leaves of *Dodonaea angustifolia*. Curr Bioact Compd 13:268–273
74. Tewabe Y, Assefa S (2018) Anti-malarial potential of the leaf exudate of *Aloe macrocarpa* Todaro and its major constituents against *Plasmodium berghei*. Clin Exp Pharmacol 8:1
75. Gemechu W, Bisrat D, Asres K (2014) Anti-malarial anthrone and chromone from the leaf latex of Aloe *debrana Christian*. Ethiop Pharm J 30:1–9

76. Batista R, De Jesus Silva A Jr, de Oliveira A (2009) Plant derived antimalarial agents: new leads and efficient phytomedicines. Part II. Non-alkaloidal natural products. Molecules 14(8): 3037–3072
77. Muthaura CN, Keriko JM, Mutai C et al (2015) Antiplasmodial potential of traditional phytotherapy of some remedies used in treatment of malaria in Meru-Tharaka Nithi County of Kenya. J Ethnopharmacol 175:315–323
78. Philip K, Elizabeth M, Cheplogoi P, Samuel K (2017) Ethnobotanical survey of antimalarial medicinal plants used in Butebo county, Eastern Uganda. Eur J Med Plants 21(4):1–22
79. Lacroix D, Prado S, Kamoga D et al (2011) Antiplasmodial and cytotoxic activities of medicinal plants traditionally used in the village of Kiohima. Uganda J Ethnopharmacol 133 (2):850–855
80. Muganga R, Angenot L, Tits M, Frédérich M (2010) Antiplasmodial and cytotoxic activities of Rwandan medicinal plants used in the treatment of malaria. J Ethnopharmacol 128(1):52–57
81. White NJ (2007) Cardiotoxicity of antimalarial drugs. Lancet Infect Dis 7(8):549–558
82. Price RN, Uhlemann AC, Brockman A et al (2004) Mefloquine resistance in *Plasmodium falciparum* and increased pfmdr1 gene copy number. Lancet 364(9432):438–447
83. Essien D, Obiajunwa-Otteh JI, Akuodor GC, Essien AD (2014) Evaluation of the antimalarial potential of *Icacina senegalensis* Juss (Icacinaceae). Asian Pac J Trop Med 7(1):S469–S472
84. Sonibare MA, Okorie PN, Aremu TO, Adegoke A (2015) Ethno-medicines for mosquito transmitted diseases from South-western Nigeria. J Nat Rem 15(1):33–42
85. Nigussie G, Wale M (2022) Medicinal plants used in traditional treatment of malaria in Ethiopia: a review of ethnomedicine, anti-malarial and toxicity studies. Malar J 21:262
86. Okello D, Kang Y (2019) Exploring antimalarial herbal plants across communities in Uganda based on electronic data. Evid Based Complement Alternat Med 3057180:1–27
87. Asase A, Asafo-Agyei T (2011) Plants used for treatment of malaria in communities around the Bobiri Forest Reserve in Ghana. J Herbs Spices Med Plants 17(2):85–106
88. Baird JK, Caneta-Miguel E, Masba S, Bustos DG, Abrenica JA, Layawen AV, Calulut JM, Leksana B, Wignall FS (1996) Survey of resistance to chloroquine of falciparum and vivax malaria in Palawan, The Philippines. Trans R Soc Trop Med Hyg 90:413–414
89. Bruce-Chwatt LJ (1982) Qinghaosu: a new antimalarial. Br Med J (Clin Res Ed) 284:767–768
90. Mokuolu OA, Okoro EO, Ayetoro SO, Adewara AA (2007) Effect of artemisinin based treatment policy on consumption pattern of antimalarials. Am J Trop Med Hyg 76:7–11
91. Odugbemi TO, Akinsulire OR, Aibinu IE, Fabeku PO (2007) Medicinal plants useful for malaria therapy in Okeigbo, Ondo State, Southwest Nigeria. Afr J Trad CAM 4(2):191–198
92. Anywar G, van't Klooster CIEA, Byamukama R et al (2016) Medicinal plants used in the treatment and prevention of malaria in Cegere sub-county, northern Uganda. Ethnobot Res Appl 14:505–516
93. Nwabuisi C (2002) Prophylactic effect of multi-herbal extract Agbo-Iba on malaria induced in mice. East Afr Med J 79:343–346
94. Strangeland T, Alele PE, Katuura E, Lye KA (2011) Plants used to treat malaria in Nyakayojo sub-county, Western Uganda. J Ethnopharmacol 137(1):154–166
95. Deharo E, Bourdy G, Quenevo C, Munoz V, Ruiz G, Sauvain M (2001) A search for natural bioactive compounds in Bolivia through a multidisciplinary approach. Part V. Evaluation of the anti-malarial activity of plants used by the Tacana Indians. J Ethnopharmacol 77:91–98
96. Epidi JO, Izah SC, Ohimain EI, Epidi TT (2016) Phytochemical, antibacterial and synergistic potency of tissues of *Vitex grandifolia*. Biotechnol Res 2(2):69–76
97. Epidi JO, Izah SC, Ohimain EI (2016) Antibacterial and synergistic efficacy of extracts of *Alstonia boonei* tissues. Br J Appl Res 1(1):0021–0026
98. Izah SC, Odubo TC (2023) Effects of *Citrus aurantifolia* fruit juice on selected pathogens of public health importance. ES Food Agrofor 11:829. https://doi.org/10.30919/esfaf829
99. Enaregha EB, Izah SC, Okiriya Q (2021) Antibacterial properties of *Tetrapleura tetraptera* pod against some pathogens. Res Rev Insights 5:1–5

100. Izah SC, Chandel SS, Etim NG, Epidi JO, Venkatachalam T, Devaliya R (2019) Potency of unripe and ripe express extracts of long pepper (*Capsicum frutescens* var. baccatum) against some common pathogens. Int J Pharm Phytopharmacol Res 9(2):56–70
101. Izah SC, Etim NG, Ilerhunmwuwa IA, Silas G (2019) Evaluation of crude and ethanolic extracts of *Capsicum frutescens* var. minima fruit against some common bacterial pathogens. Int J Complement Altern Med 12(3):105–108
102. Izah SC, Etim NG, Ilerhunmwuwa IA, Ibibo TD, Udumo JJ (2019) Activities of express extracts of *Costus* afer Ker–Gawl. [family COSTACEAE] against selected bacterial isolates. Int J Pharm Phytopharmacol Res 9(4):39–44
103. Izah SC, Uhunmwangho EJ, Dunga KE, Kigigha LT (2018) Synergy of methanolic leave and stem-back extract of *Anacardium occidentale* L. (cashew) against some enteric and superficial bacteria pathogens. MOJ Toxicol 4(3):209–211
104. Izah SC, Uhunmwangho EJ, Dunga KE (2018) Studies on the synergistic effectiveness of methanolic extract of leaves and roots of *Carica papaya* L. (papaya) against some bacteria pathogens. Int J Complement Altern Med 11(6):375–378
105. Izah SC, Uhunmwangho EJ, Etim NG (2018) Antibacterial and synergistic potency of methanolic leaf extracts of *Vernonia amygdalina* L. and *Ocimum gratissimum* L. J Basic Pharmacol Toxicol 2(1):8–12
106. Izah SC, Zige DV, Alagoa KJ, Uhunmwangho EJ, Iyamu AO (2018) Antibacterial efficacy of aqueous extract of *Myristica fragrans* (common nutmeg). EC Pharmacol Toxicol 6(4):291–295
107. Izah SC, Uhunmwangho EJ, Eledo BO (2018) Medicinal potentials of *Buchholzia coriacea* (wonderful kola). Med Plant Res 8(5):27–43
108. Izah SC (2018) Some determinant factors of antimicrobial susceptibility pattern of plant extracts. Res Rev Insight 2(3):1–4
109. Izah SC, Aseibai ER (2018) Antibacterial and synergistic activities of methanolic leaves extract of lemon grass (*Cymbopogon citratus*) and rhizomes of ginger (*Zingiber officinale*) against Escherichia coli, Staphylococcus aureus and Bacillus subtilis. Acta Sci Microbiol 1(6): 26–30
110. Kigigha LT, Izah SC, Uhunmwangho EJ (2018) Assessment of hot water and ethanolic leaf extracts of *Cymbopogon citratus* Stapf (Lemon grass) against selected bacteria pathogens. Ann Microbiol Infect Dis 1(3):1–5
111. Kigigha LT, Selekere RE, Izah SC (2018) Antibacterial and synergistic efficacy of acetone extracts of Garcinia kola (bitter kola) and *Buchholzia coriacea* (wonderful kola). J Basic Pharmacol Toxicol 2(1):13–17
112. Kigigha LT, Biye SE, Izah SC (2016) Phytochemical and antibacterial activities of *Musanga cecropioides* tissues against *Escherichia coli*, *Pseudomonas aeruginosa Staphylococcus aureus, Proteus* and *Bacillus* species. Int J Appl Res Technol 5(1):100–107
113. Kigigha LT, Izah SC, Okitah LB (2016) Antibacterial activity of palm wine against *Pseudomonas, Bacillus*, *Staphylococcus, Escherichia,* and *Proteus* spp. Point J Bot Microbiol Res 2(1):046–052
114. Kigigha LT, Apreala A, Izah SC (2016) Effect of cooking on the climbing pepper (*Piper nigrum*) on antibacterial activity. J Environ Treat Tech 4(1):6–9
115. Kigigha LT, Izah SC, Ehizibue M (2015) Activities of *Aframomum melegueta* seed against *Escherichia coli*, *Staphylococcus aureus* and *Bacillus* species. Point J Bot Microbiol Res 1(2): 23–29
116. Izah SC, Chandel SS, Epidi JO, Venkatachalam T, Devaliya R (2019) Biocontrol of Anopheles gambiae larvae using fresh ripe and unripe fruit extracts of *Capsicum frutescens* var. baccatum. Int J Green Pharm 13(4):338–342
117. Izah SC (2019) Activities of crude, acetone and ethanolic extracts of *Capsicum frutescens* var. minima fruit against larvae of Anopheles gambiae. J Environ Treat Tech 7(2):196–200

118. Seiyaboh EI, Seiyaboh Z, Izah SC (2020) Environmental control of mosquitoes: a case study of the effect of *Mangifera Indica* root-bark extracts (family Anacardiaceae) on the larvae of Anopheles gambiae. Ann Ecol Environ Sci 4(1):33–38
119. Seiyaboh EI, Odubo TC, Izah SC (2020) Larvicidal activity of *Tetrapleura tetraptera* (Schum and Thonn) Taubert (Mimosaceae) extracts against *Anopheles gambiae*. Int J Adv Res Microbiol Immunol 2(1):20–25
120. Youkparigha FO, Izah SC (2019) Larvicidal efficacy of aqueous extracts of *Zingiber officinale* Roscoe (ginger) against malaria vector, *Anopheles gambiae* (Diptera: Culicidae). Int J Environ Agri Sci 3(1:020):1–6
121. Willcox ML, Bodeker G (2004) Traditional herbal medicines for malaria. BMJ 329:1156–1159
122. Lu YC, Lin Q, Luo GS, Dai YY (2006) Solubility of berberine chloride in various solvents. J Chem Eng Data 51:2
123. Duan HG, Wei YH, Li BX, Qin HY, Wu XA (2009) Improving the dissolution and oral bioavailability of the poorly water-soluble drug aloe-emodin by solid dispersion with polyethylene glycol 6000. Drug Dev Res 70:7
124. Battu SK, Repka MA, Maddineni S, Chittiboyina AG, Avery MA, Majumdar S (2010) Physicochemical characterization of berberine chloride: a perspective in the development of a solution dosage form for oral delivery. AAPS Pharm SciTech 11:1466–1475
125. Shia CS, Hou YC, Tsai SY, Huieh PH, Leu YL, Chao PD (2010) Differences in pharmacokinetics and ex vivo antioxidant activity following intravenous and oral administrations of emodin to rats. J Pharm Sc 99:21852195

15 Mechanistic Approaches of Herbal Medicine in the Treatment of Arthritis

Rupesh Kumar Pandey, Lubhan Singh, Sokindra Kumar, Manish Pathak, Amit Kumar, Sachin Kumar Jain, Priyanka Pandey, Shiv Shankar Shukla, Ravindra Kumar Pandey, and Kratika Daniel

Contents

R. K. Pandey (✉) · L. Singh · S. Kumar
Department of Pharmacology, Kharvel Subharti College of Pharmacy, Swami Vivekanand Subharti University, Meerut, Uttar Pradesh, India

M. Pathak
Departments of Pharmaceutical Chemistry, Kharvel Subharti College of Pharmacy, Swami Vivekanand Subharti University, Meerut, Uttar Pradesh, India

A. Kumar
Kharvel Subharti College of Pharmacy, Swami Vivekanand Subharti University, Meerut, Uttar Pradesh, India

S. K. Jain
IPS Academy College of Pharmacy, Oriental College of Pharmacy and Research, Oriental University, Indore, Madhya Pradesh, India

P. Pandey
Departments of Pharmaceutics, NKBR College of Pharmacy, Meerut, Uttar Pradesh, India

S. S. Shukla
Department of Pharmaceutical Quality Assurance, Columbia Institute of Pharmacy, Tekari, Raipur, Chhattisgarh, India

R. K. Pandey
Columbia Institute of Pharmacy, Tekari, Raipur, Chhattisgarh, India

K. Daniel
Oriental College of Pharmacy, Indore, India

S. C. Izah et al. (eds.), *Herbal Medicine Phytochemistry*, Reference Series in Phytochemistry, https://doi.org/10.1007/978-3-031-43199-9_17

Abstract

Arthritis is a global concern, affecting the global population, especially older people. The allopathic treatments proved their efficacy but toxicities too, so the need is to establish the treatment which should be productive and safe too. Many herbal-based medicines are used in the traditional medicine system, proving their effectiveness in the disease. The need of the hour is to establish their scientific, mechanical approach to arthritis treatment, which can improve the quality of life of these patients. In recent years, scientific research has shown that plant bioactive compounds are efficacious in disease treatment. This chapter aims to establish the knowledge of herbal medicines in disease treatment through the available scientific research data and their applicability. The plant-related drugs and disease knowledge are essential for treating disease. Ayurveda is the best and gold standard in terms of efficacy. Ancient literature like Ayurvedic Pharmacopeia and Chark Samhita have not only classified but also provided evidence-based knowledge on many diseases. In this chapter, we have conceptualized the various bioactive compounds from herbal drugs in their role in disease management treatment. The plant's primary and secondary metabolites have massive potential for treating arthritis. We have discussed the targeted sites for the disease management and applicability of herbal drugs in arthritis through available scientific-based evidence.

Keywords

Arthritis · Herbal medicines · Bioactive compounds · Mechanism of action

Abbreviations

AIA	Adjuvant induced arthritis
Akt	Ak strain transforming
CIA	Collagen-induced arthritis
ERK	Extracellular signal-regulated kinase
FLS	Fibroblast-like synovial cells
MAPK	Mitogen activated protein kinases
NSAIDs	Non-steroidal anti-inflammatory drugs
PI3K	Phosphatidylinositol 3-kinase
RA	Rheumatoid arthritis

1 Introduction

Arthritis is a most worrying problem across the globe. Arthritis has prehistoric existence in the world medical background. According to ancient literature, Charka Susruta explained the kind of arthritis in well-known standard books like Charka and Susruta Samhitas. The symptoms of the disease include pain obstruction and stiffness of joint, bone, and muscle pain, which sometimes leads to paralysis in some cases. The term for arthritis is rheumatism, a joint disorder that reflects the existing blend of humoral and biomedical control globally.

Herbal drugs are defined as drugs having active constituents from natural plant parts like leaves, roots, flowers, and fruits. Herbal drugs have been used for the treatment of a variety of diseases from ancient times, and it is not an embellishment in the direction of the utilization of herbal drugs in the condition. Herbal medicines are amalgamated through the therapeutic knowledge of a generation of working physicians on an ancient system of medicine for more than long years. These days, investigators showed their enormous curiosity in therapeutic compounds that are consequent from plant origin since these drugs have the least toxicities and are not very expensive compared to available treatments. Herbal drugs have significant action in many disease treatments. Many researchers have explained the applicability of herbal medicine and their active constituents in managing arthritis treatment.

In this chapter, we have conceptualized the applicability of bioactive compounds in the treatment lane. The disorders start with joints, ligaments, muscles, tendons, bones, and nerves. Available literature suggests that the conventional medicine system incorporates complete body components. According to Charka, the rainy season, its moist, cold storms, and polluted water supply produce disturbances in wind, the primary reason for rheumatic problems. Diet, age, heredity, and lifestyle are important factors in disease development. The primary symptom is pain, which leads further to inflammation.

Pain is the sensation that leads to so many problems in day-to-day life. Pain can be tolerated or not depending on the person when it affects aged people, especially older ones, creating discomfort in daily activities. The primary problem in the case of pain is it is not measurable. Clinicians use a scale known as VAS (visual analogue scale) to treat the patients, which works on the perception of the patients. The recent evidence of chronic toxicities with the available allopathic treatment provided a path toward new research.

1.1 Inflammation

Inflammation is a foremost problem in arthritis, distinguished by redness, heat, inflammation, and failure of function. Arthritis symptoms are stiffness, pain, inflammation, and loss of joint tasks. The diverse type of arthritis contains distinguishing indication, diagnosis, and disease management [1]. The types of arthritis are osteoarthritis and rheumatoid arthritis.

A chronic inflammation-related disease that mainly involves the periarticular tissue, cartilage leads to deformity, functional disabilities, and significant disability as a result of many inflammatory transmitters secreted through macrophages, such as eicosanoids like prostaglandins, cytokines, leukotrienes, and reactive oxygen species [2].

RA is the diverse process involved in synovial cell proliferation, fibrosis with cartilage degeneration, and bone attrition progression reconciled by an inter-reliant complex of cytokines, proteolytic enzymes, and prostanoids.

The cytokines, precisely tumor necrosis factor and interleukin-1, have prominent functions in disease. The increased level of IL-1 in the synovial fluid can be seen in disease patients; the concentrations of IL-1 in the plasma reported by researchers correlate with the disease functioning [3].

1.2 Rheumatoid Arthritis

The different forms of the disease are as follows:

Palindromic rheumatoid arthritis
Rheumatoid spondylitis
Juvenile rheumatoid arthritis

1.3 Other Kinds of Arthritis

- Osteoarthritis
- Primary osteoarthritis – Affects older age population
- Secondary osteoarthritis – Affects all ages of populations
- Ankylosing spondylitis
- Infectious arthritis
- Gout

1.4 Epidemiology

The occurrence of the disease is 0.3–1% across the globe, majorly in developed nations [4]. The disease prominently affects women more than men (3:1), and the progression of the disease is mainly seen in ages 30–55 years. It affects 0.5–1.0% of adults, and 0.75% of disease cases are reported approximately in India [5]. Arthritis is highest among adults due to lack of physical activity compared to those who are inadequately active or have biological activity. The prevalence of disease may be due to concomitant diseases like cardiovascular, obesity, and diabetes. It is expected that the number of cases of the disease may increase with the highest percentage in adults due to current food habits and other factors.

1.5 Disease Etiology

The disease etiology is not known. Genetic propensity to environmental factors is probably responsible for the disease progression.

1. **Environmental factors**: The disease has been linked to mineral oils, silica exposure, diet factors, and blood transfusion.
2. **Role of sex hormones**: According to the prevalence of disease, it has been seen that disease prominently occurs in women. In female, it is observed that it generally starts at the perimenopausal, perinatal, and gestation period.
3. **Genetic**: The genetic mechanism of this disease is not precise; the genetic level's appearance compared to healthy patients strongly suggests that genetic inheritance is the factor that resembles the key for future-based discoveries for diagnosis and treatment. Some kinds of literature reported that persons with selective human leukocyte antigen genes have a greater risk for disease prevalence than those in which human leukocyte is not available.

1.6 Pathophysiology

The pathophysiology can be understood through Fig. 1. According to the disease pathophysiology of rheumatoid arthritis, it starts from the synovium. The level of cytokines plays a role in disease progression through cartilage destruction. The destruction of collagen leads to the narrowing of joint space and the destruction of the bone. The pannus formation occurs due to the exudation of fluid in the synovium. Activation of CD4 T cells releases the cytokines that cause the formation of fibroblasts by MMP and RANK ligand, leading to the activation of osteoclast and destroying the tissues, promoting the destruction of joints.

There are two chief mechanisms for the inflammation process: B cells and T cells activation by the immune system, which have significant functions linked with the disease. T cells are essential in distinguishing the antigen as "non-self," which produces cytokines. After that, the part B cells start to secrete and reproduce the

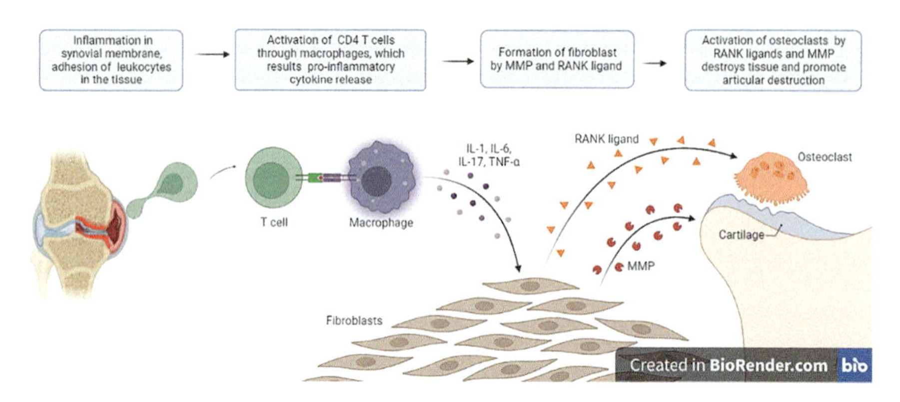

Fig. 1 Pathophysiology of rheumatoid arthritis

antibodies into the systemic circulation to identify antigens and activate the inflammation. Then, the various signaling pathways' role comes, which releases further pro-inflammatory mediators and causes the progression of the disease.

1.7 Responsibility of NF-κB/IκB-α in Inflammation Associated with RA

The disease pathophysiology entails the continual appearance of pro-inflammatory cytokines, chemokines that contain various sites like the presence of transcription factor inside their promoter, signifying the NF-κB involvement in the progression of the disease. The activation of tumor necrosis factor-alpha triggers NF-kB, which results in an increased level in the joints of rheumatoid arthritis patients. Furthermore, hyper (NF-κB) activity can also be seen in the macrophages of RA patients.

Classical and alternative pathways are the two essential signaling pathways that play a role in stimulating NF-kB. In these pathways, one common thing is IkB kinase complex activation that has the catalytic kinase subunits with a regulatory non-enzymatic protein named IKK. These NF-kB dimers are activated due to IKK-mediated phosphorylation-induced protein degradation of the IκB-α inhibitor, which liberates the active NF-κB transcription factor subunits to translocate to the nucleus and induce target gene expression. Without activating the signal, IkB sequesters NF-kB subunits and inhibits its translocation, consequently terminating transcriptional activity. In the classical signaling pathways, ligand binding happens on cell surface receptors such as a member of the Toll-like receptor superfamily, leading to the recruitment of adaptors to the receptor's cytoplasmic domain. In turn, these adaptors recruit the IKK complex, leading to phosphorylation and IB inhibitor degradation.

1.8 Role of Immunostimulants or Immunomodulators

The immune system assists in homeostasis through the active network that engages the applicability of numerous molecular basis cells capable of differentiation and reorganizing foreign particles from the self-cells. There are two types of immunity: innate and adaptive. Innate immunity works as the primary defense involving roles of cellular and molecular mechanisms that instantly respond to infection when any antigen attacks [6]. The natural immune system contains physical, chemical, and other cellular barriers. The group of cells like macrophages, neutrophils, and many more having foreign cells from self-act as cellular barriers [7]. If any pathogen or foreign cell can escape the first line of defense, immune system cells become the second line of defense or adaptive immunity [8].

1.9 NSAIDs: Other Inhibitors of Pro-inflammatory Mediators

The non-steroidal anti-inflammatory drugs treatment has been used for a long time to assuage arthritic pain and inflammation. Aspirin and Ibuprofen are drugs used extensively by clinicians to treat disease. Many NSAID products have become available for treating illness, but their toxicities are also high and concern global people. In the past, many painkillers were withdrawn from the pharmaceutical market, like rofecoxib and valdecoxib, due to cardiac toxicities. So, the safety of patients is a prime concern with efficacy.

2 Herbal Drugs and Applicability of Their Bioactive Compounds in the Treatment of Arthritis

Ayurveda has proven its efficacy in the disease since old times. Ayurveda originated from the Sanskrit words ayus and veda, which means denotation of life and knowledge. Ayurveda is a discipline of life focused on man and cures for diseases through the comprehension of medicinal plants [9]. The primary metabolites include that material that strengthens plant organization, in addition to the energy metabolism of the plant, e.g., proteins, carbohydrates, and fats as primary bioactive compounds. Secondary metabolites are the chemicals not utilized by plants for their organization or purpose. They are moderately available in small quantities with diverse application to protect plants from various microbes and insects, which also exhibits growth-regulating action in the plants. The scientific-based evidence proved the therapeutic applicability of these phytochemicals in disease management. Herbal plants used for the treatment of arthritis are summarized in Table 1. The review of available data suggests that herbal plants with bioactive compounds like flavonoids, triterpenoids, and phenolic compounds can be used in the treatment of arthritis. We have performed studies related to the plant-based drug for arthritis and found that researchers reported their work through the use of various animal models for

Table 1 Bioactive compounds of plant-based medicines for arthritis

S. no	Botanical name of plant	Family	Plant bioactive compounds	Reference
1.	*Abrus precatorius*	Fabaceae	Triterpenoid	[10, 11]
2.	*Acacia confusa*	Leguminosae	Flavonoids, melanoxetin	[12, 13]
3.	*Acacia hydaspica*	Leguminosae	Gallic acid, caffeic acid, rutin, catechin	[14, 15]
4.	*Aconitum vilmorinianum*	Ranunculacae	Diterpinoids, vilimorine A-D	[16, 17]
5.	*Albizia procera*	Fabaceae	Biochanin-A	[18, 19]
6.	*Allium sativum*	Liliaceae	Diallyl disulfide	[20, 21]
7.	*Alstonia boonei*	Apocynaceae	Phenolic acids, rutin, isoquercetin, flavonoliganes	[22, 23]
8.	*Alternanthera bettzickiana*	Amaranthaceae	Gallic acid, catechin	[24, 25]
9.	*Asystasia dalzelliana*	*Acanthaceae*	Steroids, flavonoids, alkaloids, tannins	[26]
10.	*Azadirachta indica*	Meliaceae	Nimbolide	[27, 28]
11.	*Berberis orthobotrys*	Berberidaceae	Flavonoids, phenolic compounds	[29]
12.	*Cissus quadrangularis*	Vitaceae	Russelioside B	[30, 31]
13.	*Capparis spinosa* L.	*Capparidaceae*	Flazin, guanosine, capparine	[32, 33]
14.	*Caesalpinia sappan*	Caesalpiniaceae	Sappanol, episappanol	[34, 35]
15.	*Calophyllum inophyllum*	Calophyllaceae	Triterpenoids	[36, 37]
16.	*Calotropis gigantea*	Apocynaceae	Lupeol	[38, 39]
17.	*Caltha palustris*	Ranunculaceae	Polysaccharide B	[40, 41]
18.	*Cayaponia tayuya*	Cucurbitaceae	Dihydrocucurbitacin B	[42, 43]
19.	*Celastrus*	Celastraceae	Celastrol	[44, 45]
20.	*Capparis erythrocarpos*	*Capparaceae*	Proteins, polyphenols	[46, 47]
21.	*Curcuma longa* L.	Zingiberaceae	Curcuminoids	[48, 49]
22.	*Chloranthus serratus*	Chloranthaceae	Terpenoids	[50, 51]
23.	*Clematis orientalis*	Ranunculaceae	Flavonoids and glycosides	[52, 53]
24.	*Clerodendrum serratum*	Verbenaceae	Terpenoids, steroids, flavonoids, and phenolics	[54, 55]
25.	*Cuscuta reflexa*	Convolvulaceae	Beta-sitosterol	[56, 57]
26.	*Drynaria quercifolia*	Polypodiaceae	Squalene, gamma tocopherol, N-hexadecanoic acid	[58, 59]
27.	*Eriobotrya japonica*	Rosaceae	Corosolic acid, oleanolic acid, and ursolic acid	[60, 61]
28.	*Ephedra gerardiana*	Ephedraceae	Alkaloids, flavonoids	[62, 63]
29.	*Euphorbia neriifolia*	Euphorbiaceae	Steroid and terpenoid	[64, 65]
30.	*Ficus bengalensis*	*Moraceae*	Flavonoids, terpenoid, sterols	[66, 67]
31.	*Fagopyrum cymosum*	Polygonaceae	Eugenol	[68, 69]
32.	*Ginkgo biloba*	Ginkgoaceae	Terpene	[70, 71]
33.	*Glycosmis pentaphylla*	*Rutaceae*	Sulfur-containing amides	[72, 73]

(continued)

Table 1 (continued)

S. no	Botanical name of plant	Family	Plant bioactive compounds	Reference
34.	*Glycyrrhiza glabra*	*Leguminosae*	Licochalcone A (flavonoid)	[74, 75]
35.	*Hemidesmus indicus*	Apocynaceae	Triterpenoids	[76, 77]
36.	*Ipomoea batatas*	Convolvulaceae	Polyphenols, 3-epifriedelinol	[78, 79]
37.	*Jatropha isabellei*	Euphorbiaceae	Alkaloid	[80, 81]
38.	*Justicia gendarussa Burm F*	*Acanthaceae*	Sterols, flavonoids	[82, 83]
39.	*Kadsura heteroclita*	Schisandraceae	Lignans and triterpenoids	[84, 85]
40.	*Lantana camara*	Verbenaceae	Triterpenoids	[86, 87]
41.	*Laportea bulbifera*	Urticaceae	Neochlorogenic acid, cryptochlorogenic acid, and chlorogenic acid	[88, 89]
42.	*Lawsonia innermis*	*Lythraceae*	Lawsochylin A and lawsonaphthoate, luteolin, apigenin	[90, 91]
43	*Leucas aspera*	*Lamiaceae*	Epicatechin, beta epicatechin, procyanidin, beta-sitosterol	[92, 93]
44.	*Linum usitatissimum*	Linaceae	Alpha linolenic acid	[94, 95]
45.	*Litsea cubeba*	Lauraceae	Litsecols	[96, 97]
46.	*Lonicerae japonica*	Caprifoliaceae	Lonicerin	[98, 99]
47.	*Melastoma malabathricum*	Melastomataceae	*B*-Sitosterol, melastomic acid	[100, 101]
48.	*Monotheca buxifolia*	Sapotaceae	Flavonoids, triterpenoids, vitamin E, phytol,	[102, 103]
49.	*Moringa rivae*	Moringaceae	Fatty acids, vitamin E	[104]
50.	*Olea europaea*	Oleaceae	Oleuropein, ligstroside	[105–107]
51.	*Panax ginseng*	Araliaceae	Ginsenosides Rg3, Rk1, and Rg5	[108, 109]
52.	*Phyllanthus amarus*	Phyllanthaceae	Phyllanthin and hypophyllanthin	[110, 111]
53.	*Pinus maritime*	Pinaceae	Flavangenol	[112, 113]
54.	*Piper betle*	Piperaceae	Hydroxychavicol	[114, 115]
55.	*Pistia stratiotes*	Araceae	Triterpenes, flavonoids	[116, 117]
56.	*Premna serratifolia*	Lamiaceae	Phenolic compounds and flavonoids	[118, 119]
57.	*Punica granatum*	Punicaceae	Tannins and anthocyanins	[120, 121]
58.	*Rhus verniciflua*	Anacardiaceae	Flavonol, fisetin	[122, 123]
59.	*Ribes alpestre*	Grossulariaceae	Phenolic and flavonoid	[124]
60.	*Ribes orientale*	Grossulariaceae	Polyphenolic compounds, flavonoid	[125]
61.	*Ruta graveolens*	Rutaceae	Alkaloids, polyphenols	[126, 127]
62.	*Salacia reticulata*	Celastraceae	Polyphenols,	[128, 129]
63.	*Salix nigra*	Salicaceae	Sterols, terpenes, flavonoids	[130, 131]

(continued)

Table 1 (continued)

S. no	Botanical name of plant	Family	Plant bioactive compounds	Reference
64.	*Saraca asoca*	Fabaceae	Polyphenols	[132, 133]
65.	*Sargassum wightii*	Sargassaceae	Alginic acid	[134, 135]
66.	*Saussurea lappa*	Asteraceae	Polyphenols, flavonoids	[136, 137]
67.	*Semecarpus anacardium*	Anacardiaceae	Flavonoids	[138, 139]
68.	*Sida rhombifolia*	Malvaceae	Rhombifoliamide, ß-sitosterol	[140–142]
69.	*Sinomenium acutum*	Menispermaceae	Sinomenine	[143, 144]
70.	*Solanum nigrum*	Solanaceae	Polyphenols, flavonoids	[145, 146]
71.	*Sophora flavescens*	Fabaceae	Flavonoids and alkaloids	[147, 148]
72.	*Strobilanthus callosus*	Acanthaceae	Lupeol, 19 alpha-h-lupeol	[149, 150]
73.	*Strychnos potatorum*	Loganiaceae	Alkaloids, ß-sitosterol	[151, 152]
74.	*Syzygium aromaticum*	Myrtaceae	Eugenol, terpenes	[153, 154]
75.	*Torilis japonica*	Apiaceae	Torilin	[155, 156]
76.	*Toxicodendron pubescens*	Anacardiaceae	Quercetin and rutin	[157–159]
77.	*Tragia involucrata*	*Euphorbiaceae*	Rutin, quercetin	[160, 161]
78.	*Trewia polycarpa*	*Euphorbiaceae*	Terpenoids, flavonoids	[162, 163]
79.	*Tridax procumbens*	Asteraceae	Flavonoids	[164, 165]
80.	*Trigonella foenum graecum*	Fabaceae	Steroids, alkaloids, polyphenols, flavonoids	[166, 167]
81.	*Urtica pilulifera*	Urticaceae	Phenolic compounds	[168, 169]
82.	*Vernonia cinerea*	Asteraceae	Steroids, flavonoids	[170, 171]
83.	*Vitex negundo*	Verbenaceae	Phenylnaphthalene-type lignans	[172, 173]
84.	*Wendlandia heynei*	Rubiaceae	Terpenoids, B-carotene	[174, 175]
85.	*Withania somnifera*	Solanaceae	Phenolic compounds, withanolides	[176–178]
86.	*Xanthium strumarium*	Asteraceae	Sesquiterpenoids, phenylpropenoids, lignanoids	[179, 180]
87.	*Yucca schidigera*	Asparagaceae	Resveratrol, spirostanol	[181, 182]

arthritis, which gives a direction in the treatment of disease and also attracts the phytoconstituents' role.

3 Bioactive Compounds and Their Targets in Disease

Bioactive compounds are the secondary metabolites isolated from herbal medicinal plants, animals, fungus, and microorganisms. Moreover, they are having pharmacological or toxicological effects on organisms, foremost to utilization in food and pharmaceutical industries. In the drug discovery process, the new property of these compounds gave the direction in the treatment of disease.

Treatment of any disease requires the knowledge of the targets for the drug delivery. Bioactive compounds of plant-based medicines have played a vital role in diseases, so the need of the hour is to explore these compounds with the moa that can give direction in disease management. As per the disease, etiology immune system plays a vital role, so targeting the cytokines in arthritis can give us the direction for treatment. In this section, we have highlighted that the various pathways like PI3K and Akt signaling, MAPK signaling, NF-κB, STAT signaling, and NRF2 signaling responsible for disease progression with the applicability of primary and secondary metabolites from herbal drugs in the lane of treatment reported by researchers.

3.1 Flavonoids

Flavonoids are successive compounds, and the structure of these compounds suggests that they are made up of phenyl rings and phenolic hydroxyl groups throughout their carbon chains [183]. Flavonoids have significant efficacy in treating arthritis and have been reported by many researchers that can be correlated with Table 1. Flavonoids also significantly influence therapy through inhibition of PI3K and Akt signaling pathways. This pathway plays a vital role in disease progression through fibroblast-like synovial cells that may secret an abundance of cytokines that incessantly excite FLS cells by unrestrained propagation of cells. PI3K and Akt signaling path is believed to be a connection between the propagation and programmed cell death of FLS cells. The studies on arthritis suggest that these signaling are highly articulated and reflect the consequence of the extreme movement of fibroblast-like synovial cells. Many pro-inflammatory mediators like IL-17 and IL-21 can support the inflammatory propagation of fibroblast-like synovial cells by stimulating and activating PI3K [184].

Some researchers reported that LY294002, a PI3K inhibitor, can improve overgrowth and inflammation in synovial joints in animal studies [185].

Another critical action of flavonoids is the inhibition of the MAPK signaling pathway, which is also responsible for disease progression through stimulation of various kinase systems like ERK, p38, c-Jun N-terminal kinase, and ERK5, which control many enzymes cytokines and chemokines causes persistent inflammation and abnormal hyperplasia [186].

Some researchers revealed that a flavonoid isolated from *Glycyrrhiza uralensis* can considerably block the IL-1β-induced rheumatoid arthritis fibroblast-like synovial cell propagation by slow-down through the kinase system [187].

Flavonoids also inhibit the NF-κB STAT signaling pathways in disease. The NF-κB and STAT lead to shifting in the nucleus that causes inflammation, cell propagation, and apoptosis. Some studies suggested that flavonoid Genkwanin extracted from *Daphne genkwa* showed anti-arthritic activity by slowing down the p-NF-κB p-STAT3 in AIA mice [188].

Several researchers reported through their research on NRF2 signaling that it plays an essential role in arthritis through oxidative stress concerned with etiological and diagnostic parameters. The activation of the Nrf2 signaling leads a range of enzymes like heme oxygenase-1 HO-1 to be released, having antioxidant properties that regulate the oxidative stress condition in disease [189]. Some studies implicated that epigallocatechin 3-gallate has anti-arthritic potential up-regulation of Nrf2 and HO-1 in joints of CIA rats [190].

Moreover, a flavonoid calycosin isolated from the *Astragali radix* can inhibit the pro-inflammatory mediators in RA-FLS by up-regulation of Nrf2 and HO1, as reported by some researchers [191].

3.2 Lignans

Some researchers reported that lignans with antioxidant potential can inhibit the ROS through the NF-κB path, eventually decreasing the appearance of inflammatory cytokines and pro-inflammatory enzymes. Activating the AMPK and Nrf2 path increases antioxidant-related genes and promotes the release of anti-inflammatory cytokines [192].

3.3 Anthraquinone

The reported work by authors suggests that rhein is an anthraquinone, which demonstrated cell viability and differentiation without toxicities at humans' physiological levels but considerably decreased cytokine levels in urate crystal-activated macrophages [193].

3.4 ß-Sitosterol

ß-sitosterol structure is similar to the cholesterol group. The research work reported by authors revealed that ß-sitosterol showed marked inhibition on synovial angiogenesis through suppression and proliferation of endothelial cells that result in alleviated joint inflammation and bone erosion in CIA mice [194].

3.5 Terpenoids

Terpenes are an important class of organic compounds produced mainly by herbal plants. The researchers reported that terpenes moderate the immunologic response and cartilage destruction by inhibiting various cytokines responsible for disease progression [195].

4 Challenges with Herbal-Based Medicines

The major challenge with herbal-based medicine is standardization and quality control. Globally, various guidelines have been approved for standard, quality-related parameters of herbal drugs. WHO approved the guidelines for taking care of the standardization parameters of herbal medicines [196, 197]. This process involves the phytochemical investigation of raw material of plant-based treatment related to features like collection and management of plant material, therapeutic and safety profile measurement of the processed products with proper documentation based on knowledge, and proviso of produce products to consumers [198]. Growth and accomplishment of the guidelines of traditional or herbal drugs within diverse divisions of the globe are frequently tackled by quite a few confrontations [199, 200]. The problems often encountered in many countries are those connected to regulatory standings, evaluation of effectiveness with safety, and insufficient information concerning traditional and plant-related drugs in nationwide drug regulations. Many herbal medicines have inadequate information on their mechanism of action in disease treatment, and the adverse drug reactions are not known anymore. The interaction with other drugs data is not available.

The various parameters related to standardization and quality control are as follows.

4.1 Collection and Authentication

People use many herbal drugs on their traditional knowledge, causing problems in the collection, adversely affecting the yield and quality of chemical constituents, and errors in experiments. Authentication is the basic process before the experimentation of any herbal drugs.

4.2 Presence of Other Organic Materials

It's essential to collect, identify, and remove the other foreign organic materials during the collection because they can lead to errors in the experimental process and the purity of drugs too.

4.3 Ash Values Determination

Ash values are an essential method for the purity of crude drugs. The specific test is performed for the identification of the total ash value. These values also have significance with quality principles. The ash value also gives an idea of finding the suitable part for the extraction process.

4.4 Moisture in Plant-Related Drugs

The moisture content determination helps reduce the mistake in the judgment of the actual weight of drug material, i.e., little moisture recommends improved stability adjacent to product deprivation.

4.5 Extractive Yield

These are the investigative weights of the extractive yield of bioactive compounds of herbal drugs in different solvents. These give the knowledge of the relation between the solvent and plant metabolites, which can be utilized for isolation and impact the cost of herbal medicine.

4.6 Qualitative Profile Evaluation

Identifying and categorizing crude drugs with reverence to phytochemicals is essential in photo-constituent processing. It provides work for diverse investigative methods to recognize and separate the active ingredient. Bioactive compounds showing plans provide botanical characteristics, solvent selection, purification, and categorization of the active constituents, which are essential parameters for pharmaceutical utilization.

4.7 Chromatography

The detection of raw drugs is based on using major chemical constituents as markers. Chromatographic methods also play an essential role in identifying active plant-based constituents, which helps describe the relation with treatment. The process should be accurate and standardized.

4.8 Quantitative Evaluation of Chemicals

The plant phytoconstituents are one of the critical parameters for evaluating its efficacy in disease management and the applicability of major constituents, such as in dosage determination parameters.

4.9 Toxicological Screening

The toxicological screening is essential for determining toxic residues present in the preparation, which can lead to toxicities. Hence, it is essential to perform these studies per the available guidelines of the OECD. The processes are based on international guidelines based on scientific evidence.

Standardization is one of the critical parameters and needs for herbal drugs. Much literature is available on the uses of herbal medicines, but the problem is to prove the same through standardization parameters. Some issues may arise if someone is unaware of these drugs' therapeutic dosage. More work is to be needed in the standardization part of this medication.

Apart from these, additional factors affect the standardization and quality control of herbal drug's efficacy parameters for processing herbal medications.

4.10 Adulteration in Drugs

Adulteration affects efficacy as well as causes toxicities all over the world. There is a need to develop regulatory guidelines for adulteration of herbal drugs since quality testing is an important parameter.

4.11 Faulty Preparations

The irrational combination of drugs is a leading cause of drug-to-drug interaction. There is a need for strict guidelines for herbal preparations since the herbal market is increasing in terms of volume and consumption day-by-day. Faulty herbal preparation leads to adverse effects and reduced efficacy.

4.12 Storage

The ancient and current literature implicated the storage conditions of herbal drugs. Some factors like temperature, light, moisture, and air are essential parameters for primary and secondary metabolites as they directly affect the metabolites.

4.13 Dosage and Time

Many countries still use their traditional medicine system to treat various diseases. The Indian people use many herbal medicines and speak about dosage and time through their traditional knowledge. The ancient standard literature also implicates the therapeutic applicability of the same, so it is necessary to relate the traditional knowledge and scientific evidence to overcome the toxicities.

5 Prospects of Herbal Medicines in the Disease

The challenge for herbal medicine in arthritis management is the applicability of these formulations in disease management. In recent years, the surge in curiosity concerning the endurance of Ayurvedic formulations increased. From a comprehensive point of view, there is a transfer toward plant-based medicine, as the threats and the inadequacy of contemporary medicine have ongoing getting more perceptible, and the preponderance of Ayurvedic drugs are processed from herbal sources. In many countries, many herbal formulations are available for arthritis treatment, and people use these medicines without any prescriptions. Many small towns have some orthodox people who do not have proper qualifications and knowledge about these drugs but are selling them drugs. The need is to provide the treatment with the support of healthcare professionals, which is essential to create awareness among them for prescribing herbal medicine. It is the liability of the countries' regulators to ensure that the consumers get good quality medicine that is pure, effective, and safe for treatment. The regulatory setups are required for the clinical trials related to herbal drugs. The repositories should be available online for patient compliance, which can give a clear-cut idea for reporting adverse drug reactions with herbal medicines. The quality standards of raw materials related to plant sources of herbal medicines are a supreme consequence in mitigating their satisfactoriness in the current structure of the medical system. Most of the time, problems encountered by herbal drug manufacturers are the non-availability of raw materials for the quality control profile of herbal material and their formulations. Herbal product usage is increasing due to noise made by manufacturing companies through advertisements. Systems like Ayurveda still require awareness among the people and experiential support from contemporary medical sciences to make them plausible and tolerable for all [201]. The inventive investigation endeavors to describe the benefit of conventional systems of medicine with reverence to their safety profiles, which will possibly affect improved exploitation of these systems of medicine. Globally, several pharmacopoeias have offered herbal drug information through monographs describing the constraints and standards of many herbal medicines and several products completed by these standards, which are safe and productive. The value of herbal medicines in treating disease is increasing due to less toxicity when compared with allopathic medication. The herbal drugs market is expanding in terms of the volume of people who use these drugs as per their traditional knowledge. The acceptance of these drugs is increasing due to their safety margins. Many Pharma organizations have started manufacturing facilities for herbal medicines globally. Technological advancement is needed as primary and secondary metabolites of the drug have a prominent role in disease management. Herbal medicines can be a good choice compared to available drug treatments and can improve the daily life of these patients. As arthritis disease mainly affects older patients and has some concomitant treatment that increases chances of drug-to-drug interaction, Herbal medicines can be a good option for disease management as their safety margin is satisfactory. Many herbal formulations are available in the market for the treatment of arthritis and for improving the quality of life of these patients. Ethno-pharmacological studies are increasing regularly. The need is to relate traditional medicine knowledge with scientific evidence for appropriate formulations for patients' safety. Primary and secondary metabolites can be utilized for the future-based treatment of arthritis, as these metabolites can reduce the dosage due to their efficacy at minimum concentration. The time is to establish the ethanol-pharmacology concept on a broader base for the herbal medicines and changes needed in available manufacturing processes. The time is to show the ethanol-pharmacology idea on a wider base for the herbal medicines and modifications needed in available manufacturing processes. Labeling and product information for patients is also required to change from time to time for the acceptance purpose in disease treatment.

Almost three-fourths of the herbal medicines used wide-reaching were expected to be exposed after local treatment. As per the available data from WHO, the minimum percentage of contemporary drugs come from herbal sources used traditionally [202]. In the present scenario, almost all countries focus on the herbal medicine market due to its safety and efficacy. The growing demand for herbal medicine requires attention in terms of quality and methods for improving product processing and thought processes to get the maximum active bioactive compounds from the plant's raw material. The cost of processing bioactive compounds is high, and the need to develop methods like solvent selection is required. Many studies also focused on how that combination of water and alcohol can achieve better extractive yield. The modernization in processing methods not only affects the work but will also reduce the cost of medicine for the patients. All countries should focus on their native herbal drugs and explore their applicability in disease management. Every country should have an awareness program for their citizens to establish knowledge of herbal medicine. The various programs should be the plan for arthritis management for the stakeholders of the disease management. The time is to find the drugs that have not been explored well yet; many plants are not identified and stored in a database worldwide. Aquatic plants are another excellent option for treating arthritis in the future drug discovery process. The arthritic patients need to develop a formulation that can relieve pain and should have anti-inflammatory action. Patient tolerability to dosage is also a significant concern because many patients have some concomitant treatment.

6 Conclusion

Arthritis is increasing consistently, so it requires attention in diagnosing and treating the disease. The available allopathic treatment toxicities cannot be ignored. The need is to establish a treatment that is safe and effective. In this chapter, we have explained the mechanistic approaches of plant-related drugs with scientific evidence that can be utilized for future drug discovery processes. Herbal medicine can be an appropriate drug candidate for the management of arthritis. Perhaps more studies are needed to explore their potential. As per the etiology of the disease, cytokines play a crucial role in disease progression. We have highlighted the plant-based bioactive compounds such as flavonoids, lignans, anthraquinone, ß-sitosterol, and terpenoids that can be an option in future drug discovery as it has therapeutic values in the disease. The need of the hour is more research focusing on metabolites for getting a drug candidate and molecular basis for drug targeting.

The worldwide recognition and utilization of herbal medicines plus associated products continue to presume and increase. The problems involving adverse drug reactions at the current time are too much more dramatic, increasing in occurrence and no longer controversial since the previous delusions of concerning or else classifying herbal medicinal products as safer as they are extracted from natural origin. The truth is that saying safer and origin from nature are not the same; therefore, regulatory guidelines for herbal medicines must standardize with fortified approaches on the international level. The applicable regulatory requirements in diverse nations of the globe necessitate being proactive and enduring in the direction of suitable actions toward looking after public health by ensuring that every single herbal medicine permitted for sale is safe and of appropriate quality. Quality education specifically related to herbal medicine is needed for a better understanding. Globally, the guidelines should meet equal parameters. Through this, quality herbal medicine will be available for the patients. The pharmacist should give all the information related to drugs, like drug-to-drug interactions, for the betterment of the disease patients.

The contributors of plant-based drugs, like doctors, nurses, and pharmacists, have minute training in understanding the effects of herbal medicines on the health of their patients. Several are inadequately informed about these products, and how to use them is a big question. Sufficient training and teaching are required because patients use different types of prescription and counter-product drugs. The fact is that healthcare professionals have to play a vital role in expressions of their precious assistance to safety assessments of medications with the significance of the same. All contributors to herbal medicines should be adequately allowed to observe the safety of herbal medicines. It can be achieved through partnerships with conventional healthcare professionals. The need of the hour is to create such an environment for smooth functioning and adequate knowledge about the use and safety of herbal medicines. The qualification of every doctor, nurse, pharmacist, and patient linked to herbal drugs is essential for treating and understanding the misuse of herbal medicines. To improve the quality of life of arthritis patients, it is required to develop the formulation based on natural sources.

References

1. Colburn N (2012) Review of rheumatology. Springer, New York/London. https://doi.org/10.1007/978-1-84882-093-7.1-156
2. Ebringer A (2012) Rheumatoid arthritis and proteus, vol X. Springer, London/New York, pp 1–233
3. Choy EHS, Panayi GS (2001) Cytokine signaling pathways involved in inflammatory arthritis. Pathogenesis of RA. N Engl J Med 344:907–916
4. Kvien TK, Uhlig T, Odegard S, Heiberg MS (2006) Epidemiological aspects of rheumatoid arthritis: the sex ratio. Ann N Y Acad Sci 1069:212–222
5. Malaviya A, Kapoor S, Singh RR, Kumar A, Pande I (1993) Prevalence of rheumatoid arthritis in the adult Indian population. Rheumatol Int 13(4):131–134
6. Turvey SE, Broide DH (2010) Innate immunity. J Allergy Clin Immunol 125(2 Suppl 2): S24–S32
7. Kuby, Immunology 6th edition
8. Schmid D, Münz C (2007) Innate and adaptive immunity through autophagy. Immunity 27(1): 11–21
9. Micozzi MS (2002) Culture, anthropology, and the return of "complementary medicine". Med Anthropol Quarterly 16(4):398–403. https://doi.org/10.1525/maq.2002.16.4.398
10. Anam EM (2001) Anti-inflammatory activity of compounds isolated from the aerial parts of *Abrus precatorius* (Fabaceae). Phytomedicine 8(1):24–27. https://doi.org/10.1078/0944-7113-00001

11. Qian H, Wang L, Li Y, Wang B, Li C et al (2022) The traditional uses, phytochemistry and pharmacology of *Abrus precatorius* L.: a comprehensive review. J Ethnopharmacol 296: 115463. https://doi.org/10.1016/j.jep.2022.115463
12. Wu JH, Tung YT, Chien SC, Wang SY, Kuo YH et al (2008) Effect of phytocompounds from the heartwood of *Acacia confusa* on inflammatory mediator production. J Agric Food Chem 56(5):1567–1573. https://doi.org/10.1021/jf072922s
13. Lin HY, Chang TC, Chang ST (2018) A review of antioxidant and pharmacological properties of phenolic compounds in *Acacia confusa*. J Tradit Complement Med 8(4):443–450. https://doi.org/10.1016/j.jtcme.2018.05.002
14. Afsar T, Khan MR, Razak S, Ullah S, Mirza B (2015) Antipyretic, anti-inflammatory and analgesic activity of *Acacia hydaspica* R. Parker and its phytochemical analysis. BMC Complement Altern Med 15:136
15. Afsar T, Razak S, Shabbir M, Khan MR (2018) Antioxidant activity of polyphenolic compounds isolated from ethyl-acetate fraction of *Acacia hydaspica* R. Parker. Chem Cent J 12(1): 5. https://doi.org/10.1186/s13065-018-0373-x
16. Li M, He J, Jiang LL, Ng ES, Wang H et al (2013) The anti-arthritic effects of *Aconitum vilmorinianum*, a folk herbal medicine in southwestern China. J Ethnopharmacol 147(1): 122–127. https://doi.org/10.1016/j.jep.2013.02.018
17. Yin TP, Cai L, Fang HX, Fang YS, Li ZJ et al (2015) Diterpenoid alkaloids from *Aconitum vilmorinianum*. Phytochemistry 116:314–319. https://doi.org/10.1016/j.phytochem.2015.05.002
18. Sangeetha M, Chamundeeswari D, Saravana Babu C, Rose C, Gopal V (2020) Attenuation of oxidative stress in arthritic rats by ethanolic extract of *Albizia procera* benth bark through modulation of the expression of inflammatory cytokines. J Ethnopharmacol 250:112435. https://doi.org/10.1016/j.jep.2019.112435
19. Somwong P, Theanphong O (2021) Quantitative analysis of triterpene lupeol and anti-inflammatory potential of the extracts of traditional pain-relieving medicinal plants *Derris scandens*, *Albizia procera*, and *Diospyros rhodocalyx*. J Adv Pharm Technol Res 12(2): 147–151. https://doi.org/10.4103/japtr.JAPTR_13_21
20. Chen Y, Xue R, Jin X, Tan X (2018) Antiarthritic activity of diallyl disulfide against Freund's adjuvant-induced arthritic rat model. J Environ Pathol Toxicol Oncol 37(4):291–303. https://doi.org/10.1615/JEnvironPatholToxicolOncol.2018027078
21. El-Saber Batiha G, Magdy Beshbishy A, Wasef L, Elewa YHA, Al-Sagan A et al (2020) Chemical constituents and pharmacological activities of garlic (*Allium sativum* L.): a review. Nutrients 12(3):872. https://doi.org/10.3390/nu12030872
22. Mollica A, Zengin G, Sinan KI, Marletta M, Pieretti S et al (2022) A study on chemical characterization and biological abilities of *Alstonia boonei* extracts obtained by different techniques. Antioxidants (Basel) 11(11):2171. https://doi.org/10.3390/antiox11112171
23. Akinnawo OO, Anyasor GN, Osilesi O (2017) Aqueous fraction of *Alstonia boonei* de wild leaves suppressed inflammatory responses in carrageenan and formaldehyde induced arthritic rats. Biomed Pharmacother 86:95–101. https://doi.org/10.1016/j.biopha.2016.11.145
24. Manan M, Saleem U, Akash MSH, Qasim M, Hayat M et al (2020) Antiarthritic potential of comprehensively standardized extract of *Alternanthera bettzickiana*: *in vitro* and *in vivo* studies. ACS Omega 5(31):19478–19496. https://doi.org/10.1021/acsomega.0c01670
25. Manan M, Saleem U, Ahmad B, Aslam N, Anwar A et al (2022) Anti-arthritic and toxicological evaluation of ethanolic extract of *Alternanthera bettzickiana* in rats. Front Pharmacol 13:1002037. https://doi.org/10.3389/fphar.2022.1002037
26. Kumar S, Kumar VR (2011) Invitro antiarthritic activity of isolated fractions from methanolic extract of *Asystasia dalzelliana* leaves. Asian J Pharmaceutical Clin Res
27. Israr M, Naseem N, Akhtar T, Aftab U, Zafar MS et al (2023) Nimbolide attenuates complete Freund's adjuvant induced arthritis through expression regulation of toll-like receptors signaling pathway. Phytother Res 37(3):903–912. https://doi.org/10.1002/ptr.7672

28. Cui X, Wang R, Bian P, Wu Q, Seshadri VDD et al (2019) Evaluation of antiarthritic activity of nimbolide against Freund's adjuvant induced arthritis in rats. Artif Cells Nanomed Biotechnol 47(1):3391–3398. https://doi.org/10.1080/21691401.2019.1649269
29. Alamgeer UAM, Hasan UH (2017) Anti-arthritic activ*ity of aqueous-methano*lic extract and various fractions of *Berberis orthobotrys* Bien ex Aitch. BMC Complement Altern Med 17:371
30. El-Shiekh RA, El-Mekkawy S, Mouneir SM, Hassan A, Abdel-Sattar E (2021) Therapeutic potential of russelioside B as anti-arthritic agent in Freund's adjuvant-induced arthritis in rats. J Ethnopharmacol 270:113779. https://doi.org/10.1016/j.jep.2021.113779
31. Bafna PS, Patil PH, Maru SK, Mutha RE (2021) *Cissus quadrangularis* L: a comprehensive multidisciplinary review. J Ethnopharmacol 279:114355. https://doi.org/10.1016/j.jep.2021.114355
32. Zhou H, Jian R, Kang J, Huang X, Li Y et al (2010) Anti-inflammatory effects of caper (*Capparis spinosa* L.) fruit aqueous extract and the isolation of main phytochemicals. J Agric Food Chem
33. Kdimy A, El Yadini M, Guaadaoui A, Bourais I, El Hajjaji S et al (2022) Phytochemistry, biological activities, therapeutic potential, and socio-economic value of the caper bush (*Capparis spinosa* L.). Chem Biodivers 19(10):e202200300. https://doi.org/10.1002/cbdv.202200300
34. Mueller M, Weinmann D, Toegel S, Holzer W, Unger FM et al (2016) Compounds from *Caesalpinia sappan* with anti-inflammatory properties in macrophages and chondrocytes. Food Funct 7(3):1671–1679. https://doi.org/10.1039/c5fo01256b
35. Yuanting J, Ruikang H, Yang L, Hanqiao L (2022) Two new cassane-type diterpenoids from the seeds of *Caesalpinia sappan*. Nat Prod Res 36(8):2078–2084. https://doi.org/10.1080/14786419.2020.1849196
36. Perumal SS, Ekambaram SP, Dhanam T (2017) In vivo antiarthritic activity of the ethanol extracts of stem bark and seeds of *Calophyllum inophyllum* in Freund's complete adjuvant induced arthritis. Pharm Biol 55(1):1330–1336. https://doi.org/10.1080/13880209.2016.1226346
37. Li YZ, Li ZL, Yin SL, Shi G, Liu MS et al (2010) Triterpenoids from *Calophyllum inophyllum* and their growth inhibitory effects on human leukemia HL-60 cells. Fitoterapia 81(6):586–589. https://doi.org/10.1016/j.fitote.2010.02.005
38. Saratha V, Subramanian SP (2012) Lupeol, a triterpenoid isolated from *Calotropis gigantea* latex ameliorates the primary and secondary complications of FCA induced adjuvant disease in experimental rats. Inflammopharmacology 20(1):27–37. https://doi.org/10.1007/s10787-011-0095-3
39. Sivapalan S, Dharmalingam S, Venkatesan V, Angappan M, Ashokkumar V (2023) Phytochemical analysis, anti-inflammatory, antioxidant activity of *Calotropis gigantea* and its therapeutic applications. J Ethnopharmacol 303:115963. https://doi.org/10.1016/j.jep.2022.115963
40. Suszko A, Obmińska-Mrukowicz B (2013) Influence of polysaccharide fractions isolated from *Caltha palustris* L. on the cellular immune response in collagen-induced arthritis (CIA) in mice. A comparison with methotrexate. J Ethnopharmacol 145(1):109–117. https://doi.org/10.1016/j.jep.2012.10.038
41. Bhandari P, Gray AI, Rastogi RP (1987) Triterpenoid saponins from *Caltha palustris*. Planta Med 53(1):98–100. https://doi.org/10.1055/s-2006-962634
42. Escandell JM, Recio MC, Máñez S, Giner RM, Cerda-Nicolas M et al (2006) Dihydrocucurbitacin B, isolated from *Cayaponia tayuya*, reduces damage in adjuvant-induced arthritis. Eur J Pharmacol 532(1–2):145–154. https://doi.org/10.1016/j.ejphar.2005.12.028
43. Aquila S, Giner RM, Recio MC, Spegazzini ED, Rios JL (2009) Anti-inflammatory activity of flavonoids from *Cayaponia tayuya* roots. J Ethnopharmacol 121(2):333–337. https://doi.org/10.1016/j.jep.2008.11.002

44. Nanjundaiah SM, Venkatesha SH, Yu H, Tong L, Stains JP et al (2012) Celastrus and its bioactive celastrol protect against bone damage in autoimmune arthritis by modulating osteoimmune cross-talk. J Biol Chem 287(26):22216–22226. https://doi.org/10.1074/jbc.M112.356816
45. Li G, Liu D, Zhang Y, Qian Y, Zhang H et al (2013) Celastrol inhibits lipopolysaccharide-stimulated rheumatoid fibroblast-like synoviocyte invasion through suppression of TLR4/NF-κB-mediated matrix metalloproteinase-9 expression. PLoS One 8(7):e68905. https://doi.org/10.1371/journal.pone.0068905
46. Danquah CA, Woode E, Boakye-Gyasi E (2011) Anti-arthritic effects of an ethanolic extract of *Capparis erythrocarpos* roots in Freund's adjuvant-induced arthritis in rats. J Pharmacol Toxicol
47. Twumasi MA, Tandoh A, Mante PK, Ekuadzi E, Boakye-Gyasi ME et al (2019) Leaves and stems of *Capparis erythrocarpos*, more sustainable than roots, show antiarthritic effects. J Ethnopharmacol 238:111890. https://doi.org/10.1016/j.jep.2019.111890
48. Funk JL, Oyarzo JN, Frye JB, Chen G, Lantz RC et al (2006) Turmeric extracts containing curcuminoids prevent experimental rheumatoid arthritis. J Nat Prod Mar 69(3):351–355. https://doi.org/10.1021/np050327j
49. Wang Z, Jones G, Winzenberg T, Cai G, Laslett LL et al (2020) Effectiveness of *Curcuma longa* extract for the treatment of symptoms and effusion-synovitis of knee osteoarthritis: a randomized trial. Ann Intern Med 173(11):861–869. https://doi.org/10.7326/M20-0990
50. Sun S, Li S, Du Y, Cai G, Laslett LL et al (2020) Anti-inflammatory effects of the root, stem and leaf extracts of *Chloranthus serratus* on adjuvant-induced arthritis in rats. Pharm Biol 58(1):528–537. https://doi.org/10.1080/13880209.2020.1767159
51. Zhang M, Iinuma M, Wang JS, Oyama M, Ito T et al (2012) Terpenoids from *Chloranthus serratus* and their anti-inflammatory activities. J Nat Prod 75(4):694–698. https://doi.org/10.1021/np200968p
52. Karimi E, Ghorbani NM, Habibi M et al (2018) Antioxidant potential assessment of phenolic and flavonoid rich fractions of *Clematis orientalis* and *Clematis ispahanica* (Ranunculaceae). Nat Prod Res 32(16):1991–1995. https://doi.org/10.1080/14786419.2017.1359171
53. Hasan UH, Alamgeer SM, Ebrahimi M, Mehrafarin A et al (2019) Inhibitory effects of *Clematis orientalis* aqueous ethanol extract and fractions on inflammatory markers in complete Freund's adjuvant-induced arthritis in Sprague-Dawley rats. Inflammopharmacology 27(4):781–797. https://doi.org/10.1007/s10787-018-0543-4
54. Tiwari RK, Chanda S, Udayabanu M, Singh M, Agarwal S (2021) Anti-inflammatory and anti-arthritic potential of standardized extract of *Clerodendrum serratum* (L.) moon. Front Pharmacol 12:629607. https://doi.org/10.3389/fphar.2021.629607
55. Patel JJ, Acharya SR, Acharya NS (2014) *Clerodendrum serratum* (L.) moon – a review on traditional uses, phytochemistry and pharmacological activities. J Ethnopharmacol 154(2): 268–285. https://doi.org/10.1016/j.jep.2014.03.071
56. Niazi SG, Uttra AM, Qaiser MN, Ahsan H (2017) Appraisal of anti-arthritic and nephroprotective potential of *Cuscuta reflexa*. Pharm Biol 55(1):792–798. https://doi.org/10.1080/13880209.2017.1280513
57. Aung TTT, Xia MY, Hein PP, Tang R, Zhang DD et al (2020) Chemical constituents from the whole plant of *Cuscuta reflexa*. Nat Prod Bioprospect 10(5):337–344. https://doi.org/10.1007/s13659-020-00265-x
58. Modak D, Paul S, Sarkar S, Thakur S, Bhattacharjee S (2021) Validating potent anti-inflammatory and anti-rheumatoid properties of *Drynaria quercifolia* rhizome methanolic extract through in vitro, in vivo, in silico and GC-MS-based profiling. BMC Complement Med Ther 21(1):89. https://doi.org/10.1186/s12906-021-03265-7
59. Saravanan S, Mutheeswaran S, Saravanan M, Chellappandian M, Gabriel Paulraj M et al (2013) Ameliorative effect of *Drynaria quercifolia* (L.) J. Sm., an ethnomedicinal plant, in arthritic animals. Food Chem Toxicol 51:356–363. https://doi.org/10.1016/j.fct.2012.10.020

60. Kuraoka-Oliveira ÂM, Radai JAS, Leitão MM, Lima Cardoso CA, Silva-Filho SE et al (2020) Anti-inflammatory and anti-arthritic activity in extract from the leaves of *Eriobotrya japonica*. J Ethnopharmacol 249:112418. https://doi.org/10.1016/j.jep.2019.112418
61. Zhu X, Wang L, Zhao T, Jiang Q (2022) Traditional uses, phytochemistry, pharmacology, and toxicity of *Eriobotrya japonica* leaves: a summary. J Ethnopharmacol 298:115566. https://doi.org/10.1016/j.jep.2022.115566
62. Uttra AM, Alamgeer SM, Shabbir A, Jahan S (2018) Ephedra gerardiana aqueous ethanolic extract and fractions attenuate Freund complete adjuvant induced arthritis in Sprague Dawley rats by downregulating PGE2, COX2, IL-1β, IL-6, TNF-α, NF-kB and upregulating IL-4 and IL-10. J Ethnopharmacol 224:482–496. https://doi.org/10.1016/j.jep.2018.06.018
63. Kumar K, Sharma YP, Manhas RK, Bhatia H (2015) Ethnomedicinal plants of Shankaracharya Hill, Srinagar, J&K, India. J Ethnopharmacol 170:255–274. https://doi.org/10.1016/j.jep.2015.05.021
64. Palit P, Mandal SC, Bhunia B (2016) Total steroid and terpenoid enriched fraction from *Euphorbia neriifolia* Linn offers protection against nociceptive-pain, inflammation, and in vitro arthritis model: an insight of mechanistic study. Int Immunopharmacol 41:106–115. https://doi.org/10.1016/j.intimp.2016.10.024
65. Mali PY, Panchal SS (2017) *Euphorbia neriifolia* L.: review on botany, ethnomedicinal uses, phytochemistry and biological activities. Asian Pac J Trop Med 10(5):430–438. https://doi.org/10.1016/j.apjtm.2017.05.003
66. Patil VV, Patil VR (2010) A comparative evaluation of anti-inflammatory activity of the bark of *Ficus bengalensis* in plants of different age. J Basic Clin Pharm 1(2):107–113
67. Rajdev K, Jain S, Bhattacharaya SK (2018) Antinociceptive effect of *Ficus bengalensis* bark extract in experimental models of pain. Cureus 10(3):e2259. https://doi.org/10.7759/cureus.2259
68. Shen L, Wang P, Guo J, Du G (2013) Anti-arthritic activity of ethanol extract of Fagopyrum cymosum with adjuvant-induced arthritis in rats. Pharm Biol 51(6):783–789. https://doi.org/10.3109/13880209.2013.766892
69. Zhao J, Jiang L, Tang X, Peng L, Li X et al (2018) Chemical composition, antimicrobial and antioxidant activities of the flower volatile oils of Fagopyrum esculentum, Fagopyrum tataricum and Fagopyrum Cymosum. Molecules 23(1):182. https://doi.org/10.3390/molecules23010182
70. Han Y (2005) Ginkgo terpene component has an anti-inflammatory effect on Candida albicans-caused arthritic inflammation. Int Immunopharmacol 5(6):1049–1056. https://doi.org/10.1016/j.intimp.2005.02.002
71. Wang H, Shi M, Cao F, Su E (2022) Ginkgo biloba seed exocarp: a waste resource with abundant active substances and other components for potential applications. Food Res Int 160: 111637. https://doi.org/10.1016/j.foodres.2022.111637
72. Nian H, Xiong H, Zhong F, Teng H, Teng H et al (2020) Anti-inflammatory and anti-proliferative prenylated Sulphur-containing amides from the leaves of Glycosmis pentaphylla. Fitoterapia 146:104693. https://doi.org/10.1016/j.fitote.2020.104693
73. Khandokar L, Bari MS, Seidel V, Haque MA (2021) Ethnomedicinal uses, phytochemistry, pharmacological activities and toxicological profile of Glycosmis pentaphylla (Retz.) DC.: a review. J Ethnopharmacol 278:114313. https://doi.org/10.1016/j.jep.2021.114313
74. Su X, Li T, Liu Z, Huang Q, Liao K et al (2018) Licochalcone a activates Keap1-Nrf2 signaling to suppress arthritis via phosphorylation of p62 at serine 349. Free Radic Biol Med 115:471–483. https://doi.org/10.1016/j.freeradbiomed.2017.12.004
75. Yang F, Su X, Pi J, Liao K, Zhou H et al (2018) Atomic force microscopy technique used for assessment of the anti-arthritic effect of licochalcone a via suppressing NF-κB activation. Biomed Pharmacother 103:1592–1601. https://doi.org/10.1016/j.biopha.2018.04.142
76. Mehta A, Sethiya NK, Mehta C, Shah GB (2012) Anti-arthritis activity of roots of Hemidesmus indicus r.Br. (Anantmul) in rats. Asian Pac J Trop Med 5(2):130–135. https://doi.org/10.1016/S1995-7645(12)60011-X

77. Nandy S, Mukherjee A, Pandey DK, Ray P, Dey A (2020) Indian sarsaparilla (Hemidesmus indicus): recent progress in research on ethnobotany, phytochemistry and pharmacology. J Ethnopharmacol 254:112609. https://doi.org/10.1016/j.jep.2020.112609
78. Majid M, Nasir B, Zahra SS, Khan MR, Mirza B et al (2018) Ipomoea batatas L. lam. Ameliorates acute and chronic inflammations by suppressing inflammatory mediators, a comprehensive exploration using in vitro and in vivo models. BMC Complement Altern Med 18(1):216. https://doi.org/10.1186/s12906-018-2279-5
79. Majid M, Farhan A, Baig MW, Khan MT, Kamal Y et al (2022) Ameliorative effect of structurally divergent Oleanane triterpenoid, 3-Epifriedelinol from *Ipomoea batatas* against BPA-induced gonadotoxicity by targeting PARP and NF-κB signaling in rats. Molecules 28(1):290. https://doi.org/10.3390/molecules28010290
80. Silva CR, Fröhlich JK, Oliveira SM, Cabreira TN, Rossato MF et al (2013) The anti-nociceptive and anti-inflammatory effects of the crude extract of Jatropha isabellei in a rat gout model. J Ethnopharmacol 145(1):205–213. https://doi.org/10.1016/j.jep.2012.10.054
81. Fröhlich JK, Stein T, da Silva LA, Biavatti MW, Tonussi CR et al (2017) Antinociceptive and anti-inflammatory activities of the Jatropha isabellei dichloromethane fraction and isolation and quantitative determination of jatrophone by UFLC-DAD. Pharm Biol 55(1):1215–1222. https://doi.org/10.1080/13880209.2017.1295999
82. Paval J, Kaitheri SK, Potu BK, Govindan S, Kumar RS et al (2009) Anti-arthritic potential of the plant Justicia gendarussa Burm F. Clinics (Sao Paulo) 64(4):357–362. https://doi.org/10.1590/s1807-59322009000400015
83. Kavitha SK, Viji V, Kripa K, Helen A (2011) Protective effect of Justicia gendarussa Burm.f. on carrageenan-induced inflammation. J Nat Med 65(3–4):471–479. https://doi.org/10.1007/s11418-011-0524-z
84. Yu H, Zeng R, Lin Y, Tasneem S, Yang Z et al (2019) Kadsura heteroclita stem suppresses the onset and progression of adjuvant-induced arthritis in rats. Phytomedicine 58:152876. https://doi.org/10.1016/j.phymed.2019.152876
85. Wang M, Jiang S, Yuan H, Zafar S, Hussain N et al (2021) A review of the phytochemistry and pharmacology of Kadsura heteroclita, an important plant in Tujia ethnomedicine. J Ethnopharmacol 268:113567. https://doi.org/10.1016/j.jep.2020.113567
86. Wu P, Song Z, Wang X, Li Y, Li Y et al (2020) Bioactive triterpenoids from Lantana camara showing anti-inflammatory activities in vitro and in vivo. Bioorg Chem 101:104004. https://doi.org/10.1016/j.bioorg.2020.104004
87. Ghisalberti EL (2000) Lantana camara L. (Verbenaceae). Fitoterapia 71(5):467–486. https://doi.org/10.1016/s0367-326x(00)00202-1
88. Tang J, Zhang Q, Wu D, Chen SY, Chen Y et al (2022) Zhongguo Zhong Yao Za Zhi 47(17): 4755–4764. https://doi.org/10.19540/j.cnki.cjcmm.20220609.201
89. Lu X, Zhao Y, Li B, Feng W, Qi J et al (2022) Phytochemical, chemotaxonomic and bioinformatics study on Laportea bulbifera (Urticaceae). Chem Biodivers 19(7): e202200070. https://doi.org/10.1002/cbdv.202200070
90. Kore KJ, Shete RV, Desai NV (2011) AntiArthritic activity of hydroalcoholic extract of Lawsonia Innermis. Int J Drug Dev Res 3(4):217–224
91. Liou JR, El-Shazly M, Du YC, Tseng CN, Hwang TL et al (2013) 1,5-Diphenylpent-3-en-1-ynes and methyl naphthalene carboxylates from Lawsonia inermis and their anti-inflammatory activity. Phytochemistry 88:67–73. https://doi.org/10.1016/j.phytochem.2012.11.010
92. Kripa KG, Chamundeeswari D, Thanka J, Uma Maheswara Reddy C (2011) Modulation of inflammatory markers by the ethanolic extract of Leucas aspera in adjuvant arthritis. J Ethnopharmacol 134(3):1024–1027. https://doi.org/10.1016/j.jep.2011.01.010
93. Prajapati MS, Patel JB, Modi K, Shah MB (2010) Leucas aspera: a review. Pharmacogn Rev 4(7):85–87. https://doi.org/10.4103/0973-7847.65330
94. Kaithwas G, Majumdar DK (2010) Therapeutic effect of Linum usitatissimum (flaxseed/linseed) fixed oil on acute and chronic arthritic models in albino rats. Inflammopharmacology 18(3):127–136. https://doi.org/10.1007/s10787-010-0033-9

95. Akter Y, Junaid M, Afrose SS, Nahrin A, Alam MS et al (2021) A comprehensive review on Linum usitatissimum medicinal plant: its phytochemistry, pharmacology, and ethnomedicinal uses. Mini Rev Med Chem 21(18):2801–2834. https://doi.org/10.2174/1389557521666210203153436
96. Guo Q, Bai RF, Su GZ, Zhu ZX, Zeng KW et al (2016) Chemical constituents from the roots and stems of Litsea cubeba. J Asian Nat Prod Res 18(1):51–58. https://doi.org/10.1080/10286020.2015.1118063
97. Lin B, Zhang H, Zhao XX, Rahman K, Wang Y et al (2013) Inhibitory effects of the root extract of Litsea cubeba (lour.) pers. on adjuvant arthritis in rats. J Ethnopharmacol 147(2): 327–334. https://doi.org/10.1016/j.jep.2013.03.011
98. Lee JH, Han Y (2011) Antiarthritic effect of lonicerin on Candida albicans arthritis in mice. Arch Pharm Res 34(5):853–859. https://doi.org/10.1007/s12272-011-0520-6
99. Shang X, Pan H, Li M, Miao X, Ding H (2011) Lonicera japonica Thunb.: ethnopharmacology, phytochemistry and pharmacology of an important traditional Chinese medicine. J Ethnopharmacol 138(1):1–21. https://doi.org/10.1016/j.jep.2011.08.016
100. Kumar V, Bhatt PC, Rahman M, Patel DK, Sethi N et al (2016) Melastoma malabathricum Linn attenuates complete freund's adjuvant-induced chronic inflammation in Wistar rats via inflammation response [retracted in: BMC complement med Ther. 2022 Sep 20;22(1):246]. BMC Complement Altern Med 16(1):510. https://doi.org/10.1186/s12906-016-1470-9
101. Manzoor-I-Khuda MM, Chowdhury SA, Reza T, Chowdhury AK (1981) Chemical investigation on Melastoma Malabathricum. Part 1: isolation of Melastomic acid and Beta-sitosterol from the roots. J Bangladesh Acad Sci 5:55–59
102. Akhtar MF, Khan K, Saleem A, Baig MMFA, Rasul A et al (2021) Chemical characterization and anti-arthritic appraisal of Monotheca buxifolia methanolic extract in complete Freund's adjuvant-induced arthritis in Wistar rats. Inflammopharmacology 29(2):393–408. https://doi.org/10.1007/s10787-020-00783-7
103. Ali JS, Azeem M, Mannan A, Zia M (2022) Chemical composition, antibacterial and antioxidative activities of *Monotheca buxifolia* (Falc.) A. DC leaves essential oil. Nat Prod Res 36(22):5848–5851. https://doi.org/10.1080/14786419.2021.2018591
104. Saleem A, Saleem M, Akhtar MF, Shahzad M, Jahan S (2020) Moringa rivae leaf extracts attenuate Complete Freund's adjuvant-induced arthritis in Wistar rats via modulation of inflammatory and oxidative stress biomarkers [published correction appears in Inflammopharmacology. 2019 Sep 4]. Inflammopharmacology 28(1):139–151. https://doi.org/10.1007/s10787-019-00596-3
105. Hong YH, Song C, Shin KK, Choi E, Hwang SH et al (2021) Tunisian Olea europaea L. leaf extract suppresses Freund's complete adjuvant-induced rheumatoid arthritis and lipopolysaccharide-induced inflammatory responses. J Ethnopharmacol 268:113602. https://doi.org/10.1016/j.jep.2020.113602
106. Song C, Hong YH, Park JG, Kim HG, Jeong D et al (2019) Suppression of Src and Syk in the NF-κB signaling pathway by Olea europaea methanol extract is leading to its anti-inflammatory effects. J Ethnopharmacol 235:38–46. https://doi.org/10.1016/j.jep.2019.01.024
107. Charoenprasert S, Mitchell A (2012) Factors influencing phenolic compounds in table olives (*Olea europaea*). J Agric Food Chem 60:7081–7095. https://doi.org/10.1021/jf3017699
108. Kim KR, Chung TY, Shin H, Son SH, Park KK et al (2010) Red ginseng saponin extract attenuates murine collagen-induced arthritis by reducing pro-inflammatory responses and matrix metalloproteinase-3 expression. Biol Pharm Bull 33(4):604–610. https://doi.org/10.1248/bpb.33.604
109. Kiefer D, Pantuso T (2003) Panax ginseng. Am Fam Physician 68(8):1539–1542
110. Mali SM, Sinnathambia A, Kapasea CU, Bodhankar SL, Mahadik KR (2011) Anti-arthritic activity of standardised extract of *Phyllanthus amarus* in Freund's complete adjuvant induced arthritis. Biomed Aging Pathol 1:85–190
111. Sharma A, Singh RT, Handa SS (1993) Estimation of phyllanthin and hypophyllanthin by high performance liquid chromatography in *Phyllanthus amarus*. Phytochem Anal 4:226–229. https://doi.org/10.1002/pca.2800040507

112. Tsubata M, Takagaki K, Hirano S, Iwatani K, Abe C (2011) Effects of flavangenol, an extract of French maritime pine bark on collagen-induced arthritis in rats. J Nutr Sci Vitaminol (Tokyo) 57(3):251–257. https://doi.org/10.3177/jnsv.57.251
113. Gabaston J, Leborgne C, Waffo-Téguo P, Pedrot E, Richard T et al (2020) Separation and isolation of major polyphenols from maritime pine (Pinus pinaster) knots by two-step centrifugal partition chromatography monitored by LC-MS and NMR spectroscopy. J Sep Sci 43(6): 1080–1088. https://doi.org/10.1002/jssc.201901066
114. Pandey A, Bani S, Dutt P, Suri KA (2010) Modulation of Th1/Th2 cytokines and inflammatory mediators by hydroxychavicol in adjuvant induced arthritic tissues. Cytokine 49(1):114–121. https://doi.org/10.1016/j.cyto.2009.08.015
115. Biswas P, Anand U, Saha SC, Kant N, Mishra T et al (2022) Betelvine (Piper betle L.): a comprehensive insight into its ethnopharmacology, phytochemistry, and pharmacological, biomedical and therapeutic attributes. J Cell Mol Med 26(11):3083–3119. https://doi.org/10.1111/jcmm.17323
116. Kyei S, Koffuor GA, Boampong JN (2012) Antiarthritic effect of aqueous and ethanolic leaf extracts of Pistia stratiotes in adjuvant-induced arthritis in Sprague-Dawley rats. J Exp Pharmacol 4:41–51. https://doi.org/10.2147/JEP.S29792
117. Singh B, Sahu PM, Sharma MK (2002) Anti-inflammatory and antimicrobial activities of triterpenoids from Strobilanthes callosus nees. Phytomedicine 9:355–359
118. Rajendran R, Krishnakumar E (2010) Anti-arthritic activity of Premna serratifolia Linn., wood against adjuvant induced arthritis. Avicenna J Med Biotechnol 2(2):101–106
119. Woo SY, Hoshino S, Wong CP, Win NN, Awouafack MD et al (2019) Lignans with melanogenesis effects from Premna serratifolia wood. Fitoterapia 133:35–42. https://doi.org/10.1016/j.fitote.2018.12.008
120. Ahmed S, Wang N, Hafeez BB, Cheruvu VK, Haqqi TM (2005) Punica granatum L. extract inhibits IL-1beta-induced expression of matrix metalloproteinases by inhibiting the activation of MAP kinases and NF-kappaB in human chondrocytes in vitro. J Nutr 135(9):2096–2102. https://doi.org/10.1093/jn/135.9.2096
121. Shukla M, Gupta K, Rasheed Z, Khan KA, Haqqi TM (2008) Consumption of hydrolyzable tannins-rich pomegranate extract suppresses inflammation and joint damage in rheumatoid arthritis. Nutrition 24(7–8):733–743. https://doi.org/10.1016/j.nut.2008.03.013. Epub 2008 May 19
122. Lee JD, Huh JE, Jeon G, Yang HR, Woo HS et al (2009) Flavonol-rich RVHxR from Rhus verniciflua Stokes and its major compound fisetin inhibits inflammation-related cytokines and angiogenic factor in rheumatoid arthritic fibroblast-like synovial cells and in vivo models. Int Immunopharmacol 9(3):268–276. https://doi.org/10.1016/j.intimp.2008.11.005
123. Kim SH, Huh CK (2022) Isolation and identification of Fisetin: an antioxidative compound obtained from *Rhus verniciflua* seeds. Molecules 27(14):4510. https://doi.org/10.3390/molecules27144510
124. Hassan UH, Alamgeer SM, Jahan S, Saleem M et al (2019) Amelioration of adjuvant induced arthritis in Sprague Dawley rats through modulation of inflammatory mediators by Ribes alpestre Decne. J Ethnopharmacol 235:460–471. https://doi.org/10.1016/j.jep.2019.02.025
125. Uttra AM, Alamgeer SM, Jahan S, Bukhari IA et al (2019) Ribes orientale: a novel therapeutic approach targeting rheumatoid arthritis with reference to pro-inflammatory cytokines, inflammatory enzymes and anti-inflammatory cytokines. J Ethnopharmacol 237:92–107. https://doi.org/10.1016/j.jep.2019.03.019
126. Ratheesh M, Shyni GL, Sindhu G, Helen A (2010) Protective effects of isolated polyphenolic and alkaloid fractions of Ruta graveolens L. on acute and chronic models of inflammation. Inflammation 33(1):18–24. https://doi.org/10.1007/s10753-009-9154-y
127. Freire RB, Borba HR, Coelho CD (2010) Ruta graveolens L. toxicity in Vampirolepis nana infected mice. Indian J Pharmacol 42(6):345–350. https://doi.org/10.4103/0253-7613.71898
128. Sekiguchi Y, Mano H, Nakatani S, Shimizu J, Kobata K et al (2012) Anti-proliferative effects of Salacia reticulata leaves hot-water extract on interleukin-1β-activated cells derived from the

synovium of rheumatoid arthritis model mice. BMC Res Notes 5:198. https://doi.org/10.1186/1756-0500-5-198

129. Sekiguchi Y, Mano H, Nakatani S, Shimizu J, Wada M (2010) Effects of the Sri Lankan medicinal plant, Salacia reticulata, in rheumatoid arthritis. Genes Nutr 5(1):89–96. https://doi.org/10.1007/s12263-009-0144-3
130. Sharma S, Sahu D, Das HR, Sharma D (2011) Amelioration of collagen-induced arthritis by Salix nigra bark extract via suppression of pro-inflammatory cytokines and oxidative stress. Food Chem Toxicol 49(12):3395–3406. https://doi.org/10.1016/j.fct.2011.08.013
131. Tawfeek N, Mahmoud MF, Hamdan DI, Sobeh M, Farrag N et al (2021) Phytochemistry, pharmacology and medicinal uses of plants of the genus *Salix*: an updated review. Front Pharmacol 12:593856. https://doi.org/10.3389/fphar.2021.593856
132. Mukhopadhyay MK, Nath D (2011) Phytochemical screening and toxicity study of *Saraca asoca* bark methanolic extract. Int J Phytomed 3:498–505
133. Ahmad F, Misra L, Tewari R, Gupta P, Mishra P et al (2016) Anti-inflammatory flavanol glycosides from Saraca asoca bark. Nat Prod Res 30(4):489–492. https://doi.org/10.1080/14786419.2015.1023728
134. Sarithakumari CH, Renju GL, Kurup GM (2013) Anti-inflammatory and antioxidant potential of alginic acid isolated from the marine algae, Sargassum wightii on adjuvant-induced arthritic rats. Inflammopharmacology 21(3):261–268. https://doi.org/10.1007/s10787-012-0159-z
135. Ramu S, Murali A, Jayaraman A (2019) Phytochemical screening and toxicological evaluation of *Sargassum wightii* Greville in Wistar rats. Turk J Pharm Sci 16(4):466–475. https://doi.org/10.4274/tjps.galenos.2018.68442
136. Gokhale AB, Damre AS, Kulkami KR, Saraf MN (2002) Preliminary evaluation of anti-inflammatory and anti-arthritic activity of S. lappa, A. speciosa and A. aspera. Phytomedicine 9(5):433–437. https://doi.org/10.1078/09447110260571689
137. Saleem TS, Lokanath N, Prasanthi A, Madhavi M, Mallika G et al (2013) Aqueous extract of Saussurea lappa root ameliorate oxidative myocardial injury induced by isoproterenol in rats. J Adv Pharm Technol Res 4(2):94–100. https://doi.org/10.4103/2231-4040.111525
138. Ramprasath VR, Shanthi P, Sachdanandam P (2006) Immunomodulatory and anti-inflammatory effects of Semecarpus anacardium LINN. Nut milk extract in experimental inflammatory conditions. Biol Pharm Bull 29(4):693–700. https://doi.org/10.1248/bpb.29.693
139. Chakraborty M, Asdaq SM (2011) Interaction of Semecarpus anacardium L. with propranolol against isoproterenol induced myocardial damage in rats. Indian J Exp Biol 49(3):200–206
140. Gupta SR, Nirmal SA, Patil RY, Asane GS (2009) Anti-arthritic activity of various extracts of Sida rhombifolia aerial parts. Nat Prod Res 23(8):689–695. https://doi.org/10.1080/14786410802242778
141. Sireeratawong S, Lertprasertsuke N, Srisawat U, Thuppia A, Suwanlikhid N et al (2008) Acute and subchronic toxicity study of the water extract from root of *Sida rhombifolia* Linn. In rats. Songklanakarin J Sci Technol 30:729–737
142. Kamdoum BC, Simo I, Wouamba SCN, Tchatat Tali BM, Ngameni B et al (2022) Chemical constituents of two Cameroonian medicinal plants: *Sida rhombifolia* L. and *Sida acuta* Burm. f. (Malvaceae) and their antiplasmodial activity. Nat Prod Res 36(20):5311–5318. https://doi.org/10.1080/14786419.2021.1937156
143. Liu L, Buchner E, Beitze D, Schmidt-Weber CB, Kaever V et al (1996) Amelioration of rat experimental arthritides by treatment with the alkaloid sinomenine. Int J Immunopharmacol 18(10):529–543. https://doi.org/10.1016/s0192-0561(96)00025-2
144. Yamasaki H (1976) Pharmacology of sinomenine, an anti-rheumatic alkaloid from Sinomenium acutum. Acta Med Okayama 30(1):1–20
145. Alamgeer SA, Ambreen MU, Umme Habiba H (2019) Alkaloids, flavonoids, polyphenols might be responsible for potent antiarthritic effect of Solanum nigrum. J Tradit Chin Med 39(5):632–641
146. Hameed A, Akhtar N (2018) Comparative chemical investigation and evaluation of antioxidant and tyrosinase inhibitory effects of Withania somnifera (L.) Dunal and Solanum nigrum (L.) berries. Acta Pharma 68(1):47–60. https://doi.org/10.2478/acph-2018-0007

147. Jin JH, Kim JS, Kang SS, Son KH, Chang HW et al (2010) Anti-inflammatory and anti-arthritic activity of total flavonoids of the roots of Sophora flavescens. J Ethnopharmacol 127(3):589–595. https://doi.org/10.1016/j.jep.2009.12.020
148. He X, Fang J, Huang L, Wang J, Huang X (2015) Sophora flavescens Ait.: traditional usage, phytochemistry and pharmacology of an important traditional Chinese medicine. J Ethnopharmacol 172:10–29. https://doi.org/10.1016/j.jep.2015.06.010
149. Agarwal RB, Rangari VD (2003) Anti-inflammatory and anti-arthritic activities of Lupeol and 19a-H Lupeol isolated from *Strobilanthus callosus* and *Strobilanthus ixiocephala* roots. Indian J Pharmacol. 35:384–387
150. Shrinivasa B, Desu R, Elango K, Satish Kumar MN, Suresh B et al (2011) Investigation of selected medicinal plants (*Strobilanthes kunthianus, Strobilanthes cuspidatus*) and marketed formulation (Shallaki) for their anti-inflammatory and anti-osteoarthritic activity. Pharmanest 2:492–499
151. Ekambaram S, Perumal SS, Subramanian V (2010) Evaluation of antiarthritic activity of Strychnos potatorum Linn seeds in Freund's adjuvant induced arthritic rat model. BMC Complement Altern Med 10:56. https://doi.org/10.1186/1472-6882-10-56
152. Yadav KN, Kadam PV, Patel JA, Patil MJ (2014) Strychnos potatorum: phytochemical and pharmacological review. Pharmacogn Rev 8(15):61–66. https://doi.org/10.4103/0973-7847.125533
153. Grespan R, Paludo M, Lemos Hde P, Barbosa CP, Bersani-Amado CA et al (2012) Anti-arthritic effect of eugenol on collagen-induced arthritis experimental model. Biol Pharm Bull 35(10):1818–1820. https://doi.org/10.1248/bpb.b12-00128
154. Batiha GE, Alkazmi LM, Wasef LG, Beshbishy AM, Nadwa EH et al (2020) Syzygium aromaticum L. (Myrtaceae): traditional uses, bioactive chemical constituents, pharmacological and toxicological activities. Biomolecules 10(2):202. https://doi.org/10.3390/biom10020202
155. Endale M, Lee WM, Kwak YS, Kim NM, Kim BK et al (2013) Torilin ameliorates type II collagen-induced arthritis in mouse model of rheumatoid arthritis. Int Immunopharmacol 16: 232–242
156. Rahimpour Y, Doorandishan M, Dehsheikh AB, Sourestani MM, Mottaghipisheh J (2023) A review on Torilis japonica: ethnomedicinal, phytochemical, and biological features. Chem Biodivers 20(5):e202201071. https://doi.org/10.1002/cbdv.202201071
157. Patil CR, Rambhade AD, Jadhav RB, Patil KR, Dubey VK et al (2011) Modulation of arthritis in rats by Toxicodendron pubescens and its homeopathic dilutions. Homeopathy 100(3): 131–137. https://doi.org/10.1016/j.homp.2011.01.001
158. Patil CR, Gadekar AR, Patel PN, Rambhade A, Surana SJ et al (2009) Dual effect of Toxicodendron pubescens on carrageenan induced paw edema in rats. Homeopathy 98(2): 88–91. https://doi.org/10.1016/j.homp.2009.01.003
159. Jadhav HP, Chaudhari GG, Patil DD, Jadhav RB, Reddy NM et al (2016) Standardization of homeopathic mother tincture of Toxicodendron pubescens and correlation of its flavonoid markers with the biological activity. Homeopathy 105(1):48–54. https://doi.org/10.1016/j.homp.2015.08.003
160. Vigneshwaran S, Maharani K, Sivasakthi P, Selvan PS, Saraswathy SD et al (2023) Bioactive fraction of Tragia involucrata Linn leaves attenuates inflammation in Freund's complete adjuvant-induced arthritis in Wistar albino rats via inhibiting NF-κB. Inflammopharmacology 31(2):967–981. https://doi.org/10.1007/s10787-023-01154-8
161. Pallie MS, Perera PK, Kumarasinghe N, Arawwawala M, Goonasekara CL (2020) Ethnopharmacological use and biological activities of *Tragia involucrata* L. Evid Based Complement Alternat Med 2020:8848676. https://doi.org/10.1155/2020/8848676
162. Chamundeeswari D, Vasantha J, Gopalakrishnan S, Sukumar E (2003) Free radical scavenging activity of the alcoholic extract of Trewia polycarpa roots in arthritic rats. J Ethnopharmacol 88(1):51–56. https://doi.org/10.1016/s0378-8741(03)00143-0
163. Chamundeeswari D, Vasantha J, Gopalakrishnan S, Sukumar E (2004) Anti-inflammatory and antinociceptive activities of Trewia polycarpa roots. Fitoterapia 75(7–8):740–744. https://doi.org/10.1016/j.fitote.2004.07.001

164. Jain DK, Patel NS, Nagar H, Patel A, Chandel HS (2012) Anti-arthritic activity of *Tridax procumbens* ethanolic extract of leaves. J Pharm Sci 2:80–86
165. Al-Bari MAA, Hossain S, Mia U, Al Mamun MA (2020) Therapeutic and mechanistic approaches of *Tridax Procumbens* flavonoids for the treatment of osteoporosis. Curr Drug Targets 21(16):1687–1702. https://doi.org/10.2174/1389450121666200719012116
166. Sindhu G, Ratheesh M, Shyni GL, Nambisan B, Helen A (2012) Anti-inflammatory and antioxidative effects of mucilage of Trigonella foenum graecum (fenugreek) on adjuvant induced arthritic rats. Int Immunopharmacol 12(1):205–211. https://doi.org/10.1016/j.intimp.2011.11.012
167. Nagulapalli Venkata KC, Swaroop A, Bagchi D, Bishayee A (2017) A small plant with big benefits: fenugreek (Trigonella foenum-graecum Linn.) for disease prevention and health promotion. Mol Nutr Food Res 61(6). https://doi.org/10.1002/mnfr.201600950
168. Abudoleh S, Disi A, Qunaibi E, Aburjai T (2011) Anti-arthritic activity of the methanolic leaf extract of *Urtica pilulifera* L. on albino rats. Am J Pharmacol Toxicol 6:27–32
169. Ozen T, Cöllü Z, Korkmaz H (2010) Antioxidant properties of Urtica pilulifera root, seed, flower, and leaf extract. J Med Food 13(5):1224–1231. https://doi.org/10.1089/jmf.2009.1303
170. Latha RM, Geetha T, Varalakshmi P (1998) Effect of Vernonia cinerea less flower extract in adjuvant-induced arthritis. Gen Pharmacol 31(4):601–606. https://doi.org/10.1016/s0306-3623(98)00049-4
171. Pandey A, Bani S, Satti NK, Gupta BD, Suri KA (2012) Anti-arthritic activity of agnuside mediated through the down-regulation of inflammatory mediators and cytokines. Inflamm Res 61(4):293–304. https://doi.org/10.1007/s00011-011-0410-x
172. Zheng CJ, Zhao XX, Ai HW, Lin B, Han T et al (2014) Therapeutic effects of standardized Vitex negundo seeds extract on complete Freund's adjuvant induced arthritis in rats. Phytomedicine 21(6):838–846. https://doi.org/10.1016/j.phymed.2014.02.003
173. Jing R, Ban Y, Xu W, Nian H, Guo Y et al (2019) Therapeutic effects of the total lignans from Vitex negundo seeds on collagen-induced arthritis in rats. Phytomedicine 58:152825. https://doi.org/10.1016/j.phymed.2019.152825
174. Maryam S, Khan MR, Shah SA, Zahra Z, Batool R et al (2019) Evaluation of anti-inflammatory potential of the leaves of Wendlandia heynei (Schult.) Santapau & Merchant in Sprague Dawley rat. J Ethnopharmacol 238:111849. https://doi.org/10.1016/j.jep.2019.111849
175. Maryam S, Khan MR, Shah SA, Zahra Z, Majid M et al (2018) *In vitro* antioxidant efficacy and the therapeutic potential of *Wendlandia heynei* (Schult.) Santapau & Merchant against bisphenol A-induced hepatotoxicity in rats. Toxicol Res (Camb) 7(6):1173–1190. https://doi.org/10.1039/c7tx00322
176. Gupta A, Singh S (2014) Evaluation of anti-inflammatory effect of Withania somnifera root on collagen-induced arthritis in rats. Pharm Biol 52(3):308–320. https://doi.org/10.3109/13880209.2013.835325
177. Hussain A, Aslam B, Muhammad F, Faisal MN, Kousar S et al (2021) Anti-arthritic activity of *Ricinus communis* L. and *Withania somnifera* L. extracts in adjuvant-induced arthritic rats *via* modulating inflammatory mediators and subsiding oxidative stress. Iran J Basic Med Sci 24(7):951–961. https://doi.org/10.22038/ijbms.2021.55145.12355
178. Kumar S, Singh R, Gajbhiye N, Dhanani T (2018) Extraction optimization for phenolic- and Withanolide-rich fractions from *Withania somnifera* roots: identification and quantification of Withaferin a, 12-Deoxywithastromonolide, and Withanolide a in plant materials and marketed formulations using a reversed-phase HPLC-photodiode Array detection method. J AOAC Int 101(6):1773–1780. https://doi.org/10.5740/jaoacint.18-0081
179. Lin B, Zhao Y, Han P, Ma XQ, Rahman K et al (2014) Anti-arthritic activity of Xanthium strumarium L. extract on complete Freund′s adjuvant induced arthritis in rats. J Ethnopharmacol 155(1):248–255. https://doi.org/10.1016/j.jep.2014.05.023

180. Fan W, Fan L, Peng C, Wang L, Li L, Wang J et al (2019) Traditional uses, botany, phytochemistry, pharmacology, pharmacokinetics and toxicology of *Xanthium strumarium* L: a review. Molecules 24(2):359. https://doi.org/10.3390/molecules24020359
181. Cheeke PR, Piacente S, Oleszek W (2006) Anti-inflammatory and anti-arthritic effects of Yucca schidigera: a review. J Inflamm (Lond) 3:6. https://doi.org/10.1186/1476-9255-3-6
182. Qu L, Wang J, Ruan J, Huang P, Wang Y et al (2018) Spirostane-type Saponins obtained from Yucca schidigera. Molecules 23(1):167. https://doi.org/10.3390/molecules23010167
183. Eleonora C, Patrizia F, Piero G, Gubbiotti R, Samperi R et al (2011) Flavonoids: chemical properties and analytical methodologies of identification and quantitation in foods and plants. Nat Prod Res 25(5):469–495. https://doi.org/10.1080/14786419.2010.482054
184. Feng FB, Qiu HY (2018) Effects of Artesunate on chondrocyte proliferation, apoptosis and autophagy through the PI3K/AKT/mTOR signaling pathway in rat models with rheumatoid arthritis. Biomed Pharmacother 102:1209–1220
185. Qi W, Lin C, Fan K, Liu L, Feng X et al (2019) Hesperidin inhibits synovial cell inflammation and macrophage polarization through suppression of the PI3K/AKT pathway in complete Freund's adjuvant-induced arthritis in mice. Chem Biol Interact 306:19–28
186. Kim EK, Choi EJ (2010) Pathological roles of MAPK signaling pathways in human diseases. Biochim Biophys Acta 1802(4):396–405
187. Zhai KF, Duan H, Cui CY, Cao YY, Si JL et al (2019) Liquiritin from Glycyrrhiza uralensis attenuating rheumatoid arthritis via reducing inflammation, suppressing angiogenesis, and inhibiting MAPK signaling pathway. J Agric Food Chem 67(10):2856–2864
188. Bao Y, Sun YW, Ji J, Gan L, Zhang CF et al (2019) Genkwanin ameliorates adjuvant-induced arthritis in rats through inhibiting JAK/STAT and NF-κB signaling pathways. Phytomedicine 63:Article 153036
189. Chu J, Wang X, Bi H, Li L, Ren M, Wang J (2018) Dihydromyricetin relieves rheumatoid arthritis symptoms and suppresses expression of pro-inflammatory cytokines via the activation of Nrf2 pathway in rheumatoid arthritis model. Int Immunopharmacol 59:174–180
190. Karatas A, Dagli AF, Orhan C, Gencoglu H, Ozgen M et al (2020) Epigallocatechin 3-gallate attenuates arthritis by regulating Nrf2, HO-1, and cytokine levels in an experimental arthritis model. Biotechnol Appl Biochem 67(3):317–322
191. Su X, Huang Q, Chen J, Wang M, Pan H et al (2016) Calycosin suppresses expression of pro-inflammatory cytokines via the activation of p62/Nrf2-linked heme oxygenase 1 in rheumatoid arthritis synovial fibroblasts. Pharmacol Res 113(Pt A):695–704
192. Osmakov DI, Kalinovskii AP, Belozerova OA, Andreev YA, Kozlov SA (2022) Lignans as pharmacological agents in disorders related to oxidative stress and inflammation: chemical synthesis approaches and biological activities. Int J Mol Sci 23(11):6031
193. Chang WC, Chu MT, Hsu CY, Wu YJ, Lee JY et al (2019) Rhein, an Anthraquinone drug, suppresses the NLRP3 Inflammasome and macrophage activation in urate crystal-induced gouty inflammation. Am J Chin Med 47(1):135–151. https://doi.org/10.1142/S0192415X19500071
194. Qian K, Zheng X-X, Wang C, Liu XB, Xu SD et al (2022) β-Sitosterol inhibits rheumatoid synovial angiogenesis through suppressing VEGF signaling pathway. Front Pharmacol 12: 816477. https://doi.org/10.3389/fphar.2021.816477
195. Carvalho AMS, Heimfarth L, Santos KA, Guimaraes AG, Picot L et al (2019) Terpenes as possible drugs for the mitigation of arthritic symptoms - a systematic review. Phytomedicine 57:137–147. https://doi.org/10.1016/j.phymed.2018.10.028
196. Ekor M (2014) The growing use of herbal medicines: issues relating to adverse reactions and challenges in monitoring safety. Front Pharmacol 4:177. https://doi.org/10.3389/fphar.2013.00177
197. Bandaranayake WM (2006) Quality control, screening, toxicity, and regulation of herbal drugs. In: Ahmad I, Aqil F, Owais M (eds) Modern phytomedicine. Turning medicinal plants into drugs. Wiley-VCH GmbH & Co. KGaA, Weinheim, pp 25–57. https://doi.org/10.1002/9783527609987.ch2

198. Studdert D, Eisenberg D, Miller F, Curto DA, Kaptchuk TJ et al (1998) Medical malpractice implications of alternative medicine. J Am Med Assoc 280:1610–1615. https://doi.org/10.1001/jama.280.18.1610
199. Brevort P (1998) The booming us botanical market: a new overview. Herbal Gram 44:33–48
200. Raynor DK, Dickinson R, Knapp P, Long AF, Nicolson DJ (2011) Buyer beware? Does the information provided with herbal products available over the counter enable safe use? BMC Med 9:94. https://doi.org/10.1186/1741-7015-9-94
201. Petrovska BB (2012) Historical review of medicinal plants' usage. Pharmacogn Rev 6(11):1–5. https://doi.org/10.4103/0973-7847.95849
202. Yuan H, Ma Q, Ye L, Piao G (2016) The traditional medicine and modern medicine from natural products. Molecules 21(5):559. https://doi.org/10.3390/molecules21050559

Assessment of the Phytochemical Constituents and Metabolites in the Medicinal Plants and Herbal Medicine Used in the Treatment and Management of Respiratory Diseases

16

Prasann Kumar, Lalit Saini, and Monika Sharma

Contents

P. Kumar · L. Saini · M. Sharma (✉)
Department of Agronomy, School of Agronomy, Lovely Professional University, Phagwara, Punjab, India

S. C. Izah et al. (eds.), *Herbal Medicine Phytochemistry*, Reference Series in Phytochemistry, https://doi.org/10.1007/978-3-031-43199-9_19

Abstract

Worldwide, chronic obstructive pulmonary disease, asthma, lung cancer, and cystic fibrosis are some of the top causes of death and morbidity. As a result, these illnesses place a significant burden on healthcare systems, economies, and societies in many nations. Chronic respiratory illnesses impact the respiratory airways, lung parenchyma, and pulmonary vasculature, among other parts of the respiratory system. Respiratory disorders are a fairly frequent diagnosis in children, adolescents, and adults in the era of air pollution. Both upper and lower respiratory system disorders can have a significant negative impact on overall health as well as economic and psychological burdens. Because they are eco-friendly and have few side effects, plant-based solutions are receiving a lot of attention as a means of treating and preventing health issues. It is believed that active plant contact and exposure to nature are good for both physical and mental health. The immunological and cardiovascular systems are the main targets of plant-based medications. The medicinal properties of both terrestrial and marine botanicals are an effective control of a variety of disorders. Biologically active compounds with distinct values may be isolated from both. Therefore, the main aim of this chapter is to know the potential of pulmonary diseases and to provide the knowledge of plant therapy or some medicinal herbs for curing.

Keywords

Chronic pulmonary disease · Lung disease · Respiration · Immune · Psychological · Airborne · Botanicals · Medicinal herbs · Anti-inflammation

Abbreviations

AYUSH	Ayurveda, Yoga and Naturopathy, Unani, Siddha, and Homeopathy
ACE2	*Angiotensin-converting enzyme*
AKT	Ak strain transforming
BAX	Bcl-2-associated X protein
CD	*Cluster of differentiation*

CDK	Cyclin-dependent kinases
COPD	Chronic obstructive pulmonary disease
COVID-19	*Corona virus disease – 2019*
CXCR4	C-X-C motif chemokine receptor 4
DMDS	Dimethyl disulphide
ET	Endotracheal Tube
FEV	Forced expiratory volume
GSH	Glutathione
HIV	Human immunodeficiency virus
HUVEC	*Human umbilical vein endothelial cells*
LOX	Lipoxygenase
MCT	Medium chain triglycerides
MDA	*Malondialdehyde*
NOS	Nitric oxide synthase
PAP	Pulmonary artery pressure
PASMC	*Pulmonary artery smooth muscle cells*
PDE	Phosphodiesterase enzyme
RNA	Ribonucleic acid
ROS	Reactive oxygen species
RVH	Right ventricular hypertrophy
RVSP	*Right Ventricular Systolic Pressure*
SARS CoV-2	Severe acute respiratory syndrome coronavirus 2
SOD	Superoxide dismutases
TGF	Transforming growth factor
TMS	Transcranial magnetic stimulation
VEGF	*Vascular endothelial growth factor*
WBC	White blood cells

1 Introduction

Medicinal herbs plays a critical role in the well-being of rural, remote, and indigenous people. Plants with medicinal properties are used to cure a wide range of illnesses in both humans and animals. Scientists are interested in the ethnomedical and nutritional uses of locally sourced plant products, prompting a search for bioactive chemicals. Carbohydrates, protein, and fat are all found in medicinal plants. Physiological, metabolic, and morphological processes rely on these substances, making them essential for human survival [1]. Medicines, dietary supplements, and other healthcare items often contain ingredients originating from plants. Antioxidants, hypoglycaemics, and hypolipidaemics are just a few examples of the types of phytochemicals found in plants that play a significant role in the search for new and improved therapeutic components. Many pharmaceuticals have their origins in plants, either as direct or indirect sources [2]. Humans rely on plants for a wide variety of things, including food, shelter, fibre, and medicine. Herbal remedies are the primary choice for health care among rural residents. Even in places where conventional medicine is readily available, there has been a recent uptick in curiosity about herbal remedies. Minerals and plant-based chemicals found in medicinal plants are two examples of the types of physiologically active components that have been shown to have an array of impacts on human physiology [2]. As they are an abundant source of several bioactive chemicals, essential oils are becoming more and more popular. They are preferred over chemical antioxidants due to their antioxidant and antibacterial capabilities as well as their GRAS (generally recognised as safe) status [3].

Botanicals are used as ingredients in medications, cosmetics, dietary supplements, and herbal teas, among other healthful products, and have seen a surge in popularity in recent years [3]. Plants used for medical purposes have been found to have bioactive substances that may reduce the risk of developing serious illnesses like cancer, cardiovascular disease, and diabetes. Medicinal herbs are essential in the treatment of oral health problems such halitosis, gums bleeding, dental caries, ulcers of mouth, and gingivitis due to their high effectiveness and low risk of adverse effects. Many plant species have unique mechanisms for producing secondary metabolites [3]. A lack of certain vitamins and minerals in the diet has been linked to a variety of potentially serious health problems, including malnutrition and a host of diseases caused by micronutrient deficiencies. Historically, all medicines were derived from plants, either in the form of a crude extract mixture or a simpler sample of plant parts. The primary advantage of using plant-based drugs is that they are non-toxic, far safer than their synthetic counterparts, while also providing significant therapeutic benefits and being much cheaper.

The last several decades have seen a surge of curiosity about medicinal plants and their traditional applications across the Indian subcontinent. Many young people nowadays are leaving rural communities to pursue higher education and work in urban centres, and with them, there has been a decline in the value placed on traditional medicines. The rich medicinal plant knowledge of the country is in danger due to the present trends of deforestation, environmental degradation, and modernization in the area. Hence, herbal knowledge is only accessible by the elderly, and only a select few have access to conventional therapies. Hence, by recording indigenous people's traditional plant medicine, we can close the generational gap in herbal knowledge. So, the goal of this chapter is to examine the big picture of the various herbal plants, formulations, and plant constituents that are currently being utilized to treat respiratory ailments.

2 Phytoconstituents Contributed in Pulmonary Disease Treatment

Plants contain molecules called phytochemicals, which have protective and disease-fighting qualities. Phytochemical components such as triterpenoids, quercetin, resveratrol, etc., have shown significant clinical efficacy against cardiovascular and respiratory problems [4]. In Table 1, we have outlined just a handful of the plant-based chemicals that are effective in treating respiratory disorders.

Table 1 Plant-based chemicals in treating respiratory disorders

Phytochemicals	Disease
Androsin	Bronchial asthma
Apple polyphenol	High blood pressure in the lungs
Curcumin	Bronchitis, emphysema, cystic fibrosis
Ligustrazine	High blood pressure in the lungs
Quercetin	Asthma, COPD, pneumonia, pulmonary hypertension
Resveratrol	Asthma, COPD, pneumonia, pulmonary hypertension, pneumonia, sarcoidosis
Salidroside	High blood pressure in the lungs
Triterpenoids	Pulmonary hypertension, chronic bronchitis

3 Medicinal Plants as Beneficial Agents

The reduction in blood pressure diastolic and activity of sympathetic nervous system and the induction of more relaxed, calmed, and natural sentiments are only a few of how nature and active plant engagement are thought to benefit physical and mental health [5]. Plants are the most common form of life and are widely believed to be the most emblematic component of a natural setting. It is important to remember that indoor plants have their unique challenges that cannot be faced in the outdoors. As a result, houseplants can be read in two ways [6] either as a representation of nature or as a product of human engagement with the natural world. The presence of even a small number of houseplants has been linked to fewer sick days taken at work and a greater sense of satisfaction, comfort, and well-being. The visual density of urban tree coverage has been found to positively correlate with both stress recovery [7] and landscape choice [8] in various studies. There should be more interest and use of energy-efficient indoor plants that can improve air quality and moderate local climate. There are plant species and cultivars that are particularly effective at removing volatile organic chemicals [9]. Diffusion of volatile organic molecules into a plant's leaf depends on several factors, including the leaf's stomatal features, wax layer, and hair growth, all of which vary from plant to plant [10]. Members of the Araliaceae family tended to have intermediate to high removal rates of organic compounds which are highly volatile, as compared to the family of Araceae which have demonstrated low [11] rates of removal. Ferns, followed by herbs, had the highest formaldehyde elimination rates, according to research by [10]. Dry deposition and atmospheric dispersion are two mechanisms that vegetation can use to its advantage to improve air quality [12]. The study of dry deposition, which analyses the procedure by which pollutants are at least momentarily removed from the ambient air, takes into account interception, sedimentation, capture, biological processes, and other factors. Pollution levels can be lowered locally by the use of green infrastructure. Many studies have found that persons who reside in more peaceful environments (rural areas, big green spaces) report higher levels of happiness [13]. Vegetation in urban areas might mitigate some of the adverse effects of noise pollution caused by human activity [14]. Certain green areas can provide thermally comfortable circumstances because they permit deeper water penetration into the ground, increasing the amount of water available for evaporation and so aiding in the further cooling of the surrounding air [15]. The Ministry of AYUSH (Government of India) has occasionally issued distinct guidelines for the treatment of respiratory issues dependent on different medicine systems of India. A respiratory infection called bronchitis is marked by a rise in mucus production and airway inflammation [10]. It is challenging to separate from other illnesses, such as influenza, the common cold, etc., that frequently cause cough. In the upper respiratory, mild viral infection system is referred to as the "common cold." Each year, separate influenza outbreaks, a viral infection of the respiratory system, occur. According to Rawal et al. [16], acute respiratory distress syndrome is a fatal inflammatory lung injury that develops along with the acute failure of the respiratory as a consequence of clearly identified non-pulmonary and pulmonary insults [17]. Emphysema is characterized by persistent breathlessness and an inability to tolerate physical exercise due to damage to the respiratory membrane. There is evidence of a rise in neutrophils, macrophages, T-lymphocytes, and eosinophils in the terminal air spaces and emphysematous tissue of the pulmonary parenchyma. A deteriorating lung condition called COPD is marked by airway narrowing. The inflammatory reaction causes emphysema in the small air cells but chronic bronchitis in the larger airways [18]. These various methods, which the hospitals use following their areas of expertise and modern medicine primarily as adjuvants and may be pertinent for boosting immunity, treating fevers, or providing other respiratory treatments, are as follows:

3.1 Aush Kwath

The AYUSH Ministry encourages AYUSH kwath used as a pre-made mixture for promoting the general public's health. *Ocimum sanctum* leaves, *Cinnamomum verum* stem barks, *Zingiber officinale* rhizomes, and Piper nigrum fruits make up the mixture. Ayush Kwath, Ayush Kudineer, and Ayush Joshanda are a few of the titles under which the formulation is marketed in the market. The market offers it in tablet and powder shapes. These herbs are effective treatments for different viral diseases and are said to boost immunity [19, 20].

3.2 Samshamani Vati

Ayurvedic medicine, that is, Guduchi ghana vati (Samshamani vati), is used to treat all kinds of fevers. It is also used as an anti-inflammatory and painkiller [21]. Samshamani vati, an aqueous extract of Tinospora cordifolia, is thought to have immunomodulatory activities as a result of the synergistic effects of the many chemicals present [22]. Additionally, it works well for many infectious illnesses.

Hareetaki Agasthaya, a popular "Avaleha kalpana," Agastya Haritaki Rasayana contains more than 15 herbal components and is used to treat a variety of respiratory infections. The majority of its components demonstrated antiviral, anti-asthmatic, anti-inflammatory, and immunomodulatory properties [23–25].

3.3 Anuthaila

Arn. and Retz. (Leptadenia reticulate) Wight has been used in the therapy of throat, bronchitis, and asthma issues. Anuthaila Anuthaila contains about twenty ingredients. Similarly, *Ocimum sanctum* and *Sesamum indicum* oil are suggested for a variety of ailments, including dry cough, asthma, migraines, and respiratory illnesses [26]. There are accounts of *S. indicum* seeds combined with *Tachyspermum ammi* seeds that are being used to treat colds, dry coughs, and lung conditions [27].

3.4 Roghan-e-Baboona

A traditional Unani medicine called roghan-e-baboona is used to manage inflammation and asthma. The primary component of Roghan-e-Baboona is Matricaria chamomilla L. flowers. It is made from *M. chamomilla* flowers, which have been proven to be effective for sore throats and acute viral nasopharyngitis [28].

3.5 Sharbat-e-Sadar

It is a formulation polyherbal syrup of unani region plants and is extensively used for the common cold, cough, and respiratory diseases. A key component, *Trachyspermum ammi*, was found to promote the growth of B-cells and to block the neutralization of antibodies to the Japanese encephalitis virus [29]. According to Singh et al. [30], *Adhatoda vasica* inhibits the HIV protease, and *Bombyx mori* was found to boost immune reactions against viral infection. Other components with antiviral and immunomodulatory properties include *Ziziphus jujuba*, *Glycyrrhiza glabra*, *Onosma bracteatum*, and *Ficus carica*.

3.6 Asgandh Safoof

Withania somnifera Dunal, also known as Asgandh Safoof Asgand, is a widely used medicinal herb in India. In the Unani school of medicine, the root powder is used as an immunomodulator. According to reports, the extract of roots substantially raises counts of CD8+ and CD4+ as well as blood profile measurements, particularly WBC and platelet counts (Bani et al., 2006). By disrupting the connections between the host ACE2 receptor and the viral S-protein receptor binding domain, aqueous suspension prevented SARS-CoV-2 entrance and demonstrated strong inhibitory action against the mitogen-induced proliferative response of T lymphocytes [31].

4 PH Treatment with Medicinal Plants

4.1 Allium Sativum

The plant species Garlic (*Allium sativum* L.) has been utilized in traditional medicine for centuries. Cysteine sulfoxides and other sulphur-containing compounds are primarily responsible for the therapeutic effects of *A. sativum*. The category includes allicin, which has the potential to be converted as thiosulfinates [32]. It has been shown that *A. sativum* has therapeutic effects, including as lowering the level of glucose in the blood, decreasing oxidative stress, fighting cancer, as well as reducing inflammation, boosting the immune system, warding off infections, and protecting the heart [33]. *A. sativum* may have the ability to hyperpolarize the membrane of smooth muscle cells, eNOS activity increases, and thereby relax vascular smooth muscles. According to research by Fallon et al. (1998), the vasorelaxant action of a fine extract in the powder form of *A. sativum* can slow the pH development in rats exposed to hypoxia. These are Bunge Allium macrostemon. Traditional Chinese medicine makes use of Bunge Allium macrostemon, a member of the family having Amaryllidaceae, for its many pharmacological activities, including those of being antitumourous, antiasthmatic, antioxidant, and immune-enhancing immunity [34]. The primary component of *A. macrostemon* is dimethyl disulphide (DMDS), which has several biological functions including melanin synthesis regulation, anti-hypertensive and anti-inflammatory, and is thus utilized for myocardial ischemia treatment [35].

4.2 Allium Ursinum

Also known as bear's garlic or wild garlic, it is a plant that is endemic to Eurasia. It is an Amaryllidaceae family member. It has a long history of application in alternative medical practices [36]. *A. ursinum* is capable of producing a miscellaneous set of chemical structures, which are ultimately responsible for the organism's biological activity. The aroma, which is reminiscent of garlic, is caused by flavonoid glycosides and sulphur components, which are two of the most important chemicals. Several of the chemicals produced by *A. ursinum*, such as phytosterols and galactolipids, are unique to the species [37]. On human platelets it may be antiaggregatory effects, inhibition effect on 5-LOX (5-lipoxygenase), and synthase of prostaglandin-endoperoxide (PTGS) enzymes in conditions of in vitro, high antioxidant activity and biosynthesis of cholesterol inhibition in vitro are few essential pharmacological activities that are exhibited by *A. ursinum*. Other important pharmacological activities include the following. In addition to this, its activity has been linked to blood pressure reduction as well as an inhibition of the PDE 5A (phosphodiesterase enzyme). *A. ursinum* contains flavonoids and saponins, both of which have the potential to suppress the activity of PDE5A [38]. There have been reports of an upregulation of PDE5 in PH [39]. Remodelling of vascular pulmonary and RVH can be described through the rise of RV/(LV + S) ratio and thickness of pulmonary arterial wall, respectively. In contrast to untreated hypertensive rats which are untreated, rats having pulmonary hypertension treated with *A. ursinum* are shown to be remodelling the pulmonary vascular and RVH [40] (Table 2).

4.3 Mimosa Pigra

Throughout Africa, the Americas, and Asia, the *Mimosa pigra* L. is widely used as a medicine due to its status as a member of the Fabaceae family. This plant is used against a variety of conditions, including heart problems, gastrointestinal issues, and infections [41]. In a research on the pharmaceutical potential of *Mimosa pigra* against hypoxia-induced PH in rats, Rakotomalala et al. (2013) discovered that *Mimosa pigra* can upregulate NO generation, downregulate PAP and RV/(LV + S) and p38 MAPK expression and phosphorylation in lung tissue. In rats with pulmonary hypertension, P38 MAPK is activated [42]. This research suggests that *M. pigra* may have a beneficial effect on PH outcomes.

4.4 Trifolium Pratense

Trifolium pratense, or Red clover, is a member of the Fabaceae family of plants. Its original range includes West Asia, Europe, and Northwest Africa. *T. pratense* relies heavily on several different flavonoids [43] for its health benefits. *T. pratense* contains isoflavones, which a recent study on animal models found to inhibit delicate muscle contractions in the uterus, ileum, and bladder [44]. Jiang and Yang (2016) showed that feeding broiler chickens with PH an extract of T. *pratense* containing isoflavones decreased ET-1 in the hens' serum and lungs while simultaneously increasing NOS secretion. To a greater extent, ET-1 can stimulate MMP-2 expression, which is associated with respiratory vascular remodelling [45]. *T. pratense* can raise serum levels of NOS and NO due to its phytoestrogen activity [46]. These results imply that Trifolium pratense isoflavones may be useful for treating PH in avian models, where it has been observed to increase NO and decrease ET-1 expression.

4.5 Tulsi (Ocimum Sanctum)

A sacred Hindu herb that has a very large history of medical usage and is now being studied for its potential to alleviate symptoms of bronchial asthma and bronchitis. Tulsi's main ingredient, Eugenol, has a variety of therapeutic effects, including antimicrobial, analgesic, antispasmodic, and adaptogenic. Symptoms of bronchitis and the duration of the illness were reduced in randomized, double-blind, placebo-controlled clinical trials with the combination of primrose root tincture and thyme

Table 2 Medicinal plants used for various diseases

Name of the plant	Scientific name and family	Methods used	Parts used	Purpose of use
Spiny amaranth	*Amaranthus spinosus* Amaranthaceae	Decoction	Foliage	Used for treatment of cough and asthma
Mango	*Mangifera indica* Anacardiaceae	Freshly consumed and decoction	Foliage and fruits	Wound cuts and diabetes can be cured
Soursop	*Annona muricata* Annonaceae	Freshly consumed and decoction	Foliage and fruits	Kidney infection, diabetes, hypertension, and wound cuts
Custard apple	*Annona squamosa* Annonaceae	Poultice	Foliage	Stomachache and headache
Indian snakeroot/ Sarpagandha	*Rauwolfia serpentina* Apocynaceae	Decoction	Foliage	Cure cold and cough
Bagawak	*Philodendron lacerum* Araceae	Poultice	Foliage	Sprain and stomachache
Aloe	*Aloe vera* Asphodelaceae	Extraction	Foliage	Burning of the skin
Common mugwort	*Artemisia vulgaris* Asteraceae	Decoction and poultice	Foliage	Diseases of the lung, blood vomiting, ringworm, and cough
Sambung	*Blumea balsamifera* Asteraceae	Decoction	Foliage	Cold, headache, flu, fever, urinary tract infection, and stomachache
African marigold	*Tagetes erecta* Asteraceae	Pounding and poultice	Leaves	Wound and cuts
Fukien tea tree	*Carmona retusa* Boraginaceae	Poultice	Leaves	Cough and body pain
Indian cherry	*Carmona dichotoma* Boraginaceae	Decoction	Leaves	Post-delivery fever
Papaya	*Carica papaya* Caricaceae	Pounding and extraction	Leaves	Malaria and dengue
Rangoon creeper	*Quisqualis indica* Combretaceae	Decoction	Leaves	Kidney infection
Sweet potato	*Ipomea batatas* Convolvulaceae	Decoction	Leaves	Diabetes and cut and wound infection

(continued)

Table 2 (continued)

Name of the plant	Scientific name and family	Methods used	Parts used	Purpose of use
Bitter-melon	*Momordica charantia* Cucurbitaceae	Cooking and extraction	Leaves and fruits	Diabetes and cut and infection of the wound
Alingaro	*Elaeagnus alingaro* Elaeagnaceae	Decoction	Roots	Headache, stomachache, Kidney infection, muscle pain, gall bladder stone and liver disease
Chinese laurel	*Antidesma bunius* Phyllanthaceae	Decoction	Foliage	Arthritis and kidney infection
Croton	*Codiaeum variegatum* Euphorbiaceae	Poultice	Foliage	Body pain
Snake weed	*Euphorbia hirta* Euphorbiaceae	Decoction	Stem, leaves and roots	Malaria and dengue
Physic nut	*Jatropha curcas* Euphorbiaceae	Poultice	Foliage	Sprain and stomachache
Bitter cassava	*Manihot esculenta* Euphorbiaceae	Decoction	Foliage	Diabetes, wound and cuts
Quick stick	*Gliricidia sepium* Euphorbiaceae	Pounding	Foliage	Vomiting, diarrhoea, and kidney infection
Lima bean	*Phaseolus lunatus* Euphorbiaceae	Cooking	Seeds	Diabetes and wound infection
Glorybower	*Clerodendrum intermedium* Fabaceae/ Lamiaceae	Poultice	Foliage	Sprain and stomachache
Spearmint	*Mentha spicata* Lamiaceae	Decoction	Foliage	Cold, cough, and headache
Common basil	*Ocimum basilicum* Lamiaceae	Decoction	Foliage	Cough and cold
Oregano	*Origanum vulgare* Lamiaceae	Decoction	Foliage	Cold, cough, headache,
Cat's whiskers	*Orthosiphon aristatus* Lamiaceae	Decoction	Foliage and roots	Kidney infection
Painted nettle	*Plectranthus scutellarioides* Lamiaceae	Poultice	Foliage	Kidney infection, cold, and cough

(continued)

Table 2 (continued)

Name of the plant	Scientific name and family	Methods used	Parts used	Purpose of use
Fragrant premna	*Premna adorata* Lamiaceae	Extraction and poultice	Foliage and stem	Skin disease and allergies
Chinese chase tree	*Vitex negundo* Lamiaceae	Decoction	Foliage	Cough and cold
Cinnamon	*Cinnamomum mercadoi* Lauraceae	Decoction	Foliage	Vomiting, nausea, and hypertension
Avocado	*Persea Americana* Lauraceae	Decoction	Foliage	Kidney infection, stomachache, hypertension, and diabetes
Queen's flower	*Lagerstroemia speciosa* Lythraceae	Decoction	Foliage	Kidney infection
Drumstick tree	*Moringa oleifera* Moringaceae	Eating in a fresh state and decoction	Fruits and foliage	Diabetes and wound Infection
Guava	*Psidium guajava* Myrtaceae	Decoction	Foliage	Diarrhoea, wounds and cuts Infection
Java plum	*Syzygium cumini* Myrtaceae	Eating in a fresh state and decoction	Fruits and foliage	Diabetes and wound infection
Bilimbi	*Averrhoa bilimbi* Oxalidaceae	Decoction	Foliage	Fever
Screw pine	*Pandanus tectorius* Lythraceae	Decoction	Foliage	Fever and cold, stomachache, flu, kidney infection, wounds cuts infection, and diabetes
Pepper elder	*Peperomia pellucida* Moringaceae	Decoction	Foliage	Body pain and hypertension
Betel pepper	*Piper betle* Myrtaceae	Poultice	Foliage	Sprain
Love grass	*Chrysopogon aciculatus* Poaceae	Decoction	Roots	Stomachache, headache
Cogongrass	*Imperata cylindrica* Poaceae	Decoction	Roots	Muscle pain, headache, stomachache, liver disease, kidney infection, and gall bladder stone
Calamondin	*Citrofortunella microcarpa* Piperaceae	Extraction	Fruits	Cold and cough

(continued)

Table 2 (continued)

Name of the plant	Scientific name and family	Methods used	Parts used	Purpose of use
Star apple	*Chrysophyllum cainito* Sapotaceae	Eating in a fresh state and decoction	Fruits and foliage	Diabetes and wound infection
Bell pepper	*Capsicum annuum* Poaceae	Poultice	Foliage	Boil and wounds
Burr bush	*Triumfetta bartramia* Tiliaceae	Decoction	Roots	Muscle pain, headache, stomachache, kidney infection, gall bladder stone and liver disease
Turmeric	*Curcuma longa* Rutaceae	Decoction	Rhizome	Stomachache

fluid extract [47]. Some research [48] also shows that ivy (*Hedera helix*) leaf extract helps reduce chronic bronchitis symptoms. In contrast to thymoquinone, a different phytochemical, nigellone, derived from black cumin was reported to display anti-spasmodic and facilitative effects on pulmonary clearance [49].

4.6 Argemone Mexicana (Papaveraceae)

Brahmadandi, the Indian plant *A. mexicana*, is a regular sight along highways there. The most active compounds identified from *A. mexicana* for the curing of asthma are berberine and protopine [50]. Native Andhra Pradesh residents have found success in treating asthma by taking a powder made from the seeds two times a day for 2 weeks [51]. At a concentration of 50 mg/kg, aqueous extracts from the stem of *A. mexicana* have been shown to reduce milk-induced leukocytosis and eosinophilia, two hallmarks of the allergic response [52].

4.7 Piper Longum (Piperaceae)

Piperine extracted from *P. longum* (long pepper) fruit has shown antiasthmatic and anti-bronchitis action (Hussain et al., 2011), and the plant is native to India. Alkaloids, steroids, glycosides, and flavonoids are all present in *Piper longum* fruit. The guinea pig's bronchospasm caused by histamine was successfully reduced by the fruit [47].

4.8 Tamarindus Indica (Fabaceae)

Breathing disorders, vaginal and uterine symptoms, and inflammation are just a few of the many conditions that *T. indica* (Tamarind), a traditional tropical medicine, is renowned for treating in folk medicine. Researchers have found that an extract of *T. indica* leaves has potent anti-asthmatic activity due to its bronchodilating, anti-histaminic (H1-antagonist), and anti-inflammatory properties [53].

4.9 Ageratum Conyzoides

The New World's tropical regions are home to the annual herb *Ageratum conyzoides*, abbreviated as *A. conyzoides*, which is a member of the Asteraceae family (previously the Compositae). Yet, it is also found throughout the rest of the world's tropical and subtropical climates. This plant's hydro-alcoholic leaf extract has been shown to have significant antihistaminic activity and to prevent clonidine-induced catalepsy in rats when given orally at doses of 250–1000 mg/kg [54].

5 Phytochemicals Used for PH Treatment

The Orchidaceae genus includes the *Eulophia macrobulbon* (Corduroy orchid) species. In Asia, Europe, and America, conventional medicine frequently makes use of orchids. Narkhede et al. (2016) state that they also treat cancer, diabetes, autoimmune, and inflammatory illnesses, as well as pulmonary and other diseases. PDE5 inhibitors, flavonoids, and terpenoids are present in the ethanolic extract of *E. macrobulbon*, indicating that it may be a pulmonary vasodilator [55]. A treatment with an ethanolic extract of *E. macrobulbon* decreases calcium chloride 2–induced in vitro contraction of PA rings and lowers the ratio (RV/LV + S), according to research on MCT-induced PH in rats by Wisutthathum et al. [56]. Pigs with high cholesterol have their coronary arteries relaxed by *E. macrobulbon* extracts because of stilbenes and their compounds. As a result, it has been proposed that *E. macrobulbon* extract may regulate pH and cause PAs (like coronary arteries) to relax.

5.1 Apple Polyphenol

Especially treat conditions like gastric mucosal injury, cancer, inflammation, and cardiovascular disease, doctors often prescribe apple polyphenol, a chemical extracted from several apple fruits (*Malus* spp.) [57]. Apple polyphenol can influence hemodynamic indicators such as mPAP and PVR, which may reduce the contraction of pulmonary vascular rings, as discovered by Hua et al. [58] in their investigation of hypoxia-induced PH in rats. It reduces cytosolic Ca2+ in PASMCs in vitro. Apple polyphenol increases eNOS expression, boosts NO levels, and reduces caspase-3 and iNOS expression in PAEC at the genetic level. Our results demonstrate that apple polyphenol has therapeutic potential as a treatment for PH, as it reduces contraction and increases NO levels to bring about its therapeutic impact.

5.2 Asiaticoside

The saponin asianoside, isolated from the plant *Centella asiatica* (L.) Urb., has a wide variety of biological activities that have led to its widespread use in traditional medicine. These include effects on temporary cerebral ischemia and reperfusion that are antioxidant, anti-inflammatory, antihepatofibrotic [59], and neuroprotective. In research of hypoxia-induced PH, [60] discovered that asiaticoside therapy decreased RVH and mPAP while simultaneously increasing cGMP and NO concentrations and decreasing ET-1 concentrations in the blood. Not only does this chemical prevent apoptosis in endothelial cells, but it also stimulates nitric oxide synthase (NOS) and activates AKT. Evidence suggests that asiaticoside can influence NO production, which slows the development of PH via phosphorylating and activating the serine/threonine protein kinase/eNOS pathway. According to research by Wang X. B. et al. [61] on hypoxia-induced PH in rats and PASMCs, treatment with asiaticoside raises RVH and mPAP, lowers the production of TGF-bR2, TGF-b1, and TGF-bR1, and inhibits the phosphorylation of Smad2/3 in lung tissue of animals. Smad2/3 phosphorylation is shown to increase when PH develops. Asiaticoside decreases both the quantity and migration of PASMCs, as well as the expression of TGF-b1. Thus, asiaticoside inhibits cell growth and encourages cell death through apoptosis to deliver its therapeutic benefits [61] (Fig. 1).

5.3 Astragalus Polysaccharides

The polysaccharides in *Astragalus membranaceus* Fisch. ex Bunge is the actual chemical responsible for the plant's many pharmacological effects, including its anti-inflammation and antioxidant properties [62]. According to Yuan et al. (2017), therapy with astragalus polysaccharides improves hemodynamic indicators

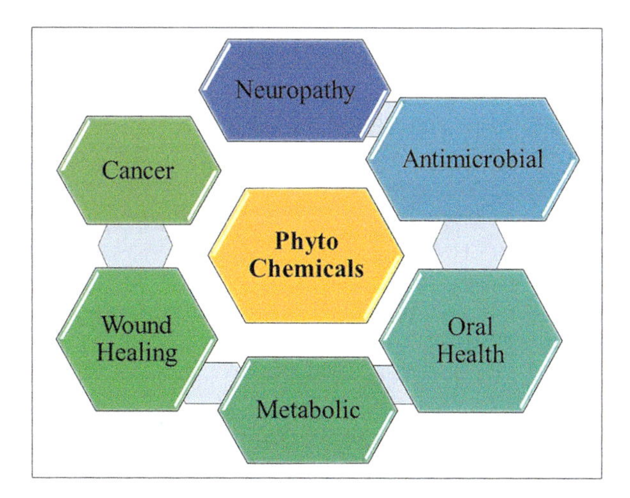

Fig. 1 Medicinal properties of phytochemicals

including PVR, RVH, mPAP, and medial wall thickness in rats with MCT-induced PH. Increased levels of NO and overexpression of eNOS were also noted in the therapy groups. Astragalus polysaccharides were effective in reducing pro-inflammatory mediator levels. The pho-IkBa expression level was also shown to be decreasing. Astragalus polysaccharides, from the research, can be an effective treatment for PH by inhibiting inflammation and by boosting NO levels, which improves hemodynamic indicators. Pho-IkBa is a marker for the evaluation of inflammation.

5.4 Baicalin

Flavonoid baicalin is extracted from the roots of the Scutellaria baicalensis Georgi plant and has several pharmacological effects, including antioxidant, anticancer, anti-inflammatory, and apoptosis-inducing [59]. Because of its anti-fibrotic effect, baicalin has been shown to reduce collagen I expression. It was shown by Wang et al. [61] that hypoxic conditions play a role in PH via pulmonary arterial collagen. In their investigation of hypoxia-induced PH, Liu et al. (2015) found that baicalin treatment improved hemodynamic parameters such as mean arterial pressure (mPAP), mean systemic arterial pressure (mSAP), and right ventricular (RV)/(LV + S) percent. Collagen I is a marker of PH development in lung tissue, and baicalin inhibits its protein and mRNA expression. Baicalin has a much larger effect on protein expression, collagen I mRNA than it does on collagen III, which is worth noting. Baicalin reduces collagen I expression by increasing ADAMTS1 expression. As baicalein inhibits collagen I synthesis and improves hemodynamic parameters, it may be useful as a therapy for PH [63].

5.5 Rhodiola Tangutica

Golden root, also known as *Rhodiola tangutica*, is a significant plant used in Qinghai and Tibetan ancient medicine. This plant is used to cure a range of diseases, most notably to avoid colds brought on by high altitude [64]. Phenylethanols, flavonoids, and terpenoids are a few of the *R. tangutica* enrich fraction's physiologically active compounds [65]. According to Yuan et al. (2017), therapy with astragalus polysaccharides improves hemodynamic indicators including PVR, RVH, mPAP, and medial wall thickness in rats with MCT-induced PH. These three proteins are essential for the transition of the cell cycle from G0/G1 phase to S phase. A regulator of CDKs is P27kip1. As a result, controlling these variables can reduce cell proliferation, which in turn reduces medial vessel thickness and contributes to structural alterations in pulmonary hypertension. In a lab study on isometric tension changes brought on by a force transducer in rat's pulmonary artery verified the vasorelaxant activity of *R. tangutica* [63]. As a result, these investigations suggest that *R. tangutica* may be used as a successful pH treatment.

5.6 Andrographis Somnifera

In particular in India, ashwagandha, a plant belonging to the Solanaceae family, is utilised as a herbal remedy to treat a number of diseases [66]. HPLC study identifies withanolide A and withaferin A as the most important active components in *W. somnifera* root powder extract. *W. somnifera* is used medicinally to treat a number of ailments, such as cancer, Alzheimer's, ageing, inflammation, and bone-building [67]. It is also used to lower blood pressure and cholesterol levels [66]. Kaur et al. (2015) discovered that therapy with *W. somnifera* can lessen RVP and RVH as well as PCNA expression in a study on MCT-induced pulmonary hypertensive mice. Procaspase-3 expression rises as a result of the *W. somnifera* therapy, inducing apoptosis in the pulmonary vessels. Additionally, *W. somnifera* therapy might lower the ROS concentrations in pulmonary tissue. Its anti-inflammatory impact is related to the ability of it to raise IL-10 levels while lowering NFkB and TNF-a levels. Withanolides discovered that *W. somnifera* has an anti-inflammatory impact by inhibiting TNF-a, which in turn inhibits the activation of NFkB [68]. *W. somnifera* also alters the expression of HIF-1a and eNOS in lung tissue (HIF-1a increases in hypoxia conditions). Increased eNOS action may result in higher NO levels, which relax the blood vessels. *W. somnifera* antioxidant, vasorelaxant, and anti-inflammatory properties make it a promising herb for the treatment of PH.

5.7 Luteolin, Apin, and Rhoifolin

Luteolin, Apiin, and Rhoifolin are three kinds of flavonoids used in traditional medicine. Studies have shown that the bulk of plants and veggies contain flavonoids. These flavonoids have biochemical and pharmaceutical effects that include anti-inflammatory, anti-allergic, and spasmolytic properties. The slow cardiac fibres of dogs may be inhibited by luteolin, apiin, and rhoifolin, which can also lower aortic pressure and capillary respiratory pressure in healthy animals. In a research on hypoxia-induced hypertension in dogs, Occhiuto and Limardi [69] showed that structural changes in the flavonoids (apiin, rhoifolin, and luteolin) significantly alter their antihypertensive activity. They showed that rhoifolin decreases aortic pressure but has no effect on pulmonary vascular resistance, whereas apiin and luteolin exhibit antihypertensive characteristics.

5.8 Resveratrol

Resveratrol has a polyphenolic composition and is a potent antioxidant. Due to its numerous biological effects, it has a variety of medicinal uses, including anti-inflammatory, antioxidant, and anti-aging properties [70]. Because it prevents ACT, which promotes apoptosis in human endometrial cancer cells [71], resveratrol possesses anticancer characteristics. Guan et al.'s [72] study on hypoxic pulmonary artery smooth muscle cells (PASMCs) shows that resveratrol administration may inhibit PASMC migration and proliferation, which in turn might reduce p-AKT and AKT expression in PAMSCs. Aside from that, MMP-2 and MMP-9 production is decreased by resveratrol. Due to MMPs' activation of the PI3K/AKT signalling pathway, PASMC migration and proliferation are enhanced [73]. Resveratrol may boost NO production, reduce oxidative stress, inflammation, and cell proliferation, according to some research. As a result, it might enhance PH and pulmonary artery endothelium performance. Resveratrol can also stop the contraction and inflammation that lead to PAH in the rat pulmonary artery dysfunctions and cardiac myocyte hypertrophy [74]. In a study on hypoxia-induced PH in vitro and in vivo, Xu et al. (2016) showed that resveratrol has the ability to improve PH and reduce inflammatory mediators in the lungs of rats. Additionally, resveratrol treatments cause a decrease in H2O2 in the lungs of rats. According to Prabhakar and Semenza [75], ROS and NO boost HIF-1a expression whereas resveratrol's antioxidant activity decreases HIF-1a expression. HIF-1a is a crucial gene in the evolution of PH. Resveratrol boosted the expression of the proteins Trx-1 and Nrf-2. Antioxidant proteins like Trx-1 are balanced by the antioxidant sensor Nrf2. Resveratrol boosts SOD and GSH production, Trx-1 and Nrf-2 protein levels, and PASMC growth. It also reduces ROS, H2O2 production, and HIF-1a expression. These indicators demonstrate the potency of resveratrol as an antioxidant. In contrast to resveratrol, a synthetic substance called trimethoxystilbene (TMS) is produced by methylating resveratrol and has more activities [31]. Increasing the involvement of apoptosis and reducing cell proliferation, TMS is more effective than resveratrol in treating pH, according to a 2012 research by Gao et al. on PASMCs. Resveratrol is therefore proposed to be a potent treatment for PH because of its antioxidant and anti-proliferative properties.

5.9 Carvacrol

The leaves of the herbs thyme (*Thymus* spp.) and oregano (*Origanum* spp.) are the source of the phenolic compound carvacrol. Pharmacologically, carvacrol has been shown to have antiviral, antibacterial, anticancer, and antioxidant effects. In a study of hypoxia-induced PH in rats, Zhang et al. (2016) found that carvacrol decreased RVH and pulmonary artery thickness. By boosting GSH and SOD activity and lowering MDA levels in pulmonary artery vascular smooth muscle cells, it can also mitigate oxidative stress. Carvacrol treatment decreases Bcl-2 (anti-apoptotic protein) protein expression and mRNA, enhances caspase-3 expression, and decreases Procaspase-3 expression. Apoptosis is inhibited and cell survival is promoted by ERK1/2 and Akt. The ERK1/2 and PI3K/Akt pathways are also down-regulated by carvacrol therapy. These results show that carvacrol can reduce oxidative stress and increase apoptosis in PASMCs, both of which are important in reducing PH.

5.10 Nobiletin

Flavonoid nobiletin originates in the rinds of citrus fruits. In light of its wide range of biological actions, including antioxidant, anti-inflammatory, and anti-tumour effects, this molecule has several potential uses [76]. According to Cheng et al. (2017) in a research of MCT-induced pulmonary hypertensive rats, treatment with nobiletin enhances hemodynamic markers such as RVSP and RVH, suggesting that it has a positive impact on hypoxia-induced respiratory vascular remodelling. Nobiletin prevents the phosphorylation of Src, STAT3, and PDGF-BB in vitro, which prevents the expression of Pim1 and NFATc2, two Src and STAT3 target genes. In the end, this chain of genes acts to halt further cell division. Therefore, nobiletin's ability to block the proliferation-promoting Src/STAT3 axis means it can slow the development of pH.

5.11 Panax Notoginseng Saponins

The biological and therapeutic effects of *Panax notoginseng* Saponins (PNS) are used in Therapeutic Potential and Cellular Mechanisms (TCM). They include antioxidant, vasorelaxant, and inhibitory impact on vascular smooth muscle receptor of the voltage-gated Ca2+ channels [35]. Injecting PNS into rats with hypoxia-induced pulmonary hypertension improved hemodynamic indicators like mean pulmonary arterial pressure (mPAP), mean capillary permeability (mCAP), and right ventricular to left systolic ratio (RV/(LV + S)), and thus decreased p38MAPK mRNA expression in lung tissue of these rats. The expression of P38MAPK, a key indicator of physiological stress, rises as PH worsens. Research suggests that PNS may be an option for treating PH.

5.12 Quercetin

In addition to its anti-inflammatory, antioxidant, anti-tumour, anti-proliferative, inducing apoptosis, anti-metastatic, vasodilator, and blood pressure-lowering properties, the flavonoid quercetin is found in a wide variety of plants and fruits. Quercetin may also slow the development of pH in rats, according to certain studies. Quercetin treatment improves hemodynamic markers such PAH and RVH, as shown in research of hypoxia-induced PH in rats by He et al. (2015). By blocking the TrkA/AKT signalling axis, quercetin suppresses PASMC growth and triggers apoptosis in the cells. Increased expression of BAX and decreased expression of Bcl-2 were both found to occur after quercetin administration. Protein expression of cyclin D1 is upregulated by quercetin, while cyclin B1 and Cdc2 are downregulated. Cell migration regulators including MMP2, MMP9, CXCR4, integrin b1, and integrin a5 are under their control because of their ability to suppress their expression. All medicinal effects of herbs and phytochemicals are triggered by one of six mechanisms:

(a) **Antiproliferation:** Antiproliferative effects can be achieved through the use of plant extracts and phytochemicals such as *W. somnifera*, *R. tangutica*, genistein, berberine, magnesium lithospemate B, isohynchophylline, nobiletin, and quercetin all of which work by inhibiting the expression of genes or suppressing pathways involved in cell proliferation. Reducing the expression of CDK 4 and 6, cyclin D1, PDGF-BB, PP2AC, and PCNA while simultaneously increasing the expression of the p27kip1 (cyclin inhibitor) marker, which suppresses the ACT/GSK3b and Trk A/ACT pathways, leads to decreased cell proliferation [65].

(b) **Antioxidant:** The antioxidant effects of *B. vulgaris*, *C. rhipidophylla*, *K. odoratissima*, *M. oleifera*, *T. pratense*, carvacrol, *T. arjuna*, oxymatrine, magnesium lithospemate B, and resveratrol in the treatment of PH occur via increases in GPx, SOD, CAT, and GSH and decreases in ROS, H2O2, and MDA. These metrics point to less oxidative stress and less tissue damage overall [72].

(c) **Antivascular remodelling:** It has previously been explained that vascular remodelling is a crucial event in the development of PH. Several plant extracts and phytochemicals have their effects by inhibiting the process of vascular remodelling by first reducing ET-1 expression and then MMP-9 and MMP-2 production. As an example, *K. odoratissima*, *C. rhipidophylla*, *T. pratense*, *S. miltiorrhiza*, baicalin, asiaticoside, punicalagin, ginsenoside Rb1, and resveratrol are all examples of plant extracts and phytochemicals that have a vascular anti-remodelling effect and thus inhibit the progression of PH [60, 72].

(d) **Vasodilator:** Increased blood vessel constriction and subsequent PH is a major contributor to PH. The major vascular relaxation inducers reduce TXA2 and cytosolic Ca2+ levels while boosting NO levels by increasing eNOS expression. *A. sativum, A macrostemon, K. odoratissima, E. macrobulbum, S. miltiorrhiza, M. pigra, T. pratense, S. securidaca, W. somnifera*, Apple polyphenols, polydatin, and resveratrol are all examples of plant extracts and phytochemicals with these characteristics [54].

(e) **Apoptosis inducer:** Wsomnifera, T. arjuna, baicalin, apple polyphenols, isohynchophylline, carvacrol, quercetin, punicalagin, salidroside, and resveratrol are only a few phytoextracts and phytochemicals that have been demonstrated to enhance apoptosis and hence improve PH. They carry out their function by suppressing the activity of ERK1/2/STAT3 and PI3K/ACT, limiting the action of growth factors like HIF-1a, TGF-b1, and AhR, or increasing the expression of apoptosis-inducing genes like bax and caspase-3 while concurrently decreasing the expression of BCL2 (an anti-apoptosis factor). By causing lung tissue cells, PAECs, and PASMCs to undergo apoptosis, these chemicals stop the progression of PH [58].

(f) **Anti-inflammation:** Increased inflammation is a significant mechanism in PH. It has been demonstrated that a variety of phytochemicals and plant extracts, such as Astragalus polysaccharides, *Withania somnifera*, punicalagin, oxymatrine, and resveratrol, can reduce levels of inflammatory mediators such IL-6, IL-10, TNF-a, IL-1b, NFkB, VEGF, and pho-IkB, which in turn can reduce chronic inflammation. As a result, they can halt the inflammatory process, which in turn slows the development of PH [70].

6 Pulmonary Disease Treatment Using Indian Herbal Medicine

6.1 Asthma

The candlenut tree (Aleurites moluccana) may be found all over the United States and has long been used as a folk remedy for a variety of ailments, including bronchial asthma, joint discomfort, fever, and migraines. This is because it has several beneficial qualities, including anti-viral, anti-nociceptive, anti-hypersensitivity, and anti-microbial [77]. Studies on a semisolid herbal medication made from *A. moluccana* leaf extract show promise as a phytomedicine for the treatment of asthma [78]. This extract showed anti-inflammatory, analgesic, and wound-healing properties in pre-clinical experiments. The therapeutic benefits of *Nigella sativa* (Black cumin) extracts are extensive and include but are not limited to the following: anti-inflammatory, anti-oxidant, antihistaminic, anti-allergic, immunomodulatory, antitussive, and bronchodilatory effects. Patients' asthma symptoms were reduced when they were given *N. sativa* seed, boiling extract, or oil, according to clinical trials [79]. Mountain knotgrass, or *Aerva lanata*, is a widespread weed that thrives in the country's warmer plains. The ethanol extract of this plant has anti-asthmatic properties due to its ability to cause catalepsy and degranulation of mast cells [80]. One additional medicinal herb that can stabilize mast cells is *Bacopa monnieri* (Brahmi). Extracts from medicinal plants like *Bauhinia variegate* (Rakta Kanchnar), *Cassia sophera* (Kasaunda), *Casuarina equisetifolia* (Whistling Pine), and *Clerodendrum serratum* (Bharangi) have the potential to treat asthma by inhibiting immune responses like bronchial and tracheal constriction brought on by histamine and clonidine.

6.2 Chronic Obstructive Pulmonary Disease

In addition to its usage as a Siddha medicine, the leaves and fruits of the *Solanum nigrum* (Manathakkali) plant are commonly utilized as a culinary item throughout most of India. Inflammation-induced oedema can be reduced by using extracts from its leaves. By blocking the ability of AP-1 and NF-kB to bind to DNA, a glycoprotein isolated from *S. nigrum* was discovered to reduce the production of inflammatory chemicals. Reduced levels of superoxide and nitric oxide, both of which have a great value in the disease physiology of chronic obstructive pulmonary disease (COPD), were observed in those who took an ethanolic extract of *Boerhavia diffusa* (Punarnava). In addition to being able to relax muscles and prevent infection, it also protects cells from damage [81]. Vasadi syrup and Shwasaghna dhuma, two Indian herbal medicines, have been shown to significantly improve COPD symptoms and forced expiratory volume in 1 second (FEV1%) in clinical trials involving patients with COPD. Equal parts of the following ingredients make up the Vasadi syrup: *Curcuma longa* (Haridra), *Justicia adhatoda* (Vasa), *Clerodendrum serratum* (Bharangi), *Coriandrum sativum* (Dhanyaka), *Zingiber officinale* (Shunthi), *Tinospora cordifolia* (Guduchi), *Piper longum* (Pippali), and *Solanum virginianum* (Kantakari). Contents of Shwasaghna dhuma are dry leaves of *Datura stramonium* (Dhatura), powder of seeds of *Solanum virginianum* (Kantakari), seeds of *Hyoscyamus niger* (Khurasani ajwain), *Trachyspermum ammi* (Ajwain), *Curcuma longa* (Haridra), Potassium Nitrate (Kalmi shora), and *Cannabis sativa* (Bhanga) in equal parts [82].

6.3 Bronchial Asthma

Ocimum sanctum (Tulsi), an ancient medicinal herb, is effective against both acute bronchitis and chronic bronchial asthma. Tulsi's main ingredient, Eugenol, has a wide range of therapeutic effects, including anti-microbial, analgesic, antispasmodic, and adaptogenic [53]. In double-blind, randomised, placebo-controlled clinical studies, the combination of thyme fluid extract and primrose root tincture decreased bronchitis symptoms and the length of the disease [47]. Some research [48] also shows that ivy (*Hedera helix*) leaf extract helps reduce chronic bronchitis symptoms. In contrast to thymoquinone, a different phytochemical, nigellone, derived from *Nigella sativa* (Black cumin) was reported to display antispasmodic and facilitative effects on respiratory clearance [49].

6.4 Lung Cancer

In Indian cuisine, *Curcuma longa* (Turmeric) is a staple ingredient. In vitro studies using the A549 lung cancer cell line revealed that *C. longa* extract inhibited telomerase activity and exhibited lethal effects in a dose-dependent manner. The phytochemical conferone, which was isolated from the Ferula species, was found to have weak cytotoxic effects against the A549 lung cancer cell line [76]. The cytotoxic action of *Annona muricata* (Mamaphal) can be seen in the A549 lung carcinoma cell line, and this can be traced back to the plant's constituent chemicals, annomuricin A and B [83]. Using a mouse model, researchers showed that alcoholic extracts of *Andrographis paniculata* (Kiryat) had chemopreventive benefits by increasing lung levels of superoxide dismutase, catalase, and DT-diaphorase [30]. An extract of *Phyllanthus urinaria* (Jaramla) demonstrated anti-angiogenic efficacy in a mouse model in which Lewis lung carcinoma cells were implanted by inhibiting neovascularization in the tumour cells as well as the migration of HUVEC [51]. Other medicinal plants that are used to treat lung cancer symptoms include *Glycyrrhiza glabra* (Liquorice), *Zingiber officinale* (Ginger), *Ocimum sanctum* (Tulsi), *Adhatoda vasica* (Malabar nut), and *Terminalia chebula* (Myrobalan) [52].

6.5 Pneumonia

The antibacterial properties of the *Verbascum* species, better known as Mullein, have led to its usage in the treatment of pneumonia as part of traditional herbal medicine.

Verbascum fruticulosum extract showed strong antibacterial activity against the multidrug-resistant *Streptococcus pneumoniae* strain, suggesting its potential for application in the treatment of pneumonia. Similar to the results seen with *V. fruticulosum* [84], the extract of *Urtica urens* (Dwarf nettle) also showed anti-microbial action against *S. pneumoniae* in the same investigation. Beetroots, or *Beta vulgaris*, are a popular vegetable. Antibacterial activity, as measured by a concentration-dependent expansion of the zone of inhibition against Klebsiella pneumonia, was present in alcoholic leaf extracts of *B. vulgaris* fractionated by n-hexane and chloroform [85]. *Nepeta glutinosa* (Benth), *Ficus racemosa* (Gular fig), *Terminalia chebula* (Myrobalan), *Ricinus communis* (Castor bean), and *Vitex negundo* (Chaste tree) are some other medicinal plants that have been used to cure pneumonia [86].

6.6 COVID-19

The medicinal plants' essence, such as *Astragalus membranaceous* (Katira), *Cullen corylifolium* (Scurfy Pea), *Agastache rugosa* (Indian mint), *Cassia alata* (Candlebush), *Mollugo cervine* (Carpetweed), *Gymnema Sylvestre* (Gurmar), *Tinospora cordifolia* (Gurjo), and *Quercus infectoria* (Manjakani), is capable of inhibiting the action of coronavirus (Benarba and Pandiella (2020). Due to their ability to inhibit the SARS-CoV-2 mechanism of action by obstructing the activity of spike protein, the primary protease, ACE-2, and their receptor, phytochemicals like curcumin, quercetin, luteolin, withaferin A, apigenin, amaranthin, and gallic acid have the potential to be used as drug candidates [34]. An Ayurveda treatment plan was successful in curing COVID-19 symptoms in a 43-year-old guy, according to a case study. Sudarsana Churna was used to reduce fever, Talisadi Churna was used to restore a sense of taste, and Dhanwantara Gutika was used to help with breathing difficulties [87].

7 Pharmaceutical Industry and Herbal Plants

The plant world naturally contains phytochemicals, which have demonstrated their therapeutic value as a source of a broad range of biological activity [88]. Various active components found in plants allow us to speculate about their possible therapeutic value. Herbivores, microbes, and plants use the structures of secondary metabolites as protection systems within the cell by interfering with molecular targets [89]. For the creation of plant-based medicines and pharmaceuticals, it is extremely beneficial to identify as many chemical units and groups as possible, such as phenol and organic acids, carbohydrates, and amino acids [90]. Terrestrial or marine botanicals can yield biologically active substances, and the value of these substances relies on their characteristics [91]. Plants' nutritional and therapeutic properties lead to effective disease management for a variety of conditions. Plants demonstrate the importance of their flexible ecomorphology for environments. Animal, human, insect, and overall ecosystem well-being are all directly influenced by plants. Farming and herbal gardening are increasingly being studied in depth.

Climate, soil, and geographic variables all have an indirect or direct impact on plant development [92]. Unfortunately, many medicinal plants are in danger of going extinct due to overharvesting, habitat destruction, a lack of sustainable agriculture, and conservation practices. However, by controlling their accumulated secondary compounds, plants can get around many obstacles (physical, biological, and chemical) [93, 94]. Only 6% of the estimated 215,000–500,000 higher plant species in the globe have had their biological activity screened [94]. Economic commerce in more than 2000 different medicinal and aromatic plant species threatens 10% of those species in at least one European country [95]. Over 82% of drug substances are directly derived from natural products, and over 50% of drug substances come from chemicals separated from plants or herbs [96]. According to Markets 2019, the plant extracts industry is expected to grow from its current estimated value of USD 23.7 billion to USD 59.4 billion by 2025. The use of medications to relieve some of the most prevalent respiratory disease symptoms, such as a sore tongue, cough, nasal congestion, etc., is supported by extensive data [14]. In the accessible literature, reports of individual or combined herbal sources have been made. Homeopathy is widely used, and using it has been linked with positive outcomes.

8 Conclusion

In conclusion, medicinal plants have been used for centuries to cure respiratory diseases and provide relief from respiratory symptoms. They contain bioactive compounds that have antiviral, antibacterial, anti-inflammatory, and immune-modulatory properties, which make them effective in treating respiratory infections and diseases. Studies have shown that various medicinal plants, such as ginger, turmeric, eucalyptus, garlic, liquorice, and thyme, can be used to treat respiratory infections and diseases like asthma, bronchitis, and pneumonia. However, it is essential to note that the use of medicinal plants as a treatment for respiratory diseases should not replace conventional medicine but should be used as a complementary therapy. In addition, more research is needed to determine the safety and efficacy of these medicinal plants in the treatment of respiratory diseases. Nonetheless, their use can offer a natural and cost-effective way to manage respiratory diseases and improve overall respiratory health. Therefore, the integration of medicinal plants into respiratory disease management programs can be a promising approach to improving respiratory health outcomes. Medicinal plants have been used for centuries as a natural remedy for respiratory diseases. They offer a variety of therapeutic benefits, including anti-inflammatory, antioxidant, and expectorant properties, which make them an effective treatment option. Many of these plants have been studied extensively, and their efficacy in treating respiratory conditions has been proven by scientific research. Some of the commonly used medicinal plants for respiratory diseases include liquorice root, ginger, eucalyptus, thyme, and peppermint. These plants can be used in different forms such as teas, tinctures, and extracts and they can be easily incorporated into daily routines. While medicinal plants offer a natural and safe alternative to conventional medicine, it is essential to seek medical advice before using them, especially if you have pre-existing medical conditions or are taking medication. Additionally, it is crucial to purchase these plants from a reputable source to ensure their quality and purity. Overall, medicinal plants remain an important component of traditional medicine, and their role in the treatment of respiratory diseases cannot be ignored. With further research and exploration, these plants may provide new avenues for the development of effective and sustainable treatments for respiratory diseases.

Acknowledgment We would like to extend our sincere appreciation and gratitude to the Department of Agronomy at the School of Agriculture at Lovely Professional University, Punjab, 144411, India.

Conflicts of Interest A conflict of interest has not been identified.

References

1. Cheeke P (2009) Applications of saponins as feed additives in poultry production. In: Proceedings of the 20th Australian poultry science symposium, Sydney, Australia, p 50, December
2. Dagli N, Dagli R, Mahmoud RS, Baroudi K (2015) Essential oils, their therapeutic properties, and implication in dentistry: a review. J Int Soc Prev Community Dent 5(5):335
3. Kumar M, Changan S, Tomar M, Prajapati U, Saurabh V, Hasan M et al (2021) Custard apple (Annona squamosa L.) leaves: nutritional composition, phytochemical profile, and health-promoting biological activities. Biomol Ther 11(5):614
4. Jasemi SV, Khazaei H, Aneva IY, Farzaei MH, Echeverría J (2020) Medicinal plants and phytochemicals for the treatment of pulmonary hypertension. Front Pharmacol 11:145
5. Lee MS, Lee J, Park BJ, Miyazaki Y (2015) Interaction with indoor plants may reduce psychological and physiological stress by suppressing autonomic nervous system activity in young adults: a randomized crossover study. J Physiol Anthropol 34(1):1–6
6. Bringslimark T, Hartig T, Patil GG (2009) The psychological benefits of indoor plants: a critical review of the experimental literature. J Environ Psychol 29(4):422–433
7. Jiang B, Li D, Larsen L, Sullivan WC (2016) A dose-response curve describing the relationship between urban tree cover density and self-reported stress recovery. Environ Behav 48(4):607–629
8. Jiang B, Larsen L, Deal B, Sullivan WC (2015) A dose–response curve describing the relationship between tree cover density and landscape preference. Landsc Urban Plan 139: 16–25
9. Dela Cruz M, Christensen JH, Thomsen JD, Müller R (2014) Can ornamental potted plants remove volatile organic compounds from indoor air? – a review. Environ Sci Pollut Res 21: 13909–13928
10. Kim KJ, Jeong MI, Lee DW, Song JS, Kim HD, Yoo EH et al (2010) Variation in formaldehyde removal efficiency among indoor plant species. HortScience 45(10):1489–1495
11. Yang DS, Pennisi SV, Son KC, Kays SJ (2009) Screening indoor plants for volatile organic pollutant removal efficiency. HortScience 44(5):1377–1381
12. Janhäll S (2015) Review on urban vegetation and particle air pollution–deposition and dispersion. Atmos Environ 105:130–137
13. Shepherd D, Welch D, Dirks KN, McBride D (2013) Do quiet areas afford greater health-related quality of life than noisy areas? Int J Environ Res Public Health 10(4):1284–1303
14. Van Renterghem T, Forssén J, Attenborough K, Jean P, Defrance J, Hornikx M, Kang J (2015) Using natural means to reduce surface transport noise during propagation outdoors. Appl Acoust 92:86–101

15. Coccolo S, Pearlmutter D, Kaempf J, Scartezzini JL (2018) Thermal comfort maps to estimate the impact of urban greening on the outdoor human comfort. Urban For Urban Green 35:91–105
16. Rawal G, Yadav S, Kumar R (2018) Acute respiratory distress syndrome: an update and review. J Transl Intern Med 6(2):74–77
17. Confalonieri M, Salton F, Fabiano F (2017) Acute respiratory distress syndrome. Eur Respir Rev 26(144):160116
18. Chalmers JD (2017) Management of chronic airway diseases: what can we learn from real-life data? COPD: J Chron Obstruct Pulmon Dis 14(sup1):S1–S2
19. Alsuhaibani S, Khan MA (2017) Immune-stimulatory and therapeutic activity of Tinospora cordifolia: double-edged sword against salmonellosis. J Immunol Res 2017:1787803
20. Bhalla G, Kaur S, Kaur J, Kaur R, Raina P (2017) Antileishmanial and immunomodulatory potential of Ocimum sanctum Linn. and Cocos nucifera Linn. in murine visceral leishmaniasis. J Parasit Dis 41:76–85
21. Patgiri B, Umretia BL, Vaishnav PU, Prajapati PK, Shukla VJ, Ravishankar B (2014) Anti-inflammatory activity of Guduchi Ghana (aqueous extract of Tinospora Cordifolia Miers.). Ayu 35(1):108
22. More P, Pai K (2011) Immunomodulatory effects of Tinospora cordifolia (Guduchi) on macrophage activation. Biol Med 3(2):134–140
23. Kumar S, Kamboj J, Sharma S (2011) Overview for various aspects of the health benefits of Piper longum linn. fruit. J Acupunct Meridian Stud 4(2):134–140
24. Lampariello LR, Cortelazzo A, Guerranti R, Sticozzi C, Valacchi G (2012) The magic velvet bean of Mucuna pruriens. J Tradit Complement Med 2(4):331–339
25. Jiang ZY, Liu WF, Zhang XM, Luo J, Ma YB, Chen JJ (2013) Anti-HBV active constituents from Piper longum. Bioorg Med Chem Lett 23(7):2123–2127
26. Cohen MM (2014) Tulsi-Ocimum sanctum: a herb for all reasons. J Ayurveda Integr Med 5(4):251
27. Patil GG, Mali PY, Bhadane VV (2008) Folk remedies used against respiratory disorders in Jalgaon district, Maharashtra. Indian J Nat Prod Resour 7(4):354–358
28. Kyokong O, Charuluxananan S, Muangmingsuk V, Rodanant O, Subornsug K, Punyasang W (2002) Efficacy of chamomile-extract spray for prevention of post-operative sore throat. J Med Assoc Thai 85:S180–S185
29. Roy P, Abdulsalam FI, Pandey DK, Bhattacharjee A, Eruvaram NR, Malik T (2015) Evaluation of antioxidant, antibacterial, and antidiabetic potential of two traditional medicinal plants of India: Swertia cordata and Swertia chirayita. Pharm Res 7(5s):S57–S62
30. Singh KP, Upadhyay B, Pra R, Kumar A (2010) Screening of Adhatoda vasica Nees as a putative HIV-protease inhibitor. J Phytology 2(4):78–82
31. Balkrishna A, Pokhrel S, Singh J, Varshney A (2020) Withanone from Withania somnifera may inhibit novel coronavirus (COVID-19) entry by disrupting interactions between viral S-protein receptor binding domain and host ACE2 receptor. https://doi.org/10.21203/rs.3.rs-17806/v1
32. Horníčková J, Kubec R, Cejpek K, Velíšek J, Ovesná J, Stavělíková H (2010) Profiles of S-alk (en) ylcysteine sulfoxides in various garlic genotypes. Czech J Food Sci 28(4):298–308
33. Lanzotti V, Scala F, Bonanomi G (2014) Compounds from allium species with cytotoxic and antimicrobial activity. Phytochem Rev 13:769–791
34. Bhuiyan FR, Howlader S, Raihan T, Hasan M (2020) Plants metabolites: possibility of natural therapeutics against the COVID-19 pandemic. Front Med 7:444
35. Chu CC, Wu WS, Shieh JP, Chu HL, Lee CP, Duh PD (2017) The anti-inflammatory and vasodilating effects of three selected dietary organic sulfur compounds from allium species. J Funct Biomater 8(1):5
36. Sobolewska D, Podolak I, Makowska-Wąs J (2015) Allium ursinum: botanical, phytochemical and pharmacological overview. Phytochem Rev 14:81–97
37. Oszmianski J, Kolniak-Ostek J, Wojdyło A (2013) Characterization and content of flavonol derivatives of Allium ursinum L. plant. J Agric Food Chem 61(1):176–184

38. Lines TC, Ono M (2006) FRS 1000, an extract of red onion peel, strongly inhibits phosphodiesterase 5A (PDE 5A). Phytomedicine 13(4):236–239
39. Kass DA, Champion HC, Beavo JA (2007) Phosphodiesterase type 5: expanding roles in cardiovascular regulation. Circ Res 101(11):1084–1095
40. Bombicz M, Priksz D, Varga B, Kurucz A, Kertész A, Takacs A et al (2017) A novel therapeutic approach in the treatment of pulmonary arterial hypertension: Allium ursinum liophylisate alleviates symptoms comparably to sildenafil. Int J Mol Sci 18(7):1436
41. Rosado-Vallado M, Brito-Loeza W, Mena-Rejon GJ, Quintero-Marmol E, Flores-Guido JS (2000) Antimicrobial activity of Fabaceae species used in Yucatan traditional medicine. Fitoterapia 71(5):570–573
42. Welsh D, Mortimer H, Kirk A, Peacock A (2005) The role of p38 mitogen-activated protein kinase in hypoxia-induced vascular cell proliferation: an interspecies comparison. Chest 128(6): 573S–574S
43. Bouea SM, Wiese TE, Nehls S, Burow ME, Elliott S, Carter-Wientjes CH, Shih BY, Mclachlan JA, Cleveland TE (2003) Evaluation of the estrogenic effects of legume extracts containing phytoestrogens. J Agric Food Chem 51:2193–2199
44. Wang LD, Qiu XQ, Tian ZF, Zhang YF, Li HF (2008) Inhibitory effects of genistein and resveratrol on Guinea pig gallbladder contractility in vitro. World J Gastroenterol: WJG 14(31): 4955
45. Tan X, Chai J, Bi SC, Li JJ, Li WW, Zhou JY (2012) Involvement of matrix metalloproteinase-2 in medial hypertrophy of pulmonary arterioles in broiler chickens with pulmonary arterial hypertension. Vet J 193(2):420–425
46. Simoncini T, Fornari L, Mannella P, Caruso A, Garibaldi S, Baldacci C, Genazzani AR (2005) Activation of nitric oxide synthesis in human endothelial cells by red clover extracts. Menopause 12(1):69–77
47. Gruenwald J, Graubaum HJ, Busch R (2005) Efficacy and tolerability of a fixed combination of thyme and primrose root in patients with acute bronchitis. Arzneimittelforschung 55(11):669–676
48. Guo R, Pittler MH, Ernst E (2006) Herbal medicines for the treatment of COPD: a systematic review. Eur Respir J 28(2):330–338
49. Wienkötter N, Höpner D, Schütte U, Bauer K, Begrow F, El-Dakhakhny M, Verspohl EJ (2008) The effect of nigellone and thymoquinone on inhibiting trachea contraction and mucociliary clearance. Planta Med 74(02):105–108
50. Singh RP, Banerjee S, Rao AR (2001) Modulatory influence of Andrographis paniculata on mouse hepatic and extrahepatic carcinogen metabolizing enzymes and antioxidant status. Phytother Res 15(5):382–390
51. Huang ST, Yang RC, Lee PN, Yang SH, Liao SK, Chen TY, Pang JHS (2006) Anti-tumor and anti-angiogenic effects of Phyllanthus urinaria in mice bearing Lewis lung carcinoma. Int Immunopharmacol 6(6):870–879
52. Garodia P, Ichikawa H, Malani N, Sethi G, Aggarwal BB (2007) From ancient medicine to modern medicine: ayurvedic concepts of health and their role in inflammation and cancer. J Soc Integr Oncol 5(1):25–37
53. Prakash PAGN, Gupta N (2005) Therapeutic uses of Ocimum sanctum Linn (Tulsi) with a note on eugenol and its pharmacological actions: a short review. Indian J Physiol Pharmacol 49(2):125
54. Kurashima K, Takaku Y, Ohta C, Takayanagi N, Yanagisawa T, Kanauchi T, Takahashi O (2017) Smoking history and emphysema in asthma–COPD overlap. Int J Chron Obstruct Pulmon Dis 12:3523–3532
55. Temkitthawon P, Changwichit K, Khorana N, Viyoch J, Suwanborirux K, Ingkaninan K (2017) Phenanthrenes from Eulophia macrobulbon as novel phosphodiesterase-5 inhibitors. Nat Prod Commun 12(1):1934578X1701200121
56. Wisutthathum S, Demougeot C, Totoson P, Adthapanyawanich K, Ingkaninan K, Temkitthawon P, Chootip K (2018) Eulophia macrobulbon extract relaxes rat isolated

pulmonary artery and protects against monocrotaline-induced pulmonary arterial hypertension. Phytomedicine 50:157–165
57. Espley RV, Butts CA, Laing WA, Martell S, Smith H, McGhie TK et al (2014) Dietary flavonoids from modified apple reduce inflammation markers and modulate gut microbiota in mice. J Nutr 144(2):146–154
58. Hua C, Zhao J, Wang H, Chen F, Meng H, Chen L et al (2018) Apple polyphenol relieves hypoxia-induced pulmonary arterial hypertension via pulmonary endothelium protection and smooth muscle relaxation: in vivo and in vitro studies. Biomed Pharmacother 107:937–944
59. Dong MS, Jung SH, Kim HJ, Kim JR, Zhao LX, Lee ES et al (2004) Structure-related cytotoxicity and anti-hepatofibric effect of asiatic acid derivatives in rat hepatic stellate cell-line, HSC-T6. Arch Pharm Res 27:512–517
60. Wang X, Cai X, Wang W, Jin Y, Chen M, Huang X et al (2018) Effect of asiaticoside on endothelial cells in hypoxia-induced pulmonary hypertension. Mol Med Rep 17(2):2893–2900
61. Wang XB, Wang W, Zhu XC, Ye WJ, Cai H, Wu PL et al (2015) The potential of asiaticoside for TGF-β1/Smad signaling inhibition in prevention and progression of hypoxia-induced pulmonary hypertension. Life Sci 137:56–64
62. Auyeung KK, Han QB, Ko JK (2016) Astragalus membranaceus: a review of its protection against inflammation and gastrointestinal cancers. Am J Chin Med 44(01):1–22
63. Li G, Gai X, Li Z, Chang R, Qi Y, Zhaxi D et al (2016) Preliminary study of active component and mechanism of Rhodiola algida var. tangutica on inducing rat pulmonary artery vasorelaxation. J Qin Med Coll 1:40–45
64. Li HX, Sze SCW, Tong Y, Ng TB (2009) Production of Th1-and Th2-dependent cytokines induced by the Chinese medicine herb, Rhodiola algida, on human peripheral blood monocytes. J Ethnopharmacol 123(2):257–266
65. Nan X, Su S, Ma K, Ma X, Wang X, Zhaxi D et al (2018) Bioactive fraction of Rhodiola algida against chronic hypoxia-induced pulmonary arterial hypertension and its anti-proliferation mechanism in rats. J Ethnopharmacol 216:175–183
66. Mohanty IR, Arya DS, Gupta SK (2008) Withania somnifera provides cardioprotection and attenuates ischemia–reperfusion induced apoptosis. Clin Nutr 27(4):635–642
67. Ojha SK, Arya DS (2009) Withania somnifera Dunal (Ashwagandha): a promising remedy for cardiovascular diseases. World J Med Sci 4(2):156–158
68. Ichikawa H, Takada Y, Shishodia S, Jayaprakasam B, Nair MG, Aggarwal BB (2006) Withanolides potentiate apoptosis, inhibit invasion, and abolish osteoclastogenesis through suppression of nuclear factor-κB (NF-κB) activation and NF-κB–regulated gene expression. Mol Cancer Ther 5(6):1434–1445
69. Occhiuto F, Limardi F (1994) Comparative effects of the flavonoids luteolin, apiin and rhoifolin on experimental pulmonary hypertension in the dog. Phytother Res 8(3):153–156
70. Yeung AWK, Tzvetkov NT, Balacheva AA, Georgieva MG, Gan RY, Jozwik A et al (2020) Lignans: quantitative analysis of the research literature. Front Pharmacol 11:37
71. Testa JR, Bellacosa A (2001) AKT plays a central role in tumorigenesis. Proc Natl Acad Sci 98 (20):10983–10985
72. Guan Z, Shen L, Liang H, Yu H, Hei B, Meng X, Yang L (2017) Resveratrol inhibits hypoxia-induced proliferation and migration of pulmonary artery vascular smooth muscle cells by inhibiting the phosphoinositide 3-kinase/protein kinase B signaling pathway. Mol Med Rep 16(2):1653–1660
73. Shivakrupa R, Bernstein A, Watring N, Linnekin D (2003) Phosphatidylinositol 3′-kinase is required for growth of mast cells expressing the kit catalytic domain mutant. Cancer Res 63(15): 4412–4419
74. Paffett ML, Lucas SN, Campen MJ (2012) Resveratrol reverses monocrotaline-induced pulmonary vascular and cardiac dysfunction: a potential role for atrogin-1 in smooth muscle. Vasc Pharmacol 56(1–2):64–73

75. Prabhakar NR, Semenza GL (2012) Adaptive and maladaptive cardiorespiratory responses to continuous and intermittent hypoxia mediated by hypoxia-inducible factors 1 and 2. Physiol Rev 92(3):967–1003
76. Kooti W, Servatyari K, Behzadifar M, Asadi-Samani M, Sadeghi F, Nouri B, Zare Marzouni H (2017) Effective medicinal plant in cancer treatment, part 2: review study. J Evid Based Complementary Altern Med 22(4):982–995
77. Clarke R, Lundy FT, McGarvey L (2015) Herbal treatment in asthma and COPD–current evidence. Clin Phytoscience 1(1):1–7
78. Cesca TG, Faqueti LG, Rocha LW, Meira NA, Meyre-Silva C, De Souza MM et al (2012) Antinociceptive, anti-inflammatory and wound healing features in animal models treated with a semisolid herbal medicine based on Aleurites moluccana L. Willd. Euforbiaceae standardized leaf extract: semisolid herbal. J Ethnopharmacol 143(1):355–362
79. Gholamnezhad Z, Shakeri F, Saadat S, Ghorani V, Boskabady MH (2019) Clinical and experimental effects of Nigella sativa and its constituents on respiratory and allergic disorders. Avicenna J Phytomed 9(3):195
80. Kumar D, Prasad DN, Parkash J, Bhatnagar SP, Kumar D (2009) Antiasthmatic activity of ethanolic extract of Aerva lanata Linn. Pharmacologyonline 2:1075–1081
81. Ram A, Balachandar S, Vijayananth P, Singh VP (2011) Medicinal plants useful for treating chronic obstructive pulmonary disease (COPD): current status and future perspectives. Fitoterapia 82(2):141–151
82. Sharma PK, Johri S, Mehra BL (2010) Efficacy of Vasadi syrup and Shwasaghna Dhuma in the patients of COPD (Shwasa Roga). Ayu 31(1):48
83. Desai AG, Qazi GN, Ganju RK, El-Tamer M, Singh J, Saxena AK et al (2008) Medicinal plants and cancer chemoprevention. Curr Drug Metab 9(7):581–591
84. Gupta VK, Kaushik A, Chauhan DS, Ahirwar RK, Sharma S, Bisht D (2018) Anti-mycobacterial activity of some medicinal plants used traditionally by tribes from Madhya Pradesh, India for treating tuberculosis related symptoms. J Ethnopharmacol 227:113–120
85. Hussain Z, Mohammad P, Sadozai SK, Khan KM, Nawaz Y, Perveen S (2011) Extraction of anti-pneumonia fractions from the leaves of sugar beets Beta vulgaris. J Pharm Res 4(12):4783–4785
86. Asadbeigi M, Mohammadi T, Rafieian-Kopaei M, Saki K, Bahmani M, Delfan M (2014) Traditional effects of medicinal plants in the treatment of respiratory diseases and disorders: an ethnobotanical study in the Urmia. Asian Pac J Trop Med 7:S364–S368
87. Girija PLT, Sivan N (2022) Ayurvedic treatment of COVID-19: a case report. J Ayurveda Integr Med 13(1):100329
88. Kasote DM, Katyare SS, Hegde MV, Bae H (2015) Significance of antioxidant potential of plants and its relevance to therapeutic applications. Int J Biol Sci 11(8):982
89. Pan SY, Zhou SF, Gao SH, Yu ZL, Zhang SF, Tang MK et al (2013) New perspectives on how to discover drugs from herbal medicines: CAM's outstanding contribution to modern therapeutics. Evid Based Complementary Altern Med 2013:627375
90. Pye CR, Bertin MJ, Lokey RS, Gerwick WH, Linington RG (2017) Retrospective analysis of natural products provides insights for future discovery trends. Proc Natl Acad Sci 114(22): 5601–5606
91. Mann J (2002) Natural products in cancer chemotherapy: past, present and future. Nat Rev Cancer 2(2):143–148
92. Kebaili Z, Hameurlaine S, Fellah O, Djermane M, Gherraf N (2019) Assessment of alkaloid content and antibacterial activity of Hyoscyamus albus and Hyoscyamus muticus collected in two different climatic regions in Algeria. J Biochem Technol 10(1):1
93. Hanif A, Juahir H, Lananan F, Kamarudin MKA, Adiana G, Azemin A, Yusra AI (2018) Spatial variation of Melaleuca cajuputi powell essential oils. J Fundam Appl Sci 10(1S):139–155

94. Fellah O, Hameurlaine S, Gherraf N, Zellagui A, Ali T, Abidi A et al (2018) Anti-proliferative activity of ethyl acetate extracts of grown at different climatic conditions in Algeria. Acta Sci Nat 5(2):23–31
95. Sharrock S, Jones M (2009) Conserving Europe's threatened plants: progress towards target 8 of the global strategy for plant conservation. BGCI, Richmond
96. Maridass M, De Britto AJ (2008) Origins of plant derived medicines. Ethnobot Leafl 2008(1):44

Assessment of the Phytochemical Constituents and Metabolites in Medicinal Plants and Herbal Remedies Used in the Treatment and Management of Reproductive Diseases: Polycystic Ovary Syndrome

17

Prasann Kumar, Subham Saurabh, and Khushbu Sharma

Contents

Abstract

The present study explores the phytochemical constituents and metabolites found in medicinal plants and herbal remedies traditionally used for treating and managing polycystic ovary syndrome (PCOS). This study's primary aims involve identifying, isolating, and characterizing bioactive compounds derived from natural sources. Subsequently, an assessment of their bioactivity will be conducted. Comparative analyses are undertaken to identify the most favorable approaches for managing polycystic ovary syndrome (PCOS), encompassing a thorough evaluation of safety and toxicity considerations. The study aims to establish standardized formulations and conduct clinical trials to assess these natural remedies' practical effectiveness. The primary objective is to develop evidence-based recommendations for

P. Kumar (✉) · S. Saurabh · K. Sharma
Department of Agronomy, School of Agriculture, Lovely Professional University, Phagwara, Punjab, India

S. C. Izah et al. (eds.), *Herbal Medicine Phytochemistry*, Reference Series in Phytochemistry, https://doi.org/10.1007/978-3-031-43199-9_20

responsibly utilizing medicinal plants and herbal remedies within the polycystic ovary syndrome (PCOS) treatment framework.

Keywords

Antioxidants · Herbal · Infertility · PCOS · Lumen · Reproductive system · No poverty · Zero hunger

Abbreviations

ANT	Anti-Mullerian hormone
DHEA	Dehydroepiandrosterone
ERs	Estrogen receptors
EV	Estradiol valerate
FSH	Follicle-stimulating hormone
IL	Interleukin
IVF	In vitro fertilization
LDLC	Low-density lipoprotein cholesterol
LH	Luteinizing hormone
LPA	lysophosphatidic acid
MDA	Malondialdehyde
NO	Nitric oxide
OC	Ovarian cancer
PCOS	Polycystic ovarian syndrome
TAC	Total antioxidant capacity
TC	Total cholesterol
TG	Triglyceride

1 Introduction

Traditional Asian medicine and plants have been used to treat various illnesses since ancient times. South Korea, Japan, and the tribal and agricultural populations of Bangladesh, China, India, Nepal, and Vietnam [22] for their profitable and user-friendly nature. Due to socioeconomic and geographic circumstances, ethnic people in many Asian nations rely heavily on medicinal plants for their primary healthcare in rural areas. Seventy to eighty percent of the population in developing nations depend on medicinal plants for primary healthcare, and ethnomedicine is becoming more popular in developed nations because it has almost no side effects. Ethnic people rely on the medicinal plants nearby to learn about the economic values and medicinal qualities of many plants and traditional medicine based on their needs, observations, and prior experiences [30]. Research on ethnomedicinal herbal plants led to the discovery of 75% of herbal drugs, and about 25% of modern medicines are derived from sources of widely used medicinal plants. Worldwide, the World Health Organization (WHO) has identified over 21,000 plant types for medicinal purposes.

Ayurvedic medicine, also known as Ayurveda, is a form of complementary and alternative treatment originating in India [6]. Ancient medicine used to treat issues with male reproduction: The herbal remedy made from *Centella asiatica, Hemidesmus indicus,* and *Hibiscus rosa* is used to cure spermaturia. *Helianthus fraternus, Dracaena terniflora, Sinensis*, and *Cuminum cyminum* are employed [14]. *Ocimum sanctum* and *Evolvulus alsinoides* are used to boost sperm concentration. To cure impotence, whole *Withania somnifera* is used (Table 1). Traditional medicine used to treat issues with feminine reproduction: *Celastrus paniculatus* and *Hibiscus rosa-sinensis* were used to cure leukorrhea. *Clerodendrum viscosum* and *Premna latifolia* have both been used to treat menorrhagia. Several plants were found to be effective in treating abortion, including *Ensete superbum*, *Mirabilis jalapa*, *Securinega leucopyrus*, *Celastrus paniculatus*, *Gardenia gummifera*, *Ziziphus oenoplia*, *Erythrina indica*, *Ixora coccinea*, and *Ziziphus rugosa.* To address dysmenorrhea, *Ocimum basilicum* and *Tabernaemontana divaricata* were used [16, 20, 21]. It was asserted that *Wrightia tinctoria* and *Diospyros montana* were usually helpful for treating different menstrual disorders and irregularities

Table 1 Plant species involved in disease prevention

Disease	*Plant species*	Part used	Use	Admin
1.	*Manihot esculenta*	Tuber, fresh	Vaginal infection, Vaginal discharge	Oral
2.	*Caesalpinia spinosa* (Molina) Kuntze	Seeds pods	Inflammation of ovaries	Oral
3.	*Ingo edulis* C. Martius	Huaba, Pacae, Guava, Pacai Flowers, fresh	Hair growth	Oral
4.	*Pelargonium roseum* Willd	Geranio	Flowers Uterus pain	Topical
5.	*Illicium verum* Hook.	Seeds, dried	Expel residues of feces in the stomach of newborn babies	Oral
6.	*Origanum vulgare* L.	Leave	Menstrual cramps	Oral
7.	*Salvia officinalis*	Whole plant	Control and regulate the menstrual cycle	Topical
8.	*Malva sylvestris* L.	Leave, stem	Pain	Topical
9.	*Mirabilis jalapa* L.	Root, fresh	Prostate, Pre-prostate cancer	Oral
10.	*Cynodon dactylon* L.	Stems, dried	Cysts of the ovary, Cysts of the uterus	Oral
11.	*Laccopetalum giganteum*	Leaves, fresh	Fertilization (Heat Ovaries)	Oral
12.	*Cinchona officinalis* L	Bark, dried	Fertility, Sexual potency	Oral
13.	*Cestrum strigilatum* R	Flowers, stem	Control and regulate the menstrual cycle	Oral
14.	*Typha angustifolia* L	Stems, dried	Prostate	Topical
15.	*Pilea microphylla* (L.)	Whole plant,	Prostate, Cysts	Oral
16.	*Lantana scabiosiflora*	Leaves, stem	Cold of the ovaries, Menstruation	Oral

(References: [14], [16], [20], [21])

(Table 1). The botanical names of the plants used to treat male and female reproductive disorders are listed alphabetically by disease. Additionally given are the components used in the drug preparation process, the dosage, and the recommended treatment time. Where accessible (Table 1), information about the other ingredients, if any, is also provided. The use of ethnomedicine as complementary and alternative medicine is growing in popularity despite the absence of scientific validation of the efficacy and safety of medicinal plants. It is also because it is cost-effective and has few side effects [34]. Pharmaceutical firms and scholars have recently concentrated on medicinal plants as a lucrative field. Polycystic ovary syndrome (PCOS), a prevalent and multifaceted endocrinological and metabolic disorder in women, is associated with infertility or subfertility. Initially referred to as the Stein-Leventhal syndrome, PCOS manifests as a severe condition characterized by ovarian enlargement and numerous small, undeveloped follicles [37]. It is a common female disorder with a prevalence rate of 2.2–26% globally and 11.96% among Indian teenagers [13]. Clinical manifestations of PCOS include menstrual irregularities or absence, abdominal obesity, acanthosis nigricans, and signs of androgen excess, such as acne or seborrhea, and insulin intolerance. Over the long term, PCOS increases the risk of endometrial cancer, type 2 diabetes, dyslipidemia, hypertension, and cardiovascular diseases. Notably, endometrial malignancy, cardiovascular disease, dyslipidemia, and type 2 diabetes mellitus are more prevalent in women with PCOS. The pathophysiology of PCOS primarily stems from abnormalities in the hypothalamic-pituitary axis, insulin secretion and action, and ovarian function. Obesity and insulin intolerance are closely linked to PCOS. Elevated insulin levels prompt the ovaries to produce testosterone, leading to anovulation, a significant cause of infertility. An altered luteinizing hormone/follicle-stimulating hormone (LH/FSH) relationship in patients with PCOS, with higher LH levels relative to FSH secretion, leads to increased androgen production by theca cells, resulting in irregular hormonal regulation, higher estrogen levels, erratic menstruation, and infertility [7]. Individuals with PCOS also exhibit an abnormal lipoprotein profile, characterized by elevated plasma triglyceride (TG) levels, slightly increased low-density lipoprotein cholesterol (LDL-C), and reduced high-density lipoprotein cholesterol (HDLC) levels. This lipid profile is often associated with heightened hepatic lipase activity, increased triglyceride content in circulation, insulin resistance, and elevated concentrations of smaller LDL particles. Current treatments for PCOS predominantly involve lifestyle modifications and pharmaceutical interventions. Lifestyle changes include dietary adjustments and supervised exercise to facilitate weight reduction. Pharmaceutical options like oral clomiphene citrate and metformin are employed for PCOS management; however, they may be accompanied by side effects such as nausea, vomiting, and gastrointestinal disturbances [5]. To mitigate these side effects, herbal medications are combined with pharmaceutical treatments to address PCOS symptoms effectively. This study aims to conduct a comprehensive analysis of the phytochemical constituents and bioactivity of medicinal plants and herbal remedies traditionally employed for managing reproductive diseases. The objectives encompass the identification and characterization of these compounds, the assessment of their bioactivity, the execution of comparative analyses to ascertain the most efficacious alternatives, the evaluation of safety and toxicity, the development of standardized formulations, the initiation of clinical trials, and the establishment of usage guidelines. The primary objective of this study is to provide evidence-based healthcare interventions that can improve the overall well-being of individuals impacted by reproductive health conditions.

2 Exploring Medicinal Plants and Herbal Remedies for Managing Reproductive Diseases with a Focus on Polycystic Ovary Syndrome (PCOS)

Polycystic ovarian syndrome (PCOS) is a complex metabolic disorder that significantly impacts the lives of numerous women during their reproductive years. This intricate condition poses challenges and potential implications for long-term health [44, 45]. Polycystic ovary syndrome (PCOS) can present a significant obstacle to a woman's desire to become a mother, as infertility frequently emerges as a prominent clinical manifestation of the condition. Nevertheless, the extensive ramifications of polycystic ovary syndrome (PCOS) transcend reproductive health, significantly impacting the psychosocial welfare of individuals affected by this condition. Polycystic ovary syndrome (PCOS) has attracted considerable attention due to its complex interconnections with other medical conditions. Polycystic ovary syndrome (PCOS) is often linked to diabetes, a metabolic disorder characterized by irregular blood glucose levels. Women diagnosed with PCOS face an increased risk of developing type 2 diabetes, which adds an extra health concern [37, 38].

Furthermore, the prevalence of obesity poses a significant concern in the lives of numerous women affected by polycystic ovary syndrome (PCOS). The presence of hormonal imbalances in this condition frequently leads to increased body weight and challenges in effectively managing weight [44, 45]. Consequently, this phenomenon exacerbates the susceptibility to diabetes and other metabolic complications. Polycystic ovary syndrome (PCOS) has been associated with hypertension, also known as high blood pressure. The co-occurrence of this condition seems to be prevalent, posing difficulties in the holistic management of individuals affected by this syndrome. Moreover, it is worth noting that polycystic ovary syndrome (PCOS) frequently leads to the grave outcome of heart disease, which is recognized as the primary cause of mortality in women.

Polycystic ovary syndrome (PCOS) is characterized by dysregulation of hormones, insulin resistance, and excessive body weight, increasing the likelihood of cardiovascular complications [26]. Consequently, women with PCOS should exercise prudence and vigilance in monitoring their cardiovascular health, emphasizing proactive interventions and adjustments to their daily habits. The range of lipid abnormalities presents an increasingly diverse range of health hazards linked to PCOS. Dyslipidemia, a medical condition characterized by aberrant levels of lipids in the bloodstream, is a common comorbidity in individuals with PCOS [28–30]. Elevated levels of low-density lipoprotein (LDL) cholesterol and triglycerides, in conjunction with diminished levels of high-density lipoprotein (HDL) cholesterol, have been implicated in the pathogenesis of atherosclerosis and the progression of cardiovascular disease. The increased prevalence of lipid abnormalities in individuals with PCOS adds a layer of complexity to the already intricate health profile observed in women affected by this condition. Autoimmune thyroiditis, characterized by chronic thyroid gland inflammation, represents an additional element within the intricate framework of PCOS [31, 32–34]. PCOS demonstrates an increased susceptibility to thyroid disorders, specifically autoimmune thyroiditis, a significant concern. This phenomenon adds to the array of difficulties that need to be tackled in the holistic administration of the syndrome. Maintaining optimal thyroid function is a crucial priority in the pursuit of overall health and well-being. The correlation between PCOS and malignancies exacerbates the overall impact of the condition. While PCOS has no direct causal relationship with cancer, existing evidence indicates a correlation between PCOS and an elevated long-term susceptibility to certain malignancies. Breast cancer, the most common malignancy among women, and endometrial cancer in the uterine lining are two noteworthy and worrisome health hazards [35–38]. The relationship between hormonal imbalances and PCOS is characterized by increased estrogen levels, which appears to contribute to an increased risk of developing cancer. Cancer is a poignant reminder of the importance of timely detection, cancer screenings, and preventive measures for women managing PCOS. The precise etiology of PCOS remains uncertain, highlighting the intricate interaction between genetic and environmental elements. The phenomenon is hypothesized to arise from a complex combination of innate genetic characteristics and external environmental influences. A genetic component in PCOS suggests a familial predisposition, as indicated by the high prevalence of the syndrome among close relatives. However, the activation of latent potential in PCOS is significantly influenced by environmental factors, ultimately resulting in its manifestation. The complex interplay between genetic and environmental factors gives rise to a syndrome that is a multifaceted puzzle that is still not fully comprehended. Extensive research has been undertaken to tackle PCOS's intricate challenges and comprehend, manage, and mitigate its effects. Various management strategies have been developed to improve the quality of life for individuals who self-identify as women and experience the effects of PCOS [39–41]. These strategies encompass various approaches, including adjustments to an individual's lifestyle, pharmaceutical interventions, and alternative therapies. Using medicinal plants has garnered considerable attention as a potential therapeutic approach, offering a holistic and nature-based strategy for managing PCOS. Throughout history, medicinal herbs have played a crucial role in traditional medicine systems, highly esteemed for their therapeutic properties and ability to effectively treat diverse health conditions. Botanical remedies have been employed within PCOS to alleviate the specific symptoms associated with this condition. PCOS has two significant symptoms in women: oligo/amenorrhea and hyperandrogenism. These symptoms have considerable implications for reproductive health and affect individuals' overall physical and mental well-being. Exploring medicinal plants as a potentially effective alternative therapy for PCOS represents a promising avenue for scholarly inquiry. Plants often demonstrate the presence of bioactive compounds that possess therapeutic properties, capable of addressing the hormonal imbalances and metabolic disruptions inherent to PCOS. By incorporating these natural remedies, which have strong roots in historical healing practices, individuals can adopt an integrative methodology to manage the syndrome proficiently. While it is not within the scope of this paragraph to conduct an extensive analysis of individual medicinal herbs and their corresponding properties, it is important to highlight the prevailing inclination toward natural and complementary treatments in the management of PCOS. By embracing these traditional botanical treatments, women diagnosed with PCOS can access various resources to rebalance hormone levels, regulate menstrual cycles, and alleviate the distressing symptoms commonly linked to this condition [42, 43]. To summarize, PCOS is a multifaceted metabolic disorder that substantially impacts women's health in the reproductive-age bracket. The wide-ranging health challenges associated with this phenomenon include diabetes, obesity, hypertension, heart disease, lipid abnormalities, and autoimmune thyroiditis. The ripple effect of this phenomenon extends to these various health conditions. The existence of malignant neoplasms, such as breast and endometrial carcinomas, underscores the severity of this syndrome [44, 45].

3 Unraveling the Enigmatic Nature of PCOS: Traditional Herbal Remedies as a Ray of Hope for Management and Well-Being

The enigmatic etiology of polycystic ovary syndrome (PCOS), intricately influenced by genetic and environmental factors, is a poignant testament to the unresolved enigmas within medical science [14]. Nevertheless, amidst these challenges, there is a glimmer of hope with the emergence of research on managing polycystic ovary syndrome (PCOS) and using medicinal herbs [16]. These traditional remedies, rooted in longstanding customs, offer an alternative approach to managing the complex array of symptoms associated with polycystic ovary syndrome (PCOS), thereby augmenting the overall well-being of individuals affected by this condition [20, 21]. As we navigate the dynamic terrain of PCOS research and treatment, using medicinal plants is a testament to the enduring nature and flexibility of healthcare methodologies across diverse societies and historical periods (Table 1).

4 Compounds from Plants for PCOS

Ecklonia cava: In addition, the administration of *Ecklonia cava* extract resulted in the normalization of various hormone levels, such as testosterone, estrogen, luteinizing hormone (LH), and follicle-stimulating hormone. Moreover, it demonstrated the ability to alleviate anti-Mullerian hormone (AMH) levels. The results of this study indicate that the extract in question has the potential to modulate multiple factors related to the development of ovarian follicles. The histological analysis provided additional evidence to support the assertion that rats administered with

Ecklonia cava extract demonstrated a notable decrease in symptoms associated with polycystic ovary syndrome (PCOS), resembling the characteristics observed in the ovaries of healthy individuals. This well-established natural remedy mitigates the effects of allergies and inflammation [21].

***Glycyrrhiza glabra*:** Glycyrrhiza glabra, commonly utilized in Korean herbal formulas, is recognized for its efficacy in addressing various health conditions such as inflammation, immune system modulation, hepatic failure, spasms, and metabolic disorders, specifically in women. A study conducted in 2018 by Lee et al. examined the impact of an ethanol extract of *Glycyrrhiza glabra* on female Sprague Dawley rat models that displayed symptoms similar to polycystic ovary syndrome (PCOS) induced by letrozole. The experimental group that received a combination of licorice and letrozole demonstrated a marked increase in follicle-stimulating hormone (FSH) recovery and a significantly reduced ratio of luteinizing hormone (LH) to FSH. According to Lee et al. [26], licorice extract demonstrated significant efficacy in mitigating polycystic ovary syndrome (PCOS) symptoms. These symptoms encompassed the thickening of the theca layer, the thinning of the granulosa layer of antral follicles, a decrease in the quantity of antral follicles, and the development of follicular cysts.

***Aegle marmelos*:** The present study investigated the potential antiandrogenic effects of a hydroalcoholic preparation obtained from *Aegle marmelos* in the management of polycystic ovary syndrome (PCOS) induced by letrozole in a female Wistar rat model. The present study investigated the impact of the hydroalcoholic extract derived from *Aegle marmelos* on ovarian histopathology and a range of biochemical and physical parameters. The research findings indicated significant elevations in blood luteinizing hormone (LH) levels and reductions in serum follicle-stimulating hormone (FSH) levels. An increased occurrence of corpus luteum, developing follicles, and oocytes enveloped by granulose cells [17].

***Bougainvillea spectabilis*:** In a 2018 study examining a rat model of polycystic ovary syndrome (PCOS) induced by estradiol valerate (EV), Ebrahim N.A. and colleagues conducted an assessment of the protective potential of *Bougainvillea spectabilis* leaves (BSL) extract on ovarian folliculogenesis. The research revealed that the group treated with estradiol valerate and *Bougainvillea spectabilis* extract exhibited significantly reduced blood levels of LH, estrogen, and glucose while also displaying elevated serum levels of FSH and antioxidants. Furthermore, this group's ovarian cortex featured corpora lutea, preantral, and antral follicles [18].

***Matricaria chamomilla*:** Various studies have explored the potential of natural remedies in alleviating symptoms and complications associated with polycystic ovary syndrome (PCOS), a complex metabolic disorder affecting women. For instance, research by Heidary et al. in 2018 examined the effects of chamomile on lipid and hormonal parameters in reproductive-age PCOS women. The study found oral chamomile capsule administration significantly decreased overall testosterone levels in PCOS women. However, it did not significantly affect dehydroepiandrosterone sulfate levels, LH/FSH ratio, or lipid factors [20].

***Cinnamomum zeylanicum*:** In 2018, Maryam Rafraf and her team conducted a study investigating how cinnamon supplements could affect women with PCOS.

They examined various factors, including blood total antioxidant capacity (TAC), malondialdehyde (MDA), and serum lipids. The women who took cinnamon supplements experienced significant improvements in their serum levels of total cholesterol (TC), low-density cholesterol, and high-density cholesterol. Additionally, the antioxidant capacity of their blood increased, indicating better protection against oxidative stress. They also noticed decreased malondialdehyde levels, a positive sign for their overall health.

Galega officinalis: The potential protective effects of *Galega officinalis* in a rodent model of polycystic ovary syndrome (PCOS) induced by estradiol valerate. The study unveiled several encouraging results. The researchers noted noteworthy decreases in fasting blood sugar, insulin, testosterone, luteinizing hormone (LH), and follicle-stimulating hormone (FSH) concentrations, alongside substantial elevations in serum estrogen and aromatase levels. Additionally, the experimental group that received *Galega officinalis* demonstrated a significant augmentation in both preantral and antral follicles, accompanied by a notable reduction in cystic follicles. The results of this study indicate that *Galega officinalis* possesses potential advantages attributed to its antioxidant and insulin-related characteristics, as evidenced in a historical context tracing back to 1190 A.D.

Moringa oleifera: This study investigates the impact of different doses of *Moringa oleifera* leaf extract on a polycystic ovary syndrome (PCOS) model that exhibits insulin resistance. The researchers made noteworthy observations during their investigation. Initially, a conspicuous decline in the accumulation of body weight was observed – a significant augmentation in folliculogenesis in the polycystic ovary syndrome (PCOS) ovary model. Additionally, androgen levels were declining in the cohort that underwent treatment with *Moringa oleifera*.

Nigella sativa: The study encompassed a sample size of ten female participants clinically diagnosed with polycystic ovary syndrome (PCOS) and oligomenorrhea. Following the administration of *Nigella sativa*, noteworthy enhancements were observed in multiple health parameters, encompassing a considerable decrease in serum cholesterol levels, triglycerides, fasting blood sugar, insulin, aspartate aminotransferase (AST), and LH levels.

Vitis: This study investigates the impact of grape seed extract (GSE) on lipid and inflammatory markers in Wistar rats with polycystic ovary syndrome (PCOS) induced by estradiol valerate. The study primarily examined triglyceride (TG), total cholesterol (TC), high-density lipoproteins cholesterol (HDL-C), low-density lipoproteins cholesterol (LDL-C), and interleukin 6 (IL-6). Significantly, the cohort that was administered grape seed extract demonstrated noteworthy decreases in LDL-C, TC, and IL-6 concentrations. Administering specific doses of GSE had a beneficial effect on rats with polycystic ovary syndrome (PCOS). This effect was observed through the mitigation of dyslipidemia and reduction of inflammation, improving the overall systemic symptoms of the rats.

Bambusa bambos: In recent studies, scholars have investigated a range of natural interventions to mitigate the difficulties associated with polycystic ovary syndrome (PCOS). This multifaceted metabolic disorder primarily affects women. The investigations above have produced encouraging results in mitigating particular symptoms and complications linked to polycystic ovary syndrome (PCOS). In a study conducted in 2018, the effects of processed *Nigella sativa* on menstrual irregularities associated with polycystic ovary syndrome (PCOS) were investigated. The findings of this study demonstrated noteworthy enhancements in cholesterol, triglycerides, blood sugar, insulin, AST, and LH levels after the treatment. The potential of grape seed extract (GSE) was examined in order to improve lipid profiles and mitigate inflammation in rats with polycystic ovary syndrome (PCOS). The findings of this investigation revealed favorable outcomes. Furthermore, the study conducted by examined the effects of bamboo seed oil on metabolic symptoms commonly associated with polycystic ovary syndrome (PCOS). This research revealed significantly reduced glucose, cholesterol, and triglyceride levels. The findings above highlight the potential of utilizing natural remedies to alleviate complications associated with polycystic ovary syndrome (PCOS) and improve the overall quality of life for individuals affected by this condition.

***Commiphora wightii*:** A study conducted in 2016 examined the effects of *Commiphora wightii* on hyperandrogenism in a polycystic ovary syndrome (PCOS) model. The findings of this study demonstrated a decrease in morphological abnormalities of ovarian follicles and a restoration of hormone levels to normal. Furthermore, a study conducted in 2015 examined the efficacy of hazelnut oil as a treatment for polycystic ovary syndrome (PCOS) in rats. The findings of this study indicated favorable results, including weight loss, restoration of hormonal equilibrium, and an augmentation in uterine weight. Moreover, a study conducted in 2016 revealed the potential of curcumin in regulating testosterone levels, mitigating insulin resistance, and preventing diabetic complications in rats with polycystic ovary syndrome (PCOS). These studies highlight the potential efficacy of natural remedies, including *Commiphora wightii*, hazelnut oil, and curcumin, in mitigating hormonal imbalances and metabolic disturbances commonly observed in individuals with polycystic ovary syndrome (PCOS) ([16, 24])

***Corylus avellana*:** The study's results indicated a significant reduction in body weight, restoring LH and FSH levels to normal, and increasing uterine weight. Furthermore, the researchers observed a substantial elevation in estrogen and progesterone levels and a decrease in testosterone. These findings highlighted the antioxidant properties of *Corylus avellana* oil and its efficacy in treating PCOS by regulating gonadotropins, steroids, and serum lipid parameters [16].

***Curcumin*:** Curcumin effectively restored serum testosterone levels to normal and successfully elevated progesterone levels, addressing hormonal imbalances associated with PCOS. Notably, curcumin can prevent insulin resistance and diabetic complications when administered orally, as evidenced by its capacity to halt the rise in HbA1c levels. Furthermore, the research revealed that curcumin was successful in eliminating cysts. These findings underscore the potential of curcumin as a promising medication for the treatment of both pathological and clinical abnormalities in PCOS.

Palm pollen: Palm pollen extract reveals reduced LH, estrogen levels, and cystic follicles, coupled with increased FSH, progesterone, and corpus luteum [23]. Similarly, research by Shetty et al. in 2015 found that *Vitex negundo* effectively lowered excessive androgen levels in rats with PCOS, accompanied by decreased glucose and testosterone levels. Additionally, in 2014, Sadrefozalayi et al. explored the renoprotective effects of the aqueous extract of *Foeniculum vulgare* in experimental PCOS rats, observing improvements in serum urea levels and kidney function. These findings underscore the potential of natural remedies, such as palm pollen, *Vitex negundo*, and *Foeniculum vulgare*, in alleviating symptoms and complications associated with PCOS.

***Vitex negundo*:** The effects of *Vitex negundo* on female Sprague Dawley rats that had experienced hyperandrogenism due to PCOS. Through daily administration of the hydroalcoholic extract of *Vitex negundo*, they observed a significant reduction in elevated androgen levels. Additionally, the nirgundi extract led to a substantial decrease in serum glucose and testosterone levels. Notably, the group that received the nirgundi extract displayed the lowest incidence of follicular cysts and lesions.

***Foeniculum vulgare*:** The research revealed that PCOS-afflicted rats exhibited lower serum urea levels. Notably, the administration of *Foeniculum vulgare* led to significant improvements in Bowman's space and acute tubular necrosis, restoring these parameters to a more normal state.

***Allium cepa*:** The influence of an ethanolic extract derived from *Allium cepa* seeds on apoptosis modulation in rats with experimentally induced PCOS. The findings pointed to elevated levels of antioxidant capacity, a reduction in cyst formation, and a decrease in apoptotic granulosa cells, signifying the potential advantages of *Allium cepa* in managing PCOS. In addition, in 2010, Laxmipriya Nampoothiri and her colleagues investigated the efficacy of an Aloe vera gel formulation in a PCOS rat model induced by letrozole. The administration of Aloe vera gel resulted in enhanced ovarian health, the restoration of hormonal balance, and the maintenance of estrus cyclicity, underscoring its protective role against a PCOS phenotype [19].

***Aloe barbadensis Mill*:** Aloe vera, scientifically classified as Liliaceae, was the focal point of research conducted by Laxmipriya Nampoothiri et al. in 2010. This study aimed to assess the effectiveness of a specific formulation of Aloe vera gel in a rat model of polycystic ovary syndrome (PCOS) induced by letrozole. Significant results were noted after applying Aloe vera gel, including a lack of weight increase and a decrease in ovarian atretic cysts. In addition, the oral administration of a 1 ml daily dosage of Aloe vera gel formulation for 45 days was crucial in restoring the subjects' estrus cyclicity, improving glucose sensitivity, and altering important steroidogenic activity. As a result, it served as a protective factor against the PCOS phenotype [33].

5 Exploring Natural Remedies for Enhancing Reproductive Health: A Focus on PCOS and Fertility

Polycystic ovary syndrome (PCOS) is a commonly encountered, heterogeneous, endocrinological, and metabolic disorder affecting women of reproductive age, often resulting in infertility or subfertility [14, 26]. Due to their diverse phytoconstituents, medicinal plants have been explored for their potential to manage various ailments, including PCOS [30]. Various herbal extracts following the onset of PCOS have been associated with reduced levels of blood testosterone and LH, alongside elevated blood progesterone and FSH levels [16]. Notably, these treatments have been observed to induce the development of various follicles at different growth phases, encompassing primary, antral, preantral, large oocytes, and corpora lutea in the treatment groups. Research suggests that many herbal extracts exhibit efficacy in treating PCOS, normalizing blood and sex hormone levels, resolving cysts, reducing insulin levels, and ameliorating other PCOS symptoms [21, 22]. Furthermore, some studies have emphasized the safety of these PCOS-targeted plants, attributing their favorable effects to the presence of phytoestrogens [23]. These compounds, characterized as weak estrogen antagonists, tend to exhibit more pronounced estrogenic effects when the body's estrogen levels are low in PCOS patients, rendering them safe for widespread utilization [24]. The medicinal plants discussed in this chapter can potentially enhance fertility through their interactions with the hypothalamic-pituitary-gonadal (HPG) axis and modulation of estrogen receptors (ERs). Additionally, these plants can provide an environment conducive to regulating ovulation, implantation, uterine embryo tolerance, and fetal development. They also show promise in averting reproductive transmitted bacterial, viral, and fungal infections, mitigating inflammatory reactions, hypersensitivity, and autoimmune disorders [2].

Estrogen receptors (ERs) mediate physiological responses to estrogen within specific tissues. While ERs are present in ovarian tissues, bone, and blood vessels, ERs are expressed in breast and uterine tissues. These ERs act as ligand-activated nuclear transcription factors, falling under the category of nuclear hormone receptors. Several plants contain phytoestrogens, which exhibit estrogenic effects in mammals. Notable examples of these substances with a strong affinity for estrogen receptors include formononetin, genistein, daidzein, and biochanin A. It is important to note that the concentration required for induction by these derivatives is significantly higher than anticipated despite their stronger binding to ER compared to ER [8]. These phytoestrogens contribute to the reduction in bone loss by lowering the levels of uric acid deoxypyridinoline. However, they do not impact osteogenic markers such as alkaline phosphatase and osteocalcin. Additionally, their vasoconstrictive properties make them beneficial in treating menopausal symptoms, such as reducing hot flashes, vaginismus, and dyspareunia [12]. Furthermore, bioactive compounds present in medicinal plants, including flavonoids and isoflavones, exhibit the potential to improve lipid profiles by reducing triglycerides and LDL cholesterol while increasing HDL cholesterol, which is crucial for reproductive health. Moreover, numerous studies have demonstrated that plants rich in polyphenols can impede the development of breast tumors by blocking pathways like IGF-1 (insulin-like growth factor 1)/P13K/Akt and ERK1/2 MAPK-Bax. Research into the effects of plant polyphenols, such as isoflavones, on the skin has shown that these compounds enhance the synthesis of hydroxyproline, a marker of collagen and elastic fiber synthesis, leading to vasodilation vascular endothelial growth factor (VEGF) in cutaneous blood vessels, regulation of sweat and sebaceous gland secretion, and control of collagen synthesis in the growth of hair layers, particularly in the context of menopause and infertility-related symptoms.

***Punica granatum* (Pomegranate):** Pomegranates, identified by their scientific name *Punica granatum*, have garnered significant attention and are cultivated across various geographical areas, including Asia, the Middle East, and the Mediterranean [31]. Pomegranates possess abundant vital constituents, encompassing vitamin C, water, and flavonoids, namely anthocyanins, punicalagin, ellagic acid, and gallic acid. The pomegranate seeds contain phytoestrogens of significance, such as genistein, daidzein, coumestrol, and glutamic and aspartic acids [9]. Significantly, the utilization of pomegranate extract, containing these valuable phytoestrogens, has demonstrated the ability to regulate and alleviate symptoms commonly associated with polycystic ovary syndrome (PCOS) in animal studies involving rats affected by PCOS. The botanical extract in question plays a role in the augmentation of uterine wall thickness and the stimulation of mucus secretion by facilitating increased uterine blood flow through vasodilation. The increased production of mucosal secretions, facilitated by anti-inflammatory mechanisms, has been linked to a higher successful implantation rate. Mohammadzadeh et al. [32] conducted a rigorous clinical trial employing a triple-blind, randomized, controlled design, which involved a sample of 110 women who were in good health. The study yielded a noteworthy finding. The researchers discovered that the presence of calcium and tannins in pomegranate peel was associated with an enhancement in sexual satisfaction among women and a simultaneous reduction in the symptoms of inflammation and infection in their reproductive canal. Furthermore, a comprehensive examination of diverse human cell lines, such as ovarian (SKOV3), cervical (SiHa, HeLa), endometrial (HEC-1A), breast (MCF-7, MDA MB-231), and abnormal breast fibroblast (MCF-10A) cells, has unveiled the noteworthy characteristics of pomegranate extract as a selective estrogen receptor modulator (SERM). The observed effect of this activity can be attributed to its capacity to interact with estrogen receptors (ERs), suppressing cell line proliferation in laboratory and animal models, specifically in mice undergoing ovariectomy. Additionally, the study conducted a randomized controlled, triple-blind parallel trial to investigate the effects of pomegranate fruit extract on testosterone levels in the serum of 23 women diagnosed with polycystic ovary syndrome (PCOS). The findings revealed that the administration of pomegranate fruit extract reduced testosterone levels, improving the participants' lipid profile. Furthermore, a research study conducted on rats with polycystic ovary syndrome (PCOS) found that administering pomegranate fruit extract increased serum estrogen levels and a simultaneous improvement in related symptoms. The study reported significant findings after 81 days [22].

***Matricaria chamomilla* (Chamomile):** *Matricaria chamomilla*, commonly known as chamomile, is a member of the chicory family and possesses a range of beneficial compounds, including flavonoids and antioxidants such as apigenin, choline, camelina, farnesene, matricin, and gallic acid. Numerous studies have demonstrated that chamomile extracts can substantially impact various physiological processes. The investigation of the growth and maturation of isolated mouse ovarian follicles within a three-dimensional culture system revealed the significant contribution of chamomile extracts. The extracts above exhibited the ability to enhance progesterone, 17-estradiol, and dehydroepiandrosterone levels in the culture medium while mitigating oxidative stress and reducing indicators of follicular diameter and antrum formation. Furthermore, the influence of chamomile resulted in an extended duration of oocyte survival. Applying chamomile extract has also exhibited the capacity to augment estrogen-dependent sexual characteristics in mice undergoing gonadectomy. The phenomenon above can be observed in facilitating modifications such as enhanced hair growth, body temperature fluctuations, and managing the menstrual cycle. Conducted a double-blinded clinical trial with a sample size of 80 postpartum pregnant women, all of whom had a gestational age of 40 weeks or more. The researchers observed that the administration of chamomile extract-containing capsules for 1 week resulted in a facilitation of labor onset. Furthermore, it is worth noting that this particular intervention resulted in a decreased duration of labor compared to the control group. The effects of chamomile plant extract on lactation have been elucidated through human research. According to it was observed that the administration of chamomile plant extract exhibited the potential to induce lactogenesis in women who were lactating. The observed mechanism can be attributed to the interaction between chamomile and dopamine receptors, which promotes galacta production. Furthermore, conducted a pilot randomized trial with a sample size of 56 women diagnosed with idiopathic hyperprolactinemia. The study findings suggest that chamomile might potentially play a role in regulating prolactin secretion in women. The findings of this study indicate that female participants who received a dosage of 5 ml of chamomile syrup twice daily experienced a reduction in prolactin levels after 4 weeks, as opposed to the group that received a placebo. In addition, the aroma of chamomile has been linked to increased intensity of uterine contractions during labor, as demonstrated in a controlled trial that included a sample of 130 female participants. Chamomile extract has the potential to provide supplementary advantages in comparison to chemical medications, such as mefenamic acid and nonsteroidal anti-inflammatory drugs (NSAIDs), in the prevention of postpartum pain among women. The inhibitory effect on COX-2 has been attributed to this phenomenon.

Vitex agnus-castus (Verbenaceae): Historically, *Vitex agnus-castus*, a plant belonging to the Verbenaceae family, has been employed for the management of acne and various menstrual disorders associated with corpus luteum insufficiency, including spasmodic dysmenorrhea, premenstrual symptoms, certain menopausal conditions, and insufficient lactation. The analysis of Verbenaceae plant extracts using liquid chromatography-electrospray ionization/mass spectrometry (LC-ESI/MS) confirmed the presence of various compounds. These compounds include orientin, catechin, rutin, rosmarinic acid glycoside, quercetagetin trimethyl ether, biochanin A, genistein, syringetin C glycoside, agnuside, kaempferol-7-*O*-glucuronide, luteolin-7-*O*-glucoside, homorientin, and isovitexin. Most of these compounds are isoflavones and flavonoids, which benefit women's reproductive health due to their strong binding affinity for estrogen ERs. The plant's flavones have been found to exert an inhibitory effect on the release of prolactin and FSH hormones by modulating the hypothalamic-pituitary-gonadal (HPG) axis. Additionally, its flavonoids have been shown to enhance the release of nitric oxide (NO) and cyclic guanosine monophosphate (cGMP) from vascular endothelial cells, thereby promoting increased blood flow to the endometrium ([4]. Approximately one in six couples is affected by the condition of infertility. When a couple cannot achieve pregnancy after 1 year, they are clinically diagnosed with infertility. The term "female infertility" describes situations in which the female partner is identified as the primary cause of infertility. Approximately 50% of infertility cases can be attributed to factors associated with female infertility, with approximately one-third of all cases solely attributed to female infertility. The term "unexplained sterility" is commonly employed in the context of reproductive processes. The term "unexplained infertility" describes cases where no identifiable cause for the inability to conceive has been determined. In general, the outcomes become apparent following the completion of all available tests by a couple. According to Beal [10], approximately 25% of couples experiencing infertility are diagnosed with unexplained infertility. The topic of discussion pertains to the issue of female infertility. The topic under discussion pertains to signs. Alterations in the menstrual cycle and ovulatory patterns among females can potentially indicate a medical condition associated with impaired fertility. One of the signs that can be observed is the presence of unusual menstrual cycles. Variations in menstrual patterns: bleeding that deviates from the typical flow in increased or decreased intensity. The duration between consecutive menstrual cycles fluctuates every month. The individual experiences recurring episodes of pain. The occurrence of cramps, pelvic discomfort, and back pain is possible. In certain instances, hormonal imbalances may contribute to infertility among women. In this particular scenario, it is plausible that supplementary indicators may be present. In addition to alterations in sexual drive and longing, according to Smith et al. [44], individuals may experience various skin alterations such as heightened acne, weight gain, hair loss or thinning, and the emergence of dark hair on the chin, torso, and lips. The diagnosis of female infertility refers to identifying and assessing factors contributing to a woman's inability to conceive or carry a pregnancy to term. The female individual and her significant other must undergo diagnostic examinations as a component of the inquiry into potential infertility. The woman has the potential to undergo the subsequent examinations: comprehensive physical examination, encompassing a thorough review of the patient's medical history. Laparoscopy, a minimally invasive surgical procedure involving inserting a specialized instrument through a small incision in the abdominal region, is employed to examine the reproductive organs. Additionally, blood tests are conducted to assess the levels of hormones associated with ovulation, while ultrasound scans are employed to detect the presence of fibroids. According to Lloyd and Hornsby [33], infertility can be attributed to various factors, including an underlying medical condition that adversely affects the fallopian tubes, inhibits ovulation, or disrupts hormonal balance. The medical conditions include endometriosis, polycystic ovarian syndrome, premature ovarian failure, pelvic inflammatory disease, uterine tumors, and polycystic ovary syndrome. Addressing any preexisting medical conditions that may contribute to reproductive challenges is imperative before initiating infertility treatment. Various treatment options can be considered if fertility restoration does not occur. The topic of interest pertains to the relationship between plants and the process of conception. Herbal medicines have historically been employed in the treatment of fertility issues. Historical records indicate that using herbal remedies to enhance male and female fertility dates back to approximately 200 A.D. The herbal fertility treatments encompass distinct botanical species and their corresponding extracts, which are believed to have advantageous effects on the reproductive system, hormonal equilibrium, and sexual drive. Both individuals seeking to enhance their chances of conception and couples experiencing fertility challenges utilize these medications. According to Agha-Hosseini et al. [1], herbal reproductive treatments have demonstrated potential in addressing hormonal imbalances, irregular menstrual cycles, erectile dysfunction, and impaired sperm motility. Throughout human history, individuals across various cultures and regions have explored and used herbal medicine for countless centuries. Infertility treatment is becoming more prevalent in Western societies among males and females. Nevertheless, the American Society for Reproductive Medicine asserts that there is limited scientific substantiation for the notion that herbal remedies can enhance fertility. Nevertheless, it is possible to provide credible evidence by utilizing evidence-based herbal medicine, specifically through clinical trials [11]. The emotional impact of infertility is well understood by women who undergo this condition. The conditions above encompass anxiety, isolation, prolonged depressive states, and difficulties concentrating on routine activities. Many infertility women may initiate their treatment by considering in vitro fertilization (IVF) techniques. Nevertheless, implementing these techniques incurs significant costs, rendering them financially inaccessible for certain women. Various natural and herbal remedies have been identified as potential interventions that may contribute to addressing the root causes of infertility and enhancing the likelihood of successful conception. According to Akhondzadeh et al. [3], incorporating a nutritious diet and engaging in consistent physical activity are lifestyle factors that can yield positive outcomes.

Ashwagandha: Ashwagandha, scientifically referred to as *Withania somnifera*, is an herbal remedy recognized as Indian ginseng. It has been observed to offer notable advantages for females encountering difficulties conceiving. This botanical remedy effectively promotes optimal functioning of the reproductive organs and helps maintain hormonal balance. Pomegranate has been found to possess properties that contribute to the toning of the endometrium in individuals who experience recurrent miscarriages. Pomegranate has been found to enhance female fecundity. This intervention facilitates the enhancement of uterine blood circulation and the thickening of the uterine lining, thereby reducing the likelihood of experiencing a miscarriage. Furthermore, it supports the optimal development of the fetus [15].

Cinnamon: Cinnamon has been found to promote healthy ovarian function, thereby potentially mitigating sterility. This therapeutic effect is associated with the potential to derive five distinct benefits from cinnamon consumption. Female infertility is a prominent factor contributing to infertility in individuals diagnosed with polycystic ovary syndrome (PCOS). According to a study published in the *Fertility and Sterility* journal in 2007, including cinnamon in the diet has been shown to potentially improve the regularity of menstrual cycles in individuals with polycystic ovary syndrome (PCOS). Additionally, it encompasses various factors that can hinder a woman's reproductive capacity, such as endometriosis, uterine fibroids, and amenorrhea, characterized by the absence of menstrual periods. Furthermore, it has been suggested that cinnamon possesses potential benefits in preventing yeast infections.

Chasteberry: Chasteberry, scientifically referred to as Vitex has been identified as an effective remedy for addressing infertility caused by hormonal imbalances within the body. The herb's ability to enhance ovulation is attributed to its elevated prolactin levels. Additionally, it combats polycystic ovary syndrome (PCOS).

Dates: The consumption of dates has been associated with potential enhancements in fertility due to the presence of various beneficial nutrients. According to these substances possess a rich concentration of essential nutrients such as vitamins A, E, D, and B, as well as iron and other minerals that are crucial for facilitating a woman's ability to conceive and successfully carry a pregnancy to full term. The maca root is a highly effective botanical remedy that has demonstrated efficacy in addressing both male and female infertility. This botanical specimen facilitates the synthesis of hormonally balanced substances and confers particular benefits to females afflicted with hypothyroidism due to its ability to enhance thyroid functionality. Moreover, it serves as a noteworthy reservoir of diverse nutrients that enhance fertility.

6 Ovarian Cancer and Its Prevention Through Herbal Medicines

Ovarian cancer (OC) is a prevalent form of cancer that affects a significant number of women annually, accounting for approximately 244,000 cases worldwide. This represents approximately 2% of all reported cancer cases globally. Nevertheless, obsessive-compulsive disorder (OC) exhibits notable variations across different geographical areas. In 2017, Fiji exhibited the highest prevalence of OC, with an estimated rate of 15 cases per 100,000 females. In contrast, China and specific regions of Africa reported a considerably lower incidence, with only four cases per 100,000 females. The incidence rate of ovarian cancer in women is approximately 1 in 70. Regrettably, when OC is concomitant with other gynecological malignancies, it emerges as the primary contributor to cancer-related mortality, leading to approximately 150,000 deaths globally annually. The elevated mortality rate can be ascribed to delayed detection, as symptoms frequently remain undetectable during the early stages and only become apparent during stages III and IV. Various therapeutic modalities have been developed to impede the dissemination of cancer cells, encompassing chemotherapy, laser therapy, radiation, gene therapy, hyperthermia, and surgical interventions. Medical interventions encompass a multifaceted approach, incorporating radiation therapy, chemical treatments, and surgical procedures. Although these methods possess certain benefits, it is essential to acknowledge their inherent limitations and drawbacks. Therefore, the scientific community is currently working to develop a more efficient treatment to tackle the obstacles presented by ovarian cancer.

Previous generations have used medicinal plants to address various human afflictions, including cancer treatment. Numerous botanical species have demonstrated various pharmacological attributes, encompassing antioxidative, antimicrobial, anticancer, and antidiabetic properties. One notable advantage associated with the utilization of plants in the field of medicine lies in their extensive collection of diverse phytochemicals, which possess the ability to specifically address a wide range of medical conditions. Searching for novel pharmaceuticals to combat cancer is contingent mainly upon identifying biologically significant phytochemicals [41]. Using biomolecules derived from plants has become a well-regarded and promising approach with considerable potential. Significantly, there has been a notable increase in the incorporation of pharmacologically beneficial botanicals in the process of pharmaceutical advancement during recent decades. Considerable research efforts have been undertaken over the past two decades to investigate the potential of natural products in enhancing the efficacy of cancer treatment, with a particular focus on numerous countries.

7 Other Therapeutic Plants' Ability to Prevent Ovarian Cancer

Africa is home to various indigenous medicinal plants traditionally employed for therapeutic purposes. Some of these plants include *Aframomum arundinaceum*, *Aframomum alboviolaceum*, *Aframomum kayserianum*, *Aframomum polyanthum*, *Echinops giganteus*, *Xylopia aethiopica*, *Piper capense*, and *Imperata cylindrica*. *Additionally*, *Gladiolus quartinianus*, *Vepris soyauxii [25]*, *Polygonum limbatum*, *Polyscias fulva*, *Beilschmiedia acuta*, *Crinum zeylanicum*, *Dioscorea bulbifera*, *Elaoephorbia drupifera*, *Solanum aculeastrum*, *Albizia schimperiana*, *Zanthoxylum gilletii*, and *Strychnos usambarensis* have demonstrated their efficacy in the treatment and management of various malignancies, including cancer [16]. The observed plants have demonstrated a notable cytotoxic impact, particularly in regions with a high cancer incidence, and have exhibited resistance to pharmaceutical interventions. As an illustration, in the context of in vitro investigations involving Korean medicinal plants to examine their potential anticancer properties, it was observed that the ethyl acetate fraction derived from the methanolic extract of *Lespedeza cuneata* exhibited a noteworthy cytotoxic impact on A2780 human ovarian carcinoma cells. The IC50 value, indicating the concentration required to inhibit 50% cell growth, was determined to be 77.25 ± 2.05 μg/mL. Similarly, the compound known as (+)-9′-*O*-(α-l-rhamnopyranosyl) lyoniresinol, derived from this particular plant, demonstrated in vitro antiproliferative activity against A2780 cells, with an IC50 value of 77.24 ± 2.05 μM. Significantly, a study involving athymic mice and human SKOV3 ovarian cancer xenografts observed that the tea seeds (*Camellia sinensis*), known to contain saponins, exhibited cancer chemopreventive effects. In addition, it has been observed that lysophosphatidic acid (LPA), a biologically active lipid, can enhance the invasive properties of cancer cells and promote their migration. This effect is accompanied by the phosphorylation of STAT3, which triggers the secretion of interleukin-6 (IL-6) and interleukin-8 (IL-8). Curcumin, a polyphenolic compound derived from the plant species *Curcuma longa*, possesses a range of curcuminoids and exhibits potential therapeutic efficacy in ovarian cancer. It is worth noting that traditional practices in various regions of Nigeria have employed various plant species, such as *Kigelia africana*, *Pistia stratiotes*, *Chenopodium africanum*, *Nymphaea lotus*, *Parquetina nigrescens*, *Nicotiana tabacum*, Nigrescentongensis, *Elaeis guineensis*, *Piper guineense*, *Aframomum melegueta*, and *Petiveria alliacea*, for the treatment of cancer, including ovarian cancer. In particular, *Securidaca longipedunculata* has gained recognition as an effective remedy for ovarian cancer [42]. Although there have been dedicated studies on the Ijebu people, an ethnic group within the Yoruba community, similar practices have been observed in various regions, including Ogun State in northwest Nigeria.

8 Conclusion

This evaluation examines the phytochemical constituents and metabolites in medicinal plants and herbal remedies utilized to treat and control reproductive ailments, specifically emphasizing conditions such as polycystic ovary syndrome (PCOS) and ovarian cancer. This analysis underscores the considerable potential of plant-based therapies in advancing women's reproductive health. This study examines various medicinal plants and compounds, such as *Ecklonia cava*, *Glycyrrhiza glabra*, and *Punica granatum*, and their potential to alleviate reproductive diseases and prevent ovarian cancer. These substances demonstrate promising properties in this regard. The complex nature of polycystic ovary syndrome (PCOS) and ovarian cancer necessitates a comprehensive approach to both treatment and prevention. These herbal remedies can alleviate the symptoms associated with these conditions and tackle the root causes, thereby providing a natural and complementary alternative to conventional medical interventions. Furthermore, this study elucidates the potential therapeutic efficacy of less prevalent botanicals such as *Aegle marmelos*, *Bambusa bambos*, and *Corylus avellana*, which have historically been employed in various regions across the globe. The assessment above highlights the significance of ongoing research on the effectiveness and safety of these botanical treatments. Scientific validation and clinical trials are crucial in establishing the optimal dosages, potential adverse effects, and potential interactions with other medications. This will assist healthcare professionals in making well-informed decisions and providing recommendations to patients who are seeking alternative treatment options. Moreover, it is imperative to underscore that herbal remedies should not be regarded as substitutes for conventional medical interventions but as adjunctive modalities. Patients should seek guidance from healthcare professionals and herbalists to develop personalized treatment strategies, considering their unique circumstances and requirements. Within women's reproductive health, investigating botanical remedies and their constituents holds considerable potential for forthcoming scholarly inquiry and practical implementation. As further knowledge is acquired regarding these therapeutic interventions, their incorporation into holistic healthcare approaches can enhance treatment efficacy and improve the overall well-being of individuals grappling with reproductive disorders and ovarian malignancies. This assessment incentivizes the broader medical and scientific community to engage in collaborative efforts, conduct further research on these botanical agents, and advocate for their responsible utilization in treating and preventing reproductive diseases.

Acknowledgments We would like to extend our sincere appreciation and gratitude to the Department of Agronomy at the School of Agriculture at Lovely Professional University, Punjab, 144411, India.

Conflicts of Interest A conflict of interest has not been identified

References

1. Agha-Hosseini M, Kashani L, Aleyaseen A, Ghoreishi A, Rahmanpour HAEH, Zarrinara AR, Akhondzadeh S (2008) *Crocus sativus* L. (saffron) in the treatment of premenstrual syndrome: a double-blind, randomized, and placebo-controlled trial. BJOG 115(4):515–519
2. Akbari Bazm M, Khazaei M, Khazaei F, Naseri L (2019) *Nasturtium officinale* L. hydroalcoholic extract improved oxymetholone-induced oxidative injury in mouse testis and sperm parameters. Andrologia 51(7):e13294
3. Akhondzadeh S, Malek-Hosseini M, Ghoreishi A, Raznahan M, Rezazadeh SA (2008) Effect of ritanserin, a 5HT2A/2C antagonist, on negative symptoms of schizophrenia: a double-blind randomized placebo-controlled study. Prog Neuropsychopharmacol Biol Psychiatry 32(8): 1879–1883
4. Amégbor K, Metowogo K, Eklu-Gadegbeku K, Agbonon A, Aklikokou KA, Napo-Koura G, Gbeassor M (2012) Preliminary evaluation of the wound healing effect of *Vitex doniana* sweet (Verbenaceae) in mice. Afr J Tradit Complement Altern Med 9(4):584–590
5. Arentz S, Abbott JA, Smith CA (2014) Herbal medicines for the management of PCOS and associated oligo/amenorrhea and hyperandrogenism: a review of the laboratory evidence for effects with corroborative clinical findings. BMC Complement Altern Med 14:511
6. Baliga MS (2010) Triphala, Ayurvedic formulation for treating and preventing cancer: a review. J Altern Complement Med 16:1301–1308
7. Barbosa G, Arbex A, Rocha D (2016) Polycystic ovary syndrome and fertility. Obs Gyn Rep Med 6:58–65
8. Barnes S, Kim H, Darley-Usmar V, Patel R, Xu J, Boersma B, Luo M (2000) Beyond ERα and ERβ: estrogen receptor binding is only part of the isoflavone story. J Nutr 130(3):656S–657S
9. Battineni JK, Boggula N, Bakshi V (2017) Phytochemical screening and evaluation of anti-emetic activity of *Punica granatum* leaves. Eur J Pharm Med Res 20(4):4
10. Beal MW (1998) Women's use of complementary and alternative therapies in reproductive health care. J Nurse Midwifery 43(3):224–234
11. Bhattacharya S, Johnson N, Tijani HA, Hart R, Pandey S, Gibreela A (2010) Female infertility. BMJ Clin Evid 2010:0819
12. Cheng G, Wilczek B, Warner M, Gustafsson JÅ, Landgren BM (2007) Isoflavone treatment for acute menopausal symptoms. Menopause 14:468–473
13. Choudhary N, Padmalatha V, Nagarathna R (2011) Prevalence of polycystic ovarian syndrome in Indian adolescents. J Pediatr Adolesc Gynecol 24(2):223–227
14. Dash MK, Joshi N, Dwivedi LK (2016) Effects of Pushpadhanwa rasa on psychological imbalances in PCOS patients and its management. Ayur Pharm Int J Ayur Alli Sci 5(7):90–97
15. DeBenedectis C, Ghosh E, Lazarus E (2015) Pitfalls in imaging of female infertility. Semin Roentgenol 50(4):273–283

16. Demirel MA, Ilhan M, Suntar I (2015) Activity of *Corylus avellana* seed oil in letrozole-induced PCOS model in rats. Rev Bras 26:83–88
17. Dhivya C, Dhanalakshmi S, Chitra V (2018) Alleviation of polycystic ovarian syndrome by hydroalcoholic leaf extract of *Aegle marmelos* L. Correa in letrozole-induced rat model. Drug Invent Today 10(7):1246–1250
18. Ebrahim NA, Badawi MA, Ahmed BS (2018) The possible protective effect of *Bougainvillea spectabilis* leaves extract on estradiol valerate-induced polycystic ovary syndrome in rats (biochemical and histological study). Eur J Anat 22(6):461–469
19. Ghasemzadeh A, Farzadi L, Khaki A (2013) Effect of *Allium cepa* seeds ethanolic extract on experimental polycystic ovary syndrome (PCOS) apoptosis induced by estradiol valerate. Life Sci 10(4s):170–175
20. Heidary M, Yazdanpanahi Z, Dabbaghmanesh MH (2018) Effect of chamomile capsule on lipid and hormonal-related parameters among women of reproductive age with polycystic ovary syndrome. J Res Med Sci 23(33):1–7
21. Hong EJ, Lee HW, Kim HJ (2018) Therapeutic effect of *Ecklonia cava* extract in letrozole-induced polycystic ovary syndrome rats. Front Pharmacol 9:1325
22. Hossen MJ (2015) Alternative medicine: health safety and therapeutic potentialities. EC Vet Sci 869(1):28–29
23. Jashni HK, Jahromi KH, Bagheri Z (2016) The effect of Palm pollen extract on PCOS in rats. Int J Med Res Health Sci 5(5s):317–321
24. Kavitha A, Narendra Babu A, Sathish Kumar M (2016) Evaluation of effects of *Commiphora wightii* in Dehydroepiandrosterone (DHEA) induced PCOS in rats. Pharmacophore 4(1):47–55
25. Kuete V, Fankam AG, Wiench B, Efferth T (2013) Cytotoxicity and modes of action of the methanol extracts of six Cameroonian medicinal plants against multidrug-resistant tumor cells. Evid Based Complement Alternat Med 2013:285903
26. Lee HW, Yang H, Kim HJ (2018) Licorice ethanol extract improves symptoms of polycystic ovary syndrome in Letrozole-induced female rats. Integr Med Res 7:264–270
27. Lloyd KB, Hornsby LB (2009) Complementary and alternative medications for women's health issues. Nutr Clin Pract 24(5):589–608
28. Luciano AA, Lanzone A, Goverde AJ (2013) Management of female infertility from hormonal causes. Int J Gynecol Obstet 123(Suppl 2):S9–S17
29. Malarvizhi D, Anusooriya P, Meenakshi P, Sowmya S, Perumal PC, Oirere EK, Gopalakrishnan VK (2015) Antioxidant properties and analysis of bioactive compounds present in n-hexane root extract of *Zaleya decandra*. Int J Pharm Sci Rev Res 34:118–123
30. Malla B, Gauchan DP, Chhetri RB (2015) An ethnobotanical study of medicinal plants used by ethnic people in Parbat district of Western Nepal. J Ethnopharmacol 165:103–117
31. Melgarejo P, Nunez-Gomez D, Legua P, Martínez-Nicolás JJ, Almansa MS (2020) Pomegranate (*Punica granatum* L.) a dry pericarp fruit with fleshy seeds. Trends Food Sci Technol 102: 232–236
32. Mohammadzadeh F, Babazadeh R, Salari R, Afiat M, Heidarian Miri H (2019) The effect of pomegranate peel gel on orgasm and sexual satisfaction of women in reproductive age: a triple-blind, randomized, controlled clinical trial. Iran J Obstet Gynecol Infertil 22(7):66–76
33. Nampoothiri L, Nagar PS, Maharjan R (2010) Effect of *Aloe barbadensis* Mill. Formulation on Letrozole-induced polycystic ovarian syndrome rat model. J Ayurveda Integr Med 1(4): 273–279
34. Nasir S, Batool M, Hussain SM (2015) Bioactivity of oils from medicinal plants against immature stages of Dengue mosquito *Aedes aegypti* (Diptera: Culicidae). Int J Agric Biol 17: 843–847
35. Ndefo UA, Eaton A, Green MR (2013) Polycystic ovary syndrome: a review of treatment options with a focus on pharmacological approaches. P T 38(6):336–355
36. Olujimi OO, Bamgbose O, Arowolo T, Steiner O, Goessler W (2014) Elemental profiles of herbal plants commonly used for cancer therapy in Ogun State, Nigeria. Part I. Microchem J 117:233–241

37. Pachiappan S, Matheswaran S, Saravanan PP (2017) Medicinal plants for PCOS: a review of phytomedicine research. Int J Herbal Med 5(2):78–80
38. Palanirajan A, Raj CA, Perumal PC, Sundaram S, Balu V, Prabhakaran P, Gopalakrishnan VK (2015) Screening of novel CXC chemokine receptor 4 inhibitors from ethyl acetate extract of *Alpinia purpurata* using GC-MS analysis and its molecular docking studies. Int J Pharmacognosy Phytochem Res 7:480–488
39. Palanisamy CP, Ashafa AOT (2018) Analysis of novel CXC Chemokine receptor type 4 (CXCR4) inhibitors from hexane extract of *Euclea crispa* (Thunb.) Leaves by chemical fingerprint identification and molecular docking analysis. J Young Pharm 10(2):173
40. Perry TE, Hirshfeld-Cytron J (2013) Role of complementary and alternative medicine to achieve fertility in uninsured patients. Obstet Gynecol Surv 68(4):305–311
41. Ruibin J, Bo J, Danying W, Chihong Z, Jianguo F, Linhui G (2017) Therapy effects of wogonin on ovarian cancer cells. Biomed Res Int 2017:9381513
42. Segun PA, Ogbole O, Ajaiyeoba EO (2018) Medicinal plants used in the management of cancer among the Ijebus of southwestern Nigeria. J Herbal Med 14:68–75
43. Singh A (2015) Observations on the wild medicinal flora of Banaras Hindu University main campus, India. Int J Mod Biol Med Chem 6:1–21
44. Smith CA, Armour M, Ee C (2016) Complementary therapies and medicines in reproductive medicine. Semin Reprod Med 34(2):67–73
45. Soumya V, Muzile YI, Venkatesh P (2019) A novel method of extraction of bamboo seed oil (*Bambusa bambos* Druce) and its promising effects on metabolic symptoms of experimentally induced polycystic ovarian disease. Indian J Pharmacol 48(2):162–167

Assessment of the Phytochemical Constituents and Metabolites in the Medicinal Plants and Herbal Medicine Used in the Treatment and Management of Skin Diseases

18

Prasann Kumar, Nirmala Karam, and Debjani Choudhury

Contents

P. Kumar
Department of Agronomy, School of Agriculture, Lovely Professional University, Phagwara, Punjab, India

N. Karam · D. Choudhury (✉)
Department of Plant Pathology, School of Agriculture, Lovely Professional University, Phagwara, Punjab, India

S. C. Izah et al. (eds.), *Herbal Medicine Phytochemistry*, Reference Series in Phytochemistry, https://doi.org/10.1007/978-3-031-43199-9_21

Abstract

The most significant organ present in the human body is the skin. It helps in protecting the body from different microbial organisms. Skin also helps regulate body temperature and enables one to sense touch, heat, and cold. A variety of skin diseases are widespread and frequently occurring health problems that affect people of all ages, from newborns to elders. These diseases can cause harm in several different ways. It is essential to maintain healthy skin to maintain a healthy body. People can suffer from various skin diseases that affect their skin, causing skin cancer, herpes, cellulitis, acne, and eczema. Many wild plants and their parts having medicinal properties are widely available. Those plants are frequently applied to treat these diseases and their symptoms. Plants have been used by humanity since the beginning of time. In addition to being cheap, natural treatments are also claimed to be safe. The raw material can also be used to produce novel synthetic agents, as it is suitable for this purpose. In this chapter, you will find a review of some plants that can be used to treat skin diseases and different symptoms produced due to these diseases and a summary of the recent technological advancements in this area during the past few years.

Keywords

No poverty · Zero hunger · Skin disease · Eczema · Herpes · Cancer · Health · Symptoms

Abbreviations

COLIPA	European Cosmetic, Toiletry, and Perfumery Association (Now Cosmetic Europe)
CPE	Catechin polyphenols

DPPH	2, 2-Diphenyl-1-picrylhydrazyl
HICV-RT	Human Immunodefeciency Virus-Reverse Transcriptase
HuDe	Human Dermal
IL	Interleukin
LCMS	Liquid Chromatography-Mass Spectrometry
NO	Nitrous oxide
PCPSO	Pomegranate Cold-Pressed Seed Oil
PFJ	Pomegranate Fermented Juice
PGE2	Prostaglandin E2
UHPLC	Ultra-High-Performance-Liquid Chromatography
UV	Ultraviolet
UVA	Ultraviolet Aging
UVB	Ultraviolet burning

1 Introduction

Since immemorial times, a wide range of plants of medicinal value have been utilized as a medicine to treat different diseases. More attention has been gained due to their remedial effects [1]. Other traditional herbs from various habitats can be a new strategy to protect against skin infections. The antimicrobial properties of those plants are now in demand due to their remedial properties. Due to a lack of awareness regarding phytomedicines and deforestation, the value of medicinal herbs is declining. These modern medicines come with many challenges, creating side impacts and toxicity [2]. Mother Nature has provided us with many herbal plants contributing to the health care system. These plant-based medicines are beneficial against various diseases and disorders. Due to lower side effects, there has been an increase in the adaptability and acceptability of these phytomedicines in the recent era [3]. The plants having herbal remedies are efficacious against various skin disorders, skin cancers, eczema, fungal and bacterial infections, acne, etc. [4]

Human skin is considered the largest organ of the human body and is the first defense line to protect the body from many diseases and disorders. Skin disorders occur worldwide, affect all age groups, and present a significant health burden in developed and developing countries. Healthy skin can only be achieved through the use of herbal medicines. The demand for herbal medicines is increasing progressively every day due to their low price and the hazardous effects of modern medications. These medicines have gained a vital contribution to health care [5] due to the presence of various bioactive components like glycosides, tannins, alcohols, aldehydes, etc. Different therapeutic agents can be discovered from these bioactive components and be a future general asset. Cosmetics have also been prepared to look beautiful. In the olden days, many homemade remedies were prepared with different plants and their parts, like banana, neem, sugar beet, *aloe vera*, saffron, papaya, carrot, neem, tulsi, marigold, etc. All these plants are used as medicinal herbs and various homemade remedies to manage different skin diseases. Polyherbal formulations, plant extracts, powders, oils, face packs, soaps, etc., can be prepared with these herbs. People use various synthetic products as cosmetics, but these products may lead to side effects such as skin reactions, inflammations, swellings, irritations, etc., use herbal skin products that are safe for the skin to avoid various side effects and allergies. Various synthetic products are used for oily and dry skin, but those products are not up to the mark. However, the homemade herbal remedies used in different combinations had a marked effect when applied on the skin with few side effects. Dental care products and perfumes also come under cosmetics to prevent tooth decay and body odor. There are different types of skin: oily skin and dry skin. Applying facewash, lotion, and sunscreen for protection from UV radiation and using moisturizing creams for cold climates. All these skin care products should be of herbal origin for better care. Among young stars, acne is the biggest problem that spoils most people's look, appearance, and beauty. It happens due to access sebum secretion, which increases oils in the skin. This can be treated with acne-treating products using gels and creams.

Due to stress and anxiety, many youngsters face aging experiences, which can be prevented by applying herbal skin care products and taking good care of their mental health. Anti-aging helps skin firmness and smoothness, improving brightening and a young appearance.

With the help of various herbal preparations containing fruit pulp extracts, blackheads and black spots can be cleared as these products supply a good amount of antioxidants, which helps give a toned skin. Herbal plants also help brighten skin and provide a good look, enhancing the complexities of the skin and giving it a shiny appearance. Many people face skin dehydration, which ultimately dries up, and the skin is subject to peeling. Various herbal cosmetics preparations can be applied to the skin to increase its moisturizing capacity.

Hair fall has become a significant problem for many people due to dandruff. Most men are facing such issues. So, protecting the hair from dandruff is very important to prevent baldness. The application of herbal anti-dandruff shampoo can do this. Various hair oils can also be provided to control hair fall. Growing hair follicles on the face and all over the body is a significant problem many women face. This can be overcome by applying various herbal preparations to prevent hair growth [6]. The focus of this chapter is to assess the role of phytomedicine in skin infection management.

2 Skin Layers and Their Functions

Skin is an organ of tissues that work together to perform unique and critical functions that provide overall protection. The skin consists of multiple layers, of which epidermal tissue is the most superficial layer attached to the dermis and external to the hypodermis. The epidermal layer is keratinized and protects the skin cells. The dermis is the vascular layer that provides flexible tissues. Additionally, it comprises several connective tissues, sweat glands, hair follicles, and connective tissues and helps provide support and flexibility to the tissues. It also helps to regulate body temperature and provides physical protection against injury. It is a barrier against the external environment, preventing the entry of bacteria and other harmful substances. The innermost layer of the hypodermis is one of the body's most vascularized layers, consisting of loose connective tissues that serve as energy reserves and maintain the body's temperature (Fig. 1). It also contains sweat glands, hair follicles, and oil glands. The hypodermis also plays a vital role in the body's immune system by providing a physical barrier to keep out infectious agents.

Absorption of the drug using the skin is a route through which the drug enters the skin. Drug absorption depends on factors like concentration, time duration, medicine stability, skin condition, and body part exposure. Skin absorption is the transport of chemicals using the skin's outer surface as well as into the circulation. While transporting herbal medication into the dermal region for the treatment of skin diseases, a series of steps are involved in releasing herbal medicines in the dosage form. These steps include formulating the dosage forms, selecting the appropriate delivery system, and ensuring the release of the medicine at the desired rate. Additionally, the stability and safety of the drug must be considered to provide the highest quality of treatment. After a therapeutic photothermal constituent penetrates through the epidermal layer of the skin, it diffuses into the dermal region of the body. The light energy then activates the photothermal branch and produces photothermal effects, leading to the desired therapeutic outcome. The photothermal effects can be monitored and controlled by modulation of light energy. As soon as the herbal constituent reaches the circulation system, it is absorbed by the body. Figure 2 below depicts a graph illustrating how drugs penetrate the body. The drug is then

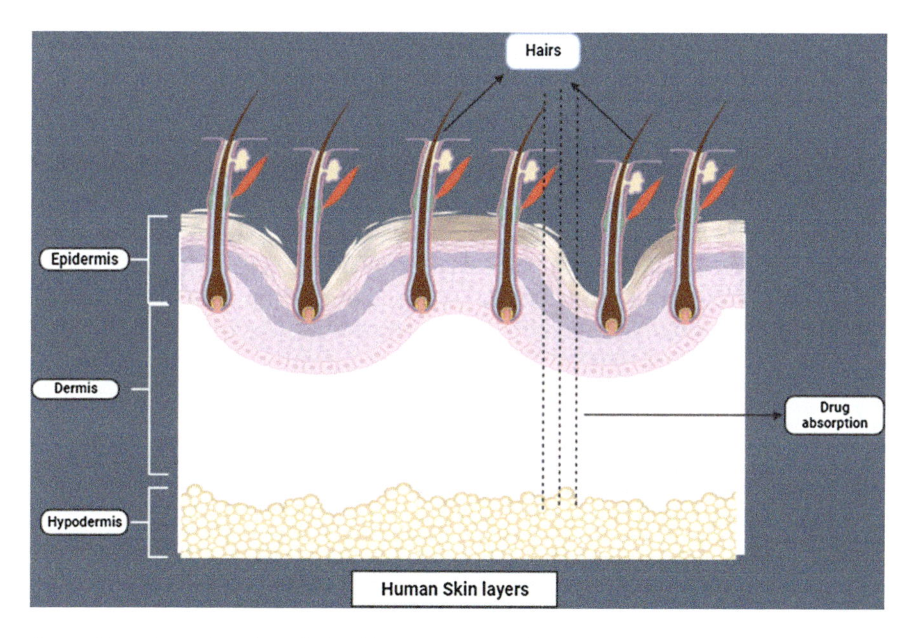

Fig. 1 Drug absorption through the skin that transport the chemicals through outer surface of the skin and through blood circulation

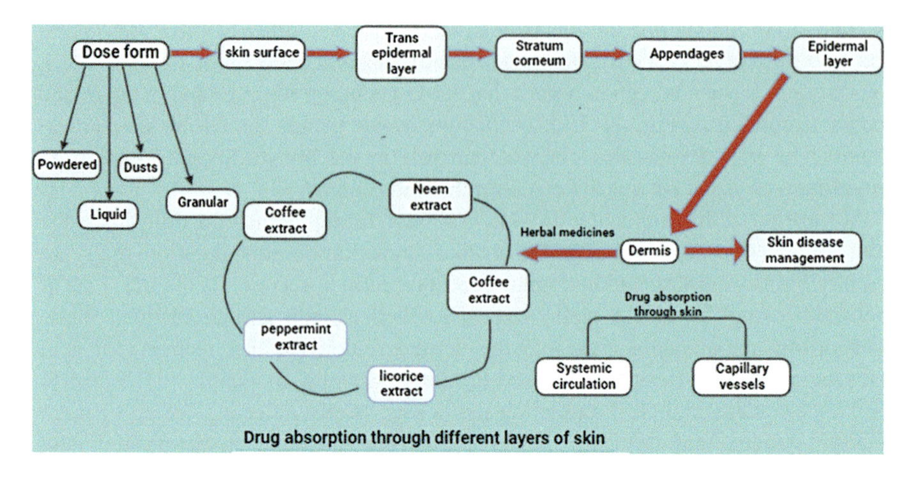

Fig. 2 Drug absorption through different layers of skin, by means of systemic circulation and capillary vessels

distributed through the body, achieving the desired therapeutic effect. The light energy can then be turned off, ending the photothermal effect. The drug is then metabolized and eliminated from the body.

3 Diverse Skin Conditions and Diseases: Characteristics, Effects, and Pathogen Involvement

3.1 Skin Disease

It includes various types of conditions that create irritation or inflame in the skin. Skin diseases also cause rashes or other changes in skin appearance. All the skin diseases mentioned below are not due to pathogens. The list of pathogens occurring in skin diseases is mentioned in Table 1. Various skin conditions can affect individuals, each with distinct characteristics and effects. These conditions include acne, characterized by oil clogging in skin follicles and the formation of bacteria and dead skin in pores; alopecia areata, which leads to hair loss in small patches; atopic dermatitis (eczema), causing itchy skin, dryness, swelling, and cracking; psoriasis, resulting in scaly, swollen, and hot skin; Raynaud's phenomenon, involving periodic reduced blood flow in fingers, toes, and other body parts with associated changes in skin color and numbness; rosacea, typically presenting with thickened skin and pimples on the face; skin cancer, marked by uncontrolled abnormal development of skin cells; vitiligo, leading to pigment loss in skin patches; prurigo, causing itchy skin when exposed to the sun; argyria, involving changes in skin color as the body accumulates a silvery buildup; chromhidrosis, a pigmentation disorder; epidermolysis bullosa, a connective tissue-related disorder resulting in fragile skin prone to blisters and tears; Harlequin ichthyosis, characterized by the presence of thick, hard patches on the skin at birth;

Table 1 List of pathogens causing skin infections

Pathoges	Skin infections	References
Bacterial pathogens		
Propionibacterium acnes	Acne, dendruss, psoriasis, and blepharitis	[9]
Microsporum audouinii., Trichophytion spp.	Tinea capitis	[87]
Staphylococcus aureus	Atopic dermatitis and skin and soft tissue infections	[9]
Fungi		
Malassezia dermatis, M. furfur, M. globose, M. japonica, M. obtuse, M. restricta, M. slooffiae, M. sympodialis, and M. yamatoensis	Tinea versicolor, psoriasis, and atopic dermatitis	[88]

lamellar ichthyosis, which leads to the development of a waxy skin layer, revealing scaly and red skin; and necrobiosis lipoidica, a condition where rashes on the lower legs may progress to ulcers.

3.2 Cause of Skin Disease

Certain lifestyle factors for developing skin disease include (a) trapping of bacteria in the hair follicles or pores, (b) sun rays, (c) viruses, (d) parasites on the skin, e.g., any conditions affecting a person's immune system, (e) certain medicines, bowel infections, (f) kidney diseases, (g) thyroid problems, (h) diabetes, etc.

3.3 Diagnosis of Skin Disease

Medical professionals employ various diagnostic methods to assess skin conditions. These methods include performing a biopsy to examine a small-skin sample under a microscope, conducting a culture to identify the presence of fungi, bacteria, or viruses on the skin, administering a skin patch test to detect allergic reactions to chemicals, utilizing a Wood's light test involving UV light to examine the skin, employing diascopy to assess skin patches by applying pressure with a microscope, using dermoscopy for inspecting skin lesions and conducting a Tzanck test to examine fluid from a blister (Fig. 3).

3.4 Various Skin Diseases Caused by Pathogens

Table 1 presents the list of pathogens causing skin infections

3.4.1 Bacterial Infections

The skin's external surface is mainly affected by bacteria. Skin acts as a medium for bacterial growth. It also allows commensal bacteria to grow on the skin to protect it

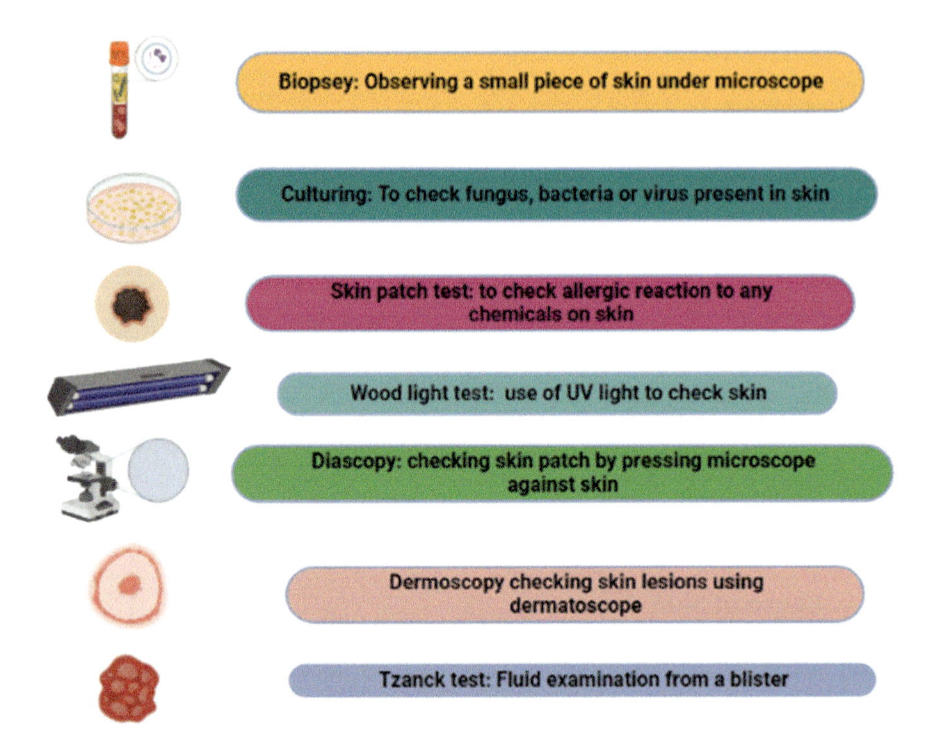

Fig. 3 Methods for diagnosing skin disease

from disease-causing bacteria. Some gram-positive bacteria like corynebacterium, staphylococcus, micrococcus, and streptococcus are notorious bacterial pathogens for the skin. A pathogenicbacteria allows bacteria to adhere, grow, and invade the skin's surface. With an increase in bacteria, disruption of the integumentary barrier, invasion by these colonizing bacteria emanates, and skin and soft tissue infections develop.

3.4.2 Viral Infections

It is believed that viruses can penetrate the stratum corneum, which is the top layer of the skin and is susceptible to infection. The stratum corneum is a natural barrier to illness, but viruses can pass through the skin, leading to disease. Viruses can also be transmitted through contact with contaminated surfaces or objects. Therefore, it is essential to take precautions such as washing your hands often and avoiding contact with people who are sick. As well as infecting the epidermis, it also infects the subcutaneous layer. To prevent infection, avoiding touching your face is essential, as this is an easy way for viruses to enter your body. Additionally, it is necessary to regularly clean and disinfect surfaces and objects that are frequently touched. There are a variety of viral skin infections, such as measles, warts, chickenpox, etc., which are not healable by antibiotics [7]. Therefore, practicing good hygiene and wearing protective clothing is essential to avoid contracting the virus. Vaccination is another effective way to reduce the risk of viral skin infections.

3.4.3 Fungal Infections

Fungal infection occurs in the depth of the skin tissue. But sometimes, the conditions may occur superficially, like skin, nails, or hair, called athlete's foot infection or ringworm infection, where trial antibiotics are needed to be taken, which in the long-term cause severe disorders [7].

3.4.4 Tumors and Skin Cancers

The lack of melanin causes skin cancer, which ensues in vitiligo, an auto-immune disorder. The scarcity of melanin cells causes albinism, which may cause skin cancer when exposed to UV radiation. If detected, there is a chance for the survival of patients. [7].

3.4.5 Skin Pigmentation

With the increase in the melanin content of the body, hyperpigmentation occurs, which causes skin irritation and premature aging [7].

Trauma: The burns, cuts, or skin injury leads to a pathogenic attack known as trauma [7].

3.4.6 Rashes

Inflamed-skin conditions can cause pain, itching, and discomfort, leading to scarring and psychological distress. Treatment of such conditions can include topical medications, antibiotics, and lifestyle changes. It is essential to seek professional medical advice when dealing with skin inflammation.

3.4.7 Psoriasis

Infections such as psoriasis that affect the skin can harm the patient's quality of life and negatively affect the patient's outlook on life. Having psoriasis can be physically and emotionally draining. Symptoms can be uncomfortable and embarrassing; flare-ups can cause pain, itchiness, and inflammation. These symptoms can interfere with everyday activities and make it difficult for the patient to enjoy their life. Excessive proliferation of the skin can result from either hereditary factors or hyperproliferation of the skin. Psoriasis is an autoimmune disorder that causes the body to mistakenly attack healthy skin cells, resulting in a rapid overproduction of skin cells that accumulate on the skin's surface. This leads to the characteristic red, scaly patches of psoriasis. In addition, psoriasis can lead to psychological distress, as patients may feel embarrassed or self-conscious about their appearance. Three types of psoriasis are known to exist: nail psoriasis, which causes pitting and yellowing of the nails along with severe hyperkeratosis; skin psoriasis, which causes yellow pus to appear on the skin and blisters to appear near the joints; and psoriatic arthritis, in which bone erosion occurs near the joints [8]. Psychological distress can be exacerbated by the physical discomfort caused by the condition and the fact that psoriasis can be visible to the public. Covering nail or skin psoriasis with clothing is often tricky, and

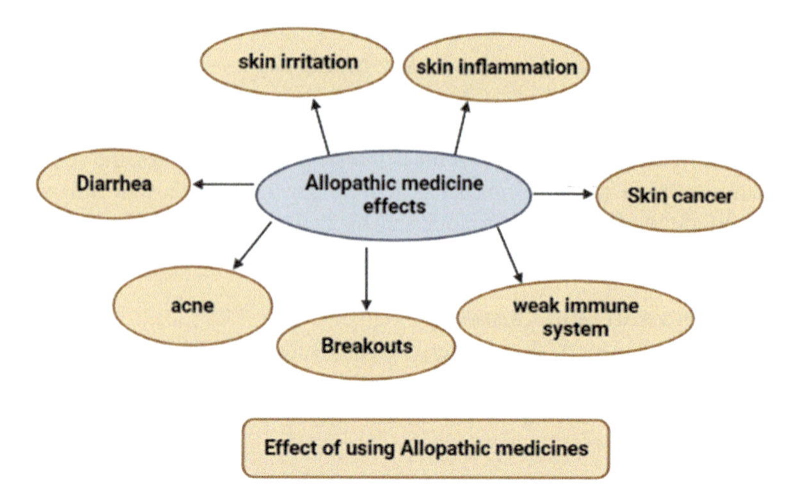

Fig. 4 Side effects of allelopathic medicines

psoriatic arthritis can cause severe pain and joint stiffness. All of these can lead to social isolation and a decrease in self-esteem.

3.5 Allopathic Approaches for Skin Disorders

Skin disorders treatment with allopathic medicines causes severe ripple effects and can be diminished by phytomedicines. Numerous hazardous effects occur due to these allopathic medicines, like skin inflammation, irritation, breakouts, and diarrhea. Immunosuppressants lead to skin cancer and a weak immune system [5] (Fig. 4).

4 Role of Phytomedicines for Skin Infection

4.1 *Azadirachta indica*

Neem, a member of the family Meliaceae, plays a significant role in the promotion of better health due to the presence of a high content of antioxidants, azadirachtin, glycerides, polyphenols, triterpenes, limonoids, beta-sitosterol, quercetin, catechins, carotenes, and vitamin C that contribute to its health-promoting effects. It has been used in Chinese, Unani, and Ayurvedic medicine for thousands of years. It has been used for centuries as a traditional medicine due to its anti-inflammatory, antibiotic, antifungal, and antiviral properties. It is also a key ingredient in many natural skin care and hair care products. The paste prepared from leaves helps treat acne, eczema, scabies, skin allergies, and psoriasis. Neem is also known to have antibacterial and anti-ulcer effects, making it an excellent tool for boosting the overall health of individuals. Additionally, it has been shown to help reduce cholesterol levels, improve digestion, and regulate blood sugar levels. Neem is also known for its antioxidant, anti-inflammatory, and anti-cancer properties that help boost immunity and relieve stress and anxiety. Neem is known to have antiviral and antifungal properties. The findings from earlier reports have confirmed that the role of neem and its constituents play an essential role in the pathogenesis of many diseases [86]. It treats various illnesses, such as malaria and intestinal parasites. In addition, neem has also been used to treat eye infections, gastrointestinal disorders, and fever. The bark of this plant helps maintain healthy gums, increases oral hygiene, and can also be used as an antiviral against chicken pox due to its antiviral properties. It can also reduce swelling, promote healing, and has contraceptive effects that can help manage neurodegenerative disorders and diabetes. As far as treating eczema, scabies, and other skin disorders is concerned, neem paste has a variety of beneficial effects. The product contains essential fatty acids that provide moisture and texture to the skin during the healing process by inhibiting the growth of inflammatory cells. Several studies have shown that the oil extracted from neem can treat ringworm, psoriasis, and scrofula diseases [10]. Oil extracted from neem is also one of the best herbal remedies for treating various skin disorders, allergies, infections, inflammations, and other skin disorders [11, 12]. The oil extracted from neem also reduces wrinkles, dark spots, and discoloration, excellently making it a tremendous natuan for minor skin irritations and infections. Applying neem and turmeric as lotion during night hours helps prevent scthe abies, eczema, and various logic problems. Neem can also be taken orally, either as a tea, in caps as a tea, or in capsule form to boost the immune system. Neem is also a natural insect repellent, making it useful for insects from gardens and homes away.

4.2 *Glycyrrhiza glabra*

It is a sweet and aromatic herb that comes from the licorice family. This herb's active metabolite is glycyrrhizin, an antiviral, anti-diabetic, antifungal, expectorant, demulcent, mild laxative, antioxidant, and skin-whitening substance. It is important to note that liquiritin, an antioxidant, is the anotis component of licorice that contributes to skin lightening. It also contains glycyrrhizin, which is known to have antibacterial properties that can help keep skin acne-free. It is believed to have many health benefits and has been used in traditional herbal medicine for centuries. Licorice has anti-inflammatory, antibacterial, and antioxidant properties that improve overall health and skin health. It also has a mild sedative effect, making it a popular choice for people looking for natural sleep aids. Despite its ethnopharmacological value, licorice has been explored worldwide due to its ethnopharmacological properties. Research has also suggested that it may positively affect cardiovascular health. Glycyrrhizin also shows potential in treating liver disorders, chronic bronchitis, and ulcers. In addition, research suggests that licorice may be beneficial in supporting cognitive function and mental health. Additionally, it is also a good source of valuable flavonoids and saponins as well. It may also help to reduce the risk of cancer and heart disease. The benefits of this plant provide a place for research and discovery of other benefits and potent effects that the plant has to offer. Licorice has also been used in traditional medicine for centuries, making it a trusted and reliable source of health and wellness. In addition, it is widely available and affordable, making it an attractive option for those seeking natural remedies. Glabridin is another constituent of *G. glabra,* which can lighten the skin's pigment due to its ability to inhibit the activity of tyrosine B16 in melanoma cells. Due to its anti-inflammatory and antioxidant properties, it reduces the appearance of wrinkles and improves skin texture. It has also been found to have beneficial effects on the skin, such as reducing redness and irritation. Additionally, licorice can also help protect skin from UV radiation and reduce melanin production, which can help with skin whitening. As a cosmetic, the herbal extracts of this plant can be used to depigment the skin and protect it from UV rays [13]. This can benefit those who want to lighten their skin tone or reduce the appearance of dark spots. According to various reports, root extracts of *G. glabra* have been proven effective in treating acne, eczema, and psoriasis. Furthermore, its anti-inflammatory properties can help reduce redness, pain, and swelling caused by skin conditions [14]. Licorice gel can control atopic dermatitis in patients with asthma and hay fever in their family history. Additionally, licorice helps to protect against UV radiation. It can also be used to treat cold sores and heal wounds. In combination with polyethene glycol, this gel can be applied at night along with polyethene glycol to reduce itching on the face [14]. Licorice gel can thus be used to reduce the severity of atopic dermatitis and for anti-aging purposes. It is important to note that licorice gel should be used in moderation and with the advice of a healthcare professional. In addition, some studies have shown that licorice extract also reduced pigmentation in the skin of black mollyfish by inhibiting tyrosine synthesis due to the dispersal of melanin (which contributes to skin tanning) from its location in the body. However, caution must be taken when using licorice gel as it can cause irritation and other skin conditions. It is, therefore, recommended to do a patch test before using the gel. As a result of the nutritional properties of licorice extract, new possibilities exist for the development of beautifying preparations that are safe, effective, and provide effective treatment for allergic dermatitis [15]. Such practices should be developed using carefully selected, high-quality raw materials and tested adequately in pre-clinical and clinical trials. This way, licorice gel can be used safely and effectively for its intended purpose.

4.3 *Ocimum sanctum*

It has become apparent that the "queen of herbs," *Ocimum sanctum* [16], has become a drug of choice for many people. Studies have found that this herb has anti-inflammatory, antibacterial, antifungal, and antiviral properties. Some essential constituents of basil are eugenol, ursolic acid, linalool, carvacrol, and rosmarinic acid, which are believed to have anti-inflammatory and antioxidant properties [16]. It has been used to treat various health conditions, ranging from colds and coughs to skin diseases and digestive issues. It is also known to boost immunity and fight oxidative stress. As it has been reported, it is used in several countries to treat various skin diseases and disorders, including allergies, rashes, insect bites, snake bites, scars, leucoderma, blisters, chickenpox, and roundworm infections. In addition, it has been used in traditional medicine to treat cancer, diabetes, and cardiovascular diseases. It is thought to reduce inflammation and boost the immune system. There is also evidence that the phyto extracts of basil are helpful in the reduction of itchy skin that occurs due to acne, pimples, and scars [17]. Studies have shown that basil effectively treats various conditions, including anxiety, inflammation, and skin irritation. Therefore, *O. sanctum* would be a striking plant for developing cosmetics. Its strong antibacterial and antifungal properties make it a viable option for producing natural skin care products. With its numerous therapeutic benefits, *O. sanctum* has the potential to be a valuable ingredient in the cosmetic industry. The first report of anti-aging activity has been investigated using collagenase, elastase, and hyaluronidase activity inhibition under in vitro conditions. These results suggest that basil could help reduce the visible signs of aging. Thus, basil could be a powerful and natural ingredient for skin care products. Some studies also confirmed the presence of rosmarinic acid in basil, which possesses potent antioxidant and anti-inflammatory activity that can protect the skin from damage caused by free radicals. Moreover, it can stimulate collagen production, which in turn can help reduce wrinkles and fine lines. *Extracts of basil* have inhibitory activities against oxidation, inflammation, collagenase, elastase, and hyaluronidase [18]. This indicates that *O. sanctum* extracts have the potential for use in cosmetics and dermatological applications due to their anti-aging properties. It can also be used to boost the immune system and reduce stress levels. Another report from a study includes the beneficial effects of herbal extracts as safe, effective, potent antibacterial, antiviral, and antifungal agents in controlling skin blisters, treatment of wound infections, chickenpox, and roundworm infections by modifying lifestyle [19]. Many skin disorders can be managed using tulsi plants, which have been found to have good herbal potential. Basil extract formulations and base (containing no extracts) were applied to the cheeks of 11 healthy human volunteers to determine the presence of erythema and melanin. Significant results were observed in the base product, while erythema decreased in the formulation. The study concluded that tulsi plants have the potential to manage skin disorders. Even though the melanin content of the skin increased significantly with the application of the base, it decreased with the application of the formulation [14]. The results indicate that the formulation containing the basil extract had a significant anti-inflammatory effect on the skin and a skin-lightening outcome. This suggests that basil extract could be helpful in skin care products. In addition, the essential oil from basil leaves has been used to treat headaches and other ailments. The compounds present in tulsi are believed to contribute to the therapeutic effects of basil, although further research is needed to understand the mechanisms of action fully.

4.4 *Andrographis paniculata*

This plant has been used for decades in the subcontinents of India to treat various diseases. It contains diterpenes, lactones and flavonoids, alkanes, keto aldehydes, and andrographolide. It helps treat pyrexia, dyspepsia, malaria, colic pains, dysentery, and diarrhea with aerial parts of the plants, roots, leaves, and whole plants of *Andrographis paniculata*. It has also been used to treat common colds and sore throats and improve the immune system. The plant is known for its antioxidant, anti-inflammatory, and anti-cancer properties. In this plant, there is a bioactive compound called andrographolide that is highly bioavailable and has anti-inflammatory, antioxidant, and other activities that can be used as a therapeutic agent to treat inflammatory skin diseases [20]. Andrographolide also exhibits various pharmacological activities, such as anti-tumor, anti-arthritis, and anti-diabetic. Furthermore, it can also be used as an analgesic and anti-diabetic agent. *A. paniculata* helps cure primary skin infections like eczema and leukoderma. Moreover, it can be mixed with oil and applied to treat wounds as a paste on the skin [21]. This compound also helps reduce inflammation and itching while promoting the healing of wounds. It is also known to have antibacterial, antifungal, and antiviral properties.

4.5 *Emblica officinalis*

As a medicinal herb, Emblica officinalis, also known as Indian gooseberry or Amla, is used for the treatment of various skin disorders and diseases. Among the chemical constituents, it contains tannins, polyphenols, gallic acid, alkaloids, emblicanin A and B, phyllembein, quercetin, ascorbic and ellagic acids, vitamins, and minerals. It is also believed to have anti-inflammatory and antioxidant properties, which can help the body fight against diseases. Amla can be eaten raw or taken as a supplement and has been used in traditional medicine for centuries. It has been found that the fruits of this plant have immense pharmacological and medicinal applications. It also boosts immunity, improves digestion, and reduces cholesterol levels. Amla has been used to treat diabetes, high blood pressure, and other health conditions. Amla is highly valued in Ayurveda for its therapeutic properties and is used in various herbal preparations. It is also known to be a rich source of antioxidants, essential for fighting free radicals and maintaining overall good health. In addition to its antioxidant properties, it also has anti-inflammatory and wound-healing properties. Amla is also rich in vitamin C, which helps boost the immune system and protect against infectious diseases. Furthermore, it is known to have anti-aging benefits, helping to keep skin looking youthful and healthy. Several conventional therapeutic agents can be replaced with remarkable herbs due to their outstanding efficacy and lack of side effects [22]. Amla is one such herb that provides a wide range of health benefits. Ayurveda, an ancient Indian medical practice, highly recommends amla and rasayan due to their antioxidant properties as part of its daily healthcare regimen that can help protect the body from the damaging effects of free radicals. They can also help to promote healthy skin, hair, and nails. According to various reports, amla contains high amounts of vitamin C, making it an effective hair tonic and treatment for warts and skin infections [23]. Amla is also rich in minerals like calcium, phosphorus, iron, and carotene, essential for healthy bones, teeth, and muscles. Rasayan provides various health benefits, including promoting better digestion, boosting immunity, and relieving stress [22]. In addition, amla is known to have anti-inflammatory properties, which can help reduce inflammation and pain associated with arthritis. It has been suggested to help lower cholesterol levels. Oral consumption of amla helps improve the condition in females, and drinking the extract improves the skin's elasticity and reduces the appearance of wrinkles [24]. Additionally, this fruit is known to be full of vitamins and minerals that are essential for overall health and well-being. It inhibits microbial adhesion, inactivates enzymes, suppresses influenza A virus, prevents viral absorption, and treats HIV by inhibiting HIV-RT [22]. Furthermore, it has been used to treat various skin conditions such as acne, eczema, and psoriasis.

4.6 *Aloe vera*

It is one of the most valuable and oldest herbs belonging to the family Liliaceae, indigenous to India, and belongs to the genus Aloe vera. *Aloe vera* has traditionally been used to treat various ailments, including skin conditions, digestive issues, and inflammation. It contains many nutrients, including folic acid and vitamins A, C, E, and B12. It also contains calcium, manganese, zinc, and magnesium minerals. The most investigated active compounds are aloe-emodin, aloin, aloesin, emodin, and acemannan [28]. Several species of Aloe have been used to prevent and treat skin-related problems for a long time. *Aloe vera* also contains a range of anti-inflammatory compounds, which have been found to have antibacterial, antifungal, and antiviral properties. It is a natural ingredient in many skin care, hair care, and health products. Depending on the individual application, aloe materials can be used for cosmetic, radiation, cancer, wound healing, and antimicrobial purposes [25]. It is also known to aid digestion, reduce constipation, and help regulate blood sugar levels. *Aloe vera* can also be taken orally as a juice or in capsule form. Additionally, *aloe vera* can be applied directly to the skin as a topical remedy. There is a derivative of *aloe vera* called aloesin, which helps to lighten dark patches, age spots, and acne marks. *Aloe vera* is also known to possess anti-inflammatory and antioxidant properties, which can help to reduce inflammation and protect the skin from environmental damage. It can even help to reduce the signs of aging, making skin look smoother and more youthful. As one of the most potent skin healers on the planet, this herb is used to treat sunburns, insect bites, wound infections, cuts, itching, and swelling in people worldwide. *Aloe vera* is also beneficial for treating skin conditions such as psoriasis, eczema, and acne. It is also used to reduce the appearance of scars, stretch marks, and dark spots. As a result of its antioxidant, antiaging, and anti-wrinkle properties, it purifies and detoxifies the skin. Aqueous extract from this plant is crucial in repairing damaged skin cells, reducing wrinkles, curing acne, reducing pimples, treating eczema, and working as a herbal gel for glowing skin [26]. Kumar et al. [27] published an eminent report regarding the application of *aloe vera* to wounds and burns. *Aloe vera* also helps treat psoriasis and other skin problems due to its anti-inflammatory and antibacterial properties [28]. It is also used in the treatment of skin cancer [29].

4.7 *Carica papaya*

Carica papaya is a herbaceous succulent plant, popularly known as pawpaw, belongs to the Caricaceae family and is considered the powerhouse of all vital nutrients for medicinal use. The roots, pulp, bark, seeds, and peels can be utilized for various ailments due to abundant vitamins and enzymes. The active components in papaya are alkaloids, glycosides, tannins, saponins, and flavonoids, responsible for medicinal activity. Platelet count can be increased if leaf juices are given to dengue fever patients [30]. The principal constituent of papaya is papain, which has antioxidant, antifungal, and antiviral effects. It is one of the safest and most widely promoted nutraceuticals against diseases like eczema, warts, and skin infections, preventing premature aging and scabies and reducing itching and wrinkles. Due to its inflammatory activity, papaya extract has wound-healing properties. Papaya peels can be used as potent sunscreen and, when mixed with honey, can be used as a skin-lightening and moisturizing agent. Adding lemon juice to papaya pulp can be used as anti-dandruff, and blending essential oils with papaya can be a good muscle relaxant [31].

4.8 *Jatropha curcus*

Jatropha is a drought-resistant, perennial plant belonging to the family Euphorbiaceae, famous for the production of biodiesel along with several medical applications. It contains terpenoids, coumarins, alkaloids, phenols, steroids, flavonoids, tannins, lignoides, and coumarins. Latex of Jatropha contains Jatrophine, Jatropha, and curcumin with anti-cancerous properties that help heal central skin lesions with wound healing properties and have purgative, anti-helminthic, and abortifacient effects. The oil produced from seeds help manage parasitic skin infections, which validates the plant in minimizing skin infections by promoting the healing of wounds with potent anti-inflammatory and antimicrobial processes [32]`.

4.9 *Zingiber officinalis*

Ginger belongs to the Zingiberaceae family, which can be used as a medicine and flavoring agent. Among the many essential components of ginger, gingerol *helps form new blood vessels at the site of damaged skin, promoting the repairing of skin cells and reducing the pain the patient feels.* Gingerol has potent medicinal properties like anti-inflammatory, antioxidant, and anti-tumor properties and can help protect against bacteria and viruses. It also contains carbohydrates, lipids, terpenes, and phenolic compounds, a popular ingredient in many Asian cuisines. It is also known for herbal ointments that can treat primary skin lesions. These ointments usually contain natural ingredients such as *aloe vera*, tea tree oil, and lavender oil, which can help reduce inflammation and promote healing. They are gentle and typically well-tolerated by people with sensitive skin. It has been shown that the oil obtained from seeds can be used to treat parasitic skin infections through its ability to promote wound healing and wound care and its potent anti-inflammatory and antimicrobial properties [33]. It also has antifungal and antiviral properties, making it an ideal choice for treating skin infections. Furthermore, it can help to protect the skin from further damage and irritation. Ginger can be consumed in various forms, including fresh, powdered, dried, and as an oil or extract. A ginger extract can help improve the structure and function of the skin, as well as reduce the formation of wounds that do not heal and make skin susceptible to disease. It also helps to improve the production of collagen, which is essential for skin health and increases the body's antioxidant capacity. Ginger may also help reduce wrinkles, age spots, and other signs of aging. It helps reduce inflammation, which is a significant cause of skin damage and can help reduce the risk of skin cancer. In addition, ginger can help to regulate oil production in the skin, helping to keep it balanced and healthy. In a study conducted on corticosteroid-treated rats, pre-treatment with topical ginger and curcumin extract improved the healing of induced abrasion skin wounds [34]. The study concluded that ginger and curcumin extract have a protective effect on the skin and can help wound healing and reduce inflammation. It also helps to reduce pain and can improve digestion. Gingerol can also help improve circulation, lower cholesterol levels, and improve immunity. It has also effectively combat bacterial, fungal, and viral infections. *The root extracts of ginger have been explored for their activity against various skin problems.* Ginger can also be a natural remedy for headaches, nausea, arthritis, and menstrual cramps. It can also treat multiple respiratory issues such as cough, cold, and sore throat. *The herbal extracts of ginger help in wound healing by improving skin texture and structure.* Additionally, ginger can relieve digestive issues such as indigestion and bloating. *The combination of ginger and turmeric helps to speed up the healing process of wounds.* It has been observed that when ginger extract, which is also an anti-inflammatory agent, is applied to mice ears, croton oil-induced edema can be inhibited by inhibition of cytokines cyclooxygenase and nitric acid production. In addition, ginger extract has been found to reduce inflammation and pain in cases of rheumatoid arthritis. Due to its antioxidant properties, ginger helps to protect against oxidative stress and tissue damage. There has also been evidence that wound-healing mechanisms have been observed in humans when these results have been simulated [34]. These findings suggest that ginger and turmeric can help to reduce inflammation and support the healing of wounds. In addition, applying these natural ingredients can reduce the need for other medications, such as antibiotics.

4.10 *Eucalyptus globulus*

It is estimated that there are more than 600 species of eucalyptus in the large genus Eucalyptus within the family Myrtaceae. Eucalyptus is also a famous ornamental tree, grown for its exciting bark and the shade it provides. It contains sideroxylonal C, leptospermone, grandinol, cypellogins A, B, and C, (+)-oleuropeic acid, cypellocarpins A, B, and C, and isoleptospermone. Many of these species are native to Australia and are used for their medicinal and aromatic properties in many cultures worldwide. Eucalyptus provides a wide variety of herbal by-products, food additives, antioxidants, healing properties of wounds, and protection against many fungal infections. They are also known to have anti-inflammatory, antiseptic, and antiviral properties. It is also used as an insect repellent, and its leaves and bark can be used to make tea. There are reports of different species of eucalyptus, like *Eucalyptus globulus, E. maculate*, and *E. viminalis,* in managing acne, scabies, and skin infections. Also, it offers a cooling sensation to significant cuts and wounds at the site of application, thereby reducing inflammation at the site of application [35].

4.11 *Euphorbia hirta*

The Euphorbia hirta *plant is a pantropical weed and is part of the Euphorbiaceae family.* It contains alkaloids, flavonoids, terpenes, chromenes, and sterols from all the plant parts. Quercetin is the active metabolite in euphorbia with potent antioxidant and anti-inflammatory properties that can be used to treat severe wounds, boils, and swellings. It is one of the oldest herbal medicines used today and is an effective remedy for acne, warts, and cancer. *E. hirta* is also known for its anti-inflammatory, antiseptic, and antifungal properties, making it a popular choice for treating skin ailments. It can also treat colds, fevers, and stomach pain. If you mix the extract with warm coconut oil and turmeric, it can be used to relieve the itching of boils and sores that are caused by bacteria. Roots of *E. hirta* are also adequate food nutrients for infants and breast-feeding mothers and are effective against snakebites. It has sedative properties and treats urinary tract infections [36]. The extract of *E. hirta* is also used to treat fever, diarrhea, and dysentery. It also improves digestion and treats respiratory problems such as asthma. It has analgesic properties and is used to relieve pain. Not only that, but it is also used to treat fever, cough, and other respiratory ailments. In addition, it is an effective anti-inflammatory and antiseptic. It can be administered topically to relieve pain and reduce swelling.

4.12 *Ficus carica*

In temperate climates, the fig tree (*Ficus carica* L.) is one of the world's oldest cultivated trees and one of the most popular edibles. It produces a sweet, nutritious, and versatile fruit enjoyed in fresh and dried forms. Figs are a high source of fiber, vitamins, and minerals, making them a great addition to any diet. It also contains phenolic acids and flavonoids. They are also known for their anti-inflammatory and antioxidant properties. There are two types of figs: fresh and dried. Fresh figs are available in the summer and early fall, while dried figs are open year-round. Figs can be enjoyed in various dishes, from salads to desserts. They can also be eaten as a snack on their own. Fresh figs are widely eaten near production areas because they are highly perishable. Figs have several nutrients, including fiber, potassium, calcium, and iron. Dried figs are a good source of dietary fiber and calcium. They are also a good source of minerals like magnesium, manganese, and copper. Figs are a good source of antioxidants, which can help protect the body from damage caused by free radicals. Several health benefits can be gained from the fruits, including treating skin ulcers, eczema, acne, skin dryness, and wounds, and preventing the aging process. It has also been found to have cholesterol-lowering properties and can aid in weight loss. Figs may also help reduce the risk of certain types of cancer. It has been shown that when applied to human cheeks for 8 weeks to investigate melanin, sebum, and trans-epidermal edema, the fruits reduced hyperpigmentation, acne, and wrinkles after 8 weeks of application. Figs may also have anti-inflammatory properties and can help protect the skin from environmental damage. Furthermore, they are a good source of dietary fiber and other essential nutrients. It has been suggested that the herbal extract of ficus can be used as a source for future skin care products [37]. Figs are a rich source of antioxidants, which can help to repair and protect the skin from oxidative stress. They also contain polyphenols, which can help to reduce the appearance of wrinkles and fine lines. Overall, figs can be a powerful source of nutrition for skin health.

4.13 *Allium cepa*

Since ancient times, onions have been used in the traditional Ayurvedic system of medicine. It contains Kaempferol, Myricetin, Quercetin, Quercitrin, and Allyl sulfide. For managing skin allergies and anti-aging, fruits and bulbs contain essential constituents such as pyruvic acid and quercetin. Onion helps as a remedy for various infections, including healing wounds, scarring, and cancers. Onions are rich in antioxidants, which help to reduce inflammation and fight infection. They also contain sulfur compounds that help to kill bacteria, fungi, and viruses. Onion juice is often used as a topical treatment for skin infections, and the poultice helps reduce scars. Raw or cooked onions can also be consumed for various health benefits, including improved digestion and cardiovascular health. In addition, onion extract has also been found to help prepare new herbal skin care products due to its anti-aging properties [38]. Onion extract is also rich in antioxidants, which help protect the skin from environmental damage, eczema, and psoriasis and reduce inflammation. It can also help reduce wrinkles and make the skin more elastic. Onion extract can also lighten the skin, reduce acne, and tighten facial pores. It can even help reduce dark circles under the eyes. The onion can be a skin irritant because it stimulates blood circulation, thus causing the skin to itch. Onion extract is also an effective antiseptic, helping to reduce the risk of infection and soothe sunburns and other minor skin irritations. Finally, it can also act as a natural skin toner. The poultice can also treat boils and warts and produce heat to speed up healing. Due to its strong scent, the onion can also be used as a natural insect repellent. It can also be used as an antiseptic to help protect wounds from infection. It inhibits ultraviolet B (UV-B) Inhibits free radical scavengers.

4.14 *Abrus percatorius*

It is also known as a jequirity bean or rosary pea containing a potent poison known as abrin. It also has luteolin, orientin, isoorientin, glycyrrhizin, abrectorin, abrusoside A to D, abrusogenin desmethoxycentaviridin-7-o-rutinoside, and abruquinones D, E, and F. The leaves with lime can be applied to scars, sores, blisters, and skin allergies. It helps stimulate blood circulation, prevents gray hairs, and heal wounds and leukoderma [40]. It also helps in controlling inflammation and cure rabies. The seeds also have insecticidal properties and can control worm infections [39].

4.15 *Daucus carota*

Carrots are the oldest cultivated vegetables and produce different colored varieties with different nutrients and flavors. It contains beta carotene, fiber, vitamin K1, potassium, antioxidants, and lutein. The presence of rich beta carotene maintains immunity, vision, and skin health by converting it into vitamin A in the body. Carrot oil is considered the bootstrapper of rich available vitamins A and E, which is beneficial for skin and hair and prevent wrinkles and aging [41]. It also produces ample vitamin C, producing collagen in the body. It is an essential protein for making the skin elastic and preventing wrinkles. It is also an excellent moisturizer with antiseptic and calming effects. It helps in detoxification to treat different skin problems. When mixed with sunflower, the roots of this plant can be a good anti-tanning agent. A study was conducted to determine the SPF of carrot oil as it contains high B-carotene concentration. Four oils, i.e., carrot seed oil, wheat germ oil, jojoba oil, and olive oil, are selected for B-carotene, a vitamin A precursor. It also contains vitamin B1, B2, B5, B6, and B9. Another study was executed with *Aloe vera*, *Cucumis sativus* and *Daucus carota* to prepare a polyherbal cosmetic cream. Oil in water preparations is incorporated with stearic acid and cetyl alcohol. The results of the studies showed no redness, edema, irritation, or inflammation on the skin of individuals, indicating that the formulation is stable and safe for the skin [42].

4.16 *Menta piperata*

Peppermint, an aromatic plant hybrid mint, is a combination of the flavors of spearmint and wild mint. It contains essential oils that contain menthone and carboxyl esters, mainly menthyl acetate, and dried peppermint contains volatile oil containing 1,8-cineol, menthone, menthol, menthofuran, and methyl acetate. Since ancient times, this herbal remedy has been used for various ailments, such as digestive problems, headaches, colds, and skin conditions. It helps reduce dandruff and the growth of hair lice due to its antimicrobial and antiseptic effects. It increases blood circulation in the scalp and can be used in the preparation of creams, shampoos, and various hair and skin products and also helps in preventing hair loss [43]

4.17 *Beta vulgaris*

Beetroot, the root vegetable, is deep red. It contains flavonoids, betalains (betaxanthins and betacyanins), saponins, polyphenols, and inorganic nitrate. It also contains potassium, zinc, sodium, copper, phosphorus, calcium, magnesium, iron, and manganese.

It contains flavonoids, betalains (e.g., betacyanins and betaxanthins), saponins, polyphenols, and inorganic nitrate. It also contains potassium, zinc, sodium, copper, phosphorous, calcium, magnesium, iron, and manganese. It has many health benefits, such as boosting stamina, reducing inflammation, and protecting cells from damage. Beetroot can be eaten raw, roasted, boiled, or juiced. Beetroot juice is rich in antioxidants and nutrients which improve skin health. When the phytoconstituents were investigated from their aqueous and methanolic extracts, they were rich, especially in quercetin, kaempferol, and phenolic compounds. Ointments were prepared at different concentrations and treated against acne and psoriasis, which showed a healing effect for acne. It was also observed that dried leaf extracts were more predominant than fresh leaf extracts. The dose prepared in the solution was more effective than ointment. The same result was obtained in the case of psoriasis, but in this case, the psoriasis ointment dose was more efficacious [44].

4.18 *Musa paradisiaca*

Beauty treatments, especially for skin and hair, are most effective with bananas. It contains Leucocyanidin (flavonoids). A small piece of banana skin placed on the wart with the yellow side out can treat warts. Banana peel can also help treat poison ivy rashes by rubbing the peel over the rashes. Rubbing peel on sore skin can increase discomfort. It helps reduce skin allergies, warts, poison ivy rashes, irritation, and inflammation. In such cases, a thick paste can be applied, reducing irritation and inflammation. Drinking smashed bananas in milk 2–3 times a day can treat skin allergies [23].

4.19 *Persea americana*

The avocado is an evergreen tree called an alligator or avocado pear. Due to the presence of vitamin C, vitamin E, carotenoids, and polyphenols, these compounds have an antioxidant effect, protecting cells from free-radical damage. It contains flavonoids, tannins, phenolic compounds, saponins and alkaloids, hydroxybenzoic acid, ferulic acid, catechin, caffeic acid, coumaric acid, chlorogenic acid, and triterpenoid glycosides. It also helps in the treatment of psoriasis, wrinkles, and stretch marks [45, 46].

4.20 *Calendula officinalis*

An annual and perennial herb with medicinal properties. Calendula CO2 extract acts as a well-standard cosmetic preparation, especially for the preparation of ointments for different skin diseases as well as for the preparation of cosmetics. It contains adinol, carotenoids, isorhamnetic, saponins, triterpenes, sesquiterpenoids, scopoletin, flavonoids, quercetin, and kaempferol. Anti-inflammatory principles of marigold by CO2 extraction are due to pentacyclic triterpene alcohols such as faradiol-3-monoesters and taxa sterols, which are the essential ingredients [47]. The extract was proven to be an effluent scavenger of hydrogen peroxide radicals under in vitro studies within the mitochondria of rat cardiac muscles. The section shows excellent antioxidant activity due to carotenoids, flavonoids, and polyphenols [48]. Water-alcohol extracts of marigolds were checked for their antimicrobial, antifungal, and antiviral properties against acne and dermal problems among young stars at the beauty and edicationsl centre "Top Beauty" in Sofia [49].

4.21 *Coffea arabica*

Coffee has been cultivated and consumed for centuries and contains caffeine, a stimulant that boosts energy and chlorogenic acid, which reduces redness when exposed to UV. It also has antioxidants and other beneficial compounds. Coffee can produce collagen and elastin and protect the skin from moisture loss. Coffee can protect against photo aging. According to the study by researchers, when hairless mice were exposed to UVB rays, and then caffeine applied to them 3 times/week for 11 weeks, they showed limited photodamage. This may be due to apoptosis (programmed cell death) of UVB, which removes damaged skin cells before they cause aging and skin cancer. An innovative facial cream has been designed that includes ingredients that help in improving the appearance of wrinkles, firmness, redness, and texture. According to different researchers, coffee bean extract rejuvenates skin aging. There were significant improvements in the formation of wrinkles, firmness, redness, and consistency of facial skin, and it also helps to prevent the development of aging and inflammatory skin disorders [50–53].

4.22 *Cucumis sativus*

It is a creeping plant widely cultivated for its edible fruit, which contains 95% water, 4% carbohydrates, and 1% protein with negligible fat content. It contains alkaloids, flavonoids, and a cyanogenic glycoside. *Cucumis sativus* extract mixed with *Glycyrrhiza glabra* extract and almond oil was evaluated for redness, edema, inflammation, and irritation. It was observed to be safe to use for skin [54]. The extracts of *Cucumis sativus* were investigated for their antimicrobial and antioxidant properties, and their cosmetic value in the treatment of acne shows the presence of phenolic and flavonoid compounds in fresh and dried plant extracts. An herbal cream prepared with fresh cucumber extract mixed with tea tree oil and linseed oil showed good spreadability, consistency, ease of removal, and no evidence of phase separation [55]. Herbal cream made of cucumber fruits, *Cyperus rotundus* roots, and almond oil were tested, showing no edema, redness, inflammation, or irritation during irritancy studies. These plant formulations are safe for use on the skin [56]. An emulsion of 3% cucumber extract and lemon oil was prepared to adjust the odor. The formulation was then evaluated to see the effects on sebum, erythema, melanin, moisture, and water loss on the skin of healthy human volunteers for 4 weeks. The odor was adjusted with a few drops of lemon oil. Over time, the smell disappeared due to the volatilization of lemon oil. The formulation shows eloquent effects on skin sebum secretion in which there was an increase in erythema while skin dehydration level and melanin in the skin decreased [57].

4.23 *Allium sativum*

It is an aromatic herb known for its biological properties and is a good source of antioxidants. It contains allicin, alliin, S-allyl cysteine, and methiin. When administered orally, it affects ultraviolet protection, cancer treatment, and immunologic properties.

Topically, garlic extract is effective against viral and fungal infections, alopecia areata, skin aging, psoriasis, keloid scar, leishmaniasis, and cutaneous horn. It also helps in skin rejuvenation and wound healing. However, the effectiveness of oral and topical garlic extract application is not yet appropriately explored [58]. An herbal shampoo was prepared with aqueous extract of *Allium sativum,* coconut oil, castor oil, and olive oil. The results of this combination help produce stable herbal shampoo with good foam formation, foam quality, and proper retention [59].

4.24 *Vitis vinifera*

Grapes are a rich source of antioxidants due to the presence of valuable phenolic compounds. It also contains flavonols, proanthocyanidins, flavon-3-ols, myricetin, flavonoids, peonidin, resveratrol, tannins, quercetin, anthocyanins, kaempferol, ellagic acid, and cyaniding. These phenolic compounds have anti-aging, anti-inflammatory, and antimicrobial actions and can permeate through the skin's barriers. The phenolic compounds present in grapes are the source of natural ingredients for cosmetics [60], fighting against acne [61], skin tightening and healing [62], reduction of dark circles surrounding eyes, hydration, anti-aging [63, 64], and UVB radiation protection [65, 66]. The active ingredients extracted from grapes can be used as cosmetics like skin-conditioning, antioxidants, flavoring agents, and colorants. The safety of concentration and use of these active ingredients from grapes has been confirmed by The Cosmetic Ingredient Review Expert Panel [67].

4.25 *Punica granatum*

The pomegranate is a fruit with a great deal of nutritional value, and its juice is an excellent source of antioxidants, vitamin C, and polyphenols, which may have health benefits in the future. It also contains gallic acid, ellagic acid, punicalagin's, and punicalins. Studies have shown that drinking pomegranate juice can improve heart health, reduce inflammation, and even boost memory. It is also a rich source of fiber, potassium, and vitamin K. A pomegranate extract, which is made from the peel, seeds, or flowers of the fruit, has been used in traditional medicine systems for a long to treat inflammations, infections, and cancers, among other conditions. An evaluation was carried out in a study in which creams were formulated with neem oil, jamul powder, and carrot powder at different concentrations based on the antioxidant properties of the phytoextracts and evaluated [68]. Upon evaluation of the extract and seed oil, it contained high levels of ellagic acid (peel, juice, seed extract) and punic acid (seed oil). Aside from these components, pomegranate derivatives contain other polyphenols that make them practical and valuable as anti-inflammatory and anti-aging agents for topical application. There are several advantages for using pomegranate seed oil, including its moisturizing and nourishing properties. Additionally, pomegranate seed oil can help reduce wrinkles and signs of aging and promote skin elasticity and firmness. It can also help to protect the skin from sun damage and environmental stressors. As a result, it is excellent for mature and aging skin, dry and cracked skin, and irritated and sunburnt skin. Pomegranate seed oil is also non-greasy and has anti-inflammatory properties, making it suitable for all skin types. It can also be a natural remedy to reduce acne, diminish blemishes and scars, and balance oily skin. This oil is used for treating acne, eczema, psoriasis, and rosacea and assisting in several specialist facials and body treatments. Additionally, pomegranate seed oil is a potent antioxidant, rich in fatty acids and vitamin C, which helps reduce inflammation and protect against free radical damage. It also helps to improve the skin's elasticity and can help to reduce wrinkles and fine lines. The combination of pomegranate and oil extracted from seed is highly effective in reducing and preventing fine wrinkles and firming up the epidermis [69]. It can also provide intense hydration and help to reduce inflammation while providing antioxidant protection. Pomegranate seed oil is also known to be a natural sun protectant and can help reduce age spots' appearance. Fermented juice and seed oil flavonoids of pomegranate were studied for antioxidant and eicosanoid enzyme inhibition properties. Pomegranate fermented juice (PFJ) and cold-pressed seed oil (PCPSO) showed vigorous antioxidant activity. A study of total phenols in PCPSO showed the presence of punicic acid (65.3%) along with palmitic acid, stearic acid, oleic acid, linoleic acid, and three unidentified peaks from which two peaks are probably isomers of punicic acid [70]. A study was conducted to check the sunscreen activity of combined *Pongamia pinnata* leaf extract and *Punica granatum* peel extract. To determine the sun protection factor, absorption spectroscopy, transmission spectroscopy, and the COLIPA standard method were used. Absorption spectroscopy measures the amount of UV radiation absorbed by the sunscreen, while transmission spectroscopy measures the amount of UV radiation transmitted through the sunscreen. The COLIPA standard method measures the amount of UV radiation reflected by the sunscreen. Combining the results of these three methods allows the sun protection factor to be accurately determined. Evidence suggests that sunscreens do a better job of protecting against UVA and UVB rays following the study by Patil and Fegade [71]. This is because the COLIPA standard method is designed to measure the radiation reflected by the sunscreen rather than the amount absorbed. This allows for a more accurate measurement of the sun protection factor, vital in protecting skin from the harmful effects of UVA and UVB rays.

4.26 *Cucurbita maxima*

The pumpkin is a fruit rich in nutrients, antioxidants, and fiber and has various health benefits, especially for the eyes, heart, skin, and immune system. Beta-carotene, an antioxidant in pumpkin, helps protect the cells from damage from free radicals and can reduce inflammation. It also contains vitamin A, which helps maintain healthy vision, and vitamin C, which helps strengthen the immune system. It contains peptides, fixed oils, carotenoids, polysaccharides, para-aminobenzoic acid, proteins, γ-aminobutyric acid, and sterols. Additionally, due to the presence of fiber in pumpkin, it helps to keep the gut healthy and can reduce the risk of certain diseases. Pumpkin also contains zinc, phosphorus, iron, magnesium, calcium, ferulic acid, alpha-linolenic acid, vitamin A, and vitamin B. The vitamins and minerals in pumpkins are essential for maintaining healthy vision, skin, cardiovascular, and immune health. The vitamins A and B are necessary for healthy skin and eyes, while zinc, magnesium, iron, phosphorus, and calcium are essential for bone, muscle, and heart health. Ferulic acid and alpha-linolenic acid's antioxidant and anti-inflammatory properties also benefit overall health. In addition to these nutrients, it contains other components such as copper, niacin, thiamin, riboflavin, salicylic acid, ferulic acid, unsaturated oils, retinol, antioxidants, and beta-carotene. As a powerful antioxidant and mild alternative to retinoic acid, pumpkin is an exfoliation accelerator. Zinc in pumpkin helps in wound healing by boosting the immune system and promoting the production of collagen in the skin, which helps keep the skin elastic and reduces the appearance of wrinkles. Beta-carotene and retinol are both antioxidants that help to protect the skin from damage caused by free radicals. The salicylic acid and unsaturated oils help to exfoliate the skin and reduce the appearance of pores. In addition to being rich in vitamin A derivatives, pumpkin also contains retinoids that hold a high affinity for the retinoic acid receptors in the skin, activating the cellular turnover process. In addition to its keratolytic properties, pumpkin enzymes are an excellent non-abrasive preparation for much stronger exfoliations while at the same time reducing the oxidative and free radical stress associated with chemical exfoliants [72]. Retinoids are a form of vitamin A that helps to reduce wrinkles, improve skin firmness, and even out skin tone. Pumpkin's retinoids target the retinoic acid receptors in the skin, which helps to promote skin cell turnover and reduce oxidative and free radical damage. This makes pumpkin enzymes an ideal choice for gentle exfoliation that can produce results similar to more potent chemical exfoliants.

4.27 *Camellia sinensis*

Green tea is one of the richest sources of antioxidants and catechins. Other compounds like theaflavins, Quercetin, L-Theanine, GABA, statin, and aromatic volatiles are also present. Furthermore, it has astringent properties as well. The raw material of green tea has a wide range of applications as a raw material. The catechins in green tea are known to have anti-inflammatory and antibacterial properties and are great for cardiovascular health. The astringent properties help to reduce inflammation, soothe skin, and even help to reduce wrinkles. Additionally, the raw material of green tea can be used in various ways, from skin care products to herbal teas. It also contains anti-inflammatory, antibacterial, and antiviral properties capable of boosting the immune system and protecting the body from infections [73]. The astringent properties of green tea work to tighten and tone the skin, which can reduce the appearance of wrinkles and other signs of aging. The antibacterial and anti-inflammatory properties can help reduce redness and inflammation, while the antiviral properties can help protect against viruses and other pathogens. As a result of the presence of polyphenols and other ingredients in the product, it may also provide some skin benefits. The polyphenols in green tea can neutralize free radicals, which are molecules that can cause cellular damage. This can lead to both anti-aging and skin protective benefits. Additionally, green tea can help to reduce inflammation and redness from skin conditions such as acne or rosacea. The antibacterial and antiviral properties of green tea can also help to reduce the risk of bacterial or viral infections on the skin. It has been reported that the formulations of green tea have been evaluated regarding the corneum transepidermal water loss status: water content, viscoelastic-to-elastic ratio (UV/UE), and skin microrelief. According to the results revealed from applying green tea extract for an extended period, this extract has a prolonged moisturizing effect on the skin. This can be attributed to the polyphenol and proanthocyanidin compounds present in green tea, which have antioxidant properties that help to protect the skin from environmental damage. Additionally, the polysaccharides in green tea help increase the skin's water content and create a protective barrier that helps lock in moisture. The skin microrelief improved significantly due to the topical application of the formulation [74]. The moisturizing effect of the green tea extract was due to its ability to increase the water content of the stratum corneum and reduce the transepidermal water loss. This resulted in a higher viscoelastic-to-elastic ratio (UV/UE) and improved skin microrelief. These effects were long-lasting, indicating the potential of green tea extract to be used as a moisturizing ingredient.

4.28 *Curcuma longa*

Curcumin is one of the main components of turmeric, a common spice known for its antioxidant and anti-inflammatory properties, which can help with various ailments such as arthritis and digestive issues. It also contains saponin, alkaloid, sterol, tannin, flavonoid, phytic acid, and phenol. It is also believed to have cancer-fighting properties. An evaluation of the effect of a cream prepared from turmeric extract on skin sebum secretion in humans was conducted using a sebumeter to measure the impact of the cream on human volunteers. The results showed that the cream had an anti-sebum effect and improved the skin's elasticity, which is beneficial in reducing wrinkles and other signs of aging. The anti-inflammatory and antioxidant properties of curcumin were also demonstrated, as the cream improved skin hydration and decreased the appearance of wrinkles. The DPPH method was used to determine the extract's antioxidant activity. In the 6th week after applying the cream, there was a significant increase in the observed sebum values. This suggests that the curcumin extract was able to penetrate deep into the skin and have a positive effect on the sebum levels, which helps to maintain healthy skin and reduce wrinkles. Additionally, the DPPH method is often used to measure antioxidant activity, and the findings suggest that the curcumin extract had a significant positive effect [75].

4.29 *Crocus sativus*

The spice of saffron is a precious spice, an expensive and rare spice that has been used in medicine for centuries. There are three main constituents of saffron, namely, croccin, picrocrocin, and crocetin, as well as safranal. It is harvested from the crocus flower and has many medicinal properties. It treats various illnesses, such as depression, anxiety, and insomnia. Saffron is also used in cooking to flavor and color dishes. Saffron is also known to have inflammatory and antioxidant properties, making it a voguish component in traditional medicines. It is also used in perfumes and cosmetics and is widely used in the Middle East and India. It is used to evaluate the antioxidant properties of saffron by using the DPPH (2, 2-Diphenyl-1-picrylhydrazyl) method (2, 2-Diphenyl-1-picrylhydrazyl). Using a hexameter, a water-in-oil (w/o) formulation of this plant extract was evaluated in human volunteers for the amount of melanin and erythema in their skins. A significant depigmentation and anti-erythematic effect were observed on human skin due to these results. In addition, applying the saffron formulation may also be beneficial in treating melanoma in humans. Further studies are necessary to assess the effect of *C. sativus* extract on melanoma. Animal studies can also be conducted to understand the mechanism of action better. Additionally, the safety of long-term use of the extract should be determined [76].

4.30 *Fragaria vesca*

Strawberries are not actual berries but are the richest source of vitamin C, manganese, folate, potassium, antioxidants, polyphenols, anthocyanins, and vitamins. Anthocyanins, ellagic acid, flavonols. The potential photoprotective capacity of different strawberry-based formulations was assessed in human dermal fibroblasts (HuDe) exposed to UV radiation, which showed noticeable photoprotection in HuDe, increasing cell viability due to the presence of Coenzyme Q10 (antioxidant) [77]

4.31 *Passiflora incarnata*

It is a flowering tropical vine which gained a lot of attention as it is a powerful source of antioxidants. Seed Oil is a superior salve, rich in vitamin C, calcium, riboflavin, niacin, iron, and phosphorus. Due to the high amount of fatty acids and liquid texture, a wide array of personal care products to nourish oily and dry skin are prepared from the seed oil of passion fruit. It increases vitality and improves hair growth. It can also be included in bath care products as it has a calming, sedating effect, and promotes relaxation. The oil is included in massage applications due to its anti-inflammatory, anti-spasmodic, and sedative properties [78].

4.32 *Petroselinum crispum*

This plant is generally considered a versatile herb that provides a concentrated source of nutrients. Rice in vitamins, minerals, and antioxidants helps boost the immune system and protect against illness. It also contains Flavonols (kaempferol and quercetin) and flavones (apigenin and luteolin). Additionally, it has anti-inflammatory properties that help in reducing puffiness and swelling. Many cosmetic industries use parsley to freshen the face, minimize discoloration, wrinkles, dark age spots, and freckles, prevent red sites, and strengthen hair. Several in vitro studies have shown that the herb has oil-controlling properties [79]. It can also be used as an antiseptic and can be used to treat acne. Parsley also has antimicrobial properties, which can help to protect the skin from infection.

4.33 *Theobroma cacao*

Cocoa beans are well-known, as they are rich in polyphenols. It contains carboxylic acid, phenolic acid, fatty acid, flavonoids (flavonol and flavones), stilbenoids, and terpenoids. Cocoa beans are a rich source of antioxidants and phytonutrients, linked to reducing inflammation, improving heart health, and promoting healthy blood sugar levels. They are also a good source of fiber and minerals such as magnesium, manganese, and zinc. To measure the antioxidant capacity of cocoa pods, a UHPLC extraction method was used to extract from cocoa pods that shows the quantification of the total polyphenols and flavonoids present in the cocoa pods. This method allowed researchers to determine cocoa beans' antioxidant capacity and their beneficial effects on the human body. To assess the impact of pod extract on skin whitening, elastase assays, collagenase assays, and sun screening tests were conducted to determine if they would inhibit the action of skin-degrading enzymes, and mushroom tyrosinase assays were performed to determine if they would impede the movement of skin-degrading enzymes. The elastase, collagenase, and sun screening assays showed that the pod extract had an inhibitory effect on skin-degrading enzymes, indicating that it could have a whitening effect. The results of the mushroom tyrosinase assay showed that the pod extract inhibited the activity of the mushroom tyrosinase, which further supported the idea that it could potentially have a whitening effect. The study results revealed that cocoa extract had antioxidant properties and showed a more decisive action against enzymes such as elastase and collagenase than other extracts. This is important because these enzymes can break down collagen and elastin, essential proteins that keep skin looking firm and youthful. By inhibiting the activity of these enzymes, the cocoa extract could help to protect the skin and reduce the signs of aging. Despite its limited performance as a UVA sunscreen agent, CPE also showed potential as a UVB sunscreen. The cocoa extract contains a type of polyphenol known as catechin polyphenols (CPE). Polyphenols have been shown to inhibit certain enzymes that break down collagen and elastin. By inhibiting these enzymes, CPE helps protect the skin from damage and reduce the signs of aging. In addition, CPE also has UVB protection properties, although its performance in this area is limited. CPE was found to contain carboxylic acid, phenolic acid, fatty acid, flavonoids (flavonol and flavones), stilbenoids, and terpenoids when LCMS analysis was conducted [80]. There are scientific validations for using cocoa-derived phytoconstituents as a practical approach for skin protection [81]. Cocoa butter is a common ingredient in different skin moisturizers, but cocoa's beneficial effects on the skin may extend beyond its topical agent. According to some studies, it has been found that cocoa flavonoids protect the skin from UV damage caused by the sun's UV rays [82]. Cocoa flavonoids contain antioxidants that have been found to reduce inflammation and improve skin elasticity. The appearance of dark spots and wrinkles is also reduced. In addition, these flavonoids can help protect against further skin damage caused by UV rays.

4.34 *Psidium guajava*

Guava is a tropical fruit full of nutrients and contains fibers, vitamin C, and folic acid. The leaves have inflammatory action and potent antibacterial activity. Acne and black spots can be eliminated with guava leaf extract. Guava leaves blended with water can be used as a scrub to remove blackheads. Due to its antioxidant properties, it protects the skin from damage and improves texture, skin tone, and skin tightening. The leaves can also give an instant cure for itchiness as they contain allergy-blocking compounds. An evaluation was done with the essential oil of guava against four dermatophytes, namely, *Microsporium canis, Trichophyton rubrum*, *T. tonsurans,* and *T. verusossum,* that were isolated from the infected part of the nail, skin, scalp, and genital organs of patients from the district hospital of Bareilly. Excellent inhibitory effects were given by the oil against all the pathogenic dermatophytes [83]. It has also been reported that aqueous extracts of guava leaves inhibit the growth of pathogenic microbes like *Proteus mirabilis,*

Streptococcus pyogenes, Escherichia coli, Staphylococcus aureus, and *Pseudomonas aureginosa,* which is more effective than the organic extracts. These results also suggest that aqueous extracts can be used as a natural antibacterial agent, providing a potential alternative to synthetic antibiotics. More research is necessary to confirm the efficacy of this method. Furthermore, the extract acne can also be used to formulate oral antibacterial drugs that are used to treat surgical and skin infections [84]. Extracts from medicinal plants can treat many bacterial infections and may provide an effective alternative to synthetic antibiotics. Further research is needed to confirm the efficacy of this method. Various herbal plants containing different metabolites and their uses are mentioned below in Tables 2, 3, and 4.

5 Conclusion

Phytoherbal remedy is an art of healing with plant-based medicine, penetrating deeply into Indian culture. With new emerging technology and modern medicinal practices, people from every part of the world depend on herbal healing. It is increasing daily due to its admiration, lower cost, less toxicity, few side effects, and widespreadness in our natural ecosystem. Moreover, to fulfill the needs of the growing human population and their deteriorating health conditions, nature is contributing an immense amount of herbs for the survival of human beings and other organisms on the planet. Phytomedicines provide tremendous options for curing various skin infections. Different herbal medicines have been tested against skin infections like rashes, acne, eczema, psoriasis, pigmentation, fungal, viral and bacterial infections, skin cancers, inflammation, cuts, wounds, etc. In these presented chapters, there is a mention of different plant-based products, from tulsi, neem, jatropha, ginger, garlic, cucumber, grapes, pomegranate, cucumber, pumpkin, turmeric, tea, passion fruit, papaya, etc., to treat various skin problems. There are mentions of multiple phytoconstituents from those plants that help treat different skin remedies. Consequently, the demand for herbal cosmetics has been increasing tremendously. Healthy and natural skin maintenance has become vital as it may lead to many severe conditions. Cosmetics can be prepared from a single plant or combinations of plant parts and polyherbal formulations. These formulations can be further explored against harmful microorganisms that cause severe health hazards. A comprehensive screening of these natural herbs can be a new commanding and valuable target for developing herbal-based industries that tie up with pharmaceutical companies. The positive effects of herbal medicines may help researchers reduce skin infections and screen out newer leading molecules, appropriate approaches and new processes in drug discovery.

Table 2 Applications of herbal phytomedicines in skin disorders

Sl no	Common name	Botanical name and family	Useful against skin infections	Parts used
1	Neem	*Azadirachta indica*, Meliaceae	Leaves paste in treating acne, eczema, scabies, skin allergies, psoriasis	Leaves, barks
2	Licorice	*Glycyrrhiza glabra*, Leguminosae	Skin lightening, anti-inflammatory, depigmentation, atopic dermatitis, protect skin from UV radiation, and reduce melanin production, allergic dermatitis	Roots
3	Tulsi	*Ocimum santhum*, Labiatae	Cure rashes, scars, leucoderma, allergy, blisters, reduce inflammation, and boost immune system, cure chickenpox, insect bites, round worm infections	Leaves
4	Bitterweed	*Andrographis paniculata*, Acanthaceae	Eczema, leucoderma, cuts, and wounds	Aerial parts, roots, and leaves
5	Amla	*Embillica officinalis*, Euphorbiaceae	Treatment of warts, skin infections, and prevents premature aging, reduces itching, wrinkles, and scabies	Fruit
6	*Aloe vera*	*Aloe barbadensis*, Lilaceae	Helps in healing wounds, burns, sunburns, insect bites, cuts, itching swelling, scabies, and dandruff	Leaves
7	Papaya	*Carica papaya*, Caricaceae	Eczema, warts	Roots, pulp, bark, seeds, and peels
8	American purging nut	*Jathropha curcus*, Euphorbiaceae	Major skin lesions, wound healing	Fruits, seeds, and roots
9	Ginger	*Zingiber officinalis*, Zingiberaceae	Skin problems	Root and rhizomes
10	Eucalyptus	*Eucalyptus globulus*, Myrtaceae	Heals fungal infection, skin problems, and healing of wounds	Leaves
11	Asthma plant	*Euphorbia hirta*, Euphorbiaceae	Major wounds, boils, swelling, and acne	Root
12	Common fig	*Ficus carica*, Moraceae	Hyperpigmentation, wrinkles, itchy skin and skin irritation, eczema, skin ulcers, and acne	Fruits
13	Onion	*Allium cepa*, Amaryllidaceae	Skin allergy, anti-aging, scars, and stimulates blood circulation	Bulbs
14	Rosary Pea	*Abrus precatorius*, Fabaceae	Leucoderma, gray hair prevention, and wound healing property	Leaves
15	Carrot	*Daudus carota*, Apiaceae	Hair and skin moisturizer, eczema, and warts	seeds

(continued)

Table 2 (continued)

Sl no	Common name	Botanical name and family	Useful against skin infections	Parts used
16	Peppermint	*Menta piperata,* Labitae	Reduces dandruff and lice	Seeds
17	Beetroot	*Beta vulgaris,* Chinopodiaceae	Acne, psoriasis	Roots and leaves
18	Banana	*Musa paradisiaca,* Musaceae	Skin allergies, warts, poison ivy rashes, irritation, and inflammation	Peel and fruit
19	Avocado	*Persea americana,* Lauraceae	Psoriasis, wrinkles, and stretch marks	Fruit
20	Pot marigold	*Calendula officinalis,* Asteraceae	Anti-inflammatory	Flowers and leaves
21	Coffee	*Coffee arabica, C. canephora*	Wrinkles, firmness, and redness	Leaves
22	Cucumber	*Cucumis sativus,* Cucurbetaceae	Photoaging, skin disorders	Fruit
23	Garlic	*Allium sativum,* liliaceae	psoriasis, alopecia areata, keloid scar, and wound healing	Bulb
24	Grape	*Vinis vitifera,* vitaceae	Removal of fungal and viarl infections, leishmaniasis, control aging of skin, and rejuvenates skin	Berry
25	Pomegranate	*Punica granatum,* Lythraceae	Antioxidant	Seeds
26	Pumpkin	*Cucurbita maxima,* Cucurbetaceae	Aid in the healing process and exfoliators	Fruit
27	Tea	*Camellia sinensis,* Theaceae	Antiaging, antioxidant and astringent activity, and skin roughness (62)	Leaves
28	Turmeric	*Curcuma longa,* Zingiberaceae	Skin sebum secretion	Rhizomes
29	Saffron	*Crocus sativus,* Iridaceae	Melanin and erythema, management of melanoma	Flower
30	Strawberry	*Fragaria ananassa,* Rosaceae	Antioxidant, photoprotection	Fruit
31	Passion fruit	*Passiflora edulis,* Passifloraceae	Hair growth, treating dry skin, treating muscular aches, and swelling	Fruit seed
32	Parsley	*Petroselinum crispum,* Apiaceae	Removes dark age spots, face discoloration, freckles, wrinkles, and prevents red spot occurence, growing and strengthening hair, control oils	Leaves

(continued)

Table 2 (continued)

Sl no	Common name	Botanical name and family	Useful against skin infections	Parts used
33	Cocoa Bean	*Theobroma cacao,* Malvaceae	Sunscreen, antioxidant properties,	Pods
34	Guava	*Psidium guajava* Myrtaceae	Acne, dark spots, and blackheads	Leaves

Source: [5, 6]

Table 3 Constituents and mechanism of action of herbal plants [5, 6]

Sl no	Common name	Chief constituents	Mechanism of action
1	Neem	Azadirachtin, limonoides, vitamin C, glycerides, polyphenols, triterpenes, quercetin, chtechins, carotenes, and beta-sitesterol	The product contains essential fatty acids and during healing add moisture and texture to the akin thereby inhibiting the inflammatory cell growth
2	Licorice	Glycerrihizin, liquiritin	Tyrosin B16 inhibition of melanoma cells by dispersed tyrosin B16
3	Tulsi	Contanis ursolic acid, rosmarinic acid, oleanolic acid, linalool, carvacrol, β-caryophyllene, and eugenol	Blocks cylcoxygenase lipoxygenase pathways of arachidonic acid metabolism and immune response enhancement
4	Bitterweed	Diterpenes, lactones flavonoids, alkenes, ketonaldehydes, and andrographolide	It inhibits the inflammatory mediators nitrous oxide release (NO), interleukin (IL), and prostaglandin E2 (PGE2)
5	Amla	Ascorbic acid, polyphenols (ellagic acid, chebulinic acid, gallic acid, chebulagic acid, apeigenin, quercetin, corilagin, leutolin, etc.)	Inhibits adhesion of microbes, causes inactivation of enzymes. Influenza A virus suppression and prevention of viral absorption, treats HUV by inhibiting HIV-RT
6	*Aloe vera*	Anthraquinones, phenolics, flavonoids, chromones, anthrones, steroids, alkaloids and tannins anthraquinones, phenolics, chromones, anthrones, flavonoids, steroids, alkaloids, and tannins	The proliferation cells of skin, blood cells production, enhancement of elastin and collagen production, suppresses the immunity of skin, self-hydration of skin, and removal of dead skin cells due to penetration
7	Papaya	Papain, phenolic acids, chymopapain, cyanogenic glucosides, tocopherol, cystatin, glucosinolates, and vitamin C	Increases prothrombin, removes necrotic tissues
8	American purging nut	Tannins, flavonoids, phenols, saponins, lignoides, steroids, coumarins, terpenoides, and alkaloids	Phagocytosis of bacterial cells, during proliferative phase release of cytokine can cause the migration and division of cells

(continued)

Table 3 (continued)

Sl no	Common name	Chief constituents	Mechanism of action
9	Ginger	Carbohydrates, lipids, terpenes, and phenolic compounds	Cytokinin cyclooxygenase and nitric acid is inhibited, inhibits nuclear factor kappa B (NF-Kb) cells
10	Eucalyptus	Sideroxylonal C, (+)-oleuropeic acid, cypellocarpins A, B, and C, cypellogins A, B, and C, leptospermone, isoleptospermone, grandinol	Inhibits gram-negative and gram-positive bacteria by interaction of trichosolan with enzymes of the fatty acid pathway
11	Asthma plant	Alkaloides, flavonoids, terpenes, chromenes, and sterols from all the plant parts	Serum cholesterol triglycerides and urea reduction in alloxan-induced m ice, due to antioxidant and free radical scavenger
12	Common fig	Phenolic acid and flavonoides	Inhibits tyrosin, reduces melanin content
13	Onion	Kaempferol, muricetin, quercetin, quercitrin, and allyl sulfide	Inhibits ultraviolet B (UV-B), inhibits free radical scavenger.
14	Rosary Pea	Luteolin O-rutinoside, orientin, abrectorin, isoorientin, desmethoxycentaviridin-7-, abrusoside A to D, abrusogenin, glycyrrhizin, and abruquinones D, E, and F	Collagen synthesis enhancement through elevated hydroxyproline
15	Carrot	Beta carotene, fibrem vitamin K1, potassium, antioxidants, and lutein	Acts on epidermal cells
16	Peppermint	Essential oils of peppermint contains menthone and carbaryl esters, mainly menthyl acetate while dried peppermint contains volatile oil containing menthyl acetate, menthone, menthofuran, menthol, and 1,8-cineol	Improvement in the circulation of bold in scalp
17	Beetroot	Contains betalains (e.g., betaxanthins and betacyanins), saponins, flavonoids, inorganic nitrate, and polyphenols. It also contains sodium, potassium, phosphorus, calcium, magnesium, copper, iron, zinc, and manganese	–
18	Banana	Leucocyanidin (flavonoids)	Reduce skin irritation and inflammation
19	Avocado	Flavonoides, saponins, phenolics, tannins, alkaloids, hydroxybenzoic acid, catechin, chlorogenic acid, ferulic acid, coumaric acid, caffeic acid, and triterpenoid glycosides	Protect cells from free radical damage
20	Pot marigold	Adinol, carotenoids, isorhamnetin, saponins, triterpenes, sesquiterpenoids, scopoletin, flavonoids, quercetin, and kaempferol.	Effective scavenger of H2O2 radicals, change in malondialdehyde level, superoxide dismutase glutathione, ascorbic acid, and catalase

(continued)

Table 3 (continued)

Sl no	Common name	Chief constituents	Mechanism of action
21	Coffee	Chlorogenic acid	Reduce redness on skin due to ultraviolet exposure, increase apoptosis (programmed cell death)
22	Cucumber	Alkaloids, flavonoids, and a cyanogenic glycoside	Skin sebum secretion
23	Garlic	Alliin, methiin and S-allylcysteine, and allicin	Immunologic properties, cutaneous microcirculation, and protection against UVB
24	Grape	Presence of phenolic acids, flavon-3-ols, myrcetin, flavonols, flavonoids, resveratrol, quercetin, tannins, ellagic acid, kaempferol, anthocyanins, cyaniding, and proanthocyanidins peonidin	Skin protection against UVB radiation, antioxidant, antiaging, and skin tightening
25	Pomegranate	Gallic acid, ellagic acid, punicalagins, and punicalins	Antioxidant potential
26	Pumpkin	Carotenoides, polysaccharides, paraaminobenzoic acid, fixed oils, peptides, proteins, sterols, and γ- aminobutyric acid.	Exfoliation accelerator, activation of cell turnover
27	Tea	Catechins, theaflavins, quercetin, L-Theanine, GABA, statin, and aromatic volatiles	Antioxidant and astringent activity
28	Turmeric	Saponin, alkaloid, sterol, tannin, flavonoid, phytic acid, and phenol.	Skin sebum secretion
29	Saffron	Crocin, Picrocrocin, and safranal	Change in the levels of skin erythema and melanin, management of melanoma
30	Strawberry	Vitamin C, manganese, folate, potassium, antioxidants, polyphenols, anthocyanins, and vitamins. Anthocyanins, ellagic acid, and flavonols	Photoprotective capacity, increasing cell viability
31	Passion fruit	Riboflavin, niacin, iron, and phosphorus	Anti-inflammatory, anti-spasmodic, and sedative properties
32	Parsley	Flavonols (kaempferol and quercetin), flavones (apigenin and luteolin)	Control oil secretion on the skin
33	Cocoa bean	Carboxylic acid, flavonoids (flavonol and flavones), stilbenoids, fatty acid, phenolic acid, and terpenoides	Higher inhibition toward tyrosinase enzyme, shows better action against collagenase and elastase enzymes
34	Guava	Fibers and vitamin C, Folic acid	Destroy the free radicals that damage skin

Source: [5, 6]

Table 4 List of some other phytomedicines

Sl no	Skin care products	Scientific name and family	Phytoconstituents	uses	Reference
1	Coconut oil	*Cocos nucifera*, Arecaceae	Glycerides and lower chain fatty acids	Baking and cooking, used as hair oil	[89]
2	Sunflower oil	*Helianthus annuus*, Asteraceae	Lecithin, tocopherols, carotenoids, and waxes	Products prepared for face and body	[90]
3	Jojoba oil	*Simmondsia chinenesis*, Simmondsiaceae	Linear liquid wax esters, alcohols, fatty acids, sterols, and vitamins	Used as moisturizer exotic fragrances carrier oil, removes odor	[91]
4	Olive oil	*Oleaeur opaea*, Oleaceae	Triolein,tripalmitin, trilinolein, tristearate, monosterate, triarachidin,squalene, β- sitosterol, and tocopherol	Skin and hair conditioners like lotions, shampoos, etc.	[85]
5	Rhodiola rosea	*Lignum rhodium,* Crassulaceae		Enhance physical endurance, increase in work productivity, longevity, resistance to high altitude sickness, treatment of fatigue, anxiety, anemia, impotence, gastrointestinal problems, skin disorders and infections, and nervous system disorders	[92]
6	Ginkgo	*Ginkgo biloba*, Ginkgoaceae	Glycosides, mostly quercetin and kaempferol derivatives, and terpenes	Antioxidant, anti-inflammatory, reduce UBV sunrays	[93]
7	Henna	*Lawsonia inermis*, Lythraceae	Gallic acid, resin, mannitol, fats, mucilage, glucose, and alkaloid traces	Henna powders are produced for hair application and mehendi designs on hands and legs	[94]
8	Shikakai	*Acacia concinna*, Leguminosae	Saponins, flavonoids, tannin, alkaloid, sugar, and anthraquinone glycosides	Hair cleaning and washing, hair growth improvement, purgative, expectorant, and emetic	[95]
9	Tamarind	*Tamarindu syndical*, Fabaceae	Rich source of sugars, contain minerals and excellent source of vitamin B, contains high antioxidant phenolic compounds	Improves skin health	[96]

Acknowledgment We would like to extend our sincere appreciation and gratitude to the Department of Agronomy & Plant Pathology at the School of Agriculture at Lovely Professional University, Punjab, 144411, India.

Conflicts of Interest A conflict of interest has not been identified

References

1. Yuan H, Ma Q, Ye L, Piao G (2016) The traditional medicine and modern medicine from natural products. Molecules 21(5):559
2. Bhatia M, Siddiqui N, Gupta S (2013) *Abrus precatorius* (L.): an evaluation of traditional herb. J Pharm Res 3:3296–3315
3. Parekh J, Jadeja D, Chanda S (2005) Efficacy of aqueous and methanol extracts of some medicinal plants for potential antibacterial activity. Turk J Biol 29(4):203–210
4. Gupta P, Kumar A, Sharma N, Patel M, Maurya A, Srivastava S (2017) A review on phytomedicines used in treatment of most common skin diseases. Indian J Drugs 5(4):150–164
5. Ahuja A, Gupta J, Gupta R (2021) Miracles of herbal phytomedicines in treatment of skin disorders: natural healthcare perspective. Infect Disord Drug Targets (Formerly Current Drug Targets-Infectious Disorders) 21(3):328–338
6. Chandrasekar R (2020) A comprehensive review on herbal cosmetics in the management of skin diseases. Res J Top Cosmet Sci 11(1):32–44
7. Lawrence HS, Nopper AJ (2018) Superficial bacterial skin infections and cellulitis. In: Principles and practice of pediatric infectious diseases. Elsevier, pp 436–444
8. Aghmiuni AI, Khiavi AA (2017) Medicinal plants to calm and treat psoriasis disease. In: Aromatic and medicinal plants–back to nature. IntechOpen, pp 1–28. 2016
9. Byrd Allyson L, Yasmine B, Segre Julia A (2018) The human skin microbiome [J]. Nat Rev Microbiol 16:143–155
10. Bhowmik D, Chiranjib YJ, Tripathi KK, Kumar KS (2010) Herbal remedies of *Azadirachta indica* and its medicinal application. J Chem Pharm Res 2(1):62–72
11. Satralkar S, Zagade TB (2019) Effectiveness of application of neem paste on face acne among teenagers in selected area of Sangli, Miraj and Kupwad Corporation. Int J Sci Res 8(6): 1387–1396
12. Lakshmi T, Krishnan V, Rajendran R, Madhusudhanan N (2015) *Azadirachta indica*: a herbal panacea in dentistry–an update. Pharmacogn Rev 9(17):41
13. Damle M (2014) Glycyrrhiza glabra (Liquorice)-a potent medicinal herb. Int J Herb Med 2(2): 132–136
14. Akhtar N, Khan M S, Iqbal A, Khan BA, Bashir S (2011) Glycyrrhiza glabra extract cream: effects on skin pigment 'Melanin'. In: 2011 international conference on bioscience, biochemistry and bioinformatics IPCBEE, 5
15. Radhakrishnan N, Gnanamani A, Sadulla S (2005) Effect of licorice (*Glycyhrriza glabra* Linn.), a skin-whitening agent on Black molly (*Poecilia latipinnaa*). J Appl Cosmetol 23(4):149
16. Nagarajan N, Kumar MS (2022) Queen of herb – *Ocimum sanctum* Linn (Tulsi) and its medicinal importance – a review. AIP Conf Proc 2473(1). https://doi.org/10.1063/5.0096483
17. Kulkarni KV, Adavirao BV (2018) A review on: Indian traditional shrub Tulsi (*Ocimum sanctum*): the unique medicinal plant. J Med Plants Stud 6(2):106–110
18. Chaiyana W, Anuchapreeda S, Punyoyai C, Neimkhum W, Lee KH, Lin WC, Lue SC, Viernstein H, Mueller M (2019) *Ocimum sanctum* Linn. as a natural source of skin anti-ageing compounds. Ind Crop Prod 127:217–224
19. Cohen MM (2014) Tulsi-Ocimum sanctum: a herb for all reasons. J Ayurveda Integr Med 5(4):251

20. Bayazid ALB, Jang YA (2021) The role of andrographolide on skin inflammations and modulation of skin barrier functions in human keratinocyte. Biotechnol Bioprocess Eng 26: 804–813
21. Okhuarobo A, Falodun JE, Erharuyi O, Imieje V, Falodun A, Langer P (2014) Harnessing the medicinal properties of Andrographis paniculata for diseases and beyond: a review of its phytochemistry and pharmacology. Asian Pac J Trop Dis 4(3):213–222
22. Khurana SK, Tiwari R, Khan S, Yatoo MI, Gugjoo MB, Dhama K (2019) *Emblica officinalis* (Amla) with a particular focus on its antimicrobial potentials: a review. J Pure Appl Microbiol 13(4):1–18
23. Kumar KPS, Bhowmik D, Duraivel, Umadevi M (2012) Traditional and medicinal uses of banana. J Pharmacogn Phytochem 1(3):51–63
24. Uchiyama T, Tsunenaga M, Miyanaga M, Ueda O, Ogo M (2019) Oral intake of lingonberry and amla fruit extract improves skin conditions in healthy female subjects: a randomized, double-blind, placebo-controlled clinical trial. Biotechnol Appl Biochem 66(5):870–879
25. Svitina H, Swanepoel R, Rossouw J, Netshimbupfe H, Gouws C, Hamman J (2019) Treatment of skin disorders with Aloe materials. Curr Pharm Des 25(20):2208–2240
26. Rajeswari R, Umadevi M, Rahale CS, Puspha R, Kumar S, Bhowmik D (2012) Aloe vera: the miracle plant its medicinal and traditional uses in India. J Pharmacog Phytochem 1(4):118–124
27. Kumar KPS, Debjit B (2010) Aloe vera: a potential herb and its medicinal importance. J Chem Pharm Res 2(1):21–29
28. Sánchez M, González-Burgos E, Iglesias I, Gómez-Serranillos MP (2020) Molecules: pharmacological update properties of aloe vera and its major active constituents 25(6):1324. https://doi.org/10.3390/molecules25061324
29. Manirakiza A, Irakoze L, Manirakiza S (2021) Aloe and its effects on cancer: a narrative literature review. East Afr Health Res J 5(1):1–16
30. Singh SP, Kumar S, Mathan SV, Tomar MS, Singh RK, Verma PK, Kumar A, Kumar S, Singh RP, Acharya A (2020) Therapeutic application of *Carica papaya* leaf extract in the management of human diseases. DARU J Pharm Sci 28:735–744
31. Aravind G, Bhowmik D, Duraivel S, Harish G (2013) Traditional and medicinal uses of *Carica papaya*. J Med Plants Stud 1(1):7–15
32. Thomas R, Sah NK, Sharma PB (2008) Therapeutic biology of *Jatropha curcas*: a mini review. Curr Pharm Biotechnol 9(4):315–324
33. Esimone CO, Nworu CS, Jackson CL (2008) Cutaneous wound healing activity of a herbal ointment containing the leaf extract of *Jatropha curcus* L. (Euphorbiaceae). Int J Appl Res Nat Prod 1(4):1–4
34. Kazerouni A, Kazerouni O, Pazya N (2013) Effects of Ginger (*Zingiber officinale*) on skin conditions: a non quantitative review article. J Turk Acad Dermatol 7:137–222
35. Takahashi R, Kukubo SM (2004) Antimicrobial activities of *Eucalyptus maculate*. Appl Microbiol 39:60–64
36. Kumar S, Malhotra R, Kumar D (2010) Euphorbia hirta: its chemistry, traditional and medicinal uses, and pharmacological activities. Pharmacogn Rev 4(7):58
37. Khan H, Akhtar N, Ali A (2014) Effects of cream containing *Ficus carica* L. fruit extract on skin parameters: in vivo evaluation. Indian J Pharm Sci 76(6):560
38. Upadhyay RK (2016) Nutraceutical, pharmaceutical and therapeutic uses of *Allium cepa*: a review. Int J Green Pharm (IJGP) 10(1):46–64
39. Kirtikar KR, Basu BD (1956) Indian medicinal plants. International Book Distributors, Dehradun. Lisa M, Pomegranate power. Time Magazine, December 2003. http://www.morretec.com/files/Pomegrante.pdf
40. Kuo SC, Chen SC, Chen LH, Wu JB, Wang JP, Teng CM (1995) Potent antiplatelet, anti-inflammatory and antiallergic isoflavanquinones from the roots of *Abrus precatorius*. Planta Med 61(04):307–312
41. Jyotsna A, Suryawanshi S (2016) In-vitro determination of sun protection factor and evaluation of herbal oils. Ijpr 6(1):37–43

42. Aswal A, Kalra M, Rout A (2013) Preparation and evaluation of polyherbal cosmetic cream. Pharm Lett 5(1):83–88
43. Sharma R, Hooda M (2017) Essential oils against dandruff: an alternative treatment. Int J Pharm Chem Res 3:248–251
44. Nidhal KM, Thukaa ZAJ, Anas TA, Hassan AJ (2014) Phytochemical study of the Iraqi beta vulgaris leaves and its clinical applications for the treatment of different dermatological diseases. WJPPS 3(8):05–19
45. Dreher ML, Davenport AJ (2013) Hass avocado composition and potential health effects. Crit Rev Food Sci Nutr 53(7):738–750
46. Oliveira APde, Franco EdeS, Barreto RR, Cordeiro DP, Melo RGde, Aquino CMFde et al (2013) Effect of semisolid formulation of Persea Americana Mill (Avocado) oil on wound healing in rats. Evid Based Complement Alternat Med 1–8. https://doi.org/10.1155/2013/472382
47. May P, Quirin KW (2014) Supercritical marigold flower CO2-extract-evergreen in evidence based cosmetic application. Cosmet Sci Technol 1:19–25
48. Bernatoniene J, Masteikova R, Julija D, Peciura R, Gauryliene R, Bernatoniene R et al (2011) Topical application of *Calendula officinalis* (L.). Formulation and evaluation of hydrophilic cream with antioxidant activity. J Med Plants Res 5(6):868–877
49. Ibrahim Z, Dimitrova Z, Georgiev S, Madzharov V, Titeva S (2010) use of cosmetic products containing extract of marigold (calendula officinalis) in cases of acne problem at the educational and beauty center "top beauty"-sofia. Trakia J Sci 8
50. Del Carmen Velazquez Pereda M, de Campos Dieamant G, Eberlin S, Nogueira C, Colombi D, Di Stasi LC, De Souza Queiroz ML (2009) Effect of green *Coffea arabica* L. seed oil on extracellular matrix components and water-channel expression in in vitro and ex vivo human skin models. J Cosmet Dermatol 8(1):56–62
51. Kitagawa S, Yoshii K, Morita S, Teraoka R (2011) Efficient topical delivery of chlorogenic acid by an oil-in-water microemulsion to protect skin against UV-induced damage. Chem Pharm Bull 59(6):793–796
52. Koo S, Hirakawa S, Fujii S, Kawasumi M, Nghiem P (2007) Protection from photodamage by topical application of caffeine after ultraviolet irradiation. Br J Dermatol 156(5):957–964
53. Palmer DM, Kitchin JS (2010) A double-blind, randomized, controlled clinical trial evaluating the efficacy and tolerance of a novel phenolic antioxidant skin care system containing *Coffea arabica* and concentrated fruit and vegetable extracts. J Drugs Dermatol: JDD 9(12):1480–1487
54. Singh M, Sharma S, Khokra SL, Sahu RK, Jangde R (2011) Preparation and evaluation of herbal cosmetic cream. Pharmacologyonline 2:1258–1264
55. Budhirajaa H, Guptaa RK, Pratibha N (2014) Formulation and characterization of *Cucumis sativus* extract in the treatment of acne. WJPPS 3(12):1043–1057
56. Rajvanshi A, Sharma S, Khokra SL, Sahu RK, Jangde R (2011) Formulation and evaluation of *Cyperus rotundus* and *Cucumis sativus* based herbal face cream. Pharmacologyonline 2(1): 1238–1244
57. Akhtar N, Mehmood A, Khan BA, Mahmood T, Muhammad H, Khan S, Saeed T (2011) Exploring cucumber extract for skin rejuvenation. Afr J Biotechnol 10(7):1206–1216
58. Pazyar N, Feily A (2011) Garlic in dermatology. Dermatol Rep 3(1):5–7
59. Saraf S, Hargude SM, Kaur CD, Saraf S (2011) Formulation and evaluation of herbal shampoo containing extract of *Allium sativum*. Res J Top Cosmet Sci 2(1):18–20
60. Soto ML, Falqué E, Domínguez H (2015) Relevance of natural phenolics from grape and derivative products in the formulation of cosmetics. Cosmetics 2(3):259–276
61. Aburjai T, Natsheh FM (2003) Plants used in cosmetics. Phytother Res 17(9):987–1000
62. Khanna S, Venojarvi M, Roy S, Sharma N, Trikha P, Bagchi D, Bagchi M, Sen CK (2002) Dermal wound healing properties of redox-active grape seed proanthocyanidins. Free Radic Biol Med 33(8):1089–1096
63. Luque-Rodríguez JM, De Castro MDL, Pérez-Juan P (2005) Extraction of fatty acids from grape seed by superheated hexane. Talanta 68(1):126–130

64. Spiers SM, Cleaves FT (1999) Topical treatment of the skin with a grape seed oil composition. Google Patents 60/002,553
65. Perde-Schrepler M, Chereches G, Brie I, Tatomir C, Postescu ID, Soran L, Filip A (2013) Grape seed extract as photochemopreventive agent against UVB-induced skin cancer. J Photochem Photobiol B Biol 118:16–21
66. Katiyar SK (2015) Proanthocyanidins from grape seeds inhibit UV–radiation-induced immune suppression in mice: detection and analysis of molecular and cellular targets. Photochem Photobiol 91(1):156–162
67. Fiume MM, Bergfeld WF, Belsito DV, Hill RA, Klaassen CD, Liebler DC, Marks JG Jr, Shank RC, Slaga TJ, Snyde PW (2014) Safety assessment of *Vitis vinifera* (Grape)-derived ingredients as used in cosmetics. Int J Toxicol 33(3_suppl):48S–83S
68. Matangi SP, Mamidi SA, Gulshan MD, Raghavamma STV, Nadendla RR (2014) Formulation and evaluation of anti aging poly herbal cream. Int J Pharm Sci Rev Res 24(2):133–136
69. Lisa M (2003) Pomegranate power. Time Magazine, December. http://www.morretec.com/files/Pomegrante.pdf
70. Schubert SY, Lansky EP, Neeman I (1999) Antioxidant and eicosanoid enzyme inhibition properties of pomegranate seed oil and fermented juice flavonoids. J Ethnopharmacol 66(1): 11–17
71. Patil S, Fegade B, Zamindar U, Bhaskar VH (2015) Determination of sun protection effect of herbal sunscreen cream. WJPPS 4(08):1554–1565
72. Allison RR (2013) Pumpkin E serum–vital nutrient treatment. White Paper 4(1):1–2
73. Jacek Arct BB, Anna Oborska KP, Poland BHC (2003) The tea ano its cosmetic application. J Appl Cosmetol 21:117–127
74. Gianeti MD, Mercurio DG, Maia Campos PMBG (2013) The use of green tea extract in cosmetic formulations: not only an antioxidant active ingredient. Dermatol Ther 26(3):267–271
75. Zaman SUZ, Akhtar N (2013) Effect of turmeric (*Curcuma longa* Zingiberaceae) extract cream on human skin sebum secretion. Trop J Pharm Res 12(5):665–669
76. Akhtar N, Khan HMS, Ashraf S, Mohammad IS, Ali F (2014) Skin depigmentation activity of crocus sativus extract cream. Trop J Pharm Res 13(11):1803–1808
77. Gasparrini M, Forbes-Hernandez TY, Afrin S, Alvarez-Suarez JM, Gonzàlez-Paramàs AM, Santos-Buelga C et al (2015) A pilot study of the photoprotective effects of strawberry-based cosmetic formulations on human dermal fibroblasts. Int J Mol Sci 16:17870–17884
78. http://www.naturalsourcing.com
79. Sahasrabuddhe S (2015) To study oil control property of parsley leaf extract in cosmetic products. Adv Biol Biomed 11(1):1–6
80. Abdul Karim A, Azlan A, Ismail A, Hashim P, Abd Gani SS, Zainudin BH, Abdullah NA (2014) Phenolic composition, antioxidant, anti-wrinkles and tyrosinase inhibitory activities of cocoa pod extract. BMC Complement Altern Med 14(1):1–13
81. Scapagnini G, Davinelli S, Di Renzo L, De Lorenzo A, Olarte HH, Micali G, Cicero AF, Gonzalez S (2014) Cocoa bioactive compounds: significance and potential for the maintenance of skin health. Nutrients 6(8):3202–3213
82. Williams S, Tamburic S, Lally C (2009) Eating chocolate can significantly protect the skin from UV light. J Cosmet Dermatol 8(3):169–173
83. Bhushan G, Sharma SK, Kumar S, Tandon R, Singh AP (2014) In-vitro antidermatophytic activity of essential oil of *Psidium guajava* (Linn.). Indian J Pharm Biol Res 2(2):57
84. Abubakar EMM (2009) The use of *Psidium guajava* Linn. in treating wound, skin and soft tissue infections. Sci Res Essays 4(6):605–611
85. Venkatachalam D, Thavamani SB, Varghese VK, Vinod KR (2019) Review on herbal cosmetics in skin care. Indo Am J Pharm Sci 6(1):781–789
86. Mohammad AA (2016) Therapeutics role of *Azadirachta indica* (Neem) and their active constituents in diseases prevention and treatment. Evid Based Complement Alternat Med 2016:7382506. https://doi.org/10.1155/2016/7382506. Epub 2016 Mar 1

87. Cohen BA (2022) Introduction to pediatric dermatology. In: Pediatric dermatology, 5th edn. pp 1–13. https://doi.org/10.1016/B978-0-7020-7963-4.00010-6
88. Dolenc-Voljč M (2014) Diseases caused by Malassezia species in human beings. In: The microbiology of skin, soft tissue, bone and joint infections. pp 77–91. https://doi.org/10.1016/B978-0-12-811079-9.00005-7
89. Athar M, Syed MN (2005) Taxonomic perspective of plant species yielding vegetable oils used in cosmetics and skin care products. Afr J Biotechnol 4:36–44
90. Gediya SK, Mistry RB, Patel UK, Blessy M, Jain HN (2011) Herbal plants: used as cosmetics. J Nat Prod Plant Resour 1:24–32
91. Rabasco AAM, Gonzalez RML (2000) Lipids in pharmaceutical and Cosmetic preparations. Grasas Aceites 51:74–96
92. Brown RP, Gerbarg PL, Ramazanov Z (2002) *Rhodiola rosea* A phytomedical overview. Herbal Gram. J Am Bot Counc 56:40–52
93. Dixit SN, Srivastava HS, Tripathi RD (1980) Lawsone, the antifungal antibiotic from leaves of *Lawsonia inermis* and some aspects of its mode of action. Indian Phytopathol 31:131–133
94. Chaudhary G, Goyal S, Poonia P (2010) *Lawsonia inermis* Linnaeus: a phytopharmacological review. Int J Pharm Sci Drug Res 2:91–98
95. Rao Diwan PV (2001) Herbal formulation useful as therapeutic and cosmetic applications for the treatment of general skin disorders. US6200570 B1
96. Khanpara K, Renuka V, Shukla J, Harsha CR (2012) A Detailed Investigation of shikakai (*Acacia concinna* Linn) fruit. J Curr Pharm Res 9:06–10

Metabolites and Phytochemicals in Medicinal Plants Used in the Management and Treatment of Neurological Diseases

19

Okon Godwin Okon

Contents

Abstract

In medicine, neurological disorders can be well defined as disorders which affect the brain and nerves found all through the body of humans as well as the spinal cord. Structural, biochemical or electrical abnormalities in terms of biochemistry,

O. G. Okon (✉)
Department of Botany, Faculty of Biological Sciences, Akwa Ibom State University, Uyo, Akwa Ibom State, Nigeria
e-mail: okonokon@aksu.edu.ng

S. C. Izah et al. (eds.), *Herbal Medicine Phytochemistry*, Reference Series in Phytochemistry, https://doi.org/10.1007/978-3-031-43199-9_22

electrical or structure in the brain, spinal cord or other nerves can lead to a range of symptoms. Some good examples of such symptoms include poor or loss of coordination and sensation, seizures, muscle weakness, paralysis, pain, confusion and an altered sense of consciousness. There are many causative factors when it comes to neurological disorders, this may include lifestyle, genetics, congenital abnormalities, brain injury, infections and nerve and spinal cord injuries. Neurological disorders include but are not limited to the following: Alzheimer's disease, brain aneurysm, epilepsy and seizures, meningitis, multiple sclerosis, muscular dystrophy, Parkinson's disease, stroke, migraine, headaches and encephalitis. Plants used in the treatment and management of neurological disorders include *Andrographis paniculata*, *Areca catechu*, *Centella asiatica*, *Sesamum indicum*, *Cinnamomum cassia*, *Euphorbia tirucalli*, *Piper methysticum*, *Areca catechu*, *Angelica sinensis*, *Centella asiatica*, *Aegle marmelos*, *Gardenia jasminoides*, *Tripterygium wilfordi*, *Xylopia aethiopi*, *Curcuma longa*, *Panax ginseng*, *Garcinia kola*, *Glycine max*, *Schizandrae chinensis*, *Ginkgo biloba*, *Piper nigrum*, *Withania somnifera*, *Cyperus rotundus*, *Suctellaria baicalensis*, *Silybinisus laborinum*, etc. these plants possess some key phytochemical groups such as Phenolics, alkaloids, saponins, flavonoids, triterpene, glyco withanolides, sesquiterpenes, lignans, neolignans, etc. While the active constituents/metabolites include curcumin, wogonin, flavonoid, silibinin, stilbenoid/resveratrol, hyperforin, rutin, ferulic acid, paeoniflorin, safranal, crocin, naringenin, berberine, chrysin, thymoquinone, nardostachysin, jatamansic acid, jatamansone, etc. These phytochemicals have been reported by many researchers to possess and exhibit neuroprotective properties. The plants discussed in this chapter as well as their active constituents provide great insight into affordable means of treating and managing neurological disorders as well as a basis for pharmaceutical companies searching for novel sources for the production of neuroprotective pharmacotherapy drugs.

Keywords

Brain · Curcuma longa · Curcumin · Metabolites · Naringenin · Neurological disorder · Panax ginseng · Phytochemicals

1 Introduction

Neurological diseases or disorders can be described as diseases/disorders which particularly affect or disturb the brain, spinal cord as well as nervous coordination which run throughout the human body. Distortions or abnormalities which can be biochemical, structural or electrical as the case may be in the brain, nerves or spinal cord can create an assortment of symptoms. These symptoms include seizures, poor coordination, paralysis, confusion, muscle weakness, altered levels of consciousness, loss of sensation and pain [1].

There are a great number range of specific causes of neurological diseases/disorders. The prominent of these causes are genetic disorders, contagions, deformities or disorders, malnutrition which could be linked to the health of one's environment, lifestyle, serious injuries to the brain, nerves as well as the spinal cord are also amongst the common causes of neurological diseases.

Let's also note that there's a big difference between neurological disorders and mental disorders (which could also be referred to as psychiatric illnesses, mainly deformities of ones way of thinking, behaviour as well as feeling ultimately resulting in anguish or diminishing of function). On the other hand, there are numerous neurological diseases, very few are rare ones while others are very common [1]. Figures from the US National Library of Medicine put the number of neurological disorders at 600.

Disabilities associated with neurological diseases/disorders are numerous, including cerebral palsy, brain tumours, epilepsy, autism, learning disabilities, ADD and neuromuscular disorders. A few of these disorders appear from childbirth thus making them congenital. Aside neurological diseases being congenital, some are actually caused by trauma, physical defects, tumours, degeneration and contagions all of which causes injury to the nervous system which will ultimately determine the magnitude of vision, communication, cognition, movement and hearing impairment [1].

The use of several synthetic agents and medication with neuroprotective properties has been properly documented. However, there are some discomforting side effects associated with these synthetic neuroprotective medications which include fretfulness or nervousness, fatigue, dry mouth, tiredness, drowsiness, struggle with balance, etc. However, the use of plants and herb with neuroprotective properties in recent times has gained prominence globally. This has led many plant scientists globally to focus on and research more into the metabolites and phytochemicals present in the said plants and their mechanism of neurological moderation and neuroprotective conferment against neurodegeneration.

Plants used in the treatment and management of neurological diseases and disorders include but are not limited to *Andrographis paniculata*, *Areca catechu*, *Centella asiatica*, *Sesamum indicum*, *Cinnamomum cassia*, *Euphorbia tirucalli*, *Piper methysticum*, *Areca catechu*, *Angelica sinensis*, *Centella asiatica*, *Aegle marmelos*, *Gardenia jasminoides*, *Tripterygium wilfordi*, *Xylopia aethiopi*, *Curcuma longa*, *Panax ginseng*, *Garcinia kola*, *Glycine max*, *Schizandrae chinensis*, *Ginkgo biloba*, *Piper nigrum*, etc. Some key phytochemical groups in the plants listed above include Phenolics, alkaloids, saponins, flavonoids, triterpene, glyco withanolides, sesquiterpenes, lignans, neolignans, etc. While the active constituents/metabolites include curcumin, wogonin, flavonoid, silibinin, stilbenoid/resveratrol, hyperforin, rutin, ferulic acid, paeoniflorin, safranal, crocin, naringenin, berberine, chrysin, thymoquinone, nardostachysin, jatamansic acid, jatamansone, etc.

According to Phani et al. [2], saponins, phenolic acids, flavonoids, polyphenols, terpenoids, etc. are regarded as the most essential phytochemicals from dietary sources. These phytochemicals work by decreasing or by retrogressive cellular injury and by decelerating advancement of neuronal cell loss.

This chapter with explicitly discuss the numerous medicinal plants that have been used in the management of neurological diseases and disorders globally already. It will also highlight the phytochemical groups and bioactive constituents/metabolites and their mechanism of neurological moderation and neuroprotective conferment against neurodegeneration as well as providing the basis for pharmaceutical companies searching for novel sources for the potential production of neuroprotective pharmacotherapy drugs.

2 Types of Neurological Diseases

Several disorders of the nervous system exist, which require professional healthcare physicians. Some of these disorders include: Alzheimer's disease, brain aneurysm, epilepsy and seizures, meningitis, multiple sclerosis, muscular dystrophy, Parkinson's disease, stroke, migraine, headaches, encephalitis, etc.

2.1 Alzheimer's Disease

According to John Hopkins Medicine [3], over 5.2 million people in America are affected with Alzheimer's disease particularly those above 65 years of age as well as thousands under the said age as well. Alzheimer's disease which is also regarded as the most common form of dementia affects about two-thirds of women as well.

The motor function of Alzheimer's patients is not affected when compared to other forms of dementia until in the later stages. Common behaviours associated with Alzheimer's disease include [3]: language descent, misperception, diminished memory, behaviour and thinking, agitation, compromised judgement, diminished ability to communicate, changes in behaviour and personality, impaired awareness, incapability to think of following directions and emotional indifference.

2.2 Brain Aneurysm

This (brain, intracranial or cerebral aneurysm) can be defined as a ballooning effect or swelling which occurs as a result of the weakening in a particular area specifically in the wall of the brain's blood vessel. Brain aneurysms usually do not show any symptoms and are insignificant in size (10 mm) and are usually not noticed until rupturing occurs. The smaller the aneurysm, the lower the risk and chances of rupturing [3].

More so, sentinel haemorrhage which occurs sparingly may show symptoms before rupture. This is as a result of a small quantity of blood leakage into the brain. These aneurysms symptoms arise sometimes because of the pressing on other adjourning structures like the eye nerves which results in a loss of vision and suspension of eye movement [3]. Some of the symptoms of an unruptured brain aneurysm include change in vision, reduced eye movement, headaches and pains in the eyes.

2.3 Epilepsy and Seizures

Epilepsy happens to be the utmost common neurological disorder which affects patients regardless of background, race or age. According to John Hopkins Medicine [3], over

2.2 million people in America live with epilepsy. Epilepsy is a nervous system disorder that involves the brain, thus making the patients vulnerable to repeated seizures.

One is described as epileptic or having epilepsy if he/she has two or more seizures. However, seizures may be caused by several factors that may disturb the networks between the brain and nerve cells such as concussion of the brain, alcohol, drug withdrawal, high/low blood sugar, high fever, etc. a few of the major causes of epilepsy include strokes, tumours, brain damage caused by either injury or illness, and disproportion in neurotransmitters.

2.4 Meningitis

An inflammation or swelling of the meninges results in the disease meningitis. Meninges here are the three layers that surround and protect the brain and spinal cord. It is usually, an infection that results in the swelling of the meninges [3].

Meningitis caused by viruses is more severe and more of a threat to life than those caused by less common bacterial. Meningitis caused by viruses is usually transmitted via different forms such as pitiable hygiene, sneezing, coughing and sparingly through common insects like ticks and mosquitoes [3]. Some symptoms of meningitis include misperception, seizures, stiffness of the neck, sleepiness, extreme temperature, headache, vomiting, photophobia and pains in body joints.

Symptoms observed in children include: fever, difficulty waking, backache, vomiting, loss of appetite, agitated, loud cry and change in skin colour to pale.

2.5 Parkinson's Disease

Parkinson's disease is a very serious neurodegenerative disorder that is progressive and shreds away all motor capabilities consequently, reducing the balance in individuals usually above 65 years of age. Parkinson's disease is sometimes regarded as hereditary or occurs extemporaneously [3]. However, it is known that this disease may occur as a result of dead brain cells in the "substantia nigra" resulting in the termination of dopamine production, hence loss of muscle movement or control [3].

Recent researches are currently targeted at investigating genes responsible for Parkinson's disease as well as discovering new complex biochemical pathways which is involved in the disease for targeted therapy.

2.6 Stroke

Stroke happens as a result of the stoppage of blood flow to the brain which put the patient in an emergency state. Brain functions appropriately when there is adequate nutrient and oxygen supply, hence, stoppage of blood supply notwithstanding duration may create a lot of complications [3].

Loss of brain functions due to dead brain cells may result in the patient's inability to carry out the following functions [3]: movement, eating, making speech, remember or think, regulate bowel and bladder movement and regulate additional dynamic body functions.

Strokes occur unannounced without prior notice. Strokes are of two types: haemorrhagic and ischaemic stroke.

Risk factors of stroke include: 140/90 or higher blood pressure, chronic smokers, usage of contraceptives, patients with heart diseases, diabetes, elevated blood cholesterol, drug abuse, obesity, excessive alcohol intake, not exercising the body, old age (from 55 years), genetics and males are more at risk of having a stroke than women.

2.7 Migraine

The intensity of migraines can range from modest intensity to tremendously severe ones which are usually characterized by an excruciating beating feel. Often time migraines occur on one part but can also occur on a whole or either part of the head, face or neck [3]. Photophobia and sensitivity to smell and noise are associated with migraines while vomiting is one of the most common symptoms. Migraines can occur as a result of the following: excessive alcohol consumption, insomnia, loss of water due to dehydration, menstruation, grinding of tooth, extreme weather changes and starvation.

2.8 Encephalitis

This is caused by swelling of the brain's active tissues either caused by an autoimmune reaction or infection leading to serious swelling in the brain with resultant effects like photophobia, rigid neck, headaches, seizures as well as mental misperception [3].

Observed symptoms of encephalitis include photophobia, sound sensitivity, headaches, high body, temperature, fever, unconsciousness and stiffness of the neck.

3 Medicinal Plants Used in the Treatment of Neurological Disorders

Medicinal plant can be any plant in which one or more of its parts contains bioactive substances which can be used for therapeutic purposes or which are antecedents for the synthesis of useful drugs. Right from time immemorial, natural plant products have occupied a place of prime importance in the field of medicine. They continue to be vital to many who do not have access to modern medicines or cannot afford synthetic drugs. Medicinal plants, e.g., herbs, and trees contain chemical compounds known to modern and ancient civilizations for their bioactive properties. Until the development of chemistry, particularly, the synthesis of secondary metabolites in the nineteenth century, medicinal plants and herbs have been the sole source of bioactive compounds capable of healing human ailments. Researchers believe that if medicinal herbs are taken in the appropriate doses and forms, they will be more effective than modern pharmaceutical drugs (synthetic drugs).

Plant products contain secondary metabolites that heal a variety of ailments and conditions in humans and animals. Over three-quarters of the world's population rely mainly on plants and their products for primary healthcare. About 30% of the entire plant species worldwide, at one time or the other was used for medicinal values [4].

It has been estimated that in developed countries like the USA, plant drugs constitute as much as 25% of the total drugs. Fast developing countries such as China and Singapore, their contributions are as much as 80%. Therefore, the economic uses of medicinal plants are much more in countries such as China than the rest of the world. These countries provide two-third of the plants used in modern systems of medicine and the health care delivery of rural population depends on indigenous systems of medicine. Of the 250,000 higher plant species on earth, more than 80,000 have medicinal purposes [5].

Current reviews and research have shown that there are a lot of medicinal plants with several phytochemicals which have been reported to decline neuropathy in animal models of CIPN. Curcumin, rutin, quercetin, matrine, euphol, thioctic acid, rosmarinic acid and cannabinoids are the most relevant natural products with therapeutic effects in CIPN [6] (Table 1).

4 Medicinal Plants, Bioactive Compounds/Phytochemicals and Mechanisms of Action

4.1 *Panax ginseng*

This medicinal plant is well known for its numerous healing properties and it is indigenous to the Chinese/Koreans. The plant has been credited with the ability to aid patients resist sicknesses by boosting the patient's immune system and has been used in the treatment of several illnesses including neurodegenerative disorder, diabetes, cancer, hypertension, etc. According to a research by Nah et al. [51], the active compound found in *Panax ginseng* ginsenoside, which is a saponin, has the potential to impede voltage-dependent Ca^{2+} channels linked by a G protein receptor sensitive to the Ca^{2+} which modulates neuronal Ca^{2+}. The neuroprotective effect of ginsenoside occurs when it acts on dopaminergic neurons by impeding the promotion of the nigral iron level, thus dropping expression of divalent metal transporter (DMT1), elevating ferroportin (FP1) expression in Parkinson's disease's patient [52].

4.2 *Ginkgo biloba*

Ginkgo biloba is widely used as a plant supplement with great build-up of bioactive compounds such as bilobalide, ginkgolides A, B, and C, quercetin,

Table 1 List of medicinal plants in the treatment of neurological disorders

Botanical names	Common Names	Family	Bioactive compounds/ phytochemicals	Mechanism of action/effects	References
Andrographis paniculata	King of bitters	Acanthaceae	Diterpene lactone and Andrographolide	Decreases of aggregation of platelet. Increases endothelial nitric oxide synthase (eNOS) activity	[7]
Areca catechu	Areca, Betel-nut Palm, Pinang	Arecaceae	Alkaloids, Arecoline, Guvacine, Arecaidine	Involvement in the binding activity of muscarinic (M2)	[8, 9]
Sesamum indicum Linn	Sesame	Pedaliaceae	Sesamol	By improving cognitive functions and the reduction of oxidative stress	[10, 11]
Cinnamomum cassia	Chinese cinnamon or cassia	Lauraceae	Coumarin	Reduces cold allodynia	[12]
Euphorbia tirucalli	Milk bush	Euphorbiaceae	Euphol	Reduces persistent and mechanical hypersensitivity	[13]
Piper methysticum	Kava	Piperaceae	Kavain, Yangonin, Kavalactones, Methysticin	Modulation of γ-aminobutyric acid type A (GABAA) receptors	[14]
Angelica sinensis	Garden angelica	Apiaceae	Ferulic acid	Involvement in antioxidant action	[15]
Centella asiatica	Indian pennywort	Apiaceae	Asiatic acid/triterpene	Protects against dysfunction caused by 3-nitropropionic acid (3-NP). Decreases acetylcholinesterase (AChE) and involvement in antioxidant activity	[16, 17]
Aegle marmelos	Bilwa or bael	Rutaceae	Alkaloid-amide and Aegeline	Reduces reactive oxygen species (ROS)	[18]
Gardenia jasminoides	Common gardenia or cape jasmine	Rubiaceae	Crocin/carotenoid	Decreases acetylcholinesterase (AChE) and oxidative stress	[19]

Tripterygium wilfordii	Thunder god vine	Celastraceae	Triptolide	Increases cognitive function, complemented by reduction in neuroinflammation and Aβ deposition as well as the improvement of memory	[20, 21]
Xylopia aethiopica	Spice tree	Annonaceae	Xylopic acid	Reduces inert motorized hyperalgesia	[22]
Curcumin longa	Turmeric	Zingiberaceae	Curcumin	Upsurge manifestation of brain neurotropic factor by the reversal of hippocampal neurogenesis caused by stress	[23, 24]
Panax ginseng	Ginseng	Araliaceae	Ginsenoside and triterpene glycosides	Promotion of hippocampal gamma-aminobutyric acid (GABA) level in the brain through the increased enzyme activities of glutamate decarboxylase. Reduces Ca^{2+}	[25]
Garcinia kola	Bitter kola	Clusiaceae	Kolaviron	Reduces necrotic cell death and myeloperoxidase (MPO)	[26]
Glycine max	Soybean	Fabaceae	Genistein and flavonoid	Reduces monoamine oxidase (MAO) and inflammation	[27, 28]
Schisandrae chinensis	Chinese magnolia vine	Magnoliaceae	Schizandrin	Plays a part in acetylcholinesterase (AChE) and γ-Aminobutyric acid type A (GABAA) receptors	[29]
Ginkgo biloba	Maidenhair trees	Ginkgoaceae	Ginkgolides, Gastrodin, Vanillin, Bilobalide	By promotion of hippocampal gamma-aminobutyric acid (GABA) levels.	[30]
Piper nigrum	Black pepper	Piperaceae	Piperine	Suppresses the actions of N-methyl-D-aspartate (NMDA) receptor	[31]
Withania somnifera	Indian Winter cherry	Solanaceae	Withanolides/Withanols/ Withamides/Withanone/Amido compounds	Stabilization of mood	[32]
Cyperus rotundus	Nut grass	Cyperaceae	α-Cyperone	Decreases of aggregation of platelet	[33]

(continued)

Table 1 (continued)

Botanical names	Common Names	Family	Bioactive compounds/ phytochemicals	Mechanism of action/effects	References
Scutellaria baicalensis	Huang Qin or Chinese skullcap	Lamiaceae	Wogonin/flavonoid	Decrease in monoamine oxidase A (MOA-A) and synthesis of thrombin	[34]
Silybinisus laborinum		Fabaceae	Silibinin	Decreases acetylcholinesterase (AChE) and reactive oxygen species (ROS)	[34, 35]
Vitis vinifera	Common grape	Vitaceae	Stilbenoid/Resveratrol	Decreases beta-amyloid plaques	[36]
Hypericum perforatum	St. John's wort	Hypericaceae	Hyperforin and Rutin	Decrease in monoamine oxidase A (MOA-A) and monoamine oxidase B	[37]
Paeonia lactiflora	Common garden peony	Paeoniaceae	Paeoniflorin	Controlling adrenergic systems	[38]
Crocus sativus	Saffron crocus	Liliaceae	Safranal, Crocin	By promotion of hippocampal gamma-aminobutyric acid (GABA) levels	[39]
Citrus paradise	Grapefruit	Rutaceae	Naringenin	Reduces the inflammation of cytokines	[40]
Citrus sinensis	Orange	Rutaceae	Naringenin	Reduces the inflammation of cytokines	[40]
Berberis genus	Barberry	Berberidaceae	Berberine	Decreases acetylcholinesterase (AChE)	[41]
Hypericum afrum	–	Hypericaceae	Flavonoid	Decrease in monoamine oxidase A (MOA-A)	[27]
Cytisus villosus	Brooms	Fabaceae	Chrysin	Decrease in monoamine oxidase A (MOA-A)	[27]
Nigella sativa	Black seed	Ranunculaceae	Thymoquinone	Upsurge in GABAergic tone via opioid receptor	[42]
Ficus platyphylla	Broad leaf fig	Moraceae	Saponin	Overturns excitatory and inhibitory synaptic traffic. Works by the suppression of excited synaptic traffic	[43]
Nardostachys jatamansi	Balachara	Valirenaceae	Nardostachysin, Jatamansic acid, Jatamansone	Enhances dopamine	[44]

Biota orientalis	Oriental arborvitae	Cupressaceae	Pinusolides	Confers neuroprotection	[45]
Camellia sinensis	Green tea	Theaceae	Epigallocatechin gallate/catechin	Decreases acetylcholinesterase (AChE) and reduces oxidative stress	[19, 46]
Valeriana officinalis	Garden heliotrope	Caprifoliaceae	Hesperidin/flavonoid	Reduces oxidative stress	[35]
Angelica gigas	Korean angelica	Apiaceae/ Umbelliferae	Decursin/pyranocoumarin	Decrease in monoamine oxidase A (MOA-A)	[47]
Galanthus nivalis	Snowdrop	Amaryllidaceae	Galanthamine	Inhibits acetylcholinesterase AChE	[48]
Gastrodia elata	Tian ma	Orchidaceae	Gastrodin	GABA transaminase inhibition	[30]
Ziziphus jujube	Common jujube	Rhamnaceae	Spinosin, Jujubosides	Reduction of hippocampal hyper-action	[49]
Allium sepa	Onions, shallots	Alliaceae	Allium and Allicin	Activation of sulphides containing allyl conveys neuroprotective properties by the initiation of mitochondrial uncoupling proteins	[50]
Allium sativum	Garlic	Alliaceae	Allium and Allicin	Activation of sulphides containing allyl conveys neuroprotective properties by the initiation of mitochondrial uncoupling proteins	[50]
Ocimum sanctum	Tulsi	Lamiaceae	Essential oils	Involvement in dopamine functions	[2]
Rubia cordifolia	Common madder	Rubiaceae	Terpenoids	Increases gamma-aminobutyric acid (GABA) levels	[2]
Cymbopogon citratus	Lemongrass	Gramineae	Essential oils	Increases gamma-aminobutyric acid (GABA) neurotransmission	[2]
Salvia lavandulaefolia	Spanish sage	Lamiaceae	α-Pinene, 1,8-cineole β-Pinene	Inhibits acetylcholinesterase AChE	[2]

isorhamnetin, kaempferol, etc. *Ginkgo biloba* has been reported to have positive effects on neuropathy (vincristine-induced) [53]. Research also shows the impeding properties of *Ginkgo biloba* through the nitrosative- and oxidative-induced neural damage on diabetic neuropathy in animal model [54]. Reports by Taliyan and Sharma [55] also show ginkgolide B regulation of neural damage via cellular pro-apoptotic pathways.

4.3 *Curcuma longa*

Curcuma longa also known as turmeric is known for its numerous benefits and medicinal properties. In relation to neurodegenerative disorder, curcumin found in *Curcuma longa* has been reported to be beneficial in management and treatment of the disorder. Xu et al. [56] reported curcumin benefits which includes betterment of behavioural deficits and protects the neurons against ischemic cell death as observed in animal models. Other researches have also reported that curcumin has been used in in vitro treating of age-related neurodegenerative diseases. Lingering stress-induced damage of hippocampal neurogenesis has been reportedly upturned by curcumin as well as the potential to increase the manifestation of brain-derived neurotrophic factor (BDNF) [56, 57].

4.4 *Allium sativum* and *A. sepia*

Allium sativum (garlic) and *A. sepia* (onions) belongs to the family Alliaceae. They possess high quantity of allium and allicin (organosulphur compounds) which have been reported to have positive effects on neurodegenerative disorders and free radical foraging actions. Activation of sulphides containing allyl conveys neuroprotective properties by the initiation of mitochondrial uncoupling proteins [50].

4.5 Camellia sinensis

Camellia sinensis also referred to as tea, in which the popular black and green teas are produced from depending on their degree of fermentation. This plant is a great source of polyphenolics. Phytochemicals and other bioactive compounds found in *Camellia sinensis* include flavonoids (aflavins) and catechins (epicatechin, gallocatechin, epigallocatechin-3-gallate (EGCG), epigallocatechin) all of which drops chemotherapy-associated dose-preventive side effects in the early stage of therapy [58]. ECCG has been reported in neurological disturbances studies in many models as a result of its antioxidant and anti-inflammatory properties [59, 60].

5 Chemical Structures of Bioactive Compounds/Phytochemicals Found in Plants Used in the Treatment and Management of Neurological Disorders

See Figs. 1, 2, 3, 4, 5, 6, 7, 8, 9, 10, 11, 12, 13, 14, 15, 16, 17, 18, 19, 20, 21, 22, 23, 24, 25, 26, 27, and 28.

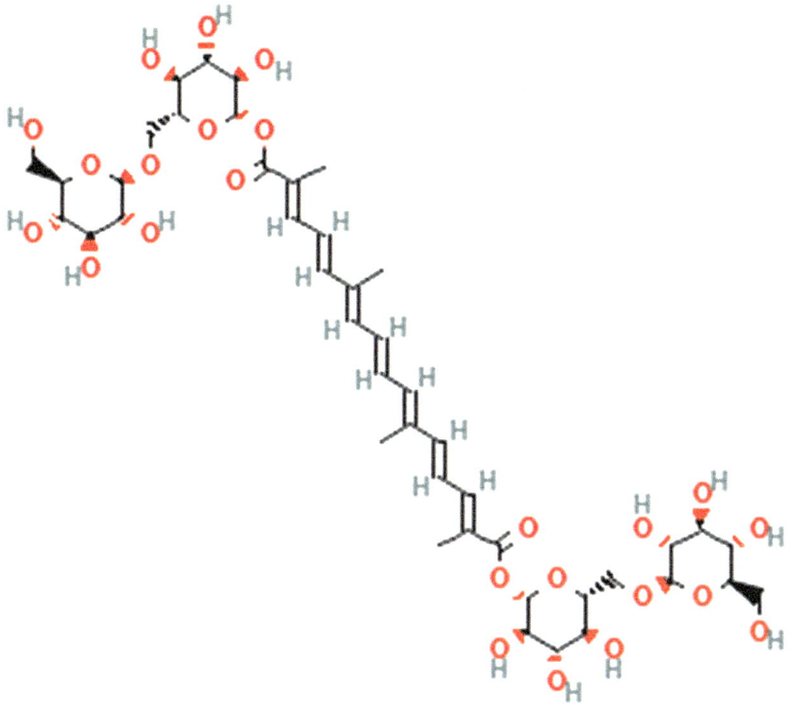

Fig. 1 Crocin (*Crocus sativus*)

Fig. 2 Naringenin (*Citrus sinensis*)

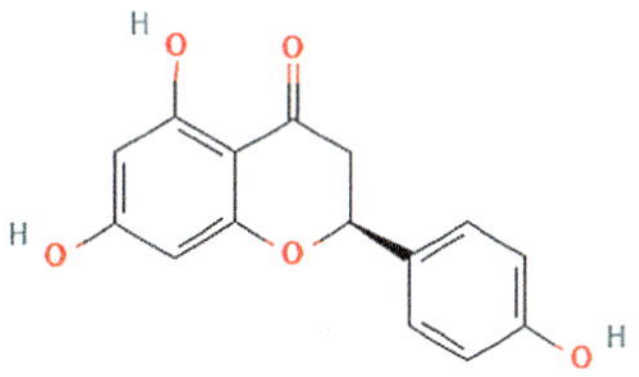

Fig. 3 Silibinin (*Silybinisus laborinum*)

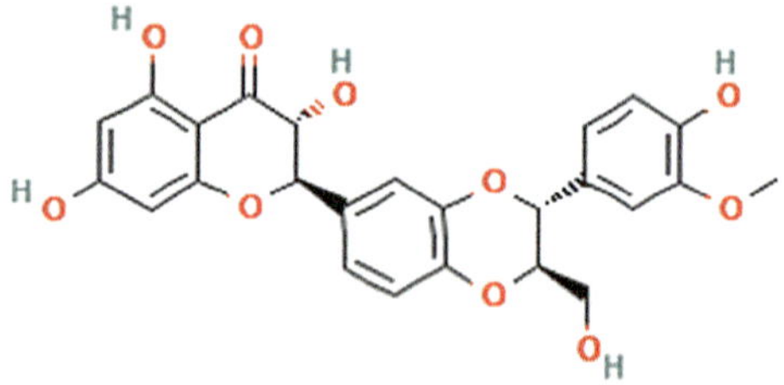

Fig. 4 Curcumin (*Curcumin longa*)

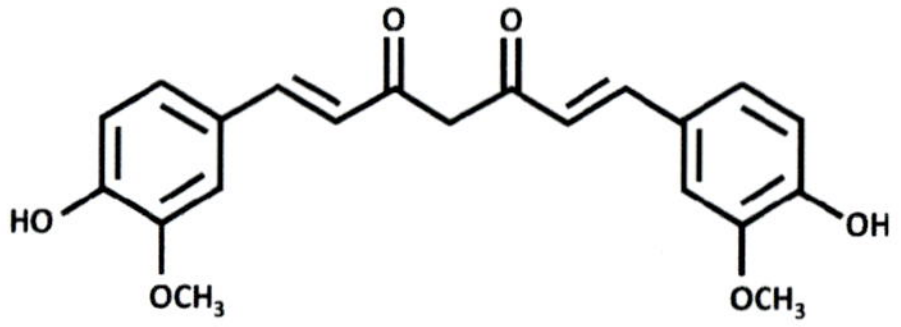

Fig. 5 Spinosin (*Ziziphus jujube*)

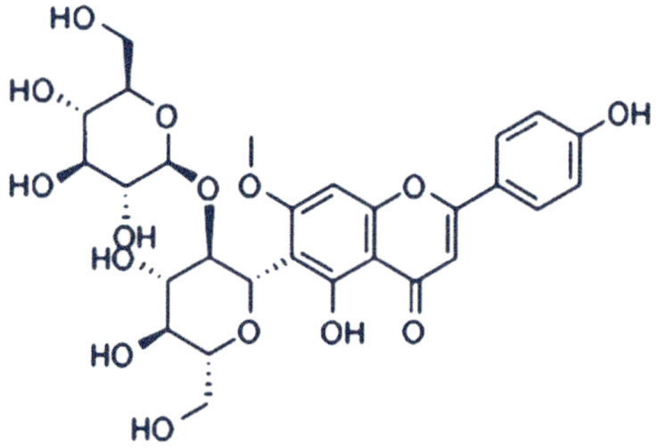

Fig. 6 Arecaidine (*Areca catechu*)

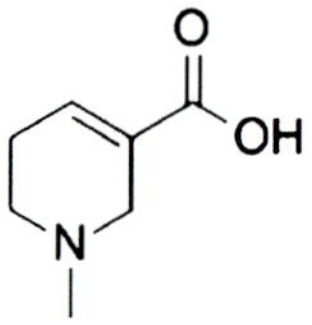

Fig. 7 Gastrodin (*Gastrodia elata*)

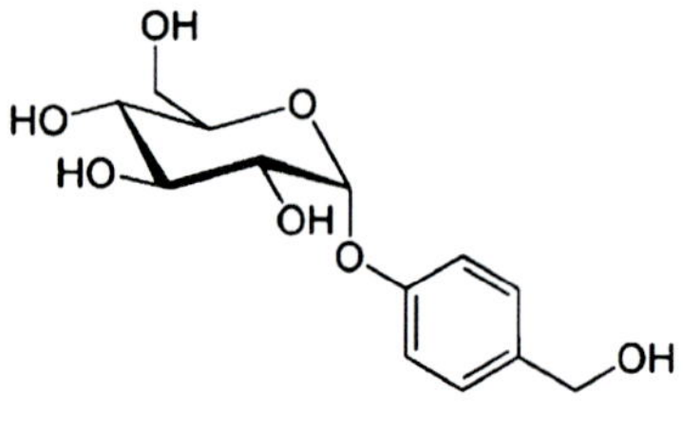

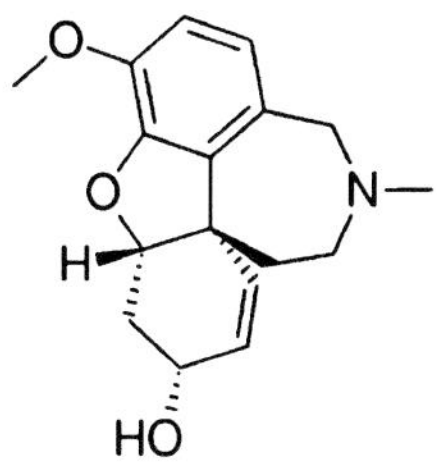

Fig. 8 Galanthamine (*Galanthus nivalis*)

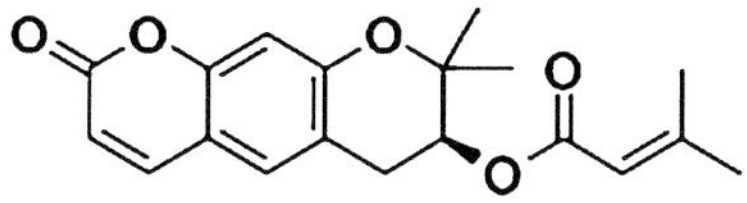

Fig. 9 Decursin (*Scutellaria baicalensis*)

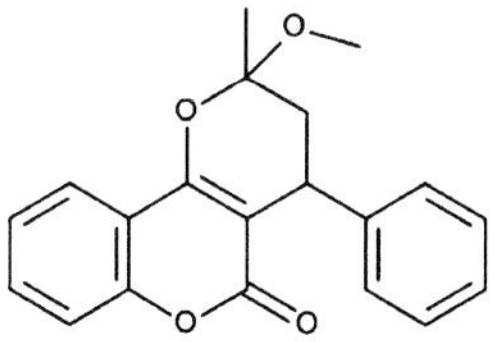

Fig. 10 Pyranocoumarin (*Scutellaria baicalensis*)

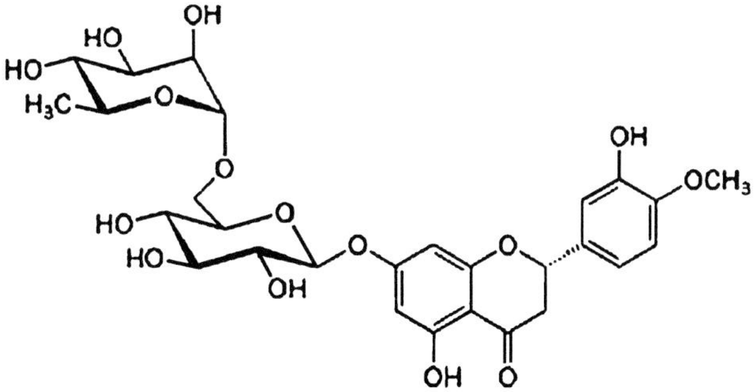

Fig. 11 Hesperidin (*Valeriana officinalis*)

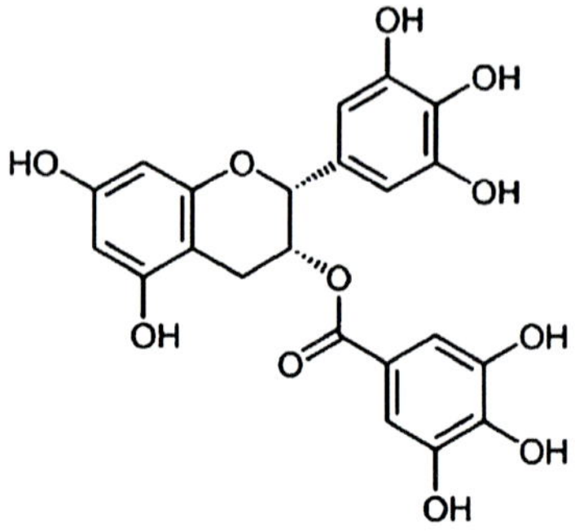

Fig. 12 Epigallocatechin gallate (*Camellia sinensis*)

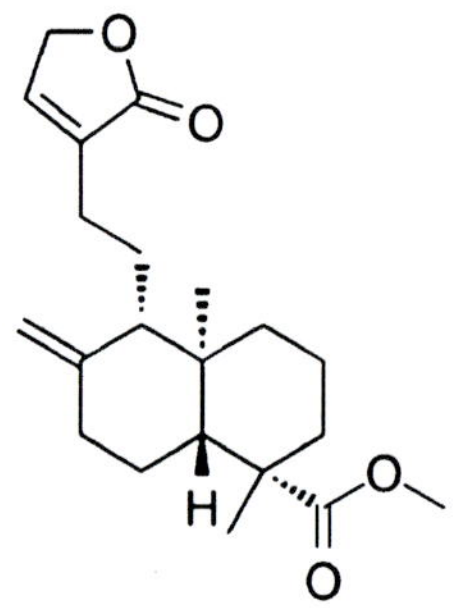

Fig. 13 Pinusolides (*Biota orientalis*)

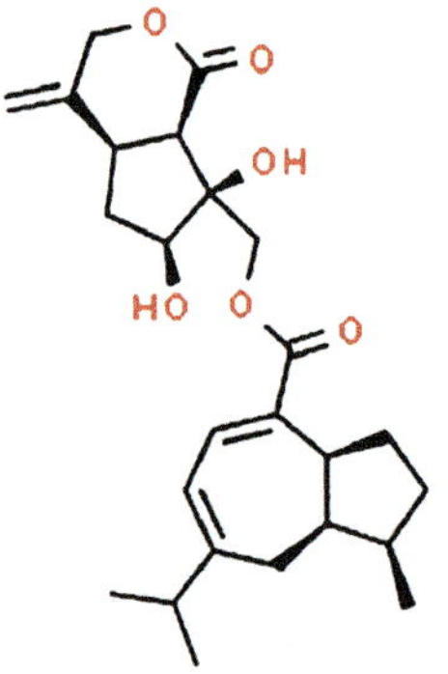

Fig. 14 Nardostachysin (*Nardostachys jatamansi*)

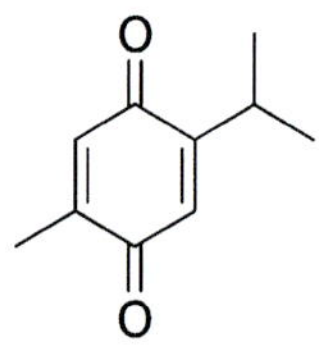

Fig. 15 Thymoquinone (*Nigella sativa*)

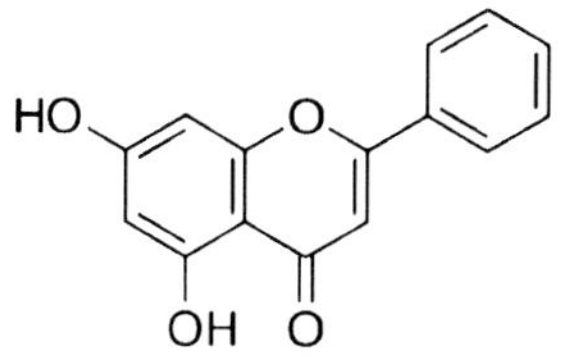

Fig. 16 Chrysin (*Cytisus villosus*)

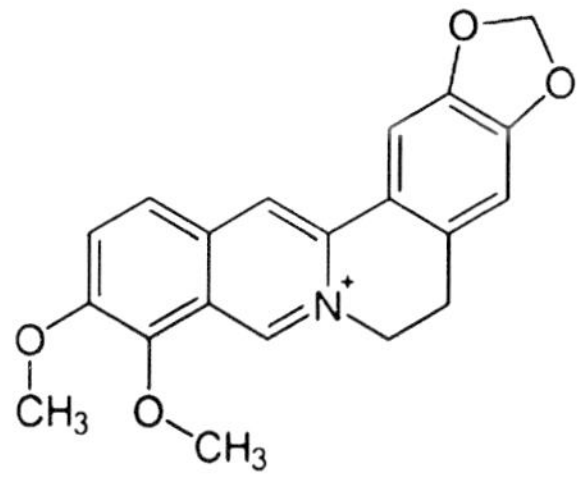

Fig. 17 Berberine (*Berberis* genus)

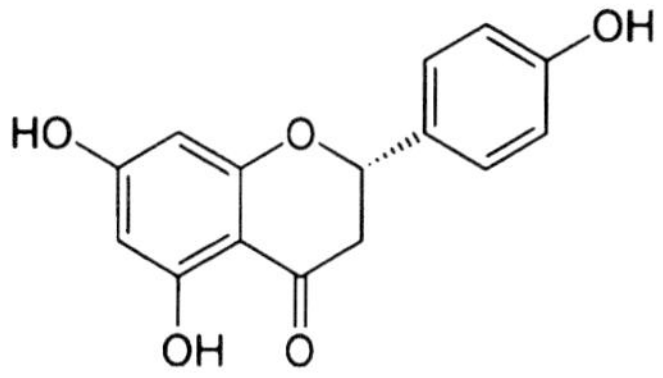

Fig. 18 Naringenin (*Citrus paradise*)

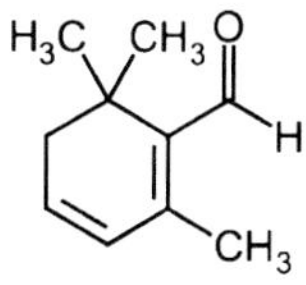

Fig. 19 Safranal (*Crocus sativus*)

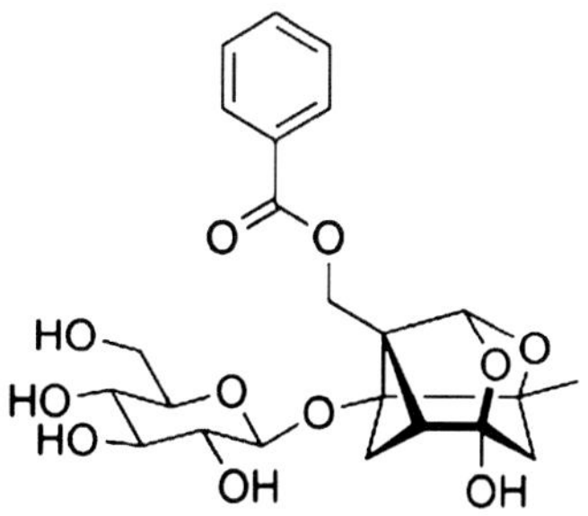

Fig. 20 Paeoniflorin (*Paeonia lactiflora*)

Fig. 21 Rutin (*Hypericum perforatum*)

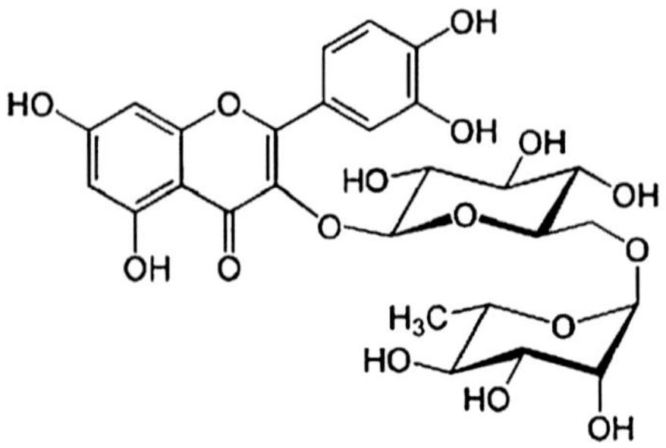

Fig. 22 Stilbenoid (*Vitis vinifera*)

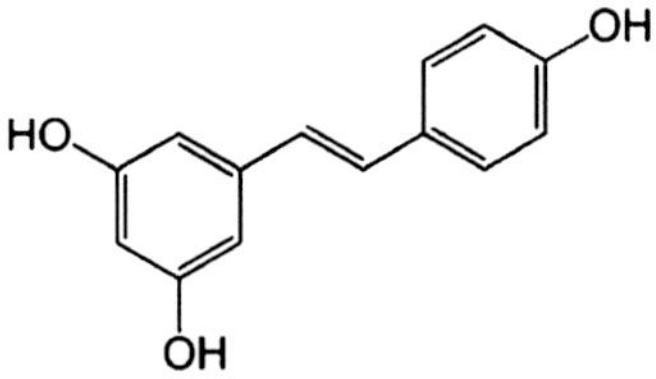

Fig. 23 Wogonin (*Suctellaria baicalensis*)

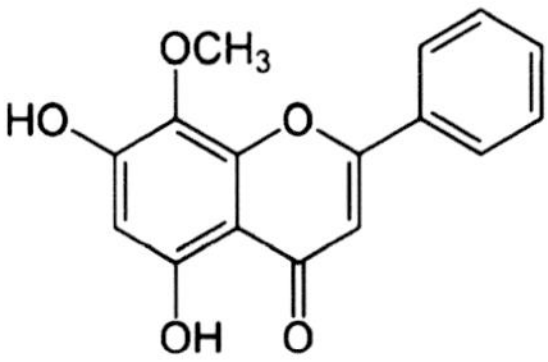

Fig. 24 Withanolide (*Withania somnifera*)

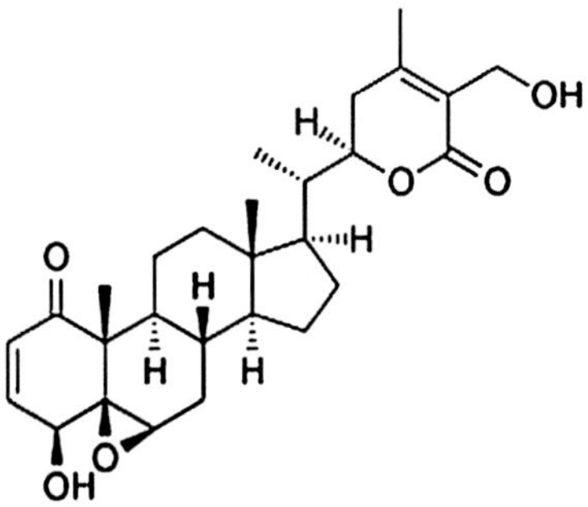

Fig. 25 Piperine (*Piper nigrum*)

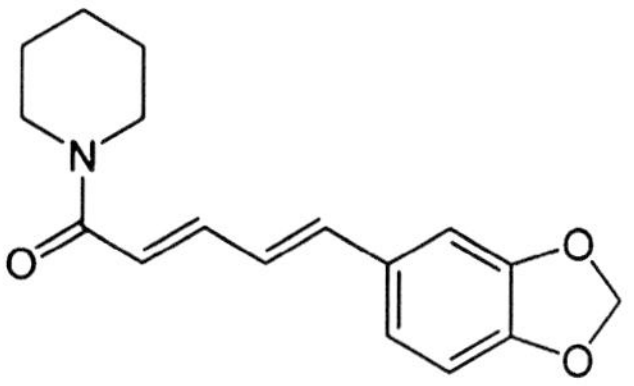

Fig. 26 Ginkgolides (*Ginkgo biloba*)

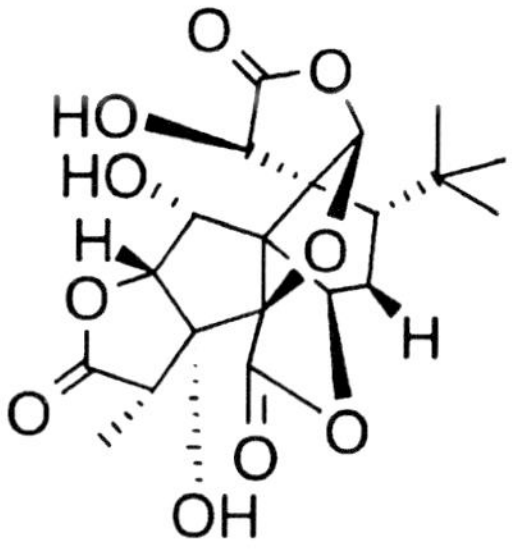

Fig. 27 Kolaviron (*Garcinia kola*)

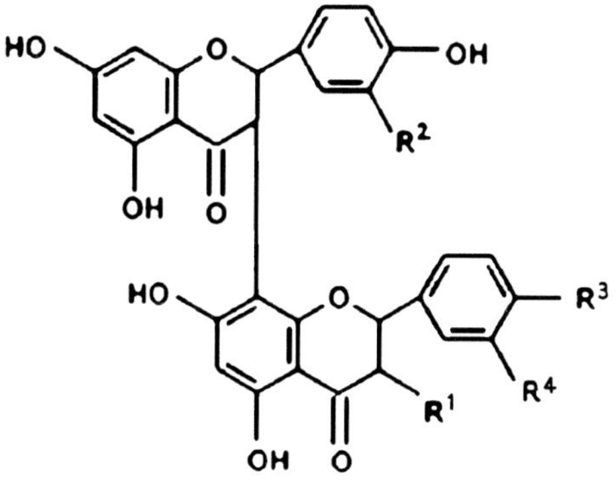

6 Conclusion

Neurological diseases or disorders can be described as diseases/disorders which particularly affect or disturb the brain, spinal cord as well as nervous coordination which run throughout the human body. There are a great number of specific causes of neurological diseases/disorders. The prominent of these causes are genetic disorder, contagions, deformities or disorders, malnutrition which could be linked to the health of one's environment, lifestyle, serious injuries to the brain, nerves as well as the spinal cord are also amongst the common causes of neurological diseases. The use of several synthetic agents and medication with neuroprotective properties has

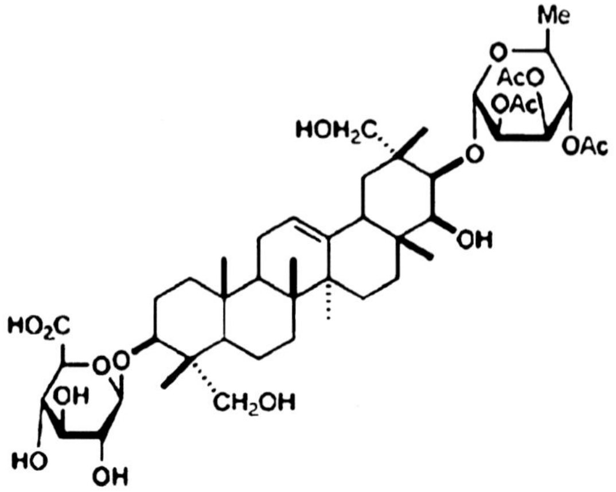

Fig. 28 Triterpene glycosides (*Panax ginseng*)

been properly documented. But there are some discomforting side effects associated with these synthetic neuroprotective medications which include fretfulness or nervousness, fatigue, dry mouth, tiredness, drowsiness, struggle with balance, etc. This chapter has reviewed a lot of medicinal plants with several phytochemicals and their mechanism of actions which have been reported to decline neuropathy. The plants discussed in this chapter as well as their active constituents provide a great insight into affordable means of treating and managing neurological disorders as well as a basis for pharmaceutical companies searching for novel sources for the production of neuroprotective pharmacotherapy drugs.

References

1. Montana DPHHS (2022) Neurological disorders. Accessed at: https://dphhs.mt.gov/schoolhealth/chronichealth/neurologicaldisorders. Retrieved: 15 March 2022
2. Phani KG, Anilakumar KR, Naveen S (2015) Phytochemicals having neuroprotective properties from dietary sources and medicinal herbs. Pharm J 7(1):1–17
3. John Hopkins Medicine (2022) Neurological disorders. Accessed at: https://www.hopkinsmedicine.org/health/conditions-and-diseases/neurological-disorders. Retrieved: 15 March 2022
4. Salehi B, Kumar NV, Sener B, Sharifi-Rad M, Kılıç M, Mahady GB, Vlaisavljevic S, Iriti M, Kobarfard F, Setzer WN (2018) Medicinal plants used in the treatment of human immunodeficiency virus. Int J Mol Sci 19:1459
5. Hu R, Lin C, Xu W, Liu Y, Long C (2020) Ethnobotanical study on medicinal plants used by Mulam people in Guangxi. China J Ethnobiol Ethnomed 16:1–50
6. Vahideh O, Mahboobe R, Roodabeh B, Farnaz E, Roja R, Rozita N, Tarun B, Hari PD, Zahra A, Mohammad HF (2019) Medicinal plants and their isolated phytochemicals for the management of chemotherapy-induced neuropathy: therapeutic targets and clinical perspective. DARU J Pharmaceut Sci 27:389–406
7. Yang CH, Yen TL, Hsu CY, Thomas PA, Sheu JR, Jayakumar T (2017) Multi-targeting andrographolide, a novel NF-kappaB inhibitor, as a potential therapeutic agent for stroke. Int J Mol Sci 18:1638

8. Dar A, Khatoon S (1997) Antidepressant activities of Areca catechu fruit extract. Phytother Res 4(1):41–45
9. Choudhary S, Kumar P, Malik J (2013) Plants and phytochemicals for Huntington's disease. Pharm Rev 7(14):81–91
10. Shinomol GK, Muralidhara M (2008) Prophylactic neuroprotective property of *Centella asiatica* against 3-nitropropionic acid induced oxidative stress and mitochondrial dysfunctions in brain regions of prepubertal mice. Neurotoxicology 29(6):948–957
11. Hsu DZ, Wan CH, Hsu HF, Lin YM, Liu MY (2008) The prophylactic protective effect of sesamol against ferric–nitrilotriacetate-induced acute renal injury in mice. Food Chem Toxicol 46(8):2736–2741
12. Chen XC, Zhu YG, Zhu LA, Huang C, Chen Y, Chen LM, Fang F, Zhou YC, Zhao CH (2003) Ginsenoside Rg1 attenuates dopamine-induced apoptosis in PC12 cells by suppressing oxidative stress. Eur J Pharmacol 473(1):1–7
13. De Jong WH, Borm PJ (2008) Drug delivery and nanoparticles: applications and hazards. Int J Nanomedicine 3(2):133–149
14. Feltenstein MW, Lambdin LC, Ganzera M, Ranjith H, Dharmaratne W, Nanayakkara NP, Khan IA, Sufka KJ (2003) Anxiolytic properties of *Piper methysticum* extract samples and fractions in the chick social-separation-stress procedure. Phytother Res 17(3):210–216
15. Yan JJ, Cho JY, Kim HS, Kim KL, Jung JS, Huh SO, Suh HW, Kim YH, Song DK (2001) Protection against beta-amyloid peptide toxicity in vivo with long-term administration of ferulic acid. Br J Pharmacol 133(1):89–96
16. Nataraj J, Manivasagam T, Justin Thenmozhi A, Essa MM (2017) Neurotrophic effect of asiatic acid, a triterpene of *Centella asiatica* against chronic 1-methyl 4-phenyl 1, 2, 3, 6-tetrahydropyridine hydrochloride/probenecid mouse model of Parkinson's disease: The role of MAPK, PI3K-Akt-GSK3beta and mTOR signalling pathways. Neurochem Res 42: 1354–1365
17. Nataraj J, Manivasagam T, Justin Thenmozhi A, Essa MM (2017) Neuroprotective effect of asiatic acid on rotenone-induced mitochondrial dysfunction and oxidative stress-mediated apoptosis in dierentiated SH-SYS5Y cells. Nutr Neurosci 20:351–359
18. Derf A, Sharma A, Bharate SB, Chaudhuri B (2018) Aegeline, a natural product from the plant *Aegle marmelos*, mimics the yeast SNARE protein Sec22p in suppressing alpha-synuclein and Bax toxicity in yeast. Bioorg Med Chem Lett 29:454–460
19. Zang CX, Bao XQ, Li L, Yang HY, Wang L, Yu Y, Wang XL, Yao XS, Zhang D (2018) The protective effects of *Gardenia jasminoides* (Fructus Gardenia) on amyloid-beta-Induced mouse cognitive impairment and neurotoxicity. Am J Chin Med 46:389–405
20. Newman DJ, Cragg GM (2016) Natural products as sources of new drugs from 1981 to 2014. J Nat Prod 79:629–661
21. Chiu HFK, Zhang M (2000) Dementia research in China. Int J Geriatr Psychiatry 15:947–953
22. Ameyaw EO, Woode E, Boakye-Gyasi E, Abotsi WKM, Kyekyeku JO, Adosraku RK (2014) Anti-Allodynic and antihyperalgesic effects of an ethanolic extract and xylopic acid from the fruits of *Xylopia aethiopica* in murine models of neuropathic pain. Pharm Res 6(2):172–179
23. Bertoncello KT, Aguiar GPS, Oliveira JV, Siebel AM (2018) Micronization potentiates curcumin's anti-seizure effect and brings an important advance in epilepsy treatment. Sci Rep 8:2645
24. Seo EJ, Fischer N, Efferth T (2018) Phytochemicals as inhibitors of NF-kappaB for treatment of Alzheimer's disease. Pharmacol Res 129:262–273
25. Nabavi SF, Sureda A, Habtemariam S, Nabavi SM (2015) Ginsenoside Rd and ischemic stroke: a short review of literatures. J Ginseng Res 39:299–303
26. Akinmoladun AC, Akinrinola BL, Olaleye MT, Farombi EO (2015) Kolaviron, a *Garcinia kola* biflavonoid complex, protects against ischemia/reperfusion injury: pertinent mechanistic insights from biochemical and physical evaluations in rat brain. Neurochem Res 40:777–787
27. Larit F, Elokely KM, Chaurasiya ND, Benyahia S, Nael MA, Leon F, Abu-Darwish MS, Efferth T, Wang YH, Belouahem-Abed D et al (2018) Inhibition of human monoamine oxidase

A and B by flavonoids isolated from two Algerian medicinal plants. Phytomed Int J Phytother Phytopharm 40:27–36
28. Dey A, Bhattacharya R, Mukherjee A, Pandey DK (2017) Natural products against Alzheimer's disease: pharmaco-therapeutics and biotechnological interventions. Biotechnol Adv 35:178–216
29. Zhang JT (2002) New drugs derived from medicinal plants. Therapie 57(2):137–150
30. An H, Kim IS, Koppula S, Kim BW, Park PJ, Lim O, Choi WS, Lee KH, Choi DK (2010) Protective effects of *Gastrodia elata* Blume on MPP+-induced cytotoxicity in human dopaminergic SH-SY5Y cells. J Ethnopharmacol 130(2):290–298
31. D'Hooge R, Pei YQ, Roes A, Lebrun P, van Bogaert PP, De Deyn PP (1996) Anticonvulsant activity of piperine on seizures induced by excitatory amino acid receptor agonists. Arzneim Forsch 46(6):557–560
32. Bhattacharya SK, Bhattacharya A, Kumar A, Ghosal S (2000) Antioxidant activity of *Bacopa monniera* in rat frontal cortex, striatum and hippocampus. Phytother Res 14(3):174–179
33. Pirzada AM, Ali HH, Naeem M, Latif M, Bukhari AH, Tanveer A (2015) *Cyperus rotundus* L.: traditional uses, phytochemistry, and pharmacological activities. J Ethnopharmacol 174:540–560
34. Chen C, Yang FQ, Zhang Q, Wang FQ, Hu YJ, Xia ZN (2015) Natural products for antithrombosis. Evid Based Complement Altern Med 2015:876426
35. Santos G, Giraldez-Alvarez LD, Avila-Rodriguez M, Capani F, Galembeck E, Neto AG, Barreto GE, Andrade B (2016) SUR1 receptor interaction with hesperidin and linarin predicts possible mechanisms of action of *Valeriana officinalis* in Parkinson. Front Aging Neurosci 8:97
36. Velmurugan BK, Rathinasamy B, Lohanathan BP, Thiyagarajan V, Weng CF (2018) Neuroprotective role of phytochemicals. Molecules 23:2485
37. Cervo L, Rozio M, Ekalle-Soppo CB, Guiso G, Morazzoni P, Caccia S (2002) Role of hyperforin in the antidepressantlike activity of *Hypericum perforatum* extracts. Psychopharmacology 164(4):423–428
38. Ohta H, Matsumoto K, Watanabe H, Shimizu M (1993) Involvement of beta-adrenergic systems in the antagonizing effect of paeoniflorin on the scopolamine-induced deficit in radial maze performance in rats. Jpn J Pharmacol 62(4):345–349
39. Hosseinzadeh H, Khosravan V (2002) Anticonvulsant effects of aqueous and ethanolic extracts of *Crocus sativus* L. stigmas in mice. Arch Iran Med 5(1):44–47
40. Shal B, Ding W, Ali H, Kim YS, Khan S (2018) Anti-neuroinflammatory potential of natural products in attenuation of Alzheimer's disease. Front Pharm 9:548
41. Kaufmann D, Kaur Dogra A, Tahrani A, Herrmann F, Wink M (2016) Extracts from traditional Chinese medicinal plants inhibit acetylcholinesterase, a known Alzheimer's disease target. Molecules (Basel Switzerland) 21:1161
42. Hosseinzadeh H, Parvardeh S (2004) Anticonvulsant effects of thymoquinone, the major constituent of *Nigella sativa* seeds, in mice. Phytomedicine 11(1):56–64
43. Chindo BA, Anuka JA, McNeil L, Yaro AH, Adamu SS, Amos S, Connelly WK, Lees G, Gamaniel KS (2009) Anticonvulsant properties of saponins from *Ficus platyphylla* stem bark. Brain Res Bull 78(6):276–282
44. Rao VS, Rao A, Karanth KS (2005) Anticonvulsant and neurotoxicity profile of *Nardostachys jatamansi* in rats. J Ethnopharmacol 102(3):351–356
45. Koo KA, Sung SH, Kim YC (2002) A new neuroprotective pinusolide derivative from the leaves of *Biota orientalis*. Chem Pharm Bull 50(6):834–836
46. Zhang H, Bai L, He J, Zhong L, Duan X, Ouyang L, Zhu Y, Wang T, Zhang Y, Shi J (2017) Recent advances in discovery and development of natural products as source for anti-Parkinson's disease lead compounds. Eur J Med Chem 141:257–272
47. Lee HW, Ryu HW, Kang MG, Park D, Lee H, Shin HM, Oh SR, Kim H (2017) Potent inhibition of monoamine oxidase A by decursin from *Angelica gigas* Nakai and by wogonin from *Scutellaria baicalensis* Georgi. Int J Biol Macromol 97:598–605

48. Woodruff-Pak DS, Vogel RW, Wenk GL (2003) Mecamylamine interactions with galantamine and donepezil: effects on learning, acetylcholinesterase, and nicotinic acetylcholine receptors. Neuroscience 117(2):439–447
49. Shou C, Feng Z, Wang J, Zheng X (2002) The inhibitory effects of ujuboside A on rat hippocampus *in vivo* and *in vitro*. Planta Med 68(9):799–803
50. Sabogal-Guaqueta AM, Osorio E, Cardona-Gomez GP (2016) Linalool reverses neuropathological and behavioural impairments in old triple transgenic Alzheimer's mice. Neuropharmacology 102:111–120
51. Nah SY, Park HJ, Mccleskey EW (1995) Pharmacology A trace component of ginseng that inhibits Ca21 channels through a pertussis toxin-sensitive G protein. Proc Natl Acad Sci 92: 8739–8743
52. Wang J, Xu HM, Yang HD, Du XX, Jianq H, Xie JX (2009) Rg1 reduces nigral iron levels of MPTP- treated C57BL6 mice by regulating certain iron transport proteins. Neurochem Int 54 (1):43–48
53. Park HJ, Lee HG, Kim YS, Lee JY, Jeon JP, Park C et al (2012) *Ginkgo biloba* extract attenuates hyperalgesia in a rat model of vincristine induced peripheral neuropathy. Anesth Analg 115(5): 1228–1233
54. Yang X, Zheng T, Hong H, Cai N, Zhou X, Sun C et al (2018) Neuroprotective effects of *Ginkgo biloba* extract and Ginkgolide B against oxygen-glucose deprivation/reoxygenation and glucose injury in a new in vitro multicellular network model. Front Med 12(3):307–318
55. Taliyan R, Sharma PL (2012) Protective effect and potential mechanism of *Ginkgo biloba* extract EGb 761 on STZ-induced neuropathic pain in rats. Phytother Res 26(12):1823–1829
56. Xu Y, Ku B, Cui L, Li X, Barish PA, Foster TC (2007) Curcumin reverses impaired hippocampal neurogenesis and increases serotonin receptor 1A mRNA and brain-derived neurotrophic factor expression in chronically stressed rats. Brain Res 1162:9–18
57. Oi YM, Imafuku C, Shishido Y, Kominato S, Nishimura K (1999) Allyl containing sulfides in garlic increase uncoupling protein content in brown adipose tissue, and noradrenaline and adrenaline secretion in rats. J Nutr 129(2):336–342
58. Lee JS, Kim YT, Jeon EK, Won HS, Cho YS, Ko YH (2012) Effect of green tea extracts on oxaliplatin-induced peripheral neuropathy in rats. BMC Complement Altern Med 12:124
59. Chen SQ, Wang ZS, Ma YX, Zhang W, Lu JL, Liang YR et al (2018) Neuroprotective effects and mechanisms of tea bioactive components in neurodegenerative diseases. Molecules 23(3)
60. Kakuda T (2002) Neuroprotective effects of the green tea components theanine and catechins. Biol Pharm Bull 25(12):1513–1518

Diabetes Treatment and Prevention Using Herbal Medicine

20

Babalola Ola Yusuf, Rukayat Abiola Abdulsalam, and Saheed Sabiu

Contents

Abstract

Diabetes is a cluster of prevalent hormonal conditions that display abnormally high blood glucose level. Diabetes is caused by the improper utilization of insulin by body cells or absence of pancreatic synthesis of insulin. Diabetes management involves measures used in maintaining diabetic condition to be close to normoglycaemia. This includes the use of medications such as insulin therapy and oral antidiabetic agents, exercise, weight loss, diet, and natural remedies. 65–80% of people worldwide use herbal medicine for their main medical care, which entails using medicinal plants to treat, diagnose, and prevent illnesses. Due

B. O. Yusuf · R. A. Abdulsalam · S. Sabiu (✉)
Department of Biotechnology and Food Science, Faculty of Applied Sciences, Durban University of Technology, Durban, South Africa
e-mail: BabalolaY@dut.ac.za; 22175907@dut4life.ac.za; sabius@dut.ac.za

S. C. Izah et al. (eds.), *Herbal Medicine Phytochemistry*, Reference Series in Phytochemistry, https://doi.org/10.1007/978-3-031-43199-9_43

to their accessibility and alleged absence of adverse effects, most people, especially those who live in developing nations, prefer to treat their diabetes with medicinal plants including fenugreek, cinnamon, bitter lemon, and *Gymnema sylvestre*. Since many of these plants inhibit amylase, encourage insulin secretion, and aid in glucose uptake, they have similar antidiabetic effects to those of common oral antidiabetic drugs. There have been efforts to isolate bioactive components in antidiabetic plants as well as comprehending how they work due to the recent increase in interest in medicinal herbs and technological advances.

Keywords

Diabetes · Diabetes management · Herbal treatment · Traditional medicine · Medicinal herbs

1 Introduction

Diabetes is an enduring metabolic ailment branded by high blood glucose levels [1]. There are approximately 422 million diabetics in the world, and this figure is steadily rising [1]. Either insufficient pancreatic insulin synthesis or inappropriate insulin use by body cell is to blame for the disorder [2]. It is regarded as the foremost causes of illness and mortality [3] and can cause several consequences, such as cardiovascular disease, renal failure, blindness, as well as amputation [2]. Dietary vicissitudes, insulin treatment, and oral hypoglycaemic drugs are typical management options for diabetes [1]. Herbal medicine has been utilized for ages to treat diabetes and has lately grown in favour of supplemental therapy [1]. The utilization of phytotherapy in the management and prevention of diabetes is examined in this chapter. Diabetes has long been treated with herbal medicines in conventional healthcare systems of Chinese (TCM) and Indian (Ayurveda) [4]. Herbal remedies work in a variety of ways, including by improving insulin sensitivity and glucose absorption [4].

2 Herbal Medicine

2.1 The Herbal/Traditional Medicine Structure

Traditional medicine, commonly known as complementary and alternative medicine (CAM), includes an array of procedures and approaches, including the use of therapeutic exercises, spirituality, and medicinal herbs for treatment, prevention, and maintenance of health [5]. Due to regional variations in historical, economic, and lifestyle circumstances, these behaviours, which are grounded on ideas and knowledge, fluctuate significantly from one place to another [6]. Traditional folk medicine has been practised and improved upon since the Stone Age [7]. The theory and practice of traditional medicine are said to work together to diagnose, treat, and prevent ailments. It is based on earlier information and discoveries that were frequently shared verbally through oral tradition, oral lessons from ancestors, or more recently, written down [7, 8]. The most prevalent kind of treatment in contemporary medical settings is the usage of medicinal plants or herbal medications [9].

Herbal remedies are frequently used in conventional medicine, which emphasizes their important place in ethnomedicine [10]. The investigation of pharmacognosy and the use of herbal plants is the foundation of conventional medicine [11, 12], which is known as herbal medicine also referred to as phytomedicine or phytotherapy [13]. Herbal medicine, phytotherapy, and phytomedicine are additional names for the research on pharmacognosy and the use of herbal plants, which is the cornerstone of contemporary medicine [14]. The WHO claims that herbal remedies are pharmaceuticals that have undergone processing and labelling and are utilized to cure illness or enhance life's quality. They include active compounds found in above- (e.g. seed, flower, leaf, and bark) or below-ground plant components (e.g. root, rhizome) or certain combinations of plant parts [15, 16]. Traditional Chinese herbal medicine and Ayurveda are the most well-known and widely used traditional healthcare systems [17, 18]. Among these systems are the following:

- System of Traditional African Medicine

  Traditional African medicine encompasses a range of conventional medical practices, such as native herbalism and African theology, and often involves the involvement of shamans, midwives, and herbalists [19]. Shamans, midwives, and herbalists are frequently involved in traditional African medicine, which includes a variety of Western medicinal techniques such native herbalism and African religion [19]. Ancient healing magic may be more common than Western medicine [7] and has a longer history and wider prevalence on the African continent than other traditional medical practices [20]. Different specialties of African traditional medicine, which is a complete healthcare system, include divination, spiritualism, and herbalism. However, in some circumstances there might be some overlap [7, 21].

  An individual who practises medical care in their tribe by utilizing methods such as herbs, minerals, animal parts, and incantations, among others, in accordance with their community's traditions and beliefs is known to be a traditional therapist/healer [7]. They should be considered as skilled, experienced, adaptable, and trustworthy [22]. Depending on the culture, there may be various classifications of traditional healers, including high priests, herbalists, midwives, priestesses, diviners, seers/spiritualists, and witch doctors [7]. In certain societies in Nigeria, these individuals are identified by their native names, such as *Babalawo*, *Buka*, or *Dibia* [23]. It is believed in ancient African medicine that certain aspects of the practice are only shared with initiates. Therefore, to gain access to these components, initiation into a cult is usually required [7]. According to this form of medicine, illnesses are caused by spiritual forces rather than random chance. As a result, the prescribed remedy typically involves a plant-based treatment that is believed to possess therapeutic qualities as well as symbolic and spiritual significance [7].

- Traditional Chinese Medical System

  Traditional Chinese medicine (TCM) is a complementary medical methodology that originates from Chinese traditional medicine [24]. TCM encompasses a wide range of health and therapeutic practices that are typically viewed as incompatible. These practices include traditional beliefs, intellectual theories, Confucian ideology, herbal treatments, dietary interventions, physical activities, various medical specialties, and different schools of thought [25]. Traditional Chinese medicine (TCM) is based on the philosophy of yinyangism, which combines the *Yin Yang* theory and the Five Phases hypothesis [26]. This philosophy applies the concept of *yin* and *yang* to the human body, where the top and back are considered as *yang*, and the bottom is viewed as *yin* [27]. *Yin* and *yang* are also used to describe various biological processes and disease symptoms. For example, sensations of cold and heat are considered *yin* and *yang* symptoms [27]. Traditional Chinese medicine (TCM) includes a list of medications that are believed to address specific symptom clusters by enhancing either *yin* or *yang* [29]. Additionally, strict guidelines have been established for the order of actions, countermeasures, and other interactions among the Five Phases theory [28]. The *zàng-fû* concept is closely linked to each of these Five Phases tenets and significantly affects the TCM body model [28]. The Five Phases concept is frequently utilized in both diagnosis and treatment within TCM [28] (Fig. 1, Table 1).

  In traditional Chinese medicine, the vital energy of the body, known as *ch'i* or *qi*, is believed to flow through meridians that are linked to multiple organ systems and physical functions through forks [31]. Rather than focusing on structural

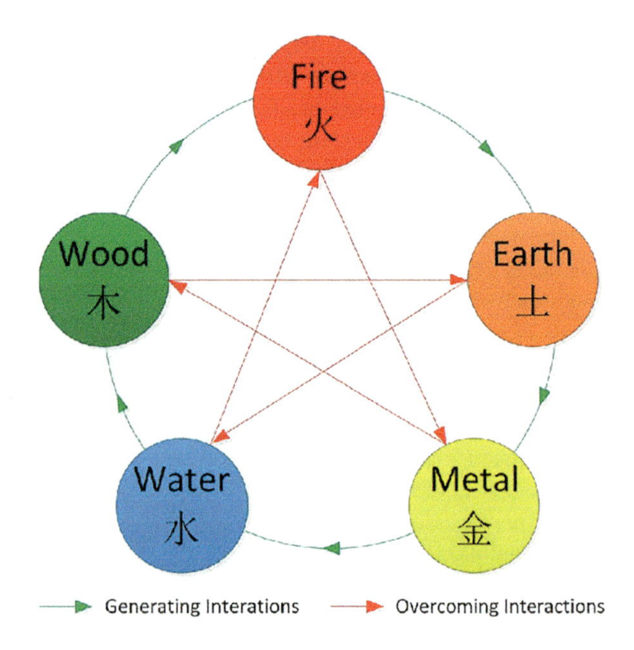

Fig. 1 Relations between *Wu-xing* (the Five Phases of TCM) [29]

Table 1 The arrangement of things according to the Five Phases character and *záng-fû* organ as well as distribution

Phenomenon	Wood	Fire	Earth	Metal	Water	References
Direction	East	South	Centre	West	North	[29]
Colour	Green or violet	Red or purple	Yellow or pink	White	Black	[30]
Climate	Wind	Heat	Damp	Dryness	Cold	[28]
Taste	Sour	Bitter	Sweet	Acrid	Salty	[28]
Záng organ	Liver	Heart	Spleen	Lung	Kidney	[29]
Fû organ	Gall bladder	Small intestine	Stomach	Large intestine	Bladder	[29]
Sense organ	Eye	Tongue	Mouth	Nose	Ears	[30]
Facial part	Above bridge of nose	Between eyes and lower part of the eyes	Bridge of nose	Between eyes and middle part	Below cheekbone	[30]
Eye part	Iris	Inner/outer corner of eye	Upper and lower lid	Sclera	Pupil	[30]

details, traditional Chinese medicine places greater emphasis on physiological processes such as digestion, respiration, and temperature regulation [32]. The perception of illness in this system is that it reflects an imbalance or disharmony in the operations or relationships of various elements, including *yin*, *yang*, *qi*, *xuê*, *zàng-fû*, meridians, and their interplay with the surroundings and the human body [27]. Traditional Chinese medicine (TCM) considers recognizing a "pattern of disharmony" to be a critical step in the diagnostic process, as it forms the basis of therapy [28, 33]. However, identifying various patterns, which is also known as "pattern discrimination," is considered to be the most challenging aspect to master within TCM [34].

- Traditional Indian Medical System

  Ayurveda is a type of complementary therapy, being a 2000-year-old tradition originated from India [35]. Although it has changed over time, its beliefs and methods, which include herbal medicines, dietary programmes, meditation, yoga, massage, enemas, laxatives, and medicinal oils, are frequently regarded as pseudoscientific [36, 37]. Samkhya and Vaisheshika are two philosophical ideas that served as the foundation for Ayurveda, as well as Buddhist and Jain principles [38]. Maintaining balance is the major goal of Ayurveda, and it is thought that stifling natural desires might make one sick [39]. In fact, according to Ayurveda, even suppressing a sneeze may result in shoulder ache [40]. Ayurveda places a strong emphasis on a number of elements, including healthy eating practices, enough sleep, and appropriate sexual behaviour [39]. The human body, according to Ayurveda, is made up of three components: *doshas*, which are humoral substances; *malas*, which are waste products; and *dhatus*, which are the body's tissues [40]. Ayurveda recognizes three primary physiological humours or *doshas*, namely, *Pitta*, *Kapha*, and *Vata* [41]. These *doshas*, collectively known as *tridosha*, are often associated with the nervous system by contemporary writers [41]. An individual's mental state can be influenced by the *doshas* in Ayurveda [42]. The dominance of any *dosha* impacts an individual's bodily composition (prakriti) and character, as each dosha possesses distinct qualities and performs specific tasks in the body and mind [42, 43] (Fig. 2).

Ayurveda considers an imbalance in both mental and physical *doshas* as a significant cause of illnesses [45]. To diagnose illnesses, Ayurveda employs eight diagnostic methods, which include examining the urine, stool, speech, touch, vision, tongue, appearance, and pulse [45]. In Ayurveda, physicians rely on the five senses to diagnose patients, and for example, they use hearing to observe breathing and speech [46]. Ayurveda emphasizes the study of the *marman* or *marma* points, which are points in the body that could be potentially lethal [46]. Contemporary Ayurveda primarily focuses on improving the body's metabolism, promoting effective elimination of substances, and enhancing longevity, while two of the original eight branches of Ayurveda concentrate on surgical procedures [46]. There is also a significant emphasis on exercise, yoga, and meditation in this approach [47]. To maintain good health, Ayurveda subscribes to the Dinacharya school of thought, which emphasizes following the natural rhythms of the brain for activities such as waking, sleeping, and working [48]. The Sattvic diet is recommended as a part of the dietary regimen. In addition, important hygiene practices include regular showers, teeth brushing, tongue scrubbing, oil pulling, caring for the skin, and cleaning the eyes [48].

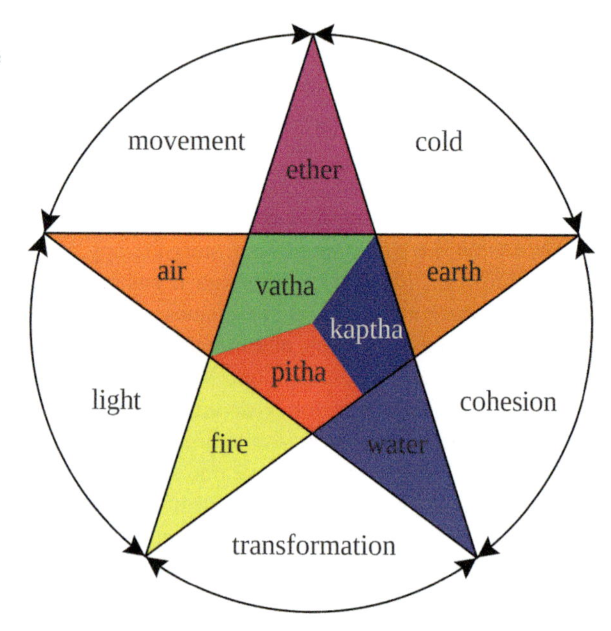

Fig. 2 The three *doshas* in Ayurveda are made up of five elements [44]

2.2 Herbal Preparation

There are different methods for preparing herbal remedies that vary based on regional and cultural practices. It is permissible to use both fresh and dried parts of plants, and a specific method is chosen based on experience to maximize effectiveness and minimize potential harm [7]. In general, there are several techniques for preparing herbal remedies, including:

(a) Extraction: The extraction method involves using a solvent in a specific weight-to-volume ratio. Sometimes, the solvent may evaporate and transform into a jelly-like substance [7].
(b) Infusion: To create an infusion, the plant materials are soaked in hot or cold water for a little while. Honey or another addition can be added to stop spoilage [7, 49].
(c) Decoction: It is created by making woody part of wood boil for a fixed period and then filtering out the solids. Potash can be used both as a preservative and to aid in the extraction process [49].
(d) Tincture: Tinctures, which are alcohol infusions, can be diluted before use if they are too concentrated [50].
(e) Ashing: Ashing involves burning the dry ingredients until they turn to ash, filtering the ashed samples, and then mixing them with water or food for consumption [7].

    There are several other types of herbal preparations. Lotions are thin treatments applied to the body, while liniments are surface treatments that come in watery, semi-liquid, or oily forms and contain bioactive ingredients [7]. Poultices are made from freshly macerated plant parts that retain their liquid and are applied to the skin [49]. Snuffs, which are made from powdered dried herbs, are burned to create charcoal and inhaled through the nose [7]. Gruels are grain-based cereals or porridges that are mixed with pulverized dried vegetables or their ash and consumed. Sometimes, mixtures of multiple plants are used to enhance the combined effects of the substances [49].

Aside from the usual methods of taking herbal remedies, such as orally, rectally, topically, or nasally, there are other ways to consume them. These include passive inhalation or smoking a poorly constructed cigar made of dried plant materials [7]. Another method involves inhaling the volatile oils released from heating or burning the plant matter [7]. These techniques can be helpful in relieving congestion, headaches, and respiratory issues. Sitz baths, however, are specifically used for treating piles [7, 51].

2.3 Prevalence of Use

Herbal treatments are a popular form of primary medical treatment worldwide, with estimates indicating that 65–80% of people use them [52]. Conventional medicine is responsible for about 40 percentage of medical services in China [53], and about 80% of African utilize conventional remedies to assist with healthcare difficulties [54]. As of 2018, the cases of herbal medicine usage varied across several cultural groups [55], and people's perceptions of herbal remedies are shaped by ancient traditions and lifestyle [56, 57]. For instance, the significant utilization of spices and herbs in Australian culture is recognized as a way to preserve food [57, 58]. According to data from the WHO, the use of phytomedicine was highest in European countries and Southeast Asia in 2019, at a rate of 91%, followed by Africa at 87%, and the lowest usage was in the America at 80% [6].

People who suffer from chronic illnesses such as HIV/AIDS [59], diabetes, asthma, end-stage kidney disease [60], and cancer [61] are more likely to use herbal remedies. Furthermore, several factors, including sex, age, race, level of education, and socioeconomic status, have been linked to the frequency of natural remedy usage [62]. Additional factors include arthritis, depression, anxiety, hypertension [63], and various chronic conditions [64, 65].

3 Diabetes Management

The principal aim of managing diabetes is to regulate sugar level as close to normoglycaemia, without causing it to drop too low [66]. To achieve this goal, individuals may need to make changes to their diet, engage in regular exercise, lose weight, and use medications such as insulin, oral medications, or herbal remedies [67].

Being well-informed about diabetes and vigorously partaking in treatment are imperative, as people who manage their blood sugar well have low risk of complication [68, 69]. Managing diabetes typically involves modifying one's diet, engaging in physical activity, losing weight, and using appropriate medications (insulin, oral, and herbal) [67].

The American College of Physicians recommends maintaining a target glycated haemoglobin (HbA1c) level of 7–8% for diabetes treatment [70]. Health conditions including high blood pressure, obesity, metabolic syndrome, as well as physical idleness can exacerbate the negative effects of diabetes [71].

The same principles for controlling diabetes can be applied to the general population, but interventions may need to be tailored to specific populations. Further research is necessary to determine if self-monitoring is effective for individuals with serious mental illnesses as well as type 2 diabetes and if the outcomes are comparable to those seen in the general population [72].

3.1 Medication

To be effective, most treatments for diabetes work by reducing glucose levels in different ways [73]. It is widely accepted that maintaining tight control over blood glucose levels helps people with diabetes to avoid additional health problems such as kidney or eye issues [74, 75]. The kind of medications used here include the following:

- Oral antidiabetic medications: Type 2 diabetes is frequently managed with oral antidiabetic medicines [76]. Metformin, the first-line treatment, is so due to its proven effectiveness in reducing mortality [76]. Metformin reduces glucose production by the liver [78]. There are several classes of pharmaceuticals, mostly administered orally, that can aid in managing hyperglycaemia in individual having type 2 diabetes such as insulin-stimulating production (like sulfonylureas), reduce glucose uptake from the gastrointestinal system (like acarbose), block the metabolic function of the enzyme DPP-4 which inhibits incretins like GLP-1 and GIP (like sitagliptin), improve the body's insulin responsiveness (like thiazolidinediones), and increase glucose excretion from the body (like SGLT2 inhibitors) [78] (Table 2).
- Insulin Therapy: Insulin is an anti-glycaemic agent drug that lowers high sugar levels, especially in people with type 1 diabetes and, to a lesser extent, in people with type 2 diabetes as well as associated consequences [86]. The World Health Organization recognizes insulin as an essential medication that should be widely available and accessible to people globally [6, 87]. Type 1 diabetes results from a deficiency of insulin, and insulin is commonly used to manage this condition [88]. Subcutaneous injections or insulin pumps are the typical methods of administering insulin, but some types can also be given through intravenous or intramuscular injections [86, 88]. The speed at which insulin is processed in the body differs, resulting in various peak times and lengths of effect [88]. Insulins that act more quickly often have extended peak durations and remain in the body for longer periods [89]. In the management of type 2 diabetes, a sustained-release form of insulin is usually introduced initially, while continuing with oral medications [77]. Insulin dosages are then increased until the desired glucose levels are achieved [90].

  It is important to monitor patients regularly and provide comprehensive education on insulin therapy, as improper administration can have serious consequences. For example, if food intake is reduced, the insulin dosage may need to be reduced as well [91]. Physical exercise affects the absorption of glucose by cells in the body, which is regulated by insulin with varying onset times and durations. This means that exercise can reduce the need for insulin [92]. However, insulin therapy also carries risks due to the inability to consistently monitor an individual's blood glucose level and adjust insulin infusion accordingly [93].
- Herbal Medication: For centuries, people have used herbal medicine to treat several illnesses, including diabetes [94]. Diabetes is a chronic condition that results in high blood sugar levels, which can lead to severe complications if not managed effectively. Despite the availability of conventional treatments like oral hypoglycaemic drugs and insulin therapy, numerous individuals with diabetes opt for complementary and alternative therapies, such as herbal medicines [94]. Fenugreek, cinnamon, bitter melon, and *Gymnema sylvestre* are some of the herbs that have been employed in the past to treat diabetes. These herbs are believed to help

Table 2 The antidiabetic drugs that can be taken orally have varying modes of action and potential side effects

Oral agent	Mechanism of action	Side effect	References
Sulfonylurea, e.g. gliclazide, glibenclamide, glipizide, glimepiride	Sulfonylureas act on pancreatic beta cells by binding to and blocking the KATP channels, which are sensitive to ATP and located in the cellular membrane. This action prevents potassium from leaving the cell, leading to depolarization. As a result, voltage-gated channels for Ca^{2+} open, causing an increase in cytosolic Ca^{2+}. This increase in Ca^{2+} levels results in a greater integration of insulin granules with the phospholipid bilayer and consequently, an increase in the production of mature insulin	Hypoglycaemia, weight gain, gastrointestinal upset, headache, and hypersensitivity reactions	[79, 80]
Postprandial glucose regulators, e.g. repaglinidine, nateglinide	Postprandial glucose control medications work by interacting with KATP channels in the cellular membrane of pancreatic beta cells, which are sensitive to ATP. These medications block the postprandial glucose control mechanism by preventing potassium from leaving the cell, leading to depolarization. This depolarization causes the opening of the Ca^{2+} channels which is voltage-gated. This allows an influx of Ca^{2+} ion into the cell increasing the concentration. The increased Ca^{2+} concentration facilitates the interaction of the membrane of the cells and insulin granules, increasing the synthesis of mature insulin. However, the effects of these medications are short-	Hypoglycaemia and weight gain	[81]

(continued)

Table 2 (continued)

Oral agent	Mechanism of action	Side effect	References
	lived with a peak of two to 4 h after administration and a duration of less than 6 h, as they are rapidly absorbed and eliminated from the body		
Biguanide, e.g. metformin	Biguanides work by activating the AMP-activated protein kinase (AMPK). The breakdown and synthesis of fats and carbohydrates, insulin signalling pathways, as well as the maintenance of overall ATP homeostasis heavily depend on this protein. When AMPK is activated, expression of the small heterodimer partner is increased, which leads to the reduction in the synthesis of genes partaking in glucose production in the liver, particularly those responsible for phosphoenolpyruvate carboxykinase as well as glucose-6-phosphatase	Gastrointestinal upset and lactic acidosis	[82, 83]
Thiazolidinedione, e.g. pioglitazone, rosiglitazone	Thiazolidinediones (TZDs) are a class of drugs that work by activating PPARs, which are specific nuclear receptors also known as PPAR-gamma or PPARG. They are categorized as PPARG agonists. Free fatty acids (FFAs) and eicosanoids are the main ligands for these receptors. When the PPAR receptor is activated in combination with the retinoid X receptor (RXR), it can affect gene expression, suppressing some genes and increasing the transcription of others. The transcription of certain genes mainly causes adipose cells to store more fatty	Weight gain and oedema	[84, 85]

(continued)

Table 2 (continued)

Oral agent	Mechanism of action	Side effect	References
	acids, leading to a decreased transported of fatty acids. Consequently, energy is generated through the breakdown of sugars, especially glucose, to produce energy for various biological processes		
Alpha-glucosidase inhibitors, e.g. acarbose	Alpha-glucosidase inhibitors work by hindering the conversion of oligosaccharides into monosaccharides, which is carried out by alpha-glucosidase enzymes in the small intestine. Acarbose, a type of alpha-glucosidase inhibitor, lowers postprandial blood glucose levels by competing with dietary polysaccharides for alpha-glucosidase enzymes, which reduces the absorption of glucose	Gastrointestinal upset	[81]

regulate blood sugar levels by boosting insulin production, reducing sugar cravings, enhancing insulin sensitivity, and slowing down carbohydrate absorption [95]. Various herbal remedies, including *Acacia arabica*, *Allium sativum*, *Aegle marmelos* L., *Zingiber officinale*, *Aloe barbadensis*, and *Eucalyptus globulus*, are naturally insulin-mimetic or insulin secretagogues [93]. Some herbs, such as *Juniperus communis*, *Salix planifolia*, and *Pinus banksiana*, help in the efficient absorption of glucose [95], while others serve as α-amylase inhibitors, such as *Eugenia cumini* seeds (632 ρg/ml) [96], *Aloe barbadensis* (30 ρg/ml) [96], *Morus alba* (1440 ρg/ml) [97], *Pterocarpus marsupium* (0.9 ρg/ml) [98], and *Costus pictus* (2510 ρg/ml) [99], among others.

3.2 Current Research on Chinese and Ayurvedic Medicines

One of the first methods of treating and managing illness dates back to the Chinese and Ayurveda medicine [100]. Chinese and Ayurvedic medicine make up more than 65% of all herbal on the market today [101]. It is crucial to analyse the conclusions the most recent research has produced.

Emblica officinalis's antidiabetic properties were evaluated by [102]. According to their findings, 200 mg/kg b.w. of the fruit extract significantly lessened sugar level in the blood. In their study on *Salacia reticulata*, [103] found that compounds isolated from the plant, such as Salcinol and kotalanol, had the capacity to inhibit -glucosidase. Combinations of *S. reticulata* and *Catharanthus roseus* L. showed reduced level of fat and glucose in rats induced to diabetes. Furthermore, they say that in comparison to controls, herbs lowered level of glucose in the blood of diabetic-induced rats. Inhibition of the enzymes aldose, glucosidase, as well as pancreatic lipase is what mostly lowers blood glucose levels. Research carried out in vitro employing a rat pancreatic cell line RINm5F have revealed that the fractions of *Tinospora cordifolia* stem rich in isoquinoline alkaloids exhibit properties of insulin release and mimicry. Similarly, in vivo experiments have also confirmed these effects. These fractions include palmatine, jatrorrhizine, and magnoflorine [104]. Another isoquinoline alkaloid, called berberine, has been the subject of research and has been proven effective in both experimental and human diabetes. According to reports, berberine can reduce high blood sugar levels just as effectively as metformin [105]. Furthermore, the stem and root of *Tinospora cordifolia* contain neosporin, isocolumbin, palmatine, tinocordiside, cordioside, and sitosterol compounds, which are believed to possess properties such as antidiabetic, anti-hyperlipidemic, and antioxidant properties [103]. *Aegle marmelos* (L.) Corrêa demonstrated a considerable drop in HbA1c, PPBG, and BMI, while *Allium sativum* L. showed a significant decrease in PPBG and DBP; *Aloe vera* L. was found to significantly reduce BMI, and *Anethum graveolens* L. was found to suggestively lower fasting insulin, insulin resistance, LDL, and other parameters, according to [106] systematic review and meta-analysis from 2022. They arrive at the conclusion that Ayurveda botanicals have favourable effects on outcomes connected to type 2 diabetes [107]. conducted a research on people who have been diagnosed of type 2 diabetes, where they evaluated the impact of a hydroalcoholic extract of *Crocus sativus* L. on various factors such as HbA1c, lipid profile, fasting blood glucose, and renal and liver function tests. According to the study findings, the extract was potent in reducing HbA1c, fasting blood glucose, serum cholesterol concentration, and LDL-c levels, by improving insulin sensitivity and enhancing antioxidant enzyme activity. However, there was no noticeable difference observed in the liver and renal function tests. The study conducted by [108] on *Azadirachta indica* leaves and twigs revealed that the aqueous extract had a significant effect in reducing fasting blood sugar, post-prandial glucose level, insulin resistance, as well as HbA1c when compared to a placebo. Moreover, neem was found to be more effective than the placebo in terms of improving endothelial function, reducing oxidative stress, and systemic inflammation. The extract was effective across all doses, but it had no impact on lipid profile or platelet aggregation. As per [109], supplementing with *Anethum graveolens* (dill) powder led to significant reductions in total cholesterol, malondialdehyde, and low-density lipoprotein cholesterol levels in the group treated. The homeostatic typical valuation of insulin resistance and mean serum levels of insulin also showed significant improvement. The intervention group had considerably greater average of serum total antioxidant capacity and high-density lipoprotein compared to baseline assessment. The frequency of digestive complaints related to colonic motility was the only one that significantly decreased with supplementation.

The treated groups had considerably lower mean scores than the control group when malondialdehyde, low-density lipoprotein cholesterol, insulin, and total cholesterol levels were considered. Additionally, the treatment group's mean improvements in high-density lipoprotein were substantially higher in comparison to that of the control group. In an experiment conducted by [110], rats are being nourished diet heavy in fat and receiving streptozotocin (STZ/HFD) treatment, which resulted in distorted insulin resistance (IR) pointers across the three IR valuation methods used in the study. Their insulin and sugar levels were much higher than normal throughout the experiment. The study used EmbliQur, a treatment that comprises *Emblica officinalis* and *Curcuma longa*, and it is verified to be effective in reducing the elevated BGL caused by HFD/STZ and improving IR in all animals subjected to the IR test. According to a study conducted by [111] on the results of using separated compounds and unrefined components from *Striga orobanchioides* Benth on diabetic rats induced by streptozotocin, pentacyclic triterpenoids were found to be present (81.5% w/w) in the ethylacetate-methanolic fraction of the ethanol extract. Betulin was discovered through spectroscopic analysis of the isolated chemical. The bioactive fraction and betulin demonstrated substantial inhibition of digestive enzymes in vitro, and the group receiving betulin at a dose of 40 mg/kg showed substantial improvement in their level of diabetes. Gene expression studies disclosed that betulin at a dosage of 40 mg kg^{-1} improves liver glucose usage, sinks hepatic inflammation, and surges insulin making. In silico tests revealed that the plant compound has a strong ability to impede the enzymes DPP-IV, α-glucosidase, GK, and α-amylase.

According to [112], JQ-R, a Chinese herbal medicine preparation, consists of three active components, namely, the polyhydric alcohols of *L. japonica*, *A. membranaceus* saponin, and the whole alkaloid fraction of *C. chinensis*. These fractions have been verified to possess antidiabetic effects in diabetic KK^{Ay} mice and in vitro. The aforementioned components efficiently control the metabolism of glucose, reduce oxidative stress, and regulate related inflammatory responses. By promoting glucose absorption through the insulin-dependent PI3K-AKT signalling pathway, JQ-R was effective in improving glucose metabolism partially in diabetic KK^{Ay} mice. According to a study by [113], mulberry leaves were found to have potential in treating diabetes through in vivo and in vitro tests. Aqueous extract of the leaves which revealed alkaloids and flavonoids as the main hypoglycaemic components was given to diabetic mice, producing rates of inhibition of $87.29 \pm 1.32\%$ and $86.12 \pm 1.79\%$, respectively, on glucosidase activity. When given to diabetic rats, aqueous mulberry leaf extract improved glucose tolerance by 19.02% and lowered blood sugar levels by 28.17%. The extract also had positive effects on gut flora, reduced kidney damage, and lowered blood levels of insulin, tumour necrosis factor-α (TNF-α), glycosylated serum protein, and free fatty acid. These results imply that the beneficial interactions between the different components of the leaves of mulberry may be responsible for their hypoglycaemic actions. According to [114], the study found that the polyphenols of mulberry leaf, which consist of various compounds such as rutin, benzoic acid, hyperoside, and chlorogenic acid, effectively suppressed glucose transport. The degree of inhibition increased over time and with higher concentrations of MLPs. The blood glucose lowering impact of *Dendrobium officinale* and the related mechanisms were demonstrated in a study by [115]. The study found that *Dendrobium officinale* had an impact on the cAMP-PKA and Akt/FoxO1 signalling pathways conciliated by glucagon. This resulted in the encouraged synthesis of glycogen as well as inhibition of glycogen breakdown and gluconeogenesis in the liver. In a study conducted by [116], it was discovered that polyphenols derived from *Azadirachta indica* have the ability to enhance glucose absorption in muscle tissue and decrease the functioning of the proteins responsible for breaking down polysaccharides. As per [117], *Cordyceps militaris* extract exhibits a blood sugar-lowering effect. In addition, it has the ability to improve and repair damaged pancreatic islet cells, increase the thymus index, reduce pro-inflammatory cytokines release, and limit the release of cytokines. This indicates that the immunomodulatory and inflammatory-reducing qualities of *Cordyceps militaris* are closely associated. The study also showed that quercetin has a curative impact on diabetic retinopathy in rats with the disease. Furthermore, the study found a correlation between the treatment outcome and the level of expression of HO-1. In a study conducted by [118], a diabetic rat was treated with a microcapsule encapsulating quercetin, and the results on enteric nerves were assessed. The outcomes showed that by reducing oxidative stress, this therapy may mitigate the nerve damage brought on by diabetes [119]. conducted a study in 2022 that found ZSP to significantly increase the PI3K/AKT pathway's activity through increasing the expression of Akt1 and Akt2 proteins and the copying of the 4 catalytic PI3K subunit genes pik3ca, pik3cb, pik3cd, and pik3cg. The phosphorylated forms of these proteins were also considered due to their role in PI3K and AKT activation. The outcomes disclosed that ZSP enhanced PI3K/AKT signalling at both translational and transcriptional levels when likened to the model group. However, there was a discrepancy in the normal group's PI3K and AKT mRNA and protein expression, which could be due to possible differences in post-transcriptional regulation between wt/wt and db/db animals. Additionally, the high insulin levels in db/db mice may donate to the elevated PI3K and AKT expression as a result of ongoing insulin stimulation.

3.3 Several Traditional Herbal Medicines' Modes of Action

Plants exhibit antidiabetic properties by reactivating pancreatic cells to increase insulin production, inhibiting glucose absorption in the gastrointestinal tract, or promoting metabolites involved in insulin-dependent activities. The existence of certain compounds, such as polyphenols, flavonoids, terpenoids, and coumarins, in medicinal plants is believed to be responsible for their antidiabetic effects [120]. Different herbal remedies have diverse methods of demonstrating their antidiabetic function. Some of these include the following:

- Ginseng (*Panax* spp.): The greatest study has been done on this medicinal herb's hypoglycaemic effects. Ginseng's geographic origin, amount, dispensation, and diabetes types have the main effects on its potency and effectiveness [121]. *Panax ginseng*, sometimes recognized as Chinese or Korean ginseng, possess the uppermost medicinal potential [121]. Ginseng diminishes blood sugar by regulating several different mechanisms (Fig. 3). *Panax ginseng* at a concentration of 0.1–1.0 g/mL was found to significantly increase the amount of insulin released from pancreatic islets of rats which was isolated when glucose is present at an amount of 3.3 mmol/L [122]. Oral ingestion for 20 days of 100 mg/kg b.w. of heat-processed American ginseng was taken and was found to decrease serum glucose levels, glycosylated proteins, and glycated haemoglobin (HbA1c) in rats with diabetes caused by streptozotocin (STZ). Additionally, the treatment increased values of creatinine clearance and reduced the storage of N (ϵ)-(carboxymethyl) lysine and its receptors in the kidneys [123]. Ginsenosides are one of the numerous constituents that have been isolated from ginseng and are regarded to be the active ingredient. Polysaccharides, peptides, and poly acetylenic alcohols are additional components [124]. The water-soluble extract of ginseng root has the ability to reduce diabetes when compared to its fat-soluble components. The ginsenosides present in both American and Panax ginseng are thought to improve insulin resistance and its sensitivity at the target tissues [121]. There are some foods that don't appear to assist lower blood sugar levels since they have low ginsenoside contents [125].

- Bitter lemon (*Momordica charantia*): Research has indicated that bitter melon exhibits hypoglycaemic effects in cell culture, animal experiments, as well as clinical trials involving humans [126, 127]. The bitter melon's antidiabetic ingredients include alkaloids, vicine, charantin, polypeptide-P, and other

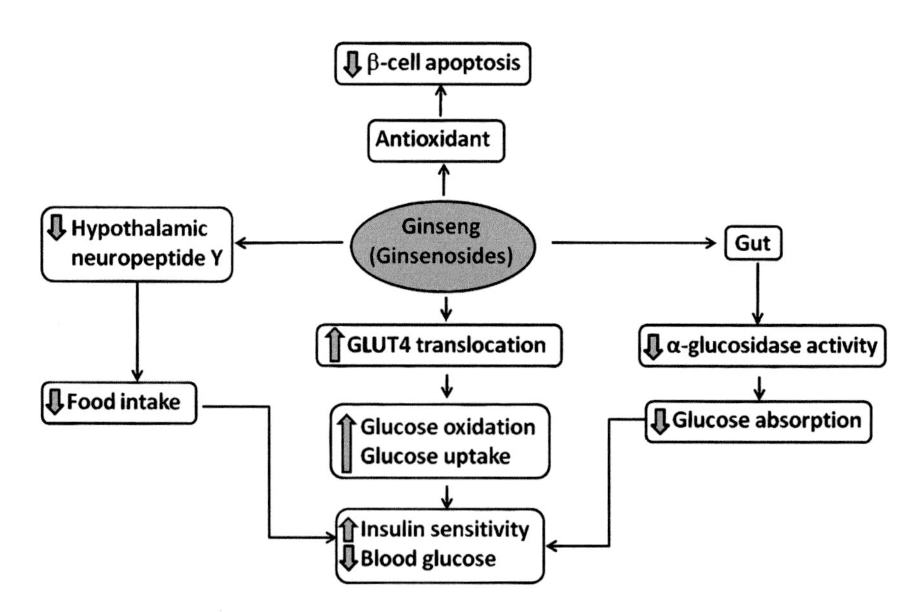

Fig. 3 The mechanism by which ginseng increases glucose metabolism [121]

non-specific bioactive antioxidants [121]. Bitter melon's oleanolate-glycoside was found to enhance glucose tolerance in individuals with type 2 diabetes by inhibiting the absorption of sugar in the intestines [121]. Bitter melon has been verified to reduce levels of glucose in the blood in mice with diabetes caused by STZ by preventing cell peroxidation and apoptosis, increasing the amount of liver and muscle glycogen, and stimulation of the enzymes in the liver such as phosphofructokinase, glucokinase, and hexokinase activity [128]. In diabetic rats caused by STZ, the inactivity of liver fructose-1,6-bisphosphatase and glucose-6-phosphatase activities resulted in decreased gluconeogenesis. However, overexpression of glucose-6-phosphate dehydrogenase (G6PDH) in red blood cells and hepatocytes enhanced glucose oxidation [129]. In summary, research conducted on both humans and animals has demonstrated that bitter melon plays a role in regulating glucose and lipid metabolism (as depicted in Fig. 4). Although the exact mode of action is still being investigated, bitter melon can be utilized as a dietary supplement or herbal medicine for managing diabetes and/or metabolic conditions [130].

- Fenugreek (*Trigonella foenum-graecum*): A type of medicinal product made from a plant's dried seed that is indigenous to North Africa, China, and India [121]. Fenugreek was used in Ayurveda for millennia as both a demulcent (an agent that soothes inflammation of the throat, nose, mouth, or skin) and a laxative [121]. Fenugreek seed powder users apparently experience decreased postprandial glucose levels [132]. This can be due to a slowdown in gastrointestinal glucose absorption and a bulk laxative action. Fenugreek, however, also appears to boost insulin release since it contains 4-isoleucine [131, 133]. Research has demonstrated that fenugreek seeds are effective in improving glycaemic regulator and reducing resistance of insulin in people with type 2 diabetes that is moderate. Furthermore, fenugreek seed supplementation has a positive effect on hypertriglycaeridemia and increases HDL-C levels. Supplementation with

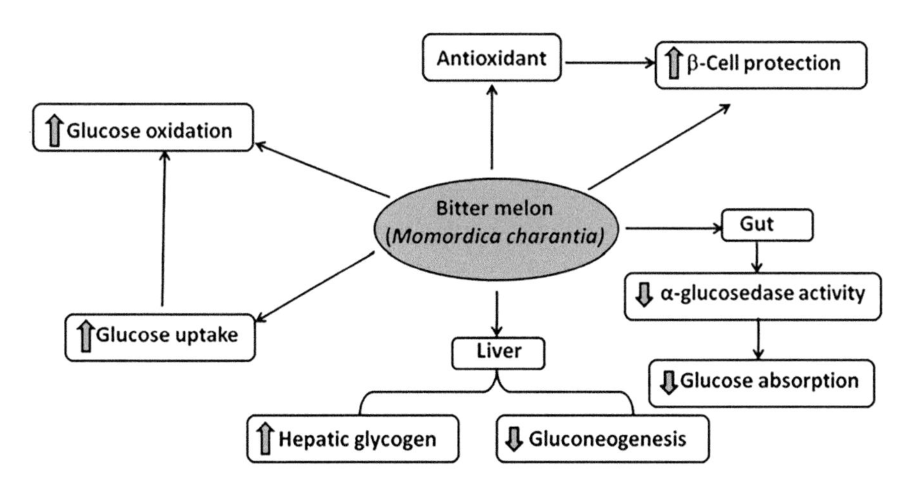

Fig. 4 Bitter melon's mechanism of action in lowering blood sugar [121]

fenugreek leaves improves body weight and liver glycogen levels, and its impact on glucose metabolism is significant, similar to the effects of glibenclamide [134]. Studies conducted on experimental diabetes indicate that fenugreek leaf powder has the ability to reduce oxidative stress. In rats with diabetes, administration of fenugreek has been found to decrease lipid peroxidation and significantly enhance the antioxidant system [135].

- *Coptis chinensis*: In China, *Coptis chinensis* is widely utilized for the management of diabetes. The active component of the *Coptis chinensis* plant is an isoquinoline alkaloid known as berberine. The plant stems, bark, roots, and rhizomes contain it [121]. Giving diabetic rats berberine through intragastric administration at doses of 100 and 200 mg kg^{-1} resulted in a reduction of fasting blood sugar levels, total cholesterol, triglycerides, and low-density lipoprotein cholesterol (LDL-C) in the blood. At the same time, there was an increase in high-density lipoprotein cholesterol (HDL-C), nitric oxide (NO) levels, and prevention of the elevation of glutathione peroxidase (GSH-px) and superoxide dismutase (SOD) [136]. Berberine can bring about a reduction in body weight and an increase in insulin response, which may be due to various mechanisms. In MIN6 cells and rat islets, it decreases glucose-stimulated insulin secretion (GSIS), boosts AMPK activity, and enhances the translocation of GLUT4 in adipocytes and myotubes [137, 138]. In addition, it raises the synthesis of PPAR α/δ/γ/ proteins in the liver and enhances the expression of insulin receptors in the liver and skeletal muscle cells, which stimulates the use of glucose by cells in the presence of insulin [139]. Furthermore, in rats with diabetes induced by STZ, berberine significantly lowers the activity of disaccharidases and α-glucuronidase in the intestine [140]. A berberine derivative called dihydroberberine (dhBBR) displayed advantageous benefits in vivo in animals fed a high-fat diet [141]. Figure 5 provides an overview of the several ways that berberine lowers blood sugar levels.
- Banaba (*Lagerstroemia speciosa*): Banaba belongs to the same family as the banaba plant. Banaba leaf extracts are highly favoured in the Philippines and Southeast Asia, and they are frequently employed by diabetics in North America. Banaba comprises two significant components that have two significant impacts on individuals with diabetes [121]. Corosolic acid and ellagitannins are two components found in banaba extracts that appear to function similarly to insulin and stimulate the activation of insulin receptors [142]. The main reasons for the above-mentioned effect are thought to be either the inhibition of tyrosine phosphatase or the stimulation of tyrosine kinase insulin receptor. Initial clinical investigations indicate that individuals suffering from type 2 diabetes who take a special banaba extract (named glucosol) for 2 weeks experience a notable 10% drop in their the levels of sugar in the blood as opposed to those who receive a placebo [143]. Research has shown that banaba possesses a hypoglycaemic impact; however, additional investigation is required to assess its potential long-term effects [121].
- *Agaricus* mushroom (*Agaricus* spp.): Despite having its roots in Brazil, *Agaricus* is now readily accessible in Japan, China, and other Asian countries. It has carbohydrates like betaglucans, which, like those in other mushrooms, seem to boost immune function markers and serve as an immunostimulant [121]. *Agaricus*

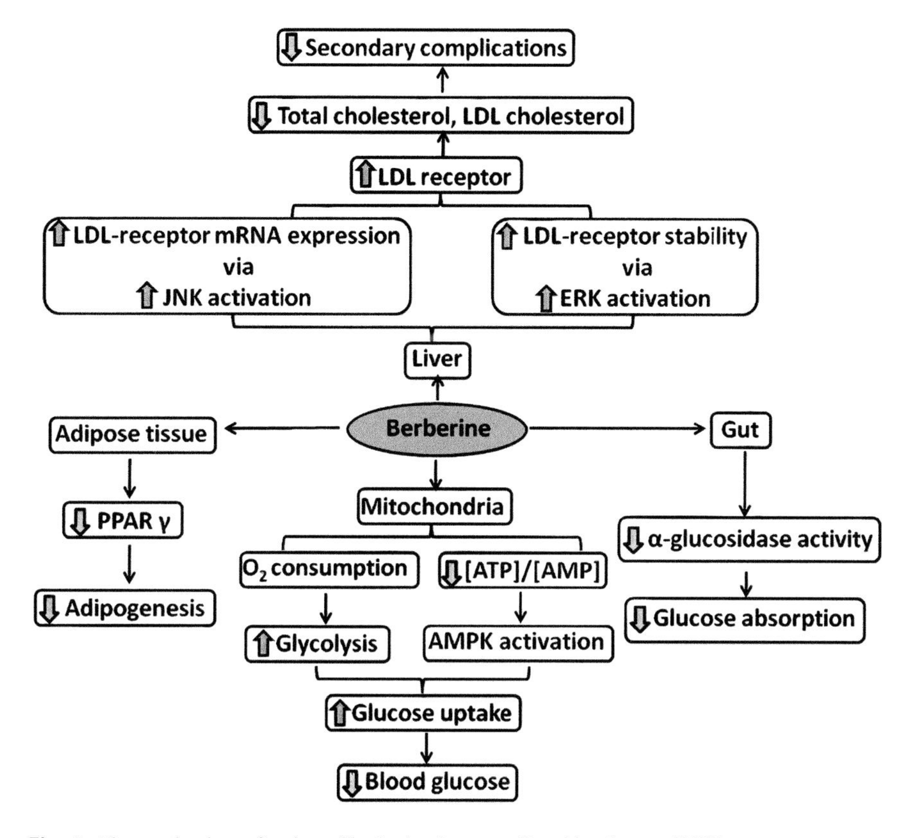

Fig. 5 The mechanism of action of berberine in controlling blood sugar [121]

mushrooms contain certain compounds that can enhance sensitivity of insulin as well as lowering resistance of insulin in individuals with type 2 diabetes. Additionally, these compounds may elevate adiponectin levels, which could potentially decrease insulin resistance [121]. Clinical studies have shown that individuals with type 2 diabetes who augmented their regular treatment with half gram of *Agaricus* mushroom extract thrice daily had decreased overnight levels of insulin in comparison to others who abstained from eating the mushroom [144]. For the majority of people, the *Agaricus* mushroom extract appears to be harmless when used for a period of up to 12 weeks. However, it can cause potentially harmful drops in blood sugar levels (hypoglycaemia) [144].

3.4 Surgery

If type 1 diabetics experience severe complications from their illness, such as end-stage kidney disease that requires a kidney transplant, a pancreas transplant may be considered as an uncommon alternative [145, 146].

Weight loss surgery, which is often used to treat obesity, can also be a successful therapy for obese patients that are diabetic [147]. Following surgery, many patients are competent to maintain normoglycaemia without the need for medication, which can reduce the risk of long-term mortality [148, 149]. However, it is unclear what BMI thresholds should be used to determine if surgery is appropriate [149]. It is recommended that individuals who are unable to control their blood sugar levels and weight should consider this option [150].

3.5 Self-Management and Support

In countries like the United Kingdom that have a general practitioner system, diabetes treatment is typically provided outside of hospitals. Hospital-based care is usually only necessary in cases of complications, difficulty in controlling blood sugar levels, or for research purposes. General practitioners and specialists may work together on occasion to provide care. Additionally, managing diabetes through telehealth from the comfort of one's own home can be a successful strategy [151].

Computer-based self-management interventions are utilized to gather data for educational programmes that aim to aid self-care in type 2 diabetics [152]. Nonetheless, there is insufficient proof to back the effects on various physiological, psychological, and emotional outcomes such as blood pressure, cholesterol, weight, dietary changes, depression, depression-related quality of life, and other related factors [152, 153].

- Glycaemic surveillance and management: Glycaemic monitoring and control refers to the process of measuring and managing blood glucose levels in individuals with diabetes [154]. Blood glucose monitoring using a glucometer is a common method employed for this purpose. If necessary, adjustments are made to insulin dosages, dietary habits, and exercise routines to ensure that blood glucose levels remain within a predetermined target range [155, 156]. Regular monitoring is critical in reducing short-term and long-term issues associated with diabetes, including hypoglycaemia, renal damage, and cardiovascular disease. It is recommended to consult with a healthcare professional to determine the most appropriate target range and monitoring frequency for your individual needs [153]. Additionally, measuring haemoglobin A1c levels every three to 6 months offers an average of blood glucose levels over the past two to 3 months and is vital for effective monitoring [157]. Managing blood glucose levels effectively in addition to monitoring blood glucose also requires a healthy diet, regular exercise, and appropriate medication use, including insulin and oral hypoglycaemic agents [158]. Individuals with diabetes should work with a healthcare professional to develop a personalized plan for monitoring and regulating their blood glucose levels.
- Diabetes education: Diabetes education pertains to educating people on the essential expertise as well as ability to treat their disease, which involves understanding the causes; recognizing warning signs, symptoms, and possible

complications; as well as learning preventative and treatment strategies [159]. This education encompasses several topics, such as comprehending the effects of diabetes on the body, monitoring blood glucose levels, making necessary insulin adjustments, creating healthy meal plans, engaging in physical activity and exercise, understanding medication usage, identifying and avoiding potential complications, managing emotional well-being and stress, and many more [159]. Diabetes education can be offered in various settings, including hospitals, clinics, community centres, and online [159]. Self-management lessons, workshops, or one-on-one counselling sessions may be included, and it can be provided either in a group or individual context [87, 160]. Proper diabetes education and receiving appropriate care can lead to improved overall quality of life, decreased chances of complications, and more fulfilling lives for individuals with diabetes. By collaborating with healthcare professionals to design a customized education programme and attending diabetes education sessions on a regular basis, individuals with diabetes can stay informed about new developments in science and treatment methods [159].

4 Prevention of Diabetes

Type 1 diabetes has no known mitigating measures [161]. However, healthy lifestyle choices, such as diet and exercise modifications, can often prevent type 2 diabetes from developing or postpone it [162], and it constitutes approximately 85–90% of all global instances [163].

4.1 Lifestyle

Numerous healthy lifestyle practices have been established to assist in the prevention of diabetes. It is believed that smoking cigarettes significantly increases the risk of developing diabetes [164]. Stopping smoking is a critical preventive measure as it increases the chances of developing diabetes and its associated consequences [165]. According to a 2015 study, there is a clear link between the amount of exposure to cigarette smoke along with the chance of developing type 2 diabetes, with the risk increasing as the amount of exposure to cigarette smoke increases [164]. The 2014 Surgeon General's Report highlights the position of promoting smoking termination as a crucial public health strategy to combat the global diabetes epidemic, stating that people who smoke frequently have 30–40% higher likelihood of developing type 2 diabetes than individuals who do not [166].

4.2 Alcohol

In 2010, approximately 41% of adults worldwide consumed alcohol based on available data [167]. The WHO reported that on average, individuals above

15 years of age consume 6.2 L of unadulterated alcohol per year, corresponding to 13.5 g [168]. A 2015 report indicated that 86.4% of individuals 18 years or older in the United States reported consuming alcohol [169]. The consumption of alcohol has the potential to increase the likelihood of developing diabetes. Furthermore, extreme alcohol ingestion portends various chronic health conditions, including diabetes mellitus and its associated problems [169]. Multiple studies have revealed different associations between alcohol ingestion and diabetes risk, including positive, null, U-shaped, or J-shaped relationships [170]. While liquor ingestion may have various effects on blood sugar levels, studies have found that prolonged and regular drinking can cause upsurge insulin resistance [171]. To prevent the negative effects of alcohol on insulin resistance and blood sugar, it is recommended to minimize alcohol intake to a maximum of two drinks per day. Additionally, it is advised to consume alcohol with food or a snack, rather than on an empty stomach [172]. Average alcohol consumption is linked to a minimized chance of developing type 2 diabetes mellitus (T2DM) compared to those who drink minimally or not at all, while consuming excessive amounts of alcohol carries the same danger as not drinking [169]. Assessment of cohort research as a whole conducted in various countries has provided convincing evidence that consuming fewer than 30 g of alcohol in a day is linked with reduced risk of developing T2DM in the populace [173]. Another study suggests that the highest risk reduction for developing T2DM is 18% for individuals who consume between 10 and 14 g of alcohol per day [174]. Research has indicated that alcohol consumption may prevent diabetes from occurring in men and women by delaying its onset [175, 176]. According to research, the consumption of approximately 24 g and 22 g of alcohol per day for women and men, respectively, is linked to the highest level of protection against developing diabetes. However, doses exceeding 50 g for women and 60 g for men daily are deemed harmful [177]. The frequency of alcohol use can also affect the level of risk [174]. Binge drinking, which involves consuming large amounts of alcohol in one go, has been linked with amplified risk of heart disease and other negative health consequences when compared to individuals who consume the same amount of alcohol but with a more frequent consumption pattern spread out across three to 4 days per week [174, 177].

According to a study that followed participants over time, drinking alcohol may advance the chances of getting type 2 diabetes mellitus [178]. Appraisal of multiple studies found that women who drink a lot of alcohol daily are more susceptible to type 2 diabetes development, but this correlation isn't seen as strongly in men [177]. Another study shows that excessive alcohol intake can prevent the body from controlling glucose levels, putting both men and women at higher risk of developing type 2 diabetes [179]. A summary of numerous studies that looked at participants over time found no considerable contrast in the likelihood of getting type 2 diabetes mellitus between those who drink heavily and those who consume less than 48 g of alcohol per day [180]. However, consuming large amounts of alcohol can worsen issues with managing blood glucose levels, which can promote the likelihood of having type 2 diabetes and its related complications [178]. There is evidence that drinking alcohol can lower insulin resistance [181]. Alcohol intake can affect how the body processes glucose, whether someone has diabetes or not. Drinking alcohol on an empty stomach can surge the risk of hypoglycaemia by reducing the making of glucose and breakdown of stored glucose, especially when combined with certain medications and low glycogen levels [182]. Explanations for how alcohol can add to the progression of diabetes includes decreased activity in beta cells, increased insulin resistance, and a reduction in glucose tolerance. Studies have found a connection between higher blood glucose levels and decreased beta cell activity, as well as increased alcohol consumption [183]. Frequent and excessive alcohol intake can make it harder for the body to maintain healthy blood glucose levels, which can cause problems with glucose tolerance and increase insulin resistance [184]. Alcohol consumption can lead to weight increase, an important trigger for type 2 diabetes, according to data [184, 185].

4.3 Diet and Exercise Modification

Combining a healthy diet with regular exercise can reduce your chances of developing diabetes, help you lose weight, lower your cholesterol levels, and reduce your blood pressure [160]. The American Diabetes Association (ADA) endorses maintaining a healthy weight, engaging in at least two and a half hours of physical activity per week (even just taking long walks can be enough), consuming foods with moderate amounts of fat (about 80% of caloric intake should come from healthy fats), and getting enough fibre from sources like whole grains [186]. Getting more exercise can help lower type 2 diabetes risk, particularly after eating a high-carbohydrate meal that causes a spike in blood sugar levels [187, 188].

Increase your intake of whole grains, fibre, and the polyunsaturated fats in fish, nuts, and vegetable oils [172, 188]. Red meat and other foods high in saturated fat should be avoided so as to lessen the likelihood of diabetes development [189].

5 Conclusion

Diabetes has long been treated and prevented with herbal therapy. Herbs work in a variety of ways, including by boosting glucose absorption, lowering oxidative stress, and improving insulin sensitivity. It is encouraging that herbal medicine can be used to treat and prevent diabetes. However, despite these functions, there need to be more research on identifying bioactive constituents of plants and elucidating their mechanism of action.

References

1. Sun J, Ren J, Hu X, Hou Y, Yang Y (2021) Therapeutic effects of Chinese herbal medicines and their extracts on diabetes. Biomed Pharmacother 142(111977):1–12
2. Nithya V, Sangavi P, Srinithi R, Nachammai KT, Kumar SG, Prabu D (2023) Diabetes and other comorbidities: macrovascular and macrovascular diseases: diabetes and cancer. In: Noor R (ed) Advances in diabetes research and management, 1st edn. Springer, Singapore, pp 21–29

3. Zhu M, Li J, Li Z, Luo W, Dai D, Weaver SR, Stauber C, Luo R, Hu H (2013) Mortality rates and the causes of death related to diabetes mellitus in Shanghai Songjiang District: an 11-year retrospective analysis of death certificates. BMC Endocr Disord 15(45):1–8
4. Wang Z, Wang J, Chan P (2013) Treating type 2 diabetes mellitus with traditional Chinese and Indian medicinal herbs. Evid Based Complement Alternat Med 2013(343594):1–17
5. Gunjan M, Naing TW, Saini RS, Ahmad A, Naidu JR, Kumar I (2015) Marketing trends and future prospects of herbal medicine in the treatment of various disease. World J Pharm Res 4 (9):132–155
6. WHO (2019) WHO global report on traditional and complementary medicine 2019. World Health Organization, Geneva, pp 15–28
7. Ozioma EJ, Nwamaka-Chinwe OA (2019) Herbal medicine in African traditional medicine. In: Builders PF (ed) Herbal medicine. Intech Open, United Kingdom, pp 191–214
8. Mokgobi MC (2014) Understanding traditional African healing. Afr J Phys Health Educ Recreat Dance 20(Suppl 2):24–34
9. Liu WJH (2011) Introduction to traditional herbal medicines and their study. In: Liu WJH (ed) Traditional herbal medicine research methods: identification, analysis, bioassay and pharmaceutical and clinical studies. Wiley, Hoboken, pp 1–2
10. Awodele O, Fajemirokun O, Oreagba I, Samuel TA, Popoola TD, Coulidaty AV, Emeka PM (2018) Herbal medicines use: remedies or risks. West Afr J Pharm 29(2):1–18
11. El-Naggar S, Ayyad EK, Khandil R, Salem M (2021) Phoenix dactylifera seed extract ameliorates the biochemical toxicity induced by silver nanoparticles in mice. Int J Can Biomed Res 5(4):73–84
12. Vinod M, Yogesh T, Shrishail G, Suyash I, Avinash B, Gajanand N, Moulaui S, Saili M, Smeeta P, Zainab B, Aishwaryar S (2021) A brief review on herbal medicines. Res J Pharmaco Phytochem 13(2):101–102
13. Yasmin AR, Chia SL, Look QH, Omar AR, Noordin MM, Ideris A (2020) Herbal extracts as antiviral agents. In: Florou-Paneri P, Christaki E, Giannenas I (eds) Feed additives: aromatic plant and herbs in animal nutrition and health. Elsevier Science & Technology, San Diego, pp 115–132
14. Tsay H, Shyur L, Agrawal D, Wu Y, Wang S (2016) Medicinal plants – recent advances in research and development. Springer, Singapore
15. Parveen A, Parveen B, Parveen R, Ahmad S (2015) Challenges and guidelines for clinical trial of herbal drugs. J Pharm Bioallied Sci 7(4):329–323
16. World Health Organization (2008) Media centre: Traditional medicine. Geneva, Switzerland, p 20
17. Che CT, George V, Ijinu TP, Pushpangadan R, Andrae-Marobela K (2017) Traditional medicine. In: Badal S, Delgoda R (eds) Pharmacognosy: fundamentals, application and strategies. Academic Press, Boston, pp 15–30
18. D'souza MR (2021) Traditional Indian herbs for the management of diabetes mellitus and their herb-drug interaction potentials: an evidence-based review. In: Chen H, Zhang M (eds) Structure and health effects of natural products on diabetes mellitus. Springer, Singapore, pp 279–296
19. Kanu IA (2013) The dimensions of African cosmology. Filisofia Theoretica: J Afr Philos Cult Relig 2(2):533–555
20. Asakitikpi A (2020) African indigenous medicines: towards holistic healthcare system in Africa. Afr Identities 20(14):365–375
21. Mahomoodally MF (2013) Traditional medicine in Africa: an appraisal of ten potent African medicinal plants. Evid Based Complement Alternat Med 2013:1–14
22. Alem A (2019) African system of thought: whether they fit scientific knowledge. J Philos Cult 7(2):14–25
23. Abdullahi AA (2011) Trends and challenges of traditional medicine in Africa. Afr J Tradit Complement Altern Med 8(Suppl 1):115–123

24. Eigenschink M, Dearing L, Dablander TE, Maier J, Sitter HH (2020) A critical examination of the main premises of traditional Chinese medicine. Weiner Klinnische Wochenschrift 132(9–10):260–273
25. Andrews B (2013) The making of modern Chinese medicine. UBC Press, Vancouver, BC, pp 10–17
26. Xue R, Fang Z, Zhang M, Yi Z, Wen C, Shi T (2013) TCMID: traditional Chinese medicine integrative database for herb molecular mechanism analysis. Nucleic Acids Res 41(Database issue):D1089–D1096
27. Law KMY, Kesti M (2014) Yin Yang and organisational performance: Five elements for improvement and success, 1st edn. Springer, London, pp 1–55
28. Ergil MC, Ergil KV (2009) Pocket atlas of Chinese medicine. Thieme, Stuttgart, p 15
29. Maciocia G (2015) The five elements. In: 3rd (ed) The foundations of Chinese medicine: A comprehensive text. Elsevier, China, pp 34–37
30. Deng T (2005) Practical diagnosis in traditional Chinese medicine, 5th reprint. Churchill Livingstone, pp 30–32
31. Kassiola JJ (2022) Zhang Zai's cosmology of Qi/qi and the refutation of arrogant anthropocentrism: Confucian green theory illustrated. Environ Values 31(5):533–554
32. Xu J, Xia Z (2019) Traditional Chinese medicine (TCM) – does its contemporary business booming and globalization really reconfirm its medical efficacy & safety. Med Drug Discov 1 (100003):1–5
33. Cheung H, Doughty H, Hinsley A, Hsu E, Lee TM, Milner-Gulland EJ, Possingham HP, Biggs D (2020) Understanding traditional Chinese medicine to strengthen conservation outcomes. People Nat 3:115–128
34. Armour M, Betts D, Roberts K, Armour S, Smith CA (2021) The role of research in guiding treatment for women's health: a qualitative study of traditional Chinese medicine acupuncturists. Int J Environ Res Public Health 18(834):1–11
35. Karousatos CM, Lee JK, Braxton DR, Fong TL (2021) Case series and review of ayurvedic medication induced liver injury. BMC Complement Med Ther 21(91):1–11
36. Beall J (2018) Scientific soundness and the problem of predatory journals. In: Kaufman AB, Kaufman JC (eds) Pseudoscience: The conspiracy against science. 293. MIT Press, Cambridge
37. Kumari S, Gheena S, Premavathy D (2020) Effects of Ayurvedic medicines on COVID-19 affected patients. Eur J Mol Clin Med 7(1):276–284
38. Sebastian CD (2022) Ayurveda and the medical knowledge in ancient India: shadows and realities. Indian J Med Ethics 8(1):8–15
39. Mathpati MM, Albert S, Porter JDH (2020) Ayurveda and medicalisation today: the loss of important knowledge and practice in health? J Ayurveda Integr Med 11(1):89–94
40. Majumder B, Ray S (2022) Studies on functions of vata dosha – a novel approach. Int J Ayurveda Pharma Res 10(Suppl 2):117–124
41. Jat R, Meena P, Saini S, Meena P (2023) A critical review on rakta as chaturtha dhosa or dhasu. J Ayurveda Holistic Med 11(2):1–8
42. Payyappallimana U, Venkatasubramanian P (2016) Exploring Ayurvedic knowledge on food and health for providing innovative solutions to contemporary healthcare. Front Public Health 4(57):57
43. Shipla S, Venkatesha MCG (2011) Understanding personality from Ayurvedic perspective for psychological assessment: a case study. Ayu Institute for Post Graduate Teaching & Research in Ayurveda 32(1):12–19
44. Khan R, Ullah M, Shafi B, Orakzai UB, Shafi S (2020) Knocking at alternative doors: Survey of android-based smartphones applications for alternative medicines. In: Mobile devices and smart gadgets in medical science. IGI Global, pp 1–13
45. Bhadresha K, Patel M, Brahmbhatt J, Jain N, Rawal R (2022) Ayurgenomics: A brief note on ayurveda and their cross kingdom genomics. Int Assoc Biol Comput Digest 1(2):254–258
46. Chopra AS (2003) Ayurveda. In: Selin H (ed) Medicine across cultures: history and practice of medicine in non-western cultures. Kluwer Academic, pp 75–83

47. Upadhay D (2014) An approach to healthy life through yoga in Ayurveda. Int J Res 1(3):40–44
48. Underwood EA, Rhodes P (2008) History of medicine. In: Encyclopaedia of Britannica. 2008 edition
49. Haruna NA, Erhabor O, Erhabor T, Adias TC (2021) Review of some herbs with jaemato-therapeutic properties in use in Nigeria. Sokoto J Med Lab Sci 6(4):65–81
50. Hudz N, Makowitz E, Shanaida M, Białon M, Jasicka-Masiak I, Yezerska O, Svydenko L, Wieczorek PP (2020) Phytochemical evaluation of tinctures and essential oil obtained from *Satureja montana* herb. Molecules 25(4763):1–20
51. Baig Z, Abu-Omar N, Harington M, Gill D, Ginther DN (2022) Be kind to your behind: a systematic review of the habitual use of bidets in benign perianal disease. Evid Based Complement Alternat Med 2022(1633965):1–9
52. World Health Organization (2015) Traditional medicine. Fact sheet number 134. www.who.int/mediacentre/factsheets/fs134/en/
53. Sen S, Chakraborty R, De B (2011) Challenges and opportunities in the advancement of herbal medicine: India's position and role in a global context. J Herb Med 1(3–4):67–75
54. Laelago T, Yohannes T, Lemango F (2016) Prevalence of herbal medicine use and associated factors among pregnant women attending antenatal care at public health facilities in Hossana town, Southern Ethiopia: facility based cross sectional study. Arch Public Health 74(7):1–8
55. Harris PE, Cooper KL, Reltan C, Thomas KJ (2012) Prevalence of complementary and alternative medicine (CAM) use by the general population: a systematic review and update. Int J Clin Pract 66(10):924–939
56. Akhagba OM (2017) Cultural influence in the consumption of herbal medicine among Nigerian women: a theoretical exploration. Misc Anthropol Sociol 18(May):193–206
57. Máthé A, Bandoni A (2021) Medicinal and aromatic plants of South America Vol. 2: Argentina, Chile and Uruguay, pp 3–48
58. Richmond R, Bowyer M, Vuong Q (2019) Australian native fruits: potential uses as functional food ingredients. J Funct Foods 62(103547):1–16
59. Mbali H, Sithole JJK, Nyondo-Mipando AL (2021) Prevalence and correlates of herbal medicine use among anti-retroviral therapy (ART) clients at Queen Elizabeth Central Hospital (QECH), Blantyre, Malawi: a cross-sectional study. Malawi Med J 33(3):153–158
60. Ginwala R, Bhavsar R, Chigbu DGI, Jain P, Khan ZK (2019) Potential role of flavonoids in treating chronic inflammatory diseases with a special focus on the anti-inflammatory activity of *Apigenin*. Antioxidants 8(35):1–28
61. Asiimwe JB, Nagendrappa PB, Atukurda EC, Kamatenesi MM, Nambozi G, Tolo CU, Ogwang PE, Sarki AM (2021) Prevalence of the use of herbal medicine among patients with cancer: a systematic review and meta-analysis. Evid Based Complement Alternat Med 2021:1–18
62. Ashraf H, Salehi A, Sousani M, Sharifi MH (2021) Use of complementary alternative medicine and the associated factors among patients with depression. Evid Based Complement Alternat Med 2021(6626394):1–8
63. Peltzer K, Pengpid S (2022) The use of herbal medicines among chronic disease patients in Thailand: a cross-sectional survey. J Multidiscip Healthc 12:573–582
64. Nur N (2010) Knowledge and behaviours related to herbal remedies: a cross-sectional epidemiological study in adults in middle Anatolia, Turkey. Health Soc Care Community 18 (4):389–395
65. Peltzer K, Huu TN, Ngoc NB, Pengpid S (2017) The use of herbal remedies and supplementary products among chronic disease patients in Vietnam. Stud Ethno-Med 1(2):137–145
66. Holt RIG, de Vries JH, Hess-Fischi A, Hirsch IB, Kirkman MS, Klupa I, Ludwig B, Nørgaad K, Pettus J, Renard E, Skyler JS, Snoek FJ, Weinstock RS, Peters AL (2021) The management of type 1 diabetes in adults. A consensus report by the American Diabetes Association (ADA) and the European Association for the Study of diabetes (EASD). Diabetes Care 44(11):2589–2625

67. Toumpanakis A, Turnbull T, Alba-Barba I (2018) Effectiveness of plant-based diets in promoting well-being in the management of type 2 diabetes: A systematic review. BMJ Open Diabetes Res Care 6(1):e000534, 1–10
68. Cornelius J, Doran F, Jefford E, Salehi N (2020) Patient decision aids in clinical practice for people with diabetes: a scoping review. Diabetol Int 11:344–359
69. Awuchi GC, Echeta CK, Igwe VS (2020) Diabetes and the nutrition and diets for its prevention and treatment: a systematic review and dietic perspective. Health Sci Res 6(1):5–19
70. Qaseem A, Wilt TJ, Kansagara D, Horwitch C, Barry MJ, Forciea MA (2015) Hemoglobin A1c targets for glycemic control with pharmacologic therapy for non pregnant adults with type 2 diabetes mellitus: a guidance statement update from the American College of Physicians. Ann Intern Med 168(8):569–576
71. NICE (2008) Clinical guideline 66: Type 2 diabetes. https://nice.org.uk/guidance/cg66
72. McBain A, Mulligan K, Haddad M, Flood C, Jones J, Simpson A (2016) Self-management interventions for type 2 diabetes in adult people with severe mental illness. Cochrane Database Syst Rev 4:CD011361. 1–74
73. Chaudhuryy A, Duvoor C, Dendi VSR, Kraleti S, Chada A, Ravilla R, Marco A, Shekawat NS, Montales MT, Kuriakose K, Sasapu A, Beebe A, Patil N, Musham CK, Lohani GP, Mirza W (2017) Clinical review of antidiabetic drugs: implications for type 2 diabetes mellitus management. Front Endocrinol 8(6):1–12
74. MacIsaac RJ, Jerums G, Ekinci G (2018) Glycemic control as primary prevention for diabetic kidney disease. Adv Chronic Kidney Dis 25(2):141–148
75. Rosberger DF (2013) Diabetic renopathy: current concepts and emerging therapy. Endocrinol Metab Clin N Am 42(4):721–745
76. NICE (2015) Type 1 diabetes in adults: Diagnosis and management. https://www.nice.org.uk/guidance/ng17
77. Yang Y, Liu Y, Chen S, Cheong KL, Tang B (2020) Carboxymethyl β-cyclodextrin grafted carboxymethyl chitosan hydrogel-based microparticles for oral insulin delivery. Carbohydr Polym 246(116617):1–9
78. Feingold KR (2022) Oral and injectable (non-insulin) pharmacological agents for the treatment of type 2 diabetes. In: Feingold KR, Anawalt B, Blackman MR (eds) Endotext (internet). MDText.com, Inc., South Darmouth (MA). https://www.ncbi.nlm.nih.gov/books/NBK279141
79. Wei L, Xianqing W, Qian X, Wencong L (2020) Mechanism and characteristics of sulfonylureas and glinides. Curr Top Med Chem 20(1):37–56
80. Garcin L, Mericq V, Fauret-Amsellem A, Cave H, Polak M, Beltrand J (2020) Neonatal diabetes due to potassium channel mutation: Response to sulfonylurea according to the genotype. Pediatr Diabetes 2020(21):932–941
81. Campbell I (2007) Oral antidiabetic drugs: Their preparations and recommended use. Drug Rev 18(6):56–74
82. Di Magno L, Di Pastena F, Bordone R, Coni S, Canettieri G (2022) The mechanisms of biguadines: New answers to a complex questions. Cancer 14(3220):1–32
83. Xin G, Ming Y, Ji C, Wei Z, Liu S, Morris-Natschke SL, Zhange X, Yu K, Li Y, Zhang B, Xing Z, He Y, Chen Z, Yang X, Niu H, Lee KH, Huang W (2020) Novel patient antiplatelet thrombotic agent derived from biguanide for ischemic stroke. Eur J Med Chem 200:112462. 1–14
84. Eggleton JS, Jialal I (2022) Thiazolidinediones. StatPearls Publishing, Treasure Island
85. Lebovitz HE (2019) Thiazolidinediones: The forgotten diabetes medications. Curr Diab Res 19(12):151
86. Thota S, Akbar A (2022) Continuing education activity. StatPearls Publishing. https://ncbi.nlm.nih.gov/books/NBK560688
87. WHO (2021) World Health Organization model list of essential medicines: 22nd list (2021). World Health Organization, Geneva. https://who.int/publications/i/item/WHO-MHP-HPS-EML-2021.02

88. Powers MA, Bardsley J, Cypress M, Duker P, Funnel MM, Fischi AH, Maryniuk MD, Siminerio L, Vivian E (2016) Diabetes self-management education and support in type 2 diabetes: a joint position statement of the American Diabetes Association, the American Association of Diabetes Educators and the academy of nutrition and dietetics. Clin Diabetes 34 (2):70–80
89. Ching NL, Gupta M (2020) Transdermal drug delivery systems in diabetes management: a review. Asian J Pharm Sci 15(1):13–25
90. Zhang J, Xu J, Lim J, Nolan JK, Lee H, Lee CH (2021) Wearable glucose monitoring and implantable drug delivery systems for diabetes management. Adv Healthc Mater 10(17):1–23
91. Trief PM, Cibula D, Rodriguez E, Aket B, Weinstock RS (2016) Incorrect insulin administration: a problem that warrants attention. Clin Diabetes 34(1):25–33
92. Sgrò P, Emerenziani GP, Antinozzi C, Sacchetti M, di Luigi L (2021) Exercise as a drug for glucose management and prevention in type 2 diabetes mellitus. Curr Opin Pharmacol 59:95–102
93. McCall AL, Farhy LS (2013) Treating type 1 diabetes: from strategies for insulin delivery to dual hormonal control. Minerva Endocrinol 38(2):145–163
94. Sharma RK, Bhatia SR, Sharma PR (2013) Herbal drugs in the treatment of diabetes mellitus: a review. J Diabetes Metab Disord 12(1):1
95. Rana SK, Sharma RK, Sharma PR (2013) Herbal drugs in diabetes: a review. J Diabetes Metab Disord 12(1):1
96. Abu Soud RS, Hamdan II, Afifi FU (2004) Alpha amylase inhibitory activity of some plant extracts with hypoglycemic activity. Sci Pharm 72:25–33
97. Sudha P, Zinjarde SS, Bhargava SY, Kumar AR (2011) Potent α-amylase inhibitory activity of Indian Ayurvedic medicinal plants. BMC Complement Altern Med 11(5):1–10
98. Gulati V, Harding IH, Palombo E (2012) Enzyme inhibitory and antioxidant activities of traditional medicinal plants: potential application in the management of hyperglycemia. BMC Complement Altern Med 12(1):1–9
99. Rege A, Ambaye R, Chowdhary A (2014) Effect of *Costus pictus* d. Don on carbohydrate hydrolysing enzymes. Int J Pharm Pharm Sci 6:278–280
100. Yuan H, Ma Q, Ye L, Piao G (2016) The traditional medicine and modern medicine from natural products. Molecules 21(5):559
101. Sen S, Chakraborty R (2017) Revival, modernization and integration of Indian traditional herbal medicine in clinical practice: importance, challenges and future. J Tradit Complement Med 7(2):234–244
102. Garg R, Sachdev K, Dharmendra A (2017) A scientific evaluation of Ayurvedic drugs in the management of diabetes mellitus type 2: an evidence based review. Int J Ayurveda Pharma Res 5(11):21–27
103. Rajendra A, Sudeshraj R, Sureshkumar S (2018) Potential antidiabetic activity of medicinal plants – a short review. J Phytopharmacol 7(5):456–459
104. Sharma R, Ruknuddin G, Prajapati P, Amon H (2015) Antidiabetic claims of *Tinospora cordifolia* (wild.) Miers: critical appraisal and role in therapy. Asian Pac J Trop Biomed 5 (1):68–78
105. Sharma R, Bolleddu R, Maji JK, Rukuddin G, Prajapati PK (2021) In-vitro α-amylase, α-glucosidase inhibitory activities and in-vivo anti-hyperglycemic potential of different dosage forms of Guduchi (*Tinospora cordifolia* [wild.] Miers) prepared with Ayurvedic Bhavana process. Front Pharmacol 12:642300
106. Chattopadhay K, Wang H, Kaur J, Nalbant G, Almaqhawi A, Kundakci B, Panniyammakal J, Heinrich M, Lewis SA, Greenfield SM, Tandon N, Biswas TK, Kinra S, Leonardi-Bee J (2022) Effectiveness and safety of Ayurvedic medicines in type 2 diabetes mellitus management: a systematic review and meta-analysis. Front Phatmacol 13(821810):1–31
107. Aleali AM, Amani R, Shahbazian H, Namjooyan F, Latifi SM, Cheraghian B (2019) The effect of hydroalcoholic saffron (*Crocus sativus* L.) extract on fasting plasma glucose, HbA1c, lipid

profile, liver and renal function tests in patients with type 2 diabetes mellitus: a randomized double-blind clinical trial. Phytother Res 2019:1–10

108. Pingali U, Ali MA, Gundagani S, Nutalapati C (2020) Evaluation of the effect of an aqueous extract of *Azadirachta indica* (neem) leaves and twigs on glycemic control, endothelial dysfunction and systematic inflammation in subjects with type 2 diabetes mellitus – a randomized, double-blind, placebo-controlled clinical study. Diabetes Metab Syndr Obes 13: 4401–4412
109. Haidari F, Zakerkish M, Borazjani F, Angali KA, Foroushani GA (2020) The effects of *Anethum graveolens* (dill) powder supplementation on clinical and metabolic status in patient with type 2 diabetes. Trials 21(483):1–11
110. Panda V, Deshmukh A, Singh S, Shah T, Hingorani L (2021) An Ayurvedic formulation of *Emblica officinalis* and *Curcuma longa* alleviates insulin resistance in diabetic rats: involvement of curcuminoids and polyphenolics. J Ayurveda Integr Med 12(3):506–513
111. Vikhe S, Kunkulol R, Raut D (2022) Antidiabetic and hyperlipidemic effects of crude fractions and isolated compound from *Striga orobanchioides* Benth on streptozotocin-induced diabetic rats. J Ayurveda Integr Med 13(3):1–13
112. Liu Q, Liu S, Gao L, Sun S, Huan Y, Li C, Wong Y, Guo N, Shen Z (2017) Anti-diabetic effects and mechanisms of action of a Chinese herbal medicine preparation JQ-R *in vitro* and in diabetic KK^{Ay} mice. Acta Pharm Sin B 7(4):461–469
113. Han X, Sung C, Feng X, Wang Y, Meng T, Li S, Bai Y, Du B, Sun Q (2020) Isolation and hypoglycemic effects of water extracts from mulberry leaves in Northeast China. Food Funct 11(4):3112–3125
114. Li Q, Wang G, Liu F, Hu T, Shen W, Li E, Liao S, Zou Y (2020) Mulberry leaf polyphenols attenuated postprandial glucose absorption via inhibition of disaccharides activity and glucose transport in Caco-2 cells. Food Funct 11(2):1835–1844
115. Liu Y, Yang L, Zhang Y, Liu X, Wu Z, Gilbert RG, Deng B, Wan K (2020) *Dendrobium officinale* polysaccharide ameliorates diabetic hepatic glucose metabolism via glucagon-mediated signalling pathways and modifying liver-glucagon structure. J Ethnopharmacol 248(112308):1–12
116. Liu RM, Dai R, Luo Y, Xiao JH (2019) Glucose-lowering and hypolipidemic activities of polysaccharides from *Cordyceps taii* in streptozotocin-induced diabetic mice. BMC Complement Altern Med 19(230):1–10
117. Chai GR, Liu S, Yang HW, Chen XL (2021) Quercetin protects diabetic retinopathy in rats by inducing heme oxygenase-1 expression. Neural Regen Res 16(7):1344–1350
118. Saher-Sierakowski CC, Viera-Frez FC, Hermes-Uliana C, Martins HA, Bossolani G, Lima MM, Blegniski FP, Guarnier FA, Baracat MM, Perks J, Zanoni JN (2021) Protective effects of quercetin-loaded microcapsules on the enteric nervous system of diabetic rats. Auton Neurosci 230:1–33
119. Wu Y, Sun B, Guo X, Wu L, Hu Y, Qin L, Yang T, Li M, Qin T, Jiang M, Liu T (2022) Zishen pill alleviates diabetes in Db/db mice via activation of PI3K/AKT pathway in the liver. Chin Med 17(128):1–13
120. Patel DK, Prasad SK, Kumar R, Hemalatha S (2012) An overview on antidiabetic medicinal plants having insulin mimetic property. Asian Pac J Trop Biomed 2(4):320–330
121. Prabhakar PK, Doble M (2011) Mechanism of action of natural products used in the treatment of diabetes mellitus. Chin J Integr Med 17(8):563–574
122. Kim K, Kim HY (2008) Korean red ginseng stimulates insulin release from isolated rat's pancreatic islets. J Ethnopharmacol 120:190–195
123. Kim EK, Kwon KB, Lee JH, Park BH, Park JW, Lee HK (2007) Inhibition of cytokine-mediated nitric oxide synthase expression in rat insulinoma cells by scoparone. Biol Pharm Bull 30:242–246
124. Karmazyn M, Gan XT (2021) Chemical components of ginseng, their biotransformation products and their potential as treatment of hypertension. Mol Cell Biochem 476:333–347

125. Huang L, Ren C, Li HJ, Wu TC (2021) Recent Progress on processing technologies, chemical components, and bioactivities of Chinese red ginseng, American red ginseng, and Korean red ginseng. Food Bioprocess Technol 15:47–71
126. Choudhary U, Sabikhi L, Kapila S (2021) Double emulsion-based mayonnaise encapsulated with bitter gourd extract exhibits improvement *in vivo* anti-diabetic action in STZ-induced rats. 3 Biotech 11:363. 1–14
127. Kim SK, Jung J, Jung JH, Yoon NA, Kang SS, Roh GS, Hahm JR (2020) Hypoglycemic efficacy and safety of *Momordica charantia* (bitter melon) in patients with type 2 diabetes mellitus. Complement Ther Med 52(102524):1–5
128. Rathi SS, Grover JK, Vikrant V, Biswas N (2007) Prevention of experimental diabetic cataract by Indian Ayurvedic plant extracts. Phytother Res 16:774–777
129. Shibib BA, Khan LA, Rahman R (1993) Hypoglycemic activity of *Coccinia indica* and *Momordica charantia* in diabetic rats: depression of the hepatic glucogenic enzymes glucose-6-phosphatase and fructose-1,6-bisphosphatase and evaluation of both liver and red-cell shunt enzyme glucose-6-phosphate dehydrogenase. Biochem J 299(pt 1):267–270
130. Bhaskaran K, Ting OJ, Tengli T (2019) Management of diabetes and hyperlipidemia by natural medicines. In: Dhosh D, Mokherjee PK (eds) Natural medicines, 1st edn. CRC Press, Boca Raton, pp 1–23
131. Rao AS, Hegde S, Pacioretty LM, de Benedetto J, Babish JG (2020) *Nigella sativa* and *Trigonella foenum-graecum* supplemented chapatis safely improve HbA1c, body weight, waist circumference, blood lipids, and fatty liver in overweight and diabetic subjects: a twelve-week safety and efficacy study. J Med Food 23(9):905–919
132. Semwal DK, Kumar A, Aswal S, Chauhan A, Semwal RB (2020) Protective and therapeutic effects of natural products against diabetes mellitus via regenerating pancreatic β-cells and restoring their dysfunction. Phytother Res 35(3):1218–1229
133. Wickramasinghe ASD, Kalansuriya P, Attanayake AP (2021) Herbal medicines targeting the improved β-cell functions and β-cell regeneration for the management of diabetes mellitus. Evid Based Complement Alternat Med 2021(2920530):1–32
134. Devi BA, Kamalakkannan N, Prince PS (2003) Supplementation of fenugreek leaves to diabetic rats: effect on carbohydrate metabolic enzymes in diabetic liver and kidney. Phytother Res 17:1231–1233
135. Annida B, Stanely MP (2005) Supplementation of fenugreek leaves reduce oxidative stress in streptozocin-induced diabetic rats. J Med Food 8:382–385
136. Yu R, Hui H, Shlomo M (2005) Insulin secretion and action. In: Endocrinology, 2nd edn. Hamana Press, pp 311–319
137. Lee YS, Kim WS, Kim KH, Yoon MJ, Cho HJ, Shen Y (2006) Berberine, a natural plant product, activates AMP-activated protein kinase with beneficial metabolic effects in diabetic and insulin-resistant states. Diabetes 55:2256–2284
138. Zhou L, Wang X, Shao L, Yang Y, Shang W, Yuan G (2008) Berberine acutely inhibits insulin secretion from beta-cells through 3′, 5′-cyclic adenosine 5′-monophosphate signalling pathway. Endocrinology 149:4510–4518
139. Zhou JY, Zhou SW, Zhang KB, Tang JL, Guang LX, Ying Y (2008) Chronic effects of berberine on blood, liver glucolipid metabolism and liver PPARs expression in diabetic hyperlipidemic rats. Biol Pharm Bull 31:1169–1176
140. Liu WH, Hei ZQ, Nei H, Tang FT, Huang HQ, Li XJ (2008) Berberine ameliorates renal injury in streptozotocin-induced diabetic rats by suppression of both oxidative stress and aldose reductase. Clin Med J (Engl) 121:706–712
141. Turner N, Li JY, Gosby A, To SW, Cheng Z, Miyoshi H (2008) Berberine and its more biologically available derivative, dihydroberberine, inhibit mitochondrial respiratory complex: a mechanism for the action of berberine to activate AMP-activated protein kinase and improve insulin action. Diabetes 57:1414–1418

142. Derosa G, D'Angelo A, Maffiola P (2022) The role of selected nutraceuticals in the management of prediabetes and diabetes: an updated review of the literature. Psychother Res 36(10): 3709–3765
143. Judy WV, Hari SP, Stoggdill WW, Judy JS, Naguio YM, Passwater R (2002) Antidiabetic activity of a standardized extract (Glucosol™) from *Lagerstroemia speciosa* leaves in type II diabetes: a dose-dependent study. J Ethnopharmacol 87:115–117
144. Hsu CH, Liao YL, Lin SC, Hwang KC, Chou P (2007) The mushroom *Agaricus blazei* Mirill in combination with metformin and gliclazide improves insulin resistance in type 2 diabetes: a randomized, double-blinded and placebo-controlled trial. J Altern Complement Med 13:97–102
145. Jiang AT, Rowe N, Sener A, Luke P (2014) Simultaneous pancreas-kidney transplantation: the role in the treatment of type 1 diabetes and end-stage renal disease. Can Urol Assoc J 8(3–4): 135–138
146. Pérez-Sáez MJ, Pascual J (2015) Kidney transplantation in the diabetic patients. J Clin Med 4 (6):1265–1280
147. Affinati AH, Esfandiari NH, Oral EA, Kraftson AT (2019) Bariatric surgery in the treatment of type 2 diabetes. Curr Diab Rep 19(156):1–10
148. O'Kane M, Parretti HM, Pinkney J, Welbourn R, Hughes CA, Mok J, Walker N, Thomas D, Devin J, Coulman KD, Pinnock G, Batterham RL, Mahawar KK, Sharma M, Blakemore AI, McMillan I, Barth JH (2020) British obesity and metabolic surgery society guidelines on perioperative and postoperative biochemical monitoring and micronutrient replacement for patients undergoing bariatric surgery – 2020 update. Obes Rev 21(11):1–23
149. Wiggins T, Guidozzi N, Welbourn R, Ahmed AR, Markar SR (2020) Association of bariatric surgery with all-cause mortality and incidence of obesity-related disease at a population level: a systematic review and meta-analysis. PLoS Med 17(7):e1003206. 1–18
150. Dixon JB, le Roux CW, Rubino F, Zimmet P (2012) Bariatric surgery for type 2 diabetes. Lancet 379(9833):2300–2311
151. Crossen S, Raymond J, Neinstein A (2020) Top 10 tips for successfully implementing a diabetes telehealth program. Diabetes Technol Ther 22(12):920–928
152. Pal K, Eastwood SV, Michie S, Farmer AJ, Barnard ML, Peacock R (2013) Computer-based diabetes self-management interventions for adults with type 2 diabetes mellitus. Cochrane Database Syst Rev 2013(3):CD008776. 1–114
153. Carpenter R, di Chiacchio T, Barker K (2019) Interventions for self-management of type 2 diabetes: an integrative review. Int J Nurs Sci 6(1):70–91
154. Mathew TK, Tadi P (2022) Blood glucose monitoring (online). StatPearls [internet]. Treasure Island. https://pubmed.ncbi.nlm.nih.gov/32310436/
155. Czupyniak L, Barhai L, Bolgarska S, Bronisz A, Broz A, Cupryk K, Honka M, Janez A, Krnic M, Lalic H, Matinka E, Rahelic D, Roman G, Tankosa T, Vákonyi T, Welnik B, Zherdova N (2014) Self-monitoring of blood glucose in diabetes from evidence to clinical reality in central and Eastern Europe – recommendations from the international central-eastern European expert group. Diabetes Technol Ther 16(7):460–478
156. Pleus S, Freckmann G, Schavefr S, Heinemann L, Ziegler R, Ji L, Mohan V, Calliari LE, Hinzmann R (2022) Self-monitoring of blood glucose as an integral part in the management of people with type 2 diabetes mellitus. Diabetes Ther 13:829–846
157. Ehrmann BJ (2014) Diabetes. A comprehensive guide to geriatric rehabilitation, 3rd edn. Churchill Livingstone/Elsevier, Edinburgh, pp 329–336
158. Davies MJ, D'Alesio DA, Fradkin J, Kernan WN, Mathieu C, Mingrone G, Rossing P, Tsapas A, Wexler DJ, Buse JB (2018) Management of hyperglycemia in type 2 diabetes, 2018. A consensus report by the American Diabetes Association (ADA) and the European Association for the Study of diabetes (EASD). Diabetes Care 41(12):2669–2701
159. Nazar CMJ, Bojerenu MM, Safdar M, Marwat J (2016) Effectiveness of diabetes education and awareness of diabetes mellitus in combating diabetes in the United Kingdom: a literature review. J Nephropharmacol 5(2):110–115

160. Sushko K, Menezes HT, Strachan P, Butt M, Sherifali D (2021) Self-management education among women with pre-existing diabetes in pregnancy: a scoping review. Int J Nurs Stud 117 (103883):1–10
161. World Health Organization (2016) Diabetes fact sheet 2013
162. Balk EM, Earley A, Raman G, Avendano EA, Pitta AG, Remington PL (2015) Combined diet and physical activity promotion programs to prevent type 2 diabetes among persons at increased risk: a systematic review the community preventive service takes force. Ann Intern Med 163(6):437–451
163. Alustiza E, Perales A, Mateo-Abad M, Ozcoidi I, Aizpuru G, Albaina O, Vergara I (2020) Tackling risk factors for type 2 diabetes in adolescent: pre-START study in Euskadi (translated). Ann Pediatr 3017:1–12
164. Pan A, Wang Y, Hu FB TM, Wu T (2015) Relation of active, passive and quitting smoking with incident type 2 diabetes: a systematic review and meta-analysis. Lancet Diabetes Endocrinol 3:958–966
165. Mutyambizi C, Pavlova M, Hongoro C, Groot W (2020) Inequalities and factors associated with adherence to diabetes self-care practices amongst patients at two public hospitals in Gauteng, South Africa. BMC Endocr Disord 20(15):1–10
166. Maddatu J, Anderson-Baucum E, Evans-Molina C (2017) Smoking and the risk of type 2 diabetes. Transl Res 184:101–107
167. Shield KD, Rylett M, Gmel G, Kehoe-Chan TAK (2010) Global alcohol exposure estimates by country territory and region for 2005 – a contribution to the comparative risk assessment for the 2010. Global burden of disease study. Addiction 108(5):912–922
168. World Health Organization (2018) Global status report on alcohol and health, 2018. World Health Organization, Geneva, Switzerland, pp 14–19, MIT Press, Cambridge
169. Volaco A, Ercolano CA (2018) Alcohol consumption and its relationship to diabetes mellitus: friend or foe? Endocrinol Metab Int J 6(1):32–35
170. Polsky S, Akturk HK (2017) Alcohol consumption, diabetes risk and cardiovascular disease within diabetes. Curr Diab Rep 17(136):1–12
171. Steiner JL, Crowell KT, Lang CH (2015) Impact of alcohol on glycemic control and insulin action. Biomol Ther 5(4):2223–2248
172. Asif M (2014) The prevention and control the type 2 diabetes by changing lifestyle and dietary pattern. J Educ Health Promot 3(1):1–8
173. Li X, Yu F, Zhou Y, He J (2016) Association between alcohol consumption and the risk of incident type 2 diabetes: a systematic review and dose-response meta-analysis. Am J Clin Nutr 103(3):818–829
174. Knott C, Bell S, Briton A (2018) Alcohol consumption and the risk of type 2 diabetes: a systematic review and dose-response meta-analysis of more than 1.9 million individuals from 38 observational studies. Diabetes Care 38(9):1804–1812
175. Holst C, Becker U, Jørgensen ME, Grønbæk M, Tolstrup JS (2017) Alcohol drinking patterns and risk of diabetes: a cohort study of 70,551 men and women from the general Danish population. Diabetologia 60(10):1941–1950
176. Huang J, Wang X, Zhang Y (2017) Specific types of alcoholic beverage consumption and risk of type 2 diabetes: a systematic review and meta-analysis. J Diabetes Investig 8(1):56–68
177. Baliunas DO, Taylor BJ, Irving H, Roerecke M, Patra J (2009) Alcohol as a risk factor for type 2 diabetes: a systematic review and meta-analysis. Diabetes Care 32(11):2123–2132
178. Wake AD (2021) Alcohol consumption and the incidence of type 2 diabetes mellitus: "issue that requires more attention". Endocrinol Metab Int J 8(6):143–147
179. Cullmann M, Hillding A, Östenson CG (2012) Alcohol consumption and risk of pre-diabetes and type 2 diabetes development in a Swedish population: alcohol consumption and risk of type 2 diabetes and pre-diabetes. Diabet Med 29:441–452
180. Koppes LLJ, Dekker JM, Hendriks HFJ (2005) Moderate alcohol consumption lowers the risk of type 2 diabetes: a meta-analysis of prospective observational studies. Diabetes Care 28:719–725

181. Kawamoto R, Kohara K, Taraba Y (2009) Alcohol consumption is associated with decreased insulin resistance independent of body mass index in Japanese community-dwelling men. Tohoku J Exp Med 218:331–337
182. Kalaria T, Ko YL, Issuree KKJ (2021) Literature review: drug and alcohol-induced hypoglycaemia. J Lab Precis Med 6(21):1–16
183. Choi S, Lee G, Kang J (2020) Association of change in alcohol consumption on fasting serum glucose, insulin resistance and beta cell function among Korean men. Alcohol 885:127–133
184. Sakurai Y, Kubota N, Yamauchi T, Kadowaki T (2021) Role of insulin resistance in MAFLD. Int J Mol Sci 22(8):4156. 1-26
185. Ley SH, Korat AVA, Sun Q, Tobias DK, Zhang C, Qi L, Willett WC, Manson JW, Hu FB (2016) Contribution of the nurses' health studies to uncovering risk factors for type 2 diabetes: diet, lifestyle, biomarkers, and genetics. Am J Public Health 106:1624–1630
186. Diabetes Prevention Program Research Group, Knowler WC, Fowler SE, Hamman RF, Christophi CA, Hoffman HJ, Brenneman AT, Brown F, Goldberg R, Venditti E, Nathan DM (2009) 10-year follow-up of diabetes incidence and weight loss in the diabetes prevention program outcomes study. Lancet 374(9702):1677–1686
187. Aune D, Norat T, Leitzmann M, Vatten LJ (2015) Physical activity and the risk of type 2 diabetes: a systematic review and dose–response meta-analysis. Eur J Epidemiol 30:529–542
188. Balk EM, Earley A, Raman G, Avendano EA, Pittas AG, Remington PL (2015) Combined diet and physical activity promotion programs to prevent type 2 diabetes among persons at increased risk: a systematic review for the community preventive services task force. Ann Intern Med 136(6):1–18
189. Ley SH, Hardy O, Mohan V, Hu FB (2014) Prevention and management of type 2 diabetes: dietary components and nutritional strategies. Lancet 383(9933):1999–2007

Therapeutic Applications of Herbal Medicines for the Prevention and Management of Cancer

21

Manish Pathak, Rupesh Kumar Pandey, Lubhan Singh, Amit Kumar, Ganesh Prasad Mishra, Ravindra Kumar Pandey, Bina Gidwani, and Shiv Shankar Shukla

Contents

Abstract

Plants serve as a revered base for the natural products that support human health. There is growing evidence that Ayurvedic medicinal herbs and the secondary metabolites they produce can be used to treat cancer. Numerous research have documented the value of herbal plants in immune system modulation, cancer

M. Pathak
Departments of Pharmaceutical Chemistry, Kharvel Subharti College of Pharmacy, Swami Vivekanand Subharti University, Meerut, Uttar Pradesh, India

R. K. Pandey · L. Singh · G. P. Mishra
Department of Pharmacology, Kharvel Subharti College of Pharmacy, Swami Vivekanand Subharti University, Meerut, Uttar Pradesh, India

A. Kumar
Kharvel Subharti College of Pharmacy, Swami Vivekanand Subharti University, Meerut, Uttar Pradesh, India

R. K. Pandey
Columbia Institute of Pharmacy, Tekari, Raipur, Chhattisgarh, India

B. Gidwani · S. S. Shukla (✉)
Department of Pharmaceutical Quality Assurance, Columbia Institute of Pharmacy, Tekari, Raipur, Chhattisgarh, India

S. C. Izah et al. (eds.), *Herbal Medicine Phytochemistry*, Reference Series in Phytochemistry, https://doi.org/10.1007/978-3-031-43199-9_23

patient survival, and quality of life. For the treatment and prevention of cancer, Ayurvedic herbal remedies are included into mainstream therapy. Humans are becoming more and more reliant on herbal remedies because they are safer and more potent than manufactured pharmaceuticals. The value of medicinal plants is rising due to its potential and uses. The use of Ayurvedic medicinal herbs for the prevention and treatment of various malignancies is examined in this chapter's review work. This chapter also discusses a report on the potential use of phytoconstituents in the treatment of cancer.

Keywords

Ayurvedic medicinal plants · Cancer · Phytoconstituents

Abbreviations

DNA	Deoxyribonucleic acid
S.Nr	Serial number
WHO	World Health Organization

1 Introduction

Herbalism is a term that literally refers to the use of plants to treat a variety of human illnesses. The word "herb" derives from a Latin word that designates a plant's vegetative and reproductive parts that are used for therapeutic purposes. The use of medicinal plants as a source of drugs to treat or prevent various diseases may be dated back about five millennia in recorded accounts from India's early civilisations. The natural products that enhance human health are derived from plants. The Plantae might contain a range of therapeutically useful physiologically active compounds, as well as unique templates for synthesising analogues and stimulating tools for better understanding biological processes [1–8].

According to the WHO in 2011, approximately 88% of the world's population relies on herbal medications that have therapeutic efficacy. Long before the prehistoric era, plants were employed for medical purposes. Chinese, Egyptian, and Unani authors of old or ancient writings all agreed that items derived from medicinal plants are utilised as areas of medicine. Over a thousand years ago, Vaids and Hakim from various civilisations used medicinal herbs for therapeutic purposes [9–20].

1.1 Traditional Medicine System

Traditional systems of drugs still are being widely practised in various ways. The use of synthetic drugs and its negative impact and high cost leads to faith of human beings on medicinal plants. Utilisation of plant materials in the form of herbal medicines progressed because of improper supply of medicines and increase in population. The diversified and verdant plant kingdom is renowned for their therapeutic value [21–32]. More utilisation of plants for medicinal purposes enhanced because of development of varied herbal therapies. According to the Turkish Journal, significant value of medicinal plants or herbal plants increased importance over allopathic medicines because of presence of many narcotic compounds. Medicinal plants bestowed with varied compounds having biotic origin and showing medicinal and therapeutic values. Medicinal plants are exceptionally a magnificent possibility for curing human diseases because of presence of analgesic, anti-inflammatory, anti-tumour, and antiviral properties. Medicinal plants have various better properties over chemical medicines because plant-derived compounds are more tolerant, non-toxic, and beneficial for the human body [33–48].

India is the birth place and vast repository of various plants having medicinal value. Among many ancient and primitive civilisations, Indian civilisation has been evidenced for medicines assessed from herbs. Flora of India provides path for processing of herbal drugs. Nearly eight thousand herbal medicines are documented in India by AYUSH. Various types of medications are accepted by Indian culture which was assessed by Unani, Siddha, and Ayurvedic systems [49–62].

80% of the world's population, according to the WHO (2012), relies on herbal medicines for health reasons. Around 21,000 plant species with therapeutic properties that are used globally for basic health were also documented by the WHO (2012). More than 30% of a plant species' total biomass was employed for therapeutic reasons. The contribution from quickly developing nations like China and India is at most 80%. India and China contributed more towards the herbal medication. Rural population of these countries mainly rely on indigenous medication system. Treatment is associated with natural system having maximum positive impact and minimal negative impact. Remedies based on nature are good for human beings and medicinal plants play very vital role in herbal treatment. Medicinal plants are restrained as ironic possessions of components which can be utilised for drug formation. Therapeutic value of plants mainly depends on medicinal properties of plants [63–85].

The biologically active chemicals found in the plant kingdom constitute a largely untapped source of potential medicines as well as useful tools and special templates for studying biological processes. In India, the use of plant composites for therapeutic purposes has grown over time [80–82, 84–88]. Traditional medicines have played a significant part in addressing the world's healthcare requirements in the past, are still doing so today, and will continue to do so in the future. The Ayurvedic structure in India has described an enormous number of those medicines buttressed plants or plant material and therefore the determination of their morphological and pharmacological or pharmacognostical characters. It can offer a better considerate of their active principles and mode of accomplishment. Young people nowadays are completely dependent on allopathy, therefore using medicinal plant mixtures to halt the disease's sneaky effects is an alternative to Western medications, which often have terrible side effects. Secondary metabolites found in medicinal plants, such as flavonoids, flavones, anthocyanins, lignans, and coumarins, have been shown to have anticancer activities. Items are made from herbs, especially in Ayurveda. The use of traditional remedies in India was emphasised in the ancient Vedas and other writings [89–101]. Ayurveda has two primary goals: first, to promote health and second, to treat disease. Ayus and Veda, two Sanskrit terms that translate to "life" and "knowledge," respectively, are the origin of Ayurveda. Literally, it signifies understanding of and research into life. Ayurveda, the science of life, may be a part of India's illustrious Vedic legacy. Ayurveda, the oldest system of medicine in India, uses plant extracts to combat cancer-causing chemicals and eradicate tumours associated with cancer. Three major expositions on Ayurveda – the Charaka Samhita (a work on ancient Indian medicine), the Sushruta Samhita (a text on ancient Indian surgery), and the Kashyapa Samhita – are well known (text on gynaecology and child health). Ayurveda which prescribes a variety of medications for preserving health, vitality, youthfulness, and longevity. Ayurveda considers health as a equilibrium of delight or happiness of the soul, sense, and mind. Medicines prepared by Ayurvedic mechanisms well proved for cancer treatment having anti-cancerous properties [102–114]. Ayurvedic medicines have an ability to fight against the cancerous tumours. Preparations of medicines by Ayurvedic method act as an ancillary or a concomitant treatment alongside radiotherapy or chemotherapy. Negative impact of chemotherapy and radiotherapy can be reducing by usage of Ayurvedic medicines. Ayurvedic medicines assist to detain the cancerous cells where radiotherapy and chemotherapy have significant side effects. The ancient therapeutic art and science of Ayurveda focuses on enhancing, sustaining, and balancing mind-body health. It also takes a holistic approach to health and wellness. Meditation is the cornerstone of Ayurvedic medicine's approach to mind-body healing. Ayurvedic medicines that are sold in many parts of the world have undergone substantial research about their anti-cancer properties. Both Charaka and Sushruta Samhita referred to granthi and arbuda as the equivalents of cancer (malignant or major neoplasm). Based on the doshas involved, both are frequently inflammatory or non-inflammatory. The Ayurveda definition of health is the harmonious with coordination of vata, pitta, and kapha system within the body [115–131].

1.2 Cancer and Its Molecular Basis

Cancer is characterised by uncontrolled and unrestrained cell proliferation. It appears as a growth in a clinical context. A malformed mass of tissue known as neoplasm grows erratically and keeps growing long after the factors that first triggered its growth have been removed. A tumour is described as benign if its characteristics are believed to be essentially harmless, which implies that it wouldn't spread to neighbouring or distant regions, could be surgically removed without difficulty, and didn't constitute a severe threat to the patient's life [132–141]. Malignant tumours are collectively referred to as cancers, which are derived from the Latin word for crab, because they have a crab-like effect on the tissues. A malignant tumour has the ability to spread to distant areas (metastasize), harm nearby structures, and ultimately lead to death. Not all cancers result in death; with early detection and appropriate care, certain cancers can be cured [142–178].

Cells that proliferate expand quickly.

- Nucleotide sequence changes due to genetic mutation
- Agents that cause cancer, or carcinogens

Such genetically modified crops, rising cigarette use, fast food, alteration, or mutation may be inherited through the germ line or acquired through the action of environmental forces. Environmental agents Rapid development, neighborhood invasion, and the ability to metastasize far away are characteristics of malignant neoplasms. The ability of a cancer cell to proliferate in the absence of growth signals, resistance to signals that would normally stop a cell from dying, ability to form new blood vessels, ability to spread to other tissues, ability to metastasize to distant organs, and failure to repair damaged DNA are just a few of the key changes that occur in a cancer cell [179–202].

1.3 Common Cancers

Men are most likely to have lung, prostate, colorectal, stomach, and liver cancers worldwide. The most often diagnosed cancers in women are those of the breast, cervix, colorectum, lung, and stomach [203–217]. Oral, lung, stomach, and colorectal cancers are becoming prevalent among men in India. The most prevalent cancers in women are breast, cervix, colorectal, ovarian, and oral cancers [213, 216, 218–230]. According to the geographic region, prevalent social customs, and socioeconomic strata, different malignancies have different incidence rates. For instance, oral malignancies are more common in the Indian subcontinent than in Western nations. This is a result of increased use of chewable tobacco products including gutkha, paan, paan masala, khaini, and supari, among others. Due to inadequate genital cleanliness, cervical cancer is more prevalent in women from lower socioeconomic groups. Those who consume more fatty foods and fewer fibre in their diets are more likely to develop colorectal cancer [231–242].

1.4 Common Causes of Cancer

There are many agents who causes cancer or increases the risk of cancer. Some major agents are as follows (Fig. 1). Also some risk factors of cancer are shown in Table 1.

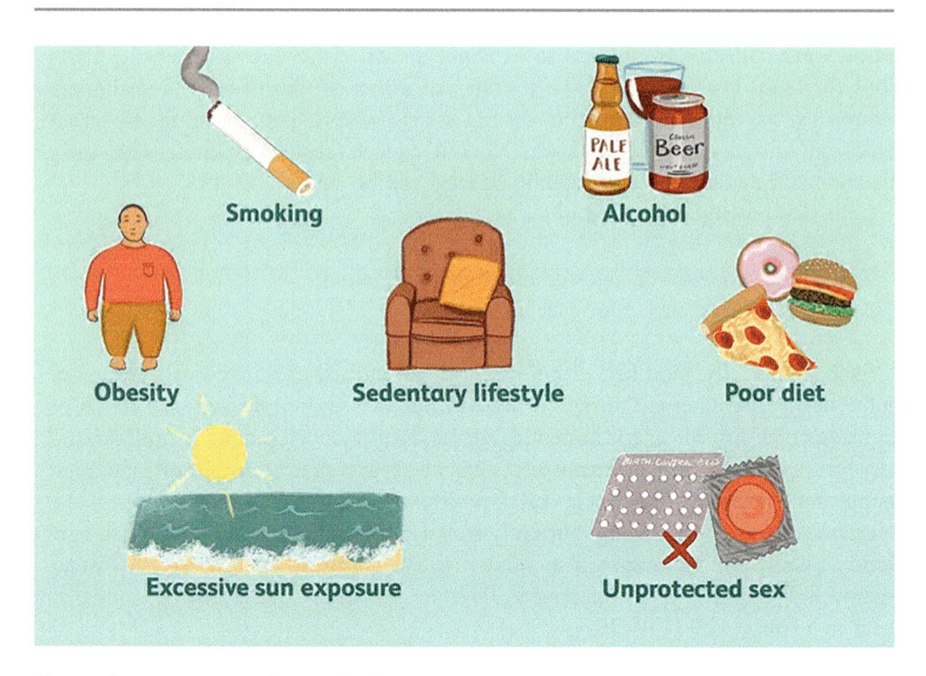

Fig. 1 Common causes of cancer [100]

Table 1 Some common conditions and risk factors of certain cancer

Tobacco	The single most significant preventable risk factor for cancer death worldwide is tobacco use. It is thought to be responsible for 22% of cancer deaths annually, according to the WHO. The vast majority of lung cancers are linked to smoking. Those who don't smoke have also been linked to malignancies by passive smoking. Danger rises as smoking frequency increases. The usage of smokeless tobacco is more common in the Indian subcontinent. They include using raw tobacco, betel quid, gutkha, pan masala, masheri, and other substances
Alcohol	Drinking alcohol increases your risk of acquiring a number of cancers that are related to alcohol. It has a synergistic impact when taken with smoking that significantly raises the risk of cancer development compared to when either chemical is consumed alone
Obesity	People with BMIs over 30 are considered obese. They are more likely to get cancer, diabetes, and heart disease. Regular exercise and a healthy, balanced diet are crucial for battling the obesity pandemic
Areca nut	It is also referred to as supari in India. It is called as Pan or betel quid and can be chewed either alone or in combination with betel leaf, catechu, and slaked lime. It is linked to a condition called submucous fibrosis, which gradually narrows the chewer's mouth aperture and is a sign of malignancy
Pollution	This refers to the environmental pollution of the air, water, and soil caused by the usage of carcinogenic substances. Both ingestion and air exposure to these compounds, which are known as carcinogens, have the potential to cause cancer

(continued)

Table 1 (continued)

Occupational exposure	Numerous compounds, such as asbestos, cadmium, ethylene oxide, benzopyrene, silica, ionising radiation, including radon, tanning booths, aluminium and coal manufacturing, and the manufacture of iron and steel, are known to cause various cancers. Understanding these is essential because the majority of these cancers are preventable with the correct knowledge and safety precautions
Radiation	Exposure to ionising radiation has been associated with a greater risk of acquiring a number of cancers. UV radiation has been related to skin cancers such basal cell carcinoma, squamous cell carcinoma, and melanoma
Biological agents	A wide range of germs, parasites, and viruses that raise people's chance of developing cancer. Oropharyngeal and cervical cancers have been linked to the human papilloma virus (HPV). Hepatitis B and C are associated with liver cancer. Schistosomiasis, a parasite infection, is associated with an increased risk of bladder cancer. *H. pylori* infection may result in stomach cancer
Lifestyle factors	High body mass index is associated with cancers of the oesophagus, colorectum, breast, endometrium, etc. Consuming red meat increases your chances of colon and rectal cancer. A nutritious diet that includes fruits, salads, and vegetables has been found to reduce the risk
Unprotected sex	Unprotected sex also causes cancer in males or females

2 Importance of Traditional Plant-Based Drugs in Treatment of Cancer

Plant-based drugs are becoming more and more common all around the world. Recent studies on herbal plants or medicine have made substantial progress in the pharmacological study of several plants used in traditional medical systems. The secondary metabolites or compounds found in therapeutic plants include tannins, terpenoids, alkaloids, and flavonoids. These compounds determine the therapeutic effectiveness of the plants, especially their antimicrobial activity. The use of plant-derived remedies as an indigenous treatment in traditional systems of medicine has been related to the advent of plant-derived medications in contemporary medicine. It has been discovered that several of the plants have potent anticancer properties. Cancer is treated with medicines made from plants [211, 212, 243–248]. The traditional herbal plants used to treat cancer are listed in the table below.

2.1 List of Traditional Plants Used to Treat Cancer

There are so many plants that possess anticancer properties. Some of these plants and their chemical constituents are shown in Table 2.

Table 2 Plants with anticancer potentials

Sr. No.	Botanical name	Family	Chemical constituents	Types of cancer
1.	*Ferula assafoetida*	Apiaceae	Coumarin compounds	Liver cancer [9]
2.	*Thymus vulgaris*	Lamiaceae	Thymol	Prostate cancer [10], head cancer [11]
3.	*Thymbra spicata*	Lamiaceae	Thymol	Lung cancer [12]
4.	*Taverniera spartea*	Fabaceae	Isoflavonoid compounds	Breast and prostate cancer [13]
5.	*Peganum harmala*	Nitrariaceae	Alkaloids	Breast cancer [14], cervical cancer [13, 22]
6.	*Viola tricolor*	Violaceae	Flavonoids	Cervical cancer [23]
7.	*Achillea wilhelmsii*	Asteraceae	Phenolic compounds	Colon cancer [24]
8.	*Mentha pulegium*	Lamiaceae	Piperitone	Blood cell cancer [25]
9.	*Ammi visnaga*	Apiaceae	Visnadine	Breast cancer [26]
10.	*Camellia sinensis*	Theaceae	Epicatechin	Breast cancer [27]
11.	*Avicennia marina*	Acanthaceae	Flavonoids	Breast and larynx cancer [15, 16]
12.	*Silybum marianum*	Asteraceae	Flavonoids	Colorectal cancer [17], breast cancer [18]
13.	*Artemisia absinthium*	Asteraceae	Artemisinin, isorhamnetin	Colon cancer [19], blood cancer [20]
14.	*Curcuma longa*	Zingiberaceae	Curcumin	Breast [21], prostate [28], cervix [3], larynx cancer [29]
15.	*Crocus sativus*	Iridaceae	Phenolic compounds	Sarcoma, oral cancer [30, 31]
16.	*Zingiber officinale*	Zingiberaceae	Flavonoids	Urinary cancer [32], prostate cancer [33]
17.	*Olea europaea*	Oleaceae	Oleic acid	Breast cancer [34], colon cancer[35]
18.	*Taxus baccata*	Taxaceae	Taxol	Prostate cancer [36], pancreatic cancer [37], blood cell cancer [38]
19.	*Nigella sativa*	Ranunculaceae	Thymoquinone	Prostate [39]
20.	*Allium sativum*	Amaryllidaceae	Allicin	Cervical cancer [40], breast [41, 42] colon [43, 44], larynx [45]
21.	*Lepidium sativum*	Brassicaceae	Vitamins (A, B, C, and E)	Breast cancer [46]

(continued)

Table 2 (continued)

Sr. No.	Botanical name	Family	Chemical constituents	Types of cancer
22.	*Trigonella foenum-graecum*	Fabaceae	Alkaloids	Breast cancer [47, 48]
23.	*Glycyrrhiza glabra*	Fabaceae	Glycyrrhizin	Kidney cancer [48]
24.	*Physalis alkekengi*	Solanaceae	Physalins	Cervical cancer [50, 51]
25.	*Lagenaria siceraria* Standl	Cucurbitaceae	Vitamins B and C	Lung cancer [52], breast cancer [53]
26.	*Ferula gummosa*	Apiaceae	Sesquiterpenes	Lung cancer [54], skin cancer, stomach cancer [55]
27.	*Urtica dioica* L.	Urticaceae	Phenolic compounds	Prostate cancer [56–58]
28.	*Ammi majus*	Apiaceae	Coumarin compounds	Breast cancer [59]
29.	*Rosa damascene*	Rosaceae	Phenolic compounds	Lung cancer [60], breast cancer, cervix [61]
30.	*Astragalus cystosus*	Fabaceae	Lectins, flavonoids	Lung cancer [62]
31.	*Myrtus communis*	Myrtaceae	Myrtucommulone	Breast cancer [63–65]
32.	*Vinca rosea*	Apocynaceae	Vindoline	Breast cancer, larynx cancer [64]
33.	*Citrullus colocynthis*	Cucurbitaceae	Quercetin	Liver cancer [66]
34.	*Polygonum aviculare*	Polygonaceae	Flavonoids and alkaloids	Breast cancer [67], cervical cancer [68, 69]
35.	*Astrodaucus orientalis*	Apiaceae	A-thujene and a-copaene	Breast cancer [70]
36.	*Actinidia chinensis*	Actinidiaceae	Polysaccharide	Hepatic cancer [71]
37.	*Aegle marmelos*	Rutaceae	Lupeol	Breast cancer and leukaemia
38.	*Agave americana*	Agavaceae	Steroidal and alkaloid	Cancerous tumour
39.	*Aloe vera*	Asphodelaceae	Emodin	Leukaemia, stomach cancer [72]
40.	*Alpinia galanga*	Zingiberaceae	Pinocembrin	Leukaemia
41.	*Amoora rohituka*	Meliaceae	Amooranin	Lymphocytic [73]
42.	*Andrographis paniculata*	Acanthaceae	Flavonoids	Colon cancer [74]

(continued)

Table 2 (continued)

Sr. No.	Botanical name	Family	Chemical constituents	Types of cancer
43.	*Annona muricata*	Annonaceae	Acetogenins	Lung cancer and liver cancer [75]
44.	*Apis mellifera*	Apidae	Protein	Lymphoid cancer [76]
45.	*Ananas comosus*	Bromeliaceae	Bromelain	Leukaemia [77], gastrointestinal carcinoma [78]
46.	*Angelica sinensis*	Apiaceae	Polysaccharide	Cervical cancer [79], brain tumour [80], glioblastoma multiforme [81]
47.	*Annona species*	Annonaceae	Acetogenins	Nasopharyngeal cancer [82]
48.	*Arctium lappa*	Asteraceae	Arctigenin [85]	Prostate cancer [83, 84]
49.	*Artemisia asiatica*	Asteraceae	Iso-liquiritigenin	Liver tumour
50.	*Astragalus membranaceus*	Fabaceae	Swainsonine	Gastrointestinal cancers and liver cancer [85]
51.	*Azadirachta indica*	Meliaceae	Limonoids and nimbolide	Leukaemia [78]
52.	*Bauhinia variegata*	Caesalpiniaceae	Cyanidin glucoside	Breast cancer [79]
53.	*Berberis vulgaris*	Berberidaceae	Berbamine and berberine	Prostate cancer, oral cancer [80]
54.	*Betula alba*	Betulaceae	Betulinic acid	Prostate cancer [82], thyroid [86], breast cancer [84], colon cancer [85]
55.	*Betula utilis*	Betulaceae	Ursolic acid and betulin	Liver cancer, breast cancer [78]
56.	*Bolbostemma paniculatum*	Cucurbitaceae	Tubeimoside-5	Glioblastoma cancer [79]
57.	*Cannabis sativa*	Cannabaceae	Cannabinoids	Breast cancer, colorectal cancer [80, 81]
58.	*Catharanthus roseus*	Apocynaceae	Vincristine and vinblastine	Non-Hodgkin's lymphoma and Hodgkin's disease
59.	*Cinnamomum cassia*	Lauraceae	Coumarin	Promyelocytic leukaemia and basal cell carcinoma [84]
60.	*Colchicum luteum*	Liliaceae	Colchicines	Hodgkin lymphoma

(continued)

Table 2 (continued)

Sr. No.	Botanical name	Family	Chemical constituents	Types of cancer
61.	*Combretum caffrum*	Combretaceae	Combretastatin	Colon cancer and lung cancer [85]
62.	*Coriandrum sativum*	Apiaceae	Quercetin	Adenocarcinoma [87, 88]
63.	*Daphne mezereum*	Thymelaeaceae	Mezerein	Lung cancer and lymphocytic leukaemia [89]
64.	*Echinacea angustifolia*	Asteraceae	Arabinogalactan	Oesophagus cancer [90]
65.	*Emblica officinalis*	Phyllanthaceae	Emblicanin a and b	Liver cancer
66.	*Fagopyrum esculentum*	Polygonaceae	Inhibitor-1 protein,	Lymphoblastic leukaemia [91]
67.	*Ginkgo biloba*	Ginkgoaceae	Ginkgolides	Invasive oestrogen-receptor breast cancer
68.	*Glycine max*	Fabaceae	Genistein and daidzein	Cancer in oral
69.	*Gossypium hirsutum*	Malvaceae	Gossypol	Breast cancer and melanoma [92]
70.	*Indigofera tinctoria*	Fabaceae	Tannins and phenols	Lung cancer [84]
71.	*Lentinus edodes*	Polyporaceae	Terpenoids and lentinan	Colon cancer and lung carcinoma [93]
72.	*Linum usitatissimum*	Linaceae	Lignans	Breast cancer [80, 94]
73.	*Nothapodytes foetida*	Icacinaceae	Scopoletin, camptothecin	Colon cancer [85, 95]
74.	*Ochrosia elliptica*	Apocynaceae	9-methoxy ellipticine	Breast, kidney cancer [85, 96]
75.	*Ocimum sanctum*	Lamiaceae	Ursolic acid	Fibrosarcoma
76.	*Oldenlandia diffusa*	Rubiaceae	Stigmasterol	Stomach cancer [88, 97]
77.	*Origanum vulgare*	Lamiaceae	Rosmarinic acid	Lung cancer [85]
78.	*Panax ginseng*	Araliaceae	Polyacetylene	Stomach cancer [85]
79.	*Pfaffia paniculata*	Amaranthaceae	Cytotoxic substances	Oestrogen positive breast cancer [88]
80.	*Picrorhiza kurroa*	Plantaginaceae	Picrosides-i, ii, iii	Liver cancer [90]
81.	*Plumbago zeylanica*	Plumbaginaceae	Plumbagin	Fibrosarcoma cancers [90]
82.	*Podophyllum hexandrum*	Berberidaceae	Podophyllotoxin	Leukaemia [87]

(continued)

Table 2 (continued)

Sr. No.	Botanical name	Family	Chemical constituents	Types of cancer
83.	*Prunella vulgaris*	Lamiaceae	Oleanolic acid	Oral cavity [88]
84.	*Psoralea corylifolia*	Fabaceae	Psoralidincorylfolinin	Osteosarcoma [88]
85.	*Viscum album*	Santalaceae	Phenolic compounds	Testis cancer [89]

3 Conclusion

Cancer is the unchecked growth of malignant cells or tissues within the body. A cancerous growth is referred to as a malignant tumour or malignancy. A growth that is not cancerous is referred to as a benign tumour. Each of the sequential, connected processes that restrict the rate of cancer spread is sequential and interconnected. The traditional medical systems that are currently in use each have their own linked concepts and historical settings. While some of them, like Ayurveda traditional medicines, are spreading over the globe, some of these, like Tibetan traditional medicine, are still mostly exclusive to the country of origin. Many compounds found in plants have been demonstrated in clinical studies to have chemoprotective effects. Inhibiting angiogenesis is a novel approach to cancer treatment. The proper selection and application of this plant could very well aid in the management of cancer through antiangiogenic therapy. Plant-derived anticancer medicines are in high demand because they effectively inhibit cancer cell lines. In order to meet demand and be sustainable, the exploitation of these agents needs to be controlled. The attempt has been made to compile a list of the many different plant species found throughout the world that are either regularly utilised or being researched for their potential anticancer properties.

References

1. Coseri S (2009) Natural products and their analogues as efficient anticancer drugs. Mini-Rev Med Chem 9(5):560–571
2. Kaur R, Kaur H (2010) The antimicrobial activity of essential oil and plant extracts of Woodfordia fruticosa. Arch Appl Sci Res 2:302–309
3. Fan W, Johnson KR, Miller MC (1998) In vitro evaluation of combination chemotherapy against human tumor cells. Oncol Rep 5(5):1035–1042
4. Wamidh HT (2011) Anticancer and antimicrobial potential of plant-derived natural products. In: Phytochemicals – bioactivities and impact on health. IntechOpen, London, pp 142–158
5. Kinghorn AD, Farnsworth NR, Soejarto DD et al (2003) Novel strategies for the discovery of plant-derived anticancer agents. Pharm Biol 41:53–67
6. Sakarkar DM, Deshmukh VN (2011) Ethnopharmacological review of traditional medicinal plants for anticancer activity. Int J PharmTech Res 3:298–308

7. Nussbaumer S, Bonnabry P, Veuthey JL, Sandrine F (2011) Analysis of anticancer drugs: a review. Talanta 85:2265–2289
8. Patel S, Gheewala N, Suthar A, Shah A (2009) In-vitro cytotoxicity activity of Solanum nigrum extracts against Hela cell- line and Vero cell-line. Int J Pharm Pharm Sci 1(1):38–46
9. Sadooghi SD, Nezhad-Shahrokh-Abadi K, Zafar Balanezhad S, Baharara J (2013) Investigating the cytotoxic effects of ethanolic extract of Ferula Assa-foetida resin on HepG2 cell line. Feyz 17:323–330
10. Keramati K, Sanai K, Babakhani A, Rakhshan M, Vaezi G, Haeri A (2011) Effect of hydroalcoholic extract Thymus vulgaris induced prostate cancer injection DMBA in Wistar rats. J Pazhuhesh 35:135–140
11. Sertel S, Eichhorn T, Plinkert PK, Efferth T (2011) Cytotoxicity of Thymus vulgaris essential oil towards human oral cavity squamous cell carcinoma. Anticancer Res 31:81–87
12. Sabzali S, Arman R, Panahi J, Havasian MR, Haghani K, Bakhtiyari S (2014) Investigation on the inhibitory effects of hydro-alcoholic extract of Thymbra spicata on the growth of lung cancer cell line SK-Mes-1. J Ilam Univ Med Sci 22:153–158
13. Khalighi-Sigaroodi F, Jeddi-Tehrani M, Ahvazi M et al (2014) Cytotoxicity evaluation of Taverniera spartea on human cancer cell lines. J Med Plants 2:114–128
14. Ayoob I, Hazari YM, Lone SH, Khuroo MA, Fazili KM, Bhat KA et al (2017) Phytochemical and cytotoxic evaluation of Peganum harmala: structure activity relationship studies of Harmine. ChemistrySelect 2(10):2965–2968
15. William PJ, Lobo-Echeverri T, Mi Q et al (2005) Antitumour activity of 3-chlorodeoxylapachol, a naphthoquinone from Avicennia germinans collected from an experimental plot in southern Florida. J Pharm Pharmacol 57:1101–1108
16. Ramasamy K, Agarwal R (2008) Multitargeted therapy of cancer by silymarin. Cancer Lett 269(2):352–362
17. Shariatzadeh SM, Hamta A, Soleimani M, Fallah Huseini H, Samavat S (2014) The cytotoxic effects of Silymarin on the 4T1 cell line derived from BALB/c mice mammary tumors. J Med Plants 4:55–65
18. Chen HH, Zhou HJ, Fang X (2003) Inhibition of human cancer cell line growth and human umbilical vein endothelial cell angiogenesis by artemisinin derivatives in vitro. Pharm Res 48: 231–236
19. Zhou HJ, Wang WQ, Wu GD, Lee J, Li A (2007) Artesunate inhibits angiogenesis and down regulates vascular endothelial growth factor expression in chronic myeloid leukemia K562 cells. Vasc Pharmacol 47:131–138
20. Ooko E, Kadioglu O, Greten HJ, Efferth T (2017) Pharmacogenomic characterization and isobologram analysis of the combination of ascorbic acid and curcumin-two main metabolite of Curcuma longa-in cancer cells. Front Pharmacol 8:38
21. Mohammad P, Nosratollah Z, Mohammad R, Abbas A, Javad R (2010) The inhibitory effect of Curcuma longa extract on telomerase activity in A549 lung cancer cell line. Afr J Biotechnol 9: 912–919
22. Forouzandeh F, Salimi S, Naghsh N, Zamani N, Jahani S (2014) Evaluation of anti-cancer effect of Peganum harmala L hydroalcoholic extract on human cervical carcinoma epithelia cell line. J Shahrekord Univ Med Sci 16:1–8
23. Mortazavian SM, Ghorbani A, Ghorbani HT (2012) Effect of hydro-alcoholic extracts of Viola tricolor and its fractions on proliferation of cervix carcinoma cells. Iran J Obstet Gynecol Infertil 15:9–16
24. Dalali Isfahani L, Monajemi R, Amjad L (2013) Cytotoxic effects of extract and essential oil leaves of Achillea wilhelmsii C. Koch on colon cancers cells. Exp Anim Biol 1:1–6
25. Aslani E, Naghsh N, Ranjbar M (2014) Cytotoxic effect of Mentha pulegium plants before flowering on human chronic myelogenous leukemia K562 cancer category. J Arak Univ Med Sci 16: 1–10 (2014); Mohammed ZY, Nada SM, Al-Halbosiy MM, Abdulfattah SY, Abdul-Hameed B. Cytotoxic effects of Ammi visnaga volatile oil on some cancer cell lines. J Biotechnol Res Center 8:5–7

26. Kumari M, Pattnaik B, Rajan SY, Shrikant S, Surendra SU (2017) EGCG-A promis anticancer. Phytochemistry 3(2):8–10
27. Momtazi Borojeni A, Behbahani M, Sadeghi-Aliabadi H (2011) Evolution of cytotoxic effect of some extracts of Avicennia marina against MDA-MB231 human breast cancer cellline. Pharm Sci 16:229–238
28. Anand P, Sundaram C, Jhurani S, Kunnumakkara AB, Aggarwal BB (2008) Curcumin and cancer: an "old-age" disease with an "age old" solution. Cancer Lett 267:133–164
29. Bakshi HA, Sam S, Anna F, Zeinab R, Ahmad SG, Sharma M et al (2009) Crocin from Kashmiri saffron (Crocus sativus) induces in vitro and in vivo xenograft growth inhibition of Dalton's lymphoma (DLA) in mice. Asian Pac J Cancer Prev 10:887–890
30. Mousavi M, Baharara J, Asadi-Samani M (2014) Anti-angiogenesis effect of Crocus sativus L. extract on matrix metalloproteinase gene activities in human breast carcinoma cells. J Herb Med Pharmacol 3:101–105
31. Aung HH, Wang CZ, Ni M, Fishbein A, Mehendale SR, Xie JT (2007) Crocin from Crocus sativus possesses significant anti-proliferation effects on human colorectal cancer cells. Exp Oncol 29:175–180
32. Kurapati KR, Samikkannu T, Kadiyala DB et al (2012) Combinatorial cytotoxic effects of Curcuma longa and Zingiber officinale on the PC-3M prostate cancer cell line. J Basic Clin Physiol Pharmacol 23:139–146
33. Hosain Zadegan H, Ezzet Por B, Abdollah Por F, Motamedy M, Rashidipor M (2010) Study of cytotoxic activity of olive and green tea extracts on breast tumor cell line. J Ardabil Univ Med Sci 10:287–294
34. Fini L, Hotchkiss E, Fogliano V et al (2008) Chemopreventive properties of pinoresinol-rich olive oil involve a selective activation of the ATM-p53 cascade in colon cancer cell lines. Carcinogenesis 29:139–146
35. De Bono JS, Oudard S, Ozguroglu M, Hansen S, Machiels JP, Kocak I et al (2010) Prednisone plus cabazitaxel or mitoxantrone for metastatic castration-resistant prostate cancer progressing after docetaxel treatment: a randomised open-label trial. Lancet 376:1147–1154
36. Dieras V, Limentani S, Romieu G, Tubiana-Hulin M, Lortholary A, Kaufman P et al (2008) Phase II multicenter study of larotaxel (XRP9881), a novel taxoid, in patients with metastatic breast cancer who previously received taxane-based therapy. Ann Oncol 19(7):1255–1260
37. Sadeghi-Aliabadi H, Alavi M, Asghari G, Mirian M (2013) Cytotoxic evaluation of different extracts of Taxus baccata against MDA-MB-468, HeLa and K562 cancer cell lines. J Isfahan Med Sch 31:1508–1517
38. Tu LY, Pi J, Jin H, Cai JY, Deng SP (2016) Synthesis, characterization and anticancer activity of kaempferol-zinc (II) complex. Bioorg Med Chem Lett 26(11):2730–2734
39. Karmakar S, Roy Choudhury S, Banik N, Ray S (2011) Molecular mechanisms of anti-cancer action of garlic compounds in neuroblastoma. Anti Cancer Agents Med Chem 11:398–407
40. Nakagawa H, Tsuta K, Kiuchi K, Senzaki H, Tanaka K, Tsubura A (2001) Growth inhibitory effects of diallyl disulfide on human breast cancer cell lines. Carcinogenesis 22:891–897
41. Colic M, Vucevic D, Kilibarda V, Radicevic N, Savic M (2002) Modulatory effects of garlic extracts on proliferation of T-lymphocytes in vitro stimulated with concanavalin A. Phytomedicine 9:117–124
42. Arunkumar A, Vijayababu MR, Srinivasan N, Aruldhas MM, Arunakaran J (2006) Garlic compound, diallyl disulfide induces cell cycle arrest in prostate cancer cell line PC-3. Mol Cell Biochem 288:107–113
43. Robert V, Mouille B, Mayeur C, Michaud M, Blachier F (2001) Effects of the garlic compound diallyl disulfide on the metabolism, adherence and cell cycle of HT-29 colon carcinoma cells: evidence of sensitive and resistant sub-populations. Carcinogenesis 22:1155–1161
44. Hadjzadeh MAIR, Tavakol Afshari J, Ghorbani A, Shakeri MT (2006) The effects of aqueous extract of garlic (Allium sativum L.) on laryngeal cancer cells (Hep-2) and L929 cells in vitro. J Med Plants 2(18):41–48

45. Mahassni SH, Al-Reemi RM (2015) Apoptosis and necrosis of human breast cancer cells by an aqueous extract of garden cress (Lepidium sativum) seeds. Saudi J Biol Sci 20: 131–139 (2013); Aslani E, Naghsh N, Ranjbar M. Cytotoxic effects of hydro- alcoholic extracts of cress (Lepidium sativum)—made from different stages of the plant—on k562 leukemia cell line. Hormozgan Med J 18:411–419
46. Amin A, Alkaabi A, Al-Falasi S, Daoud SA (2005) Chemopreventive activities of Trigonella foenum graecum (fenugreek) against breast cancer. Cell Biol Int 29:687–694
47. Zhang YY, Huang CT, Liu SM, Wang B, Guo J, Bai JQ et al (2016) Licochalcone A exerts antitumor activity in bladder cancer cell lines and mice models. Trop J Pharm Res 15(6):1151–1157
48. Luo C-F, Kong J, Wang H, Liu H, Chen C, Shao S-H (2012) Ethanol extracts of Physalis alkekengi Linn induced the apoptosis of human esophageal carcinoma cells. Chin J Clin Lab Sci 12:21
49. Li X, Zhao J, Yang M et al (2014) Physalins and withanolides from the fruits of Physalis alkekengi L var franchetii (Mast) Makino and the inhibitory activities against human tumor cells. Phytochem Lett 10:95–100
50. Shokrzadeh M, Parvaresh A, Shahani S, Habibi E, Zalzar Z (2013) Cytotoxic effects of Lagenaria siceraria Standl. Extract on cancer cell line. J Mazandaran Univ Med Sci 23:225–230
51. Ghosh K, Chandra K, Ojha AK, Sarkar S, Islam SS (2009) Structural identification and cytotoxic activity of a polysaccharide from the fruits of Lagenaria siceraria (Lau). Carbohydr Res 344:693–698
52. Valiahdi SM, Iranshahi M, Sahebkar A (2013) Cytotoxic activities of phytochemicals from ferula species. DARU 21:39–45
53. Gharaei R, Akrami H, Heidari S, Asadi MH, Jalili A (2013) The suppression effect of Ferula gummosa Boiss. Extracts on cell proliferation through apoptosis induction in gastric cancer cell line. Eur J Integr Med 5:241–247
54. Durak I, Biri H, Devrim E, Sozen S, Avcı A (2004) Aqueous extract of Urtica dioica make significant inhibition on adenosine deaminase activity in prostate tissue from patients with prostate cancer. Cancer Biol Ther 3:855–857
55. Konrad L, Muller HH, Lenz C, Laubinger H, Aumuller G, Lichius JJ (2000) Antiproliferative effect on human prostate cancer cells by a stinging nettle root (Urtica dioica) extract. Planta Med 66:44–47
56. Safarinejad MR (2005) Urtica dioica for treatment of benign prostatic hyperplasia: a prospective, randomized, double-blind, placebo-controlled, crossover study. J Herb Pharmacother 5: 1–11
57. Nemati F, Eslami Jadidi B, Talebi DM (2013) Investigation cytotoxic effects of Ammi majus extract on MCF-7 and Hela cancer cell line [in Persian]. J Anim Biol 5:59–66
58. Zu Y, Yu H, Liang L et al (2010) Activities of ten essential oils towards Propionibacterium acnes and PC-3, A-549 and MCF-7 cancer cells. Molecules 15:3200–3210
59. Zamiri-Akhlaghi A, Rakhshandeh H, Tayarani-Najaran Z, Mousavi SH (2011) Study of cytotoxic properties of Rosa damascena extract in human cervix carcinoma cell line. Avicenna J Phytomed 1:74–77
60. Cassileth BR, Rizvi N, Deng G et al (2009) Safety and pharmacokinetic trial of docetaxel plus an Astragalus-based herbal formula for non-small cell lung cancer patients. Cancer Chemother Pharmacol 65:67–71
61. Ogur R (2014) Studies with Myrtus communis L.: anticancer properties. J Intercult Ethnopharmacol 3:135–137
62. Sumbul S, Ahmad MA, Asif M, Akhtar M (2011) Myrtus communis Linn—a review. Indian J Nat Prod Resour 2:395–402
63. Mothana RA, Kriegisch S, Harms M, Wende K, Lindequist U (2011) Assessment of selected Yemeni medicinal plants for their in vitro antimicrobial, anticancer, and antioxidant activities. Pharm Biol 49:200–210

64. Ayyad S-EN, Abdel-Lateff A, Alarif WM, Patacchioli FR, Badria FA, Ezmirly ST (2012) In vitro and in vivo study of cucurbitacins-type triterpene glucoside from Citrullus colocynthis growing in Saudi Arabia against hepatocellular carcinoma. Environ Toxicol Pharmacol 33:245–251
65. Habibi RM, Mohammadi RA, Delazar A et al (2011) Effects of Polygonum aviculare herbal extract on proliferation and apoptotic gene expression of MCF-7. DARU 19:326–331
66. Banazadeh H, Delazar A, Habibi Roudkenar M, Rahmati Yam-chi M, Sadeghzadeh Oscoui B, Mehdipour A (2012) Effects of knotweet or Polygonum aviculare herbal extract on proliferation of HeLa cell line. Med J Mashhad Univ Med Sci 54:238–241
67. Mohammad R, Hossein B, Davood F, Farnaz T, Ali F, Yuse R (2011) The apoptotic and cytotoxic effects of Polygonum aviculare extract on Hela-S cervical cancer cell line. Afr J Biochem Res 5:373–378
68. Abdolmohammadi MH, Fouladdel S, Shafiee A, Amin G, Ghaffari SM, Azizi E (2009) Antiproliferative and apoptotic effect of Astrodaucus orientalis (L.) Drude on T47D human breast cancer cell line: potential mechanisms of action. Afr J Biotechnol 8:4265–4276
69. Fang T, Hou J, He M, Wang L, Zheng M, Wang X, Xia J (2019) Actinidia chinensis planch root extract (acRoots) inhibits hepatocellular carcinoma progression by inhibiting EP3 expression. J Ethnopharmacol 251:112–529
70. Baliga MS, Thilakchand KR, Rai MP (2013) Aegle marmelos (L.)Correa (Bael) and its phytochemicals in the treatment and prevention of cancer. Integr Cancer Ther 12(3):187–196
71. Pecere T, Gazzola MV, Micignat C et al (2000) Aloe-emodin is a new type of anticancer agent with selective activity against neuro-ectodermal tumors. Cancer Res 60:2800–2804
72. Osman NHA, Said UZ, El-Waseef AM, Ahmed ESA (2015) Luteolin supplementation adjacent to aspirin treatment reduced dimethyl hydrazine-induced experimental colon carcinogenesis in rats. Tumour Biol 36:1179–1190
73. Yajid AI, Rahman HSA, Wong MPK, Zain WZW (2018) Potential benefits of Annona muricata in combating cancer: A review. Malays J Med Sci 25(1):5–15
74. Premratanachai P, Chanchao C (2014) Review of the anticancer activities of bee products. Asian Pac J Trop Biomed 4(5):337–344
75. Sastri BN (1949) The wealth of India A dictionary of Indian raw materials and industrial products. Ind Med Gaz 84(10):476–477
76. Amini A, Ehteda A, Moghaddam SM, Akhter J, Pillai K, Morris DL (2013) Cytotoxic effects of bromelain in human gastrointestinal carcinoma cell lines (MKN45, KATO-III, HT29-5F12, and HT29-5M21). Onco Targets Ther 6:403–409
77. Herdwiani W, Soemardji A, Elfahmi TM (2016) A review of cinnamon as a potent anticancer drug. Asian J Pharm Clin Res 9(3):8–13
78. Lauritano H, Andersen JH, Hansen E, Albrigtsen M, Escalera L, Esposit F et al (2016) Bioactivity screening of microalgae for antioxidant, anti-inflammatory, anticancer, anti-diabetes, and antibacterial activities. Front Mar Sci 3:68
79. Tang ELH, Rajarajeswaran J, Fung SY, Kanthimathi MS (2013) Antioxidant activity of Coriandrum sativum and protection against DNA damage and cancer cell migration. BMC Complement Altern Med 13:347
80. Nithya TG, Sumalatha D (2014) Evaluation of in vitro anti-oxidant and anticancer activity of Coriandrum Sativum against human colon cancer Ht-29 cell lines. Int J Pharm Pharm Sci 6(2): 421–424
81. Sovrlić MM, Manojlović NT (2017) Plants from the genus Daphne: A review of its traditional uses, phytochemistry, biological and pharmacological activity. Serbian J Exp Clin Res 18(1): 69–79
82. Jean B (1993) Pharmacognosy, phytochemistry medicinal plants. Lavoisier Publisher, p 151
83. Lim TK (2013) Edible medicinal and non-medicinal plants: fruits, vol 5. Springer
84. Gilbert NE, Reilly JE, Chang CJ, Lin YC, Brueggemeier RW (1995) Antiproliferative activity of gossypol and gossypolone on human breast cancer cells. Life Sci 57:61–67
85. Ladanyi A, Timar J, Lapis K (1993) Effect of lentinan on macrophage cytotoxicity against metastatic tumor cells. Cancer Immunol Immunother 36:123–126

86. Lim TK (2013) Edible medicinal and non-medicinal plants, vol 5. Springer
87. Mizuno TS (1995) Lentinus edodes: functional properties for medicinal and food purposes. Food Rev Int 11:111–128
88. Mizuno T, Saito H, Nishitoba T, Kawagishi H (1995) Antitumor-active substances from mushrooms. Food Rev Int 11:23–61
89. Serraino M, Thompson LU (1991) The effect of flaxseed supplementation on early risk markers for mammary carcinogenesis. Cancer Lett 60:135–142
90. Lampe JW, Martini MC, Kurzer MS, Adlercreutz H, Slavin JL (1994) Urinary lignan and isoflavonoid excretion in premenopausal women consuming flaxseed powder. Am J Clin Nutr 60:122–128
91. Dixit S, Ali H (2010) Anticancer activity of medicinal plant extract-a review. J Chem Chem Sci 1:79–85
92. Gupta S, Zhang D, Yi J, Shao J (2004) Anticancer activities of Oldenlandia diffusa. Herb Pharmacother 4(1):21–33
93. Balusamy SR, Perumalsamy H, Huq MA, Balasubramanian BK (2018) Anti-proliferative activity of Origanum vulgare inhibited lipogenesis and induced mitochondrial mediated apoptosis in human stomach cancer cell lines. Biomed Pharmacother 108:1835–1844
94. Dou YQ, Yang MH, Wei ZM, Xiao CD, Yang XH (2020) The study of early application with Dixiong decoction for non-small cell lung cancer to decrease the incidence and severity of radiation pneumonitis: a prospective, randomized clinical trial. Chin J Integr Med 16(5):411–416
95. Mvondo MA, Trésor M, Kamgaing W, Mvondo MA, Kamgaing MTW (2021) Aqueous extract of Dacryodes edulis (Burseraceae) leaves inhibited tumor growth in female Wistar rats with 7,12-Dimethylbenz[a]anthracene-induced breast cancer. Oxidative Med Cell Longev 2021:234–240
96. Akkol EK, Bardakci H, Barak TH (2022) Herbal ingredients in the prevention of breast cancer: comprehensive review of potential molecular targets and role of natural products. Oxidative Med Cell Longev 2022:124–129
97. Ali M, Wani SUD, Salahuddin M, Manjula SN (2023) Recent advance of herbal medicines in cancer- a molecular approach. Heliyon 9(2):e13684
98. Fayed L (2021) Causes and risk factors of cancer. Verywell health 2021 September 07
99. Dheda K, Barry CE, Maartens G (2016) Tuberculosis. Lancet 387(10024):1211–1226
100. Comas I, Coscollá M, Luo T, Borrell S, Holt K, Kato-maeda M et al (2013) Out-of-Africa migration and Neolithic co-expansion of mycobacterium tuberculosis with modern humans. Nat Genet 45:1176–1182
101. Ryan F (1994) The forgotten plague: how the battle against tuberculosis was won—and lost. Little Brown & Co, Boston. WHO, 2020, Global tuberculosis report
102. Raviglione M, Harries A, Msiska R, Wilkinson D (1997) Nunn PJA tuberculosis and HIV: current status in Africa. AIDS 11(Suppl B):S115–S123
103. Falzon D, Schünemann HJ, Harausz E, González-Angulo L, Lienhardt C, Jaramillo E, Weyer K (2017) World health organization treatment guidelines for drug-resistant tuberculosis, 2016 update. Eur Respir J 49(3):1602308
104. Nachega JB, Chaisson RE (2003) Tuberculosis drug resistance: a global threat. Clin Infect Dis 36(Suppl_1):S24–S30
105. Arya V, Baba A, Singh A, Memorial JS (2011) A review on anti-tubercular plants. Int J PharmTech Res 3(2):872–880
106. Copp BR (2003) Antimycobacterial natural products. Nat Prod Rep 20(6):535–557
107. Mangwani N, Singh PK, Kumar V (2019) Medicinal plants: adjunct treatment to tuberculosis chemotherapy to prevent hepatic damage. J Ayurveda Integr Med 11(4):522–528
108. Newton SM, Lau C, Wright CW (2000) A review of antimycobacterial natural products. Phytother Res 14(5):303–322
109. Okunade AL, Elvin-Lewis MP, Lewis WH (2004) Natural antimycobacterial metabolites: current status. Phytochemistry 65(8):1017–1032

110. Sanusi SB, Abu Bakar MF, Mohamed M, Sabran SF, Mainasara MM (2017) Southeast Asian medicinal plants as a potential source of antituberculosis agent. Evid Based Complement Alternat Med 2017:7185649
111. Chinsembu KC (2016) Tuberculosis and nature's pharmacy of putative anti-tuberculosis agents. Acta Trop 153:46–56
112. León-Díaz R, Meckes M, Said-Fernández S, Molina-Salinas GM, Vargas-Villarreal J, Torres J et al (2010) Antimycobacterial neolignans isolated from Aristolochia taliscana. Mem Inst Oswaldo Cruz 105(1):45–51
113. Navarro-García V, Luna-Herrera J, Rojas-Bribiesca G, Álvarez-Fitz P, Rios GY (2011) Antibacterial activity of Aristolochia brevipes against multidrug-resistant mycobacterium tuberculosis. Molecules 16:7357–7364
114. Jiménez-Arellanes A, León-Díaz R, Meckes M, Tapia A, Molina-Salinas G, Luna-Herrera J, Yépez-Mulia L (2012) Antiprotozoal and Antimycobacterial activities of pure compounds from Aristolochia elegans rhizomes. Evid Based Complement Alternat Med 2012:593403
115. Jyoti MA, Nam KW, Jang WS, Kim YH, Kim SK, Lee BE, Song HY (2016) Anti-mycobacterial activity of methanolic plant extract of Artemisia capillaris containing ursolic acid and hydroquinone against Mycobacterium tuberculosis. J Infect Chemother 22(4):200–208
116. Ntutela S, Smith P, Matika L, Mukinda J, Arendse H, Allie N et al (2009) Efficacy of artemisia afra phytotherapy in experimental tuberculosis. Tuberculosis 1(Suppl 1):S33–S40
117. María GMS, Borquez J, Ardiles A, Said-Fernández S, San-Martín A et al (2010) Bioactive metabolites from the Andean flora. Antituberculosis activity of natural and semisynthetic azorellane and mulinane diterpenoids. Phytochem Rev 9:271–278
118. Lenta BN, Chouna JR, Nkeng-Efouet PA, Sewald N (2015) Endiandric acid Derivatives and other constituents of plants from the Genera Beilschmiedia and Endiandra (Lauraceae). Biomol Ther 5(2):910–942
119. Aponte J, Estevez Y, Gilman R, Lewis W, Rojas R, Sauvain M et al (2008) Anti-infective and cytotoxic compounds present in Blepharodon nitidum. Planta Med 74:407–410
120. Torres-romero D, Jiménez I, Rojas R, Gilman R, López M, Bazzocchi I (2011) Dihydro-β-agarofuran sesquiterpenes isolated from Celastrus vulcanicola as potential anti-Mycobacterium tuberculosis multidrug-resistant agents. Bioorg Med Chem 19:2182–2189
121. Rukachaisirikul T, Prabpai S, Champung P, Suksamrarn A (2002) Chabamide, a novel piperine dimer from stems of Piper chaba. Planta Med 68(9):853–855
122. Acevedo L, Martínez E, Castañeda P, Franzblau S, Timmermann B, Linares E, Mata R (2000) New phenylethanoids from Buddleja cordata subsp. cordata. Planta Med 66:257–261
123. Serkani JE, Isfahani BN, Safaei HG, Kermanshahi RK, Asghari G (2012) Evaluation of the effect of Humulus lupulus alcoholic extract on rifampin-sensitive and resistant isolates of Mycobacterium tuberculosis. Res Pharm Sci 7(4):235–242
124. Inui T, Wang Y, Nikolić D, Smith D, Franzblau S, Pauli G (2010) Sesquiterpenes from Oplopanax horridus. J Nat Prod 73:563–567
125. Limmatvapirat C, Sirisopanaporn S, Kittakoop P (2004) Antitubercular and Antiplasmodial constituents of Abrus precatorius. Planta Med 70:276–278
126. Macabeo APG, Lee CA (2014) Sterols and triterpenes from the non-polar antitubercular fraction of Abutilon Indicum. Pharmacognosy J 6(4):49–52
127. Phongpaichit S, Vuddhakul V, Subhadhirasakul S, Wattanapiromsakul C (2006) Evaluation of the antimycobacterial activity of extracts from plants used as self-medication by AIDS patients in Thailand. Pharm Biol 44(1):71–75
128. Elkington BG, Sydara K, Newsome A, Hwang CH, Lankin DC, Simmler C et al (2014) New finding of an anti-TB compound in the genus Marsypopetalum (Annonaceae) from a traditional herbal remedy of Laos. J Ethnopharmacol 151(2):903–911
129. Mohamad S, Zin NM, Wahab HA, Ibrahim P, Sulaiman SF, Zahariluddin ASM, Noor SSM (2011) Antituberculosis potential of some ethnobotanically selected Malaysian plants. J Ethnopharmacol 133(3):1021–1026

130. Macabeo APG, Vidar WS, Chen X, Decker M, Heilmann J, Wan B et al (2011) Mycobacterium tuberculosis and cholinesterase inhibitors from Voacanga globosa. Eur J Med Chem 46 (7):3118–3123
131. Hasan N, Osman H, Mohamad S, Chong WK, Awang K, Zahariluddin ASM (2012) The chemical components of Sesbania grandiflora root and their antituberculosis activity. Pharmaceuticals 5(8):882–889
132. Radji M, Kurniati M, Kiranasari A (2015) Comparative antimycobacterial activity of some Indonesian medicinal plants against multi-drug resistant Mycobacterium tuberculosis. Pharm Biol 5(1):019–022
133. Mulyani Y, Sukandar E, Adnyana I (2012) Petiveria alliacea: new alternative for the treatment of sensitive and multi-resistant Mycobacterium tuberculosis. J Pharmacogn Phytother 4(7):91–95
134. Jiangseubchatveera N, Liawruangrath B, Liawruangrath S, Korth J, Pyne SG (2015) The chemical constituents and biological activities of the essential oil and the extracts from leaves of Gynura divaricata (L) DC growing in Thailand. J Essent Oil Bear Plants 18(3):543–555
135. Saludes JP, Garson MJ, Franzblau SG, Aguinaldo AMJ, Derivatives TE (2002) Antitubercular constituents from the hexane fraction of Morinda citrifolia Linn(Rubiaceae). Phytother Res 16 (7):683–685
136. Jang WS, Jyoti MA, Kim S, Nam K-W, Ha TKQ, Oh WK, Song H-Y (2016) In vitro antituberculosis activity of diterpenoids from the Vietnamese medicinal plant Croton tonkinensis. J Nat Med 70(1):127–132
137. Chaisson RE, Bishai WR (2017) Overview of tuberculosis. In: Handbook of tuberculosis. Adis, London, pp 155–158; Flynn JL, Chan J (2003) Immune evasion by Mycobacterium tuberculosis: living with the enemy. Curr Opin Immunol 15(4): 450–455
138. Xiang L, Li Y, Xu DX, Kosanovic D, Schermuly RT, Li X (2018) Natural plant products in treatment of pulmonary arterial hypertension. Pulm Circ 8(3):1–20
139. Simonneau G, Montani D, Celermajer DS, Denton CP, Gatzoulis MA, Krowka M, Williams PG, Souza R (2019) Haemodynamic definitions and updated clinical classification of pulmonary hypertension. Eur Respir J 53:1801913
140. Sánchez-Gloria JL, Osorio-Alonso H, Arellano-Buendía AS, Carbó R, Hernández-Díazcouder A, Guzmán-Martín CA, Rubio-Gayosso I, Sánchez-Muñoz F (2020) Nutraceuticals in the treatment of pulmonary arterial hypertension. Int J Mol Sci 21:4827
141. Jasemi SV, Hosna K, Ina YA, Mohammad HF, Javier E (2020) Medicinal plants and phytochemicals for the treatment of pulmonary hypertension. Front Pharmacol 11:145
142. Jasemi SV, Khazaei H, Aneva IY, Farzaei MH, Echeverria J (2020) Medicinal plants and phytochemicals for pulmonary hypertension. Front Pharmacol 11:145
143. Schermuly RT, Ghofrani HA, Wilkins MR, Grimminger F (2011) Mechanisms of disease: pulmonary arterial hypertension. Nat Rev Cardiol 8(8):443
144. Yeager ME, Halley GR, Golpon HA, Voelkel NF, Tuder RM (2001) Microsatellite instability of endothelial cell growth and apoptosis genes within plexiform lesions in primary pulmonary hypertension. Circ Res 88(1):E2–E11
145. Dorfmüller P, Perros F, Balabanian K, Humbert M (2003) Inflammation in pulmonary arterial hypertension. Eur Respir J 22(2):358–363
146. Balabanian K, Foussat A, Dorfmüller P, Durand-Gasselin I, Capel F, Bouchet-Delbos L, Portier A, Marfaing-Koka A, Krzysiek R, Rimaniol AC, Simonneau G, Emilie D, Humbert M (2002) CX(3)C chemokine fractalkine in pulmonary arterial hypertension. Am J Respir Crit Care Med 165(10):1419–1425
147. Waxman AB, Zamanian RT (2013) Pulmonary arterial hypertension: new insights into the optimal role of current and emerging prostacyclin therapies. Am J Cardiol 111(5):1–16
148. Wong CM, Bansal G, Pavlickova L, Marcocci L, Suzuki YJ (2013) Reactive oxygen species and antioxidants in pulmonary hypertension. Antioxid Redox Signal 18(14):1789–1796
149. Hui-li G (2011) The management of acute pulmonary arterial hypertension. Cardiovasc Ther 29(3):153–175

150. Reyes RV, Castillo-Galán S, Hernandez I, Herrera EA, Ebensperger G, Llanos AJ (2018) Revisiting the role of TRP, orai, and ASIC channels in the pulmonary arterial response to hypoxia. Front Physiol 9:486
151. Benza RL, Miller DP, Gomberg-Maitland M, Frantz RP, Foreman AJ, Coffey CS et al (2010) Predicting survival in pulmonary arterial hypertension: insights from the registry to evaluate early and long-term pulmonary arterial hypertension disease management (REVEAL). Circulation 122(2):164–172
152. Galie N, Hoeper MM, Humbert M, Torbicki A, Vachiery J-L, Barbera JA et al (2009) Guidelines for the diagnosis and treatment of pulmonary hypertension: the task force for the diagnosis and treatment of pulmonary hypertension of the European Society of Cardiology (ESC) and the European Respiratory Society (ERS), endorsed by the International Society of Heart and Lung Transplantation (ISHLT). Eur Heart J 30(20):2493–2537
153. Hussain A, Bennett RT, Tahir Z, Isaac E, Chaudhry MA, Qadri SS, Loubani M, Morice AH (2019) Differential effects of atrial and brain natriuretic peptides on human pulmonary artery: an in vitro study. World J Cardiol 11(10):236–243
154. Mintah SO, Asafo-Agyei T, Archer MA, Atta-Adjei Junior P, Boamah D, Kumadoh D, Appiah A, Ocloo A, Boakye YD, Agyare C (2019) Medicinal plants for treatment of prevalent diseases. Intechopen
155. Voswinckel R, Reichenberger F, Enke B et al (2008) Acute effects of the combination of sildenafil and inhaled treprostinil on haemodynamics and gas exchange in pulmonary hypertension. Pulm Pharmacol Ther 21:824–832
156. Hoeper MM, Barberà JA, Channick RN et al (2009) Diagnosis, assessment, and treatment of non–pulmonary arterial hypertension pulmonary hypertension. J Am Coll Cardiol 54:85–96
157. Njila NIM, Mahdi E, Lembe MD, Nde Z, Nyonseu D (2017) Review on extraction and isolation of plant secondary metabolites. In: 7th International conference on agricultural, chemical, biological and environmental sciences, vol 2017, pp 67–72
158. Fabricant DS, Farnsworth NR (2001) The value of plants used in traditional medicine for drug discovery. Environ Health Perspect 109:69–75
159. Cosa P, Vlietinck AJ, Berghe DV, Maes L (2006) Anti-infective potential of natural products: how to develop a stronger in vitro 'proof-of-concept'. J Ethnopharmacol 106:290–302
160. Zhang Q, Lin GL, Ye CW (2018) Techniques for extraction and isolation of natural products: a comprehensive review. Chin Med 13(20):1–26
161. Rasul MG (2018) Extraction, isolation and characterization of natural products from medicinal plants. Int J Basic Sci Appl Comput 2(6):1–6
162. Handa SS, Khanuja SPS, Longo G, Rakesh DD (2008) Extraction technologies for medicinal and aromatic plants. ICS UNIDO, Trieste
163. Vidyadhar S, Saidulu M, Gopal TK, Chamundeeswari D, Rao U, Banji D (2010) In vitro anthelmintic activity of the whole plant of Enicostemma littorale by using various extracts. Int J Appl Biol Pharm Technol 1(3):1119–1125
164. Fan XH, Cheng YY, Ye ZL, Lin RC, Qian ZZ (2006) Multiple chromatographic fingerprinting and its application to the quality control of herbal medicines. Anal Chim Acta 555:217–224
165. Cannell RJP (1998) Natural products isolation. Human Press Inc., Totowa, pp 165–208
166. Tonthubthimthong SC, Douglas P, Luewisutthichat W (2011) Supercritical CO_2 extraction of nimbin from neem seeds an experimental study. J Food Eng 47(4):2
167. Eberhardt TL, Li X, Shupe TF, Hse CY (2007) Chinese tallow tree (Sapium sebiferum) utilization: characterization of extractives and cell-wall chemistry. Wood Fiber Sci 39:319–324
168. Horníčková J, Kubec R, Cejpek K, Velíšek J, Ovesna J, Stavělíková H (2010) Profiles of S-alk (en)ylcysteine sulfoxides in various garlic genotypes. Czech J Food Sci 28(4):298–308
169. Fallon MB, Abrams GA, Abdel-Razek TT, Dai J, Chen S-J, Chen Y-F et al (1998) Garlic prevents hypoxic pulmonary hypertension in rats. Am J Physiol Lung Cell Mol Physiol 275(2):L283–L287

170. Tan X, Chai J, Bi SC, Li JJ, Li WW, Zhou JY (2012) Involvement of matrix metalloproteinase2 in medial hypertrophy of pulmonary arterioles in broiler chickens with pulmonary arterial hypertension. Vet J 193(2):420–425
171. Chu MP, Xu ZX, Zhang XL (2007) Changes of expression of ICAM-1 in autoimmune myocarditis in mice and the interfering effect of Panax notoginseng saponin. J Wenzhou Med Coll 37:203–205
172. Behnam A, Majid K, Seyed MH, Zia UR, Farmanullah F, Mohammad HK, LiGuo Y (2019) Hawthorn (Crataegus oxyacantha) flavonoid extract as an effective medicinal plant derivative to prevent pulmonary hypertension and heart failure in broiler chickens. Kafkas Univ Vet Fak Derg 25(3):321–328
173. Himanshu M, Pankaj P, Soheb AM, Sandeep S, Milind PH, Sanjay KB, Sudheer A, Ruma R, Subir KM (2016) Beneficial effects of aqueous extract of stem bark of Terminalia arjuna (Roxb.), an Ayurvedic drug in experimental pulmonary hypertension. J Ethnopharmacol. https://doi.org/10.1016/j.jep.2016.07.029
174. Masoumeh P, Reza M, Saber A, Nasrollah N, Siamak N, Fariba B (2017) Medicinal plants used for hypertension. J Pharm Sci Res 9(5):537–554
175. Jiang YB, Yang YR (2015) Trifolium pratense isoflavones improve pulmonary vascular remodelling in broiler chickens. J Anim Physiol Anim Nutr. https://doi.org/10.1111/jpn.12424
176. Landazuri P, Chamorro NL, Cortes BP (2017) Medicinal plants used in the management hypertension. J Anal Pharm Res 5(2):00134
177. Gan H (2010) The management of acute pulmonary arterial hypertension. Cardiovasc Ther 29 (2011):153–175
178. Ritu S, Sakshi J, Jyoti KB, Parul S (2016) The use of Rauwolfia serpentina in hypertensive patients. IOSR J Biotechnol Biochem 2(5):28–32
179. Lili X, Ying L, Xu D, Djuro K, Ralph TS, Xiaohui L (2018) Natural plant products in treatment of pulmonary arterial hypertension. Pulm Circ 8(3):1–20
180. Meresa A, Fekadu N, Degu S, Tadele A, Geleta B (2017) An ethno botanical review on medicinal plants used for the management of hypertension. Clin Exp Pharmacol 7:228
181. Lakshmi T, Anitha R, Durgha K, Manjusha V (2011) Coping with hypertension using safer herbal medicine – a therapeutic review. Int J Drug Dev Res 3(3):31–57
182. Qilian H, Xingmei N, Silin L, Shanshan S, Ke M, Zhanqiang L, Dianxiang L, Rili G (2018) Tsantan sumtang alleviates chronic hypoxia-induced pulmonary hypertension by inhibiting proliferation of pulmonary vascular cells. Biomed Res Int 2018:9504158
183. Zhouye W, Zhaoqing X (2020) Compound xiebai capsule alleviates pulmonary vascular remodeling in monocrotaline-induced pulmonary arterial hypertension in rats. Trop J Pharm Res 19(1):107–114
184. Tingting W, Jun H, Wenjing X, Yaolei Z, Longfu Z, Li Y, Xiaoqiang Y, Xin C, Yonghe H (2020) Chinese medicinal plants for the potential management of high-altitude pulmonary oedema and pulmonary hypertension. Pharm Biol 58(1):815–827
185. Manish A, Nandini D, Vikas S, Chauhan NS (2010) Herbal remedies for treatment of hypertension. IJPSR 1(5):1–21 49; Thomas FMK, Michael H, Heinrich W, Hagen K, Raimund E (2005) Clinical and pharmacological testing of patients with pulmonary hypertension for treatment decision. Herz 30(2010): 286–295
186. Rakotomalala G, Agard C, Tonnerre P, Tesse A, Derbré S, Michalet S, Hamzaoui J, Rio M, Cario-Toumaniantz C, Richomme P, Charreau B, Loirand G, Pacaud P (2013) Extract from Mimosa pigra attenuates chronic experimental pulmonary hypertension. J Ethnopharmacol 148(1):106–116
187. Bai MK, Guo Y, Bian BD et al (2011) Integripetal rhodiola herb attenuates high altitude induced pulmonary arterial remodeling and expression of vascular endothelial growth factor in rats. Sheng Li Xue Bao 63:143–148
188. Huang XY, Fan R, Lu YY, Lin QD (2011) The protective effect of salidroside on cor pulmonale rats induced by chronic hypoxia in normal pressure. Chin Arch Tradit Chin Med 29:1868–1871

189. Zhang S, Gao W, Xu K et al (1999) Early use of Chinese drug rhodiola compound for patients with post-trauma and inflammation in prevention of ALI/ARDS. Zhonghua Wai Ke Za Zhi 37:238–240
190. Wang XQ, Wang BH, Li YH, Zhao YL, Wang YY, Wang Y (2012) Regulating effect and its mechanism of Tibet Rhodiola crenulate on blood pressure in spontaneous hypertension rat. Chin J Exp Tradit Med Formulae 18:150–154
191. Sui XL, Yang F, Chen RH et al (2006) Inhibition of rhodiola on the growth of EVC-304 cell line. Xi Bao Yu Fen Zi Mian Yi Xue Za Zhi 22:524–525
192. Yao BW, Chen W (2005) Research advances in the pharmacological effects of Ginkgo biloba leaves (in Chinese). Zhejiang J Integr Tradit Chin West Med 15:192–193
193. Reddy KN, Reddy CS, Trimurthulu G (2006) Ethnobotanical survey on respiratory disorders in eastern Ghats of Andhra Pradesh. Ethnobot Leafl 1:16
194. Taur DJ, Patil RY (2011) Antiasthmatic activity of Ricinus communis L. roots. Asian Pac J Trop Biomed 1(1):S13–S16
195. Taur DJ, Patil RY (2011) Mast cell stabilizing and antiallergic activity of Abrus precatorius in the management of asthma. Asian Pac J Trop Med 4(1):46–49
196. Taur DJ, Patil RY (2011) Some medicinal plants with anti-asthmatic potential: a current status. Asian Pac J Trop Biomed 1(5):413–418
197. Haq I (2004) Safety of medicinal plants. Pak J Med Res 43(4):203–210
198. Singh P, Shukla R, Kumar A, Prakash B, Singh S, Dubey NK (2010) Effect of Citrus reticulata and Cymbopogon citratus essential oils on aspergillus flavus growth and aflatoxin production on Asparagus racemosus. Mycopathologia 170(3):195–202
199. Ali SI, Qaiser M (1986) A phytogeography analysis of the phanerogames of Pakistan and Kashmir. Proc R Soc Edinb 89:89–101
200. Shinwari ZK, Qaiser M (2011) Efforts on conservation and sustainable use of medicinal plants of Pakistan. Pak J Bot 43:5–10
201. Haq I (1983) Medicinal plants. Hamdard Foundation Press, Karachi 10. Karou D, Nadembega WMC, Ouattara L, Ilboudo DP, Canini A, Nikiéma JB, Simpore J, Colizzi V, Traore AS (2007) African ethnopharmacology and new drug discovery. Med Aromat Plant Sci Biotechnol 1(1): 61–69
202. Hussain K, Shahazad A, Zia-ul-Hussnain S (2008) An ethnobotanical survey of important wild medicinal plants of Hattar district Haripur, Pakistan. Ethnobot Leafl 1:5
203. Hamayun M, Khan SA, Sohn EY, Lee IJ (2006) Folk medicinal knowledge and conservation status of some economically valued medicinal plants of district swat, Pakistan. Lyonia 11(2): 101–113
204. Bremer K, Anderberg AA (1994) Asteraceae: cladistics and classification. Timber Press, Portland
205. Jeffrey C (2006) Introduction with key to tribes. In: Kadereit JW, Jeffrey C (eds) The families and genera of vascular plants, vol. 8, flowering plants. Eudicots: Asterales. Springer, Berlin, pp 61–87
206. Barker MS, Kane NC, Matvienko M, Kozik A, Michelmore RW, Knapp SJ, Rieseberg LH (2008) Multiple paleopolyploidizations during the evolution of the Compositae reveal parallel patterns of duplicate gene retention after millions of years. Mol Biol Evol 25(11):2445–2455
207. Cronquist A (1982) An integrated system of classification of flowering plants. Columbia University Press, New York
208. Adekenov SM (1995) Sesquiterpene lactones from plants of the family Asteraceae in the Kazakhstan flora and their biological activity. Chem Nat Compd 31(1):21–25
209. Uysal T, Ozel E, Bzkurt M, Ertuğrul K (2012) Genetic diversity in threatened populations of the endemic species Centaurea lycaonica Boiss. Heldr (Asteraceae). Res J Biol 2(3):110–116
210. Stepp JR, Moerman DE (2001) The importance of weeds in ethnopharmacology. J Ethnopharmacol 75(1):19–23
211. Kumar A, Singh L (2022) Pharmacognostic study and development of quality control parameters for whole plant of acampe papillosa lindl. Neuroquantology 15(20):5614–5622

212. Kumar A, Rastogi P et al (2023) Phyto-chemical potential of Sida cordifolia leaves for antinociceptive, anti-inflammatory and antioxidant activity. Eur Chem Bull 12(si6): 2245–2256
213. Sulaiman MR, Moin S, Alias A, Zakaria ZA (2008) Antinociceptive and anti-inflammatory effects of S. rhombifolia L. in various animal models. Res J Pharmacol 2:13–16
214. Venkatesh S, Reedy YSM, Suresh B, Reedy BM, Ramesh M (1999) Antinociceptive and anti-inflammatory activity of *Sida rhombifolia*. J Ethopharmacol 67(29):279–232
215. Zakaria ZS, Rahman NLA, Sulaiman MR (2009) Antinociceptive and anti-inflammatory activities of the chloroform extract of Bauhinia purpurea L. in animal models. Int J Trop Med 4:140–145
216. Franzotti EM, Santos CVF, Rodrigues HMSL, Mourao MR, Andrade AR, Antoniolli. (2000) Anti-inflammatory, analgesic activity and acute toxicity of S. cordifolia L. J Ethnopharmacol 72:273–278
217. Ahmed S, Sultana M, Hasan MM, Azhar I (2011) Analgesic and antiemetic activity of cleome viscose L. Pak J Bot 43:119–122
218. Antoniolli AR, Franzotti EM, Santos CVF, Rodrigues HMSL, Mourao RHV, Andrade MR (2000) Anti-inflammatory, analgesic activity and acute toxicity of Sida cordifolia L. (Malva-branca). J Ethopharmacol 72:273–278
219. Rahman MA, Paul LC, Solaiman M, Rahman AA (2011) Analgesic and cytotoxic activities of S. rhombifolia Linn. Pharmacology 2:707–714
220. Pawa RS, Jain A, Sharma P, Chaurasiya PK, Singour PK (2011) In vitro studies on S. cordifolia Linn for anthelmintic and antioxidant properties. Chin Med 2:47–52
221. Antoniolli AR, Franzotti EM, Santos CVF, Rodrigues HMSL, Mourao RHV, Andrade MR (2000) Anti-inflammatory, analgesic activity and acute toxicity of S. Cordifolia. J Ethopharmacol 72:273–278
222. Singh A, Dhariwal S, Navneet. (2018) Traditional uses, antimicrobial potential, pharmacological properties and phytochemistry of S. rhombifolia Linn. Int J Innov Pharm Sci Res 6(02): 54–68
223. Franco CIF, Morasis LCSL, Quintas LJ, Antoniolli AR (2005) CNS pharmacological effects of the hydroalcoholic extract of S. cordifolia leaves. J Ethnopharmacol 98:275–279
224. Gupta SR, Nirmal SA, Patil RY, Asana GS (2009) Anti-arthritic activity of various extracts of S. rhombifolia aerial parts. Nat Prod Res 23(8):689–695
225. Narendhirakannan RT, Limmy TP (2010) In vitro antioxidant studies on ethanolic extracts of leaf, stem and root of S. rhombifolia L. Int J Pharm Bio Sci 2:105–106
226. Rao KS, Mishra SH (1997) Anti-inflammatory and hepatoprotective activities of S. rhombifolia Linn. Indian J Pharmacol 29(1):110–116
227. Shekhar TC, Bahuguna YM, Vijender S (2010) Anti-inflammatory activity of ethanolic stem extracts of Rubi cordifolia Linn. in rats. IJRAP 1:126–130
228. Narendhirakannan RT, Limmy TP (2011) Anti-inflammatory and antioxidant properties of S. rhombifolia stems and roots in adjuvant induced arthritic rats. Immunopharmacol Immunotoxicol 34(2):326–336
229. Khalil MS, Sperotto SJ, Manfron PM (2006) Anti-inflammatoryactivity of the hydroalcoholic extract of leaves S. rhombifoli L. (malvaceae). Bonaerense 25:260
230. Daehler CC (1998) The taxonomic distribution of invasive angiosperm plants: ecological insights and comparison to agricultural weeds. Biol Conserv 84(2):167–180
231. Ahmed AA, Bassuony NI (2009) Importance of medical herbs in animal feeding. World J Agric Sci 5(4):456–465
232. Khan SM, Page S, Ahmad H, Harper D (2013) Identifying plant species and communities across environmental gradients in the Western Himalayas: method development and conservation use. Eco Inform 14:99–103
233. Kumar D, Trivedi N, Dixit RK (2015) Herbal medicines used in the traditional Indian medicinal system as a therapeutic treatment option for diabetes management: a review. World J Pharm Pharm Sci 4(4):368–385

234. Koffi N, Marie-Solange T, Emma AA, Noel ZG (2009) Ethnobotanical study of plants used to treat arterial hypertension in traditional medicine, by abbey and Krobou population of Agboville (Cote d'ivoire). Eur J Sci Res 35:85–98
235. Adeyemi OO, Akindele AJ, Nwumeh KI (2013) Acute and subchronic toxicological assessment of Byrsocarpus coccineus Schum. and Thonn. (Connaraceae) aqueous leaf extract. Int J Appl Res Nat Prod 3(2):1–11
236. Ghorbani A (2005) Studies on pharmaceutical ethnobotany in the region of Turkmen Sahra, north of Iran: (part 1): general results. J Ethnopharmacol 102(1):58–68
237. Bhat JA, Kumar M, Bussmann RW (2013) Ecological status and traditional knowledge of medicinal plants in Kedarnath wildlife sanctuary of Garhwal Himalaya, India. J Ethnobiol Ethnomed 9(1):b5
238. Sosnowska J, Balslev H (2013) American palm ethnomedicine: a meta-analysis. J Ethnobiol Ethnomed 5(1):43
239. Adnan M, Begum S, Latif A, Tareen AM, Lee LJ (2012) Medicinal plants and their uses in selected temperate zones of Pakistani Hindukush-Himalaya. J Med Plant Res 6(24):4113–4127
240. Adnan M, Ullah I, Tariq A, Murad W, Azizullah A, Khan AL, Ali N (2014) Ethnomedicine use in the war affected region of Northwest Pakistan. J Ethnobiol Ethnomed 10(1):16
241. Manish P, Lubhan S, Ganesh P, Kamal D, Sagarika M (2021) Medicinal plants used in treatment of bronchitis. In: Medicinal plants for lung diseases a pharmacological and immunological perspective. Springer. https://doi.org/10.1007/978-981-33-6850-716
242. Manish P, Prateek P, Habibullah K et al (2021) Green synthesis of silver nanoformulation of Scindapsus officinalis as potent anticancer and predicted anticovid alternative: exploration via experimental and computational methods. Biocatal Agric Biotechnol 35:02072
243. Manish P, Prateek P et al (2022) Attenuation of hepatic and breast cancer cells by Polygonatum verticillatum embedded silver nanoparticles. Biocatal Agric Biotechnol 30:101863
244. Manish P, Priyanshu T et al (2019) Cytotoxic action of silver nanoparticles synthesized from Phyllanthus fraternus on hepatic and breast cancer cell lines: A green approach. Int J Green Pharm 13(3):1
245. Rupesh K, Manish P et al (2022) Plant bioactive compounds and their mechanistic approaches in the treatment of diabetes: a review. Future J Pharm Sci 8:52
246. Manish P, Lubhan S et al (2019) Antimicrobial activity of Lagenaria siceraria fruit extract against pathogenic bacterial strains. EC. Pharmacol Toxicol 7(8):825–830
247. Amit K, Manish P et al (2022) Pharmacognostical studies and quality control parameters of Sida Rhombifolia. IJBPAS 11(2):662–672
248. Kumar A, Mishra PG, Tomar SD, Pathak M, Pandey KR, Singh L (2023) Phyto-therapeutic potential of aerial part of Sida rhombifolia for anti-inflammatory, Antinociceptive, and antioxidant activity. Int J Pharm Qual Assur 14(1):91–95

Herbs and Herbal Formulations for the Management and Prevention of Gastrointestinal Diseases

22

Nitu Singh, Urvashi Sharma, Bharat Mishra, Ashish M. Kandalkar, and Sachin Kumar Jain

Contents

Abstract

Diseases affecting the digestive tract's activities are referred to as GIT disorders. Stomach or abdominal pain, dyspepsia, bloating, diarrhea, constipation, dysentery, vomiting, and gastroenteritis are typical gastrointestinal illnesses. Patients with gastrointestinal issues are routinely treated with herbal or alternative treatments. In several regions of the world, herbs have historically been trusted for the management of numerous illnesses. They are considered vital bases of pharmaceutical items, especially herbal remedies, and their use considerably impacts the delivery of basic health care. Because they have minimal side effects when used

N. Singh (✉) · B. Mishra
R.G.S College of Pharmacy, Lucknow, India

U. Sharma
Faculty of Pharmacy, Medi-Caps University, Indore, India

A. M. Kandalkar
Vidyaniketan College of Pharmacy, Anjangaon, India

S. K. Jain
Oriental College of Pharmacy and Research, Oriental University, Indore, India

S. C. Izah et al. (eds.), *Herbal Medicine Phytochemistry*, Reference Series in Phytochemistry, https://doi.org/10.1007/978-3-031-43199-9_24

appropriately, herbal medications are becoming more and more popular with Westerners. In recent years, a widespread trend has been there toward a resurgence of interest in a conventional system of medical care. According to the WHO, traditional medicine has a place in the primary healthcare system. Medicinal plants remain to be a significant source of treatment in developing nations. The majority of people in developing nations, or about 88% of them, are thought to rely on a traditional system of medicine for their medical care. When treating GI problems, evidence-based, rational phytotherapy which employs herbs and herbal preparations to provide the desired therapeutic effect plays a significant role. Given the variety of these disorders, it stands to reason that a wide range of herbal products can be utilized in both prevention and treatment. This chapter aims to review the use of various herbs and herbal preparations for managing and preventing gastrointestinal diseases as well as their safety and efficacy.

Keywords

Gastrointestinal disorders · Herbal medicines · Herbs · Herbal formulation · Safety

Abbreviations

5-HT	5-hydroxytryptamine
BMI	Body mass indices
DFHE	Dill Fruit Hydroalcoholic Extract
DGL	Deglycyrrhizinated licorice
DNA	Deoxyribonucleic acid
ESCOP	European Scientific Cooperative on Phytotherapy
FD	Functional Dyspepsia
FGIDs	Functional gastrointestinal disorders
GERD	Gastroesophageal Reflux Disease
GI	Gastrointestinal
GLP-1	Glucagon-like peptide 1
H2	Histamine Type 2
IBS	Irritable Bowel Syndrome
IL 10	Interleukin 10
LES	Lower esophageal sphincter
L-NAME	L-N^G-Nitro arginine methyl ester
MMP	Matrix metalloproteinase
NERD	Nonerosive reflux disease
PDS	Postprandial distress syndrome
PKS	Polyketide synthases
PMO	Peppermint Oil
POAL	Probiotics originating from Aloe leaf)
PPIs	Proton pump inhibitors
QOL	Quality of Life
RCT	Randomized control trial

TCM	Traditional Chinese medicine
TID	Thrice a day
VEGF	Vascular endothelial growth factor
WHO	World Health Organization
XO	Xanthine oxidase

1 Introduction

The gastrointestinal tract (GIT), also known as the human digestive system, is made up of a hollow, twisting muscular tube that starts from the oral cavity, where food is taken in the mouth and travels through the throat, esophagus, stomach, and intestines before ending at the rectum and anus, from where digested food is discharged. Solid organs of the digestive system cover the gallbladder, pancreas, and liver. GI tract bacteria, also called gut flora or microbiome, aid digestion. Bacteria are the most prevalent and diverse group of microbial eukaryotes (fungi, protozoa, archaea, and viruses) that make up the human gastrointestinal microbiota. Its bacterial ecosystem is extensive and complex, with up to 1014 bacteria present – more than ten times as numerous as somatic human cells [1]. The GI tract's primary functions include digestion, absorption, excretion, and protection. Such tasks are carried out by a few specialized organs extending from the mouth to the anus. The stomach and small intestine are majorly responsible for digestion and absorption due to physical mechanisms like retropulsion along with chemical factors such as the presence of bile and enzymes in the small intestine. Desiccating and compacting waste, which is then kept in the rectum and sigmoid colon until it is evacuated, are the main jobs of the large intestine [2]. Ailments affecting the digestive tract's activities, like absorption of food and liquid, digestion, or excretion, are referred to as gastrointestinal disorders [3]. These conditions are brought on by infections with different bacterial, viral, and parasite species [4, 5].

1.1 Types of Gastrointestinal Disorders

There are two types: functional and structural.

Functional gastrointestinal disorders (FGIDs) are ailments where the digestive system appears normal but does not operate as it should. Functional disorders can affect anyone at any age, including adults, teens, and children. Yet, rather than being the outcome of an illness or infection, FGIDs cause sensitivity and a variety of GI symptoms due to improper functioning. The three main traits of FGIDs are dysfunction in the brain-gut axis, sensation, and motility.

Motility is the term for the muscular activity of the GI tract. The sensation is produced by the GI tract's nerves in response to stimuli (e.g., digestion of food). Functional GI illnesses may have nerves that are so responsive that even normal contractions can be painful or uncomfortable. A failure in communication between the brain and the GIT is known as a brain-gut dysfunction. FGIDs may impair the regulatory pathway between brain and gut function. Some of the indications of FGIDs are nausea, stomach pain, constipation, bloating, burping, flatulence, diarrhea, vomiting, and indigestion [6].

Structural gastrointestinal disorder refers to conditions where the colon seems aberrant and is not functioning properly. Sometimes the structural anomaly must be surgically removed. GI structural illnesses include inflammatory bowel disease, colon cancer, stenosis, colon polyps, hemorrhoids, and diverticular disease. Untreated structural GI problems usually aggravate symptoms and lead to life-threatening consequences [6].

1.2 General Signs of Digestive Disorders

Clearly, the conditions and symptoms of digestive illnesses are unique and vary from individual to individual. Nonetheless, most digestive problems exhibit certain indications and symptoms. Figure 1 illustrates typical indications [7].

The few commonest gastrointestinal conditions are:

1. Celiac Disease
2. Irritable Bowel Syndrome (IBS)
3. Lactose Intolerance

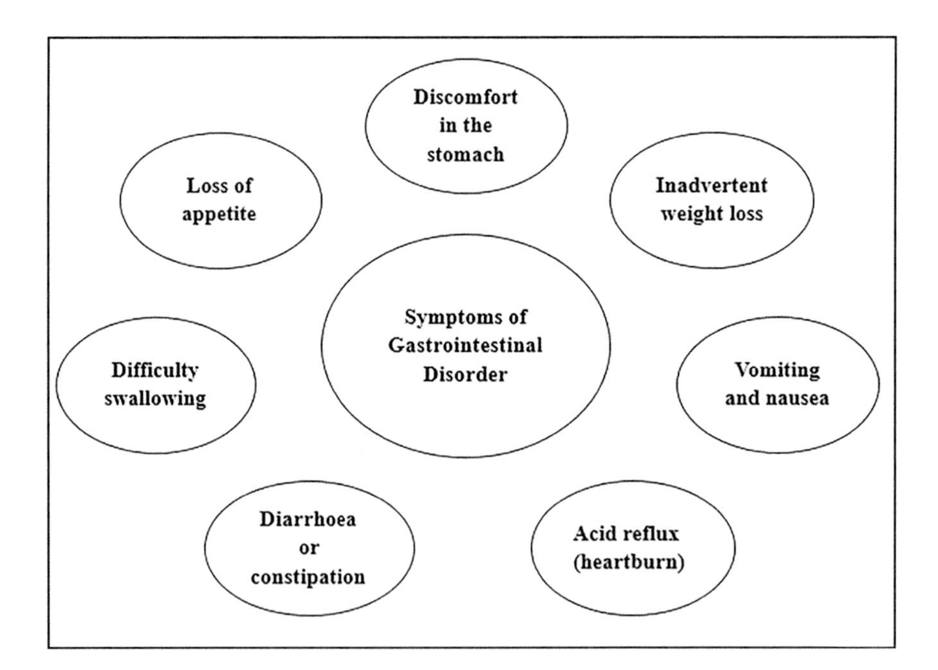

Fig. 1 Common symptoms of gastrointestinal disorders

4. Chronic Diarrhea
5. Constipation
6. Gastroesophageal Reflux Disease (GERD)
7. Peptic Ulcer Disease
8. Crohn's Disease
9. Ulcerative Colitis
10. Gallstones
11. Chronic and Acute Pancreatitis
12. Liver Disease
13. Diverticulitis [7]

1.3 Synthetic Medicines for Gastrointestinal Disorders and Their Adverse Effects

Gastrointestinal disorders are typically treated with the following class of medications.

- **Antacids**

  An antacid is a blend of many components having various salts of calcium, magnesium, and aluminum as their active substances. The antacid functions by neutralizing the acid present in the stomach by inhibiting the action of the proteolytic enzyme pepsin. The elderly and newborn populations are particularly vulnerable to negative consequences. It is not advised to take antacids frequently in this group for reasons of safety. Table 1 lists the therapeutic applications and negative effects of antacids [8].
- **Histamine H2-receptor antagonists (Histamine Blockers)**

  For a range of gastrointestinal disorders, including those listed in Table 1, stomach acid-suppressing drugs also called H2 receptor blockers or H2 receptor antagonists (H2RAs) are routinely prescribed. H2RAs prevent the endogenous ligand histamine from binding and showing its action by reversibly attaching themselves to receptors available on stomach parietal cells, hence reducing stomach acid output. Antagonists of the H2 receptor are often well tolerated. Headache, weariness, drowsiness, fatigue, abdominal pain, constipation, or diarrhea are examples of mild side effects. There is a correlation between the usage of H2RAs and some side effects, as shown in Table 1, in individuals who are over 65 or have hepatic or renal impairment [9].
- **Proton pump inhibitors**

  Proton pump inhibitors (PPIs) prevent the final step of gastric acid secretion by blocking the hydrogen/potassium adenosine triphosphatase (H^+/K^+ ATPase) enzyme, also known as the "proton pump," in the parietal cells of the stomach. The following conditions associated with stomach acid secretion are advised for the prevention and treatment of PPIs. In general, doctors do not worry about PPIs having serious side effects when used over a brief amount of time say 2 weeks,

Table 1 Reported adverse effects of synthetic medications

S.No.	Category of medications	Therapeutic use	Adverse effects	References
1	Antacids	Heartburn symptoms of GERD Gastric and duodenal ulcers Gastritis due to stress Non-ulcer dyspepsia Bile acid caused diarrhea Biliary reflux Constipation	Constipation Hypophosphatemia Hypercalcemia Osteopenia Microcytic anemia Fecal impaction Nausea Vomit Stomach cramps	[8]
2	Histamine H2-receptor antagonists (Histamine Blockers)	GERD Ulcer of duodenum or GIT, hypersecretion of gastric juices, and indigestion and heartburn	Delirium, confusion, hallucinations, or slurred speech	[9]
3	Proton pump inhibitors	Gastro-esophageal ulcers, peptic ulcer disease, non-steroidal anti-inflammatory drug-associated ulcers, Zollinger-Ellison syndrome, and eradication of *Helicobacter pylori*	Dementia, fracture risk, hepatocellular carcinoma, nosocomial pneumonia, chronic kidney disease, nutritional deficiencies, vitamin B12, iron and calcium deficiency, bacterial overgrowth, gastric polyps	[10]
4	Prokinetics	Functional dyspepsia	Parkinson-like symptoms, Dystonia, Akathisia, Galactorrhoea, and breast engorgement	[11]

but reports of adverse effects arise with the growing use of these drugs on long-term use. According to current studies, PPIs can be consumed only for a shorter duration of time in the lowest possible dose which is effective enough because using them for a longer time may pose adverse effects like reduced nutritional absorption, dementia, infections, kidney disease, and hypergastrinemia. The therapeutic uses and side effects of chronic PPI usage are listed in Table 1 [10].

- **Prokinetics**

  Dopamine antagonists (metoclopramide, domperidone, levosulpiride, and itopride) and serotonin (5-HT) receptor agonists for example cisapride and mosapride make up the majority of the prokinetic medications used in clinical settings. Although the effectiveness of all prokinetic medications for the treatment of gastrointestinal hypomotility disorders is well recognized, these medications are also linked to a number of negative side effects, which are mentioned in Table 1 [11].

2 Herbal Therapy for the Management and Prevention of Gastrointestinal Diseases

Herbal therapy has significantly improved human health because of its capacity to promote health, heal illnesses, and treat and rehab patients. Herbal medicine, also known as phytomedicine, or phytotherapy, comprises herbs, herbal components, or formulations, which have plant-based active chemicals or other compounds [12]. Numerous studies indicate that traditional people around the world frequently employ plant-based therapies or herbal medications to address problems in the digestive system. People can treat several digestive-related illnesses simply by regularly incorporating herbal substances into their diet for improving the performance of the digestive system. These herbal treatments function in a variety of ways, either by easing the bowel movements, enhancing the digestive process, raising the bowel movement frequency, detoxifying the body, eliminating toxins, minimizing gas, settling an upset stomach, and decreasing the bloating and digestive discomfort. All societies have employed medicinal plants throughout history as a type of medicine for the treatment of various human maladies, and they still have a crucial role in our current technology-based civilization. Approximately half of all medications currently used in clinical settings are plant-derived components or natural drug products or their derivatives [13]. For their health care, almost 80% of individuals in underdeveloped nations mostly use traditional herbal remedies that contain plant extracts or their active ingredients [14]. Secondary metabolites, also known as phytochemicals, are abundant in medicinal plants, and their derivatives display a wide range of biological characteristics, including anti-plasmodial, anti-inflammatory, anticancer, and antioxidant effects [13, 15, 16].

This review's objective is to provide some illustrations of studies that have been successful in establishing the efficacy of various gastrointestinal illnesses treated with plants.

2.1 Herbs Used in Gastrointestinal Diseases

2.1.1 *Aloe barbadensis* Miller

Aloe barbadensis Miller is one of more than 400 species of aloe that are native to arid tropical and subtropical weathers, including the southern United States, also referred to as *Aloe vera* (Fig. 2a). They are all members of the Liliaceae family and originated in South Africa [17]. Only a few aloe species have recently been thought to be important commercially, with *A. vera* being of utmost strength and thus the maximum well-liked herb in the research community [18].

Secondary metabolites produced by *A. vera* are recognized to be beneficial [17, 19]. The two most important secondary metabolites are anthraquinones and tricyclic aromatic quinines, both of which are widely distributed. Chrysophanol and Aloe-emodin are the two main anthraquinone derivatives that are found in nature [20]. It has been suggested that the type III polyketide biosynthesis pathway is used

Fig. 2 (**a**) Leaves of *Aloe barbadensis Miller* (**b**) Dried ripe fruits of *Anethum graveolens L* (**c**) Dried ripe fruits of Carum carvi L. (**d**) Fruits of Emblica officinalis (**e**) Dried ripe fruits of *Foeniculum vulgare* Mill. (**f**) Dried roots of Glycyrrhiza glabra (**g**) Leaves of Mentha piperita (**h**) Seeds of Plantago ovata Forsk (**i**) Fruits of Trachyspermum ammi L. (**j**) Seeds of Trigonella foenum-graecum L (**k**) Rhizome of Zingiber officinale

to generate the tricyclic aromatic quinines found in aloe [21]. The important secondary metabolites in *A. vera* gel are aloesin, aloin, and aloe-emodin (an oxidative derivative of aloin) [22].

For the treatment of stomach infections due to *H. pylori*, *aloe vera* gel acts as a special and effective natural remedy. It exhibits antibacterial action against both susceptible and resistant *H. pylori* strains. The cytoprotective properties of *Aloe vera* leaf extracts are explored for their benefits in digestion and peptic ulcer treatment [23]. *Aloe vera* latex is frequently used to treat constipation because it contains anthraquinone glycosides, which are known to have laxative effects [24, 25]. In a double-blind, randomized controlled trial (RCT) with 28 healthy individuals, it is discovered that aloin had a greater laxative effect than the stimulant laxative phenolphthalein [26]. The *L. brevis* strains were isolated from naturally fermented *A. vera* gel and called POAL (probiotics from aloe leaf) strains because they displayed exclusionary resistance to a variety of antibiotics while preventing the growth of various dangerous enteropathogens in the gut [27]. The colonic flora converts the aloin included in the gel into reactive aloe-emodin, which has a purgative effect. Aloe-emodin, a compound derived from *A. vera*, via decreasing MMP-2/9 reduces the migration of colon cancer cells and suppresses vascular endothelial growth factor (VEGF) and Ras homolog gene family member B through lowering the nuclear translocation and binding of DNA to NF-KB, assuring the effectiveness of aloe-emodin tumor via multiple target sites [28].

2.1.2 *Anethum graveolens* Linn

Dill, *Anethum graveolens* L. (Apiaceae family), is a well-known herb that is frequently used as a spice and also produces essential oil (Fig. 2b). It has been utilized in Ayurveda treatment since ancient times. It is a flavorful annual herb. In Ayurvedic medicine, dill is well known in the treatment of gastrointestinal pain, colic, and to aid with digestion. In Unani medicine, it is utilized for colic, digestive issues, and even for diarrhea [29].

Anethum is a component found in gripe water and used for the treatment of colic pain in infants and flatulence in young children [30]. The seed is stomachic, somewhat diuretic, stimulant, and galactagogue with a pleasant aroma [31, 32]. Dill oil from grapes contains the monoterpenes carvone and limonene as their primary ingredients [33]. Three different compounds phellandrene, dill ether, and myristicin combine to give the dill herb its distinctive aroma [34, 35]. The essential oil from the seeds decreases intestinal cramps and spasms, thereby calming colic [36, 37]. The volatile oil acts as an appetite booster and relieves gas aiding in digestion. Chewing the seeds helps to reduce bad breath [38]. The mucosa is strongly protected by *A. graveolens* L. seed extracts, that possess antisecretory, anti-ulcer, and anti-HCl and ethanol-induced stomach ulcers in mice [39]. The seeds of *A. graveolens* L. were used to extract two flavonoids, quercetin and isorhamnetin, which have antioxidant qualities and may be utilized to fight free radicals. This outcome might help avoid peptic ulcers [40, 41]. The HCl extract of dill fruit strongly relaxed the contractions caused by a range of spasmogens in rat ileum, supporting its importance in traditional medicine for digestive problems [42].

In another study, the effects of methanolic extract of dill on both spontaneous contractions and even those induced by acetylcholine were explored. Cumulative dosages of dill methanolic extract significantly reduced ($p < 0.01$) the spontaneous rat ileum contractions. The extract also reduced the contraction induced by acetylcholine in a dose-dependent manner ($p < 001$). The findings confirmed that the methanolic dill extract had a relaxing effect on the isolated rat gut. Dill extract prevented both naturally occurring and acetylcholine-induced ileum contractions [43].

2.1.3 *Carum carvi* Linn

C. carvi also known as Caraway is an essential medicinal plant belonging to the Apiaceae family (Fig. 2c). It has been grown for a very long time in Egypt, Australia, China, and Iran, as well as in the north and center of Europe [44]. It has been used traditionally as a spice in food and drinks as well as an herbal alternative for the treatment of GI diseases such as dyspepsia, diarrhea, various spasmodic symptoms, bloating, and flatulent colic [45].

Two important natural chemicals are limonene and carvone, having mucoprotective and antiulcerogenic actions on gastroduodenitis and duodenal peptic ulcers [46]. Caffeic acids, quercetin, kaempferol [47, 48], fatty acids (10–18%) (linoleic and oleic acids, petroselinic), essential oil (3–7%), protein (20%), and carbohydrates (15%) are the main constituents of caraway fruits [49]. As per European Pharmacopoeia, caraway fruit possesses essential oil up to 3%, the primary component is D-carvone which is up to 50–65%, (+)-limonene (45%), and carveol and dihydrocarveol (<1.5%). The odor is due to its main constituent D-carvone [50, 51].

An alcoholic extract of caraway has been found to have antispasmodic properties that block the spasmogens acetylcholine and histamine-induced smooth muscle spasms [52, 53]. Caraway oil demonstrated significant levels of selectivity in a study on 12 intestinal bacteria, suppressing the growth of possible pathogens to levels that did not affect the study's beneficial bacteria. This outcome was related to the efficacy and value of caraway oil in traditional medicine for the management of dysbiosis, which is associated with a variety of gastrointestinal and systemic disorders [54]. Caraway extracts reduced acid and leukotriene production, increased mucin secretion, and released prostaglandin E2 preventing indomethacin-induced stomach ulcers in a dose-dependent manner. Histological testing also supported the antiulcerogenic action, which was linked to the flavonoid content and free radical scavenging abilities of the food [55]. In one clinical trial, patients with functional dyspepsia (FD), according to Rome III criteria who were on a placebo with their regular medications, such as PPIs, anticonvulsants, antihistamines, antidepressants / TCAs, beta-blockers, pain modulators, H2RAs, and antacids, were compared with those who were taking enteric-coated capsules encapsulating caraway oil and menthol. The results of the study showed that the symptomatic reduction in the intervention group was quantitatively greater than that in the placebo group. Clinical Global Impressions were improved by intervention and placebo treatments in 61% and 49% of patients, respectively ($p = 0.23$). The treated group experienced no significant side effects [56]. The effectiveness of hot or cold olive oil poultices in curing IBS for 3 weeks, followed by 2-week "wash-out" periods, was compared to the efficiency of hot poultices created with caraway poultices in a cross-over RCT. Two-thirds of patients who used a caraway oil poultice on their skin rated the treatment as satisfactory or very good. The overall IBS-QOL (Quality of Life) score showed a statistically significant difference between caraway oil and the other oils. Comparatively, Caraway oil reported adequate alleviation at 51.8%, compared to hot olive oil 23.5% and cold olive oil 25.8% [57].

2.1.4 Emblica Officinalis

Emblica officinalis holds a revered place in Ayurveda, an ancient medical system practiced in India (Fig. 2d). It is a member of the Euphorbiaceae family. The Indian gooseberry or *Phyllanthus emblica* is another name [58, 59]. The juice of this fruit contains Vitamin C, i.e. 4.78 mg mL^{-1} [60, 61] in larger quantities. Several different acids that are present include gallic acid, ellagic acid, chebulinic acid, chebulagic acid, 1-O-galloyl-beta-D-glucose, quercetin, 3,6-di-O-galloyl-D-glucose, 1, 6-di-O-galloyl beta-D glucose, corilagin, 3-ethylgallic acid (3-ethoxy-4 [62]. Kaempferol-3-O-alpha-L-(6″- ethyl)-rhamnopyranoside and kaempferol-3-O-alpha-L-(6″-methyl)-rhamnopyranoside are flavonoids found in *P. emblica* [63].

P. emblica can improve meal digestion, absorption, and assimilation when used regularly. People that use it discover they prefer the taste of food more. Additionally, it enhances iron absorption for healthy blood. Stomach acids are balanced. For reducing mild to severe hyperacidity, *P. emblica* is the best choice. It is administered medicinally to alleviate diarrhea. In cases of dysentery, it is provided by the locals as a fruit decoction blended with sour milk. Fenugreek seeds infused with leaves are used to treat chronic diarrhea [64].

Polyphenols from *P. emblica* L. are widely used for gastrointestinal organ protection. Since *Helicobacter pylori* are responsible for stomach ulcers, the bioactive substances of amla act as one of the potent inhibitors of these strains in-vitro [65].

The *Phyllanthus emblica* extract was investigated for antiulcer and anti-secretory activities using several models to prompt GI ulcers in mice such as pylorus ligation, administration of indomethacin or necrotizing agent (25% sodium chloride, 0.2 M sodium hydroxide, and 80% ethanol), and hypothermia induction are all used to cause gastrointestinal ulcers [66]. In all the test models employed, oral treatment of extract at doses of 250 mg/kg and 500 mg/kg significantly reduced the development of stomach lesions. In a different experiment, mice with pancreatitis caused by L-arginine had their blood levels of lipase and IL-10 reduced when *P. emblica* L. (rich in tannins and gallic acid) in 200 mg/100 g was administered [67]. The research also showed that the group of rats in *P. emblica* had greater amounts of nucleic acids, pancreatic protein, DNA synthesis rate, and pancreatic amylase. The histological changes in the colon of mice with colitis brought on by acetic acid were also lessened by the methanol fruit extract of *P. emblica* [68]. The daily intake of 500 mg/tablet of *P. emblica* extract daily twice decreased the frequency and intensity of vomiting and indigestion when compared with the placebo group [69].

2.1.5 *Foeniculum vulgare* Mill.

A biennial plant from the Apiaceae family, *Foeniculum vulgare* Mill. (Fig. 2e) has both aroma and therapeutic properties. Due to its flavor, it is a medicinal herb that humans have been aware of and used since ancient times. It was grown in practically every nation [70]. Due to the valuable nutritional makeup of fennel, which includes the presence of important fatty acids, a diet containing the right amount of fennel may have potential health benefits [71].

Fennel water also possesses qualities just like anise and dill water. And when combined with syrup and sodium bicarbonate, this collection forms the domestic "gripe water," that is given to treat newborn flatulence. By adding boiling water to a teaspoonful of smashed fennel seeds, fennel tea can be made which is also used as a carminative [72]. From *F. vulgare*, flavonoids such quercetin-3-rutinoside, rosmarinic acid, and eriodictyol-7-rutinoside have been extracted [73]. The most prevalent flavonoids in *F. vulgare* include quercetin-3-glucuronide, isoquercitrin, kaempferol-3-glucuronide, quercetin-3-arabinoside, isorhamnetin glucoside, and kaempferol-3-arabinoside [74]. There have also been reports of aqueous extract of *F. vulgare* found to include quercetin-3-O-galactoside, kaempferol-3-O-rutinoside, quercetin-3-O-galactoside, and kaempferol-3-O-glucoside [75]. While the methanolic extract was used to isolate quercetin 3-O-rutinoside, quercetin 3-O-glucoside, and kaempferol 3-O-rutinoside, the ethyl acetate extract also yielded quercetin, kaempferol, and isorhamnetin 3-O-rhamnoside. These flavonoids have outstanding anti-inflammatory and antinociceptive properties [76]. Additionally, it was claimed that the antioxidants quercetin, rutin, and isoquercitrin had immunomodulatory properties [77]. Chlorogenic acids and rosmarinic acid (6.8% and 14.9% respectively) are the main phenolic components obtained in the methanolic extract of fennel seeds, while the main flavonoids are quercetin and apigenin (17.1% and 12.5%, respectively) [78].

Fennel essential oil reduces intestinal gas while also controlling the intestine's smooth muscle movement. *Foeniculum vulgare* is approved for the treatment of several symptoms of dyspepsias such as gastrointestinal atony, a feeling of heaviness in the stomach, and many others. It can be taken alone or in conjunction with other herbal remedies for the treatment of these conditions. Fennel can be added to treatments that contain anthraquinone ingredients to lessen the frequency of stomach pain that is frequently related to this kind of laxative [79]. The significant anti-ulcerogenic effect was shown by an aqueous extract of *F. vulgare* on a rat stomach on which lesions were brought on by ethanol. It was discovered that pre-treating with an aqueous extract considerably lessened the harm caused by ethanol to the stomach. When compared to the control group of animals, the effect of the aqueous extract was greatest and statistically significant ($P < 0.001$) in the 300 mg kg^{-1} group of mice. Aqueous extract of *F. vulgare* also markedly decreased levels of malondialdehyde in whole blood while markedly raising levels of Vitamin C, nitrite, Vitamin A, nitrate, and beta-carotene. As a result, rats' stomach mucosal lesions caused by ethanol were prevented by using aqueous fruit extract of *F. vulgare* [80].

2.1.6 *Glycyrrhiza glabra* Linn.

Commonly known as **Licorice**, it belongs to the Leguminosae family and has the scientific name *Glycyrrhiza glabra* (Fig. 2f). Both in parts of Europe and Asia, this therapeutic herb can be found [81].

The triterpenoid saponin glycyrrhizin is the primary constituent of licorice root, also referred to as glycyrrhizinic acid or glycyrrhizic acid, and is typically present in between 6% and 10%. The aglycone molecule (glycyrrhetinic acid) is produced by the hydrolysis of glycyrrhizin due to intestinal flora and a sugar moiety, causing both to be absorbed [82].

By removing the glycyrrhizin molecule, deglycyrrhizinated licorice (DGL) a processed licorice extract is produced that is used to treat peptic and aphthous ulcers. Flavonoids are those parts of DGL that are active. In animal trials, these compounds showed remarkable resistance to the development of chemical-induced ulcers [83]. A few more components of licorice include isoflavonoids (like, licoricone, isoflavonol, glabrol, and kumatakenin), coumarins like herniarin and umbelliferone, chalcones, triterpenoids, lignins, amino acids, amines, sterols, gums, and volatile oils [84].

Even though, glycyrrhetinic acid was the first medication to be shown to speed up the healing process in duodenal and gastric ulcers [85], many doctors who used to use licorice to treat peptic ulcers have started using DGL. As it has been proven to be superior and efficient to glycyrrhetinic acid with no side effects [86]. Over the years, numerous clinical investigations have revealed DGL to be a potent drug for ulcers. It has proven to be incredibly successful to treat stomach ulcers [87–91]. In one trial, DGL (760 mg, thrice a day) or a placebo was given to 33 individuals with stomach ulcers for 1 month. The DGL group (78%) experienced a considerably larger decrease in ulcer size than the placebo group (34%) [89].

DGL was found to be superior to cimetidine (Tagamet), ranitidine (Zantac), and antacids in multiple head-to-head comparison studies, both for the short-term treatment of peptic ulcers and for maintenance therapy [87, 88, 91]. For example, a hundred volunteers were given DGL (760 mg, between the meals, TID) or Tagamet in a head-to-head comparison with 200 mg of Tagamet thrice a day, with a bedtime dose of 400 mg [88]. Despite the fact that Tagamet has several serious adverse effects, using DGL is very safe. Since it was demonstrated in human tests to minimize the gastrointestinal bleeding brought on by aspirin, DGL is strongly advised in patients needing treatment for longer duration with ulcerogenic medications including aspirin, NSAIDs, and corticosteroids, to prevent gastric ulcers [90]. Duodenal ulcers can benefit from DGL as well. DGL was used to treat 40 patients who had more than six relapses the year prior and had chronic duodenal ulcers that had been present for four to 12 years. For 8 weeks, a daily dose of 4.5 g of DGL was given to half of the patients, while the other half had 3 g daily for the same period. Typically, within 5 to 7 days, all 40 patients had significant improvement, and no one needed surgery at the time of the one-year follow-up [92].

2.1.7 *Mentha piperita* Linn.

The earliest medicinal herb utilized in both Eastern and Western traditional medical systems is *Mentha piperita* (Fig. 2g) belonging to the family Lamiaceae. Herbalists have employed *M. piperita* as an emmenagogue, antiseptic, antispasmodic, astringent, carminative, antipruritic, antiemetic, diaphoretic, antibacterial, rubefacient, anticatarrhal, and analgesic [93]. It has historical usage as a spasmolytic for a variety of digestive issues, including colic in newborns, flatulence, diarrhea, dyspepsia, nausea, and vomiting, as well as to lessen gas and cramps [94]. In addition to menthol, peppermint oil also contains limonene, menthone, isomethone, cineole, isopulegol, pulegone, menthofuran, caron, and menthyl acetate [95]. Menthol is the main component of peppermint oil. According to the British Herbal Compendium [96], the leaves of piperita were used to cure biliary problems, indigestion, and intestinal colic. Its usage as a cholagogue and carminative in conditions affecting the gastrointestinal system, gallbladder, and bile ducts was advised by the German Commission, European Scientific Cooperative on Phytotherapy (ESCOP).

The major component of Dabur's "Pudina Hara," an Ayurvedic medicine that is used for the treatment of several stomach-related issues like gas, indigestion, acidity, etc., is *M. Piperita* oil. Mostly the oil is employed for treating the conditions like IBS, liver issues, ulcerative colitis, biliary tract ailments, and gallbladder problems. The gastric emptying rate is increased following *M. piperita* oil administration in a blinded, controlled research in patients with dyspepsia and normal gastrointestinal function [97]. It has a statistically significant impact on gynecological surgery patients who experience nausea [98]. Research indicates that *M. piperita* causes intestinal smooth muscle relaxation by lowering the calcium flux to the jejunum and large intestine [99]. Menthol and similar terpenes exhibit choleretic effects in in-vivo investigations and are useful in the treatment of a patient with bile duct and gallbladder cholesterol stones [100]. Although, piperita is acknowledged to be safe and is approved by Food and Drug Administration (FDA), still people with GI reflexes must use it cautiously because it might cause gallbladder discomfort and bile duct blockage [94].

2.1.8 *Plantago ovata* Forsk

The psyllium plant, *Plantago ovate* (Fig. 2h) of the family Plantaginaceae, is currently grown all over the world, including India [101]. It is originally from Persia. Several members of the *Plantago* genus are collectively referred to as "psyllium," and their seeds are a fantastic source of both soluble and insoluble fiber and are used commercially to manufacture mucilage [102]. The term "mucilage"refers to a class of transparent, colorless gelling substances that are produced by the seed coat and are frequently referred to as the husk or psyllium husky. For ailments such as hypercholesterolemia, colitis, diabetes, colon cancer, and inflammatory bowel disease, *P. ovate* is being utilized in many cultures as a home remedy [103]. Mucilaginous polysaccharide, mostly constitutes rhamnose, xylose, galactose, galacturonic acid, and arabinose, is the main component of the seeds and husk. Fatty acids are also present in the seeds, namely linoleic, oleic, and linolenic acids in concentrations of about 40.6%, 39.1%, and 6.9% respectively making up the majority. Dietary fiber of *P. ovata* may be extracted and has medical benefits, making it a useful source for making low-calorie foods [102].

Chronic constipation, digestion, hemorrhoids, moderate diarrhea, relief of duodenal ulcer, colon cleaning, IBS, demulcent, GERD, and other disorders are regularly treated with psyllium husk as a therapeutic agent [104]. Being an herbal, non-irritating laxative, which helps promote digestion, psyllium husk has been suggested. It transforms into a gelatinous, viscous substance that absorbs water to carry out a specific function when submerged in water. Fibers of husk are usually used to treat constipation and are frequently taken as a supplement. Clinical studies have consistently indicated the raised moisture level in stool, causing it to become heavier and softer while facilitating bowel movements [105]. On the other side, water uptake by psyllium has been demonstrated to impede colon transit and gastric emptying, which is advantageous for people who have diarrhea or stools that are incompatible with liquid stools. Both the number of bowel motions and the strain on the gastrointestinal walls are increased. Hemorrhoids and diverticulitis can be treated with psyllium husk [106]. Because of the accompanying health benefits, the FDA has authorized using psyllium husk in eatable items [107].

According to numerous research, using psyllium husk for medicinal purposes is risk-free and has few negative effects [103]. However, to validate its pharmacological qualities and postulate mechanisms of action for improvements to human health, further research is essentially required.

2.1.9 *Trachyspermum ammi* Linn.

Trachyspermum ammi L. (Fig. 2i), a drug of the Apiaceae family, is a highly prized seed spice with significant therapeutic value [108]. Ajwain fruits consist of brownish essential oil of 2–4%, of which thymol makes up the majority (35%–60%) [109].

Carvacrol, dipentene, α and β-pinenes, α-terpinene, and γ-terpenine as well as para-cymene are all components of the nonthymol fraction thymene [110]. The plant also contains trace quantities of myrcene, caphene, and α-3-carene. Alcoholic extracts possess a saponin that is extremely hygroscopic. A steroid-like compound and a yellow, crystalline flavone have been extracted from the fruits, along with a glucoside, 6-O-β-glucopyranosyloxythymol [111], and 25% of oleoresin that contains volatile oil [112]. One of the main ingredients of *T. ammi* oil is carvone which is 46% [113].

The fruits of *T. ammi* have been used medicinally for amoebiasis, expectorant, stomachic, carminative, and antiseptic purposes. The investigation of calcium channel blockage, which is revealed to facilitate the spasmolytic activities of herbs, supports the conventional usage of Ajwain in gastrointestinal ailments, like colic and diarrhea, as well as in hypertension [114]. Various ulcer models were used to demonstrate the antiulcer activity of *Trachyspermum ammi* fruit. The ulcer index and the proportion of ulcer prevention in mice pre-treated with ethanolic extract considerably lowered in all models. According to the findings, the extract significantly reduced ulcerative lesions, indicating protection ($p < 0.001$) in comparison to the animals in the control group [115]. When *T. ammi* was added in infusion, the gastric acid amount rose because it would cause the secretion of gastric acid to increase. *T. ammi* enhanced gastric acid output by about four times [114].

T. ammi decreased meal transit time in in-vivo experiments with rats, hence boosting the production of bile acids, and/or increasing digestive enzymes activity [116]. One of the studies found that the essence made from the seeds of *T. ammi* considerably decreased ($P < 0.05$) the contractions caused by the neurotransmitter acetylcholine (10^{-4} M). Upon adding the essence of *T. ammi* (0.01%) to the organ bath, it quickly began to have an inhibitory impact on acetylcholine-induced contraction; it reached its peak within 0.5 min and remained for at least 5 min. On the other hand, the same doses of *T. ammi* demonstrated a substantial antispasmodic activity [$P < 0.05$] with cumulative doses of 10^{-5}–10^{-2} M acetylcholine. The highest and lowest inhibitory effect on contraction of about 96.16% and 88.17% were elicited by concentrations of 0.01% and 0.002% *T. ammi*, respectively, in presence of 10^{-3} M acetylcholine with the statistically significant [$P < 0.05$] variations in the anti-spasmodic activity of both the concentrations [117].

2.1.10 *Trigonella foenum-graecum* Linn.

Trigonella foenum-graecum L. (Fabaceae) is also known as fenugreek (Fig. 2j). It has been a valuable spice for generations [118, 119]. One of the many chemical components of fenugreek is steroid sapogenins. It has been established that the fenugreek embryo, which is oily, contains diosgenin. In the stem alkaloids such as Nicotinic acid, trigocoumarin, trigonelline, and trimethyl coumarin are present. Seeds contain mucilage that stands out in specific [29]. The stem contains volatile oil, mucilage up to 28%, fixed oil of about 5% with a great aroma and a bitter taste, alkaloids namely choline and trigonelline, 22% proteins, and a substance that gives the stem its yellow color [120]. Fenugreek has about 58% carbohydrates, 23–26% protein, 6% fat, and 25% dietary fiber [121].

Fenugreek seeds have been found to lessen the GI adverse effects of tablets of metformin in people having diabetes type 2. Fenugreek seeds enhanced and controlled the intestinal microbiota, increased mucus secretion, and selectively stimulated intestinal bacteria. Fenugreek seeds work by decreasing the GI adverse effects of metformin tablets and increasing acceptance of them. This allows for sustained use of the pills as a treatment [122]. Fenugreek seed extract, oil, and powder have been found to hasten the healing of mucosal wounds in mice. In their research, mucilage-containing fenugreek seeds reduced intestinal and stomach swelling by creating a thin, supple covering on the fortifications of the affected organs [123, 124].

Similar to ranitidine, fenugreek seeds are used to treat gastroesophageal reflux [125]. It lessens bloating as well. The number of bacteria was reported to increase by polysaccharides isolated from fenugreek seeds, just like or even superior to inulin in behavior [126]. Fenugreek seeds have antiulcer properties. Seeds of fenugreek have a comparable effect as a popular proton pump inhibitor omeprazole, used to treat digestive problems such as gastroesophageal reflux disease, gastric and duodenal ulcers, and gastritis [127]. The hydro-extract and a fraction of gel extracted from seeds of fenugreek demonstrated notable ulcer-protective properties in a rat model of alcohol-provoked stomach ulcers, which are due to its actions on mucosal glycoproteins and anti-secretory activity. Additionally, fenugreek seed extract prevents alcohol-caused peroxidation of lipids and following mucosa damage likely through increasing the antioxidant capability of the stomach mucosa [127]. An aqueous trigonella seed extract has also been proven by researchers to have an antioxidant capacity [128], and it is well-known that giving antioxidants to rats can prevent ethanol-induced stomach damage [129]. Additionally, it has been discovered that soluble gel from fenugreek seeds prevents gastric lesion formation more effectively than omeprazole. The gastro-protective and anti-secretory properties of seeds of fenugreek may be caused by the polysaccharide content of the gel or the flavonoids [127]. This theory may be true because fenugreek seeds are acknowledged to have great quantities of flavonoids and it was proved that flavonoids guard the mucosa by reducing the growth of certain necrotic agent-induced lesions [130].

2.1.11 Zingiber officinale

Ginger (Family: Zingiberaceae) is a rhizome (Fig. 2k) that is utilized as a spice all over the world. Depending on the growth, harvesting, and processing conditions, this rhizome includes an extensive range of volatile and non-volatile chemicals in varying amounts [131]. Ginger is a crucial dietary component that has a carminative action, eases intestinal cramping, reduces strain on the lower esophageal sphincter [LES], and prevents bloating, gas, and dyspepsia [132–134]. It is a key ingredient in Asian medicine and has been utilized for a long time to treat a number of ailments, including digestive disorders like heartburn, gas, colic, vomiting, and diarrhea including pain in the abdomen, nausea, and cramping [135, 136]. The majority of the components of ginger that have been extracted by various analytical techniques are gingerols [137].

It improved the excretion of digestive enzymes. Clinical trials have demonstrated that ginger is just as active as dimenhydrinate [138] as well as being more efficient than vitamin B6 [139] in alleviating morning sickness and nausea during pregnancy. In addition, it has been suggested in the management of chemotherapy-related vomiting [140], gynecological laparoscopy [141], and postoperative preventive antiemetics [142]. Additionally, ginger is also suggested as a remedy for motion and sea sickness [143]. Ginger's exact mode of action is uncertain. It may, however, block receptors of serotonin and have antiemetic effects on the central nervous and gastrointestinal tract [144]. In a clinical trial, 100 mg ginger extract or 2 g of rhizome for 2 days was given and their effect on GI motility was examined [145]. The findings showed that the intervention group's gastrointestinal motility significantly improved when compared to the control group. Ginger has been demonstrated to quicken stomach emptying and increase antral contractions in healthy people [146]. In research, no variations were observed in the length of the fundus, GI symptoms, or the levels of the gut peptides like ghrelin, motilin, and GLP-1 of serum in individuals with functional dyspepsia [147].

2.2 The Herbal Formulations Used in Gastrointestinal Disorders

2.2.1 Triphala (India)

An Ayurvedic herbal formulation Triphala is made up of three herbs as shown in Table 2 [148, 149]. Tannins, ellagic acid, gallic acid, and chebulinic acid are their major constituents. These powerful antioxidants may be responsible for some of the formula's immunomodulatory effects [150–152]. Other bioactive substances found in Triphala include saponins, flavonoids like luteolin and quercetin, anthraquinones, fatty acids, amino acids, and several types of carbohydrates [151]. The human gut microbiome also converts triphala-derived polyphenols like chebulinic acid into active metabolites, possibly in vitro to reduce oxidative damage [153]. The usage of Triphala for gastrointestinal health, in general, is likely its most well-known benefit. Triphala extracts made from both aqueous and alcohol have been reported to stop diarrhea in animal trials [154]. Triphala also has enteroprotective properties, which are most likely brought on by its high antioxidant content, at least in part. In a mouse model, Triphala restored reduced protein, glutathione, and phospholipid amounts in the intestinal villi of the brush border while concurrently lowering myeloperoxidase and XO levels in the epithelium of the intestine [155]. Triphala had gastro-protective effects on stress-induced ulcers in rats [156]. Triphala was tested in a human clinical experiment on individuals with gastrointestinal diseases, and it was found to improve the occurrence, yield, and constancy of stools while reducing constipation, mucus, abdominal pain, hyperacidity, and flatulence [157]. The antioxidant properties and great quantities of flavonoids in Triphala were responsible for the healing effect, which also decreased colitis in mouse screening methods [158].

Triphala contains phytochemicals like quercetin and gallic acid that are known to encourage the evolution of Bifidobacteria and *L. bacillus* species by suppressing the

Table 2 Herbal formulations used in gastrointestinal disorders with their constituents

S.no	Formulation Name	Herbs used in the formulation	Indication	References
1	Triphala	Fruits of *Terminalia chebula* (family Combretaceae), *Terminalia bellerica* (family Combretaceae), and *Phyllanthus emblica* (Phyllanthaceae)	Appetite stimulation, gastric hyperacidity reduction, constipation, abdominal pain, and flatulence	[148, 149, 155]
2	STW-5 (IBEROGAST®)	*Iberis amara* (family Brassicaceae), Roots of *Angelica Sinensis* (family Apiaceae), flowers of *Matricaria chamomilla* L. (Family Asteraceae), fruit of *Carum carvi* L. (family Apiaceae), fruits of *Silybum marianum* (family Asteraceae), leaves of *Melissa officinalis* (family Lamiaceae), leaves of *Mentha x piperitae* (family Lamiaceae), *Chelidonium majus L.* (family Papaveraceae), and roots of *Glycyrrhiza glabra* (family Fabaceae)	Functional dyspepsia Irritable bowel syndrome	[166–171]
3	Peppermint oil (PMO)	*Mentha piperata* (family Lamiaceae)	Digestive tract smooth muscle relaxant Diarrhea Irritable bowel syndrome Antispasmodic	[172, 174, 177–179]
4	Zhizhu Kuanzhong	*Rhizome of Atractylodes macrocephala* (family Asteraceae), fruits of *Citrus aurantium* (family Rutaceae), roots of *Bupleurum* (family Apiaceae), and fruits of *Crataegus pinnatifida* (family Rosaceae)	Functional dyspepsia	[180, 181]
5	Rikkunshito RKT	Tuber of *Pinellia ternata* (family Araceae), roots of *Panax ginseng* (family Araliaceae), rhizome of *Atractylodis lanceae* (family Asteraceae) *Pachyma hoelen* (family Polyporaceae) *Aurantii nobilis pericarpium*	Dyspepsia Stress-induced gastric hypersensitivity Epigastric fullness, heartburn, burp, and nausea.	[184, 185, 194, 195]

(continued)

Table 2 (continued)

S.no	Formulation Name	Herbs used in the formulation	Indication	References
		(family Rutaceae), fruits of *Zizyphi fructus* (family Rhamnaceae)		
6	DA-9701	Seeds of *Pharbitis nil* (family Convolvulaceae) tubers of *Corydalis turtschaninovii* (family Papaveraceae)	Functional dyspepsia	[197, 198, 200–202]
7	Pudin Hara	Pudin Hara Pearls: leaves of *Mentha piperata* (family Lamiaceae) and *Mentha spicata* (family Lamiaceae) Pudin Hara Liquid: *Mentha piperata* (family Lamiaceae)	Indigestion, functional dyspepsia	[204]

formation of unwanted gut flora like *E. coli* [159–162]. Additionally, lactic acid bacteria have enzymes (such as tannase) that can break down plant tannins like the gallic acid in Triphala [163–165]. For instance, chebulinic acid, a polyphenol derived from Triphala, is metabolized by the human gut microbiota into products called urolithins that is able to fend off oxidative injury.

2.2.2 STW-5 (Iberogast®) (Germany)

A liquid concoction known as STW-5 is created by combining extracts from nine well-known herbs as indicated in Table 2, that were obtained using alcohol in a certain ratio. It is marketed in Europe as an over-the-counter drug and has been clinically used for many years in German-speaking nations [166, 167]. The concoction has distinctive ingredients, like extracts of well-known herbs produced by distinct extraction methods with stable quantities of components [167].

STW-5 demonstrated spasmolytic effects in an experiment using a guinea pig ileal muscle strip by decreasing acetylcholine and spasms brought on by histamine. Because of *Iberis amara* the basal resting tone is increased and the atonic segment is contracted. These double actions make it spasmolytic and tonic and were reliant on the basal tone of the gut. The duodenum, jejunum, and colon also experienced these effects [168].

315 individuals with functional dyspepsia as indicated using the Rome II criteria were included in a randomly, double-blind, placebo-controlled research. The human volunteers were randomly assigned to receive either a placebo or STW 5, 3×20 drops/day for an 8-week treatment after a washout period of a week. Six months were spent monitoring the patients. The gastrointestinal symptoms' (GIS's) evolution during the treatment period served as the primary endpoint. In both groups, the GIS improved. Following 4 and 8 weeks of treatment, the STW 5 group showed a

clinically meaningful improvement that was more than the placebo. Global evaluations by researchers and patients reaffirmed STW 5's appreciable superiority [169, 170].

The effectiveness and security of STW 5 in treating IBS were examined in one random, double-blind, and placebo-controlled research [171]. The test used a gastroenterological expert panel-developed abdominal symptom score, which contained eight specific IBS indications assessed on a 4-point Likert scale, as one of its major effectiveness endpoints. After a 1-week washout period, for four weeks, 300 patients were randomized to receive STW 5, two research medications, or a placebo. In comparison to the placebo, the IBS symptom score dropped by 1.5 points because of STW 5, which was clinically relevant and statistically significant ($p < 0.0004$). Similar results were found for the stomach pain, which was assessed using four-point Likert scales for each of the four abdominal quadrangles. STW 5 once more demonstrated a statistically significant advantage over placebo [171].

2.2.3 Peppermint Oil

Steam distillation is used to create peppermint oil (PMO) through plant leaves of *Mentha piperata* as indicated in Table 2 [172]. Its numerous constituents (>80) include 35–55% of menthol, 20–31% of menthone, menthyl acetate in 3–10%, cineol, and several other volatile oils [173–175]. Menthol, which naturally occurs as a pure stereoisomer, appears to be its main component and active element [174].

There are numerous lines of evidence that point to PMO's potential role as a digestive tract smooth muscle relaxant [174]. In guinea pig ileal smooth muscle, PMO and its ingredient menthol inhibit calcium channels, which aids in relaxing intestinal smooth muscle, according to in vitro studies [176]. A different investigation discovered that PMO greatly decreased the guinea pig coli's sensitivity to acetylcholine, 5-hydroxytryptamine, and histamine [99]. By limiting the uptake of extracellular calcium ions via voltage-dependent channels, they hypothesized that the PMO relaxes GI smooth muscle [99, 175]. PMO reduced stomach spasms when administered orally or topically [177, 178].

According to Rome III criteria, in a recent study, patients with mixed IBS or IBS with diarrhea participated in a randomized controlled experiment of 4 weeks to test the effectiveness and tolerability of a new PMO formulation intended to sustain the drug release within the small intestine. A solid-state customized enteric-coating matrix designed to deliver peppermint oil to the small intestine that is triple-coated with prolonged release and fewer potential side effects was used in their research. At the end of the experiment, the PMO group had a 40% reduction from baseline in the whole IBS scale scores when compared with the placebo group 24.3% ($p = 0.0246$). In addition, compared to the placebo, there was a superior perfection in numerous distinct GI symptoms and also in rigorous or intolerable symptoms [179].

2.2.4 Zhizhu Kuanzhong (China)

Since 2002, a proprietary fixed combination formulation known as Zhizhu Kuanzhong (ZZKZ) has been sold commercially in China [180]. Four herbs as indicated in Table 2 combine to make it. The primary ingredients Shan Zha and

Bai Zhu have been investigated in animal pharmacological investigations, and it has been found that they speed up intestinal propulsion and gastric emptying when compared to control treatments [181, 182]. Zhi Shi is frequently used as a TCM treatment for FD on its own. Zhi Shi demonstrated an inhibitory effect in in vitro tests on the impulsive retrenchment of the pyloric circular smooth muscle strip [183]. It has been demonstrated that chai hu has an anti-anxiety and antidepressant effect in addition to speeding up small intestinal transit time and gastric fluid emptying [180].

ZZKZ, a patented formula, enhanced intestinal propulsion and motility as well as stomach emptying in rats [180, 181]. ZZKZ was demonstrated to be highly efficacious than placebo for 392 patients meeting Rome III postprandial distress syndrome criteria when taken at a dose of 3 $\times$ 2 capsules daily for functional dyspepsia. This particularly enhanced satiety and fullness after meals. Since no serious side effects had been documented in any of the RCTs or post-marketing monitoring data, it suggests that this medication is safe [180, 181].

ZZKZ was shown to be non-inferior to cisapride in FD symptom control and gastric emptying in a limited Phase II, RCT with 403 FD patients [182]. In a study, St. John's wort and domperidone combined with ZZKZ at the same dosage of three capsules per day had comparable benefits on FD patients' scores for depression and anxiety [183].

2.2.5 Rikkunshito (RKT) (Japan)

As stated in Table 2 [184, 185], RKT is an extracted granule for oral use made up of a dried mixture of eight unprocessed herbs. RKT has been shown to improve GE, antral contractions, and fundic relaxation in addition to reducing GI symptoms like dyspepsia or reflux [186]. It has been proposed that each component of RKT exhibits synergistic effects through a complicated interacting route [187, 188].

By triggering the esophageal acid clearance mechanisms, RKT was able to minimize clinical symptoms and the amount of time that acid was present in the distal esophagus during 24-hour esophageal pH monitoring tests in a trial with eight children who had symptoms of gastroesophageal reflux [189]. In a pilot survey, RKT treatment for 8 weeks enhanced clearance of the esophagus by lessening LES residual pressure throughout swallowing and elevating the entire frequency of bolus transit and peristaltic contraction rate in esophageal multichannel impedance and manometry in 30 patients from PPI-refractory NERD [190]. Rikkunshito produced a relaxing effect on the stomach smooth muscles of a rat, which was brought about by the Ca2+-stimulated K+ channel being activated [191]. RKT is demonstrated to improve adaptive relaxation of the guinea pig stomach when it is isolated [192]. RKT demonstrated a high bile salt binding capacity and may guard against bile acid exposure harming the esophageal mucosa. Refractory GERD and duodenogastroesophageal reflux can both benefit from it [193]. RKT can improve stomach accommodation and reduce stress-induced gastric hypersensitivity [194]. In Japan, during a placebo-controlled, randomized trial with 42 patients having chronic idiopathic dyspepsia, the acetaminophen absorption technique revealed that RKT possesses prokinetic activity for enhancing gastric emptying, it also decreases upper gastrointestinal symptoms such as nausea, heartburn, epigastric fullness, and burp in comparison with placebo [195]. According to a randomized, parallel comparison experiment in Japan with 104 patients with GERD who were resistant to PPIs, the effectiveness of RKT with a standard dose of PPI on decreasing acid-related dysmotility symptom and reflux symptoms was comparable to that of a double dose of PPI [196].

2.2.6 DA-9701 (South Korea)

The Korean Food and Drug Administration approved the novel herbal medication DA-9701, which was created in South Korea, in May 2011 [197]. It contains ethanolic extracts of plants indicated in Table 2. These two herbs are extensively utilized as an herbal medication in Korea, China, and Japan since ancient times to treat gastrointestinal as well as gynecological disorders. Pharbitidis semen is employed as an intestinal peristalsis activator and an analgesic for the stomach [198]. Tuber of Corydalis is prescribed as an analgesic or an antispasmodic for the gastrointestinal tract [199]. The active ingredients in DA-9701 are corydaline and tetrahydropalmatine in Corydalis tuber and chlorogenic acid in Pharbitidis semen. The pharmacological investigation has shown that DA-9701 exhibits dopamine D2 antagonistic, 5-HT1A agonist, adrenergic a2 agonist, and 5-HT4 agonist action [197, 198].

According to the Rome II criterion, 462 Korean FD patients participated in a randomized, controlled, multi-center trial that revealed the effects of 30 mg DA9701 TID and 50 mg itopride hydrochloride TID on FD symptoms [200]. At 4 weeks, the itopride group had a responder rate of 36%, whereas the DA-9701 group had a responder rate of 37%. DA-9701 with itopride significantly decreased from baseline, and all consumer complaints related to gastrointestinal. In short research, FD patients according to Rome III definitions and the presence of H. pylori, the group having DA-9701 presented better indication recovery rates (73.3%) at 12 weeks as compared to the elimination group (60%). Due to the short sample size, there was no statistically significant difference [201]. In a subclass examination, DA-9701 greatly lowered the reflux symptom score in participants 65 years of age or older when compared to the placebo [202]. After taking 30 mg of DA-9701 TID for 24 days, 27 patients who satisfied Rome III's parameters for functional constipation showed improved involuntary movement of the bowel, stool form, and subjective symptoms of constipation [203].

2.2.7 Pudin Hara (India)

Two Ayurvedic preparations manufactured by Dabur India Ltd. called Pudin Hara Pearls and Pudin Hara Liquid include two Mentha species as shown in Table 2 and are intended for indigestion. Along with other ingredients, preservatives, and colors, each 180 mg of Pudin Hara Pearls contains 0.174 ml of Mentha oil obtained from the aerial part of *Mentha piperata*, and 0.034 ml of Spearmint oil (*Mentha spicata*, Aerial part, Oil). In addition to other ingredients, preservatives, and colors, each 1 ml or 20 drops of Liquid Pudin Hara contain 0.0337 ml of *Mentha piperata* oil.

One capsule of Pudin Hara Pearls (thrice a day) or ten drops of Liquid Pudin Hara was given to 79 individuals with functional dyspepsia after mealtimes for 5 days or until signs subsided, whichever came first. Based on improvements in indications such as nausea, acid regurgitation, heartburn, stomach discomfort, and upper abdominal bloating as well as the overall assessment of the investigator and patient therapeutic efficacy was determined from the baseline. Safety was evaluated based on the emergence of negative events and modifications to the hematological and biochemical profiles. Liquid Pudin Hara and Pearls of Pudin Hara dramatically reduced the symptoms of functional dyspepsia, including the above-mentioned conditions. Although no statistically significant variations were observed in the research products while comparing the effectiveness in treating the symptoms of dyspepsia Pudin Hara Pearls were more effective in treating bloating of the upper abdomen, heartburn, acid regurgitation, and stomach pain, while Liquid Pudin Hara was more effective at treating nausea and vomiting as well as the feeling of acidity. In any of the study subjects, no adverse events were recorded during the investigation. Pudin Hara Pearls and Pudin Hara Liquid could be used safely and effectively in treating functional dyspepsia [204].

3 Prospects and Challenges of Herbal Medicine Formulation

A single herbal remedy or medicinal plant may include hundreds of natural ingredients, and a polyherbal formulation has many times as many as one. Single active ingredient may practically be hard to identify in such a study, especially when a herbal product is a blend of two or more herbs [205]. Raw material quality is influenced by environmental factors, farming practices, and good collection techniques, such as plant selection and cultivation, in addition to intrinsic (genetic) aspects. It is challenging to conduct quality controls on the raw materials for herbal medicines due to a variety of factors. Correct identification of medicinal plant species, specific storage, and unique cleaning processes for diverse materials are significant requirements for the quality control of raw materials, according to good manufacturing practice (GMP). The main difficulties come in the quality control of final herbal medicines, particularly herbal mixtures [205]. As a result, compared to other pharmaceuticals, the overall specifications and procedures for quality control of final herbal products continue to be far more complicated. The World Health Organization (WHO) continues to support the implementation of quality assurance and control measures, such as National Quality Specifications and Standards for Herbal Materials, GMP, labeling, and licensing schemes for manufacturing, to assure the safety and efficacy of herbal medicines. The use of incorrect plant species, adulteration of herbal products, contamination, overdosage, misuse of herbal medications by consumers or healthcare professionals, and the combination of herbal medications with other medications are just a few of the causes of adverse effects resulting from the consumption of herbal medicines. Most manufacturers of herbal medicines are not properly informed about the value of taxonomic botany and documentation, and this presents specific obstacles during the identification and collection of medicinal plants used in herbal remedies. It is vital to use the most widely used binomial names for medicinal plants to avoid confusion caused by the common names. Botanists, phytochemists, pharmacologists, and other key players must work effectively together to monitor herbal medicine [205].

In human health care systems around the world, medicinal herbs have been crucial not only in treating illnesses but also as a potential resource for maintaining good health. The herbal industry has the potential to significantly impact the world economy. Future international labeling guidelines should effectively address quality issues given the rising use of herbal products. To comprehend the use of herbal medicines, standardization of methods and quality control data on safety and efficacy are required. The formation of businesses based on medicinal plants in undeveloped countries has been severely hampered by a lack of understanding of the potential social and economic benefits of doing so. Much research is being done on herbal medications to incorporate them into brand-new drug delivery systems [206]. Innovative approaches to reducing drug instability, like the use of suspension, biodegradable cellulose, therapeutic proteins, nanoparticles, and emulsifiers, can be especially helpful. Preventing drug deterioration brought on by environmental factors is a major issue when it comes to medication stability [207]. To maintain medication stability, certain procedures, including packing and container standards, must be followed. Pharmacokinetics and ADME research are essential for ensuring the efficacy, toxicity, and safety profile of herbal therapies because the side effects and toxicity of herbal treatments have been well-documented, including kidney damage, the development of stones, acute neuropathy, and newborn mortality. The validation of efficacy, safety, and toxicity levels occurs at the most crucial stage of developing herbal medications, which includes clinical trials and ethical considerations [207]. Such validation is necessary to compete with existing allopathic medicines and survive in the global medication market. For herbal medicine producers to maintain the quality of their products, competent authority regulations for quality and safety criteria are essential [208].

4 Conclusion

Numerous studies show that traditional people around the world frequently employ plant-based therapies or herbal medications to address digestive system problems. Different ethnic groups have employed a wide variety of herbs to cure different digestive issues from generation to generation. Herbal drugs are becoming more and more well-liked among Westerners because, when used properly, they have few negative effects. Recently, there has been a general trend toward a recovery of interest in a traditional healthcare system. People can treat a range of digestive-related ailments by regularly including herbal medicines in their diets, which can enhance how well the digestive system works. Numerous secondary metabolites or phytochemicals found in plants can have a part in the management and inhibition of certain gastrointestinal problems. These herbal treatments for digestion function in a variety of methods, counting encouraging effortless bowel movements, enhancing digestion, growing the incidence of bowel movements, purifying the body and eliminating contaminants, settling an upset stomach, and minimizing gas, bloating, and discomfort in digestion. Unfortunately, not enough research has been done on the vast majority of plant species utilized in indigenous traditional medicine in many underdeveloped nations. Triphala is a potent polyherbal blend with a wide range of medicinal benefits for preserving homeostasis as well as illness prevention and therapy. Numerous scientific investigations have documented evidence-based support for Triphala's many traditional uses. It may be possible to explain STW 5 (Iberogast) observed therapeutic effectiveness in treating hypotonic and spastic dysmotility symptoms of functional dyspepsia and irritable bowel syndrome, at least in part, by the pharmacological evidence that points to a dual-action principle. The practice of STW-5 as a prospective alternate medicine for functional gastrointestinal disorders, particularly functional dyspepsia and irritable bowel syndrome, will be made possible by further research on their involvement in the pathophysiology of FGIDs. Clinical experiments and in vitro investigations have shown that peppermint oil appears to reduce IBS indications, particularly stomach pain, by soothing smooth muscles in the stomach. It is necessary to have pertinent data derived from thorough studies to support the efficacy and security of PMO. Rats' intestinal propulsion and motility as well as stomach emptying were improved by the patented formula Zhizhu Kuanzhong. It was discovered to be more effective at treating FD when used alone or in combination with conventional Western therapy. RKT enhances stomach motor activity, together with gastric accommodation and emptying, and boosts esophageal clearance and motility, according to basic research and clinical investigations. DA-9701 increases stomach emptying and antral motility in healthy and sick situations, according to a number of preclinical and human investigations. Furthermore, it was asserted that DA-9701 might lessen constipation and GERD in the particular group. By using Pudin Hara Pearls and Pudin Hara Liquid, functional dyspepsia symptoms such as acid regurgitation, heartburn, a sense of acidity, stomach pain, and upper abdominal bloating were dramatically reduced. Both Pudin Hara Liquid and Pudin Hara Pearls may be used safely and successfully to treat functional dyspepsia. For the establishment of the scientific basis for activity and to more accurately assess the superiority, efficiency, and security of these formulations, ongoing examination of traditional plant medicines is necessary. Well-designed clinical trials will offer the required proof of efficacy to validate herbal medicines. In a small number of clinical trials, the safety and acceptability of conventional and herbal medicines used to treat digestive system illnesses have been examined. The findings generally demonstrate that there have been minimal side effects observed. Recent research on conventional plant-based therapeutics has produced evidence that supports additional studies in the hope that new treatments for digestive illnesses may be created.

Acknowledgments The Chancellor of R G S College of Pharmacy is acknowledged by the writers for his gracious assistance in giving the essential resources and inspiration for the work to be completed successfully.

References

1. Ley RE, Turnbaugh PJ, Klein S, Gordon JI (2006) Microbial ecology: human gut microbes associated with obesity. Nature 444:1022–1023
2. Cheng LK, O'Grady GDP, Egbuji JU, Windsor JA, Pullan AJ (2010) Gastrointestinal system. Wiley interdisciplinary reviews. Syst Biol Med 2:65–79
3. Neamsuvan O, Tuwaemaengae T, Bensulong F, Asae A, Mosamae K (2012) A survey of folk remedies for gastrointestinal tract diseases from Thailand's three southern border provinces. J Ethnopharmacol 144:11–21
4. Mathabe MC, Nikolova RV, Lall N, Nyazema NZ (2006) Antibacterial activities of medicinal plants used for the treatment of diarrhoea in Limpopo Province, South Africa. J Ethnopharmacol 105:286–293
5. Karki A, Tiwari BR (2007) Prevalence of acute diarrhoea in Kathmandu valley. JNMA J Nepal Med Assoc 46:175–179
6. Jeremy S (2022) Brief note on functional and structural gastrointestinal disorder. J Hepatol Gastroint Dis 8
7. 13 Most Common Gastrointestinal Conditions and What to Do About Them (2020) Available via DIALOG https://www.centurymedicaldental.com/13-most-common-gastrointestinal-conditions-and-what-to-do-about-them/
8. Salisbury BH, Terrell JM (2022) Antacids. In: StatPearls [Internet]. StatPearls Publishing, Treasure Island
9. Nugent CC, Falkson SR, Terrell JM (2022) H2 blockers. In: StatPearls [Internet]. StatPearls Publishing, Treasure Island
10. Yibirin M, De Oliveira D, Valera R, Plitt AE, Lutgen S (2021) Adverse effects associated with proton pump inhibitor use. Cureus 13:e12759
11. Biswas M, Singh KNM, Shetty YC, Koli PG, Ingawale S, Bhatia SJ (2019) Prescription pattern & adverse drug reactions of prokinetics. Indian J Med Res 149:748–754
12. World Health Organization (WHO) (1993) Summary of WHO guidelines for the assessment of herbal medicines. Herbal Gram 28:13–14
13. Cheema HS, Prakash O, Pal A, Khan F, Bawankule DU, Darokar MP (2014) Glabridin induces oxidative stress-mediated apoptosis-like cell death of malaria parasite *Plasmodium falciparum*. Parasitol Int 63:349–358
14. Kashyap S, Rao PB, Mishra P, Supriya (2019) Antioxidant potential and activity of aerial parts of eight medicinal plants of Uttarakhand, India. Bangladesh J Bot 48:265–270
15. Gaur R, Cheema HS, Yadav DK, Singh SP, Darokar MP, Khan F, Bhakuni RS (2015) In vitro antimalarial activity and molecular modeling studies of novel artemisinin derivatives. RSC Adv 5:47959–47974
16. Boniface PK, Verma S, Shukla A, Cheema HS, Srivastava SK, Khan F, Darokar MP, Pal A (2015) Bioactivity guided isolation of antiplasmodial constituents from *Conyza sumatrensis* (Retz.) E.H. Walker. Parasitol Int 64:118–123
17. Reynolds T, Dweck AC (1999) *Aloe vera* leaf gel: a review update. J Ethnopharmacol 68:3–37
18. Eshun K, He Q (2004) *Aloe vera*: a valuable ingredient for the food, pharmaceutical and cosmetic industries – a review. Crit Rev Food Sci Nutr 44:91–96
19. Boudreau MD, Beland FA (2006) An evaluation of the biological and toxicological properties of *Aloe barbadensis* (miller), *Aloe vera*. J Environ Sci Health C Environ Carcinog Ecotoxicol Rev 24:103–154
20. Tan Z, Li F, Xing J (2011) Separation and purification of *Aloe* anthraquinones using PEG/salt aqueous two-phase system. Sep Sci Technol 46:1503–1510
21. Mizuuchi Y, Shi SP, Wanibuchi K et al (2009) Novel type III polyketide synthases from Aloe arborescent. FEBS J 276:2391–2401
22. Rajasekaran S, Ravi K, Sivagnanam K, Subramanian S (2006) Beneficial effects of *Aloe vera* leaf gel extract on lipid profile status in rats with streptozotocin diabetes. Clin Exp Pharmacol Physiol 33:232–237

23. Babaee N, Zabihi E, Mohseni S, Moghadamnia AA (2012) Evaluation of the therapeutic effects of *Aloe vera* gel on minor recurrent aphthous stomatitis. Dent Res J (Isfahan) 9: 381–385
24. Chapman DD, Pittelli JJ (1974) Double-blind comparison of alophen with its components for cathartic effects. Curr Ther Res Clin Exp 16:817–820
25. de Witte P (1993) Metabolism and pharmacokinetics of anthranoids. Pharmacology 47:86–97
26. Ulbricht C, Armstrong J, Basch E et al (2007) An evidence-based systematic review of *Aloe vera* by the natural standard research collaboration. J Herb Pharmacother 7:279–323
27. Kang MC, Kim SY, Kim YT et al (2014) In vitro and in vivo antioxidant activities of polysaccharide purified from *Aloe vera* (*Aloe barbadensis*) gel. Carbohydr Polym 99: 365–371
28. Suboj P, Babykutty S, Valiyaparambil Gopi DR, Nair RS, Srinivas P, Gopala S (2012) Aloe emodin inhibits colon cancer cell migration/angiogenesis by downregulating MMP-2/9, RhoB and VEGF via reduced DNA binding activity of NF-kappaB. Eur J Pharm Sci 45:581–591
29. Khare CP (2004) Indian herbal remedies: rational western therapy, ayurvedic and other traditional usages, botany. Springer, Berlin
30. Pulliah T (2002) Medicinal plants in India. Regency Publications, New Delhi
31. Hornok L (1992) Cultivation and processing of medicinal plants. Wiley, New York
32. Sharma R (2004) Agrotechniques of medicinal plants. Daya Publishing House, New Delhi
33. Santos AG, Figueiredo AC, Lourenco PM, Barrosa JG, Pedro LG (2002) Hairy root cultures of *Anethum graveolens* (dill): establishment, growth, time-course study of their essential oil and its comparison with parent plant oils. Biotechnol Lett 24:1031–1036
34. Blank I, Grosch W (1991) Evaluation of potent odorants in dill seed and dill herb (*Anethum graveolens* L.) by aroma extract dilution analysis. J Food Sci 56:63–67
35. Bonnlander B, Winterhalter P (2000) 9-Hydroxypiperitone beta-D-glucopyranoside and other polar constituents from dill (*Anethum graveolens* L.) herb. J Agric Food Chem 48:4821–4825
36. Duke JA (2001) Handbook of medicinal herbs. CRC Press, London
37. Fleming T (2000) PDR for herbal medicines. Medical Economics Company, Montvale
38. Nair R, Chanda S (2007) Antibacterial activities of some medicinal plants of the western region of India. Turk J Biol 31:231–236
39. Hosseinzadeh H, Karimi GR, Ameri M (2002) Effects of *Anethum graveolens* L. seed extracts on experimental gastric irritation models in mice. BMC Pharmacol 2:21–25
40. Mahran GH, Kadry HA, Thabet CK, Al-Azizi M, Liv N (1992) GC/MS analysis of volatile oil of fruits of *Anethum graveolens*. Int J Pharmacog 30:139–144
41. Mohele B, Heller W, Wellmann E (1985) UV-induced biosynthesis of quercetin 3-o-beta-d-glucuronide in dill *Anethum graveolens* cell cultures. Phytochem 24:183–185
42. Naseri-Gharib MK, Heidari A (2007) Antispasmodic effect of *A. graveolens* fruit extract on rat ileum. Int J Pharm 3:260–264
43. Gocmanac-Marija I, Dusanka K, Milica R et al (2015) Spasmolytic effect of *Anethum graveolens* l. Methanol extract on isolated rat ileum. Acta Med Median 54:5–10
44. Salehi Surmaghi MH, Amin GR, Kaveh S (2002) *Carvi fructus*. In: Iranian herbal pharmacopeia scientific committee, 1st edn. Iranian Ministry of Health & Medical Education Publications, Tehran
45. Johri RK (2011) *Cuminum cyminum* and *Carum carvi*: an update. Pharmacogn Rev 5:63–72
46. Zheng GQ, Kenney PM, Lam LK (1992) Anethofuran, carvone, and limonene: potential cancer chemopreventive agents from dill weed oil and caraway oil. Planta Med 58:338–341
47. ESCOP (2003) Monographs: the scientific foundation for herbal medicinal products. Thieme, Stuttgart
48. Sachan AK, Das DR, Kumar M (2016) *Carum carvi*-an important medicinal plant. J Chem Pharm Res 8:529–533
49. Olennikov DN, Kashchenko NI (2014) Polysaccharides. Current state of knowledge: an experimental and scientometric investigation. Khimiya rastitel'nogo syr'ya 1:5–26

50. Ravid U, Putievsky E, Katzir I, Weinstein V, Ikan R (1992) Chiral GC analysis of (S)(+)- and (R)(−)-carvone with high enantiomeric purity in caraway, dill and spearmint oils. Flavour Fragr J 7:289–292
51. de Carvalho CCCR, da Fonseca MMR (2006) Carvone: why and how should one bother to produce this terpene. Food Chem 95:413–422
52. Forster HB, Niklas H, Lutz S (1980) Antispasmodic effects of some medicinal plants. Planta Med 40:309–319
53. Al-Essa MK, Shafagoj YA, Mohammed F, Afifi FU (2010) Relaxant effect of ethanol extract of *Carum carvi* on dispersed intestinal smooth muscle cells of the Guinea pig. Pharm Biol 48: 76–80
54. Hawrelak JA, Cattley T, Myers SP (2009) Essential oils in the treatment of intestinal dysbiosis: a preliminary in vitro study. Altern Med Rev 14:380–384
55. Khayyal MT, el-Ghazaly MA, Kenawy SA et al (2001) Antiulcerogenic effect of some gastrointestinally acting plant extracts and their combination. Arzneimittelforschung 51: 545–553
56. Chey WD, Lacy BE, Cash BD, Epstein M, Shah SM (2017) Randomized controlled trial to assess the efficacy & safety of caraway oil/L-menthol plus usual care polypharmacy vs. placebo plus usual care polypharmacy for functional dyspepsia. Gastroenterology 152: S306
57. Lauche R, Janzen A, Lüdtke R, Cramer H, Dobos G, Langhorst J (2015) Efficacy of caraway oil poultices in treating irritable bowel syndrome – a randomized controlled cross-over trial. Digestion 92:22–31
58. Bhandari PR, Kamdod MA (2012) *Emblica officinalis* (Amla): a review of potential therapeutic applications. Int J Green Pharm 6:257–269
59. Khan KH (2009) Roles of *Emblica officinalis* in medicine – a review. Bot Res Int 2:218–228
60. Khattak KF (2013) Proximate composition, phytochemical profile and free radical scavenging activity of radiation processed *Emblica officinalis*. Int Food Res J 20:1125–1131
61. Jain SK, Khurdiya DS (2004) Vitamin C enrichment of fruit juice based ready-to-serve beverages through blending of Indian gooseberry (*Emblica officinalis* Gaertn.) juice. Plant Foods Hum Nutr 59:63–66
62. Zhang LZ, Zhao WH, Guo YJ, Tu GZ, Lin S, Xin LG (2003) Studies on chemical constituents in fruits of Tibetan medicine *Phyllanthus emblica*. Zhongguo Zhong Yao Za Zhi 28:940–943
63. Rehman H, Yasin KA, Choudhary MA et al (2007) Studies on the chemical constituents of *Phyllanthus emblica*. Nat Prod Res 21:775–781
64. Singh E, Sharma S, Pareek A, Dwivedi J, Yadav S, Sharma S (2011) Phytochemistry, traditional uses and cancer chemopreventive activity of Amla (*Phyllanthus emblica*): the sustainer. J Appl Pharm Sci 2:176–183
65. Mehrotra S, Jamwal R, Shyam R et al (2011) Anti-helicobacter pylori and antioxidant properties of *Emblica officinalis* pulp extract: a potential source for therapeutic use against gastric ulcer. J Med Plants Res 5:2577–2583
66. Al-Rehaily AJ, Al-Howiriny TS, Al-Sohaibani MO, Rafatullah S (2002) Gastroprotective effects of "Amla" *Emblica officinalis* on in vivo test models in rats. Phytomedicine 9:515–522
67. Sidhu S, Pandhi P, Malhotra S, Vaiphei K, Khanduja KL (2011) Beneficial effects of *Emblica officinalis* in L-arginine-induced acute pancreatitis in rats. J Med Food 14:147–155
68. Deshmukh CD, Bantal V, Pawar A (2010) Protective effect of *Emblica officinalis* fruit extract on acetic acid-induced colitis in rats. J Herb Med Toxicol 4:25–29
69. Karkon Varnosfaderani S, Hashem-Dabaghian F, Amin G et al (2018) Efficacy and safety of amla (*Phyllanthus emblica* L.) in non-erosive reflux disease: a double-blind, randomized, placebo-controlled clinical trial. J Integr Med 16:126–131
70. Muckensturm B, Foechterlen D, Reduron JP, Danton P, Hildenbrand M (1997) Phytochemical and chemotaxonomic studies of *Foeniculum vulgare*. Biochem Syst Ecol 25:353–358

71. Barros L, Carvalho AM, Ferreira ICFR (2010) The nutritional composition of fennel (*Foeniculum vulgare*): shoots, leaves, stems and inflorescences. LWT- Food Sci Technol 43: 814–818
72. Agarwal R, Gupta SK, Agarwal SS, Srivastava S, Saxena R (2008) Oculohypotensive effects of Foeniculum vulgare in experimental models of glaucoma. Indian J Physiol Pharmacol 52:77–83
73. Faudale M, Viladomat F, Bastida J, Poli F, Codina C (2008) Antioxidant activity and phenolic composition of wild, edible, and medicinal fennel from different Mediterranean countries. J Agric Food Chem 56:1912–1920
74. Kunzemann J, Herrmann K (1977) Isolation and identification of flavon(ol)-O-glycosides in caraway (*Carum carvi* L.), fennel (*Foeniculum vulgare* Mill.), anise (*Pimpinella anisum* L.), and coriander (*Coriandrum sativum* L.), and of flavon-C-glycosides in anise – I. Phenolics of spices. Z Lebensm Unters Forsch 164:194–200
75. Parejo I, Jauregui O, Sánchez-Rabaneda F, Viladomat F, Bastida J, Codina C (2004) Separation and characterization of phenolic compounds in fennel (*Foeniculum vulgare*) using liquid chromatography-negative electrospray ionization tandem mass spectrometry. J Agric Food Chem 52:3679–3687
76. Nassar MI, Aboutabl EA, Makled YA, ElKhrisy EA, Osman AF (2010) Secondary metabolites and pharmacology of *Foeniculum vulgare* Mill. Subsp. Piperitum. Rev Latinoam de Química 38:103–112
77. Cherng J, Chiang W, Chiang L (2008) Immunomodulatory activities of common vegetables and spices of Umbelliferae and its related coumarins and flavonoids. Food Chem 106:944–950
78. Roby MHH, Sarhan MA, Selim KA, Khalel KI (2013) Antioxidant and antimicrobial activities of essential oil and extracts of fennel (*Foeniculum vulgare* L.) and chamomile (*Matricaria chamomilla* L.). Ind Crops Prod 44:437–445
79. Chakŭrski I, Matev M, Koĭchev A, Angelova I, Stefanov G (1981) Treatment of chronic colitis with an herbal combination of *Taraxacum officinale, Hipericum perforatum, Melissa officinalis, Calendula officinalis*, and *Foeniculum vulgare*. Intern Dis 20:51–54
80. Birdane FM, Cemek M, Birdane YO, Gülçin I, Büyükokuroğlu E (2007) Beneficial effects of vulgare ethanol-induced acute gastric mucosal injury in rat. World J Gastroenterol 13:607–611
81. Fiore C, Eisenhut M, Ragazzi E, Zanchin G, Armanini D (2005) A history of the therapeutic use of liquorice in Europe. J Ethnopharmacol 99:317–324
82. Hattori M, Sakamoto T, Kobashi K et al (1983) Metabolism of glycyrrhizin by human intestinal flora. Planta Med 48:38–42
83. Yamamoto K, Kakegawa H, Ueda H et al (1992) Gastric cytoprotective anti-ulcerogenic actions of hydroxyl chalcones in rats. Planta Med 58:389–393
84. Chandler RF (1985) Licorice, more than just a flavour. Can Pharm J 118:421–424
85. Doll R, Hill I, Hutton C et al (1962) Clinical trial of a triterpenoid liquorice compound in gastric and duodenal ulcer. Lancet 2:793–796
86. Wilson JA (1972) A comparison of carbenoxolone sodium and deglycyrrhizinated liquorice in the treatment of gastric ulcer in the ambulant patient. Br J Clin Pract 26:563–566
87. Morgan AG, Pacsoo C, McAdam WA (1985) Maintenance therapy. A two-year comparison between Caved-S and cimetidine treatment in the prevention of symptomatic gastric ulcer. Gut 26:599–602
88. Morgan AG, McAdam WA, Pacsoo C et al (1982) Comparison between cimetidine and Caved-S in the treatment of gastric ulceration, and subsequent maintenance therapy. Gut 23: 545–551
89. Turpie AG, Runcie J, Thomson TJ (1969) Clinical trial of deglycyrrhizinised liquorice in gastric ulcer. Gut 10:299–303
90. Rees WD, Rhodes J, Wright JE et al (1979) Effect of deglycyrrhizinated liquorice on gastric mucosal damage by aspirin. Scand J Gastroent 14:605–607
91. Kassir ZA (1985) Endoscopic controlled trial of four drug regimens in the treatment of chronic duodenal ulceration. Ir Med J 78:153–156

92. Tewari SN, Wilson AK (1973) Deglycyrrhizinated liquorice in duodenal ulcer. Practitioner 210:820–823
93. Rita P, Animesh DK (2011) An Updated overview of Peppermint (*Mentha Piperita* L.). Int Res J Pharm 2:1–10
94. Grigoleit HG, Grigoleit P (2005) Peppermint oil in irritable bowel syndrome. Phytomedicine 12:601–606
95. Loolaie M, Moasef N, Rasouli H, Adibi H (2017) Peppermint and its functionality: a review. Arch Clin Microbiol 4:54
96. Shah PP, D'Mello PM (2004) A review of medicinal uses and pharmacological effects of *Mentha piperita*. Nat Prod Rad 3:214–221
97. Dalvi SS, Nadkarni PM, Pardesi R, Gupta KC (1991) Effect of peppermint oil on gastric emptying in man: a preliminary study using a radiolabelled solid test meal. Indian J Physiol Pharmacol 35:212–214
98. Tate S (1997) Peppermint oil: a treatment for postoperative nausea. J Adv Nurs 26:543–549
99. Hill JM, Aronson OI (1991) The mechanism of action of peppermint oil on gastrointestinal smooth muscle an analysis using patch clamp electrophysiology and isolated tissue pharmacology in rabbit and Guinea pig. Gastroenterology 101:55–65
100. Somerville KW, Ellis WR, Whitten BH, Balfour TW, Bell GD (1985) Stones in the common bile duct: experience with medical dissolution therapy. Postgrad Med J 61:313–316
101. Verma A, Mogra R (2015) Psyllium (*Plantago ovata*) husk: a wonder food for good health. IJSR 4:1581–1585
102. Theuissen EAM (2008) Water soluble dietary fibers and cardiovascular disease. Physiol Behav 94:285–292
103. Purohit P, Rathore HS (2019) Isabgol: a herbal remedy. World J Pharm Res 8:579–585
104. Shabbir S (2019) Psyllium the hidden superfood of all times. J Nutraceuticals Food Sci 4:1–2
105. Garg P (2017) Psyllium husk should be taken at higher dose with sufficient water to maximize its efficacy. J Acad Nutr Diet 117(5):681
106. Chaplin MF, Chaudhury S, Dettmer PW, Sykes J, Shaw AD, Davies GJ (2000) Effect of ispaghula husk on the faecal output of bile acids in healthy volunteers. J Steroidal Biochem Mol Biol 72:283–292
107. Leeds AR (2009) Dietary fiber; role in nutrition management of disease. In: Caballero B (ed) Guide to nutritional supplements. Academic Press, USA
108. Ayurvedic Pharmacopoeia of India (1999–2011) Government of India, Ministry of Health and Family Welfare Department of Ayush
109. Ishikawah T, Sega Y, Kitajima J (2001) Water-soluble constituents of ajowan. Chem Pharm Bull 49:840–844
110. Chopra RN (1982) Chopra's indigenous drug of India, 2nd edn. Academic Publishers, Calcutta
111. Garg SN, Kumar S (1998) A new glucoside from *Trachyspermum ammi*. Fitoterapia 69: 511–512
112. Nagalakshmi S, Shankaracharya NB, Naik JP, Rao LJM (2000) Studies on chemical and technological aspects of ajowan (*Trachyspermum ammi* syn. *Carum copticum*). J Food Sci Technol 37:277–281
113. Choudhury S (1998) Composition of the seed oil of *Trachyspermum ammi* (L.) Sprague from Northeast India. J Essen Oil Res 10:588–590
114. Vasudevan K, Vembar S, Veeraraghavan K, Haranath PS (2000) Influence of intragastric perfusion of aqueous spice extracts on acid secretion in anesthetized albino rats. Indian J Gastroenterol 19:53–56
115. Ramaswamy S, Sengottuvelu S, Sherief S et al (2010) Gastroprotective activity of Ethanolic extract of *Trachyspermum Ammi* fruit. Int J Pharm Biosci 1:1–15
116. Patel K, Srinivasan K (2001) Studies on the influence of dietary spices on food transit time in experimental rats. Nutr Res 21:1309–1314

117. Gilani AH, Jabeen Q, Ghayur MN, Janbaz KH, Akhtar MS (2005) Studies on the antihypertensive, antispasmodic, bronchodilator and hepatoprotective activities of the *Carum copticum* seed extract. J Ethnopharmacol 98:127–135
118. Basch E, Ulbricht C, Kuo G, Szapary P, Smith M (2003) Therapeutic applications of fenugreek. Altern Med Rev 8:20–27
119. Parthasarathy VA, Kandinnan K, Srinivasan S (2008) Fenugreek in: organic spices. New India Publishing Agencies, New Delhi
120. Grieve M (1984) A modern herbal: the medicinal, culinary, cosmetic and economic properties, cultivation and folklore of herbs, grasses, fungi, shrubs and trees with all their modern scientific uses. Savvas Publishing
121. U.S. Department of Health and Human Services (2012) National Institutes of Health Website
122. Verma N, Usman K, Patel N et al (2016) A multicenter clinical study to determine the efficacy of a novel fenugreek seed (*Trigonella foenum-graecum*) extract (FenfuroTM) in patients with type 2 diabetes. Food Nutr Res 60:32382
123. Helmy H (2011) Study the effect of fenugreek seeds on gastric ulcers in experimental rats. World J Dairy Food Sci 6:152–158
124. Bhat BG, Sambaiah K, Chandrasekhara N (1985) The effect of feeding fenugreek and ginger on bile composition in the albino rat. Nutr Rep Int 32:1145–1151
125. DiSilvestro RA, Verbruggen MA, Offutt EJ (2011) Anti-heartburn effects of a fenugreek fiber product. Phytother Res 25:88–91
126. Haghshenas B, Nami Y, Haghshenas M et al (2015) Effect of addition of inulin and fenugreek on the survival of microencapsulated *Enterococcus durans* 39C in alginate-psyllium polymeric blends in simulated digestive system and yogurt. Asian J Pharm Sci 10:350–361
127. Pandian RS, Anuradha CV, Viswanathan P (2002) Gastroprotective effect of fenugreek seeds (*Trigonella foenum-graecum*) on experimental gastric ulcer in rats. J Ethnopharmacol 81: 393–397
128. Anuradha CV, Ravikumar P (1998) Anti-lipid peroxidative activity of seeds of fenugreek (*Trigonella foenum-graecum*). Med Sci Res 26:317–321
129. Ligumsky M, Sestieri M, Okon E, Ginsburg I (1995) Antioxidants inhibit ethanol-induced gastric injury in the rat. Role of manganese, glycine, and carotene. Scand J Gastroenterol 30: 854–860
130. Saurez J, Herrera MD, Marhuenda E (1996) Hesperidine and neohesperidin dihydrochalcone on different experimental models of induced gastric ulcer. Phytother Res 10:616–618
131. Haniadka R, Saldanha E, Sunita V, Palatty PL, Fayad R, Baliga MS (2013) A review of the gastroprotective effects of ginger (*Zingiber officinale* Roscoe). Food Funct 4:845–855
132. Ali BH, Blunden G, Tanira MO, Nemmar A (2008) Some phytochemical, pharmacological and toxicological properties of ginger (*Zingiber officinale* Roscoe): a review of recent research. Food Chem Toxicol 46:409–420
133. Chrubasik S, Pittler MH, Roufogalis BD (2005) Zingiberis rhizoma: a comprehensive review on the ginger effect and efficacy profiles. Phytomedicine 12:684–701
134. Lohsiriwat S, Rukkiat M, Chaikomin R, Leelakusolvong S (2010) Effect of ginger on lower esophageal sphincter pressure. J Med Assoc Thai 93:366–372
135. Johns Cupp M (2000) Toxicology and clinical pharmacology of herbal products. Humana Press, Totowa
136. Capasso F, Gaginella TS, Grandolini G, Izzo AA (2003) Phytotherapy. A quick reference to herbal medicine. Springer, Heidelberg
137. Koh EM, Kim HJ, Kim S et al (2009) Modulation of macrophage functions by compounds isolated from Zingiber officinale. Planta Med 75:148–151
138. Pongrojpaw D, Somprasit C, Chanthasenanont A (2007) A randomized comparison of ginger and dimenhydrinate in the treatment of nausea and vomiting in pregnancy. J Med Assoc Thail 90:1703–1709

139. Chittumma P, Kaewkiattikun K, Wiriyasiriwach B (2007) Comparison of the effectiveness of ginger and vitamin B6 for treatment of nausea and vomiting in early pregnancy: a randomized double-blind controlled trial. J Med Assoc Thail 90:15–20
140. Sharma SS, Gupta YK (1998) Reversal of cisplatin-induced delay in gastric emptying in rats by ginger (*Zingiber officinale*). J Ethnopharmacol 62:49–55
141. Chaiyakunapruk N, Kitikannakorn N, Nathisuwan S, Leeprakobboon K, Leelasettagool C (2006) The efficacy of ginger for the prevention of postoperative nausea and vomiting: a meta-analysis. Am J Obstet Gynecol 194:95–99
142. Phillips S, Ruggier R, Hutchinson SE (1993) *Zingiber officinale* (ginger)-an antiemetic for day case surgery. Anaesthesia 48:715–717
143. Mowrey DB, Clayson DE (1982) Motion sickness, ginger, and psychophysics. Lancet 1: 655–657
144. DerMarderosian A, Beutler JA (2006) The review of natural products. Wolters Kluwer, St. Louis
145. Giacosa A, Morazzoni P, Bombardelli E, Riva A, Bianchi Porro G, Rondanelli M (2015) Can nausea and vomiting be treated with ginger extract? Eur Rev Med Pharmacol Sci 19: 1291–1296
146. Wu KL, Rayner CK, Chuah SK et al (2008) Effects of ginger on gastric emptying and motility in healthy humans. Eur J Gastroenterol Hepatol 20:436–440
147. Hu ML, Rayner CK, Wu KL et al (2011) Effect of ginger on gastric motility and symptoms of functional dyspepsia. World J Gastroenterol 17:105–110
148. Rani B, Prasad M, Kumar R, Vikram Y, Kachhawa GR, Sharma S (2013) Triphala: a versatile counteractive assortment of ailments. Int J Pharm Chem Sci 2:101–109
149. Kumar NS, Nair AS, Nair AM, Murali M (2016) Pharmacological and therapeutic effects of Triphala – a literature review. J Pharmacogn Phytochem 23:23–27
150. Lu K, Chakroborty D, Sarkar C et al (2012) Triphala and its active constituent chebulinic acid are natural inhibitors of vascular endothelial growth factor-a mediated angiogenesis. PLoS One 7:e43934
151. Belapurkar P, Goyal P, Tiwari BP (2014) Immunomodulatory effects of Triphala and its individual constituents: a review. Indian J Pharm Sci 76:467–475
152. Lee HS, Won NH, Kim KH, Lee H, Jun W, Lee KW (2005) Antioxidant effects of aqueous extract of *Terminalia chebula* in vivo and in vitro. Biol Pharm Bull 28:1639–1644
153. Olennikov DN, Kashchenko NI, Chirikova NK (2015) In vitro bioaccessibility, human gut microbiota metabolites and hepatoprotective potential of chebulic ellagitannins: a case of Padma Hepaten® formulation. Nutrients 7:8456–8477
154. Biradar YS, Singh R, Sharma K, Dhalwal K, Bodhankar SL, Khandelwal KR (2007) Evaluation of anti-diarrhoeal property and acute toxicity of Triphala Mashi, an Ayurvedic formulation. J Herb Pharmacother 7:203–212
155. Nariya M, Shukla V, Jain S, Ravishankar B (2009) Comparison of enteroprotective efficacy of Triphala formulations (Indian herbal drug) on methotrexate-induced small intestinal damage in rats. Phytother Res 23:1092–1098
156. Nariya MB, Shukla VJ, Ravishankar B, Jain SM (2011) Comparison of gastroprotective effects of Triphala formulations on stress-induced ulcer in rats. Indian J Pharm Sci 73:682–687
157. Pulok K, Mukherjee SR, Bhattacharyya S et al (2005) Clinical study of 'Triphala' – a well-known phytomedicine from India. Iran J Pharmacol Ther 5:51–54
158. Rayudu V, Raju AB (2014) Effect of Triphala on dextran sulphate sodium-induced colitis in rats. Ayu 35:333–338
159. Carlsen MH, Halvorsen BL, Holte K et al (2010) The total antioxidant content of more than 3100 foods, beverages, spices, herbs and supplements used worldwide. Nutr J 9:1–11
160. Yadav S, Gite S, Nilegaonkar S, Agte V (2011) Effect of supplementation of micronutrients and phytochemicals to fructooligosaccharides on growth response of probiotics and *E. coli*. Biofactors 37:58–64

161. Tabasco R, Sánchez-Patán F, Monagas M et al (2011) Effect of grape polyphenols on lactic acid bacteria and bifidobacteria growth: resistance and metabolism. Food Microbiol 28: 1345–1352
162. Boto-Ordóñez M, Urpi-Sarda M, Queipo-Ortuño MI, Tulipani S, Tinahones FJ, Andres-Lacueva C (2014) High levels of Bifidobacteria are associated with increased levels of anthocyanin microbial metabolites: a randomized clinical trial. Food Funct 5:1932–1938
163. Jimenez N, Esteban-Torres M, Mancheno JM et al (2014) Tannin degradation by a novel tannase enzyme present in some *Lactobacillus plantarum* strains. Appl Environ Microbiol 80: 2991–2997
164. Matoba Y, Tanaka N, Noda M, Higashikawa F, Kumagai T, Sugiyama M (2013) Crystallographic and mutational analyses of tannase from *Lactobacillus plantarum*. Proteins 8: 2052–2058
165. Jimenez N, Curiel JA, Reveron I et al (2013) Uncovering the *Lactobacillus plantarum* WCFS1 gallate decarboxylase involved in tannin degradation. Appl Environ Microbiol 79: 4253–4263
166. Malfertheiner P (2017) STW 5 (Iberogast) therapy in gastrointestinal functional disorders. Dig Dis:3525–3529
167. Pilichiewicz AN, Horowitz M, Russo A et al (2007) Effects of Iberogast on proximal gastric volume, antropyloroduodenal motility and gastric emptying in healthy men. Am J Gastroenterol 102:1276–1283
168. Ammon HP, Kelber O, Okpanyi SN (2006) Spasmolytic and tonic effect of Iberogast (STW 5) in intestinal smooth muscle. Phytomedicine 13:67–74
169. von Arnim U, Peitz U, Vinson B, Gundermann KJ, Malfertheiner P (2007) STW 5, a phytopharmacon for patients with functional dyspepsia: results of a multicenter, placebo-controlled double-blind study. Am J Gastroenterol 102:1268–1275
170. von Arnim U, Vinson BR, Malfertheiner P et al (2008) Functional dyspepsia: are relapse rates influenced by active treatment. Gastroenterology:134
171. Madisch A, Holtmann G, Plein K (2004) Treatment of irritable bowel syndrome with herbal preparations: results of a double-blind, randomized, placebo-controlled, multi-centre trial. Aliment Pharmacol Ther 19:271–279
172. Kearns GL, Chumpitazi BP, Abdel-Rahman SM, Garg U, Shulman RJ (2015) Systemic exposure to menthol following administration of peppermint oil to pediatric patients. BMJ Open 5:e008375
173. Haber SL, El-Ibiary SY (2016) Peppermint oil for treatment of irritable bowel syndrome. Am J Health Syst Pharm 73:22–26
174. Grigoleit HG, Grigoleit P (2005) Pharmacology and preclinical pharmacokinetics of peppermint oil. Phytomedicine 12:612–616
175. Blumenthal M, Goldberg A, Brinckmann J (2000) Herbal medicine: expanded commission E monographs. Integrative Medicine Communications, Boston
176. Hawthorn M, Ferrante J, Luchowski E, Rutledge A, Wei XY, Triggle DJ (1988) The actions of peppermint oil and menthol on calcium channel dependent processes in intestinal, neuronal and cardiac preparations. Aliment Pharmacol Ther 2:101–118
177. Mizuno S, Kato K, Ono Y et al (2006) Oral peppermint oil is a useful antispasmodic for double-contrast barium meal examination. J Gastroenterol Hepatol 21:1297–1301
178. Imagawa A, Hata H, Nakatsu M et al (2012) Peppermint oil solution is useful as an antispasmodic drug for esophagogastroduodenoscopy, especially for elderly patients. Dig Dis Sci 57: 2379–2384
179. Cash BD, Epstein MS, Shah SM (2016) A novel delivery system of peppermint oil is an effective therapy for irritable bowel syndrome symptoms. Dig Dis Sci 61:560–571
180. Xiao Y, Li Y, Shu J et al (2019) The efficacy of oral Zhizhu Kuanzhong, a traditional Chinese medicine, in patients with postprandial distress syndrome. J Gastroenterol Hepatol 34: 526–531

181. Wen M, Zhang F, Wang Y (2019) Effect of Zhizhu Kuanzhong capsules on treatment of functional dyspepsia: a meta-analysis of randomized controlled trials. Chin J Integr Med 25: 625–630
182. Wei Z, Ai L, Chen X et al (2019) Comparative studies on the regulatory effects of raw and charred hawthorn on functional dyspepsia and intestinal flora. Trop J Pharm Res 18:333–339
183. Wu Z, Zhang S, Li P, Lu X, Wang J, Zhao L, Wang Y (2016) Effect of *Aurantii Fructus* Immaturus Flavonoidon the contraction of isolated gastric smooth muscle strips in rats. Evid Based Complement Alternat Med 2016:1–7
184. Cremonini F (2014) Standardized herbal treatments on functional bowel disorders: moving from putative mechanisms of action to controlled clinical trials. Neurogastroenterol Motil 26: 893–900
185. Tominaga K, KatoM TH et al (2014) A randomized, placebo-controlled, double-blind clinical trial of rikkunshito for patients with non-erosive reflux disease refractory to proton-pump inhibitor: the G-PRIDE study. J Gastroenterol 49:1392–1405
186. Kusunoki H, Haruma K, Hata J et al (2010) Efficacy of Rikkunshito, a traditional Japanese medicine (Kampo), in treating functional dyspepsia. Intern Med 49:2195–2202
187. Tominaga K, Kido T, Ochi M et al (2011) The traditional Japanese medicine rikkunshito promotes gastric emptying via the antagonistic action of the 5-HT(3) receptor pathway in rats. Evid Based Complement Alternat Med 2011:248481
188. Kitagawa H, Munekage M, Matsumoto T et al (2015) Pharmacokinetic profiles of active ingredients and its metabolites derived from rikkunshito, a ghrelin enhancer, in healthy Japanese volunteers: a cross-over, randomized study. PLoS One 10:e0133159
189. Kawahara H, Kubota A, Hasegawa T et al (2007) Effects of rikkunshito on the clinical symptoms and esophageal acid exposure in children with symptomatic gastroesophageal reflux. Pediatr Surg Int 23:1001–1005
190. Odaka T, Yamato S, Yokosuka O (2017) Esophageal motility and rikkunshito treatment for proton pump inhibitor-refractory non-erosive reflux disease: a prospective, uncontrolled, open-label pilot study trial. Curr Ther Res Clin Exp 84:37–41
191. Kito Y, Suzuki H (2010) Properties of Rikkunshi-to (TJ-43)-induced relaxation of rat gastric fundus smooth muscles. Am J Physiol Gastrointest Liver Physiol 298:G755–G763
192. Hayakawa T, Arakawa T, Kase Y et al (1999) Liu- Jun-Zi-tang, a kampo medicine, promotes adaptive relaxation in isolated Guinea pig stomachs. Drugs Exp Clin Res 25:211–218
193. Araki Y, Mukaisho KI, Fujiyama Y, Hattori T, Sugihara H (2012) The herbal medicine rikkunshito exhibits strong and differential adsorption properties for bile salts. Exp Ther Med 3:645–649
194. Shiratori M, Shoji T, Kanazawa M et al (2011) Effect of rikkunshito on gastric sensorimotor function under distention. Neurogastroenterol Motil 23:323–329, 155–156
195. Tatsuta M, Iishi H (1993) Effect of treatment with liu-jun-zi-tang (TJ-43) on gastric emptying and gastrointestinal symptoms in dyspeptic patients. Aliment Pharmacol Ther 7:459–462
196. Tominaga K, Iwakiri R, Fujimoto K et al (2012) Rikkunshito improves symptoms in PPI-refractory GERD patients: a prospective, randomized, multicenter trial in Japan. J Gastroenterol 47:284–292
197. Kwon YS, Son M (2013) DA-9701: a new multi-acting drug for the treatment of functional dyspepsia. Biomol Ther 21:181–189
198. Jin M, Son M (2018) DA-9701 (Motilitone): a multi-targeting botanical drug for the treatment of functional dyspepsia. Int J Mol Sci 19:4035
199. Jung JW, Kim JM, Jeong JS et al (2014) Pharmacokinetics of chlorogenic acid and corydaline in DA-9701, a new botanical gastroprokinetic agent, in rats. Xenobiotica 44:635–643
200. Choi MG, Rhee PL, Park H et al (2015) Randomized, controlled, multicenter trial: comparing the safety and efficacy of DA-9701 and itopride hydrochloride in patients with functional dyspepsia. J Neurogastroenterol Motil 21:414–422

201. Park JY, Kim JG, Hong SJ et al (2019) A randomized double-blind comparative study of the efficacy of helicobacter pylori eradication therapy and motilitone® for functional dyspepsia. Korean J Helicobacter Up Gastrointest Res 19:106–114
202. Park CH, Kim HS, Lee SK (2014) Effects of the new prokinetic agent DA-9701 formulated with corydalis tuber and pharbitis seed in patients with minimal change esophagitis: a bi-center, randomized, double-blind, placebo-controlled study. J Neurogastroenterol Motil 20:338–346
203. Kim SY, Woo HS, Kim KO et al (2017) DA-9701 improves colonic transit time and symptoms in patients with functional constipation: a prospective study. J Gastroenterol Hepatol 32: 1943–1948
204. Sastry JLN, Vats A, Vedula S, Kumar S (2016) Symptomatic management of functional dyspepsia: evaluation of efficacy and safety of pudin hara pearls and pudin hara liquid. Int J Res Ayurveda Pharm 7:65–69
205. Bhardwaj S, Verma R, Gupta J (2018) Challenges and future prospects of herbal medicine. Int Res Med Health Sci 1:12–15
206. Saggar S, Mir PA, Kumar N, Chawla A, Uppal J, Shilpa, Kaur A (2022) Traditional and herbal medicines: opportunities and challenges. Pharmacogn Res 14:107–114
207. Prabhakar P, Mamoni B (2021) Technical problems, regulatory and market challenges in bringing herbal drug into mainstream of modern medicinal practices. Res J Biotechnol 16:3
208. Lee JY, Jun SA, Hong SS, Ahn YC, Lee DS, Son CG (2016) Systematic review of adverse effects from herbal drugs reported in randomized controlled trials. Phytother Res 30: 1412–1419

Herbal Medicine and Pregnancy 23

Priyanka Devi and Prasann Kumar

Contents

Abstract

The usage of herbal medications is increasingly commonplace worldwide, and this trend is also noticeable among pregnant women. Pregnant women may use herbal medicine to improve their general health. Although self-diagnosis and self-dosing are common in herbal medicine, they are unsafe and highly discouraged. It is advisable to use herbs for pregnancy management only under the guidance of a midwife, physician, herbalist, naturopathic, or homeopathic doctor because each pregnancy is unique. This chapter discusses the use of herbal remedies during pregnancy and the types of herbal remedies used. Herbal medicines include herbs, herbal preparations, and finished products made from herbs, and they contain active compounds from plants with known or suspected therapeutic advantages.

P. Devi · P. Kumar (✉)
Department of Agronomy, School of Agriculture, Lovely Professional University, Phagwara, Punjab, India

S. C. Izah et al. (eds.), *Herbal Medicine Phytochemistry*, Reference Series in Phytochemistry, https://doi.org/10.1007/978-3-031-43199-9_25

To ensure the best outcome for pregnant women and their babies, herbal medicine products should only be recommended for use by individuals who are well-versed in the types of herbs, their parts (e.g., roots, leaves, etc.), dosage, and how they could be used (e.g., capsules, tonics, teas) before consuming herbs during pregnancy. Most pregnant women who use herbal medicine conceal their usage from their attending conventional medical professionals. However, most women were encouraged by their families and friends to utilize herbal medications because are considered safer, more effective, and had fewer adverse side effects than modern medicine. During pregnancy, herbs that are considered safe to use are usually those used for food and other related purposes. The most common form of these supplements is a tablet, a tea, or an infusion.

Keywords

Pregnancy · Tonics · Baby · Roots · Leaves · Herbs · Naturopathic · No poverty · Zero hunger

Abbreviations

ACOG	American College of Obstetricians and Gynecologists
AEDs	Antiepileptic drugs
FDA	Food and Drug Administration
HMs	Herbal medicine
UDP	Glucuronosyltransferase

1 Introduction

In many societies across the globe, the use of medicinal plants to promote a healthy pregnancy and facilitate a smooth delivery has been a long-standing tradition. In the past three decades, there has been a noticeable surge in the availability of comprehensive literature on plant-based therapies in Western countries. Despite the medical community lacking substantial evidence to deem homeopathic remedies unsafe during pregnancy, this perception still endures. It is worth noting, however, that most natural remedies employed during pregnancy and labor have not undergone rigorous scientific scrutiny. While anecdotal evidence may offer some support for their use, it is imperative to establish scientific validation regarding their safety and efficacy.

Nevertheless, women in North America and elsewhere continue to rely on botanical remedies they have come to know through word of mouth or natural birthing literature. This study draws from a range of sources, including HerbClips, a resource provided by the American Botanical Council, and various herbal and pregnancy literature. These sources shed light on the subject [16].

Pregnancy triggers a series of significant physiological changes in a woman's body, often leading to a variety of discomforts such as nausea, vomiting, indigestion, and diarrhea. It is a common response for expectant women, including the use of natural remedies, to self-medicate to alleviate these conditions [54]. This trend has contributed to the substantial increase in the global utilization of natural remedies by pregnant women [14, 54].

The prevalence of natural medication use during pregnancy is influenced by various factors, including geographic region, ethnicity, cultural practices, and socio-economic status. To investigate this phenomenon, researchers conducted a cross-sectional study involving 9459 pregnant women across 23 countries in Europe, encompassing Western Europe, Northern Europe, Eastern Europe, North and South Americas, and Australia. Among the participants, 28.9% reported the use of herbal medicines during their pregnancies. Notably, the countries with the highest reported usage of natural medications were Russia (60.0%), Poland (49.8%), and Australia (43.8%) [57].

Furthermore, Hwang et al. [37] found that countries like Iran (49.2%), Egypt (27.31%), Bangladesh (70%), Iraq (53.7%), Palestine (40%), and Taiwan (33.6%) also exhibited a substantial prevalence of natural medication use among pregnant women.

While previous studies indicate the widespread use of natural medicines during pregnancy in sub-Saharan Africa [6], there remains a scarcity of data on the prevalence of herbal medicine usage among expectant women in numerous sub-Saharan African nations. This study provides insights into the reasons, potential adverse effects, and effectiveness of the most commonly used herbal medicines during pregnancy in sub-Saharan Africa and the varying frequencies of herbal drug utilization across different countries in the region.

The utilization of herbal medicine (HM) during pregnancy is a common practice, with the majority of users being female, and many continue using HMs even during pregnancy [24]. However, the prevalence of HM use during pregnancy can vary widely depending on factors such as region, race, culture, and socioeconomic status, ranging from 7% to 55% [24, 114]. For example, in Australia, 34% of pregnant women use HMs [27], while in the European Union, the figure is as high as 50% [36, 62]. In contrast, in the United States and Canada, 6–9% of pregnant women use HMs [64].

Pregnant women often turn to herbal products to support their health and address non-life-threatening conditions, viewing them as a safe and natural alternative to conventional medications, particularly for issues like nausea and constipation [27, 38]. Notably, some HMs, such as black cohosh (*Actaea racemosa*) and blue cohosh (*Caulophyllum thalictroides*), have a well-documented history of use as uterine tonics and labor inducers [27]. However, despite reports of associated risks [109], there remains a lack of comprehensive information regarding their safety and efficacy during pregnancy.

One key issue is the widespread availability of HMs without needing a prescription in many countries. This can lead to challenges when pregnant women self-prescribe HMs alongside traditional medications recommended by their healthcare providers [99]. Patients may not always disclose their use of HMs to their healthcare providers, often because they do not perceive them as potentially hazardous substances [1].

This chapter serves as a comprehensive exploration of the multifaceted landscape of herbal medicine use during pregnancy. Its primary objective is providing valuable education to medical practitioners and expectant mothers. It seeks to shed light on the prevalence of herbal medicine use among pregnant women and, more crucially, to highlight the potential interactions between herbal medicines (HMs) and conventional medications employed to manage underlying medical conditions or pregnancy-related issues.

Physiological changes that occur during pregnancy can trigger or exacerbate various medical conditions. For instance, hypertensive disorders, encompassing chronic hypertension, fetal hypertension, and preeclampsia, contribute to approximately 10% of pregnancy-related complications in the United States. Some expectant mothers grapple with preexisting conditions like epilepsy or asthma, with approximately 9% reporting a history of asthma. Moreover, the use of antiepileptic drugs (AEDs) during pregnancy has been associated with potential risks, including adverse effects like miscarriage, antepartum, and postpartum hemorrhage.

Within this chapter, we delve into the demographics of pregnant herbal medicine users and explore the risks associated with herbal medicine usage during pregnancy. By highlighting the key medical conditions that herbal remedies aim to address, we aim to provide a comprehensive understanding of the nuances surrounding this practice.

In essence, the use of herbal medicine during pregnancy is a topic that requires thoughtful consideration, well-informed decision-making, and the active engagement of healthcare professionals. By examining prevalence, risks, and interactions with conventional medications, we endeavor to equip both medical practitioners and expectant mothers with the knowledge necessary to make informed choices regarding herbal medicine use during pregnancy. This knowledge ultimately plays a pivotal role in ensuring the best possible outcomes for both mother and child.

2 Navigating Pregnancy: A Guide to Herbal Medicine and Medicinal Plants

Pregnancy is a remarkable period marked by profound physical and emotional transformations as women's bodies adapt to nurture and develop a new life. Pregnancy is a profound and intricate life-altering journey, encompassing a multitude of physical, emotional, and social transformations. This process typically spans 9 months and culminates in the birth of a child. This journey often brings about a host of typical symptoms, such as morning sickness, sleep disturbances, and heartburn. In response to these challenges, many individuals turn to herbal remedies and medicinal plants as an alternative or complementary approach to conventional medical care.

The various facets of the pregnancy voyage are summarized in Table 1. It's imperative to acknowledge each pregnancy's uniqueness and its individualized experiences. Consultation with healthcare professionals is indispensable for

Table 1 Key stages and aspects in pregnancy

Topic	Key points
Conception	Pregnancy begins with sperm fertilizing an egg.
	Zygote forms in the fallopian tube.
	Zygote implants in the uterine wall.
Prenatal development	Zygote evolves into an embryo.
	Divided into three trimesters.
	Each trimester is marked by changes and milestones.
Pregnancy symptoms	Includes morning sickness, fatigue, mood swings, and appetite changes.
	Symptoms can vary among individuals.
Prenatal care	Essential for a healthy pregnancy.
	Involves routine check-ups, prenatal vitamins, and screening for complications.
Diet and nutrition	Balanced diet crucial for fetal development.
	Avoidance of harmful substances like alcohol and specific medications.
Physical changes	Pregnancy results in weight gain, breast modifications, and abdominal expansion.
	Hormonal changes can lead to skin alterations like stretch marks.
Emotional changes	Pregnancy-related emotional adjustments due to hormonal shifts and anticipation of parenthood.
	Effective communication and emotional support are vital.
Maternal health	Monitoring the health of the expectant mother is paramount.
	Conditions like preeclampsia and gestational diabetes may require specific care.
Labor and delivery	Culmination of pregnancy with uterine contractions leading to birth.
	Various delivery methods, including vaginal birth and cesarean section.
Postpartum period	A phase of physical and emotional recuperation post-childbirth.
	Involves caring for the newborn and adapting to parenthood.
Challenges	Some pregnancies may face complications like ectopic pregnancies, miscarriages, or preterm birth.
	Awareness and timely medical intervention are crucial.
Parenting and childcare	Pregnancy initiates the lifelong journey of parenthood.
	Considerations include childcare, parenting approaches, and the well-being of both child and parents.
Support system	Quality of pregnancy and postpartum period depends on a strong support system including partners, family, friends, and healthcare professionals.
Cultural and societal factors	Cultural and societal factors influence an individual's pregnancy experience.
	Wide variations exist across different geographic regions and communities.
Fertility and reproductive health	Vital concepts for those trying to conceive.
	Infertility concerns and assisted reproduction options warrant discussion.

(continued)

Table 1 (continued)

Topic	Key points
Family planning	Decisions about family size and birth spacing significantly impact pregnancy discussions.
	Timing and methods of contraception are part of family planning.
Legal and ethical issues	Complex topics like surrogacy, adoption, and abortion are discussed in the context of pregnancy.
	These matters involve moral, legal, and psychological considerations.

personalized guidance and addressing specific concerns or medical conditions during pregnancy. Additionally, emotional support and open communication are indispensable components for expectant parents to navigate this transformative journey effectively.

Herbal medicine, an ancient practice deeply ingrained in human culture, involves using plants and their derivatives for therapeutic purposes [7, 8, 48]. Medicinal plants have been significant in various cultural traditions, serving as remedies for various health conditions [40–47]. The use of medicinal herbs and plants during pregnancy has garnered considerable attention due to the potential relief they may offer for the common discomforts experienced by expectant mothers. However, it also raises substantial concerns regarding their effectiveness and safety.

The use of herbal remedies and medicinal plants during pregnancy necessitates a careful evaluation, as it entails a delicate balance between potential benefits and safety, always with the overarching goal of safeguarding the well-being of both the expectant mother and the developing fetus. This discussion endeavors to provide participants with a comprehensive understanding of the subject, enabling them to make well-informed choices concerning their health and well-being during this transformative period, benefiting themselves and their unborn children.

This chapter section lays the foundation for a comprehensive exploration of the use of herbal medicine and medicinal plants during pregnancy. The associated potential risks that underscore the importance of seeking expert medical advice, identifying known safe herbs, discussing appropriate dosage considerations, exploring potential benefits, and navigating the legal aspects are summarized in Table 2. Furthermore, we will consider the varying responses of individuals to herbal remedies and complementary therapies, emphasizing the necessity for personalized guidance and informed decision-making.

In summary, safety should be a paramount concern for pregnant individuals, and consultation with healthcare providers is essential. While some medicinal plants and herbal remedies may be used cautiously to address common discomforts, they should complement rather than replace conventional medical care during pregnancy. Informed and open discussions with healthcare professionals are vital for making well-informed decisions regarding the use of herbal remedies during pregnancy.

Table 2 Aspects of herbal medicine and medicinal plants in pregnancy

Topic	Key points
Safety concerns	Limited scientific data on medicinal plants' safety and fetal effects during pregnancy.
Consultation with healthcare professionals	Pregnant individuals should always seek advice from healthcare providers before using herbal remedies.
Known safe herbs	Some herbs are considered safe during pregnancy when used in moderation.
	Examples: ginger, peppermint, and red raspberry leaf (in the third trimester with supervision).
Exercise caution with some herbs	Avoid herbs like black cohosh, pennyroyal, and blue cohosh, which can induce preterm labor or miscarriage.
Regulation and quality	Quality of herbal products varies; use products from reputable sources to ensure safety and purity.
Dosage and duration	Adhere to recommended dosages and usage durations for safe herbal product use during pregnancy.
Potential benefits	Some herbs may offer advantages for common pregnancy discomforts, but use cautiously and under supervision.
Herbal treatments for common pregnancy symptoms	Examples of herbal treatments for common discomforts.
Individual variability	Response to herbs varies between individuals, emphasizing the need for personalized guidance.
Alternative therapies	In some cases, alternative therapies like acupuncture or aromatherapy can provide relief without potential risks.

3 The Herbal Approach to Pregnancy Care: Tradition, Safety, and Well-Being

The use of herbal plants in traditional medicine has a long-standing history, spanning centuries, and this practice extends to the realm of pregnancy care. Incorporating herbal remedies into pregnancy-related healthcare is a venerable tradition observed across diverse cultures worldwide. These remarkable botanicals, possessing a multitude of therapeutic properties, provide pregnant women with natural solutions to alleviate various discomforts and enhance overall well-being, all while minimizing potential risks associated with pharmaceutical medications. When employed judiciously and under the guidance of knowledgeable practitioners, herbal medicine during pregnancy can serve as a valuable and holistic approach to maintaining a healthy pregnancy and addressing common maternal and fetal health concerns.

Pregnancy is a unique and transformative phase in a woman's life, characterized by physiological, emotional, and hormonal changes. Many expectant mothers contend with a range of common discomforts, including morning sickness, heartburn, insomnia, anxiety, and swollen ankles. Herbal medicine offers a diverse array of herbal remedies that can be harnessed to safely and effectively manage these symptoms [81, 82]. One of the most renowned herbs for alleviating pregnancy-related nausea and vomiting is ginger (Zingiber officinale), with a history of use spanning centuries, and its efficacy in curbing morning sickness is supported by modern scientific research. Ginger can be consumed as tea, capsules, or candied ginger.

Another commonly used herb during pregnancy is chamomile (*Matricaria chamomilla*). Chamomile tea is a soothing and gentle remedy for anxiety, insomnia, and digestive discomfort, rendering it a valuable herb for expectant mothers. The calming effects of chamomile can help reduce stress and promote better sleep, which is especially beneficial during pregnancy when sleep disruptions are expected.

Beyond managing specific discomforts, herbal medicine during pregnancy encompasses herbs recognized for their uterine toning and nourishing properties. For example, raspberry leaf (*Rubus idaeus*) is a well-regarded herb for strengthening uterine muscles and preparing the body for labor when consumed as tea in the second and third trimesters. Nettle (*Urtica dioica*), celebrated for its rich nutritional profile, provides essential vitamins and minerals, including iron and folic acid, crucial for both the developing fetus and the expectant mother's health. Nettle tea is often recommended to combat anemia, improve circulation, and alleviate swelling, which are common concerns during pregnancy.

Herbal medicine also addresses the emotional and psychological aspects of pregnancy. Herbs like St. John's Wort (*Hypericum perforatum*) and lavender (*Lavandula angustifolia*) can offer relief from anxiety and mood swings. St. John's Wort, known for its antidepressant and anti-anxiety properties, can be safely used in moderation. Lavender essential oil, when diluted and applied topically or diffused, can help soothe nerves and promote relaxation.

However, it is essential to underscore the need for caution when considering herbal remedies during pregnancy. Safety is paramount, as certain herbs may be contraindicated during pregnancy due to potential adverse effects. Pregnant women should consult a qualified herbalist, naturopathic doctor, or midwife experienced in herbal medicine to select appropriate herbs and dosages tailored to their unique needs. Additionally, expectant mothers should be aware of herbs to avoid during pregnancy. Herbs such as black cohosh (*Actaea racemosa*), pennyroyal (*Mentha pulegium*), and tansy (*Tanacetum vulgare*) have been associated with adverse pregnancy outcomes and should be strictly avoided. Herbs that stimulate uterine contractions, like blue and black cohosh, should not be used without proper supervision during pregnancy.

Regulation and quality control of herbal products are crucial considerations when incorporating herbal medicine during pregnancy. Pregnant women should obtain herbs from reputable suppliers to ensure their purity, free from contaminants, and that they have been responsibly harvested and processed. Choosing organically grown herbs whenever possible is essential to minimize exposure to pesticides and other potentially harmful chemicals.

It is essential to emphasize that herbal medicine should not replace standard prenatal care, including regular check-ups with healthcare providers, ultrasound scans, and appropriate screenings. Instead, it should complement conventional medical care by addressing specific concerns and promoting overall well-being, providing a holistic approach to pregnancy care.

The importance of informed and cautious herbal medicine use during pregnancy cannot be overstated. Each pregnancy is unique, and what may be safe and beneficial for one expectant mother may not be suitable for another. This underlines the necessity of individualized care and guidance from a qualified herbalist or healthcare practitioner. Such an approach ensures that the chosen herbs are appropriate for the expectant mother's health and pregnancy. Herbal medicine offers a holistic and natural approach to supporting maternal and fetal health during pregnancy. When used with caution and under the guidance of experienced practitioners, herbal remedies can relieve common discomforts, nurture the mother's overall well-being, and even prepare the body for labor and delivery. However, pregnant women need to approach herbal medicine with a discerning eye, seeking professional guidance to ensure the safety and efficacy of their chosen herbs [98]. With the right approach, herbal medicine can be a valuable and harmonious addition to a healthy and happy pregnancy journey. Numerous studies have indicated that pregnant women have a range of options for consuming herbal medicines (HMs) during pregnancy. These options include unprocessed herbal mixtures, herbal extracts, finished and branded herbal medicines, as well as nutritional supplements containing specific HMs, vitamins, and minerals [57]. In Table 3, you can find a list of medicinal plants commonly used during pregnancy and their intended purposes.

Pregnant women across the globe are increasingly turning to herbal remedies as they seek natural solutions to common discomforts associated with pregnancy. Among the most widely embraced herbal options are ginger, chamomile, peppermint, Echinacea, cranberry, Vaccinium, and garlic. These herbal choices transcend geographical boundaries, recognized for their potential to alleviate typical pregnancy-related ailments.

For instance, ginger is highly regarded for its ability to combat morning sickness, while chamomile and peppermint are sought after for their calming properties. Cranberry and Echinacea are known for their respective benefits in promoting urinary health and bolstering the immune system. Garlic is appreciated for its versatile medicinal properties.

Regional preferences for specific herbal remedies may vary. In Eastern and Western Europe, uva ursi has traditionally been employed for its purported urinary tract benefits. Conversely, in Eastern European nations, herbal options like motherwort, centaurium erythraea, and lovage have taken precedence in the repertoire of pregnant women. These regional variations are deeply rooted in historical practices and cultural traditions.

Understanding the popularity and regional disparities in herbal choices among pregnant women is of paramount importance for both healthcare professionals and expectant mothers. It facilitates informed decision-making, underscoring the significance of effectiveness and safety. This global perspective serves as an invaluable resource for promoting the well-being of both mothers and their unborn children as herbal medicine continues to play a role in pregnancy care.

In Sweden, expectant mothers often incorporate dietary supplements such as Floradix® IRON + HERBS, ginseng, and valerian into their routines [35]. The decision to use herbal medicines during pregnancy can yield benefits for both the

Table 3 Different medicinal plants used by pregnant women in their different physiological conditions of pregnancy

Physiological condition	Medicinal plants	Scientific names
Cold and flu	Echinacea	*Echinacea purpurea*
	Elderberry	*Sambucus nigra*
Pain (Gastralgia and other types of pain)	Chamomile	*Matricaria chamomilla*
	Peppermint	*Mentha × piperita*
	Lavender	*Lavandula angustifolia*
Gastrointestinal disorders, constipation, flatulence	Fennel	*Foeniculum vulgare*
	Dandelion	*Taraxacum officinale*
Anxiety, stress	Valerian	*Valeriana officinalis*
	Lemon balm	*Melissa officinalis*
Urinary tract infection	Cranberry	*Vaccinium macrocarpon*
	Uva Ursi	*Arctostaphylos uva-ursi*
Fetal health promotion	Raspberry leaf	*Rubus idaeus*
	Nettle leaf	*Urtica dioica*
Labor preparation, facilitation, and induction	Black cohosh	*Actaea racemosa*
	Blue cohosh	*Caulophyllum thalictroides*
Milk production and secretion	Fenugreek	*Trigonella foenum-graecum*
	Blessed thistle	*Cnicus benedictus*
Nausea and vomiting	Ginger	*Zingiber officinale*
Stress and anxiety	*Matricaria chamomilla* (chamomile)	*Matricaria chamomilla*
Digestive discomfort	Peppermint	Mentha × piperita
Immune support	Echinacea	*Echinacea purpurea*
Uterine tonic for late pregnancy	*Rubus idaeus* (raspberry leaf)	*Rubus idaeus*
Sleep and relaxation	*Lavandula angustifolia* (lavender)	*Lavandula angustifolia*
Nutrient support	*Urtica dioica* (stinging nettle)	*Urtica dioica*
Preparing for labor	*Rubus idaeus* (red raspberry leaf)	*Rubus idaeus*

mother and the unborn child, and they are occasionally employed to manage specific health issues. Some of the most frequently mentioned uses of medicinal plants during pregnancy include addressing concerns like nausea and vomiting, urinary tract infections (UTIs), childbirth preparation and comfort, common colds and flu, digestive issues, pain management, improving fetal outcomes, preventing abortion, managing anxiety, maintaining general health, and alleviating swelling [33, 57].

4 Global Perspectives on Herbal Medicine Use During Pregnancy: Practices, Prevalence, and Safety Considerations

The utilization of HMs is not confined to specific nations; however, certain HM practices are more prevalent in particular regions. For instance, expectant mothers in India and Ghana display a higher tendency to use HMs for the prevention of abortions and to enhance the health of their unborn children, compared to their counterparts in more industrialized regions [5, 15].

These practices are often rooted in a variety of factors, some of which are directly related to the well-being of the fetus. For example, the prevention of newborn hyperbilirubinemia, protection of the unborn child from injury, promotion of increased prenatal weight, and support for optimal brain development are among the commonly cited reasons [23, 105]. Surprisingly, certain indications also strongly influence societal perceptions of pregnant women. In places like Lesotho and Zimbabwe, it is customary for pregnant women to use HMs, and adherence to cultural standards often mandates this practice [72, 94, 95, 103].

The use of regional medicinal plants for prenatal care is widespread in developing nations, especially in Africa and Asia [68, 96, 97, 112]. These medicinal herbs are consumed in various forms, including decoctions, medications, and soups. For example, in South Africa, Zulu women in their final trimester frequently consume a concoction known as "Isihlambezo." This mixture, which comprises several medicinal herbs, is believed to promote a favorable pregnancy and facilitate uncomplicated labor [17]. Some of the most commonly used ingredients in this mixture include *Agapanthus africanus* (L.) Hoffmanns, *Pentanisia prunelloides* (Klotzsch ex Eckl. & Zeyh.), *Clivia miniata* (Lindl.) Verschaff. Walp., and *Gunnera perpensa* L. [17].

The specific formulations employed often depend on several factors, including the preferences of the expectant mother, the traditional healer's recommendations, and the woman's current health status. It's worth noting that the uterine toning effects of these herbal mixtures are activated upon consumption, but excessive doses leading to heightened vaginal stimulation may result in adverse effects. For example, the "Isihlambezo" mixture has been linked to low newborn birth weights, embryonic meconium spotting (a potential sign of prenatal distress and oxygen deprivation), and uterine breach [17].

Malawi, situated in southeastern Africa, is renowned for its picturesque landscapes, diverse cultural heritage, and the long-standing practice of utilizing herbal medicine and medicinal plants. Traditional healers and local communities in Malawi have relied on the healing properties of indigenous plants to address a wide array of health concerns. This practice is deeply ingrained in Malawian culture and operates in tandem with modern medical approaches.

The realm of Malawian herbal medicine encompasses a plethora of plant species, each possessing distinctive therapeutic attributes. Notably, the Moringa tree (*Moringa oleifera*) stands as a symbol of nutrition and offers a potential solution to combat malnutrition, a prevailing challenge in the nation. Additionally, Malawi's abundant flora includes plants like *Artemisia annua*, known for their antimalarial properties, providing a natural alternative in the fight against this prevalent disease. In Malawi, the "Mwanamphepo" mixture is commonly used to induce labor and consists of various herbs, including *Ampelocissus obtusata* (Welw. ex Baker) Planch., *Cyphostemma hildebrandtii*, and *Cissus verticillata*, often consumed with oatmeal just before childbirth (Maliwichi-Nyirenda and Maliwichi 2010).

Medicinal plants are integral to the Malawian healthcare landscape. The country boasts a diverse array of plant species renowned for their medicinal applications. For example, Senna (*Cassia abbreviata*) is employed for its laxative qualities, while the African potato (*Hypoxis hemerocallidea*) is sought after for its potential to address a spectrum of health issues.

This exploration of herbal medicine and medicinal plants in Malawi seeks to shed light on the profound interplay between the nation's rich biodiversity and its healthcare traditions. We delve into the cultural significance, traditional healing methodologies, and the potential integration of these herbal therapies into the broader healthcare framework. Furthermore, we address the vital issue of sustainability concerning medicinal plants and their role in delivering accessible and cost-effective healthcare solutions to the people of Malawi.

In Korea, the herbal combination "Anjeonicheon-tang," which includes therapeutic plants like Atractylodes macrocephala Koidzumi white rhizoma, *Glycyrrhiza glabra* L., and Panax ginseng C.A.Mey, is used to address pregnancy-related symptoms, particularly hyperemesis gravidarum. In Taiwan, a tea known as "An-Tai-Yin," composed of medicinal plants aimed at inducing labor, is commonly consumed by expectant mothers [20]. However, mothers who consumed An-Tai-Yin were found to have an increased risk of connective tissue, joint, and ocular abnormalities in their babies [20].

Research conducted by Pallivalapila et al. [102] suggests that the first and third trimesters witnessed the highest prenatal use of herbal medicines. During the first trimester, herbal remedies were often used to manage early pregnancy symptoms like morning sickness, nausea, and gastrointestinal issues. In contrast, the third trimester use aimed at aiding in childbirth and preparing the cervix for delivery [79, 97]. Nevertheless, it was not uncommon for women to use herbal medicines throughout their entire pregnancy with the goal of promoting both maternal and fetal health. Notably, no consistent trends have been observed by scholars [110], and some women continued using these herbal remedies until they felt better [77].

According to Kennedy et al. [58], married women with higher levels of education were more inclined to use complementary and alternative medicine (CAM) that included herbal medicines during pregnancy (Forster et al. 2006; [2, 57, 58]). This demographic of herbal medicine consumers was typically middle class, Caucasian, non-smoking males over the age of 30 who had prior experience with herbal remedies. On the other hand, in Eastern European countries, pregnant women under the age of 20 were more likely to use herbal medicines [57].

In contrast, in Africa and Asia, women who incorporated herbal medicines during childbirth tended to be under the age of 30, possess lower socioeconomic status, lack a high school diploma, and reside in remote areas with limited access to public healthcare facilities [54, 113]. These pregnant women often sought advice from friends, family, traditional birth attendants, local plant vendors, and herbalists before deciding to use herbal medicines [54, 57, 113].

Media coverage played a role in influencing the decision to use herbal medicines during pregnancy in industrialized nations, as reported by Hall et al. [33]. In Eastern Europe, pregnant women typically followed their doctor's recommendations regarding herbal medicine use [57].

The practice of using both herbal medicines and traditional remedies during pregnancy is common worldwide [36, 79, 110, 111]. It is worth noting that when herbal medicines are used in conjunction with conventional drugs, they can either enhance or diminish the effects or side effects of these medications. However, many pregnant women refrain from disclosing their herbal medicine usage to healthcare providers due to the stigma attached to this practice.

Overall, a significant proportion of pregnant women turn to herbal medicines and traditional remedies for support during pregnancy, labor, and postpartum, with the prevalence of herbal medicine use varying based on demographic factors such as geographical location, cultural background, and educational level. This highlights the urgent need for research into the efficacy and safety of herbal medicines and traditional plant remedies in pregnant individuals, and potentially, in humans. Table 4 summarizes the global Perspectives on Herbal Medicine Use During Pregnancy: Practices, Prevalence, and Safety Considerations.

5 Herbal Medicine in Pregnancy: Nurturing Maternal and Fetal Health

The utilization of herbal plants in traditional medicine has a rich historical tradition that spans centuries and extends its influence to prenatal care. Across diverse cultures globally, the incorporation of herbs into pregnancy-related healthcare has been a time-honored practice. These botanical wonders, with their diverse therapeutic attributes, offer expectant mothers natural remedies to alleviate a variety of discomforts and enhance overall well-being, all while mitigating potential risks associated with pharmaceutical medications.

When used judiciously and under the guidance of knowledgeable practitioners, herbal medicine during pregnancy can be a valuable and holistic approach to fostering a healthy gestation period and addressing common maternal and fetal health concerns. Pregnancy is a remarkable and transformative phase marked by physiological, emotional, and hormonal changes. Many expectant mothers grapple with a range of discomforts, including morning sickness, heartburn, insomnia, anxiety, and swollen ankles. Herbal medicine provides a spectrum of botanical solutions that can effectively manage these symptoms safely.

One of the most renowned herbs for alleviating pregnancy-related nausea and vomiting is ginger (*Zingiber officinale*). With a history of use spanning centuries, modern scientific research substantiates ginger's efficacy in curbing morning

Table 4 Global perspectives on herbal medicine use during pregnancy: practices, prevalence, and safety considerations

Topic	Key findings
Regional variation in HM practices	Herbal medicine (HM) practices during pregnancy vary across regions. In India and Ghana, there's a higher tendency to use HMs for pregnancy health compared to industrialized regions. Cultural factors often influence this choice.
Prevalent reasons for HM use	Reasons for using HMs during pregnancy include preventing newborn hyperbilirubinemia, protecting the unborn child from injury, promoting increased prenatal weight, and supporting optimal brain development.
HM use in developing nations	Developing nations, particularly in Africa and Asia, widely use regional medicinal plants for prenatal care. These medicinal herbs are consumed in various forms, but some formulations may have adverse effects.
Country-specific HM practices	Specific examples include South African Zulu women using "Isihlambezo," Korean "Anjeonicheon-tang" for hyperemesis gravidarum, and Taiwanese "An-Tai-Yin" to induce labor (with reported risks to babies).
Trimester-specific HM use	HM use peaks during the first and third trimesters. In the first trimester, it's used to manage early pregnancy symptoms, while the third trimester use aims to aid in childbirth and prepare the cervix for delivery.
Demographic variations	Demographic factors like education and age influence HM use. Higher education levels correlate with increased CAM usage, while in Eastern Europe, younger women are more likely to use HMs.
Socioeconomic and cultural factors	In Africa and Asia, women with lower socioeconomic status, lacking a high school diploma, and residing in remote areas tend to use HMs. Societal perceptions and cultural practices often influence HM use.
Media influence	Media coverage affects HM use in industrialized nations. In Eastern Europe, pregnant women typically follow their doctor's recommendations regarding herbal medicine use.
HM use alongside conventional medicine	Many pregnant women use both herbal medicines and conventional remedies worldwide. The combined use can enhance or diminish the effects of conventional drugs, but there's a stigma associated with HM use.
Need for further research	The prevalence and safety of herbal medicines and traditional plant remedies during pregnancy call for further research and exploration, taking into account regional, cultural, and demographic differences.

sickness. It can be incorporated into one's diet in various forms, such as ginger tea, capsules, or candied ginger.

Chamomile (*Matricaria chamomilla*) is another herb frequently employed during pregnancy. Chamomile tea is a soothing and gentle remedy for anxiety, insomnia, and digestive discomfort, making it a valuable herbal option for expectant mothers.

Its calming effects can help reduce stress and facilitate improved sleep, which is especially valuable during pregnancy when sleep disruptions are common.

However, it is crucial to exercise caution when considering herbal remedies during pregnancy, with safety being of utmost importance. Some herbs can be contraindicated due to their potential to induce adverse effects. Pregnant women should consult with qualified herbalists, naturopathic doctors, or experienced midwives to ensure the selection of appropriate herbs and determine suitable dosages tailored to their specific needs.

In addition to professional consultation, expectant mothers should familiarize themselves with herbs that must be avoided during pregnancy. Herbs like black cohosh (*Actaea racemosa*), pennyroyal (*Mentha pulegium*), and tansy (*Tanacetum vulgare*) have been associated with adverse pregnancy outcomes and should be categorically avoided. Furthermore, herbs known to stimulate uterine contractions, such as blue and black cohosh, should not be used without appropriate supervision during pregnancy.

The regulation and quality control of herbal products are paramount considerations when incorporating herbal medicine during pregnancy. Pregnant women should source their herbs from reputable suppliers to ensure that these botanicals are free from contaminants and have been harvested and processed responsibly. Opting for organically grown herbs whenever possible is essential to minimize pesticide exposure and potential exposure to other harmful chemicals.

It is essential to reiterate that herbal medicine should not replace standard prenatal care. Regular check-ups with healthcare providers, ultrasound scans, and appropriate screenings are non-negotiable components of comprehensive pregnancy care. Herbal medicine should complement conventional medical practices by addressing specific concerns and enhancing overall well-being.

When approached with prudence, herbal medicine during pregnancy offers a holistic and natural methodology for supporting maternal and fetal health throughout this unique journey. When carefully selected and monitored under the guidance of experienced practitioners, these herbal remedies can effectively alleviate common discomforts, nurture maternal well-being, and prepare the body for labor and delivery. Pregnant women, however, need to approach herbal medicine with a discerning eye, seeking professional guidance to ensure the safety and efficacy of their chosen herbs. With a thoughtful and informed approach, herbal medicine can be an invaluable and harmonious addition to a healthy and joyful pregnancy experience.

6 Raspberry Leaves in Pregnancy: A Natural Supportive Remedy

Various herbal plants play a role in pregnancy support, particularly tonic herbs that are deemed safe for regular consumption during pregnancy. These herbs serve a dual purpose: they are non-toxic and not highly therapeutic, making them suitable for expectant mothers, while also nourishing the mother and fortifying the cervix in preparation for a smooth childbirth. Among the herbs falling into this category, we find raspberry leaves (*Rubus idaeus*), partridge berries (*Mitchella repens*), and stinging nettles (*Urtica dioica*).

Raspberry leaves have a long history of use by pregnant women in China, Europe, and North America. Both fresh and dried raspberry leaves can be steeped in hot water to create a flavorful tea with high nutritional value, believed to promote the well-being of the pregnant woman's cervix. Additionally, women at risk of miscarriage have been encouraged for centuries to incorporate raspberry leaf tea into their daily routines throughout pregnancy, as it is believed to aid in carrying the pregnancy to full term. This recommendation underscores the significance of proper nourishment in reducing the likelihood of pregnancy and childbirth complications, such as miscarriage, postpartum bleeding, and premature or delayed labor [3].

Raspberry leaves are a rich source of essential nutrients, including vitamins A, B complex, C, E, as well as minerals like calcium, iron, phosphate, and potassium [4]. Furthermore, they contain magnesium and iron [10]. The herb's bitterness, invigorating properties, and soothing effects can be attributed to its tannins, polypeptides, and antioxidants. Of particular importance is the alkaloid fragarine, discovered in 1941, which is known to inhibit uterine contractions [12]. Some reports suggest that fragarine strengthens the muscles of the uterus and pelvis, potentially easing the process of childbirth. While animal studies have shown that raspberry leaf can both stimulate and relax the uterus, human research in this regard has been limited. However, a recent observational study involving raspberry leaf tea in women of reproductive age indicated a reduced risk of preterm or delayed labor and a decreased need for medical intervention during labor. Notably, the plant has not been linked to adverse pregnancy outcomes, and there is no evidence to suggest it is toxic or teratogenic. While raspberry leaf tea has been promoted for use throughout pregnancy, its significant impact on the endometrium is sometimes recommended only in the third trimester [25]. Additionally, chewing on small fragments of raspberry leaf tea ice cubes may provide a moist tongue, which, in turn, could facilitate uterine contractions.

7 Herbal Remedies for Pregnancy: Efficacy and Safety

Approximately 25% of pregnancies end prematurely, and the causes can vary, including the mother's immune system rejecting the embryo, having an incompetent cervix, or carrying a nonviable fetus. Miscarriage is a distressing issue, and women at risk of losing their pregnancies often face limited options. Early pregnancy symptoms, like spotting and cramps, typically lead to bed rest until the threat either subsides or results in a miscarriage [119].

However, two North American native plants, black haw (*Viburnum prunifolium*) and false unicorn (*Chamaelirium luteum*), offer potential for miscarriage prevention.

Black Haw: Historically, Eclectic Physicians used both black haw and its close relative, cramp bark (*Viburnum opulus*), for similar purposes, such as preventing and treating abortion, preparing for labor, and alleviating false labor and postpartum symptoms [118]. Black haw has been found effective when used in small, gradual amounts as a preventative measure or in larger doses for imminent situations, according to Eclectic practices [120]. As a result of widespread use in both alternative and conventional medicine, we now have a better understanding of the pharmacological effects of black haw. Its active compounds, salicin and scopoletin, act as efficient uterine antispasmodics, soothing and relaxing uterine muscles. In modern Chinese therapy, black haw is employed to safeguard pregnancies that might otherwise end in abortion due to cervical laxity or uterine instability. It remains unclear whether the herb's efficacy varies depending on its form, whether consumed as a pill, tea, or extract. Although clinical studies are limited, black haw and cramp bark are generally considered safe for use throughout pregnancy [116]. However, due to the potential risk of syncope (fainting), it is advisable for pregnant or nursing women to avoid high doses or regular usage of black haw. There is also anecdotal evidence suggesting that cramp root might be a viable alternative.

False Unicorn: False unicorn, commonly used in herbal preparations to enhance fertility, reduce the risk of abortion, and prepare the body for childbirth, is associated with various benefits. This herb is particularly known for its ability to address uterine incontinence and strengthen a cervix that has loosened due to recurrent pregnancies. It is believed that taking false unicorn before pregnancy may enhance a woman's fecundity [101]. In certain situations, women turn to false unicorn to either prevent miscarriage or alleviate the symptoms of an impending loss, such as bleeding and cramps [65].

While there is limited rigorous scientific testing on the safety of false unicorn during pregnancy, it is generally considered safe by most herbalists. Some women have used false unicorn even after the onset of abortion, taking frequent doses to successfully carry their pregnancies to full term. There is some indication that the plant may lead to increased levels of human chorionic gonadotropin in the blood, potentially averting abortion. Additionally, cases have been reported where false unicorn and other herbs effectively halted excessive bleeding associated with abortion.

Despite the absence of clinical studies regarding the pharmacological effects of false unicorn, it is commonly believed to possess qualities that restore ovarian and uterine health while acting as an emmenagogue. Contrary to popular belief, it does not stimulate the uterus during pregnancy, but it may contribute to hormone regulation and strengthen the liver and intestinal system [19, 78, 80]. False unicorn can be consumed in various forms, including as a drink, extract, or pill. The specific active components in these different forms are not well documented due to limited research on the plant's mechanisms. Given its acrid flavor, false unicorn is typically preferred in pill or extract form rather than being brewed into a beverage.

Labor Induction: In North America, women often turn to various herbs for labor induction, as highlighted by the research of Moshi and Mhame [71]. Some commonly used herbs for this purpose include blue cohosh (*Caulophyllum thalictroides*), black cohosh (*Cimicifuga racemosa*), and even beetroot (*Trillium erectum*).

Blue Cohosh, a plant native to North America, holds a long history of use by Native Americans and Eclectic Physicians for its medicinal properties, particularly the roots and tubers. It was employed to hasten labor in cases where it had stalled due to maternal frailty, exhaustion, or a lack of reproductive energy [56, 94, 95]. During the 1800s, it gained popularity in both Western and African-American traditional medicine as a means of inducing childbirth and increasing menstruation flow. In modern herbal medicine, blue cohosh is highly regarded for its uterus-stimulating, antispasmodic, and emmenagogue properties [39, 91, 100]. This plant owes its oxytocic effects to flavonoids such as caulosaponin and caulophyllosaponin, making it one of the most potent natural labor inducers. It is often blended with other herbs, particularly black cohosh, in a 1:1 ratio to expedite labor. However, it's important to note that some of blue cohosh's active components are not water soluble, making it more effective as a preparation rather than a tea.

Recent concerns regarding the safety of blue cohosh have emerged. There have been reports of increased fetal distress, as evidenced by meconium-stained amniotic fluid or fetal arrhythmia, and elevated fetal heart rates associated with its use. Some case reports highlight severe health issues in newborns whose mothers used blue cohosh, particularly when used alongside black cohosh. One study [49] described a case where a woman took an unknown quantity of blue and black cohosh to induce labor, resulting in the newborn exhibiting symptoms of illness, necessitating hospitalization and medical intervention for convulsions, renal injury, and artificial respiration. This has raised concerns about the safety of blue cohosh.

Blue cohosh contains compounds like caulosaponin, which researchers believe may contribute to these health problems by constricting cardiac blood vessels and inducing harmful myocardial responses. Some experts have argued that our understanding of these plants' dangers is based on studies conducted on individual components in animals, often using unrealistically high doses. The toxic impact of caulosaponin in animal studies was observed at levels far exceeding what a human would typically consume and could be fatal.

Furthermore, there have been documented cases of nausea, severe gastrointestinal discomfort, and other adverse reactions associated with blue cohosh use. Additionally, blue cohosh contains at least ten alkaloids, including the teratogenic-methylcytosine and extremely embryotoxic sparteine. There's also concern regarding anagyrine, a substance found in the plant's roots, which has been linked to congenital disabilities, such as twisted calf syndrome. Although there are no direct evidence of such issues in humans, a case report connected a similar genetic malformation in a human to the consumption of anagyrine-contaminated goat's milk. However, teratogenic effects of anagyrine have not been observed in studies involving rat embryos.

Given the growing body of evidence pointing to potential risks, it's advisable to exercise caution when considering the use of blue cohosh. The continued use as a labor induction method in North America suggests a need for further research and education on its safety [29, 92] (▶ Chap. 59, "Research Needs of Medicinal Plants Used in the Management and Treatment of Some Diseases Caused by Microorganisms"). Conversely, black cohosh, another plant with labor-inducing potential, has a lower reputation outside North America. While no poisonous, carcinogenic, or teratogenic traits have been established, some European authorities recommend against its use during pregnancy and breastfeeding [50, 107]. Extremely high doses of black cohosh may lead to unpleasant side effects, including intense pain, vertigo, nausea, and vomiting. It is more common to use blue cohosh, but midwives often recommend a combination of these plants to expedite labor. Black cohosh is believed to complement the effects of blue cohosh by regulating the frequency and intensity of uterine contractions. Its purported properties include being antispasmodic, calming, vasorelaxant, and hypotensive.

Beetroot, a plant historically used by Native Americans in North America, particularly its rhizomes, has been employed for centuries to aid in childbirth and reduce postpartum hemorrhage. The term "beth" in beetroot likely stems from "birth" [119]. This maroon-flowered plant was commonly used by early American colonies, Shakers, and Eclectic Physicians to treat conditions such as vaginal bleeding and heavy menstrual flow [13]. While scientific investigations into beetroot's pharmacological effects are lacking, ample evidence suggests its efficacy in inducing uterine contractions. It is known for its properties as an antihemorrhagic and emmenagogue for the genitourinary system, and it also possesses expectorant, astringent, and antimicrobial qualities [87, 89, 90].

In modern herbal medicine, beetroot, often consumed as a drink or extract, is a common choice for stimulating labor that has stalled. Laboratory tests have shown that saponin glycosides present in beetroot and its close relative, *Trillium grandiflorum*, have significant antifungal properties. However, the glycosides in beetroot have not been thoroughly examined regarding their chemistry and toxicity. Clinical studies on beetroot are highly unlikely due to its limited recognition, even though they could provide valuable insights into its phytochemistry and pharmacology. While more research is needed, beetroot may be considered a potentially safer option for inducing labor compared to other alternatives [22].

8 The Efficacy and Safety of Herbal Medicine in Pregnant Women

Historically, people have traditionally relied on the wisdom and experience of their elders to ascertain the safety and effectiveness of specific medicinal plants. Over the past two to three decades, extensive preclinical and clinical research has been conducted to shed light on the efficacy and safety of various therapeutic plants and herbal combinations. Some of these studies have involved in vivo and in vitro assessments to evaluate their potential embryotoxicity.

For instance, ginsenosides, the active components in ginseng, had direct teratogenic effects on rat fetuses, as determined in an in vitro investigation [18]. This study further inferred that these herbal medicines, due to their oxytocic activity, might lead to uterine hyperstimulation, potentially resulting in fetal ischemia and premature birth, based on their findings using an isolated rat uterus model.

In another study by Mahmoudian et al. [66], the impact of valerian (*V. officinalis* L.) preparation on the development and functioning of mouse embryonic brain cells was explored. This research revealed that the use of valerian during the second trimester significantly reduced the levels of zinc in the embryonic brain, a critical component crucial for normal brain development [85, 93]. As a result, it is advisable for pregnant women to exercise caution when considering the use of valerian products.

Furthermore, research involving thermopsis lanceolata R.Br., an expectorant, indicated that pregnant rodents exposed to this herbal medicine had offspring with reduced body weights and craniocaudal dimensions by the time they reached 20 days of age [59]. While the pups did not experience an impact on their postnatal survival rate, their sensory, motor, and emotional-motor behaviors were altered. Consequently, the use of thermopsis or any product containing it for therapeutic purposes is discouraged due to these findings.

9 Interactions Between Medications and Herbal Supplements

9.1 Use of Homeopathic Medicines During Pregnancy and the Risk of Epilepsy

Women who experience frequent seizures may become pregnant, but the scientific research on herbal remedies for epilepsy is somewhat limited. Some herbal medicines (HMs) might increase the risk of seizures, cause unwanted side effects, or interact with conventional antiepileptic drugs (AEDs). An exception is a medication derived from the *Cannabis sativa* L. species that has received FDA approval for the treatment of certain types of epileptic seizures. Specifically, cannabidiol extracted from *Cannabis sativa* has shown promise in the management of certain forms of epilepsy. Interestingly, while these cannabis preparations have been available in several other countries for some time, the United States has only recently gained access to them. These products come in various forms, with some containing additional compounds like tetrahydrocannabinol.

However, it's worth noting that cannabis use during pregnancy has been associated with adverse effects on both the mother and the fetus, as pointed out by Gunn et al. [32]. Furthermore, in vitro studies have shown that cannabidiol inhibits several cytochrome enzymes, including CYP3A4, CYP3A5, and CYP2D6, while inducing CYP1A1 [52]. Additionally, these studies have demonstrated that cannabidiol activates the UDP-glucuronosyltransferases UGT1A9 and UGT2B7, as highlighted by Yamaori et al. [121–123]. These complex interactions could affect the processing of AEDs and may lead to treatment failures or adverse effects, altering the levels of these medications in the bloodstream. Notably, antiepileptic drugs such as clobazam, rufinamide, topiramate, zonisamide, and eslicarbazepine have been found to interact with cannabidiol, as reported by Gaston et al. [30].

While some herbal medicines are used during pregnancy, there is less scientific evidence to support their use compared to their application in treating epileptic patients. Additionally, the pharmacokinetics of various antiepileptic drugs are intricate, and they may interact with other drugs or herbal medicines in ways that can influence the course of treatment. Herbal medicines commonly used by individuals with epilepsy include St. John's Wort (*Hypericum perforatum* L.), *Ginkgo biloba* L., *Borago officinalis* L., *Valeriana officinalis* L., and *Cannabis sativa* L., as indicated by Liu et al. [63].

9.2 Pregnancy-Related Hypertension and the Use of Herbal Medicine

Pregnancy can lead to various forms of hypertension, including chronic, prenatal, preeclamptic, and chronic hypertension with concurrent preeclampsia, as noted by Lai et al. [61]. Pregnancy-related hypertension is one of the major causes of maternal and fetal illness and mortality during pregnancy, as highlighted by Mustafa et al. [74]. Preeclampsia, characterized by the onset of hypertension and proteinuria after 20 weeks of pregnancy, is on the rise, particularly among older women and those with underlying medical conditions [74, 115]. Townsend et al. [115] further emphasize that preeclampsia poses a significant risk for preterm birth and maternal mortality. Moreover, it can have detrimental effects on various bodily systems, including the renal, liver, neurological, cardiovascular, and nervous systems (e.g., eclampsia, cerebral blood clots, and bleeding, acute fatty liver of pregnancy, hepatic periportal inflammation, glomerular endotheliosis, and acute kidney injury).

Managing pregnancy-related hypertension is essential to reduce the risk of complications, including renal damage. One of the antihypertensive medications used for cases of mild to severe hypertension is labetalol [21]. It is important to avoid angiotensin-converting enzyme inhibitors during pregnancy, and alternative medications like methyldopa and nifedipine are recommended [21].

In 2017, the American College of Obstetricians and Gynecologists (ACOG) released updated recommendations for the management of sudden-onset significant hypertension during pregnancy based on the best available data [21]. Sublingual nifedipine with immediate release is the first-line treatment for pregnant women with abrupt-onset hypertension. In cases where immediate treatment is needed, hydralazine or labetalol administered intravenously are also options [21]. Most pregnant women are prescribed either nifedipine or labetalol.

The metabolic pathways of these medications differ. Nifedipine is metabolized in the body into pyridine and nifedipine carboxylic acid, primarily through the enzyme CYP3A4 [124]. Labetalol, on the other hand, undergoes significant metabolic changes in the human liver and intestines. According to Jeong et al. [51], glucuronidation, mainly mediated by enzymes UGT1A1 and UGT2B7, plays a central role in the elimination of labetalol. This process results in the formation of two inactive drug substances known as alcoholic glucuronide and o-phenyl glucuronide.

It's important to be aware that the use of herbal medicines by pregnant women, whether under medical supervision or not, can potentially alter the way these drugs are metabolized.

9.3 Using HM During Pregnancy and Asthma

Asthma is a respiratory condition that impacts the bronchial airways of the lungs, and it is estimated to affect approximately 4–8% of pregnant women [73, 76, 86]. Inadequate management of asthma during pregnancy has been associated with adverse outcomes for both the mother and the infant. These maternal outcomes may include conditions such as hypertension, gestational diabetes, preterm delivery, and even maternal mortality. Similarly, infants of mothers with poorly controlled asthma may experience fetal malformations, illness, and, in severe cases, even mortality. Multiple studies have provided evidence supporting these associations [11, 69].

Research conducted by Al Ghobain et al. [9] and McLaughlin et al. [69] underscores the importance of effectively managing asthma during the perinatal period, showing that this leads to better outcomes for both mothers and their children. Consequently, the US Department of Health and Human Services (HHS) recommends giving high priority to pregnancy-specific asthma treatment. The primary focus of such treatment is to keep the airways open, reduce the need for short-acting inhaled beta2-agonists, and minimize the risk of severe exacerbations, which helps protect both the mother and her infant from experiencing low oxygen levels.

Standard daily asthma management during pregnancy often involves a combination of medications. This includes the use of a long-acting beta-agonist like salmeterol or formoterol, a leukotriene receptor antagonist such as montelukast or zafirlukast, and a short-acting beta2-agonist like albuterol. In some cases, systemic corticosteroids like prednisone, prednisolone, or methylprednisolone may be prescribed. Additionally, healthcare providers may recommend the use of methylxanthines like theophylline or 5-lipoxygenase inhibitors like omalizumab or cromolyn as part of the treatment regimen [75, 76].

10 Conclusion

Pregnant women and their healthcare providers must make informed decisions regarding the use of pharmaceuticals or natural remedies. However, the safety and efficacy of many plant-based treatments during pregnancy remain poorly documented, often relying on anecdotal evidence. In some cases, herbal remedies might offer alternatives to invasive medical procedures, potentially reducing complications associated with malnutrition. Natural remedies like black haw and false unicorn could help save fetuses that might otherwise be lost. Compared to hospital inductions, herbal labor induction methods may provide pregnant women with a greater sense of control during childbirth.

The herbal remedies discussed here deserve increased attention for their potential to assist expectant mothers and their caregivers in achieving safe, healthy pregnancies and satisfying birthing experiences. Nevertheless, natural treatments should be approached with a degree of skepticism. Recommendations for any medication, whether natural or synthetic, should be founded on comprehensive laboratory research and clinical trials.

The combination of herbal medicines (HMs) or other natural products with conventional medications during pregnancy raises growing concerns, particularly due to our limited knowledge of potential interactions and associated risks. Many pregnant women use HMs without consulting their healthcare providers, despite the documented positive outcomes HMs can have on maternal and infant health before, during, and after pregnancy. The common assumption that HMs are inherently safe can lead to unintended consequences.

Both pregnant women and medical professionals would benefit from enhanced education and a shift away from the misconception that HMs are universally harmless. There is a real risk of interactions between HMs and traditional over-the-counter medications. For example, pregnant women taking aspirin and HMs with antiplatelet properties face an elevated risk of hemorrhage, which can, in severe cases, lead to fatal consequences. The lack of research in this area is a significant concern, as little is known about the effects of HMs on fetal development and the potential adverse reactions when combined with conventional medications.

Moreover, the potential for medication interactions can significantly compromise the management of certain medical conditions during pregnancy, such as hypertension, asthma, and seizures. The reluctance to conduct research during pregnancy makes it challenging to determine the optimal medication dosage for a woman in her third trimester, especially when she is concurrently using HMs that may interfere with the effectiveness of conventional treatments. Equally concerning is the lack of study into the embryotoxic effects of HMs. Clinical pharmacologists, scientists, and healthcare providers should strive to better understand the results of experimental and clinical research concerning the harmful effects of HMs on fetal development.

Acknowledgments We would like to extend our sincere appreciation and gratitude to the Department of Agronomy at the School of Agriculture at Lovely Professional University, Punjab, 144411, India.

Conflicts of Interest A conflict of interest has not been identified.

References

1. Adams J (2011) The growing popularity of complementary and alternative medicine during pregnancy and implications for healthcare providers. Expert Rev Obstet Gynecol 4(6): 365–366. https://doi.org/10.1586/eog.11.29
2. Adams J, Lui CW, Sibbritt D, Broom A, Wardle J, Homer C (2009) Women's use of complementary and alternative medicine during pregnancy: a critical review of the literature. Birth-Iss Perinat C 36(3):237–245. https://doi.org/10.1111/j.1523-536X.2009.00328.x

3. Adane F, Seyoum G, Alamneh YM, Abie W, Desta M, Sisay B (2020) Herbal medicine use and predictors among pregnant women attending antenatal care in Ethiopia: a systematic review and meta-analysis. BMC Pregnancy Childbirth 20(1):157. https://doi.org/10.1186/s12884-020-2856-8
4. Adnan M, Gul S, Batool S, Bibi F, Rehman A, Yaqoob S (2017) A review on the ethnobotany, phytochemistry, pharmacology and nutritional composition of *Cucurbita pepo* L. J Phytopharmacol 6:133–139
5. Adusi-Poku Y, Vanotoo L, Detoh EK, Oduro J, Nsiah RB, Natogmah AZ (2015) Are orthodox prescribers aware of the type of herbal medicines used by pregnant women attending an antenatal clinic in Offinso North district? Ghana Med J 49(4):227–232. https://doi.org/10.4314/gmj.v49i4
6. Ahmed SM, Nordeng H, Sundby J, Aragaw YA, de Boer HJ (2018) The use of medicinal plants by pregnant women in Africa: a systematic review. J Ethnopharmacol 224:297–313. https://doi.org/10.1016/j.jep.2018.05.032
7. Aigberua AO, Izah SC (2019) pH variation, mineral composition and selected trace metal concentration in some liquid herbal products sold in Nigeria. Int J Res Stud Biosci 7(1):14–21
8. Aigberua AO, Izah SC (2019) Macro nutrient and selected heavy metals in powered herbal medicine sold in Nigeria. Int J Med Plants Nat Prod 5(1):23–29
9. Al Ghobain MO, AlNemer M, Khan M (2018) Assessment of knowledge and education relating to asthma during pregnancy among women of childbearing age. Asthma Res Pract 4:2. https://doi.org/10.1186/s40733-017-0038-x
10. Alammar N, Wang L, Saberi B, Nanavati J, Holtmann G, Shinohara RT (2019) The impact of peppermint oil on the irritable bowel syndrome: a meta-analysis of the pooled clinical data. BMC Complement Altern Med 19(1):21. https://doi.org/10.1186/s12906-018-2409-0
11. Baghlaf H, Spence AR, Czuzoj-Shulman N, Abenhaim HA (2017) Pregnancy outcomes among women with asthma. J Matern Fetal Neonatal Med:1–7. https://doi.org/10.1080/14767058.2017.1404982
12. Barnes LAJ, Barclay L, McCaffery K, Aslani P (2018) Complementary medicine products used in pregnancy and lactation and an examination of the information sources pertaining to maternal health literacy: a systematic review of qualitative studies. BMC Complement Altern Med 18(1):229. https://doi.org/10.1186/s12906-018-2283-9
13. Batisai K (2016) Towards an integrated approach to health and medicine in Africa. SAHARA J J Soc Aspects HIV/AIDS Res Alliance 13(1):113–122. https://doi.org/10.1080/17290376.2016.1220323
14. Bayisa B, Tatiparthi R, Mulisa E (2014) Use of herbal medicine among pregnant women on antenatal care at Nekemte hospital, Western Ethiopia. Jundishapur J Nat Pharm Prod 9(4):e17368. https://doi.org/10.17795/jjnpp-17368
15. Bhatt P, Leong TK, Yadav H, Doshi A (2016) Prevalence and factors associated with the use of herbal medicines among women receiving antenatal care in hospitals-Mumbai, India. Int J Phytother 6(2):82–90
16. Blumenthal M, Goldberg A, Brinckmann J (eds) (2000) Herbal medicine: expanded commission E monographs. American Botanical Council, Austin
17. Brookes K (2018) Chemical investigation of isihlambezo or traditional pregnancy related medicine. University of Kwa-Zulu Nata, School of Laboratory Medicine and Medical Science, Physiology. http://hdl.handle.net/10413/6093
18. Chan LY, Chiu PY, Lau TK (2003) An in-vitro study of ginsenoside Rb1-induced teratogenicity using a whole rat embryo culture model. Hum Reprod 18(10):2166–2168. https://doi.org/10.1093/humrep/deg401
19. Chime AO, Aiwansoba RO, Eze CJ, Osawaru ME, Ogwu MC (2017) Phenotypic characterization of tomato *Solanum lycopersicum* L. cultivars from southern Nigeria using morphology. Malaya J Biosci 4(1):30–38
20. Chuang CH, Doyle P, Wang JD, Chang PJ, Lai JN, Chen PC (2006) Herbal medicines used during the first trimester and major congenital malformations: an analysis of data from a

pregnancy cohort study. Drug Saf 29(6):537–548. https://doi.org/10.2165/00002018-200629060-00006

21. Committee on Obstetric Practice (2017) Committee opinion no. 692: emergent therapy for acute-onset, severe hypertension during pregnancy and the postpartum period. Obstet Gynecol 129(4):e90–e95. https://doi.org/10.1097/AOG.0000000000002019
22. Dante G, Bellei G, Neri I, Facchinetti F (2014) Herbal therapies in pregnancy: what works? Curr Opin Obstet Gynecol 26(2):83–91. https://doi.org/10.1097/gco.0000000000000052
23. Dika H, Dismas M, Iddi S, Rumanyika R (2017) Prevalent use of herbs for the reduction of labor duration in Mwanza, Tanzania: are obstetricians aware? Tanzan J Health Res 19(2):1–8. https://doi.org/10.4314/thrb.v19i2.5
24. Dugoua JJ (2010) Herbal medicines and pregnancy. J Popul Ther Clin Pharmacol 3(17): e370–e378
25. Duru CB, Uwakwe KA, Chinomnso NC, Mbachi II, Diwe KC, Agunwa CC (2016) Socio-demographic determinants of herbal medicine use in pregnancy among Nigerian women attending clinics in a tertiary hospital in Imo State, South-East, Nigeria. Am J Med Stud 4(1):1–10. https://doi.org/10.12691/ajms-4-1-1
26. Enioutina EY, Salis ER, Job KM, Gubarev MI, Krepkova LV, Sherwin CM (2017) Herbal medicines: challenges in the modern world. Part 5. Status and current directions of complementary and alternative herbal medicine worldwide. Expert Rev Clin Pharmacol 10(3): 327–338. https://doi.org/10.1080/17512433.2017.1268917
27. Frawley J, Adams J, Steel A, Broom A, Gallois C, Sibbritt D (2015) Women's use and self-prescription of herbal medicine during pregnancy: an examination of 1,835 pregnant women. Womens Health Issues 25(4):396–402. https://doi.org/10.1016/j.whi.2015.03.001
28. Fruscalzo A, Salmeri MG, Cendron A, Londero AP, Zanni G (2012) Introducing routine trial of labour after caesarean section in a second level hospital setting. J Matern Fetal Neonatal Med 25(8):1442–1446. https://doi.org/10.3109/14767058.2011.640367
29. Gakuya DW, Okumu MO, Kiama SG, Mbaria JM, Gathumbi PK, Mathiu PM (2020) Traditional medicine in Kenya: past and current status, challenges and the way forward. Sci Afr 8: e00360. https://doi.org/10.1016/j.sciaf.2020.e00360
30. Gaston TE, Bebin EM, Cutter GR, Liu Y, Szaflarski JP, Program UC (2017) Interactions between cannabidiol and commonly used antiepileptic drugs. Epilepsia 58(9):1586–1592. https://doi.org/10.1111/epi.13852
31. Gilad R, Hochner H, Savitsky B, Porat S, Hochner-Celnikier D (2018) Castor oil for induction of labor in post-date pregnancies: a randomized controlled trial. Women Birth 31(1):e26–e31. https://doi.org/10.1016/j.wombi.2017.06.010
32. Gunn JK, Rosales CB, Center KE, Nunez A, Gibson SJ, Christ C (2016) Prenatal exposure to cannabis and maternal and child health outcomes: a systematic review and meta-analysis. BMJ Open 6(4):e009986. https://doi.org/10.1136/bmjopen-2015-009986
33. Hall HG, Griffiths DL, McKenna LG (2011) The use of complementary and alternative medicine by pregnant women: a literature review. Midwifery 27(6):817–824. https://doi.org/10.1016/j.midw.2010.08.007
34. Hekmatpou D, Mehrabi F, Rahzani K, Aminiyan A (2019) The effect of *Aloe vera* clinical trials on prevention and healing of skin wound: a systematic review. Iran J Med Sci 44(1):1–9
35. Holst L, Nordeng H, Haavik S (2008) Use of herbal drugs during early pregnancy in relation to maternal characteristics and pregnancy outcome. Pharmacoepidemiol Drug Saf 17(2): 151–159. https://doi.org/10.1002/pds.1527
36. Holst L, Wright D, Haavik S, Nordeng H (2011) Safety and efficacy of herbal remedies in obstetrics-review and clinical implications. Midwifery 27(1):80–86. https://doi.org/10.1016/j.midw.2009.05.010
37. Hwang JH, Kim Y-R, Ahmed M, Choi S, Al-Hammadi NQ, Widad NM (2016) Use of complementary and alternative medicine in pregnancy: a cross-sectional survey on Iraqi women. BMC Complement Altern Med 16:191–191. https://doi.org/10.1186/s12906-016-1167-0

38. Imarhiagbe O, Ogwu MC (2022) Sacred groves in the global south: a panacea for sustainable biodiversity conservation. In: Izah SC (ed) Biodiversity in Africa: potentials, threats and conservation. Sustainable development and biodiversity, vol 29. Springer, Singapore, pp 525–546. https://doi.org/10.1007/978-981-19-3326-4_20
39. Isa A, Gaya B, Saleh M, Mohammed A, Muhammad AAM (2014) The modulatory role of aqueous extract of *Vernonia amygdalina* extract on pregnant rats. IOSR J Pharm Biol Sci 9(3): 45–47. https://doi.org/10.9790/3008-09324547
40. Izah SC, Uhunmwangho EJ, Dunga KE, Kigigha LT (2018) Synergy of methanolic leave and stem-back extract of *Anacardium occidentale* L. (cashew) against some enteric and superficial bacteria pathogens. MOJ Toxicol 4(3):209–211
41. Izah SC, Uhunmwangho EJ, Dunga KE (2018) Studies on the synergistic effectiveness of methanolic extract of leaves and roots of *Carica papaya* L. (papaya) against some bacteria pathogens. Int J Complement Altern Med 11(6):375–378
42. Izah SC, Uhunmwangho EJ, Etim NG (2018) Antibacterial and synergistic potency of methanolic leaf extracts of *Vernonia amygdalina* L. and *Ocimum gratissimum* L. J Basic Pharmacol Toxicol 2(1):8–12
43. Izah SC, Zige DV, Alagoa KJ, Uhunmwangho EJ, Iyamu AO (2018) Antibacterial efficacy of aqueous extract of *Myristica* fragrans (common nutmeg). EC Pharmacol Toxicol 6(4):291–295
44. Izah SC, Uhunmwangho EJ, Eledo BO (2018) Medicinal potentials of *Buchholzia* coriacea (wonderful kola). Med Plant Res 8(5):27–43
45. Izah SC, Chandel SS, Etim NG, Epidi JO, Venkatachalam T, Devaliya R (2019) Potency of unripe and ripe express extracts of long pepper (*Capsicum* frutescens var. baccatum) against some common pathogens. Int J Pharm Phytopharmacol Res 9(2):56–70
46. Izah SC, Etim NG, Ilerhunmwuwa IA, Silas G (2019) Evaluation of crude and ethanolic extracts of *Capsicum* frutescens var. minima fruit against some common bacterial pathogens. Int J Complement Altern Med 12(3):105–108
47. Izah SC, Etim NG, Ilerhunmwuwa IA, Ibibo TD, Udumo JJ (2019) Activities of express extracts of *Costus* afer Ker–Gawl. [family COSTACEAE] against selected bacterial isolates. Int J Pharm Phytopharmacol Res 9(4):39–44
48. Izah SC, Aigberua AO, Richard G (2022) Concentration, source, and health risk of trace metals in some liquid herbal medicine sold in Nigeria. Biol Trace Elem Res 200:3009–3302
49. Jafarian A, Zolfaghari B, Parnianifard M (2012) The effects of methanolic, chloroform, and ethylacetate extracts of the *Cucurbita pepo* L. on the delay type hypersensitivity and antibody production. Results Pharma Sci 7(4):217–224
50. James PB, Wardle J, Steel A, Adams J (2018) Traditional, complementary and alternative medicine use in sub-Saharan Africa: a systematic review. BMJ Glob Health 3(5):e000895. https://doi.org/10.1136/bmjgh2018-000895
51. Jeong H, Choi S, Song JW, Chen H, Fischer JH (2008) Regulation of UDP-glucuronosyltransferase (UGT) 1A1 by progesterone and its impact on labetalol elimination. Xenobiotica 38(1):62–75. https://doi.org/10.1080/00498250701744633
52. Jiang R, Yamaori S, Takeda S, Yamamoto I, Watanabe K (2011) Identification of cytochrome P450 enzymes responsible for metabolism of cannabidiol by human liver microsomes. Life Sci 89(5–6):165–170
53. Jo J, Lee SH, Lee JM, Lee H, Kwack SJ, Kim DI (2016) Use and safety of Korean herbal medicine during pregnancy: a Korean medicine literature review. Eur J Intern Med 8(1):4–11. https://doi.org/10.1016/j.eujim.2015.10.008
54. John LJ, Shantakumari N (2015) Herbal medicines use during pregnancy: a review from the Middle East. Oman Med J 30(4):229–236. https://doi.org/10.5001/omj.2015.48
55. Jurgens TM (2003) Potential toxicities of herbal therapies in the developing fetus. Birth Defects Res B Dev Reprod Toxicol 68(6):496–498. https://doi.org/10.1002/bdrb.10050
56. Kaingu CK, Oduma JA, Kanui TI (2011) Practices of traditional birth attendants in Machakos District, Kenya. J Ethnopharmacol 137(1):495–502. https://doi.org/10.1016/j.jep.2011.05.044

57. Kennedy DA, Lupattelli A, Koren G, Nordeng H (2013) Herbal medicine use in pregnancy: results of a multinational study. BMC Complement Altern Med 13(355):1–10. https://doi.org/10.1186/1472-6882-13-355
58. Kennedy DA, Lupattelli A, Koren G, Nordeng H (2016) Safety classification of herbal medicines used in pregnancy in a multinational study. BMC Complement Altern Med 16: 102. https://doi.org/10.1186/s12906-016-1079
59. Krepkova LV, Savinova TB, Bortnikova VV, Borovkova MV, Dmitrieva OP, Sidel'nikov NI (2016) The effect of thermopsis. (thermopsis lanceolata R. BR.) dry extract on the embryogenesis of rats. Problems Biol Med Pharmaceut Chem 1:8–12
60. Kwon HL, Triche EW, Belanger K, Bracken MB (2006) The epidemiology of asthma during pregnancy: prevalence, diagnosis, and symptoms. Immunol Allergy Clin N Am 26(1):29–62. https://doi.org/10.1016/j.iac.2005.11.002
61. Lai C, Coulter SA, Woodruff A (2017) Hypertension and pregnancy. Tex Heart Inst J 44(5): 350–351. https://doi.org/10.14503/THIJ-17-6359
62. Lapi F, Vannacci A, Moschini M, Cipollini F, Morsuillo M, Gallo E et al (2010) Use, attitudes and knowledge of complementary and alternative drugs (CADs) among pregnant women: a preliminary survey in Tuscany. Evid Based Complement Alternat Med 4(7):477–486. https://doi.org/10.1093/ecam/nen031
63. Liu W, Ge T, Pan Z, Leng Y, Lv J, Li B (2017) The effects of herbal medicine on epilepsy. Oncotarget 8(29):48385–48397. https://doi.org/10.18632/oncotarget.16801
64. Louik C, Gardiner P, Kelley K, Mitchell A (2010) Use of herbal treatments in pregnancy. Am J Obstet Gynecol 5(202):e431–e439. https://doi.org/10.1016/j.ajog.2010.01.055
65. Mahmoodpoor A, Medghalchi M, Nazemiyeh H, Asgharian P, Shadvar K, Hamishehkar H (2018) Effect of *Cucurbita* maxima on control of blood glucose in diabetic critically ill patients. Adv Pharm Bull 8(2):347–351. https://doi.org/10.15171/apb.2018.040
66. Mahmoudian A, Rajaei Z, Haghir H, Banihashemian S, Hami J (2012) Effects of valerian consumption during pregnancy on cortical volume and the levels of zinc and copper in the brain tissue of mouse fetus. Zhong Xi Yi Jie He Xue Bao 10(4):424–429. https://doi.org/10.3736/jcim20120411
67. Maluma S, Kalungia AC, Hamachila A, Hangoma J, Munkombwe D (2017) Prevalence of traditional herbal medicine use and associated factors among pregnant women of Lusaka Province, Zambia. J Prev Rehabil Med 1:5–11. https://doi.org/10.21617/jprm.2017.0102.1
68. Maputle SM, Mothiba TM, Maliwichi L (2015) Traditional medicine and pregnancy management: perceptions of traditional health practitioners in Capricorn district, Limpopo Province. Stud Ethno-Med 9(1):67–75. https://doi.org/10.1080/09735070.2015.11905422
69. McLaughlin K, Foureur M, Jensen ME, Murphy VE (2018) Review and appraisal of guidelines for the management of asthma during pregnancy. Women Birth 31(6):e349–e357. https://doi.org/10.1016/j.wombi.2018.01.008
70. Mishra A, Dave N (2013) Neem oil poisoning: case report of an adult with toxic encephalopathy. Indian J Crit Care Med 17(5):321–322. https://doi.org/10.4103/0972-5229.120330
71. Moshi M, Mhame P (2013) Legislation on medicinal plants in Africa. In: Kuete V (ed) Medicinal plant research in Africa: pharmacology and chemistry. Elsevier, Amsterdam
72. Mugomeri E, Seliane K, Chatanga P, Maibvise C (2015) Identifying promoters and reasons for medicinal herb usage during pregnancy in Maseru, Lesotho. Afr J Nurs Midwifwery 17(1): 4–16. https://doi.org/10.25159/2520-5293/63
73. Murphy VE, Schatz M (2014) Asthma in pregnancy: a hit for two. Eur Respir Rev 23(131): 64–68. https://doi.org/10.1183/09059180.00008313
74. Mustafa R, Ahmed S, Gupta A, Venuto RC (2012) A comprehensive review of hypertension in pregnancy. J Pregnancy:105918. https://doi.org/10.1155/2012/105918
75. Namazy JA, Schatz M (2018) Management of asthma during pregnancy: optimizing outcomes and minimizing risk. Semin Respir Crit Care Med 39(1):29–35. https://doi.org/10.1055/s-0037-1606216

76. National Heart Lung, Blood Institute, National Asthma Education, Prevention Program Asthma, Pregnancy Working Group (2005) NAEPP expert panel report. Managing asthma during pregnancy: recommendations for pharmacologic treatment-2004 update. J Allergy Clin Immunol 115(1):34–46. https://doi.org/10.1016/j.jaci.2004.10.023
77. Nergard CS, Ho TPT, Diallo D, Ballo N, Paulsen BS, Nordeng H (2015) Attitudes and use of medicinal plants during pregnancy among women at health care centers in three regions of Mali, West-Africa. J Ethnobiol Ethnomed 11:73. https://doi.org/10.1186/s13002-015-0057-8
78. Nordeng H, Bayne K, Havnen GC, Paulsen BS (2011) Use of herbal drugs during pregnancy among 600 Norwegian women in relation to concurrent use of conventional drugs and pregnancy outcome. Complement Ther Clin Pract 17(3):147–151. https://doi.org/10.1016/j.ctcp.2010.09.002
79. Nyeko R, Tumwesigye NM, Halage AA (2016) Prevalence and factors associated with use of herbal medicines during pregnancy among women attending postnatal clinics in Gulu district, northern Uganda. BMC Pregnancy Childbirth 16(1):296. https://doi.org/10.1186/s12884-016-1095-5
80. Ogwu MC (2018) Effects of indole-3-acetic acid on growth parameters of *Citrullus lanatus* (Thunberg) Matsum and Nakai. Momona Ethiop J Sci 10(1):109–125. https://doi.org/10.4314/mejs.v10i1.8
81. Ogwu MC (2019) Lifelong consumption of plant-based GM foods: is it safe? In: Papadopoulou P, Misseyanni A, Marouli C (eds) Environmental exposures and human health challenges. IGI Global, Pennsylvania, pp 158–176. https://doi.org/10.4018/978-1-5225-7635-8.ch008
82. Ogwu MC (2019) Towards sustainable development in Africa: the challenge of urbanization and climate change adaptation. In: Cobbinah PB, Addaney M (eds) The geography of climate change adaptation in urban Africa. Springer Nature, Cham, pp 29–55. https://doi.org/10.1007/978-3-030-04873-0_2
83. Ogwu MC (2020) Value of *Amaranthus* [L.] species in Nigeria. In: Waisundara V (ed) Nutritional value of Amaranth. IntechOpen, London, pp 1–21. https://doi.org/10.5772/intechopen.86990
84. Ogwu MC (2023) Local food crops in Africa: sustainable utilization, threats, and traditional storage strategies. In: Izah SC, Ogwu MC (eds) Sustainable utilization and conservation of Africa's biological resources and environment. Sustainable development and biodiversity, vol 888. Springer, Singapore, pp 353–374. https://doi.org/10.1007/978-981-19-6974-4_13
85. Ogwu MC, Osawaru ME (2022) Traditional methods of plant conservation for sustainable utilization and development. In: Izah SC (ed) Biodiversity in Africa: potentials, threats and conservation. Sustainable development and biodiversity, vol 29. Springer, Singapore, pp 451–472. https://doi.org/10.1007/978-981-19-3326-4_17
86. Ogwu MC, Osawaru ME (2023) Disease outbreaks in ex-situ plant conservation and potential management strategies. In: Izah SC, Ogwu MC (eds) Sustainable utilization and conservation of Africa's biological resources and environment. Sustainable development and biodiversity, vol 888. Springer, Singapore, pp 497–518. https://doi.org/10.1007/978-981-19-6974-4_18
87. Ogwu MC, Osawaru ME, Chime AO (2014) Comparative assessment of plant diversity and utilization patterns of tropical home gardens in Edo State, Nigeria. Sci Afr 13(2):146–162
88. Ogwu MC, Osawaru ME, Aiwansoba RO, Iroh RN (2016) Ethnobotany and collection of west African okra [*Abelmoschus caillei* (A. Chev.) Stevels] germplasm in some communities in Edo and Delta States, Southern Nigeria. Borneo J Resour Sci Technol 6(1):25–36. https://doi.org/10.33736/bjrst.212.2016
89. Ogwu MC, Osawaru ME, Obadiaru O, Itsepopo IC (2016) Nutritional composition of some uncommon indigenous spices and vegetables. In: International conference on natural resource development and utilization (4th edition): positioning small and medium enterprises (SMEs) for optimal utilization of available resources and economic growth. Raw Materials Research and Development Council (Federal Ministry of Science and Technology). Maitama, Abuja, pp 286–300
90. Ogwu MC, Osawaru ME, Aiwansoba RO, Iroh RN (2016) Status and prospects of vegetables in Africa. In: Borokini IT, Babalola FD (eds) Conference proceedings of the joint biodiversity conservation conference of Nigeria Tropical Biology Association and Nigeria chapter of

Society for conservation biology on MDGs to SDGs: toward sustainable biodiversity conservation in Nigeria. University of Ilorin, Ilorin, pp 47–57pp

91. Ogwu MC, Osawaru ME, Obahiagbon GE (2017) Ethnobotanical survey of medicinal plants used for traditional reproductive care by Usen people of Edo State, Nigeria. Malaya J Biosci 4(1):17–29
92. Ogwu MC, Chime AO, Oseh OM (2018) Ethnobotanical survey of tomato in some cultivated regions in Southern Nigeria. Maldives Natl Res J 6(1):19–29
93. Ogwu MC, Izah SC, Iyiola AO (2022) An overview of the potentials, threats and conservation of biodiversity in Africa. In: Izah SC (ed) Biodiversity in Africa: potentials, threats and conservation. Sustainable development and biodiversity, vol 29. Springer, Singapore, pp 3–20. https://doi.org/10.1007/978-981-19-3326-4_1
94. Ogwu MC, Osawaru ME, Owie MO (2023) Effects of storage at room temperature on the food components of three cocoyam species (*Colocasia esculenta, Xanthosoma atrovirens, and X. sagittifolium*). Food Stud Interdiscip J 13(2):59–83. https://doi.org/10.18848/2160-1933/CGP/v13i02/59-83
95. Ogwu MC, Osawaru ME, Amodu E, Osamo F (2023) Comparative morphology, anatomy, and chemotaxonomy of two *Cissus* Linn. species. Rev Bras Bot. https://doi.org/10.1007/s40415-023-00881-0
96. Ogwu MC, Iyiola AO, Izah SC (2023) Overview of African biological resources and environment. In: Izah SC, Ogwu MC (eds) Sustainable utilization and conservation of Africa's biological resources and environment. Sustainable development and biodiversity, vol 888. Springer, Singapore, pp 1–31. https://doi.org/10.1007/978-981-19-6974-4_1
97. Onyiapat JL, Okafor C, Okoronkwo I, Anarado A, Chukwukelu E, Nwaneri A (2017) Complementary and alternative medicine use: results from a descriptive study of pregnant women in Udi local Government area of Enugu state, Nigeria. BMC Complem Altern M 17(1):189. https://doi.org/10.1186/s12906-017-1689-0
98. Osawaru ME, Ogwu MC (2014) Ethnobotany and germplasm collection of two genera of cocoyam (*Colocasia* [Schott] and *Xanthosoma* [Schott], Araceae) in Edo state Nigeria. Sci Technol Arts Res J 3(3):23–28. https://doi.org/10.4314/star.v3i3.4
99. Osawaru ME, Ogwu MC (2020) Survey of plant and plant products in local markets within Benin City and environs. In: Filho LW, Ogugu N, Ayal D, Adelake L, da Silva I (eds) African handbook of climate change adaptation. Springer Nature, Cham, pp 1–24. https://doi.org/10.1007/978-3-030-42091-8_159-1
100. Osawaru ME, Ogwu MC, Omoigui ID, Aiwansoba RO, Kevin A (2016) Ethnobotanical survey of vegetables eaten by Akwa Ibom people residing in Benin City, Nigeria. Uni Benin J Sci Technol 4(1):70–93
101. Pai MR, Acharya LD, Udupa N (2004) Evaluation of antiplaque activity of Azadirachta indica leaf extract gel – a 6-week clinical study. J Ethnopharmacol 90(1):99–103. https://doi.org/10.1016/j.jep.2003.09.035
102. Pallivalapila AR, Stewart D, Shetty A, Pande B, Singh R, McLay JS (2015) Use of complementary and alternative medicines during the third trimester. Obstet Gynecol 125(1):204–211. https://doi.org/10.1097/AOG.0000000000000596
103. Panganai T, Shumba P (2016) The African Pitocin – a midwife's dilemma: the perception of women on the use of herbs in pregnancy and labour in Zimbabwe, Gweru. Pan Afr Med J 25:9. https://doi.org/10.11604/pamj.2016.25.9.7876
104. Peprah P, Agyemang-Duah W, Arthur-Holmes F, Budu HI, Abalo EM, Okwei R (2019) 'We are nothing without herbs': a story of herbal remedies use during pregnancy in rural Ghana. BMC Complement Altern Med 19(1):65. https://doi.org/10.1186/s12906-019-2476-x
105. Rahman AA, Sulaiman SA, Ahmad Z, Daud WN, Hamid AM (2008) Prevalence and pattern of use of herbal medicines during pregnancy in tumpat district, Kelantan. Malays J Med Sci 15(3):40–48
106. Sahalie NA, Abrha LH, Tolesa LD (2018) Chemical composition and antimicrobial activity of leave extract of *Ocimum* lamiifolium (Damakese) as a treatment for urinary tract infection. Cogent Chem 4(1):1440894. https://doi.org/10.1080/23312009.2018.1440894

107. Saper RB, Kales SN, Paquin J, Burns MJ, Eisenberg DM, Davis RB et al (2004) Heavy metal content of ayurvedic herbal medicine products. JAMA 292(23):2868–2873. https://doi.org/10.1001/jama.292.23.2868
108. Shewamene Z, Dune T, Smith CA (2017) The use of traditional medicine in maternity care among African women in Africa and the diaspora: a systematic review. BMC Complement Altern Med 17(1):382. https://doi.org/10.1186/s12906-017-1886-x
109. Smeriglio A, Tomaino A, Trombetta D (2014) Herbal products in pregnancy: experimental studies and clinical reports. Phytother Res 28(8):1107–1116. https://doi.org/10.1002/ptr.5106
110. Tabatabaee M (2011) Use of herbal medicine among pregnant women referring to Valiasr hospital in Kazeroon, Fars, South of Iran. J Med Plants Res 1(37):96–108
111. Tamuno I, Omole-Ohonsi A, Fadare J (2010) Use of herbal medicine among pregnant women attending a tertiary hospital in Northern Nigeria. Int J Gynecol Obstet 15(2). https://doi.org/10.5580/2932
112. Tang L, Lee AH, Binns CW, Hui YV, Yau KKW (2016) Consumption of Chinese herbal medicines during pregnancy and postpartum: a prospective cohort study in China. Midwifery 34:205–210. https://doi.org/10.1016/j.midw.2015.11.010
113. Teni FS, Birru EM, Surur AS (2017) Pattern and predictors of medicine use among households in Gondar Town, northwestern Ethiopia: a community-based medicine utilization study. BMC Res Notes 10(1):357. https://doi.org/10.1186/s13104-017-2669-7
114. Tiran D (2003) The use of herbs by pregnant and childbearing women: a riskbenefit assessment. Complement Ther Nurs Midwifery 9(4):176–181. https://doi.org/10.1016/S1353-6117(03)00045-3
115. Townsend R, O'Brien P, Khalil A (2016) Current best practice in the management of hypertensive disorders in pregnancy. Integr Blood Pressure Control 9:79–94. https://doi.org/10.2147/IBPC.S77344
116. Trabace L, Tucci P, Ciuffreda L, Matteo M, Fortunato F, Campolongo P (2015) "Natural" relief of pregnancy-related symptoms and neonatal outcomes: above all do no harm. J Ethnopharmacol 174:396–402. https://doi.org/10.1016/j.jep.2015.08.046
117. Tunaru S, Althoff TF, Nüsing RM, Diener M, Offermanns S (2012) Castor oil induces laxation and uterus contraction via ricinoleic acid activating prostaglandin EP3 receptors. Proc Natl Acad Sci U S A 109(23):9179–9184. https://doi.org/10.1073/pnas.1201627109
118. Viljoen E, Visser J, Koen N, Musekiwa A (2014) A systematic review and meta-analysis of the effect and safety of ginger in the treatment of pregnancyassociated nausea and vomiting. Nutr J 13:20. https://doi.org/10.1186/1475-2891-13-20
119. Wang HP, Yang J, Qin LQ, Yang XJ (2015) Effect of garlic on blood pressure: a meta-analysis. J Clin Hypertens (Greenwich) 17(3):223–231. https://doi.org/10.1111/jch.12473
120. Xia HC, Li F, Li Z, Zhang ZC (2003) Purification and characterization of Moschatin, a novel type I ribosome-inactivating protein from the mature seeds of pumpkin (*Cucurbita moschata*), and preparation of its immunotoxin against human melanoma cells. Cell Res 13(5):369–374. https://doi.org/10.1038/sj.cr.7290182
121. Yamaori S, Ebisawa J, Okushima Y, Yamamoto I, Watanabe K (2011) Potent inhibition of human cytochrome P450 3A isoforms by cannabidiol: role of phenolic hydroxyl groups in the resorcinol moiety. Life Sci 88(15–16):730–736. https://doi.org/10.1016/j.lfs.2011.02.017
122. Yamaori S, Okamoto Y, Yamamoto I, Watanabe K (2011) Cannabidiol, a major phytocannabinoid, as a potent atypical inhibitor for CYP2D6. Drug Metab Dispos 39(11):2049–2056. https://doi.org/10.1124/dmd.111.041384
123. Yamaori S, Kinugasa Y, Jiang R, Takeda S, Yamamoto I, Watanabe K (2015) Cannabidiol induces expression of human cytochrome P450 1A1 that is possibly mediated through aryl hydrocarbon receptor signaling in HepG2 cells. Life Sci 136:87–93. https://doi.org/10.1016/j.lfs.2015.07.007
124. Zisaki A, Miskovic L, Hatzimanikatis V (2015) Antihypertensive drugs metabolism: an update to pharmacokinetic profiles and computational approaches. Curr Pharm Des 21(6):806–822. https://doi.org/10.2174/1381612820666141024151119

Herbal Medicine and Rheumatic Disorders Management and Prevention 24

Retno Widyowati, Rizki Rahmadi Pratama, Irawati Sholikhah, and Sachin Kumar Jain

Contents

R. Widyowati (✉)
Department of Pharmaceutical Science, Faculty of Pharmacy, Universitas Airlangga, Surabaya, Indonesia
e-mail: rr-retno-w@ff.unair.ac.id

R. R. Pratama
Master Program of Pharmaceutical Science, Faculty of Pharmacy, Universitas Airlangga, Surabaya, Indonesia

I. Sholikhah
Department of Chemistry, Faculty of Science and Technology, Universitas Airlangga, Surabaya, Indonesia

S. K. Jain
IPS Academy College of Pharmacy, Oriental College of Pharmacy and Research, Oriental University, Indore, Madhya Pradesh, India

S. C. Izah et al. (eds.), *Herbal Medicine Phytochemistry*, Reference Series in Phytochemistry, https://doi.org/10.1007/978-3-031-43199-9_26

Abstract

Rheumatoid disorder is a chronic inflammatory disorder of the joints that causes discomfort, inflexibility, swelling, deformity, and sometimes disability. Its cause remains a mystery. The severity of irreversible joint degradation may be mitigated with early diagnosis and therapy with disease-modifying anti-rheumatic drugs, as recommended by the American Academy of Rheumatology. Although effective treatments were quickly identified, harmful side effects often undermine their effectiveness. Herbal treatments are gaining in popularity and effectiveness for several forms of arthritis. According to a recent US poll, 90% of people with rheumatoid arthritis use complementary therapy, like herbal medications. The evidence for herbal treatments for four prevalent arthritic conditions back pain, fibromyalgia, osteoarthritis, and rheumatoid arthritis is briefly reviewed in this chapter. Some plants commonly used for the prevention or therapy of rheumatoid disorder are *Zingiber officinale*, *Piper methysticum*, *Ephedra* sp., *Oenothera biennis*, *Panax ginseng*, *Ginkgo biloba*, *Allium sativum*, *Ganoderma lucidum*, *Tripterygium wilfordii*, *Uncaria tomentosa*, *Boswellia serrata*, *Tinospora cordifolia*, *Tinospora cordifolia*, *Tribulus terrestris*, *Curcuma longa*, *Caesalpinia sappan*, *Clematis mandshurica*, *Rosmarinus officinalis*, *Elaeagnus angustifolia*, *Crocus sativus*, *Camellia sinensis*, *Matricaria chamomilla*, *Punica granatum*, *Cinnamomum cassia*, *Centella asiatica*, *Phyllanthus amarus*, *Psidium guajava*, *Moringa oleifera*, *Nigella sativa*, *Cannabis sativa*, *Andrographis paniculata*, *Artemisia annua*, and *Piper betle*. The possibility of treating rheumatic diseases using these plants as well as the chemical components that have this impact will be discussed in silico, in vitro, in vivo, and in clinical terms.

Keywords

Clinical study · Herbal medicine · In silico · In vitro · In vivo · Rheumatoid

Abbreviations

ADAMTS	Disintegrin and metalloproteinase with thrombospondin motifs
BSA	Bovine serum albumin
BW	Body weight
CAM	Complementary treatments
CFA	Complete Freund adjuvant
CIA	Collagen-induced arthritis
COX	Cyclooxygenase
CRP	C-reactive protein
DMARD	Disease-modifying anti-rheumatic drugs
DMS	Duration of morning stiffness
EAE	Experimental autoimmune encephalomyelitis
IL	Interleukin
IL-1B	Interleukin-1β
INOS	Inducible nitric oxide synthase
KG	Kilogram
LPS	Lipopolysaccharide
MAPK	Mitogen-activated protein kinase
MG	Milligram
MMP	Metalloproteinases
MSM	Methylsulfonylmethane
MTX	Methotrexate
NO	Nitric oxide
NSAID	Non-steroidal anti-Inflammatory drugs
PDB	Protein Data Bank
PGE2	Prostaglandin-E2
RA	Rheumatoid arthritis
RF	Rheumatoid factor
ROS	Reactive oxygen species
SJC	Number of swollen joints
SLE	Systemic lupus erythematosus
TJC	Number of painful joints
TLR	Toll-like receptor
TNF-α	Tumour necrosis factor-alpha
VAV	Visual analog scale
WHO	World Health Organization
WT	Wild type

1 Introduction

Rheumatic illnesses include fibromyalgia, osteoarthritis, and rheumatoid arthritis (Fig. 1). Rheumatic disorders are associated with decreased productivity, substantial disability, and diminished quality of life [1]. Conventional treatment for rheumatic

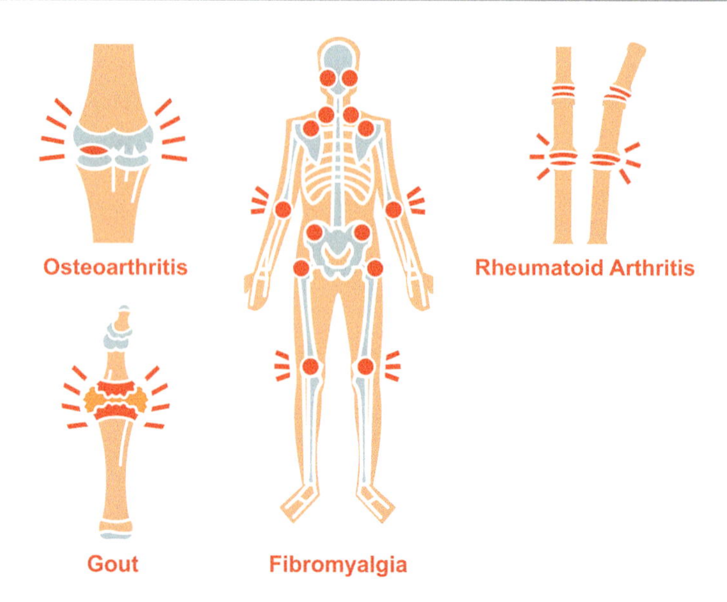

Fig. 1 Rheumatic disorders

disorders usually uses disease-modifying anti-rheumatic drugs (DMARDs), non-steroidal anti-inflammatory drugs (NSAIDs), and biologics. Some individuals may not react well to these conventional therapies. Thus, people with rheumatic illnesses usually seek alternative and complementary treatments (CAM) [2–4].

According to the World Health Organization, complementary and alternative medicine includes a wide variety of healthcare practices that are not integrated into the mainstream healthcare system and are not part of the country's tradition [5]. Under the guise of complementary and alternative medicine, a range of methods may be utilized to prevent or treat illnesses. Five groups can be used to categorize complementary and alternative medical practices, such as acupuncture, homoeopathy, herbal remedies, dietary ingredients, or additions derived from nature [6]. Around 60–90% of rheumatoid arthritis patients have employed complementary and alternative therapy, according to prior studies [7].

Prevention techniques for rheumatoid arthritis are classified as primary, secondary, or tertiary interventions. The objective of primary prevention is to prevent the onset of illness by removing particular risk factors and boosting an individual's resistance to the condition. The purpose of secondary prevention is to slow the development of a disease from its latent or asymptomatic phase to its symptomatic phase. Hence, a secondary preventive intervention aims to thwart the processes of disease development before they manifest as a disease. Cancer screening programmes such as mammography and colonoscopies are examples of this strategy. Tertiary prevention has the purpose to postpone or mitigate the effects of a disease that has already taken hold. Here is where the majority of rheumatic illnesses are being treated, with rheumatologists seeking to avoid the progression of the disease to a disability or early death when a patient presents with clinically evident disease (swollen joints in RA, or skin rash in SLE) [5].

The therapeutic objectives for rheumatoid arthritis are to lessen symptoms, stop the disease's progression, and improve quality of life. The conditions considered include analgesic relief, inflammation reduction, articular structure protection, function maintenance, and systemic involvement control before starting rheumatoid arthritis treatment [8].

This chapter examines various traditional herbs from China, India, and other Asian nations that are used as arthritis treatments. It also will go into greater detail about in silico, in vitro, in vivo, and clinical trial research on arthritis-curing plants. The data are gotten by searching the keyword (clinical study, herbal medicine, in silico, in vitro, in vivo, rheumatoid) at Google Scholar, PubMed, and other Internet sources.

2 Herbal Medicine

Herbal medicine is generally defined as a raw material or a preparation derived from plants with a therapeutic or medicinal effect that benefits human health. Herbal medicine is composed of raw materials or materials that have gone through further processing from one or more plant species. Below are some plants that have benefits as herbal medicine.

2.1 *Zingiber officinale*

Zingiber officinale (ginger) belongs to the Zingiberaceae family and has numerous advantages because of its compounds. In in silico studies, their bioactive compounds 8-gingerol, 6-gingerol, and zingerone had the potential to be inhibitors of several rheumatoid arthritis proteins [10]. These compounds carried out molecular binding on several rheumatoid arthritis protein targets, namely, cyclooxygenase-2 (COX-2), interleukin-1β (IL-1β), macrophage colony-stimulating factor (MCSF), matrix metalloproteinases-9 (MMP-9), and tumour necrosis factor-alpha (TNF-α) proteins (Fig. 2) [2]. An in vitro experiment utilizing the aqueous extract of ginger at 125 g/ml revealed that its anti-inflammatory activity was higher than diclofenac (58% vs. 52%) at the same concentration. In another study, prostaglandin-E2 (PGE2), IL-6, and IL-8 were inhibited by the 1 g/ml of ginger ethanol extract on the mRNA expression of inducible nitric oxide synthase (iNOS) and COX-2 in Caco-2 adenocarcinoma cells [11]. Rats were induced by I/R and fed once per day for 3 days in a row before the application of 6-gingerol (1, Fig. 5) from *Zingiber officinale* Roscoe extract. When administered orally at 25 and 50 mg/kg BW, the interleukin-1 and 6 and TNF-α were found to be at lower levels, as was the neutrophil count [12]. Expression of IL-12 and TGF-α in the central nervous system and serum was shown to decrease when ginger essential oil was given to experimental autoimmune encephalomyelitis (EAE) mice at 200 and 300 mg/kg [13]. In different clinical trial, it was discovered that ginger, either by itself or in combination with *Echinacea* extract (15 patients over 60 years old, 25 mg of ginger, and 5 mg of *Echinacea* extract for 30 days), considerably reduced knee discomfort [9].

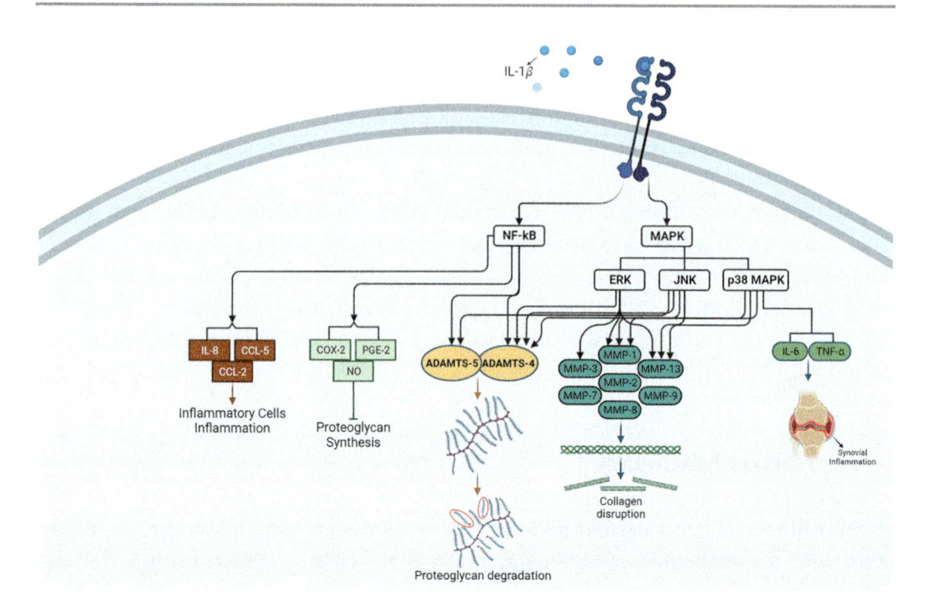

Fig. 2 Mechanism of rheumatic arthritis that is influenced by several protein targets [9]

2.2 *Piper methysticum*

Piper methysticum (kava) is a plant originating from Papua, Indonesia. Their bioactive compounds include alkaloids, flavonoids, tannins, saponins, anthraquinones, and essential oils. Based on the outcomes of molecular docking of the side chain with the morpholine ring occupying the polar domain and ring, one of the compounds from this plant showed action in suppressing iNOS protein is known as protein data bank (PDB) ID: 3E67 [10]. In in vitro investigation of nitric oxide (NO) inhibitory generation, this extract had an inhibition rate of 84.0% at a dose of 10 mM (IC_{50} 14 6.4 M) with the least amount of cytotoxicity (Fig. 3). In addition, the kavain (2, Fig. 5) in this extract significantly reduced TNF-α production induced by *E. coli* lipopolysaccharide (LPS) in primary macrophages of WT mice. In in vivo evaluation, this compound also has a significant anti-inflammatory effect on mice-induced CAIA (collagen antibody-induced arthritis) [14].

2.3 *Ephedra* Sp.

A herb known as *Ephedra* is frequently used in traditional medicine to treat fever, headaches, cough, and shortness of breath. *Ephedra alte* contains β-sitosterol (3, Fig. 5) and androstan-3-one efficiently inhibited inflammation and reduced oxidative stress by inhibiting five target proteins such as IL-6, toll-like receptor-4 (TLR4) hybrid, TNF-α, TLRs, and IL-1β. Fresh chicken egg albumin denaturation

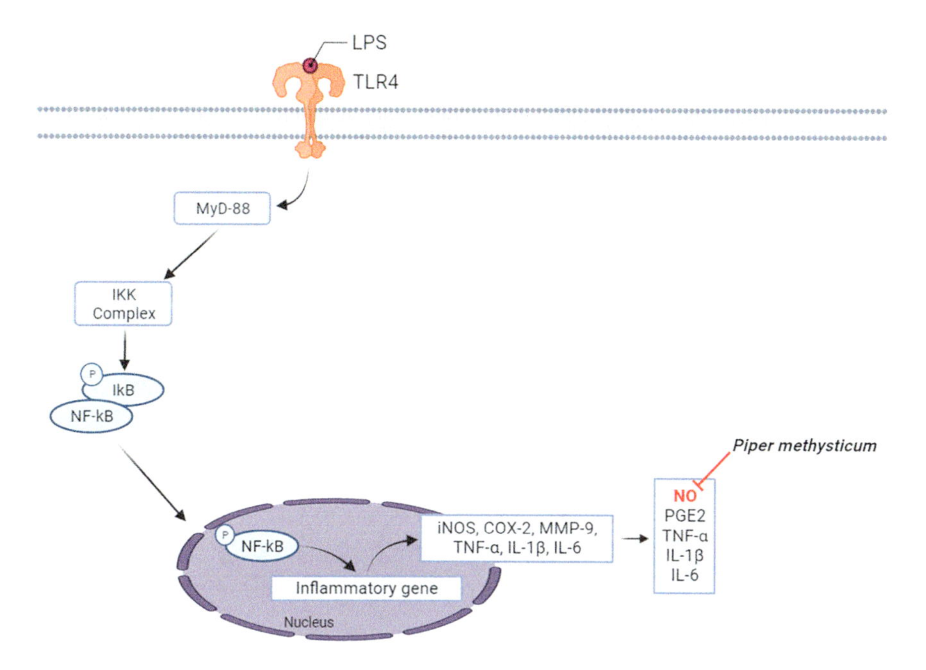

Fig. 3 Mechanism of *Piper methysticum* in inhibiting NO in RAW 264.7 cells

and membrane stabilization were performed on ethanol, ethyl acetate, n-butanol, and aqueous extracts of *Ephedra gerardiana* at 6400 μg/mL in vitro experiments involving bovine serum albumin (BSA) inhibition tests. Their aqueous extract exhibited the greatest inhibitory impact (98.5%). Then n-butanol and ethyl acetate inhibited 92.1 and 85.7%, respectively. Meanwhile, the ethanol extract with the same concentration showed an inhibitory effect of 99.1%. All extracts showed better inhibitory effect results compared to aspirin (80.5%) [15]. In other species, extracts of *Ephedra alata* and *Ephedra fragilis* inhibited NO in RAW cells 264.7 by 62% and 70% at 50 mg/mL, respectively [16]. ESP-B4 (an acidic hetero-polysaccharide) derived from *Ephedra sinica* also demonstrated reduced LPS-induced cytokine production. Moreover, pre-treatment with ESP-B4 greatly reduced the phosphorylation of mitogen-activated protein kinases (MAPKs) brought on by LPS [17]. *Ephedra gerardiana* at 200 mg/kg reduced leg volume and diameter in formaldehyde-induced rats [18]. To cure rheumatoid arthritis, it has been demonstrated that the ESP-B4 from *Ephedra sinica* inhibited the TLR4 signaling pathway, which reduces the production of inflammatory agents and cytokines [17]. This plant activity in rats given Freund's complete adjuvant was examined (FCA). Rat foot oedema can be reduced and haematological and biochemical alterations were greatly improved by this extract. Moreover, in all extracts-treated mice, the overproduction of pro-inflammatory cytokine was markedly reduced. Nonetheless, there was a significant rise in IL-4 and 10 [18].

2.4 *Oenothera biennis*

NO, TNF-α, IL-1, and thromboxane B2 (TXB2) production can be reduced by the sterol components found in *Oenothera biennis* oil or evening primrose oil (EPO). Sterols do not, however, lower PGE2 [19]. TNF-α, IL-6, and IL-1 were significantly inhibited by this extract in peritoneal macrophages (PMs), and extracellular signal-regulated ERK signaling, P38 of MAPK, and LPS-activated NF-kappaB (NF-kB) were also inhibited by *Oenothera biennis* L. [19]. A study of the arthritic group's response to EPO revealed a substantial decrease in body weight and an increase in the ankle, plasma angiopoietin-1, and TNF-α [20]. In arthritic rats, EPO increased platelet count, white blood cell count, and high ankle diameter. EPO therapy significantly reduced the levels of inflammatory biomarkers in arthritic animals, including rheumatoid factor, TNF-α, IL-1β, and IL-6. As measured by disease activity score-28 (DAS28), IL-17, tacrolimus (TAC), rheumatoid factor (RF), and C-reactive protein (CRP) values after 3 months in clinical studies on 90 rheumatoid arthritis patients who received EPO 500 mg/TDS, the results might lessen DAS28 score and severity of rheumatoid arthritis patients [21].

2.5 *Panax ginseng*

With binding affinities of −7.58 and −7.89 kcal/mol for NF-kB protein, the ginsenoside (4, Fig. 5) derived from *Panax ginseng* has an anti-inflammation medicinal agent. Moreover, it reduced the reactive oxygen species (ROS) in anti-inflammatory experiments utilizing an in vitro model of NF-kB in RAW 264.7 macrophages activated by LPS [22]. It boosted the expression of early and late-stage differentiation markers (alkaline phosphatase, collagen type-I, and osteocalcin) during osteoblast differentiation on MC3T3-E1 cells that were exposed to H_2O_2. It also decreased the overexpression of pro-inflammatory cytokines such as IL-1, NO expression, and ROS generation [23]. Ginsenoside also suppresses osteoclastogenesis in human CD14+ cells [24]. In IL-1-stimulated chondrocytes, ginsenoside Rg3 increased the expression of aggrecan and collagen type 2 (COL-II), two chondroprotective proteins, while decreasing MMP-1 and MMP-13 [25]. A mouse model of osteoarthritis was used to demonstrate that ginsenoside Rg5 dramatically ameliorated osteoarthritis symptoms by decreasing cartilage breakdown and cell death. Whereas other research demonstrated that 15 mg/kg of ginsenoside-Rg5 orally for 1 month lowered levels of IL-1, TNF-α, NO, and iNOS by 67, 54, 32, and 49%, respectively [26]. Ginsenoside Rg1 administration to osteoarthritis mice attenuated cartilage degradation and lowered the loss of type II collagen and MMP-13 levels [27]. In the joints of collagen-induced arthritis (CIA) mice treated with *Panax ginseng*, pro-inflammatory cytokines including IL-17, IL-6, TNF-α, and IL-1 were significantly reduced [28]. Fifty-two postmenopausal women with a diagnosis of degenerative arthritis of the hand and symptoms of oedema and pain in the hands had therapy with 3 g red ginseng daily for 12 weeks, then both blood oestradiol and endometrial thickness were unaffected by ginseng [28, 29]. It had considerably elevated serum OC concentrations and a dramatically decreased DPD/OC ratio, which represents bone resorption and bone formation. Twelve weeks of treatment greatly alleviated knee arthritis symptoms [30].

2.6 *Ginkgo biloba*

Ginkgo biloba is a plant from the Ginkgoaceae family and is classified as a phytopharmaceutical or medicine. It had a potential effect as an anti-inflammatory based on its ability to bind to each target protein (such as COX-2, iNOS, TNF-α, TLR-4, and IL-1β) by forming H-bonds and hydrophobic interactions [31]. In addition, *Ginkgo biloba* extract inhibited IL-1-induced chemokine synthesis, as evidenced by a decrease in THP-1 cell motility to the cell culture medium and a decrease in IL-1-stimulated iNOS expression and NO release [32]. The presence of bilobalide in *Ginkgo biloba* extract had a crucial role in suppressing the synthesis of iNOS, COX-2, and MMP-13 in ATDC5 chondrocytes stimulated by IL-1, decreasing the activity of ECM-degrading enzymes, and increasing the expression of cartilage anabolic protein [33–35]. The ginkgolide C inhibited the release of pro-apoptotic factors and stimulates the release of anti-apoptotic proteins that are produced by H_2O_2. This substance inhibited the production of MMP-3 and MMP-13, ADAMTS, and iNOS. ECM breakdowns via inhibiting Cox-2 and SOX-9 [36].

In mice models of osteoarthritis produced by ACLT, ginkgolide C had been shown to suppress joint discomfort, cartilage degradation, and aberrant subchondral bone remodeling [37]. Moreover, bilobalide inhibited the development of aberrant osteophytes in the subchondral bone and stopped the degeneration of cartilage after trauma [38]. The ginkgolide B reduced blood levels of IL-1, IL-6, MCP-1, TNF-a, MMP-3, and MMP-13 and increases IL-10 [34]. In clinical studies, *Ginkgo biloba* extract treatment for 6 weeks in 51 OA patients had much greater benefits than glucosamine/chondroitin [39, 40].

2.7 *Allium sativum*

Allium sativum is commonly used as a medicinal plant. Based on its bioactive constituents (S-allyl cysteine, polyphenols, and flavonoids), this plant has anti-inflammatory properties. In in silico testing, the presence of the pyridoxamine from *Allium sativum* inhibited all receptors except NT5C2-1497 against the four target proteins (XO, ADA, GDA, and NT5C2-1498). It indicated that the substance has the potential to be employed as an anti-gout agent [37]. Meanwhile, in a study using *Allium sativum* methanol extract, it was shown that the extract had the highest inhibition value at 100 μg/ml of 88.5% in denaturation protein egg albumin and 61.2% in bovine serum albumin [41]. This plant also showed regulation of the IL-17 gene which significantly did not influence cell proliferation and lymphocyte proliferation (T CD4+ and CD8+) [42]. In a mouse model of arthritis, treatment with

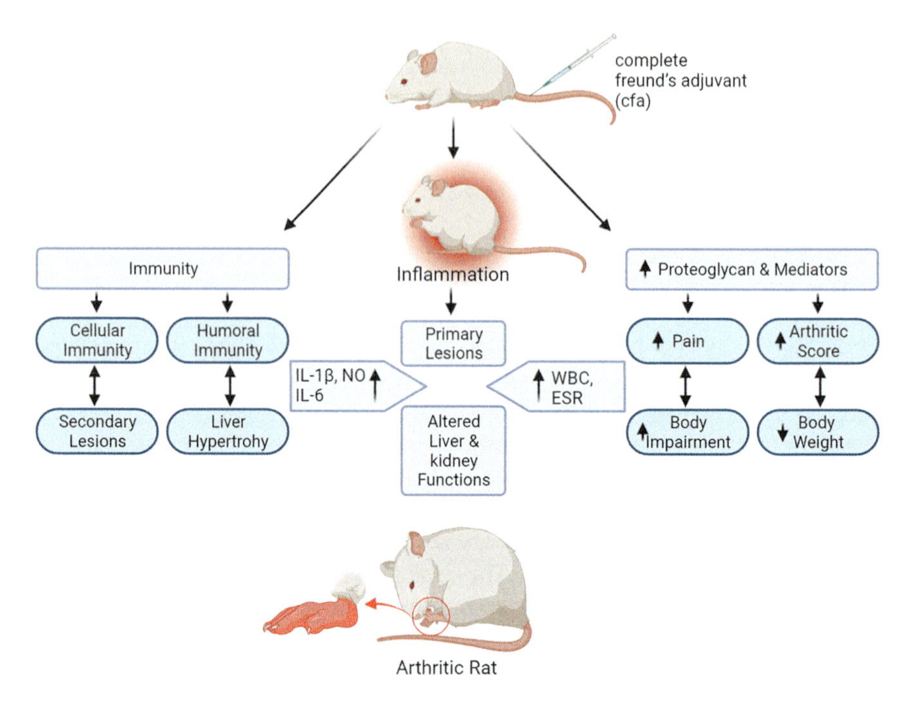

Fig. 4 Mechanism and pathway of CFA-induced arthritis [44]

100 M diallyl trisulfide suppressed the production of pro-inflammatory cytokines and regulated immunological activity by restoring the balance between Th17 and Treg cells [39]. In addition, administration of diallyl disulfide (20 and 50 mg/kg) in a CFA rat model of rheumatoid arthritis (Fig. 4) induced a significant decrease in leg volume and oedema formation; this compound also reduced inflammation, decreased white blood cell count, and increased RBC count in CFA-induced mice [40]. Seventy women with rheumatoid arthritis were separated into two groups for a clinical experiment, with one group receiving 1000 mg of garlic and the other receiving a placebo for 8 weeks. The results showed that serum C-reactive protein decreased significantly, pain intensity also decreased decreases, and the number of swollen joints also decreased [43].

2.8 *Ganoderma lucidum*

Ganoderma lucidum is a medical mushroom that contains triterpenoids, β-glucan, sterols, coumarins, mannitol, organic germanium, vitamins, flavonoids, and minerals. In silico sterol from this plant was directly bound actively to p38 and p65 to suppress their activation [45]. Furthermore, sterol has a crucial role in reducing inflammation in macrophages, as it may greatly reduce LPS-induced cell polarization and the production and expression of mRNA from pro-inflammatory mediators such as NO, TNF-α, IL-1, and 6 [46, 47]. *Ganoderma lucidum* extract's oil can prevent CIA-induced cartilage degradation in knee synovial membrane inflammation. The GLS oil group demonstrated a significant drop in knee eosinophils, and this oil lowered IL-6 mRNA expression. *Ganoderma lucidum* at 250 mg/kg BW was effective as an anti-inflammatory drug [48]. Clinically, this plant had an analgesic effect for patients with active rheumatoid arthritis, which is generally safe and well tolerated due to demonstrated activities, including significant antioxidant, anti-inflammatory, or immunomodulating properties [49].

2.9 *Tripterygium wilfordii*

Tripterygium wilfordii (thunder god vine) is commonly used in TCM to treat inflammatory conditions such as rheumatoid arthritis and autoimmune. Based on the structure of the CD2 protein, the triptolide derivative (5R)-5-hydroxy triptolide (LLDT-8) from this plant received the greatest docking score. This compound effectively blocked immune and inflammatory signaling pathways [46]. Another compound of tripterygium glycoside showed an anti-arthritis effect in rheumatoid arthritis therapy by initiating apoptosis in affected synovial lymphocytes or fibroblasts to prevent their proliferation, restrain Th17 cell differentiation, and inhibit angiogenesis [47]. Their alkaloids also had an anti-arthritis effect and significantly reduced leg swelling and suppress articular cartilage degeneration in CIA-induced mice. In addition, it was found that levels of IL-6, IL-8, NF-kB, and TNF-α in CIA serum mice decreased significantly [50]. Clinically, tripterygium glycosides significantly influenced the duration of morning stiffness (DMS), the number of painful joints (TJC), the number of swollen joints (SJC), the visual analog scale (VAS), the protein C-reactive (CRP), the erythrocyte sedimentation rate (ESR), and the rheumatoid factor (RF) in rheumatoid arthritis patients [51]. *Tripterygium wilfordii* extract at a single dose had a significant efficiency compared to MTX, while their combination gave a good significance than a single dose of MTX [52].

2.10 *Uncaria tomentosa*

The cat's claw plant (*Uncaria tomentosa*) is found in the peat forests of West Kalimantan, Indonesia. Some of the empirical benefits of this plant are used for ulcers, rheumatism, cirrhosis, gonorrhea, and female genital tract cancer. It contains alkaloids, flavonoids, polyphenols, triterpenoids, steroids, saponins, and tannins. This plant reduced osteoclast differentiation induced by RANKL without affecting cell viability [53]. Also, injection of this extract reduced ecto-nucleoside triphosphate diphosphohydrolase activity in mice models of CFA-induced arthritis but does not affect ecto-nucleoside triphosphate diphosphohydrolase and elevated erythrocyte adenosine deaminase activities in healthy animals [54]. In a separate study, administration of *Uncaria tomentosa* extracts decreased the RANKL/OPG ratio, an essential gene involved in osteoclast differentiation, reducing TRAP-positive cells lining alveolar bone and raising bone-specific alkaline phosphatase (BAP) and anabolic bone markers in plasma rat arthritic models [53]. Clinically, this plant was given to patients with advanced cancer for 8 weeks to improve their quality of life and reduce fatigue and inflammation that occurs [55]. During hemorrhagic cystitis (HC) in rats, quinovic acid glycoside was demonstrated to reduce oedema, bleeding, IL-1β levels, and neutrophil migration, exhibiting a protective impact on CH-induced urothelial damage, control of visceral discomfort, and reduction of IL-1 levels [56].

2.11 *Boswellia serrata*

Boswellia serrata (salai/salai guggul) is one of the components of the ancient and most famous herb in traditional Indian medicine, Ayurveda. This plant is beneficial for the treatment of arthritis, as well as diarrhoea, dysentery, ringworm, boils, and fever, according to empirical evidence (antipyretic). This plant is pharmacologically effective as an anti-atherosclerotic, analgesic, anti-inflammatory, and anti-arthritic. Boswellic acid (5, Fig. 5) and curcumin from this plant had the potential effect for anti-rheumatism as indicated by the high binding energy, which is between −8.66 and −9.67 [57]. The dry extract of this plant (0.1 μg/ml) showed pro-angiogenic activity, whereas the hydro enzymatic extract did not show any effect on the migratory capacity of endothelial cells [58]. Treatment of this extract for 21 days to CIA-induced Wistar rats altered numerous parameters, including MPO, LPO, GSH, and NO at doses 100 and 200 mg/kg BW. Furthermore, it resulted in a considerable reduction in numerous inflammatory mediators and a rise in IL-10 levels [59]. The injection of this extract at 180 mg/kg BW resulted in a rise in the rat's body weight and a decrease in knee diameter, leg thickness, ankle diameter, foot volume, and arthritis index, but the impact was less than that of the conventional cyclophosphamide medication. Clinically, a 12-day study of curcumin complex or its combination with boswellic acid improved pain-related symptoms in individuals with degenerative hypertrophic knee osteoarthritis. Curamin®, which contains *Curcuma longa* and *Boswellia serrata* extracts, improved the treatment of osteoarthritis due to the synergistic impact of curcumin and boswellic acid [60]. Similar to ibuprofen, the combination of *E. angustifolia* and *Boswellia thurifera* relieved pain and enhanced function in knee osteoarthritis patients [61]. Comparing methylsulfonylmethane (MSM) 5 g and boswellic acid (BA) 7.2 mg in 60 patients with knee arthritis revealed a statistically significant difference between the experimental group and the placebo group in the number of patients using anti-inflammatory medicines [62].

2.12 *Tinospora cordifolia*

Tinospora cordifolia (gurjo) is a traditional medicinal herb having anti-inflammatory properties that use to treat rheumatism, gout, bruising, fever, and stimulating hunger.

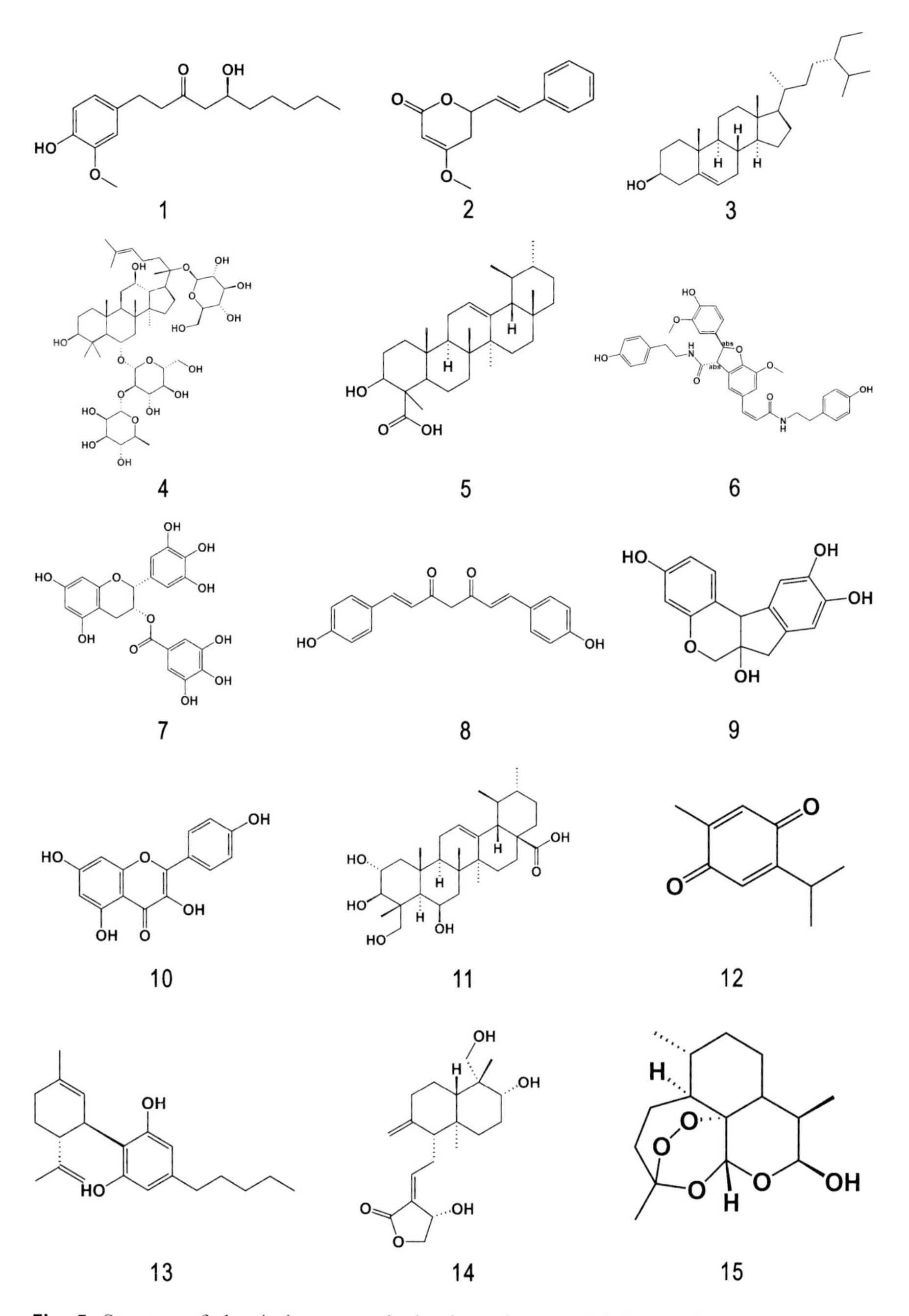

Fig. 5 Structure of chemical compounds that have the potential for anti-rheumatoid disorders therapy

It contains alkaloids, saponins, glycosides, tannins, polyphenols, flavonoids, picroretin, berberine, tinocrisposide, and columbin. This plant contains a tyramine-Fe molecule that forms a complicated binding with COX-2 (COX 2-tyramine Fe) [63]. Their extract increased the expression of genes that regulate osteoblastogenesis and mineralization, such as type I collagen, alkaline phosphatase, bone sialoprotein, osteopontin, osteonectin, osteomodulin, and osteocalcin, and has a positive effect on osteogenesis and bone formation [64]. Oral treatment of 50 mg/kg of *Tinospora cordifolia* extract to ovariectomized rats for 21 days prevented bone loss and suppressed osteoclast development [64]. This plant suppressed TNF-α and IL-1 and reduces NO generation in RAW cells 264.7 [65]. The chloroform extracts of this plant downregulated pro-inflammatory biomarkers in THP-1 macrophages by inhibiting NF-B nuclear translocation in THP cells [66]. Their extract considerably lowered levels of pro-inflammatory like IL-1, 6, and 17 in rat models of RA-induced arthritis. Hence, it may be inferred that the extract inhibited bone destruction by altering the ratio of mediators in bone remodeling (RANKL and MMP-9) that support anti-osteoclastic activity [67].

2.13 *Withania somnifera*

Withania somnifera (ashwagandha) is one of the endemic plants from India and is believed to improve physical and mental health as well as prevent various diseases, such as antimicrobial, immunostimulant, anti-inflammatory, anti-arthritis, antioxidant, and anti-diabetic. Their activity conducted cannot be separated from the role of the active compounds contained in this plant. Withanolide B present in the *Withania somnifera* had an excellent inhibition value of −7.8 kcal/mol for rheumatoid arthritis by targeting TNF-α through in silico analysis [68]. The ethanol extract of this plant inhibited 5-LOX (65%) in vitro with an IC_{50} of 0.92 mg/mL [69]. While their chloroform and water extracts were able to control the expression of TNF, IL-1, and IL-6 by reducing the production of iNOS and by downregulating NF and activator proteins. These extracts did not affect the expression of TNF-α. Moreover, the extract inhibited the migration of activated microglia by reducing MMP expression [70]. The alcoholic extract of this plant significantly decreased the diameter of the right leg's tarsal joint. In addition, the histology findings revealed that the joint membranes of mice administered this extract did not exhibit an inflammatory reaction [71]. Ashwagandha at a dose of 200 mg/kg has shown to minimize pannus production in rheumatoid arthritis joints [72]. Their extracts showed dose-dependent relief of disease severity based on reduced lipid peroxidation and ceruloplasmin, which lowered some serum levels of TNF-α, IL-1β, COX-1, and COX-2 [73]. Ashwagandha at 300 mg/kgBB increased the secretion of IL-10, lowered pro-inflammatory cytokines, and inhibited NF-kB activity [74]. The anti-inflammatory effect in adult zebrafish of comparable size and weight was determined by extracting fresh leaves of this plant with methanol and water. Due to phenolic acids and flavonoids, this extract can block TNF-α channels in zebrafish [75]. In collagen-induced rheumatic rats, administering root powder at 600 mg/kg body weight reduced the severity of arthritis by enhancing functional recovery of motor activity and radiological scores [76]. In a separate trial, this plant extract delayed morphine's analgesic impact. Moreover, it may inhibit hyperalgesia in the tail-flick test [77]. The anti-inflammatory effect of this root aqueous extract was investigated by measuring TNF-α, IL-1, 6, and 10 levels in rats with collagen-induced arthritis. Oral treatment of 300 mg/kg of their aqueous root extract reduced ROS and normalized MMP-8 levels in collagen-induced arthritic mice [74]. Clinically, the consumption of 500–1000 mg ashwagandha tablets by 77 patients for 8–12 weeks resulted in managing arthritis symptoms in these patients [78].

2.14 *Tribulus terrestris*

Tribulus terrestris is a member of the family Zygophyllaceae. This plant has been utilized in TCM and Ayurveda for centuries. Its pharmacological properties included anti-inflammatory, analgesic, hepatoprotective, and anti-cariogenic. Its action cannot be separated from the existence of active chemicals such as flavonoids, steroids, saponins, and alkaloids in this plant. Tribulusamide D and N trans-caffeoyl tyramine in the ethanol extract of *Tribulus terrestris* possessed anti-inflammatory activity, as indicated by the ability of the compounds to inhibit NO induced by LPS and PGE2 by reducing iNOS which can be induced and expression of COX-2. Tribulusamide D (6, Fig. 5) decreased the production of LPS-induced inflammatory cytokines, including IL-6, IL-10, and TNF-α, in RAW 264.7 macrophage-induced LPS [79]. These extracts were capable of repairing LPS and inducing excessive NO release and transcription of inflammatory cytokine genes. This plant inhibited LPS-induced increases in intracellular iNOS and NF-kB, which are responsible for cellular NO and cytokine production, respectively [80]. In vivo anti-arthritis action was proven by a substantial decrease in pro-inflammatory cytokines in an arthritic rat model treated with the extract at a dose of 200 and 300 mg/kg orally for 21 days [81]. In addition, this extract inhibited NO synthase 2, COX-2, TNF-α, and IL-6 in animals with MIA-induced OA [82].

2.15 *Curcuma longa*

Curcuma longa or turmeric is a plant that comes from the Zingiberaceae family. Traditionally, turmeric is often used as a treatment such as an antioxidant, antibacterial, anti-inflammatory, and antiviral. This plant includes important components such as curcumin, resin, curcuminoid, curcumin diglucoside, curcumin monoglucoside, sophoricoside, and epigallocatechin gallate (7, Fig. 5). Molecular docking scores and energy-energy ligand-protein interactions demonstrated that turmeric reduced TNF-α and IL-1 via strong molecular interactions [83]. The bisdemethoxycurcumin (8, Fig. 5) exhibited potent anti-inflammatory activity in conjunction with the suppression of NO, xanthine oxidase, and lipoxygenase. This chemical also inhibited the LPS-induced synthesis of COX-2, iNOS, IL-6, and TNF in RAW macrophages [84]. Turmeronol strongly inhibited the LPS-induced synthesis of PGE2 and NO, as well as the mRNA expression of the corresponding synthetic enzymes. Furthermore, this compound effectively reduced the LPS-induced mRNA and protein elevation of IL-1, IL-6, and TNF-α. Turmeronol derivates suppressed NF-kB nuclear translocation in a manner comparable to that of the NF-B inhibitor pyrrolidine dithiocarbamate, but not that of curcumin (another NF-kB inhibitor) [85]. Curcumin lowered inflammatory activity 6 h after zymosan-induced arthritis in rats. Curcumin at 1 mg/kg was more efficacious than prednisone in the first 6 h following arthritis onset. It did not have the same anti-inflammatory impact as prednisone after 12, 24, and 48 h. It was more effective than prednisone at decreasing inflammatory infiltration after 48 h, independent of the prednisone dosage employed [86]. Clinically, treatment of theracurmin containing curcumin 180 mg/day orally every day for 8 weeks considerably reduced reliance on celecoxib to the point where no adverse effects are noticed [87]. Compared to placebo, treatment with curcuminoids was linked with considerably higher decreases in WOMAC, VAS, and LPFI ratings. Curcuminoids are a safe and effective alternative therapy for OA [88]. *Curcuma longa* inhibited inflammation and had therapeutic benefits, such as a drop in VAS/WOMAC score level [89].

2.16 *Caesalpinia sappan*

Caesalpinia sappan known as secang is a member of the Caesalpiniaceae family. This plant is utilized in several nations, particularly in Southeast Asia. In Indonesia, secang is historically used as an immune-boosting and immune-maintenance herbal beverage. The part of this plant often used as herbal medicine is bark because it contains active compounds and is used in treating various diseases such as inflammation, diarrhoea, anti-osteoporosis, and other conditions. Phytochemical compounds in secang include brazilin, sappanone, cassane, and caesalpiniaphenol. Cassane cured rheumatoid arthritis via the TCR, TLR, VEGF, and osteoclast differentiation pathways, which are strongly associated with antigen recognition, inflammation, angiogenesis, and osteoclastogenesis [90, 91]. 3,7-dihydroxy-4H-chromen-4-one and 4-hydroxy-3,5-dimethoxybenzaldehyde exhibited potent inhibition of NO with IC_{50} values of 14.5 and 21.5 μM [92]. In CIA rats, brazilin (9, Fig. 5) at 10 mg/kg BW decreased the severity of acute foot oedema. Bone mineral density was considerably lower in CIA rats and appeared to rise proportionally with the administration of methotrexate and brazilin. The microstructural analysis demonstrates that brazilin reduced joint degradation, surface erosion, and enhanced bone growth. Brazilin significantly lowered blood levels of inflammatory cytokines in rats with CIA [93]. Sappanchalcone prevents chronic disease progression, leg oedema, arthritis severity, radiography, and histomorphometric alterations. Serum levels of pro-inflammatory cytokines were considerably lowered [94].

2.17 *Clematis mandshurica*

Clematis mandshurica roots are commonly used to treat inflammatory disorders such as gout, arthritis, and tetanus. This plant is traditionally used to relieve pain and fever in Korea. It contains hederagenin which inhibited the protein expression of iNOS, COX-2, and NF-kB at doses of 10, 30, and 100 μM triggered by LPS and generated NO, PGE2, TNF-α, IL-1β, and IL-6 on RAW 264.7 cells [95]. Feruloyl and iso-feruloyl derivatives at 50 mM exhibited a moderate inhibitory effect ranging from 29.3% to 38.7%, whereas the 3-carboxy-indole derivative exhibited a strong inhibitory effect with an IC_{50} of 7.7 to 12.4 mM [96]. The 5-O-isoferuloyl-2-deoxy-D-ribono lactone (5-IRL) stimulated inducible mRNA expression of COX-2 and iNOS [97]. This plant extract lowered arthritis scores relative to the control group in CIA-induced rats. Normal rats treated with this plant had no adverse effects on gut microbiota and SCFA metabolism [100]. This root extract at 0.25, 0.5, and 1 mg/mL inhibited protein expression LPS caused by iNOS, COX-2, NF-kB, and MAPKs (ERK, JNK, and p38) as well as the generation of TNF-α, IL-1β, IL-6, NO, and PGE2 [98]. The anti-oedema activity of hederagenin was demonstrated by the dependent reduction of LPS-induced mRNA levels of iNOS and COX-2 and the results of the carrageenan-induced rat hind limb oedema test. In addition, 30 mg/kg of hederagenin prevented carrageenan-induced increases in skin thickness, inflammatory cell infiltration, and mast cell degranulation [95]. Clematichinenoside significantly reduced leg swelling in AIA mice. It also increased levels of IL-10 and TGF-β1 released from ConA-activated PP lymphocytes, while decreased levels of IL-17 A and TNF-α [99].

2.18 *Rosmarinus officinalis*

Rosmarinus officinalis often known as rosemary is a member of the Lamiaceae family. This plant is utilized as traditional medicine like anti-diabetic, anti-inflammatory, and antioxidant. This plant contains carnosic acid, carnosol, and rosmarinic acid. Carnosic acid suppressed the expression of pro-inflammatory cytokines, including TNF-α, IL-1β, IL-6, IL-8, IL-17, and MMP-3, and downregulated the creation of RANKL. Moreover, it suppressed osteoclastogenesis and bone resorption [100]. The 70% ethanol extract of this plant exhibited oxidative damage and an enzymatic antioxidant defence system and anti-inflammatory effects [101]. In rats with arthritis, the aqueous extract of this plant reduced oxidative damage and enhanced antioxidant capacity by boosting GSH levels. Also, this extract decreased leg oedema, the number of leukocytes recruited in the femorotibial joint cavity, and the weight of lymph nodes and delayed the formation of subsequent diseases [102]. At 60 mg/kg, the methanol extract of this plant exerted a potentially effective anti-arthritic by controlling inflammation in an adjuvant-induced experimental model [103].

2.19 *Elaeagnus angustifolia*

Elaeagnus angustifolia or oleaster is a plant that comes from the Elaeagnaceae family. Kaempferol (10, Fig. 5) contained in this plant significantly inhibited IL-1-β-induced protein production from inflammatory mediators such as inducible iNOS and COX-2. It also inhibited general matrix-degrading enzymes (MMP-1, MMP-3, MMP-13, and a disintegrin), MMP with a thrombospondin-5 motif produced by IL-1β, hence preventing collagen II degradations [104]. The extract at a dose of 32 mg/kgBW significantly reduced serum concentrations of MMP-3 and MMP-13 [105] and at 1000 mg/kg effectively decreased leg oedema swelling [106]. In a randomized clinical trial conducted on 90 female osteoarthritis patients, this extract at 15 g/day for 8 weeks reduced serum levels of TNF-α and MMP-1 [107]. In addition, kaempferol reduced osteoarthritis symptoms with an efficacy comparable to that of ibuprofen [108].

2.20 *Crocus sativus*

Crocus sativus known as saffron is a member of the Iridaceae family and a traditional remedy for chronic disorders such as asthma, arthritis, and respiratory problems. Active chemicals found in saffron include phenolic and carotenoid compounds. The extract of this plant exhibited anti-inflammatory properties such as NO and IL-6 by murine RAW 264,7 cells. This extract possessed anti-osteoclastogenic properties by a full suppression of tartrate-resistant acid phosphatase (TRAP)-positive osteoclast production and a reduction in the expression of key osteoclast-related genes [109]. The extract administered orally at 25, 50, and 100 mg/kg BW to rats with arthritis for 47 days decreased TNF-a and IL-1 levels in CSE-2 and 3-group rats compared to rheumatic mice [110]. Dosages of 200–2000 mg/kg of saffron ethanol extract exhibited an anti-inflammatory effect of 18.2–27.8% against mice treated with 49.5% DEX (15 mg/kg) [111]. Histological scores of joint inflammation, bone erosion, chondrocyte mortality, cartilage surface erosion, and bone erosion in CIA mice treated with 40 mg/kg crocin were equal to those of normal mice. The protein expression levels of MMP-1, MMP-3, and MMP-13 in CIA mice treated with 40 mg/kg crocin were lowered to normal levels. In addition, crocin inhibited the expression of TNF-a, IL-17, IL-6, and CXCL8 in CIA rats' blood and ankle tissue [112, 113].

2.21 *Camellia sinensis*

Green tea also known as *Camellia sinensis* is one of the herbal plants frequently utilized in traditional medicine such as losing weight, lowering cholesterol, and overcoming an inflammation. This plant has several active compounds including flavones, flavonols, flavanones, catechins, and isoflavones. Epigallocatechin gallate in this plant had a bond to TNF-α of −90.90 and to the IL-1 target protein of

−120.75 [83]. This compound showed anti-rheumatoid arthritis efficacy by killing synovial fibroblast cells (66.1%), downregulating TNF-α (93.33%), IL-6 (87.97%), and inhibiting p-STAT3 (77.75%) [114]. A dose of 400 mg/kg/BB of *Camellia sinensis* extract had anti-arthritis effectiveness of 60% and boosted levels of superoxide dismutase, catalase, and glutathione by 34%, 59%, and 50% [115]. At 400 mg/kg/BB, this extract effectively reduced TNF-α, IL-1, and IL-6 levels in the joint tissue of rats. Compared to the arthritic and infected group, the therapy of this extract exhibited substantial reductions in cartilage degradation and pannus development, and there was no evidence of inflammation in the small intestine [116]. Clinically, administration of this extract for 4 weeks resulted in a significant difference in the average VAS pain score, total WOMAC, and WOMAC physical function between the green tea and control groups [117].

2.22 *Matricaria chamomilla*

Matricaria chamomilla (camomile) is a member of the Asteraceae family and used for treating stomach discomfort and sleeplessness, as well as for anti-inflammatory, antibacterial, and sedative effects. The compounds from this plant had comparable scores in the active site by the free binding energy ($\Delta G = 54.19$ kcal/mol) [118]. The extract at 400 mg/kg BW decreased blood calcium, phosphate ion, and magnesium levels by 54.01%, 27.73%, and 20%, respectively. In addition, this dose also decreased creatinine and alkaline phosphatase levels by 29.41% and 27.82%, respectively [118, 119]. Compared to diclofenac and placebo, chamomile oil greatly reduced the patient's requirement for acetaminophen. Nevertheless, there were no statistically significant changes in the WOMAC questionnaire's domains, and no adverse effects were reported by patients taking chamomile oil [120].

2.23 *Punica granatum*

Punica granatum is used as an anti-inflammatory, antibiotic, antibacterial, etc. It contains flavonoids, polyphenols, alkaloids, anthocyanins, tannins, etc. The ellagic acid derived from the *Punica granatum* could bind to NF-kB protein; however, the various binding efficiencies are examined based on the energy gained for binding and the number of hydrogen bonds generated. The calculated bond-free energy of the NF-kB-EA complex was 7.1 kcal/mol with two hydrogen bonds [121]. The 70% ethanol extract of this plant decreased IL-1-inducible, iNOS, COX-2, and MMP-13 protein production in primary mouse chondrocytes (PRC) [122]. At a dose of 50 mg/kg, the butanol fraction increased RBC and Hb while decreasing WBC and ESR [123]. The injection of 500 and 750 mg/kg of this ethanol extract for 4 weeks resulted in a reduction in paw volume; a rise in latency time and levels of ALT, AST, and ALP; and a substantial drop in RF and MDA activity compared to the rheumatism control group. The extract reduced the levels of IL-1 and TNF in rats with arthritis. Histological analysis revealed that therapy with the extract reversed greatly the histological alterations produced by arthritis [124]. Rosmarinic acid considerably decreased the severity of rheumatism, leg volume, joint diameter, white blood cell count, and erythrocyte sedimentation rate. Moreover, it dramatically boosted body mass, haemoglobin, and red blood cells [125]. In a mouse model of osteoarthritis caused by collagenase type II, POMx (150 mg/kg) reduced knee-associated inflammation and nociception and corrected alterations in the weight-bearing ratio of experimental mice [122]. In clinical studies, 250 mg of pomegranate enhanced disease activity and multiple blood indicators of inflammation and oxidative stress in individuals with rheumatoid arthritis [126]. Pomegranate juice enhanced physical function and stiffness, as well as it is antioxidant and MMP-13 inhibiting capabilities in osteoarthritis patients [127]. Women with knee OA who took 500 mg of pomegranate peel twice a day for 8 weeks saw a reduction in pain and an improvement in symptoms [128].

2.24 *Cinnamomum cassia*

Cinnamomum cassia is a plant that belongs to the Lauraceae family and is traditionally used to treat gastritis, blood circulation abnormalities, dyspepsia, and inflammatory conditions. This plant contains several active compounds such as cinnamic aldehyde, coumarin, tannin, methyl dihydromelilotoside derivative, hierochin B, balanophonin, quercetin, and luteolin that were able to interact with JAK2, mPEGS-1, COX-2, IL-1β, and PPARγ target proteins [129]. This plant possessed antipyretic, anti-inflammatory, antibacterial, anti-diabetic, and anticancer properties. The ethyl acetate fraction of this plant inhibited carrageenan-induced leg swelling in rheumatoid arthritis and LPS-induced NO generation in murine immune cells [130]. Cinnamaldehyde from this plant dramatically reduced synovial inflammation in rats with arthritis. It was related to the inhibition of IL-1β. This compound was also capable of reducing HIF-1α activity by preventing succinate build-up in the cytoplasm. This reduction inhibits IL-1β synthesis because HIF-1α promotes the expression of NLRP3, which is important in the assembly and processing of IL-1β by the inflammation. In addition, cinnamaldehyde suppressed the expression of the GPR91 succinate receptor, hence inhibiting HIF-1α activation even more [131]. Cinnamomi lactone inhibited MMP-3, IL-1β, and MMP-1 gene expression in FLS cells [132]. Stems of this plant considerably lowered MDA levels and joint swelling, accompanied by a rise in GSH levels. This extract dramatically decreased joint swelling, IL-1, and TNF levels in rats with arthritis caused by CFA. The expression of TNF-α receptors was reduced in mice treated with indomethacin or *Cinnamomum cassia* stem extract [133].

2.25 *Centella asiatica*

Centella asiatica is a plant that is often used in wound healing and contains saponins, asiaticoside, asiatic acid, and carotenoids. The 2000 g/ml of *C. asiatica* extract inhibited protein denaturation and membrane stabilization in the greatest percentages of 89.76% and 94.97%, respectively [134]. Madecassoside and madecassic acid (11, Fig. 5) directly inhibited the release of pro-inflammatory cytokines and IL-10 from CD4+ CD25+ T cells [135]. The extract dramatically reduced mechanical allodynia and hyperalgesia, spontaneous pain, and motor changes. Their efficacy was greater than or equivalent to that of the reference medication, triamcinolone acetonide (100 μg intramuscularly). The histological study of MIA+14G1862-treated rats revealed the improvement of several morphological parameters, with significantly positive effects on joint space and fibrin deposition [135]. The methanol fraction of *Centella asiatica* (150 and 250 mg/kg) considerably mitigated the severity of CIA and decreased synovial inflammation, cartilage erosion, and bone erosion, as demonstrated by histological and radiological data. The increase in plasma levels of TNF-α, IL-1, IL-6, IL-12, and NO in CIA mice was significantly attenuated by treatment with *Centella asiatica*. Serum levels of anti-collagen type-II antibodies were significantly lower in the treatment group than in the rheumatic group in rats given *Centella asiatica* (150 and 250 mg/kg) [136].

2.26 *Phyllanthus amarus*

Phyllanthus amarus is a medicinal plant that is often used to treat pain and includes the Phyllanthaceae family. Based on its pharmacological activity, this plant is often used to treat diabetes, malaria, analgesics, arthritis, and hypertension. This plant contains compounds such as filantin, hypophylantine, filantenol, nirantin, and Quercetin. In a cartilage explant model, *P. amarus* extract (PAE) and its major components, including phyllanthin and hypophyllanthin, were evaluated. Following 4 days of incubation, the DMMB binding assay and zymography were used to assess the release of sulfated glycosaminoglycans (sGAGs) and MMP-2 activity, respectively, in a culture medium. The chondroprotective potential of PAE and its major components against IL-1β-induced degradation of cartilage explants was demonstrated by decreased levels of sGAGs and MMP-2 activity in culture media, which were consistent with increased uronic acid and proteoglycan content in explants, compared to IL-1β treatment [137]. Phyllanthin considerably inhibited the compound's response to LPS-stimulated prostaglandin overproduction via a mechanism connected to the modification impact of COX-2 protein and gene expression. Moreover, phyllanthin dramatically prevented the release of mRNA and the production of IL-1ß and TNF-α. In a concentration-dependent way, phyllanthin also dramatically downregulated IκBα, NF-kB (p65), and IKKα/β phosphorylation and inhibited JNK, ERK, p38MAPK, and Akt activation [138]. In addition, the extract suppressed the synthesis of pro-inflammatory (TNF-α, IL-1β, PGE2) and the expression of COX-2 protein in LPS-stimulated U937 human macrophages. Pre-treatment of this plant also substantially inhibited the pro-inflammatory marker mRNA transcription in U937 macrophages [139].

Phyllanthus amarus extract was able to modulate the development of physical parameters (e.g. weight loss, increase in body temperature, reduction in hind limb volume, and severity of arthritis), bone mineral density, haematological and biochemical disturbances, serum cytokine production, and levels of MMPs together with their inhibitors in synovial fluid. Examination of the knee joint's histopathology demonstrated that the extract efficiently lowered synovitis, pannus development, bone resorption, and cartilage disintegration [140]. The extracts and leaves of two *Phyllanthus amarus* and *Phyllanthus fraternus* elevated the threshold for mechanical withdrawal (mechanical hyperalgesia) to return to the original reaction. The percentage of maximal achievable effectiveness and the percentage of suppression of hyperalgesia further demonstrate the efficiency of *Phyllanthus* extract [141]. In 40 OA patients, *Phyllanthus amarus* extract nanoparticles administered through phonophoresis showed a substantial rise in the visual analog scale (VAS) and 6-MWT following therapy in both groups. The phonophoresis (PP) group had a more substantial reduction in VAS pain levels and an increase in 6-MWT than the ultrasonic therapy (UT) group [142].

2.27 *Psidium guajava*

Psidium guajava is a plant that is known for its many benefits in Indonesia. The phytochemical analysis of guava leaves revealed the presence of flavonoids, saponins, tannins, phenols, carotenoids, and quercetin. In addition, epicatechin-3-O-gallate shows tremendous promise as a selective anti-inflammatory COX-2 inhibitor due to its strong bond strength and stability, as well as its low bond energy when bound to the target protein (6-COX) [143]. Guava leaf extract significantly inhibited LPS-induced production of NO and PGE2 in RAW264.7 macrophages by inhibiting the expression and activity of iNOS and COX-2 which was partially induced by the downregulation of ERK1/2 activation [144]. The ethanol extract of this leaf considerably suppressed leg oedema in the chronic phase, as measured by an increase in oedema beginning on day 20. The most beneficial dose of EEPG for lowering arthritis scores was 250 mg/kg. Histopathological analysis revealed that the knee's synovial membrane and cartilage have been repaired [145]. Extract of this leaf was also capable of reducing the number of lymphocytes and leukocytes. The group that received extract dosages of 250 and 750 mg/kg BW showed a substantial reduction in leukocytes both before and after treatment. Following administration of 250, 500, and 750 mg/kg BW of ethanol extract of *P. guajava* leaf, the number of lymphocytes was also reduced. This leaf extract at 750 mg/kg BW suppressed leukocytes and lymphocytes most effectively [146]. In the 12th week of the clinical experiment, knee pain and stiffness (a subgroup of the JKOM score) were considerably lower in the guava group than in the placebo group. Utilizing a visual analog scale (VAS) to measure knee pain, a significant correlation was seen between treatment effect and test length, with the guava group having a lower VAS score after 12 weeks than the placebo group. In conclusion, continual use of guava leaf extract can alleviate knee discomfort, indicating a possible preventative benefit against OA symptoms [147].

2.28 *Moringa oleifera*

Moringa oleifera or commonly known by the local name "moringa" is a type of plant that has various benefits. One of the potentials possessed by moringa based on phytochemical analysis (flavonoids, saponins, and polyphenols) is as an anti-inflammatory. The moringa extract inhibited the production of inflammatory cytokines (IL-1β, TNF-α, and IL-6) and the degradation of cytoplasmic IB- and nuclear translocation of p65 protein, resulting in reduced NF-kB transactivation. The moringa extract also inhibited NFk-B activation in RAW 264.7 cells, hence reducing the levels of pro-inflammatory mediators like NO, IL-1β, TNF-α, and IL-6 [148]. In vivo investigation of moringa leaf extract at 250 and 500 mg/kg in rats with arthritis showed higher inhibitory action against inflammatory leg oedema with fewer toxicity and adverse effects than the indomethacin-treated group [149]. Oral treatment of *Moringa oleifera* (25, 50, and 100 mg/kgBW) in rats induced with CFA significantly reduced joint inflammation as evidenced by reductions in joint diameter, rheumatic scores, and inflammatory cell infiltration [150]. Oral administration of *Moringa oleifera* at doses between 250 and 500 mg/kg lowered the rheumatic index and sole oedema. It is believed that moringa can reduce serum pro-inflammatory cytokines when administered [151].

2.29 *Nigella sativa*

Nigella sativa (black cumin) is one of the herbal plants that are frequently employed as a traditional medicine due to the presence of its active constituents. One of the pharmacological activities of this plant is as an anti-inflammatory. This plant produces thymoquinone (12, Fig. 5) which acts as an anti-inflammatory and anti-proliferative in osteoarthritis. This compound inhibited apoptosis-regulated signaling kinase 1 (ASK1) in response to TNF-induced activation of phospho-p38 and phospho-JNK. Moreover, TNF-α promotes ASK1 phosphorylation at the Thr845 site in RA-FLS selectively [152]. Application of this plant oil (1.82 mL/kg) to arthritic rats dramatically decreased the onset of arthritis by 56%. The oil possessed anti-inflammatory, anti-rheumatic, and analgesic properties [153]. Thymoquinone derivative considerably lowered the total leukocyte count, inflammatory clinical score, and blood urea and serum creatinine levels [154]. Clinically, therapy with *Nigella sativa* caused an increase in the number of swollen joints, as compared to the initial and placebo groups. This plant treatment led to a drop in CD8 + and a rise in the proportion of CD4+ CD25+ T cells and the CD4+/CD8+ ratio. In the NS group, a significant negative connection was identified between CD8+ change and CD4+ CD25+ T cell change [155]. *Nigella sativa* lowered the DAS28 score considerably compared to the placebo group. Serum levels of IL-10, TNF-, MDA, SOD, catalase, TAC, and NO did not alter significantly from the placebo group [156].

2.30 *Cannabis sativa*

Cannabis sativa is a plant that comes from the Cannabidaceae family and is often used as an antidepressant, for fatigue, and mental disorders. The active compounds of this plant include cannabinoids, tryptophan, and cannabidiol (13, Fig. 5). Tryptophan had an inhibition value of −5.6 against TNF-α protein compared to ascorbic acid and linoleic acid [157]. Cannabidiol contributed to the reduction of cell viability, proliferation, and RASF IL-6/IL-8 production. Moreover, pre-treatment with TNF enhanced intracellular calcium and absorption of cationic PoPo3 viability dyes in RASF [158]. *Cannabis sativa* oil decreases the survival rate of MH7A cells and induces apoptotic cell death in a dose- and time-dependent manner. In MH7A cells treated with HO, lipid accumulation and intracellular ROS increase. Then co-treatment with the antioxidant tyrone substantially negated the harmful impact of H_2O_2 on MH7A cells. The result was that ROS levels decreased, cell viability was restored, and apoptotic cell death was significantly decreased [159]. Cannabidiol increased the amount of early apoptotic B cells at the expense of viable cells and decreased the production of TNF and IL-10 when independently stimulating T cells. CBD boosted IL-10 synthesis in PBMC when B cells were triggered by T cells but decreased TNF levels when T cells were activated independently. CBD inhibited IL-10 production in PBMC/rheumatoid synovial fibroblast co-cultures when B cells were stimulated independently of T cells [160]. In a dose-dependent manner, administration of cannabidiol in a rat model of OA decreased joint oedema, limb posture score as a spontaneous pain rating, immune cell infiltration, and synovial membrane thickening. Immunohistochemical study of bone marrow (CGRP, OX42) and dorsal root ganglia (TNF-α) demonstrated a dose-dependent decrease of pro-inflammatory biomarkers, with 6.2 and 62 mg/day being effective levels [160]. Cannabidiol resulted in improvement in patient-reported outcome measures, including pain on the visual analog scale; arm, shoulder, and hand disabilities; and single-assessment numerical evaluation scores, compared to the control group in a clinical trial involving 10 patients with arthritis of the thumb. Physical metrics associated with a range of motion, grip, and pinch strength were found to be comparable [161, 162].

2.31 *Andrographis paniculata*

Andrographis paniculata popularly known as sambiloto is one of the nine most extensively used medicinal herbs in Indonesia, as well as other nations. This plant demonstrated antithrombotic, antiviral, immunosuppressive, and anti-inflammatory properties. Andrographolide is one of the primary active chemicals in this plant. This compound dose-dependently inhibited IL-1β and IL-6 release from TNF-α-stimulated RA-SFs [163]. It was effective at alleviating arthritis symptoms by reducing neutrophil infiltration and necrosis in the ankle joint, relieving systemic inflammation, accelerating LPS-activated neutrophil apoptosis, and inhibiting extracellular-dependent autophagy-dependent neutrophil formation [164]. Andrographolide (14, Fig. 5) was also capable of dose-dependently suppressing

MMP-1, 3, and 9 inhibited by these substances [165]. In in vivo experiments, andrographolide reduced blood levels of anti-CII, TNFα, IL-1β, and IL-6 to considerably reduce the severity of arthritis and joint destruction [163]. By directly targeting NF-κB, andrographolide inhibited the inducible activation of iNOS and COX-2 as indicated by immunoblot analysis [166]. Compared to mice administered dexamethasone, animals treated with andrographolide had a reduction in leg oedema, cell cytotoxicity, and NO generation [167]. Moreover, the administration of this molecule enhanced the anti-rheumatic efficacy of MTX. Combining AD and MTX decreased inflammatory symptoms in CFA mice additively. The hepatoprotective effect of AD and MTX combination treatment was characterized by an improvement in blood indicators, potentially related to antioxidant activity, and verified by histological alterations in the liver. Moreover, combination treatment substantially decreased blood levels of TNF-α, IL-6, and IL-1β [168]. On days 28, 56, and 84, individuals treated with 300 and 600 mg/day of paractin experienced significantly less discomfort than the control group. Compared to the placebo group, stiffness, physical function, and tiredness scores improved significantly in both paractin-treated groups [169].

2.32 *Artemisia annua*

Artemisia annua is a medicinal plant originating from sub-tropical regions such as China. This plant has several properties including being antimalarial, anti-inflammatory, and effective in treating autoimmune diseases. These benefits and properties cannot be separated from the role of their active compounds such as artemisinin, alkaloids, saponins, sterols, triterpenes, tannins, and coumarins. In RAW 264.7 cells, casticin and chrysosplenol D decreased the LPS-induced production of IL-1β, IL-6, and MCP-1, limited cell motility, and lowered the LPS-induced phosphorylation of IB and c-JUN. In addition, it was a JNK SP600125 inhibitor that prevents the inhibition of cytokine release by chrysosplenol D. Application of casticin and chrysosplenol D inhibited ear oedema produced by croton oil and dexamethasone in rats [170]. The dihydroartemisinin (15, Fig. 5) contained in *Artemisia annua* extract was able to inhibit bone loss in ovariectomized rats so that osteoclast formation will also be inhibited [157, 171, 172]. The clinical administration of *Artemisia annua* extract treatment at 150 and 300 mg twice a day orally for 12 weeks resulted in a substantial decrease in pain ratings and joint swelling [10].

2.33 *Piper betle*

Betel leaf (*Piper betle*) is one of the plants used in traditional medicine by the community. Betel leaves contain secondary metabolites including flavonoids, tannins, saponins, and alkaloids. Moreover, betel leaves have the potential for different therapeutic properties, including anti-inflammatory properties. The 4-allyl-1,2-diacetoxybenzene exhibited stronger interactions with less binding energy to RA targets than MMP-1(−6,4 kcal/mol), TGF(−6,9 kcal/mol), and IL-1(−5,4 kcal/mol) [173]. Betel leaf methanol extract (PBME) was able to suppress protein denaturation more efficiently than normal ibuprofen, with an IC_{50} concentration of 20.5 μg/ml compared to 17.5 μg/ml for ibuprofen [173]. Moreover, the recovery of rat joint injury demonstrates the anti-arthritic action of PBME (250 and 500 mg/kg BW). The reduction of leg oedema and rheumatic scores, backed by radiographic and histological alterations, demonstrates that combined treatment with allylpyrocatechol (APC) and MTX dramatically improved arthritic parameters. This combination inhibited the production of pro-inflammatory cytokines, such as TNF-α and IL-6. In addition, the APC-MTX combination reduced cachexia, splenomegaly, and oxidative stress associated with these conditions [174]. The use of lower (200 mg/kg) and higher (400 mg/kg) dosages of *Piper betle* extract showed a dose-dependent reduction in SGPT, SGOT, and ALP, leg volume, foot diameter, and increased body weight as compared to the rheumatism control group. Serum rheumatoid factor was significantly decreased in all treated animals. In addition, the anti-arthritic effect was validated by histology of the synovial joints of treated rats, which revealed decreased constriction of the inter-tarsal joint gaps. The acquired findings are comparable to the standard [175–179].

Medicinal plants have activity as anti-rheumatoid disorders through various mechanisms depending on their secondary metabolites. Table 1 summarizes the mechanism of anti-rheumatoid from the selected plants in this chapter of the book. Some are through the inhibition of PGE2, IL-6, IL-8, and iNOS, etc. In more detail the mechanism of action of medicinal plants as anti-rheumatoid disorders can be seen in Table 1.

3 Conclusion

Arthritis disorders are associated with reduced productivity, substantial disability, and decreased quality of life. Conventional therapies include disease-modifying anti-rheumatic drugs (DMARDs) and non-steroidal anti-inflammatory drugs (NSAIDs).

Table 1 The mechanisms of anti-rheumatoid disorders form the selected plants

Plant	Mechanism
Zingiber officinale	Aqueous extract of ginger at 125 g/ml revealed that its anti-inflammatory activity was higher than diclofenac Ethanol extract of ginger inhibit PGE2, IL-6, IL-8, and iNOS Essential oil of ginger decreasing expression of IL-12 and TGF-α
Piper methysticum	Extract of *Piper methysticum* inhibit nitric oxide (NO) Kavain compound reduce production of TNF-α
Ephedra sp.	β-Ssitosterol and androstan-3-one inhibit IL-6, TLR4, TNF-α, and IL-1β Extract of *Ephedra alata* and *Ephedra fragilis* inhibited NO *Ephedra sinica* inhibit production of cytokine

(continued)

Table 1 (continued)

Plant	Mechanism
Oenothera biennis	*Oenothera biennis* inhibit TNF-α, IL-6, and IL-1 Evening primrose oil inhibit TNF-α and reduce IL-1β, and IL-6
Panax ginseng	*Panax ginseng* reduce of reactive oxygen species (ROS), IL-1, NO expression, and ROS. Ginsenoside compound suppresses osteoclastogenesis and increased the expression of aggrecan and collagen type 2 (COL-II) Ginsenoside decreasing of MMP-1 and MMP-13
Ginkgo biloba	*Ginkgo biloba* extract inhibit IL-1, iNOS, NO, MMP-13 Ginkgolide B reduces IL-1, IL-6, MCP-1, TNF-a, MMP-3, and MMP-13 and increases IL-10
Allium sativum	*Allium sativum* regulate of the IL-17 gene
Ganoderma lucidum	Sterol compound reduce pro-inflammatory mediators such as NO, TNF-α, IL-1, and 6
Tripterygium wilfordii	Compound of tripterygium glycoside prevent their proliferation, restrain Th17 cell differentiation and inhibit angiogenesis Alkaloid compound from *Tripterygium wilfordii* decrease levels of IL-6, IL-8, NF-kB, and TNF-α
Uncaria tomentosa	*Uncaria tomentosa* extracts decreased the RANKL/OPG ratio Quinovic acid glycoside was demonstrated to reduce oedema, bleeding, IL-1β levels, and neutrophil migration
Boswellia serrata	*Boswellia serrata* extract reduce IL-10 levels *Boswellia serrata* extract altered numerous parameters, including MPO, LPO, GSH, and NO
Tinospora cordifolia	*Tinospora cordifolia* suppressed TNF-α and IL-1 and reduces NO generation Chloroform extract *Tinospora cordifolia* downregulated pro-inflammatory biomarkers in THP-1 *Tinospora cordifolia* extract lowered levels of pro-inflammatory like IL-1, 6, and 17
Withania somnifera	Chloroform and water extracts *Withania somnifera* were able to control the expression of TNF, IL-1, and IL-6, COX-1, and COX-2
Tribulus terrestris	Ethanol extract of *Tribulus terrestris* inhibit NO induced by LPS and PGE2 Tribulusamide D compound decreased the production of LPS-induced inflammatory cytokines, including IL-6, IL-10, and TNF-α, in RAW 264.7
Curcuma longa	Bisdemethoxycurcumin compound suppresses NO, xanthine oxidase, and lipoxygenase and inhibits COX-2, iNOS, IL-6, and TNF Turmeronol compound inhibited the LPS-induced synthesis of PGE2, NO, IL-1, IL-6, and TNF-α
Caesalpinia sappan	Cassane compound cured rheumatoid arthritis via the TCR, TLR, VEGF, and osteoclast differentiation pathway 3,7-dihydroxy-4H-chromen-4-one and 4-hydroxy-3,5-dimethoxybenzaldehyde exhibited potent inhibition of NO
Clematis mandshurica	Hederagenin inhibited the protein expression of iNOS, COX-2, and NF-kB Clematichinenoside compound significantly increase level of IL-10 and TGF-β1
Rosmarinus officinalis	Carnosic acid compound suppressed the expression of TNF-α, IL-1β, IL-6, IL-8, IL-17, and MMP-3.

(continued)

Table 1 (continued)

Plant	Mechanism
Elaeagnus angustifolia	Kaempferol compound significantly inhibited IL-1β, MMP-1, MMP-3, MMP-13
Crocus sativus	*Crocus sativus* extract inhibit NO and IL-6 *Crocus sativus* extract decreased TNF-a and IL-1 levels Crocin compound inhibited the expression of TNF-a, IL-17, IL-6, and CXCL8
Camellia sinensis	Epigallocatechin gallate compound killing synovial fibroblast, downregulating TNF-α, IL-6 and inhibiting p-STAT3
Matricaria chamomilla	*Matricaria chamomilla* extract decreased blood calcium, phosphate ion, and magnesium levels
Punica granatum	70% ethanol extract of *Punica granatum* decreased IL-1-inducible, iNOS, COX-2, and MMP-13 Ethanol extract *Punica granatum* reduced the levels of IL-1 and TNF
Cinnamomum cassia	Compound from *Cinnamomum cassia* were able to interact with JAK2, mPEGS-1, COX-2, IL-1β, and PPARγ target proteins Cinnamaldehyde inhibit IL-1β, reducing HIF-1α Cinnamomi lactone inhibited MMP-3, IL-1β, and MMP-1
Centella asiatica	Madecassoside and madecassic acid directly inhibited the release of pro-inflammatory cytokines and IL-10 *Centella asiatica* extract reduce TNF-α, IL-1, IL-6, IL-12, and NO
Phyllanthus amarus	*Phyllanthus amarus* extract decreased levels of sGAGs and MMP-2 Phyllanthin compound reduce IL-1ß and TNF-α
Psidium guajava	Guava leaf extract inhibited LPS-induced production of NO and PGE2
Moringa oleifera	The moringa extract inhibited the production of inflammatory cytokines (IL-1β, TNF-α, and IL-6) The moringa extract also inhibited NFk-B and reducing NO, IL-1β, TNF-α, and IL-6
Nigella sativa	Thymoquinone compound inhibited apoptosis-regulated signaling kinase 1 (ASK1) in response to TNF-induced activation of phospho-p38 and phospho-JNK *Nigella sativa* led to a drop in CD8+ and a rise in the proportion of CD4+ CD25+ T cells and the CD4+/CD8+ ratio
Cannabis sativa	Cannabidiol compound contributed to the reduction of cell viability, proliferation, and RASF IL-6/IL-8 production Cannabidiol decreased the production of TNF and IL-10 Cannabidiol boosted IL-10 synthesis
Andrographis paniculata	Andrographolide compound inhibited IL-1β and IL-6 Andrographolide suppressing MMP-1, 3, and 9 Combination of andrographolide and methotrexate substantially decreased blood levels of TNF-α, IL-6, and IL-1β
Artemisia annua	Casticin and chrysosplenol D compound decrease IL-1β, IL-6, and MCP-1
Piper betle	4-Allyl-1,2-diacetoxybenzene exhibited stronger interactions with MMP-1, TGF, and IL-1 Allylpyrocatechol (APC) and methotrexate (MTX) inhibit the production of pro-inflammatory cytokines, such as TNF-α and IL-6.

However, this therapy has several disadvantages related to gastrointestinal, cardiovascular, and nephrotoxic effects. Therefore, alternative medicine, such as the consumption of herbal medicine, is highly beneficial for complementary therapy. However, more information is required about plants or compounds that potentially be used to treat rheumatic disorders. In this book chapter, we identified 33 plants containing more than 15 compounds for anti-rheumatoid arthritis activity, osteoarthritis, gout, etc. The articles mentioned in the chapter book apply several study designs, including in silico, in vitro, and in vivo techniques. The findings of this review indicate that these plants have an effective mechanism for inhibiting multiple protein targets in rheumatic disorders, including TNF-α, interleukins, NO, iNOS, ROS, MMP, NF-kB, COX-1, COX-2, PGE2, MCSF, and others. Regarding the activity of several plants and compounds on arthritis protein targets, however, additional research is still required so that plants with potential anti-arthritis disorder activity can be identified.

References

1. Sangha O (2000) Epidemiology of rheumatic diseases. Rheumatology (Oxford) 39(2):3–12
2. Smolen JS, Aletaha D, Barton A, Burmester GR, Emery P, Firestein GS et al (2018) Rheumatoid arthritis. Nat Rev Dis Prim 4(1):18001. https://doi.org/10.1038/nrdp.2018.1
3. Panush RS (2000) American College of Rheumatology Position Statement: complementary and alternative therapies for rheumatic disease. Rheum Dis Clin N Am 26(1):189–192. https://www.sciencedirect.com/science/article/pii/S0889857X05701319
4. Rao JK, Mihaliak K, Kroenke K, Bradley J, Tierney WM, Weinberger M (1999) Use of complementary therapies for arthritis among patients of rheumatologists. Ann Intern Med 131(6):409–416
5. WHO (2004) WHO guidelines on developing consumer information on proper use of traditional, complementary and alternative medicine. World Health Organization, Geneva. https://apps.who.int/iris/handle/10665/42957
6. Barnes PM, Powell-Griner E, McFann K, Nahin RL (2004) Complementary and alternative medicine use among adults: United States, 2002. Adv Data 343:1–19
7. Struthers GR, Scott DL, Scott DG (1983) The use of alternative treatments by patients with rheumatoid arthritis. Rheumatol Int 3(4):151–152
8. Heidari B (2011) Rheumatoid arthritis: early diagnosis and treatment outcomes. Casp J Intern Med 2(1):161–170
9. Molnar V, Matišić V, Kodvanj I, Bjelica R, Jeleč Ž, Hudetz D et al (2021) Cytokines and chemokines involved in osteoarthritis pathogenesis. Int J Mol Sci 22(17):9208. https://doi.org/10.3390/ijms22179208
10. Murugesan S, Venkateswaran MR, Jayabal S, Periyasamy S (2020) Evaluation of the antioxidant and anti-arthritic potential of *Zingiber officinale* Rosc. by in vitro and in silico analysis. S Afr J Bot 130:45–53
11. Kim GJ, Lee JY, Choi HG, Kim SY, Kim E, Shim SH et al (2017) Cinnamomulactone, a new butyrolactone from the twigs of *Cinnamomum cassia* and its inhibitory activity of matrix metalloproteinases. Arch Pharm Res 40(3):304–310
12. Li Y, Xu B, Xu M, Chen D, Xiong Y, Lian M et al (2017) 6-Gingerol protects intestinal barrier from ischemia/reperfusion-induced damage via inhibition of p38 MAPK to NF-κB signalling. Pharmacol Res 119:137–148
13. Jafarzadeh A, Ahangar-Parvin R, Nemat M, Taghipour Z, Shamsizadeh A, Ayoobi F et al (2017) Ginger extract modulates the expression of IL-12 and TGF-β in the central nervous

system and serum of mice with experimental autoimmune encephalomyelitis. Avicenna J Phytomed 7(1):54–65

14. Rondanelli M, Riva A, Morazzoni P, Allegrini P, Faliva MA, Naso M et al (2017) The effect and safety of highly standardized ginger (*Zingiber officinale*) and Echinacea (*Echinacea angustifolia*) extract supplementation on inflammation and chronic pain in NSAIDs poor responders. A pilot study in subjects with knee arthritis. Nat Prod Res 31(11): 1309–1313
15. Yang M, Guo MY, Luo Y, Yun MD, Yan J, Liu T et al (2017) Effect of *Artemisia annua* extract on treating active rheumatoid arthritis: a randomized controlled trial. Chin J Integr Med 23(7): 496–503
16. Tang X, Amar S (2015) Kavain inhibition of LPS-induced TNF-α via ERK/LITAF. Toxicol Res (Camb) 5(1):188–196
17. Uttra AM, Alamgeer. (2017) Assessment of anti-arthritic potential of *Ephedra gerardiana* by in vitro and in vivo methods. Bangladesh J Pharmacol 12(4):403–409
18. Soumaya B, Yosra E, Rim BM, Sarra D, Sawsen S, Sarra B et al (2020) Preliminary phytochemical analysis, antioxidant, anti-inflammatory and anticancer activities of two Tunisian Ephedra species: *Ephedra alata* and *Ephedra fragilis*. S Afr J Bot 135:421–428. https://doi.org/10.1016/j.sajb.2020.09.033
19. Wang Q, Shu Z, Xing N, Xu B, Wang C, Sun G et al (2016) A pure polysaccharide from *Ephedra sinica* treating on arthritis and inhibiting cytokines expression. Int J Biol Macromol 86:177–188. https://doi.org/10.1016/j.ijbiomac.2016.01.010
20. Uttra AM, Alamgeer SM, Shabbir A, Jahan S (2018) *Ephedra gerardiana* aqueous ethanolic extract and fractions attenuate Freund Complete Adjuvant induced arthritis in Sprague Dawley rats by downregulating PGE2, COX2, IL-1β, IL-6, TNF-α, NF-kB and upregulating IL-4 and IL-10. J Ethnopharmacol 224:482–496. https://doi.org/10.1016/j.jep.2018.06.018
21. Montserrat-De LPS, Fernández-Arche Á, Ángel-Martín M, García-Giménez MD (2012) The sterols isolated from evening primrose oil modulate the release of proinflammatory mediators. Phytomedicine 19(12):1072–1076. https://doi.org/10.1016/j.phymed.2012.06.008
22. El-Sayed RM, Moustafa YM, El-Azab MF (2014) Evening primrose oil and celecoxib inhibited pathological angiogenesis, inflammation, and oxidative stress in adjuvant-induced arthritis: novel role of angiopoietin-1. Inflammopharmacology 22(5):305–317
23. Ma R, Chen Q, Li H, Wu S, Lian M, Jin X et al (2020 Jul) Extract of *Oenothera biennis* L. stem inhibits LPS-induced inflammation by regulating MAPK and NF-κB signaling pathways. Pak J Pharm Sci 33(4):1473–1481
24. Abd-Nikfarjam B, Abbasi M, Memarzadeh M, Farzam SA, Jamshidian A, Dolati-Somarin A (2022) Therapeutic efficacy of *Urtica dioica* and evening primrose in patients with rheumatoid arthritis: a randomized double-blind, placebo-controlled clinical trial. J Herb Med 32:100556. https://doi.org/10.1016/j.hermed.2022.100556
25. Ahn S, Siddiqi MH, Noh HY, Kim YJ, Kim YJ, Jin CG et al (2015) Anti-inflammatory activity of ginsenosides in LPS-stimulated RAW 264.7 cells. Sci Bull 60(8):773–784. https://doi.org/10.1007/s11434-015-0773-4
26. Kang S, Siddiqi MH, Yoon SJ, Ahn S, Noh HY, Kumar NS et al (2016) Therapeutic potential of compound K as an IKK inhibitor with implications for osteoarthritis prevention: an in silico and in vitro study. Vitr Cell Dev Biol Anim 52(9):895–905. https://doi.org/10.1007/s11626-016-0062-9
27. Choi YS, Kang EH, Lee EY, Gong HS, Kang HS, Shin K et al (2013) Joint-protective effects of compound K, a major ginsenoside metabolite, in rheumatoid arthritis: in vitro evidence. Rheumatol Int 33(8):1981–1990. https://doi.org/10.1007/s00296-013-2664-9
28. So MW, Lee EJ, Lee HS, Koo BS, Kim YG, Lee CK et al (2013) Protective effects of ginsenoside Rg3 on human osteoarthritic chondrocytes. Mod Rheumatol 23(1):104–111
29. Zhang P (2017) Ginsenoside-Rg5 treatment inhibits apoptosis of chondrocytes and degradation of cartilage matrix in a rat model of osteoarthritis. Oncol Rep 37(3):1497–1502

30. Cheng W, Jing J, Wang Z, Wu D, Huang Y (2017) Chondroprotective effects of ginsenoside Rg1 in human osteoarthritis chondrocytes and a rat model of anterior cruciate ligament transection. Nutrients 9(3):263
31. Jhun J, Lee J, Byun JK, Kim EK, Woo JW, Lee JH et al (2014) Red ginseng extract ameliorates autoimmune arthritis via regulation of stat3 pathway, th17/treg balance, and osteoclastogenesis in mice and human. Mediat Inflamm 2014:351856
32. Kim HI, Chon SJ, Seon KE, Seo SK, Choi YR (2021) Clinical effects of Korean red ginseng in postmenopausal women with hand osteoarthritis: a double-blind, randomized controlled trial. Front Pharmacol 12:1–8
33. Cho SK, Kim D, Yoo D, Jang EJ, Jun JB, Sung YK (2018) Korean red ginseng exhibits no significant adverse effect on disease activity in patients with rheumatoid arthritis: a randomized, double-blind, crossover study. J Ginseng Res 42(2):144–148. https://doi.org/10.1016/j.jgr.2017.01.006
34. Jung SJ, Oh MR, Lee DY, Lee YS, Kim GS, Park SH et al (2021) Effect of ginseng extracts on the improvement of osteopathic and arthritis symptoms in women with osteopenia: a randomized, double-blind, placebo-controlled clinical trial. Nutrients 13(10):3352
35. Kury ALT, Dayyan F, Shah FA, Malik Z, Khalil AAK, Alattar A et al (2020) *Ginkgo biloba* extract protects against methotrexate-induced hepatotoxicity: a computational and pharmacological approach. Molecules 25(11):2540
36. Ho LJ, Hung LF, Liu FC, Hou TY, Lin LC, Huang CY et al (2013) *Ginkgo biloba* extract individually inhibits JNK activation and induces c-Jun degradation in human chondrocytes: potential therapeutics for osteoarthritis. PLoS One 8(12):1–12
37. Ma T, Lv L, Yu Y, Jia L, Song X, Xu XY et al (2022) Bilobalide exerts anti-inflammatory effects on chondrocytes through the AMPK/SIRT1/mTOR pathway to attenuate ACLT-induced post-traumatic osteoarthritis in rats. Front Pharmacol 13:1–14
38. Xie C, Jiang J, Liu J, Yuan G, Zhao Z (2020) Ginkgolide B attenuates collagen-induced rheumatoid arthritis and regulates fibroblast-like synoviocytes-mediated apoptosis and inflammation. Ann Transl Med 8(22):1497–1497
39. Al-Rekabi MD (2014) Comparative study between the clinical effects of glucosamine/*Ginkgo biloba* & glucosamine/chondroitin in treatment of knee osteoarthritis. Iraqi Acad Sci J 2:1–8
40. Ma T, Jia L, Zhao J, Lv L, Yu Y, Ruan H et al (2022) Ginkgolide C slows the progression of osteoarthritis by activating Nrf2/HO-1 and blocking the NF-κB pathway. Front Pharmacol 13: 1–16
41. Lestari AR, Batubara I, Wahyudi ST, Ilmiawati A, Achmadi SS (2022) Bioactive compounds in garlic (*Allium sativum*) and black garlic as antigout agents, using computer simulation. Life 12(8):1131
42. Zhao Z, Liu Y, Lu Y, Hou M, Shen X, Yang H et al (2022) *Gingko biloba* inspired lactone prevents osteoarthritis by activating the AMPK-SIRT1 signaling pathway. Arthritis Res Ther 24(1):1–17
43. Liang JJ, Li HR, Chen Y, Zhang C, Chen DG, Liang ZC et al (2019) Diallyl trisulfide can induce fibroblast-like synovial apoptosis and has a therapeutic effect on collagen-induced arthritis in mice via blocking NF-κB and Wnt pathways. Int Immunopharmacol 71:132–138
44. Chen Y, Xue R, Jin X, Tan X (2018) Antiarthritic activity of diallyl disulfide against Freund's adjuvant-induced arthritic rat model. J Environ Pathol Toxicol Oncol Off Organ Int Soc Environ Toxicol Cancer 37(4):291–303
45. Kalpana T, Chaitanya B, Mounica P, Rama ML, Sri ML, Sireesha PS (2018) In-vitro anti-arthritic activity of methanolic extracts of *Allium sativum*. World Journal of Pharmaceutical Research 7(16):955–959
46. Moutia M, Seghrouchni F, Abouelazz O, Elouaddari A, Al Jahid A, Elhou A et al (2016) *Allium sativum* L. regulates in vitro IL-17 gene expression in human peripheral blood mononuclear cells. BMC Complement Altern Med 16(1):1–10. https://doi.org/10.1186/s12906-016-1365-9

47. Moosavian SP, Paknahad Z, Habibagahi Z, Maracy M (2020) The effects of garlic (*Allium sativum*) supplementation on inflammatory biomarkers, fatigue, and clinical symptoms in patients with active rheumatoid arthritis: a randomized, double-blind, placebo-controlled trial. Phyther Res. 34(11):2953–2962
48. Majid M, Nasir B, Zahra S, Khan M, Mirza B, Haq I (2018) *Ipomoea batatas* Lam. ameliorates acute and chronic inflammations by suppressing inflammatory mediators, a comprehensive exploration using in vitro and in vivo models. BMC Complement Altern Med 18:216. https://doi.org/10.1186/s12906-018-2279-5
49. Xu J, Xiao C, Xu H, Yang S, Chen Z, Wang H et al (2021) Anti-inflammatory effects of Ganoderma lucidum sterols via attenuation of the p38 MAPK and NF-κB pathways in LPS-induced RAW 264.7 macrophages. Food Chem Toxicol 150:112073. https://www.sciencedirect.com/science/article/pii/S027869152100106X
50. Zhang Y, Wang H, Mao X, Guo Q, Li W, Wang X et al (2018) A novel circulating miRNA-based model predicts the response to Tripterysium glycosides tablets: moving toward model-based precision medicine in rheumatoid arthritis. Front Pharmacol 9:1–11
51. Xu X, Li QJ, Xia S, Wang MM, Ji W (2016) Tripterygium glycosides for treating late-onset rheumatoid arthritis: a systematic review and meta-analysis. Altern Ther Health Med 22(6):32–39
52. Tiyah SW, Ratnaningtyas NI, Wibowo ES, Mumpuni A, Ekowati N (2023) *Ganoderma lucidum* as anti-inflammatory agent on the level of albumin and globulin in rat (*Rattus norvegicus*) rheumatoid arthritis (RA) model. In: Proceeding ICMA-SURE 5th International Conference on Multidisciplinary Approaches Sustainable Rural Development, vol 2(1). http://jos.unsoed.ac.id/index.php/eprocicma/article/view/7808
53. Li EK, Tam LS, Wong CK, Li WC, Lam CWK, Wachtel-Galor S et al (2007) Safety and efficacy of *Ganoderma lucidum* (lingzhi) and san Miao san supplementation in patients with rheumatoid arthritis: a double-blind, randomized, placebo-controlled pilot trial. Arthritis Rheum 57(7):1143–1150
54. Zhang Y, Xu W, Li H, Zhang X, Xia Y, Chu K et al (2013) Therapeutic effects of total alkaloids of *Tripterygium wilfordii* Hook f. on collagen-induced arthritis in rats. J Ethnopharmacol 145(3):699–705
55. Zheng W, Mei Y, Chen C, Cai L, Chen H (2021) The effectiveness and safety of *Tripterygium wilfordii* glycosides combined with disease-modifying anti-rheumatic drugs in the treatment of rheumatoid arthritis: a systematic review and meta-analysis of 40 randomized controlled trials. Phytother Res 35(6):2902–2924
56. Chen WJ, Li TX, Wang XY, Xue ZP, Lyu C, Li HZ et al (2020) Meta-analysis of RCT studies on clinical efficacy of single administration of Tripterygium glycosides tablets or combined administration with methotrexate against rheumatoid arthritis. China J Chin Mater Med 45(4):791–797
57. Lima V, Melo IM, Taira TM, Buitrago LYW, Fonteles CSR, Leal LKAM et al (2020) *Uncaria tomentosa* reduces osteoclastic bone loss in vivo. Phytomedicine 79:153327. https://doi.org/10.1016/j.phymed.2020.153327
58. Castilhos LG, Rezer JP, Ruchel JB, Thorstenberg MLML, dos Jaques JAS, Schlemmer JB et al (2015) Effect of *Uncaria tomentosa* extract on purinergic enzyme activities in lymphocytes of rats submitted to experimental adjuvant arthritis model. BMC Complement Altern Med 15(1): 1–11. https://doi.org/10.1186/s12906-015-0694-4
59. de Paula LCL, Fonseca F, Perazzo F, Cruz FM, Cubero D, Trufelli DC et al (2015) *Uncaria tomentosa* (cat's claw) improves quality of life in patients with advanced solid tumors. J Altern Complement Med 21(1):22–30
60. Dietrich F, Martins JP, Kaiser S, Silva RBM, Rockenbach L, Edelweiss MIA et al (2015) The quinovic acid glycosides purified fraction from *Uncaria tomentosa* protects against hemorrhagic cystitis induced by cyclophosphamide in mice. PLoS One 10(7):1–15
61. Raghunath S, Kumar GR, Rashmiranjan B, Sekhar NS (2013) Computational study of some bowsellic acid and curcumin derivatives as potential anti-arthritis compounds. Ijbr 6(1):52–59
62. Chen Y, Liu P, Yu Z, Tan J (2014) Synthesis of nanosized NaY molecular sieve under dynamic hydrothermal crystallization. Key Eng Mater 609–610:288–297

63. Umar S, Umar K, Sarwar AHMG, Khan A, Ahmad N, Ahmad S et al (2014) *Boswellia serrata* extract attenuates inflammatory mediators and oxidative stress in collagen induced arthritis. Phytomedicine 21(6):847–856. https://www.sciencedirect.com/science/article/pii/S0944711314000750
64. Haroyan A, Mukuchyan V, Mkrtchyan N, Minasyan N, Gasparyan S, Sargsyan A et al (2018) Efficacy and safety of curcumin and its combination with boswellic acid in osteoarthritis: a comparative, randomized, double-blind, placebo-controlled study. BMC Complement Altern Med 18(1):1–16
65. Karimifar M, Soltani R, Hajhashemi V, Sarrafchi S (2017) Evaluation of the effect of *Elaeagnus angustifolia* alone and combined with *Boswellia thurifera* compared with ibuprofen in patients with knee osteoarthritis: a randomized double-blind controlled clinical trial. Clin Rheumatol 36(8):1849–1853
66. Notarnicola A, Tafuri S, Fusaro L, Moretti L, Pesce V, Moretti B (2011) The "mESACA" study: methylsulfonylmethane and boswellic acids in the treatment of gonarthrosis. Adv Ther 28(10):894–906
67. Widodo WT, Widyarti S, Sumitro SB, Santjojo DH (2021) In silico study of Tyramine-Fe complex in Brotowali (*Tinospora crispa*) as anti-inflammatory. In: Proceedings of the 11th annual international conference on industrial engineering and operations management. WHO, pp 3473–3480
68. Abiramasundari G, Gowda CMM, Pampapathi G, Praveen S, Shivamurugan S, Vijaykumar M et al (2017) Ethnomedicine based evaluation of osteoprotective properties of *Tinospora cordifolia* on in vitro and in vivo model systems. Biomed Pharmacother 87:342–354. https://doi.org/10.1016/j.biopha.2016.12.094
69. Ghatpande NS, Misar AV, Waghole RJ, Jadhav SH, Kulkarni PP (2019) *Tinospora cordifolia* protects against inflammation associated anemia by modulating inflammatory cytokines and hepcidin expression in male Wistar rats. Sci Rep 9(1):1–11
70. Philip S, Tom G, Balakrishnan NP, Sundaram S, Velikkakathu VA (2021) *Tinospora cordifolia* chloroform extract inhibits LPS-induced inflammation via NF-κB inactivation in THP-1 cells and improves survival in sepsis. BMC Complement Med Ther 21(1):1–13
71. Sannegowda KM, Venkatesha SH, Moudgil KD (2015) *Tinospora cordifolia* inhibits autoimmune arthritis by regulating key immune mediators of inflammation and bone damage. Int J Immunopathol Pharmacol 28(4):521–531
72. Zaka M, Sehgal SA, Shafique S, Abbasi BH (2017) Comparative in silico analyses of *Cannabis sativa*, *Prunella vulgaris* and *Withania somnifera* compounds elucidating the medicinal properties against rheumatoid arthritis. J Mol Graph Model 74:296–304. https://doi.org/10.1016/j.jmgm.2017.04.013
73. Madhusudan M, Zameer F, Naidu A, Nagendra Prasad MN, Dhananjaya BL, Hegdekatte R (2016) Evaluating the inhibitory potential of *Withania somnifera* on platelet aggregation and inflammation enzymes: an in vitro and in silico study. Pharm Biol 54(9):1936–1941
74. Gupta M, Kaur G (2016) Aqueous extract from the *Withania somnifera* leaves as a potential anti-neuroinflammatory agent: a mechanistic study. J Neuroinflammation 13(1):193
75. Hasan HF, Khazal KF, Luaibi OK (2019) The effect of crude alcoholic extract of *Withania somnifera* leaves in experimentally induced arthritis in mice. Univ Thi-Qar J Sci 4(2):45–52
76. Khaled H, Hanna J, Shoukry NMM, Darwesh A, Fares N (2022) Therapeutic potential of *Withania somnifera* extract on experimental model of arthritis in rats: histological study. Front Sci Res Technol 4:82–90
77. Khaled HE, Ayad J, Kamel H, Bahaa A, Darwesh E, Hassan N et al (2022) Anti-inflammatory activity of *Withania somnifera* root extract in complete Freund's adjuvant-induced arthritis in male albino rats. Egypt J Hosp Med 89(2):7997–8003
78. Khan MA, Ahmed RS, Chandra N, Arora VK, Ali A (2018) In vivo, extract from Withania somnifera root ameliorates arthritis via regulation of key immune mediators of inflammation in experimental model of arthritis. Antiinflamm Antiallergy Agents Med Chem 18(1): 55–70

79. Sivamani S, Joseph B, Kar B (2014) Anti-inflammatory activity of *Withania somnifera* leaf extract in stainless steel implant induced inflammation in adult zebrafish. J Genet Eng Biotechnol 12(1):1–6. https://www.sciencedirect.com/science/article/pii/S1687157X14000031
80. Gupta A, Singh S (2014) Evaluation of anti-inflammatory effect of *Withania somnifera* root on collagen-induced arthritis in rats. Pharm Biol 52(3):308–320
81. Orrù A, Marchese G, Casu G, Casu MA, Kasture S, Cottiglia F et al (2014) *Withania somnifera* root extract prolongs analgesia and suppresses hyperalgesia in mice treated with morphine. Phytomedicine 21(5):745–752
82. Kanjilal S, Gupta AK, Patnaik RS, Dey A (2021) Analysis of clinical trial registry of India for evidence of anti-arthritic properties of *Withania somnifera* (Ashwagandha). Altern Ther Health Med 27(6):58–66
83. Lee HH, Ahn EK, Hong SS, Oh JS (2017) Anti-inflammatory effect of Tribulusamide D isolated from *Tribulus terrestris* in lipopolysaccharide-stimulated RAW264.7 macrophages. Mol Med Rep 16(4):4421–4428
84. Zhao WR, Shi WT, Zhang J, Zhang KY, Qing Y, Tang JY et al (2021) *Tribulus terrestris* L. extract protects against lipopolysaccharide-induced inflammation in RAW 264.7 macrophage and zebrafish via inhibition of Akt/MAPKs and NF-B/iNOS-NO signaling pathways. Evid Based Complement Altern Med 2021(2 Supplementary):1–11
85. Mishra NK, Biswal GS, Chowdary KA, Mishra G (2013) Anti-arthritic activity of *Tribulus terrestris* studied in Freund's adjuvant induced arthritic rats. J Pharm Educ Res 4(1):41–46
86. Park YJ, Cho YR, Oh JS, Ahn EK (2017) Effects of *Tribulus terrestris* on monosodium iodoacetate-induced osteoarthritis pain in rats. Mol Med Rep 16(4):5303–5311
87. Xu S, Peng H, Wang N, Zhao M (2018) Inhibition of TNF-α and IL-1 by compounds from selected plants for rheumatoid arthritis therapy: in vivo and in silico studies. Trop J Pharm Res 17:277–285
88. Gouthamchandra K, Sudeep HV, Chandrappa S, Raj A, Naveen P, Shyamaprasad K (2021) Efficacy of a standardized turmeric extract comprised of 70% bisdemothoxy-curcumin (Reverc3) against lps-induced inflammation in raw264.7 cells and carrageenan-induced paw edema. J Inflamm Res 14:859–868
89. Okuda-Hanafusa C, Uchio R, Fuwa A, Kawasaki K, Muroyama K, Yamamoto Y et al (2019) Turmeronol A and Turmeronol B from: Curcuma longa prevent inflammatory mediator production by lipopolysaccharide-stimulated RAW264.7 macrophages, partially via reduced NF-κB signaling. Food Funct 10(9):5779–5788
90. Nonose N, Pereira JA, Machado PRM, Rodrigues MR, Sato DT, Martinez CAR (2014) Oral administration of curcumin (*Curcuma longa*) can attenuate the neutrophil inflammatory response in zymosan-induced arthritis in rats. Acta Cir Bras 29(11):727–734
91. Nakagawa Y, Mukai S, Yamada S, Matsuoka M, Tarumi E, Hashimoto T et al (2014) Short-term effects of highly-bioavailable curcumin for treating knee osteoarthritis: a randomized, double-blind, placebo-controlled prospective study. J Orthop Sci Off J Jpn Orthop Assoc 19(6):933–939
92. Panahi Y, Rahimnia AR, Sharafi M, Alishiri G, Saburi A, Sahebkar A (2014) Curcuminoid treatment for knee osteoarthritis: a randomized double-blind placebo-controlled trial. Phytother Res 28(11):1625–1631
93. Srivastava S, Saksena AK, Khattri S, Kumar S, Dagur RS (2016) Curcuma longa extract reduces inflammatory and oxidative stress biomarkers in osteoarthritis of knee: a four-month, double-blind, randomized, placebo-controlled trial. Inflammopharmacology 24(6): 377–388
94. Wang Y, Hu B, Peng Y, Xiong X, Jing W, Wang J et al (2019) In silico exploration of the molecular mechanism of cassane diterpenoids on anti-inflammatory and immunomodulatory activity. J Chem Inf Model 59(5):2309–2323

95. Chu MJ, Wang YZ, Itagaki K, Ma HX, Xin P, Zhou XG et al (2013) Identification of active compounds from *Caesalpinia sappan* L. extracts suppressing IL-6 production in RAW 264.7 cells by PLS. J Ethnopharmacol 148(1):37–44. https://doi.org/10.1016/j.jep.2013.03.050
96. Min BS, Cuong TD (2013) Phenolic compounds from *Caesalpinia sappan* and their inhibitory effects on LPS-induced NO production in RAW264.7 cells. Nat Prod Sci 19(3):201–205
97. Jung EG, Il HK, Hwang SG, Kwon HJ, Patnaik BB, Kim YH et al (2015) Brazilin isolated from *Caesalpinia sappan* L. inhibits rheumatoid arthritis activity in a type-II collagen induced arthritis mouse model. BMC Complement Altern Med 15(1):1–11
98. Jung EG, Il HK, Kwon HJ, Patnaik BB, Kim WJ, Hur GM et al (2015) Anti-inflammatory activity of sappanchalcone isolated from *Caesalpinia sappan* L. in a collagen-induced arthritis mouse model. Arch Pharm Res 38(6):973–983. https://doi.org/10.1007/s12272-015-0557-z
99. Lee CW, Park SM, Zhao R, Lee C, Chun W, Son Y et al (2015) Hederagenin, a major component of *Clematis mandshurica* Ruprecht root, attenuates inflammatory responses in RAW 264.7 cells and in mice. Int Immunopharmacol 29(2):528–537. https://doi.org/10.1016/j.intimp.2015.10.002
100. Fu Q, Chen J, Yuan HM, Ma Y, Yu T, Zou L (2016) Alkaloids and phenolic glycosides from *Clematis mandshurica* and their inhibitory effects against NO production in LPS-induced RAW 246.7 macrophages. Phytochem Lett 17:238–241
101. Dilshara MG, Lee KT, Lee CM, Choi YH, Lee HJ, Choi IW et al (2015) New compound, 5-O-isoferuloyl-2-deoxy-D-ribono-γ-lacton from *Clematis mandshurica*: anti-inflammatory effects in lipopolysaccharide-stimulated BV2 microglial cells. Int Immunopharmacol 24(1):14–23
102. Lee CW, Park SM, Kim YS, Jegal KH, Lee JR, Cho IJ et al (2014) Biomolecular evidence of anti-inflammatory effects by *Clematis mandshurica* Ruprecht root extract in rodent cells. J Ethnopharmacol 155(2):1141–1155. https://doi.org/10.1016/j.jep.2014.06.048
103. Xiong Y, Ma Y, Han W, Kodithuwakku ND, Liu LF, Li FW et al (2014) Clematichinenoside AR induces immunosuppression involving Treg cells in Peyers patches of rats with adjuvant induced arthritis. J Ethnopharmacol 155(2):1306–1314. https://doi.org/10.1016/j.jep.2014.07.028
104. Liu M, Zhou X, Zhou L, Liu Z, Yuan J, Cheng J et al (2018) Carnosic acid inhibits inflammation response and joint destruction on osteoclasts, fibroblast-like synoviocytes, and collagen-induced arthritis rats. J Cell Physiol 233(8):6291–6303
105. Amaral GP, de Carvalho NR, Barcelos RP, Dobrachinski F, de Portella RL, da Silva MH et al (2013) Protective action of ethanolic extract of *Rosmarinus officinalis* L. in gastric ulcer prevention induced by ethanol in rats. Food Chem Toxicol 55:48–55. https://doi.org/10.1016/j.fct.2012.12.038
106. De Almeida GG, De Sá-Nakanishi AB, Comar JF, Bracht L, Dias MI, Barros L et al (2018) Water soluble compounds of *Rosmarinus officinalis* L. improve the oxidative and inflammatory states of rats with adjuvant-induced arthritis. Food Funct 9(4):2328–2340
107. Wei T, Liu Y, Li M (2021) Anti-inflammatory and anti-arthritic activity of rosmarinic acid isolated from *Rosmarinus officinalis* in an experimental model of arthritis. Indian J Pharm Educ Res 55(2):507–516
108. Huang X, Pan Q, Mao Z, Wang P, Zhang R, Ma X et al (2018) Kaempferol inhibits interleukin-1ß stimulated matrix metalloproteinases by suppressing the MAPK-associated ERK and P38 signaling pathways. Mol Med Rep 18(3):2697–2704
109. Heydari NM, Parsivand M, Mohammadi N, Asghari MN (2022) Comparison of *Elaeagnus angustifolia* L. extract and quercetin on mouse model of knee osteoarthritis. J Ayurveda Integr Med 13(2):100529. https://doi.org/10.1016/j.jaim.2021.10.001
110. Motevalian M, Shiri M, Shiri S, Shiri Z, Shiri H (2017) Anti-inflammatory activity of *Elaeagnus angustifolia* fruit extract on rat paw edema. J Basic Clin Physiol Pharmacol 28(4):377–381

111. Nikniaz Z, Ostadrahimi A, Mahdavi R, Ebrahimi AA, Nikniaz L (2014) Effects of *Elaeagnus angustifolia* L. supplementation on serum levels of inflammatory cytokines and matrix metalloproteinases in females with knee osteoarthritis. Complement Ther Med 22(5):864–869
112. Panahi Y, Alishiri GH, Bayat N, Hosseini SM, Sahebkar A (2016) Efficacy of *Elaeagnus angustifolia* extract in the treatment of knee osteoarthritis: a randomized controlled trial. EXCLI J 15:203–210
113. Orabona C, Orecchini E, Volpi C, Bacaloni F, Panfili E, Pagano C et al (2022) *Crocus sativus* L. petal extract inhibits inflammation and osteoclastogenesis in RAW 264.7 cell model. Pharmaceutics 14(6):1290
114. Rathore B, Jaggi K, Thakur S, Mathur A (2015) Anti-inflammatory activity of *Crocus sativus* extract in experimental arthritis. Int J Pharm Sci Res 6(4):1473–1478
115. Zamani TRS, Sahebari M, Mahmoudi Z, Hosseinzadeh H, Haghmorad D, Tabasi N et al (2015) Inhibitory effect of *Crocus sativus* L. ethanol extract on adjuvant-induced arthritis. Food Agric Immunol 26(2):170–180
116. Liu W, Sun Y, Cheng Z, Guo Y, Liu P, Wen Y (2018) Crocin exerts anti-inflammatory and anti-arthritic effects on type II collagen-induced arthritis in rats. Pharm Biol 56(1):209–216. https://doi.org/10.1080/13880209.2018.1448874
117. Sahebari M, Heidari H, Nabavi S, Khodashahi M, Rezaieyazdi Z, Dadgarmoghaddam M et al (2021) A double-blind placebo-controlled randomized trial of oral saffron in the treatment of rheumatoid arthritis. Avicenna J Phytomed 11(4):332–342
118. Misra S, Ikbal AMA, Bhattacharjee D, Hore M, Mishra S, Karmakar S et al (2022) Validation of antioxidant, antiproliferative, and in vitro anti-rheumatoid arthritis activities of epigallocatechin-rich bioactive fraction from *Camellia sinensis* var. assamica, Assam variety white tea, and its comparative evaluation with green tea fraction. J Food Biochem 46(12):e14487. https://doi.org/10.1111/jfbc.14487
119. Tanwar A, Chawla R, Ansari MM, Neha TP, Chakotiya AS et al (2017) In vivo anti-arthritic efficacy of *Camellia sinensis* (L.) in collagen induced arthritis model. Biomed Pharmacother 87:92–101. https://doi.org/10.1016/j.biopha.2016.12.089
120. Tanwar A, Chawla R, Basu M, Arora R, Khan HA (2017) FRI0032 Curative effect of *Camellia sinensis* (CS) against opportunistic infection in vulnerable animal model of rheumatoid arthritis. Ann Rheum Dis 76(2):491. http://ard.bmj.com/content/76/Suppl_2/491.2.abstract
121. Hashempur MH, Sadrneshin S, Mosavat SH, Ashraf A (2018) Green tea (*Camellia sinensis*) for patients with knee osteoarthritis: a randomized open-label active-controlled clinical trial. Clin Nutr 37(1):85–90. https://doi.org/10.1016/j.clnu.2016.12.004
122. Raja A, Singh GP, Fadil SA, Elhady SS, Youssef FS, Ashour ML (2022) Prophylactic anti-osteoporotic effect of *Matricaria chamomilla* L. flower using steroid-induced osteoporosis in rat model and molecular modelling approaches. Antioxidants 11(7):1316
123. Ortiz MI, Cariño-Cortés R, Ponce-Monter HA, González-García MP, Castañeda-Hernández G, Salinas-Caballero M (2017) Synergistic interaction of *Matricaria chamomilla* extract with diclofenac and indomethacin on carrageenan-induced paw inflammation in rats. Drug Dev Res 78(7):360–367
124. Shoara R, Hashempur MH, Ashraf A, Salehi A, Dehshahri S, Habibagahi Z (2015) Efficacy and safety of topical *Matricaria chamomilla* L. (chamomile) oil for knee osteoarthritis: a randomized controlled clinical trial. Complement Ther Clin Pract 21(3):181–187. https://www.sciencedirect.com/science/article/pii/S1744388115000493
125. Khan MA, Rabbani G, Kumari M, Khan MJ (2022) Ellagic acid protects type II collagen induced arthritis in rat via diminution of IKB phosphorylation and suppression IKB-NF-kB complex activation: in vivo and in silico study. Inflammopharmacology 30(5):1729–1743. https://doi.org/10.1007/s10787-022-01022-x
126. Lee CJ, Chen LG, Liang WL, Hsieh MS, Wang CC (2018) Inhibitory effects of punicalagin from *Punica granatum* against type II collagenase-induced osteoarthritis. J Funct Foods 41:216–222. https://www.sciencedirect.com/science/article/pii/S1756464617307491

127. Gautam RK, Sharma S, Sharma K, Gupta G (2018) Evaluation of antiarthritic activity of butanol fraction of *Punica granatum* Linn. Rind extract against Freund's complete adjuvant-induced arthritis in rats. J Environ Pathol Toxicol Oncol 37(1):53–62
128. Li Z, Gai S (2019) Effect of ethanol extract of *Punica granatum* against Freund's complete adjuvant-induced arthritis in rats yanming wang, tao he1. Trop J Pharm Res 18(3):591–595
129. Gautam RK, Gupta G, Sharma S, Hatware K, Patil K, Sharma K et al (2019) Rosmarinic acid attenuates inflammation in experimentally induced arthritis in Wistar rats, using Freund's complete adjuvant. Int J Rheum Dis 22(7):1247–1254
130. Ghavipour M, Sotoudeh G, Tavakoli E, Mowla K, Hasanzadeh J, Mazloom Z (2017) Pomegranate extract alleviates disease activity and some blood biomarkers of inflammation and oxidative stress in rheumatoid arthritis patients. Eur J Clin Nutr 71(1):92–96
131. Ghoochani N, Karandish M, Mowla K, Haghighizadeh MH, Jalali MT (2016) The effect of pomegranate juice on clinical signs, matrix metalloproteinases and antioxidant status in patients with knee osteoarthritis. J Sci Food Agric 96(13):4377–4381
132. Rafraf M, Hemmati S, Jafarabadi MA, Moghaddam A, Haghighian MK (2017) Pomegranate (*Punica granatum* L.) peel hydroalcoholic extract supplementation reduces pain and improves clinical symptoms of knee osteoarthritis: a randomized double-blind placebo controlled study. Iran Red Crescent Med J 19(1):1–8
133. Zhang Q, Li R, Liu J, Peng W, Gao Y, Wu C et al (2019) In silico screening of anti-inflammatory constituents with good drug-like properties from twigs of *Cinnamomum cassia* based on molecular docking and network pharmacology. Trop J Pharm Res 18(10):2125–2131
134. Lee JS, Lim S (2021) Anti-inflammatory, and anti-arthritic effects by the twigs of *Cinnamomum cassia* on complete Freund's adjuvant-induced arthritis in rats. J Ethnopharmacol 278:114209. https://doi.org/10.1016/j.jep.2021.114209
135. Liu P, Wang J, Wen W, Pan T, Chen H, Fu Y et al (2020) Cinnamaldehyde suppresses NLRP3 derived IL-1β via activating succinate/HIF-1 in rheumatoid arthritis rats. Int Immunopharmacol 84:106570. https://doi.org/10.1016/j.intimp.2020.106570
136. Kim Y, Kim DM, Kim JY (2017) Ginger extract suppresses inflammatory response and maintains barrier function in human colonic epithelial Caco-2 cells exposed to inflammatory mediators. J Food Sci 82(5):1264–1270
137. Sharma H, Chauhan P, Singh S (2018) Evaluation of the anti-arthritic activity of *Cinnamomum cassia* bark extract in experimental models. Integr Med Res 7(4):366–373. https://www.sciencedirect.com/science/article/pii/S2213422018301872
138. Chippada SC, Volluri S, Bammidi SR, Vangalapati M (2011) In vitro anti-arthritic activity of methanolic extract of *Centella asiatica*. Biosci Biotechnol Res Asia 8(1):337–340
139. Wang T, Wei Z, Dou Y, Yang Y, Leng D, Kong L et al (2015) Intestinal interleukin-10 mobilization as a contributor to the anti-arthritis effect of orally administered madecassoside: a unique action mode of saponin compounds with poor bioavailability. Biochem Pharmacol 94(1):30–38. https://doi.org/10.1016/j.bcp.2015.01.004
140. Sharma S, Gupta R, Thakur SC (2014) Attenuation of collagen induced arthritis by *Centella asiatica* methanol fraction via modulation of cytokines and oxidative stress. Biomed Environ Sci 27(12):926–938
141. Pradit W, Chomdej S, Nganvongpanit K, Ongchai S (2015) Chondroprotective potential of *Phyllanthus amarus* Schum. & Thonn. in experimentally induced cartilage degradation in the explants culture model. Vitr Cell Dev Biol Anim 51(4):336–344
142. Harikrishnan H, Jantan I, Haque MA, Kumolosasi E (2018) Phyllanthin from *Phyllanthus amarus* inhibits LPS-induced proinflammatory responses in U937 macrophages via downregulation of NF-κB/MAPK/PI3K-Akt signaling pathways. Phyther Res 32(12): 2510–2519
143. Harikrishnan H, Jantan I, Haque MA, Kumolosasi E (2018) Anti-inflammatory effects of *Phyllanthus amarus* Schum. & Thonn. Through inhibition of NF-KB, MAPK, and PI3K-Akt

signaling pathways in LPS-induced human macrophages. BMC Complement Altern Med 18(1):1–13

144. Alam J, Jantan I, Kumolosasi E, Nafiah MA, Mesaik MA (2018) Suppressive effects of the standardized extract of *Phyllanthus amarus* on Type II Collagen-induced Rheumatoid Arthritis in Sprague Dawley Rats. Curr Pharm Biotechnol 19(14):1156–1169
145. Chopade AR, Sayyad FJ (2014) Antifibromyalgic activity of standardized extracts of *Phyllanthus amarus* and *Phyllanthus fraternus* in acidic saline induced chronic muscle pain. Biomed Aging Pathol 4(2):123–130. https://www.sciencedirect.com/science/article/pii/S2210522014000069
146. Pinkaew D, Kiattisin K, Wonglangka K, Awoot P (2020) Phonophoresis of *Phyllanthus amarus* nanoparticle gel improves functional capacity in individuals with knee osteoarthritis: a randomized controlled trial. J Bodyw Mov Ther 24(1):15–18. https://doi.org/10.1016/j.jbmt.2019.04.013
147. Ahsana D, Andika A, Nashihah S (2021) Molecular docking study of flavonoid compounds in the guava leaves (Psidium guajava L.) which has potential as anti-inflammatory COX-2 inhibitors. Lumbung Farm J Ilmu Kefarmasian 2(2):67
148. Jang M, Jeong SW, Cho SK, Ahn KS, Lee JH, Yang DC et al (2014) Anti-inflammatory effects of an ethanolic extract of guava (*Psidium guajava* L.) leaves in vitro and in vivo. J Med Food 17(6):678–685
149. Baroroh HN, Utami ED, Achmad A (2016) Psidium guajava leaves decrease arthritic symptoms in adjuvant-induced arthritic rats. Univ Med 34(3):197
150. Baroroh HN, Lesty R, Utami ED (2022) *Psidium guajava* leaves extract decreased leukocytes and lymphocytes count in Complete Freund's Adjuvant-induced arthritis rats. Acta Pharm Indo 10:1–7
151. Kakuo S, Fushimi T, Kawasaki K, Nakamura J, Ota N (2018) Effects of *Psidium guajava* Linn. leaf extract in Japanese subjects with knee pain: a randomized, double-blind, placebo-controlled, parallel pilot study. Aging Clin Exp Res 30(11):1391–1398. https://doi.org/10.1007/s40520-018-0953-6
152. Lee HJ, Jeong YJ, Lee TS, Park YY, Chae WG, Chung IK et al (2013) Moringa fruit inhibits LPS-induced NO/iNOS expression through suppressing the NF-κB activation in RAW264.7 cells. Am J Chin Med 41(5):1109–1123
153. Mahdi HJ, Khan NAK, Asmawi MZB, Mahmud R, A/L Murugaiyah V (2018) In vivo anti-arthritic and anti-nociceptive effects of ethanol extract of *Moringa oleifera* leaves on complete Freund's adjuvant (CFA)-induced arthritis in rats. Integr Med Res 7(1):85–94. https://doi.org/10.1016/j.imr.2017.11.002
154. Pandey P, Bhatt PC, Kumar V (2017) *Moringa oleifera* Lam ameliorates adjuvant induced arthritis via inhibition of inflammatory mediators and down-regulation of MMP3 and MMP9 proteins. Lupus Sci Med 4(1):A32–A32. https://lupus.bmj.com/content/4/Suppl_1/A32.1
155. Mohan CV, Chatterjee S, Jha DK (2022) Pharmacological evaluation of *Moringa oleifera* on collagen-induced arthritis in rats. J Pharm Res Int 34(54):11–32
156. Umar S, Hedaya O, Singh AK, Ahmed S (2015) Thymoquinone inhibits TNF-α-induced inflammation and cell adhesion in rheumatoid arthritis synovial fibroblasts by ASK1 regulation. Toxicol Appl Pharmacol 287(3):299–305
157. Nasuti C, Bordoni L, Fedeli D, Gabbianelli R (2019) Effect of *Nigella sativa* oil in a rat model of Adjuvant-Induced Arthritis. Proceedings 17:16
158. Faisal R, Shinwari L, Jehangir T (2015) Comparison of the therapeutic effects of thymoquinone and methotrexate on renal injury in pristane induced arthritis in rats. J Coll Phys Surg Pak 25(8):597–601
159. Kheirouri S, Hadi V, Alizadeh M (2016) Immunomodulatory effect of *Nigella sativa* oil on T lymphocytes in patients with rheumatoid arthritis. Immunol Investig 45(4):271–283
160. Hadi V, Kheirouri S, Alizadeh M, Khabbazi A, Hosseini H (2016) Effects of *Nigella sativa* oil extract on inflammatory cytokine response and oxidative stress status in patients with

rheumatoid arthritis: a randomized, double-blind, placebo-controlled clinical trial. Avicenna J Phytomed 6(1):34–43
161. Ge X, Chen Z, Xu Z, Lv F, Zhang K, Yang Y (2018) The effects of dihydroartemisinin on inflammatory bowel disease-related bone loss in a rat model. Exp Biol Med 243(8):715–724
162. Lowin T, Tingting R, Zurmahr J, Classen T, Schneider M, Pongratz G (2020) Cannabidiol (CBD): a killer for inflammatory rheumatoid arthritis synovial fibroblasts. Cell Death Dis 11(8):1–11
163. Jeong M, Cho J, Il SJ, Jeon YJ, Kim JH, Lee SJ et al (2014) Hempseed oil induces reactive oxygen species- and C/EBP homologous protein-mediated apoptosis in MH7A human rheumatoid arthritis fibroblast-like synovial cells. J Ethnopharmacol 154(3):745–752. https://doi.org/10.1016/j.jep.2014.04.052
164. Lowin T, Laaser SA, Kok C, Bruneau E, Pongratz G (2023) Cannabidiol: influence on B cells, peripheral blood mononuclear cells, and peripheral blood mononuclear cell/rheumatoid arthritis synovial fibroblast cocultures. Cannabis Cannabinoid Res 8(2):321–334
165. Hammell DC, Zhang LP, Ma F, Abshire SM, McIlwrath SL, Stinchcomb AL et al (2016) Transdermal cannabidiol reduces inflammation and pain-related behaviours in a rat model of arthritis. Eur J Pain 20(6):936–948
166. Heineman JT, Forster GL, Stephens KL, Cottler PS, Timko MP, DeGeorge BR (2022) A randomized controlled trial of topical cannabidiol for the treatment of thumb basal joint arthritis. J Hand Surg Am 47(7):611–620. https://www.sciencedirect.com/science/article/pii/S0363502322001332
167. Li Z-Z, Tan J-P, Wang L-L, Li Q-H (2017) Andrographolide benefits rheumatoid arthritis via inhibiting MAPK pathways. Inflammation 40(5):1599–1605
168. Li X, Yuan K, Zhu Q, Lu Q, Jiang H, Zhu M et al (2019) Andrographolide ameliorates rheumatoid arthritis by regulating the apoptosis–NETosis balance of neutrophils. Int J Mol Sci 20(20):5035
169. Li GF, Qin YH, Du PQ (2015) Andrographolide inhibits the migration, invasion and matrix metalloproteinase expression of rheumatoid arthritis fibroblast-like synoviocytes via inhibition of HIF-1α signaling. Life Sci 136:67–72. https://doi.org/10.1016/j.lfs.2015.06.019
170. Gupta S, Mishra KP, Singh SB, Ganju L (2018) Inhibitory effects of andrographolide on activated macrophages and adjuvant-induced arthritis. Inflammopharmacology 26(2):447–456
171. Gupta S, Mishra KP, Kumar B, Singh SB, Ganju L (2020) Andrographolide attenuates complete Freund's adjuvant induced arthritis via suppression of inflammatory mediators and pro-inflammatory cytokines. J Ethnopharmacol 261:113022. https://doi.org/10.1016/j.jep.2020.113022
172. Li F, Li H, Luo S, Ran Y, Xie X, Wang Y et al (2018) Evaluation of the effect of andrographolide and methotrexate combined therapy in complete Freund's adjuvant induced arthritis with reduced hepatotoxicity. Biomed Pharmacother 106:637–645
173. Hancke JL, Srivastav S, Cáceres DD, Burgos RA (2019) A double-blind, randomized, placebo-controlled study to assess the efficacy of *Andrographis paniculata* standardized extract (ParActin®) on pain reduction in subjects with knee osteoarthritis. Phyther Res. 33(5):1469–1479
174. Li YJ, Guo Y, Yang Q, Weng XG, Yang L, Wang YJ et al (2015) Flavonoids casticin and chrysosplenol D from *Artemisia annua* L. inhibit inflammation in vitro and in vivo. Toxicol Appl Pharmacol 286(3):151–158
175. Feng MX, Hong JX, Wang Q, Fan YY, Yuan CT, Lei XH et al (2016) Dihydroartemisinin prevents breast cancer-induced osteolysis via inhibiting both breast cancer cells and osteoclasts. Sci Rep 6(1):19074. https://doi.org/10.1038/srep19074
176. Zhou L, Liu Q, Yang M, Wang T, Yao J, Cheng J et al (2016) Dihydroartemisinin, an antimalaria drug, suppresses estrogen deficiency-induced osteoporosis, osteoclast formation, and RANKL-induced signaling pathways. J Bone Miner Res Off J Am Soc Bone Miner Res 31(5):964–974

177. Murugesan S, Ravichandran D, Lakshmanan DK, Ravichandran G, Arumugam V, Raju K et al (2020) Evaluation of anti rheumatic activity of *Piper betle* L. (Betelvine) extract using in silico, in vitro and in vivo approaches. Bioorg Chem 103:104227. https://doi.org/10.1016/j.bioorg.2020.104227
178. De S, Kundu S, Chatterjee U, Chattopadhyay S, Chatterjee M (2018) Allylpyrocatechol attenuates methotrexate-induced hepatotoxicity in a collagen-induced model of arthritis. Free Radic Res 52(6):698–711
179. Hegde K, Emani A, Shrijani JK, Shabaraya AR (2020) Anti arthritic potentials of *Piper betle*: a preclinical study. Indian J Pharm Pharmacol 5(1):21–28

Evidence from the Use of Herbal Medicines in the Management and Prevention of Common Eye Diseases

25

Prasann Kumar, Aman Khokhar, and Shipa Rani Dey

Contents

Abstract

The fact that botanical compounds have been extensively used to cure various diseases and ailments throughout history is well known. Several of these compounds have been reported to be powerful antioxidants, anti-inflammatory agents, and antiapoptotic agents. Furthermore, these mechanisms have also been involved in ocular diseases, such as age-related macular degeneration (AMD), glaucoma, diabetic retinopathy, cataracts, and retinitis pigmentosa. Recent studies have demonstrated that several epidemiological and clinical studies have been performed that have shown the benefit of plant-derived compounds on these ocular pathologies, such as curcumin, lutein, zeaxanthin, ginseng, and many others in recent years. Cell cultures and animal models have shown promising results when these drugs have been used to treat eye diseases in the laboratory. It is important to note that although there appear to be many significant correlations between these botanical compounds, further investigation is necessary to determine their mechanistic pathways. This will facilitate widespread pharmaceutical application and the development of non-invasive alternatives for significant eye diseases to provide non-invasive solutions to those diseases.

P. Kumar · A. Khokhar · S. R. Dey (✉)
Department of Agronomy, School of Agriculture, Lovely Professional University, Phagwara, India

S. C. Izah et al. (eds.), *Herbal Medicine Phytochemistry*, Reference Series in Phytochemistry, https://doi.org/10.1007/978-3-031-43199-9_27

Keywords

Eye · Glaucoma · Cataracts · Diabetic retinopathy · Macular degeneration · Lutein

Abbreviations

AMD	Age-related macular degeneration
AMPK	AMP-activated protein kinase
AREDS	Age-related eye disease study
CB1	Cannabinoid receptor type 1
CBD	Cannabidiol
CGRP	Calcitonin gene-related peptide
COX-2	Cyclooxygenase-2
DR	Diabetic retinopathy
EGCG	Epigallocatechin gallate
fERG	Flash electroretinogram
GBE	*Ginkgo biloba* extract
GO	Graves' ophthalmopathy
HSP72	Heat shock proteins
HSV	Herpes simplex virus
HUVECs	Human umbilical vein endothelial cells
IGF-I	Insulin growth factor-1
IOP	Intraocular pressure
JNK	c-Jun N-terminal kinase
KRG	Korean red ginseng
LDLs	Low-density lipoproteins
LIRD	Light-induced retinal degeneration
MCP	Monocyte chemoattractant protein
MDA	Malondialdehyde
NF-kappa B	Nuclear factor kappa B
NO	Nitric oxide
NOS	Nitric oxide synthase
NTG	Normal tension glaucoma
PACAP	Pituitary adenylate cyclase-activating peptide
PCG-1alpha	Peroxisome proliferator-activated receptor gamma coactivator 1alpha
POAG	Primary open-angle glaucoma
Rb1	Retinoblastoma protein 1
RCTs	Randomized controlled trials
Rg3	Retinoblastoma protein 3
RGC-5	Retinal ganglion cell
ROS	Reactive oxygen species
RP	Retinitis pigmentosa
RPE	Retinal pigment epithelium

THC	Delta-9-tetrahydrocannabinol
TNF-alpha	Tumor necrosis factor-alpha
UV	Ultraviolet
VF	Visual field
WHO	World Health Organization

1 Introduction

The World Health Organization (WHO) estimates that approximately 34 million blind people live worldwide [10]. The eye has several natural defense mechanisms that it can use to protect itself from injury or infection. Tears, for instance, maintain a moist environment around the eye and physically remove any foreign particles, such as dust or microorganisms, that may have settled there. They also keep the eye from drying out. Moreover, tears have been found to contain several anti-infective substances, such as lysozymes and interferon. The eyelashes and eyelids keep the eye's surface moist while serving as a barrier between the ocular surface and the outside world. On the other hand, these protective mechanisms can become compromised occasionally, leading to eye inflammation. Eye infections are frequently and commonly reported in the South African province known as the Eastern Cape. These infections are brought on by exposure to bacteria, fungi, viruses, and other types of microorganisms. Traditional healers and herbalists recommend plant-based remedies for treating conditions like these [34, 56]. According to the various ancient and modern Chinese pharmacopoeia, numerous herbal treatments have been utilized for a significant amount of time to treat different eye conditions. They effectively treat night blindness, cataracts, floaters, and glaucoma, among other eye conditions. The usefulness of these treatments is deduced from anecdotal evidence and specific clinical results rather than going through the rigorous process of modern systematic scientific research, which would involve testing on animals and participating in clinical trials. In addition, the efficacies have a reputation for having a large amount of variation due to herbal remedies needing to be standardized in terms of their sources, qualities, combinations, or preparation methods [59, 60, 62, 63, 65, 66]. This is because these aspects of herbal remedies can affect their effectiveness. Despite this, the historical treatments documented for a long time are a goldmine for discovering natural products. Herbal extracts and herbal molecules that have been defined offer a superior alternative for the continued research and development of herbal medicines, making it possible to circumvent the challenges posed by variations. Because it has such a high metabolic rate, the eye is constantly damaged by the light's photooxidation. Damage caused by oxidative agents can quickly occur in the tissue cells of the eye. Even though there are several pathological factors associated with ocular diseases, such as age, inflammation, oxidative stress, genetic predisposition, tumorigenesis, diabetes, angiogenesis, and ischemia, it appears that oxidative stress plays a pivotal role in the pathogenesis of ocular diseases. Any of these contributors can cause ocular diseases, but they all work together.

The Aim and Objective of the Chapter

- To analyze the effectiveness of herbal medicines in managing glaucoma symptoms and preventing progression.
- To investigate the potential of herbal remedies in reducing intraocular pressure for glaucoma patients.
- To assess the impact of herbal treatments on retinal health in diabetic retinopathy cases.
- To examine the role of herbal supplements in preventing cataract formation and progression.
- To explore the antioxidant properties of herbal extracts for protecting against age-related macular degeneration.
- To evaluate the safety and efficacy of herbal therapies in alleviating dry eye symptoms.
- To investigate the mechanisms by which herbal medicines can improve visual function in eye diseases.
- To identify potential herbal interventions for reducing ocular inflammation in various eye conditions.
- To examine the impact of herbal supplements on improving ocular blood flow in glaucoma.
- To assess the clinical outcomes of using herbal treatments in conjunction with conventional therapies for eye diseases.
- To determine the neuroprotective effects of specific herbal compounds in glaucoma management.
- To analyze the anti-inflammatory properties of herbal medicines for reducing eye discomfort and redness.
- To investigate the role of herbal remedies in preventing or delaying the onset of cataracts.
- To explore the potential of herbal therapies in addressing oxidative stress in diabetic retinopathy.
- To evaluate the safety and tolerability of herbal supplements as adjunct treatments in ophthalmology.
- To examine the impact of herbal interventions on improving contrast sensitivity in eye diseases.
- To assess the effectiveness of herbal treatments in preventing vision loss associated with age-related eye conditions.
- To investigate the mechanisms of action of herbal compounds in preserving retinal health.
- To explore the synergistic effects of combining herbal and traditional glaucoma medications.

Several different risk factors have been associated with age-related macular degeneration, also known as AMD, including getting older, smoking cigarettes, being exposed to sunlight, and consuming a diet that is high in fat and low in antioxidants, in addition to having specific genetic variants (such as the polymorphism of complement factor H), has been shown to increase the risk of developing cardiovascular disease. There is a correlation between these potential dangers and oxidative pressure, ultimately damaging retinal tissue. This includes damage to retinal pigment epithelium, photoreceptor cells, and choroidal capillaries. Glaucoma causes progressive and irreversible death of retinal ganglion cells, damage to the optic nerve and loss of vision, even without increased intraocular pressure. This is because glaucoma causes pressure to build up inside the eye. Oxidative stress is responsible for cytotoxic effects, a degeneration of the retinal ganglion cells, and damage to the trabecular meshwork, all of which contribute to an obstruction of the flow of an increase in intraocular pressure as well as the production of aqueous humor [44, 58, 68, 69]. The gradual clouding of the lens that occurs because of prolonged exposure to light is the underlying cause of cataracts. This photooxidation is responsible for the breakdown of biological components of the lens, most notably crystallin. Molecules that possess crystalline structures and contain oxidized thiol groups tend to establish disulfide bonds, resulting in the agglomeration of crystals and the consequent onset of cataract formation. Oxidative stress is increased due to diabetes, which leads to damage to capillary cells and the microvasculature of the retina. An impairment in glyceraldehyde-3-phosphate dehydrogenase is another factor that contributes to the pathogenesis of diabetic retinopathy brought on by diabetes. This factor incorrectly activates significant biochemical pathways [46]. The medical condition known as Graves ophthalmopathy, or GO, is characterized by an abnormal eyeball bulging and is partially brought on by hyperthyroidism. It affects anywhere between 25% and 50% of Graves' patients. Roughly 5% of patients being treated have a condition that could harm their vision. Numerous pieces of evidence indicate that the pathogenesis of GO may involve oxidative stress [2]. The acceleration of essential metabolic processes and cellular oxidative metabolism in mitochondria results from excessive thyroid hormone production, resulting in an elevation in the generation of reactive oxygen species (ROS). The presence of an oxidative environment has the potential to induce cellular proliferation in fibroblasts located within the eye, which can lead to eye protrusion. Some research showed that active GO patients had higher levels of 8-hydroxy-20-deoxyguanosine (8-OHdG) in their urine and that GO orbital fibroblasts had higher levels of 8-OHdG, superoxide anions, malondialdehyde (MDA), and hydrogen peroxides. Among the elderly population, blindness ranks as the second most significant fear, following death. Common causes of blindness include ocular diseases like glaucoma, cataracts, and age-related macular degeneration (AMD). The development of cataracts is a significant challenge faced by the elderly population and has a considerable bearing on the cost of medical care. The lens is the component of the eye that is particularly vulnerable to oxidative damage. Cataracts are caused by the exposure of underlying epithelial cells to both exogenous and endogenous reactive oxygen species, in addition to cellular death and degeneration. The phenomenon above results in the cross-linking and aggregation of the crystallin proteins present in the lens, leading to the development of cataracts. The formation of cataracts is one of the age-related processes that is significantly impacted by oxidative stress. Because the type of cataract and its pigmentation are determined by the intensity and form of oxidative stress, this factor makes preventing cataracts more difficult. Dry eye syndrome is a prevalent ocular condition. It can cause eye pain, vision problems, and possible corneal surface damage. The fact that more and more people are going to ophthalmology clinics with symptoms of dry eye syndrome shows how important etiology is. It also shows how important it is to find out what causes dry eyes and stop doing those things. Surgeons should identify and assess preoperative risk factors while taking measures to prevent factors that may exacerbate this condition. Given its established status as a recognized syndrome in periorbital surgery, it is imperative to avoid this condition. The patient will be able to have a more straightforward recovery process if all potential risk factors associated with dry eye syndrome have been identified and considered. This will reduce the likelihood of the patient developing this complication. In 2007, according to the International Dry Eye Work Shop, dry eye disease was a condition that impacted both the tear film and the ocular surface. Symptoms of dry eye disease include pain, blurred vision, unstable tear film, and eye surface damage. Symptoms include a foreign body sensation, burning, itching, blurred vision, pain, and photophobia; scleral injection; ocular discomfort; and blurred vision. Non-healing epithelial defects, corneal thinning, infections, corneal perforation, and permanent vision impairment are some visual complications that can result from all these symptoms.

2 Different Types of Eye Infections

Bacterial eye infections: Microorganisms such as *Haemophilus influenza*, *Staphylococcus aureus*, *Streptococcus pneumonia*, *Pseudomonas aeruginosa*, *Escherichia coli*, *Bacillus cereus*, *Neisseria gonorrhoeae*, *Staphylococcus epidermidis*, *and Chlamydia trachomatis* are responsible for the eye infection [29]. The bacteria *Staphylococcus aureus* and *Staphylococcus epidermidis* are the most common causes of external eye infections. *C. trachomatis* is the pathogen that leads to trachoma. This condition is the primary contributor to ocular morbidity and is the foremost infectious etiology of blindness globally [83]. According to the World Health Organization's estimation, the global population of individuals affected by trachoma is approximately 146 million. Some symptoms associated with bacterial eye infections include irritation, burning, tearing, and typically a mucopurulent or purulent discharge. It is possible to wake up with your eyelids stuck together. This is especially common in the morning. Even though bacterial eye infections are typically considered self-limiting, if they are allowed to go untreated, they can progress into more severe conditions threatening the patient's vision.

Fungal eye infections: *Fusarium oxysporum*, *Aspergillus niger*, *Fusarium solani*, *Candida albicans*, *Aspergillus flavus*, and *Penicillium notatum* are some of the fungal species that can cause infections in the eye [27]. The treatment of these infections poses a significant challenge and has the potential to lead to blindness. Redness, blurred vision, and sensitivity to light are some of the symptoms. Amphotericin B and Natamycin are two antifungal agents that can be applied topically to the eye to treat fungal infections [11, 19].

Viral eye infections: Coxsackieviruses, Herpes simplex virus-1, and adenoviruses are the three most common culprits behind viral infections of the eye [40]. In developed countries, ocular infection caused by HSV-1 is the primary etiology of visual impairment. Herpes simplex type 1 accounts for more than 95% of ocular herpes infections. Currently, 51 human adenovirus serotypes are known, which can be broken down into 6 species (A–F). In particular, the eyes can become infected with species D. Eye infections caused by viruses are highly contagious and are passed on through direct contact, most commonly through the sharing of items that have been in connection with the eye secretions of an infected person. For instance, the virus can be passed on when someone who already has individuals come into contact with contaminated surfaces, such as door handles, after touching their eyes. Additionally, sharing objects in connection with an individual's eyes, such as towels or pillowcases, between two people can also facilitate transmission. Antiviral medications like acyclovir, famciclovir, and valacyclovir can shorten the infection's duration and lessen the severity of the symptoms. Glaucoma is a common form of ocular disease that affects people all over the world. The demise of retinal ganglion cells characterizes glaucoma, the formation of a hollow at the optic nerve head, and, typically, a rise in intraocular pressure (IOP). Although elevated, the etiology of ganglion cell degeneration and optic disc cupping has long been attributed to intraocular pressure (IOP). However, research on low-tension glaucoma has provided evidence that challenges this notion, indicating that IOP may not be a sufficient or necessary factor in developing the disease. The cause of this condition is still unknown. Other mechanisms involving the neurosensory retina have been proposed. Recent research has brought attention to the possible neurotoxic function of glutamate. It has indicated that apoptotic mechanisms could be implicated in the deterioration of ganglion cells [38]. Glaucomatous eyes of dogs, monkeys, and humans have been found to have abnormally high levels of glutamate in the vitreous. Many of its morphological features resemble those of human congenital glaucoma, including an expanded retinal surface, an atypical corneal endothelium exhibiting cellular degeneration, insufficiently differentiated cells with a collapsed trabecular meshwork, and an adhesion of the anterior aspect of the iris to the posterior cornea. It is believed that several stimuli can start or start the process of apoptosis in glaucoma. Therefore, elevated intraocular pressure (IOP) and the accumulation of organelles at the optic nerve head may be able to obstruct axoplasmic flow, thereby impeding the circulation of trophic factors [30]. Mitosis, conjunctival hyperemia, and the inflammatory response in the eye are comprised of the breakdown of the blood-aqueous barrier, resulting in the subsequent leakage of protein into the aqueous humor. The nature, duration, and strength of the noxious stimulus determine the extent of these reactions. Substance P and calcitonin gene-related peptide (CGRP) are two examples of C-fiber neurotransmitters that have been demonstrated to play an essential role in the ocular response to injury. Both neurotransmitters are involved in healing [26]. Recent research has identified pituitary adenylate cyclase-activating peptide (PACAP), a C-fiber neuropeptide that has been shown to play a role in ocular inflammation in rabbits. Any transmitter removed from local nerve fibers will diffuse into the anterior chamber because there is no barrier between the iris, the ciliary body, and the anterior chamber. This is because the anterior chamber is directly connected to the ciliary body. Because of this, research on the release of transmitters benefits significantly from using the eye as a model. Nitric oxide (NO) is a molecule with a short half-life that participates in various biological processes. Recent observations suggest that NO may significantly regulate ocular function in its physiological or pathophysiological aspects. Therefore, the presence of NOS activity in the anterior uvea of rabbits has been demonstrated. At the same time, NOS immunoreactivity has been observed by immunizing nerve fibers in the uvea of rat eyes. Both findings can be found in the literature. It was discovered that administering L-NAME to the rabbit via intravenous injection caused a reduction in the regional blood flow in the uvea. It is interesting to note that the activation of ocular C-fibers in response to a minor injury (infrared irradiation of the iris) appears to involve the participation of nitric oxide (NO).

Ocular injuries – Exposure to industrial chemicals like acid and alkali, radiation energy like ultraviolet light, and physical trauma all contribute to ocular injuries in the workplace. Although injuries to the eye caused by plant saps are uncommon, they can have serious consequences [42].

Conjunctivitis – Conjunctivitis is a medical condition characterized by mucous membrane inflammation that lines the eyelid's inner surface. Various things can cause conjunctivitis, such as viruses, bacteria, allergies, injuries, or other foreign substances. The following are some symptoms: itchiness, watery eyes, bloodshot eyes, pain, and occasionally blurred vision. Depending on what caused conjunctivitis in the first place, there are a few different treatments available. In most cases, antibiotics are prescribed for the treatment of bacterial infections. Eyewashes can also be made from several herbal combinations [9].

3 Major Eye Diseases Share Common Mechanistic Pathways

The most common causes of blindness can be broken down into four categories: age-related macular degeneration (AMD), cataracts, glaucoma, and other retinal diseases such as retinitis pigmentosa (RP) and diabetic retinopathy (DR) [74]. According to the findings of an epidemiologic study that the Eye Diseases Prevalence Research Group carried out, it is estimated that by the year 2020, there will be 30.1 million people in the United States with cataracts. Furthermore, it was estimated that 4.1 million individuals aged 40 and above in the United States would have diabetic retinopathy [15]. Approximately 2.95 million individuals in the United States would receive a diagnosis of age-related macular degeneration (AMD), and about 2.2 million individuals would receive a diagnosis of glaucoma. It is interesting to note that all these diseases are linked to aging and that the mechanisms that underlie their respective etiologies and pathophysiologies are similar in several respects. The pathways above encompass apoptotic factors, inflammation, and oxidative stress, offering valuable insights into areas that may be amenable to targeted interventions. Oxidative stress is induced by reactive oxygen or nitrogen species, and lipid peroxidation is responsible for the death of ocular cells in many eye diseases. Various pathological pathways are associated with inflammatory factors, including nuclear factor kappa B (NF-kappa B) and tumor necrosis factor-alpha (TNF-alpha). It is interesting to note that the mechanisms of these pathways often intersect with those of numerous botanical compounds. Oxidative stress causes the generation of reactive oxygen species, leading to their interaction with mitochondria, subsequently triggering the activation of the JNK pathway, ultimately resulting in the cell's death. This review will concentrate on AMD, DR, RP, and glaucoma as pathologies and the possible benefits of botanical compounds in preventing and treating these eye diseases. This is because all these conditions have a significant impact on populations all over the world.

Age-related macular degeneration (AMD): Age-related macular degeneration is a persistent retinal condition commonly observed in individuals aged 50 years or above. The degeneration of photoreceptor and RPE cells in the macula results in a loss of central vision, both of which are necessary to provide sharp and clear vision. Smoking, one's racial or ethnic background, and one's family medical history are various factors that have been identified as potential contributors to the development of age-related macular degeneration, a prevalent cause of blindness among elderly individuals on a global scale, according to numerous epidemiologic studies. Oxidative stress and inflammation are essential mechanistic pathways to the pathology [31]. As per the results of the National Eye Institute's Age-Related Eye Disease Study (AREDS), the consumption of dietary supplements that comprise antioxidants and zinc has the potential to diminish the likelihood of advanced AMD onset and consequent vision impairment to a considerable extent [4]. This research opens the door for discovering potentially beneficial natural compounds for managing and prophylaxis of age-related macular degeneration (AMD). The investigation of the active ingredients of various plants has revealed their potential as antioxidants and anti-inflammatory agents. These characteristics are crucial in managing and avoiding age-related macular degeneration (AMD) and other ocular ailments which have already been present.

Glaucoma: Glaucoma is a term that refers to a set of ocular disorders that may interrupt the transmission of visual stimuli from the eye to the brain [3]. Most glaucoma cases are attributed to optic nerve damage resulting from the apoptosis of retinal ganglion cells [78]. The medical term for the heightened force within the eye is intraocular pressure (IOP), frequently used in clinical settings. Glaucoma patients typically have their intraocular pressure (IOP) lowered using oral medications, eye drops, or surgery [47]. This is the primary treatment method for the disease. Even though the elevation of intraocular pressure (IOP) is considered a leading etiological factor in the development of glaucoma, many cases continue to worsen even after the eye pressure has been brought down to the normal range. Developing novel and creative approaches to damage prevention and mitigation and reducing intraocular pressure is essential in such situations. The role of apoptosis in the pathogenesis of glaucoma has been well established. Therefore, investigations into the efficacy of neuroprotective agents may yield promising results. Various chemical components are found in plants that exhibit neuroprotective qualities and have the potential to be helpful in the treatment and prevention of glaucoma.

4 Diabetic Retinopathy (DR) and Retinitis Pigmentosa (RP)

Individuals diagnosed with either type 1 or type 2 diabetes are susceptible to the development of diabetic retinopathy. Diabetic retinopathy (DR) is considered one of the primary causes of blindness worldwide, in addition to age-related macular degeneration and glaucoma. It is estimated that DR affects 140 million people [85]. Changes in the retinal blood vessels bring on the pathology in this case. There is a possibility that blood vessels will swell or leak, and the emergence of anomalous blood vessels on the retina's surface is a possibility. Four stages comprise the DR pathology: severe no proliferative retinopathy, moderate no proliferative retinopathy, mild no proliferative retinopathy, and proliferative retinopathy. Proliferative retinopathy is the final stage. No treatments are necessary for the first three stages of the disease; even so, patients in these phases are required to regulate their blood cholesterol, blood pressure, and blood sugar [21]. This is done to prevent the disease from progressing further in these stages. Laser surgery is required to stem fluid flow out of the eye when proliferative retinopathy is present. However, in severe bleeding, a vitrectomy may be necessary to remove blood from the macula (the central part of the retina) [85]. The laser treatment may help reduce the size of the abnormal blood vessels, but a vitrectomy is necessary in these circumstances. Another disease that affects the retina is called retinitis pigmentosa, or RP for short. This condition causes rod and cone cells in the retina to become damaged, resulting in impaired vision and, in severe cases, total blindness. Retinitis pigmentosa is a rare disease affecting only 1 in 4000 people in the United States [75]. Genetic predisposition is the primary factor that puts a person at risk for developing this condition. Because there are currently no treatments available for retinitis pigmentosa, the state is a significant source of concern, even though it is not as common as other eye diseases. However, results from earlier experiments conducted on mice suggested that high doses of antioxidants, such as vitamin A palmitate, may slow the progression of the disease. Numerous plant extracts have the potential to have significant effects on the treatment as well as the prevention of both RP and DR. However, because no studies have been conducted in a clinical setting, there are plenty of questions regarding the potential advantages of taking such supplements.

Cataract: A cataract is an eye disease that commonly affects older people and typically begins to manifest in people's eyes around the age of 60. The lens becomes cloudy due to this pathology because of the breakdown of proteins, which causes the lens to be affected. Damage to the lens causes vision to become blurry, as well as a reduction in the sensitivity to colors and shapes; the lens is an essential component for achieving focus on objects situated at varying distances, whether they are nearby or distant. It is known that several risk factors, including smoking, exposure to ultraviolet (UV) light, diabetes, eye injury, and family history, can increase the likelihood of developing cataracts [91]. Patients who suffer from cataracts can easily be cured by undergoing surgery. The cloudy lens causing the condition will be surgically removed during the procedure, and an intraocular lens will be inserted. It is estimated that cataracts cause 51% of the world's blindness. However, surgery can restore sight in most patients who undergo it; several nations lack access to sophisticated eye care services. Fortunately, new research suggests that limiting protein degradation in the eye by eating botanical compounds rich in antioxidants may help reduce the risk of cataracts.

5 Therapeutic and Adverse Effects of Herbal Remedies

5.1 Ginkgo *(Ginkgo biloba)*

GBE, an extract derived from the *Ginkgo biloba* plant, has been extensively documented in various sources and is made by processing the leaves of a tree estimated to have originated in China more than 250 million years ago. For centuries, *Ginkgo biloba* extract has been utilized in traditional Chinese medicine and is thought to help improve cognitive function. It is widely available in supplement form and is popular among people looking to improve their memory and focus. As we all know, *Ginkgo biloba* extract (GBE) is predominantly comprised of flavonoids and terpenoids, with a total of more than 60 distinct bioactive compounds, of which 30 are exclusive to GBE and cannot be found anywhere else in nature [16]. GBE is a safe and effective natural supplement with a variety of applications. It has been employed to treat various medical conditions, including depression, memory loss, anxiety, and fatigue. Research suggests that GBE can also help improve mental clarity and focus. In addition, it is an herbal supplement most frequently used by individuals who have reached the age of 65 or above [13]. GBE exhibits a high degree of tolerability and presents a limited incidence of adverse effects. Therefore, it can be safely used as a supplement for elderly patients looking for natural remedies for their health issues. It has been found that the pathogenesis of glaucomatous optic nerve degeneration remains largely unknown. However, evidence suggests that neuroinflammation, oxidative stress, and optic nerve ischemia may be involved in its development. GBE can act as an antioxidant and reduce these oxidative stress levels, thus potentially slowing down the progression of glaucoma. It also has anti-inflammatory properties, which can help reduce the inflammation in the optic nerve and improve patients' vision. According to research, gamma-butyrobetaine, or GBE, is a potent antioxidant that can protect tissues against damage caused by free radicals, just like other antioxidants, like vitamins C and E, do. These properties of GBE make it a promising option for treating glaucoma. GBE supplementation is safe and effective in early studies. Additional investigation is required to validate its effectiveness in the treatment of glaucoma. By contrast, GBE operates at the organelle level by providing molecular stabilization to the mitochondria. The fact that it has this property sets it apart from the other agents. During oxidative stress, it has been observed that alterations in mitochondrial function can render retinal ganglion cells more susceptible to the deleterious impact of oxidative stress [23, 49]. GBE has been found to increase the stability of the mitochondria and reduce the damage caused by oxidative stress, suggesting that it may be effective in treating glaucoma. Further research is needed to confirm its efficacy.

According to the findings of one study, the application of GBE has the potential to decrease the quantity of reactive oxygen species within cultured neuronal cells and protect the mitochondrial membrane. The vasodilatory properties of GBE have been discovered to have potential benefits for enhancing coronary and peripheral circulation. Additionally, GBE has been found to have rheological effects that could improve blood viscosity [33]. Both findings were made possible by the fact that GBE has vasodilatory properties. Additionally, GBE can lessen the number of active cells (such as glial cells) in low-grade inflammation [12]. GBE is regarded as a neuroprotective agent because of its antioxidant, vast regulatory, and anti-inflammatory properties. As a result, there has been a suggestion that it could serve as a viable treatment option for individuals with glaucoma. Researchers have conducted four randomized controlled trials (RCT) to investigate GBE's impact on glaucoma. The first study was a double-blind, placebo-controlled crossover experiment on patients with normal tension glaucoma (NTG) [73]. The participants were allocated randomly to either of the two intervention groups: 4 weeks of GBE administration followed by 8 weeks of placebo or the opposite. Even though GBE supplementation led to statistically better visual field (VF) indices compared to baseline, the observed VF indices did not exhibit sustained improvement after the washout period, and no statistically significant alterations in intraocular pressure (IOP) were noted. In contrast, [25], who conducted a similar cross-over study, the study did not observe any statistically significant variations in the visual field (VF) indices or intraocular pressure (IOP). The treatment order and duration were identical in both studies; however, the second study only included patients recently diagnosed with NTG and taking topical hypotensive agents. The fact that the first study was carried out on a population from Europe and the second study was carried out on a population from Asia both suggest that differences in patient populations' ethnicities might be another factor that contributed to the findings. Another randomized controlled trial has demonstrated that using GBE has been observed to elicit a favorable impact on blood circulation in individuals diagnosed with NTG. This proposition implies that GBE can potentially serve as a therapeutic intervention for NTG. Further research is needed to confirm the findings and assess the safety and efficacy of GBE as a treatment. Specifically, following 4 weeks of *Ginkgo biloba* extract (GBE) supplementation compared to a placebo, there were notable enhancements in ocular blood volume, blood flow, and blood velocity due to GBE intake. The results of this study indicate that *Ginkgo biloba* extract (GBE) could serve as a viable therapeutic option for individuals with normal tension glaucoma (NTG). However, further clinical trials are needed to confirm these results. Further studies should also be conducted to understand the long-term impact of GBE supplementation on NTG patients. In addition 2016, Dewi Sari and colleagues conducted a study on individuals diagnosed with primary open-angle glaucoma (POAG). The study involved the administration of GBE to the participants, and the researchers observed noteworthy enhancements in various parameters, including VF indexes, superior and inferior retinal nerve fiber layer thickness, malondialdehyde (a plasma-derived oxidative stress marker), and glutathione peroxidase (an antioxidant enzyme). These improvements were observed after 6 months of GBE administration compared to the placebo group. The study concluded that GBE could be a potential treatment option for POAG. It is noteworthy to mention that there were no reported negative consequences associated with the administration of GBE. Nevertheless, it is unknown whether, statistically speaking, there was a significant disparity in the level of improvement observed between the treatment group and the baseline group. Additional investigation is required to validate the efficacy of *Ginkgo biloba* extract (GBE) as a therapeutic intervention for primary open-angle glaucoma (POAG). Different dosage levels and treatment durations should be investigated to determine the optimal treatment protocol. Clinical trials should be conducted on the safety and effectiveness of *Ginkgo biloba* extract (GBE) in managing primary open-angle glaucoma (POAG). There are numerous benefits associated with Ginkgo. Studies have demonstrated the potential of Ginkgo to enhance memory, mitigate the likelihood of developing dementia, and ameliorate circulation. The safety and tolerability of this substance are widely acknowledged among the general population. Furthermore, Ginkgo has few known side effects and is widely available. GBE is typically well-tolerated, although its adverse effects should be considered, which are the most severe related to the fact that GBE acts as an antithrombotic. This means, it can increase the risk of bleeding, especially in those taking other medications that thin the blood. Therefore, it is important to consult a physician before taking Ginkgo to ensure there are no potential interactions with any other medications. There have been several case reports in which ocular complications have been identified, such as retinal hemorrhage and hyphema. In contrast, systemic effects have been identified, like subdural hematoma and subarachnoid hemorrhage have been reported in other case reports. Ginkgo should not be taken without consulting a doctor, as it may have adverse effects. These can range from ocular complications to systemic effects like subarachnoid hemorrhage. Following the doctor's advice on dosage and usage when taking Ginkgo is important. Nevertheless, it has been found that the incidence of bleeding is not significantly different between patients who take GBE and those who take a placebo (based on the results of the two randomized controlled trials (RCTs) that examined the effect of GBE on elderly patients [17].

Bilberry *(Vaccinium myrtillus)*: Since the sixteenth century, people have used bilberry fruit for its medicinal properties. About half of the glaucoma patients who responded to a survey about using herbal remedies for treatment said they took bilberry [77]. Anthocyanin, a type of flavonoid, is the bilberry component responsible for its medicinal properties [76]. Like GBE, bilberry's proposed therapeutic effect is based on the evidence that a neurodegenerative process characterizes glaucoma, and this substance has the potential to act as a neuroprotective agent and combat that process. This statement pertains to the augmentation of retinal ganglion cell density, the fortification of optic nerve architecture, the amelioration of retinal ganglion cell resilience against mechanical or ischemic perturbations, and the mitigation of neuroinflammatory processes. Several proposed mechanisms of action exist for bilberry, including 1) antioxidative properties, 2) decreased capillary fragility, 3) collagen stabilization and promotion of collagen production, and 4) prevention of the production and release of proinflammatory compounds. Research into the use of bilberry as a treatment for glaucoma has been conducted due to the positive effects that bilberry has on antioxidation and blood circulation. This study is a retrospective investigation examining the controlled impact of bilberry supplementation and GBE for 6 to 59 months. It found that those who took the supplements improved their visual acuity, while those in the control group saw a decline in their visual acuity [80]. Additionally, VF improved, but the statistical analysis revealed no significant differences in VF between the treatment and control groups. There was no discernible impact on IOP from the use of bilberry supplements. A randomized controlled trial lasted 24 months, and individuals who were administered anthocyanin supplements exhibited a decrease in VF mean deviation. Nevertheless, the reduction in mean deviation was notably inferior in the group receiving treatment compared to the group receiving placebo. In the studies discussed above, the use of bilberry was not associated with any adverse effects. Conversely, an overdose of bilberry can result in symptoms such as icterus, anemia, and cachexia.

Marijuana *(Cannabis sativa)*: Marijuana, also referred to as cannabis, has a long history of use for therapeutic purposes dating back thousands of years; however, its acceptance in Western medicine has a tumultuous past [1]. In 1937, the United States passed a law that made marijuana use illegal. The use of marijuana for medicinal purposes was made legal in California in 1996, making it the first state to do so. Since then, a cumulative count of 32 states and the District of Columbia have enacted legislation similar to California's, legalizing the medicinal use of marijuana. The increasing legalization of marijuana and a shift in social attitudes toward its use have led to a greater likelihood of glaucoma patients utilizing marijuana as a therapeutic alternative. Cannabidiol (CBD) and delta-9-tetrahydrocannabinol (THC) are the chemical compounds responsible for marijuana's physiological effects. Marijuana is made up of over 400 different compounds (CBD). There is a lack of knowledge regarding the mechanism of action responsible for the ocular effect that marijuana produces. The endocannabinoid system relies on the crucial G-protein receptors, cannabinoid receptor type 1 (CB1) and cannabinoid receptor type 2 (CB2). The CB1 receptor is situated in the ciliary body, muscles, trabecular meshwork, and Schlemm's canal, indicating its potential influence on the production of aqueous humor and the outflow of trabecular and uveoscleral fluids. On the other hand, the CB2 receptor is known to regulate cytokine release in the immune system [6]. It is hypothesized that marijuana possesses a neuroprotective property, among other mechanisms, reducing free radicals and inhibiting apoptosis. The effects of marijuana smoking on a limited number of healthy participants were investigated by Hepler and Frank, revealing a reduction of 30% in intraocular pressure (IOP). On the other hand, only about 65% of people would feel this [24]. The pressure-lowering effect lasts for only a short time – between 3 and 4 hours on average. THC can be administered through various routes, including inhalation, oral ingestion, sublingual administration, intravenous injection, and topical application. The topical application of cannabinoid extracts is not an optimal route of administration for treating glaucoma due to the low aqueous solubility and highly lipophilic nature of these extracts, which result in poor ocular penetration. This is contrary to what one might assume to be the case. The study conducted did not reveal any noticeable variation in the intraocular pressure (IOP) among the subjects who were administered with topical 1% THC or light mineral oil (control) within a short duration [36]. However, WIN55212-2, a more recently developed topical synthetic cannabinoid, lowers intraocular pressure (IOP) by 20–30% in glaucoma patients, but the maximal effect takes place after an hour [72]. The study demonstrated that synthetic cannabinoids, namely, BW29Y and BW146Y, when orally administered to patients with glaucoma, reduced intraocular pressure (IOP). The findings indicated that BW146Y was equally effective in reducing IOP compared to the consumption of THC through smoking marijuana. However, BW29Y did not exhibit any significant impact on IOP. Notably, CBD increased intraocular pressure (IOP) in mouse eyes at intervals of 1 and 4 hours following topical administration, in contrast to THC [48]. In a study conducted on humans, analogous outcomes about elevated intraocular pressure (IOP) were noted after the sublingual administration of greater cannabidiol (CBD) after a lapse of 4 hours. Marijuana use has been associated with various negative outcomes, including psychotropic effects such as cognitive impairment, euphoria, dysphoria, decreased coordination, decreased short-term memory, time distortion, and sleepiness. Cardiovascular effects such as tachycardia, hypotension, conjunctival hyperemia, and decreased lacrimation have also been linked to the use of this substance. Moreover, the potential hypotensive impact of marijuana on intraocular pressure (IOP) may be counteracted by the concomitant reduction in blood pressure, resulting in inadequate blood supply to the optic nerve and consequent ischemic alteration [50]. If you smoke marijuana, you risk developing emphysema and possibly lung cancer. Topical THC can cause corneal injury and local irritation in some people. These are some of the side effects specific to the route of administration. There is a need for additional concern regarding the possible addictive properties and the level of tolerance. This is a particularly concerning finding because for patients to attain round-the-clock intraocular pressure (IOP) regulation through marijuana consumption, they must engage in smoking activities approximately six to eight times within 24 hours. However, this frequency of use is likely to lead to the development of cannabinoid use disorder. The monthly cost is anticipated to be approximately $690 to keep this dose, a sizeable increase from the currently prescribed medications.

Curcumin: Curcumin, also called *Curcuma longa*, commonly known as turmeric, is a widely utilized spice in South Asian cuisine. Curcumin is a naturally occurring compound obtained from the *Curcuma longa* plant. It has been extensively used for its anti-inflammatory properties in managing various inflammatory disorders [23]. Curcumin is a polyphenol with lipophilic properties, rendering it insoluble in aqueous solutions; however, it can maintain its integrity in environments with an acidic pH, such as the human stomach [49]. Interaction with various molecular targets for inflammation is required for curcumin's anti-inflammatory effects. For instance, this is achieved through the regulation of enzymatic activity, including but not limited to nitric oxide synthase (iNOS), cyclooxygenase-2 (COX-2), and lipoxygenase. This is how it works to regulate the inflammatory response. Inflammatory cytokines such as monocyte chemoattractant protein (MCP), TNF, interleukins, and migration inhibitory protein are all produced in lower quantities when curcumin is present [18]. This study investigates the impact of curcumin, an anti-inflammatory and antioxidative agent, on light-induced retinal degeneration (LIRD) in rat models and retina-derived cell lines. Studies have demonstrated the anti-inflammatory and antioxidative properties of curcumin. Retinal neuroprotection was noted in rats supplemented with 0.2% curcumin for 2 weeks. Through the inhibition of NF-kB activation and the downregulation of inflammatory gene expression, curcumin protected the retina from the effects of LIRD. When used in experiments as a pretreatment for the utilization of retina-derived cell lines, namely, 661W and ARPE-19, curcumin has been observed to confer cellular protection against hydrogen peroxide (H_2O_2)-induced cell death. This protective effect is attributed to activating cellular protective enzymes, including HO-1 and thioredoxin [33]. These results were found in studies. The study revealed that the cytoprotective properties of curcumin against oxidative stress induced by H_2O_2 were augmented when human retinal cells were incubated with 15 M curcumin. This effect was attributed to the upregulation of HO-1 expression, which decreased reactive oxygen species (ROS) levels [12]. Curcumin can also change the expression of NF-B, AKT, NRF2, and growth factors, which stop inflammation and protect cells. Studies with real-life rat models have demonstrated that curcumin directly impacts ocular disorders. Curcumin has been shown to have protective effects. The present study investigates the impact of oral administration on Wistar albino rats with streptozotocin-induced diabetic retinopathy in recent DR studies. Studies have demonstrated the potential of curcumin to inhibit the development of cataracts in rat models induced by selenite ingestion, galactose ingestion, naphthalene exposure, and diabetes [25]. The inclusion of curcumin in the dietary regimen of diabetic rats was found to confer protection to their lens against cataractogenesis. This was attributed to preserving the chaperone-like function of lens-crystallin [14]. Curcumin has been linked to several unwanted side effects since it was first studied in 1976, even though it has shown some promise as a possible natural treatment. Alterations to chromosomes and DNA are the primary types that can occur at higher doses of curcumin. Curcumin has been identified as the preferred compound for managing and prophylaxis of age-related macular degeneration (AMD), cataracts, and diabetic retinopathy (DR). However, it is noteworthy that curcumin possesses antioxidative and anti-inflammatory properties that may result in adverse effects at elevated dosages.

Lutein and zeaxanthin: Understanding that macular pigments comprise carotenoids such as lutein and zeaxanthin is important. They are found in the macula, the part of the retina responsible for the sharpest central vision, and are essential for healthy vision. Lutein and zeaxanthin are believed to act as "sunglasses for the macula," protecting it from the negative effects of blue light. Generally speaking, these carotenoid pigments are found in higher concentrations in the macula and retina of humans. Carotenoids are present in various fruits and vegetables, including but not limited to kale, spinach, and maize. They are frequently consumed in the form of dietary supplements. Research has suggested that augmenting the consumption of lutein and zeaxanthin can potentially mitigate the likelihood of age-related macular degeneration. In addition, they can be found in a wide variety of fruits and vegetables, including red grapes, kale, spinach, corn, kiwi, and kiwifruit [37]. Lutein and zeaxanthin are also found in egg yolks, oranges, and mangoes. They may also have a protective effect on the eyes from UV light. In addition, carotenoids have the potential to exhibit anti-inflammatory and antioxidative characteristics. Multiple epidemiological studies have demonstrated an association between decreased likelihood of age-related macular degeneration and elevated levels of lutein and zeaxanthin in the bloodstream. Notably, this relationship has been observed in studies such as the Age-Related Eye Disease Study (AREDS) and the Case-Control Study Group for Eye Diseases in the United States. This suggests that consuming foods rich in lutein and zeaxanthin may help reduce the risk of AMD [52–54, 55, 57]. Furthermore, supplementation with lutein and zeaxanthin has also reduced the risk of AMD. There is still some debate about the exact mechanism by which lutein and zeaxanthin work, but it has been hypothesized that carotenoids have the potential to safeguard the macula and outer retinal segments of photoreceptors against oxidative stress through the activation of an antioxidant cascade, which subsequently deactivates reactive oxygen species [24]. Furthermore, it has been proposed that lutein and zeaxanthin play a significant role in the metabolic processes of photoreceptor cells, thereby contributing to the preservation of optimal visual health. Research has indicated that lutein and zeaxanthin possess the potential to safeguard the macula against harm caused by blue light. For this hypothesis to be accepted, it is necessary to consider that lutein and zeaxanthin possess antioxidant properties. These antioxidants help to scavenge free radicals and protect the macula from oxidative damage. Thus, lutein and zeaxanthin play a vital role in maintaining healthy vision. In addition, lutein and zeaxanthin also perform the function of light filters within the eye, absorbing blue light before it reaches the retina, thereby protecting the retina from damage. Lutein and zeaxanthin can be found in many fruits, vegetables, and supplements, making them easy to obtain. A balanced diet with these nutrients is essential for maintaining healthy vision. As a result, the retina is protected from the harmful effects of LIRD, which can be caused by acute light exposure and exposure to highlight levels. The consumption of lutein and zeaxanthin has been associated with a potential decrease in the likelihood of developing age-related macular degeneration and cataracts. Taking these nutrients regularly can help improve vision and eye health. A study on cultured ARPE-19 cells indicates that providing lutein and zeaxanthin led to decreased photooxidative damage and suppressed gene expression linked to inflammation in RPE cells. The results of this study indicate that lutein and zeaxanthin can safeguard the eyes against oxidative stress and inflammation commonly linked to the deterioration of vision due to aging. This can significantly improve individuals' eye health and vision as they age. Furthermore, several epidemiological studies have revealed an inverse correlation between the amount of macular pigment and the likelihood of age-related macular degeneration (AMD) among older individuals. Therefore, augmenting the dietary consumption of lutein and zeaxanthin can potentially mitigate the likelihood of acquiring age-related macular degeneration (AMD). Supplementation with these two carotenoids has also been beneficial for reducing the risk of AMD. As per a recent clinical study that was published in an American Academy of Ophthalmology journal, the administration of lutein and zeaxanthin supplements exhibited an improvement in visual function and deceleration in the advancement of AMD pathology among patients diagnosed with early AMD [48]. Additionally, the research above revealed a positive correlation between the intake of lutein and zeaxanthin supplements and increased macular pigment optical density. This suggests that the intake of lutein and zeaxanthin supplements has been suggested to potentially mitigate the likelihood of age-related macular degeneration (AMD) by augmenting the optical density of the macular pigment. Zeaxanthin has also been shown to enhance the visual function of older male patients with AMD in clinical studies that were similar to this one. This indicates that lutein and zeaxanthin supplementation may have a protective effect against AMD and may also help improve the visual function of people already suffering from AMD. The consumption of lutein and zeaxanthin supplements may be a useful choice for preventing macular degeneration in the future [50]. Additional research is required to ascertain the precise dosage and duration necessary to elicit the protective effects of lutein and zeaxanthin supplementation. It is also important to consider the potential side effects of supplementing these carotenoids. Finally, diet is a key factor in determining an individual's risk for AMD and should also be considered.

Saffron: Because of its ability to reduce toxicity, saffron is one of the spices frequently used in traditional medical practices since it reduces toxicity. Saffron is recognized for its antioxidative and anti-inflammatory characteristics. It has been employed to treat diverse medical conditions, from dermatological disorders to gastrointestinal ailments. Additionally, it is believed to have mood-boosting effects. To provide it with its antiapoptotic properties, crocin and crocetin are the active ingredients that are present in them. These two compounds work together to reduce oxidative stress, thus providing saffron with its health benefits. Studies have also suggested that saffron has been found to possess potential health benefits in reducing the incidence of specific types of cancer and cardiovascular disease. The antioxidant carotenoids crocin and crocetin are known to prevent the onset of apoptosis by protecting cells from the effects of reactive oxygen species. Saffron also has anti-inflammatory effects, and consumption of certain foods may have a beneficial effect on reducing inflammation and promoting overall health. Additionally, saffron can help to reduce the risk of depression and improve cognitive function, concentration, and memory. There is evidence that crocin inhibits apoptosis in serum-starved PC12 cells exposed to hypoxic conditions, peroxidation of membrane lipids, and caspase-3 activation in serum-starved PC12 cells. Furthermore, research has indicated that saffron may have potential therapeutic benefits in addressing oxidative stress and inflammation in brain tissue, thereby indicating its potential utility in treating Alzheimer's disease. In addition, it also increases the levels of glutathione (GSH) in the body. It prevents the activation of the JNK pathway, which plays a crucial role in the downstream signaling cascade initiated by ceramide [45]. These findings suggest that saffron could be a potential therapy for Alzheimer's and other neurological diseases. Further research is needed to validate these findings and explore saffron's full potential as a therapeutic agent. It has been demonstrated that low GSH levels in the body lead to enhanced sensitivity of the cells to agents that induce apoptosis; dietary supplementation with saffron may potentially safeguard cells against damage and death by preserving glutathione (GSH) levels within the body

[51]. This suggests that saffron may exhibit a safeguarding impact against cellular damage and mortality caused by oxidative stress. Further research will be necessary to confirm these findings and to elucidate the exact mechanisms by which saffron exerts its protective effects. The addition of crocin to the RGC-5 retinal ganglion cell line inhibited oxidative stress in the cells, as shown by a decrease in the production of caspase-3 and -9, thereby preventing the death of the cells [84]. The findings of this investigation propose that saffron exhibits promising prospects in the prophylaxis and management of neurodegenerative ailments. Additional investigation is required to establish the effectiveness and safety of saffron for the aforementioned medical conditions. These findings were gleaned from research conducted on the RGC-5 cell line. Numerous studies have examined the potential impact of saffron as a dietary supplement, including clinical trials conducted on both human and rodent subjects. The effects of prolonged exposure to bright light were significantly mitigated in rats given saffron supplements. Falsini et al. [20] found that when saffron was supplemented at a dose of 20 milligrams per day for 90 days, the study observed notable enhancements in the macular photopic flash electroretinogram (FERG) parameters, specifically in terms of amplitude and modulation threshold [20]. This supports the idea that saffron may be effective as a natural remedy for treating eye diseases caused by prolonged exposure to strong light. Further research is needed better to understand the effects of saffron on eye health. The findings above were obtained through clinical trials on human subjects diagnosed with early age-related macular degeneration (AMD). The administration of saffron through oral supplementation resulted in a noteworthy enhancement in macular function. However, the precise mechanisms underlying the advantageous impacts of saffron on photoreceptors and bipolar cells remain unclear. Studies on rats have shown evidence of saffron's ability to inhibit cell death in situations where the rats were exposed to intense light. Based on the findings of these preclinical investigations, it can be inferred that saffron exhibits neuroprotective properties. These findings suggest the possibility of saffron as a prospective therapeutic intervention for addressing retinal degenerative pathologies. Additional investigation is required to comprehensively comprehend the impacts of saffron on ocular health and its prospective therapeutic applications. According to the findings of the clinical trials, adding saffron to one's diet may temporarily improve one's retinal function; age-related macular degeneration can manifest in its early stages. As part of the research, it has been demonstrated that the inhibitory effect of saffron on the proteolysis of water-soluble protein fractions within the lens can prevent the formation of selenite-induced cataracts in Wistar rats. This is attributed to saffron's antiapoptotic and antioxidative properties. Additional investigation is required to comprehend the precise mechanisms underlying the impact of saffron on ocular well-being and whether its benefits can be sustained in the long term. Additionally, it is imperative to consider all possible adverse effects of consuming saffron supplements. This is accomplished through the saffron's ability to prevent cell death. Even though the studies do not provide conclusive evidence for the potential neuroprotective properties of saffron in age-related macular degeneration (AMD) and its potential role in the prevention of cataracts, despite the need for further investigation, the data above exhibit potential

for the advancement of prophylactic and remedial applications of nutritional supplements in combatting the ailments above. Further research is needed to understand the exact mechanisms of saffron's neuroprotective effects and validate the efficacy of saffron as a therapeutic or prophylactic agent in treating the aforementioned ocular conditions.

Catechin – Catechin is a polyphenolic antioxidant frequently discovered in green tea [28]. Epigallocatechin gallate (EGCG), found in the most significant quantity in green tea, is a catechin that has powerful antioxidative effects. Researchers have shown in the past that previous research has demonstrated that the administration of EGCG in conjunction with sodium nitroprusside via intraocular injections resulted in a safeguarding impact on the retinal photoreceptors, which suggests that EGCG may have a beneficial effect on patient individuals who experience ocular ailments characterized by oxidative stress [86]. This protective effect was further confirmed in a recent study that found that EGCG improved the visual acuity of individuals diagnosed with age-related macular degeneration. Thus, EGCG is a promising therapeutic agent for ocular diseases. Catechins have a few different mechanisms of action, including mitigating glutamate-induced toxicity via lipid peroxidation and protein modification, eliminating reactive oxygen species, and the oxidative changes of low-density lipoproteins. The oral administration of EGCG has been shown to the findings of the study indicate that the administration of the substance in rat models may have a potential therapeutic effect in mitigating light-induced retinal neuronal degeneration, thereby potentially serving as a prophylactic measure against photoreceptor cell death [87]. Similarly, when catechin was given to Sprague-Dawley rats with cataracts caused by N-methyl-N-nitrosourea, the cataract-caused apoptosis in the lens epithelium was stopped. This may help treat or prevent cataracts in people. Additionally, RPE cell migration and adhesion can be inhibited by EGCG, which offers a potential preventative action against AMD. As a result, catechin is a compound that should be considered for treating and preventing diseases like AMD.

Ginseng – The Panax ginseng plant's root, commonly called ginseng in traditional Chinese medicine, has been widely utilized in this field for an extended period. It is believed to help boost energy levels and improve overall well-being. It is also thought to possess antioxidant and anti-inflammatory properties, which may help to protect the body from disease. There are several active components in ginseng, but the most important are ginsenosides. Ginsenosides are thought to be responsible for many of the purported health benefits of ginseng, as they appear to have antioxidant, anti-inflammatory, and immune-modulating effects. Empirical evidence suggests that these substances positively impact cognitive abilities and physical aptitude. Ginsenosides, a class of steroidal saponins, have demonstrated the ability to impede glutamate-induced neurotoxicity, lipid peroxidation, and calcium influx into cells. These effects may contribute to preventing apoptosis [78]. The ginsenoside saponins Rb1 and Rg3 have been found to inhibit apoptotic cascades in cells. There has been evidence that retinoblastoma protein 1 (Rb1) and retinoblastoma protein 3 (Rg3) inhibit NMDA glutamate receptor activity, thus suppressing tumor necrosis factor-alpha production and providing neuroprotective effects to cultured cortical cells, which have been shown to have a wide range of pharmacological effects [7], targeting a wide range of tissues. This suggests that Rb1 and Rg3 could be potential drug targets for neurodegenerative diseases. Further research on their effects on other tissues is warranted. The medicinal use of ginseng dates back several millennia. In recent years, studies have found that ginseng may benefit the brain, including improved cognition and memory. It is thought that ginseng may work by increasing levels of Rb1 and Rg3, which could provide new insights into how it may be used as a treatment. As a result of oral administration of Korean Red Ginseng (KRG) to patients with glaucoma during clinical trials, there have been significant increases in the amount of blood flowing to the retina in the temporal peripapillary areas of the eye. The potential outcome of this could be a decrease in the intraocular pressure and a postponement in the advancement of the ailment. Further research is needed to confirm the efficacy of KRG in treating glaucoma. To prevent glaucoma, augmenting the blood flow to the retina in the initial phases of the ailment could potentially yield advantageous outcomes [39]. Further research should also focus on the effects of KRG on other ocular diseases. Additionally, further studies are needed to identify the long-term efficacy and safety of KRG for the treatment of glaucoma. It is commonly thought that glaucoma is characterized by the enlargement of blood vessels and diminished blood circulation, which are significant risk elements that may lead to optic nerve damage in individuals with glaucoma. The efficacy of KRG in lowering intraocular pressure has been demonstrated, potentially mitigating the likelihood of optic nerve impairment. Further research should provide more evidence to support this hypothesis. The evidence indicates that ginsenosides derived from ginseng may potentially contribute to preventing age-related macular degeneration (AMD) due to their ability to inhibit TNF-a. This could suggest that KRG can be used to reduce the risk of AMD. Further studies are needed to confirm this hypothesis. Undoubtedly, inflammation is a principal risk factor associated with age-related macular degeneration (AMD). Therefore, further research into the effects of KRG as an anti-inflammatory agent could be beneficial in reducing the risk of AMD. Additionally, research into the long-term safety of KRG would be beneficial for determining its potential as an AMD prevention strategy. It is believed that patients suffering from conditions such as age-related macular degeneration, glaucoma or cataracts may benefit from ginseng's antiapoptotic and antioxidant properties. This could potentially assist in the retardation of the advancement of the ailment, mitigation of symptoms, and enhancement of the general visual capacity. Further investigation is required to establish the effectiveness of ginseng as a treatment for age-related macular degeneration (AMD). It is noteworthy that extensive research has been conducted on the advantageous impacts of ginseng for individuals with diabetes, encompassing the mitigation of blood glucose levels, regulation of weight gain, and augmentation of insulin secretion. Therefore, ginseng may have similar benefits for those suffering from AMD. It is also important to consider the safety of ginseng when using it as an AMD treatment. Further research is needed to assess the potential risks of using ginseng for AMD treatment. In a study that was conducted very recently, it was shown that ginseng treatment significantly reduced the amount of retinal oxidative stress that was caused by diabetes in mouse models, which were associated with diabetes. Nonetheless, it remains ambiguous whether this phenomenon holds for human beings. More studies need to be conducted to understand the long-term effects of ginseng on AMD. Many of the findings in this study were based on the fact that ginseng has antioxidative properties. Ginseng also has anti-inflammatory effects, which could help reduce the symptoms of AMD. Additional investigation is required to comprehensively comprehend the potential efficacy of ginseng as a treatment for age-related macular degeneration. Even though researchers have given more attention to the effects of ginseng on AMD and diabetes than to the effects of ginseng on cataracts caused by selenite in rats, ginseng has also been shown to have the ability to reduce the severity of cataracts caused by selenite. This suggests that ginseng may have potential benefits for people with cataracts. However, more studies are needed to confirm if ginseng can be used as a potential cataract treatment. Furthermore, a team of Korean researchers led by Lee et al. [43] have identified the non-saponin component of ginseng as one of the key agents responsible for preventing cataracts. Their research found that the non-saponin component of ginseng could block the damage caused by oxidative stress on the proteins in the eyes and ultimately reduce the risk of cataracts. Further research is required to validate the effectiveness of ginseng as a plausible remedy for cataracts. As a supplement, ginseng is an extremely promising compound for further investigation into the treatment of macular degeneration, diabetic retinopathy, and even cataracts in the future. In addition, ginseng has also been linked to improved visual acuity and contrast sensitivity, indicating potential benefits for treating age-related vision loss. Further research is needed to determine the full potential of ginseng for eye care. This is because ginseng is renowned for its potent antioxidant properties. Ginseng also contains compounds that are thought to have therapeutic effects on the eyes and can help protect the eyes from oxidative stress. It is also believed to reduce inflammation and improve blood circulation, benefiting eye health.

Resveratrol – Researchers have suggested that red wine's protective effects on human health could explain why France has a lower mortality rate than other countries due to the lack of cardiovascular diseases in the country. Researchers suggest this is linked to the French consuming much more red wine than other countries. This has led to studies into the potential health benefits of moderate red wine consumption. The findings indicate that red wine may confer health advantages, including a decreased cardiovascular disease risk. In red wines, large amounts of polyphenols can be found in them. Polyphenols possess antioxidative and anti-inflammatory characteristics, potentially mitigating the susceptibility to specific ailments. However, it is important to remember that these benefits are only achievable if red wine is consumed in moderation. It is widely believed that polyphenols are a class of compounds that are capable of exhibiting a variety of properties, including this phenomenon that involves the suppression of platelet aggregation, the production of eicosanoids that promote inflammation and blood clotting, and the hindrance of endothelin synthesis which triggers vasoconstriction. These properties can reduce the risk of coronary heart disease and other cardiovascular diseases. However, drinking too much red wine can have unwanted side effects, including an increased risk of liver disease, high blood pressure, and obesity. Additionally, too much alcohol can lead to impaired judgment, poor impulse control, and long-term health issues. Therefore, it is important to consume red wine responsibly in moderation. It is well known that red wines have particular health benefits for the cardiovascular system. Therefore, it is important to consume red wine in moderation and combine it with a balanced and healthy diet. Regular physical activity and other lifestyle factors can also help to maximize the health benefits of drinking red wine. The administration of resveratrol in mice that were subjected to a high-fat diet resulted in a noteworthy improvement in insulin sensitivity, a reduction in insulin-like growth factor-1 (IGF-I) levels, and an elevation in the activity of AMP-activated protein kinase (AMPK) and peroxisome proliferator-activated receptor gamma coactivator 1alpha (PCG-1alpha), which in turn has been shown to result in significantly longer survival and health [8]. The results of this study indicate that resveratrol could potentially and significantly impact the preservation of metabolic equilibrium and the mitigation of metabolic disorders. Further studies are needed to understand resveratrol's effects on metabolic health better. Additionally, it is important to consider the potential side effects of the long-term use of resveratrol supplements. Furthermore, resveratrol has also been shown to reduce inflammation and improve mitochondrial health. Thus, it is possible that resveratrol could be beneficial for overall health and well-being. However, more research is needed before any definitive conclusions can be made. The researchers also found that supplementation with resveratrol reduced oxidative stress, vascular endothelial growth factor, and the development of early vascular lesions in rat and mouse models of diabetes [32]. However, there is still debate over the efficacy of resveratrol as a supplement. Additionally, more research is needed to determine the potential long-term effects of resveratrol supplementation. In patients with diabetic retinopathy, this quality may be helpful due to its ability to delay or even stop the death of cells. Despite these potential benefits, resveratrol must be studied further before it can be recommended as a treatment for any condition. Further research is needed to confirm the potential benefits of resveratrol supplementation. As a result of inhibiting CHOP and IRE1 expression, resveratrol prevented injury-induced capillary degeneration and endoplasmic reticulum stress in rats. However, further studies are needed to determine the long-term effects of resveratrol supplementation. Additionally, studies should be conducted to assess the efficacy of resveratrol in humans. It was demonstrated by the fact that it could protect the body against both conditions. Thus, it is likely that resveratrol could be used as a therapeutic agent to treat conditions associated with capillary degeneration and endoplasmic reticulum stress. Further research is needed to explore the potential of this promising compound further. There is some evidence that resveratrol may be able to be a novel drug in terms of treating vascular dysfunction in the retina [71]. However, animal studies have shown conflicting results. More research is needed to confirm the effectiveness of resveratrol in humans. Additionally, further investigation is required to assess the compound's safety over an extended period. This is because retinal ischemia plays a significant role in the development of closed-angle glaucoma and diabetic retinopathy, as well as other types of eye diseases. Thus, it is imperative to evaluate the effectiveness of resveratrol and appraise the possible hazards linked to its prolonged consumption. Further research is needed to clarify the effects of resveratrol on eye health. There is no doubt that resveratrol is an efficient antioxidant and that the potential benefit of this substance lies in its ability to mitigate the oxidation of low-density lipoproteins (LDLs), thereby conferring cellular protection against the deleterious effects of free radicals. However, more research is needed to fully understand the potential risks of the long-term use of resveratrol on eye health. Additionally, further studies should be conducted to determine the optimal dosage for humans. Additionally, the compound resveratrol has exhibited potential in promoting increased blood flow, which may be a preventative measure against vessel damage and apoptosis of optic nerve cells in individuals with glaucoma. The results of these studies could help guide further research into the potential therapeutic uses of resveratrol for eye health. Additionally, these findings could be used to develop new treatments for glaucoma and other eye diseases. It has been shown that resveratrol, the induction of heme-oxygenase-1 and the inhibition of pro-oxidant intracellular heme effects, neuronal cell cultures following strokes were observed to exhibit a mitigated response, provide neuroprotection and demonstrate a novel pathway for cellular neuroprotection [90]. These findings provide evidence of the potential for resveratrol to treat glaucoma by protecting the neural cells from oxidative damage. Further research is needed to investigate the cellular mechanisms of resveratrol-mediated neuroprotection further. The administration of resveratrol as a supplement to rats resulted in a reduction of malonyl dialdehyde levels and an increase in glutathione levels in the lens. This study also revealed that resveratrol supplementation effectively suppressed selenite-induced oxidative stress and the development of cataracts. Further studies are needed to uncover which pathways are regulated by resveratrol and how to optimize the dose and timing of resveratrol for maximal neuroprotection. Additionally, human safety and efficacy trials are needed to confirm these findings. Because oxidative stress plays a crucial role in the pathophysiology of significant ocular ailments, including cataracts, age-related macular degeneration, glaucoma, and diabetic retinopathy. The characteristics of resveratrol may hold significant potential in creating innovative therapeutic approaches and preventative measures for these conditions. Further research is needed to explore the full potential of resveratrol as a therapeutic and preventive agent for major ocular diseases. Studies should focus on the long-term administration of resveratrol and its effects on ocular health. Ultimately, these studies may lead to novel treatments and preventive interventions. *Salvia miltiorrhiza*, also referred to as Asian red sage or Danshen, comprises salvianolic acid B. This potent polyphenolic antioxidant is water-soluble and possesses anti-inflammatory properties. Danshen is also known as *Salvia miltiorrhiza*. A diabetic retinopathy mouse model was used in a recent investigation conducted to examine the impact of intravenous administration of Danshen. Ischemia of the blood capillaries is one of the most serious complications that can arise from DR. This condition is characterized by a change in structure; the thickening of the basement membrane of the capillaries induces the condition. Under such circumstances, the prompt elimination of oxygen radicals following ischemia is hindered, which destroys the permeability membrane and the formation of edema due to the destruction of nerve cells due to lipid peroxidation. This can permanently damage the capillaries, decreasing the organ's oxygen supply. Furthermore, this can cause organ failure and other serious health issues. The administration of Danshen to hypoxic and ischemic retinal tissues may facilitate the restoration of blood-oxygen transportation, enhance the absorption of retinal hemangiomas, and consequently avert visual impairment if it is injected into the tissues of the retina that are suffering from hypoxia and ischemia. There is also evidence that Danshen may help patients with diabetes maintain healthy blood sugar levels by scavenging free radicals [88]. There has also been evidence that Danshen is also beneficial for treating glaucoma, with intravenous use of Danshen after glaucoma treatment resulting in a reduction of retinal ganglion cell damage as a result of intravenous use of Danshen. Danshen has been suggested to have the potential to stabilize the visual field during the middle to late stages of glaucoma based on the outcomes of prior clinical trials. In other research, in rabbits, Danshen has demonstrated the ability to prevent the mortality of retinal ganglion cells and suppress the activation of nuclear factor kappa B (NF-kB) by the inflammatory protein TNF-[89]. Like ginseng, Danshen's neuroprotective mechanism may operate through the inhibition of NMDA receptor antagonist activity, similar to how ginseng protects neurons. This suggests that Danshen could be a prospective therapeutic intervention for neurodegenerative disorders, including but not limited to Alzheimer's and Parkinson's disease. Additional investigation is required to ascertain the effectiveness and safety of Danshen for human consumption. The results of preclinical studies conducted on patients suffering from ocular diseases involving oxidative stress have shown promising results for Danshen. Clinical trials conducted in humans are now needed to explore the therapeutic potential of Danshen further. If results prove successful, Danshen could provide an effective and safe treatment for neurodegenerative diseases. Among the most common diseases are diabetic retinopathy, age-related macular degeneration, and cataracts. Danshen has been shown to reduce inflammation in the eye's blood vessels. It also helps to improve vision and protect the retinal cells from further damage caused by oxidative stress. Furthermore, it could potentially mitigate the likelihood of ocular disease-induced visual impairment.

Quercetin – Researchers have paid the most attention to quercetin as one of the flavonoids, and a diverse array of plant-based sources, including but not limited to Brassica vegetables, black and green teas, and various types of berries, contain this particular nutrient, which are all great sources of this flavonoid. Quercetin is acknowledged for its anti-inflammatory characteristics and has been associated with a decreased susceptibility to specific categories of cancer and cardiovascular ailments. Additionally, it may also have the potential to improve cognitive performance. Researchers found that populations with diets rich in flavonoids, such as those located in the Mediterranean region, tend to have lower rates of cardiovascular disease and are more likely to live longer than those with diets low in flavonoids [22], which sparked an interest in flavonoids. Studies have suggested that flavonoids can improve memory, focus, and attention. Moreover, flavonoids have also been linked to improved mood, enhanced cognitive function, and reduced cognitive decline. Much research has been conducted on quercetin's anti-inflammatory and antioxidative properties [82]. However, there are currently no medications based on quercetin; these products have received approval from the regulatory body known as the Food and Drug Administration. Quercetin is a type of flavonoid that has been found to have many health benefits. More research is needed to understand the potential benefits of quercetin and how it can be used to treat various illnesses. Research on the protective effects of quercetin was demonstrated in the retinal cell line ARPE-19 by demonstrating its ability to inhibit proinflammatory molecules and directly inhibit the intrinsic apoptosis pathway of the cells. Further research is needed to understand the effects of quercetin on human cells and its potential use as a therapeutic agent. Currently, quercetin can be found in various supplements and dietary sources. The findings of an in vitro study involving RF/6A rhesus choroids-retinal endothelial cells indicate that the administration of quercetin led to a dose-dependent reduction in the migration and tube formation of the cells. These results suggest the potential of quercetin to be used as an anti-angiogenic agent in the treatment of retinal diseases. Further studies need to be conducted to evaluate the effects of quercetin on human cells in vivo. Those two processes are crucial to the phenomenon of retinal angiogenesis, which is a distinctive hallmark of age-related macular degeneration and is the subject of inquiry. Therefore, quercetin could be a promising therapeutic agent for treating age-related macular degeneration. Further research should also be done to evaluate the efficacy and safety of quercetin for other retinal diseases. The administration of quercetin after oxidative damage exhibited a dose-dependent reduction in cellular damage and senescence. The findings of this investigation indicate that quercetin exhibits promise as a viable therapeutic intervention for managing age-related macular degeneration and other retinal pathologies. Additional investigation is required to comprehensively comprehend the effectiveness and safety of quercetin for these medical conditions. Additional research involving human RPE cells grown in culture further demonstrated that this was the case. Quercetin also showed protective effects on the human RPE cells against oxidative stress, indicating its potential as a therapeutic agent for retinal diseases. Further research should be conducted to investigate the effects of quercetin on human RPE cells in vivo. It has also been shown that quercetin (along with other natural flavonoids) significantly decreased the production of reactive oxygen species (ROS) in retinal cell cultures that were subjected to oxidative stress as a result of ascorbate and Fe+2 [5]. This suggests that quercetin may be useful in treating and preventing retinal diseases caused by oxidative stress. Further research should explore the potential of quercetin to reduce ROS production in vivo. According to a study, human umbilical vein endothelial cells (HUVECs) were used to determine the effect of quercetin in reducing human umbilical vein endothelial cell viability following exposure to oxidants. The study's findings indicate that quercetin can mitigate oxidative stress and cellular apoptosis in HUVECs. The results of this study indicate that quercetin may possess the ability to act as a potent agent in mitigating oxidative stress and safeguarding the retina against harm. Researchers have also studied the effects of quercetin on laser-induced choroidal neovascularization in vivo, finding that quercetin significantly reduced the size of the choroidal neovascularization compared to untreated controls when compared to untreated controls. These results demonstrate the potential for quercetin to be an effective treatment for reducing the risk of vision loss caused by laser-induced choroidal neovascularization. Further research is needed to determine the long-term efficacy of quercetin in clinical settings. Quercetin was observed to impede the expression of heat shock proteins, particularly HSP72, in rat models of glaucoma, thereby hindering the neuroprotective properties of these proteins in the models above. This suggests that quercetin may have a detrimental effect on the vision of glaucoma patients. Therefore, further research is necessary to understand healthfully quercetin's implications on vision. Even though there is some evidence that quercetin may have anticataract effects, scientists have not yet understood the exact mechanisms by which this may be achieved [35, 61, 64, 65, 67, 70]. Hence, additional investigation is required to explore the potential of quercetin in safeguarding visual function. Additionally, more studies should be conducted to evaluate the safety of quercetin in long-term use. The compound also acts as an antioxidant, modulating glycation, epithelial cellular signaling, calpain protease activity, and sorbitol-aldose reductase activity [79]. Additional research should be conducted to explore the other potential benefits of quercetin. Furthermore, the impact of quercetin on different age groups should be studied to determine its efficacy in different populations. In fact, due to its ability to inhibit aldose reductase, quercetin served as a gold standard for cataract inhibition in 2011 when Gacche and Dhole tested the abilities of other flavonoids that also showed similar abilities. Further research on the efficacy of quercetin for cataract inhibition is needed to determine the full extent of its potential to improve vision. Long-term studies should be conducted to assess the impact of quercetin on different age groups over time. The precise mechanisms underlying the impact of quercetin on cataract formation and its effect on the eye's lens remain unclear, despite its status as the most prevalent flavonoid in the human diet and the existence of multiple plausible pathways for its physiological activity [41, 81]. More research is needed to understand how quercetin affects eye health and vision. In particular, more research is needed to understand its long-term effects on different age groups. Also, studies should be conducted to determine whether quercetin can prevent or treat cataracts. Although there is significant evidence associating quercetin and other flavonoids with medical benefits, particularly regarding ocular health, additional research is necessary to establish the suitability of this promising compound as a therapeutic intervention for ocular inflammatory diseases and cataracts. Additional research is warranted to explore the impact of quercetin on various ocular ailments. It is recommended that clinical trials be undertaken to assess the safety and efficacy of quercetin in the treatment of various ocular conditions.

6 Conclusion

Finding evidence to support the use of complementary and alternative medications is difficult. Still, nutritional supplements and herbal remedies are easily marketed in popular culture as promoting health benefits. While these medications may not have been proven effective, many people take them to improve their health. However, it is essential to understand the risks associated with taking these medications and the potential for interactions with other medicines. As a result of patients with eye diseases frequently using complementary and alternative medicine (CAM) products, ophthalmologists must be knowledgeable about the appropriate use of these products in common eye diseases. They should also be aware of potential adverse events or interactions associated with using CAM products. Furthermore, healthcare providers should be mindful of the potential for additive effects when CAM products are used concomitantly with traditional medications. Several studies have shown that specific antioxidant vitamins and zinc can significantly reduce the progression of age-related macular degeneration (AMD) in patients by slowing retina deterioration. Therefore, healthcare providers should be knowledgeable about the potential benefits of CAM products and should consider recommending them to patients who may benefit from them. Additionally, healthcare providers should educate patients on adequately using the potential risks associated with using complementary and alternative medicine (CAM) products. As a result of the reviewed studies, antioxidant vitamins and mineral supplements are not currently supported by evidence in the literature about their effectiveness in preventing or treating cataracts in healthy individuals. Therefore, healthcare providers should discourage their use, as they may be a waste of money and resources. Patients should be aware that using said products carries potential hazards, and there is a dearth of substantiating evidence regarding their effectiveness. In both diabetic retinopathy and glaucoma research, these products demonstrate no benefit. Patients should be encouraged to discuss with their healthcare provider any alternative treatments that are evidence-based and offer a better chance of success. Furthermore, healthcare providers should stay current on the latest research to make the best patient decisions. The challenges associated with later studies on the effects of antioxidants on cataract development and progression are attributed to several factors. These include the variability within and between study populations, inadequate definition of subjects' lifetime exposure to antioxidants, reliance on self-reported dietary intake, imprecise assessment of antioxidant nutrients, variations in the doses and formulations of vitamins used in different studies, and the duration of the follow-up period required to evaluate cataract development and progression. Therefore, it is difficult to draw definitive conclusions from studies on the association between antioxidant vitamins and cataract risk. Further research is needed to understand the potential role of antioxidants in cataract prevention. In terms of determining whether herbal remedies can be used in the treatment of eye diseases, currently, there exists a dearth of observational studies or prospective clinical trials that satisfy the prerequisites for level I or level II evidence. Thus, more research is necessary to assess the efficacy and safety of herbal remedies in treating eye diseases. Further studies should also evaluate the role of antioxidants in cataract prevention. Consequently, it presents a challenge to formulate any definitive inferences regarding the effectiveness of herbal remedies in addressing ocular ailments. Given the lack of reliable evidence, caution should be exercised when using herbal remedies to treat eye diseases. Further research should focus on the potential short- and long-term effects of herbal remedies on eye health.

Additionally, more research is needed to explore the potential of herbal remedies in preventing eye diseases.

Acknowledgment We would like to extend our sincere appreciation and gratitude to the Department of Agronomy at the School of Agriculture at Lovely Professional University, Punjab, 144411, India.

Conflicts of Interest A conflict of interest has not been identified.

References

1. Abuhasira R, Shapiro L, Landschaft Y (2018) Medical use of cannabis and cannabinoids containing products – regulations in Europe and North America. Eur J Intern Med 49:2–6. https://doi.org/10.1016/j.ejim.2018.01.001
2. Ademoglu E, Ozbey N, Erbil Y, Tanrikulu S, Barbaros U, Yanik BT (2006) Determination of oxidative stress in thyroid tissue and plasma of patients with Graves' disease. Eur J Intern Med 17:545–550
3. Anderson DR (2009) The optic nerve in glaucoma. In: Tasman W, Jaeger EA (eds) Duane's ophthalmology. Lippincott Williams & Wilkins
4. AREDS2 Research Group, Chew EY, Clemons T et al (2012) The age-related eye disease study 2 (AREDS2): study design and baseline characteristics (AREDS2 report number 1). Ophthalmology 119(11):2282–2289
5. Areias FM, Rego AC, Oliveira CR, Seabra RM (2001) Antioxidant effect of flavonoids after ascorbate/Fe2+−induced oxidative stress in cultured retinal cells. Biochem Pharmacol 62(1): 111–118
6. Atakan Z (2012) Cannabis is a complex plant with different compounds and effects on individuals. Ther Adv Psychopharm 2(6):241–254. https://doi.org/10.1177/2045125312457586
7. Attele AS, Wu JA, Yuan C (1999) Ginseng pharmacology: multiple constituents and multiple actions. Biochem Pharmacol 58(11):1685–1693
8. Baur JA, Pearson KJ, Price NL et al (2006) Resveratrol improves health and survival of mice on a high-calorie diet. Nature 444(7117):337–342
9. Bevans N, Alford M, Guanchez F, Aregullin M, Rodriguez E (2001) J Undergrad Study Independent Res 2:20–24
10. Bourne RR, Flaxman SR, Braithwaite T, Cicinelli MV, Das A, Jonas JB et al (2017) Magnitude, temporal trends, and projections of the global prevalence of blindness and distance and near vision impairment: a systematic review and meta-analysis. Lancet Glob Health 5:e888–e897
11. Cesaro S, Toffolutti T, Messina C, Calore E, Alaggio R, Cusinato R, Pillon M, Zanesco L (2004) Eur J Hematol 73:50–55
12. Cybulska-Heinrich AK, Mozaffarieh M, Flammer J (2012) Ginkgo biloba: an adjuvant therapy for progressive normal and high-tension glaucoma. Mol Vis 18:390–402
13. de Souza Silva JE, Santos Souza CA, da Silva TB, Gomes IA, Brito GC, de Souza Araujo AA et al (2014) Use of herbal medicines by elderly patients: a systematic review. Arch Gerontol Geriatr 59(2):227–233. https://doi.org/10.1016/j.archger.2014.06.002
14. DeKosky ST, Williamson JD, Fitzpatrick AL, Kronmal RA, Ives DG, Saxton JA et al (2008) Ginkgo biloba for prevention of dementia: a randomized controlled trial. JAMA 300(19): 2253–2262. https://doi.org/10.1001/jama.2008.683
15. Denning DW (1998) J Clin Infect Dis 26:781–803
16. Diamond BJ, Shiflett SC, Feiwel N, Matheis RJ, Noskin O, Richards JA et al (2000) Ginkgo biloba extract: mechanisms and clinical indications. Arch Phys Med Rehabil 81(5):668–678

17. Dodge HH, Zitzelberger T, Oken BS, Howieson D, Kaye J (2008) A randomized placebo-controlled trial of Ginkgo biloba for the prevention of cognitive decline. Neurology 70(19 Pt 2): 1809–1817. https://doi.org/10.1212/01.wnl.0000303814.13509.db
18. Eckert A, Keil U, Scherping I, Hauptmann S, Müller WE (2005) Stabilization of mitochondrial membrane potential and improvement of neuronal energy metabolism by Ginkgo biloba extract EGb 761. Ann N Y Acad Sci 1056:474–485
19. Fabiana BM, Marangon DM, Joann AG, Eduardo CF (2004) Am J Ophthalmol 137:820–825
20. Falsini B, Piccardi M, Minnella A et al (2010) Influence of saffron supplementation on retinal flicker sensitivity in early age-related macular degeneration. Investig Ophthalmol Vis Sci 51(12):6118–6124
21. Fong DS, Aiello L, Gardner TW et al (2003) Diabetic retinopathy. Diabetes Care 26(1): S99–S102
22. Formica JV, Regelson W (1995) Review of the biology of quercetin and related bioflavonoids. Food Chem Toxicol 33(12):1061–1080
23. Geyman LS, Suwan Y, Garg R, Field MG, Krawitz BD, Mo S et al (2018) Noninvasive detection of mitochondrial dysfunction in ocular hypertension and primary open-angle glaucoma. J Glaucoma 27(7):592–599. https://doi.org/10.1097/ijg.0000000000000980
24. Green K (1998) Marijuana smoking vs cannabinoids for glaucoma therapy. Arch Ophthalmol 116(11):1433–1437
25. Guo X, Kong X, Huang R, Jin L, Ding X, He M et al (2014) Effect of Ginkgo biloba on visual field and contrast sensitivity in Chinese patients with normal tension glaucoma: a randomized, crossover clinical trial. Investig Ophthalmol Vis Sci 55(1):110–116. https://doi.org/10.1167/iovs.13-13168
26. Hakanson R, Wang ZY (1996) Neurogenic inflammation (Geppetti P, Holzer P, eds). CRC Press, pp 131–140
27. Hedayati MT, Pasqualotto AC, Warn PA, Bowyer P, Denning DW (2007) J Microbiol 153: 1677–1692
28. Higdon JV, Frei B (2003) Tea catechins and polyphenols: health effects, metabolism, and antioxidant functions. Crit Rev Food Sci Nutr 43(1):89–143
29. Hirotoshi IB, Takashi S, Yoshiaki K, Kiyofumi O, Yasushi I, Wei Z, Mohammad MS, Yoshihiro KF, Yuichi O, Takayuki E (2006) J Diagn Microbiol Infect Dis 56:297–303
30. Hollander H, Makarov F, Stefani FH, Stone J (1995) Ophthalmic Res 27:296–309
31. Hollyfield JG, Bonilha VL, Rayborn ME et al (2008) Oxidative damage-induced inflammation initiates age-related macular degeneration. Nat Med 14(2):194–198
32. Hua J, Guerin KI, Chen J et al (2011) Resveratrol inhibits pathologic retinal neovascularization in Vldlr-/- mice. Investig Ophthalmol Vis Sci 52(5):2809–2816
33. Huang SY, Jeng C, Kao SC, Yu JJ, Liu DZ (2004) Improved hemorrheological properties by Ginkgo biloba extract (Egb 761) in type 2 diabetes mellitus complicated with retinopathy. Clin Nutr 23(4):615–621
34. Imarhiagbe O, Ogwu MC (2022) Sacred groves in the global south: a panacea for sustainable biodiversity conservation. In: Izah SC (ed) Biodiversity in Africa: potentials, threats and conservation, Sustainable development and biodiversity, vol 29. Springer, Singapore, pp 525–546. https://doi.org/10.1007/978-981-19-3326-4_20
35. Imarhiagbe O, Onyeukwu II, Egboduku W, Mukah FE, Ogwu MC (2022) Forest conservation strategies in Africa: historical perspectives, status and sustainable avenues for progress. In: Izah SC (ed) Biodiversity in Africa: potentials, Threats and conservation, Sustainable development and biodiversity, vol 29. Springer, Singapore, pp 547–572. https://doi.org/10.1007/978-981-19-3326-4_21
36. Jay WM, Green K (1983) Multiple-drop study of topically applied 1% delta 9-tetrahydrocannabinol in human eyes. Arch Ophthalmol 101(4):591–593
37. Katz J, Costarides AP (2019) Facts vs fiction: the role of cannabinoids in the treatment of glaucoma. Curr Ophthalmol Rep 7(3):177–181
38. Kerrigan LA, Zack DJ, Quigley HA, Smith SD, Pease ME (1997) Arch Ophthalmol 115: 1031–1035

39. Kim NR, Kim JH, Kim CY (2010) Effect of Korean red ginseng supplementation on ocular blood flow in patients with glaucoma. J Ginseng Res 34(3):237–245
40. Kojaoghlanian T, Flomenberg P, Horwitz MS (2003) Rev Med Virol 13:155–171
41. Kosoe EA, Achana GTW, Ogwu MC (2023) Meta-evaluation of the one health implication on food systems of agrochemical use. In: Ogwu MC, Chibueze Izah S (eds) One health implications of agrochemicals and their sustainable alternatives, Sustainable development and biodiversity. Springer, Singapore, pp 387–409. https://doi.org/10.1007/978-981-99-3439-3_14
42. Lam TSK, Wong OF, Leung CH, Fung HT (2009) Hong Kong J Emerg Med 16(4):267–270
43. Lee SM, Sun JM, Jeong JH et al (2010) Analysis of the effective fraction of sun ginseng extract in selenite-induced cataract rat model. J Korean Ophthalmol Soc 51:733–739
44. Liu Q, Ju WK, Crowston JG, Xie F, Perry G, Smith MA et al (2007) Oxidative stress is an early event in hydrostatic pressure-induced retinal ganglion cell damage. Investig Ophthalmol Vis Sci 48:4580–4589
45. Maccarone R, Di Marco S, Bisti S (2008) Saffron supplement maintains morphology and function after exposure to damaging light in mammalian retina. Investig Ophthalmol Vis Sci 49(3):1254–1261
46. Madsen-Bouterse SA, Kowluru RA (2008) Oxidative stress and diabetic retinopathy: pathophysiological mechanisms and treatment perspectives. Rev Endocr Metab Disord 9:315–327
47. Mandelcorn E, Gupta N (2009) Lens-related glaucomas. In: Tasman W, Jaeger EA (eds) Duane's ophthalmology. Lippincott Williams & Wilkins
48. Miller S, Daily L, Leishman E, Bradshaw H, Straiker A (2018) Delta9-tetrahydrocannabinol and cannabidiol differentially regulate intraocular pressure. Investig Ophthalmol Vis Sci 59(15): 5904–5911
49. Mirzaei M, Gupta VB, Chick JM et al (2017) Age-related neurodegenerative disease associated pathways identified in retinal and vitreous proteome from human glaucoma eyes. Sci Rep 7(1):12685
50. Novack GD (2016) Cannabinoids for treatment of glaucoma. Curr Opin Ophthalmol 27(2): 146–150
51. Ochiai T, Soeda S, Ohno S, Tanaka H, Shoyama Y, Shimeno H (2004) Crocin prevents the death of PC-12 cells through sphingomyelinase-ceramide signaling by increasing glutathione synthesis. Neurochem Int 44(5):321–330
52. Ogwu MC (2019) Lifelong consumption of plant-based GM foods: is it safe? In: Papadopoulou P, Misseyanni A, Marouli C (eds) Environmental exposures and human health challenges. IGI Global, Pennsylvania, pp 158–176. https://doi.org/10.4018/978-1-5225-7635-8.ch008
53. Ogwu MC (2019) Towards sustainable development in Africa: the challenge of urbanization and climate change adaptation. In: Cobbinah PB, Addaney M (eds) The geography of climate change adaptation in urban Africa. Springer Nature, Switzerland, pp 29–55. https://doi.org/10.1007/978-3-030-04873-0_2
54. Ogwu MC (2019) Understanding the composition of food waste: an "-omics" approach to food waste management. In: Gunjal AP, Waghmode MS, Patil NN, Bhatt P (eds) Global initiatives for waste reduction and cutting food loss. IGI Global, Pennsylvania, pp 212–236. https://doi.org/10.4018/978-1-5225-7706-5.ch011
55. Ogwu MC (2023) Local food crops in africa: sustainable utilization, threats, and traditional storage strategies. In: Izah SC, Ogwu MC (eds) Sustainable utilization and conservation of Africa's biological resources and environment, Sustainable development and biodiversity, vol 888. Springer, Singapore, pp 353–376. https://doi.org/10.1007/978-981-19-6974-4_13
56. Ogwu MC, Osawaru ME (2022) Traditional methods of plant conservation for sustainable utilization and development. In: Izah SC (ed) Biodiversity in Africa: potentials, threats and conservation, Sustainable development and biodiversity, vol 29. Springer, Singapore, pp 451–472. https://doi.org/10.1007/978-981-19-3326-4_17
57. Ogwu MC, Osawaru ME (2023) Disease outbreaks in ex-situ plant conservation and potential management strategies. In: Izah SC, Ogwu MC (eds) Sustainable utilization and conservation of

Africa's biological resources and environment, Sustainable development and biodiversity, vol 888. Springer, Singapore, pp 497–518. https://doi.org/10.1007/978-981-19-6974-4_18
58. Ogwu MC, Osawaru ME, Ahana CM (2014) Challenges in conserving and utilizing plant genetic resources (PGR). Int J Genet Mol Biol 6(2):16–22. https://doi.org/10.5897/IJGMB2013.0083
59. Ogwu MC, Osawaru ME, Aiwansoba RO, Iroh RN (2016) Ethnobotany and collection of West African Okra [*Abelmoschus caillei* (A. Chev.) Stevels] germplasm in some communities in Edo and Delta States, Southern Nigeria. Borneo J Resour Sci Technol 6(1):25–36. https://doi.org/10.33736/bjrst.212.2016
60. Ogwu MC, Osawaru ME, Obahiagbon GE (2017) Ethnobotanical survey of medicinal plants used for traditional reproductive care by Usen people of Edo State, Nigeria. Malaya J Biosci 4(1):17–29
61. Ogwu MC, Izah SC, Iyiola AO (2022) An overview of the potentials, threats and conservation of biodiversity in Africa. In: Izah SC (ed) Biodiversity in Africa: potentials, threats and conservation, Sustainable development and biodiversity, vol 29. Springer, Singapore, pp 3–20. https://doi.org/10.1007/978-981-19-3326-4_1
62. Ogwu MC, Osawaru ME, Owie MO (2023) Effects of storage at room temperature on the food components of three cocoyam species (*Colocasia esculenta, Xanthosoma atrovirens, and X. sagittifolium*). Food Stud Interdiscip J 13(2):59–83. https://doi.org/10.18848/2160-1933/CGP/v13i02/59-83
63. Ogwu MC, Osawaru ME, Amodu E, Osamo F (2023) Comparative morphology, anatomy, and chemotaxonomy of two *Cissus* Linn. species. Braz J Bot 46:397. https://doi.org/10.1007/s40415-023-00881-0
64. Ogwu MC, Iyiola AO, Izah SC (2023) Overview of African biological resources and environment. In: Izah SC, Ogwu MC (eds) Sustainable utilization and conservation of Africa's biological resources and environment, Sustainable development and biodiversity, vol 888. Springer, Singapore, pp 1–31. https://doi.org/10.1007/978-981-19-6974-4_1
65. Osawaru ME, Ogwu MC (2014) Ethnobotany and germplasm collection of two genera of cocoyam (*Colocasia* [Schott] and *Xanthosoma* [Schott], Araceae) in Edo State Nigeria. Sci Technol Arts Res J 3(3):23–28. https://doi.org/10.4314/star.v3i3.4
66. Osawaru ME, Ogwu MC (2014) Conservation and utilization of plant genetic resources. In: Omokhafe K, Odewale J (eds) Proceedings of 38th Annual Conference of The Genetics Society of Nigeria. Empress Prints Nigeria Limited, pp 105–119
67. Osawaru ME, Ogwu MC (2020) Survey of plant and plant products in local markets within Benin City and environs. In: Filho LW, Ogugu N, Ayal D, Adelake L, da Silva I (eds) African handbook of climate change adaptation. Springer Nature, Switzerland, pp 1–24. https://doi.org/10.1007/978-3-030-42091-8_159-1
68. Osawaru ME, Ogwu MC, Ahana CM (2013) Current status of plant diversity and conservation in Nigeria. Niger J Life Sci 3(1):168–178
69. Osawaru ME, Ogwu MC, Omologbe J (2014) Characterization of three Okra [*Abelmoschus* (L.)] accessions using morphology and SDS-PAGE for the basis of conservation. Egypt Acad J Biol Sci 5(1):55–65
70. Osawaru ME, Ogwu MC, Omoigui ID, Aiwansoba RO, Kevin A (2016) Ethnobotanical survey of vegetables eaten by Akwa Ibom people residing in Benin City, Nigeria. Univ Benin J Sci Technol 4(1):70–93
71. Osborne NN, Casson RJ, Wood JPM, Chidlow G, Graham M, Melena J (2004) Retinal ischemia: mechanisms of damage and potential therapeutic strategies. Prog Retin Eye Res 23(1):91–147
72. Porcella A, Maxia C, Gessa GL, Pani L (2001) The synthetic cannabinoid WIN55212-2 decreases the intraocular pressure in human glaucoma resistant to conventional therapies. Eur J Neurosci 13(2):409–412

73. Quaranta L, Bettelli S, Uva MG, Semeraro F, Turano R, Gandolfo E (2003) Effect of Ginkgo biloba extract on preexisting visual field damage in normal tension glaucoma. Ophthalmology 110(2):359–362
74. Quigley H, Broman AT (2006) The number of people with glaucoma worldwide in 2010 and 2020. Br J Ophthalmol 90(3):262–267
75. Radu RA, Yuan Q, Hu J et al (2008) Accelerated accumulation of lipofuscin pigments in the RPE of a mouse model for ABCA4-mediated retinal dystrophies following vitamin A supplementation. Investig Ophthalmol Vis Sci 49(9):3821–3829
76. Rhee DJ, Katz LJ, Spaeth GL, Myers JS, Steinmann WC (2001) Complementary and alternative medicine for glaucoma. Surv Ophthalmol 46(1):43–55
77. Rhee DJ, Spaeth GL, Myers JS, Augsburger JJ, Shatz LJ et al (2002) Prevalence of the use of complementary and alternative medicine for glaucoma. Ophthalmology 109(3):438–443
78. Ritch R (2007) Natural compounds: evidence for a protective role in eye disease. Can J Ophthalmol 42(3):425–438
79. Shetty AK, Rashmi R, Rajan MGR, Sambaiah K, Salimath PV (2004) Antidiabetic influence of quercetin in streptozotocin-induced diabetic rats. Nutr Res 24(5):373–381
80. Shim SH, Kim JM, Choi CY, Kim CY, Park KH (2012) Ginkgo biloba extract and bilberry anthocyanins improve visual function in patients with normal tension glaucoma. J Med Food 15(9):818–823
81. Stefek M, Karasu C (2011) Eye lens in aging and diabetes: effect of quercetin. Rejuvenation Res 14(5):525–534
82. Stewart LK, Soileau JL, Ribnicky D et al (2008) Quercetin transiently increases energy expenditure but persistently decreases circulating markers of inflammation in C57BL/6J mice fed a high-fat diet. Metabolism 57(1):S39–S46
83. Taylor KI, Taylor HR (1999) Br J Ophthalmol 83:134–135
84. Yamauchi M, Tsuruma K, Imai S et al (2011) Crocetin prevents retinal degeneration induced by oxidative and endoplasmic reticulum stresses via inhibition of caspase activity. Eur J Pharmacol 650(1):110–119
85. Yau JWY, Rogers SL, Kawasaki R et al (2012) Global prevalence and major risk factors of diabetic retinopathy. Diabetes Care 35(3):556–564
86. Zhang B, Osborne NN (2006) Oxidative-induced retinal degeneration is attenuated by epigallocatechin gallate. Brain Res 1124(1):176–187
87. Zhang B, Rusciano D, Osborne NN (2008) Orally administered epigallocatechin gallate attenuates retinal neuronal death in vivo and light-induced apoptosis in vitro. Brain Res 1198: 141–152
88. Zhang L, Dai SZ, Nie XD, Zhu L, Xing F, Wang LY (2013) Effect of Salvia miltiorrhiza on retinopathy. Asian Pac J Trop Med 6(2):145–149
89. Zhu M, Cai F (1993) Evidence of compromised circulation in the pathogenesis of optic nerve damage in chronic glaucomatous rabbit. Chin Med J 106(12):922–927
90. Zhuang H, Kim Y, Koehler RC, Doré S (2003) Potential mechanism by which resveratrol, a red wine constituent, protects neurons. Ann N Y Acad Sci 993:276–286
91. Zigler JS Jr, Datiles MB III (2011) Pathogenesis of cataracts. In: Tasman W, Jaeger EA (eds) Duane's ophthalmology, vol 2. Lippincott Williams & Wilkins

Assessment of the Phytochemical Constituents and Metabolites of Some Medicinal Plants and Herbal Remedies Used in the Treatment and Management of Injuries

26

Arinze Favour Anyiam, Ejeatuluchukwu Obi, and Onyinye Cecilia Arinze-Anyiam

Contents

A. F. Anyiam (✉)
Department of Medical Laboratory Science, Thomas Adewumi University, Oko, Nigeria
e-mail: arinze.anyiam@tau.edu.ng

E. Obi
Toxicology Unit, Department of Pharmacology and Therapeutics, College of Health Sciences, Nnamdi Azikiwe University, Awka, Nigeria
e-mail: e.obi@unizik.edu.ng

O. C. Arinze-Anyiam
Department of Medical Laboratory Science, Thomas Adewumi University, Oko, Nigeria
e-mail: onyinye.arinze@tau.edu.ng

S. C. Izah et al. (eds.), *Herbal Medicine Phytochemistry*, Reference Series in Phytochemistry, https://doi.org/10.1007/978-3-031-43199-9_57

Abstract

This chapter discusses the phytochemical constituents and metabolites in some medicinal plants and herbal remedies used in the treatment and management of injuries. A meta-analysis was conducted to identify the studies that investigated the chemical composition and biological activity of these natural remedies. The medicinal plants and herbal medicine included in the review were chosen for their traditional use in wound healing and were found to contain a range of phytochemicals with known medicinal properties, such as flavonoids, terpenoids, alkaloids, and saponins. These phytochemicals have been proven to exhibit antioxidant, anti-inflammatory, and wound-healing activity in various in vitro and in vivo studies. In addition, the review identified several metabolites produced by these medicinal plants and herbal medicine that may contribute to their therapeutic effects. Overall, the results of this review suggest that medicinal plants and herbal medicine may be valuable sources of phytochemicals and metabolites with medicinal properties, and encourage their usage in the treatment and management of wounds. However, further research is needed to fully understand the mechanisms of action and potential therapeutic effects of these compounds. This involves research into the best doses and combinations of natural remedies for various sorts of wounds, along with research into their safety and effectiveness in clinical settings.

Keywords

Wound healing · Antioxidant activity · Traditional medicine · Therapeutic effects · In vitro · Flavonoids

Abbreviations

5-LOX	5-lipoxygenase
ALT	Alanine transferase
AMPK	Activated protein kinase
AST	Aspartate aminotransferase
BC	Before Christ
bFGF	Basic fibroblast growth factor

CADs	Cichoric acids
CC	Creative Commons
CFE	Calendula flower alcohol extracts
CFS	Chronic fatigue syndrome
COX-2	Cyclooxygenase-2
DCQA	Dicaffeoyl quinic acid
DPPH	2,2-Diphenyl-1-picrylhydrazyl
EGF	Epidermal growth Factor
ET-1	Endothelin-1
GABA	Gamma amino butyric acid
HB-EGF	Heparin-binding epidermal growth factor-like growth factor
HPA axis	Hypothalamic–pituitary–adrenal axis
IFN-γ	Interferon gamma
IGF-1	Insulin-like growth factor-1
IL-1β	Interleukin-1 beta
IL-6	Interleukin-6
KGF	Keratinocyte growth factor
MCF	Monocyte chemoattractant protein
MIP	Macrophage inflammatory protein
MMP	Matrix metalloproteinase
MPP	1-Methyl-4-phenylpyridinium
MTT	(3- [4,5-Dimethylthiazol-2-yl]-2,5 diphenyl tetrazolium bromide)
NF-yB	Nuclear factor y B
NF-κB	Nuclear factor kappa B
NK cells	Natural killer cells
NO	Nitric oxide
NrF2	Nuclear factor erythroid 2-related factor
O/W	Oil-in-water
PCL	Polycaprolactone
PDGF	Platelet-derived growth factor
PGE2	Prostaglandin E2
ROS	Reactive oxygen species
STAT3	Signal transducer and activator of transcription 3
TGFs	Transforming growth factors
TGF-α	Transforming growth factor alpha
TGF-β	Transforming growth factor beta
TIMPs	Transforming inhibitors of metalloproteinases
TNF-α	Tumour necrosis factor alpha
TRPA 1	Transient receptor potential ankyrin 1
TRPV 1	Transient receptor potential cation channel subfamily V member 1
UVB rays	Ultraviolet B rays
VEGF	Vascular endothelial growth factor
VPF	Vascular permeability factor

1 Introduction

The largest organ in the body is the skin. The protection against physical harm from the environment, control of body temperature, and sensitivity to stimuli are just a few of the important jobs that skin performs. It serves different functions, and any harm or injury makes the organism more vulnerable to other biological and physical dangers, leading to the wound. A wound is a type of injury to the human body which forces the skin to break and subsequently damages the affected underlying tissue [1].

Herbal medicines, also known as phytomedicine, include herbs (such as seeds, leaves, bark, and roots), herbal materials (such as essential oils, and fresh juices), herbal preparations (such as tinctures, extracts, and fatty oils), and completed herbal products. It is a chemical compound combination from multiple chemical classes, the bioactivity of which combine to produce effects as a result of combining activities of the individual components [2, 3]. Additionally, other inert compounds serve as ballast or fillers. The medicine in question is therefore a complex mixture, both chemically and in relation to how different bioactivities interact to generate a particular effect. Contemporary synthetic drugs are very expensive, sometimes dangerous, and have withdrawal symptoms that come back. Nonetheless, because they come from a natural source and already have been utilized from the beginning of time, herbs are typically claimed to be safe [2]. Herbal medicine has been utilized for centuries to treat an array of ailments [4, 5], and it remains a popular choice for people seeking natural remedies [6]. At the locations of injury, wounds are the rupture of the anatomical and functional continuity of cells and tissues. They may be brought on by mechanisms that damage the tissue physically, chemically, microbiologically, or immunologically [7]. Each year, 14 million persons are projected to suffer from burns and wounds, with more than 80% living in countries with low and moderate incomes [7]. Nevertheless, this might be underreported because wounds are an everyday event which can befall anybody any time, and burns and minor wounds are not always treated medically.

The skin protects our bodies against both chemical and physical harm, as well as loss of water, and assists in maintaining homeostasis in the body. In everyday life, skin damage occurs when the skin fails in its defensive role, resulting in the emergence of an injury on the surface of the skin. The skin defends our internal body system against the outside world, and thus regulates temperature and maintains homeostasis. To sustain its functions after damage, the skin's integrity must be restored [8]. A complex circulatory and neural network connects the outer epidermis, dermis, and lower subcutaneous layer of human skin. The epidermis, located on the outermost layer, is primarily made up of keratinocytes, which establish an airtight barrier for protection alongside Langerhans cells, melanocytes, and Merkel cells [9]. The dermis is located beneath the epidermis and is connected to it by the basement membrane, which is a slim layer of extracellular matrix (ECM) composed primarily of collagen IV, integrins, perlecan, laminins, and nidogen [10]. The dermis has a complex composition that differs quite significantly from the epidermis [10]. It is made up of a slim layer of extracellular matrix (ECM), which serves as a framework for hair follicles, fibroblasts, sweat glands, blood vessels, and other mesenchymal cells. Additionally, it contains chemicals that modulate homeostasis, such as enzymes and growth factors [11]. The dermis is further divided into multiple layers, with the papillary layer nearest to the basement membrane and made up of poorly organised thin collagen fibres that house a dense population of fibroblasts [12]. The skin's complexity makes it extremely challenging to duplicate in a laboratory environment.

The tear and exposure of the skin could end up in illnesses and lesions, which could give rise to severe diseases and possibly fatality. As a result, one of the main subjects in medical research on wound treatment is how to repair these wounds in as little time as possible and with the lowest possible level of effect on the wound in regards to appearance [13, 14].

The aim of this chapter is to explore the phytochemical constituents and metabolites present in medicinal plants and herbal remedies used in the treatment and management of injuries, with a focus on their traditional use in wound healing. The objectives of this chapter are to identify the various phytochemicals present in these natural remedies, examine their biological activity, and evaluate their potential therapeutic effects on wounds.

To achieve the aims and objectives of the chapter, the authors have conducted a meta-analysis of studies that have investigated the chemical composition and biological activity of medicinal plants and herbal remedies used in wound healing. The traditional use of these natural remedies and various phytochemicals and metabolites present in these remedies that may contribute to their therapeutic effects have also been identified and examined. In addition, the in vitro and in vivo studies that support the antioxidant, anti-inflammatory, and wound-healing activity of these phytochemicals have been discussed.

The chapter is organized into different sections that discuss the phytochemicals and metabolites present in medicinal plants and herbal remedies used in wound healing. Each section focuses on a specific class of phytochemicals, such as flavonoids, terpenoids, alkaloids, and saponins. The biological activity of these compounds and their potential therapeutic effects on wounds have also been evaluated.

The information used in this chapter was sourced from various peer-reviewed journals, scientific databases, and books. The information presented in this chapter is supported by a meta-analysis of studies that have investigated the chemical composition and biological activity of medicinal plants and herbal remedies used in wound healing. The authors have also relied on traditional knowledge and practices, as well as modern scientific research, to support the use of these natural remedies in wound healing.

1.1 Herbal Medicine Classifications

According to Balap and Gaikwad [2], there are three main classifications of herbal medicine:

- Chinese herbal medicine
- Western herbal medicine
- Traditional Indian medicine
- Chinese herbal medicine: The oldest traditional treatise known to exist on the topic was written by Huang-ti, the Yellow monarch (2697–2597 BC), who is responsible for coordinating and standardizing Chinese herbal remedies. Due to the health benefits of both, the majority of Chinese herbalists combine one principal herb with a diet. The Indian subcontinent employs herbalism in the same manner. In Chinese medicine, there are four main ways to combine herbs. When two or more herbs are used together to complement one another's effects on the body, complementary herbs are used. Herbs that are helpful have different uses and effects. When the other herb gives supportive catalysis, one herb takes the initiative. Herbs of fright lessen the risk of wound by reducing the ferocity and power of partner herbs in an admixture. Cancelling herbs, that is, a mixture of laxative and astringent herbs, are employed in reducing constipation caused by the latter. Herbs used to reduce constipation caused by such injury are utilized to remove unwanted side effects [2].
- Western herbal medicine: Before the twentieth century, Western herbal medicine predominated the medical landscape; however, conventional or inorganic medicine founded on scientific medical models has since supplanted it. The general public's attitude towards conventional medicine has changed recently, though. The bilious, sanguine, choleric, and melancholy temperaments, which determine the makeup of the patient, the level of cold, wet, heat, or dry administered to sickness conditions and herbs, and Hippocrates' sense of humour were the foundations of European herbalism until the seventeenth century.
- Traditional Indian medicine: Traditional Indian medicine, also regarded as Ayurvedic medicine, is an ancient Indian health system that focuses on general healthcare, medicine, dietary guidance, and health education. Ayurveda medicinal system is founded on the three distinct biotic humours of Vata, which are Pitta (fire), (air), and Kapha (water). These tridoshas are present in everyone, but one element is constitutionally dominant. The primary factor defines the constitutional strengths and limitations. Vata is known for its lack of power and stamina, as well as breathlessness, indigestion, bloating, and circulation problems [15].

1.2 Traditional Use of Herbal Remedies in Wound Healing

Herbs are frequently used in combination in traditional medicine to produce a synergetic effect. It implies that the effects of numerous plants combined together can be larger when compared to the effects of each individual herb alone. This is known as a 'herbal preparation' or 'herbal formula' and is a regular technique in traditional medicine.

Traditional healers in various cultures possess a plethora of knowledge regarding the application of medicinal herbs in wound healing. They have established specialized procedures and methods for producing and applying herbal treatments, which they frequently combine with other therapies like massage, dietary modifications, or acupuncture.

One advantage of employing herbal therapies for healing wounds is that they may be adjusted to the needs of the individual. Practitioners of traditional medicine frequently consider the patient's general state of health and welfare, as well as specific wound features such as severity, location, and type. This personalized approach can aid in the process of healing and encourage a faster recovery.

An additional benefit of medicinal herbs is that they are frequently less expensive and easier to obtain than prescription drugs or medical treatments. A lot of herbs can be cultivated in the garden or bought at agricultural markets or natural food shops. This could make them an especially enticing choice for people who prefer a natural healing approach or those who do not have access to medical services.

The field of science has begun to understand the phytochemical components and metabolites that give these plants their medicinal properties [16]. Studies have demonstrated that substances such as polysaccharides, flavonoids, triterpenoids, and curcumin have anti-inflammatory and wound-healing properties [16]. With this knowledge, medicinal plants can be used effectively to treat and manage wounds, which helps to improve the healing process.

2 Wound Classification

Wounds are often categorized based on the underlying aetiology of wound development. They include the following.

2.1 Acute Wounds

Acute wounds are those that recover completely within 5–10 days, and at most, within 30. Acute wounds typically happen during surgery or due to acute tissue injury. Trauma (such as burns or injuries), a subsequent disease condition (such as leg ulcers), or a mix of both can result in acute skin lesions [17]. These wounds heal on their own using a prompt and systematic healing process that returns the skin's structural and functional integrity [18]. In contrast to chronic wounds, which take months to fully recuperate, acute wounds take just a few weeks. A recent injury that has not gone through the wound-healing stages is characterized as an acute wound. These might be superficial wounds that involve both the dermis and the epidermis, or extend beyond these layers and cause damage to the skin that compromises the subcutaneous layer [16]. Tissue damage/injury in an acute wound typically takes place throughout a systematic, phase of time correction that successfully restores functional and anatomical integrity. Cuts or surgical incisions frequently result in acute wounds [19].

2.2 Closed Wounds

Closed wounds allow blood to exit the circulatory system, but it still remains within the human body and includes contusions or abrasions, hematomas or blood tumours, and crush injuries [20]. Bumps and bruises make it obvious [21].

2.3 Open Wounds

In this scenario, blood leaks out with obvious bleeding from the person. It is further classified as follows: incised wounds, laceration or tear wounds, puncture wounds, abrasions or superficial wounds, and penetration wounds.

2.3.1 Incised Wounds

This wound has minimal tissue damage and no tissue loss. Sharp instruments like scalpels and knives are the major culprits.

2.3.2 Laceration or Tear Wounds

This non-surgical injury, when combined with other types of stress, causes tissue damage and loss.

2.3.3 Puncture Wounds

They are brought about by an object puncturing the skin, like a needle or a nail. Infection is common in these wounds because dirt can seep deep into them.

2.3.4 Abrasions or Superficial Wounds

Abrasion results from slipping on a rough surface. The top layer of skin, or epidermis, is scraped off during this process, exposing nerve endings and causing a painful injury.

2.3.5 Penetration Wounds

Penetration wounds are generally the result of an item, for instance, a knife, passing through the skin.

2.3.6 Gunshot Wounds

They often result from a bullet or similar projectile that enters the body.

2.4 Chronic Wounds

These wounds do not pass via the typical healing phases and have instead progressed to a level of pathological inflammation. They require a longer time to heal or they become recurrent [22] (Fig. 1).

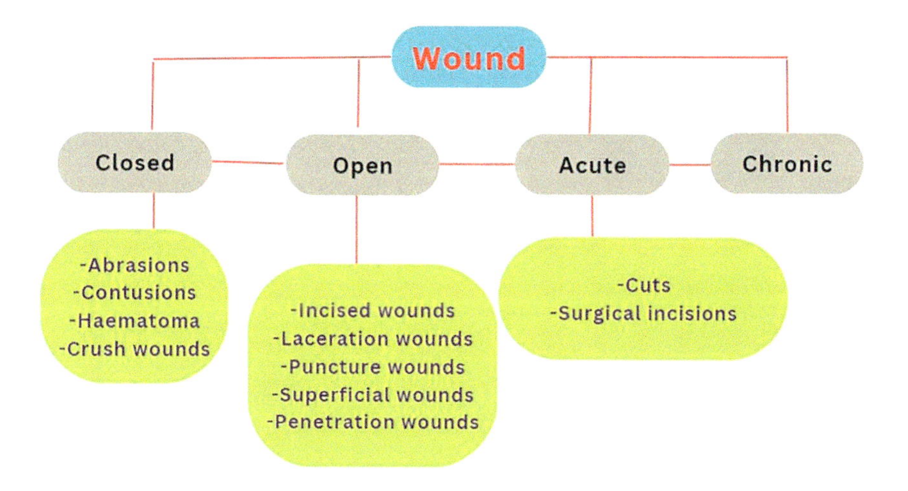

Fig. 1 Types of wound summarized

3 Wound Healing

Wound healing is a complicated process which comprises the interaction of various cell types, inflammatory mediators, and extracellular matrix components, and also several signalling pathways [23]. It is defined by a well-coordinated series of activities aimed to restore the skin's mechanical integrity and barrier function [24]. Tissue damage initiates healing, and this has four time-dependent phases: haemostasis, inflammatory, proliferative, and remodelling phases [25] (Fig. 2).

3.1 Haemostasis Phase

Coagulation and haemostasis occur immediately after injury to prevent bleeding [18, 19]. These mechanisms necessitate a series of interconnected procedures to secure blood vessels and establish matrixes for cell influx required in the latter stages of the healing process [20, 21]. Extrinsic and intrinsic routes trigger the coagulation cascade, resulting in platelet aggregation and the formation of clots [19]. During haemostasis, three steps occur in rapid succession. The first reaction is vascular contraction, which happens when blood vessels constrict to reduce the loss of blood. In the second step, platelets attach within a few seconds of a blood vessel's epithelial wall rupturing to establish a temporary plug over the vessel wall breach. Collagen, when exposed at the injury location, enhances platelet adhesion to the wound site. The third and last phase is blood clotting, often known as coagulation. The coagulation cascade, also regarded as secondary haemostasis, is defined as a set of stages

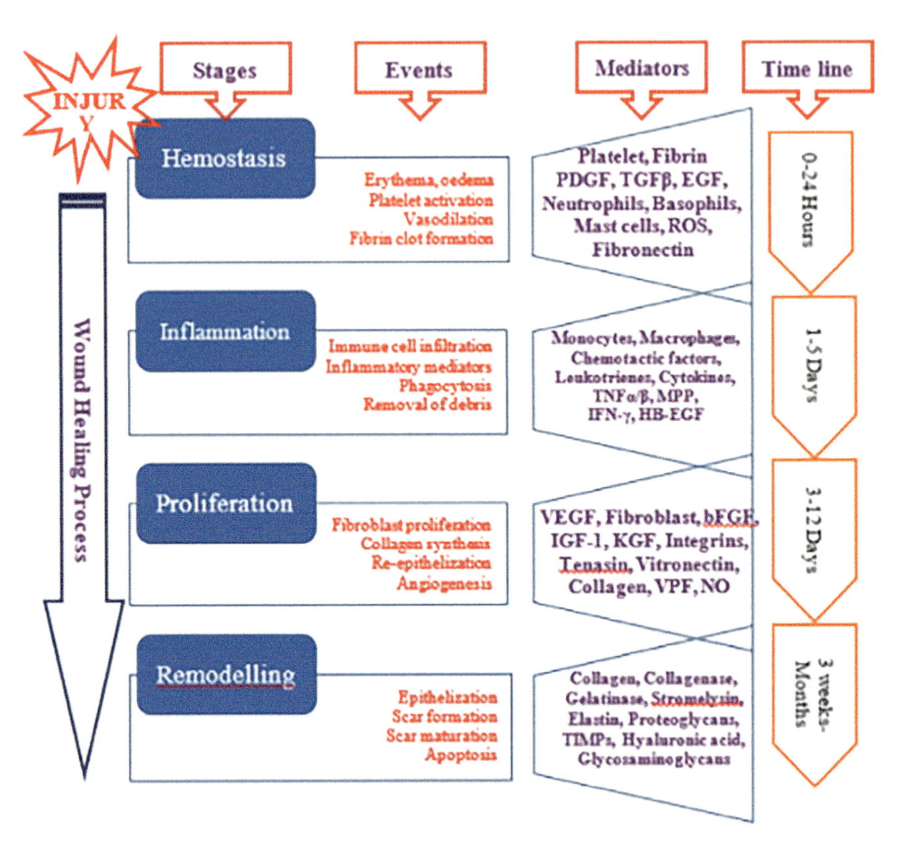

Fig. 2 Schematic representation of wound-healing phases and mediators involved [26]. (Adapted from Khaire et al. [26]. "An Insight into the Potential Mechanism of Bioactive Phytocompounds in the Wound Management" by Manisha Khaire is licensed underCCBY4.0)

that occurs as a result of tissue injury-induced bleeding, with each step activating the next and eventually producing blood clots. Coagulation strengthens the haemostatic plug by forming fibrin threads, which act as 'molecular glue'.

It takes around 60 s for the first set of fibrin strands to entangle with the wound. The fibrin-rich platelet block is completely established after a few minutes [19–22]. Growth factors derived from platelets, also known as platelet-derived growth factors, with cytokines also aid wound healing by activating/triggering and recruiting neutrophils, endothelial cells, fibroblasts, and macrophages [17, 21].

The phytochemicals and metabolites of medicinal herbs have been found to possess different effects on the different stages of wound healing. For example, flavonoids exhibit anti-inflammatory and antioxidant effects, which can promote the early wound-healing stages [23]. Triterpenoids, on the other hand, have been found to have qualities that reduce inflammation and promote healing [25]. Man can heal wounds in situ via continuous tissue regeneration and repair [27]. Curing chronic and acute wounds, along with burn wounds, follow the same basic steps: haemostasis, inflammation, propagation, fibroplasia, deposition of collagen, re-epithelialization, contraction, remodelling, and maturation [28]. The skin's ability to regenerate and restore the barrier function to the external environment after tissue damage is critical for survival. Healing of wounds is crucial following surgical incisions, burns, and chronic ulcers [29].

3.2 Inflammatory Phase

This is divided into two stages: the early inflammatory phase and the late inflammatory phase [30]. The early inflammatory phase/stage initiates molecular mechanisms which result in the infiltration of neutrophils at the site of the wound, whose primary objective is to avert infection [31]. Several chemoattractive drugs begin to attract neutrophils to the site of the wound within 24–36 h of damage. Neutrophils are responsible for the removal of foreign substances, microbes, and injured tissue from wounds by producing oxygen-derived free radicals and proteolytic enzymes [32].

The neutrophils are evacuated from the area once the task is finished. Then macrophages arrive at the injury site and carry on with the phagocytosis before moving on to the next healing phase in the course of the late inflammatory phase. Macrophages are crucial regulatory cells that store a considerable number of powerful growth factors in subsequent stages of the inflammatory response [31]. Lymphocytes are the last cells to arrive at the wound area (in the course of the late inflammatory phase), and they perform a vital role in the control of collagenase they are later on necessary for remodelling of collagen, extracellular matrix component formation, and breakdown [30].

3.3 Proliferative Phase

The proliferative phase includes the key healing mechanisms and begins three days following the injury, and goes on for about a fortnight. After the injury, myofibroblasts and fibroblasts multiply for the first three days in the surrounding tissue [33] prolyl moving into the wound, drawn there by substances like platelet-derived growth factors (PDGFs) and transforming growth (TGFs), which are generated by platelets and inflammatory cells [34]. When fibroblasts enter the wound, they proliferate rapidly and begin to deposit the new extracellular matrix by generating matrix proteins which include type 1 and 3 procollagen, proteoglycans, fibronectin, and hyaluronan. The synthesis of collagen is a critical event. Moreover, collagen is a critical part of all wound-healing stages because it helps to maintain wound integrity [16].

3.4 Remodelling Phase

Scar tissue production occurs through wound remodelling (Fig. 3d, above) and is finished in roughly one year or more [35]. Once the lesion heals, capillary growth ceases, fibroblast and macrophage density is decreased by apoptosis, and blood supply and metabolic process diminish [35]. The result is a developed scar with fewer blood vessels, cells, and a strong tensile strength [30]. Though the majority of injuries are the consequence of basic wounds, local and systemic factors can change and slow down the finely regulated process of repair, resulting in injuries whose healing is not orderly and timely, and grow into non-healing, chronic wounds. Consequently, they are classed as either chronic or acute wounds subject to the tissue's capacity to mend the injury [16].

4 Some Facts About Chronic and Acute Wounds

Though chronic wounds take several months to heal, acute wounds require just a few weeks to recover entirely. A fresh wound that has not yet developed via successive wound-healing stages is classified as an acute wound. These might be superficial wounds that involve both dermis and epidermis or full-thickness damage to the skin that compromises the hypodermal layer. Surgical incisions, abrasions, lacerations, and thermal wounds are all examples of acute wounds. They heal promptly via the regular inflammatory process, new tissue formation, and remodelling. Growth factors, together with cytokines that are released close to the injury site, regulate the healing of the acute wound [36].

On the other hand, chronic wounds do not heal normally and are not treated in a timely and organized manner [37]. The healing process is incomplete and disrupted by different circumstances that cause haemostasis, inflammation, propagation, or remodelling to be delayed by one or more stages [16]. Tissue hypoxia, infection, exudate, necrosis, and high inflammatory cytokine levels are instances of such causes [38]. The injury's continuous inflammatory state leads to a response by the tissue in a cascade format which, when combined, keeps the injury from healing. These wounds usually relapse because the procedure of healing is disorganized and the anatomical and functional outcomes are poor [37]. Different factors, like neuropathic disorders, venous and arterial insufficiency, high blood pressure, vasculitis, and/or burns, cause chronic wounds [39] (Fig. 4).

5 Herbs Utilized in Herbal Medicine for the Treatment of Wounds

Plants have been used for therapeutic purposes by humans throughout the course of time. Herbal medicine, also referred to as phytotherapy, is an ancient practice which treats numerous health ailments with plants and their extracts. Accidents, burns, and surgeries can all result in wounds.

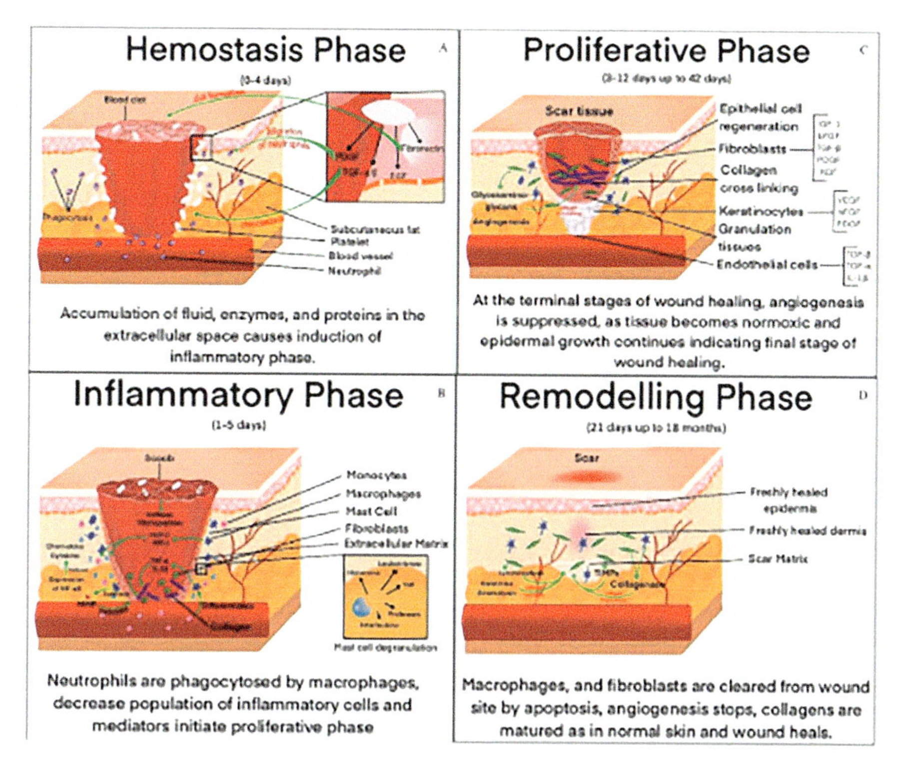

Fig. 3 Healing of wounds: pathophysiology and the mechanisms of different mediators involved. (Adapted from Khaire et al. [26]. "An Insight into the Potential Mechanism of Bioactive Phytocompounds in the Wound Management" by Manisha Khaire is licensed underCCBY4.0.) (**a**) Haemostasis phase, (**b**) inflammatory phase, (**c**) proliferative phase, and (**d**) remodelling phase. bFGF: basic fibroblast growth factor; EGF: epidermal growth factor; IGF-1: insulin-like growth factor-1; IL-1β: interleukin-1 beta; KGF: keratinocyte growth factor; MCF: monocyte chemoattractant protein; MIP: macrophage inflammatory protein; MMP: matrix metalloproteinase; PDGF: platelet-derived growth factor; NF-κB: nuclear factor kappa B; TGF-α: transforming growth factor alpha; TGF-β: transforming growth factor beta; TIMPs: inhibitors of metalloproteinases; VEGF: vascular endothelial growth factor. Haemostasis (**a**): a clot forms, creating a momentary obstacle to loss of fluid and the entrance of pathogens, which act as a storehouse for antimicrobials and bioactive substances, producing a functional extracellular matrix that supports the infiltration and movement of immunological cells, and initiating tissue healing pathway. (**b**) Inflammatory phase: An initial step that entails activating damage-associated molecular patterns, followed by the generation of free radicals, and the creation of reactive biomolecule-recruiting immune cells; then antimicrobial species are released, followed by immune cells which infiltrate and produce amplifying alarmin signals (which are intrinsic, expressed constitutively, chemotactic, and proteins/peptides that are immune activation which are released as a consequence of degranulation, cell injury or even death, or in reaction to immunological stimulation), plus keratinocyte and fibroblast activation. (**c**) Proliferation phase: this includes fibroblast, keratinocyte, and endothelial migration plus multiplication; the resolution of inflammation; extracellular matrix and collagen production, as well as a reduction in vascular permeability; the neovascularization of new lymphatic vessels and capillaries; re-epithelialization; and the new development of granular tissue. (**d**) Extracellular matrix/collagen turnover (production and breakdown); remodelling of the extracellular matrix and readjustment; contraction of the extracellular matrix; apoptosis of the fibroblast and endothelia; repigmentation [16]

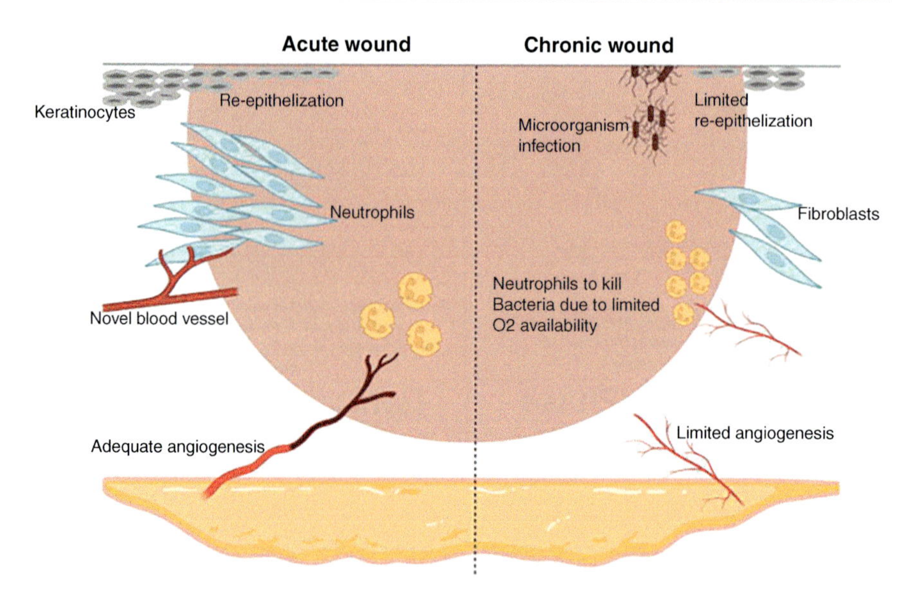

Fig. 4 Processes that are engaged in both acute and chronic wounds. Appropriate angiogenesis accompanied by re-epithelialization promotion, fibroblast propagation, and then neutrophil anti-infection activity within acute wounds (left side). Chronic wounds as seen on the right: featuring persistent local bacterial contagions, and failure of new blood vessel development. The proliferation of fibroblasts is decreased, and neutrophil anti-infection capabilities are limited by inadequate angiogenesis. (Adapted from Vitale et al. [16]. "Phytochemistry and Biological Activity of Medicinal Plants in Wound Healing: An Overview of Current Research" by Vitale et al. is licensed underCCBY4.0)

Although some wounds can be managed through simple first-aid procedures like disinfecting and dressing, others necessitate more specialized medical attention. Herbal medicine is a natural method of wound treatment that can facilitate healing while also lowering the likelihood of infection. Wounds, whether mild or serious, are common medical conditions that can benefit immensely from herbal treatment. Many herbs have been discovered to possess anti-inflammatory, antimicrobial, antioxidant, and wound-healing properties, making them effective in wound treatment. Herbs with antimicrobial properties can help decrease the chances of infection by preventing viruses, bacteria, and fungi from growing and reproducing. This is especially vital for open wounds, which can become infected. Herbs with anti-inflammatory properties can assist in reducing inflammation and swelling, enabling the body's built-in healing mechanisms to take place and so improving wound healing. Herbs with antioxidant properties help protect the body from harmful free radical and oxidative stress. Herbs with wound-healing properties can aid in the growth and renewal of healthy tissue, necessary for the repair of injured skin and tissue. In this chapter, we will look at a few of the herbs that are often used in herbal therapy to treat wounds. It is crucial to emphasize, however, that herbal medication should not be used in place of professional medical treatment.

One area where herbal medicine has been demonstrated to be particularly effective is in wound management and treatment. A diverse range of therapeutic plants and herbs are utilized in standard wound treatment, and scientists are just beginning to comprehend the phytochemical components and metabolites that give these plants their therapeutic effects.

Plants that are traditionally used for wound treatment include *Aloe vera* [7], *Calendula officinalis* [40], *Centella asiatica* [41], and turmeric [42].

6 Phytochemical Composition of Selected Plants Used in the Treatment and Management of Wounds and Infections Resulting from Wounds

Phytochemical analysis, the investigation of the chemical makeup of plants, has helped identify the specific compounds which give these plants their medicinal properties. All plants naturally produce chemical molecules as by-products of their primary metabolites carbohydrates and fats, and their supplementary metabolites, sugars, and sterols [43], which are only present in a few plant genera or species. The pigments, as well as secondary metabolites, could be converted into drugs that have therapeutic effects on people. As a result, this has led to the growth of standardized extracts and isolated compounds, which can be used, that is, wound treatment with increased safety and efficacy.

6.1 *Aloe vera*

A plant from the tropics with a wide geographic range that thrives in hot, arid climates is *Aloe vera*. The gel can be extracted from an *A. vera* leaf's mucilaginous centre. For several thousand years, the plant has been utilized and remains a key component in numerous commercial burn and skin injury care products [44]. Along with polysaccharides, enzymes, amino acids, minerals, and carbohydrates, the gel of *Aloe vera* also possesses vitamins B, A, E, and C and anthraquinone.

According to research, the glycoprotein fraction taken from *A. vera* encourages the growth and migration of cells [45].

6.2 *Calendula officinalis*

Calendula officinalis originates from the Asteraceae family and is a widely cultivated herb used for medicinal purposes in India, the United States, China, and Europe. It is known by a variety of common names, like pot marigold and marigold. *C. officinalis* has historically been applied externally to heal minor burns, other skin issues, and wounds [46, 47]. Herpes, wounds, scars, and products for the skin and hair can all be treated with *C. officinalis* infusions, tinctures, liquid extracts, creams, or ointments [48]. Chemical and pharmacological studies conducted over the past ten years have revealed that *C. officinalis* contains a vast array of additional metabolites with several pharmaceutical activities that attest to its medicinal use [16]. Triterpenoids, both esterified and free, flavonoids, quinones, carotenoids, coumarins, volatile oil, polyunsaturated fatty acids, along with amino acids, are the most active substances [40]. Studies using in vitro fibroblast and keratinocyte cells along with in vivo rodent animal models made up the majority of the studies on the roles of *C. officinalis* in the healing of acute wounds [49–52]. Most clinical research has primarily focused on chronic wounds [40], while only one clinical trial examined acute wounds [53]. In vitro, tests of calendula flower alcohol extracts (CFE) showed the presence of anti-inflammatory and anti-oedematous properties [25]. They also promoted angiogenesis in the chorioallantois membrane model, increased human keratinocyte and fibroblast migration and proliferation, and decreased collagenase activity. Transcription factor NF-kB was activated by CFE, which then affected the inflammatory stage in keratinocytes, whereas when it had a negligible impact on keratinocyte migration, it enhanced interleukin-8 chemokine levels at both the translational and transcriptional levels.

6.3 *Centella asiatica*

Centella asiatica (Gotu Kola, or Asian pennywort) is utilized to accelerate wound healing. *Centella asiatica* aerial section extracts are reported to help with healing chronic ulcers about their size, depth, and location. A punch-type wound has been demonstrated to benefit from the impact of asiaticoside, a compound isolated from *Centella asiatica*. Isolated triterpenes from *Centella asiatica* enhance glycosaminoglycan synthesis and collagen remodelling. Additionally, oral medication administration of the madecassoside in Gotu kola enhances angiogenesis and collagen synthesis at the wound site [54]. It contains several compounds like Asiatic acid, madecassoside, madecassic acid, triterpenoids, flavonoids, and asiaticoside which are responsible for its wound-healing abilities by stimulating collagen formation [41].

6.4 *Hypericum perforatum*

The herb *Hypericum perforatum*, also referred to as St. John's Wort, for millennia, has been utilized in traditional European herbal medicine and is widely distributed in Asia, Europe, and Africa [55]. It has also become naturalized in America. Because of this species' burn/wound healing and anti-inflammatory abilities, its medicinal qualities have been known since Hippocrates' time [56]. A compilation of prescriptions from a Cypriot monastery mentions the herb's use in treating cuts. Tinctures and oils from the herb are used to cure a variety of diseases, including myalgia, minor burns, bruises, abrasions, contusions, and ulcers [57]. Pharmacological research has shown that the presence of a variety of biologically active compounds is what gives *H. perforatum* its medicinal benefits. These compounds include flavonoids (quercetin, hyperoside, rutin, and quercitrin), naphtodianthrones (hypericin and pseudohypericin), xanthones (Norathyriol), prenylated acylphloroglucinols (hyperforin and hyperforin), and sesquiterpenes-rich essential oil [57]. According to research, these substances exhibit anti-inflammatory, antiseptic, antiviral, and wound-healing properties [58].

6.5 Turmeric

Turmeric, or *Curcuma longa*, a common spice, has been discovered to possess the active compound curcumin which possesses wound-healing and anti-inflammatory qualities. Since it is an active component present in the ginger family and in the *Curcuma longa* root, curcumin has been used for both medicinal and culinary purposes for many years. Ayurvedic traditional medicine practitioners utilize curcumin in treating diabetes, liver disorders, airway disorders, skin injuries, and asthma [59, 60]. Curcumin is utilized in conventional Chinese medicine as a common treatment for stomach pain. It is a widely researched nutraceutical, with frequent usage for decades by various ethnic groups. An extremely pleiotropic chemical has now been shown to correspond with important cellular pathways at the DNA replication, translation, and post-translational levels. Pro-inflammatory cytokines, adhesion cell molecules, NF-yB, apoptosis, prostaglandin E2, cyclooxygenase 2, STAT3, 5-LOX, and β-transforming growth factor, phosphorylase kinase, triglycerides, heme oxygenase-1, ET-1, AST, creatinine, and ALT which are observed in the goal pathways. Several of the helpful effects of curcumin are produced by changing the pericellular and extracellular matrix, according to experimental results from numerous in vitro and in vivo studies [21]. Therefore, it may not come as a surprise that curcumin promotes fibroblast growth, granulation tissue development, and collagen build-up during the healing of cutaneous wounds [61].

6.6 Comfrey

Comfrey is a herb that has been used for centuries to heal broken bones and wounds. It contains compounds like Allantoin, which promotes cell growth and repair.

It helps to stimulate the growth of new skin cells, which can hasten the recovery process. Rosmarinic acid has anti-inflammatory properties, Mucilage creates a protective barrier above a wound, helping to keep it moist and promoting healing, Saponins have antimicrobial properties, inulin, a polysaccharide located in the comfrey root, is considered to promote the growth of new cells and helps in healing [62], and *Symphytum officinale* is a pyrrolizidine alkaloid believed to be responsible for comfrey's capacity to heal wounds, but which has been linked to liver damage and has been banned in several countries [63]. In an experiment, comfrey leaf extract was employed to determine whether it could aid in wound healing. The topical compositions of glycerol alcohol solution, carbomer gel, and oil-in-water (O/W) emulsion were investigated. O/W emulsions have proven to be the most effective at accelerating the process of recovery. From the 3rd to the 28th day, inflammatory cell infiltration decreased from 3% to 46% while collagen deposition rose from 40% to 240%. However, the best outcomes were obtained with an 8% extract prepared as an emulsion [64].

6.7 *Echinacea*

Echinacea is a herb known for its immune system-boosting properties [65]. This was shown by its ability to stimulate the monocytes and natural killer cells, which are often deployed in innate immunity against infections [66]. This antimicrobial property makes it suitable for use to treat burns, minor cuts, and skin irritations. The plant *Echinacea* and its component chemical echinacoside look to have beneficial anti-inflammatory and tissue-healing properties [43]. Scientific research has proven the anti-hyaluronidase properties of echinacoside, an *Echinacea* coffee oil conjugate. The authors in [67] studied the significance of *Echinacea* on the hyaluronidase enzyme in pig vocal cords along with the herb's relevance in wound healing. *Echinacea* dosages of 300, 600, and 1200 mg were applied topically to the affected side at random. The findings demonstrated that anti-hyaluronidase therapy was successful in accelerating vocal cord recovery.

6.8 *Arnica montana*

Arnica montana is used topically to treat some ailments such as joint pain, bruises, wound healing, muscle aches, sprains, superficial phlebitis, insect bite swelling, and inflammation from bone breaks. Recent research indicates that it might be beneficial in treating scalds [68, 69]. Increased total phenolic acid levels in *Arnica montana* L. roots were also correlated with mycorrhization [70]. *A. montana* contains 150 therapeutically active substances, which include sesquiterpene lactones such as helenalin and 11 alpha, 13-dihydohelenalin, together with their carbonic acid ester short chains (0.1–0.5% of dry matter in leaves, 0.3–1% in flower shoots), flavonoids (0.6–1.7%) using capillary electrophoresis [71] in the form of flavonoid glucuronides, flavonoid glycosides, and flavonoid aglycones.

Other components of *A. montana* are arnidiol; carotenoids; diterpenes; coumarins (umbelliferone and scopoletin); polyacetylenes; pyrrolizidine alkaloids (tussilagine and isotussilagine) [72]; phenolic acids (Heriguard, limonin, 1.0–2.2%) and caffeic acid [26]; dicaffeoyl quinic acid (DCQA) derivatives (cynarin, ZINC13451209, and 4,5 dicaffeoyl quinic acid); oligosaccharides; and lignans [73]. It consists of sesquiterpene lactone derivatives like tigloyl, methacryloyl, metacryl, isobutyryl, and isovaleryl helenalin [74]. The composition and quantity of phytochemicals such as helenalin esters, phenolics, caffeic acid derivatives, and dihydrohelenalin esters which exist in the flower shoots are different, depending on climate and latitudinal variations, according to a phytobiological study of *A. montana.* Several researchers have discovered that the plant's flowers are mostly rich in active components [73, 74]. Sesquiterpene lactone quantity and nature vary with latitude. Flowers gathered from high-latitude grasslands primarily possess helenalin esters, whereas flowers collected from lower-latitude pastures primarily contain dihydrohelenalin esters. Another study looked at the significance of environmental conditions on the degree of xanthotoxin lactones in 10 German grasslands. Xanthotoxin lactone concentration was higher (0.59–1.10%) [75].

6.9 Chickweed

Chickweed or *Stellaria media* Linn. is a herb that has already been traditionally utilized in treating skin irritations and wounds. It contains compounds that can decrease inflammation and promote cell growth [76]. Chickweed has a high concentration of lipids [77], triterpenoid [78], C-glycosyl flavonoids [79], phenolic compounds and flavonoids [76], sitosterol, phlobatannins, pentasaccharide and alkaloids [80], and saponins [81]. Furthermore, (S)-2-aminohexanedioic acid, 6,7 dimethylheptacosane, oxalic acid, 5-acetoxydotetracont-3-en-1-ol, saponarin, and saccharopine are bioactive metabolites identified from chickweed aerial parts [82]. Another phytochemical investigation discovered active biological metabolites in several regions of chickweed and its preparations [76]. Flavonoids are mostly saponarin, apigenin 6-C-glucoside (isovitexin), apigenin glucoside arabinoside, isovitexin 7,2-di-O-glucoside, isovitexin 7-O-galactosyl-2-O-glucoside, apigenin 6-C-α- apigenin 6-C-β-D- galactosyl-8-C-β-L-arabinoside, quercetin 3-O-rutinoside (rutin), apigenin 6,8-di-C-α- L-arabinoside, L-arabinosyl-8-C-β-D-galactoside, apigenin 6-C-β-D-glycosyl-8-C-β-D- galactoside, coumarins, likewise ascorbic acid [83]. Luteolin, kaempferol, quercetin, rutin, sinapic acid, protocatechuic acid, vanillic acid, catechin hydrate, and resorcylic acid are phenolic chemicals found in chickweed leaf extracts [84]. Other chemicals extracted through the aerial section of the chickweed include 6-methylheptyl-3′-hydroxy-2′-methylpropanoate, 2, 2, 4-trimethyloctan-3-one and 2, 4, 5, 7-tetramethyloctane [76]. These vital metabolites exhibit anti-inflammatory properties and aid in weight loss [85].

Hydroxycinnamic acids, catechins [86], saponosides, ascorbic acid, mucilage, polyphenols, silicon, and carotene are important antioxidant active components [87]. Chickweed possesses a high concentration of phenolic acid, flavonoids, phlobatannins, alkaloids, tocopherol, and Vitamin C, all of which play important antioxidant roles in the human body [76].

6.10 Lavender

Lavender is a herb that has been traditionally utilized in treating minor cuts, scalds, and other derma irritations. It has antiseptic and anti-inflammatory properties, which can assist in lowering the risk of infection and promote healing [88]. Before flowering, lavender flowers (*Lavandulae flos*) are gathered to make the medicinal raw material. Lavender's principal biologically active agents are phenolic compounds, triterpenes, essential oil components, and sterols [89]. Lavender is mostly recognized for its essential oil, which is available in concentrations ranging from 2% to 3% [43]. By steam distillation or hydrodistillation, it is extracted from the petals. Linalyl acetate and linalool are the essential oil's two main components, with concentrations ranging from 1.2% to 59.4% and 9.3% to 68.8% respectively [43]. Both the high concentration of linalyl acetate and linalool and their relative proportions affect the quality of lavender essential oil [90]. Terpenes such as limonene, borneol, eucalyptol, camphene, camphor, −ocimene, 1,8-cineol, fenchone, acetate, α-terpineol, lavandulol, β-caryophyllene, α-pinen and geraniol, and also non-isoprenoid cyclic components such as octenol, octanon, octanol, and octenyl acetate are the major chemicals [89]. Polyphenols are another significant class of chemical elements found in lavender flowers. These are plant secondary metabolites that possess diverse biological characteristics. Flowers, leaves, stalks, fruits, and seeds are just a few of the plant's components where they are found [91]. Almost 8000 polyphenolic chemicals have so far been discovered [92]. They are identified through single or multiple cyclic rings and various numbers of hydroxy groups in a molecule about chemical structure [93]. The shikimate pathway is used in the biosynthesis of polyphenols [94]. They could be separated into numerous different classes, including lignans, coumarins, stilbenes, flavonoids, and phenolic acids [94]. The majority of phenol compounds are discovered in mixtures with organic acids, sugars, and esters [93]. The hydroxybenzoic acid derivatives and the 3-coumaric acid derivatives are the two categories of phenolic acids [95]. The most prevalent hydroxycinnamic acid in the *Lavandula* genus is rosmarinic acid [96]. The principal participants in this group are 3-hydroxybenzoic acid, benzoic acid, syringic acid, 4-hydroxybenzoic acid, vanillic acid, gallic acid, protocatechuic acid, homovanillic acid, and homoprotocatechuic acid [97]. The research of [98] in a rat model showed that topical lavender oil treatment dramatically improved the synthesis of collagen by fibroblasts, which was associated with elevated TGF- β expression in wound abrasions and increases the production of granulation tissue by synthesizing collagen, tissue remodelling by switching from type III to type I collagen, and wound shrinking.

6.11 *Panax ginseng*

This is a popular medicinal plant in Japan, China, Eastern Siberia, and Korea [21]. Recollection is also thought to boost immunity, improve physical agility, and lessen weariness [21]. Hence, the herb is utilized in treating anxiety, chronic fatigue syndrome (CFS), and depression. It has been demonstrated that *Panax ginseng* has vasodilatory, blood lipid-controlling, anticancer, anti-inflammatory, antibacterial, anti-allergic, anti-ageing, and immunomodulatory properties [99]. Several bioactive compounds make up *Panax ginseng*, but a type of saponin is its most effective active component (also known as ginsenosides by Asian scientists and panaxosides by Russian researchers) [21]. Preparations of the root of *Panax ginseng* are attested to shield the skin from severe UVB rays and significantly accelerate the healing process after laser burns and excision wound damage [21]. According to studies, *Panax ginseng* extracts boost human dermal fibroblasts' ability to proliferate, promote keratinocyte migration, and produce collagen ex vivo. However, ginsenoside Rb2, which is separated from *Panax ginseng*, has been revealed to stimulate the growth of the epidermis in raft kelp culture by boosting fibronectin and the receptor, epidermal growth factor, and receptor expression, as well as collagenase I and keratin, which are all essential/critical for wound healing [100]. According to Park and Park [101], *Panax ginseng* may promote the repair of full-thickness skin lesions and also has a great effect on skin regeneration, suggesting its usage as a wound dressing and development as a food product.

6.12 Neem

Neem (*Azadirachta indica*) is renowned for its antifungal, anti-ulcer, anticancer, antiviral, antibacterial, and antioxidant properties in wound dressing. The anti-oxidation of nitric oxide in RAW 264.7 cell lines was evaluated by [102]. Inside wells coated with integrated collagen and 1000 g/mL of *Azadirachta indica* extract, the nitric oxide level was determined to be 10 g/mL. The biocomposite film possesses effective nitric oxide antioxidant and anti-inflammatory properties. In additional studies, the scientists used the *Azadirachta indica* -incorporated collagen film cell lines to perform the antioxidant activities and the biocompatibility test. The combined collagen films of its extract (400 g/mL) demonstrated an 80% improvement in DPPH scavenging ability and had at least 80% cell viability using the MTT assay. Research shows that the herb contains azadirone, azadirachtin, nimbin, nimbidin, and nimbinin [103].

6.13 German Chamomile (*Chamomilla recutita*)

For German chamomile, scientists investigated the effectiveness of PCL/PS electrospinning nanofiber membranes as active wound dressings that contain chamomile. Quercetin, apigenin, luteolin, patuletin, and their glucosides are some of the specific phenolics and flavonoids that provide the Asteraceae plant, C. *recutita* (L.) Rauschert, its therapeutic properties [21]. The most uncommon flavonoid discovered in chamomile plants, apigenin, has a notable impact on how quickly wounds heal [104].

Nanofibers were demonstrated to be efficient against microorganisms, *C. albicans* (fungi) and *S. aureus*, showing inhibitory zones measuring around 7.6 mm in width, according to studies of antifungal and antibacterial in vitro [104]. The mesenchymal stem cells' survival and ex vivo cell adherence to the nanofibers were demonstrated by the MTT assay. The authors' research revealed that the nanofibers, mixed with about 15% of *Chamomilla recutita* extract, could cure a large proportion of the injury 14 days following therapy, which was verified using a rat injury model. This injury evaluation revealed that collagen accumulated in the tissue's dermis together with re-epithelization as well as the absence of necrosis [104] (Table 1 and Fig. 5).

7 Conclusion

This review is a significant addition to natural product research and wound therapy. As such, it has the potential to influence both research and development and clinical treatment in this field. Going forward, the outcomes of this review could be utilized to direct future studies into the individual phytochemicals and metabolites that are most beneficial in wound treatment and management. This might facilitate the invention of novel chemicals with wound-healing characteristics, which could then be turned into new patient medicines. Furthermore, the study could have practical implications for the use of herbal medicine in wound treatment. Clinicians and patients could benefit from a more evidence-based approach to selecting herbal treatments for the care of wounds if specific plants and metabolites that are beneficial in treating wounds are defined. Overall, the literature evaluation has the potential to assist and guide future research and clinical practice in natural products research and wound healing.

The evaluation of phytochemical components and metabolites in medicinal plants and herbal medicine used in wound therapy and management is an important area of research. Traditional medicine usage for wound healing has been applied for centuries and continues to be significant in modern times. Many medicinal plants contain phytochemicals such as phenolic compounds, flavonoids, terpenoids, and alkaloids, which have been demonstrated to have wound-healing properties. Furthermore, by the growing need for natural and safe alternatives to traditional therapy, using herbal medicine to treat wounds has attracted increased attention in recent times. The assessment of phytochemical elements and metabolites in medicinal plants used for wound remedy has shown encouraging results, and the discovery of active chemicals in these plants may lead to the development of novel and effective wound-healing medications.

Finally, the evaluation of phytochemical components and metabolites in herbal medicine and medicinal plants used in the treatment and management of wounds has the potential to provide a better understanding of these plants' modes of action, in addition to helping to create novel and effective wound treatments.

Table 1 Plants and their constituents used in wound repair

Scientific name	Common name	Plant part and formulation	Phytochemical compositions	Biological function	Mechanism of action	References
Aloe vera* (L.) *Burm. f. anagallis	Aloe	Corpulent leaves, gel	Flavonoids- (aleosin, aloin, rhein, emodin) Polysaccharides- (glucomannan, acemannan, and acetylated polymannan)	Antibacterial, anti-inflammatory, epithelialization process	Regulates inflammatory response; regulates the autophosphorylation of signalling proteins; enhances collagen deposition and angiogenesis; boosts the propagation of fibroblasts and mildly stimulates the migration of keratinocytes.	[105–107]
Calendula officinalis	Pot marigold	Flowers	Triterpenoids, Flavonoids- (5-O-caffeoylquinic acid, caffeic acid, isorhamnetin-isorhamnetin-3-O-rutinoside, rosmarinic acid, medicocarpin, nicotiflorin, venoruton, and isoquercetin) Quinones – coumarins	Re-epithelialization process, anti-inflammatory	Increases the appearance of inflammatory mediators; enhances keratinocyte and fibroblast proliferation; boosts collagen biosynthesis and angiogenesis; suppresses lipoxygenase activity; decreases glutathione levels.	[26, 52, 108, 109]
Centella asiatica	Gotu kola	Leaf	Triterpenoids/ triterpenes Flavonoids (Asiatic acid, madecassoside, madecassic acid, and asiaticoside	Antioxidant, antimicrobial, anti-inflammatory, neuroprotective, and wound healing	Collagen depositing and re-epithelialization in a punch-type wound Collagen repair and biosynthesis of glycosaminoglycans Synthesis of collagen and angiogenesis at the wound site	[110, 111]
Hypericum perforatum	St. John's wort	Flower	Hyperforin, hyperoside, hypericin, quercetin,	Antioxidant, antiatherogenic activity, antidepressant, oestrogen-	Inhibition of lipoxygenase activity and prostaglandin E reduction; shorten the interval of	[112–115]

(continued)

Table 1 (continued)

Scientific name	Common name	Plant part and formulation	Phytochemical compositions	Biological function	Mechanism of action	References
			flavonoids, tannins, and saponins	mimetic activity anti-inflammatory, analgesic, gastroprotective	inflammation, stop fibroblast migration during the process of wound healing	
Curcuma longa	Turmeric	Rhizomes	Diarylheptanoids/ curcuminoids Curcumin, demethoxycurcumin, bisdemethoxycurcumin, 5′-methoxycurcumin, dihydrocurcumin, Sesquiterpenes turmerone, zingiberene, arturmerone, fatty acids, sesquiterpenoids, monoterpenoids, proteins	Antioxidant, gastroprotective, anti-inflammatory, antiseptic, antibacterial, antimicrobial, antimutagenic, anticancer, antifungal, hepatoprotective, cardioprotective, hypoglycaemic, antiarthritic, antidepressant	Prevent local activation and appearance of NF-κB genes which mediate joint inflammation and destruction, reduce histamine production and prolong the impact of cortisol against inflammation, reduce in 9,2-benzanthrazen-induced DNA adducts and consequent reduction of carcinogenesis; granulation and tissue formation, increase fibroblast migration, increase TGF-β production; collagen deposition; increase fibroblast proliferation; inhibition of membrane phospholipid peroxidation and surge in liver lipid metabolism; inhibition of the increase in thiobarbituric acid-reactive substance and protein carbonyl and reversing changed activities of antioxidant enzymes	[25, 116–121]
***Symphytum officinale* L.**	Comfrey	Leaves	Tannins, terpenoids, triterpenoids, amino acids,	Anti-inflammatory, analgesic, antioxidant,	Decreasing inflammation, apoptosis, and oxidant stress,	[122–126]

			flavonoids, reducing agents, flavones, saponins, alkaloids, allantoin, polysaccharides, phenolic acids (caffeic, rosmarinic, and caffeoylquinic acids), glycosides, pyrrolizidine phytosterols, and triterpene saponins	antitumour, antiapoptotic, neuroprotective, antibacterial, hepatoprotective, antihistamine	through the NF-kB, Nrf2, and caspase-3 pathways; correcting cholinergic disorders, correcting the antioxidant-oxidative imbalance, delaying the occurrence of the pro-inflammatory event in primary human epithelial tissue	
Echinacea	Purple coneflower	Roots, flowers, leaves	Phenolics – echinacoside, caftaric acid, chlorogenic acid and cichoric acids (CADs), cynarin Alkylamides – polysaccharides/ glycoproteins Arabinogalactans	Antioxidant, antibacterial, anxiolytic, anti-inflammatory, anti-hyaluronidase, analgesic, tissue healing, antimicrobial, antifungal, cyclooxygenase, and lipoxygenase enzyme activities, immunostimulant, vitamin C	Stimulate phagocytosis, T-cell production, lymphocytic activity, cellular respiration (anti-oxidation) action against tumour cells, inhibit hyaluronidase enzyme secretion	[127–131]
Arnica montana	Mountain tobacco	Flowers	Sesquiterpene Flavonoids – flavonoid glucuronides, flavonoid glycosides, and flavonoid aglycones Lactones – Helenalin,11a,13-dihydohelenalin Essential oils – monoterpenes, fatty acids, thymol derivatives, and sesquiterpene; carotenoids;	Antiphlogistic, inotropic, anti-inflammatory, antibiotic, uterotonic, immunomodulatory, antiplatelet, analgesic and anti-rheumatic, antibacterial, antifungal, antioxidant	Increase blood circulation to the affected area, which can improve the delivery of nutrients and oxygen to the injury site, aiding in the healing process; reduce inflammation, stimulate the immune system, promote tissue regeneration, and scale down development of scar tissue	[71, 73, 74, 132]

(continued)

Table 1 (continued)

Scientific name	Common name	Plant part and formulation	Phytochemical compositions	Biological function	Mechanism of action	References
			a triterpene – arnidiol; diterpenes; necine bases (isotussilagine and tussilagine); coumarins (scopoletin and umbelliferone); polyacetylenes; phenolic acids (caffeic acid, chlorogenic acid, and cynarin, 1.0–2.2%); dicaffeoyl quinic derivatives (cynarin, ZINC13451209, and 4,5 dicaffeoyl quinic acid); lignans; and oligosaccharides			
***Stellaria media* (Linn.)**	Chickweed	Whole plant, leaves, flowers, roots, stems	Flavonoids, saponins, isoflavonoids, tannins, alkaloids, triterpenoids, phenolic acids, anthraquinone, and phenolic compounds	Anti-anxiety, antiobesity, antifungal, antidiabetic, antibacterial, anti-inflammatory, anti-leishmanial, and antitoxicity	Promote the propagation and movement of fibroblasts; produce collagen and other extracellular matrix components; stimulate the immune system, support angiogenesis, and accelerate the removing the wound's debris and dead tissue	[76–78, 80, 81]
Lavandula angustifolia	Lavender	Flowers	Flavonoids, coumarins, terpinen-4-ol, camphor,	Anti-inflammatory, analgesic, anxiolytic,	Inhibit pro-inflammatory cytokines and enzyme production, e.g.,	[90, 133, 134]

			cineole, linalyl acetate, linalool	antioxidant, antimicrobial, sedative	cyclooxygenase-2 (COX-2), TNF-α, (IL-6), and prostaglandin E2 (PGE2); interact with pain receptors in the body, such as the transient receptor potential cation channel subfamily A member 1 (TRPA1) and subfamily V member 1 (TRPV1); regulate the activity of gamma-aminobutyric acid (GABA) neurotransmitter in the brain; disrupt the cell membrane of microbes; scavenge free radicals and reduce oxidative stress; activate GABA receptors in the brain and reduce the activity of the system for the histamine neurochemical	
Panax ginseng	Korean ginseng	Roots	Ginsenosides, phenolic compounds (flavonoids and phenolic acids); polysaccharides; essential oils; polyacetylenes; peptides	Anti-inflammatory, antioxidant, adaptogenic, immunomodulatory, antidiabetic, antitumour, cognitive enhancing	Inhibit the manufacture of enzymes and proinflammatory cytokines, such as interleukin-6 (IL-6), TNF-α, COX-2, and PGE2; regulate the sympathetic nervous system and hypothalamic–pituitary–adrenal (HPA) axis; scavenge free radicals and reduce oxidative stress; stimulate the production of natural killer (NK) cells and T cells; induce apoptosis, deposit collagen, enhanced proliferation and migration of epidermal cells	[101, 135–138]

(continued)

Table 1 (continued)

Scientific name	Common name	Plant part and formulation	Phytochemical compositions	Biological function	Mechanism of action	References
					as well as keratinocytes; enhance angiogenesis, increase proliferation and trafficking of fibroblasts and boost the protein expression of collagen	
Azadirachta indica	Neem	Leaves, bark, seeds	Azadirachtin, nimbin, nimbinin, quercetin, salannin, beta-sitosterol, kaempferol, azadirone, meliacin, nimbidin, tiglic acids, fatty acids	Anti-pyretic, anti-inflammatory; antiviral; antioxidant; immunomodulatory	Inhibit the manufacture of pro-inflammatory cytokines; activated protein kinase (AMPK) activation, an enzyme required in regulating the metabolism of glucose; inhibit the replication of viral particles by blocking the viral protease activity; scavenges free radicals and reduces oxidative stress; inhibit the assimilation of cholesterol in the intestines	[102, 103, 139]
Matricaria chamomilla	German chamomile	Flowers	Chamazulene, apigenin, bisabolol; matricin; coumarins (herniarin, umbelliferone); quercetin; luteolin; patuletin	Anti-inflammatory; antispasmodic; antioxidant; sedative	Inhibit the making of pro-inflammatory cytokines and enzymes such as COX-2; free radicals scavenging and reduction of oxidant stress; bind to benzodiazepine receptors in the brain; relax smooth muscles; promote an immune response	[104, 111]

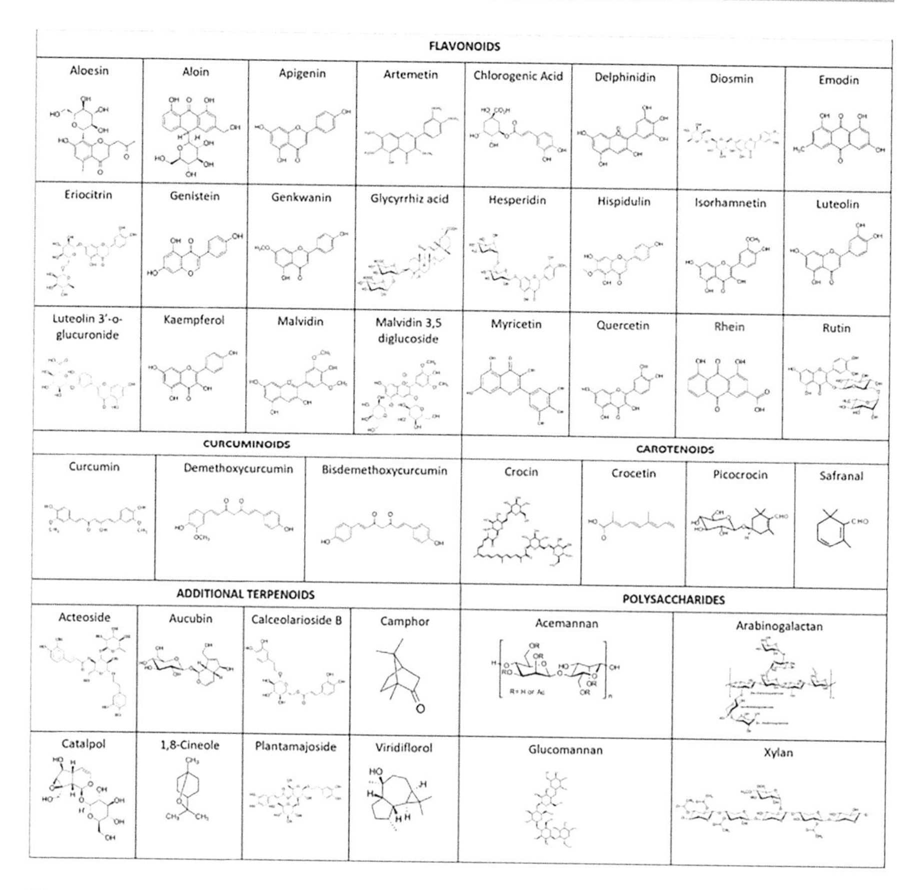

Fig. 5 The structures of selected primary phytochemical metabolites involved in wound healing. (Adapted from Vitale et al. [16]. "**Phytochemistry and Biological Activity of Medicinal Plants in Wound Healing: An Overview of Current Research**" by Vitale et al. is licensed under CC BY 4.0)

References

1. Ernst E (2007) Herbal medicines: balancing benefits and risks. Novartis Found Symp 282: 154–172. https://doi.org/10.1002/9780470319444.CH11
2. Balap A, Gaikwad A (2021) Challenges, advances and opportunities of herbal medicines in wound healing: a review. Int J Pharm Sci Rev Res 71:125–136. https://doi.org/10.47583/ijpsrr.2021.v71i01.015
3. Kurian A (2012) Health benefits of herbs and spices. Handb Herbs Spices, 2nd ed 2:72–88
4. Alasbahi RH, Groot MJ (2020) Evaluation of the wound healing activity of twelve herbal and non-herbal remedies used in Sana'a-Yemen for the treatment of wounds and burns. J Med Herbs Ethnomed:90–116. https://doi.org/10.25081/jmhe.2020.v6.6379
5. Cauilan BY, Lungan MHT (2021) Wound healing activity of ointment infused with aqueous extracts of plants used by the agtas of Peñablanca and Malauegs of Rizal in the province of

wound healing activity of ointment infused with aqueous extracts of plants used by the Agtas of Peñablanca. Int J Biol Sci. https://doi.org/10.12692/ijb/19.4.168-174
6. World Health Organization Injuries and violence. https://www.who.int/news-room/fact-sheets/detail/injuries-and-violence. Accessed 20 Jan 2023
7. Namunana S, Lutoti S, Nyamaizi G, Agaba G, Apun I, Ssebunnya C, Tenywa GM, Wangalwa R, Kaggwa B, Fadhiru Kamba P, Musoke-Muweke D, Ogwang Engeu P (2018) Formulation, development and validation of a wound healing herbal ointment from extracts of Bidens pilosa and Aloe barbadensis. J Pharm Pharmacol Res 02. https://doi.org/10.26502/jppr.0008
8. Mousavi SM, Nejad ZM, Hashemi SA, Salari M, Gholami A, Ramakrishna S, Chiang WH, Lai CW (2021) Bioactive agent-loaded electrospun nanofiber membranes for accelerating healing process: a review membranes (Basel) 11: https://doi.org/10.3390/MEMBRANES11090702
9. Malabadi RB, Kolkar KP, Acharya M, Br N, Chalannavar RK (2022) Wound healing: role of traditional herbal medicine treatment. Int J Innov Scient Res Rev 4(12):3686–3691. https://doi.org/10.1128/JMP.40.8.2919-2921.2002
10. Murray RZ, West ZE, Cowin AJ, Farrugia BL (2019) Development and use of biomaterials as wound healing therapies. Burn trauma 7. https://doi.org/10.1186/S41038-018-0139-7
11. Napavichayanun S, Pienpinijtham P, Reddy N, Aramwit P (2021) Superior technique for the production of agarose dressing containing sericin and its wound healing property. Polymers (Basel) 13. https://doi.org/10.3390/POLYM13193370
12. Patil TV, Patel DK, Dutta SD, Ganguly K, Randhawa A, Lim KT (2021) Carbon nanotubes-based hydrogels for bacterial eradiation and wound-healing. Applications Appl Sci 11. https://doi.org/10.3390/APP11209550
13. Muhammad HS, Muhammad S (2005) The use of Lawsonia inermis linn. (henna) in the management of burn wound infections. Afr J Biotechnol 4:934–937
14. Lazarides MK, Giannoukas AD (2007) The role of hemodynamic measurements in the management of venous and ischemic ulcers. Int J Low Extrem Wounds 6:254–261. https://doi.org/10.1177/1534734607306878
15. Dwivedi G, Dwivedi S (2007) Sushruta – the Clinician – Teacher par Excellence. Hist Med 49: 243–244
16. Vitale S, Colanero S, Placidi M, Di Emidio G, Tatone C, Amicarelli F, D'Alessandro AM (2022) Phytochemistry and biological activity of medicinal plants in wound healing: an overview of current research. Molecules 27:27. https://doi.org/10.3390/molecules27113566
17. Ather S, Harding KG (2019) Wound management and dressings. Adv Text Wound Care:3–19. https://doi.org/10.1533/9781845696306.1.3
18. Velnar T, Bailey T, Smrkolj V (2009) The wound healing process: an overview of the cellular and molecular mechanisms. J Int Med Res 37:1528–1542. https://doi.org/10.1177/147323000903700531
19. Wilkinson HN, Hardman MJ (2020) Wound healing: cellular mechanisms and pathological outcomes: cellular mechanisms of wound repair. Open Biol 10. https://doi.org/10.1098/RSOB.200223
20. Strodtbeck F (2001) Physiology of wound healing. Newborn Infant Nurs Rev 1:43–52. https://doi.org/10.1053/NBIN.2001.23176
21. Sharma A, Khanna S, Kaur G, Singh I (2021) Medicinal plants and their components for wound healing applications. Futur J Pharm Sci 7. https://doi.org/10.1186/s43094-021-00202-w
22. Nagori BP, Solanki R (2011) Undefined role of medicinal plants in wound healing. Res J Med Plant 5(4):392–405. https://scialert.net/abstract/?doi=rjmp.2011.392.405
23. Gonzalez ACDO, Andrade ZDA, Costa TF, Medrado ARAP (2016) Wound healing – a literature review. An Bras Dermatol 91:614–620
24. Janis JE, Harrison B (2016) Wound healing: Part I. Basic science. Plast Reconstr Surg 138:9S–17S. https://doi.org/10.1097/PRS.0000000000002773

25. Mahmoud MF, Monti DM, Vitale S, Colanero S, Placidi M, Di Emidio G, Tatone C, Amicarelli F, Maria D'alessandro A (2022) Phytochemistry and biological activity of medicinal plants in wound healing: an overview of current research. Molecules 27:27. https://doi.org/10.3390/molecules27113566
26. Khaire M, Bigoniya J, Bigoniya P (2023) An insight into the potential mechanism of bioactive phytocompounds in the wound management. Pharmacogn Rev 17:43–68. https://doi.org/10.5530/097627870153
27. Owen JA, Punt J, Stranford SAJP (2013) Kuby immunology 7th Edition 2013 by Judy Owen, 7th edn. W.H. Freeman and Company, New York
28. Wallace HA, Zito PM (2022) Wound healing phases. StatPearls Publishing
29. Tsioutsiou EE, Miraldi E, Governa P, Biagi M, Giordani P, Cornara L (2017) Skin Wound Healing: From Mediterranean Ethnobotany to Evidence based Phytotherapy. Athens J Sci 4: 199–212. https://doi.org/10.30958/ajs.4-3-2
30. Hart J (2013) Inflammation 1: its role in the healing of acute wounds. 11:205–209. https://doi.org/10.12968/JOWC.2002.11.6.26411
31. Hess C (2000) Skin care basics. Adv Skin Wound Care 13:127–128
32. Flanagan M (2013) The physiology of wound healing. J Wound Care 9:299–300. https://doi.org/10.12968/JOWC.2000.9.6.25994
33. Witte MB, Barbul A (1997) General principles of wound healing. Surg Clin North Am 77: 509–528. https://doi.org/10.1016/S0039-6109(05)70566-1
34. Goldman R (2004) Growth factors and chronic wound healing: past, present, and future. Adv Skin Wound Care 17:24–35
35. Broughton G, Janis JE, Attinger CE (2006) Wound healing: an overview. Plast Reconstr Surg 117:1e-S-32e-S. https://doi.org/10.1097/01.PRS.0000222562.60260.F9
36. Li J, Chen J, Kirsner R (2007) Pathophysiology of acute wound healing. Clin Dermatol 25:9–18. https://doi.org/10.1016/J.CLINDERMATOL.2006.09.007
37. Robson MC, Steed DL, Franz MG (2001) Wound healing: biologic features and approaches to maximize healing trajectories. Curr Probl Surg 38:A1–140. https://doi.org/10.1067/MSG.2001.111167
38. Vanwijck R (2001) Biologie chirurgicale de la cicatrisation. Bull Mem Acad R Med Belg 156:185
39. Komarcević A (2000) The modern approach to wound treatment. Med Pregl 53:363–368
40. Givol O, Kornhaber R, Visentin D, Cleary M, Haik J, Harats M (2019) A systematic review of *Calendula officinalis* extract for wound healing. Wound Repair Regen 27:548–561
41. Arribas-López E, Zand N, Ojo O, Snowden MJ, Kochhar T (2022) A systematic review of the effect of *Centella asiatica* on wound healing. Int J Environ Res Public Health 19:3266
42. Barchitta M, Maugeri A, Favara G, San Lio RM, Evola G, Agodi A, Basile G (2019) Nutrition and wound healing: An overview focusing on the beneficial effects of curcumin. Int J Mol Sci 20:1119
43. Shankar S, Kumar S, Saha S, Singh K, Patel A, Pathak D (2022) Review of phytomedicine and its potential contribution to wound healing. Int J Scient Develop Res 7(12):109–120.
44. Torfs H, Poels J, Detheux M, Dupriez V, Van Loy T, Vercammen L, Vassart G, Parmentier M, Vanden Broeck J (2002) Recombinant aequorin as a reporter for receptor-mediated changes of intracellular Ca_2+−levels in Drosophila S2 cells. Invertebr Neurosci 4:119–124. https://doi.org/10.1007/s10158-001-0013-2
45. Sánchez M, González-Burgos E, Iglesias I, Gómez-Serranillos MP (2020) Pharmacological update properties of aloe vera and its major active constituents. Molecules 13;25(6):1324. https://doi.org/10.3390/molecules25061324. PMID: 32183224; PMCID: PMC7144722
46. Basch E, Bent S, Foppa I, Haskmi S, Kroll D, Mele M, Szapary P, Ulbricht C, Vora M, Yong S (2009) Marigold (*Calendula officinalis L.*). 6:135–159. https://doi.org/10.1080/J157V06N03_08
47. Jan N, Iqbal Andrabi K, John R (2017) *Calendula officinalis* – an important medicinal plant with potential biological properties. Proc Indian Natn Sci Acad 83:769–787. https://doi.org/10.16943/ptinsa/2017/49126

48. Muley BP, Khadabadi SS, Banarase NB (2009) Phytochemical constituents and pharmacological activities of *Calendula officinalis Linn* (Asteraceae): a review. Trop J Pharm Res 8: 455–465. https://doi.org/10.4314/tjpr.v8i5.48090
49. Maria Leal Parente L, de Souza Lino RJ, Manrique Faustino Tresvenzol L, Clare Vinaud M, Realino de Paula J, Margarida Paulo N (2012) Wound healing and anti-inflammatory effect in animal models of *Calendula officinalis L.* growing in Brazil. 2012:375671. https://doi.org/10.1155/2012/375671
50. Shafeie N, Naini AT, Jahromi HK (2015) Comparison of different concentrations of calendula officinalis gel on cutaneous wound healing. Biomed Pharmacol J 8:979–992. https://doi.org/10.13005/BPJ/850
51. Fronza M, Heinzmann B, Hamburger M, Laufer S, Merfort I (2009) Determination of the wound healing effect of Calendula extracts using the scratch assay with 3T3 fibroblasts. J Ethnopharmacol 126:463–467. https://doi.org/10.1016/J.JEP.2009.09.014
52. Nicolaus C, Junghanns S, Hartmann A, Murillo R, Ganzera M, Merfort I (2017) In vitro studies to evaluate the wound healing properties of Calendula officinalis extracts. J Ethnopharmacol 196:94–103. https://doi.org/10.1016/J.JEP.2016.12.006
53. Eghdampour F, Jahdie F, Kheyrkhah M, Taghizadeh M, Naghizadeh S, Hagani H (2013) The impact of aloe vera and calendula on perineal healing after episiotomy in primiparous women: a randomized clinical trial. J Caring Sci 2:279–286. https://doi.org/10.5681/jcs.2013.033
54. Babu M, Prasad O, Murthy T (2011) Comparison of the dermal wound healing of *Centella asiatica* extract impregnated collagen and crosslinked collagen scaffolds. J Chem Pharm Res 3:353–362
55. Matić IZ, Ergün S, Đorđić Crnogorac M, Misir S, Aliyazicioğlu Y, Damjanović A, Džudžević-Čančar H, Stanojković T, Konanç K, Petrović N (2021) Cytotoxic activities of *Hypericum perforatum* L. extracts against 2D and 3D cancer cell models. Cytotechnology 73:373–389. https://doi.org/10.1007/S10616-021-00464-5/METRICS
56. Mikail HG (2016) Pleiotropic activity of *Hypericum perforatum* L. ~ 256 ~. J Med Plants Stud 4:256–258. https://doi.org/10.1007/S10298-010-therapie2010
57. Jarić S, Kostić O, Mataruga Z, Pavlović D, Pavlović M, Mitrović M, Pavlović P (2018) Traditional wound-healing plants used in the Balkan region (Southeast Europe). J Ethnopharmacol 211:311–328. https://doi.org/10.1016/j.jep.2017.09.018
58. Nayak SB, Isik K, Marshall JR (2017) Wound-healing potential of oil of hypercium perforatum in excision wounds of male sprague dawley rats. Adv Wound Care 6:401–406. https://doi.org/10.1089/wound.2017.0746
59. Gopinath D, Ahmed MR, Gomathi K, Chitra K, Sehgal PK, Jayakumar R (2004) Dermal wound healing processes with curcumin incorporated collagen films. Biomaterials 25:1911–1917. https://doi.org/10.1016/S0142-9612(03)00625-2
60. Saraswathy N, Rohit R, Shanmugam K, Sozheeswari C, Ramalingam P (2012) A preliminary investigation of turmeric-agar composite film as bioactive wound dressing material on excision wound in rat model. Indian J Nat Prod Resour 3:237–241
61. Sharifi-Rad J, El Rayess Y, Rizk AA, Sadaka C, Zgheib R, Zam W, Sestito S, Rapposelli S, Neffe-Skocińska K, Zielińska D, Salehi B, Setzer WN, Dosoky NS, Taheri Y, El Beyrouthy M, Martorell M, Ostrander EA, Suleria HAR, Cho WC, Maroyi A, Martins N (2020) Turmeric and its major compound curcumin on health: bioactive effects and safety profiles for food, pharmaceutical, biotechnological and medicinal applications. Front Pharmacol 11:1021
62. Schmidt M (2022) Accelerated wound healing with comfrey herb cream. Phytotherapie 2022-innovativ. HERBResearch. Germany. https://doi.org/10.13140/RG.2.2.19978.34249
63. Stickel F, Seitz HK (2000) The efficacy and safety of comfrey. Public Health Nutr 3:501–508
64. Araújo LU, Reis PG, Barbosa LCO, Saúde-Guimarães DA, Grabe-Guimarães A, Mosqueira VCF, Carneiro CM, Silva-Barcellos NMS (2012) In vivo wound healing effects of *Symphytum officinale* L. leaves extract in different topical formulations. Pharmazie 67:355–360. https://doi.org/10.1691/ph.2012.1563

65. Sharifi-Rad M, Mnayer D, Morais-Braga MFB, Carneiro JNP, Bezerra CF, Coutinho HDM, Salehi B, Martorell M, del Mar CM, Soltani-Nejad A, Uribe YAH, Yousaf Z, Iriti M, Sharifi-Rad J (2018) Echinacea plants as antioxidant and antibacterial agents: From traditional medicine to biotechnological applications. Phyther Res 32:1653–1663. https://doi.org/10.1002/ptr.6101
66. Mistríková I, Vaverková Š (2007) Morphology and anatomy of Echinacea purpurea, E. angustifolia, E. pallida and Parthenium integrifolium. Biologia (Bratisl) 62:2–5. https://doi.org/10.2478/s11756-007-0006-7
67. Rousseau B, Tateya I, Lim X, Munoz-del-Rio A, Bless DM (2006) Investigation of anti-hyaluronidase treatment on vocal fold wound healing. J Voice 20:443–451. https://doi.org/10.1016/j.jvoice.2005.06.002
68. Olioso D, Marzotto M, Bonafini C, Brizzi M, Bellavite P (2016) Arnica montana effects on gene expression in a human macrophage cell line. Evaluation by quantitative real-time PCR. Homeopathy 105:131–147. https://doi.org/10.1016/j.homp.2016.02.001
69. Kriplani P (2017) *Arnica montana L.* – a plant of healing: review. 69:925–945. https://doi.org/10.1111/jphp.12724
70. Balm L, Engel R, Szabó K, Abrankó L, Rendes K, Füzy A (2016) Effect of arbuscular mycorrhizal fungi on the growth and the polyphenol profile of marjoram, lemon balm, and marigold. J Agric Food Chem 64:3733–3742
71. Ganzera M, Egger C, Zidorn C, Stuppner H (2008) Quantitative analysis of flavonoids and phenolic acids in *Arnica montana L.* by micellar electrokinetic capillary chromatography. Anal Chim Acta 614:196–200. https://doi.org/10.1016/J.ACA.2008.03.023
72. Paßreiter CM (1992) Co-occurrence of 2-pyrrolidineacetic acid with the pyrrolizidines tussilaginic acid and isotussilaginic acid and their 1-epimers in Arnica species and Tussilago farfara. Phytochemistry 31:4135–4137. https://doi.org/10.1016/0031-9422(92)80428-H
73. Aeschbach R, Löliger J, Scott BC, Murcia A, Butler J, Halliwell B, Aruoma OI (1994) Antioxidant actions of thymol, carvacrol, 6-gingerol, zingerone and hydroxytyrosol. Food Chem Toxicol 32:31–36. https://doi.org/10.1016/0278-6915(84)90033-4
74. Kriplani P, Guarve K, Baghael US (2017) *Arnica montana L.* – a plant of healing: review. J Pharm Pharmacol 69:925–945. https://doi.org/10.1111/JPHP.12724
75. Petrova M, Zayova E, Vassilevska-Ivanova R, Vlahova M (2012) Biotechnological approaches for cultivation and enhancement of secondary metabolites in *Arnica montana L.* Acta Physiol Plant 34:1597–1606. https://doi.org/10.1007/S11738-012-0987-X
76. Singh R, Chaudhary M, Chauhan ES (2022) Stellaria media Linn.: A comprehensive review highlights the nutritional, phytochemistry, and pharmacological activities. J. HerbMed Pharmacol 11:330–338
77. Harish B, Pankaj S (2022) Lesser-known edible plants of Karsog Valley. In: Rethinking Himalaya: Its Scope and Protection, 1st edn. Blue Rose Publishers, Noida, pp 24–39
78. Hu YM, Wang H, Ye WC, Qian L (2010) New triterpenoid from Stellaria media (L.) Cyr. Nat Prod Res 23:1274–1278. https://doi.org/10.1080/14786410701642532
79. Hu Y, Ye W, Li Q, Tian H, Wang H, Du H (2006) C-glycosylflavones from Stellaria media. Chin J Nat Med 4:420–424
80. Rogowska M, Lenart M, Srečec S, Ziaja M, Parzonko A, Bazylko A (2017) Chemical composition, antioxidative and enzyme inhibition activities of chickweed herb (Stelaria media L., Vill.) ethanolic and aqueous extracts. Ind Crop Prod 97:448–454. https://doi.org/10.1016/J.INDCROP.2016.12.058
81. Böttger S, Melzig MF (2011) Triterpenoid saponins of the Caryophyllaceae and Illecebraceae family. Phytochem Lett 4:59–68. https://doi.org/10.1016/J.PHYTOL.2010.08.003
82. Mithril C, Dragsted LO (2012) Safety evaluation of some wild plants in the New Nordic Diet. Food Chem Toxicol 50:4461–4467. https://doi.org/10.1016/J.FCT.2012.09.016
83. Michael KJ, Tomczyk WM Flavonoids of the caryophyllaceae. https://doi.org/10.1007/s11101-021-09755-3

84. Augspole I, Duma M, Ozola B, Cinkmanis I (2017) Phenolic profile of fresh and frozen nettle, goutweed, dandelion and chickweed leaves. Latvia University of AgricultureFaculty of Food Technology. 11th Baltic Conference onFood Science and Technology. Latvia. In: FOODBALT, pp 36–39
85. Arora D, Sharma A (2014) Isolation and characterization of the chemical constituents of stellaria media linn. Int J Pharm Sci Res 5:3669–3673. https://doi.org/10.13040/IJPSR.0975-8232.5(9)
86. Vodoslavskyi V (2017) The quantitative content of the phenolic compounds in the Stellaria media herb. Pharma Innov J 6:174–175
87. Miere F, Teusdea AC, Laslo V, Fritea L, Moldovan L, Costea T, Uivaroşan D, Vicas SI, Pallag A (2019) Natural Polymeric Beads for Encapsulation of Stellaria media Extract with Antioxidant Properties. Mater Plast 56:671–679. https://doi.org/10.37358/mp.19.4.5252
88. Moradi M, Niazi A, Mazloomi E, Mousavi SF, Lopez V (2020) Effect of lavender on episiotomy wound healing and pain relief: a systematic review. Evid Based Care J 10:61–69
89. EMA (2012) Assessment report on *Lavandula angustifolia* Miller, aetheroleum and *Lavandula angustifolia* Miller. Comm Herb Med Prod 44:1–46
90. Prusinowska R, Śmigielski KB (2014) Composition, biological properties and therapeutic effects of lavender (*Lavandula angustifolia L*). A review. Herba Pol 60:56–66. https://doi.org/10.2478/HEPO-2014-0010
91. Hussain T, Tan B, Yin Y, Blachier F, Tossou MCB, Rahu N (2016) Oxidative stress and inflammation: what polyphenols can do for us? Oxidative Med Cell Longev 2016:1. https://doi.org/10.1155/2016/7432797
92. Dobros N, Zawada KD, Paradowska K (2022) Phytochemical profiling, antioxidant and anti-inflammatory activity of plants belonging to the *Lavandula* genus. Molecules 28. https://doi.org/10.3390/MOLECULES28010256
93. Crozier A, Jaganath IB, Clifford MN (2009) Dietary phenolics: chemistry, bioavailability and effects on health. Nat Prod Rep 26:1001–1043. https://doi.org/10.1039/B802662A
94. Francenia Santos-Sánchez N, Salas-Coronado R, Hernández-Carlos B, Villanueva-Cañongo C (2019) Shikimic acid pathway in biosynthesis of phenolic compounds. Plant Physiol Asp Phenolic Compd. https://doi.org/10.5772/INTECHOPEN.83815
95. Tsao R (2010) Chemistry and biochemistry of dietary polyphenols. Nutrients 2:1231–1246. https://doi.org/10.3390/NU2121231
96. Turgut A, Emen F, Canbay H, Demirdogen RE, Cam N, Kiliç D, Yeşilkaynak T (2016) Chemical characterization of *Lavandula angustifolia* Mill. which is a phytocosmetic species and investigation of its antimicrobial effect in cosmetic products. J Turk Chem Soc 4:283–298. dergipark.org.tr
97. Bajkacz S, Baranowska I, Buszewski B, Kowalski B, Ligor M (2018) Determination of flavonoids and phenolic acids in plant materials using SLE-SPE-UHPLC-MS/MS Method. Food Anal Methods 11:3563–3575. https://doi.org/10.1007/S12161-018-1332-9
98. Mori HM, Kawanami H, Kawahata H, Aoki M (2016) Wound healing potential of lavender oil by acceleration of granulation and wound contraction through induction of TGF-β in a rat model. BMC Complement Altern Med 16:1–11. https://doi.org/10.1186/S12906-016-1128-7/FIGURES/6
99. Xiong Y, Chen L, Man J, Hu Y, Cui X (2019) Chemical and bioactive comparison of Panax notoginseng root and rhizome in raw and steamed forms. J Ginseng Res 43:385–393. https://doi.org/10.1016/j.jgr.2017.11.004
100. Lee JS, Hwang HS, Ko EJ, Lee YN, Kwon YM, Kim MC, Kang SM (2014) Immunomodulatory activity of red ginseng against influenza a virus infection. Nutrients 6:517–529. https://doi.org/10.3390/nu6020517
101. Park KS, Park DH (2019) The effect of Korean red ginseng on full-thickness skin wound healing in rats. J Ginseng Res 43:226–235. https://doi.org/10.1016/J.JGR.2017.12.006

102. Viji Chandran S, Amritha TS, Rajalekshmi G, Pandimadevi M (2015) A preliminary in vitro study on the bovine collagen film incorporated with azadirachta indica plant extract as a potential wound dressing material. Int J PharmTech Res 8:248–257
103. Osunwoke EA, Olotu EJ, Allison TA, Onyekwere JC (2013) The wound healing effects of aqueous leave extracts of *Azadirachta Indica* on Wistar Rats. J Nat Sci Res 3:181–186
104. Motealleh B, Zahedi P, Rezaeian I, Moghimi M, Abdolghaffari AH, Zarandi MA (2014) Morphology, drug release, antibacterial, cell proliferation, and histology studies of chamomile-loaded wound dressing mats based on electrospun nanofibrous poly(ε-caprolactone)/polystyrene blends. J Biomed Mater Res – Part B Appl Biomater 102:977–987. https://doi.org/10.1002/jbm.b.33078
105. Oryan A, Mohammadalipour A, Moshiri A, Tabandeh MR (2016) Topical application of aloe vera accelerated wound healing, modeling, and remodeling. Ann Plast Surg 77:37–46. https://doi.org/10.1097/SAP.0000000000000239
106. Hormozi M, Assaei R, Boroujeni MB (2017) The effect of aloe vera on the expression of wound healing factors (TGFβ1 and bFGF) in mouse embryonic fibroblast cell: In vitro study. Biomed Pharmacother 88:610–616. https://doi.org/10.1016/J.BIOPHA.2017.01.095
107. Wahedi HM, Jeong M, Chae JK, Do SG, Yoon H, Kim SY (2017) Aloesin from Aloe vera accelerates skin wound healing by modulating MAPK/Rho and Smad signaling pathways in vitro and in vivo. Phytomedicine 28:19–26. https://doi.org/10.1016/J.PHYMED.2017.02.005
108. Dinda M, Dasgupta U, Singh N, Bhattacharyya D, Karmakar P (2015) PI3K-mediated proliferation of fibroblasts by Calendula officinalis Tincture: implication in wound healing. Phyther Res 29:607–616. https://doi.org/10.1002/PTR.5293
109. Dinda M, Mazumdar S, Das S, Ganguly D, Dasgupta UB, Dutta A, Jana K, Karmakar P (2016) The water fraction of Calendula officinalis hydroethanol extract stimulates in vitro and in vivo proliferation of dermal fibroblasts in wound healing. Phyther Res 30:1696–1707. https://doi.org/10.1002/PTR.5678
110. Babu MK, Prasad OS, Murthy TEGK (2011) Comparison of the dermal wound healing of *Centella asiatica* extract impregnated collagen and crosslinked collagen scaffolds. J Chem Pharm Res 3:353–362
111. Sharma A, Khanna S, Kaur G, Singh I (2021) Medicinal plants and their components for wound healing applications. Futur J Pharm Sci 7(1):1–13. https://doi.org/10.1186/S43094-021-00202-W
112. Müller WE, Singer A, Wonnemann M (2001) Hyperforin – antidepressant activity by a novel mechanism of action. Pharmacopsychiatry 34(Suppl 1). https://doi.org/10.1055/S-2001-15512
113. Behnke K, Jensen GS, Graubaum HJ, Gruenwald J (2002) Hypericum perforatum versus fluoxetine in the treatment of mild to moderate depression. Adv Ther 19:43–52. https://doi.org/10.1007/BF02850017
114. Bezáková L, Pšenák M, Kartnig T (1999) Effect of dianthrones and their precursors from *Hypericum perforatum* L. on lipoxygenase activity. Pharmazie 54:711–711
115. Sánchez-Mateo CC, Bonkanka CX, Hernández-Pérez M, Rabanal RM (2006) Evaluation of the analgesic and topical anti-inflammatory effects of *Hypericum reflexum* L. fil. J Ethnopharmacol 107:1–6. https://doi.org/10.1016/J.JEP.2006.01.032
116. Memarzia A, Khazdair MR, Behrouz S, Gholamnezhad Z, Jafarnezhad M, Saadat S, Boskabady MH (2021) Experimental and clinical reports on anti-inflammatory, antioxidant, and immunomodulatory effects of Curcuma longa and curcumin, an updated and comprehensive review. Biofactors 47:311–350. https://doi.org/10.1002/BIOF.1716
117. Wang X, Shen K, Wang J, Liu K, Wu G, Li Y, Luo L, Zheng Z, Hu D (2020) Hypoxic preconditioning combined with curcumin promotes cell survival and mitochondrial quality of bone marrow mesenchymal stem cells, and accelerates cutaneous wound healing via PGC-1α/SIRT3/HIF-1α signaling. Free Radic Biol Med 159:164–176. https://doi.org/10.1016/J.FREERADBIOMED.2020.07.023

118. Pandey VK, Ajmal G, Upadhyay SN, Mishra PK (2020) Nano-fibrous scaffold with curcumin for anti-scar wound healing. Int J Pharm 589:119858. https://doi.org/10.1016/J.IJPHARM.2020.119858
119. Rathinavel S, Korrapati PS, Kalaiselvi P, Dharmalingam S (2021) Mesoporous silica incorporated PCL/Curcumin nanofiber for wound healing application. Eur J Pharm Sci 167:106021. https://doi.org/10.1016/J.EJPS.2021.106021
120. Rathinavel S, Indrakumar J, Korrapati PS, Dharmalingam S (2022) Synthesis and fabrication of amine functionalized SBA-15 incorporated PVA/Curcumin nanofiber for skin wound healing application. Colloids Surf A Physicochem Eng Asp 637:128185. https://doi.org/10.1016/J.COLSURFA.2021.128185
121. Suryanarayana P, Saraswat M, Mrudula T, Krishna TP, Krishnaswamy K, Reddy GB (2005) Curcumin and turmeric delay streptozotocin-induced diabetic cataract in rats. Investig Ophthalmol Vis Sci 46:2092–2099. https://doi.org/10.1167/iovs.04-1304
122. Le V, Dolganyuk V, Sukhikh A, Babich O, Sciences SI-A (2021) U (2021) phytochemical analysis of symphytum officinale root culture extract. Appl Sci 11:1–16. https://doi.org/10.3390/app11104478
123. Staiger C (2013) Comfrey root: From tradition to modern clinical trials. Wien Med Wochenschr 163:58–64. https://doi.org/10.1007/S10354-012-0162-4/FIGURES/1
124. Segev A, Badani H, Kapulnik Y, Shomer I, Oren-Shamir M, Galili S (2010) Determination of polyphenols, flavonoids, and antioxidant capacity in colored chickpea (*Cicer arietinum L.*). J Food Sci 75. https://doi.org/10.1111/J.1750-3841.2009.01477.X
125. Neagu E, Paun G, Radu GL, Păun G, Radu LG (2011) Phytochemical study of some Symphytum officinalis extracts concentrated by membranous procedures UPB. Sci Bull Ser B Chem Mater Sci 73(3): 66–74
126. Paun G, Neagu E, Moroeanu V, Ungureanu O, Cretu R, Ionescu E, Elena Tebrencu C, Ionescu R, Stoica I, Radu GL (2017) Phytochemical analysis and in vitro biological activity of Betonica officinalis and Salvia officinalis extracts. Rom Biotechnol Lett 22(4):12751–12761
127. Cruz I, Cheetham JJ, Arnason JT, Yack JE, Smith ML (2014) Alkamides from Echinacea disrupt the fungal cell wall-membrane complex. Phytomedicine 21:435–442. https://doi.org/10.1016/J.PHYMED.2013.10.025
128. Shin DM, Choi KM, Lee YS, Kim W, Shin KO, Oh S, Jung JC, Lee MK, Lee YM, Hong JT, Yun YP, Yoo HS (2014) Echinacea purpurea root extract enhances the adipocyte differentiation of 3T3-L1 cells. Arch Pharm Res 37:803–812. https://doi.org/10.1007/S12272-013-0251-Y
129. Parsons J, Cameron S et al (2018) undefined (2018) Echinacea biotechnology: advances, commercialization and future considerations. Taylor Fr 56:485–494. https://doi.org/10.1080/13880209.2018.1501583
130. Zagumennikov VB, Molchanova AV, Babaeva EY, Petrova AL (2015) Accumulation of ascorbic acid in fresh echinacea purpurea plants and their processing products. Pharm Chem J 48:671–674. https://doi.org/10.1007/S11094-015-1168-1
131. Barrett B (2003) Medicinal properties of Echinacea: a critical review. Phytomedicine 10:66–86. https://doi.org/10.1078/094471103321648692
132. Oberbaum M, Galoyan N, Lerner-Geva L, Singer SR, Grisaru S, Shashar D, Samueloff A (2005) The effect of the homeopathic remedies Arnica montana and Bellis perennis on mild postpartum bleeding – a randomized, double-blind, placebo-controlled study – preliminary results. Complement Ther Med 13:87–90. https://doi.org/10.1016/J.CTIM.2005.03.006
133. Smigielski K, Prusinowska R, Raj A, Sikora M, Woliñska K, Gruska R (2013) Effect of drying on the composition of essential oil from *Lavandula angustifolia*. J Essent Oil Bear Plants 14: 532–542. https://doi.org/10.1080/0972060X.2011.10643970
134. Lawrence B (2015) Progress in essential oils. Perfum Flavorist 40:42–52
135. Eming SA, Krieg T, Davidson JM (2007) Inflammation in wound repair: molecular and cellular mechanisms. J Invest Dermatol 127:514–525. https://doi.org/10.1038/SJ.JID.5700701

136. Kim YG, Sumiyoshi M, Sakanaka M, Kimura Y (2009) Effects of ginseng saponins isolated from red ginseng on ultraviolet B-induced skin aging in hairless mice. Eur J Pharmacol 602: 148–156. https://doi.org/10.1016/J.EJPHAR.2008.11.021
137. Yu L, Xie J, Xin N, Wang Z (2015) Panax notoginseng saponins promote wound repair of anterior cruciate ligament through phosphorylation of PI3K, AKT and ERK. Int J Clin Exp Pathol 8:449
138. Kimura Y, Sumiyoshi M, Kawahira K, Sakanaka M (2006) Effects of ginseng saponins isolated from red ginseng roots on burn wound healing in mice. Br J Pharmacol 148:860–870. https://doi.org/10.1038/SJ.BJP.0706794
139. Thakurta P, Bhowmik P, Mukherjee S, Hajra TK, Patra A, Bag PK (2007) Antibacterial, antisecretory and antihemorrhagic activity of Azadirachta indica used to treat cholera and diarrhea in India. J Ethnopharmacol 111:607–612. https://doi.org/10.1016/J.JEP.2007.01.022

Phytochemicals and Overview of the Evolving Landscape in Management of Osteoarthritis

27

Falak Bamne, Nikhat Shaikh, Ahmad Ali, Munira Momin, and Tabassum Khan

Contents

F. Bamne · A. Ali
University Department of Life Sciences, University of Mumbai, Mumbai, India
e-mail: ahmadali@mu.ac.in

N. Shaikh
Regional Research Institute of Unani Medicine, Mumbai, India

M. Momin
Department of Pharmaceutics, SVKM's Dr. Bhanuben Nanavati College of Pharmacy, Mumbai, India
e-mail: munira.momin@bncp.ac.in

T. Khan (✉)
Department of Pharmaceutical Chemistry & Quality Assurance, SVKM's Dr. Bhanuben Nanavati College of Pharmacy, Mumbai, India
e-mail: tabassum.khan@bncp.ac.in

S. C. Izah et al. (eds.), *Herbal Medicine Phytochemistry*, Reference Series in Phytochemistry, https://doi.org/10.1007/978-3-031-43199-9_55

Abstract

Osteoarthritis (OA), characterized by degeneration of the joints, is forecast to be the major cause of disability for older people by 2030. The primary pathology of osteoarthritis are cartilage and bone rebuilding. Osteoarthritis management consists of pharmacotherapy, physiotherapy, and surgery. OA should traditionally be treated using herbal medications in various medical systems around the world in addition to steroid use. These provide symptomatic relief from pain and inflammation, but minimally reduce joint destruction. Structural damage needs to be effectively prevented and should be the main goal of new therapeutic regimens. Phytoconstituents especially belonging to phenolics, flavonoids, steroids, terpenoids, sterols, sesquiterpenes, triterpenes, and phytoalexins class exhibit antioxidant, anti-inflammatory, anti-arthritic effects and prove beneficial in OA. These phytochemicals modulate different targets like cyclooxygenases, lipoxygenases, and other signaling pathways involved in the pathophysiology of OA, thereby providing immense benefits. These also serve as potential leads for the development of new chemical entities as potential anti-inflammatory molecules of diverse utility. A number of clinical trials have found that herbal remedies and their components are essential to mitigate osteoarthritis. Reflex plus™ is a patented combination of collagen hydrolysate and rosehip extract. Ingredients in the blend have been thoroughly researched for their efficacy in arthritis. Collagen hydrolysate tends to accumulate within cartilage and assists in repairing the damage caused by osteoarthritis and other cartilage alterations. The rose hip works on the anti-inflammatory pathway to reduce joint discomfort and inflammation. The most important pro-inflammatory cytokines implicated in the pathogenesis of osteoarthritis are TNF-α and IL-1β, which act on synoviocytes and chondrocytes by specific interactions with cytokine receptors at the cell surface, assisting in the construction of cartilage for the effective functioning of the joints and the relief of pain associated with osteoarthritis and also help in the protection of cartilage. The purpose of this chapter is to incorporate the impact of phytochemicals on various molecular targets of inflammation into the overall management of osteoarthritis.

Keywords

Osteoarthritis · Anti-inflammatory · Herbal medicine · Phytochemicals · Rose hip

Abbreviations

BMD	Bone mineral density
CH	Collagen hydrolysate
COX-1	Cyclooxygenase -1
COX-2	Cyclooxygenase -2
FDA	Food and Drug Administration

Fib3-1	Fibrin-3-1
Fib3-2	Fibrin-3-2
FSTL1	Fibulin-3 peptide and follistatin-like protein-1
GMPs	Good Manufacturing Practices
GOPO	Galactolipid
GS	Glucosamine sulfate
IL-1	Interleukin-1
IL-6	Interleukin-6
IL-8	Interleukin-8
LDL	Low-density lipoproteins
MMP-13	Matrix metalloproteinase-13
NSAIDs	Non-steroidal anti-inflammatory drugs
OA	Osteoarthritis
PMN	Polymorphonuclear
PUFAs	Polyunsaturated fatty acids
RONS	Reactive oxygen and nitrogen species
ROS	Reactive oxygen species
TNF- α	Tumor necrosis factor alpha
USDA	United States Department of Agriculture
UV-B	Ultraviolet B

1 Introduction

Osteoarthritis (OA) is the most prevalent joint disease. It is the most common type of arthritis, with a higher chance of reduced mobility (defined as assistance in walking or climbing stairs) in individuals over 65 years old who have affected knees [1]. OA is a multifactorial condition with disparities in structure and function throughout the joint. The changes that occur in articular cartilage as a result of aging can lead to the degeneration of tissue and a pathological process characterized by a gradual and typically progressive deterioration of tissue structure and function [2]. In a typical knee joint, the cartilage remains intact and undamaged, preserving the joint's health and functionality. Similarly, the meniscus, a crucial component for shock absorption and stability, remains healthy and fully functional. There are no signs of osteophytes, which are bony outgrowths that can affect joint movement [3]. Moreover, the subchondral bone and ligaments exhibit no abnormalities, maintaining their normal state. Conversely, in a knee joint affected by osteoarthritis, there is a breakdown of the cartilage, leading to compromised joint integrity. The meniscus also suffers damage, reducing its ability to effectively cushion and stabilize the joint [4]. Osteophytes, which are indicators of degeneration, become present, potentially hindering smooth joint movement. Additionally, the subchondral bone undergoes remodeling and sclerosis, contributing to the altered joint structure. Ligaments may also be affected by these changes (Table 1). All the diseases associated with aging, osteoarthritis is perhaps the most strongly correlated and causes significant mobility impairment. The aging process results in significant changes to the structure, composition of the matrix, and mechanical properties of articular cartilage. Most

Table 1 Difference in normal knee and osteoarthritis knee

Normal knee	Osteoarthritis knee
No cartilage damage	Cartilage breaking down
Healthy meniscus	Damaged meniscal
No osteophytes	Presence of osteophytes
Normal subchondral bone and ligaments	Bone remodeling and sclerosis

individuals with fibrillation of the articular surface do not suffer from joint pain or dysfunction. Therefore, the age-related fibrillation of articular cartilage does not necessarily lead to the progressive degeneration of articular cartilage that results in osteoarthritis [5, 6].

Pain and stiffness in the affected joints are the most common signs of osteoarthritis. Discomfort usually intensifies when the joint is moved or toward the end of the day. Joints may become stiff after a period of rest, but this typically subsides relatively quickly once movement resumes. Symptoms of the condition often fluctuate without any obvious cause, with periods of discomfort lasting for several weeks or months, followed by periods of relief. On occasion, the affected joint may become swollen. Swelling in the joints, particularly in the fingers, can take on a knobby and hardened appearance due to the formation of additional bone tissue, or a soft and puffy appearance caused by the thickening of the joint lining and the accumulation of extra fluid in the joint capsule (Fig. 1).

The joint may produce grinding or crackling noises during regular movement, which is known as crepitation. At times, the muscles around the joint may appear weak or atrophied. Engaging in exercises that strengthen the muscles supporting the joints can prevent this from happening [7]. The exact cause of osteoarthritis remains uncertain. It is more than just a result of regular "wear and tear," and the likelihood of developing osteoarthritis is influenced by various factors [8], such as:

1. Age: Osteoarthritis typically emerges in the late forties, possibly due to age-related bodily changes like muscle weakness, weight gain, and reduced efficiency in the body's healing ability.
2. Gender: Osteoarthritis is typically more prevalent and severe in women in most joints.
3. Obesity: Being overweight, particularly in weight-bearing joints like the hip and knee, is a common cause of osteoarthritis [9].
4. Joint Injury: Arthritis can develop in a joint later in life due to a severe injury or surgery. Osteoarthritis is not caused by regular activity or exercise; however, the risk may increase with very strenuous or repetitive activities or physically demanding occupations.
5. Abnormalities in the joints: If person were born with abnormalities, OA may begin early and be more severe than usual [10].
6. Weather and Diet: Several individuals with osteoarthritis may notice that changes in weather worsen their pain, especially when the atmospheric pressure drops before it rains. However, weather changes do not cause osteoarthritis; they can

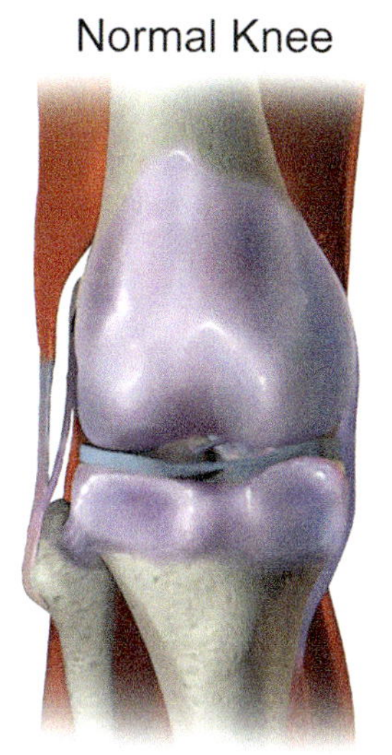

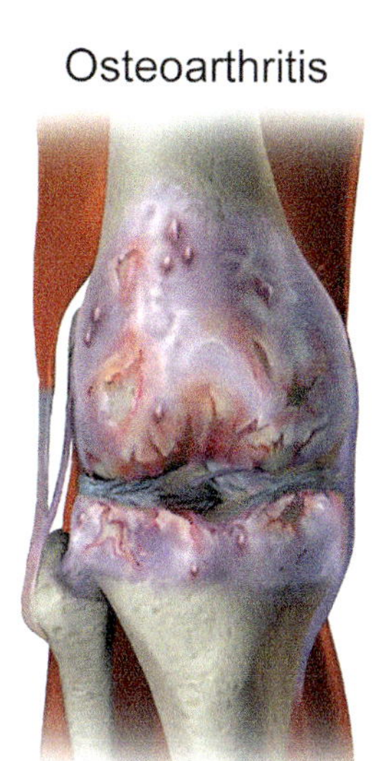

Fig. 1 Normal and osteoarthritis knee

only influence the symptoms. However, your risk of developing osteoarthritis is influenced by your weight somewhat than other specific dietary factors [11].

Accurate diagnosis of arthritis is fundamental to its treatment. There are numerous types of arthritis, and each may require different treatments. Typically, diagnosing osteoarthritis involves evaluating an individual's symptoms, medical history, the progression of their symptoms, and the physical examination of the affected joint [12]. Osteoarthritis impacts various joints across the body, leading to discomfort and reduced mobility. The affected joints encompass the neck, shoulders, lower back, hips, elbows, base of the thumbs, fingertips, knees, ankles, and the base of the toes. In these areas, the cartilage gradually deteriorates, causing pain, stiffness, and limited joint movement [13]. Osteoarthritis commonly targets these regions, contributing to challenges in daily activities and prompting the need for management strategies to alleviate its effects. Physical examination of joints involves screening for tenderness over the joint, Crepitus, also known as joint grating or creaking, bone swelling, excessive fluid, difficulty in movement, instability of joint, weakness of muscles or thinness that supports the joint.

Blood tests such as CBC, ESR, and rheumatoid factor are usually normal in OA. X-rays may not always be useful in diagnosing osteoarthritis or determining the appropriate treatment options. However, they can detect the presence of calcium deposits in the affected joint.

In rare cases, an MRI scan of the knee may prove to be useful. High-frequency radio waves can also be used to create images of soft tissue and bone changes that would not be evident on an X-ray. Its primary function is to uncover other potential joint and bone abnormalities that may be causing the symptoms [14].

2 Pathophysiology of OA

OA is deemed an organ disease that affects all joint structures. Subchondral osteosclerosis, articular marginal osteophytes, and moderate, chronic, nonspecific synovial inflammation are all associated with progressive loss of articular cartilage in synovial joints [15, 16].

2.1 Articular Cartilage

Despite the involvement of multiple joint tissues, osteoarthritis has long been characterized mainly by interruption of the repair method in damaged cartilage as a consequence of biochemical and biomechanical changes in the joint. The normal maintenance of physiological homeostasis in articular cartilage is controlled by chondrocytes, which generate a structural matrix consisting mainly of type II collagen and proteoglycans [17]. Chondrocytes can produce less effective collagen (type I collagen), smaller and less dense proteoglycans, more degradable enzymes, and nitric oxide, which produces different inflammatory mediators, including IL-1, when subjected to mechanical stress and interleukin (IL)-1. The imbalance between breakdown and synthesis of the extracellular matrix leads to a vicious cycle, resulting in the loss of articular cartilage. This process does not cause clinical symptoms until the innervated tissue is affected, as articular cartilage is avascular [18].

2.2 Osteophytes

Osteophytes near the end of the joint can expand at this point. Osteophytes are cartilage-covered bony growths. Osteophytes are classified into three types: traction spines, where ligaments and tendons adhere to bone, inflammatory spines from vertebral bodies, and osteochondrocytes, which generate cartilage by synovial metaplasia. They can cause discomfort in spinal osteoarthritis but are advantageous in lower extremity osteoarthritis because they support the joints [19, 20].

2.3 Synovial Inflammation

The synovial membrane is essential for appropriate functioning of joints by providing nourishment to chondrocytes via synovial fluid and joint space, as well as removing waste products and breakdown products from the matrix. Synovial mucosal cells produce hyaluronic acid and lubricin, which help to protect and maintain articular cartilage [21]. In osteoarthritis, synovial inflammation contributes to various clinical symptoms, such as pain, and correlates with the progression of the disease. In addition, synovitis is a main element in the pathophysiology of osteoarthritis through the action of several soluble mediators. The association between arthroscopically assessed synovitis and the fraction of functional impairment or pain insights continues to be debatable [22].

3 Molecular Targets of OA

Cytokines, proteinases, lipid mediators, and reactive oxygen species stimulate chondrocytes to produce enzymes that break down cartilage. In patients with osteoarthritis, levels of IL-6, IL-8, and MMP-13 are significantly higher in osteophyte-derived osteoblasts [23]. The expression of IL-6 and IL-8 genes in osteoblasts was found to increase in response to non-physical mechanical stress, and the degree of stress was found to be directly proportional to the magnitude of this increase. This suggests that inflammatory factors play an even greater role in the formation of osteophytes.

Furthermore, IL-6 stimulates the production and expression of MMP-13 in osteoblasts derived from osteophytes and those located in the subchondral region affected by osteoarthritis. Increased IL-8 and MMP-13 expression can cause cartilage degradation through chondrocyte enlargement [24]. Growth factors, which are responsible for the synthesis of the normal matrix, including insulin-like growth factor-1, bone morphogenetic protein, platelet-derived growth factor, and transforming growth factor, play a crucial role in inhibiting the activity of pro-inflammatory cytokines and supporting repair. Osteoarthritis can cause cartilage damage. They enhance anabolic activity and proteoglycan production in chondrocytes while inhibiting catabolic activity [25].

Currently, there are no biomarkers that can be considered reliable tools for diagnosing and predicting the progression of osteoarthritis in routine clinical practice. Fibulin-3 peptide and follistatin-like protein-1 (FSTL1), on the other hand, are extracellular proteins with promise as osteoarthritis biomarkers. Fibrin-3 is frequently distributed in several kinds of tissues and blood vessels of different sizes, and can prevent vascular encroachment and angiogenesis [26]. In addition, it also improves the cartilage in osteoarthritis. An osteoarthritis patient's urine and serum have a larger proportion of their two fibrin-3 fragments (Fib3-1 and Fib3-2) than controls. Increased levels of Fib3-1 were found to be associated with aging and hormone levels, while Fib3-2 levels were not related to gender, age, or menopause. FSTL1, which is activated by ischemic stress and inflammatory mediators, is expressed in human tissues. It is important to note that this information is paraphrased and not intended to be used as medical advice. Serum FSTL1 levels have been found to correlate with inflammatory status, making it a potential biomarker for rheumatoid arthritis and other autoimmune diseases. FSTL1 is thought to play a role in the etiology of arthritis and is activated by ischemic stress and inflammatory mediators in human tissues. The levels of serum FSTL1 were significantly higher in patients with osteoarthritis than in healthy individuals, and in women, they were correlated with the severity of the disease and the expansion of joint space [27].

4 Management of OA

Although there is still no cure for osteoarthritis, there are treatments that can relieve symptoms shown in Fig. 2. These include: 1. Changes in lifestyle, such as increased physical activity, muscle-building exercise, and weight loss (where relevant); 2. Pain-killers, e.g., pain-relieving creams or gels or steroid injections; 3. Physical therapies, e.g., application of heat, splints, or other aids; 4. Surgery; 5. Complementary treatment and supplements [28].

4.1 Physical Activity

Many patients are concerned that exercising would aggravate their discomfort and create further joint injury. Physical activity can help protect your joints by strengthening the muscles that support them, as well as control pain, relieve stress, and decrease excess weight. If the discomfort makes it difficult to move, take a pain medication like acetaminophen and warm up the bothersome joint beforehand, or apply an ice pack [29]. There are three types of workouts:

Movement exercises: Exercises that involve movement are beneficial to posture, strength, and flexibility. The exercises softly and gently loosen up the joints to a range of motion that feels pleasant to the touch [30]. This coupled with low intake and low serum levels of vitamin D are linked to increased risk of OA progression [31].

Strengthening workouts: These are exercises that use water, light weights, or resistance bands to provide resistance and strengthen the muscles that support and move your joints. Because of a natural dread of pain, many individuals become less

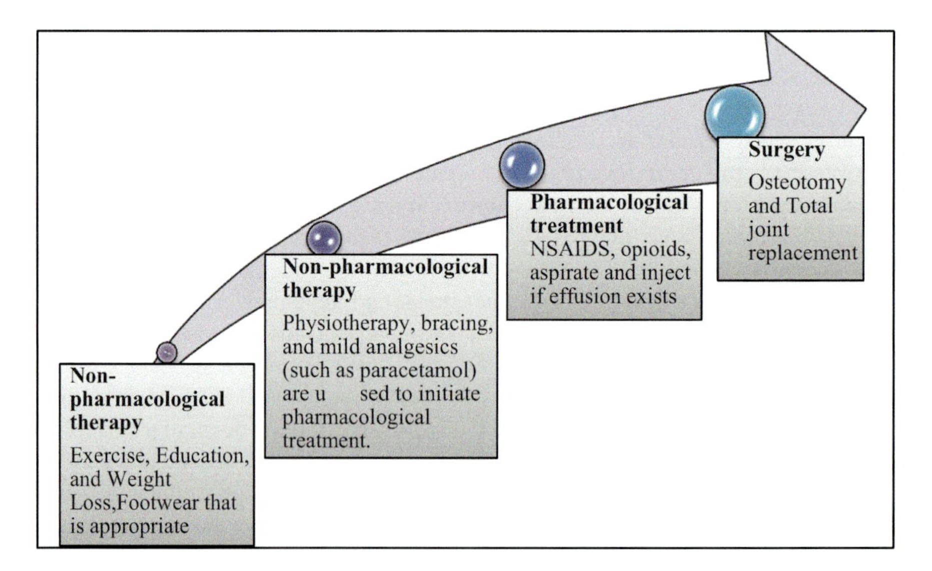

Fig. 2 Management approaches of osteoarthritis

active when they get arthritis, which can lead to increased damage causing muscles to weaken that can cause muscles to weaken and even become noticeably thinner, making it even more difficult for joints to function normally to retain [32].

Aerobic exercise: For individuals experiencing joint pain, walking, cycling, and swimming are all great forms of exercise. You could also consider using a stationary bike or elliptical machine. Walking in the shallow end of a pool is an effective way to improve leg muscle strength. Pools designed for hydrotherapy or aquatic therapy are typically warmer than regular swimming pools. Water provides a comforting and pain-relieving effect, and the buoyancy helps in weight support, making it easier to perform some stretching exercises while waiting for daily activities like cooking or doing household chores [33, 34].

Weight loss: Even a small reduction in weight can have a significant impact on symptoms, particularly for those who are overweight. A well-balanced diet is the most effective way to achieve weight loss. To reduce calorie intake, one should avoid consuming foods that are high in fat and sugar. At the same time, it's important to have a balanced diet that includes all major food groups. Gradual increment in physical activity can also assist with weight loss. There are various ideas regarding what is good and harmful to eat when you have arthritis, but no single diet has been shown to assist. Consulting with a nutritionist is recommended to determine the most appropriate dietary plan. Some studies suggest that including fatty fish or fish-derived oils in the diet may be beneficial in alleviating symptoms of certain types of arthritis [35, 36].

4.2 Medication

The medications that are broadly taken for OA do not influence the actual condition. However, they can help reduce the side effects of inflammation and tension. NSAID gels and creams: Topical creams, gels, and patches containing nonsteroidal anti-inflammatory drugs (NSAIDs) can be applied directly to the skin for pain relief. Drug stores and stores sell ibuprofen and diclofenac gels over the counter. Others, such as ketoprofen, are only available with a doctor's prescription [37]. Creams, gels, and fixes can be effective for specific joints, most notably the knees and hands, but they cannot treat joints that are deeper under the skin, such as the ankles and hips. They are quite safe and are typically used as the first line of defense in drug therapy [38, 39].

Capsaicin cream: Capsaicin cream is a powerful pain reliever derived from the pepper plant (capsicum). It is very beneficial for knee and hand osteoarthritis, as are NSAID creams and gels. It should be used on a daily basis. When most people first take capsaicin, they feel a heat or burning sensation, which usually goes away within a few days. Capsaicin is extremely safe, just like NSAID creams, gels, or fixes [40, 41].

Paracetamol: The safest form of pain treatment to attempt first is paracetamol. It is best to take them before the pain becomes unbearable. Spending extra money on expensive paracetamol brands, which can be obtained without a prescription at drugstores and supermarkets, does not provide any advantages [42, 43].

Nonsteroidal anti-inflammatory drugs (NSAIDs): Nonsteroidal anti-inflammatory drugs (NSAIDs) are a more effective pain relief option compared to acetaminophen. Ibuprofen is the most commonly used NSAID and can be purchased in low to medium doses at pharmacies and supermarkets. If necessary, the doctor may advise high doses of ibuprofen, such as naproxen or diclofenac. Celecoxib and Etoricoxib, two newer forms of NSAIDs, are said to reduce the incidence of stomach ulcers and discharges [44, 45].

Steroid injections: Long-acting steroid injections are occasionally delivered straight into a sore joint. The injection usually begins to work within a day or two and can relieve pain for several weeks or even months, particularly in the knee or thumb. Steroid injections are frequently utilized for managing painful osteoarthritis or abrupt, intense pain triggered by crystals in the joint. The doctor may also provide a steroid injection if the arthritis could interfere with an important event [46, 47].

4.3 Physical Therapies

Heat and cold: Heat applied to a painful joint might help ease pain. To prevent skin damage, use a hot water bottle covered in a towel or a microwaveable wheat bag. Another option is to soak your hands or feet in a bowl of warm water for a short period. In case of joint inflammation, an ice pack covered in a towel can help relieve swelling and discomfort. Apply ice for up to 20 minutes every few hours [48].

Splints and other supports: Orthoses, braces, and other supports are available to assist with problematic joints. When osteoarthritis has changed the orientation of a joint, these can be incredibly helpful. It is best to seek expert counsel from a competent specialist or physiotherapist before making your decision [49, 50].

Shoes: Choosing comfortable, durable shoes influences not just your feet, but also your knees, hips, and spine. The perfect shoe would have a thick but delicate bottom, a delicate upper, and plenty of space at the toe and arch. To relieve pressure on areas of the foot not affected by osteoarthritis, the doctor or podiatrist may offer orthotics for various foot problems [51, 52].

Crutches: Some people use a cane. Holding it in the other hand can also mitigate knee or hip pain. It is best to obtain the counsel of a knowledgeable medical professional because your reasons for utilizing a stick will dictate which side it should be used on. It is crucial that a walking stick is adjusted properly according to the body size, a good provider should do this. Check that the handle feels comfortable in the hand if there is arthritis in the hands [53, 54].

Posture: Posture affects joint alignment, among other things. This in turn can lead to joints or joint parts being subjected to greater stress than usual [55].

4.4 Surgery

The majority of patients with osteoarthritis do not require surgery. However, if the arthritis is severe and the symptoms have a significant influence on quality of life, the doctor may recommend surgery. Every year, thousands of hip and knee replacement procedures are performed for osteoarthritis, and other joint replacement surgeries are becoming more popular [56]. There are additional surgical options than joint replacement. Washing out loose bone fragments and other tissue from the knee can sometimes be accomplished using keyhole surgery. This is known as arthroscopic lavage. Fusion can be quite effective for some tiny joints. This means that the bones in a joint are surgically fixed, preventing movement and, as a result, pain. Although the motion loss may appear weird at first, fusion frequently produces favorable outcomes [57].

4.5 Complementary Treatment and Supplements

Complementary therapies can help ease some of the symptoms of arthritis, such as pain and stiffness, as well as some of the adverse effects of medicine. Relaxation practices, for example, can improve overall safety. Some of the most well-known treatments:

1. Acupuncture involves the insertion of thin needles into specific points on the body, which is said to help restore the natural balance of health by correcting imbalances in energy flow. There is evidence that it can help with osteoarthritis symptoms. NICE does not currently recommend this treatment for osteoarthritis.
2. The Alexander Technique entails becoming more conscious of your posture and moving more efficiently. It has been proven to be very efficient for back pain but is less marked for osteoarthritis.
3. Aromatherapy uses vegetable oils to support health and well-being. The oil can be vaporized, inhaled, exploited in showers and burners, or used as portion of a scented healing back massage. Aromatherapy Is Effective for Osteoarthritis Symptoms, according to Research No, yet some people enjoy the following advantages, such as relaxation support.
4. Although there is limited evidence that it is beneficial in alleviating osteoarthritis symptoms, a good massage can be relaxing and rejuvenating.
5. Osteopathy and chiropractic involve manipulating the body's alignment manually and applying pressure to soft tissues. The objective is to correct structural and mechanical irregularities, improve mobility, reduce pain, and facilitate the body's natural healing process. Chiropractice may provide relief in spine or extremity OA associated pain as reported in aretrospective study conducted to understand the impact of chiropractic management in patients with OA and spine pain. There is no specific scientific evidence that osteopathy is effective in the treatment of osteoarthritis [63].

Use of supplements: Arthritic patients frequently experiment with a range of supplements, including herbal medicines, vitamins, minerals, and supplements. There is minimal evidence that they improve arthritis or its negative effects in many patients. Before taking any supplements: 1. Research the supplements you intend to use; not all of them are suitable for everyone. 2. Consult your doctor or pharmacist to ensure if it is safe to combine with prescription or over-the-counter treatments. After taking medication, keep an impact diary to see if the supplement (s) are having an effect [58].

Glucosamine: Glucosamine is a substance that is present in the body, mainly in tissues like cartilage, tendons, and ligaments. It can also be obtained from the shells of crustaceans such as shrimp, crab, and lobster. Some studies have suggested that taking dietary supplements of glucosamine, particularly in the knee, may be useful in treating osteoarthritis. There seems to be proof that there is no such thing as instant pain relief. It will take several months to visualize if it works. If it doesn't work after 2 months, it possibly won't work [59, 60].

Chondroitin: Chondroitin is normally present in our body and is said to help make ligaments more versatile. Animal studies indicate that it may help slow cartilage disintegration. To a certain extent. Expect no progress for at least 2 months. Assuming the ligaments are badly damaged, chondroitin cannot be used [61].

Fish oils: Fish oil and fish liver oil are generally regarded good for the joints. There isn't sufficient information to say if it's effective. Fish liver supplements are often high in vitamin A, so it's important not to overeat. Fish oil supplements tend to be lower in vitamin A, so it's safer if you really desire to take benefit of the high fish oil content [62].

5 Herbal Medicines in OA

Herbal medicine refers to the purpose of different sections of medicinal plants. Herbology has a long practice of use outside of conventional medicine. It is the collection of information, skills, and practices that are used to preserve health as well as to prevent, diagnose, improve, or treat health. Many physical and mental diseases. The earliest written validation of the help of medicinal plants to formulate medicines was observed on his 5000-year-old Sumerian clay tablets at Nagpur. It includes 12 recipes and references over 250 plant ingredients, including alkaloids like poppy, henbane, and mandrake. The sacred Vedic publications of India pertain to healing with the help of abundant plants in the country. Pen TSao, a Chinese treatise on the utilization of roots and plants, was authored in 2500 BC by Emperor Shen Nong. Author of 365 remedies (dried bits of medicinal herbs). Many of them are still in existence today [64, 65].

The risks connected with herbal medications are introduced onto the market without a mandated safety or toxicological evaluation of their efficacy. Many of these nations also lack efficient procedures to control natural medicine manufacturing and quality laws [65]. Many countries face problems with regulatory status, safety and efficacy evaluation, quality control, safety monitoring, and insufficient or erroneous information about traditional supplementary items [66].

The difficulties in assessing safety and efficiency stem from the fact that the requirements, inquiry procedures, standards, and methodologies necessary to evaluate the safety and efficacy of natural medicines are more involved than those required for traditional medications and pharmaceuticals [67, 68]. The point is, no one can deny that it's a lot more perplexing. Routine medical care. A single home-grown pharmaceutical or medicinal plant may have hundreds of natural elements, and natural therapeutic items combined may include many more. This is essentially unimaginable when evaluating individual dynamic components in this way, especially if the natural product is a mix of at least two spices [69].

5.1 Rose Hip

Rose hip oil is extracted from dried *Rosa canina (R. canina*) fruits acquired from Hyben Vital in Langeland, Denmark. The fruit of the Rosaceae (rosebush) rose family *Rose hip* ranges in hue from red to orange. This species is indigenous to Europe and is widely dispersed there [70]. In fact, it is the most abundant rose species in Central Europe. Furthermore, the dog rose is endemic to portions of North Africa and Central Asia. European invaders transported *R. canina* to the Americas, South Australia, New Zealand, and southern Africa hundreds of years ago, where it escaped cultivation, naturalized, and now thrives in the wild [71].

R. canina is a big shrub or small tree with an arched trunk that grows to a height of 2.74 meters. The five-petaled blossoms, which range in color from white to pink and bloom alone or in tiny clusters in June and July, bloom alone or in tiny clusters. In September and October, the blooms are followed by persistent brilliant red berries, which are often referred to as fake berries (hips). The true fruit is a tiny hairy mass containing seeds in the rump. Dog roses are propagated through spreading, sucking, layering, and sowing, which is most commonly done by loin-eating birds. *R. canina* is tough to control due to its reproductive habits and massive size. Some countries have put in place control measures to prevent their spread and categorize them as invasive species [72].

Rose hip has long been utilized for its astringent flavor. It was claimed to relieve diarrhea, dysentery, thirst, coughing, and "blood vomiting." Nicholas Culpeper, an English herbalist, botanist, physician, and astrologer from the seventeenth century, observed that ripe hips were rendered temperate (usually a mixture made with sugar or honey), which was described as "gently binding the belly and diverting from humor and fluid flow." He also said that the dry, powdery loins dissolve stones, make urinating easier, and alleviate colic.

Rose hips are utilized in various formulations to treat a wide range of ailments such as colds, diabetes, diarrhea, edema, fever, gastritis, gout, stiffness, sciatica, renal and lower urinary tract diseases. They are also used as a diuretic and laxative, and for blood cleansing purposes. Additionally, they are used to enhance the flavor of several food items including jams, jellies, teas, soups, infusions, syrups, beverages, pies, bread, and wine, owing to their vitamin C content [73, 74].

In the northern region of Portugal, people consume the fruit in its raw form, and extracts or brandy are made by steeping it for its diuretic, anti-diarrheal, anti-rheumatic, anti-inflammatory, respiratory, decongestant, energizing, and tonic properties, as well as for treating skin infections. The young shoots and fruit are eaten uncooked, and the fruit is utilized to make spirits and cakes. Similarly, it is also produced in rural Cantabria, which is located in northern Spain [75, 76].

Although the FDA has not officially recognized *R. canina* as generally recognized as safe (GRAS) for use in food, extracts and oils obtained from other species of *Rose hips* such as *R. alba*, *R. centifolia*, *R. damascena*, *R. gallica*, and their variants have been approved for use in cosmetics, foods, and medicines. However, the USDA considers *Rose hips* from *R. canina* or *R. rugosa* as a food [77, 78]. The use of *Rose hip* or *Rose hip* extract is allowed as an ingredient in dietary supplements, provided that the manufacturer complies with current Good Manufacturing Practices (cGMPs) and notifies the FDA within 30 days of marketing if they make a structure-function claim. In the FDA's cGMPs for regulating dietary supplements, the term "*rosehips*" is mentioned multiple times in the preamble [79, 80]. The FDA provides detailed specifications and labeling requirements for different types of dietary supplements that contain rosehips, such as "*rosehips* dietary supplement" and "vitamin C from *rosehips* dietary supplement." [81].

5.2 Phytochemistry of *R. canina*

With the aid of tandem mass spectrometry, gas chromatography, high-performance liquid chromatography, thin-layer chromatography, diode array detection, and thin-layer chromatography, a new study was started to discover the chemicals found in *R. canina* fruits. These studies allowed for the differentiation of various fruit combinations from *R. canina* [82]. Future investigation should focus on finding additional substances that might be responsible for some of the plant's bioactive qualities that have been found in recent medical studies. Quantitative component studies, which are also lacking, might point to the creation of legal *R. canina* breeding and post-harvest control methods as well as support for therapeutic agents [83].

Triterpene

The terpene family is one of the most diverse and variable groups of natural phytochemicals. Over 4000 different complex molecules, many of which are typically outside the realm of chemical synthesis, are among them. Simple triterpenes can operate as signaling molecules, whereas complex glycosylated triterpenes or saponins provide pest and microbe protection. It can be put to a lot of different uses. Triterpenes, which are precursors to sterols, are made by both plants and animals. Sterols are crucial membrane structural underpinnings and, as steroid hormones, they also contribute to cell signaling. Triterpenes, however, are not thought to be essential for healthy development and growth. Among the components of rosehip, both simple and conjugated triterpenes are highly regarded [84].

Flavonoids

Phytochemicals called flavonoids are part of the phenylpropanoid family of compounds. Two aromatic rings are connected by three carbon bridges to form the fundamental backbone of a flavonoid. This skeletal structure is produced by two different biosynthetic pathways. One aromatic ring and a three-carbon extension are obtained via the malonic acid pathway via phenylalanine, and the other is obtained through the condensation of three acetic acid-derived units. The most prevalent substituents for flavonoids, which typically exist as glycosides, are sugars. However, flavonoids can have a variety of substituents [85].

Anthocyanins, flavones, flavanols, isoflavones, flavanols, and flavanones are the first six subclasses of flavonoids. The anthocyanins, which are beta-glycosides with a sugar in the third position, are the largest group of colored flavonoids. Anthocyanins work as attractants, enticing animals (pollinators) to flowers and fruits through visual cues. Another essential attribute of flavonoids is their capacity to protect cells from ultraviolet B (UV-B) exposure. While permitting visible wavelengths to pass through, they gather in the epidermal layers of the skin and capture harmful UV-B radiation [86, 87]. Compared to anthocyanins and carotenoids, flavones and flavanols trap light at fewer frequencies. Flavones and flavanols cannot be seen by the human eye because of this. Because they are linked to the UV patterns of flowers known as "Nectar Guides," they can appear to insects that perceive UV tones in a range of light. It has been demonstrated that isoflavones have a variety of biological functions, such as antibacterial and insecticidal qualities [88].

The in vitro findings from flavonoid bioactivity screens were supported by in vivo research on the resistance of plasma and lipoproteins to ex vivo oxidation after ingestion of flavonoid-rich meals. A lower incidence of infectious diseases, cancer, inflammation, heart disease, ischemic stroke, atherosclerosis, and other chronic diseases, is linked to consuming foods high in flavonoids, particularly fruits and vegetables. Quercetin, one of the few flavonoids, has shown fascinating bioactive properties in both in vitro and some in vivo investigations [89, 90]. Rutin and its glycosides (rutin and quercitrin, for example) have shown anti-inflammatory activity in gastrointestinal inflammation models, perhaps through downregulation of nuclear factor kappa-beta signaling. Quercetin may also influence eicosanoid biosynthesis, protect low-density lipoproteins (LDL) from oxidation, reduce clotting, and relax smooth muscle in the heart. Interestingly, the flavonoid tiliroside (kaempferol 3-O-d-(6-p-coumaryl)-glycopyranoside) discovered in *R. canina* dramatically reduced the oxidation of human LDL in vitro. It has been shown in humans to have anti-obesity properties as well as antioxidant, cytotoxic, and anti-complementary properties [91, 92].

Carotenoids

Tetraterpenoids called carotenoids absorb light with wavelengths of 400–500 nm. Carotenoids, which give fruits and flowers their colors, are therefore found in plants as red, orange, and yellow pigments. Prevents photochemical processes from occurring, particularly oxidation (radical scavenger) [93, 94]. Animals must obtain their carotenoids from their diet because they are unable to synthesize them. A 40-carbon structure with an isoprene base unit makes up the majority of carotenoids. There are typically two subgroups of carotenoids. (1) Oxygen-containing xanthophylls (lutein, zeaxanthin, cryptoxanthin, etc.); and (2) non-hydroxylated hydrocarbons (i.e., −carotene, −carotene, and lycopene). The structure of a carotenoid (the number of conjugated double bonds and the presence or absence of oxygen) has a direct impact on the compound's color [95, 96].

Dietary carotenoids have been linked to the prevention of angina pectoris, apoptosis induction, and mammary cell proliferation. It has also been proposed that carotenoids may aid in human prostate cancer prevention [97, 98]. The positive effects of carotenoids that have been discovered may be the result of this discovery, either because carotenoids are better suited to serve as indicators of high consumption of fruits and vegetables or because the effects of epidemiological studies have been attributed to substances unrelated to carotenoids [99, 100]. However, some carotenoids have reportedly been found to have intriguing physiological properties. According to human research, lutein and zeaxanthin, which are found in high concentrations in the macula of the human retina, may protect the outer segment of the macula and the retinal photoreceptors from oxidative stress are seen [101].

Diets high in lutein and zeaxanthin are associated with a lower prevalence of nuclear cataracts in older women and a lower risk of age-related macular degeneration. Additionally, it has been suggested that lycopene, when given in the form of tomato paste or extract, can treat non-eosinophilic airway inflammation and prostate cancer. Because it is a result of administering a combination of compounds, it is problematic [102, 103].

Vitamins

Vitamins are organic chemicals that are produced by plants and some lower animals. Vitamins play a variety of activities in the body, including coenzyme activity, precursor activity, antioxidant activity, calcium and phosphorus intake regulation, and coagulation regulation (blood coagulation). Numerous illnesses and diseases in humans are caused by vitamin deficiencies [104].

Vitamins that dissolve in water, like vitamin C and the B vitamins, cannot be stored by the body and must be consumed continuously through food. Water-soluble vitamins that aren't used are excreted. Recent studies have concentrated on the antioxidant activity of vitamins C and E [105].

Apart from scurvy prevention, which is attributed to vitamin C's role in collagen synthesis, vitamin C is involved in a number of significant enzymatic syntheses. It is crucial for the production of bile acids and the conversion of cholesterol to these acids. Vitamin E, which is assumed to work largely as an antioxidant, is thought to protect PUFAs of plasma membrane phospholipids and plasma lipoproteins. According to recent studies, vitamin E reduces protein kinase C activity, although the physiological significance of this impact is unknown [106].

Osteoblasts, which promote bone formation, and osteoclasts, which encourage the breakdown of mineralized tissue (breakdown of bone structure), control bone formation. Osteoclasts produce ROS (reactive oxygen species), and ROS encourage the breakdown of collagen chains, which is crucial for bone remodeling. Proteolytic enzymes like elastase and metalloproteinases can be activated by elevated oxidants, further harming the bone extracellular matrix. Antioxidant supplements may therefore aid in restoring bone mineral density (BMD) [107].

Ascorbic acid, a component of vitamin C, is crucial for procollagen and stable collagen secretion, both of which are necessary for the development of connective and bone tissue. *Rose hips* beneficial effects as an osteoporosis treatment are further supported by their high polyphenol content. Oxidative stress-related bone loss is known to be lessened by it. Strong evidence supports the idea that combining the seeds and skins of the *R. canina* subspecies helps people feel less pain and function better every day. It is evident that both the peel and the seed are required because a study on pure peel powder found no impact on patients' reported pain and discomfort [85, 108].

The beneficial effects of *R. canina* have not yet been fully explained in terms of their biochemical underpinnings, though. According to this perspective, the only molecule, a galactolipid by the name of GOPO that has anti-inflammatory properties, appears to be significant. Collagen plays a significant role in the healing process and the high natural vitamin C content in *R. canina* is readily absorbed. Linoleic acid which contributes to the COX-1 and COX-2 inhibitory effects found in rose hips may also offer an explanation in addition to triterpene acids [109, 110].

At the molecular level, OA involves an inflammatory process that produces excessive levels of RONS (reactive oxygen and nitrogen species). RONS generated by chondrocytes and associated with OA include superoxide, hydrogen peroxide, nitric oxide, and peroxynitrite [111]. Despite the fact that RONS is required as a second messenger in the control of normal physiological processes and as a destructive agent in the immune system's defense against pathogens, aberrant chondrocyte metabolism reduces its physiological buffering capacity [112]. An overabundance of anything causes oxidative stress. Proteins, lipids, nucleic acids, and extracellular matrix constituents can all be damaged by RONS overproduction. Induced cell damage and RONS signaling both amplify the inflammatory response at the same time. These comprise transcription factors, matrix-degrading enzymes (such as metalloproteinases MMP-3 andMMP-13), inflammatory cytokines (interleukins IL-1 and IL-6), and this process is closely linked to permanent cell cycle exit and chondrocyte senescence (also known as the "senescence-associated secretory phenotype") [113, 114].

TNF-α, IL-1, and IL-6 are pro-lipolytic inflammatory cytokines produced by adipocytes. Moreover, it prevents the synthesis of lipids. Lipids function as a negative feedback mechanism by suppressing the synthesis of type II collagen and proteoglycan in addition to promoting the production of prostaglandins, matrix metalloproteinases (MMPs), and other cytokines. In this method, pro-inflammatory cytokines cause bone resorption and cartilage matrix disintegration [115, 116].

5.3 Clinical Trial Study on Rose Hip Extract

This study aims to evaluate Reflex Plus's impact on people with osteoarthritis (OA). The study will last 12 or 24 weeks, not including the roughly 1-week screening period. Reflex Plus, a health supplement made of collagen hydrolysate and rosehip extract, is the test product under investigation. The supplement is believed to assist in the development of cartilage, which is essential for proper joint function and reducing arthritic pain. It is also helpful in managing joint discomfort and safeguarding cartilage. The primary aim of the study is to assess the difference in knee pain between the start and end of the trial, as determined by the WOMAC osteoarthritis index sum score, when comparing the supplement and placebo.

Study Phase: Phase 4

Condition: Osteoarthritis

Intervention: The dietary supplement contains 5 g of Reflex Plus Collagen Hydrolysate and 0.55 g of Rose hip Aqueous Extract. The placebo includes Fructose, Orange Flavor, and Sucralose.

Study: During the study duration, the Active Comparator Reflex Plus should be taken once a day before breakfast, using sachets.

Interventions: During the course of the study, the Reflex Plus placebo comparator dietary supplements placebo sachets should be consumed once daily before breakfast. Dietary Supplement is Placebo.

Status: Completed.

Rose hip extract contains galactolipids.-1,2-di-O-[(9Z,12Z,15Z)-octadeca-9,12,15trienoyl]-3-O-b-D-galactopyranosylglycerol, betulinic acid, oleanolic acid, ursolic acid, vitamin C, vitamin E, beta-carotene, lycopene and linoleic acid are all found in this supplement. Also contains monogalactosyldiglyceride and digalactosyldiglyceride. These active components control the inflammatory response and stop cartilage from being destroyed. Additionally, *Rose hip* extract has been shown to lower levels of the acute phase protein CRP and serum creatinine while also inhibiting peripheral blood polymorphonuclear leukocytes (PMN). Additionally, it was noted that *Rose hip* use lessens pain, stiffness, and disease severity while COX inhibits −2 and contributes anti-inflammatory properties. Rose hip powder was proven to be an effective dietary supplement for her OA knee patient. However, more research is needed to determine the therapeutic relevance of the reported good experimental outcomes.

6 Conclusion

OA is a frequent condition among the elderly. The first line of management consists of lifestyle changes and physical therapy, which is followed by analgesics and NSAIDs. However, these medications only treat symptoms and have no impact on how the disease develops naturally. Nutraceuticals refer to dietary supplements that are believed to modify the inflammatory process and, as a result, impact the natural progression of osteoarthritis. Collagen peptides are similarly effective in treating OA, although there are various formulations available on the market, each with its unique pharmacological effects. As a result, standardizing collagen peptide formulation is required before utilizing it to treat OA. Supplements containing collagen hydrolysate (CH) and glucosamine sulfate (GS) have been shown to be safe, however reports on their effects have been inconsistent. Despite lacking sufficient research, the European Society for Clinical and Economic Aspects of Osteoporosis and Osteoarthritis has recommended the use of nutraceuticals as a first-line treatment option for managing OA. Rose hip powder is considered a valuable nutraceutical in the treatment of OA patients due to its anti-inflammatory, chondroprotective, and immune-modulatory properties. However, more extensive research is necessary to establish its efficacy in the future.

R. canina, also known as rosehip, is a remarkable and gorgeous medicinal plant. It is a very potent source of antioxidants and anti-inflammatory compounds and can grow in very rocky soils. Additionally, it contains a lot of vitamins, particularly vitamin C. *Rose hips* are very effective at treating osteoarthritis in humans and some animal models, according to a number of clinical studies. It is crucial to note that a dose response, one of the pillars of the evaluation of prescription drugs, has been confirmed for *Rose hip* subspecies. There is proof that *Rose hip* treatment may be beneficial for conditions like rheumatoid arthritis, cardiovascular disease, and skin conditions like wrinkles. However, the amount of active ingredients in *Rose hips* varies. Different types of rose seeds exist. To identify the species that offer the greatest potential, a careful comparative analysis is all that is necessary. It is interesting to note that the *R. canina* variant of osteoarthritis was able to reduce the use of emergency medications by up to 50%, particularly when discussing anti-inflammatory diets and anti-inflammatory plants. As a result, you can alter your diet and possibly reduce the cost of your prescription drugs.

References

1. Martin JA, Buckwalter JA (2001) Roles of articular cartilage aging and chondrocyte senescence in the pathogenesis of osteoarthritis. Iowa Orthop J 21:1–7
2. Buckwalter JA, Mankin HJ (1997) Articular cartilage I. Tissue design and chondrocyte-matrix interactions. J Bone Joint Surg 79A(4):600–611
3. Buckwalter JA, Goldberg V, Woo SL-Y (1993) Musculoskeletal soft-tissue aging: impact on mobility. American Academy of Orthopaedic Surgeons, Rosemont, p 423
4. Buckwalter JA, Lane NE (1996) Aging, sports and osteoarthritis. Sports Med Arthrosc Rev 4: 276–287
5. Mow VC, Setton LA, Guilak F, Ratcliffe A (1995) Mechanical factors in articular cartilage and their role in osteoarthritis. In: Kuettner KE, Goldberg VM (eds) Osteoarthritic disorders. Rosemont, American Academy of Orthopaedic Surgeons, pp 147–171
6. Hunter DJ, Felson DT (2006) Osteoarthritis. BMJ 332(7542):639–642. https://doi.org/10.1136/bmj.332.7542.639. PMID: 16543327; PMCID: PMC1403209
7. Stewart HL, Kawcak CE (2018) The importance of subchondral bone in the pathophysiology of osteoarthritis. Front Vet Sci. PMID: 30211173
8. Chen D, Shen J, Zhao W, Wang T, Han L, Hamilton JL (2017) Osteoarthritis: toward a comprehensive understanding of pathological mechanism. Bone Res 5:16044. https://doi.org/10.1038/boneres.2016.44. PMID: 28149655; PMCID: PMC5240031
9. Felson DT, Anderson JJ, Naimark A et al (1988) Obesity and knee osteoarthritis. The Framingham study. Ann Intern Med 109:18–24

10. Radin EL (2004) Who gets osteoarthritis and why? J Rheumatol Suppl 70:10–15
11. Zhang Y, Jordan JM (2010) Epidemiology of osteoarthritis. Clin Geriatr Med 26(3):355–369. https://doi.org/10.1016/j.cger.2010.03.001. Erratum in: Clin Geriatr Med. 2013 May;29(2):ix. PMID: 20699159; PMCID: PMC2920533
12. Sen R, Hurley JA (2022) Osteoarthritis. [Updated 2022 May 1]. In: StatPearls [Internet]. StatPearls Publishing, Treasure Island. https://www.ncbi.nlm.nih.gov/books/NBK482326/
13. Ackerman IN, Cavka B, Lippa J, Bucknill A (2017) The feasibility of implementing the ICHOM standard set for hip and knee osteoarthritis: a mixed-methods evaluation in public and private hospital settings. J Patient Rep Outcomes 2:32
14. De Laroche R, Simon E, Suignard N et al (2018) Clinical interest of quantitative bone SPECT-CT in the preoperative assessment of knee osteoarthritis. Medicine (Baltimore) 97(35):e11943
15. Berenbaum F (2008) Osteoarthritis. B. Pathology and pathogenesis. In: Klippel JH, Stone JH, Crofford LJ, White PH (eds) Primer on the rheumatic diseases. Springer Science+Business Media, LLC, New York, pp 229–234
16. Symmons D, Mathers C, Pfleger B (2012) Global burden of osteoarthritis in the year 2000. World Health Organization website. www.who.int/healthinfo/statistics/bod_osteoarthritis.pdf
17. Bijlsma JWJ, Berenbaum F, Lafeber FPJG (2011) Osteoarthritis: an update with relevance for clinical practice. Lancet 377:2115–2126
18. Altman RD (2011) Osteoarthritis in the elderly population. In: Nakasato Y, Yung RL (eds) Geriatric rheumatology. A comprehensive approach. Springer Science+Business Media, LLC, New York, pp 187–196
19. Sakao K, Takahashi KA, Arai Y et al (2009) Osteoblasts derived from osteophytes produce interleukin-6, interleukin-8, and matrix metalloproteinase-13 in osteoarthritis. J Bone Miner Metab 27:412–423
20. Menkes C-J, Lane NE (2004) Are osteophytes good or bad? Osteoarthr Cartil 12(suppl A): S53–S54
21. Sellam J, Berenbaum F (2010) The role of synovitis in pathophysiology and clinical symptoms of osteoarthritis. Nat Rev Rheumatol 6:625–635
22. Scanzello CR, Goldring SR (2012) The role of synovitis in osteoarthritis pathogenesis. Bone 51:249–257
23. Goldring MB (2006) Update on the biology of the chondrocyte and new approaches to treating cartilage diseases. Best Pract Res Clin Rheumatol 20:1003–1025
24. Mobasheri A (2012) Osteoarthritis 2012 year in review: biomarkers. Osteoarthr Cartil 20: 1451–1464
25. Henrotin Y, Gharbi M, Mazzucchelli G et al (2012) Fibulin 3 peptides Fib3-1 and Fib3-2 are potential biomarkers of osteoarthritis. Arthritis Rheum 64:2260–2267
26. Wang Y, Li D, Xu N et al (2011) Follistatin-like protein 1: a serum biochemical marker reflecting the severity of joint damage in patients with osteoarthritis. Arthritis Res Ther 13: R193
27. Li D, Wang Y, Xu N et al (2011) Follistatin-like protein 1 is elevated in systemic autoimmune diseases and correlated with disease activity in patients with rheumatoid arthritis. Arthritis Res Ther 13:R17
28. Puranen J, Ala-Ketola L, Peltokallio P et al (1975) Running and primary osteoarthritis of the hip. Br Med J 2(5968):424–425
29. McAlindon TE, Wilson PW, Aliabadi P et al (1999) Level of physical activity and the risk of radiographic and symptomatic knee osteoarthritis in the elderly: the framing ham study. Am J Med 106(2):151–157
30. Newton PM, Mow VC, Gardner TR et al (1997) Winner of the 1996 Cabaud award. The effect of lifelong exercise on canine articular cartilage. Am J Sports Med 25(3):282–287
31. McAlindon TE, Felson DT, Zhang Y et al (1996) Relation of dietary intake and serum levels of vitamin D to progression of osteoarthritis of the knee among participants in the Framingham study. Ann Intern Med 125(5):353–359
32. Christensen R, Bartels EM, Astrup A, Bliddal H (2007) Effect of weight reduction in obese patients diagnosed with knee osteoarthritis: a systematic review and meta-analysis. Ann Rheum Dis 66:433–439. https://doi.org/10.1136/ard.2006.065904

33. Messier SP, Loeser RF, Miller GD, Morgan TM, Rejeski WJ, Sevick MA et al (2004) Exercise and dietary weight loss in overweight and obese older adults with knee osteoarthritis: the arthritis, diet, and activity promotion trial. Arthritis Rheum 50:1501–1510
34. Felson DT, Zhang Y, Anthony JM et al (1992) Weight loss reduces the risk for symptomatic knee osteoarthritis in women. The Framingham study. Ann Intern Med 116(7):535–539
35. Engstrom G, De Verdier MG, Nilsson PM et al (2009) Incidence of severe knee and hip osteoarthritis in relation to dietary intake of antioxidants beta-carotene, vitamin C, vitamin E and selenium: a population-based prospective cohort study. Arthritis Rheum 60:s235–s236
36. Fang W, Wu P, Hu R et al (2003) Environmental Se-Mo-B deficiency and its possible effects on crops and Keshan-Beck disease (KBD) in the Chousang area, Yao County, Shaanxi Province, China. Environ Geochem Health 25(2):267–280
37. Maillefert JF, Hudry C, Baron G, Kieffert P, Bourgeois P, Lechevalier D et al (2001) Laterally elevated wedged insoles in the treatment of medial knee osteoarthritis: a prospective randomized controlled study. Osteoarthr Cartil 9:738–745
38. Stewart M, Cibere J, Sayre EC, Kopec JA (2018) Efficacy of commonly prescribed analgesics in the management of osteoarthritis: a systematic review and meta-analysis. Rheumatol Int 38(11):1985–1997
39. Tugwell PS, Wells GA, Shainhouse JZ (2004) Equivalence study of a topical diclofenac solution (pennsaid) compared with oral diclofenac in symptomatic treatment of osteoarthritis of the knee: a randomized controlled trial. J Rheumatol 31(10):2002–2012
40. Bariguian Revel F, Fayet M, Hagen M (2020) Topical diclofenac, an efficacious treatment for osteoarthritis: a narrative review. Rheumatol Ther 7(2):217–236. https://doi.org/10.1007/s40744-020-00196-6
41. Yu SP, Hunter DJ (2015) Managing osteoarthritis. Aust Prescr 38(4):115–119. https://doi.org/10.18773/austprescr.2015.039
42. Pincus T, Swearingen C, Cummins P, Callahan LF (2000) Preference for nonsteroidal anti-inflammatory drugs versus acetaminophen and concomitant use of both types of drugs in patients with osteoarthritis. J Rheumatol 27:1020–1027
43. Yusuf E (2016) Pharmacologic and non-pharmacologic treatment of osteoarthritis. Curr Treat Options Rheumatol 2:111–125. https://doi.org/10.1007/s40674-016-0042-y
44. Hochberg MC, Dougados M (2001) Pharmacological therapy of osteoarthritis. 15(4):583–593. https://doi.org/10.1053/berh.2001.0175
45. Wolfe F, Zhao S, Lane N (2000) Preference for nonsteroidal antiin¯ammatory drugs over acetaminophen by rheumatic disease patients. Arthritis Rheum 43:378–385
46. Katz WA (1996) Pharmacology and clinical experience with tramadol in osteoarthritis. Drugs 52(Suppl 3):39–47
47. Knotkova H, Fine PG, Portenoy RK (2009) Opioid rotation: the science and the limitations of the equianalgesic dose table. J Pain Symptom Manag 38:426–439. https://doi.org/10.1016/j.jpainsymman.2009.06.001
48. Brosseau L, Yonge KA, Robinson V, Marchand S, Judd M, Wells G, Tugwell P (2003) Thermotherapy for treatment of osteoarthritis. Cochrane Database Syst Rev 2003(4): CD004522. https://doi.org/10.1002/14651858.CD004522
49. Hecht PJ, Backmann S, Booth RE, Rothman RH (1983) Effects of thermal therapy on rehabilitation after total knee arthroplasty: a prospective randomized study. Clin Orthop Relat Res 178:198–201
50. Kjeken I, Smedslund G, Moe RH, Slatkowsky-Christensen B, Uhlig T, Hagen KB (2011) Systematic review of design and effects of splints and exercise programs in hand osteoarthritis. Arthritis Care Res 63:834–848. https://doi.org/10.1002/acr.20427
51. Beaudreuil J (2016) Orthoses for osteoarthritis: a narrative review. Ann Phys Rehabil Med: S1877065716305310. https://doi.org/10.1016/j.rehab.2016.10.005
52. Paterson KL, Bennell KL, Wrigley TV, Metcalf BR, Campbell PK, Kazsa J, Hinman RS (2018) Footwear for self-managing knee osteoarthritis symptoms: protocol for the Footstep randomised controlled trial. BMC Musculoskelet Disord 19(1):219. 10.1186/s12891-018-2144-1
53. Carbone LD, Satterfield S, Liu C, Kwoh KC, Neogi T, Tolley E, Nevitt M, Health ABC Study (2013) Assistive walking device use and knee osteoarthritis: results from the health, aging and

body composition study (Health ABC study). Arch Phys Med Rehabil 94(2):332–339. https://doi.org/10.1016/j.apmr.2012.09.021
54. White DK, Tudor-Locke C, Zhang Y, Fielding R, LaValley M, Felson DT, Gross KD, Nevitt MC, Lewis CE, Torner J, Neogi T (2014) Daily walking and the risk of incident functional limitation in knee osteoarthritis: an observational study. Arthritis Care Res 66(9):1328–1336. https://doi.org/10.1002/acr.22362
55. Lv X, Ta N, Chen T, Zhao J, Wei H (2022) Analysis of gait characteristics of patients with knee arthritis based on human posture estimation. BioMed Res Int 2022:1–8. https://doi.org/10.1155/2022/7020804
56. Katz JN, Earp BE, Gomoll AH (2010) Surgical management of osteoarthritis. Arthritis Care Res 62(9):1220–1228. https://doi.org/10.1002/acr.20231
57. de l'Escalopier N, Anract P, Biau D (2016) Surgical treatments for osteoarthritis. Ann Phys Rehabil Med:S1877065716300355. https://doi.org/10.1016/j.rehab.2016.04.003
58. de Andrade P, Antônio M, de Campos TVO, de Abreu-e-Silva GM (2015) Supplementary methods in the nonsurgical treatment of osteoarthritis. Arthroscopy 31(4):785–792. https://doi.org/10.1016/j.arthro.2014.11.021
59. Towheed TE, Maxwell L, Anastassiades TP, Shea B, Houpt J, Robinson V et al (2005) Glucosamine therapy for treating osteoarthritis. Cochrane Database Syst Rev 2:CD002946
60. Henrotin Y, Marty M, Mobasheri A (2014) What is the current status of chondroitin sulfate and glucosamine for the treatment of knee osteoarthritis? Maturitas 78:184–187. https://doi.org/10.1016/j.maturitas.2014.04.015
61. Hill CL, March LM, Aitken D et al (2016) Fish oil in knee osteoarthritis: a randomised clinical trial of low dose versus high dose. Ann Rheum Dis 75:23–29
62. Peanpadungrat P (2015) Efficacy and safety of fish oil in treatment of knee osteoarthritis. J Med Assoc Thai Chotmaihet thangphaet 98(Suppl 3):S110–S114
63. Shengelia R, Parker SJ, Ballin M, George T, Reid MC (2013) Complementary therapies for osteoarthritis: are they effective? Pain Manag Nurs 14(4):e274–e288. https://doi.org/10.1016/j.pmn.2012.01.001
64. Bhardwaj S, Verma R, Gupta J (2018) Challenges and future prospects of herbal medicine. Int Res Med Health Sci 1(1):12–15
65. World Health Organization Traditional medicine strategy: 2014–2023 (2013) SAR, World Health Organization, Hong Kong
66. Kelly K (2009) History of medicine. Facts on File, New York, pp 29–50
67. Wiart C (2006) Etnopharmacology of medicinal plants. Humana Press, New Jersey, pp 1–50
68. U.S. Food and Drug Administration (2011) Regulatory framework of DSHEA of 1994. http://www.fda.gov/NewsEvents/Testimony/ucm115163.htm
69. Rodrigues E, Barnes J (2013) Pharmacovigilance of herbal medicines: the potential contributions of ethnobotanical and ethnopharmacological studies. Drug Saf 36:1–2
70. Ahmad N (2016) Essential oils in food preservation, flavor and safety||rose hip (Rosa canina L.) oils:667–675. https://doi.org/10.1016/b978-0-12-416641-7.00076-6
71. Jürgens AH, Seitz B, Kowarik I (2007) Genetic differentiation of *Rosa canina* (L.) at regional and continental scales. Plant Syst Evol 269(1–2):39–53
72. Cheikh-Affene ZB, Haouala F et al (2013) Pomological description and chemical composition of rose hips gathered on four *Rosa* species section Caninae growing wild in Tunisia. Int J Agric Sci Technol 1(3):43–50
73. Hunter I (2014) Declared plant policy under the natural resources management act 2004: dog rose (*Rosa canina*). Gov South Australia 2014:4
74. FairWild Foundation (2016) FairWild-certified ingredients under production. FairWild Foundation website. FairWild Foundation, Cambridge. www.fairwild.org/certification-overview
75. Cohen M (2012) Rose hip: an evidence based herbal medicine for inflammation and arthritis. Aust Fam Physician 41(7):495–498
76. Wichtl M (ed), Brinckmann JA, Lindenmaier MP (trans) (2004) Herbal drugs and phytopharmaceuticals, 3rd edn. Medpharm GmbH Scientific Publishers, Stuttgart

77. British Herbal Medicine Association (1996) British herbal pharmacopoeia. British Herbal Medicine Association, Dorset
78. Carvalho AM, Barros L, Ferreira ICFR, Frazão-Moreira A (2009) Folk medicine of Trás-Os-Montes (Portugal): traditional uses and bioactive compounds of six common medicinal species. Sesin 4 – Fitoterapia y Plantas Medicinales. 5th international congress of ethnobotany, San Carlos de Bariloche, Argentina
79. Blumenthal M, Busse WR, Goldberg A et al (eds) (1998) The complete German commission E monographs – therapeutic guide to herbal medicines. American Botanical Council/Integrative Medicine Communication, Austin/Boston
80. US Environmental Protection Agency (2015) Part 180-tolerances and exemptions for pesticide chemicals in food. In: Code of federal regulations, 40 CFR. US Government Printing Office, Washington, DC, pp 386–738
81. US Food and Drug Administration (2015) §182.20 essential oils, oleoresins (solvent-free), and natural extractives (including distillates). In: Code of federal regulations, 21 CFR. US FDA, Washington, DC, p 477
82. Ercisli S (2007) Chemical composition of fruit in some rose (Rosa ssp.) species. Food Chem 104:1379–1384
83. Zlatanov MD (1999) Lipid composition of Bulgarian chokeberry, black current and rose hip seed oils. J Sci Food Agric 79:1620–1624
84. Thimmappa R, Geisler K, Louveau T, O'Maille P, Osbourn A (2014) Triterpene biosynthesis in plants. Annu Rev Plant Biol 65:225–257
85. Fecka I (2009) Qualitative and quantitative determination of hydrolysable tannins and other polyphenols in herbal products from meadowsweet and dog rose. Phytochem Anal 20: 177–190
86. Buchanan BB, Gruissem W, Jones RL (eds) (2000) Biochemistry and molecular biology of plants. John Wiley & Sons, New York
87. Dennis DT, Turpin DH, Lefebvre DD, Layzell DB (eds) (1997) Plant metabolism, 2nd edn. SPB Academic Publishing, The Hague
88. Lotito SB, Frei B (2006) Consumption of flavonoid-rich foods and increased plasma antioxidant capacity in humans: cause, consequence, or epiphenomenon? Free Radic Biol Med 41(12):1727–1746
89. Chung KT, Wong TY, Wei CI, Huang YW, Lin Y (1998) Tannins and human health: a review. Crit Rev Food Sci Nutr 38:421–464
90. Comalada M, Camuesco D, Sierra S et al (2005) In vivo quercetin anti-inflammation effect involves release of quercetin, which inhibits inflammation through down-regulation of the NF-kappaB pathway. Eur J Immunol 35(2):584–592
91. Formica JV, Regelson W (1995) Review of the biology of quercetin and related bioflavonoids. Food Chem Toxicol 33(12):1061–1080
92. Rao YK, Geethanqili M, Fang SH, Tzeng YM (2007) Antioxidant and cytotoxic activities of naturally occurring phenolic and related compounds: a comparative study. Food Chem Toxicol 45(9):1770–1776
93. Kiokas S, Varzakas T, Oreopoulou V (2008) In vitro activity of vitamins, flavonoids, and natural phenolic antioxidants against the oxidative deterioration of oil-based systems. Crit Rev Food Sci Nutr 48:78–93
94. Sumantran VN, Zhang R, Lee DS, Wicha MS (2000) Differential regulation of apoptosis in normal versus transformed mammary epithelium by lutein and retinoic acid. Cancer Epidemiol Biomark Prev 9:257–263
95. Borek C (2004) Dietary antioxidants and human cancer. Integr Cancer Ther 3:333–341
96. Wigle DT, Turner MC, Gomes J, Parent ME (2008) Role of hormonal and other factors in human prostate cancer. J Toxicol Environ Health B Crit Rev 11(3–4):242–259
97. Bjelakovic G, Nikolova D, Gluud LL, Simonetti RG, Gluud C (2007) Mortality in randomized trials of antioxidant supplements for primary and secondary prevention: systematic review and meta-analysis. JAMA 297(8):842–857

98. Jansen MC, Van Kappel AL, Ocké MC et al (2004) Plasma carotenoid levels in Dutch men and women, and the relation with vegetable and fruit consumption. Eur J Clin Nutr 58:1386–1395
99. Semba RD, Dagnelie G (2003) Are lutein and zeaxanthin conditionally essential nutrients for eye health? Med Hypotheses 61(4):465–472
100. Hodisan T, Socaciu C, Ropan I, Neamtu G (1997) Carotenoid composition of Rosa canina fruits determined by thin-layer chromatography and high-performance liquid chromatography. J Pharm Biomed Anal 16:521–528
101. Krinsky NI, Landrum JT (2003) Bone RA. Biologic mechanisms of the protective role of lutein and zeaxanthin in the eye. Annu Rev Nutr 23:171–201
102. Giovannuci E, Ascherio A, Rimm EB, Stampfer MJ, Colditz GA, Willet WC (1995) Intake of carotenoids and retinol in relation to risk of prostate cancer. J Natl Cancer Inst 87(23): 1767–1776
103. Giovannucci E (2002) A review of epidemiologic studies of tomato, lycopene and prostate cancer. Exp Biol Med 227(10):852–859
104. Halvorsen BL, Holte K, Myhrstad MC et al (2002) A systematic screening of total antioxidants in dietary plants. J Nutr 132(3):461–471
105. Ozcan M (2002) Nutrient composition of rose (Rosa canina L.) seed and oils. J Med Food 5(3): 137–140
106. Hemilä H (2006) Do vitamin C and E affect respiratory infection? University of Helsinki, Helsinki
107. Banfi G, Iorio EL, Corsi MM (2008) Oxidative stress, free radicals and bone remodeling. Clin Chem Lab Med 46:1550–1555
108. Scheweita S, Khoshhal K (2007) Calcium metabolism and oxidative stress in bone fractures: role of antioxidants. Curr Drug Metab 8:519–525
109. Kharazmi A, Winther K (1999) Rose hip inhibits chemotaxis and chemiluminesence of human peripheral blood neutrophils in vitro and reduces certain inflammatory parameters in vivo. Inflammopharmacology 7(4):377–386
110. Marstrand K, Warholm L, Pedersen F, Winther K (2017) Dose dependent impact of rose-hip powder in patients suffering from osteoarthroitis of the hip and or knee – a double blind, randomized, placebo controlled, parallel group, phase III study. Int J Complementa Altern Med 7(1):1–9
111. Larsen E, Kharazmi A, Christensen LP, Christensen SB (2003) An anti-inflammatory galactolipid from Rose hip (Rosa canina) that inhibits chemotaxis of human peripheral blood neutrophils in vitro. J Nat Prod 66:994–995
112. Winther K, Kharazmi A, Hansen ASV, Falk-Rønne J (2012) The absorption of natural vitamin C in horses and anti-oxidative capacity: a randomized, controlled study on trotters during a three month intervention period. Comp Exerc Physiol 8(3/4):195–201
113. Jäger AK, Petersen KN, Thomasen G, Christensen SB (2008) Isolation of linoleic and alpha-linolenic acids as COX-1 and COX-2 inhibitors in rose hip. Phytother Res 22:982–984
114. Saaby L, Jäger AK, Moseby L, Hansen EW, Christensen SB (2011) Isolation of immunomodulatory triterpene acids from a Staqndardized rose hip powder (Rosa canina L). Phytother Res 25:195–201
115. Gruenwald J, Uebelhack R, Moré MI (2019) Rosa canina – rose hip pharmacological ingredients and molecular mechanics counteracting osteoarthritis – a systematic review. Phytomedicine 152958. https://doi.org/10.1016/j.phymed.2019.152958
116. Vaishya R, Agarwal AK, Shah A, Vijay V, Vaish A (2018) Current status of top 10 nutraceuticals used for Knee Osteoarthritis in India. J Clin Orthop Trauma 9(4):338–348. https://doi.org/10.1016/j.jcot.2018.07.015. Erratum in: J Clin Orthop Trauma. 2020 Nov–Dec;11(6):1175. PMID: 30449982; PMCID: PMC6224802

Physiological and Biochemical Outcomes of Herbal Medicine Use in the Treatment of Hypertension

28

Annaletchumy Loganathan and Natalia Shania Francis

Contents

Abstract

Complementing herbs with anti-hypertensive drugs is a standard health treatment for those with uncontrolled high blood pressure. This review compares the effects of completing different types of herbs with anti-hypertensive drugs on the physiological and biochemical outcomes such as blood pressure, body mass index

A. Loganathan (✉) · N. S. Francis
Department of Allied Health Sciences, Faculty of Science, Universiti Tunku Abdul Rahman (UTAR), Kampar, Malaysia
e-mail: annal@utar.edu.my; fnata90@1utar.my

S. C. Izah et al. (eds.), *Herbal Medicine Phytochemistry*, Reference Series in Phytochemistry, https://doi.org/10.1007/978-3-031-43199-9_44

(BMI), waist-hip circumference, glucose levels, hemoglobin level, and lipid profiles. Ambulatory blood pressure measurements (ABPMs) help detect blood pressure variations and serve as a cardiovascular risk marker for hypertension patients. In the clinical practice guidelines, ABPMs are included in diagnosing and managing isolated office hypertension among pregnant women with masked or labile hypertension, as the figure denotes blood pressure readings over 24 h. Point-of-care (PoC) testing devices provide rapid results and information at the location and time of care, the measuring procedure of which can even be done at home. This practice eliminates negative biases, and PoC equipment is capable of chalking up measurements as accurate as those of laboratory-based biochemical analyzers. PoC reduces cost, waiting time, and the necessity for multiple visits improves quality of life (QoL) and increases accessibility of people living in remote areas to medical testing. However, taking ABPMs in actual clinical settings and PoC technology are grossly underutilized in the Asian regions. Hence, this chapter will review evidence of complementing herbal medicine with anti-hypertension therapies and the physiological and biochemical outcomes using ABPMs and PoC devices.

Keywords

Herbal medicine · Traditional herbal medicine · Hypertension · Ambulatory blood pressure monitor · POC · Blood pressure · Blood lipids · Blood glucose · Hemoglobin

Abbreviations

ABPM	ambulatory blood pressure measurements
ACE	American Council on Exercise
BMI	body mass index
DBP	diastolic blood pressure
HDL-C	high-density lipoprotein cholesterol
LDL-C	low-density lipoprotein cholesterol
NO	nitric oxide
PoC	point of care
QoL	quality of life
SBP	systolic blood pressure
T&CM	traditional and complementary medicine
TC	total cholesterol
TCM	traditional Chinese medicine
TG	triglycerides

1 Introduction

Hypertension is a known global health issue that affects people regardless of age, race, and socio-economic background; it is a leading cause of premature death. Hypertension is defined as persistent elevations of systolic blood pressure (SBP) of

140 mm/Hg or greater and diastolic blood pressure (DBP) of 90 mm/Hg or more significant, taken at least twice on two separate occasions [1]. The prevalence of hypertension varies worldwide but is more common in low- and middle-income countries, and ironically, the number of affected people being undertreated is proportionally higher [2]. In Malaysia, the number of people diagnosed with hypertension from 2015 to 2021 increased by 14.1% among individuals above the age of 30 years [3].

The anti-hypertensive medication is often the first-line treatment; however, many people complement medication with herbal medicine to treat hypertension. Herbal medicine could be raw plants that are dried, minimally processed, mildly boiled, infused or herbal materials, herbal preparations, and finished herbal products that contain active ingredients, parts of plants, or other plant materials, or combinations [4, 5]. However, the safety and efficacy of concurrently taking herbal medicine and anti-hypertensive drugs have yet to be validated; the same goes for the potential risks and benefits. A study conducted by Asgary et al. [6] reported that anti-hypertensive drugs prescribed with *Vaccinium myrtillus* efficiently controlled hypertension. Besides blood pressure, beneficial evidence was also observed on parameters like BMI, glucose, and lipid when herbs such as *Vaccinium arctostaphylos* (Whortleberries) were complemented with anti-hypertensive drugs in treating blood pressure [7].

Subsequently, previous studies have also reported adverse effects associated with complementing herbal medicine and anti-hypertensive drugs. For instance, when *Nigella sativa* was taken with ACE inhibitors and diuretics in a study carried out in Indonesia, side effects such as dyspepsia, nausea, and constipation were observed among the participants [8]. Moreover, musculoskeletal pain, fatigue, breathlessness, itching, and even epigastric discomfort were exhibited among the patients who concomitantly consumed amlodipine with *Phyllanthus emblica* extracts [9]. Therefore, monitoring the physiological and biochemistry outcomes of using herbal medicine concurrently with any anti-hypertensive drug is imperative.

Globally, point-of-care (PoC) devices are increasingly popular monitoring instruments for measuring physiological and biochemistry outcomes near patients' bedside. PoC devices are handy in several settings, including remote areas. PoC-measured values obtained immediately within 5 min were comparable to those obtained in laboratories [10]. One of the prime PoC devices that measures physiological outcomes is the ambulatory blood pressure monitor (ABPM), a non-invasive tool that monitors repeated and continuous blood pressure over 24 h. This is useful as blood pressure varies throughout the day, and a single measurement may not be accurate enough to diagnose a person with hypertension [11]. Previous studies have reported that ABPM is the gold standard as it provides consistency and is an excellent diagnostic tool for hypertension [12].

Moreover, portable PoC devices can measure biochemistry parameters of lipid profiles such as cholesterol, triglycerides, glucose, and hemoglobin levels. PoC devices can provide producible physiological and biochemistry outcomes at a reduced cost, provide timely information regarding patients' health status, and reduce the necessity for multiple visits [12]. A home blood glucose monitoring device is essential for diabetes patients to monitor their glucose levels. In accordance, the ABPM and other PoC devices are extremely useful to patients taking herbal medicine anti-hypertensive drugs concurrently; they can make better decisions by observing the changes in blood pressure as well as the biochemistry level after ingesting both in different scenarios at a time, or at an additional time, in the short term or the long term, respectively. In addition, the monitoring devices must be user-friendly and suitable for individuals of all age groups suffering from chronic ailments.

A limitation was reported after using an ABPM in a study: less tolerability as patients may face difficulties self-fixing the machine. However, the ABPM blood pressure measurements have many advantages as they can be saved, downloaded, or printed and shown to medical personnel at clinical visits [13]. Moreover, a study conducted in Korea also elucidated similar beneficial outcomes of a PoC device known as the GCare lipid analyzer, which can measure lipid and glucose levels in a single machine. Other limitations of PoC devices reported are unreliable, unusable, and biased measurements compared to laboratory analyzers. Nevertheless, the PoC-produced measurements using capillary blood were in the acceptable range [14].

There are divergent outcomes of using PoC devices to gauge the concurrent effects of taking herbal medicine and conventional anti-hypertensive drugs. Thus, a thorough investigation among patients with hypertension is warranted. This chapter reviews the physiological and biochemistry outcomes of concurrently taking herbal medicine and anti-hypertension drugs among hypertensive adults.

2 Epidemiology of Herbal Medicine Use with Anti-hypertensive Drugs

Globally, the concurrent use of herbal and anti-hypertensive drugs to treat hypertension is a prevalent trend. In a study carried out among the South African communities, 21% of the study population was reported to have used traditional herbal medicine with anti-hypertensive drugs to treat their hypertension [15]. On the other hand, Jordan said an astounding 51.4% of the population uses local herbal medicine with anti-hypertensive medications, a trend similar to those in other countries like China and Iraq [16]. This could be attributed to the affordability and availability of these various local medicinal herbs and the strong beliefs of the communities in these countries to consume herbs as an effective means of curbing hypertension [16].

In Tanzania, 22.1% of the study population on anti-hypertensive drugs to manage hypertension used herbal medicine such as garlic, ginger, beetroots, papaya, moringa seeds, and Chinese medicine [17]. In a study carried out by Joachimdass [18] in a sub-urban setting of Malaysia, 30.6% of the study population was found to complement their medical treatment with self-prepared herbs such as *Centella Asiatica*, *Momordica charantia*, *Carica papaya*, *Cucumis sativus*, and *Parkia speciosa* with the highest usage. The previous study reported these factors were associated with complementing herbal medicine with conventional anti-hypertensive drugs: being females, having higher education levels, having high socio-economic backgrounds, and dealing with multiple co-morbidities [19].

Even though increasingly hypertensive patients are taking herbal medicine and conventional anti-hypertensive drugs concurrently, the physiological and biochemistry outcomes vary from patient to patient. In a previous study conducted in Tehran, Iran, a local herb, *V. arctostaphylos* berry, was complemented at different doses among subjects who were on prescribed anti-hypertensive drugs, the combinations of which were reported to control both systolic and diastolic readings [20]. The phytochemical compounds contained in the herbal plant, such as chlorogenic acid, anthocyanin, and others, produce no side effects in the study subjects [4, 7]. Similar findings were also observed in another study in Egypt; ginger slices were taken by the study population on ACE inhibitors [21]. However, in a previous study conducted among hypertension patients who used ginger powder alone did complain of gastrointestinal discomfort and diarrhea [22].

3 Evidence-Based Herbal Medicine Use in the Treatment of Hypertension

Even though many pharmaceutical drugs are readily available to treat hypertension, some people may prefer herbal medicine as an alternative and complementary therapy. Below are some evidence-based herbal medicines studied for their potential benefits and effectiveness in combating hypertension.

The first example is garlic (*Allium sativum)*, a renowned herb used since ancient times for its blood-pressure-lowering effects. In a clinical trial in Australia, the participants consumed two garlic capsules rich in a biochemical constituent named allicin over a 12-week intervention period [23]. The results showed that 11.8 mm/Hg significantly reduced the study participants' mean systolic blood pressure; specific blood lipid parameters like LDL and total cholesterol also registered improvements.

Besides, *hibiscus sabdariffa* is another herb that potently affects blood pressure readings. Based on a study published by Mckay and colleagues [24], it was discovered that drinking hibiscus tea for 6 weeks lowered systolic and diastolic readings by an average of 7.2 mm/Hg and 3.1 mm/Hg, respectively. In addition, the supplementation of hibiscus also brings about hypocholesterolemic effects because of its antioxidant properties, regardless of the age and gender of the consumers.

Furthermore, *Olea europaea* L. (olive leaf) extracts also exhibit blood-pressure-lowering effects in a clinical trial. It was reported that positive outcomes could be seen in blood pressure and lipid profiles due to the active phytochemistry constituent, namely, oleuropein, found in the olive leaf extracts. This is because oleuropein, a polyphenolic compound, inhibits angiotensin-converting enzyme (ACE) through its hypotensive actions [25]. *Momordica charantia* L. (bitter gourd) is commonly used to prevent hypertension. A meta-analysis study reported no significant difference after consumption of bitter gourd; however, reductions in systolic and diastolic measurements were observed among younger adults when used on a short-term basis [26].

Hypertensive adults should always be cautious when consuming herbal medicine and be more transparent about their usage to healthcare providers, as there could be some adverse effects, which may lead to poor hypertension management.

4 Evidence on Herbal Medicine Interaction with Anti-hypertensive Drugs

Nowadays, herbal medicine as an alternative or complementary therapy is rising. However, this herbal complementation to medical treatment may lead to potential interactions between the two, producing adverse effects on the body. Moreover, the effectiveness of conventional medication could also be impacted negatively because of the herbs' complementation. Hence, the following information sheds light on the interactions between herbal remedies and anti-hypertensive drugs.

On the other hand, Hawthorn is an excellent example of an herbal remedy that may improve blood pressure readings. On the other hand, there have been reports of explicit interactions between hawthorn and two classes of conventional first-line drugs that are typically used to treat high blood pressure. The drugs include beta-blockers, for instance, atenolol and propranolol, as well as calcium channel blockers like amlodipine. Both drugs work by dilating the blood vessels; concurrent intake of hawthorn may notably lead to a spike in the effectiveness of the drugs, which in turn causes repercussions like excessive lowering of blood pressure [27].

Panax ginseng (ginseng), which possesses ginsenoside-Rg3, is a steroidal saponin compound that provides anti-hypertensive effects. These outcomes are likely mediated by myogenic response inhibitions on blood vessels [28]. Nevertheless, interactions between ginseng and anti-hypertensive medications such as calcium channel blockers (amlodipine) have been noticed. A study by Jeon et al. [29] elucidated that this interaction may lead to elevated blood pressure readings and dire consequences as the drug's potency may be diminished. Similar outcomes of high blood pressure readings due to decreased effectiveness of drugs were also discovered when *Hypericum perforatum* (St. John's wort) interacted with calcium channel blockers, digoxin, and warfarin [27].

Furthermore, *Glycyrrhiza glabra* (liquorice root) is commonly used in beverages and as a herbal antidote. Research done by Deutch et al. [30] sheds light on the interfering effects of liquorice with hypertensive medication, such as ACE inhibitors. The active phytochemistry constituent, namely, glycyrrhizin, which is found in liquorice, causes interactions with anti-hypertensive drugs that are metabolized by Cytochrome P450 3A4 (CYP3A4), an enzyme found in the liver and intestine, which increases blood pressure. Conversely, *Salvia miltiorrhiza* (Danshen) can interact with diuretics and lead to uncontrolled hypotension outcomes [27].

Table 1 contains the types of herbs and drugs that lead to improvements when used concomitantly, as well as the phytochemical constituents and mechanism of action that lead to specific outcomes.

Table 1 Examples of herbal medicine with anti-hypertensive properties used in clinical studies

Author, year	Number of subjects	Type of herbs	Results and implications	Country
[20]	100	400 mg *V. arctostaphylos* berry leaf extract capsule	• The *V. arctostaphylos* berry leaf extract capsule significantly reduced systolic and diastolic blood pressure due to the presence of gallic acid, chlorogenic acid, and anthocyanin • Another study further justified that the anti-hypertensive effect of this leaf extract was indeed due to the presence of anthocyanin, chlorogenic acid, and quercetin • These constituents mediated actions such as endothelial nitric oxide synthase activation, ACE inhibition, as well as calcium antagonistic action [7]	Tehran, Iran
[21]	40	Ginger slices	• Blood pressure readings were controlled and found to be less than 120/80 mm/Hg over 1 month when ginger slices were complemented Improvements were attributed to the phytochemical constituents found in ginger such as gingerols when taken concurrently with conventional medicine When 1.2 g of powdered ginger was used instead of slices, reductions in blood pressure were still observed although not significantly. Overall, ginger is still deemed to possess hypotensive effects as it can inhibit ACE and work as an antagonist [22]	Menoufia University Hospital, Egypt

5 Evidence on the Integration of Herbal Medicine into the Conventional Medical Practices

Some types of complementary and herbal medicine practices directly integrated into the conventional medical methods in Malaysia, Shanghai, and Beijing are traditional Chinese medicine, acupuncture, and traditional massages. Herbal medicine practices are integrated with the mainstream of conventional medicine for various ailments, namely, chronic pain, post-stroke and cancer treatments, hypertension, and even for postpartum mothers who are in recovery. In Malaysia, 15 hospitals have started integrating traditional and complementary medicine (T&CM) into modern healthcare systems, including Malay, Chinese, and Indian traditional treatment practices, homeopathy, chiropractic, osteopathy, and Islamic medicine [31]. The T&CM system integration is similar to the application in Shanghai and Beijing, respectively; to receive herbal medicine, the patients must first seek help from Western medicine practitioners and obtain a referral letter with indications of their health problems.

In Malaysia, with a referral letter from a general practitioner, the patients can be treated by doctors who have majored in traditional and complementary medicine and assisted by trained nurses [31]. However, in Shanghai and Beijing, certified Traditional Chinese clinicians will treat patients with a referral [32]. It was reported that one-third of the patients who visited Yuetan Chinese Community Health Service (Yuetan CHS) in Beijing were referred to traditional Chinese medicine practitioners (TCM), as the conventional treatments were unsuccessful. Table 2 contains information on the types of T&CM provided and the specific types of diseases treated in these hospitals [31–34].

Integrating T&CM into modern medical systems is still an area of ongoing research, particularly in treating hypertension and other diseases at hospitals. Limited concrete evidence is available on the types of herbs that can be complemented with drugs and their potency in treating various disorders. Thus, it is of utmost importance to continually and extensively evaluate the efficacy of integrating herbal medicine into conventional medicine practices; it would be helpful for the relevant authorities to develop comprehensive regulatory frameworks for the usage and practice. Moreover, more quality research should be done to overcome the methodological limitations of integration in hospitals; adequate training and education should be provided to medical personnel on safely implementing the integration [35].

6 Anthropometry Measurements

Anthropometry is the quantitative measurement of the physical body using non-invasive tools such as height-measuring devices and weighing scales. The measures comprise height, weight, body mass index (BMI), head, waist, and hip circumferences, skin fold thickness, and waist/hip ratios, all of which can be utilized to assess the risk of cardiovascular diseases such as hypertension [36]. The measurements of waist/hip and the resultant ratios determine body fat distribution, which is correlated

Table 2 Examples of conventional medical practices with the integration of herbal treatments worldwide

Location	T&CM provided	Ailments treated	References
Cheras Rehabilitation Hospital, Kuala Lumpur, Malaysia	Basti therapy, Shirodhara, traditional massage and acupuncture	Chronic pain and post-stroke	[31]
National Cancer Institute, Putrajaya, Malaysia	Herbal therapy for cancer patients and acupuncture	Chronic pain, post-stroke, chemotherapy-induced side effects	[33]
Sungai Buloh Hospital, Selangor	Traditional Indian medicine (varmam therapy)	Pain management in knee osteoarthritis, shoulder, cervical and lumbar	[31]
Shanghai Yueyang Integrated TCM and Western Medicine Hospital	Traditional Chinese and Western medicine, massage disciplines such as tuina and acupuncture as well as dermatology	Spine, stroke, heart rehabilitation, and several disorders	[34]
Yuetan Chinese Community Health Service (CHS), Beijing	Acupuncture and herbal remedies for (Management of common and chronic diseases and physical rehabilitation)	Knee and back pain, headaches, menopausal symptoms, infertility, hypertension, gastrointestinal disorders, urinary problems, and emotional/mental problems	[32]

with hypertension and insulin resistance. In contrast, changes in fat mass contribute to raised blood pressure readings [37]. Therefore, anthropometry measurements taken at health facilities will facilitate identifying patients at risk for hypertension.

6.1 Body Mass Index (BMI)

Higher BMI levels are often associated with co-morbidities such as hypertension, diabetes, and elevated cholesterol readings. The following are some of the studies that examined the BMI outcomes of hypertensive individuals who consumed herbal products and conventional medicine at the same time.

A randomized controlled trial investigated the effects of *Cynara scolymus* (Artichoke) combined with the anti-hypertensive drug captopril on hypertensive patients' blood pressure and BMI readings. After an eight-week intervention period, the patients with co-morbidities such as obesity and diabetes were observed to show improvements in blood pressure and BMI levels [38]. Artichoke is rich in inulin and oligofructose phytochemistry bioactive compounds, which help maintain a balance in gut microbiota. This decreases the amount of inflammation in the body and promotes the modulation of hormones in the gastrointestinal tract, leading to BMI improvements [39].

Another study reported that green tea contains catechin polyphenols, which increase the hormone leptin, which maintains average body weight in the long term [40]. However, the reported results are inconsistent; it is uncertain whether the consumption of green tea improves weight loss and BMI levels. A study by Bogdanski and colleagues [41] reported no significant changes in the BMI of obese hypertensive participants when they complemented green tea extracts with anti-hypertensive drugs for 3 months. Hence, green tea consumption improves weight loss, inflammatory markers, and glucose homeostasis, but its effects on BMI require more research, mainly when used with anti-hypertensive drugs [40].

6.2 Fat Mass

The correlation between body composition and hypertension risk is inconsistent in the previous studies. Ye et al. [42] revealed a positive association between excess fat mass and elevated blood pressure among adults. Direct fat mass measurements require sophisticated equipment, such as machines with dual-energy X-ray imaging technologies and densitometry. Fat mass measurements are indirectly taken using the skinfold method and a bioimpedance machine [43]. The American Council on Exercise (ACE) has developed a guideline for men and women; men with fat mass percentages of above 25% and women above 32% are classified as obese. The following paragraphs explain the outcome of fat mass measurements for concomitant herbal medicine usage with anti-hypertensive drugs among hypertensive adults.

In a study by Boix-Castej'on et al. [44], hypotensive effects and the outcomes of fat mass percentage were observed for consuming two capsules of herbal medicine containing *Lippia citriodora* and *Hibiscus sabdariffa* extract, the participants of which were adults with stage-1 hypertension and were obese. Significant reductions were noted in fat mass percentage and systolic blood pressure readings. Moreover, the improvements in weight loss and BMI were also attributed to the participant's compliance with a regimen of proper exercise and diet. The combination of polyphenolic compounds such as verbascoside and hibiscetin from *Lippia citriodora* and *Hibiscus sabdariffa* was another probable reason for the reduction in fat mass percentage.

Another study reported effects on fat mass percentage among hypertensive adults who consumed flaxseed. The flaxseed, which contains lignans of secoisolariciresinol diglucoside, reportedly can lessen hypertension and diabetes; it can control obesity and bring about weight reductions due to the anti-inflammatory properties and presence of antioxidants [45]. The study participants with hypertension consumed the flaxseed for over 12 weeks, showing a significant reduction in BMI levels and overall fat mass percentage. A plausible reason could be the high amounts of soluble and insoluble fibers in the flaxseed, which help lose weight. Moreover, lifestyle factors such as regular exercise and a healthy diet contribute to the improvements [46]. However, more research is needed to properly implement self-management of body fat mass using herbal medicine and prescribed drugs.

6.3 Waist/Hip Circumferences

The waist/hip circumference (w/h) ratio is high among people with raised blood pressure and abdominal obesity [47]. However, there are plausible studies on the w/h ratio outcomes of hypertensive adults, particularly those who consume herbal medicine and anti-hypertensive drugs concurrently. The following studies detail the evidence obtained on the w/h ratios:

In a study by Dicks et al. [48], adults with hypertension and type 2 diabetes consumed flavanol-rich cocoa powder with multiple anti-hypertensive drugs for 12 weeks; the results showed a significant reduction of w/h ratios. The consumption of cocoa powder and anti-hypertensive drugs concomitantly reduced endothelin-1 system activation and inhibition of the ACE enzyme [49].

Reportedly, hypertensive adults who consumed herbal supplements containing hibiscus, olive leaf, and green tea with conventional medication experienced reductions in the w/h ratio and BMI levels, in addition to improvements in blood pressure readings [50]. Similarly, in another study, saffron consumed concurrently with anti-hypertensive drugs was reported to improve the w/h ratio and other anthropometry parameters due to the presence of carotenoids [51]. Carotenoids are a known antioxidant compound that prevents the damaging effects of singlet oxygen on human lymphocytes, reducing the risk of hypertension and obesity [52].

All in all, extensive research is still needed to investigate the outcomes of concurrent consumption of herbs and anti-hypertensive drugs in depth. Nevertheless, based on past studies of adults with co-morbidities such as hypertension or obesity, concomitant usage of herbal remedies and medicines seems to yield positive changes in w/h ratios.

7 Blood Pressure

Blood pressure is the pressure exerted on the walls of the arteries due to the force of the blood pushing against the walls when the heart pumps blood to all parts of the human body. It is measured in millimeters of mercury (mm/Hg) by an office blood pressure machine or ambulatory blood pressure monitor (ABPM); it is written using two figures determining a person's blood pressure levels. Systolic blood pressure is the first value measuring the pressure in the arteries when the heart beats. Subsequently, the second figure denotes the diastolic blood pressure. It is indicative of the pressure when the heart rests between beats. An ideal blood pressure reading would be 120/80 mm/Hg; when an individual has readings above 140/90 mm/Hg, that person would be classified as suffering from hypertension. It is vital to realize that the tasks may vary throughout the day depending on a person's schedule, stress levels, and several other factors that may affect the quality of life and the likelihood of being predisposed to other co-morbidities. Thus, monitoring blood pressure readings is essential and should be undertaken regularly at a health facility or at home [1].

7.1 Ambulatory Blood Pressure Monitor (ABPM)

Blood pressure measurements via an ambulatory blood pressure monitor have elucidated the effectiveness of detecting blood pressure variations and the prevalence of cardiovascular risk markers in hypertensive patients. The device can be used as a treatment guide for detecting treated but uncontrolled hypertension, as it can measure the blood pressure readings throughout 24 h. On the other hand, in the clinical practice guidelines, ABPM is included to diagnose and manage isolated office hypertension among pregnant women, masked and labile hypertension [11]. ABPM is the gold standard as it provides consistency; it is a better predictor for hypertension as repeated measurements can be done, leading to more excellent reproducibility of average blood pressure taken with increased precision [12].

In a clinical trial in Poland, the hypertensive patients consumed standardized tomato extract with diuretics and acetylsalicylic acid concomitantly over a 4-week duration; the average blood pressure reading measured using ABPM was 127/78 mm/Hg. The readings were repeated every 15–30 min over 24 h. When combined with the drugs, the presence of lycopene from the extract brought about a synergistic hypotensive effect that significantly improved both systolic and diastolic blood pressure readings [53].

Similarly, another study in China reported that the average 24-h ABPM reading was 131/84 mm/Hg, and the measurements were repeated every 20–30 min at the baseline appointment and after 8 weeks. The Chinese herbal formula, Songling Xuemaikang capsule (SXC), was used with losartan, which improved overall blood pressure readings. The changes could be attributed to puerarin, the active ingredient of Pueraria lobate found in SXC, as it can upregulate mRNA expression of ACE-2 and reduce plasma angiotensin II levels [54].

7.2 Office Blood Pressure Monitor

Since the olden days, sphygmomanometers, or office blood pressure monitors (OBP), have been the fundamental tool for measuring blood pressure. The measurements obtained via the OBP make diagnosing and managing hypertension easier as patients could be referred for treatment to prevent raised blood pressure readings [55]. However, readings obtained using OBP machines may come with errors due to masked hypertension. Thus, to avoid such mistakes and increase the accuracy of tasks, the machine should be calibrated by comparing the readings with those obtained manually, and the measurements should be triplicated to acquire an average assignment [56].

Moreover, some other steps could be taken to prevent errors: ensure the patients are seated comfortably without talking and have rested for a few minutes before beginning with the measurements [56]. In a study carried out in Thailand among hypertensive adults who consumed an array of herbs with anti-hypertensive drugs, the average blood pressure reading was 147/81 mm/Hg as measured by an office blood pressure machine. Two readings were taken at 15 intervals and then averaged

[57]. Thus, compared to readings obtained via ABPM across varying studies, the OBP measurement seems to give elevated blood pressure outcomes.

Lai and colleagues [54] explain two significant reasons for the higher blood pressure readings obtained when measured using an office blood pressure machine, compared with those obtained using the ABPM in this study and other previous studies. The first reason could be attributed to the phenomenon known as white-coat hypertension. When an office blood pressure machine measures an individual's blood pressure, the person tends to be less relaxed due to the presence of the medical personnel or equipment. However, several steps have been taken to mitigate the situation. The following reason could be this: office blood pressure measurements are commonly accepted in the morning before the patients consume their prescribed medicine, and readings tend to be higher during the day. In contrast, ABPM readings are taken many times at regular intervals over 24 h, even while a person is asleep at night; besides, there are no external factors that may cause a spike in the readings.

8 Point-of-Care Technology Evaluation and Biochemical Outcomes

Point-of-care (PoC) devices and their utility for patients' self-management of chronic disease states have been well known since the onset of the COVID-19 pandemic. In clinical settings, PoC devices provide rapid laboratory diagnostic results; they are handy monitoring tools for managing patients and titrating medication doses [58]. Several situations have prompted the transition to using PoC devices: the convenience of self-monitoring at home, further improving disease management and healthcare cost-effectiveness.

Many devices are available for home use to manage chronic disorders such as diabetes mellitus, hypertension, congestive heart failure, and anticoagulation. Many of these devices have built-in software capabilities, enabling patients to share important health information with healthcare providers via a computer. Official visits to health facilities revealed long turnaround times for patients to receive results and attend follow-up consultations; in comparison, results are immediately generated using a PoC device, and appropriate actions can be taken as soon as possible. The frequency of HbA1c produced by the PoC device is higher than in laboratories. However, another study on PoC devices reported varied sensitivities for detecting covid-19 antiCOVID-19 lower for IgM, and higher for IgG compared with those of the gold standard, automated chemiluminescent immunoassay [59]. In another study, the newly developed SARS-CoV-2 PoC device was claimed to provide higher sensitivities of neutralizing antibody production. Even though the acquisition of immunoglobulin and neutralizing antibodies varies according to vaccine type, age, days after vaccination, pain degree after immunization, and immunization diseases, this PoC device can be used to give clinical recommendations; it is a decision tool to advise a person whether to receive additional therapy through the immediate rapid testing [58]. A multi-centered cross-sectional study conducted at nine hospitals in Guangzhou reported the risk of having dyslipidemia by examining blood lipids using a PoC device [60]. An analysis of the PoC survey was conducted remotely in Australia; the findings showed timely access to pathology results and subsequent follow-up of patients for treatment, which was practicable despite the long distances to the nearest laboratory [61].

Utilizing PoC devices could be a quick and helpful way to assess health outcomes when herbal medicine is used as a means of self-management, which can then evaluate the statement that "herbal medicines are all safe." It is feasible to introduce PoC devices in China; the widespread use might raise dyslipidemia awareness, improve treatment, and facilitate better control.

9 Blood Glucose Level

Patients with diabetes and hypertension may be high-risk candidates for heart and brain disorders. Blood glucose levels must be adequately controlled along with other cardiovascular risk factors.

9.1 Fasting Blood Glucose Level

For hypertensive patients with metabolic syndrome, consuming Yiqi Huaju Formula, a traditional Chinese herbal medicine mixture, in addition to standard care, significantly improved fasting blood glucose and other assessed parameters [62]. In a different trial, patients with metabolic syndrome who took Pycnogenol® maritime pine bark extract for 6 months had a reduction in fasting blood sugar levels, going from a baseline of 123 mm/Hg to 106.4 mm/Hg after 3 months and eventually to 105.3 mm/Hg at the end of the study [63].

A unique green aquatic plant called Wolffia globosa mankai is a dietary source of high-quality protein. In a randomized crossover study with participants who were abdominally obese, the intervention group was given a shake containing Wolffia globosa to replace dinner. In contrast, the control group was assigned a shake containing yogurt. The study participants showed lower fasting blood sugar levels the following morning (83.2 mg/dL to 86.6 mg/dL) [64].

In India, diabetes and hypertension are common. Fenugreek seeds are frequently taken to treat diabetes and hypertension. Patients with diabetes got 10 gm of fenugreek seeds in hot water daily in a randomized, controlled study, while the control group did not. The patients had type 2 diabetes mellitus, managed by diet, exercise, and insulin or oral hypoglycemic drugs. According to statistical analysis, the study group's fasting blood glucose levels significantly decreased in the fifth month, whereas the study group's HbA1C significantly reduced in the sixth month [65]. On the other hand, treatment with 162 mg/d of onion skin extract quercetin decreased ABP in hypertension patients [66]. Prehypertensive males who consume an olive leaf extract high in phenols do not significantly affect fasting blood glucose levels or other measures [67].

9.2 2-Hour Post-Prandial Blood Glucose Level

Clinical investigations have indicated various advantages of herbal medications in lowering blood glucose. A herbal product containing nettle leaf 20% (w/w), berry leaf 10% (w/w), onion and garlic 20% (w/w), fenugreek seed 20% (w/w), walnut leaf 20% (w/w), and cinnamon bark 10% (w/w) was fed to study participants who failed to control blood glucose with two oral medications and were unwilling to inject insulin for 12 weeks in a double-blind, controlled study by Parham et al. [68]. In tests conducted during fasting and two hours after eating, the glucose level was dramatically reduced after consuming the herbal medicine mixture. Additionally, glycated hemoglobin A1c (HbA1c) dropped from 0.33% to 0.20%. Finally, after using the herbal treatment, the degree of insulin resistance dramatically decreased from 1.9 to 1.4.

An oral glucose tolerance test has improved due to the effects of olive leaf extracts. According to a study, borderline diabetics' blood glucose levels dramatically lowered at 30 and 60 minutes when 1 g of olive leaf was supplied with 300 g of white rice. In both cases, it was believed that the polyphenols in olive leaf extracts inhibited the activity of intestinal and salivary amylases. However, it is also possible that the aglycones in olive leaf extracts compete with the glucose released from food in the gut for glucose receptors, reducing absorption. According to the study, the use of olive leaf extracts by prehypertensive men revealed no significant effects on fasting glucose, insulin, or other indices [69].

10 Lipid Profile

Total cholesterol (TC), triglycerides (TG), high-density lipoprotein cholesterol (HDL-C), and low-density lipoprotein cholesterol (LDL-C) are lipid profile measures in the blood that are frequently evaluated. Increased TC, TG, LDL-C, and decreased HDL-C are signs of dyslipidemia. Dyslipidemia is a known modifiable risk factor for hypertension, ischemic heart disease, and stroke. Participants with dyslipidemia received treatment with the Unani polyherbal medicine Jawarish Falafili (JF) in a prospective randomized research. When JF was ingested, significant decreases in mean blood TC, TG, and LDL-C were seen at baseline compared to the control group who took atorvastatin for dyslipidemia. However, neither group's mean serum HDL-C levels significantly changed after treatment [65]. Compared to the control group in a different trial, consumption of olive leaf extract was also linked to physiologically significant decreases in TC, LDL-C, and TAG, with no change in the effect on HDL-C [70]. The TC and LDL-C reductions seen using olive leaf extracts and JF could be equivalent to decreased total CVD risk when considering earlier statin trials [69]. It is unknown what processes underlie the lipid-lowering effects of herbal therapy.

The consumption of phenolic components of herbal medicine, on the other hand, appears to reduce the activities of critical cholesterol-regulatory enzymes, 3-hydroxy-3-methylglutaryl-CoA (HMG-CoA) reductase (the main target of statins), and acetyl-CoA cholesterol acyltransferase (ACAT), leading to a reduction in cholesterol biosynthesis [71]. Contrarily, neither in the entire trial group nor in the subgroup of hypertension patients did quercetin supplementation for 6 weeks impact fasting serum concentrations of total cholesterol, LDL cholesterol, or HDL cholesterol. As opposed to this, Lee et al. [72] discovered that supplementing male smokers with quercetin (100 mg/d) for 10 weeks considerably decreased fasting serum concentrations of total cholesterol and LDL cholesterol and significantly elevated serum HDL-C. The length of quercetin administration, the study participants' baseline characteristics, and the dosage of quercetin taken internally may be to blame for these conflicting outcomes.

Traditional medicine uses *Phyllanthus emblica*, a plant high in antioxidants, tannins, and vitamin C, to cure various illnesses. Patients with essential hypertension were randomly assigned to receive a *Phyllanthus emblica* capsule (500 mg) or a placebo twice daily, in addition to their usual medications, over 12 weeks in a randomized controlled experiment. In the beginning, the treatment and placebo groups were comparable. Patients with essential hypertension have a satisfactory safety profile when taking *Phyllanthus emblica*. However, using *Phyllanthus emblica* failed to improve oxidant status, antioxidant capacity, or lipid profile, and it did not result in any further drops in systolic and diastolic blood pressure levels [9].

In Egypt, the management of hypertension has traditionally involved the use of *Hibiscus sabdariffa* and *Olea europaea*. When captopril 25 mg, *Hibiscus sabdariffa*, and *Olea europaea* were administered to Egyptian patients with grade 1 essential hypertension for 8 weeks, the subjects' mean triglyceride levels significantly decreased [73].

11 Hemoglobin Level

Some signs of rising hemoglobin (Hb) levels may cause a rise in systolic and diastolic blood pressure. A HemoCue Hb 201+ analyzer (HemoCue AB, Ngelholm, Sweden) routinely analyses Hb levels in finger stick capillary samples. In research by Atsma et al. [74], both male and female adults with hypertension showed a positive rise of Hb with both SBP and DBP. Several scientific explanations for the connection between Hb and blood pressure have been proposed. First, pulse wave velocity, a measure of arterial stiffness, is closely correlated with Hb, which raises SBP and DBP [75]. Furthermore, Hb may relax smooth muscle cells, change peripheral resistance, and thus modify blood pressure by scavenging nitric oxide (NO) generated in the endothelial cells that line blood arteries. Increased free Hb might bind to NO, which tightens the blood vessels and raises blood pressure [76]. Third, elevated Hb concentrations would result in elevated blood viscosity. Previous studies have shown that increasing hematocrit and hemoglobin levels cause blood to become more viscous, possibly damaging cardiovascular function by affecting blood pressure. Iron content in herbal medicine and its ability to treat anemia make it a popular drug. Hematopoiesis is significantly influenced by iron. However, since other elements such as alkaloids, flavonoids, saponins, tannins, calcium, zinc, and vitamins C and K contribute to the body's ability to absorb iron, the therapeutic potential of the herbs cannot be determined only based on their availability of iron. For instance, vitamin C helps the body absorb iron [77]. Some herbal medicines were beneficial for treating iron deficiency anemia, but studies that are more likely to be accurate have primarily focused on this influence on blood pressure control.

12 Conclusions

Blood pressure, blood glucose levels, blood lipids, and hemoglobin levels can all be affected, albeit to differing degrees, by the various herbal medicine protocols. Other confounding factors, most likely conventional medicines, smoking, exercise, co-morbidities, and food, have been the primary contributors to the changes in hypertension treatment. Certain targeted herbal medicine elements have the potential to complement hypertension treatment. However, it is essential to evaluate each patient's pharmacological therapy individually.

References

1. World Health Organization (2021) Hypertension. Available via https://www.who.int/health-topics/hypertension/. Accessed 15 Dec 2021
2. Schutte AE, Venkateshmurthy NK, Mohan S, Prabhakaran D (2021) Hypertension in low- and middle-income countries. Circ Res 128:808–826
3. Zaki NAM, Ambak R, Othman F, Wong NI, Man CS, Morad MFA et al (2021) The prevalence of hypertension among Malaysian adults and its associated risk factors: data from Malaysian Community Salt Study (MyCoSS). J Health Popul Nutr 40:1–8
4. Saeed S, Islahudin F, Makmor-Bakry M, Redzuan MD (2018) The practice of complementary and alternative medicine among chronic kidney disease patients. JAPER 8:30–36
5. National Center for Complementary and Integrative Health (2019) Traditional Chinese medicine: what you need to know. Available via https://www.nccih.nih.gov/health/traditional-chinese-medicine-what-you-need-to-know#. Accessed 28 Jan 2023
6. Asgary S, Kopaei MR, Sahebkar A, Shamsi F, Goli-malekabadi N (2016) Anti-hyperglycemic and anti-hyperlipidemic effects of Vaccinium myrtillus fruit in experimentally induced diabetes (antidiabetic effect of Vaccinium myrtillus fruit). J Sci Food Agric 96:764–768
7. Mohtashami R, Fallah HH, Nabati F, Hajiaghaee R, Kianbakht S (2019) Effects of standardized hydro-alcoholic extract of Vaccinium arctostaphylos leaf on hypertension and biochemical parameters in hypertensive hyperlipidemic type 2 diabetic patients: a randomized, double-blind and placebo-controlled clinical trial. Avicenna J Phytomed 29(1):44–53
8. Rizka A, Setiati S, Lydia A, Dewiasty E (2017) Effect of nigella sativa seed extract for hypertension in elderly: a double-blind, randomized controlled trial. Acta Med Indones 49: 307–313
9. Shanmugarajan D, Girish C, Harivenkatesh N, Chanaveerappa B, Prasanna LNC (2021) Antihypertensive and pleiotropic effects of Phyllanthus emblica extract as an add-on therapy in patients with essential hypertension-a randomized double-blind placebo-controlled trial. Phytother Res 35:3275–3285
10. Plüddemann A, Thompson M, Price CP, Wolstenholme J, Heneghan C (2012) Point-of-care testing for the analysis of lipid panels: primary care diagnostic technology update. Br J Gen Pract 62:596

11. Dadlani A, Madan K, Sawhney JPS (2019) Ambulatory blood pressure monitoring in clinical practice. Indian Heart J 71:91–97
12. Peixoto AJ (2015) Practical aspects of home and ambulatory blood pressure monitoring. Cardiovasc J 11:214–218
13. Shimbo D, Abdalla M, Falzon L, Townsend RR, Muntner P (2015) Role of ambulatory and home blood pressure monitoring in clinical practice: a narrative review. Ann Intern Med 163:691–700
14. Kim HN, Yoon SY (2021) Simultaneous point-of-care testing of blood lipid profile and glucose: performance evaluation of the GCare lipid analyzer. J Clin Lab Anal 35:12
15. Hughes GD, Aboyade OM, Clark BL et al (2013) The prevalence of traditional herbal medicine use among hypertensives living in South African communities. BMC Complement Altern Med 13:38
16. Al-Hadid D, Musa RJ, Al-Talhuni A, Alkrad JA (2020) Prevalence of traditional herbs and supplements use among hypertensive patients in Om Elamad Health Center. Pharm J 12: 1612–1622
17. Liwa A, Roediger R, Jaka H, Bougaila A, Smart L, Langwick S, Peck R (2017) Herbal and alternative medicine use in Tanzanian adults admitted with hypertension-related diseases: a mixed-methods study. Int J Hypertens:9
18. Joachimdass RJ, Subramaniam K, Sit NW, Lim YM, Teo CH, Ng CJ, Yusof AS, Loganathan A (2021) Self-management using crude herbs and the health-related quality of life among adult patients with hypertension living in a suburban setting of Malaysia. PLoS One 16:e0257336
19. Pearson H, Fleming T, Chhoun P, Tuot S, Brody C, Yi S (2018) Prevalence of and factors associated with utilization of herbal medicines among outpatients in primary health centres in Cambodia. BMC Complement Altern Med 18:114
20. Kianbakht S, Hashem-Dabaghian F (2019) Antihypertensive efficacy and safety of Vaccinium arctostaphylos berry extract in overweight/obese hypertensive patients: a randomized, double-blind and placebo-controlled clinical trial. Complement Ther Med 44:296–300
21. Shaban MI, El-Gahsh NFA, El-Said A, El-Sol H (2017) Ginger: It's effect on blood pressure among hypertensive patients. IOSR JNHS 1940:79–86
22. Neto JCG et al (2022) Effectiveness of ginger in reducing pressure levels in persons with diabetes: a placebo-controlled randomized clinical trial. Rev Enferm Atual In Derme:96
23. Ried K, Frank OR, Stocks NP (2013) Aged garlic extract reduces blood pressure in hypertensives: a dose-response trial. Eur J Clin Nutr 67:64–70
24. McKay DL, Chen CY, Saltzman E, Blumberg JB (2010) Hibiscus sabdariffa L. tea (tisane) lowers blood pressure in prehypertensive and mildly hypertensive adults. J Nutr 140:298–303
25. Razmpoosh E, Abdollahi S, Mousavirad M (2022) The effects of olive leaf extract on cardiovascular risk factors in the general adult population: a systematic review and meta-analysis of randomized controlled trials. Diabetol Metab Syndr:14
26. Jandari S et al (2020) Effects of *Momordica charantia* L. on blood pressure: a systematic review and meta-analysis of randomized clinical trials. Int J Food Prop 1:1913–1924
27. Mayo Clinic (2022) Consumer health. Available via https://www.mayoclinic.org/healthy-lifestyle/consumer-health/in-depth/herbal-supplements/art-20046488. Accessed 28 Mar 2023
28. Lee CH, Kim JH (2014) A review on the medicinal potentials of ginseng and ginsenosides on cardiovascular diseases. J Ginseng Res 38:161–166
29. Jeon SY et al (2022) Simultaneous analysis of a combination of anti-hypertensive drugs, fimasartan, amlodipine, and hydrochlorothiazide, in rats, using lc-ms/ms and subsequent application to pharmacokinetic drug interaction with red ginseng extract. Toxics 10:576
30. Deutch MR, Grimm D, Wehland M, Infanger M, Krüger M (2019) Bioactive candy: effects of liquorice on the cardiovascular system. Foods 8:495
31. Ministry of Health Malaysia (2023) Traditional and complementary unit. Available via https://tcm.moh.gov.my/en/index.php/integrated-hospital/. Accessed 9 Mar 2023
32. Kushner K, Yu S (2015) Integration of traditional Chinese medicine and Western medicine in a Chinese Community Health Centre. Fam Med Commun Health 3:79–83

33. Institut Kanser Negara (2023) Unit Perubatan Tradisional dan Komplementari. Available via https://nci.moh.gov.my/index.php/ms/perkhidmatan/2014-01-27-07-35-07/2014-01-27-07-55-14/unit-perubatan-tradisional-dan-komplementari. Accessed 14 May 2023
34. Wenjun C (2022) Yueyang Hospital best in nation for integrated TCM, Western medicine consumer health. Available via https://www.shine.cn/news/metro/2207117784/. Accessed 14 May 2023
35. Fong HHS (2016) Integration of herbal medicine into modern medical practices: issues and prospects. Integr Cancer Ther 1:287–293
36. Mahmoud I, Sulaiman N (2021) Significance and agreement between obesity anthropometric measurements and indices in adults: a population-based study from the United Arab Emirates. BMC Public Health 21:1605
37. Cheah WL, Chang CT, Hazmi H, Kho GWF (2018) Using an anthropometric indicator to identify hypertension in adolescents: a study in Sarawak, Malaysia. Int J Hypertens 7
38. Ardalani H, Jandaghi P, Meraji A, Hassanpour Moghadam M (2019) The effect of *Cynara scolymus* on blood pressure and BMI in hypertensive patients: a randomized, double-blind, placebo-controlled, clinical trial. Complement Med Res 27:40–46
39. Visuthranukul C et al (2022) Effects of inulin supplementation on body composition and metabolic outcomes in children with obesity. Sci Rep 12:13014
40. Huang LH et al (2018) Effects of green tea extract on overweight and obese women with high levels of low density-lipoprotein-cholesterol (LDL-C): a randomised, double-blind, and cross-over placebo-controlled clinical trial. BMC Complement Altern Med 18:294
41. Bogdanski P, Suliburska J, Szulinska M, Stepien M, Pupek-Musialik D, Jablecka A (2012) Green tea extract reduces blood pressure, inflammatory biomarkers, and oxidative stress and improves parameters associated with insulin resistance in obese, hypertensive patients. Nutr Res 32:421–427
42. Ye S, Zhu C, Wei C, Yang M, Zheng W, Gan D (2018) Associations of body composition with blood pressure and hypertension. Obesity 26:1644–1650
43. Healthline (2022) What is my ideal body fat percentage? Available via https://www.healthline.com/health/exercise-fitness/ideal-body-fat-percentage. Accessed 6 Mar 2023
44. Boix-Castejón M et al (2021) Effect of metabolaid® on pre- and stage 1 hypertensive patients: a randomized controlled trial. Foods 84:104583
45. Fazilati M, Aarabi A, Tadayon CA (2014) Investigation the effect of flaxseed bread on reduction of blood glucose in diabetic patients. J Food Technol Nutr 11:91–96
46. Toulabi T, Yarahmadi M, Goudarzi F, Ebrahimzadeh F, Momenizadeh A, Yarahmadi S (2022) Effects of flaxseed on blood pressure, body mass index, and total cholesterol in hypertensive patients: a randomized clinical trial. Explore (NY) 18:438–445
47. Kumar M et al (2022) A critical review on obesity: herbal approach, bioactive compounds, and their mechanism. Appl Sci 12:8342
48. Dicks L et al (2018) Regular intake of a usual serving size of flavanol-rich cocoa powder does not affect cardiometabolic parameters in stably treated patients with type 2 diabetes and hypertension – a double-blinded, randomized, placebo-controlled trial. Nutrients 10:1435
49. Shallenberger FA (2016) The integrative management of hypertension and associated conditions. J Hum Hypertens
50. Rahmani J, Bazmi E, Clark C, Hashemi Nazari SS (2020) The effect of saffron supplementation on waist circumference, HA1C, and glucose metabolism: a systematic review and meta-analysis of randomized clinical trials. Complement Ther Med 49:102298
51. Tapiero H, Townsend DM, Tew KD (2004) The role of carotenoids in the prevention of human pathologies. Biomed Pharmacother 58:100–110
52. Osińska AN, Begier-Krasińska B, Rzymski P, Krasińska A, Tykarski A, Krasiński Z (2017) The influence of adding tomato extract and acetylsalicylic acid to hypotensive therapy on the daily blood pressure profiles of patients with arterial hypertension and high cardiovascular risk. Kardiochir Torakochirurgia Pol 14:245–252

53. Lai X et al (2022) Efficacy and safety of Chinese herbal medicine compared with losartan for mild essential hypertension: a randomized, multicenter, double-blind, noninferiority trial. Circ Cardiovasc Qual Outcomes 15
54. Chia YC, Wan AWA, Fong AYY et al (2022) Malaysian working group consensus statement on renal denervation for management of arterial hypertension. Hypertens Res 45:1111–1122
55. Stergiou G, Kollias A, Parati G, Brien E (2018) Office blood pressure measurement. Hypertension 71:848–857
56. Thangsuk P, Pinyopornpanish K, Jiraporncharoen W, Buawangpong N, Angkurawaranon C (2021) Is the association between herbal use and blood-pressure control mediated by medication adherence? A cross-sectional study in primary care. Int J Environ Res Public Health 18:12916
57. Goble JA, Rocafort PT (2017) Point-of-care testing. J Pharm Pract 30:229–237
58. Noce A et al (2020) Serological determinants of COVID-19. Biol Direct 15:21
59. Shim HW et al (2022) Analysis of factors affecting neutralizing antibody production after covid-19 vaccination using newly developed rapid point-of-care test. Diagnostics (Basel) 12:1924
60. Zhang PD, He LY, Guo Y, Liu P, Li GX, Wang LZ, Liu YF (2015) Blood lipid profiles and factors associated with dyslipidemia assessed by a point-of-care testing device in an outpatient setting: a large-scale cross-sectional study in southern China. Clin Biochem 48:586–589
61. Spaeth BA, Shephard MD, Schatz S (2014) Point-of-care testing for haemoglobin A1c in remote Australian indigenous communities improves timeliness of diabetes care. Rural Remote Health 14(4):2849
62. Chen Y et al (2013) Effects of Chinese herbal medicine Yiqi Huaju formula on hypertensive patients with metabolic syndrome: a randomized, placebo-controlled trial. J Integr Med 11: 184–194
63. Belcaro G et al (2013) Pycnogenol® supplementation improves health risk factors in subjects with metabolic syndrome. Phytother Res 27:1572–1578
64. Zelicha H et al (2019) The effect of *Wolffia globosa* Mankai, a green aquatic plant, on postprandial glycemic response: a randomized crossover- controlled trial. Diabetes Care 42
65. Ranade M, Mudgalkar N (2017) A simple dietary addition of fenugreek seed leads to the reduction in blood glucose levels: a parallel-group, randomized single-blind trial. Ayu 38:24–27
66. Brüll V et al (2015) Effects of a quercetin-rich onion skin extract on 24 h ambulatory blood pressure and endothelial function in overweight-to-obese patients with (pre-hypertension): a randomised double-blinded placebo-controlled cross-over trial. Br J Nutr 114:1263–1277
67. Lockyer S, Rowland I, Spencer JPE, Yaqoob P, Stonehouse W (2017) Impact of phenolic-rich olive leaf extract on blood pressure, plasma lipids and inflammatory markers: a randomised controlled trial. Eur J Nutr 56:1421–1432
68. Parham M, Bagherzadeh M, Asghari M, Akbari H, Hosseini Z, Rafiee M, Vafaeimanesh J (2020) Evaluating the effect of a herb on the control of blood glucose and insulin-resistance in patients with advanced type 2 diabetes (a double-blind clinical trial). Caspian J Intern Med 11: 12–20
69. Ain Q, Nawab M, Ahmad T, Kazmi MH, Naikodi MAR (2022) Evaluating the safety and efficacy of a polyherbal Unani formulation in dyslipidaemia-a prospective randomized controlled trial. J Ethnopharmacol 10:115036
70. Mihaylova B, Emberson J, Blackwell L et al (2012) The effects of lowering LDL cholesterol with statin therapy in people at low risk of vascular disease: meta-analysis of individual data from 27 randomised trials. Lancet 380:581–590
71. Lee JS, Choi MS, Jeon SM et al (2001) Lipid-lowering and antioxidative activities of 3,4-di (OH)-cinnamate and 3,4-di(OH)- hydro cinnamate in cholesterol-fed rats. Clin Chim Acta 314: 221–229
72. Lee KH, Park E, Lee HJ et al (2011) Effects of daily quercetin-rich supplementation on cardiometabolic risks in male smokers. Nutr Res Pract 5:28–33
73. Nabil E et al (2020) Antihypertensive efficacy and safety of a standardized herbal medicinal product of Hibiscus sabdariffa and Olea europaea extracts (NW Roselle): a phase-II, randomized, double-blind, captopril-controlled clinical trial. Phyther Res 34:3379–3387

74. Femke A, Ingrid V, Wim DK, Marian VK, Pieternel PJ, Jaap D (2012) Haemoglobin level is positively associated with blood pressure in a large cohort of healthy individuals. Hypertension 60:936–941
75. Kawamoto R, Tabara Y, Kohara K, Miki T, Kusunoki T, Katoh T et al (2012) A slightly low haemoglobin level is beneficially associated with arterial stiffness in Japanese community-dwelling women. Clin Exp Hypertens 34:92–98
76. Cabrales P, Han G, Nacharaju P, Friedman AJ, Friedman JM (2011) Reversal of haemoglobin-induced vasoconstriction with sustained release of nitric oxide. Am J Physiol Heart Circ Physiol 300:49–56
77. N'guessan K, Kouassi KH, Ouattara D (2010) Plants used to treat anaemia, in traditional medicine, by Abbey and Krobou populations, in the South of Côte-d'Ivoire. J Appl Sci Res 6:1291–2197

Efficacy of Ethno-herbal Medicines with Anti-inflammatory and Wound Healing Potentiality: A Case of West Bengal, India

29

Torisa Roy, Tamal Mazumder, Tapas Nag, Jogen Chandra Kalita, Sylvester Chibueze Izah, and Hadida Yasmin

Contents

T. Roy · H. Yasmin (✉)
Immunology & Cell Biology Laboratory, Department of Zoology, Cooch Behar Panchanan Barma University, Cooch Behar, WB, India

T. Mazumder
Department of Zoology, North Bengal St. Xavier's College, Rajganj, WB, India

T. Nag
Department of Anatomy, All India Institute of Medical Sciences, New Delhi, India

J. C. Kalita
Department of Zoology, Guwahati University, Assam, India

S. C. Izah
Department of Microbiology, Faculty of Science, Bayelsa Medical University, Yenagoa, Bayelsa State, Nigeria

S. C. Izah et al. (eds.), *Herbal Medicine Phytochemistry*, Reference Series in Phytochemistry, https://doi.org/10.1007/978-3-031-43199-9_62

Abstract

Inflammation is the hallmark of several immune defence mechanisms against pathogenic as well as non-pathogenic challenges. However, uncontrolled inflammation can often lead to chronic pathological conditions such as diabetes, autoimmunity, cancer, and other cardiovascular diseases. Present anti-inflammatory drugs pose serious physiological threats in the long run, showing severe side effects and thus, there is a huge demand for alternative drugs. Several medicinal plants used by different ethnic, indigenous, and tribal communities across the world show anti-inflammatory and wound-healing capabilities. One such indigenous community of Sub-Himalayan Bengal of India is the Rajbanshi community, who have established their own folk medicine system. This chapter discusses in detail the traditional Rajbanshi healers, the medicinal plants they use, their preparation of the medicinal formulation, as well as route and dose of administration. Plants used to treat various inflammatory ailments such as allergy, common cold, diabetes, skin diseases, cut, burn, gastrointestinal problems, and wound healing properties are enlisted in this chapter. The study also evaluates seven of the plants surveyed for judging their capability to scavenge free radicals, inhibit delayed-type hypersensitivity, and stimulate antibody production in the murine model. This chapter provides an opportunity to unravel new anti-inflammatory drugs with the potential for immunotherapeutic applicability.

Keywords

Herbal Medicine · Rajbanshi · Anti-inflammatory · Anti-oxidant · Wound healing · Immunostimulatory · DTH

Abbreviations

Ab	Antibody
ACE	*Ageratum conyzoides*
Ag	Antigen
ANOVA	Analysis of variance
CAE	*Centella asiatica*
CCE	*Cajanus cajan*
CGE	*Coccinia grandis*
COPD	Chronic obstructive pulmonary disease
CPCSEA	Committee for the Purpose of Control and Supervision of Experiments on Animals
CSE	*Cleome spinosa*
DDE	*Drymaria diandra*
DNFB	2,4 dinitrofluorobenzene
DPPH	2,2-Diphenyl-1-picrylhydrazyl Radical Scavenging assay
DTH	Delayed type hypersensitivity
EDTA	Ethylenediaminetetraacetic acid
HA	Hemagglutination assay
MA	Mannitol

MDA	Malondialdehyde
MME	*Mikania micrantha*
NSAIDs	Non-steroidal anti-inflammatory drugs
ROS	Reactive oxygen species
SCID	Severe combined immunodeficiency disease
SRBC	Sheep Red Blood Cells
TBA	Thiobarbituric acid
TCA	Trichloroacetic acid

1 Introduction

Inflammation is a host immune response to protect the body against pathogenic and non-pathogenic infection and injury. As a general mechanism, inflammation gradually initiates the healing process to recover from the local tissue injury and restores the body's homeostasis. However, the case of chronic inflammation and severe hypersensitivity, where the healing process is not restored properly, can lead to chronic inflammatory conditions affecting several organs such as the heart, liver, lungs, kidney, pancreas, gastrointestinal tract, or even brain. These situations can lead to serious inflammatory-associated disorders such as diabetes, pancreatitis, atherosclerosis, Alzheimer's, rheumatoid arthritis, osteoarthritis, tuberculosis, asthma, pulmonary fibrosis, COVID-19, and cancer [7, 16–18, 20, 51, 52, 54]. Synthetic anti-inflammatory drugs are beneficial in treating inflammatory diseases but with prolonged use, they can cause severe life-threatening side effects. Non-steroidal anti-inflammatory drugs/ NSAIDs, such as aspirin, ibuprofen, fenoprofen, indomethacin, diclofenac, flurbiprofen, naproxen, piroxicam, tenoxicam, and oxaprozin which are commonly used, cause serious deleterious effects leading to cardiovascular risks, gastrointestinal and hepatotoxicity, hypertension, and even neurotoxic effects [3]. Thus, newer and alternative approaches, showing comparable effectivity with less or no deleterious side effects need to be taken into consideration. Medicinal plants contain many phytoconstituents which show comparable efficacy as synthetic drugs [57]. There are several traditional methods practised by indigenous communities to treat inflammatory illness using herbal formulations [19] that need to be explored for new and safer anti-inflammatory drugs.

Plants are one of the most important source of medicine in many parts of the world [13–15, 21–31, 35–37, 49, 50]. In spite of enormous progress in modern medicine and pharmaceutical research, the use of medicinal plants still remains an important part of daily life. With the arrival of modern chemistry in the mid-eighteenth century, the ability to isolate, and purify specific compounds and synthesize them in large quantities became the foundations of the current pharmaceutical industry. Natural plant extracts of therapeutic relevance are of paramount importance as reservoirs of structural and chemical diversity [1, 40] and are used commercially across the globe. India represents a country of diversified ethnic groups with rich knowledge of biological resources and traditional medicines. Here, Ayurveda still remains dominant, compared to modern medicine. The basis of the ancient wisdom of Ayurveda medicine was that a system as complicated as the human body could not easily be cured by single compounds, but rather to reset harmony of the spirit and body by the administration of combinations of medicines. Different indigenous communities in India treat their daily ailments based on the knowledge that they have been sharing among their community for generations [12, 33]. Rajbanshi community is one such indigenous group of people, who have been living for thousands of years in the Terai and Brahmaputra valleys of Eastern Himalaya, which is rich in biodiversity. Rajbanshis are traditionally agriculturists and at the same time being deeply rooted in mother nature they have acquired great knowledge of plants with medicinal values in due course of time [44]. Rajbanshi traditional healers, called Kabiraj, have vast knowledge regarding the use of plant species for the treatment of various ailments. The new millennium with its developed technology has provided many opportunities for validation of the uses of natural products. The present work would like to unravel the vast knowledge of the Rajbanshi people regarding their ancient traditional medicinal practices, mainly in curing inflammation and healing wounds. At the same time, we have also put efforts to validate the efficacy of some of these medicinal plants through laboratory experiments which could provide preliminary data to explore their potentiality in a larger perspective as anti-inflammatory agents. We believe this work will pave the path for further work in the field of immunotherapy against inflammatory diseases.

2 Overview of Herbal Medicine Practices in the West Bengal Area of India

West Bengal is a state of India, situated between 27°13′15″ to 21°25′24″N latitude and 85°48′20″E 89°53′04″E longitude and the total area that the state covers is 88,752 Km^2. The state has diversified landforms, with mountains, plateaus, hills, plains, and coastal regions. In the northern region, it has the Himalayas and in the south, it has the Bay of Bengal and the Sundarbans coastal area. Due to this varied geographical pattern, West Bengal is rich in biodiversity with respect to both its flora and fauna. There are several ethnic and tribal communities in West Bengal, in the southern region there are mainly the Santhals, Koras, Mundas, Mahali, Kora, Meches, Lodhas, Bhumijs, Oraon, Kherias, and in the northern districts there are mainly the Rajbanshis, Koches, Meches, Rabha, Lepcha, Gorkhas, and Adivasis [48]. All of these communities stay very close to nature, mostly nearby forests, and have enormous knowledge regarding the use of medicinal plants, which has been passed to them from one generation to another. Folk and ethnic medicine has become an integral part of their culture and tradition. Almost every house in the community has medicinal plants that are used as a regular source of remedy to deal with day-to-day health problems such as cuts, injuries, burns, fractures, fever, headache, diarrhoea, skin rash, allergy, etc. Usually, at first, they try to cure the health problem happening in the family with their own indigenous knowledge of household herbal remedies. And in case it is not being cured by them, they approach the traditional healers. The traditional healers are commonly known as Ojha, Kabiraj, and Mahan, on whom the local people depend for their treatment either due to their age-old faith or also often due to their financial constraints [2].

Every medicinal plant that is being used for treatment has been given a local name in the vernacular language by the community people. Among the varied parts of the plants being used, the aerial parts are mostly preferred in preparing the medicine [46]. A study by Sarkar and Mandal [47] mentioned that the Santal, Munda, Lodha, and Mahali tribes have their own traditional way to treat different ailments, where they use different plants such as *Shorea robusta, Andrographis paniculata, Adhatoda vasica, Caesalpinia pulcherrima, Calotropis gigantea, Butea monosperma, Cajanus cajan, Tagetes patula, Ocimum sanctum, Psidium guajava,* and *Moringa oleifera*. For curing diarrhoea, the Santal and Lodha tribe often use roots of *Shorea robusta*, leaves of *Psidium guajava,* and also bark and leaves of *Butea monosperma*. Leaves and roots of *Datura metel* and *Calotropis gigantea* are used by Santal, Munda, and Lodha community for treating dog bites. Various parts of *Cajanus cajan* is used by the Munda and Lodha community to treat Jaundice. The Santal and Lodha community use roots and leaves of *Andrographis paniculata* during snake bite. *Adhatoda vasica* is used for treating cough and cold, *Datura metel* for treating asthma by Santal, Lodha, and Munda communities. However, these tribes now believe that the practice of treating diseases with ethno-medicines is diminishing gradually due to the decline in the number of plant species in their localities and the adjoining areas. Another work was carried out by Paul [42] in the district of Purulia, which is located in the western part of West Bengal and is predominated by Toto and Birhor tribes. The study says that these tribes as well as local communities use several plants for treating fever, such as stems of *Tinospora cordifolia*, roots and leaves of *Calotropis gigantea*, roots of *Hemidesmus indicus* and *Thysanolaena agrostis* Nees, the entire plant of *Andrographis paniculata* wall, and the bark of *Randia dumetorum*. Leaves and roots of *Breynia retusa* Alston and leaves of *Ocimum canum* for treating pneumonia and cough, the entire plant of *Alangium salviifolium* for snake bite, bark and leaves of *Streblus asper* for bronchitis, and roots of *Clitoria ternatea* and *Mimosa pudica* are used for treating infertility. Indigenous people of this district also used leaves/bulbs of *Allium cepa* Linn, the entire plant of *Andrographis paniculata* wall, and the entire plant of *Polygonum plebeium* R.Br. for treating diarrhoea and dysentery.

Raj and co-workers [43] did an extensive study by interviewing indigenous people in the villages nearby Chilapatta Reserve Forest, which is situated in the sub-Himalayan region of West Bengal. The indigenous people of this area mostly suffer from stomach-related disorders such as gastroenteritis, dysentery, stomach pain, indigestion, stomach worm, diarrhoea, and ulcers, and thus, there are about 40 different plant species that are being used for stomach-related ailments. Next to that are the plant species that are used for healing different types of cuts and wounds such as *Justicia adhatoda* L., *Justicia gendarussa* Burm. f., *Alternanthera brasiliana* (L.) Kuntze, *Rauvolfia serpentina* (L.), *Thespesia populnea* (L.), *Asparagus racemosus* Willd., *Ageratina adenophora* (Spreng.), *Ageratum conyzoides* L., *Eupatorium odoratum* L., *Oroxylum indicum* (L.) Benth, *Bryophyllum pinnatum* Kurz., *Shorea robusta* Gaerth f., *Acacia catechu* (L. f.) Willd., *Cassia sophera* L, *Leucas aspera* (Willd.) Spreng, *Melastoma malabathricum* L, *Melia azedarach* L, *Moringa oleifera* L, *Datura metel* L, *Solanum nigrum* L, *Christella dentata*

(Forssk.), *Cissus repanda* Vahl, *Alpinia malaccensis* (Burm.f.) Roscoe, *Curcuma caesia* Roxb., and *Curcuma longa* L. For treating cough and cold they mainly used *Andrographis paniculata* (Burm.f.) Wall. *ex* Nees, *Justicia adhatoda*, *Centella asiatica* (L.) Urb, *Alstonia scholaris* (L.) R. Br., *Phoenix sylvestris* (L.) Roxb, *Terminalia bellirica* (Gaertn.) Roxb, *Elaeocarpus ganitrus* Roxb. ex G. Don, *Ricinus communis* L, *Ocimum sanctum* L, *Melia azedarach* L, Moringa oleifera L., *Myristica longifolia* Wall, *Syzygium cumini* (L.) Skeels, *Piper nigrum* L., and *Lantana camara* L. Another study by Dutta et al. showed that the tribal population of Cooch Behar used different herbal formulations mainly to treat abdominal disorders, jaundice, liver problems, and intestinal worms, and also to heal cuts and wounds [9].

This traditional knowledge has become the basis of primary healthcare being practised by different ethnic and indigenous communities of West Bengal. The use of traditional medicine is important for providing effective treatments that are prepared for the individual's needs. Additionally, Ayurveda, a traditional Indian system of medicine, has been shown to be effective in treating a variety of conditions, including diabetes, heart disease, and cancer. Furthermore, traditional medicine has been found to be an effective form of preventative care, in this region. Herbal formulations prepared by making powder or paste of different parts of plants are mainly used for curing diseases such as ulcers, diabetes, skin diseases, vomiting, diarrhoea, liver problems, fever, neurological disorder and gynaecological disorders.

3 Herbal Medicine Practices in West Bengal

3.1 Study Area and the Survey Work

The study area comprises north of West Bengal, India, located at 26°22′N latitude and 89°29′E longitude, covering the Cooch Behar district (Fig.1a). Local practitioners/traditional healers were interviewed through structured questionnaires. A total of 20 informants, where 13 of them were traditional healers or Kabiraj, were interviewed to obtain details about their methods of treatment. Written consent was taken from each interviewed person, who was willing to share their knowledge. The information was recorded on the basis of the names of the plants, their medicinal use, the formulation of the herbal preparation, and the methods of administration. Interviews and discussions were conducted in the local Rajbanshi language.

3.2 Leaf Extract Preparation

Most of the plant specimens were collected from the medicinal garden of traditional healers and some from nearby natural vegetation. The collected plant specimens were preserved with 2% mercuric chloride in ethanol and were deposited at the Department of Botany, University of North Bengal for identification. A unique accession number was assigned to each herbarium specimen by the Department of Botany. For extract preparation, leaves of the plants were collected, cleaned, shade

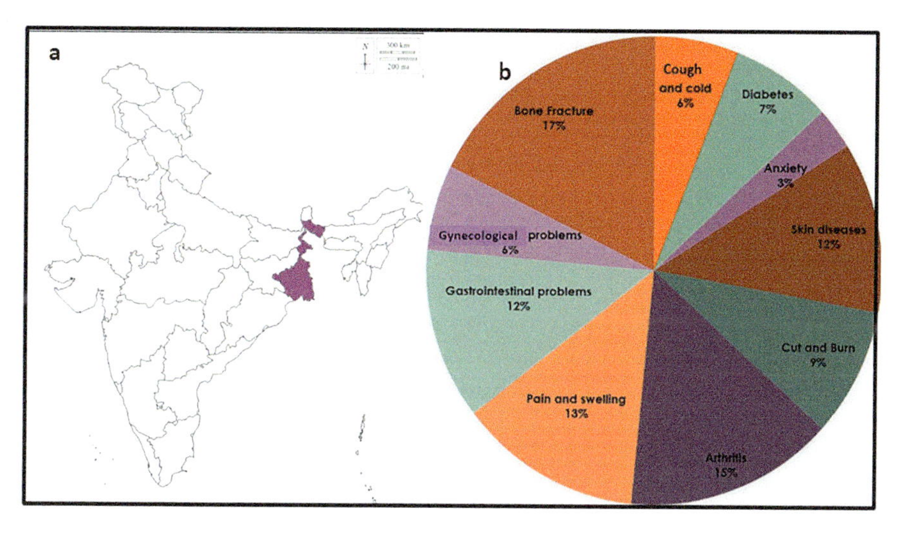

Fig. 1 (**a**) Location of West Bengal in the country map of India; (**b**) Different types of diseases cured by the Rajbanshi community with their indigenous herbal preparations

dried, and ground into powder form with a mechanical grinder. For each type of plant, 10 gm of ground leaves were then soaked in 100 ml of ethanol (80% v/v) in a 1:10 ratio and placed for 3 days in a magnetic stirrer. After 3 days, the extract was filtered first with Whatman paper (2) followed by a Millipore syringe filter (0.22 mm). The dry weight of the extracts was determined by evaporating to dryness through Rotary Vaccum (Buchi). Seven plants were chosen for further investigations, namely, *Ageratum conyzoides* (ACE), *Centella asiatica* (CAE), *Cajanus cajan* (CCE), *Mikania micrantha* (MME), *Coccinia grandis* (CGE), *Cleome spinosa* (CSE), and *Drymaria diandra* (DDE).

3.3 Free Radical Scavenging Assay

Inflammation leads to the generation of reactive oxygen species (ROS), which in due course of time causes oxidative stress and damage to the tissue. Natural compounds which are capable of quenching/scavenging ROS could be considered a potential anti-oxidant and anti-inflammatory agent. Thus, free radical scavenging assays were performed.

3.3.1 2,2-Diphenyl-1-picrylhydrazyl Radical Scavenging (DPPH) Assay

The protocol followed by Mahdi-Pour et al. [39] and Patel et al. [41] was considered for DPPH scavenging assay with some modifications. Different concentrations of 2 μl of extracts containing 0.0292 mg/ml, 0.0584 mg/ml, 0.0876 mg/ml were mixed with 800 μl of DPPH solution in methanol. The solution mixture was incubated for 30 min in the dark at room temperature. Gallic acid (GA) was used as a positive control. At 517 nm, the optical density (OD) was estimated through a spectrophotometer (HITACHI, U-2900). The free radical scavenging was calculated as a percentage of inhibition according to the following formula:

$$\text{Percentage of DPPH radical \% scavenging} = \{(\text{A0} - \text{A1})/\text{A0}\} \times 100$$

Where absorbance of the control reaction is denoted as A0 and A1 is of the extracts/ethanol.

3.3.2 Hydroxyl Radical Scavenging Assay Through Malondialdehyde (MDA) Generation

In vitro hydroxyl radical scavenging properties of different plant extracts were investigated using Fe^{2+}- ascorbate- EDTA (Ethylenediaminetetraacetic acid)- H_2O_2 system. The reaction mixture of 1 ml final volume contained L-ascorbic acid (1 Mm), 2-deoxy D-ribose (2.8 mM), ferric chloride (2 mM), hydrogen peroxide (H_2O_2) (1 mM), EDTA (1 mM) and phosphate buffer (4 mM) was incubated at 37 °C for 1 h with different concentration of leaf extract (0.146 mg/ml and 0.0584 mg/ml). TBA-TCA reagent containing 2.8% trichloroacetic acid and 1% aqueous thiobarbituric acid (TBA) was administered into the mixture. The reaction mixture was then boiled (15mins) for generating pink-coloured MDA-TBA chromogen. Once the reaction mixture had cooled the OD was estimated at 532 nm in UV- spectrophotometer (Make-Hitachi, U2900) [5, 45]. Mannitol (MA) a well-known anti-oxidant was used as a positive control to compare the level of efficacy of the plant extract. The efficacy of different plant extracts and ethanol as vehicle control was calculated assuming the level of MDA generation to be 100% in tubes deprived of extracts or ethanol and the inhibition percentage was calculated with the following formula:

$$\text{Percentage of inhibition} = \frac{\text{MDA generated in untreated tube} - \text{MDA generated in extract/ethanol treated tube}}{\text{MDA generated in untreated tube}} \times 100\%$$

3.4 Experiments on Animal Model

3.4.1 Animal Model

Swiss albino mice of 10–12 weeks (both male and female) were considered for in vivo experiments. Each experiment was repeated at least thrice taking 10 mice in each set. All experimental procedures were followed as guided by the Committee for the Purpose of Control and Supervision of Experiments on Animals (CPCSEA), Govt. of India. The Institutional Animal Ethics Committee of Cooch Behar Panchanan Barma University approved the protocols (Ref. No. CBPBU/IAEC/ZOO/010/2018).

3.4.2 Delayed Type Hypersensitivity (DTH) Assay

Delayed-type hypersensitivity (DTH) reactions represent a clinically important class of immune responses that has wide-ranging significance for various health and disease conditions [39]. Several pathogens such as mycobacteria, different proteins, as well as haptens can induce DTH reaction [34, 55]. Thus, to judge the anti-inflammatory property of the plant extract, delayed-type hypersensitivity model induced by a strong hapten, 2,4 dinitrofluorobenzene (DNFB) was considered. Swiss albino mice were injected subcutaneously with DNFB made in acetone. For primary sensitization, 0.000001% DNFB was injected in the subcutaneous region of the right foot pad of the mouse. The reinjection was done after 8 days on the left foot pad with 0.0001% DNFB. Different leaf extracts prepared in ethanol were administrated (25 μl dose) intravenously 1 h before resensitization. For each experiment, an equivalent amount of ethanol (alcohol) was administered in a separate set of animals 1 h prior to resensitization [6]. The size of the inflammation of the foot pad (diameter) was measured through a slide calliper and has been indicated as the degree of inflammation.

3.4.3 Hemagglutination Assay (HA Titre)

To understand whether the plant extract can stimulate antibody production with its treatment, HA titre was performed, where the antigen (Ag) and antibody (Ab) reaction can be observed as a visible agglutination ring at the bottom of the titre plate. Sheep erythrocyte (SRBC) was used as an antigen for immunization. Different leaf extracts/ethanol (25 μl dose) were injected intravenously 1 day before immunization. 0.1 ml of the 25% of SRBC (Ag) was injected intravenously in the lateral tail vein of the mice for immunization. The mice were sacrificed on the seventh day of immunization and serum was collected and stored at −20 °C. On the day of the experiment, the serum was heat inactivated at 56 °C for 45 min in the water bath. The heat-inactivated serum (Ab) was sterilized by passing through Millipore filters (0.45 μm, Sartorius) before experimentation. The experiment was carried out in a 96 well micro-titre plate where the first well contained 1:10 dilution of the heat-inactivated serum and the second well onward two-fold serial dilution (1:20 to 1:20,480) was made with PBS. Then in each well, 0.1% of antigen (Ag) was added before the incubation for 4–6 h at 37 °C in a humidified atmosphere. The reciprocal value of serum dilution in the last well giving positive haemagglutination reaction was considered as agglutination titre value. There will be the least agglutination in one of the wells down the row, beyond which no agglutination is visible, is considered as the end point of agglutination [4, 53].

4 Study Outcome

4.1 Outcomes of the Survey

A total of 20 informers were interviewed during the survey work. In this study we found that most of the healers are aged over 50 years old. The common diseases that were treated with different medicinal plants were cough and cold, diabetes, anxiety, skin diseases, cut and burn, pain and swelling, bone fracture, gynaecological problems, as well as gastrointestinal problems (Fig. 1b).

During this review we have recorded 37 specimens of medicinal plants used by the traditional healers, which are *Mikania micrantha, Stephania japonica,* Ageratum conyzoides, *Cleome spinosa, Cajanus cajan, Sida acuta, Eclipta prostrata,* Centella asiatica, *Ludwigia perennis, Drymaria diandra, Peperomia pellucida, Coccinia grandis, Vitex negundo, Plumbago zeylanica, Terminalia arjuna, Andrographis paniculata, Leucas zeylanica, Cissus quadrangularis, Heliotropium indicum, Calotropis procera, Scoparia dulcis, Trapa bispinosa, Ocimum sanctum, Achyranthes aspera, Euphorbia hirta, Eucalyptus globulus, Ocimum gratissimum, Abroma augusta, Mirabilis jalapa, Trichosanthes dioica, Piper nigrum, Mangifera indica, Tinospora cordifolia, Momordica charantia, Moringa oleifera, Azadirachta indica, and Curcuma longa* (Fig. 2, Tables 1 and 2).

Fig. 2 Medicinal plant used by Rajbanshi community for treating inflammatory diseases and for wound repairing and healing: *1. Mikania micrantha* 2. *Stephania japonica 3.* Ageratum conyzoides *4. Cleome spinosa 5. Cajanus cajan 6. Sida acuta 7. Eclipta prostrata 8. Centella asiatica 9. Ludwigia perennis* 10. *Drymaria diandra* 11. *Peperomia pellucida 12. Coccinia grandis 13. Vitex negundo 14. Plumbago zeylanica 15. Terminalia arjuna 16. Andrographis paniculata 17. Leucas zeylanica 18. Cissus quadrangularis 19. Heliotropium indicum 20. Calotropis procera* 21. *Scoparia dulcis 22. Trapa bispinosa 23. Ocimum sanctum 24. Achyranthes aspera 25. Euphorbia hirta 26. Eucalyptus globulus 27. Ocimum gratissimum 28. Abroma augusta 29. Mirabilis jalapa 30. Trichosanthes dioica 31. Piper nigrum 32. Mangifera indica 33. Tinospora cordifolia 34. Momordica charantia 35. Moringa oleifera 36. Azadirachta indica 37. Curcuma longa*

Table 1 Medicinal plant used by Rajbanshi community for treating various inflammatory diseases and for wound healing and repairing. The scientific name, common name, parts used, its preparation, and treatment are mentioned

Sl no.	Scientific name	Local name	Parts used	How it is being used	Ailments cured by Kabiraj
1	*Mikania micrantha*	Ashamlata	Leaves, stem	Boiled with water and decoction	Eczema, fungal infection, wound healing, stomach ache
2	*Stephania japonica*	Muchilata	Leaves, root	Mixed with other plants and boiled with water	Astringent, healing bone fracture, diarrhoea, fever, headache, urinary diseases
3	*Ageratum conyzoides*	Uchundi/ Bonpudina	Leaves, whole plant	Paste of the leaves mixed with other plants	Arthritis (joint inflammation), antifungal, bleeding control
4	*Cleome spinosa*	Hurhure	Leaves	Decoction	Bacterial infection, swelling, wound healing
5	*Cajanus cajan*	Arahar	Leaves, fruit	Leaf decoction	Jaundice; dysentery, wound healing
6	*Sida acuta*	Banmethi/ kureta	Leaves, whole plant	Decoction, juice of the leaves	Skin diseases, fever, common cold, wound healing
7	*Eclipta prostrata*	Keshraj/ kalakeshari	Whole plant	Paste with coconut oil	Promote hair growth, insomnia, dermatitis
8	*Centella asiatica*	Thankuni	Leaves	Two to three leaves were chewed daily during early morning, paste the leaves and apply 1–2 drops to the eye	Stomach problem, dysentery, conjunctivitis, wound healing
9	*Ludwigia perennis*	Bon lobongo	Leaves, roots	Boiled leaves or paste	Used to lower down fever and also for skin cut and wound healing
10	*Drymaria diandra*	Hargila	Whole plant	Plant juice	Laxative, inflammation
11	*Peperomia pellucida*	Luchipata	Whole plant	Plant juice/leaf paste	Stomach problem, cut and wounds

(continued)

Table 1 (continued)

Sl no.	Scientific name	Local name	Parts used	How it is being used	Ailments cured by Kabiraj
12	*Coccinia grandis*	Telakochu	Leaves and fruit	Fruit used as vegetable	Diabetes, osteoarthritis
13	*Vitex negundo*	Nishinda	Leaves	Mixed with different plants with mustard oil, honey, and water and directly applied to the affected area	Arthritis (joint inflammation)
14	*Plumbago zeylanica*	Agechita/ Chitrak	Leaves, roots	Powder of the leaves, paste of the leaves and roots	Piles, common cold cough, improve digestive system
15	*Terminalia arjuna*	Arjun	Bark	The bark of the tree was dried, ground, and then mixed with water/ milk	Heart-related problems, bone fracture
16	*Andrographis paniculata*	Kalmegh	Leaves	One or two washed leaves soaked in water overnight were chewed	Control diabetes, health tonic, treat intestinal worm
17	*Leucas zeylanica*	Dondokolosh	Whole plant	Used as a leafy vegetable (sauté with mustard oil), decoction	Fever, body dche, treat stomach parasite
18	*Cissus quadrangularis*	Harjora	Whole plant	Mixed with other plants with different solvents	Arthritis (joint inflammation), pain, healing of broken bone
19	*Heliotropium indicum*	Hatisur	Whole plant	Mixed with other plants with mustard oil and honey	Arthritis (joint inflammation), healing of broken bone
20	*Calotropis procera*	Akanda	Leaves	Mixed with other plants with mustard oil and honey, roasted leaves applied directly in inflamed area due to arthritis	Arthritis (joint inflammation), pain
21	*Scoparia dulcis*	Chinigur	Leaves	Juice of leaves with water taken orally	Diabetes, jaundice, skin diseases
22	*Trapa bispinosa*	Jolsingara/ paniphol	Fruit	Fresh fruits, or dry fruits made into powder	Helps to clean bowel system, treating ulcers

(continued)

Table 1 (continued)

Sl no.	Scientific name	Local name	Parts used	How it is being used	Ailments cured by Kabiraj
					and wound, flour of fruit is nutritious for health
23	*Ocimum sanctum*	Tulsi	Leaves	Juice of the leaves mixed with honey	Helps to treat cough, cold, and throat infection
24	*Achyranthes aspera*	Apang	Leaves/ roots	In the form of juice for oral consumption or for local application on skin	Consumption of juice helps to treat asthma, cough, pneumonia Local application of the juice helps to treat skin rash and itchiness Brushing tooth with the stem of helps to cure toothache
25	*Euphorbia hirta*	Dudhiya	Leaves	Juice made up of tender leaves	Used to cure respiratory diseases like asthma and bronchitis and irregular menstruation
26	*Eucalyptus globulus*	Eucalyptus	Leaves	Leaves were mixed with other plants to prepare paste	Used to treat arthritis; smoke from the burnt dried leaves helps to reduce cough
27	*Ocimum gratissimum*	Ramtulsi	Leaves	Leaves are crushed and mixed with honey	Helps to treat cough and cold
28	*Abroma augusta*	Ulotkombol	Root, bark, stem	Extracts from root, stem, and bark	Helps to cure gynaecological problems
29	*Mirabilis Jalapa*	Sandhyamalati	Leaves	Juice of the leaves	Leaf juice applied on the wound for healing

Almost all parts of medicinal plants were used for the preparation of medicinal formulations. However, leaves were used mostly compared to roots, stems, or flowers. Depending upon the disease two or more parts of the plants were used to prepare the formulation for the treatment. The mode of application of medicines was topical, orally, or by inhaling the smoke. Herbal preparations were often mixed with mustard oil, coconut oil, honey, or milk. It was observed that many of the traditional

Table 2 List of herbal formulation prepared by Rajbanshi Kabiraj for the treatment of different ailments

Sl. No.	Herbal Ingredients	Solvent	Procedure	Result
1	Leaves of *Trichosanthes dioica* *Curcuma longa* Roots of *Asparagus racemosus*	Water	Leaves were mixed with *Curcuma longa* and roots of *Asparagus racemosus*, crushed and mixed with water	The filtered portion is taken daily to cure allergic problems
3	Seeds of *Syzigium cumini* Internal white part of *Mangifera indica* seed	Honey	All of the ingredients were dried and ground in powder form. After that mixed with solvent for preparation of pills and sun dried.	This combination is very much effective for cough and cold and diabetes
4	*Andrographis paniculata* *Vitex negundo* *Glycyrrhiza glabra* *Azadirachta Indica*	Water	All the ingredients were sun dried, ground into powder form and prepared as pills	One pill every day in empty stomach will help to cure worms of abdomen and also important for healthy stomach
5	*Vitex negundo* *Azadirachta indica* *Neolamarckia cadamba* *Datura metel* *Allium sativum* *Curcuma longa*	Mustard oil	All the ingredients were boiled in mustard oil until the oil becomes black	Its helps in pain of arthritis
6	*Vetiveria zizanioides* Rice (*Oryza sativa*)	Water	Roots of the grass were mixed with rice which was soaked overnight and made into paste. The paste was taken after meal	Helps to cure Haematuria and hematemesis
7	*Centella asiatica*, *Leucas zeylanica*	Water	All the ingredients were boiled with water and then dried as pills	Pills are taken after meals and its very much effective for jaundice, body ache, abdominal dysfunction
8	*Datura metel* *Calotropis procera.* *Curcuma longa*	Mustard oil	All the ingredients were made into paste and then mixed with mustard oil and boiled for 20–30 mins. The content was then filtered and the oil collected was used.	This oil is very helpful for reducing pain and inflammation caused by fracture of bone
9	*Withania somnifera*	Water, ghee	A thick paste of all the ingredients was prepared	Its helps in arthritis and also help in joining of

(continued)

Table 2 (continued)

Sl. No.	Herbal Ingredients	Solvent	Procedure	Result
	Cissus quadrangularis *Leucas zeylanica* *Bambusa polymorpha* *Moringa oleifera* *Terminalia arjuna* *Tragia involucrata* *Piper nigrum* *Zingiber officinale* *Vitex negundo*	(clarified butter)	and applied on the area of fracture or on the area of dislocated bone	bone and also helps to reduce the pain of bone dislocation
10	Leaves of *Trichosanthes dioica* *Asparagus racemosus* *Phyllanthus emblica* *Terminalia chebula* *Curcuma longa* *Tinospora cordifolia*	Water	All the ingredients were ground and boiled with 250 ml of water for 30 min. The filtered water was taken every day on an empty stomach	Its helps to cure allergy

healers were interested in treating bone-related problems, like fractures of bone, dislocation of bones, and arthritis. For treating bone fracture they prepare a paste of fresh leaves of *Cissus quadrangularis, Vitex nigundo, Calotropis procera*, and *Heliotropicum indicum* mixed with different solvents and the paste is applied directly on the affected area. According to the local healers *Mikania micrantha, Eclipta prostrata*, and *Sida acuta* are the medicinal plants used for skin, hair-related problems, and also for the treatment of allergies. Application of paste of *Eclipta prostrata* also known as Kalakeshari/Keshraj in vernacular language, helps in the regrowth of hair, dermatitis, and curing insomnia if the leaf paste was mixed with coconut oil. For treating cough and cold, a decoction of fresh leaves of *Ocimum sanctum* (Holy basil or Tulsi) was used. *Ocimum gratissimum*, *Piper nigrum*, and *Leucas zeylanica* were also used to treat cough and cold. *Piper nigrum*, known as golmorich, and its fruits are traditionally used for flavouring food recipes. Ranjbanshi community used it for treating asthma, cough, headache, menstrual pain, gastric problems, and urinary problems too. The roots of *Vetiveria zizanioides* ground with scented-soaked rice and the paste was consumed to cure Haematuria

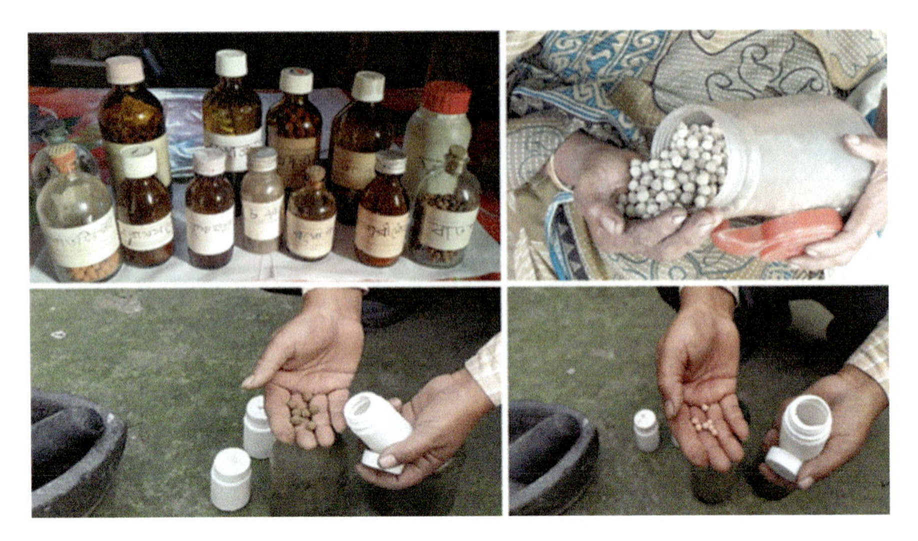

Fig. 3 Some herbal medicinal formulation in the form of syrup and pills prepared by Rajbanshi community

and Hematemesis. According to traditional medicinal practitioners *Heliotropicum indicum* was very useful in treating conjunctivitis. *Tinospora cordifolia* also known as Guloncho in vernacular language is a deciduous climbing shrub that is traditionally used to cure diabetes, reduce hypertension, and also enhance memory. The methods of application of herbal formulations are different depending on the ailment being treated and are mentioned in Tables 1 and 2. In Fig. 3, different types of medicines prepared by Kabiraj's has been shown. Seven different plants which were mainly used to treat inflammation and for healing wounds by the Rajbanshi community were chosen for comparative analysis of their anti-oxidant, anti-inflammatory, and immunostimulatory activities (Fig. 3). The dry weights of the ethanolic leaf extract and their accession number are represented in Fig. 4.

4.2 Antioxidant Property of the Different Plant Extracts

4.2.1 DPPH Scavenging

In the present investigation, three different doses (0.0292 mg/ml, 0.0584 mg/ml, and 0.876 mg/ml) of different ethanolic leaf extracts have been considered. The percentage of inhibition of DPPH generation by different extracts has been compared to the blank set, where the purple colour gradually decolorizes into pale yellow indicating DPPH scavenging. Out of seven plant extracts , DDE, CAE, CSE, and CGE showed a high percentage of inhibition in all three doses compared to other extracts. Starting from the lowest dose to the highest dose, CSE exhibited the maximum inhibition, comprising 48.80 $\pm$ 3.38 (0.0292 mg/ml), 63.82 $\pm$ 7.39 (0.0584 mg/ml), and

Fig. 4 The identification accession number and the dry weight of the ethanolic leaf extract of seven different plants used for assessing their anti-oxidant, anti-inflammatory, and immunostimulatory properties

72.26 ± 3.23 (0.876 mg/ml), whereas with DDE and CGE the percentage of inhibition ranged between 41% and 50% from lower to higher doses. The percentage of inhibition was low with the lower doses of CAE, around 34% to 36%, however, with the highest dose (0.876 mg/ml) it showed 51.38 ± 3.37% of DPPH inhibition (Fig. 5). The result of CSE at 0.876 mg/ml dose was comparable to Gallic acid, a well-known anti-oxidant which exhibited inhibition of DPPH between the range of 74% and 86% from lower to a higher dose.

4.2.2 Hydroxyl Radical Scavenging

Hydroxyl radical is one of the major reactive oxygen species, formed continuously during the process of reduction of oxygen to water in living organisms. This hydroxyl radical that is being generated can cause a breakdown of disulphide bonds in several proteins, especially fibrinogen, which can lead to abnormal protein configuration as observed during cancer, atherosclerosis, and in neurological disorders. Thus, the present study investigates the protective effect of plant extracts against damage due to hydroxyl radical generation. Two different concentrations of ethanolic leaf extract were considered in the present investigation, 0.0584 mg/ml and 0.146 mg/ml to investigate the scavenging of hydroxyl ion generation. The concentration of malonaldehyde (MDA) generated in the reaction mixture is indicative of the degree of hydroxyl radical generation. The inhibition of hydroxyl generation has been compared with the blank set (Fig. 5).

MDA generation was found significantly lower in almost all the extracts in both the concentrations (0.146 mg/ml and 0.0584 mg/ml) except CCE, where the inhibition of hydroxyl generation was not significant compared to the vehicle control (Fig. 6). The level of inhibition of hydroxyl generation was highly significant for

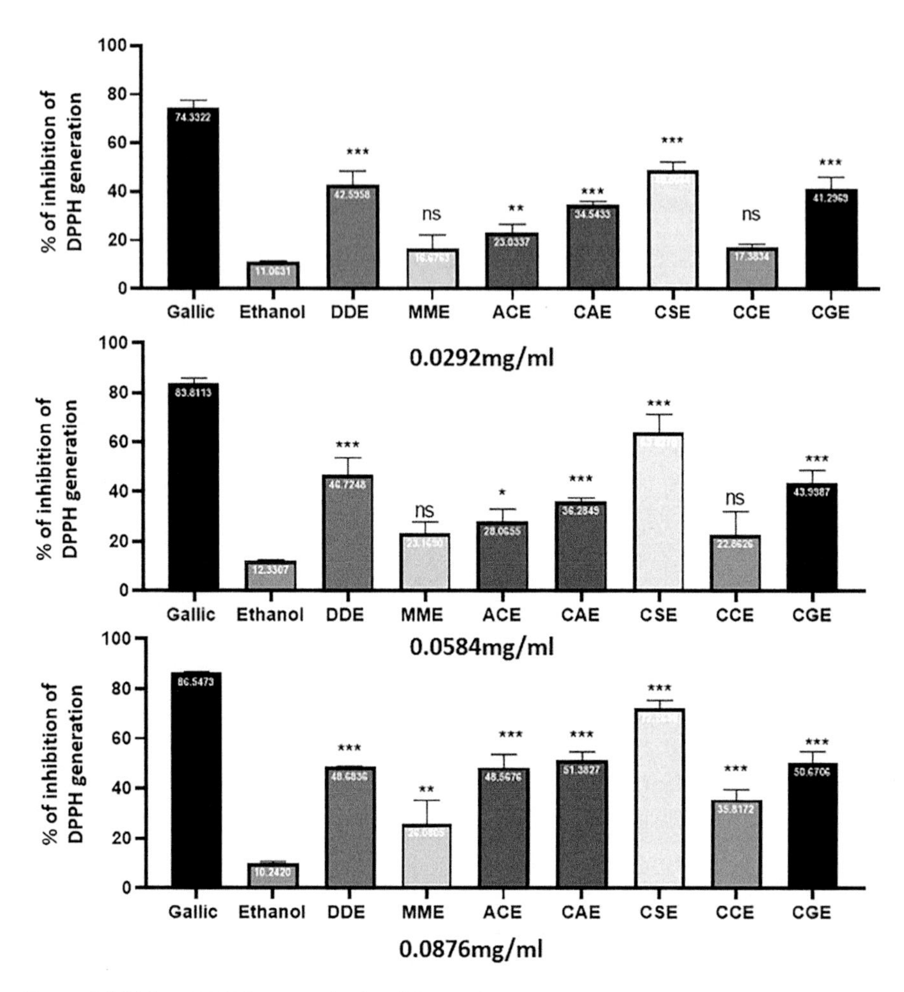

Fig. 5 Inhibition of DPPH generation by different plant extracts at 0.0.292 mg/ml, 0.0584 mg/ml, 0.0879 mg/ml. Gallic acid (Gallic) has been used as a positive control. Results were expressed as Mean ± SD ***$p < 0.001$, **$p < 0.01$, *$p < 0.05$, *ns* non significant. The significance of the antioxidant property has been estimated comparing the value of the vehicle control (ethanol)

CSE and ACE in both concentrations compared to others. However, CAE and DDE also exhibited high antioxidant properties compared to CCE and CGE. As observed in DPPH scavenging (Fig. 5), here also CSE could significantly inhibit MDA generation up to 52.84 ± 1.84% at 0.0584 mg/ml dose and 67.47 ± 3.94 at 0.146 mg/ml, dose indicating high potentiality in scavenging hydroxyl radical generation. ACE (58.21 ± 6.97%), CAE (57.55 ± 1.29%), CGE (55.44 ± 4.73%), and DDE (48.61 ± 7.24) at 0.146 mg/ml dose also showed better efficacy in inhibiting hydroxyl radical generation (Fig. 6). Mannitol (MA) has been used as a positive control to understand the potentiality of the antioxidant property of the extracts, where CSE showed to some extent similar effects with mannitol.

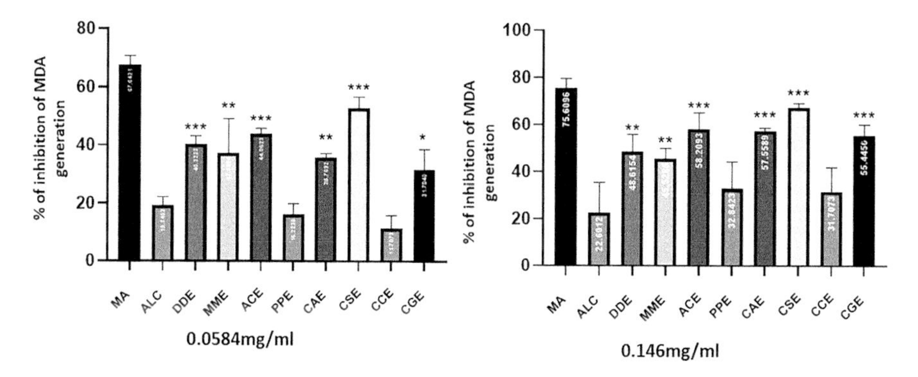

Fig. 6 Inhibition of hydr oxyl ion generation by different ethanolic leaf extracts at 0.0584 mg/ml and 0.146 mg/ml dose. Mannitol (MA) has been used as a positive control. Result expressed as Mean ± SD $***p < 0.001$, $**p < 0.01$, $*p < 0.05$, non-significant has been denoted by ns. The significance of the antioxidant property has been estimated comparing the value of vehicle control (ethanol)

4.3 Anti-inflammatory Property of Different Plant Extracts

Delayed type hypersensitivity (DTH) is a vital defence against several pathogens such as *Mycobacterium tuberculosis, Mycobacterium leprae*, *Leishmania sp*, *Histoplasma*, *Coccidioides* fungi and is also involved in tissue rejection and cancer immunity [10, 32, 38, 55, 56]. Thus, to understand the potentiality of the different ethanolic leaf extracts for curing deleterious DTH response, DNFB, a strong hapten has been used to induce DTH reaction. In case of DTH reaction in the skin, there is an early infiltration of non-specific cells during primary sensitization with the antigen, which is followed by a specific response upon resensitization with the same antigen, which is mainly T cell-mediated response. This antigen-specific hypersensitive response leads to erythema (redness of skin) and oedema at the site of the antigenic challenge in the sensitized animal [4]. The hypersensitivity response is usually delayed, showing effects after 48–72 h of resensitization. In our animal model, the DTH response exhibited within 48 h and gradually got severe. Ethanolic leaf extracts were injected intravenously 1 h before resensitization. Among all the ethanolic leaf extracts, CSE showed the best result in alleviating the DTH inflammation faster. On the 12th day of resensitization with CSE treatment the paw size was 0.295 ± 0.023, which is almost the size of a 0 h normal paw (0.276 ± 0.014). The other extracts such as CAE (0.301 ± 0.005) and ACE (0.313 ± 0.012) also showed better inhibition compared to other extracts, however, the foot pad healing process was still in continuation (Fig. 7a). In the untreated and vehicle (ethanol) control groups, the foot pad muscles as well as paw digits were also lost in the course of inflammation. However, in the plant extract-treated groups, the foot pad loss gradually recovered and the inflammation also subsided much earlier than in the control groups. In the treatment groups, none of the mice exhibited digit loss (Fig. 7b). In the case of the control groups it took more than 3 weeks to subside

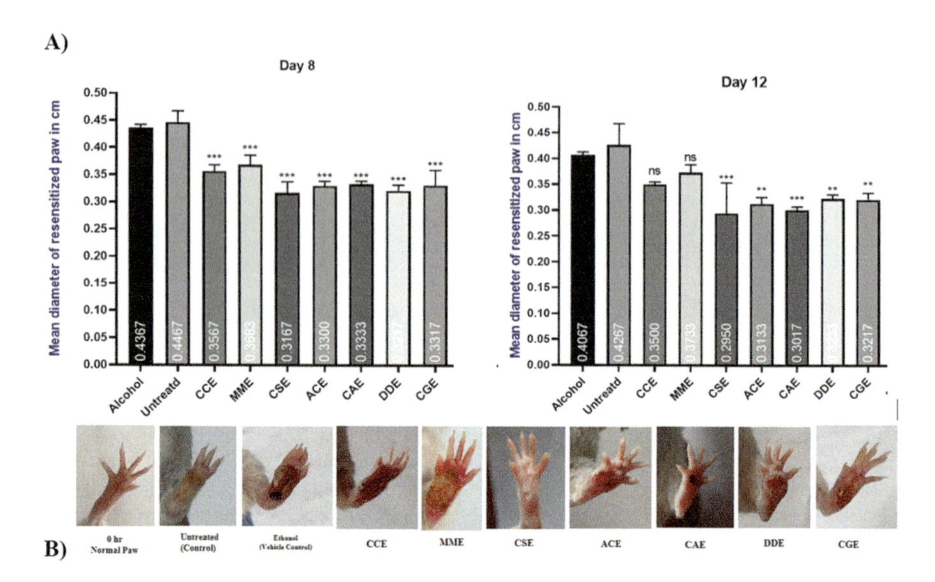

Fig. 7 (**a**) DTH (*delayed type of hypersensitivity*) reaction of resensitized paw in treated, untreated (control), and ethanol (vehicle control) at 8th and 12th days of DTH inflammation, represented by mean diameter in cm, considering the day of resensitization as day 0. Results are expressed as mean ± SD, $***p < 0.001$, $**p < 0.01$ and $*p < 0.05$, non-significant value has been denoted by ns. (**b**) Photograph shown are of the resensitized paw on the 12th day. The far left photograph shows the normal paw (0 h, before the resensitization)

the DTH inflammation, although the vehicle control groups (ethanol) recovered a bit earlier than the untreated one. CSE-treated mice never exhibited digit loss and the inflamed foot pad eventually healed back to its normal size within 12 days of resensitization, suggesting its strong anti-inflammatory and wound-healing properties. The DTH paw treated with CAE and ACE extract resumed back to normal paw diameter in 13–14 days.

4.4 Immunostimulatory Property of the Different Plant Extracts

The antibody or humoral-mediated immune response during primary immunization was investigated through a haemagglutination titre (HA) assay, which measures the kinetics of primary immunity. The mode of plant extract treatment is similar to the DTH assay, where the mice were injected intravenously with different leaf extracts, 1 h before the immunization with sheep RBC (SRBC). It was interesting to observe that almost all the plant extracts were capable enough to boost antibody (Ab) production higher than the ethanol-treated and untreated mice, however, the level of stimulation varied (Fig. 8). In the serum collected from the primary immunized mice, the highest antibody (IgM) titre value was shown with CSE (*Cleome spinosa*) treatment which was 1:2560, followed by CCE (*Cajanus cajan*) 1:1280 and then MME (*Mikania micrantha*), ACE (*Ageratum conyzoides*), CAE (*Centella*

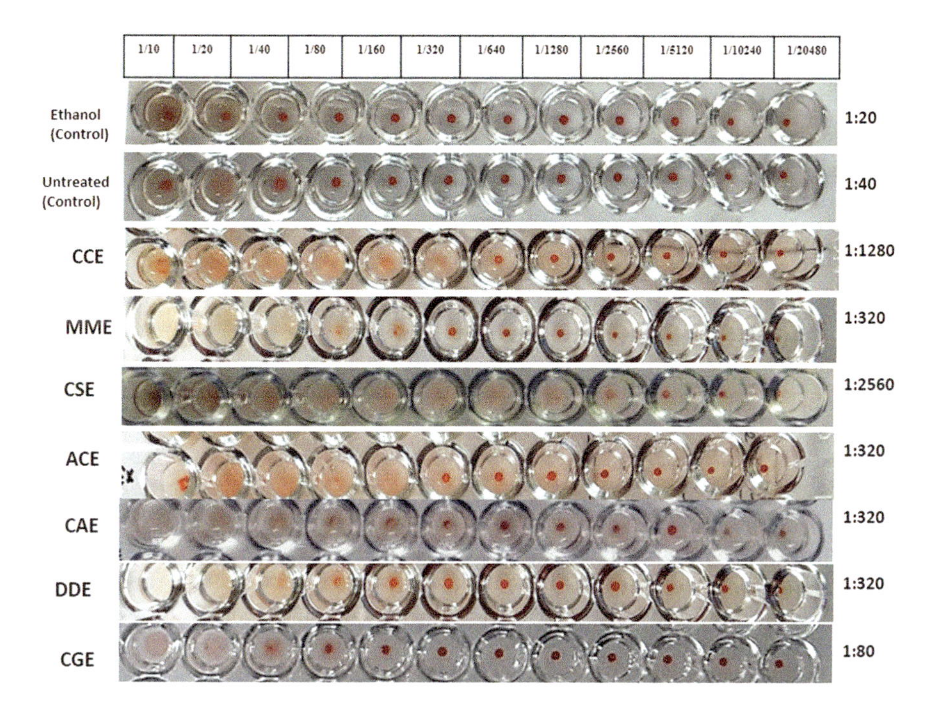

Fig. 8 Antibody response after the treatment of different plant extract. Hemagglutinin titre (HA) value shows the level of antibody production during primary immune response, mostly IgM (immunoglobulin M). The titre value mentioned on the extreme right indicates the last dilution of serum capable of agglutinating antigen

asiatica), DDE (*Drymaria diandra*) where all showed 1:320 titre value, and CGE (*Coccinia grandis*) showed the least with 1:80. Whereas, the Ab titre value was 1:20 with ethyl alcohol and 1:40 in the untreated group.

5 Discussion of Findings

Among the several ethnic communities in West Bengal, Rajbanshi community are the ancient tribe of Koch kingdom, belonging to the Indo-Mongoloid race. This chapter documents the ethnomedicinal knowledge shared by the Rajbanshi community of the Cooch Behar district of West Bengal, India. Cooch Behar is a historic location rich in culture and heritage, marked as a princely state since the sixteenth century, and is known to be the last ruling Hindu dynasty of Bengal before British rule. The Rajbanshi community inhabits mostly the northern part of West Bengal which comes under the sub-Himalayan region. The river system here plays an important role in its civilization by providing a source of water for irrigation and for drinking purposes. Torsa, Tista, Mahananda, Karatowa, and their tributaries are important rivers of this region. This Terai region of Bengal is covered with dense forest, inhabiting a huge number of exotic and wild species of flora and fauna. This area is often subjected to heavy rainfall and flash floods every year. These two important geographical features having rivers and forests makes them suitable for agricultural practices and at the same time makes them vulnerable to disease outbreak, due to drinking contaminated water during the rainy season and the dense forest that harbours parasitic diseases. The diseases that have always been prevailing in this region are dysentery, diarrhoea, cholera, malarial fever, smallpox, goitre, whooping cough, common cold, and skin diseases. In spite of the fact that the people were affected with serious diseases, the healthcare facility was never up to the mark. Due to a lack of proper healthcare facilities, the community had to be dependent on their indigenous traditional herbal medicine. In almost every household in this community, a herbal medicinal garden exists, this way they also contribute in conserving the species. Traditional medicine has been utilized by indigenous communities for centuries and the practice is still deeply rooted in the Rajbanshi community to treat a variety of ailments. Though with due course of time, the availability of modern healthcare services improved, the community still continues to rely on traditional medicine, highlighting the importance of plants in the treatment of illnesses. The results of the survey illustrated the continued significance of traditional medicine in the lives of these indigenous people in the study area. The present study documented the ancient practice of traditional medicine by the Rajbanshi community of India. About 37 medicinal plants were documented that are used to treat inflammation and for healing different types of wounds by Rajbanshis. This study highlights the plants which are focused mainly to cure inflammation, inflammation-related disorders, and healing wounds. Though some of the plants were also used for treating different types of fevers, such as *Stephania japonica, Sida acuta, Ludwigia perennis,* and *Leucas zeylanica*. For treating diarrhoea, dysentery, and for stomach-related disorders *Stephania japonica*, *Cajanus cajan*, *Mikania micrantha*, *Centella asiatica*, and *Leucas zeylanica* are used. The community used several parts of plants for preparing the herbal formulations, starting from the entire plant to its leaves, stem, roots, buds, flowers, seeds, fruits, and bark. Leaves were used the most for preparing the herbal formulation. Different modes of application were preferred for treatment such as consuming the filtrate of the leaf, applying the paste on the wound/inflamed area, inhaling the smoke of burnt leaves, or simply consuming it as cooked vegetables. The herbal preparations were utilized mostly to heal wounds and sores, abdominal pain/ache, cure jaundice, liver problems, as well as for anti-parasitic activities. However, the traditional healers of Cooch Behar district have now inclined more towards treating bone-related problems, such as breaking or dislocation of bones, and arthritis.

Almost all living beings are subject to attack by foreign pathogenic agents. Our immune system plays the most important role in combating those foreign particles and acts as a barrier in establishing the infection induced by pathogens as well as non-pathogenic agents. The non-specific immune system acts as the primary barrier by releasing several mediators and recruiting immune cells at the target site for prompt response through inflammation. Whereas, the specific immune response is target specific which responds through antibody production or through cell-mediated immune response. Immunostimulatory agents from the plant world have the potential to activate the immunological responsiveness of an organism directly at the cell level or by inducing the production of mediators. Plants and plant-derived products are safer, eco-friendly, low-cost, fast, and less toxic as compared with conventional treatment methods and act specifically on target cells without affecting normal cells when administered in proper doses. It strengthens the overall activity of the immune system including those elements that are most capable of controlling inflammatory conditions. Thus, the present study has selected seven different plant species (*Ageratum conyzoides*, *Cenetella asiatica*, *Cajanus cajan*, *Mikania micrantha*, *Coccinia grandis*, *Cleome spinosa*, and *Drymaria diandra*) which were mainly used to cure inflammatory diseases and for healing different cuts and wounds by the Rajbanshi people. Ethanolic extracts of these seven different plants were evaluated for their anti-inflammatory and immunostimulatory properties. The purpose was to look for plants with immunopharmacological properties to cure inflammatory diseases in humans. Among the seven plants investigated, *Cleome spinosa* (CSE), *Cajanus cajan* (CAE), and *Drymaria diandra* (DDE) showed high immunopharmacological properties. They exhibited anti-oxidant and anti-inflammatory properties by scavenging free radical generation, alleviating deleterious DTH response, and healing the DTH wound faster. Knowing the fact that ROS generation is one of the crucial components in aggravating inflammation and causing tissue damage, a plant product that can scavenge ROS and at the same time minimize tissue damage induced by deleterious DTH reaction and heal the wound faster without any digit loss in the murine model could turn out to be a wonderful drug.

The HA titre results were quite fascinating as with CSE and CCE treatment, the antibody titre in the host serum increased several folds higher than the control groups upon antigenic challenge. This immunostimulatory property of *Cleome spinosa* and *Cajanus cajan* has been reported here for the first time. However, other plants such as *Ageratum conyzoides* (ACE), Centella asiatica *(CAE), Mikania micrantha* (MME), and *Dymaria diandra* (DDE) also exhibited similar properties but with lower effectivity. There are certain advantages of using plants to boost the immune system as they can be utilized directly as feed and can stimulate both systemic and mucosal immunity. Our study shows that ethanolic leaf extract *Cleome spinosa* and *Cajanus cajan* plant extract can stimulate the antibody-mediated immune system in a murine model when challenged with sheep RBC. There are several pathological conditions where the immunocompromized host shows more susceptibility towards infections due to certain environmental or genetic factors [11]. One such condition is severe combined immunodeficiency (SCID) disease where there is a deficit of both humoral and adaptive immune response leading to uncontrolled damage [8]. Immunodeficiency can also arise during treatment with various agents/drugs during malignancy or for allograft rejection, leading to secondary immunodeficiency, compromising the host defence mechanism. During immunosuppression induced by pathogens, a herbal compound that has the potential to activate the humoral arm of the immune system can prove to be a wonderful agent to induce antibody production to neutralize the pathogenic effect or binds with the antigen facilitating its destruction by other cells. Thus, it will be interesting to study in detail the immunopharmacological properties of *Cleome spinosa* and *Cajanus cajan* in the future with pathogenic challenges. At the same time, we also suggest that further research on *Cleome spinosa* (CSE), *Cajanus cajan* (CAE), and *Drymaria diandra* (DDE) might provide some important leads to the search for new drugs to cure pathogenic as well as non-pathogenic inflammatory diseases.

6 Conclusion

Globally, the use of medicinal plants is widespread in the developing world, providing primary health care, livelihood improvement, and even a source of income for many people. As a result, the importance of preserving and protecting these plants is of paramount importance for the continued health of both local communities and the global population. Studies have demonstrated that medicinal plants can offer a wide range of therapeutic benefits, from treating chronic and acute illnesses to improving overall wellness. Furthermore, medicinal plants can be used to enhance nutrition, reduce the use of synthetic drugs, and provide economic opportunities for communities. As a result, the conservation of medicinal plants is imperative to ensure that their medicinal properties are not lost. Moreover, research into the potential medicinal uses of plants is an important part of the health care industry. This research can help to identify new medicines, develop safer drugs, and improve the efficiency of existing medicines. Furthermore, the use of medicinal plants can also help to reduce the cost of health care, as many of these plants are readily available and do not require the use of expensive pharmaceuticals. In this study we have documented different ethnomedicinal plants which are mainly used by the Rajbanshi community. Through ethnobotanical studies, it is important to document the indigenous knowledge associated with the plants of a certain region. This knowledge can be used to help conserve and utilize the biological resources of the region, as well as to uncover potential bio-cultural benefits. In addition, the collection of local names and traditional uses of the plants can provide important insights into the culture of the indigenous people of the area.

Strategies related to drug discovery are taking a new turn by including the knowledge gained from traditional and alternative system of medicines. The world still holds enormous medicinal plants and still there are many indigenous ethnic communities that use plants as the important source of treatment. Both need to be explored extensively, for the development of new anti-inflammatory drugs, as a majority of the plant species have not been investigated for their potentiality to cure inflammatory disorders which are non-cytotoxic. Thus, learning from the indigenous communities about their age-old traditional treatment methods and implementing it for the betterment of the mankind is the need of the hour.

Acknowledgment The present work was financially supported by Science & Technology and Biotechnology, Govt. of West Bengal, India (Memo No. 342 (sanc.)-ST/P/S&T/16 G-27/2018). The authors acknowledge the Rajbanshi community who have shared their knowledge/information that enabled the authors to put forward a well-rounded perspective on the topic. The authors also show gratitude towards Mr. Jadu Saud for assisting during the field survey.

References

1. Atanasov AG, Zotchev SB, Dirsch VM et al (2021) Natural products in drug discovery: advances and opportunities. Nat Rev Drug Discov 20:200–216. https://doi.org/10.1038/s41573-020-00114-z
2. Bamola N, Verma P, Negi C (2018) A review on some traditional medicinal plants. Int J Life Sci Scienti Res 4(1):1550–1556
3. Bindu S, Mazumder S, Bandyopadhyay U (2020) Non-steroidal anti-inflammatory drugs (NSAIDs) and organ damage: a current perspective. Biochem Pharmacol 180:114147
4. Black CA (1999) Delayed type hypersensitivity: current theories with an historic perspective. Dermatol Online J 5(1):7. PMID: 10673450
5. Chakravarty AK, Yasmin H (2005) Alcoholic turmeric extract simultaneously activating murine lymphocytes and inducing apoptosis of Ehlrich ascitic carcinoma cells. Int Immunopharmacol 5 (10):1574–1581
6. Chakravarty AK, Yasmin H (2008) Free radical scavenging and nitric oxide synthase activation in murine lymphocytes and ehlrich ascitic carcinoma cells treated with ethanolic extract of turmeric. Proc Natl Acad Sci Sect B 78(I):37–44
7. Chakravarty AK, Yasmin H, Chatterjee SN, Mazumder T (2009) Comparison of efficacy of turmeric and curcumin in immunological functions and gene expression. Int J Pharmacol 5(6): 333–345
8. Chen L, Deng H, Cui H, Fang J, Zuo Z, Deng J, Li Y, Wang X, Zhao L (2017) Inflammatory responses and inflammation-associated diseases in organs. Oncotarget 9(6):7204–7218
9. Chinn IK, Shearer WT (2015) Severe combined immunodeficiency disorders. Immunol Allergy Clin N Am 35(4):671–694. https://doi.org/10.1016/j.iac.2015.07.002
10. Datta T, Patra AK, Dastidar SG (2014) Medicinal plants used by tribal population of Coochbehar district, West Bengal, India-an ethnobotanical survey. Asian Pac J Trop Biomed 4(Suppl 1):S478–S482. https://doi.org/10.12980/APJTB.4.2014C1122
11. Disis ML, Schiffman K, Gooley TA, McNeel DG, Rinn K, Knutson KL (2000) Delayed-type hypersensitivity response is a predictor of peripheral blood T-cell immunity after HER-2/neu peptide immunization. Clin Cancer Res 6(4):1347–1350
12. Dropulic LK, Lederman HM (2016) Overview of infections in the immunocompromised host. Microbiol Spectr 4(4). https://doi.org/10.1128/microbiolspec.DMIH2-0026-2016
13. Dutta BK, Dutta PK (2005) Potential of ethnobotanical studies in North East India: an overview. Indian J Tradit Knowl 4(1):7–14
14. Enaregha EB, Izah SC, Okiriya Q (2021) Antibacterial properties of *Tetrapleura tetraptera* pod against some pathogens. Res Rev Insights 5:1–5. https://doi.org/10.15761/RRI.1000165
15. Epidi JO, Izah SC, Ohimain EI (2016) Antibacterial and synergistic efficacy of extracts of *Alstonia boonei* tissues. Br J Appl Res 1(1):0021–0026
16. Epidi JO, Izah SC, Ohimain EI, Epidi TT (2016) Phytochemical, antibacterial and synergistic potency of tissues of *Vitex grandifolia*. Biotechnol Res 2(2):69–76
17. Ferrucci L, Fabbri E (2018) Inflammageing: chronic inflammation in ageing, cardiovascular disease, and frailty. Nat Rev Cardiol 15:505–522. https://doi.org/10.1038/s41569-018-0064-2
18. Fleming TP, Watkins AJ, Velazquez MA, Mathers JC et al (2018) Origins of lifetime health around the time of conception: causes and consequences. Lancet 391(10132):1842–1852. https://doi.org/10.1016/S0140-6736(18)30312-X
19. Furman D, Campisi J, Verdin E et al (2019) Chronic inflammation in the etiology of disease across the life span. Nat Med 25:1822–1832. https://doi.org/10.1038/s41591-019-0675-0
20. Ghasemian M, Owlia S, Owlia MB (2016) Review of anti-inflammatory herbal medicines. Adv Pharmacol Sci 2016:9130979
21. Gisterå A, Hansson GK (2017) The immunology of atherosclerosis. Nat Rev Nephrol 13:368–380
22. Izah SC (2018) Some determinant factors of antimicrobial susceptibility pattern of plant extracts. Res Rev Insight 2(3):1–4

23. Izah SC (2019) Activities of crude, acetone and ethanolic extracts of *Capsicum frutescens* var minima fruit against larvae of *Anopheles gambiae*. J Environ Treat Techn 7(2):196–200
24. Izah SC, Aseibai ER (2018) Antibacterial and synergistic activities of methanolic leaves extract of lemon grass (*Cymbopogon citratus*) and rhizomes of ginger (*Zingiber officinale*) against *Escherichia coli*, *Staphylococcus aureus* and *Bacillus subtilis*. Acta Sci Microbiol 1(6):26–30
25. Izah SC, Uhunmwangho EJ, Dunga KE, Kigigha LT (2018) Synergy of methanolic leave and stem-back extract of *Anacardium occidentale* L (cashew) against some enteric and superficial bacteria pathogens. MOJ Toxicol 4(3):209–211
26. Izah SC, Uhunmwangho EJ, Dunga KE (2018) Studies on the synergistic effectiveness of methanolic extract of leaves and roots of *Carica papaya* L (papaya) against some bacteria pathogens. Int J Complement Altern Med 11(6):375–378
27. Izah SC, Uhunmwangho EJ, Etim NG (2018) Antibacterial and synergistic potency of methanolic leaf extracts of *Vernonia amygdalina* L. and *Ocimum gratissimum* L. J Basic Pharmacol Toxicol 2(1):8–12
28. Izah SC, Uhunmwangho EJ, Eledo BO (2018) Medicinal potentials of *Buchholzia coriacea* (wonderful kola). Med Plant Res 8(5):27–43
29. Izah SC, Zige DV, Alagoa KJ, Uhunmwangho EJ, Iyamu AO (2018) Antibacterial efficacy of aqueous extract of *Myristica fragrans* (Common Nutmeg). EC Pharmacol Toxicol 6(4):291–295
30. Izah SC, Chandel SS, Etim NG, Epidi JO, Venkatachalam T, Devaliya R (2019) Potency of unripe and ripe express extracts of long pepper (*Capsicum frutescens* var. baccatum) against some common pathogens. Int J Pharm Phytopharmacol Res 9(2):56–70
31. Izah SC, Etim NG, Ilerhunmwuwa IA, Silas G (2019) Evaluation of crude and ethanolic extracts of *Capsicum frutescens* var. minima fruit against some common bacterial pathogens. Int J Complement Altern Med 12(3):105–108
32. Izah SC, Etim NG, Ilerhunmwuwa IA, Ibibo TD, Udumo JJ (2019) Activities of express extracts of Costus afer Ker–Gawl. [Family COSTACEAE] against selected bacterial isolates. Int J Pharm Phytopharmacol Res 9(4):39–44
33. Janez FJ, Yasmin H, Al-Ahdal MN, Bhakta S, Kishore U (2020) Natural and trained innate immunity against *Mycobacterium tuberculosis*. Immunobiology 225(3):151951. https://doi.org/10.1016/j.imbio.2020.151951
34. Jeyaprakash K, Ayyanar M, Geetha KN et al (2011) Traditional uses of medicinal plants among the tribal people in Theni District (Western Ghats), Southern India. Asian Pac J Trop Biomed 1: S20–S25
35. Kaplan DH, Igyártó BZ, Gaspari AA (2012) Early immune events in the induction of allergic contact dermatitis. Nat Rev Immunol 12(2):114–124. https://doi.org/10.1038/nri3150
36. Kigigha LT, Biye SE, Izah SC (2016) Phytochemical and antibacterial activities of *Musanga cecropioides* tissues against *Escherichia coli*, *Pseudomonas aeruginosa Staphylococcus aureus*, *Proteus* and *Bacillus* species. Int J Appl Res Technol 5(1):100–107
37. Kigigha LT, Izah SC, Uhunmwangho EJ (2018) Assessment of hot water and ethanolic leaf extracts of *Cymbopogon citratus* Stapf (lemon grass) against selected bacteria pathogens. Ann Microbiol Infect Dis 1(3):1–5
38. Kigigha LT, Selekere RE, Izah SC (2018) Antibacterial and synergistic efficacy of acetone extracts of Garcinia kola (Bitter kola) and Buchholzia coriacea (Wonderful kola). J Basic Pharmacol Toxicol 2(1):13–17
39. Kobayashi K, Kaneda K, Kasama T (2001) Immunopathogenesis of delayed-type hypersensitivity. Microsc Res Tech 15;53(4):241–5. https://doi.org/10.1002/jemt.1090. PMID: 11340669.
40. Loveland BE, McKenzie IF (1982) Delayed-type hypersensitivity and allograft rejection in the mouse: correlation of effector cell phenotype. Immunology 46(2):313–320
41. Mahdi-Pour B, Jothy SL, Latha LY, Chen Y, Sasidharan S (2012) Antioxidant activity of methanol extracts of different parts of Lantana camara. Asian Pac J Trop Biomed 2(12):960–965

42. Mushtaq S, Abbasi BH, Uzair B, Abbasi R (2018) Natural products as reservoirs of novel therapeutic agents. EXCLI J 17:420–451. https://doi.org/10.17179/excli2018-1174
43. Patel R, Patel Y, Kunjadia P et al (2015) DPPH free radical scavenging activity of phenolics and flavonoids in some medicinal plants of India. Int J Curr Microbiol App Sci 4(1):773–780
44. Paul A (2021) Indigenous uses of ethnomedicinal plants among tribal communities of Ajodhya hill region of Purulia District, West Bengal, India. SSRG Int J Med Sci 8(7):13–19. https://doi.org/10.14445/23939117/IJMS-V8I7P102
45. Raj AJ, Biswakarma S, Pala NA et al (2018) Indigenous uses of ethnomedicinal plants among forest-dependent communities of Northern Bengal, India. J Ethnobiol Ethnomed 14:8. https://doi.org/10.1186/s13002-018-0208-9
46. Rup KB (2019) Practice of folk medicine in sub-Himalayan Bengal: a study of folk medicinal practices of the Rajbanshis in historical perspective. Abhijeet Publications. ISBN 978-93-88865-24-1
47. Sadeer NB, Montesano D, Albrizio S, Zengin G et al (2020) The versatility of antioxidant assays in food science and safety – chemistry, applications, strengths, and limitations. Antioxidants 9(8):709
48. Saha D, Sarma TK, Mukherjee SK (2016) Some medicinal plants of North 24 Parganas districts of West Bengal (India). Int J Pharm Biol Sci 6(3):191–206
49. Sarkar R, Mandal SK (2015) Role of indigenous folk medicinal plants among the tribal communities in West Bengal. Lokodarpan V–I:116–121. ISSN – 2454 – 3683
50. Sarkhel S (2014) Ethnobotanical survey of folklore plants used in treatment of snakebite in Paschim Medinipur district, West Bengal. Asian Pac J Trop Biomed 4(5):416–420. https://doi.org/10.12980/APJTB.4.2014C1120
51. Seiyaboh EI, Odubo TC, Izah SC (2020) Larvicidal activity of *Tetrapleura tetraptera* (Schum and Thonn) Taubert (Mimosaceae) extracts against *Anopheles gambiae*. Int J Adv Res Microbiol Immunol 2(1):20–25
52. Seiyaboh EI, Seiyaboh Z, Izah SC (2020) Environmental control of mosquitoes: a case study of the effect of *Mangifera indica* root-bark extracts (family Anacardiaceae) on the larvae of *Anopheles gambiae*. Ann Ecol Environ Sci 4(1):33–38
53. Taniguchi K, Karin M (2018) NF-κB, inflammation, immunity and cancer: coming of age. Nat Rev Immunol 18:309–324. https://doi.org/10.1038/nri.2017.142
54. Varghese PM, Tsolaki AG, Yasmin H, Shastri A, Ferluga J, Vatish M, Madan T, Kishore U (2020) Host-pathogen interaction in COVID-19: pathogenesis, potential therapeutics and vaccination strategies. Immunobiology 225(6):152008. https://doi.org/10.1016/j.imbio.2020.152008
55. Yasmin H (2007) Analysis of the effects of *Curcuma longa* Linn on lymphocytes and malignant cells in murine model for immunotherapy. PhD thesis, University of North Bengal, West Bengal, India
56. Yasmin H, Saha S, Butt MT, Modi RK, George AJT, Kishore U (2021) SARS-CoV-2: pathogenic mechanisms and host immune response. Adv Exp Med Biol. Springer Nature 1313:99–134. https://doi.org/10.1007/978-3-030-67452-6_6
57. Yasmin H, Varghese PM, Bhakta S, Kishore U (2021) Pathogenesis and host immune response in leprosy. Adv Exp Med Biol. Springer Nature 1313:155–177. https://doi.org/10.1007/978-3-030-67452-6_8
58. Yasmin H, Adhikary A, Al-Ahdal MN, Roy S, Kishore U (2022) Host–pathogen interaction in leishmaniasis: immune response and vaccination strategies. Immuno 2:218–254. https://doi.org/10.3390/immuno2010015
59. Yatoo MI, Gopalakrishnan A, Saxena A, Parray OR et al (2018) Anti-inflammatory drugs and herbs with special emphasis on herbal medicines for countering inflammatory diseases and disorders – a review. Recent Patents Inflamm Allergy Drug Discov 12(1):39–58

Plants Used in the Management and Treatment of Cardiovascular Diseases: Case Study of the Benin People of Southern Nigeria

30

Matthew Chidozie Ogwu, Happiness Isioma Ogwu, and Enoch Akwasi Kosoe

Contents

Abstract

There is a global increase in cardiovascular diseases (like stroke, heart attack, and heart failure), especially in developing countries. Healthcare organizations in these low- and middle-income countries are unable to handle the current and projected economic and social effects of CDVDs and are struggling to deal with other diseases and developmental challenges that disproportionately impact the region. This chapter aims to produce an inventory of plants and plant-based remedies used by Benin people, Nigeria, for the treatment and management of CDVDs. Fourteen locally identifiable key informants including professional

M. C. Ogwu (✉)
Goodnight Family Department of Sustainable Development, Appalachian State University, Boone, NC, USA
e-mail: ogwumc@appstate.edu

H. I. Ogwu
Department of Microbiology, Faculty of Life Sciences, University of Benin, Benin City, Edo State, Nigeria

E. A. Kosoe
Department of Environment and Resource Studies, SDD University of Business and Integrated Development Studies, Upper West Region, Ghana
e-mail: ekosoe@ubids.edu.gh

S. C. Izah et al. (eds.), *Herbal Medicine Phytochemistry*, Reference Series in Phytochemistry, https://doi.org/10.1007/978-3-031-43199-9_4

herbalists, traditional healers, and herbal remedy sellers were administered questionnaires to document plants and plant-based remedies used in the management and treatment of CDVD by Benin people of Southern Nigeria. From the survey, a total of 17 medicinal plants belonging to 14 plant families were identified, documented, and collected as used by Benin people for the treatment and management of CDVD. The plants included *Theobroma cacao, Aloe barbadensis, Momordica charantia, Laurus nobilis, Morinda citrifolia, Carica papaya, Allium cepa, Ficus exasperate, Elaeis guineensis, Olea europaea, Persea Americana, Moringa oleifera, Beta vulgaris, Allium sativa, Hunteria umbellate, Zingiber officinale,* and *Curcuma* species. The dominant plant families were Lauraceae, Amarylidaceae, and Zingiberaceae with two representatives each. The plants were mostly collected from home gardens, markets, and local forests and within either cultivated or wild systems. Although plants and plant-based remedies were mostly taken to address hypertension and high blood pressure, they are also used to manage stroke and to address CDVD-related chest pains without any side effects. Medicinal plants used in the management and treatment of CDVD contain natural bioactive compounds that accumulate in the plants as secondary metabolites, such as alkaloids, sterols, terpenes, flavonoids, saponins, glycosides, and tannins. The fundamental problem that most herbal medications address is the lack of standardization that may be used to classify them even though these effects usually overlap.

Keywords

Cardiovascular diseases · Plant medicine · Benin people · Ethnobotany · Heart attack · Non-communicable diseases · Healthcare systems

Abbreviations

ACE	Angiotensin-Converting Enzyme
AVG	Aloe Vera Gel
CDVD	Cardiovascular Disease(s)
NO	Nitric Oxide
NR	Not Reported
OLE	Olive Leaf Extract
OPP	Oil Palm Phenolic
SSA	Sub-Saharan African
UN	United Nations
WHO	World Health Organization

1 Introduction

There has been a considerable rise in cases of cardiovascular diseases globally [222]. CDVD is now considered among the principal cause of mortality globally. The incidences of diseases are rising at an alarming rate, especially in developing countries [102, 138, 149]. CDVD refers to any disease of the heart and blood vessels [102]. Also, cardiovascular disease was defined by Abunnaja and Sanchez [3] as "the pathologic process (usually atherosclerosis) affecting the entire arterial circulation, not just the coronary arteries" [3]. The most common CDVDs are cardiomyopathy, aortic disease, hypertension, strokes, coronary heart disease, high blood pressure, heart attack and failure, and peripheral arterial disease [102, 108, 125, 244]. CDVD causes economic, sociocultural, and health burdens to societies globally and accounts for around 31% of deaths every year [199, 236]. Globally, the World Health Organization (WHO) estimates that cardiovascular disease claims about 17.9 million lives [19, 42, 65, 122, 138, 199].

Healthcare organizations in low- and middle-income countries are unable to handle the current and projected economic and social effects of CDVDs, and the economic burden on high-income countries is unsustainable [116]. The CDVD crisis has increased the strain on the healthcare systems of sub-Saharan African (SSA) nations, which are already trying to deal with other diseases and developmental challenges that disproportionately impact the region [78, 85, 135, 139, 141, 145, 166, 183]. Also, information about region and age-specific death from CDVDs suggests SSA is disproportionately affected and so are people above 45 years [59, 227, 239]. Some causes of CDVD include unhealthy habits like smoking, and others like high cholesterol, hypertension, obesity, excess inactivity, and diabetes mellitus [149]. By 2025, the United Nations (UN) wants to reduce fatalities from noncommunicable diseases including cardiovascular disease by 25%. The current therapeutic approaches' economic costs to lower the three main CDVD risk factors – hyperlipidaemia, atherosclerosis, and hypertension – cannot be sustained on a worldwide scale [116].

Regarding the treatment of CDVD in SSA, herbal therapy is the most preferred healthcare for those affected but is also often complemented with modern medicines. It is arguably so because of sociocultural and economic motivations like affordability, availability, self-medication, and history of successful use in traditional communities [87, 128]. Herbal therapy for CDVD in SSA mostly revolves around the use of medicinal plants and spirituality because of their fewer side effects, reliability, accessibility, availability, and ability to self-medicate. Moreover, modern medicine is considered by some native people to be ineffective in the treatment and management of certain disease conditions [175]. Herbal medicines refer to the use of plant parts, materials, extracts, preparations, and products either alone or their combinations to address diverse medical uses [81, 128, 140, 143, 144, 146, 147]. Traditional medicine is a common healthcare means in developing countries of the Global South, whereas complementary or alternative medicine refers to modern medicine used in developed nations [33]. The use of traditional medicine in SSA and other parts of the Global South is often attributed to their accessibility, effectiveness, familiarity, perceived safety, naturalness, cost friendliness, and culture [29, 49, 127, 134, 178, 209].

This chapter aims to produce an inventory of the plant and plant-based remedies used by Benin people, Nigeria, for the treatment and management of CDVDs. A focused approach was adopted to select informants along with a non-experimental validation of the plants used through a literature review of the phytochemical and pharmacological information supporting the medicinal activities of the plants and plant-based remedies used in the treatment and management of CDVDs. Also, this chapter documents the traditional medicinal practices for the treatment and management of cardiovascular diseases and enumerated how these plants are used (i.e., either alone or in combination).

2 Plants and Plant-Based Remedies for the Treatment and Management of Cardiovascular Diseases in Benin City, Southern Nigeria

A survey was carried out in the six Local Government Areas that make up Benin City, Edo State (6.34° N and 5.60° E; Fig. 1) and occupied by Benin people of Southern Nigeria to document medicinal plants and traditional plant-based remedies used in the treatment and management of CDVDs. Benin City lies within the humid tropical rainforest zone of Nigeria but its vegetation has been heavily altered by

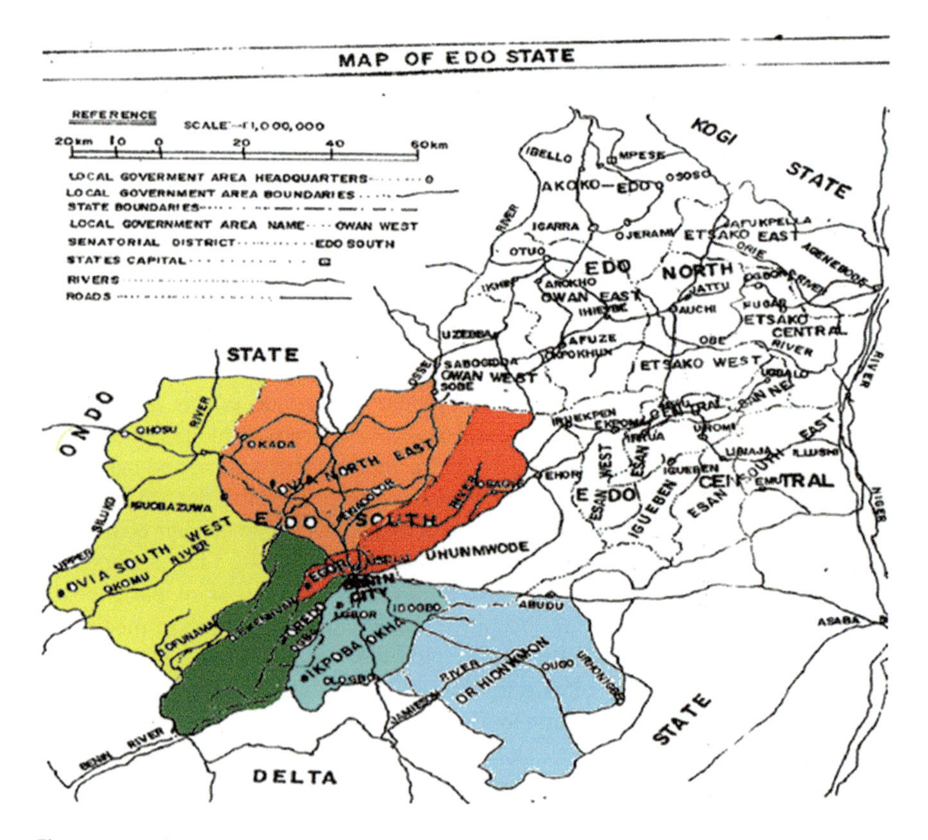

Fig. 1 Map of Edo State with the sampling area highlighted

anthropogenic activities and human population growth and is currently a mosaic of secondary forest [161–163]. The climate is monsoonal with two distinct seasons (dry and rainy season) with high rainfall (2000–3000 mm), temperature (20–40 °C), and average atmospheric humidity of 28% and with radiation of 1600 h per year [150, 152]. A detailed description of the geography, biodiversity, and soil features of the study area is present in Osawaru et al. [155–160], Osawaru and Ogwu [153, 154], Ogwu and Osawaru [142].

The survey to document traditional medicine practices and higher plants used for managing and treating CDVD by Benin people was conducted between January and June 2021 using survey sheets for information collection, a digital camera, an audio recording device for interviews, and a mobile phone-enabled position marker. The survey focused on randomly selected and locally identifiable professional herbalists, traditional healers, and herbal remedy sellers using semi-structured, open-ended questionnaires in English and/or preferred native languages. The 14 randomly selected respondents and interviewees are duly registered with all the relevant authorities (Table 1). The questionnaires were administered to the key informants (full-time traditional healers, part-time practitioners, elderly people who know the

Table 1 Demographic information of respondents

Characteristics	Frequency	Relative Percentage (%)
Marital Status		
Single	2	14.29
Married	12	85.71
Divorced	0	0.00
Unmarried but living with a partner	0	0.00
Educational Level		
Primary school	1	7.14
Secondary school	4	28.57
Technical degree	2	14.29
College/University	7	50.00
Employment Status		
Housewife	0	0.00
Student	0	0.00
Unemployed	3	21.43
Full-time employee	0	0.00
Part-time employee	11	78.57
Gender		
Male	4	28.57
Female	10	71.43
Occupation		
Trader	6	32.86
Student	3	21.43
Natural therapist	2	14.29
Pharmacist	2	14.29
Teacher	1	7.14
Tribe		
Benin	14	100.00

traditional values of the plants, relatives, and acquaintances who at one time or the other used some of these plants for the treatment and management of CDVD). The respondents animated a field survey to identify and collect plants and plant parts used for the treatment and management of CDVD. Specific information about the plants like their botanical and local names, part of the plant used, method of preparation and dosage, application, and treatment duration were asked and recorded. Collections were made from homestead farms, distant farms, and different forests present in the study area.

Data collected include the socio-demographic features of the survey respondents (e.g., marital status, educational level, employment status, gender, occupation, and tribe/ethnicity). Other pertinent data collected included plants used for CDVD treatment and/or management and preparation of the herbal remedies. Data collected were processed and analysed using Windows Microsoft Excel 2016 version software to understand inherent patterns. The therapeutic relevance of each identified species was also recorded. Based on the knowledge of the informants, a total of 17 medicinal higher plant species that are used for the treatment and management of CDVD were identified by a taxonomist at the University of Benin, Nigeria, and documented.

2.1 Medicinal Plants and Plant-Based Remedies Used by Benin People for the Treatment and Management of Cardiovascular Diseases

From the survey, a total of 17 medicinal plants belonging to 14 plant families were identified, documented, and collected as used by Benin people for the treatment and management of CDVD (Tables 2 and 3). The plants included *Theobroma cacao, Aloe barbadensis, Momordica charantia, Laurus nobilis, Morinda citrifolia, Carica papaya, Allium cepa, Ficus exasperate, Elaeis guineensis, Olea europaea, Persea Americana, Moringa oleifera, Beta vulgaris, Allium sativa, Hunteria umbellate, Zingiber officinale,* and *Curcuma* species. Some of these plants including *Allium sativum*, *Persea americana*, *Hunteria umbellate, Moringa oleifera*, and *Allium cepa* have been previously documented as useful for managing and treating CDVD [138]. The dominant plant families are Lauraceae, Amaryllidaceae, and Zingiberaceae with two representatives each. The other plant families are Malvaceae, Liliaceae, Cucurbitaceae, Rubiaceae, Caricaceae, Moraceae, Apocynaceae, Arecaceae, Moringaceae, Oleaceae, and Amaranthaceae. The habitat of these plants ranges from evergreen rainforests to warm temperate zone (Table 2). The plants were mostly collected from home gardens, markets, and local forests and within either cultivated or wild systems (Table 3). Earlier, Osawaru and Ogwu [154] confirmed the presence of plants with herbal medicine value as well as plant-based herbal remedies in local markets in Benin City and its environs. The plants and plant-based remedies were taken to address the following CDVDs – stroke, high blood pressure, hypertension, and chest pains without any side effects. It is known that plants are used to treat people with CDVDs like heart attacks, hypertension, cerebrovascular disorder, and heart failure [95, 126, 138]. Although these plants and

Table 2 Ethnobotanical documentation of plants and plant-based remedies used in the treatment and management of cardiovascular disorders

Plant and Plant-Based Remedy						
Plant name	Scientific name	Usefulness	Plant Family	Plant Habitat and Origin	Efficacy	Known Side Effects
Cocoa and Aloe Vera	*Theobroma cacao* *Aloe barbadensis*	Stroke and high blood pressure	Malvaceae Liliaceae	Evergreen tropical rainforest Tropical/subtropical forest (xerophytic environment)	Successful	None
Bitter melon/ bitter squash	*Momordica charantia*	High blood pressure	Cucurbitaceae	Terrestrial	Successful	None
Bay leaf	*Laurus nobilis*	High blood pressure	Lauraceae	Mediterranean area	Successful	None
Indian mulberry or hog apple	*Morinda citrifolia*	High blood pressure	Rubiaceae	Shady forests	Successful	None
Pawpaw	*Carica papaya*	High blood pressure	Caricaceae	Tropics of America	Successful	None
Onions	*Allium cepa*	Hypertension	Amaryllidaceae	Temperate zone	Successful	None
Sandpaper plant	*Ficus exasperata*	Hypertension	Moraceae	Moist deciduous Fringe Forest	Successful	None
Garlic, ginger, and turmeric	*Allium sativa* *Zingiber officinale* and *Curcuma* species	Hypertension	Amaryllidaceae Zingiberaceae	Terrestrial	Successful	None
Abeere	*Hunteria umbellata*	Hypertension	Apocynaceae	Terrestrial	Successful	None
Palm kernel nut Moringa seed	*Elaeis guineensis* *Moringa oleifera*	Stroke	Arecaceae Moringaceae	Tropical or subtropical region	Successful	None
Olive seed	*Olea europaea*	High blood pressure	Oleaceae	Mediterranean basin	Successful	None
Avocado pear	*Persea americana*	Chest pain	Lauraceae	Tropical/temperate region	Successful	None
Beetroot	*Beta vulgaris*	Hypertension	Amaranthaceae	Warm temperate zone	Successful	Yes, not to be taken with malaria medication

Table 3 Primary usage or methods of using these plants for treatment and management of cardiovascular disorders

Plant Remedy	Parts Used	Methods of Administration	Duration of Use	When Used	Dosage	Storage Method	Source of Plant	Wild or Cultivated	Collection Time
Cocoa Aloe Vera	Leaves and seeds	Infusion	Until recovery	Morning and evening	2 teaspoons	Canning	Garden and forest	Cultivated	Anytime
Bitter melon	Fruits and seeds	Aqueous steeping	Daily	Morning and evening	A glass	Freezing	Market and garden	Cultivated	Anytime
Bay leaf	Leaves	Aqueous steeping	Daily	Morning and evening	A glass	Bare floor or dry place	Market	Cultivated	Anytime
Indian Mulberry	Fruits and seeds	Aqueous steeping	Daily		25 ounces	Freezing	Garden	Cultivated	Anytime
Pawpaw	Fruits	Aqueous steeping	Daily	All the time	2–3 glasses	Freezing	Garden and market	Cultivated	Anytime
Onions	Whole bulb and bulb peels	Aqueous steeping	Daily	All the time	As many as possible	Freezing	Market	Cultivated	Anytime
Sandpaper plant	Leaves	Aqueous steeping	Daily	3 times daily	NR	Freezing	Garden	Wild	Anytime

Garlic Ginger Turmeric	Roots	Raw/aqueous steeping	3 weeks	Morning and evening	NR	NR	Market	Cultivated	Anytime
Abeere	Fruits and seeds	Raw/aqueous steeping	Daily	Anytime	2–3	NR	Market/ Garden	Cultivated	Anytime
Palm kernel nut Moringa seed	Seeds	Oral	Daily	Morning and evening	2 tablespoons	Canning	Market/ garden	Cultivated	Anytime
Olive seed	Seed	Infusion	Daily	3 times daily	Half tumbler	NR	Market	Cultivated	Anytime
Avocado pear	Seed	Raw	1 week	Once daily	2 tablespoons	NR	Market	Cultivated	Anytime
Beetroot	Fruit and seed	Oral	Until recovery	Morning	1 bottle	Freezing	Market	Imported	Morning

NR Not reported

their herbal remedies have active substances with pharmacological and prophylactic properties, they often have little to no side effects when taken in moderation. These bioactive compounds include phenolics, flavonoids, nitrogen, anti-inflammatory, terpenoids, antioxidants, and other secondary metabolic compounds. Other researchers [26, 101, 199] have documented and confirmed the in vitro and in vivo efficacy of *Ginseng*, *Ginkgo biloba*, *Gastrodia elata,* Nerium *oleander*, Salvia *miltiorrhiza*, *Dracocephalum moldavica, Daucus carota*, *Ganoderma lucidum*, *Tinospora cordifolia*, *Hydrocotyle asiatica, Bombax ceiba, Terminalia arjuna*, *Amaranthus viridis*, *Andrographis paniculate, Mucuna pruriens*, *Picrorhiza kurroa*, and *Gynostemma pentaphyllum in the treatment and management of CDVD.* The consumption or use of these plants reduces the risk of CDVDs like congestive heart failure, hypertension, ischaemic heart disease, and arrhythmias [96, 99, 127, 138, 179, 186].

2.2 Enumeration of Plants Used in the Management and Treatment of Cardiovascular Disease by Benin People in Southern Nigeria

2.2.1 Cocoa (*Theobroma cacao*)

The use of cocoa in traditional medicine is linked to the phytochemical constituents in the plants like flavanol, polyphenol, and procyanidin in significant concentrations [25, 41, 197]. These compounds have nutritional and pharmacological benefits. The polyphenols in cocoa have cardiovascular-protective properties by modulating different inflammatory markers connected to atherosclerosis [99]. The flavanol in cocoa also has a regulatory capacity and helps with total dietary requirements [99, 109, 229]. Cocoa consumption has been linked to flow-mediated vascular dilatation, blood pressure, total lipid profiles, and insulin resistance [74, 75, 88, 184].

Also, methylxanthines have been reported from cocoa extracts and like theobromine and caffeine help enhance the vascular and central nervous system and work in tandem with flavonols and other secondary metabolic compounds [151, 194, 197]. Both procyanidins and flavanols have been connected to diverse cardiometabolic and cardiovascular advantages in humans through the improvements of endothelium-dependent vasodilation, blood pressure regulation, inflammation, and platelet activation [40, 46, 63, 73, 75, 103, 115, 119, 165, 176, 181, 185, 188, 193, 195, 197, 214]. Also, it provides insight into the distribution, absorption, metabolism, and excretion of flavonols [5, 164, 187].

2.2.2 Aloe Vera (*Aloe barbadensis*)

Aloe vera is an important tradomedicinal plant and its innermost leaf layer contains diverse minerals and secondary metabolites [190, 200]. Aloe is used as medicine by diverse cultures and societies globally worldwide since time immemorial [105]. Aloe vera leaves contain high amounts of anthraquinone [200]. The inner layer leaf of Aloe vera is called Aloe vera gel (AVG) or aloe gel and it does not contain those essential anthraquinone compounds [190]. The plant has several beneficial properties including those lipid-lowering, immunomodulatory effects, antioxidant, anti-inflammatory, and hepatoprotective effects [190, 200]. Aloe vera contains polysaccharides like acetylated polymannan, glucomannan, acemannan, and mannose-6-phosphate that have therapeutic and pharmacological effects. AVG has direct and indirect cardioprotective impacts through hyperglycaemia, hyperlipidaemia, hypertension, and obesity effects [190]. The mineral nutrients in AVG (like iron, zinc, copper, and selenium) also participate in enzymatic, cell metabolic, and antioxidant functions with net positive effects on CDVD. Also, AVG contains a lot of antioxidant vitamins which have been linked to decreased serum levels of and risk of coronary heart disease. Both lophenol and cycloartenol are present in aloe and they can regulate blood glucose [200].

2.2.3 Bitter Melon (*Momordica charantia*)

Bitter melon offers diverse health advantages for reducing illnesses and enhancing life [191]. It is grown in most countries with tropical or sub-tropical climates and popular medicinal food in many developing countries [219]. Bitter melon's whole fruit, seeds, and leaves control oxidative stress and prevent fat build-up, and contain a variety of bioactive substances, including alkaloids, polypeptides, vitamins, and minerals. It aids in controlling blood cholesterol levels, and defending the body against cardiovascular diseases like atherosclerosis [22, 191, 203]. Also, bitter melon improves glucose and lipid metabolism [35, 208, 218]. The medicinal benefits of bitter melon are linked to its high antioxidant qualities, which are partly owing to the presence of secondary phytochemicals like isoflavones, phenols, flavonoids, anthroquinones, terpenes, and glucosinolates that give the fruit its bitter flavour [17, 47, 64]. Loss of potential health benefits may occur when active bitter components are removed by a variety of debittering techniques and selective breeding [207]. Bitter melon contains vitamins, minerals, and flavonoids [17, 117]. Because it contains a variety of bioactive components, bitter melon exhibits pharmacological effects that have been linked to improved health, including the ability to scavenge free radicals and have hypoglycaemic and hypolipidemic effects [191, 230].

2.2.4 Bay Leaf (*Laurus nobilis*)

Bay leaf can be used in alternative medicine and is effective as a blood pressure-lowering medication [14, 54, 71]. Bay leaf consumption decreases CDVD risks [98]. This plant has a track record for effectively treating disease, has few adverse effects, and is accessible. Bay leaves contain tannins, flavonoids, citrate, eugenol, and essential oils [98, 215]. Antioxidant chemicals, which include tannins and flavonoids, are present in bay leaves and have health benefits regarding blood pressure [54, 98]. The availability and prevalence of bay leaf are anticipated to aid in spreading awareness and promoting it as a herbal alternative for health. Bay leaf is a plant that is often used in the neighbourhood as a complementary medicine. It is a diuretic and has been used to treat eructation, epigastric bloating, poor digestion, and flatulence [70]. In addition to its hypoglycaemic benefits, bay leaf can improve diabetic patients' capillary function, lipid metabolism, liver, and kidney function, as well as their antioxidant status [51, 69]. The glucosidase enzyme reduces blood sugar levels, and the phenolic chemicals in bay leaves could block it [94]. The bay leaf extract may cause pancreatic cells to secrete more insulin or that it acts similarly to insulin [221]. The fact that these extracts diminish gluconeogenesis and glycogenolysis, decrease glucose absorption, and improve peripheral glucose utilization also suggests that they may have hypoglycaemic effects [51, 170].

2.2.5 Indian Mulberry (*Morinda citrifolia*)

Indian mulberry is one of the main components in several traditional formulas that are sold all over the world [53]. Mulberries and their extracts are used to treat both acute and chronic illnesses because they have powerful antimicrobial, anti-hyperlipidaemic, anti-inflammatory, and anti-cancer properties [238]. The mulberry roots, bark, leaves, fruits, and stem twigs are rich sources of vital bioactive compounds [53, 93, 167, 238]. Mulberries are a great source of phosphorus, iron, riboflavin, calcium, vitamin (C and K), potassium, and other essential minerals for our bodies [13, 93]. Additionally, they contain a sizable amount of dietary fibre as well as a variety of organic substances, such as phytonutrients, zeaxanthin, resveratrol, anthocyanins, lutein, and different polyphenolic compounds [93]. This plant's extraordinary benefits in decreasing blood cholesterol and serum glucose levels make it possible to utilize them. The presence of numerous bioactive components in this plant, including flavonoids, polyphenols, alkaloids, terpenoids, and steroids, is what gives it these qualities [93].

2.2.6 Pawpaw (*Carica papaya*)

Pawpaw has high dietary fibre, protein, carbohydrates, vitamins, and mineral nutrients. The plant is used to address diverse ailments including for treating indigestion and gastric ulcer. The consumption of pawpaw fruit helps reduce blood pressure [52, 169]. It offers numerous health advantages, including lowering blood pressure, promoting wound healing, assisting with digestion, lowering the risk of heart disease, diabetes, and cancer, and enhancing blood glucose control in diabetics [4]. Pawpaw seeds and fruits have potent antihelminthic and antiamoebic properties [11, 148]. Ascorbic acid, flavonoids, cyanogenic glucosides, and glucosinolates are just a few of the active substances found in pawpaw leaves that have been shown to lower lipid peroxidation levels and increase total antioxidant power in the blood, thus they can be used either for the prevention or treatment of CDVD [11, 196, 202].

2.2.7 Onions (*Allium cepa*)

Onions are regarded as essential vegetables [20, 30, 76, 83, 220]. Many cultures have valued onions for their culinary and therapeutic uses because they are rich sources of flavonols and organosulfur compounds [66, 131]. Also, it contains flavonoids and phenolics which have antioxidant, anti-cholesterol, anti--inflammatory, and anticancer properties. The plant is rich in protein, water, selenium, and different vitamins (B1, B2, and C) as well as potassium, diverse polysaccharides, and essential oil [225]. These phytochemicals have been linked to several health advantages and improvement in diseased conditions [20, 30, 76, 83, 86, 112, 220]. The variety of phytonutrients in onions is known to have important and extensive biological activity. Onions exhibit potent anti-atherogenic effects that are related to a variety of bioactivities [86, 112, 220]. The high amounts of flavonoids and organosulfur compounds are linked to these biological actions, which are mostly antioxidant and anti-inflammatory in nature [30, 220, 225, 235].

2.2.8 Sandpaper Plant (*Ficus exasperate*)

Sandpaper plant is a deciduous shrub or tree and extracts from the leaf have demonstrated high concentrations of secondary phytochemicals like alkaloids, tannins, flavonoids, and saponins [237]. It has been used to treat difficult childbirth, bleeding, and diarrhoea. Sandpaper leaf has hypoglycaemic potential and is utilized to manage, regulate, and/or treat CDVDs like hypertension. The leaf extract is used to treat cardiac problems and reduce blood pressure [10, 38]. Also, in the treatment of numerous degenerative diseases, the bark, leaves, fruits, and latex of sandpaper plants contain astringent, anti-oxidative, carminative, anti-inflammatory, and anticancer agents that can be used to treat diabetes, skin conditions, ulcers, dysentery, diarrhoea, stomach aches, haemorrhoids, and hypertension [12, 38, 91]. The plant's ability to treat or manage several ailments is supported by the presence of phytochemicals. Its high quantities of flavonoids and saponins, however, imply that taking the plant in excessive doses over a sustained period can cause toxicity [237].

2.2.9 Garlic (*Allium sativa*)

Garlic, the most popular herbal dietary supplement on the market, is famous for its vast health benefits, especially in the treatment and prevention of CDVDs [16, 80]. It has a lot of sulphur-based compounds and is a good source of carbohydrates as well as proteins [2, 80, 233]. More than 200 unique compounds that can shield the human body from a wide range of ailments can be found in garlic alone. Garlic's sulphur-containing components provide the body with protection by promoting the development of specific advantageous enzymes [60, 120]. Garlic contains sulphur compounds, and diverse enzymes, as well as essential minerals (like calcium, germanium, iron, copper, magnesium, potassium, zinc, and selenium), vitamins (A, B1, and C), fibre, and water [60]. Also, garlic extracts exhibit anti-inflammatory, antioxidant, and anti-microbial properties [110, 204, 234]. The consumption of garlic helps lower blood pressure, total cholesterol, body mass index, triglycerides, and inflammatory indicators [28, 34, 60, 80, 210]. Additionally, it can raise HDL-c levels and enhance several cardiovascular parameters, including carotid intima-media thickness, microcirculation, post-occlusive reactive hyperaemia, epicardial and periaortic adipose tissue, and low attenuation plaque [60, 80].

2.2.10 Ginger (*Zingiber officinale*)

Ginger is rich in antioxidant and anti-inflammatory compounds [124, 243]. Some of these compounds are effective against CDVD like hypertension and other diseases [18, 189]. The phenolic compounds in ginger include gingerols, paradol, and shogaol that have therapeutic and pharmacological potentials and are helpful in the management of inflammatory and oxidative stress diseases [114, 132, 189, 212]. In addition to its significant nutritional contributions, ginger has several medicinal and immune-boosting benefits. Additionally, they are crucial in the control of blood pressure and the treatment of cardiovascular diseases. The utilization of ginger-based products will help to advance the human health system [48, 82, 104, 114, 130].

2.2.11 Turmeric (*Curcuma* species)

Turmeric has spiritual significance and uses to treat various CDVDs and other diseases such as digestive, respiratory, and hepatic illnesses or disorders [9, 100, 106, 129, 171, 172, 192, 228]. Having a low bioavailability and pleiotropic activity, turmeric is a naturally occurring phenolic compound. Turmeric has been used as a cooking spice and a traditional treatment to heal skin conditions, cuts, and wounds since the dawn of time [92, 198]. It has been recognized as an essential dietary supplement and nutraceutical [72, 206]. There is evidence that the primary yellow bioactive component of turmeric has a variety of biological effects. These consist of its antioxidant, anti-inflammatory, anti-mutagenic, anti-carcinogenic, anti-coagulant, antifertility, antidiabetic, antibacterial, anti-fungal, anti-viral, anti-fibrotic, anti-venom, anti-ulcer, hypotensive, and hypocholesterolaemic activities [92, 192, 198, 228].

2.2.12 Moringa (*Moringa oleifera*)

Moringa tree is common in Asia and Africa and is appreciated for its therapeutic properties as the "miracle tree" [8, 19, 97, 111]. It has traditionally been utilized as both a food source and a medication to treat several illnesses like anaemia, diabetes, and infectious or cardiovascular ailments [1, 19, 62, 216]. The biggest medicinal and nutritional benefits are found in the leaves and seeds of the moringa plant. Protein, minerals, and antioxidant chemicals are plentiful in the leaves [19, 27, 77, 111]. In addition to being nutrient-rich, moringa also has anti-nutrients such as flavonoids that function as antioxidants [211, 223]. The primary phytochemical components of moringa are phenolic acids, flavonoids, saponin, tannins, alkaloids, glycosides, and glucosinolates [31, 32, 168, 226]. The bioactive components in moringa have synergistic therapeutic effects that include lowering blood sugar levels, anti-inflammatory properties, cardioprotective, anticancer, antimicrobial, neuroprotective, and immune system modulation [118, 223]. The phytochemicals N,L-rhamnopyranosyl vincosamide, isoquercetin, quercetin, quercitrin, and isothiocyanate that have been linked to functional activities related to cardiovascular disorders have been found in Moringa. The phytochemicals, such as quercetin and N,L-rhamnopyranosyl vincosamide, have molecular antioxidant, anti-inflammatory, and anti-apoptotic properties. These result in increased cardiac contractility and damage prevention for the heart's structural integrity. Additionally, these substances function as endothelium protectors and natural vasorelaxants [19, 113, 213].

2.2.13 Olive (*Olea europaea*)

Olive tree is considered one of the oldest tree species, and historically, indigenous inhabitants relied on its fruits and by-products, such as olive oil, as their primary source of nutrition [56, 67, 241]. Olives are a good source of vitamins, minerals, and carbohydrates. Numerous studies have found links between the "Mediterranean diet's" health advantages and longer life spans and lower occurrences of chronic degenerative diseases [67]. Olive leaf polyphenols and flavonoids have been shown to have anti-inflammatory, anti-carcinogenic, anti-hypertension, and antibacterial characteristics; they are therefore crucial to the outcomes that have made olive leaves significant. Olive leaves contain secoiridoid phenols, which are unique to the plant [217].

Olive seed protein hydrolysates have been shown to exhibit significant antioxidant and angiotensin-converting enzyme (ACE) inhibitor capabilities and to lower micellar cholesterol solubility [57]. Also, olive leaf extract (OLE) has lipid-lowering properties and contains significant amounts of the phenolic antioxidant known as oleuropein, far more than what is found in olive fruit or olive oil [21, 58, 68, 173, 205]. Olive oil contains at least 30 phenolic compounds, the majority of which are simple phenols (tyrosol and hydroxytyrosol), secoroids (oleuropein and ligstroside), and lignans (1-acetoxypinoresinol and pinoresinol) [42, 224]. The phenols in olive oil include straightforward compounds like vanillic, gallic, coumaric, caffeic acids, tyrosol, and hydroxytyrosol [224]. In several populations, phenolic compounds from virgin olive oil have had a positive impact on systolic and diastolic blood pressure [42, 121, 201]. A large amount of monounsaturated fat and a low amount of saturated fat are two of olive oil's key qualities. The primary monounsaturated fat, oleic acid, has several health-improving qualities, including a decrease in CDVD, neurological disorders, and cancer. Virgin olive oil is also rich in several beneficial compounds and polyphenols are likely the most important of these bioactive chemicals [42, 232]. Olive oil's phenolic components are extensively researched for their antioxidant capacity [67, 241]. It was discovered that several phenols from olive oil (like hydroxytyrosol and oleuropein) operate as radical scavengers and diminish the production of superoxide anions, neutrophil respiratory bursts, and hypochlorous acid. This may account for the decreased incidence of cancer and coronary heart disease that has been linked to the Mediterranean diet [42, 67, 231, 241]. Olive oil is resistant to heat oxidation and possesses high antioxidant potential within biological systems [42, 201, 241, 242]. Tocopherols and phenols are the principal substances enhancing stability during heat oxidation. The stability is also aided by the oil's high oleic acid content [67, 241].

2.2.14 Avocado (*Persea americana*)

The avocado pear is a widely grown and popular tropical fruit that is eaten all over the world [23, 36]. It also has significant levels of protein, potassium, and unsaturated fatty acids, which are uncommon in other fruits [23]. Its oil is high in monounsaturated fatty acids and is utilized to boost high- and low-density lipoprotein levels [137]. The chemicals included in the lipidic component of this fruit, including omega fatty acids, phytosterols, tocopherols, and squalene, have been credited with its health advantages. The health benefits of avocados include lowering cholesterol and avoiding cardiovascular illnesses [23, 137, 177]. This tree has been known to possess several therapeutic characteristics, including emmenagogue (fruit) and cicatrizant (fruit pulp). Its leaves have also been used to treat digestive disorders and skin infections and irritation. Additionally, seed oil has long been used as an ointment to soothe pain and soften the skin around wounds, as well as for the treatment of dry hair and other diseases [39, 137]. The primary bioactive substances found in avocado include tocopherols, carotenoids, polyphenols, and phytosterols. While carotenoids and tocopherols are primarily present in the avocado pulp, polyphenols can be found in the fruit's pulp, peel, seed, and leaves. Waste extracts, therefore, have a variety of biological properties, such as antimicrobial, anticancer, anti-inflammatory, antidiabetic, and antihypertensive properties [15, 89, 107, 137, 177].

2.2.15 Beetroot (*Beta vulgaris*)

Beetroot is grown in many nations throughout the world. It is consumed as part of a normal diet and is frequently used in food processing as a food colouring ingredient [37, 61, 245]. Proteins, sugar, carbs, vitamins (B complex and vitamin C), minerals, and fibre are all found in abundance in beetroot. The phenolic chemicals and antioxidants in beet include sesquiterpenoids, coumarins, triterpenes, carotenoids, and flavonoids (rhamnocitrin, tiliroside, kaempferol, astragalin, and rhamnetin) and are also present in significant amounts [37, 174]. Beetroot contains antioxidant and anti-inflammatory effects, anti-carcinogenic and anti-diabetic activities, as well as hepato-protective, hypotensive, and wound healing capabilities [37, 43, 61, 123]. Bioactive substances like dietary NO_3, betanin, antioxidants, and phenolic compounds are found in beetroot. Beetroot contains a variety of additional bioactive chemicals that may have health advantages, especially for conditions marked by chronic inflammation [37]. By enhancing nitric oxide synthesis, controlling gene expression, or modulating the activities of proteins and enzymes involved in these cellular processes, beetroot rich in nitrate and bioactive compound contents improves cardiovascular and metabolic functions [37, 45, 240]. The body uses nitric oxide (NO) for vasodilation to lower blood pressure and improve nutrients and oxygen supply within the body. According to these outcomes, beetroots may be useful in the prevention and treatment of cardiovascular disease [24, 90, 123, 180, 182, 240]. The beetroot-cereal bar exhibits high levels of antioxidant, phenolic, and nitrate components. A beetroot-cereal bar might be a good way to help individuals who have risk factors for cardiovascular disease by improving their cardiovascular parameters [44].

2.2.16 Palm Kernel Nut (*Elaeis guineensis*)

Contrary to common belief, the excess consumption of palm kernel oil does not increase the risk of having CDVD. The work of Ismail et al. [84] could not find any association but recommend their consumption in moderation to ensure good cardiometabolic health. Oil palm is abundant in West Africa and is the chief source of cooking oil [79]. Although it contains about 50% saturated fatty acid and high amounts of unsaturated fatty acids, it is not known to promote either arterial thrombosis or atherosclerosis [50]. Rather it has been documented that oil palm phenolic (OPP) found in oil palm has cardioprotective effects through mechanisms and pathways such as cholesterol biosynthesis pathway, anti-inflammatory properties, and antioxidants. OPP is recoverable from oil palm milling aqueous waste products. The CDVD effects of oil palm extolled by Benin people might likely be linked to OPP and other properties of the plant fruits.

2.2.17 Abeere (*Hunteria umbellata*)

Abeere is widely used in traditional medicine in SSA for CDVD as well as managing and treating sexually transmitted diseases, anaemia, obesity, diabetes, and stomach ulcers. The plant can be found in tropical forests all over SSA. The medical benefits of the plant are linked to its phytochemical compositions which include tannins, alkaloids, flavonoids, phenolics, and saponins, which have been implicated in lipid peroxidation, antioxidant, and free-radical scavenging activities [7, 55]. Recently, Odukoya et al. [138] implicated the use of the seed, stem, and bark of the plant in the treatment and management of CDVD. These parts of the plants contain diverse water-soluble alkaloids. It is likely that the nitric oxide synthase and arginase inhibitory effects on endothelial function contribute to the CDVD roles of the plant [136]. Another consideration is that the alkaloid fraction of *H. umbellata* helps to reduce weight gain, which otherwise may increase the risk of CDVD [6]. Hence, *H. umbellata* has a cardioprotective effect by contributing to weight loss.

3 Conclusion

Since time immemorial, plants have been used in the treatment of diverse CDVDs and recent epidemiological research has shown that eating foods derived from certain plants lowers the chance of developing cardiovascular disease. These plants will continue to play significant roles in sustaining human societal and individual well-being. Medicinal plants used in the management and treatment of CDVD contain natural bioactive compounds that accumulate in the plants as secondary metabolites, such as terpenes, flavonoids, alkaloids, sterols, glycosides, saponins, and tannins. These compounds have both therapeutic properties and pharmacological effects on humans. Traditional remedies have been utilized to treat cardiovascular illnesses for a very long time but have only started to attract the interests of researchers, government agencies, and large co-operations as they seek to explore alternative medicine, promote their usage, and search for natural raw materials and remedies for illnesses of global concern like CDVD. The fundamental problem that most herbal medications address is the lack of standardization that may be used to classify them even though these effects usually overlap. The chemical constituents of medicinal plants used for managing and treating CDVD should be further explored as potential raw materials for modern medicine as well as to understand their mode of action to improve the quality of life for individuals and society in general. Strong cardioactive glycosides are present in several plants, and they have favourable inotropic effects on the heart. The advantages of plant-based diets for the prevention of CDVD need to be investigated further. The phytochemical components found in natural goods are being researched for their potential to shield the cardiovascular and vascular systems from further harm.

References

1. Abbas RK, Elsharbasy FS, Fadlelmula AA (2018) Nutritional values of *Moringa oleifera*, total protein. Amino acid, vitamins, minerals, carbohydrates, total fat and crude fiber, under the semi-arid conditions of Sudan. J Microb Biochem Technol 10(2):56–58
2. Abe K, Hori Y, Myoda T (2020) Volatile compounds of fresh and processed garlic. Exp Therap Med 19(2):1585–1593
3. Abunnaja SS, Sanchez JA (2013) Epidemiology of cardiovascular disease. In: Maulik N (ed) Cardiovascular diseases: nutritional and therapeutic interventions. CRC Press, Boca Raton, pp 3–17
4. Abu-Saqer MM, Abu-Naser SS (2019) Developing an expert system for papaya plant disease diagnosis. J Artif Intell 1:78–85
5. Actis-Goretta L, Leveques A, Rein M, Teml A, Schafer C, Hofmann U, Li H, Schwab M, Eichelbaum M, Williamson G (2013) Intestinal absorption, metabolism, and excretion of (-)-epicatechin in healthy humans assessed by using an intestinal perfusion technique. Am J Clin Nutr 98(4):924–933
6. Adeneye AA, Crooks PA (2015) Weight losing, antihyperlipidemic and cardioprotective effects of the alkaloid fraction of *Hunteria umbellata* seed extract on normal and triton-induced hyperlipidemic rats. Asian Pac J Trop Biomed 5(5):387–394
7. Adeneye AA, Adeyemi OO, Agbaje EO, Banjo AA (2010) Evaluation of the toxicity and reversibility profile of the aqueous seed extract of *Hunteria umbellata* (K. Schum.) Hallier f. in rodents. Afr J Tradit Complement Alternat Med 7(4):350–369. https://doi.org/10.4314/ajtcam.v7i4.56704
8. Aekthammarat D, Pannangpetch P, Tangsucharit P (2019) *Moringa oleifera* leaf extract lowers high blood pressure by alleviating vascular dysfunction and decreasing oxidative stress in L-NAME hypertensive rats. Phytomedicine 54:9–16. https://doi.org/10.1016/j.phymed.2018.10.023
9. Aggarwal BB, Sundaram C, Malani N, Ichikawa H (2007) Curcumin: the Indian solid gold. In: Aggarwal BB, Surh Y-J, Shishodia S (eds) The molecular targets and therapeutic uses of curcumin in health and disease. Advances in experimental medicine and biology. Springer, New York
10. Agunloye OM, Oboh G (2018) Effect of different processing methods on antihypertensive property and antioxidant activity of sandpaper leaf (*Ficus exasperata*) extracts. J Dietary Suppl 15(6):871–883
11. Airaodion AI, Ogbuagu EO, Ekenjoku JA, Ogbuagu U, Okoroukwu VN (2019) Antidiabetic effect of ethanolic extract of Carica papaya leaves in alloxan-induced diabetic rats. Am J Biomed Sci Res 5(3):227–234
12. Ajeigbe OF, Oboh G, Ademosun AO, Oyagbemi AA (2021) Fig leaves varieties reduce blood pressure in hypertensive rats through modulation of antioxidant status and activities of arginase and angiotensin-1 converting enzyme. Comp Clin Pathol 30(3):503–513
13. Akbulut M, Özcan MM (2009) Comparison of mineral contents of mulberry (Morus spp.) fruits and their pekmez (boiled mulberry juice) samples. Int J Food Sci Nutr 60(3):231–239
14. Al Chalabi S, Majeed D, Jasim A, Al-Azzawi K (2020) Benefit effect of ethanolic extract of Bay leaves (*Laura nobilis*) on blood sugar level in adult diabetic rats induced by alloxan monohydrate. Ann Trop Med Publ Health 23(16):SP231608
15. Alagbaoso CA, Osakwe OS, Tokunbo II (2017) Changes in proximate and phytochemical compositions of *Persea americana* mill.(avocado pear) seeds associated with ripening. J Med Biomed Res 16(1):28–34
16. Alali FQ, El-Elimat T, Khalid L, Hudaib R, Al-Shehabi TS, Eid AH (2017) Garlic for cardiovascular disease: prevention or treatment? Curr Pharm Des 23(7):1028–1041. https://doi.org/10.2174/1381612822666161010124530

17. Alam MA, Uddin R, Subhan N, Rahman MM, Jain P, Reza HM (2015) Beneficial role of bitter melon supplementation in obesity and related complications in metabolic syndrome. J Lipids 2015:496169
18. Ali BH, Blunden G, Tanira MO, Nemmar A (2008) Some phytochemical, pharmacological and toxicological properties of ginger (*Zingiber officinale* Roscoe): a review of recent research. Food Chem Toxicol 46(2):409–420
19. Alia F, Putri M, Anggraeni N, Syamsunarno MRA (2022) The potency of *Moringa oleifera* Lam. as protective agent in cardiac damage and vascular dysfunction. Front Pharmacol 12:3911
20. Alissa EM, Ferns GA (2017) Dietary fruits and vegetables and cardiovascular diseases risk. Crit Rev Food Sci Nutr 57:1950–1962
21. Angelopoulos N, Paparodis RD, Androulakis I, Boniakos A, Anagnostis P, Tsimihodimos V, Livadas S (2022) Efficacy and safety of monacolin K combined with coenzyme Q10, grape seed, and olive leaf extracts in improving lipid profile of patients with mild-to-moderate hypercholesterolemia: a self-control study. Nutraceuticals 3(1):1–12
22. Anilakumar KR, Kumar GP, Ilaiyaraja N (2015) Nutritional, pharmacological and medicinal properties of *Momordica charantia*. Int J Food Sci Nutr 4(1):75–83. https://doi.org/10.11648/j.ijnfs.20150401.21
23. Arackal JJ, Parameshwari S (2017) Health benefits and uses of avocado. Rev Article 6(17): 392–399
24. Arciero PJ, Miller VJ, Ward E (2015) Performance enhancing diets and the PRISE protocol to optimize athletic performance. J Nutr Metab 2015:715859. https://doi.org/10.1155/2015/715859
25. Aron PM, Kennedy JA (2008) Flavan-3-ols: nature, occurrence and biological activity. Mol Nutr Food Res 52(1):79–104
26. Bachheti RK, Worku LA, Gonfa YH, Zebeaman M, Deepti P, D. P., & Bachheti, A. (2022) Prevention and treatment of cardiovascular diseases with plant phytochemicals: a review. Evid Based Complement Alternat Med 2022:5741198. https://doi.org/10.1155/2022/5741198
27. Baiyeri P, Akinnagbe OM (2013) Ethno-medicinal and culinary uses of *Moringa oleifera* Lam. in Nigeria. J Med Plant Res 7(13):799–804. https://doi.org/10.5897/JMPR12.1221
28. Bayan L, Koulivand PH, Gorji A (2014) Garlic: a review of potential therapeutic effects. Avicenna J Phytomed 4(1):1
29. Bekoe EO, Agyare C, Boakye YD, Baiden BM, Asase A, Sarkodie J ... Nyarko A (2020) Ethnomedicinal survey and mutagenic studies of plants used in Accra metropolis, Ghana. J Ethnopharmacol 248:112309
30. Bisen PS, Emerald M (2016) Nutritional and therapeutic potential of garlic and onion (Allium sp.). Curr Nutr Food Sci 12:190–199
31. Borgonovo G, De Petrocellis L, Schiano Moriello A, Bertoli S, Leone A, Battezzati A ... Bassoli A (2020) Moringin, a stable isothiocyanate from *Moringa oleifera*, activates the somatosensory and pain receptor TRPA1 channel in vitro. Molecules 25(4):976. https://doi.org/10.3390/molecules25040976
32. Brilhante RSN, Sales JA, Pereira VS, Castelo DDSCM, de Aguiar Cordeiro R, de Souza Sampaio CM ... Rocha MFG (2017) Research advances on the multiple uses of *Moringa oleifera*: a sustainable alternative for socially neglected population. Asian Pac J Trop Med 10 (7):621–630. https://doi.org/10.1016/j.apjtm.2017.07.002
33. Bussmann RW, Malca G, Glenn A, Sharon D, Nilsen B, Parris B ... Townesmith A (2011) Toxicity of medicinal plants used in traditional medicine in Northern Peru. J Ethnopharmacol 137(1):121–140
34. Chan JY, Yuen AC, Chan RY, Chan SW (2013) A review of the cardiovascular benefits and antioxidant properties of allicin. Phytother Res 27:637–646
35. Chaturvedi P, George S, Milinganyo M, Tripathi YB (2004) Effect of *Momordica charantia* on lipid profile and oral glucose tolerance in diabetic rats. Phytother Res 18(11):954–956

36. Chikwendu JN, Udenta EA, Nwakaeme TC (2021) Avocado pear pulp (*Persea americana*)-supplemented cake improved some serum lipid profile and plasma protein in rats. J Med Food 24(3):267–272
37. Clifford T, Howatson G, West DJ, Stevenson EJ (2015) The potential benefits of red beetroot supplementation in health and disease. Nutrients 7(4):2801–2822
38. Conrad OA, Uche AI (2013) Assessment of in vivo antioxidant properties of *Dacryodes edulis* and *Ficus exasperata* as anti-malaria plants. Asian Pac J Trop Dis 3(4):294–300
39. Dabas D, Shegog RM, Ziegler GR, Lambert JD (2013) Avocado (*Persea americana*) seed as a source of bioactive phytochemicals. Curr Pharm Des 19(34):6133–6140. https://doi.org/10.2174/1381612811319340007
40. Davison K, Berry NM, Misan G, Coates AM, Buckley JD, Howe PR (2010) Dose-related effects of flavanol-rich cocoa on blood pressure. J Hum Hypertens 24(9):568–576
41. Dillinger TL, Barriga P, Escárcega S, Jimenez M, Lowe DS, Grivetti LE (2000) Food of the gods: cure for humanity? A cultural history of the medicinal and ritual use of chocolate. J Nutr 130(8):2057S–2072S
42. Ditano-Vázquez P, Torres-Peña JD, Galeano-Valle F, Pérez-Caballero AI, Demelo-Rodríguez P, Lopez-Miranda J . . . Alvarez-Sala-Walther LA (2019) The fluid aspect of the Mediterranean diet in the prevention and management of cardiovascular disease and diabetes: the role of polyphenol content in moderate consumption of wine and olive oil. Nutrients 11 (11):2833
43. Domínguez R, Cuenca E, Maté-Muñoz JL, García-Fernández P, Serra-Paya N, Estevan MCL . . . Garnacho-Castaño MV (2017) Effects of beetroot juice supplementation on cardiorespiratory endurance in athletes. A systematic review. Nutrients 9(1):43
44. dos Santos Baião D, d'El-Rei J, Alves G, Neves MF, Perrone D, Del Aguila EM, Paschoalin VMF (2019) Chronic effects of nitrate supplementation with a newly designed beetroot formulation on biochemical and hemodynamic parameters of individuals presenting risk factors for cardiovascular diseases: a pilot study. J Funct Foods 58:85–94
45. dos Santos Baião D, Vieira Teixeira da Silva D, Margaret Flosi Paschoalin V (2021) A narrative review on dietary strategies to provide nitric oxide as a non-drug cardiovascular disease therapy: beetroot formulations – a smart nutritional intervention. Foods 10(4):859
46. Dower JI, Geleijnse JM, Gijsbers L, Schalkwijk C, Kromhout D, Hollman PC (2015) Supplementation of the pure flavonoids epicatechin and quercetin affects some biomarkers of endothelial dysfunction and inflammation in (pre)hypertensive adults: a randomized double-blind, placebo-controlled, crossover trial. J Nutr 145(7):1459–1463
47. Drewnowski A, Gomez-Carneros C (2000) Bitter taste, phytonutrients, and the consumer: a review. Am J Clin Nutr 72(6):1424–1435
48. Duarte MC, Tavares GS, Valadares DG, Lage DP, Ribeiro TG, Lage LM . . . Coelho EA (2016) Antileishmanial activity and mechanism of action from a purified fraction of *Zingiber officinalis* Roscoe against *Leishmania amazonensis*. Exp Parasitol 166:21–28
49. Eddouks M, Ajebli M, Hebi M (2017) Ethnopharmacological survey of medicinal plants used in Daraa-Tafilalet region (Province of Errachidia), Morocco. J Ethnopharmacol 198:516–530
50. Edem DO (2002) Palm oil: biochemical, physiological, nutritional, hematological, and toxicological aspects: a review. Plant Foods Human Nutr 57(3-4):319–341. https://doi.org/10.1023/a:1021828132707
51. El-Kholie E, El-Eskafy A, Hegazy N (2023) Effect of Bay Leaves (*Laurus nobilis*, L) and cardamom seeds (*Elettaria cardamomum*, L.) as anti-diabetic agents in alloxan-induced diabetic rats. J Home Econ Menofia Univ 33(1):77–88
52. Eno AE, Owo OI, Itam EH, Konya RS (2000) Blood pressure depression by the fruit juice of Carica papaya (L.) in renal and DOCA-induced hypertension in the rat. Phytother Res 14(4):235–239
53. Ercisli S, Orhan E (2007) Chemical composition of white (*Morus alba*), red (*Morus rubra*) and black (*Morus nigra*) mulberry fruits. Food Chem 103(4):1380–1384

54. Erisandi TD (2021) The difference of celery leaves and bay leaves water to decrease blood pressure among pre-elderly with primary hypertension in Public Health Center Cigugur Tengah. Jurnal Keperawatan Komprehensif (Comprehensive Nurs J) 7(2):1–11
55. Fadahunsi OS, Adegbola PI, Olorunnisola OS, Subair TI, Adepoju DO, Abijo AZ (2021) Ethno-medicinal, phytochemistry, and pharmacological importance of Hunteria umbellate (K. Schum.) Hallier f. (Apocynaceae): a useful medicinal plant of sub-Saharan Africa. Clin Phytosci 7:54. https://doi.org/10.1186/s40816-021-00287-z
56. Foscolou A, Critselis E, Panagiotakos D (2018) Olive oil consumption and human health: a narrative review. Maturitas 118:60–66
57. García MC, González-García E, Vásquez-Villanueva R, Marina ML (2016) Apricot and other seed stones: amygdalin content and the potential to obtain antioxidant, angiotensin I converting enzyme inhibitor and hypocholesterolemic peptides. Food Funct 7:4693–4701
58. Gariboldi P, Jommi G, Verotta L (1986) Secoiridoids from *Olea europaea*. Phytochemistry 25: 865–869
59. Gaziano TA (2008) Economic burden and the cost-effectiveness of treatment of cardiovascular diseases in Africa. Heart 94:140–144
60. Gebreyohannes G, Gebreyohannes M (2013) Medicinal values of garlic: a review. Int J Med Med Sci 5(9):401–408
61. Georgiev VG, Weber J, Kneschke EM, Denev PN, Bley T, Pavlov AI (2010) Antioxidant activity and phenolic content of betalain extracts from intact plants and hairy root cultures of the red beetroot Beta vulgaris cv. Detroit dark red. Plant Foods Hum Nutr 65:105–111
62. Gopalakrishnan L, Doriya K, Kumar DS (2016) *Moringa oleifera*: a review on nutritive importance and its medicinal application. Food Sci Human Wellness 5(2):49–56. https://doi.org/10.1016/j.fshw.2016.04.001
63. Grassi D, Necozione S, Lippi C, Croce G, Valeri L, Pasqualetti P, Desideri G, Blumberg JB, Ferri C (2005) Cocoa reduces blood pressure and insulin resistance and improves endothelium-dependent vasodilation in hypertensives. Hypertension 46(2):398–405
64. Grover JK, Yadav SP (2004) Pharmacological actions and potential uses of *Momordica charantia*: a review. J Ethnopharmacol 93:123–132
65. Guasch-Ferre M, Merino J, Sun Q, Fito M, Salas-Salvado J (2017) Dietary polyphenols, Mediterranean diet, prediabetes, and type 2 diabetes: a narrative review of the evidence. Oxidative Med Cell Longev 2017:6723931
66. Guercio V, Galeone C, Turati F, La Vecchia C (2014) Gastric cancer and allium vegetable intake: a critical review of the experimental and epidemiologic evidence. Nutr Cancer 66:757–773
67. Guo Z, Jia X, Zheng Z, Lu X, Zheng Y, Zheng B, Xiao J (2018) Chemical composition and nutritional function of olive (*Olea europaea* L.): a review. Phytochem Rev 17:1091–1110
68. Hadrich F, Mahmoudi A, Bouallagui Z, Feki I, Isoda H, Feve B, Sayadi S (2016) Evaluation of hypocholesterolemic effect of oleuropein in cholesterol-fed rats. Chem Int 252:54–60
69. Hamdan II, Afifi FU (2004) Studies on the in-vitro and in- vivo hypoglycemic activities of some medicinal plants used in treatment of diabetes in Jordanian traditional medicine. J Ethnopharmacol 93:117–121
70. Harismah K, Chusniatun D (2016) Pemanfaatan daun salam (*Eugenia polyantha*) Sebagai obat herbal dan rempah penyedap makanan. Warta LPM 1:110–118
71. Hartanti L, Yonas SMK, Mustamu JJ, Wijaya S, Setiawan HK, Soegianto L (2019) Influence of extraction methods of bay leaves (*Syzygium polyanthum*) on antioxidant and HMG-CoA Reductase inhibitory activity. Heliyon 5(4):e01485
72. Hay E, Lucariello A, Contieri M, Esposito T, De Luca A, Guerra G, Perna A (2019) Therapeutic effects of turmeric in several diseases: an overview. Chem Biol Interact 310: 108729
73. Heiss C, Keen CL, Kelm M (2010) Flavanols and cardiovascular disease prevention. Eur Heart J 31(21):2583–2592

74. Hooper L, Kroon PA, Rimm EB, Cohn JS, Harvey I, le Cornu KA, Ryder JJ, Hall WL, Cassidy A (2008) Flavonoids, flavonoid-rich foods, and cardiovascular risk: a meta-analysis of randomized controlled trials. Am J Clin Nutr 88:38–50
75. Hooper L, Kay C, Abdelhamid A, Kroon PA, Cohn JS, Rimm EB, Cassidy A (2012) Effects of chocolate, cocoa, and flavan-3-ols on cardiovascular health: a systematic review and meta-analysis of randomized trials. Am J Clin Nutr 95(3):740–751
76. Hricova A, Fejer J, Libiakova G, Szabov M, Gazo J, Gajdosova A (2016) Characterization of phenotypic and nutritional properties of valuable *Amaranthus cruentus* L. mutants. Turk J Agric For 40:761–771
77. Idohou-Dossou N, Diouf A, Gueye A, Guiro A, Wade S (2011) Impact of daily consumption of Moringa (*Moringa oleifera*) dry leaf powder on iron status of senegalese lactating women. Afr J Food Agric Nutr Dev 11. https://doi.org/10.4314/ajfand.v11i4.69176
78. Ikhajiagbe B, Ogwu MC, Ogochukwu OF, Odozi EB, Adekunle IJ, Omage ZE (2021) The place of neglected and underutilized legumes in human nutrition and protein security. Crit Rev Food Sci Nutr. https://doi.org/10.1080/10408398.2020.1871319
79. Ikhajiagbe B, Aituae W, Ogwu MC (2022) Morpho-physiological assessment of oil palm (*Elaeis guineensis* Jacq.) seedlings exposed to simulated drought conditions. J Oil Palm Re 34 (1):26–34. https://doi.org/10.21894/jopr.202100018
80. Imaizumi VM, Laurindo LF, Manzan B, Guiguer EL, Oshiiwa M, Otoboni AMMB … Barbalho SM (2022) Garlic: a systematic review of the effects on cardiovascular diseases. Crit Rev Food Sci Nutr 63(24):6797–6819. https://doi.org/10.1080/10408398.2022.2043821
81. Imarhiagbe O, Ogwu MC (2022) Sacred groves in the global south: a panacea for sustainable biodiversity conservation. In: Izah SC (ed) Biodiversity in Africa: potentials, threats and conservation. sustainable development and biodiversity, vol 29. Springer, Singapore, pp 525–546. https://doi.org/10.1007/978-981-19-3326-4_20
82. Imo C, Za'aku JS (2019) Medicinal properties of ginger and garlic: a review. Curr Trends Biomed Eng Biosci 18:47–52
83. Islam MA, Alam F, Solayman M, Khalil MI, Kamal MA, Gan SH (2016) Dietary phytochemicals: natural swords combating inflammation and oxidation-mediated degenerative diseases. Oxidative Med Cell Longev 2016:513743
84. Ismail SR, Maarof SK, Siedar Ali S, Ali A (2018) Systematic review of palm oil consumption and the risk of cardiovascular disease. PLoS One 13(2):e0193533. https://doi.org/10.1371/journal.pone.0193533
85. Izah SC, Ovuru KF, Ogwu MC (2022) Lassa fever in Nigeria: social and ecological risk factors exacerbating transmission and sustainable management strategies. Int J Trop Dis 5(2):65. https://doi.org/10.23937/2643-461x/1710065
86. Jaiswal N, Rizvi SI (2014) Onion extract (*Allium cepa* L.), quercetin and catechin up-regulate paraoxonase 1 activity with concomitant protection against low-density lipoprotein oxidation in male Wistar rats subjected to oxidative stress. J Sci Food Agric 94:2752–2757
87. James PB, Wardle J, Steel A, Adams J (2018) Traditional, complementary and alternative medicine use in sub-Saharan Africa: a systematic review. BMJ Glob Health 3:e000895
88. Jia L, Liu X, Bai YY, Li SH, Sun K, He C, Hui R (2010) Short-term effect of cocoa product consumption on lipid profile: a meta-analysis of randomized controlled trials. Am J Clin Nutr 92:218–225
89. Jimenez P, Garcia P, Quitral V, Vasquez K, Parra-Ruiz C, Reyes-Farias M … Soto-Covasich J (2021) Pulp, leaf, peel and seed of avocado fruit: a review of bioactive compounds and healthy benefits. Food Rev Int 37(6):619–655
90. Jonvik KL, Nyakayiru J, Pinckaers PJ, Senden JM, van Loon LJ, Verdijk LB (2016) Nitrate-rich vegetables increase plasma nitrate and nitrite concentrations and lower blood pressure in healthy adults. J Nutr 146(5):986–993. https://doi.org/10.3945/jn.116.229807
91. Joseph B, Raj SJ (2010) Phytopharmacological and phytochemical properties of three Ficus species – an overview. Int J Pharm Bio Sci 1:246–253

92. Joshi P, Joshi S, Semwal DK, Verma K, Dwivedi J, Sharma S (2022) Role of curcumin in ameliorating hypertension and associated conditions: a mechanistic insight. Mol Cell Biochem 477(10):2359–2385
93. Kadam RA, Dhumal ND, Khyade VB (2019) The Mulberry, *Morus alba* (L.): the medicinal herbal source for human health. Int J Curr Microbiol App Sci 8(4):2941–2964
94. Kalita D, Holm DG, LaBarbera DV, Petrash JM (2018) Inhibition of αglucosidase, αamylase, and aldose reductase by potato polyphenolic compounds. PLoS One 13(1):1–12
95. Kamyab R, Namdar H, Torbati M, Ghojazadeh M, Araj-Khodaei M, Fazljou SMB (2021) Medicinal plants in the treatment of hypertension: a review. Adv Pharm Bull 11(4):601–617. https://doi.org/10.34172/apb.2021.090
96. Karou SD (2011) Sub-Saharan Rubiaceae: a review of their traditional uses, phytochemistry and biological activities. PJBS 14:149–169
97. Khalil SR, Abdel-Motal SM, Abd-Elsalam M, Abd E-HN, E. and Awad, A. (2020) Restoring strategy of ethanolic extract of *Moringa oleifera* leaves against tilmicosin-induced cardiac injury in rats: targeting cell apoptosis-mediated pathways. Gene 730:144272. https://doi.org/10.1016/j.gene.2019.144272
98. Khan A, Zaman G, Anderson RA (2009) Bay leaves improve glucose and lipid profile of people with type 2 diabetes. J Clin Biochem Nutr 44(1):52–56
99. Khan N, Khymenets O, Urpí-Sardà M, Tulipani S, Garcia-Aloy M, Monagas M ... Andres-Lacueva C (2014) Cocoa polyphenols and inflammatory markers of cardiovascular disease. Nutrients 6(2):844–880
100. Kocaadam B, Şanlier N (2017) Curcumin, an active component of turmeric (*Curcuma longa*), and its effects on health. Crit Rev Food Sci Nutr 57(13):2889–2895
101. Koo YE, Song J, Bae S (2018) Use of plant and herb derived medicine for therapeutic usage in cardiology. Medicines 5(2):38. https://doi.org/10.3390/medicines5020038
102. Krisela S (2007) The heart and stroke foundation South Africa heart disease in South Africa Media data document. http://www.heartfoundation.co.za/docs/heartmonth/HeartDiseaseinSA.pdf
103. Kuebler U, Arpagaus A, Meister RE, von Kanel R, Huber S, Ehlert U, Wirtz PH (2016) Dark chocolate attenuates intracellular pro-inflammatory reactivity to acute psychosocial stress in men: a randomized controlled trial. Brain Behav Immun 57:200–208
104. Kumar A, Goyal R, Kumar S, Jain S, Jain N, Kumar P (2015) Estrogenic and anti-Alzheimer's studies of *Zingiber officinalis* as well as *Amomum subulatum* Roxb.: the success story of dry techniques. Med Chem Res 24:1089–1097
105. Kumar R, Singh AK, Gupta A, Bishayee A, Pandey AK (2019) Therapeutic potential of Aloe vera – a miracle gift of nature. Phytomedicine 60:152996
106. Kunnumakkar AB, Bordoloi D, Padmavathi G, Monisha J, Roy NK, Prasad S, Aggarwal BB (2017) Curcumin, the golden nutraceutical: multitargeting for multiple chronic diseases. Br J Pharmacol 174(11):1325–1348
107. Lara-Flores AA, Araújo RG, Rodríguez-Jasso RM, Aguedo M, Aguilar CN, Trajano HL, Ruiz HA (2018) Bioeconomy and biorefinery: valorization of hemicellulose from lignocellulosic biomass and potential use of avocado residues as a promising resource of bioproducts. In: Singhania RR, Avinash R, Praveen Kumar R, Sukumaran RK (eds) Waste to wealth energy, environment, and sustainability. pp 141–170. https://doi.org/10.1007/978-981-10-7431-8_8
108. Lee CH, Kim JH (2014) A review on the medicinal potentials of ginseng and ginsenosides on cardiovascular diseases. J Ginseng Res 38:161–166
109. Lee KW, Kim YJ, Lee HJ, Lee CY (2003) Cocoa has more phenolic phytochemicals and a higher antioxidant capacity than teas and red wine. J Agric Food Chem 51:7292–7295
110. Lee EJ, Kim KS, Jung HY, Kim DH, Jang HD (2005) Antioxidant activities of garlic (*Allium sativum* L.) with growing districts. Food Sci Biotechnol 14:123–130
111. Leone A, Spada A, Battezzati A, Schiraldi A, Aristil J, Bertoli S (2015) Cultivation, genetic, ethnopharmacology, phytochemistry and pharmacology of *Moringa oleifera* leaves: an overview. Int J Mol Sci 16(6):12791–12835. https://doi.org/10.3390/ijms160612791

112. Li W, Tang C, Jin H, Du J (2011) Effects of onion extract on endogenous vascular H2S and adrenomedulin in rat atherosclerosis. Curr Pharm Biotechnol 12:1427–1439
113. Li W, Sun C, Deng W, Liu Y, Adu-Frimpong M, Yu J, Xu X (2019) Pharmacokinetic of gastrodigenin rhamnopyranoside from Moringa seeds in rodents. Fitoterapia 138:104348. https://doi.org/10.1016/j.fitote.2019.104348
114. Li C, Li J, Jiang F, Tzvetkov NT, Horbanczuk JO, Li Y . . . Wang D (2021) Vasculoprotective effects of ginger (*Zingiber officinale* Roscoe) and underlying molecular mechanisms. Food Funct 12(5):1897–1913
115. Lin X, Zhang I, Li A, Manson JE, Sesso HD, Wang L, Liu S (2016) Cocoa flavanol intake and biomarkers for cardiometabolic health: a systematic review and meta-analysis of randomized controlled trials. J Nutr 146(11):2325–2333
116. Liwa AC, Barton EN, Cole WC, Nwokocha CR (2017) Bioactive plant molecules, sources and mechanism of action in the treatment of cardiovascular disease. In: Pharmacognosy. Elsevier, Amsterdam, pp 315–336
117. Lucas EA, Dumancas GG, Smith BJ, Clarke SL, Arjmandi BH (2010) Health benefits of Bitter Melon (*Momordica charantia*). In: Ronald Ross W, Victor RP (eds) Bioactive foods in promoting health. Academic, San Diego, pp 525–549
118. Madi N, Dany M, Abdoun S, Usta J (2016) *Moringa oleifera's* nutritious aqueous leaf extract has anticancerous effects by compromising mitochondrial viability in an ROS-dependent manner. J Am Coll Nutr 35(7):604–613. https://doi.org/10.1080/07315724.2015.1080128
119. Manach C, Williamson G, Morand C, Scalbert A, Remesy C (2005) Bioavailability and bioefficacy of polyphenols in humans. I. Review of 97 bioavailability studies. Am J Clin Nutr 81(1):230S–242S
120. Mansell P, Reckless J (1991) Effects on serum lipids, blood pressure, coagulation, platelet aggregation and vasodilation. BMJ 303:379–380
121. Martínez-González MA, Fernández-Jarne E, Serrano-Martínez M, Wright M, Gomez-Gracia E (2004) Development of a short dietary intake questionnaire for the quantitative estimation of adherence to a cardioprotective Mediterranean diet. Eur J Clin Nutr 58:1550–1552
122. Mashour NH, Lin IG, Frishman WH (1998) Herbal medicine for the treatment of 703 cardiovascular disease clinical considerations. Arch Intern Med 158(20):2225–2234
123. Mirmiran P, Houshialsadat Z, Gaeini Z, Bahadoran Z, Azizi F (2020) Functional properties of beetroot (*Beta vulgaris*) in management of cardio-metabolic diseases. Nutr Metab 17:1–15
124. Morvaridzadeh M, Fazelian S, Agah S, Khazdouz M, Rahimlou M, Agh F . . . Heshmati J (2020) Effect of ginger (*Zingiber officinale*) on inflammatory markers: a systematic review and meta-analysis of randomized controlled trials. Cytokine 135:155224
125. Mota AH (2016) A review of medicinal plants used in therapy of cardiovascular diseases. Int J Pharmacogn Phytochem Res 8:572–591
126. Motaleb MA (2011) Selected medicinal plants of Chittangong Hill tracts. IUCN, Dhaka
127. Mounanga MB, Mewono L, Angone SA, Boukandou MM, Mewono L, Aboughe AS (2015) Toxicity studies of medicinal plants used in sub-Saharan Africa. J Ethnopharmacol 174:618–627
128. Nafiu MO, Hamid AA, Muritala HF, Adeyemi SB (2017) Quality control of medicinal plants in Africa. Elsevier, Amsterdam. ISBN 9780128092866
129. Nair KPP (2013) 1 – turmeric: origin and history. In: Nair KPP (ed) The agronomy and economy of turmeric and ginger. Elsevier, Oxford, pp 1–5
130. Nasri H, Nematbakhsh M, Ghobadi S, Ansari R, Shahinfard N, Rafieian-Kopaei M (2013) Preventive and curative effects of ginger extract against histopathologic changes of gentamicin-induced tubular toxicity in rats. Int J Prev Med 4(3):316
131. Nicastro HL, Ross SA, Milner JA (2015) Garlic and onions: their cancer prevention properties. Cancer Prevent Res 8:181–189
132. Nicoll R, Henein MY (2009) Ginger (*Zingiber officinale* Roscoe): a hot remedy for cardiovascular disease? Int J Cardiol 131(3):408–409

133. Nowacka M, Tappi S, Wiktor A, Rybak K, Miszczykowska A, Czyzewski J, Drozdzal K, Witrowa-Rajchert D, Tylewicz U (2019) The impact of pulsed electric field on the extraction of bioactive compounds from beetroot. Foods 8:244
134. Ntie-Kang F, Lifongo LL, Mbaze LMA, Ekwelle N, Owono Owono LC, Megnassan E, Judson PN, Sippl W, Efange SMN (2013) Cameroonian medicinal plants: a bioactivity versus ethnobotanical survey and chemotaxonomic classification. BMC Complement Altern Med 13:147
135. Nwizugbo KC, Ogwu MC, Eriyamremu GE, Ahana CM (2023) Alterations in energy metabolism, total protein, uric and nucleic acids in African sharptooth catfish (*Clarias gariepinus* Burchell.) exposed to crude oil and fractions. Chemosphere 316:137778. https://doi.org/10.1016/j.chemosphere.2023.137778
136. Oboh G, Adebayo AA, Ademosun AO, Abegunde OA (2019) Aphrodisiac effect of *Hunteria umbellata* seed extract: modulation of nitric oxide level and arginase activity in vivo. Pathophysiology 26(1):39–47. https://doi.org/10.1016/j.pathophys.2018.11.003
137. Ochoa-Zarzosa A, Baez-Magana M, Guzman-Rodriguez JJ, Flores-Alvarez LJ, Lara-Márquez M, Zavala-Guerrero B ... López-Meza JE (2021) Bioactive molecules from native Mexican avocado fruit (*Persea americana* var. drymifolia): a review. Plant Foods Hum Nutr 76:133–142
138. Odukoya JO, Odukoya JO, Mmutlane EM, Ndinteh DT (2022) Ethnopharmacological study of medicinal plants used for the treatment of cardiovascular diseases and their associated risk factors in sub-Saharan Africa. Plants 11(10):1387. https://doi.org/10.3390/plants11101387
139. Ogwu MC (2019) Towards sustainable development in Africa: the challenge of urbanization and climate change adaptation. In: Cobbinah PB, Addaney M (eds) The geography of climate change adaptation in Urban Africa. Springer Nature, Cham, pp 29–55. https://doi.org/10.1007/978-3-030-04873-0_2
140. Ogwu MC (2020) Value of Amaranthus [L.] species in Nigeria. In: Waisundara V (ed) Nutritional value of Amaranth. IntechOpen, London, pp 1–21. https://doi.org/10.5772/intechopen.86990
141. Ogwu MC (2023) Local food crops in Africa: sustainable utilization, threats, and traditional storage strategies. In: Izah SC, Ogwu MC (eds) Sustainable utilization and conservation of Africa's biological resources and environment. Sustainable development and biodiversity, vol 888. Springer, Singapore, pp 353–376. https://doi.org/10.1007/978-981-19-6974-4_13
142. Ogwu MC, Osawaru ME (2015) Soil characteristics, microbial composition of plot, leaf count and sprout studies of cocoyam (Colocasia [Schott] and Xanthosoma [Schott], Araceae) collected in Edo State, Southern Nigeria. Sci Technol Arts Res Jl 4(1):34–44. https://doi.org/10.4314/star.v4i1.5
143. Ogwu MC, Osawaru ME (2022) Traditional methods of plant conservation for sustainable utilization and development. In: Izah SC (ed) Biodiversity in Africa: potentials, threats and conservation. Sustainable development and biodiversity, vol 29. Springer, Singapore, pp 451–472. https://doi.org/10.1007/978-981-19-3326-4_17
144. Ogwu MC, Osawaru ME, Obahiagbon GE (2017) Ethnobotanical survey of medicinal plants used for traditional reproductive care by Usen people of Edo State, Nigeria. Malaya J Biosci 4 (1):17–29
145. Ogwu MC, Iyiola AO, Izah SC (2023) Overview of African biological resources and environment. In: Izah SC, Ogwu MC (eds) Sustainable utilization and conservation of Africa's biological resources and environment. Sustainable development and biodiversity, vol 888. Springer, Singapore, pp 1–34. https://doi.org/10.1007/978-981-19-6974-4_1
146. Ogwu MC, Osawaru ME, Amodu E, Osamo F (2023) Comparative morphology, anatomy and chemotaxonomy of two Cissus Linn. species. Braz J Bot. https://doi.org/10.1007/s40415-023-00881-0
147. Ogwu MC, Osawaru ME, Owie MO (2023) Effects of storage at room temperature on the food components of three cocoyam species (*Colocasia esculenta*, *Xanthosoma atrovirens*, and

X. sagittifolium). Food Studi Interdiscip J 13(2):59–83. https://doi.org/10.18848/2160-1933/CGP/v13i02/59-83
148. Okeniyi JA, Ogunlesi TA, Oyelami OA, Adeyemi LA (2007) Effectiveness of dried Carica papaya seeds against human intestinal parasitosis: a pilot study. J Med Food 10(1):493–499
149. Olorunnisola OS, Bradley G, Afolayan AJ (2011) Ethnobotanical information on plants used for the management of cardiovascular diseases in Nkonkobe Municipality, South Africa. J Med Plant Res 5(17):4256–4260
150. Omuta GE (1980) A profile of development of Bendel state of Nigeria publication in geography 1 no. 2. Department of Geography and Regional Planning. University of Benin, Benin City
151. Oñatibia-Astibia A, Franco R, Martínez-Pinilla E (2017) Health benefits of methylxanthines in neurodegenerative diseases. Mol Nutr Food Res 61(6):1600670
152. Onwueme IC, Singh TD (1991) Field crop production in tropical Africa. CTA, Wageningen. 480p
153. Osawaru ME, Ogwu MC (2014) Ethnobotany and germplasm collection of two genera of cocoyam (Colocasia [Schott] and Xanthosoma [Schott], Araceae) in Edo State Nigeria. Sci Technol Arts Res J 3(3):23–28. https://doi.org/10.4314/star.v3i3.4
154. Osawaru ME, Ogwu MC (2015) Molecular characterization of 36 accessions of two genera of Cocoyam (Colocasia [Schott] and Xanthosoma [Schott], Araceae. Sci Technol Arts Res J 4(1): 27–33. https://doi.org/10.4314/star.v4i1.4
155. Osawaru ME, Ogwu MC, Braimah L (2013) Growth responses of two cultivated Okra species (*Abelmoschus caillei* (A. Chev.) Stevels and *Abelmoschus esculentus* (Linn.) Moench) in crude oil contaminated soil. Nig J Basic Appl Sci 21(3):215–226
156. Osawaru ME, Ogwu MC, Dania-Ogbe FM (2013) Morphological assessment of the genetic variability among 53 accessions of West African Okra [*Abelmoschus caillei* (A. Chev.) Stevels] from South Western Nigeria. Nig J Basic Appl Sci 21(3):227–238
157. Osawaru ME, Ogwu MC, Ogbeifun NS, Chime AO (2013) Microflora diversity of the phylloplane of wild Okra (*Corchorus olitorius* L. Jute). Bayero J Pure Appl Sci 6(2):136–142
158. Osawaru ME, Ogwu MC, Imarhiagbe O (2013) Biochemical characterization of some Nigerian Corchorus L. species. Bayero J Pure Appl Sci 6(2):69–75
159. Osawaru ME, Ogwu MC, Ahana CM (2013) Current status of plant diversity and conservation in Nigeria. Nigerian J Life Sci 3(1):168–178
160. Osawaru ME, Ogwu MC, Imarhiagbe O (2013) Agro-morphological characterization of some Nigerian Corchorus (L.) species. Biol Environ Sci J Trop 10(4):148–158
161. Osawaru ME, Ogwu MC, Ahana CM (2014) Study of the distinctiveness of two genera of Cocoyam (Colocasia [Schott] and Xanthosoma [Schott]) using SDS – PAGE. Nigerian Soc Exp Biol J 14(3):168–179
162. Osawaru ME, Ogwu MC, Emokpare AA (2014) Preliminary assessment of the microanatomy of Okra [Abelmoschus (L.)] wood. Egypt Acad J Biol Sci 5(1):39–54
163. Osawaru ME, Ogwu MC, Omologbe J (2014) Characterization of three Okra [Abelmoschus (L.)] Accessions using morphology and SDS-PAGE for the basis of conservation. Egypt Acad J Biol Sci 5(1):55–65
164. Ottaviani JI, Borges G, Momma TY, Spencer JPE, Keen CL, Crozier A, Schroeter H (2016) The metabolome of [2-14C](−)-epicatechin in humans: implications for the assessment of efficacy, safety, and mechanisms of action of polyphenolic bioactives. Sci Rep 6(1):29034
165. Ottaviani JI, Heiss C, Spencer JPE, Kelm M, Schroeter H (2018) Recommending flavanols and procyanidins for cardiovascular health: revisited. Mol Asp Med 61:63–75
166. Owusu IK, Acheamfour-Akowuah E (2018) Pattern of cardiovascular diseases as seen in an out-patient cardiac clinic in Ghana. World J Cardiovasc Dis 8:70–84
167. Özgen M, Serçe S, Kaya C (2009) Phytochemical and antioxidant properties of anthocyanin-rich *Morus nigra* and *Morus rubra* fruits. Sci Hortic 119(3):275–279

168. Paikra BK, Dhongade HKJ, Gidwani B (2017) Phytochemistry and pharmacology of *Moringa oleifera* Lam. J Pharmacopuncture 20(3):194–200. https://doi.org/10.3831/KPI10.3831/KPI.2017.20.022
169. Pomper KW, Layne DR, Peterson RN (1999) The pawpaw regional variety trial. Perspectives on new crops and new uses. ASHS Press, Alexandria, pp 353–357
170. Porchezhian E, Ansari SH, Shreedharan NK (2000) Antihyper-glycemic activity of *Euphrasia officinale* leaves. Fitoterapia 71(5):522–526
171. Portincasa P, Bonfrate L, Scribano ML, Kohn A, Caporaso N, Festi D ... Gasbarrini A (2016) Curcumin and fennel essential oil improve symptoms and quality of life in patients with irritable bowel syndrome. J Gastrointest Liver Dis 25(2):151
172. Prasad S, Aggarwal BB (2011) Turmeric, the golden spice. In: Herbal medicine: biomolecular and clinical aspects, 2nd edn. CRC Press, Hoboken
173. Priore P, Siculella L, Gnoni GV (2014) Extra virgin olive oil phenols down-regulate lipid synthesis in primary-cultured rat-hepatocytes. J Nutr Biochem 25:683–691
174. Punia Bangar S, Singh A, Chaudhary V, Sharma N, Lorenzo JM (2022) Beetroot as a novel ingredient for its versatile food applications. Crit Rev Food Sci Nutr 1–25
175. Rahmatullah M, Ferdausi D, Mollik MAH, Jahan R, Chowdhury MH, Haque WMA (2010) survey of medicinal plants used by Kavirajes of Chalna area, Khulna district, Bangladesh. African J Tradit Complement Altern Med 7:91–97
176. Raman G, Avendano EE, Chen S, Wang J, Matson J, Gayer B, Novotny JA, Cassidy A (2019) Dietary intakes of flavan-3-ols and cardiometabolic health: systematic review and meta-analysis of randomized trials and prospective cohort studies. Am J Clin Nutr 110(5):1067–1078
177. Ramos-Aguilar AL, Ornelas-Paz J, Tapia-Vargas LM, Ruiz-Cruz S, Gardea-Béjar AA, Yahia EM, de Ornelas-Paz J, Pérez-Martínez JD, Rios-Velasco C, Ibarra-Junquera V (2019) The importance of the bioactive compounds of avocado fruit (*Persea americana* Mill) on human health. Biotechnics 21(3):154–162. https://doi.org/10.18633/biotechnics.v21i3.1047
178. Randriamiharisoa MN, Kuhlman AR, Jeannoda V, Rabarison H, Rakotoarivelo N, Randrianarivony T, Raktoarivony F, Randrianasolo A, Bussmann RW (2015) Medicinal plants sold in the markets of Antananarivo, Madagascar. J Ethnobiol Ethnomed 11:60
179. Rastogi S, Pandey MM, Rawat AKS (2016) Traditional herbs: a remedy for cardiovascular disorders. Phytomedicine 23(11):1082–1089
180. Raubenheimer K, Hickey D, Leveritt M, Fassett R, de Zevallos O, Munoz J, Allen JD ... Neubauer O (2017) Acute effects of nitrate-rich beetroot juice on blood pressure, hemostasis and vascular inflammation markers in healthy older adults: a randomized, placebo-controlled crossover study. Nutrients 9(11):1270. https://doi.org/10.3390/nu9111270
181. Rein D, Paglieroni TG, Pearson DA, Wun T, Schmitz HH, Gosselin R, Keen CL (2000) Cocoa and wine polyphenols modulate platelet activation and function. J Nutr 130(8):2120S–2126S
182. Remington J, Winters K (2019) Effectiveness of dietary inorganic nitrate for lowering blood pressure in hypertensive adults: a systematic review. JBI Evid Synth 17(3):365–389. https://doi.org/10.11124/JBISRIR-2017-003842
183. Richard G, Izah SC, Ogwu MC (2022) Implications of artisanal crude oil refining on sustainable food production in the Niger Delta Region of Nigeria. J Environ Bioremed Toxicol 5(2):69–77. https://doi.org/10.54987/jebat.v5i2.775
184. Ried K, Sullivan TR, Fakler P, Frank OR, Stocks NP (2012) Effect of cocoa on blood pressure. Cochrane Database Syst Rev 8:CD008893
185. Ried K, Fakler P, Stocks NP (2017) Effect of cocoa on blood pressure. Cochrane Database Syst Rev 5:CD008893. https://doi.org/10.1002/14651858.cd008893.pub3
186. Rodino S, Butu M (2019) Herbal extracts – new trends in functional and medicinal beverages. In: Functional and medicinal beverages. Elsevier, Amsterdam, pp 73–108
187. Rodriguez-Mateos A, Vauzour D, Krueger CG, Shanmuganayagam D, Reed J, Calani L, Mena P, Del Rio D, Crozier A (2014) Bioavailability, bioactivity and impact on health of dietary flavonoids and related compounds: an update. Arch Toxicol 88(10):1803–1853

188. Rodriguez-Mateos A, Weber T, Skene SS, Ottaviani JI, Crozier A, Kelm M, Schroeter H, Heiss C (2018) Assessing the respective contributions of dietary flavanol monomers and procyanidins in mediating cardiovascular effects in humans: randomized, controlled, double-masked intervention trial. Am J Clin Nutr 108(6):1229–1237
189. Roudsari NM, Lashgari NA, Momtaz S, Roufogalis B, Abdolghaffari AH, Sahebkar A (2021) Ginger: a complementary approach for management of cardiovascular diseases. Biofactors 47 (6):933–951
190. Sabbaghzadegan S, Golsorkhi H, Soltani MH, Kamalinejad M, Bahrami M, Kabir A, Dadmehr M (2021) Potential protective effects of Aloe vera gel on cardiovascular diseases: a mini-review. Phytother Res 35(11):6101–6113
191. Saeed F, Sultan MT, Riaz A, Ahmed S, Bigiu N, Amarowicz R, Manea R (2021) Bitter melon (*Momordica charantia* L.) fruit bioactives charantin and vicine potential for diabetes prophylaxis and treatment. Plants 10(4):730
192. Sanidad KZ, Sukamtoh E, Xiao H, McClements DJ, Zhang G (2019) Curcumin: recent advances in the development of strategies to improve oral bioavailability. Annu Rev Food Sci Technol 10:597–617
193. Sansone R, Rodriguez-Mateos A, Heuel J, Falk D, Schuler D, Wagstaff R … Flaviola Consortium (2015) Cocoa flavanol intake improves endothelial function and Framingham Risk Score in healthy men and women: a randomised, controlled, double-masked trial: the Flaviola Health Study. Br J Nutr 114(8):1246–1255
194. Sansone R, Ottaviani JI, Rodriguez-Mateos A, Heinen Y, Noske D, Spencer JP … Heiss C (2017) Methylxanthines enhance the effects of cocoa flavanols on cardiovascular function: randomized, double-masked controlled studies. Am J Clin Nutr 105(2):352–360
195. Schroeter H, Heiss C, Balzer J, Kleinbongard P, Keen CL, Hollenberg NK, Sies H, Kwik-Uribe C, Schmitz HH, Kelm M (2006)(−)-Epicatechin mediates beneficial effects of flavanol-rich cocoa on vascular function in humans. Proc Natl Acad Sci 103(4):1024–1029
196. Seigler DS, Pauli GF, Nahrstedt A, Leen R (2002) Cyanogenic allosides and glucosides from *Passiflora edulis* and *Carica papaya*. Phytochemistry 60(8):873–882
197. Sesso HD, Manson JE, Aragaki AK, Rist PM, Johnson LG, Friedenberg G … Anderson GL (2022) Effect of cocoa flavanol supplementation for the prevention of cardiovascular disease events: the COcoa Supplement and Multivitamin Outcomes Study (COSMOS) randomized clinical trial. Am J Clin Nutr 115(6):1490–1500
198. Shah M, Murad W, Mubin S, Ullah O, Rehman NU, Rahman MH (2022) Multiple health benefits of curcumin and its therapeutic potential. Environ Sci Pollut Res 29(29):43732–43744
199. Shaito A, Thuan DTB, Phu HT, Nguyen THD, Hasan H, Halabi S, Abdelhady S, Nasrallah GK, Eid AH, Pintus G (2020) Herbal medicine for cardiovascular diseases: efficacy, mechanisms, and safety. Front Pharmacol 11:422. https://doi.org/10.3389/fphar.2020.00422
200. Shakib Z, Shahraki N, Razavi BM, Hosseinzadeh H (2019) Aloe vera as an herbal medicine in the treatment of metabolic syndrome: a review. Phytother Res 33(10):2649–2660
201. Sharifi-Rad J, Rodrigues CF, Sharopov F, Docea AO, Can Karaca A, Sharifi-Rad M … Calina D (2020) Diet, lifestyle and cardiovascular diseases: linking pathophysiology to cardioprotective effects of natural bioactive compounds. Int J Environ Res Public Health 17 (7):2326
202. Sheneni VD, Shaibu IE, Okpe JM, Omada AA (2018) In vivo biological effect of *Carica papaya* leaf extracts on P-407 induced hyperlipidemic Wistar rats. MOJ Food Process Technol 6(4):409–412
203. Shih CC, Lin CH, Lin WL, Wu JB (2009) *Momordica charantia* extract on insulin resistance and the skeletal muscle GLUT4 protein in fructose-fed rats. J Ethnopharmacol 123(1):82–90. https://doi.org/10.1016/j.jep.2009.02.039
204. Shin NR, Kwon HJ, Ko JW, Kim JS, Lee IC, Kim JC, Kim SH, Shin IS (2019) S-Allyl cysteine reduces eosinophilic airway inflammation and mucus overproduction on ovalbumin-induced allergic asthma model. Int Immunopharmacol 68:124–130
205. Silva S, Gomes L, Leitão F, Coelho AV, Boas LV (2006) Phenolic compounds and antioxidant activity of *Olea europaea* L. fruits and leaves. Food Sci Technol Int 12:385–395

206. Smith T, Kawa K, Eckl V, Morton C, Stredney R (2018) Herbal supplement sales in US increased 8.5% in 2017, topping $8 billion. HerbalGram 119:62–71
207. Snee LS, Nerurkar VR, Dooley DA, Efird JT, Shovic AC, Nerurkar PV (2011) Strategies to improve palatability and increase consumption intentions for *Momordica charantia* (bitter melon): a vegetable commonly used for diabetes management. Nutr J 10:1–11
208. Sridhar MG, Vinayagamoorthi R, Suyambunathan VA, Bobby Z, Selvaraj N (2008) Bitter gourd (*Momordica charantia*) improves insulin sensitivity by increasing skeletal muscle insulin-stimulated IRS-1 tyrosine phosphorylation in high-fat-fed rats. Br J Nutr 99(4):806–812
209. Ssegawa P, Kasenene JM (2007) Medicinal plant diversity and uses in the Sango bay area, Southern Uganda. J Ethnopharmacol 113:521–540
210. Stabler SN, Tejani AM, Huynh F, Fowkes C (2012) Garlic for the prevention of cardiovascular morbidity and mortality in hypertensive patients. Cochrane Database Syst Rev 8:CD007653
211. Stevens C, Ugese F, Otitoju G, Baiyeri K (2016) Proximate and anti- nutritional composition of leaves and seeds of *Moringa oleifera* in Nigeria: a comparative study. Agro Sci 14(2):9. https://doi.org/10.4314/as.v14i2.2
212. Suk S, Kwon GT, Lee E, Jang WJ, Yang H, Kim JH ... Lee KW (2017) Gingerenone A, a polyphenol present in ginger, suppresses obesity and adipose tissue inflammation in high-fat diet-fed mice. Mol Nutr Food Res 61(10):1700139
213. Sun C, Li W, Liu Y, Deng W, Adu-Frimpong M, Zhang H ... Xu X (2019) In vitro/in vivo hepatoprotective properties of 1-O-(4-hydroxymethylphenyl)-α-L-rhamnopyranoside from *Moringa oleifera* seeds against carbon tetrachloride-induced hepatic injury. Food Chem Toxicol 131:110531. https://doi.org/10.1016/j.fct.2019.05.039
214. Sun Y, Zimmermann D, DeCastro CA, Actis-Goretta L (2019) Dose-response relationship between cocoa flavanols and human endothelial function: a systematic review and meta-analysis of randomized trials. Food Funct 10(10):6322–6330
215. Sutrisna E, Nuswantoro Y, Said RF (2018) Hypolipidemic of ethanolic extract of Salam bark (*Syzygium polyanthum* (Wight) Walp.) from Indonesia (Preclinical study). Drug Invent Today 10(1):55–58
216. Swati S, Kaur VA, Kumari C, Ali A, Garg P, Thakur P, Attri C, Kulshrestha S (2018) *Moringa oleifera* – a never die tree: an overview. Asian J Pharm Clin Res 11(12):57. https://doi.org/10.22159/ajpcr.2018.v11i12.28049
217. Talhaoui N, Taamalli A, Gómez-Caravaca AM, Fernández Gutiérrez A, Segura-Carretero A (2015) Phenolic compounds in olive leaves: analytical determination, biotic and abiotic influence, and health benefits. Food Res Int 77:92–108
218. Tan MJ, Ye JM, Turner N, Hohnen-Behrens C, Ke CQ, Tang CP ... Ye Y (2008) Antidiabetic activities of triterpenoids isolated from bitter melon associated with activation of the AMPK pathway. Chem Biol 15(3):263–273
219. Tan SP, Kha TC, Parks SE, Roach PD (2016) Bitter melon (*Momordica charantia* L.) bioactive composition and health benefits: a review. Food Rev Int 32(2):181–202
220. Tang GY, Meng X, Li Y, Zhao CN, Liu Q, Li HB (2017) Effects of vegetables on cardiovascular diseases and related mechanisms. Nutrients 9(8):857
221. Tanko Y, Jimoh AG, Mohammed TA, Musa KY (2011) Hypoglycemic effects of the methanolic extract of aerial part of *Chrysanthellum indicum* in rats. J Nat Product Plant Resour 1:1–7
222. Tibazarwa KB, Damasceno AA (2014) Hypertension in developing countries. Can J Cardiol 30:527–533
223. Tiloke C, Anand K, Gengan RM, Chuturgoon AA (2018) *Moringa oleifera* and their phytonanoparticles: potential antiproliferative agents against cancer. Biomed Pharmacother 108:457–466. https://doi.org/10.1016/j.biopha.2018.09.060
224. Tripoli E, Giammanco M, Tabacchi G, Di Majo D, Giammanco S, La Guardia M (2007) The phenolic compounds of olive oil: structure, biological activity and beneficial effects on human health. Nutr Res Rev 18:98–112
225. Upadhyay RK (2016) Nutraceutical, pharmaceutical and therapeutic uses of *Allium cepa*: a review. Int J Green Pharm 10(1):46–64

226. Valdez-Solana MA, Mejía-García VY, Téllez-Valencia A, García-Arenas G, Salas-Pacheco J, Alba-Romero JJ, Alba-Romero JJ, Sierra-Campos E (2015) Nutritional content and elemental and phytochemical analyses of *Moringa oleifera* grown in Mexico. J Chem. https://doi.org/10.1155/2015/860381
227. Van der Sande MAB (2003) Cardiovascular disease in sub-Saharan Africa: a disaster waiting to happen. Neth J Med 61:32–36
228. Verma RK, Kumari P, Maurya RK, Kumar V, Verma RB, Singh RK (2018) Medicinal properties of turmeric (*Curcuma longa* L.): a review. Int J Chem Stud 6(4):1354–1357
229. Vinson JA, Proch J, Zubik L (1999) Phenol antioxidant quantity and quality in foods: cocoa, dark chocolate, and milk chocolate. J Agric Food Chem 47:4821–4824
230. Virdi J, Sivakami S, Shahani S, Suthar AC, Banavalikar MM, Biyani MK (2003) Antihyperglycemic effects of three extracts from *Momordica charantia*. J Ethnopharmacol 88: 107–111
231. Visioli F, Bellomo G, Galli C (1998) Free radicalescavenging properties of olive oil polyphenols. Biochem Biophys Res Commun 247:60–64
232. Vissers MN, Zock PL, Katan MB (2004) Bioavailability and antioxidant effects of olive oil phenols in humans: a review. Eur J Clin Nutr 58:955–965
233. Warade SD, Shinde KG (1998) Garlic. In: Salunkhe DK, Kadam SS (eds) Handbook of vegetable science and technology: production, composition, storage, and processing, 1st edn. Marcel Dekker, New York, pp 397–431
234. Whitmore BB, Naidu AS (2000) Thiosulfinates. In: Naidu AS (ed) Natural food antimicrobial systems. CRC Press, Boca Raton, pp 265–380
235. Wilson EA, Demming-Adams B (2007) Antioxidant, anti-inflammatory, and antimicrobial properties of garlic and onions. Nutr Food Sci 37:178–183
236. Wu M, Liu L, Xing Y, Yang S, Li H, Cao Y (2020) Roles and mechanisms of hawthorn and its extracts on atherosclerosis: a review. Front Pharmacol 11:118
237. Yahaya TO, Salisu TF, Obaroh IO, Adelabu M, Izuafa A, Danjuma JB, Sheu H, Abdulgafar IB (2022) Toxicological evaluation of phytochemicals and heavy metals in *Ficus exasperata* Vahl (Sandpaper) leaves obtained in Birnin Kebbi, Nigeria. Plant Biotechnol Persa 4(2):1–7
238. Yuan Q, Zhao L (2017) The mulberry (*Morus alba* L.) fruit: a review of characteristic components and health benefits. J Agric Food Chem 65(48):10383–10394
239. Yuyun MF, Sliwa K, Kengne AP, Mocumbi AO, Bukhman G (2020) Cardiovascular diseases in sub-Saharan Africa compared to high-income countries: an epidemiological perspective. Glob Heart 15:1–18
240. Zamani H, De Joode MEJR, Hossein IJ, Henckens NFT, Guggeis MA, Berends JE … van Breda SGJ (2021) The benefits and risks of beetroot juice consumption: a systematic review. Crit Rev Food Sci Nutr 61(5):788–804
241. Zeb A, Murkovic M (2011) Olive (*Olea europaea* L.) seeds, from chemistry to health benefits. In: Nuts and seeds in health and disease prevention. Academic, London, pp 847–853
242. Zeb A, Khan S, Khan I, Imran M (2008) Effect of temperature, UV, sun and white lights on the stability of olive oil. J Chem Soc Pak 30:790–794
243. Zhang M, Zhao R, Wang D, Wang L, Zhang Q, Wei S … Wu C (2021) Ginger (*Zingiber officinale* Rosc.) and its bioactive components are potential resources for health beneficial agents. Phytother Res 35(2):711–742
244. Zhao CN, Meng X, Li Y, Li S, Liu Q, Tang GY, Li HB (2017) Fruits for prevention and treatment of cardiovascular diseases. Nutrients 9:598
245. Zielińska-Przyjemska M, Olejnik A, Dobrowolska-Zachwieja A, Grajek W (2009) In vitro effects of beetroot juice and chips on oxidative metabolism and apoptosis in neutrophils from obese individuals. Phytophera Res 23:49–55

Nutritional Profile, Bioactive Components, and Therapeutic Potential of Edible Flowers of Chhattisgarh, India

31

Milan Hait and Nand Kumar Kashyap

Contents

M. Hait (✉) · N. K. Kashyap
Department of Chemistry, Dr. C. V. Raman University, Bilaspur, Chhattisgarh, India

S. C. Izah et al. (eds.), *Herbal Medicine Phytochemistry*, Reference Series in Phytochemistry, https://doi.org/10.1007/978-3-031-43199-9_41

Abstract

The intake of flowers as food is documented in a variety of cultures as part of customary dishes and alternative therapies. But numerous types of edible flowers are more than just a treat because they include important nutrients like protein, vitamins, carbohydrates, fats, and therapeutic potential. Flowers, with their favourable sensory and nutritional properties and the existence of bioactive components advantageous to human health, form a key sector for developing the food industry. It is essential to encourage the local use of flowers in order to maintain traditional norms, and many studies have been conducted to better understand the social and cultural contexts in which edible flowers are consumed. With the global market for natural and healthy diets, the nutritional characteristics, therapeutic advantages, active chemicals, and cooking techniques of edible species have been explored. Current techniques for extracting natural bioactive chemicals from flowers are helping to discover their constituents and produce food industry-beneficial additives. Correct taxonomy, toxicology, and a comprehensive practice guide for commercially grown flowers, processing, and preservation are still needed to promote edible flower usage. The scientific and technical information about edible flowers' nutritional profile, therapeutic effects, and phytochemical composition is evaluated and addressed to improve knowledge, dietary practices, and dietary investigation.

Keywords

Edible flowers · Chhattisgarh · Food · Therapeutic potential · Nutritional profile · Bioactive components

Abbreviations

ABTS	2,2′-azino-bis (3-ethylbenzothiazoline-6-sulfonic acid)
AGEs	Acute gastroenteritis
AMD	Age-related macular degeneration
BSA	Bovine serum albumin
BSA/fructose	Bovine serum albumin/fructose
BSA/glucose	Bovine serum albumin/glucose
CCl_4	Carbon tetra chloride
CVD	Cardio vascular disease
DMB	Dimethyl benzene
DNA	Deoxyribonucleic acid
DPPH	2,2-diphenylpicrylhydrazyl
HePG2	Human hepatoma cell
HIV	Human immunodeficiency virus
HPLC	High-performance liquid chromatography
HPLC-DAD-ESIMS	High-performance liquid chromatography equipped with photo-diode-array detection-Mass

HPLC-PDA	High-performance liquid chromatography-photo-diod-array
LPS	Lipopolysaccharide
MAP kinase	Mitogen-activated protein kinase
MCF-7	Human breast cancer cell line
MPPs	Mitochondrial processing peptidase
ORAC	Oxygen radical absorbance capacity
P13K	Phosphatidylinositol-3 kinase
T47D	Human breast cancer cell line

1 Introduction

One of the most important ecosystems in Chhattisgarh may be situated on the Indian subcontinent, which is home to a vast array of unique plant and animal species [1]. These areas in Chhattisgarh, also known as the central region of India, conserve a wide variety of plant life and provide a wealth of wood and non-timber materials, such as gums, with significant economic values [2–4]. The first step in protecting these unique natural biological components is to measure and record plant variation. We can always count on healthy ecosystems, which are beneficial to humans in a variety of ways, to be characterized by high biodiversity [5]. Food webs, food chains, bioremediation, nutrient availability, and human subsistence all rely on biodiversity to function in harmony [6]. There are approximately 6922 km^2 of forest, including deep forest and open forest, as well as mangroves. It covers around 21.5% of the nations at the geographical region level. It has several applications, including community forestry, agroforestry management, replanting, and the restoration of damaged or demolished commercial and industrial sites [7, 8]. The term "agroforestry"simply refers to a combination between agriculture and forestry that has been employed for centuries [9–11]. Agroforestry practices are also becoming a realistic means of decreasing the adverse environmental impacts of the current state of affairs [12]. Chhattisgarh has several useful and nutritional plants since its forests comprise 42% of the state's total territory and are composed of tropical moist and tropical dry deciduous trees. From a conservationist's perspective, it is crucial to keep all of these temperate forests alive and well, as they are native to a wide variety of plant varieties and serve as a safe haven for many different kinds of wildlife [6]. Hence, securing these biotic assets is critical to maintaining a healthy ecosystem. The nutritional value of food and its accessibility have long shaped the eating habits of forest-dwelling communities, and people in these areas typically harvest their own food. Roots, rhizomes, tubers, fruits, seeds, flowers, and leaves are all consumed by those who live in the forest as a means of subsistence. As a result, these edible plants are crucial to the long-term viability and food security of people who make their homes in the nation's forests [13].

The state of Chhattisgarh is located at 80° 15′ to 84°24′ E longitude and 17°46′ to 24° N latitude. The state consists of both hilly and flat areas throughout its entirety.

On average, 60 in. of precipitation fall each year. Rice is the primary crop produced in the state. Because of the state's exceptionally diverse plant and animal life, Chhattisgarh is often referred to as the "herbal state." Around 42% of the land area of the state is covered by woods. People who live in tribal communities are wholly reliant on the forest for their food and a variety of other needs [14]. Although farming is the chief source of income for indigenous communities, the forest and its resources are equally vital to the survival of indigenous and folk communities by providing for a wide range of needs, including food, medicine, fibres, and shelter. They rely heavily on farming to meet their food needs, but they also forage for wild roots, tubers, leaves, flowers, and fruits. In terms of ethnobotany, the state is not well known; the only important studies that have been written down can be found there. In this chapter, only edible flowers and their healing properties, as well as a few other ethnobotanical traits, are discussed. On the other hand, there is a lack of details concerning edible plants. This chapter discusses the wild plant species that they harvest for their edible qualities and then devour. Flowers have been eaten for a long time, but recently there has been a resurgence of interest in doing so, both for their visual beauty and any potential health benefits they may provide. New interest in edible flowers has led to a lot of research on their nutritional value, medicinal value, and phytochemical components [15]. There are many different phytonutrients in flowers, and some of them may help protect against chronic diseases. This chapter discusses the nutritional and pharmacological activities of edible flowers from Chhattisgarh, India. Whether edible flowers are to be considered functional foods or a component of our diets, however, more findings referring to their security and application are required [16, 17].

2 Edible Flowers of Chhattisgarh

Flowers are natural substances that are not only beautiful but also known for their nutritional profile and therapeutic perspective [18, 19]. People have been able to benefit in many ways from natural resources like wild edible plants. These benefits reflect long-standing interactions that have existed between plants and humans throughout different periods of history. Because of this, there is no question that people eat edible flowers that grow in the wild because the term farming has never been clearly defined. The edible flowers of Chhattisgarh originate from an extensive range of plant groups, genera, and species; however, the specific number of flowering plants that can be consumed varies from region to region. There is a wide variety of plants, but only a few are selected as edible, making accurate recognition of edible flowers crucial. A variety of edible flowers can be purchased commercially, and their popularity is growing rapidly. These blooms are most popular when consumed in moderation, but they also have many other uses in the kitchen, including in baked products, drinks, marmalade, salads, smoothies, and even as a replacement for vegetables in savoury dishes. There are so many edible flowers that it would be impossible to study them all [20, 21].

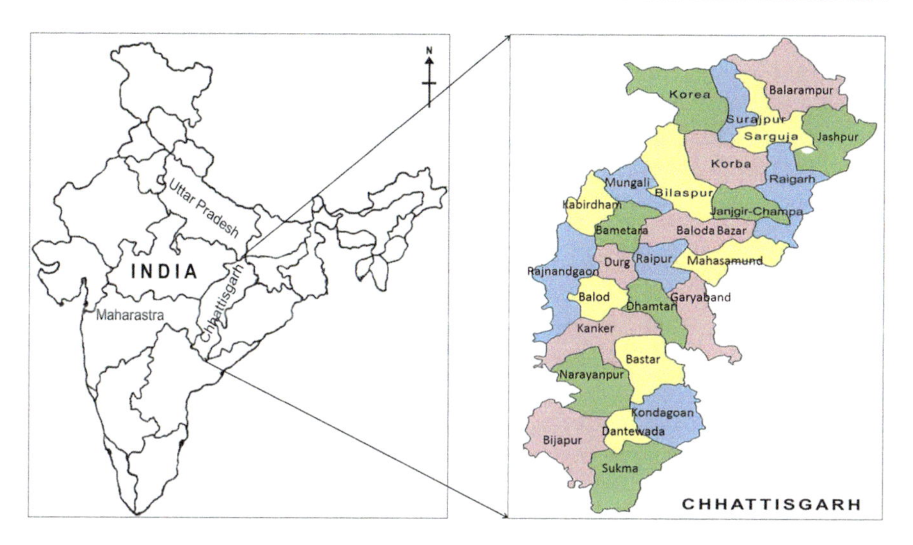

Fig. 1 Distribution of edible flowers in Chhattisgarh

3 Distribution of Edible Flowers in Chhattisgarh

An edible flower is defined as a safe, harmless, and non-toxic flower with health advantages that is taken as a component of human food. Since ancient times, they have consistently been a significant factor in supplying living organisms with various nutrients. It has been demonstrated that consuming flowers is a cultural tradition in a huge number of locations, including Raipur, Bilaspur, Raigarh, Janjgir-Champa, Baloda-Bazar, Bastar, Sarguja, Koriya, Gariyaband, Sakti, Sarangarh-Bilaigarh, Korba, Durg, and other locations of Chhattisgarh, India (Fig. 1). Edible flowers have become extremely popular around the world recently due to their attractive beauty and the therapeutic advantage associated with their specific aroma, flavour, and colouring [15]. Edible flowers have recently been at the centre of numerous investigations in the fields of food science, technology, and even pharmaceuticals. The goal of this chapter was to enhance the attractiveness of edible flowers as prospective food products by discussing their phytochemical components, health benefits, and toxicity. There is evidence that humans have been eating flowers for at least 2000 years. Several regions, including India, Pakistan, Sri Lanka, Vietnam, the USA, Australia, Kenya, and South Africa, have reported the consumption of edible flowers. Nowadays, you may eat flowers in salads, in light curries, or even by themselves as a vegetable. Certain edible flowers are suitable for stuffing or using in stir-fries [16, 22].

4 Traditional Use of Edible Flowers

The flowers are more than just decorative additions to the savoury dishes and sweet confections; rather, they offer a novel variety of tastes and boost the nutritional content of the dishes. They can be used as fresh salads, savoury dishes, soups, beverages, candies, jellies, spices, and colours. For example, *Tagetets erecta, Azadirachta indica, Rosa gallica, Cardia dichotomo, Brasica oleracea,* and *Tamarindus indicus* flowers are employed in a variety of culinary applications in dried form, powdered form, crystallized form, and foam form [23, 24].

There are many plants that have conventionally been used as edible flowers, such as *Madhuca latifolia* Roxb., *Bombax malabaricum, Butea monosperma, Holarrhena antidysenterica, Indigofera casioides, Tamarindus indicus, Holostemma rheedianum Spreng, Crotalaria juncea L., Crotalaria incana Rottl., Celastrus paniculata Willd, C. renigera (Wall) Gagnep., C. fistula L., C. arborea* Roxb., *Bauhinia racemosa Lam., Alangium salvifolium, Brassica oleracea var. botr. Linn, Couroupita guianensis Aubl, Brassica oleracea, Cordia dichotoma, Cucurbita L., Moringa oleifera, Hibiscus sabdariffa L., Azadirachta indica, Solanum xanthocarpum, Solanum xanthocarpum, Hibiscus rosa-sinensis, Helianthus annuus L., Rosa gallica, and Tagetet erecta.* The comprehensive information, however, was not provided.

The following key point describes the traditional uses of edible flowers:

- Provides a major portion of the calories and nutrients needed on a yearly basis [25].
- It is possible to use it as a main dietary replacement, especially during times of seasonal food crisis.
- For use as a salad topping, appetizer, main dish, beverage, or sweet course.
- Edible flowers have been used in a wide variety of recipes for tea, baked goods, sauces, jellies, syrups, flavoured liquors, vinegars, honey, and oils.
- Certain edible flowers that are safe to eat can be incorporated into stir-fried dishes.
- Offering an option for the application of natural dyes and colourants in the food industry.
- Salads, appetizers, soups, main dishes, confectionery, and beverages all benefit from their colourful, aromatic flavours, and refreshing aromas [26].
- Even though edible flowers are popular in their current forms, they can also be baked, frozen (as ice cubes), bottled in sweets, or stored with volatiles [15].
- Edible flower petals can be steeped in oil or vinegar to provide a floral fragrance.
- Sugar and egg whites can be employed to crystallize flowers for candy [27].
- Flowers' essential oils are applicable in aromatherapy treatments.
- Edible flowers are included in therapeutic diets.
- Edible flowers were employed as garnishes on dishes served to the nobles, particularly during formal events like feasts and festivals [15, 28]. The uses of many edible flowers in Chhattisgarh are listed in Table 1.

5 Edible Flowers as Cultural and Regional Foods

A state that is located in the centre of India is called Chhattisgarh, because the term "Chhattisgarh"is made up of two words: first "Chhattis" and second "Garh," which means 36 forts. The name Chhattisgarh is meant to represent the large group of forts

Table 1 Uses and family description of edible flower of Chhattisgarh, India

S. N.	Plant Local Name	Scientific Name	Edible parts	Family	Uses	References
1	Mahua	*Madhuca latifolia* Roxb.	Flower	Sapotaceae	The fresh as well as dry flowers are eaten along with seed	[29]
2	Semal	*Bombax malabaricum*	Flower	Bombacaceae	The young flower buds, mixed with Mahua flowers, are eaten during food scarcity	[29]
3	Palash	*Butea monosperma*	Flower	Fabaceae	The young flower buds are used as vegetable along with pickled oil	[29]
4	Koraya	*Holarrhena antidysenterica*	Flower	Apocynaceae	The flower bud is cooked as vegetable. Boiled flowers are cooked and eaten as vegetable	[29, 30]
5	Girhul	*Indigofera casioides*	Flower	Fabaceae	Flower bud is cooked as vegetable and curry	[29]
6	Imali	*Tamarindus indicus*	Flower	Fabaceae	Flowers are eaten as vegetable	[29]
7	Konga	*Holostemma rheedianum Spreng*	Flower	Asclepiadaceae	Flowers are eaten	[30]
9	Sun	*Crotalaria juncea L.*	Flower	Fabaceae	Flowers are eaten as vegetable	[30]
10	Jangli sun	*Crotalaria incana Rottl.*	Flower	Fabaceae	Flowers are cooked as vegetable	[30]
12	Kujur	*Celastrus paniculata Willd*	Flower	Celastraceae	Flowers are used as vegetable	[30]
13	Khilbiri	*C. renigera (Wall) Gagnep.*	Flower	Caesalpiniaceae	Flowers are eaten as vegetable	[30]
14	Amaltas	*C. fistula L.*	Flower	Caesalpiniaceae	Flowers are eaten as vegetable	[30]
15	Kumhi	*C. arborea* Roxb.	Flower	Lecythidacae	Flowers are eaten as vegetable	[30, 31]
16	Kachnar	*Bauhinia racemosa Lam.*	Flower	Caesalpiniaceae	Young flowering buds are used as vegetable	[30]
17	Ankol	*Alangium salvifolium*	Flower	Alangiaceae	Flowers are eaten as vegetable	[30]
18	Phoolgobhi	*Brassica oleracea var botr. Linn*	Flower	Brassicaceae	Flowers are eaten as vegetable	[14]

(continued)

Table 1 (continued)

S. N.	Plant Local Name	Scientific Name	Edible parts	Family	Uses	References
19	Kalaspati	*Couroupita guianensis Aubl*	Flower	Lecythidaceae	Flowers are eaten as vegetable	[32]
20	Broccali	*Brassica oleracea*	Flower	Boraginaceae	Flowers are eaten as vegetable	[32]
21	Bohar	*Cordia dichotoma*	Flower	Boraginaceae	Flowers are eaten as vegetable	[32]
22	Makhana	*Cucurbita pepo*	Flower	Cucurbitaceae	Flowers are eaten as vegetable	[32]
23	Munga	*Moringa oleifera*	Flower	Moringaceae	Flowers are eaten as vegetable	[32]
24	Amari patuwa	*Hibiscus sabdariffa L.*	Flower	Malvaceae	Flowers are eaten as vegetable	[32]
25	Neem	*Azadirachta indica*	Flower	Meliaceae	Flowers are eaten as vegetable	[32]
26	Bhaskatiya	*Solanum xanthocarpum*	Flower	Solanaceae	Flowers are eaten as vegetable	[32]
27	Kela	*Solanum xanthocarpum*	Flower	Musaceae	Flowers are eaten as vegetable	[32]
28	Madar	*Hibiscus rosa-sinensis*	Flower	Malvaceae	Flowers are eaten as vegetable	[32]
29	Surajmukhi	*Helianthus annuus L.*	Flower	Aesteraceae	Used as edible oil	[32, 33]
30	Gulab	*Rosa gallica*	Flower	Rosaceae	Flowers are eaten as vegetable	[32]
31	Merigold	*Tagetets erecta*	Flower	Aesteraceae	Flower are eaten as salad	[23]

that are located in the area. Almost 70% of the community is employed in agriculture, giving rise to the region's other name, the "Rice Bowl" of Central India. Rice is the major crop in the province, and it also makes up the majority of the food of the people who live here. It is abundantly clear that as the rate of urbanization rises, people's conventional dietary patterns are shifting in the direction of a culture around unhealthy food. Nevertheless, nearly 80% of the population of Chhattisgarh lives in rural areas. As a consequence of this, the state's traditional culture is still being practised, and the majority of the people's eating habits have not altered [34]. Foods from Chhattisgarh are both tasty and nutritious. Because of the state's extensive forest area, the rural and tribal people who live there supplement their diets with forest-grown roots, tubers, leaves, flowers, and fruits. They add a distinct flavour and beneficial properties of nature to traditional Chhattisgarh dishes. There are various types of leaves, tubers, roots, and flowers eaten locally here [35, 36]. This chapter delves into the typical flower-based food of Chhattisgarh and gives information on the flowers, leaves, tubers, rhizomes, and roots used in these dishes. Vegetarianism is highly prevalent in Chhattisgarh, and the state's cuisine makes excellent use of locally grown vegetables and other nutritious ingredients. Every nation's traditional cuisine has survived and is the healthiest option for its citizens. There is a worldwide trend towards continually low fruit and vegetable consumption [37]. Flowers, rhizomes, fruits, leaves, and other plant-based foods make up a significant food group that is rich in necessary minerals and vitamins; ensuring adequate consumption of this group is especially crucial in Chhattisgarh, which suffers from chronic nutritional deficits [38]. The food cultures of Chhattisgarh are the path forward towards adopting sustainable practiscs and making use of the native materials at our disposal to enhance our regular intake. Not only does the processing and preparation of ethnic foods recognize the inventiveness of the local inhabitants and their wealth of cultural and historical heritage, but it also validates their tremendous competence in supporting the existence of the environment in its entirety [39].

Dietary practices are a significant factor in determining the state of people's healthcare. The growth of the human population has been paralleled by the steady evolution of the culture surrounding local foods. It is determined by geographical position, environmental conditions, seasonal fluctuations, soil composition, water supply, forest area, agricultural practices, immigration, and the impact of conquerors, as well as the labour routines of the inhabitants in the area. The culture around traditional foods is based on an experiential context that develops over the course of generations. Food's function in human existence is not confined to simply satisfying appetite; rather, it has a far larger meaning and is an essential component in practically every facet of human existence, including the family, social relationships, festivals, and spiritual rites. The residents of a region have access to the healthiest, most nutritious food options available thanks to the traditional cuisine that has been passed down through the generations [40, 41]. Accepting unhealthy, out-of-region eating habits in the name of globalization is not only unethical but also potentially damaging to people's health. In 2015, the Global Burden of Disease study found that bad food was the largest cause of immorality worldwide, largely due to the shift from a conventional to a modern eating pattern. Each nation's traditional cuisine has stood the test of time and is the healthiest option for its citizens. While rice and vegetables are the most common foods in Chhattisgarh, pulses, green leafy vegetables (GLV), tubers, flowers, and roots are also part of a typical Chhattisgarh lunch. The rural and tribal residents of the state have acquired the ability to do with what nature provides because of the extensive forest cover in their area. The ethnic food of Chhattisgarh is thought to have beneficial health effects, which are supported by both traditional medical practices and more contemporary scientific studies. The food culture of Chhattisgarh is both abundant and extremely varied. The state's numerous dietary patterns are intricately connected to aspects of social authenticity, religious practice, and other sociocultural factors, in addition to local farming techniques and a broad variety of readily available food. Well over 50% of the state's inhabitants can acquire their meals and vegetables from the crops they grow, and edible flowers are a significant component of those crops. It is farmed or cultivated in almost every region, and about half of Chhattisgarh's population uses it as their primary source of nutrition [42].

6 Nutritional Profiles of Edible Flowers

Although the nutritional value of many forest meals is unclear, there seems to be sufficient evidence to suggest that forest foods are beneficial to one's health in terms of their nutritional content. It is of the utmost priority to support experiments on the nutritional values of foods found in forests, since this will inspire people to consume a larger amount of food and offer them a more optimal nutritional profile [14, 43]. Edible flowers have been a component of the human diet for thousands of years. Their therapeutic characteristics and subsequent health benefits have led to their classification as a type of plant cuisine. *Brassica oleracea var botr. Linn, Brassica oleracea, Cordia dichotoma, Cordia dichotoma, Moringa oleifera, Hibiscus sabdariffa L., Solanum xanthocarpum, Helianthus annuus L., Rosa gallica, Tagetes, Madhuca latifolia Roxb., Crotalaria juncea L., Bombax malabaricum, Butea monosperma, Tamarindus indicus, C. fistula L., and Bauhinia racemosa Lam.* are the many types of edible flowers that have been eaten since ancient periods. Flowers that are harmful or otherwise unsafe for humans or have poor nutritional benefits are not typically included in human food [44, 45]. Certain flower varieties include poisonous compounds that may reduce their nutritional value, such as alkaloids, oxalate, cyanogenic glycosides, nitrate, or phytate, or cause serious harm to users [15, 44, 46]. These edible flowers are sold commercially or used in human food. It is important to confirm the nutritional profile and other properties associated with human nourishment for the foods that are generally accepted as edible. Edible flowers vary widely in terms of total carbs, soluble fibre, and micronutrients, depending on the type of edible flower [15, 26, 44, 46, 47]. The majority of edible flowers are composed entirely of water molecules (more than 72%) and contain the least concentration of protein and vitamins. Edible flowers also include varying quantities of carbohydrates, crude fibre, fat, lipids, and minerals. Edible flowers include a variety of bioactive components, like pigments, carotenoids, phenolic compounds, terpenoids, flavonoids, and volatile oils, which offer a broad range of beneficial qualities. The proximate composition and mineral content concentration of many edible flowers from Chhattisgarh, India, are displayed in Tables 2 and 3, respectively.

7 Bioactive Components of Edible Flowers

Bioactive components are clinically evident, naturally produced organic molecules that do not constitute nutrients but have therapeutic properties that promote public health. They enhance the plant's appearance, aroma, and taste in addition to protecting it from illness and environmental damage [72]. Most of the bioactive chemicals in edible flowers are polyphenol compounds, carotenes, flavonoids, and anthocyanins [15, 71]. Carotenoids and flavonoids give flowers their colour and antioxidant capacity [73]. Lutein and flavonoids have been examined for their potential to treat cytotoxic cancers [73–76]. Several types of polyphenol components, including phenolic acids, flavonoids, flavonols, and anthocyanins, are found in edible flowers, which have strong antioxidant potential, act as a shield against free radicals, and have been linked to improved human metabolism [77–80]. Polyphenols are known to have antioxidant properties in in vitro and in vivo examinations [81]. Many therapeutic properties may be contributed by phenolic components, carotenes (colourants) [82], isothiocyanates, volatile oils (the primary source of the flowers' aroma) [83, 84], and cyclotides, which are circular plant peptides. The ingestion of edible flowers has substantially expanded recently due to their multiple health advantages, which are contributed in part by the abundance of phytonutrients found in edible flowers. Although many investigations have been conducted, there has not been a comprehensive literature review prepared on the nutritional profile of edible flowers and the phytonutrients found in them. Prior to recently, exploration of edible flowers focused on their nutrition, aroma, and volatile oils [85, 86]. However, phytonutrients, the primary beneficial constituents of edible flowers, have received increased attention in investigations. Phytochemicals are the active molecules found in vegetables and plant foods that significantly lower the chances of serious illnesses like tumours, fever, cancer, CVD, inflammation, and cytotoxicity [87, 88]. Figure 2 displays the various bioactive components of edible flowers from Chhattisgarh.

7.1 Anthocyanins

Anthocyanins are water-soluble phytochemicals that belong to the flavonoid group. These are the natural colour pigments that are responsible for the colouring of flowers. The pH value, metal ions, and co-pigments all play a role in determining their unique coloration. Anthocyanins have been demonstrated to have a higher potential for eliminating free radicals [89]. These molecules also play a significant role in the prevention of several illnesses, including malignancy, cardiac disease, overweight, hypertension, and diabetes. Anthocyanin can be found in high

Table 2 Proximate composition of edible flowers of Chhattisgarh India

S. N.	Plant Scientific Name	Moisture Content (g/100 g)	Total Ash (g/100 g)	Total Protein (g/100 g)	Total Fat (g/100 g)	Fibre (g/100 g)	Carbohydrate Content (g/100 g)	Total Energy (kcal)	References
1	*M. latifolia* Roxb.	79.58	1.5	6.35	1.6	10.8	–	–	[48]
2	*Bombax malabaricum*	7.4	6.55	5.3	1.21	14.4	32.47	290.31	[49, 50]
3	*Butea monosperma*	19.4	11.2	6.5	8.1	6.1	43.9	–	[51]
4	*Holarrhena antidysenterica*	79.61	4.27	7.31	7.17	6.01	53.38	561.9	[52]
5	*Indigofera casioides*	47.6	5.2	2.63	1.13	19.65	23.92	484.26	[53]
6	*Tamarindus indicus*	42.7	3.32	2.4	0.17	1.79	56.72	216.86	[54]
7	*Crotalaria juncea L.*	10.85	3.31	36.43	20.42	15.32	45.19	227.34	[55]
8	*C. fistula L.*	5.27	12.17	12.43	1.72	19.82	41.32	298.34	[56]
9	*Bauhinia racemosa lam.*	8.2	7.28	6.92	6.57	1.98	76.17	238.29	[57, 58]
10	*Alangium salvifolium*	83.9	1.61	2.71	3.8	2.95	111.67	209.13	[59]

11	*Brassica oleracea var botr. Linn*	85.58	8.02	36.15	7.23	10.17	39.5	304.42	[60]
12	*Couroupita guianensis Aubl*	–	8.94	11.23	–	6.78	–	–	[61]
13	*Cordia dichotoma*	79.29	–	2.33	–	–	31.66	–	[62]
14	*Cucurbita pepo*	69.91	9.61	11.9	2.09	3.51	36.8	226	[63]
15	*Moringa oleifera*	–	78.7	178.7	39.89	–	–	–	[64]
16	*Hibiscus sabdariffa L.*	7.13	11.09	17.29	3.73	24.51	38.59	920.21	[39, 65, 66]
17	*Solanum xanthocarpum*	91.17	1.56	1.79	0.44	–	–	32.6	[67]
18	*Hibiscus rosa-sinensis*	82.92	1.41	1.55	1.35	1.53	2.74	228.47	[68, 69]
19	*Rosa gallica*	11.18	1.83	7.94	16.09	3.71	52.23	4.22	[70]
20	*Tagetets erecta*	83.38	0.89	1.35	0.31	9.18	14.14	28. 03	[71]

Table 3 Mineral content concentration of edible flowers

S. N.	Plant Scientific Name	Ca (mg/100 g)	P (mg/100 g)	Fe (mg/100 g)	Zn (mg/100)	Cu (mg/100 g)	Co (mg/100 g)	Mg (mg/100 g)	K (mg/100 g)	References
1	*Madhuca latifolia* Roxb.	45	22	–	–	–	–	–	–	[48]
2	*Bombax malabaricum*	–	–	–	–	–	–	–	–	
3	*Butea monosperma*	44.5	–	44.4	–	–	–	–	–	[51]
4	*Holarrhena antidysenterica*	–	–	–	–	–	–	–	–	
5	*Indigofera casioides*	0.93	0.025	0.032	0.013	0.003	0.001	0.32	0.16	[53]
6	*Tamarindus indicus*	74	0.17	2.8	0.1	0.87	0.012	0.15	1.14	[54]
7	*Alangium salvifolium*	12	–	1.37	–	0.32	–	19	168	[59]
8	*Brassica oleracea var botr. Linn*	4.37	5.91	0.92	0.55	0.051	0.021	1.95	33.91	[60]
9	*Cordia dichotoma*	0.46	–	0.51	0.35	–	–	–	7.83	[62]
10	*Cucurbita pepo*	2.61	–	0.21	0.06	–	–	2.39	19.97	[63]
11	*Hibiscus sabdariffa L.*	155.57	39.5	27.6	3.85	0.71	–	38.52	23.04	[65] [40, 66]
12	*Solanum xanthocarpum*	24.2	18.7	0.79	0.67	0.35	0.42	14.47	86.9	[67]
13	*Hibiscus rosa-sinensis*	4.33	–	1.49	0.83	–	–	–	23.66	[68, 69]

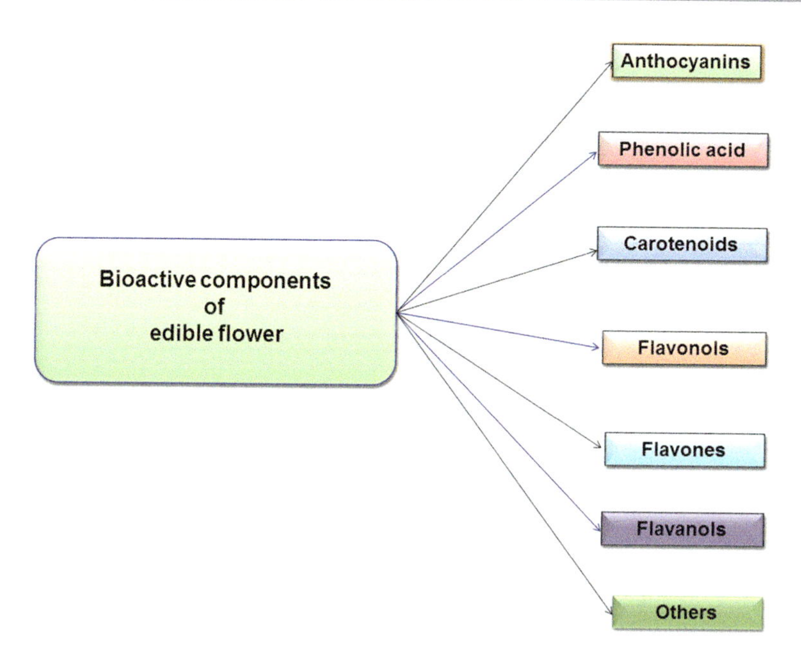

Fig. 2 Bioactive components of edible flowers

concentrations in *Rosa gallica, Solanum xanthocarpum, Hibiscus rosasinensis, Hibiscus sabdariffa L., Moringa oleifera, Cordia dichotoma, Cordia dichotoma, and Tagetets*, among other edible flowers of the plants. Kumari et al. (2017) [90] examined the anthocyanin of various Indian cultivars of *Rosa gallica*. Employing HPLC-PDA, she identified cyanidin 3,5-di-O-glucoside and pelargonidin 3,5-di-O-glucoside. The cyanidin 3,5-di-O-glucoside has been the chief anthocyanin responsible for the reddish and pinkish colour of the edible flower. Pelargonidin 3,5-di-O-glucoside has also been found as an anthocyanin in the carrot-red type of colour of the edible flower [91].

7.2 Carotenoids

Carotenoids are natural pigments found in plants. They give flowers their colours, which range from yellow to orange to red. Carotenoids contain many isoprenoid molecules [92]. The proportion of carotenoids in blossoms varies substantially across floral taxa. Carotenoids are a vital part of both human and animal nutrition and are absorbed from healthy meals and supplementation. According to the findings of numerous studies, carotenoids might decrease the development of vitamin A deficiency, age-related macular degeneration (AMD), blindness, UV exposure, malignancy, cytotoxicity, and CVD. Several studies have shown that Tagetes flowers could be a good source of carotenoids because they contain lutin and its derivatives. At a low amount, zeaxanthin, antheraxanthin, β-cryptoxanthin, β-carotene, α-carotene, zeinoxanthin, and violaxanthin have also been detected in many edible flowers like *Indigofera casioides*, *Crotalaria juncea L.*, *Bauhinia racemosa L.*, *Brassica oleracea,* etc. *Madhuca latifolia, Tamarindus indicus, Moringa oliefera, Rosa gallica, and Cardia dichotoma* flowers have been found to contain many carotenoids like lutein, flavoxanthin, β-carotene, α-carotene, lycopene, rubixanthin, zeaxanthin, α-cryptoxanthin, 13-*cis*-carotene, α-carotene, *trans*-carotene, 9-*cis*-carotene, luteoxanthin, violaxanthin, zeaxanthin, neoxanthin, and lutein epoxide [89, 92–95].

7.3 Flavonols

The group of flavonoids is referred to as flavonols. It is one of the most successful sources of flavonoids. The flavonols contain many sub-groups, such as quercetin, kaempferol, isorhamnetin, and myricetin, along with their derivatives. Flavonols were widely found in edible flowers. They were found in *Rosa gallica* [96], *Moringa oliefera, C. fistula L., Cardia dichotoma, Solonum xanthocarpum, Hibiscus rosa sinensis*, and cactus in a variety of morphologies [97]. *Tegetete*, *Rosa gallica, and Couroupita guianensis Aubl.* also noted the variety of primary flavonols that are present in phenolic flavonoids. Researchers have found that the most common flavonol aglycones in *Rosa gallica* petals are kaempferol and quercetin [91, 92, 98–100]. The majority of edible flowers contain large concentrations of flavonols in comparison to other flavonoids. Shrivastav et al. (2016) [29] reported that the edible flowers of *Butea monosperma, Bombox mealricum,* and *Madhuka latifolia* Roxb. could have isorhamnetin, quercetin, and their glycosides. Ekka and Ekka (2016) [30] also found that *Crotalaria juncea* flowers had several types of flavonols, like kaempferol, quercetin, isorhamnetin, and their glycosides. The above findings provided more evidence that flavonols could be found in many common edible flowers.

7.4 Flavones

Flavones are a subclass of flavonoid whose skeleton is 2-phenylchromen-4-one (2-phenyl-1-benzopyran-4-one). Flavones are a derivative of the Latin word "flavus," which describes a yellowish colour pigment. Flavones are widely available in edible flowers of many kinds. Acacetin, chrysoeriol, apigenin, luteolin, and their corresponding glucosides are the main flavones. There is a rich supply of flavones in a variety of edible flowers. There was a high concentration of flavones in all types of edible flowers that originated from the Asteraceae family, such as *Helianthus annuus* and *Tegetes* [16]. There is a rich supply of flavones in a variety of edible flowers. *Tamaricus indicus* flowers were discovered to contain a significant amount of luteolin-7-glucoside [29]. *Chrysanthemum* and *Tegetet* flowers contain greater amounts of flavones, including acacetin, acacetin-7-O-glucoside, apigenin,

apigenin-7-O-glucoside, luteolin, and luteolin-7-O-glucoside. These flavones have been claimed to have anti-inflammatory action. *Rosa gallica* and *Moringa olifera* both contain luteolin in their respective preparations. Apigenin glycosides were the predominant flavones identified in *Tegetet* [101].

7.5 Flavanols

Flavanols have been found in edible flowers that are made up of catechin, epicatechin, epicatechin gallate, and epigallocatechin gallate. Many researchers reported that *Rosa gallica* had a greater concentration of flavanols in edible flowers [16, 96]. Catechin, epicatechin, epicatechin gallate, and epigallocatechin gallate are the chief constituents of *Rosa gallica*. Many flavanols have been reported in *Brassica oleracea, Madhuca latifolia, Butea monosperma, and Bombox malabaricum* flowers [96, 98, 102, 103]. Many researchers reported flavanols (catechins) in *Chrysanthemum* flowers in the highest concentration. The majority of the crude extract of *Rosa gallica* flowers is composed of flavanols, specifically catechin and epigallocatechin gallate [104].

7.6 Phenolic Acids

To simplify, "phenolic acids" refer to all phenolic compounds with a single carboxylic acid group. Phenol-carboxylic acid, mostly called polyphenols, is an abundant phenolic acid found in many plants and edible flowers. They can be found in many different types of edible flowers. Another important class of secondary metabolites found in edible flowers are phenolic acids. There have been many reports of the presence of phenolic acids in *Rosa gallica* [105], *Cardia dichotoma, Alangium solvifolum, Chrysanthemum*, and *Brassica oleracea* [106]. Other phenolic acids comprise protocatechuic acid, caffeic acid, and gallic acid. An important category of phytochemicals that are identified in edible flowers is called phenolic acids. *Solanum xanthocarpum* flowers contain many phenolic compounds like coumaric acid, quercetin, rutin, ferulic acid, hydroxybenzoic acid, caffeic acid, and cinnamic acid in higher concentrations. *Indigofera cosioides* blossoms contain significant amounts of polyphenolic compounds [107].

7.7 Others

Various types of phytochemicals like carotenoid (β-carotene), phytosterols (β-sitosterol) [108], alkaloids, lignans [109–111], neolignans [112], coumarins, and bisabolo are also found in various edible flowers. The quantities of these phytonutrients were significantly low, and their functional properties were not as potent [113, 114].

8 Therapeutic Potential of Edible Flowers

The purpose of this chapter has been to examine the conventional uses, phytonutrients, therapeutic properties, and nutritional profiles of various edible flowers in Chhattisgarh, India. Flavonols, flavones, flavonoids, anthocyanins, phenolic acids, and their derivatives were discovered to be ubiquitous phytonutrients in a variety of locally eaten flowering plant species, all of which were linked to the edible flowers' reported therapeutic properties. Edible flowers have various health advantages, including antioxidant, anti-inflammatory, anticancer, anti-diabetic, anti-obesity, neuroprotective, anti-carcinogenic, anti-microbial, hepatoprotective, analgesic, and antiulcer characteristics [115–117]. These effects are due to flower phytonutrients and secondary metabolites. Advantages to health from edible flowers are also discussed [118]. The therapeutic potential in edible flowers of Chhattisgarh, India, are displayed in Fig. 3.

8.1 Anti-oxidant Activity

Oxidative stress, ageing, neurological illnesses, cerebrovascular disease, metabolic disorders, cardiovascular disease, tumours, and cytotoxemia are all caused by an excess production of free oxidative radicals. The edible flowers possess strong antioxidant properties. The antioxidant activities were investigated using both *in-vitro* (DPPH, ABTS, and ORAC) and *in-vivo* (lipid peroxidation assay)

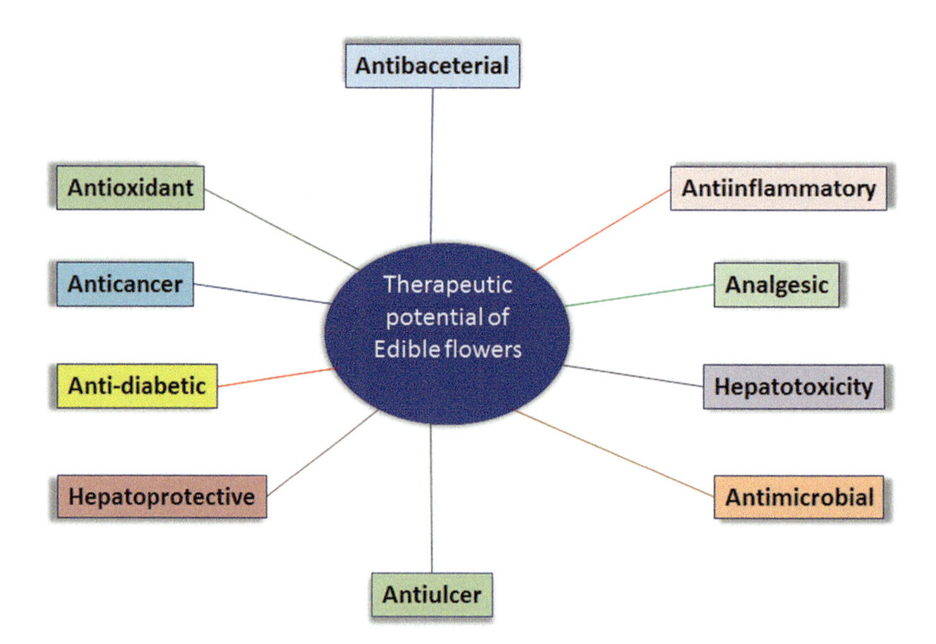

Fig. 3 Therapeutic potentials of edible flower

techniques [119]. Many researchers reported that edible flowers displayed higher antioxidant potential [120, 121]. Some edible flower extracts, like *Madhuca latifolia* Roxb., *Butea monosperma*, *Tamarindus indicus*, and *Holarrhena antidysenterica* have shown greater capacity to scavenge free radicals [122, 123]. Edible flowers have been shown to have considerable anti-oxidant potential owing to the presence of anthocyanins, flavonoids, phenolic acids, alkaloids, terpenoids, and glycosides. Edible flowers have a stronger antioxidant potential than other plant parts like leaves, roots, rhizomes, and barks, which may be due to the higher number of phenolic compounds, particularly flavonoids [124]. The phytochemical flavonoids, alkaloids, terpenoids, steroids, carbohydrates, and glycosides in edible flowers might help them fight free radicals. The biological processes that occur within the human organism generate free radicals as an unwanted consequence. Due to reactive compounds, they cause the oxidative destruction of lipids, proteins, and ultimately DNA. A well-balanced anti-oxidative inhibition system can help counteract damage. Edible flowers are known to be rich providers of a number of different natural bioactive substances, such as vitamins. Healthy food sources, including polyphenols, are of major importance because they provide better antioxidant sources. Phenolic compounds have been shown to have a good impact on chronic health issues such as CVD, diabetes, hypertension, heart disease, hepatoprotective, and neurological illnesses [125–127].

8.2 Anti-inflammatory

The inflammatory response is a necessary and helpful part of the body's physiology. Its main function is to protect the host from both poisons made by the body and microorganisms that come from outside the body. Inflammatory conditions can start off as a short-term acute condition, but they can also turn into a long-term chronic condition, which can cause irreparable harm to the host's health. The anti-inflammatory properties of *Madhuca latifolia* Roxb., *Bombax malabaricum, Butea monosperma, Holarrhena antidysenterica, Indigofera casioides, Tamarindus indicus, Crotalaria juncea L., C. fistula L., C. arborea* Roxb., *Couroupita guianensis Aubl, Cordia dichotoma, Moringa oleifera, Hibiscus sabdariffa L., and Solanum xanthocarpum* demonstrated strong inflammation action. Evaluation of the inflammation effect of edible flower extracts on chronic inflammatory diseases was conducted using cell line assays such as lipopolysaccharide (LPS)-induced RAW 264.7 macrophage activation and in vivo assays such as dimethylbenzene (DMB)-induced ear vasodilatation, acetic acid-induced fluid accumulation, and carrageenan-induced paw oedema. The results indicated that edible flower extracts reduced the severity of acute inflammation. Inflammation is a natural process that causes the immune system to work harder [128]. It is an essential part of the body's physiologic function and acts as a protective mechanism against bacterial infections, viruses, and poisons. The cell simulation method was used to investigate the characteristics of inflammation in the extract of a variety of edible flowers. Many edible flowers possess strong inflammation potential [89, 129].

8.3 Anti-cancer

Investigations in epidemiology have demonstrated that a diet rich in edible flowers, fruits, rhizomes, and vegetables is closely connected with a lower risk of developing long-term diseases [130]. Some edible flowers involve *Chrotalarea juncea* L., *Chrotalarea incana* Rottl., *Rosa gallica*, *Hibiscus sabdariffa*, and *Casea fistula* L. [131–133]. Edible flowers exhibit powerful actions against cancers of the liver, bladder, prostate, breast, and colon. The ability of edible flowers to prevent cancer is mostly attributable to phytonutrients like gallic acid, protocatechuic acids, flavonoids, and rugethanoids B [133, 134]. These edible flowers had the ability to provide cancer-fighting medications. Edible flowers have shown significant activity against a variety of different cancers. The growth and development of malignant breast tumours can be inhibited by consuming modest doses of lutein from marigold extract in their diet. In recent studies, the MTT test was used to show that the extracts of *Bauhinia tomentosa* flowers and *Tagetes erecta* petals were effective against the liver cancer cell lines HePG2 and MCF-7. Skin disorders such as skin damage and skin cancer have emerged as a topic of attention for people all over the world in recent years. The anticancer properties of *Hibiscus sabdariffa*, which is an abundant source of flavonoids, have been the subject of extensive research [135–137]. Researchers tested its cytotoxicity on T47D breast cancer cells and obtained positive results [138]. In a manner comparable to that which was addressed previously, a few studies demonstrated the anticancer action of edible rose blooms. The findings suggested, taken together, that edible flowers might be an even more effective source of anticancer drugs [139].

8.4 Anti-obesity

The build-up of excess fat in the body, the process of fatty acid synthesis, and an imbalance in energy levels all contribute to obesity, also known as overweight, which is associated with adipocyte differentiation and a higher risk of health-related issues. Substances with natural bioactive properties can be utilized to prevent obesity. Obesity or being overweight happens when a person takes in more calories than they use up, and it always leads to a number of metabolic and long-term conditions, such as trouble sleeping, high blood pressure, hypertension, high cholesterol, and diabetes [140]. Several edible flowers, including *Madhuca latifolia* Roxb., *Bombax malabaricum, Butea monosperma, Holarrhena antidysenterica, Indigofera casioides, Tamarindus indicus, Crotalaria juncea L., C. fistula L., C. arborea* Roxb., *Couroupita guianensis Aubl, Cordia dichotoma, Moringa oleifera, Hibiscus sabdariffa L., and Solanum xanthocarpum*, demonstrated an inhibiting effect on obesity [141, 142]. Various phytochemicals acted as mediators between the enzyme and biochemical pathways in lipid burning. *Rosa gallica* extract either controlled calorie consumption and expenditure by decreasing α-amylase activities and preventing carbohydrate or starch uptake or prevented adipocyte development by influencing the PI3K and MAP-kinase pathways.

8.5 Neuroprotective Activity

Alzheimer's disease and Parkinson's disease are both common forms of dementia that cause memory loss. Neurodegeneration is closely related to age, and both of these disorders are connected with ageing. Many researchers discovered, through the use of a cerebral I/R model, it has the potential to significantly enhance the neurological deficiency levels as well as the infract volume [143]. In the *Caenorhabditis elegans* model, flower extracts of *Tagetes erecta* showed potential antioxidant and neuroprotective characteristics. Polyphenols, mainly laricitin and its glycosides, were responsible for these activities and were shown to be responsible for them [144]. Degradation of dopaminergic neurons is also the root cause of ischemia, which is one of the main causes of death and chronic disability. The potent neuroprotective effect of edible flowers, such as *Chrysanthemum, Rosa gallica*, and *Honeysuckle*, was assessed in an in vivo model using glutamate, arachidonic acid, and 6-hydroxydopamine (6-OHDA)-induced serious injuries, posterior cerebral artery occlusion-induced focal cerebral ischemia, and the neurotoxin 1-methyl-4-phenylpyridinium. MPP+'s ability to cause apoptosis in human neuroblastoma cells could be stopped by the phenolic acid found in edible flowers [89].

8.6 Anti-diabetic

Metabolic disorder diabetes mellitus is associated with reduced or insufficient insulin secretion or function. Diabetes can also be referred to as "adult-onset diabetes." As a consequence, the body is unable to metabolize glucose, which ultimately leads to increased quantities of glucose in the blood. Many edible flowers have the capacity to reduce high levels of glucose in the blood, which was evaluated in comparison to insulin [145]. It was found that *Catharanthus roseus* flower ethanol extracts were better at lowering blood sugar levels than the main diagnostic medicine. Due to the liver using more glucose, hypoglycaemia has been seen [146]. Because *Butea monosperma* flowers cause a significant increase in the amount of glucose that is taken up by *Saccharomyces cerevisiae*, there is greater potential for their use in the treatment of diabetes [147]. Various phytochemicals, particularly flavonoids and terpenoids, have a stimulating impact on the body's ability to take in glucose. The anti-diabetic benefits of these flowers are caused by the suppression of enzymes involved in the metabolism of carbohydrates, the control of glucose transporters, the renewal of beta cells, and the enhancement of insulin-generating function [148].

8.7 Hepatoprotective Activity

Edible flowers and their natural active components have received an abundance of interest due to their remarkable ability to treat hepatic issues. The aqueous extract of *Madhuca latifolia* was examined, and the result indicated that it has a strong hepatoprotective reaction against carbon tetrachloride. This was a direct consequence of the serum levels of transaminases, bilirubin, alkaline phosphatase, and triglycerides regenerating themselves. Both the water and methanolic extracts of the edible flowers of *Butea monosperma* demonstrated significant hepatoprotective effects. Another aqueous extract of *Bombax ceiba* flowers showed that these flowers have hepatoprotective activity against CCl_4-induced hepatic injury [29, 149]. Abdelhafez et al. (2018) [150] investigated the possible hepatoprotective activity of *Hibiscus malvaviscus* against the hepatic injury caused by CCl_4 in mice, and findings indicated that dichloromethane and ethyl acetate fractions considerably reduced the seriousness of hepatic damage in rodents.

8.8 Antimicrobial Activity

A plant extract of *Azadirichta indica* is used in the production of natural high-antimicrobial agents. The flowers of *Brassica oleracea* and *Cardia dichotoma* plants were effective against both gram-negative and gram-positive bacteria [14]. Also, antibacterial activity has been found in both ethanolic and aqueous extracts of *Hibiscus rosa-sinensis* against a number of foodborne bacterial infections [151]. *Rosa gallica* methanolic extracts have an antimicrobial effect that is effective against a variety of bacterial and yeast strains [89].

8.9 Others

Numerous plant families and groups across Chhattisgarh, India, rely on edible flowers as a primary means of ensuring their continued access to food and means of survival. It is believed that around 800 different plant species are used as food in India, the majority of which are eaten by the indigenous people. Consuming edible flowers is vital for the eating habits of humans in several communities and is intricately tied to practically all areas of the sociocultural, spiritual, and social welfare of these humans and communities. There is a critical role for the consumption of edible flowers in satisfying the nutritional requirements of rural populations [147, 152]. There are many varieties of edible flowers, and these flowers have been used in a wide variety of ways across a variety of socioeconomic and geographic settings. In addition, there have been substantial reports that suggest that eating wild foods has been a common activity for a very long time [153]. Many of the therapeutic advantages associated with consuming edible flowers are due to the high concentrations of phytonutrients. Both *Rosa gallica* and *Hibiscus*, which are members of the Malvaceae family, have a number of medicinal properties. *Rosa gallica* was shown to be anti-cholesterol, anti-hypertensive, and anti-diabetic [16]. Both delphinidin-3-O-sambubioside and cyanidin-3-O-sambubioside were important, but the biological causes are still unknown. Hibiscus effects of on epileptic seizures and fertility suppression. Also, it helped with healthy hair, tissue repair, and immunology. Two chrysanthemum species were found to significantly reduce the formation of multiple AGEs in the bovine serum albumin (BSA)/glucose and BSA/fructose systems [154, 155]. This finding provides the possibility for the formulation of new diagnostic approaches for the prevention of diseases linked to glycosylation. In in vivo studies, the use of *Rosa gallica* extract resulted in a reduction in blood pressure in both acute and chronic situations. Anti-HIV-1 properties have been identified in rose species that could be consumed [16]. Despite substantial research into the possible health advantages of natural active components in edible flowers, surprisingly little is known about these plants.

9 Overview of Commercial Edible Flowers

Flowers have always played an important role in human culture as well as in the business and decorating worlds. Edible flowers are popular all over the world because they are pretty and healthy. Flowers can be used in many ways. They can be used to add scent, flavour, and colour, and they can also be used to decorate and make food and drinks. Before people can eat flowers, though, their biochemistry needs to be carefully studied. Roses, marigolds, hibiscus, calendula, and other flowers like these were often used in different kinds of food. The floriculture market gets a boost when flowers are grown specifically for their culinary qualities. Floriculture in India is considered an economy with tremendous growth potential. The importance of commercial floriculture with regard to international trade is growing. The liberalization of economic and trade policy created the foundation for the growth of flower production with an emphasis on exportation. Flowers are cultivated not just for their aesthetic value but also because of the nutritional and therapeutic benefits that they offer. Edible flowers have been utilized in a variety of ways throughout a wide range of international locales, including for their nutritional worth, medicinal function, flavour, form, and aesthetic appeal [15, 26]. Many flowers consumed are referred to as edible flowers. Until they are put to use, many parts of flowers, like sepals, petals, carpels, and filaments, are often plucked. Generally, edible flowers are taken in their entirety; however, depending on the kind of flower, certain sections of the flower should be ingested. In addition, it is necessary to remove some sections of certain flowers because of their bitterness. They are saved for use at a later time by employing preservation methods such as drying, freezing, or steeping in oil [156].

Regional and national preferences in food preparation and consumption are shifting. More and more consumers are learning the importance of eating well and choosing foods that meet their nutritional needs. Studies have shown that edible flowers are extremely beneficial to human health due to their abundance of beneficial chemicals, such as antioxidants, anti-carcinogens, vitamins, and many others [157]. Considering the interconnectedness of elements like effective input utilization, cost benefits, improved product quality, and ecological processes is also crucial to the enterprises' success. More people are aware of the serious decline in natural resources and the deterioration of the environment, and this has increased the pressure to make any creative activity as eco-friendly as possible [158]. Flowers have long been used as key components in a wide variety of dishes. Edible flowers are employed for both their aesthetic value and their sensory qualities [159]. The visual appeal of a meal can be affected by both the ingredients used and the presentation. Quality and aesthetics affect the appearance of meals. Edible flowers can boost a dish's attractiveness.

Edible flowers have not been grown by the floriculture industry in the past, but doing so could open up a new market and give the industry a new perspective. Certain flowers are deadly; therefore, it is important to know which ones to avoid [160]. Flowers that are often used in cooking include the *Chrysanthemum, Tegetets erecta, Madhuca latifolia* Roxb., *Bombax malabaricum, Butea monosperma, Holarrhena antidysenterica, Indigofera casioides, Tamarindus indicus, Crotalaria juncea L., C. fistula L., C. arborea* Roxb., *Couroupita guianensis Aubl., Cordia dichotoma, Moringa oleifera, Hibiscus sabdariffa L., and Solanum xanthocarpum*.

Edible flowers were a big part of traditional ways of life that are coming back thanks to globalization and better consumer knowledge. Now, more than ever, edible flowers are all the rage as a trendy new food trend in both well-known and new countries, making competition fierce in this previously untapped, unique area. Because of the COVID-19 situation right now, the flower business is looking into new options. While the number of people who grow and market edible flowers (in places like supermarkets, flea markets, and online) grows, the expectations of consumers and chefs alike continue to rise. They will be the cutting-edge components of next-generation food varieties [28]. When a flower is grown for its edible petals, it boosts the economy in a way that is hard to measure. In addition to their aesthetic value, flowers serve an essential function in agriculture: pollination. The primary factor driving the expansion of the global edible flower market is the increasing application of edible flowers as a supplement in the nutritional, food, and pharmaceutical industries. This has a beneficial effect on edible flower yield. The demand for them is rising all around the world since chefs and regular people alike see them as a healthy and secure meal option. You can also find them in a variety of dishes, drinks, decorations, baked goods, and salads. When compared to the costs of growing other crops, these ones are prohibitively expensive. The purchase of improved varieties for organic agriculture is a major cost driver. Edible flowers have a short shelf life, so they need to be eaten right away after being sold at the market. There are many geographic areas in the market for edible flowers, like Raipur, Bilaspur, Durg, Jagadalpur, and Raigarh city of Chhattisgarh, India. At the time of the evaluation, it was thought that the unique flavour and taste of edible flowers were the main reason for the growth of the regional market. Nevertheless, currently, there is no reliable evidence for the Indian edible flower market. This is due to the fact that the majority of edible flowers in India are harvested from the wild. As a result, there is a requirement to strengthen the market in economies that are experiencing economic growth [89, 161].

10 Conclusions

Edible flowers are becoming more popular in modern nutrition, mostly because of improvements in the nutritional value of a variety of foods and meals. Generally, this has a potentially therapeutic influence on public health. Edible flowers are a natural product that can be found all over the world. The vast majority of them have multiple phytochemicals that are good for your health, which is becoming more and more interesting. Numerous edible flowers have been used as a healthy diet for a long time. Recent research has shown that they are good for your health and found the active molecules and how they work to treat illness. Also, studies have been done to find out if common edible flowers are safe to eat, the best way to use them, and how much to use. Yet, due to the vast number of edible flowers that may be found across the globe, only a small portion of these flowers have been researched. We think that more research should be done to find the best ways to use edible flowers. This would make edible flowers more likely to be used as food ingredients and would also help to avoid any possible risks. Edible flowers have renewed interest in them, both because they look nice and because they may have health benefits. As their popularity has grown, more attention has been paid to their nutritional profiles, therapeutic potential, and natural bioactive components. Edible flowers have a wide range of phytonutrients. Some of these have been shown to reduce inflammation and act as antioxidants, which are two of the main reasons why we age and get sick over time. The composition and bioactivity of edible flowers have been studied. However, further data on the security and use of edible flowers is required before they can be termed functional foods or a part of our diet. Edible flowers are getting more attention in the present day as a result of their remarkable nutritional potential. Nutritionally and phytochemically, edible flowers are a great resource. The prominent phytochemicals found in edible flowers include phenolic compounds, carotenoids, terpenoids, flavonoids, and anthocyanins. Effective therapeutic potential such as anti-diabetic, anti-cancer, cytotoxicity, anti-inflammatory, antibacterial, hepatoprotective, analgesic, antimicrobial, antiulcer, and neuroprotective is present in edible flowers. The worldwide culinary and pharmaceutical industries are increasingly turning to edible flowers as a source of nutritionally rich ingredients, creating huge potential for the flower-based edibles market. Hence, the discovery of edible flowers has led to new opportunities for combating hunger, expanding agricultural production, creating new streams of revenue, and saving endangered species of wild flowers. Researchers focusing on edible flowers and their applications may find the current study a useful resource, as it covers a thorough understanding of all these issues.

References

1. Toppo P, Raj A, Harshlata (2014) Biodiversity of woody perennial flora in Badal Khole sanctuary of Jashpur district in Chhattisgarh. J Environ Biosci 28:217–221
2. Raj A (2015) Evaluation of gummosis potential using various concentration of ethephon. MSc thesis, IGAU Raipur Chhattisgarh India

3. Raj A (2015) Gum exudation in *Acacia nilotica*: effects of temperature and relative humidity. In: Proceedings of the National Expo on Assemblage of Innovative ideas/work of post graduate agricultural research scholars, Agricultural College and Research Institute, Madurai (Tamil Nadu)
4. Raj A, Haokip V, Chandrawanshi S (2015) *Acacia nilotica:* a multipurpose tree and source of Indian gum Arabic. South Indian J Biol Sci 1:66–69
5. Raj A, Toppo P (2014) Assessment of floral diversity in Dhamtari District of Chhattisgarh. J Plant Dev Sci 6:631–635
6. Jhariya MK, Raj A (2014) Effects of wildfires on flora, fauna and physico-chemical properties of soil- an overview. J Appl Nat Sci 6:887–897
7. Jhariya MK, Raj A (2014) Human welfare from biodiversity. Agrobios Newsl 12:89–91
8. Jhariya MK, Raj A, Sahu KP, Paikra PR (2013) Neem- a tree for solving global problem. Indian J Appl Res 3:66–68
9. Raj A, Jhariya MK (2016) Wasteland development through forestry. Van Sangyan 3:30–33
10. Raj A, Jhariya MK, Pithoura F (2014) Need of agroforestry and impact on ecosystem. J Plant Dev Sci 6:577–581
11. Raj A, Jhariya MK, Toppo P (2016) Scope and potential of agroforestry in Chhattisgarh state, India. Van Sangyan 3:12–17
12. Singh NR, Jhariya MK, Raj A (2013) Tree crop interaction in agroforestry system. Readers Shelf 10:15–16
13. Painkra VK, Jhariya MK, Raj A (2015) Assessment of knowledge of medicinal plants and their use in tribal region of Jashpur district of Chhattisgarh, India. J Appl Nat Sci 7:434–442
14. Lal S, Gupta DK, Dewangan B, Koreti D, Sunanda PK, Lal M, Kumar M, Dewangan B, Dewangan S, Pusphanjali TB, Thalendra KD, Tara TK, Sahu P, Kunali YS (2017) Some edible plants of Bhoramdeo wild life sanctuary Kabirdham, Chattisgarh, India. Indian J Sci Res 13: 236–247
15. Mlcek J, Rop O (2011) Fresh edible flowers of ornamental plants-a new source of nutraceutical foods. Trends Food Sci Technol 22:561–569. https://doi.org/10.1016/j.tifs.2011.04.006
16. Lu B, Li M, Yin R (2016) Phytochemical content, health benefits, and toxicology of common edible flowers: a review. Crit Rev Food Sci Nutr 56:130–148. https://doi.org/10.1080/10408398.2015.1078276
17. Egebjerg MM, Olesen PT, Eriksen FD, Ravn-Haren G, Bredsdorff L, Pilegaard K (2018) Are wild and cultivated flowers served in restaurants or sold by local producers in Denmark safe for the consumer? Food Chem Toxicol 120:129–142. https://doi.org/10.1016/j.fct.2018.07.007
18. Mahidol C, Prawat H, Prachyawarakorn V (2002) Investigation of some bioactive Thai medicinal plants. Phytochem Rev 1:287–297. https://doi.org/10.1023/A:1026085724239
19. Cruz-Garcia GS, Struik PC, Johnson DE (2011) Wild harvest: distribution and diversity of wild food plants in rice ecosystems of Northeast Thailand, NJAS-Wagen. J Life Sc 78:1–11. https://doi.org/10.1016/j.njas.2015.12.003
20. Sharifi-Rad J, Ayatollahi SA, Varoni EM (2017) Chemical composition and functional properties of essential oils from *Nepeta schiraziana* Boiss. Farmacia 65:802–812
21. Suksathan R, Rachkeeree A, Puangpradab R, Kantadoung K, Sommano SR (2021) Phytochemical and nutritional compositions and antioxidants properties of wild edible flowers as sources of new tea formulations. NFS J 24:15–25. https://doi.org/10.1016/j.nfs.2021.06.001
22. Anonymous (2023) Map of Chhattisgarh India. Google.com. https://in.pinterest.com/pin/882635226950697929/. Accessed 15 May 2023
23. Chen NH, Wei S (2017) Factors influencing consumers' attitudes towards the consumption of edible flowers. Food Qual Prefer 56:93–100. https://doi.org/10.1016/j.foodqual.2016.10.001
24. Fernandes L, Jorge A, Saraiva JAP, Susana C, Elsa R (2019) Post-harvest technologies applied to edible flowers: a review. Food Rev Intl 35:132–154. https://doi.org/10.1080/87559129.2018.1473422
25. Sasi R, Rajendran A, Maharajan M (2011) Wild edible plant diversity of Kotagiri Hills – a part of Nilgiri Biosphere Reserve, Southern India. J Res Biol 2:80–87

26. Rop O, Mlcek J, Jurikova T, Neugebauerova J (2012) Edible flowers-a new promising source of mineral elements in human nutrition. Molecules 17:6672–6683. https://doi.org/10.3390/molecules17066672
27. De LC (2020) Popular edible flowers for immunity development of individuals. Agric Obs 1: 4–9. https://doi.org/10.31782/IJCRR.2021.13901
28. Netam N (2021) Edible flower cultivation: a new approach in floriculture industry. Pharma Innov J 10:857–859. https://doi.org/10.22271/tpi.2021.v10.i3l.5896
29. Shrivastava M (2016) Study of some wild edible plants of Bastar District with special reference to Muriya tribes. Indian J Appl Pure Bio 31:23–26
30. Ekka NS, Ekka A (2016) Wild edible plants used by Tribals of north-East Chhattisgarh, India. Res J Rec Sci 5:127–131
31. Kashyap NK, Hait M, Roymohapatra G, Vaisnav MM (2022) Proximate and elemental analysis of *Careya arborea* plants root. ES Food Agrofor 7:41–47. https://doi.org/10.30919/esfa620
32. CSIR-NISCAIR (2020) Raw materials herbarium and museum Delhi (RHMD). e-bulletin 4: 1–26
33. Sumon MM (2020) Characterization, proximate composition and mineral profile analysis of different sunflower (*Helianthus annuus* L.) genotypes. MSc thesis University of Dhaka
34. Shukla A (2021) Ethnic food culture of Chhattisgarh state of India. J Ethn Food 8:1–16. https://doi.org/10.1186/s42779-021-00103-6
35. Chauhan D, Shrivastava AK, Patra S (2014) Diversity of leafy vegetables used by tribal peoples of Chhattisgarh, India. Int J Curr Microbiol App Sci 3:611–622
36. Ajay B, Sharad N, Deo S (2014) Wild edible tuber and root plants available in Bastar region of Chhattisgarh. Int J For Crop Improv 5:85–89
37. Sachdeva S, Sachdev TR, Sachdeva R (2013) Increasing fruit and vegetable consumption: challenges and opportunities. Indian J Commun Med 38:192–197. https://doi.org/10.4103/0970-0218.120146
38. Meenakshi JV (2016) Trends and patterns in the triple burden of malnutrition in India. Agric Econ 47:115–134. https://doi.org/10.1111/agec.12304
39. Singh A, Singh RK, Sureja AK (2007) Cultural significance and diversities of ethnic foods of Northeast India. Indian J Tradit Knowl 6(1):79–94
40. Pragya S, Khan M, Hailemariam H (2017) Nutritional and health importance of *Hibiscus sabdariffa*: a review and indication for research needs. J Nutr Health Food Eng 6:00212. https://doi.org/10.15406/jnhfe.2017.06.00212
41. Shukla A, Shukla A (2020) Changing food trend and associated health risks. Int J Food Nutr Sci 9:39–41. https://doi.org/10.4103/IJFNS.IJFNS_29_20
42. Vecchio MG, Paramesh EC, Paramesh H, Loganes C, Ballali S, Gafare CE, Verduci E, Gulati A (2014) Types of food and nutrient intake in India: a literature review. Indian J Pediatr 81:17–22. https://doi.org/10.1007/s12098-014-1465-9
43. FAO (1989) Forestry and nutrition- a reference manual. FAO Regional Office, Bangkok
44. Lara-Cortes E, Osorio-Diaz P, Jimenez-Aparicio A, Bautista-Banos S (2013) Nutritional content, functional properties and conservation of edible flowers. Arch Latinoam Nutr 63: 197–208
45. Alasalvar C, Pelvan E, Ozdemir KS, Kocadagh T, Mogol BA, Pash AA, Ozcan N, Ozçelik B, Gokmen V (2013) Compositional, nutritional, and functional characteristics of instant teas produced from low- and high-quality black teas. J Agric Food Chem 61:7529–7536
46. Sotelo A, López-García S, Basurto-Peña F (2007) Content of nutrients and antinutrients in edible flowers of wild plants in Mexico. Plant Foods Hum Nutr 62:133–138
47. Nnam NM, Onyeke NG (2003) Chemical composition of two varieties of sorrel (*Hibiscus sabdariffa* L.) calyces and the drinks made from them. Plant Food Hum Nutr 58:1–7
48. Pinakin DJ, Kumar V, Kumar A, Gat Y, Suri S, Sharma K (2018) Mahua: a boon for pharmacy and food industry. Curr Res Nutr Food Sci 6:1–12

49. Chauhan ES, Singh A, Tiwari A (2017) Comparative studies on nutritional analysis and phytochemical screening of *Bombax ceiba* bark and seeds powder. J Med Plant Stud 5:129–132
50. Aziz S, Nur HP, Ahmed S, Ahsan A, Siddique AB, Saha K (2016) Proximate and mineral compositions of leaves and seeds of Bangladeshi *Bombax ceiba* Linn. World J Pharm Res 5:1–13
51. Srivastava S, Yadav, Chauhan ES (2018) Nutraceutical potential of developed cake incorporated by *Butea monosperma* flower's powder. J Adv Res Appl Sci 5:608–614
52. Lilhare T, Kawale M (2019) Pharmacognostic and phytochemical studies on *Holarrhena antidysenterica* (Roth) wall. Global J Bios Biotech 8:162–167
53. Chandra S, Saklani S, Mishra AP, Rana G (2014) Nutritional, anti-nutritional profile and phytochemical screening of flowers of *Indigofera tinctoria* from Garhwal Himalaya. Inter J Herb Med 1:23–27
54. Khairunnuur FA, Zulkhairi A, Azrina A, Moklas MAM, Khairullizam S, Zamree MS, Shahidan MA (2009) Nutritional composition, in vitro antioxidant activity and *Artemia salina* L. lethality of pulp and seed of *Tamarindus indica* L. extracts. Malays J Nutr 15:65–75
55. Javed MA, Saleem M, Yamin M, Chaudri TA (1999) Lipid and protein constituents of *Crotolaria juncea* L. Nat Prod Sci 5:148–150
56. Rabha MG, Kadam AAA, Awad M, Abdel-Rahim (2021) Photochemical screening, proximate analysis and peroxide value of leaves and fruits of the tree *Cassia fistula* (L). Sudan J Sci Technol 22:1–6
57. Prabhu S, Vijayakumar S, Ramasubbu R, Praseetha PK, Karthikeyan K, Thiyagarajan G, Sureshkumar J, Prakash N (2021) Traditional uses, phytochemistry and pharmacology of *Bauhinia racemosa* Lam.: a comprehensive review. Future J Phar Sci 7:101. https://doi.org/10.1186/s43094-021-00251-1
58. Fatima M, Ahmed S, Maaz UAS, Muhammad MUH (2021) Medicinal uses, phytochemistry and pharmacology of *Bauhinia racemosa* lam. J Pharmacogn Phytochem 10:121–124. https://doi.org/10.22271/phyto.2021.v10.i2b.13972
59. Raj RD, Arun (2018) Nutritional composition of fruits of *Alangium salviifolium*, ssp Sundanum (Miq.) Bloemp.: an underutilized edible fruit plant. J Pharmacogn Phytochem 7: 3145–3148
60. Liu M-S, Ko M-H, Li H-C, Tsai S-J, Lai Y-M, Chang Y-M, Wu M-T, Long-Fang O, Chen (2014) Compositional and proteomic analyses of genetically modified broccoli (*Brassica oleracea* var. italica) harboring an agrobacterial gene. Int J Mol Sci 15:15188–15209. https://doi.org/10.3390/ijms150915188
61. Niyas N, Jasmine RK, Ally K (2021) Nutrient evaluation of *Couroupita guianensis* fruits and flowers and effect of feeding *Couroupita guianensis* flowers on growth and haemato – biochemical parameters in Wistar rats. J Vet Anim Sci 52:32–35. https://doi.org/10.51966/jvas.2021.52.1.32-35
62. Patra PA, Basak UC (2017) Nutritional and antinutritional properties of *Carissa carandas* and *Cordia dichotoma*, two medicinally important wild edible fruits of Odisha. J Basic Appl Sci Res 7:1–12
63. Biezanowska-Kopec R, Ambroszczyk AM, Piątkowska E, Leszczyńska T (2022) Nutritional value and antioxidant activity of fresh pumpkin flowers (*Cucurbita* sp.) grown in Poland. Appl Sci 12:6673. https://doi.org/10.3390/app12136673
64. Madane P, Das AK, Pateiro M, Nanda PK, Bandyopadhyay S, Jagtap P, Barba FJ, Shewalkar A, Maity B, Lorenzo JM (2008) Drumstick (*Moringa oleifera*) flower as an antioxidant dietary fibre in chicken meat nuggets. Foods 8:307. https://doi.org/10.3390/foods8080307
65. Salami SO, Afolayan AJ (2021) Evaluation of nutritional and elemental compositions of green and red cultivars of roselle: *Hibiscus sabdarifa* L. Sci Rep 11:1030. https://doi.org/10.1038/s41598-020-80433-8
66. Balarabe MA (2019) Nutritional analysis of *Hibiscus sabdariffa* L. (Roselle) leaves and calyces. Plant 7:62–65. https://doi.org/10.11648/j.plant.20190704.11

67. Fingolo CE, Braga JMA, Vieira ACM, Moura MRL, Kaplan MAC (2012) The natural impact of banana inflorescences (*Musa acuminate*) on human nutrition. An Acad Bras Cienc 84:891–898. https://doi.org/10.1590/s0001-37652012005000067
68. Bahuguna A, Vijayalaxmi KG, Suvarna VC (2018) Formulation and evaluation of fresh red Hawaiian hibiscus (*Hibiscus rosa-sinensis*) incorporated valued added products. Int J Curr Microbiol App Sci 7:4282–4290
69. Wijewardana RMNA, Nawarathne SB, Wikramasinghe I (2015) Evaluation of physicochemical and antioxidant property of dehydrated hibiscus (*Hibiscus rosa-sinensis*) flower petals and its stability in product preparation. Int J Innov Res Technol 2:179–185
70. Sivapriya T (2022) Formulation of Rosa nutballs and evaluation of its proximate principles, phytochemical components, and antioxidant property. Int J Food Nutr Sci 11:2137–2146
71. Navarro-Gonzalez I, Gonzalez-Barrio R, Garcia-Valverde V, Bautista-Ortin AB, Periago MJ (2015) Nutritional composition and antioxidant capacity in edible flowers: characterisation of phenolic compounds by HPLC-DAD-ESI/MSn. Int J Mol Sci 16:805–822. https://doi.org/10.3390/ijms16010805
72. Koche D, Shirsat R, Kawale M (2016) An overerview of major classes of phytochemicals: their types and role in disease prevention. Hislopia J 9:1–11
73. Chen GL, Chen SG, Xiao Y, Fu NL (2018) Antioxidant capacities and total phenolic contents of 30 flowers. Ind Crop Prod 111:430–445. https://doi.org/10.1016/j.indcrop.2017.10.051
74. Chkhikvishvili I, Sanikidze T, Gogia N, Enukidze M, Machavariani M, Kipian N, Rodov V (2016) Constituents of French Marigold (*Tagetes patula* L.) flowers protect jurkat t-cells against oxidative stress. Oxidative Med Cell Longev 4216285:1–16. https://doi.org/10.1155/2016/4216285
75. Moliner C, Barros L, Dias MI, Lopez V, Langa E, Ferreira IC (2018) Edible flowers of *Tagetes erecta* L. as functional ingredients: phenolic composition, antioxidant and protective effects on *Caenorhabditis elegans*. Nutrients 10:2002. https://doi.org/10.3390/nu10122002
76. Yasukawa K, Kasahara Y (2013) Effects of flavonoids from French Marigold (Florets of *Tagetes patula* L.) on acute inflammation model. Int J Inflam 53:572–578. https://doi.org/10.1155/2013/309493
77. Kaur G, Alamb MS, Jabba Z, Javed K, Athar M (2006) Evaluation of antioxidant activity of *Cassia siamea* flowers. J Ethnopharmacol 108:340–348. https://doi.org/10.1016/j.jep.2006.05.021
78. Youwei Z, Jinlian Z, Yonghong P (2008) A comparative study on the free radical scavenging activities of some fresh flowers in southern China. LWT Food Sci Technol 41:1586–1591. https://doi.org/10.1016/j.lwt.2007.10.010
79. Song L, Wang X, Zheng X, Huang D (2011) Polyphenolic antioxidant profiles of yellow camellia. Food Chem 129:351–357. https://doi.org/10.1016/j.foodchem.2011.04.083
80. Issa AY, Volate SR, Wargovich MJ (2006) The role of phytochemicals in inhibition of cancer and inflammation: new directions and perspectives. J Food Compos Anal 19:405–419
81. Chiva-Blanch G, Visioli F (2012) Polyphenols and health: moving beyond antioxidants. J Berry Res 2:63–71. https://doi.org/10.3233/JBR-2012-028
82. Niizu PY, Rodriguez-Amaya DB (2005) Flowers and leaves of *Tropaeolum majus* L. as rich sources of lutein. J Food Sci 70:605–609
83. Martinez R, Diaz B, Vasquez L, Compagnone RS, Tillet S, Canelon DJ, Torrico F, Suarez AI (2009) Chemical composition of essential oils and toxicological evaluation of *Tagetes erecta* and *Tagetes patula* from Venezuela. J Essent Oil Res 12:476–481
84. Hellinger R, Koehbach J, Fedchuk H, Sauer B, Huber R, Gruber CW, Gründemann C (2014) Immunosuppressive activity of an aqueous *Viola tricolor* herbal extract. J Ethnopharmacol 151:299–306. https://doi.org/10.1016/j.jep.2013.10.044
85. Awad AB, Fink CS (2000) Phytosterols as anticancer dietary components: evidence and mechanism of action. J Nutr 130:2127–2130

86. Bouic PJ (2001) The role of phytosterols and phytosterolins in immune modulation: a review of the past 10 years. Curr Opin Clin Nutr Metab Care 4:471–475. https://doi.org/10.1097/00075197-200111000-00001
87. Liu CL, Wang JM, Chu CY, Cheng MT, Tseng TH (2002) In-vivo protective effect of protocatechuic acid on tert-butyl hydroperoxide-induced rat hepatotoxicity. Food Chem Toxicol 40:635–641. https://doi.org/10.1016/s0278-6915(02)00002-9
88. Liu JY, Chen CC, Wang WH, Hsu JD, Yang MY, Wang CJ (2006) The protective effects of *Hibiscus sabdariffa* extract on CCl4-induced liver fibrosis in rats. Food Chem Toxicol 44:336–343. https://doi.org/10.1016/j.fct.2005.08.003
89. Kumari P, Ujala, Bhargava B (2021) Phytochemicals from edible flowers: opening a new arena for healthy lifestyle. J Funct Foods 78:104375. https://doi.org/10.1016/j.jff.2021.104375
90. Kumari P, Raju DVS, Prasad KV, Singh KP, Saha S, Arora A (2017) Quantification and correlation of anthocyanin pigments and their antioxidant activities in rose (*Rosa hybrida*) varieties. Indian J Agric Sci 87:1340–1346. https://doi.org/10.56093/ijas.v87i10.74991
91. Cendrowski A, Scibisz I, Mitek M, Kieliszek M, Kolniak-Ostek J (2017) Profile of the phenolic compounds of *Rosa rugosa* petals. J Food Qual 7941347:1–11. https://doi.org/10.1155/2017/7941347
92. Wan H, Yu C, Han Y, Guo X, Ahmad S, Tang A (2018) Flavonols and carotenoids in yellow petals of rose cultivar (Rosa 'Sun City'): a possible rich source of bioactive compounds. J Agric Food Chem 66:4171–4181. https://doi.org/10.1021/acs.jafc.8b01509
93. Bhave A, Schulzova V, Libor M, Hajslova J (2020) Influence of harvest date and post-harvest treatment on carotenoids and flavonoids composition in French marigold flowers. J Agric Food Chem 68:7880–7889. https://doi.org/10.1021/acs.jafc.0c02042
94. Ingkasupart P, Manochai B, Song WT, Hong JH (2015) Antioxidant activities and lutein content of 11 marigold cultivars (*Tagetes* spp.) grown in Thailand. Food Sci Technol 35:380–385. https://doi.org/10.1590/1678-457X.6663
95. Pavelkova P, Krmela A, Schulzova V (2020) Determination of carotenoids in flowers and food supplements by HPLC-DAD. Acta Chimica Slovaca 13:6–12. https://doi.org/10.2478/acs-2020-0002
96. Zhang J, Rui X, Wang L, Guan Y, Sun X, Dong M (2014) Polyphenolic extract from Rosa rugosa tea inhibits bacterial quorum sensing and biofilm formation. Food Control 42:125–131. https://doi.org/10.1016/j.foodcont.2014.02.001
97. Benayad Z, Martinez-Villaluenga C (2014) Phenolic composition, antioxidant and anti-inflammatory activities of extracts from Moroccan *Opuntia ficus-indica* flowers obtained by different extraction methods. Ind Crop Prod 62:412–420
98. Kumari P, Raju DVS, Singh KP, Prasad KV, Panwar S (2018) Characterization of phenolic compounds in petal extracts of rose. Indian J Hortic 75:349–351. https://doi.org/10.5958/0974-0112.2018.00060.9
99. Sarangowa O, Kanazawa T, Nishizawa M, Myoda T, Bai C, Yamagishi T (2014) Flavonol glycosides in the petal of *Rosa* species as chemotaxonomic markers. Phytochemistry 107:61–68. https://doi.org/10.1016/j.phytochem.2014.08.013
100. Wan H, Yu C, Han Y, Guo X, Luo L, Pan H (2019) Determination of flavonoids and carotenoids and their contributions to various colors of rose cultivars (*Rosa* spp.). Frontiers. Plant Sci 10:123. https://doi.org/10.3389/fpls.2019.00123
101. Ryu J, Nam B, Kim BR, Kim SH, Jo YD, Ahn JW (2019) Comparative analysis of phytochemical composition of gamma-irradiated mutant cultivars of *Chrysanthemum morifolium*. Molecules 24:3003. https://doi.org/10.3390/molecules24163003
102. Cunja V, Mikulic-Petkovsek M, Stampar F, Schmitzer V (2014) Compound identification of selected Rose species and cultivars: an insight to petal and leaf phenolic profiles. J Am Soc Hortic Sci 139:157–166. https://doi.org/10.21273/JASHS.139.2.157
103. Lee MH, Nam TG, Lee I, Shin EJ, Han AR, Lee P (2018) Skin anti-inflammatory activity of rose petal extract (*Rosa gallica*) through reduction of MAPK signaling pathway. Food Sci Nutr 6:2560–2567. https://doi.org/10.1002/fsn3.870

104. Cao X, Xiong X, Xu Z, Zeng Q, He S, Yuan Y (2020) Comparison of phenolic substances and antioxidant activities in different varieties of chrysanthemum flower under simulated tea making conditions. J Food Meas Charact 12:1–8. https://doi.org/10.1007/s11694-020-00394-4
105. Nowak R, Olech M, Pecio Ł, Oleszek W, Los R, Malm A (2014) Cytotoxic, antioxidant, antimicrobial properties and chemical composition of rose petals. J Sci Food Agric 94:560–567. https://doi.org/10.1002/jsfa.6294
106. Kao FJ, Chiang WD, Liu HM (2015) Inhibitory effect of daylily buds at various stages of maturity on nitric oxide production and the involved phenolic compounds. LWT-Food Sci Technol 61(1):130–137. https://doi.org/10.1016/J.LWT.2014.11.023
107. Lee J, Park G, Chang YH (2019) Nutraceuticals and antioxidant properties of *Lonicera japonica* Thunb. as affected by heating time. Int J Food Prop 22:630–645. https://doi.org/10.1080/10942912.2019.1599389
108. Nakamura S, Nakashima S, Tanabe G, Oda Y, Yokota N, Sakuma R, Yasikawa M (2013) Alkaloid constituents from flower buds and leaves of sacred lotus (*Nelumbo nucifera*, Nymphaeaceae) with melanogenesis inhibitory activity in B16 melanoma cells. Bioorg Med Chem Lett 21:779–787. https://doi.org/10.1016/j.bmc.2012.11.038
109. Lee DG, Lee SM, Bang MH, Park HJ, Lee TH, Kim YH, Kim JY, Baek NI (2011) Lignans from the flowers of *Osmanthus fragrans* var. aurantiacus and their inhibition effect on NO production. Arch Pharm Res 34:2029–2035. https://doi.org/10.1007/s12272-011-1204-y
110. Seo Y (2010) Antioxidant activity of the chemical constituents from the flower buds of *Magnolia denudata*. Biotechnol Bioprocess Eng 15:400–406
111. Kong CS, Lee JI, Kim JA, Seo Y (2011) In vitro evaluation on the antiobesity effect of lignans from the flower buds of *Magnolia denudata*. J Agric Food Chem 59:5665–5670. https://doi.org/10.1021/jf200230s
112. Lee CH, Kuo CY, Wang CJ, Wang CP, Lee YR, Hung CN, Lee HJ (2012) A polyphenol extract of *Hibiscus sabdariffa* L. ameliorates acetaminophen-induced hepatic steatosis by attenuating the mitochondrial dysfunction in vivo and in vitro. Biosci Biotechnol Biochem 76:646–651. https://doi.org/10.1271/bbb.110579
113. Petrulova-Poracka V, Repcak M, Vilkova M, Imrich J (2013) Coumarins of *Matricaria chamomilla* L.: aglycones and glycosides. Food Chem 141:54–59. https://doi.org/10.1016/j.foodchem.2013.03.004
114. Avonto C, Wang M, Avula B, Zhao J, Khan IA (2013) Hydroxylated bisabolol oxides: evidence for secondary oxidative metabolism in *Matricaria chamomilla*. J Nat Prod 76:1848–1853. https://doi.org/10.1021/np4003349
115. Kaisoon O, Konczak I, Siriamornpun S (2012) Potential health enhancing properties of edible flowers from Thailand. Food Res Int 46:563–571. https://doi.org/10.1016/j.foodres.2011.06.016
116. Petrova I, Petkova N, Ivanov I (2016) Fives edible flowers–valuable source of antioxidants in human nutrition. Int J Pharmacogn Phytochem Res 8:604–610
117. Skrajda-Brdak M, Dąbrowski G, Konopka I (2020) Edible flowers, a source of valuable phytonutrients and their pro-healthy effects-a review. Trends Food Sci Technol 103:179–199. https://doi.org/10.1016/j.tifs.2020.06.016
118. Moliner C, Barros L, Dias MI, López V, Langa E, Ferreira ICFR, Carlota GR (2018) Edible flowers of *Tagetes erecta* L. as functional ingredients, phenolic composition, antioxidant and protective effects on *Caenorhabditis elegans*. Nutrients 10:1–14. https://doi.org/10.3390/nu10122002
119. Xiong L, Yang J (2014) Phenolic compounds and antioxidant capacities of 10 common edible flowers from China. J Food Sci 79:517–525. https://doi.org/10.1111/1750-3841.12404
120. Adetutu A, Owoade AO (2013) Hepatoprotective and antioxidant effect of *Hibiscus* polyphenol rich extract (HPE) against carbon tetrachloride (CCl4)-induced damage in rats. J Adv Med Med Res 3:1574–1586. https://doi.org/10.9734/BJMMR/2013/3762
121. Zeng Y, Deng M, Peng Y (2014) Evaluation of antioxidant activities of extracts from 19 Chinese edible flowers. Springer Plus 3:315. https://doi.org/10.1186/2193-1801-3-315

122. Zheng J, Meenu M, Xu B (2019) A systematic investigation on free phenolic acids and flavonoids profiles of commonly consumed edible flowers in China. J Pharm Biome Anal 172:268–277. https://doi.org/10.1016/j.jpba.2019.05.007
123. Mato M, Onozaki T (2000) Flavonoid biosynthesis in white-flowered Sim carnations (*Dianthus caryophyllus*). Sci Hortic 84:333–347
124. Vandavasi SR, Ramaiah M (2015) In-vitro standardization of flowers of methanolic extract of *Dendrobium normale* Falc. for free radical scavenging activity. J Pharmacogn Phytochem 3 (5):107–111
125. Chensom S, Okumura H, Mishima T (2019) Primary screening of antioxidant activity, total polyphenol content, carotenoid content, and nutritional composition of 13 edible flowers from Japan. Prev Nutr Food Sci 24:171. https://doi.org/10.3746/pnf.2019.24.2.171
126. Peng A, Lin L, Zhao M, Sun B (2019) Identifying mechanisms underlying the amelioration effect of *Chrysanthemum morifolium* Ramat. Food Funct 10:8042–8055. https://doi.org/10.1039/c9fo01821b
127. Dantas AM, Mafaldo IM, de Lima Oliveira PM, dos Santos LM, Magnani M, Borges GDSC (2019) Bioaccessibility of phenolic compounds in native and exotic frozen pulps explored in Brazil using a digestion model coupled with a simulated intestinal barrier. Food Chem 274: 202–214. https://doi.org/10.1016/j.foodchem.2018.08.099
128. Lasselin J, Capuron L (2014) Chronic low-grade inflammation in metabolic disorders: relevance for behavioral symptoms. Neuroimmunomodulation 21:95–101. https://doi.org/10.1159/000356535
129. Harati E, Bahrami M, Razavi A, Kamalinejad M, Mohammadian M, Rastegar T (2018) Effects of viola tricolor flower hydroethanolic extract on lung inflammation in a mouse model of chronic asthma. Iran J Allergy Asthma Immunol 17:409–417. https://doi.org/10.18502/ijaai.v17i5.299
130. Davies K, Espley R (2013) Opportunities and challenges for metabolic engineering of secondary metabolite pathways for improved human health characters in fruit and vegetable crops. N Z J Crop Hortic Sci 41:154–177
131. Hu QF, Zhou B, Huang JM, Jiang ZY, Huang XZ, Yang LY, Gao XM, Yang GY, Che CT (2013) Cytotoxic oxepinochromenone and flavonoids from the flowe buds of *Rosa rugosa*. J Nat Prod 76:1866–1871. https://doi.org/10.1021/np4004068
132. Huang CN, Chan KC, Lin WT, Su SL, Wang CJ, Peng CH (2009) *Hibiscus sabdariffa* inhibits vascular smooth muscle cell proliferation and migration induced by high glucose: a mechanism involves connective tissue growth factor signals. J Agric Food Chem 57:3073. https://doi.org/10.1021/jf803911n
133. Gao XM, Shu LD (2013) Phenylethanoids from the flowers of *Rosa rugosa* and their biological activities. Bull Korean Chem Soc 34:246–248
134. Yu YG, He QT, Yuan K, Xiao XL, Li XF, Liu DM, Wu H (2011) In-vitro antioxidant activity of Bombax malabaricum flower extracts. Pharm Biol 49:569–576. https://doi.org/10.3109/13880209.2010.529614
135. Park JS, Chew BP, Wong TS (1998) Dietary lutein from marigold extract inhibits mammary tumor development in BALB/c mice. J Nutr 128:1650–1656. https://doi.org/10.1093/jn/128.10.1650
136. Solomon S, Muruganantham N, Senthamilselvi MM (2016) Anti-cancer activity of *Bauhinia tomentosa* (Flowers) against human liver cancer. J Pharmacogn Phytochem 5:287–294
137. Xavier AS, David DC (2019) In-vitro evaluation of antifungal and anticancer properties of *Tagetes erecta* petal extract. Biomed Pharmacol J 12:815–823. https://doi.org/10.13005/bpj/1705
138. Kaulika N, Febriansah R (2019) Chemopreventive activity of roselle's hexane fraction against breast cancer by in-vitro and in-silico study. In: Third international conference on sustainable innovation 2019-health science and nursing. Atlantis Press. https://doi.org/10.2991/icosihsn-19.2019.16
139. Nanda BL (2019) Antioxidant and anticancer activity of edible flowers. J Drug Deliv Ther 9: 290–295. https://doi.org/10.22270/jddt.v9i3-s.2996

140. Munoz M, Mazure R (2004) Obesidady sistema inmune. Nutr Hosp 19:319–324
141. Preuss HG, Echard B (2007) Inhibition by natural dietary substances of gastrointestinal absorption of starch and sucrose in rats and pigs: acute studies. Int J Med Sci 4:196–202
142. Kim JK, So H, Youn MJ, Kim HJ, Kim Y, Park C, Kim SJ, Ha YA, Chai KY, Kim SM, Park R (2007) *Hibiscus sabdariffa* L. water extract inhibits the adipocyte differentiation through the PI3-K and MAPK pathway. J Ethnopharmacol 114:260–267. https://doi.org/10.1016/j.jep.2007.08.028
143. Su D, Li S, Zhang W, Wang J, Wang J, Lv M (2017) Structural elucidation of a polysaccharide from *Lonicera japonica* flowers, and its neuroprotective effect on cerebral ischemia-reperfusion injury in rat. Int J Biol Macromol 99:350–357. https://doi.org/10.1016/j.ijbiomac.2017.02.096
144. Moliner C, Barros L, Dias MI, Reigada I, Ferreira IC, Lopez V (2019) *Viola cornuta* and *Viola* x *wittrockiana*: phenolic compounds, antioxidant and neuroprotective activities on *Caenorhabditis elegans*. J Food Drug Anal 27:849–859. https://doi.org/10.1016/j.jfda.2019.05.005
145. Kumar R, Janadri S, Kumar S, Swamy S (2015) Evaluation of antidiabetic activity of alcoholic extract of flower *Sesbania grandiflora* in alloxan induced diabetic rats. Asian J Pharm Pharmacol 1:21–26
146. Mishra AB, Usha T (2019) A review on pharmacological approach of the therapeutic property of *Madhuca longifolia*. Flower J Res Sid Med 2:61–68
147. Khan W, Gupta S, Ahmad S (2017) Toxicology of the aqueous extract from the flowers of *Butea monosperma* Lam. and it's metabolomics in yeast cells. Food Chem Toxicol 108:486–497. https://doi.org/10.1016/j.fct.2017.02.001
148. Salehi B, Ata A, Anil Kumar VN, Sharopov F, Ramirez-Alarcon K, Ruiz-Ortega A, Iriti M (2019) Antidiabetic potential of medicinal plants and their active components. Biomol Ther 9:551–559. https://doi.org/10.3390/biom9100551
149. Wanjari MM, Gangoria R, Dey YN, Gaidhani SN, Pandey NK, Jadhav AD (2016) Hepatoprotective and antioxidant activity of *Bombax ceiba* flowers against carbon tetrachloride-induced hepatotoxicity in rats. Hepatoma Res 2:144–150. https://doi.org/10.20517/2394-5079.2015.55
150. Abdelhafez OH, Fawzy MA, Fahim JR, Desoukey SY, Krischke M, Mueller MJ (2018) Hepatoprotective potential of *Malvaviscus arboreus* against carbon tetrachloride-induced liver injury in rats. PLoS One 13:e0202362. https://doi.org/10.1371/journal.pone.0202362
151. Mak YW, Chuah LO, Ahmad R, Bhat R (2013) Antioxidant and antibacterial activities of hibiscus (*Hibiscus rosa-sinensis* L.) and Cassia (*Senna bicapsularis* L.) flower extracts. J King Saud Univ Sci 25:275–282. https://doi.org/10.1016/j.jksus.2012.12.003
152. Bhatia H, Sharma YP, Manhas RK, Kumar K (2018) Traditionally used wild edible plants of district Udhampur, Jammu and Kashmir, India. J Ethnobiol Ethnomed 14:73–79. https://doi.org/10.1186/s13002-018-0272-1
153. Ray A, Ray R, Sreevidya EA (2020) How many wild edible plants do we eat-their diversity, use, and implications for sustainable food system: an exploratory analysis in India. Front Sustain Food Syst 4:56. https://doi.org/10.3389/fsufs.2020.00056
154. Kumar D, Agrawal P (2014) Antifertility effects of benzene extract of flowers of *Hibiscus rosa sinensis* L. on reproductive system in male albino rats. Indian J Appl Pure Biol 29:215–217
155. Kumar D, Agrawal P (2014) Contraceptive effect of *Hibiscus rosa sinensis* Corr. flower in male albino rats. Indian J Appl Pure Biol 29(2):211–214
156. Fernandes L, Ramalhosa E, Baptista P, Pereira JA, Saraiva JA, Casal SI (2019) Nutritional and nutraceutical composition of pansies during flowering. J Food Sci 84:490–498. https://doi.org/10.1111/1750-3841.14482
157. Rezende F, Sande D, Coelho AC, Oliveira G, Boaventura MA, Takahashi JA (2019) Edible flowers as innovative ingredients for future food development: anti-alzheimer, antimicrobial, and antioxidant potential. Chem Eng Trans 75:337–342. https://doi.org/10.3303/CET1975057

158. Fala NM, Contu S, Demasi S, Caser M, Scariot V (2020) Environmental impact of edible flower production: a case study. Agronamy 10:579. https://doi.org/10.3390/agronomy10040579
159. Prata GB, de Souza KO, Lopes MMA, Oliveira LS, Aragao FAS, Alves RE (2017) Nutritional characterization, bioactive compounds and antioxidant activity of Brazilian roses (*Rosa* spp). J Agric Sci Technol 19:929–941
160. Acikgoz (2017) Edible flowers. J Exp Agric Int 17:1–5
161. Fernandes L, Casal SI, Pereira JA, Saraiva JA, Ramalhosa E (2019) An overview on the market of edible flowers. Food Rev Int 1:1–18. https://doi.org/10.1080/87559129.2019.1639727

Plants Used in the Management and Treatment of Male Reproductive Health Issues: Case Study of Benin People of Southern Nigeria

32

Matthew Chidozie Ogwu and Moses Edwin Osawaru

Contents

Abstract

Through an ethnomedicinal survey, this case study presents a list of medicinal plants used for male reproductive health care by the Benin people of Southern Nigeria and their associated ethnobotanical knowledge and practices. Information was obtained from randomly selected traditional healers, tradomedical practitioners, and native midwives from the main tribal groups within the cities using a semistructured open-ended questionnaire. A total of 30 medicinal plants belonging to 22 plant families were identified, documented including *Alchornea cordifolia, Aloe vera*, *Bambusa vulgaris, Calophyllum inophyllum, Cananga odorata, Carica papaya, Cassia alata*, *Cassia mimosoides*, *Chasmanthera dependens, Cissampelos mucronata, Cissus populnea, Citrullus vulgaris, Citrus*

M. C. Ogwu (✉)
Goodnight Family Department of Sustainable Development, Appalachian State University, Boone, NC, USA
e-mail: ogwumc@appstate.edu

M. E. Osawaru
Department of Plant Biology and Biotechnology, Faculty of Life Sciences, University of Benin, Benin City, Nigeria

S. C. Izah et al. (eds.), *Herbal Medicine Phytochemistry*, Reference Series in Phytochemistry, https://doi.org/10.1007/978-3-031-43199-9_56

aurantifolia, *Combretum racemosum*, *Cyathula prostrata*, *Cynodon dactylon*, *Diodia scandens*, *Musa paradisiaca*, *Nauclea latifolia*, *Newbouldia laevis*, *Parkia biglobosa*, *Plumeria rubra*, *Rauwolfia vomitoria*, *Sansevieria trifasciata*, *Secamone afzelii*, *Sida acuta*, *Solanum nigrum*, *Tectona grandis*, *Trianthema portulacastrum*, and *Zea mays*. These plants contribute to the cultural well-being of indigenous people and are used in the treatment of venereal diseases like gonorrhea and syphilis, impotency, genito-urethra disorders, cleansing of the genital tracks, boosting sexual appetite and performance (as an aphrodisiac), boosting sperm counts to address pre-testicular, lifestyle and post-testicular causes and factors of male infertility. The dominant plant families are Apocynaceae and Poaceae with three plant representatives each and followed by Rubiaceae, Menispermaceae, and Fabaceae with two representatives each. Together these dominant plant families account for 41.4% of plants encountered in the study. The parts of the plant used in the improvement of male fertility by the Benin people of Southern Nigeria include leaves, fruits, shoot and root, flowers, leaves and root, root and flowers, seeds, stem, whole plant, root and bark, and bark. The medicinal plants documented in this chapter would benefit from research and technological innovations. The vast knowledge and practices associated with medicinal plants need to be acknowledged and conserved not only through sustained local but also global efforts. The chapter contributes toward understanding, promotion, documentation, and safeguarding of indigenous knowledge, practices, and plant resources used in addressing male reproductive issues.

Keywords

Infertile men · Sperm count · Venereal disease · Traditional medicine · Benin people

1 Introduction

Traditional medicine is the earliest form and the only health care means available to many Nigerians. This medical approach is used to treat various communicable and noncommunicable human diseases/ailments such as cough, dysentery, insomnia, cardiovascular diseases, mental illness, and infertility. Male infertility is the inability of a man to impregnate a fertile female over a minimum period of 6 months to 1 year [39, 46]. Some reversible and irreversible underlying conditions that may cause male infertility include reduced sperm motility, semen abnormalities, genital tract infection, high scrotal temperature (varicocele) and testicular disorder, excess alcohol and drug use (like anabolic steroids, ranitidine, antidepressant, and cimetidine) and abuse of illegal drugs, disturbance to the endocrine, systemic disease, genetic abnormalities, obesity, immunological factors, etc. [27, 31–33, 75]. Recently, Turner et al. [78] described male infertility as a women's and societal health issue because of stigma, unequal burden-sharing, and lack of assisted reproductive technology for diagnosis.

Male infertility has resulted in so many breakdowns in marriages across the globe. This reproductive health issue is increasing globally with little to no concerted effort [27, 72]. This may not be unconnected to the poorly understood mechanisms that are linked to male infertility like environmental, psychological, genetical, and biochemical mechanisms [8, 9]. The major means for diagnosing male infertility is through semen characterization to detect abnormality in quantity, appearance, quality, vigor or motility, and concentration. Over the last 50 years, evidence has shown that sperm count has been on the decrease as a consequence of increases in male infertility. Infertile men typically fail to impregnate fertile females due to a lack of enough or deformed sperm with excess abnormal morphology. Also, other factors such as psychological factors, which include stress, depression, guilt, and low self-esteem generally, contribute to male infertility.

This chapter aims to produce an inventory of the plants used by the residents of Benin City Nigeria for addressing male reproductive and sexual health conditions. A focused approach was adopted and carried out along with a nonexperimental validation of the plants used through a literature review of the phytochemicals and pharmacological information supporting the medicinal activities. This chapter documents the traditional medicinal practices for improving male fertility as well as in the treatment and management of male fertility issues and enumeration of how these plants are used (i.e., either alone or in combination).

2 Global Infertility and Infertility in Men

A global pandemic caused by infertility will have socioeconomic and developmental consequences besides individual and public health impacts. Infertility has been linked to individual and communal psychological, physical, spiritual, economic, and sociocultural decline mainly due to the distress, depression, and anxiety of being infertile [73, 80]. Infertility is a global health, socioeconomic, and development issue with no standardized reporting format. The problem threatens not only physical health but can lead to a serious impact on the mental and social well-being of infertile couples. At the moment, developing countries are disproportionately affected by infertility and cases continue to show an upward trend due to insufficient fertility care, caregivers, and facilities in low-income countries as well as effectiveness and affordability where available [54, 55]. On the other hand, Borumandnia et al. [12] reported a downward trajectory in infertility in high-income and developed countries due to a low tendency of reproduction and a high number of infertility care facilities. Generally, about 40–45% of infertility in men is due to the "male factor" i.e., alteration or decline of sperm concentration, motility, or morphology [31]. The concept of male factor is hinged on sperm and sperm quality, which is typically used as a surrogate measure of male fecundity. It is estimated to affect about 15% of couples or around 48.5 million couples worldwide with males solely responsible for 20–30% and 50% of cases, respectively [2]. Infertility is considered lower in men than in women

Infertility may be primary (when pregnancy has never been achieved) or secondary (when at least one prior pregnancy has been achieved). Secondary infertility is the most prevalent form in women due to reproductive tract infections whereas primary infertility is more common in men depending on several complex factors [18, 37, 79]. Male infertility may be linked to:

1. Pre-testicular causes and factors
2. Lifestyle factors and causes
3. Post-testicular causes and factors

Pre-Testicular Causes and Factors. These are conditions that impede the production and transportation of sperm and include

- Hypogonadotropin hypogonadism (low testosterone).
- Undiagnosed and untreated coeliac disease.
- Medications (particularly chemotherapies).
- Severe injury or diseases like diabetes, HIV AIDS, thyroid-linked diseases, Cushing syndrome, heart attack, liver or kidney failure, impaired sperm production, and chronic anemia.
- Genetic abnormalities such as Robertsonian translocation.
- Sexually transmitted diseases that scar the male reproductive system, impair functions, and block sperm passage like repeated *chlamydia trachomatis,* and gonorrhea infections. Human papillomavirus (genital warts).
- Poor nutrition affects sperm quality and quantity.
- Environmental factors like occupational or long-term, intensive exposure to chemicals and toxins (e.g., agrochemicals like herbicides and pesticides, bisphenol A, phthalates, and organochlorines and heavy metals like lead, cadmium, or arsenic) that reduce sperm count and quality, testicular functions, or alters the endocrine systems.
- Varicoceles (i.e., the enlargement of testicular veins which transports oxygen-depleted veins out of the scrotum resulting in the inefficient circulation of blood out of the organ), which occurs in about 15% of men and around 40% of infertile men and results in increased testicular temperature, which affects sperm production, movement, and shape [32, 33].
- Aging can reduce the count, motility, and genetic quality of sperm. This is a gradual process of male fertility decline.

Lifestyle Factors and Causes. These are factors that cause physical and emotional stress, which have the potential to temporarily reduce male fertility especially sperm count and quality and erectile dysfunction. Some issues that have been linked to low sperm count and quality like:

- Testicular overheating due to high fevers, saunas, and hot tubs.
- Excessive use of drugs like cocaine, heroin, and marijuana.
- Heavy alcohol consumption.

- Excessive cigarette smoking.
- Prolonged bicycling exposes the perineum (i.e., the region between the scrotum and the anus) to extreme shocks, vibrations, and scrotal injuries.

Post-Testicular Causes and Factors. These are conditions that decrease male fertility after testicular sperm production like

- Genital tract and ejaculation defects (premature, delayed, or retrograde ejaculation).
- Obstruction of the seminal ducts or ejaculatory duct obstruction.
- Vas deferens obstruction or lack of vas deferens.
- Infection like prostatitis.
- Retrograde ejaculation.
- Ejaculatory duct obstruction conditions like aspermia, prostatic pain, oligo asthenospermia and azoospermia, painful ejaculation, and hematospermia [34].
- Hypospadias – urethra not located at the tip of the penis.
- Erectile dysfunction or impotence.

3 Plants Used in Male Reproductive Health Care: Case Study of Benin City, Southern Nigeria

A survey was carried out in the six Local Government Areas that make up Benin City, Edo State (6.34° N and 5.60° E; Fig. 1) and is occupied by Benin people of Southern Nigeria to document medicinal plants used in the treatment and management of male fertility issues. Benin City lies within the humid tropical rainforest zone of Nigeria but its vegetation has been heavily altered by anthropogenic activities and human population growth and is currently a mosaic of secondary forest [17]. The climate is monsoonal with two distinct seasons (dry and rainy season) with high rainfall (2000–3000 mm), temperature (20–40 °C), and average atmospheric humidity of 28% and with radiation of 1600 h per year [56, 57]. A detailed description of the study area including soil characteristics can be found in Osawaru et al. [60–66], Osawaru and Ogwu [58, 59], Ogwu and Osawaru [43].

The survey to document traditional medicine practices and higher plants used for male reproductive care by Benin people was conducted between July and November 2021 using survey sheets for information collection, a digital camera, an audio recording device for interviews, and a mobile phone-enabled position marker. The survey focused on randomly selected and locally identifiable professional herbalists, traditional healers, native midwives, and herbal remedy sellers using semistructured, open-ended questionnaires in English and/or preferred native languages. The randomly selected respondents and interviewees are duly registered with all the relevant authorities. The questionnaires were administered to the key informants (full-time traditional healers, part-time practitioners, elderly people who know the traditional values of the plants, relatives, and acquaintances who at one time or the other used some of these plants for male reproductive health and fertility). The respondents

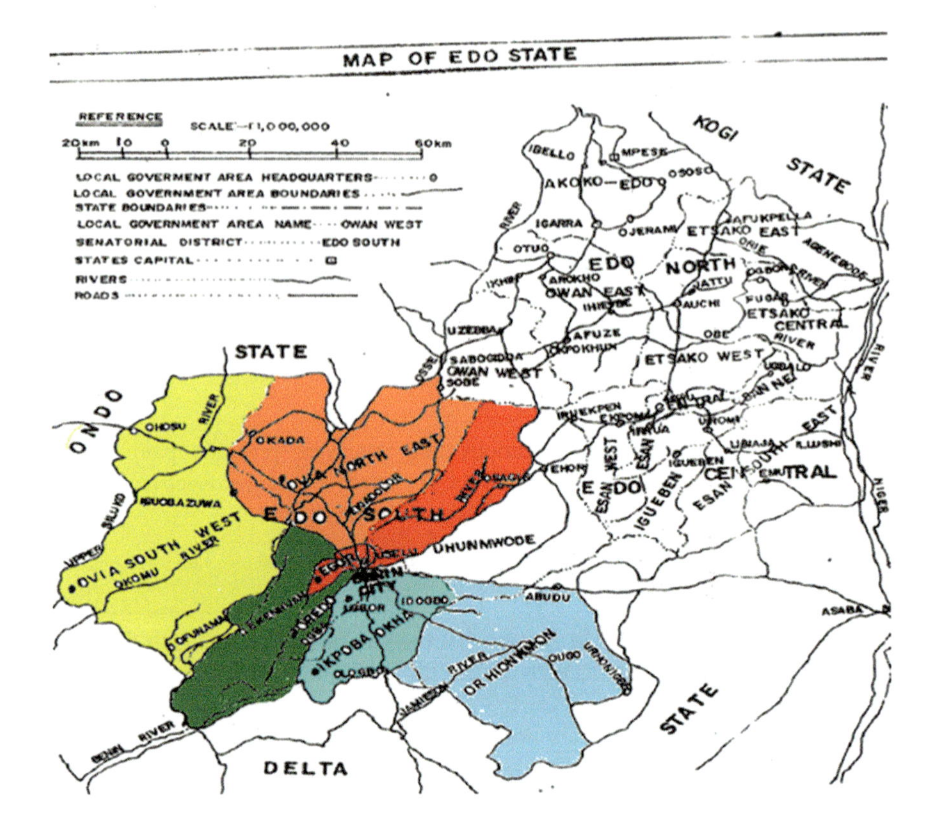

Fig. 1 Map of Edo State with the sampling area highlighted

animated a field survey to identify and collect plants and plant parts used for the treatment and management of male infertility and reproductive care. Specific questions like the botanical names, local names, plant parts used, dosage, method of preparation, application, and duration of treatment were asked during the oral interview, and the information supplied by these people was recorded. Collections were made from homestead farms, distant farms, and different forests present in the study area.

Data collected include the sociodemographic characteristics of the respondents (gender, age, marital status, educational qualification, religion, tribe/ethnic group, occupation, name of village, and languages spoken), and information on the plants used for the treatment and management of male infertility and preparation of herbal remedies for reproductive care. Data collected were processed and analyzed using Windows Microsoft Excel 2016 version software, which was also used to draw charts to understand inherent patterns. The therapeutic importance of each identified species was also recorded. Based on the knowledge of the informants, a total of 30 medicinal plants used for the treatment and management of male infertility were identified by a taxonomist at the University of Benin.

3.1 Medicinal Plants Used by Benin People for the Treatment and Management of Male Infertility and Reproductive Healthcare

From the survey, a total of 30 medicinal plants belonging to 22 plant families were identified, documented, and collected as used by Benin people for the treatment and improvement of male fertility (Tables 1 and 2). The plants included *Alchornea cordifolia, Aloe vera, Bambusa vulgaris, Calophyllum inophyllum, Cananga odorata, Carica papaya, Cassia alata, Cassia mimosoides, Chasmanthera dependens, Cissampelos mucronata, Cissus populnea, Citrullus vulgaris, Citrus aurantifolia, Combretum racemosum, Cyathula prostrata, Cynodon dactylon, Diodia scandens, Musa paradisiaca, Nauclea latifolia, Newbouldia laevis, Parkia biglobosa, Plumeria rubra, Rauwolfia vomitoria, Sansevieria trifasciata, Secamone afzelii, Sida acuta, Solanum nigrum, Tectona grandis, Trianthema portulacastrum,* and *Zea mays* (Table 1). They are used for treatment of venereal diseases like gonorrhea and syphillis, impotency, genito-urethra disorders, cleansing of the genital tracks, boosting sexual appetite and performance (as an aphrodisiac), and boosting sperm counts (Table 1). Hence, the plants can be used to address pre-testicular, lifestyle, and post-testicular causes and factors of male infertility in line with the reports of Roozbeh et al. [74] and Boroujeni et al. [11]. For greater contribution in addressing global male infertility issues, the active agents in these plants require a collaborative effort to isolate them as a prodrug candidate and elucidation at the molecular, cellular, and clinical levels to understand their mechanism of action [1, 28]. The plant families encountered in this study included Amaranthaceae, Annonaceae, Apocynaceae, Asparagaceae, Bignoniaceae, Caesalpinioideae, Calophyllaceae, Caricaceae, Combretaceae, Cucurbitaceae, Euphorbiaceae, Fabaceae, Lamiaceae, Liliaceae, Malvaceae, Menispermaceae, Musaceae, Poaceae, Rubiaceae, Rutaceae, Solanaceae, and Vitaceae (Table 2). The dominant plant families are Apocynaceae and Poaceae with three plant representatives each and followed by Rubiaceae, Menispermaceae, and Fabaceae with two representatives each (Table 2). Together these dominant plant families account for 41.4% of plants encountered in the study.

Plants possess medicinal value in addition to their nutritional importance [40–42]. From time immemorial, plants and plant extracts have been used medicinally in all parts of Nigeria especially as fertility agents without producing apparent toxic effects [39, 53]. This is backed by the report of D'Cruz et al. [16] wherein they opined that plant medicine is safe for addressing male reproductive dysfunction and other illnesses such as cancer and diabetes. Generally, there is a paucity of data and information on the safety of herbal medicine products and practices compared to modern medicine [22, 38, 69]. Where data exist, variations exist in terms of the significant toxic effect of medicinal plants on the reproductive system. The work of Nozhat et al. [38] using spearmints (Mentha spicata) suggests the absence of significant toxic effects on the reproductive system. However, toxic levels of heavy metals like cadmium, lead, arsenic, mercury, and copper have been reported in herbal medicines with significant geographical correlations in their concentrations

Table 1 Commonly used plants in the improvement of male fertility by the Benin people of Southern Nigeria

Scientific name	Family name	Common/Benin name	Parts used	Usefulness
Alchornea cordifolia	Euphorbiaceae	Christmas bush/Uwonwen	Leaves	To treat gonorrhea and other venereal diseases
Aloe vera	*Asphodelaceae*	Aloe vera/Aloe	Leaves	Treatment and management of impotency
Bambusa vulgaris	Poaceae	Bamboo/Ekpokoro	Young shoot	Treatment and management of gonorrhea and other venereal diseases
Calophyllum inophyllum	Calophyllaceae	Alexandrian laurel ball tree/ Ebe-ori	The kernel	Used in the treatment and management of gonorrhea and genito-urethra disorders
Cananga odorata	Annonaceae	Cananga tree\Erhan-uwa	Flower	Used for the treatment and management of impotency
Carica papaya	Caricaceae	Paw paw/Ughoro	Leaves and roots (used together)	For the treatment and management of venereal diseases and cleansing of the genital tracks
Cassia alata	Caesalpiniodeae	Candle bush/Asunwon	Root and flower	For the treatment and management of venereal diseases (specifically syphilis)
Cassia mimosoides	Fabaceae		Seed	It is used as an and to boost sexual performance
Chasmanthera dependens	Menispermaceae	Climbing plant/Aghoghon-ewaki	Root	A decoction made from it is taken to treat and manage gonorrhea.
Cissampelos mucronata	Menispermaceae	Abuta/Ewaki-nokhua	Leaves and roots	Leaves and roots are made into herbal juice and used as an aphrodisiac to boost sexual performance
Cissus populnea	Vitaceae	Bush mango/Iri-ogbono	Stem	Used for the treatment and management of venereal diseases
Citrullus vulgaris	Cucurbitaceae	Watermelon/Watermelon	Fruit	Fruits are eaten in large quantities as a remedy for urinary and genital tract problems arising from gravel and stone in the bladder
Citrus aurantifolia	Rutaceae	Lime/*Alimo-negiere*	Leaf	It is used to treat gonorrhea and urine retention
Combretum racemosum	Combretaceae	Christmas rose/Akoso-namwen	Root	It is used as an aphrodisiac

Cyathula prostrata	Amaranthaceae	Prostrate pasture weed/Ebe-ekperhon	Whole plant	Used for the treatment of gonorrhea
Cynodon dactylon	Poaceae	Stubborn grass/Iruvvba-ebo	Root	It is used in the treatment of secondary syphilis and irritated bladder
Diodia scandens	Rubiaceae	Not registered/Iyekegul	Leaf	It is used in treating venereal diseases
Musa paradisiaca	Musaceae	Plantain/Oghede	Root	Root juice is used to treat gonorrhea and urine retention; also, it is used in treating some other venereal diseases
Nauclea latifolia	Rubiaceae	African peach/Ero	Root and leaf	Roots and leaves are taken to improve male fertility
Newbouldia laevis	Bignoniaceae	Boundary tree/Ikhinmwin	Root with the leaves	It is used as an aphrodisiac and to boost sexual activities
Parkia biglobosa	Fabaceae	African locust bean/Ugbore	Seed	The seeds are used as an aphrodisiac
Plumeria rubra	Apocenaceae	Temple tree/Obadan-atori	The latex	Used for the treatment and management of venereal diseases
Rauwolfia vomitoria	Apocynaceae	The poison devil's pepper/Akata	The root bark	It is used as an aphrodisiac and to improve sexual performance
Sansevieria trifasciata	Asparagaceae	*Sansevieriatrifasciata*Prain/Asparagaceae	Leaf	It is used in the treatment of sexual weaknesses and to improve sexual performance
Secamone afzelii	Apocynaceae	Secamone/Iri -egile	Whole plant	It is used to cure venereal diseases like gonorrhea
Sida acuta	*Malvaceae*	Wire weed/Aranrenvbi	Seed	Used for the treatment and management of gonorrhea
Solanum nigrum	*Solanaceae*	European black nightshade/Ebeakpe	Leaf	The leaf juice is used for the treatment of gonorrhea
Tectona grandis	Lamiaceae	Teak/Ovbiakke	Bark	It is used as an aphrodisiac agent
Trianthema portulacastrum	*Aizoaceae*	Desert horse purslane/Eyen	Whole plant	Used for the treatment and management of gonorrhea and other sexual infections
Zea mays	Poaceae	Corn/oka	Seed	Roasted or boiled seeds are taken as an aphrodisiac to improve sexual performance and boost sperm count

Table 2 Distribution of plant species encountered among higher plant family

Higher plant family	Number of plant species	Relative percentage
Amaranthaceae	1	3.45
Annonaceae	1	3.45
Apocynaceae	3	10.35
Asparagaceae	1	3.45
Bignoniaceae	1	3.45
Caesalpinioideae	1	3.45
Calophyllaceae	1	3.45
Caricaceae	1	3.45
Combretaceae	1	3.45
Cucurbitaceae	1	3.45
Euphorbiaceae	1	3.45
Fabaceae	2	6.90
Lamiaceae	1	3.45
Liliaceae	1	3.45
Malvaceae	1	3.45
Menispermaceae	2	6.90
Musaceae	1	3.45
Poaceae	3	10.35
Rubiaceae	2	6.90
Rutaceae	1	3.45
Solanaceae	1	3.45
Vitaceae	1	3.45

[7, 25, 36]. The use of these plants and herbal medicine products that contain their extract poses long-term health (cancer) risks [53].

In recent times, the use of plant and plant extracts for male fertility healthcare is on the increase because of the shifting attention from synthetic drugs to natural plant products [44, 45, 47–49]. Some of these plants are used to address sexual hormonal imbalances in males and females. Phytochemical screening of medicinal plants supports the presence of bioactive materials that can affect the regulation of conception and reproduction [20, 81]. Idu and Onyibe [26] documented a total of 300 plants belonging to 274 genera and from 77 families for the treatment of various illnesses, and among those plants listed includes those used in the promotion of male fertility among the people of Edo state. Trial and error play a huge role in the initial discovery and use of these medicinal plants for the treatment and management of male fertility issues. For instance, one or more plants are first used in the treatment or management of a single ailment (e.g., gonorrhea and some other venereal diseases) and then if the patient does not get relief from the plant within a specified period, another plant is tried, secondly, if the desired plant cannot be found on a specified time, another plant is taken as a substitute. The parts of the plant used in the improvement of male fertility by the Benin people of Southern

Table 3 Frequency of plant parts used in the improvement of male fertility by the Benin people of Southern Nigeria

Plant part	Frequency	Relative percentage
Leaves	6	20
Fruits	2	6.67
Shoot and root	4	13.33
Flowers	1	3.33
Leaves and root	4	13.33
Root and flowers	2	6.67
Seeds	4	13.33
Stem	1	3.33
Whole plant	4	13.33
Root and bark	1	3.33
Bark	1	3.33

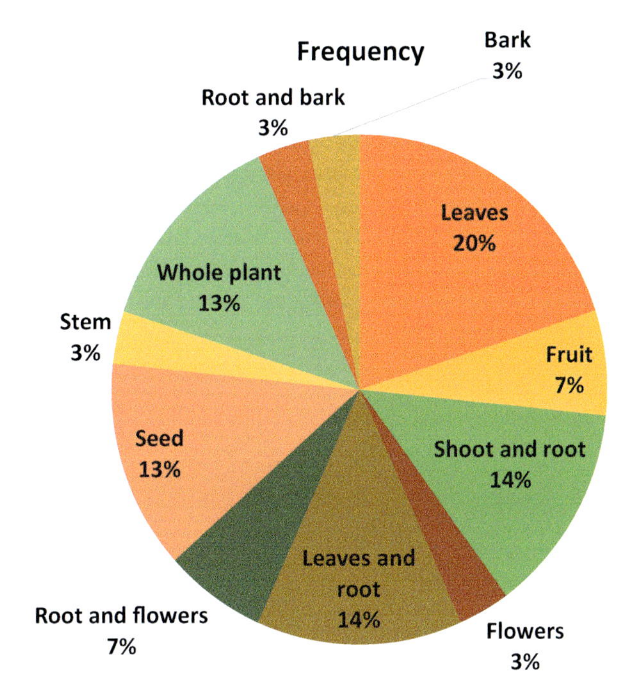

Fig. 2 Percentage of plant parts used

Nigeria include leaves, fruits, shoot and root, flowers, leaves and root, root and flowers, seeds, stem, whole plant, root and bark, and bark but the most commonly used parts are the leaves (20%), seeds (13.33%), whole plant (13.33%), leaves and root (13.33%), and shoot and root (13.33%) (Table 3 and Fig. 2). This finding is supported by the reports of Jaradat et al. [28] wherein seeds, roots, and leaves were used for infertility treatment.

3.2 Enumeration of Medicinal Plants Used by Benin People for the Treatment and Management of Male Infertility and Reproductive Healthcare

3.2.1 *Alchornia cordifolia* (Common Name – Christmas Bush and Native Name – Uwonwen)

Alchornea cordifolia, is an evergreen dioecious shrub or small tree found almost throughout tropical Africa and has fleshy seeds (Fig. 3).

PLANT PART USED: Leaves.

FOLK USE: To treat gonorrhea and other venereal diseases and is supported by the work of Boniface et al. [10]. The phytochemical constituent of the plant has significant bioactivity and a high therapeutic value.

PREPARATION OF REMEDY: It is either macerated or chewed.

DOSAGE: Not specified.

3.2.2 *Aloe vera* (*A. barbadense*) (Common Name Aloe Vera)

The plant is a cactus-like succulent perennial with strong and fibrous roots and numerous, persistent, fleshy leaves, proceeding from the upper part of the root, tapering, thick, and usually beset at the edges with spiny teeth. They can be easily cultivated especially in tropical and subtropical regions of the world. Many of the species are branching and can grow up to about 30–60 feet in height with stems as much as 10 feet in circumference (Fig. 4).

Fig. 3 *Alchornea cordifolia*

Fig. 4 *Aloe vera*

PLANT PART USED: Leaf.

FOLK USE: It is used to treat and manage impotency and to improve sperm count, motility, and other characteristics. The plant can increase testosterone levels and sperm quality and quantity [23, 52].

PREPARATION: The leaf is made into juice and taken with water.

DOSAGE: To be taken twice daily.

3.2.3 *Bambusa vulgaris* (Common Name – Bamboo, and Native Name – Ekpokoro)

The plant forms moderately loose clumps and has lemon-yellow culms (stem) with dark green leaves. The culms are not entirely straight, not easy to split, inflexible, thick-walled, and strong. The densely tufted stems can grow up to 20 m high and 10 cm thick across the base. They can be cultivated and found in the wild (Fig. 5).

PLANT PART USED: Young tender shoot.

FOLK USE: Taken to cure gonorrhea and other sexually transmitted diseases. It is used to improve reproductive vigor.

PREPARATION OF REMEDY: Shoot extract is soaked in alcohol for several days.

DOSAGE: Two shots twice a day.

3.2.4 *Calophyllum inophyllum* (Common Name – Alexandrian Laurel Ball Tree, and Native Name – Ebe-Ori)

Calophyllum inophyllum is a large evergreen plant that can grow up to 60 m tall and found in tropical regions of the world. It is a low-branching and slow-growing tree

Fig. 5 *Bambusa vulgaris*

Fig. 6 *Calophyllum inophyllum*

with a broad and irregular crown glossy oval-shaped leaves, scented flowers, and round green fruits. It usually reaches 8–20 m in height (Fig. 6).

PLANT PART USED: The fruit/kernel. Oil is produced from the kernel for use in male reproductive health.

FOLK USE: Used in the treatment of gonorrhea and genito-urethra disorders like swelling and wounds. The kernels are naturally antiseptic and anti-inflammatory [13, 70].

PREPARATION FOR USE: The kernel oil is gotten.
DOSAGE: One spoon full three times daily.

3.2.5 *Cananga odorata* (Common Name – Canaga Tree, Native Name – Erhan-Uwa)

Cananga is a fast-growing tropical tree that originates in Southeast Asia (Indonesia, Malaysia, and the Philippines) and can grow above 5 m per year. The evergreen leaves are smooth and glossy, oval, pointed, and with wavy margins and 13–20 cm long. The scented flower is drooping, long-stalked with six narrow, greenish-yellow petals (Fig. 7).

PLANT PART USED: Flowers.
FOLK USE: Cures impotency. It may be used as a male contraceptive because of its anti-spermatogenic effects [19].
PREPARATION OF REMEDY: Decoction prepared from the flowers.
DOSAGE: Two shots twice daily (morning and evening).

3.2.6 *Carica papaya* (Common Name – Paw Paw or Papaya and Native Name – Ughoro)

Papaya has a single stem growing and can reach 10 m tall, with spirally arranged leaves at the top of the trunk. The lower trunk is conspicuously scarred to mark where previous leaves and fruits were borne. The large leaves may reach 70 cm in diameter and contain a significant amount of antioxidants (Fig. 8).

Fig. 7 *Cananga odorata*

Fig. 8 *Carica papaya*

PLANT PART USED: The leaves and roots are used together.

FOLK USE: Treatment of gonorrhea and other venereal diseases. The leaves can be used as an aphrodisiac to boost libido as well as functional sterility as a contraceptive [15, 30, 35].

PREPARATION OF REMEDY: Infusion of the green leaves and roots.

DOSAGE: Three times daily (half a glass cup).

3.2.7 *Cassia alata* (Common Name – Candle Bush And Native Name – Asunwon)

Cassia alata is a shrub that can reach 4 m in height with pinnate leaves that are between 50 and 80 cm long. The leaves close in the dark while the yellow inflorescence looks like a candle. The fruits are borne in a straight pod that is approximately 25 cm long while the seeds are dispersed by water and animals (Fig. 9).

PLANT PART USED: The root and flower.

FOLK USE: For the treatment and management of venereal diseases (like syphilis). It also has hormone-boosting functions [50, 51, 77].

PREPARATION POF REMEDY: The leaves and roots are made into powder and mixed with lime.

DOSAGE: Two shots are taken twice daily.

3.2.8 *Cassia mimosoides* (Common Name – Senna Tea)

It is a low-lying to an upright diffuse shrub that can grow up to 1.5 m or more in height. The pinnate leaves are 7–10 cm long, with 40–60 pairs of narrow leaflets, and a solitary, sessile glind on the ranchis below the leaflets. Flowers grow one or two

Fig. 9 *Cassia alata*

Fig. 10 *Cassia mimosoides*

together in the axils of the leaves. Shining, small, and yellow. Pods are strap-shaped, flat, and about 5 cm long, containing rhomboid, dark-brown seeds (Fig. 10).

PLANT PART USED: Seed.

FOLK USE: Aphrodisiac to boost sexual appetite and performance and increases fertility by improving sperm characteristics.

PREPARATION OF REMEDY: Seed extract is made.
DOSAGE: Taken twice daily (half of a glass cup).

3.2.9 *Chasmanthera dependens* (Common Name – Climbing Plant And Native Name – Aghoghon-Ewaki)

The plant is a woody climber (liana) in dense evergreen or semi-deciduous lowland forest and riparian woodland of the drier Guinean zone from Sierra Leone to Nigeria, and Cameroun across Ethiopia and Somalia. They may be propagated through their seeds. The roots and leaves are considered edible (Fig. 11).

PLANT PART USED: Root.
PREPARATION OF REMEDY: Decoction prepared with root or root extracts and soaked in alcohol.
FOLK USE: The decoction is taken to treat venereal diseases like gonorrhea. It has fertility-enhancing functions and androgenic activities [21, 71].
DOSAGE: One shot is taken three times per day.

3.2.10 *Cissampelos mucronata* (Common Name – Abuta and Native Name – Ewaki-Nokhua)

Cissampelo smacronata is a liana with woody rootstock in tropical Africa. The leaves are subpeltate, ovate-cordate, and covered in grayish indumentum, mucronate at the apex, flower axillary, or in false racemes (Fig. 12).

PLANT PART USED: The leaves and roots are used.
PREPARATION OF REMEDY: The leaves and roots are made into juice for the preparation of bitter tonic.

Fig. 11 *Chasmanthera dependens*

Fig. 12 *Cissampelos macronata*

FOLK USE: For the treatment of genital-urinary diseases.
DOSAGE: Half of a glass cup twice daily.

3.2.11 *Cissus populnea* (Common Name – Bush Mango and Native Name – Iri-Ogbono)

Cissus populnea is a strong woody liana, 8–10 m long, 7 cm in diameter, dispersed generally throughout the region from the coast to the Soudanian and Sahelian woodland, of Nigeria (Fig. 13).

PLANT PART USED: Stem.
FOLK USE: Treatment of venereal diseases.
PREPARATION OF REMEDY: The decoction of the stem is mixed with the decoction prepared with *Alchornea cordifolia* root. It improves sperm quality and quantity and may be used as an aphrodisiac to enhance sexual performance [5, 50].
DOSAGE: One shot is taken twice daily (morning and evening).

3.2.12 *Citrullus vulgaris* (Common Name – Water Melon)

Citrullus vulgaris is a vine-like flowering plant originally from the Southern parts of Africa but is commonly cultivated in Central and Northern Nigeria. it is a large, sprawling oval-shaped annual plant with coarse, hairy pinnately lobed leaves and white to yellow flowers that are grown mainly for the edible fruit, which is considered a berry, botanically. The fruit is a smooth hard rind, usually green with light to dark green stripes or yellow spots, and a juicy, sweet interior flesh, usually deep red to pink, but sometimes orange, yellow, or white, with many seeds (Fig. 14).

Fig. 13 *Cissus populnea*

Fig. 14 *Citrullus vulgaris*

PLANT PART USED: Fruit.

FOLK USE: The fruit is eaten in large quantities because the pulp is considered a remedy for the urinogenital problems arising from gravel and stone in the bladder. It is considered a genital cleanser. It is believed to enhance sperm motility and kinetics [29].

PREPARATION OF REMEDY: No definite preparation. It depends on the user and is often preferred fresh or made into a juice.

DOSAGE: Not specified.

3.2.13 *Citrus aurantifolia* (Common Name – Lime and Native Name – Alimo-Negiere)

Citrus aurantifolia is a shrubby tree with many thorns. It can be grown indoors and the leaves are harvested regularly for use in the treatment and management of male reproductive issues. The trunk rarely grows straight and has many branches. The leaves are ovate and resemble orange leaves. The flowers are about 2.5 cm in diameter and are yellow-white with a light purple tinge on the margins. Flowers and fruit may be seen on the plant throughout the year, but are mostly abundant in the early rainy season in Nigeria (Fig. 15).

PLANT PART USED: Leaves.

FOLK USE: Taken to treat gonorrhea and urine retention problems. The high amount of antioxidants in the plant has a positive effect on male fertility [3].

PREPARATION OF REMEDY: Hot decoction of the leaves is prepared with water.

DOSAGE: Half a glass cup is taken three times daily.

3.2.14 *Combretum racemosum* (Common Name – Christmas Rose and Native Name – Akoso-Namwen)

It is a shrubby liane with opposite or subopposite leaves. It is common in West Africa and other wet tropical biomes (Fig. 16).

Fig. 15 *Citrus aurantifolia*

Fig. 16 *Combretum racemosum*

PLANT PART USED: The root.

FOLK USE: It is used as an aphrodisiac to improve sexual performance and appetite. Its root is known to help with infections.

PREPARATION OF REMEDY: The root is soaked in alcohol and the extract is obtained.

DOSAGE: One shot is taken three times per day.

3.2.15 *Cyathula prostata* (Common Name – Prostrate Pasture Weed and Native Name – Ebe-Ekperhon)

It is an annual to perennial erect or ascending herb that may grow up to 50 cm in height. The roots appear at the nodes while the stem is obtusely quadrangular, thickened above the nodes, often tinged with red, and covered with patent and fine hairs. The leaves are oppositely arranged, simple, and rhomboid obovate to rhomboid-oblong form and vary in size between. The inflorescence is an erect elongated raceme, terminal and in highest leaf-axis and dull pale green, hairless within and eternally clothed with patent long and white hairs. The fruit is an ellipsoid utricle, thin-walled, hairless, one-seeded, and surrounded by stiff perianth (Fig. 17).

PLANT PART USED: Whole plant.

FOLK USE: Used for the treatment of gonorrhea and other venereal diseases. It is also taken to improve sexual performance and sperm production [4]. Male sexual problems connected to obesity are treated or managed using the plant [24].

PREPARATION OF REMEDY: Decoction of the whole plant is made using water.

DOSAGE: Half a glass cup is taken three times daily.

Fig. 17 *Cyathula prostata*

3.2.16 *Cynodon dactylon* (Common Name – Stubborn Grass, Bermuda Grass, and Native Name – Iruvvba-Ebo)

The blades are gray-green and short, usually 2–5 cm long with rough edges. It is a low-statured and creeping perennial found in almost all tropical and subtropical regions. The erect stems can grow up to 30 cm tall and are slightly flattened, and often tinged purple. The seeds are produced in a cluster of six spikes at the top of the stem. It has a deep root system. The grass creeps along the ground and roots wherever a node touches the ground, forming a dense mat (Fig. 18).

PLANT PART USED: Root.

FOLK USE: It is used in the treatment of secondary syphilis and irritated bladder. It is effective in remediating stress-induced male sexual dysfunction and is believed to be a potent aphrodisiac [14].

PREPARATION OF REMEDY: Root decoction is made and soaked in water or alcohol.

DOSAGE: Two tea cups are taken per day.

3.2.17 *Diodia scandens* (Common Name – Sarmentosaswart and Native Name – Iyekegul)

Straggling, scrambling, or procumbent, often with many lateral branches from the main stem and compound leaves. The leaf blades are yellowish green, elliptic, apex, narrow to the base, and scabrid above with dense very short to longer tubercle-based hairs. Leaves are scabrid and jungle-thicket-like (Fig. 19).

Fig. 18 *Cynodon dactylon*

Fig. 19 *Diodia scandens*

PLANT PART USED: Leaf.
FOLK USE: it is used in treating venereal and urinogenital diseases.
PREPARATION OF REMEDY: Leaf decoction is prepared.
DOSAGE: Not specified.

3.2.18 *Musa paradisiaca* (Common Name – Plantain and Native Name – Oghede)

Musa paradisiaca is a large and fast-growing evergreen perennial monocotyledonous plant that can reach 9 m tall when mature. The above-ground part of the plant is a false stem or pseudostem and consists of leaves that are huge and paddle-like. Each pseudostem can produce a single flowering stem. After fruiting, the pseudostem dies, but an offshoot may develop from the base of the plant (Fig. 20).

PLANT PART USED: Root.
FOLK USE: The root juice is used to treat gonorrhea and urine retention; also, it is used in treating some other venereal diseases. It has an enhancing effect on male reproductive functions and semen quality [6].
PREPARATION OF REMEDY: Juice is made from the root.
DOSAGE: Three shots should be taken daily (morning, afternoon, and evening).

3.2.19 *Nauclea latifolia* (Common Name – African Peach and Native Name – Ero)

Nauclea latifolea is a deciduous shrub or tree with an open canopy, usually branching from low down the bole, and can reach 30 m in height depending on soil and moisture conditions. The edible fruit is gathered from the wild for local use (Fig. 21).

FOLK USE: For general male fertility and improving sexual performance.
PLANT PART USED: Root and leaves.

Fig. 20 *Musa paradisiaca*

Fig. 21 *Nauclea latifolia*

PREPARATION OF REMEDY: It is prepared by making a decoction from the root and the leaf and is typically soaked in alcohol.

DOSAGE: One glass cup is taken twice daily.

3.2.20 *Newbouldia laevis* (Common Name – Boundary Tree and Native Name – Ikhinmwin)

Newbouldia laevis is a fast-growing evergreen shrub or small tree. It only reaches a height of 8 m in the west within its range but can attain a height of up to 20 m in the east. The bole can be up to 90 cm in diameter, but it is usually less (Fig. 22).

PLANT PART USED: Root with the leaves.

FOLK USE: It is an aphrodisiac in action. It is also used to address a hormonal imbalance in men for fertility enhancement.

PREPARATION OF REMEDY: Root with leaves soaked in either palm wine or dry gin.

DOSAGE: About half a glass cup three times daily if soaked in alcohol (less than 40% alcohol) or one shot three times daily if soaked in dry gin (about 40% alcohol).

3.2.21 *Parkia biglobosa* (Common Name – African Locust Bean and Native Name – Ugbore)

Parkia biglobosa is a perennial deciduous semi-cultivated tree that can reach 20 m in height or some cases, up to 30 m. It provides a canopy for several plants that grow within its shade. The tree is considered to be fire resistant (heliophyte) and is characterized by a thick dark gray-brown bark. It has a dense spreading umbrella-

Fig. 22 *Newbouldia laevis*

Fig. 23 *Parkia biglobosa*

shaped crown and a cylindrical trunk. The pods are referred to as locust beans and are pinkish in the beginning but later turn dark brown when fully mature. Each pod can contain up to 30 seeds (Fig. 23).

PLANT PART USED: Seeds.
FOLK USE: The seeds are aphrodisiacs and improved sperm quality and quantity.
PREPARATION OF REMEDY: To be chewed.
DOSAGE: Not specified.

Fig. 24 *Plumera rubra*

3.2.22 *Plumeria rubra* (Common Name – Temple Tree and Native Name – Obadan-Atori)

Plumera rubra is a deciduous shrub or small tree that grows up to a height of 2–8 m. It has a thick succulent trunk and sausage-like branches that are covered with a thin gray bark. The branches are brittle and when broken, ooze a white latex that can be irritating to the skin and mucous membranes. The large green leaves can reach 50 cm long and are arranged alternately and clustered at the end of branches. The flowers are terminal, appearing at the end of the branches over the summer (Fig. 24).

PLANT PART USED: The latex from the bark.
FOLK USE: For the treatment of venereal diseases like syphilis and gonorrhea. The plant is believed to have a regulatory function on fertility [76].
PREPARATION OF REMEDY: The latex tapped from the bark.
DOSAGE: A very small dosage of about 0.2 g to 0.3 g is taken and not more than that is recommended for use.

3.2.23 *Rauwolfia vomitoria* (Common Name – Devil's Pepper and Native Name – Akata)

Rauwolfia vomitoria is a shrub or small tree that grows up to 8 m in height all over West Africa. The older part of the plant contains no latex, unlike the younger parts. The branches are whorled and the nodes are enlarged and lumpy. Leaves are in threes, elliptic-acuminate to broadly lanceolate. Flowers are minute and sweet-scented, and branches of the inflorescence are distinctively puberulous with hardy-free corolla lobes. Fruits are fleshy and red (Fig. 25).

Fig. 25 *Rauwolfia vomitoria*

PLANT PART USED: Root bark.
FOLK USE: It is used as an aphrodisiac. It is used to address stress-induced infertility issues in men and to improve fertility.
PREPARATION OF REMEDY: The root back is soaked in gin.
DOSAGE: One shot is taken twice daily.

3.2.24 *Sansevieria trifasciata* (Common Name – Viper's Bowstring Hemp and Native Name – Erhurhue-Ekpen)

The plant is an evergreen perennial plant and forms dense stands that are spread by way of its creeping rhizomes, which are sometimes above ground. Its stiff leaves grow vertically from a basal rosette. Mature leaves are dark green with light gray-green cross-banding and usually range between 70 cm and 90 cm long and 5 cm and 6 cm wide (Fig. 26).

PLANT PART USED: Leaf.
FOLK USE: It is used in the treatment of sexual weakness and is considered an aphrodisiac. Used to maintain sexual rigor and improve performance.
PREPARATION OF REMEDY: Decoction of the leaf with *Piper guineensis*.
DOSAGE: About half a glass cup is taken twice daily.

3.2.25 *Secamone afzelii* (Common Name – Secamone and Native Name – Iri -Egile)

It is a scandent creeping woody shrub or climber, of secondary jungle and savanna thickets, common on unkempt farmland and in boundaries. It has compound leaves (Fig. 27).

Fig. 26 *Sansevieria trifasciata*

Fig. 27 *Secamone afzelii*

PLANT PART USED: Whole plant.

FOLK USE: It is used to cure venereal diseases like gonorrhea and improve fertility.

PREPARATION OF REMEDY: Decoction is prepared using the whole plant.

DOSAGE: The whole plant is used to cook light-boiled soup and is eaten together with the soup.

Fig. 28 *Sida acuta*

3.2.26 *Sida Acuta* (Common Name – Wire Weed and Native Name – Aranrenvbi)

It is an undershrub, with mucilaginous juice, aerial, erect, cylindrical, branched, solid, and green. The leaves are alternate, simple, lanceolate to linear, rarely ovate to oblong, obtuse at the base, acute at the apex, coarsely and remotely serrate, and the petiole is shorter than the blade. The inflorescence is a racemose (Fig. 28).

PLANT PART USED: The seeds.
FOLK USE: It is used for treating venereal diseases like gonorrhea.
PREPARATION OF REMEDY: Infusion of the seed is made.
DOSAGE: Not specified.

3.2.27 *Solanum nigrum* (Common Name – European Black Nightshade and Native Name – Ebeakpe)

It is a common herb or short-lived shrub, found in many wooden areas, as well as disturbed habitats. It can reach a height of 30–120 cm while the leaves average between 4 and 7.5 cm long and about 2 and 5 cm wide and are ovate to heart-shaped, with wavy or large-toothed edges, both surface hairy or hairless. The flowers have petals greenish to whitish (Fig. 29).

PLANT PART USED: Leaves.
FOLK USE: The leaf juice is used for the treatment of gonorrhea and other venereal diseases.
PREPARATION OF REMEDY: Juice prepared with or obtained from the leaves.
DOSAGE: Half of a glass cup is taken twice daily.

Fig. 29 *Solanum nigrum*

Fig. 30 *Tectona grandis*

3.2.28 *Tectona grandis* (Common Name – Teak and Native Name – Ovbiakke)

Teak is a large deciduous tree that measures up to 40 m tall with gray to grayish brown branches. Leaves are ovate-elliptic to ovate, 15–45 cm long by 8–23 cm wide, and are held on robust petioles that are 2–4 cm long. Leaf margins are entire (Fig. 30).

PLANT PART USED: The bark.

FOLK USE: It is used as an aphrodisiac agent to improve sexual appetite and performance.

PREPARATION OF REMEDY: The bark juice is commonly extracted with and soaked in alcohol.

DOSAGE: About half a glass cup is taken daily if extracted with palm wine but one shot three times daily if extracted with gin (about 40% alcohol).

3.2.29 *Trianthema portulacastrum* (Common Name – Desert Horsepurslane and Native Name – Eyen)

It is an annual herb forming a prostrate map or clump with stems up to a meter long. It is green to red, hairless except for small lines of hairs near the leaves, and fleshy. The leaves have small round or oval blades up to 4 cm long borne on short petioles. Solitary flowers occur in leaf axils (Fig. 31).

PLANT PART USED: The whole plant.

FOLK USE: Used for the treatment of gonorrhea.

PREPARATION OF REMEDY: The whole plant is dried and pulverized for use.

DOSAGE: Not specified but it is taken with cereals like pap (ogi).

3.2.30 *Zea Mays* (Common Name – Corn, Maize, and Native Name – Oka)

It is a vigorous annual grass, varying greatly in size according to race and growth conditions, culms, stout, often with prop roots from the lower nodes, many-noded, terminated by inflorescences of the male spikelet, and with one or more female spikelet in the axils of leaves below the tassel, leaf blade expanded, usually drooping, usually green (Fig. 32).

Fig. 31 *Trianthema portulacastrum*

Fig. 32 *Zea mays*

PLANT PART USED: The seeds.

FOLK USE: Taken to boost sperm count and as an aphrodisiac to boost sexual performance and appetite.

PREPARATION OF REMEDY: Grounded seeds are taken as pap or the seeds are roasted and consumed.

DOSAGE: Not specified.

4 Conclusion

The survey reported in this case study revealed 30 medicinal plants that are used by the Benin people of Southern Nigeria for promoting male fertility. The scientific, common, and indigenous names of the plants along with the parts used, dosage (where they exist), and the methods of preparation for the treatment and management of diverse male infertility issues were documented. The plants that were identified belong to 23 families. The survey revealed that Poaceae, Apocynaceae, Rubiaceae, Menispermaceae, and Fabaceae are the dominant plant families as they have the most plant representations with at least two plants within the family used to promote male fertility among Benin people. The mode or method of administration of the plant is entirely based on oral administration either by making a juice out of the plant, by making a decoction of the plant or by infusion of the plant. The various parts of the plant used are the plant leaf, the plant stem, the plant root, the plant seed, stem bark, root bark young shoot, and fruit juice. It is worth noting that the genetic diversity of medicinal plants is continuously in decline and is increasingly threatened by extinction due to human population exploitation, unsustainable harvesting and utilization techniques, loss of habitats, and unmonitored and unregulated trade in medicinal plants. These plants contribute to the culture and well-being of indigenous people together. Medicinal plants used to treat illnesses including fertility issues would benefit from research and technological innovations. The vast knowledge and practices associated with medicinal plants need to be acknowledged and conserved not only through sustained local but also global efforts. It is important to recognize the roles played by remote communities and indigenous people as custodians of the world's medicinal plant genetic heritage.

References

1. Abarikwu SO, Onuah CL, Singh SK (2020) Plants in the management of male infertility. Andrologia 52(3):e13509. https://doi.org/10.1111/and.13509
2. Agarwal A, Mulgund A, Hamada A, Chyatte MR (2015) A unique view on male infertility around the globe. Reprod Biol Endocrinol 13:37. https://doi.org/10.1186/s12958-015-0032-1
3. Ahmadi S, Bashiri R, Ghadiri-Anari A, Nadjarzadeh A (2016) Antioxidant supplements and semen parameters: an evidence based review. Int J Reproduct Biomed 14(12):729–736
4. Ajuogu PK, Ere R, Nodu MB, Nwachukwu CU, Mgbere OO (2020) The influence of graded levels of *Cyathula prostrata* (Linn.) Blume on semen quality characteristics of adult New Zealand white bucks. Transl Anim Sci 4(2):txaa060. https://doi.org/10.1093/tas/txaa060
5. Akomolafe SF, Oboh G, Akindahunsi AA, Akinyemi AJ, Tade OG (2013) Inhibitory effect of aqueous extract of stem bark of Cissus populnea on ferrous sulphate- and sodium nitroprusside-induced oxidative stress in Rat's testes in vitro. ISRN Pharmacol 2013:130989. https://doi.org/10.1155/2013/130989
6. Alabi AS, Omotoso GO, Enaibe BU, Akinola OB, Tagoe CN (2013) Beneficial effects of low dose Musa paradisiaca on the semen quality of male Wistar rats. Niger Med J 54(2):92–95. https://doi.org/10.4103/0300-1652.110035
7. Annan K, Dickson RA, Amponsah IK, Nooni IK (2013) The heavy metal contents of some selected medicinal plants sampled from different geographical locations. Pharm Res 5(2): 103–108. https://doi.org/10.4103/0974-8490.110539
8. Assidi M (2022) Infertility in men: advances towards a comprehensive and integrative strategy for precision Theranostics. Cell 11(10):1711
9. Babakhanzadeh E, Nazari M, Ghasemifar S, Khodadadian A (2020) Some of the factors involved in male infertility: a prospective review. Int J Gen Med 13:29–41. https://doi.org/10.2147/IJGM.S241099
10. Boniface PK, Ferreira SB, Kaiser CR (2016) Recent trends in phytochemistry, ethnobotany and pharmacological significance of Alchornea cordifolia (Schumach. & Thonn.) Muell. Arg. J Ethnopharmacol 191:216–244. https://doi.org/10.1016/j.jep.2016.06.021
11. Boroujeni SN, Malamiri FA, Bossaghzadeh F, Esmaeili A, Moudi E (2022) The most important medicinal plants affecting sperm and testosterone production: a systematic review. JBRA Assist Reprod 26(3):522–530. https://doi.org/10.5935/1518-0557.20210108
12. Borumandnia N, Alavi Majd H, Khadembashi N, Alaii H (2022) Worldwide trend analysis of primary and secondary infertility rates over past decades: a cross-sectional study. Int J Reprod Biomed 20(1):37–46. https://doi.org/10.18502/ijrm.v20i1.10407
13. Cassien M, Mercier A, Thétiot-Laurent S, Culcasi M, Ricquebourg E, Asteian A, Herbette G, Bianchini JP, Raharivelomanana P, Pietri S (2021) Improving the antioxidant properties of *Calophyllum inophyllum* seed oil from French Polynesia: development and biological applications of resinous ethanol-soluble extracts. Antioxidants (Basel, Switzerland) 10(2):199. https://doi.org/10.3390/antiox10020199

14. Chidrawar V, Chitme H, Patel K, Patel NJ, Racharla V, Dhoraji N, Vadalia K (2011) Effects of Cynodon dactylon on stress-induced infertility in male rats. J Young Pharm 3(1):26–35. https://doi.org/10.4103/0975-1483.76416
15. Chinoy NJ, George SM (1983) Induction of functional sterility in male rats by low dose Carica papaya seed extract treatment. Acta Eur Fertil 14(6):425–432
16. D'Cruz SC, Vaithinathan S, Jubendradass R, Mathur PP (2010) Effects of plants and plant products on the testis. Asian J Androl 12(4):468–479. https://doi.org/10.1038/aja.2010.43
17. Dania-Ogbe FM, Adebooye OC, Bamidele JF (2001) Ethnobotany of indigenous food crops and useful plants; leafy vegetables of Southwest Nigeria; Their identification, nutritional studies and cultivation of farmer assisted selected endangered species. Paper presented at the biennial meeting of the UNU/INRA College of Research Associates, 19–20 April, 2001, Accra, Ghana
18. Deshpande PS, Gupta AS (2019) Causes and prevalence of factors causing infertility in a public health facility. J Hum Reprod Sci 12(4):287–293. https://doi.org/10.4103/jhrs.JHRS_140_18
19. Dhanapal R, Ratna JV, Sarathchandran I, Gupta M (2012) Reversible antispermatogenic and antisteroidogenic activities of Feronia limonia fruit pulp in adult male rats. Asian Pac J Trop Biomed 2(9):684–690. https://doi.org/10.1016/S2221-1691(12)60210-X
20. Edeoga HO, Okwu DE, Mbaebie BO (2005) Phytochemical constituents of some Nigerian medicinal plants. Afr J Biotechnol 4(7):685–688
21. Enenebeaku CK, Ogukwe CE, Nweke CO et al (2022) Antiplasmodial and in vitro antioxidant potentials of crude aqueous and methanol extracts of *Chasmanthera dependens* (Hochst). Bull Natl Res Centre 46:33. https://doi.org/10.1186/s42269-022-00721-3
22. Fasinu PS, Bouic PJ, Rosenkranz B (2012) An overview of the evidence and mechanisms of herb-drug interactions. Front Pharmacol 3:69. https://doi.org/10.3389/fphar.2012.00069
23. Guo X, Mei N (2016) Aloe vera: a review of toxicity and adverse clinical effects. J Environ Sci Health C Environ Carcinog Ecotoxicol Rev 34(2):77–96. https://doi.org/10.1080/10590501.2016.1166826
24. Hassan RI, Oloyede HOB, Salawu MO, Akolade JO (2021) Protective effects of Cyathula prostrata leaf extract on olanzapine-induced obese rats. J Phytopharmacol 10(2):68–79
25. Hlihor RM, Roșca M, Hagiu-Zaleschi L, Simion IM, Daraban GM, Stoleru V (2022) Medicinal plant growth in heavy metals contaminated soils: responses to metal stress and induced risks to human health. Toxics 10(9):499. https://doi.org/10.3390/toxics10090499
26. Idu M, Onyibe HI (2007) Medicinal plants of Edo state, Nigeria. Res J Med Plants 1:32–41. https://doi.org/10.3923/rjmp.2007.32.41
27. Jafari H, Mirzaiinajmabadi K, Roudsari RL, Rakhshkhorshid M (2021) The factors affecting male infertility: a systematic review. Int J Reprod Biomed 19(8):681–688. https://doi.org/10.18502/ijrm.v19i8.9615
28. Jaradat N, Zaid AN (2019) Herbal remedies used for the treatment of infertility in males and females by traditional healers in the rural areas of the West Bank/Palestine. BMC Complement Altern Med 19(1):194. https://doi.org/10.1186/s12906-019-2617-2
29. Jimoh OA, Akinola MO, Oyeyemi BF, Oyeyemi WA, Ayodele SO, Omoniyi IS, Okin-Aminu HO (2021) Potential of watermelon (*Citrullus lanatus*) to maintain oxidative stability of rooster semen for artificial insemination. J Anim Sci Technol 63(1):46–57. https://doi.org/10.5187/jast.2021.e21
30. Kaur S, Singla N, Mahal AK (2022) *Carica Papaya* modulates the organ histology, biochemicals, estrous cycle and fertility of *Bandicota bengalensis* rats. J Appl Anim Res 50(1):289–298
31. Kumar N, Singh AK (2015) Trends of male factor infertility, an important cause of infertility: a review of literature. J Hum Reprod Sci 8(4):191–196. https://doi.org/10.4103/0974-1208.170370
32. Leslie SW, Sajjad H, Siref LE (2023) Varicocele. In: StatPearls [Internet]. Treasure Island (FL): StatPearls Publishing. https://www.ncbi.nlm.nih.gov/books/NBK448113/
33. Leslie SW, Soon-Sutton TL, Khan MAB (2023) Male infertility. In: StatPearls [Internet]. Treasure Island (FL): StatPearls Publishing

34. Lira FT, Neto, Bach PV, Miranda EP, Calisto SLDS, Silva GMTD, Antunes DL, Li PS (2020) Management of ejaculatory duct obstruction by seminal vesiculoscopy: case report and literature review. JBRA Assist Reprod 24(3):382–386. Advance online publication. https://doi.org/10.5935/1518-0557.20190075
35. Lohiya NK, Goyal RB, Jayaprakash D, Ansari AS, Sharma S (1994) Antifertility effects of aqueous extract of Carica papaya seeds in male rats. Planta Med 60(5):400–404. https://doi.org/10.1055/s-2006-959518
36. Luo L, Wang B, Jiang J, Fitzgerald M, Huang Q, Yu Z, Li H, Zhang J, Wei J, Yang C, Zhang H, Dong L, Chen S (2021) Heavy metal contaminations in herbal medicines: determination, comprehensive risk assessments, and solutions. Front Pharmacol 11:595335. https://doi.org/10.3389/fphar.2020.595335
37. Masoumi SZ, Parsa P, Darvish N, Mokhtari S, Yavangi M, Roshanaei G (2015) An epidemiologic survey on the causes of infertility in patients referred to infertility center in Fatemieh Hospital in Hamadan. Iran J Reprod Med 13(8):513–516
38. Nozhat F, Alaee S, Behzadi K, Azadi Chegini N (2014) Evaluation of possible toxic effects of spearmint (Mentha spicata) on the reproductive system, fertility and number of offspring in adult male rats. Avicenna J Phytomed 4(6):420–429
39. Ogbe FMD, Eruogun OL, Uwagboe M (2009) Plants used for female reproductive health care in Oredo local government area, Nigeria. Sci Res Essay 4(3):120–130
40. Ogwu MC (2019) Towards sustainable development in Africa: the challenge of urbanization and climate change adaptation. In: Cobbinah PB, Addaney M (eds) The geography of climate change adaptation in urban Africa. Springer Nature, pp 29–55. https://doi.org/10.1007/978-3-030-04873-0_2
41. Ogwu MC (2020) Value of *Amaranthus* [L.] species in Nigeria. In: Waisundara V (ed) Nutritional value of Amaranth. IntechOpen, pp 1–21. https://doi.org/10.5772/intechopen.86990
42. Ogwu MC (2023) Local food crops in Africa: sustainable utilization, threats, and traditional storage strategies. In: Izah SC, Ogwu MC (eds) Sustainable utilization and conservation of Africa's biological resources and environment. Sustainable development and biodiversity, vol 888. Springer, Singapore, 353–376 pp. https://doi.org/10.1007/978-981-19-6974-4_13
43. Ogwu MC, Osawaru ME (2015) Soil characteristics, microbial composition of plot, leaf count and sprout studies of cocoyam (*Colocasia* [Schott] and *Xanthosoma* [Schott], Araceae) collected in Edo State, Southern Nigeria. Sci Technol Arts Res J 4(1):34–44. https://doi.org/10.4314/star.v4i1.5
44. Ogwu MC, Osawaru ME, Aiwansoba RO, Iroh RN (2016) Status and prospects of vegetables in Africa. In: Borokini IT, Babalola FD (eds) Conference proceedings of the joint biodiversity conservation conference of Nigeria Tropical Biology Association and Nigeria chapter of Society for Conservation Biology on MDGs to SDGs: toward sustainable biodiversity conservation in Nigeria. University of Ilorin, 47–57pp
45. Ogwu MC, Osawaru ME, Obadiaru O, Itsepopo IC (2016) Nutritional composition of some uncommon indigenous spices and vegetables. In: International conference on natural resource development and utilization (4th edition): positioning small and medium enterprises (SMEs) for optimal utilization of available resources and economic growth. Raw Materials Research and Development Council (Federal Ministry of Science and Technology), Maitama, pp 286–300
46. Ogwu MC, Osawaru ME, Obahiagbon GE (2017) Ethnobotanical survey of medicinal plants used for traditional reproductive care by Usen people of Edo State, Nigeria. Malaya J Biosci 4(1):17–29
47. Ogwu MC, Izah SC, Iyiola AO (2022) An overview of the potentials, threats and conservation of biodiversity in Africa. In: Izah SC (ed) Biodiversity in Africa: potentials, threats and conservation, Sustainable development and biodiversity, volume 29. Springer, Singapore, pp 3–20. https://doi.org/10.1007/978-981-19-3326-4_1
48. Ogwu MC, Iyiola AO, Izah SC (2023) Overview of African biological resources and environment. In: Izah SC, Ogwu MC (eds) Sustainable utilization and conservation of Africa's

biological resources and environment, Sustainable development and biodiversity, volume 888. Springer, Singapore, 1–34 pp. https://doi.org/10.1007/978-981-19-6974-4_1

49. Ogwu MC, Osawaru ME, Owie MO (2023) Effects of storage at room temperature on the food components of three cocoyam species (*Colocasia esculenta, Xanthosoma atrovirens, and X. sagittifolium*). Food Stud Interdiscip J 13(2):59–83. https://doi.org/10.18848/2160-1933/CGP/v13i02/59-83
50. Ogwu MC, Osawaru ME, Amodu E, Osamo F (2023) Comparative morphology, anatomy and chemotaxonomy of two *Cissus* Linn. species. Rev Bras Bot. https://doi.org/10.1007/s40415-023-00881-0
51. Oladeji OS, Adelowo FE, Oluyori AP, Bankole DT (2020) Ethnobotanical description and biological activities of *Senna alata*. Evid Based Complement Alternat Med 2020:2580259. https://doi.org/10.1155/2020/2580259
52. Olugbenga OM, Olukole SG, Adeoye AT, Adejoke AD (2011) Semen characteristics and sperm morphological studies of the West African Dwarf Buck treated with Aloe vera gel extract. Iran J Reprod Med 9(2):83–88
53. Olusola JA, Akintan OB, Erhenhi HA, Osanyinlusi OO (2021) Heavy metals and health risks associated with consumption of herbal plants sold in a major urban market in southwest, Nigeria. J Health Pollut 11(31):210915. https://doi.org/10.5696/2156-9614-11.31.210915
54. Ombelet W (2011) Global access to infertility care in developing countries: a case of human rights, equity and social justice. Facts Views Vis Obgyn 3(4):257–266
55. Ombelet W (2020) WHO fact sheet on infertility gives hope to millions of infertile couples worldwide. Facts Views Vis Obgyn 12(4):249–251
56. Omuta GE (1980) A profile of development of Bendel state of Nigeria publication in geography 1 no. 2. Department of geography and regional planning. University of Benin
57. Onwueme IC, Singh TD (1991) Field crop production in tropical Africa. CTA, Ede, the Netherlands 480p
58. Osawaru ME, Ogwu MC (2014) Ethnobotany and germplasm collection of two genera of cocoyam (*Colocasia* [Schott] and *Xanthosoma* [Schott], Araceae) in Edo State Nigeria. Sci Technol Arts Res J 3(3):23–28. https://doi.org/10.4314/star.v3i3.4
59. Osawaru ME, Ogwu MC (2015) Molecular characterization of 36 accessions of two genera of cocoyam (*Colocasia* [Schott] and *Xanthosoma* [Schott], Araceae). Sci Technol Arts Res J 4(1): 27–33. https://doi.org/10.4314/star.v4i1.4
60. Osawaru ME, Ogwu MC, Braimah L (2013) Growth responses of two cultivated Okra species (*Abelmoschus caillei* (A. Chev.) Stevels and *Abelmoschus esculentus* (Linn.) Moench) in crude oil contaminated soil. Niger J Basic Appl Sci 21(3):215–226
61. Osawaru ME, Ogwu MC, Dania-Ogbe FM (2013) Morphological assessment of the genetic variability among 53 accessions of West African Okra [*Abelmoschus caillei* (A. Chev.) Stevels] from South Western Nigeria. Niger J Basic Appl Sci 21(3):227–238
62. Osawaru ME, Ogwu MC, Ogbeifun NS, Chime AO (2013) Microflora diversity of the phylloplane of wild Okra (*Corchorus olitorius* L. Jute). Bayero J Pure Appl Sci 6(2):136–142
63. Osawaru ME, Ogwu MC, Imarhiagbe O (2013) Biochemical characterization of some Nigerian *Corchorus* L. species. Bayero J Pure Appl Sci 6(2):69–75
64. Osawaru ME, Ogwu MC, Ahana CM (2013) Current status of plant diversity and conservation in Nigeria. Niger J Life Sci 3(1):168–178
65. Osawaru ME, Ogwu MC, Imarhiagbe O (2013) Agro-morphological characterization of some Nigerian *Corchorus* (L.) species. Biol Environ Sci J Tropics 10(4):148–158
66. Osawaru ME, Ogwu MC, Ahana CM (2014) Study of the distinctiveness of two genera of cocoyam (*Colocasia* [Schott] and *Xanthosoma* [Schott]) using SDS – PAGE. Niger Soc Exp Biol J 14(3):168–179
67. Osawaru ME, Ogwu MC, Emokpare AA (2014) Preliminary assessment of the microanatomy of Okra [*Abelmoschus* (L.)] wood. Egypt Acad J Biol Sci 5(1):39–54

68. Osawaru ME, Ogwu MC, Omologbe J (2014) Characterization of three Okra [*Abelmoschus* (L.)] accessions using morphology and SDS-PAGE for the basis of conservation. Egypt Acad J Biol Sci 5(1):55–65
69. Pizzorno J (2018) Environmental toxins and infertility. Integr Med (Encinitas, Calif) 17(2):8–11
70. Pribowo A, Girish J, Gustiananda M, Nandhira RG, Hartrianti P (2021) Potential of Tamanu (*Calophyllum inophyllum*) oil for atopic dermatitis treatment. Evid Based Complement Alternat Med 2021:6332867. https://doi.org/10.1155/2021/6332867
71. Quadri AL, Yakubu MT (2017) Fertility enhancing activity and toxicity profile of aqueous extract of Chasmanthera dependens roots in male rats. Andrologia 49(10). https://doi.org/10.1111/and.12775
72. Ravitsky V, Kimmins S (2019) The forgotten men: rising rates of male infertility urgently require new approaches for its prevention, diagnosis and treatment. Biol Reprod 101(5): 872–874. https://doi.org/10.1093/biolre/ioz161
73. Rooney KL, Domar AD (2018) The relationship between stress and infertility. Dialogues Clin Neurosci 20(1):41–47. https://doi.org/10.31887/DCNS.2018.20.1/klrooney
74. Roozbeh N, Amirian A, Abdi F, Haghdoost S (2021) A systematic review on use of medicinal plants for male infertility treatment. J Family Reprod Health 15(2):74–81. https://doi.org/10.18502/jfrh.v15i2.6447
75. Shaik A, Yalavarthi PR, Bannoth CK (2017) Role of anti-fertility medicinal plants on male & female reproduction. J Complement Altern Med Res 3(2):1–22
76. Tsobou R, Mapongmetsem PM, Van Damme P (2016) Medicinal plants used for treating reproductive health care problems in Cameroon, Central Africa. Econ Bot 70:145–159. https://doi.org/10.1007/s12231-016-9344-0
77. Rowe PJ, Comhaire FH, Hargreave TB, Mahmoud AHA (2000) WHO manual for the standardized investigation and diagnosis of the infertile male, 1st edn. World Health Organization. Cambridge University Press, Cambridge, UK, 102 p
78. Turner KA, Rambhatla A, Schon S, Agarwal A, Krawetz SA, Dupree JM, Avidor-Reiss T (2020) Male infertility is a Women's health issue-research and clinical evaluation of male infertility is needed. Cell 9(4):990. https://doi.org/10.3390/cells9040990
79. Vander Borght M, Wyns C (2018) Fertility and infertility: definition and epidemiology. Clin Biochem 62:2–10. https://doi.org/10.1016/j.clinbiochem.2018.03.012
80. Walker MH, Tobler KJ (2022) Female infertility. In: StatPearls [Internet]. Treasure Island (FL): StatPearls Publishing; 2023 January. https://www.ncbi.nlm.nih.gov/books/NBK556033/
81. Yakubu MT, Akanji MA, Oladiji AT (2005) Aphrodisiac potentials of the aqueous extract of Fadogia agrestis (Schweinf. Ex Hiern) stem in male albino rats. Asian J Androl 7(4):399–404. https://doi.org/10.1111/j.1745-7262.2005.00052.x

Plants Used in the Management and Treatment of Female Reproductive Health Issues: Case Study from Southern Nigeria

33

Moses Edwin Osawaru and Matthew Chidozie Ogwu

Contents

Abstract

Through an ethnomedicinal survey, this case study presents a list of medicinal plants associated with female fertility, pre- and postnatal care among Benin and Warri residents in Southern Nigeria, and their associated ethnobotanical knowledge and practices. Information was obtained from randomly selected traditional healers, tradomedical practitioners, and native midwives from the main tribal groups within the cities using a semistructured open-ended questionnaire. A total of 24 medicinal plant species belonging to 21 families were of ethnomedicinal importance for female fertility, pre- and postnatal care of Benin and Warri

M. E. Osawaru
Department of Plant Biology and Biotechnology, Faculty of Life Sciences, University of Benin, Benin City, Edo State, Nigeria

M. C. Ogwu (✉)
Goodnight Family Department of Sustainable Development, Appalachian State University, Boone, NC, USA
e-mail: ogwumc@appstate.edu

S. C. Izah et al. (eds.), *Herbal Medicine Phytochemistry*, Reference Series in Phytochemistry, https://doi.org/10.1007/978-3-031-43199-9_5

residents in Southern Nigeria. Some of the plants include *Psidium guajava, Pennisetum purpureum, Oenothera biennis, Aloe barbadensis, Asparagus racemosus, Gossypium hirsutum, Newbouldia laevis, Phyllanthus amarus*, and *Afrormosia laxiflora*. These plants were said to have activities against menopausal symptoms, hormonal effects, fertility, and anti-fertility effects as well as abortifacients, contraceptive efficacy, aphrodisiac properties, treating reproductive tract infections, postnatal complications, relieving uterine bleeding, etc. Herbs and trees are the most used growth forms whereas leaves and seeds are the most cited plant part used in most of the tradomedical remedy preparations. Most of the remedies were prepared from a single and/or multiple plant source with other ingredients. The chapter contributes toward understanding, promotion, documentation, and safeguarding of indigenous knowledge, practices, and resources used in addressing female reproductive issues.

Keywords

Reproductive health · Female fertility · Postnatal care · Traditional healers · Menopause

1 Introduction

From time immemorial, man's life has always been dependent on plants around him, from which he draws all his basic needs [1–7]. Despite the effectiveness of chemically synthesized medicines, plants have been used medicinally in all civilizations [8, 9]. Medicinal plants contribute to the development of new pharmaceuticals to address new or existing medical issues including reproductive issues [10]. A large proportion of the Global South populace in rural and suburban areas including Nigerians rely on traditional or ethnomedicine for their primary health care due to the accessibility, availability, effectiveness, affordability, and inherent trust associated with it. This agrees with the view of CGSPS [11] that an estimated 85% of developing nations mainly use traditional health care systems. About 33% of Nigeria's population is classified as poor with women accounting for a substantial amount of the group and living in remote rural communities [12, 13]. According to WHO (2010) an estimated 25% of maternal deaths occur during pregnancy. Almost 20% of global maternal death happen in Nigeria with about 600,000 death and 900,000 near death mainly due to bleeding in pregnancy, infection, preeclampsia, eclampsia, and anemia in pregnancy [11, 14, 15]. Other issues include low access, inadequate, inefficient, and inappropriate reproductive health care system in Nigeria for the large youthful and reproductively active population [12, 14]. There is a need for high-quality care to maintain reproductive and sexual vigor as well before, during, and after pregnancy and childbirth.

Appropriate medical use of Nigeria's rich and diverse flora can contribute to addressing female reproductive health issues. For instance, *Alstonia boonei* stem bark and root and *Cymbopogon citrates* leaves that are used to treat fever and itching during pregnancy and epilepsy and *Bryophyllum pinnatum* leaves that is used to treat cough [9, 16–18]. *Ficus exasperata* is important for treating venereal diseases [16]. *Ocimum gratissimum* and *Vernonia amygdalina* are used by Usen people in Edo State Nigeria to boost immunity for mother and child after delivery and to regulate breast flow, respectively [5]. However, these medicinal plants are not isolated from the ongoing global biodiversity loss due to environmental change and anthropogenic activities as well as erosion of ethnomedicinal knowledge and practices that characterize the diverse ethnic groups of Nigeria [19–22]. The major issue connected to the erosion of ethnomedicinal knowledge and practices is the aging and death of the custodians of this knowledge, migration, and a lack of interest by the younger generation [8, 23, 24]. Put together, traditional medicine may be under threat. Confronted with illness and disease, early men discovered a wealth of useful therapeutic agents in the plant and this ageless system is still valuable today. The empirical knowledge and systems associated with medicinal substances in plants were and is still being passed on by oral tradition [5, 6, 25, 26].

This chapter aims to produce an inventory of the plants used by the residents of Benin and Warri for addressing female reproductive and sexual health conditions. A focused approach was adopted and carried out along with a nonexperimental validation of the plants used through a literature review of the phytochemical/pharmacological information supporting the medicinal activities. Here, we defined reproductive conditions and infertility according to Ogbe et al. [9] and Olooto et al. [27] as those that affect reproductive and sexual success as well as the prevention of conception and infertility as the inability to achieve pregnancy over 6–12 months despite adequate, intentional, and regular (i.e., about 3–4 times per week), unprotected sexual intercourse, respectively. Some causes of female infertility include leukorrhea, menopause, sexually transmitted infections (like chlamydia and gonorrhea), preexisting conditions (like malaria, HIV/AIDS, anemia, and malnutrition) abnormal menstrual cycles, faulty uterus and ovaries, menstrual disorder, and inappropriate lifestyles like excessive smoking and drinking that alter the hormonal pattern and ovulation [28–31]. Miscarriage (i.e., vaginal bleeding before 20 gestational weeks resulting in fetal abortion) is a common complication in pregnancies [32–34]. Prenatal health care help prepare for safe childbirth and parenthood and is beneficial for the survival of the mother, and prevents, detects, alleviates, and manages actual or potential health problems of the fetus and expectant mother. It is important to note that female genital mutilation increases the likelihood of complications during childbirth [35–37].

At present, there is little to no documentation of higher plants used for female fertility, pre- and postnatal care by Benin and Warri residents in Southern Nigeria. This chapter will document the traditional medicinal practices for female fertility, prenatal and postnatal of Benin residents and identify and evaluate how these plants are used (i.e., either alone or in combination).

2 Plants Used in Female Reproductive Health Care: Case Study of Benin and Warri Cities, Southern Nigeria

2.1 Methodology

Benin City is in Edo State (6.34° N and 5.60° E) and to the South is Delta State where Warri City is located (5.55° N and 5.77° E). Benin and Warri Cities lie within the humid tropical rainforest zone of Nigeria but their vegetation has been heavily altered by anthropogenic activities and human population growth and is currently a mosaic of secondary forest [9]. The climate is monsoonal with two distinct seasons (dry and rainy season) with high rainfall (2000 mm–3000 mm), temperature (20–40 °C) and atmospheric humidity and with radiation of 1600 h per year [38, 39]. A detailed description of the study area including soil characteristics can be found in Osawaru et al. [40]; Osawaru and Ogwu [41, 42]; Ogwu and Osawaru [43].

Survey to document traditional medicine practices and higher plants used for female fertility, prenatal and postnatal care by Benin and Warri residents was conducted between July and November 2021 using survey sheets for information collection, a digital camera, an audio recording device for interviews, and a mobile phone enabled position marker. The survey focused on randomly selected and locally identifiable professional herbalists, traditional healers, native midwives, and herbal remedy sellers using semistructured, open-ended questionnaires in English and/or preferred native languages. The randomly selected respondents and interviewees are duly registered with all the relevant authorities. The questionnaires were administered to the key informants (full-time traditional healers, part-time practitioners, elderly people who know the traditional values of the plants, relatives, and acquaintances who at one time or another used some of these plants for female fertility). The respondents animated a field survey to identify and collect plants and plant parts used for female fertility, prenatal and postnatal care. Specific questions like the botanical names, local names, plant parts used, dosage, method of preparation, application, and duration of treatment were asked during the oral interview, and the information supplied by these people was recorded. Collections were made from homesteads, distant farms, and different forests present in the study area.

Data collected include the sociodemographic characteristics of the respondents (gender, age, marital status, educational qualification, religion, tribe/ethnic group, occupation, name of village, and languages spoken), information on the plants used for female fertility, prenatal and postnatal care, and preparation of herbal remedies. Data collected were processed and analyzed using Windows Microsoft Excel 2016 version software, which was also used to draw charts and graphs to understand inherent patterns. The variables were presented as percentages. The therapeutic importance of each identified species was also recorded. Based on the knowledge of the informants, a total of 24 plants were identified and properly identified by a taxonomist at the University of Benin and through literature review.

2.2 Demographic Characteristics of Respondents

The demographic characteristic of the respondents and animators is presented in Table 1. There were more female (80%) compared to male (20%) respondents. The most dominant age category was 36–45 years (40%) and most respondents were married (66.67%) and had at least secondary school education (53.33%). Marital status, gender, culture, occupation, affordability, accessibility, and education level are major drivers for the practice, acceptance, and use of traditional medicine [44, 45]. Economic status is no longer a major determinant factor for the use of traditional medicine but techniques (like mind-body methods), spirituality, dietary practices, and comfort [46]. African traditional religion was the dominant religion of the professional herbalists, traditional healers, native midwives, and herbal remedy

Table 1 Demographic characteristics of respondents

Characteristics	Frequency	%
Gender		
Male	3	20
Female	12	80
Age range		
18–20	0	0
21–25	3	20
26–35	2	13.33
36–45	6	40
46 and above	4	26.67
Marital status		
Single	3	20
Engaged	0	0
Married	10	66.67
Divorce	2	13.33
Highest educational qualification		
Primary	3	20
Secondary	8	53.33
University	4	26.67
Religion		
Christian	8	53.33
Muslim	0	0
African traditional religion	7	46.67
Others	0	0
Occupation		
Tradomedical practitioner (healer, native midwives, nurses, etc.)	8	53.33
Government worker	1	
Trader	6	40
Private company worker	0	0
Unemployed	0	0

sellers surveyed in Benin and Warri Cities (Table 1). The religious affiliation documented in the survey is different from those reported in Curlin et al. [47] wherein it was reported that the majority of alternative medicine practitioners are most likely not to have any religious affiliation. However, they noted that these practitioners describe themselves as very spiritual and indirectly use religious beliefs in their dealings. On the other hand, higher spirituality and religious affiliation among contemporary medicine practitioners positively correlate with their acceptance to integrate alternative medicine into their practice [47]. This may be exploited in the Global South where religious affiliation is high among contemporary medicine practitioners to integrate and increase the relevance of traditional medicine practices for female fertility, pre- and postnatal care. A greater number of the preferential and concurrent use of herbal and contemporary medicine has been documented in parts of the Global South like Tamale, Northern Ghana [48]. According to Ngere et al. [49], this may be connected to the perceived sociocultural relevance of belief systems in disease etiology and diagnosis as well as caregiving and care reception in intervention formulation. In more developed countries, age and chronic diseases are the most important factors for accepting and using herbal medicines [50].

2.3 Medicinal Plants Used by Benin and Warri Residents for Female Reproductive Health Care by Benin and Warri Residents in Southern Nigeria

A total of 24 medicinal plants are used by Benin and Warri residents for the treatment and management of female reproductive issues. The 24 plants recorded from this survey are distributed into 21 families including Myrtaceae, Poaceae, Lamiaceae, Fabaceae, Apiaceae, and Lauraceae (Tables 2 and 3). The 24 plants included *Psidium guajava, Pennisetum purpureum, Oenothera biennis, Aloe barbadensis, Asparagus racemosus, Gossypium hirsutum, Newbouldia laevis, Phyllanthus amarus, Vitex agnus-castus, Portulaca oleracea, Afrormosia laxiflora, Agelaea obliqua, Alstonia boonei, Aneilema hockii, Apium graveolens, Brunfelsia splendida, Cucumeropsis mannii, Euphorbia deightonii, Heeria insignis, Macaranga barteri, Penianthus zenkeri, Persea americana, Azadirachta indica,* and *Carica papaya*. The diversity of the flora suggests the strength and diversity of reproductive health issues that may be treated with them and the wide variety of medicinal recipes [51]. Fasola [52] documented 61 plants from 32 families used in addressing female reproductive problems in Ibadan, Oyo State, Nigeria whereas Ogwu et al. [5] recorded 36 plant species belonging to 25 plant families in Usen Edo State, Nigeria. Joudi and Ghasem [53] and Ogwu et al. [5] reported the diverse uses of plants from three families – Asteraceae, Fabaceae, and Malvaceae for reproductive ailments. In the current study, Euphorbiaceae had the highest plant species representation followed by Myrtaceae. Tsobou et al. [54] recorded 70 plants from 37 families used in the treatment of reproductive ailments in Cameroon. In a similar study, Nduche et al. [55] documented 62 plant species from 41 families that are used in Ebonyi State, Nigeria to address fertility-related

Table 2 Plants used by Benin and Warri residents for the treatment of female fertility, prenatal and postnatal issues

S/No.	Collection source	Plant name	Family	Common name	Status	Cultivation status	Origin
1	Home garden	*Psidium guajava*	Myrtaceae	Guava	Not assessed	Cultivar	Exotic
2	Roadside	*Pennisetum purpureum*	Poaceae	Napier grass	Not assessed	Weed	Indigenous
3	Roadside	*Oenothera biennis*	Myrtaceae	Evening primrose	Not assessed	Weed	Exotic
4	Home garden	*Aloe barbadensis*	Asphodelaceae	Aloe vera	Endangered	Cultivar	Exotic
5	Roadside	*Asparagus racemosus*	Asparagaceae	Asparagus grass	Endangered	Weed	Exotic
6	Home garden	*Gossypium hirsutum*	Malvaceae	Cotton	Threatened	Cultivar	Exotic
7	Distant farmlands/forest	*Newbouldia laevis*	Bignoniaceae	Akoko	Not assessed	Wild	Indigenous
8	Roadside	*Phyllanthus amarus*	Euphorbiaceae	Indian gooseberry	Not assessed	Weed	Exotic
9	Distant farmlands/forest	*Vitex agnus-castus*	Lamiaceae	Chaste tree	Not assessed	Wild	Exotic
10	Home garden	*Portulaca oleracea*	Portulacaceae	Purslane plant	Not assessed	Cultivar	Exotic
11	Distant farmlands/forest	*Afrormosia laxiflora*	Fabaceae	Afrormosia	Endangered	Wild	Indigenous
12	Home garden	*Agelaea obliqua*	Connaraceae	Esura	Least concern	Cultivar	Indigenous
13	Distant farmlands/forest	*Alstonia boonei*	Apocynaceae	Stool wood	Least concern	Wild	Indigenous
14	Market	*Aneilema hockii*	Commelinaceae	Aneilema	Least concern	Weed	Indigenous
15	Home garden	*Apium graveolens*	Apiaceae	Celery	Least concern	Cultivar	Exotic
16	Home garden	*Brunfelsia splendida*	Solanaceae	Kiss me quick	Vulnerable	Cultivar	Exotic
17	Home garden	*Cucumeropsis mannii*	Cucurbitaceae	Egusi melon	Least concern	Cultivar	Indigenous
18	Roadside	*Euphorbia deightonii*	Euphorbiaceae	Ejagham	Not assessed	Weed	Indigenous
19	Distant farmlands/forest	*Heeria insignis*	Anacardiaceae	Heer plant	Not assessed	Wild	Indigenous
20	Distant farmlands/forest	*Macaranga barteri*	Euphorbiaceae	Agbasa	Not assessed	Wild	Indigenous
21	Distant farmlands/forest	*Penianthus zenkeri*	Menispermaceae	Ewe	Not assessed	Wild	Indigenous
22	Home garden	*Persea americana*	Lauraceae	Avocado	Least concern	Cultivar	Exotic
23	Home garden	*Azadirachta indica*	Meliaceae	Neem	Least concern	Cultivar	Exotic
24	Home garden	*Carica papaya*	Caricaceae	Papaya	Least concern	Cultivar	Exotic

Table 3 Distribution of the plant families of plants used for female reproductive health care by Benin and Warri residents in Southern Nigeria

Families	Number of species
Myrtaceae	2
Poaceae	1
Asphodelaceae	1
Asparagaceae	1
Malvaceae	1
Bignoniaceae	1
Euphorbiaceae	3
Lamiaceae	1
Fabaceae	1
Cucurbitaceae	1
Connaraceae	1
Apocynaceae	1
Commelinaceae	1
Apiaceae	1
Solanaceae	1
Anacardiaceae	1
Menispermaceae	1
Lauraceae	1
Meliaceae	1
Caricaceae	1
Portulacaceae	1

issues. These reports agree with the submission by Gill [56] that Nigerian flora is invaluable to the health system care of Nigerians. Some medicinal plants have multi-curative abilities where they are used in the preparation of reproductive health remedies for multiple ailments and taken as infusions, decoctions, and powdering [52, 57]. In the present study, most of the plants documented were reported to be associated with more than one female reproductive health issue as a remedy. Nonetheless, these native plants are often neglected or underutilized albeit they possess a substantial amount of undescribed benefits [58, 59]. A significant proportion of the African populace uses traditional plants and traditional healers and elders to manage health issues including the treatment of reproductive health problems [54]. The knowledge and practice are deep-rooted in the sociocultural background of the locale [60]. In some cases, a blend of indigenous and modern scientific knowledge is adopted to address certain reproductive health problems [5].

When these plants are used for female reproductive care, they are prepared and used in the form of oral tinctures, macerations, concoctions, or infusions and applied topically in rare cases [61, 62]. The survey of Benin and Warri residents revealed that the plants were mostly used alone and are commonly soaked in hot water or gin or boiled in water to extract the bioactive constituents (Table 4). In some cases, the plant seeds are dried and then grinded. These plant parts contain bioactive

Table 4 Plants used by Benin residents for female reproductive care, the parts used (whether alone or in combination with other herbal remedies), and the mode of administration is presented

S/N.	Botanical name	Part used	Decoction means	Alone or in combination with	Mode of administration
1	*Psidium guajava*	Leaves and stems	Soaking in hot water or gin	Alone	Swallowed
2	*Pennisetum purpureum*	Leaf extract	Soaking in hot water or gin	Alone	Swallowed
3	*Oenothera biennis*	Seed oil extract	Soaking in hot water or gin	Alone or in combination with *Cocos nucifera*	External application and/or swallowed
4	*Aloe barbadensis*	Leaf extract	Used fresh	Alone	Swallowed or applied externally
5	*Asparagus racemosus*	Root extract	Soaking in hot water or gin	Alone or in combination with other plant extracts	Swallowed
6	*Gossypium hirsutum*	Leaf and seed extract	Soaking in hot water or carbonated soft drink	Alone	Swallowed
7	*Newbouldia laevis*	Leaf and stem bark	Soaking in hot water or gin	Alone	Swallowed
8	*Phyllanthus amarus*	Leaf extract	Soaking in hot water or gin	Alone or in combination with yam extract	Swallowed
9	*Vitex agnus-castus*	Leaf and stem	Soaking in hot water or gin	Alone	Swallowed
10	*Portulaca oleracea*	Leaf and stem	Soaked in gin	Alone	Swallowed
11	*Afrormosia laxiflora*	Leaf extract	Soaking in hot water or gin	Alone	Swallowed
12	*Agelaea obliqua*	Leaf extract	Soaking in hot water or gin	Alone	Swallowed
13	*Alstonia boonei*	Leaf extract	Soaking in hot water or gin	Alone	Swallowed
14	*Aneilema hockii*	Leaf extract	Soaking in hot water or gin	Alone	Swallowed
15	*Apium graveolens*	Leaf and seed extract	Soaking in hot water or gin	Alone or in combination with other herbal remedies	Swallowed
16	*Brunfelsia splendida*	Leaf extract	Soaking in hot water or gin	Alone	Chewed and swallowed
17	*Cucumeropsis mannii*	Seed extract	Dried and grinded	Alone and/or in combination with other herbal remedies	Swallowed

(continued)

Table 4 (continued)

S/N.	Botanical name	Part used	Decoction means	Alone or in combination with	Mode of administration
18	*Euphorbia deightonii*	Succulent stem	Used fresh	Alone	Swallowed
19	*Heeria insignis*	Leaf extract	Soaking in hot water or gin	Alone	Swallowed
20	*Macaranga barteri*	Leaf and seed extract	Soaking in hot water or gin. Seeds are grinded.	Alone and/or in combination with other herbal remedies	Swallowed
21	*Penianthus zenkeri*	Root, stems, and leaf extract	Soaking in hot water or gin	Alone and/or in combination with other herbal remedies	Swallowed
22	*Persea americana*	Seeds	Dried and grinded	Added to meals	Swallowed
23	*Azadirachta indica*	Leaves	Boiled in water and soaked	Alone	Swallowed
24	*Carica papaya*	Seeds	Dried and grinded	Added to meals	Swallowed

polyphenolic compounds like isoflavones and flavonoids that confer beneficial reproductive health effects on women [63].

The plants used by Benin and Warri residents for female reproductive care and the parts used, whether alone or in combination with other herbal remedies, and the mode of administration is presented in Table 4. The most common plant part used is the leaf and they can be used alone or in combination with other local herbal remedies. The decoction made from the plants is more likely to be ingested than externally applied.

The commonly used plant part in the preparation of traditional medical remedies by Benin and Warri residents for the treatment and management of female reproductive health issues is presented in Fig. 1. Besides the whole plant, various plant parts are commonly collected and used in the preparation of herbal remedies including stem, bark, root, leaves, whole plant fruits, and seeds [6, 52, 64–66]. However, in the present study, leaves followed by leaves and seeds and leaves and stem bark were the most used parts, while root, stems and leaves, root, oil, and stem had the least usage. Other workers have made a similar observation regarding the plant parts that are most commonly used including Idowu et al. [67]; Thirupathy [68]; and Shosan et al. [69] wherein leaves were reported as the most used plant part for the treatment and management of reproductive issues. These leaves may be used alone or in combination with other plant parts in the preparation of traditional remedies [70]. In agreement with the finding of Tsobou et al. [54] most herbal remedies are prepared as a decoction. We also observed that all the plants were used as a decoction. The method of preparing the herbal remedy is diverse but in most cases are soaked in dry gin (approximately 40% alcohol) or boiled in hot water

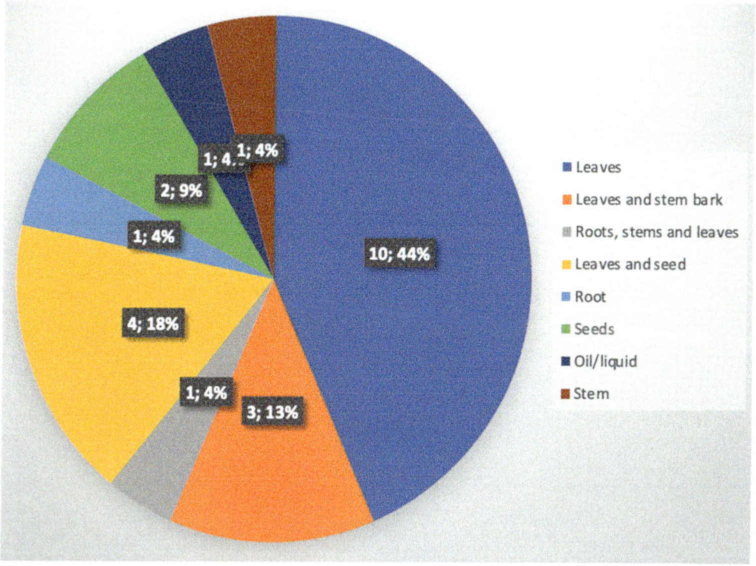

Fig. 1 Frequency of plant parts used for reproductive health care by Benin and Warri residents in Southern Nigeria

[6, 52, 71]. This work has highlighted the importance of plant-based medicine in female reproductive care, which is in line with the findings of Tsobou et al. [54] that medicinal plants are invaluable in the management of reproductive healthcare. Some of these plants used for female reproductive care have activities against menopausal symptoms, hormonal effects, fertility, and anti-fertility effects as well as abortifacients, contraceptive efficacy, aphrodisiac activity, reproductive tract infections, childbirth complications, and efficacy in relieving uterine bleeding, etc. Medicinal plants contain secondary metabolites that have important in vivo and in vitro functions to both ovarian folliculogenesis and steroidogenesis in many animal species [72].

2.4 Enumeration of Medicinal Plants Used for Female Reproductive Health Care by Benin and Warri Residents in Southern Nigeria

2.4.1 *Psidium Guajava* (Common Name - Guava)

It is a perennial evergreen tree. Flowering starts within the first 2 years and trees typically reach maturity after 5–8 years. Leaves are oppositely arranged while the fruit exudes a strong, sweet, musky odor when ripe and may be round, ovoid, or pear shape. The central pulp of the fruit is concolorous. Guava leaves are known to have attenuating effects on hypertension, hepatic steatosis, and other oxidative stress diseases because of their high oxidation activities [73–75].

Fig. 2 *Psidium guajava* leaves and fruits

Parts used: Leaves and stem bark. The leaves contain rutin, kaempferol, epicatechin, naringenin, isoflavonoids, gallic acid, and flavonoids [76].

Mode of administration: Tea made from leaves or bark is taken to regulate menstrual flow and pre-pregnancy stomach cleansing (Fig. 2).

2.4.2 *Pennisetum Purpureum* (Common Name – Elephant Grass)

A perennial grass that easily forms dense thickets with ecosystem-impacting characteristics functions especially on hydrology cycles, biophysical dynamics, and community composition. The plant has hormone-regulatory, antioxidant, and immune capacities as well as postpartum blood pressure management [77–79]. The decoction made from the plant is also used to manage menstrual pains and facilitate childbirth.

Parts used: Leaves.

Mode of administration: Young leaves are boiled and used in the treatment of venereal diseases preventing pregnancy and for superovulatory treatment as a hormonal control (Fig. 3).

2.4.3 *Oenothera Biennis* (Common Name – Evening Primrose)

A biennial plant that grows up to 1.2 m. It is not frost tender and flowers from June to September. The plant is hermaphrodite and is often polluted by Lepidoptera. The stem is hairy and often purple-tinged. Bright yellow, four-petaled flowers that open at night. The seed remains in the soil and germinates when the soil is disturbed. It is used in the treatment of mastalgia (breast pains) and hot flashes [80]. Hutcherson

Fig. 3 Mature *Pennisetum purpureum* plant

et al. [81] reported that using the plant decreases the risk of cesarean section, macrosomia, preeclampsia, and stillbirth. The plant is relevant for the cervical preparation of women about to undergo hysteroscopy [82].

Parts used: Oil extract from the plant.

Mode of administration: Used to improve sexual appetite, treat menopause symptoms, and regulate estrogen (Fig. 4).

2.4.4 *Aloe Barbadensis* (Common Name – Aloe Vera)

A succulent evergreen perennial that is stemless or short-stemmed and grows up to 60–100 cm. Leaves are thick and fleshy, green to gray-green. Leaf margins are serrated with small white teeth. The plant is used to treat internal and external injuries during pregnancy and after childbirth [83]. External application of aloe vera helps increase the sensitivity, elasticity, and functioning of reproductive organs and increases estrogen formation. Also, Maharjan et al. [84] connected the use of aloe vera gel to ovarian steroid restoration and protection.

Parts used: Leaves and leaf extract (juice).

Mode of administration: Leave and leaf extract are used to prepare juice and taken to improve ovary development, and follicle-stimulating hormones and to rejuvenate the uterus (Fig. 5).

2.4.5 *Asparagus Racemosus* (Common Name – Asparagus Grass)

Possess small pine-needle-like phylloclades that are mostly uniform and shiny green. The white flowers are short and spiky. Fruits are blackish-purple, globular berries. It has a tuberous adventitious root system that is about one meter in length and tapering at both ends. The plant is used to increase fertility and to address stress-related issues

Fig. 4 Mature *Oenothera biennis* plant with flowers

in pregnant women [85, 86]. It also has hormonal balancing effects in women, which increases the chances of pregnancy as well as lactation-boosting functions.

Parts used: Root extract.

Mode of administration: Extracts from the roots are used to balance female reproductive hormones (Fig. 6).

2.4.6 *Gossypium Hirsutum* (Common Name – Cotton)

The plant is an annual subshrub with alternate leaves, petiolate, palmately 3–5 lobed, hirsute, and cordate blades. Fruit is a dehiscent capsule (4–6 cm long), spherical smooth, light green, and bearing hairs on the epidermis strongly attached to the seed coat. Well-developed taproot with numerous lateral roots. A herbal remedy prepared from cotton regulates labor-inducing proteins and supports uterine quiescence [87]. The plant is also used to increase fertility and address health issues like fever, headaches, gastrointestinal problems, and nausea during pregnancy.

Fig. 5 Mature *Aloe barbadensis*

Fig. 6 *Asparagus racemosus*

Parts used: Leaves, seeds, and roots.

Mode of administration: Extracts from leaves and seeds are used in the treatment of menstrual disorders, symptoms of menopause, inducing labor, and afterbirth cleansing, and improve breast milk production. It is applied vaginally and on the breast surface (Fig. 7).

Fig. 7 *Gossypium hirsutum* leaves and flowers. (Source: Morris [88])

2.4.7 *Newbouldia Laevis* (Common Name – Boundary Tree, Akoko)

A perennial tree that grows up to 15 m high. The stem is woody. Each leaflet is oblanceolate 10–20 cm long and 5–10 cm broad, serrated, sessile, or with short petioles. The fruit is a long, pendulous, dehiscent capsule about 30 cm long dotted with glands. The plant leaves have uterine contractile effects and antihemorrhagic functions [89, 90]. It is also used to treat fever during pregnancy.

Parts used: Leaves and bark.

Mode of administration: Leaves are squeezed in water and applied in the virginal to stop bleeding in threatened miscarriage. The leaves are cooked and eaten by pregnant women to facilitate safe and easy delivery and milk production after birth. The barks are used to prepare a solution and used to treat stomach pains and cleanse the stomach in preparation for childbirth (Fig. 8).

Fig. 8 *Newbouldia laevis* leaves. (Source: Nwokolo [91])

2.4.8 *Phyllanthus Amarus* (Common Name – Sleeping Plant, Indian Gooseberry Plant, Seed under Leaf)

A slender annual herb with an erect stem and leafy angular branches. The leaves are very small and the minute pedicelled flowers are very numerous. The plant is used to regulate hormones and for hepatoprotection during pregnancy because of its immunomodulatory capacity and for the treatment of jaundice, diabetes, dysentery, urogenital diseases, and malaria before pregnancy [92–95]. It is also used to treat gonorrhea and other wart-like genital issues.

Parts used: The whole plant.

Mode of administration: The plant is soaked in alcohol (dry gin) and allowed to stand. The extract is taken orally to treat dysentery and stomach aches both during pregnancy and after delivery (Fig. 9).

2.4.9 *Vitex Agnus-Castus* (Common Name – Sleeping Plant, Indian Gooseberry Plant)

A deciduous hermaphrodite shrub or herb. It is pollinated by Insects and grows well in sandy and loamy soils that are well-drained even in nutritionally poor ones. Suitable pH: acid, neutral, and basic (alkaline) soils. It does not grow well in the shade. The plant is used mainly to manage and treat uterine bleeding disorders and mastodynia [96]. It is also used to improve sexual dysfunction in women undergoing menopause [97]. The plant helps with stress during pregnancy by reducing prolactin and has a hormone-stabilizing effect on prepregnant women.

Parts used: Leaves and stems.

Fig. 9 *Phyllanthus amarus* leaves. (Source: Ghosh et al. [92])

Mode of administration: The leaves and stems are soaked in dry gin and the extract is taken orally to treat premenstrual syndrome and menstrual disorder. It increases menstrual flow (Fig. 10).

2.4.10 *Portulaca Oleracea* (Common Name – Purslane Plant)

An annual or perennial prostrate succulent herb. Stem may spread on the ground or be erect. They are smooth and fleshy, more or less reddish, and could reach 40 cm. It produces mucilage in high amounts that enhance the flexibility of female reproductive organs. The plant has antiaging effects on female genitalia as well as numerous antioxidants and antiatherogenic activities. The plant is also used to fight infections and complications during and after childbirth [62].

Parts used: Leaf and stem extract.

Mode of administration: Extract is taken to improve women's fertility and libido (Fig. 11).

2.4.11 *Afrormosia Laxiflora* (Common Name – Aformosia, African Teak)

A small to medium-sized straight tree bearing crooked, drooping branches forming a disheveled crown. It is common in woodland savanna and fringing to dry dense

Fig. 10 Mature *Vitex agnus-castus* with flowers

Fig. 11 *Portulaca oleracea* leaves and flowers. (Source: SEINet Portal Network [98])

Fig. 12 *Afromosia laxiflora* leaves and flowers. (Source: Green Institute [101])

forests. It is used to boost sexual appetite and performance, cure venereal diseases, and stimulate higher milk production [99, 100].

Parts used: Leaves.

Mode of administration: Leaves are soaked and used as a laxative and stomach cleanser in preparation for pregnancy (Fig. 12).

2.4.12 *Agelaea Obliqua* (Common Name – Esura)

Scrabbling shrub with shining leaves. Flowers pure white or tinged yellow, green or pink, fragrant; fruits scarlet, in forest and secondary growth. The plant is taken to support and boost sexual functions [102].

Part used: Leaves.

Mode of administration: Leaves are consumed raw or in processed form due to their aphrodisiac properties and to treat mild vaginal infections (Fig. 13).

2.4.13 *Alstonia Boonei* (Common Name – Stool Wood)

A tree with grayish bark that exudes a bitter-tasting latex. The leaves may be elliptic or oblanceolate in shape. Whorled and angular branches covered with white lenticels. Inflorescence is repeatedly branched in whorl, with flowers in threes. Herbal decoction prepared with the plants are taken to address fever, insomnia, joint and body pains, and diarrhea. It is used to address abdominal pains, pelvic congestions, ovarian cysts, uterine fibroid, menstrual pains, and microbial infections before, during, and after pregnancy [105].

Part used: Leaves.

Fig. 13 Agelaea species. (Source: Willis [103]; Martin and Harvey [104])

Mode of administration: Leaves in combination with alligator pepper and other medicinal plants are used to prepare soup for pregnant women that are believed to allow the fetus to develop well and promote safe delivery (Fig. 14).

2.4.14 *Aneilema Hockii* (Common Name – Aneilema)

Tufted perennial with erect or straggling shoots. Roots are thick and fleshy. Leaves are spirally arranged. The large, bluish flowers readily distinguish this species.

Fig. 14 *Alstonia boonei* leaves and flowers. (Source: Ileke [106])

Capsules are generally longer, less pubescent, and have seeds in capsules that are emarginated and lack a terminal apicule. The plant is taken to increase fertility and fight venereal diseases. It is also used to treat malaria during pregnancy [107].

Part used: Leaf and root extract.

Mode of administration: Leaf and root extract are taken to prevent miscarriage and maintain the health of the mother and fetus (Fig. 15).

2.4.15 *Apium Graveolens* (Common Name – Celery)

Biennial to perennial herb. The leaves are lobed on each side of the central axis and has a row of two or more lobes on each side of the central axis. The leaves are used to increase fertility and balance hormone production. Herbal decoction prepared from the seed is used to stabilize the blood pressure of pregnant women [109]. The plant is also known to have antioxidant properties relevant for addressing oxidative stress issues in pregnant women.

Part used: Leaves and seed extract.

Mode of administration: Herbal extract prepared with *Apium* species to improve reproductive vigor in women. It is also taken to correct any sexual dysfunction in women (Fig. 16).

2.4.16 *Brunfelsia Uniflora* (Common Name Kiss Me Quick)

A medium-sized, evergreen shrub or small tree with diffuse, spreading branches. It usually grows 0.5–3 m tall. The plant is known to have antioxidant properties relevant for addressing oxidative stress issues in pregnant women.

Part used: Leaf.

Fig. 15 *Aneilema hockii.* (Source: Hyde et al. [108])

Mode of administration: The leaf is believed to increase the chances of getting pregnant in young women (Fig. 17).

2.4.17 *Cucumeropsis Mannii* (Common Name – Melon)

A climber with stems that are sparse crisped-hairy. Leaf-blade are broad ovate-cordate or reniform-cordate. Fruits are cylindrical-ellipsoid, and rounded. Seeds ovate in outline, compressed, smooth, and white. The plant is used to protect and maintain a healthy female genital system and to boost fertility [112].

Part used: Seeds.

Mode of administration: Crushed seeds are blended and mixed with other herbal remedies to improve female fertility and conception (Fig. 18).

Fig. 16 *Apium graveolens* leaves. (Source: Dennis et al. [110])

2.4.18 *Euphorbia Deightonii* (Common Name – Ejagham)

A shrub, branching from the base without a distinct trunk, to 6 m high. The plant appears to be always cultivated. It is commonly grown as a hedge-plant, the lowest branches becoming buried in the ground thus providing a thick barrier to the base. The plant contains copious white latex, which is said to be poisonous. It is an irritant in wounds. An herbal remedy prepared with the plant is used to improve endocrine pathways and treat polycystic ovary syndrome [63].

Part used: Succulent stem.

Mode of administration: Chewed and swallowed (Fig. 19).

2.4.19 *Heeria Insignis* (Common Name – Heer Plant)

A branched shrub or small tree. Leaf undersurface is silvery-white and flowers small, and creamy; fruits are glossy and black. Inflorescences of terminal and axillary, much-branched panicles. The leaves are boiled and used in the treatment of viral and microbial infections before and during pregnancy [114]. It contains important secondary metabolites. It is used to prevent and treat malaria before and during pregnancy.

Fig. 17 *Brunfelsia uniflora.* (Source: Makoto [111])

Part used: Leaves.

Mode of administration: The plant is taken to improve lactation after childbirth (Fig. 20).

2.4.20 *Macaranga Barteri* (Common Name – Macaranga, Akan)

An evergreen climbing shrub or a tree. The spiny bole have spreading and spiny branches. It is a dioecious species. Leaves and seed extracts are used to improve fertility, treat stomach ache, and fight venereal diseases like gonorrhea.

Part used: Leaf and seed extracts.

Fig. 18 Opened *Cucumeropsis mannii* fruits. (Source: Morris [113])

Fig. 19 *Euphorbia deightonii*

Mode of administration: Leaf and seed extract are taken as an abortifacient, to treat sexually transmitted infections, and stomach cleanser in preparation to get pregnant (Fig. 21).

Fig. 20 *Heeria insignis*. (Source: Tropical Plants Database [115])

2.4.21 *Penianthus Zenkeri* (Common Name – Ewe)

A dioecious, evergreen shrub or small tree. Leaves are arranged spirally, simple, and entire; with stipules absent. Seed ellipsoid, seedling with epigeal germination.

Part used: Root, stem, and leaf.

Mode of administration: The roots, stem, and leaf extract are used as an aphrodisiac, and branches, bark, and roots are so used. They are also used to treat impotence (Fig. 22).

2.4.22 *Persea Americana* (Common Name – Avocado, Pear)

It is a medium to large evergreen tree (although some varieties lose their leaves before flowering starts). The tree canopy ranges from low, dense, and symmetrical to upright and asymmetrical. Leaves are variable in shape (elliptic, oval, lanceolate). They are often pubescent and reddish when young, becoming smooth, leathery, and dark green when mature. Flowers are yellowish green, and 1–1.3 cm in diameter. The plant seed is used for hormonal management and to improve lactation [117, 118].

Part used: Leaf and seed extract.

Mode of administration: Extract from the leaf and seed are taken to improve female reproductive hormone and fertility control as well as improve maternal health (Fig. 23).

2.4.23 *Azadirachta Indica* (Common Name – Neem, Dongoyaro)

A fast-growing evergreen tree with wide and spreading branches. The leaves are opposite and pinnate with missing terminal leaflets. Petioles are mostly short with

Fig. 21 *Macaranga barteri* leaves. (Source: Willis [103]; Martin and Harvey [104])

Fig. 22 *Penianthus zenkeri* leaves and fruits. (Source: Davidson and Christoph [116])

Fig. 23 *Persea americana* leaves and fruits. (Source: Krist [119])

axillary panicle drooping flowers. The fruit endocarp contains 1–3 seeds with a brown coat. The plant is used to manage menstrual flow, improve fertility, and treat malaria during pregnancy [120, 121].

Parts used: Leaves and flowers.

Mode of administration: A decoction of the leaves is taken orally to treat malaria during pregnancy (Fig. 24).

2.4.24 *Carica Papaya* (Common Name - Pawpaw, Uhoro)

It is a tree with a stem that yields copious white latex. It has a weak, soft wooden stem crowned by a terminal cluster of large edible fruits. It is used to boost sexual performance and fertility in women through hormonal balancing and increase in estrogen production.

Parts used: Seeds and Leaves.

Mode of administration: The seeds are used to cause abortion in pregnant women, especially young girls. The leaves are used to treat malaria during pregnancy as a prenatal and postnatal measure (Fig. 25).

3 Conclusion

Though traditional reproductive health service is considered affordable and accessible, the younger generation is more interested in modern medicine, which is characteristically more expensive and less available in certain parts of Benin and Warri cities. The issue is driving the decline and neglect in the practice and resources used

Fig. 24 Azadirachta indica leaves and flowers

Fig. 25 *Carica papaya*

for female reproductive health care, respectively. This is further compounded by the process of urbanization and Western culture, which adds an extra layer to the complex and multidimensionality of female reproductive dysfunction. The threats to this indigenous practice and knowledge system require an urgent revitalization strategy to protect indigenous medical knowledge. Also, only little scientific evidence exists to prove the efficacy of the remedies and plant species employed to treat or manage reproductive disorders. Addressing these issues could lead to the confirmation of their efficacy and the discovery and development of new drugs for modern medicine. Moreover, the medicinal plants documented in this study are considered invaluable for addressing female reproduction dysfunction through traditional approaches and need conservation attention. Their conservation will help preserve the heritage and plant diversity.

References

1. Chime AO, Aiwansoba RO, Eze CJ, Osawaru ME, Ogwu MC (2017) Phenotypic characterization of tomato *Solanum lycopersicum* L. cultivars from southern Nigeria using morphology. Malaya J Biosci 4(1):30–38
2. Chime AO, Aiwansoba RO, Osawaru ME, Ogwu MC (2017) Morphological evaluation of tomato (*Solanum lycopersicum* Linn.) cultivars. Makara J Sci 21(2):97–106. https://doi.org/10.7454/mss.v21i2.7421
3. Ogwu MC (2020) Value of *Amaranthus* [L.] species in Nigeria. In: Waisundara V (ed) Nutritional value of Amaranth. IntechOpen, UK, pp 1–21. https://doi.org/10.5772/intechopen.86990
4. Ogwu MC (2023) Local food crops in Africa: sustainable utilization, threats, and traditional storage strategies. In: Izah SC, Ogwu MC (eds) Sustainable utilization and conservation of Africa's biological resources and environment, Sustainable development and biodiversity, vol 888. Springer, Singapore, pp 353–376. https://doi.org/10.1007/978-981-19-6974-4_13
5. Ogwu MC, Osawaru ME, Obahiagbon GE (2017) Ethnobotanical survey of medicinal plants used for traditional reproductive care by Usen people of Edo state, Nigeria. Malaya J Biosci 4(1):17–29
6. Ogwu MC, Osawaru ME, Onosigbere-Ohwo U (2017) Characterization of okra (*Abelmoschus* [Medik.]) accessions using dehydrogenase isozymes and protein. Maldives Natl J Res 5(1): 45–62
7. Osawaru ME, Ogwu MC (2020) Survey of plant and plant products in local markets within Benin City and environs. In: Filho LW, Ogugu N, Ayal D, Adelake L, da Silva I (eds) African handbook of climate change adaptation. Springer Nature, Switzerland, pp 1–24. https://doi.org/10.1007/978-3-030-42091-8_159-1
8. Imarhiagbe O, Ogwu MC (2022) Sacred groves in the global south: a panacea for sustainable biodiversity conservation. In: Izah SC (ed) Biodiversity in Africa: potentials, threats and conservation. Sustainable development and biodiversity, vol 29. Springer, Singapore, pp 525–546. https://doi.org/10.1007/978-981-19-3326-4_20
9. Ogbe FMD, Eruogun OL, Uwagboe M (2009) Plants used for female reproductive health care in Oredo local government area, Nigeria. Sci Res Essay 4(3):120–130
10. UNFP [United Nations Population Fund] (2020) Sexual reproduction and health. https://www.unfpa.org/sexual-reproductive-health. Accessed 17 Apr 2021
11. Center for Gender and Social Policy Studies [CGSPS] (2002) Critical issues in engendering reproductive healthcare practice in Nigeria. Obafemi Awolowo University, Ile-ife, Nigeria and the Ford Foundation, 25p
12. Obilade O, Mejiuni O (2005) Poverty alleviation through reproductive health in Nigeria exploring other non-formal alternatives. www.gla.ac.uk/centres/cradall/docs/Botswana-papers
13. Oluwole V (2022) Nigeria is no longer the poverty capital of the world but still has over 70 million people living in extreme poverty – the highest in Africa. Business Insider Africa. https://africa.businessinsider.com/local/markets/nigeria-is-no-longer-the-poverty-capital-of-the-world-but-still-has-over-70-million/2txm7g3
14. UNDP (2008) Human development report 2007/2008 – Nigeria. http://hdrstats.undp.org/countries/data_sheets/cty_ds_NGA.html
15. WHO (2007) World Health Statistics–Health status: mortality. www.who.int/whosis/whostat2007_1mortality.pdf

16. Aiyeloja AA, Bello OA (2006) Ethnobotanical potentials of common herbs in Nigeria: a case study of Enugu state. Educ Res Rev 1:6–22
17. Idu M, Onyibe HI, Timothy O, Erhabor J (2011) Ethno-medicinal flora of Otuo people of Edo state. Asian J Plant Sci 2008 7(1):9–11
18. Nwachukwu I, Obasi OO (2009) Use of modern birth control methods among rural communities in Imo State, Nigeria. Afr J Reprod Health 12(1):101–108
19. African Conservation Foundation [ACF] (2003) Overview on medicinal plants and traditional medicine in Africa. http://www.conserveafrica.org/medicinal_plants.rtf
20. Ogwu MC, Izah SC, Iyiola AO (2022) An overview of the potentials, threats and conservation of biodiversity in Africa. In: Izah SC (ed) Biodiversity in Africa: potentials, threats and conservation, Sustainable development and biodiversity, vol 29. Springer, Singapore, pp 3–20. https://doi.org/10.1007/978-981-19-3326-4_1
21. Ogwu MC, Osawaru ME, Ahana CM (2014) Challenges in conserving and utilizing plant genetic resources (PGR). Int J Genet Mol Biol 6(2):16–22. https://doi.org/10.5897/IJGMB2013.0083
22. Okoli RI, Aigbe O, Ohaju-Obodo JO, Mensah JK (2007) Medicinal herbs used for managing some common ailments among Esan people of Edo state, Nigeria. Pak J Nutr 6:490–496
23. Maregesi SM, Ngassapa OD, Pieters L, Vlietinck AJ (2007) Ethnopharmacological survey of the Bunda district, Tanzania: plants used to treat infectious diseases. J Ethnopharmacol 113: 457–470
24. Ogwu MC (2019) Towards sustainable development in Africa: the challenge of urbanization and climate change adaptation. In: Cobbinah PB, Addaney M (eds) The geography of climate change adaptation in urban Africa. Springer Nature, Switzerland, pp 29–55pp. https://doi.org/10.1007/978-3-030-04873-0_2
25. Ogwu MC, Osawaru ME, Owie MO (2023) Effects of storage at room temperature on the food components of three cocoyam species (*Colocasia esculenta, Xanthosoma atrovirens*, and *X. sagittifolium*). Food Stud: Interdiscip J 13(2):59–83. https://doi.org/10.18848/2160-1933/CGP/v13i02/59-83
26. Ogwu MC, Osawaru ME, Amodu E, Osamo F (2023) Comparative morphology, anatomy and chemotaxonomy of two *Cissus* Linn. species. Rev Bras Bot 2023:1. https://doi.org/10.1007/s40415-023-00881-0
27. Olooto WE, Amballi AA, Banjo TA (2012) A review of female infertility; important etiological factors and management. J Microbiol Biotech Res 2:379–385
28. Forster C, Abraham S, Taylor A, Llewellyn-Jones D (1994) Psychological and sexual changes after the cessation of breast feeding. Obstet Gynaecol 84(5):872–876
29. Agomuo VE, Samuel KC (2004) Fruit of the womb. Prevail Concepts Publication, Ikeja
30. Bertuccio P, Tavani A, Gallus S, Negri E, La Vecchia C (2007) Menstrual and reproductive factors and risk of non-fatal acute myocardial infarction in Italy. Eur J Obstet Gynecol Reprod Biol 134(1):67–72. https://doi.org/10.1016/j.ejogrb.2007.01.005
31. Kanyadan P, Ganti L, Mangal R, Stead T, Hahn L, Sosa M (2023) Understanding factors that influence whether a woman will seek care for reproductive health: a national survey. Health Psychol Res 11:67959. https://doi.org/10.52965/001c.67959
32. Makrydimas G, Sebire NJ, Lolis D, Vlassis N, Nicolaides KH (2003) Fetal loss following ultrasound diagnosis of a live fetus at 6–10 weeks of gestation. Ultrasound Obstet Gynecol 22:368–372. https://doi.org/10.1002/uog.204
33. Santelli J, Rochat R, Hatfield-Timajchy K, Gilbert BC, Curtis K, Cabral R, Hirsch JS, Schieve L, Unintended Pregnancy Working Group (2003) The measurement and meaning of unintended pregnancy. Perspect Sex Reprod Health 35(2):94–101. https://doi.org/10.1363/3509403
34. Sotiriadis A, Papatheodorou S, Makrydimas G (2004) Threatened miscarriage: evaluation and management. BMJ 329(7458):152–155. https://doi.org/10.1136/bmj.329.7458.152

35. Klein E, Helzner E, Shayowitz M, Kohlhoff S, Smith-Norowitz TA (2018) Female genital mutilation: health consequences and complications – a short literature review. Obstet Gynecol Int 2018:7365715. https://doi.org/10.1155/2018/7365715
36. Obiora OL, Maree JE, Nkosi-Mafutha NG (2021) "A lot of them have scary tears during childbirth..." experiences of healthcare workers who care for genitally mutilated females. PLoS One 16(1):e0246130. https://doi.org/10.1371/journal.pone.0246130
37. Tukue D, Gebremeskel TG, Gebremariam L, Aregawi B, Hagos MG, Gebremichael T et al (2020) Prevalence and determinants of modern contraceptive utilization among women in the reproductive age group in Edaga-hamus Town, Eastern zone, Tigray region, Ethiopia, June 2017. PLoS One 15(3):e0227795
38. Omuta GED (1988) The quality of urban life and the perception of livability: a case study of neighbourhoods in Benin City, Nigeria. Soc Indic Res 20:417–440. https://doi.org/10.1007/BF00302336
39. Onwueme IC, Sinha TD (1991) Field crop production in tropical Africa. CTA, Wageningen
40. Osawaru ME, Ogwu MC, Braimah L (2013) Growth responses of two cultivated okra species (*Abelmoschus caillei* (A. Chev.) Stevels and *Abelmoschus esculentus* (Linn.) Moench) in crude oil contaminated soil. Nig J Basic Appl Sci 21(3):215–226
41. Osawaru ME, Ogwu MC (2014) Ethnobotany and germplasm collection of two genera of cocoyam (*Colocasia* [Schott] and *Xanthosoma* [Schott], Araceae) in Edo state Nigeria. Sci Technol Arts Res J 3(3):23–28. https://doi.org/10.4314/star.v3i3.4
42. Osawaru ME, Ogwu MC (2015) Molecular characterization of 36 accessions of two genera of cocoyam (*Colocasia* [Schott] and *Xanthosoma* [Schott], Araceae). Sci Technol Arts Res J 4(1): 27–33. https://doi.org/10.4314/star.v4i1.4
43. Ogwu MC, Osawaru ME (2015) Soil characteristics, microbial composition of plot, leaf count and sprout studies of cocoyam (*Colocasia* [Schott] and *Xanthosoma* [Schott], Araceae) collected in Edo state, southern Nigeria. Sci Technol Arts Res J 4(1):34–44. https://doi.org/10.4314/star.v4i1.5
44. Ben-Arye E, Lev E, Keshet Y, Schiff E (2010) Integration of herbal medicine in primary care in Israel: a Jewish-Arab cross-cultural perspective. Evid Based Complement Alternat Med:2011. https://doi.org/10.1093/ecam/nep146
45. Chali BU, Hasho A, Koricha NB (2020) Preference and practice of traditional medicine and associated factors in Jimma town, Southwest Ethiopia. Evid Based Complementary Altern Med:2021. https://doi.org/10.1155/2021/9962892
46. AlRawi SN, Khidir A, Elnashar MS, Abdelrahim HA, Killawi AK, Hammoud MM, Fetters MD (2016) Traditional Arabic & Islamic medicine: validation and empirical assessment of a conceptual model in Qatar. BMC Complement Altern Med 17. https://doi.org/10.1186/s12906-017-1639-x
47. Curlin FA, Rasinski KA, Kaptchuk TJ, Emanuel EJ, Miller FG, Tilburt JC (2009) Religion, clinicians, and the integration of complementary and alternative medicines. J Altern Complement Med 15(9):987–994. https://doi.org/10.1089/acm.2008.0512
48. Kwame Ameade EP, Ibrahim M, Ibrahim S, Habib RH, Gbedema SY (2017) Concurrent use of herbal and orthodox medicines among residents of tamale, northern Ghana, who patronize hospitals and herbal clinics. Evid Based Complement Alternat Med 2018. https://doi.org/10.1155/2018/1289125
49. Ngere SH, Akelo V, Ridzon R, Otieno P, Nyanjom M, Omore R, Tippett Barr BA (2021) Traditional medicine beliefs and practices among caregivers of children under five years – the child health and mortality prevention surveillance (CHAMPS), Western Kenya: a qualitative study. PLoS One 17(11). https://doi.org/10.1371/journal.pone.0276735
50. Rashrash M, Schommer JC, Brown LM (2017) Prevalence and predictors of herbal medicine use among adults in the United States. Journal of Patient Experience 4(3):108–113. https://doi.org/10.1177/2374373517706612

51. Smith-Hall C, Overgaard LH, Pouliot M (2012) People, plants and health: a conceptual framework for assessing changes in medicinal plant consumption. J Ethnobiol Ethnomed 8:43
52. Fasola TR (2015) An ethnobotanical survey of plants used in the management and treatment of female reproductive health problems in Ibadan, southwestern Nigeria. J Biol Agric Healthc 5(3):7–11
53. Joudi L, Ghasem HB (2010) Exploration of medicinal species of Fabaceae, Lamiaceae and Asteraceae families in Ilkhji region, eastern Azerbaijan Province (northwestern Iran). J Med Plant Res 4(11):1081–1084
54. Tsobou R, Mapongmetsem PM, Van Damme P (2016) Medicinal plants used for treating reproductive health care problems in Cameroon, Central Africa. Econ Bot 70:145–159. https://doi.org/10.1007/s12231-016-9344-0
55. Nduche MU, Omosun G, Okwulehie IC (2015) Ethnobotanical survey of plants used as remedy for fertility conditions in Ebonyi state of Nigeria. Scholars Academic Journal of Biosciences 3:214–221
56. Gill LS (1992) Ethnomedicinal uses of plants in Nigeria. Uniben Press, Nigeria, p 276
57. Diame GLA (2010) Ethnobotany and ecological studies of plants used for reproductive health: a case study at Bia biosphere reserve in the Western region of Ghana. Final report submitted to the division of ecological sciences UNESCO (MAB) young scientist research award scheme Paris Cedex 15 France. UNESCO, Accra office. 125. 30p
58. Ogwu MC, Osawaru ME, Aiwansoba RO, Iroh RN (2016) Ethnobotany and collection of West African Okra [Abelmoschus caillei (A. Chev.) Stevels] germplasm in some communities in Edo and Delta States, Southern Nigeria. Borneo J Resour Sci Technol 6(1):25–36
59. Ogwu MC, Osawaru ME, Atsenokhai EI (2016) Chemical and microbial evaluation of some uncommon indigenous fruits and nuts. Borneo Sci 37(1):54–71
60. De Feo V, Senatore F (1993) Medicinal plants and Phytotherapy in the Amalfitan coast, Salerno Province, Campania, southern Italy. J Ethnopharmacol 39(1):39–51
61. Bafor EE (2017) Potentials for use of medicinal plants in female reproductive disorders – the way forward. Afr J Reprod Health 21(4):9–11. https://doi.org/10.29063/ajrh2017/v21i4.1
62. Bussmann RW, Glenn A (2010) Medicinal plants used in northern Peru for reproductive problems and female health. J Ethnobiol Ethnomed 6:30. (2010). https://doi.org/10.1186/1746-4269-6-30
63. Akbaribazm M, Goodarzi N, Rahimi M (2021) Female infertility and herbal medicine: an overview of the new findings. Food Sci Nutr 9(10):5869–5882. https://doi.org/10.1002/fsn3.2523
64. Hunde D, Asfaw Z, Kelbessa E (2004) Use and management of ethnoveterinary medicinal plants by indigenous people in 'Boosat', Welenchetti area. Ethiop J Biol Sci 3:113–132
65. Tibuti JR, Dhillion SS, Lye KA (2003) Ethnoveterinary medicines for cattle (Bos indicus) in Bulamogi county Uganda: plant species and mode of use. J Ethnopharmacol 88:279–286
66. Yinegar H, Kelbessa E, Bekele T, Lulekal E (2007) Ethnoveterinary medicinal plants in Bale Mountains National Park, Ethiopia. J Ethnopharmacol 112:55–70
67. Idowu OA, Soniran OT, Ajana O, Aworinde DO (2010) Ethnobotanical survey of antimalarial plants used in Ogun state, southwestern Nigeria. Afr J Pharm Pharmacol 4(2):55–60
68. Thirupathy S, Vaidyanathan D, Salai Senthilkumar MS, Ghouse Basha M (2013) Survey of ethno medicinal plants. Adv Appl Sci Res 4(6):90–95
69. Shosan LO, Fawibe OO, Ajiboye AA, Abeegunrin TA, Agboola DA (2014) Ethnobotanical survey of medicinal plants used in curing some diseases in infants in Abeokuta south local government area of Ogun state, Nigeria. Am J Plant Sci 5:3258–3268
70. Ayannar M, Ignacimuthu S (2011) Ethnobotanical survey of medicinal plants commonly used by the Kani tribals in Tirunelveli hills of Western Ghats, India. J Ethnopharmacol 134(3):851–864
71. Ermias L, Ensermu K, Tamrat B, Haile Y (2008) An ethnobotanical study of medicinal plants in Mana Angetu District, southeastern Ethiopia. J Ethnobiol Ethnomed 4:10

72. Mbemya GT, Viera LA, Canafistula FG, Pessoa ODL, Rodrigues APR (2017) Reports on *in vivo* and *in vitr* contribution of medicinal plants to improve the female reproductive function. Reproducao & Climeterio 32(2):109–119
73. Gutierrez-Montiel D, Guerrero-Barrera AL, Chávez-Vela NA, Avelar-Gonzalez FJ, Ornelas-García IG (2023) *Psidium guajava* L.: from byproduct and use in traditional Mexican medicine to antimicrobial agent. Front Nutr 10:1108306. https://doi.org/10.3389/fnut.2023.1108306
74. Irondi EA, Agboola SO, Oboh G, Boligon AA, Athayde ML, Shode FO (2016) Guava leaves polyphenolics-rich extract inhibits vital enzymes implicated in gout and hypertension in vitro. J Intercult Ethnopharmacol 5(2):122–130. https://doi.org/10.5455/jice.20160321115402
75. Yoshitomi H, Guo X, Liu T, Gao M (2012) Guava leaf extracts alleviate fatty liver via expression of adiponectin receptors in SHRSP.Z-Leprfa/Izm rats. Nutr Metab 9(13). https://doi.org/10.1186/1743-7075-9-13
76. Sampath Kumar NS, Sarbon NM, Rana SS, Chintagunta AD, Prathibha S, Ingilala SK, Jeevan Kumar SP, Sai Anvesh B, Dirisala VR (2021) Extraction of bioactive compounds from Psidium guajava leaves and its utilization in preparation of jellies. AMB Express 11(1):36
77. Chauhan G, Tadi P (2022) Physiology, postpartum changes. [updated 2022 Nov 14]. In: StatPearls [internet]. Treasure Island (FL): StatPearls Publishing; 2023 Jan. https://www.ncbi.nlm.nih.gov/books/NBK555904/
78. Huang PF, Mou Q, Yang Y, Li JM, Xu ML, Huang J, Li JZ, Yang HS, Liang XX, Yin YL (2021) Effects of supplementing sow diets during late gestation with Pennisetum purpureum on antioxidant indices, immune parameters and faecal microbiota. Vet Med Sci 7(4):1347–1358. https://doi.org/10.1002/vms3.450
79. Ojo OA, Oni AI, Grant S, Amanze J, Ojo AB, Taiwo OA, Maimako RF, Evbuomwan IO, Iyobhebhe M, Nwonuma CO, Osemwegie O, Agboola AO, Akintayo C, Asogwa NT, Aljarba NH, Alkahtani S, Mostafa-Hedeab G, Batiha GE, Adeyemi OS (2022) Antidiabetic activity of elephant grass (Cenchrus Purpureus (Schumach.) Morrone) via activation of PI3K/AkT signaling pathway, oxidative stress inhibition, and apoptosis in wistar rats. Front Pharmacol 13:845196. https://doi.org/10.3389/fphar.2022.845196
80. Mahboubi M (2019) Evening primrose (*Oenothera biennis*) oil in management of female ailments. J Menopausal Med 25(2):74–82. https://doi.org/10.6118/jmm.18190
81. Hutcherson TC, Cieri-Hutcherson NE, Lycouras MM, Koehler D, Mortimer M, Schaefer CJ, Costa OS, Bohlmann AL, Singhal MK (2022) Systematic review of evening primrose (*Oenothera biennis*) preparations for the facilitation of parturition. Pharmacy (Basel, Switzerland) 10(6):172. https://doi.org/10.3390/pharmacy10060172
82. Mansour Ghanaei M, Asgharnia M, Farokhfar M, Mohammad Asgari Ghalebin S, Rafiei E, Haryalchi K (2022) The effect of consuming evening primrose oil on cervical preparation before hysteroscopy: an RCT. Int J Reprod Biomed 20(7):591–600. https://doi.org/10.18502/ijrm.v20i7.11561
83. Hashemi SA, Madani SA, Abediankenari S (2014) The review on properties of aloe vera in healing of cutaneous wounds. Biomed Res Int 2015. https://doi.org/10.1155/2015/714216
84. Maharjan R, Nagar PS, Nampoothiri L (2010) Effect of Aloe barbadensis mill. Formulation on Letrozole induced polycystic ovarian syndrome rat model. J Ayurveda Integr Med 1(4):273–279. https://doi.org/10.4103/0975-9476.74090
85. Alok S, Jain SK, Verma A, Kumar M, Mahor A, Sabharwal M (2013) Plant profile, phytochemistry and pharmacology of *Asparagus racemosus* (Shatavari): a review. Asian Pac J Trop Dis 3(3):242–251. https://doi.org/10.1016/S2222-1808(13)60049-3
86. Pandey AK, Gupta A, Tiwari M, Prasad S, Pandey AN, Yadav PK, Sharma A, Sahu K, Asrafuzzaman S, Vengayil DT, Shrivastav TG, Chaube SK (2018) Impact of stress on female reproductive health disorders: possible beneficial effects of shatavari (Asparagus racemosus). Biomed Pharmacother 103:46–49. https://doi.org/10.1016/j.biopha.2018.04.003
87. Gruber CW, O'Brien M (2011) Uterotonic plants and their bioactive constituents. Planta Med 77(3):207–220

88. Morris R (2023) Cucumeropsis mannii. Photohgraphed by Ehoam Bidault. Tropical plants database, Ken Fern. tropical.theferns.info. https://tropical.theferns.info/image.php?id=Cucumeropsis+mannii. Retrieved 5 Apr 2023
89. Mudonhi N, Nunu WN (2022) Traditional medicine utilisation among pregnant women in sub-saharan African countries: a systematic review of literature. Inquiry 59: 469580221088618. https://doi.org/10.1177/00469580221088618
90. Bafor E, Sanni U (2009) Uterine contractile effects of the aqueous and ethanol leaf extracts of Newbouldia Laevis (Bignoniaceae) in vitro. Indian J Pharm Sci 71(2):124–127. https://doi.org/10.4103/0250-474X.54274
91. Nwokolo C (2021) Health benefits of Ogirisi leaves (Newbouldia laevis). https://healthguide.ng/ogirisi-leaves-newbouldia-laevis/. Accessed 5 Apr 2023
92. Bose Mazumdar Ghosh A, Banerjee A, Chattopadhyay S (2022) An insight into the potent medicinal plant *Phyllanthus amarus* Schum. and Thonn. Nucleus 65(3):437–472. https://doi.org/10.1007/s13237-022-00409-z
93. Geethangili M, Ding T (2017) A review of the phytochemistry and pharmacology of *Phyllanthus urinaria* L. Front Pharmacol 9. https://doi.org/10.3389/fphar.2018.01109
94. Jantan I, Haque MA, Ilangkovan M, Arshad L (2018) An insight into the modulatory effects and mechanisms of action of Phyllanthus species and their bioactive metabolites on the immune system. Front Pharmacol 10. https://doi.org/10.3389/fphar.2019.00878
95. Rao MV, Alice KM (2001) Contraceptive effects of Phyllanthus amarus in female mice. Phytotherapy research: PTR 15(3):265–267. https://doi.org/10.1002/ptr.735
96. Rafieian-Kopaei M, Movahedi M (2017) Systematic review of premenstrual, postmenstrual and infertility disorders of Vitex Agnus Castus. Electron Physician 9(1):3685–3689. https://doi.org/10.19082/3685
97. Heirati SFD, Ozgoli G, KabodMehri R, Mojab F, Sahranavard S, Nasiri M (2021) The 4-month effect of *Vitex agnus-castus* plant on sexual function of women of reproductive age: a clinical trial. J Educ Health Promot 10:294. https://doi.org/10.4103/jehp.jehp_63_21
98. SEINet Portal Network (2023) *Portulaca oleracea*. Photographer Max Licher. https://swbiodiversity.org/seinet/imagelib/imgdetails.php?imgid=219134 Accessed 5 Apr 2023
99. Nwodo NJ, Ibezim A, Ntie-Kang F, Adikwu MU, Mbah CJ (2015) Anti-trypanosomal activity of Nigerian plants and their constituents. Molecules (Basel, Switzerland) 20(5):7750–7771. https://doi.org/10.3390/molecules20057750
100. Chauhan S, Singh N (2019) Women and access to family planning: women's right to decide a distant reality in India. Gend Womens Stud 1(1):32–37
101. Green Institute (2018) *Afromosia laxiflora*. https://greeninstitute.ng/plants/afrormosia-laxiflor. Accessed 5 Apr 2023
102. Silvestris E, Lovero D, Palmirotta R (2019) Nutrition and female fertility: an interdependent correlation. Front Endocrinol 10:346. https://doi.org/10.3389/fendo.2019.00346
103. Willis KJ (2017) *Macaranga barteri*. Digitised palynological slide. In: European reference collection (version 1). Retrieved from globalpollenproject.org on 22 Oct 2017
104. Martin AC, Harvey WJ (2017) The global pollen project: a new tool for pollen identification and the dissemination of physical reference collections. Methods Ecol Evol 8:892–897. https://doi.org/10.1111/2041-210X.12752
105. Adotey JP, Adukpo GE, Opoku Boahen Y, Armah FA (2012) A review of the ethnobotany and pharmacological importance of Alstonia boonei De wild (Apocynaceae). ISRN Pharmacol 2012:587160. https://doi.org/10.5402/2012/587160
106. Ileke KD (2014) Cheese wood, Alstonia boonei De wild a botanical entomocides for the management of maize weevil, Sitophilus zeamais (Motschulsky) [Coleoptera: Curculionidae]. Octa J Biosci 2(2):64–68
107. Omara T (2020) Antimalarial plants used across Kenyan communities. Evid Based Complement Alternat Med 2020:4538602. https://doi.org/10.1155/2020/4538602

108. Hyde MA, Wursten BT, Ballings P, Coates Palgrave M (2023) Flora of Malawi: species information: individual images: Aneilema hockii. https://www.malawiflora.com/speciesdata/image-display.php?species_id=112780&image_id=9. Retrieved 5 Apr 2023
109. Moghadam MH, Imenshahidi M, Mohajeri SA (2013) Antihypertensive effect of celery seed on rat blood pressure in chronic administration. J Med Food 16(6):558–563. https://doi.org/10.1089/jmf.2012.2664
110. Dennis M, Grant A, Nicholaas P (2017) Plant-based ethnopharmacological remedies for hypertension in Suriname. IntechOpen, UK. https://doi.org/10.5772/intechopen.72106
111. Makoto M (2021) *Brunfelsia uniflora* (Pohl) D. Don. World Flora observation. https://identify.plantnet.org/the-plant-list/observations/1009732381. Retrieved 5 Apr 2023
112. Agu PC, Aja PM, Ekpono Ugbala E, Ogwoni HA, Ezeh EM, Oscar-Amobi PC, Asuk Atamgba A, Ani OG, Awoke JN, Nwite FE, Ukachi OU, Orji OU, Nweke PC, Ekpono Ugbala E, Ewa GO, Igwenyi IO, Egwu CO, Alum EU, Chukwu DC, Famurewa AC (2022) *Cucumeropsis mannii* seed oil (CMSO) attenuates alterations in testicular biochemistry and histology against bisphenol a-induced toxicity in male Wister albino rats. Heliyon 8(3): e09162. https://doi.org/10.1016/j.heliyon.2022.e09162
113. Morris R (2023) Tropical plants database, Ken Fern tropicalthefernsinfo 2023-04-05. tropical.theferns.info/viewtropical.php?id=Gossypium+hirsutum
114. Abubakar IB, Kankara SS, Malami I, Danjuma JB, Muhammad YZ, Yahaya H, Singh D, Usman UJ, Ukwuani-Kwaja AN, Muhammad A, Ahmed SJ, Folami SO, Falana MB, Nurudeen QO (2022) Traditional medicinal plants used for treating emerging and re-emerging viral diseases in northern Nigeria. Eur J Integr Med 49:102094. https://doi.org/10.1016/j.eujim.2021.102094
115. Tropical Plants Database (2023) Heeria insignis. https://tropicalthefernsinfo/viewtropicalphp?id=Ozoroa+insignis. Retrieved 5 Apr 2023
116. Davidson C, Christoph S (2022) Flora of the world: *Penianthus zenkeri.* https://floraoftheworldorg/taxon/flora/17592186249636 retrieved 5 April 2023
117. de Melo MFFT, Pereira DE, Moura RL, da Silva EB, de Melo FALT, Dias CCQ, Silva MDCA, de Oliveira MEG, Viera VB, Pintado MME, Dos Santos SG, Soares JKB (2019) Maternal supplementation with avocado (*Persea americana* mill.) pulp and oil alters reflex maturation, physical development, and offspring memory in rats. Front Neurosci 13:9. https://doi.org/10.3389/fnins.2019.00009
118. Orabueze IC, Babalola R, Azuonwu O, Okoko II, Asare G (2021) Evaluation of possible effects of *Persea Americana* seeds on female reproductive hormonal and toxicity profile. J Ethnopharmacol 273:113870. https://doi.org/10.1016/j.jep.2021.113870
119. Krist S (2020) Avocado oil. In: Vegetable fats and oils. Springer, Cham. https://doi.org/10.1007/978-3-030-30314-3_11
120. Gbotolorun SC, Osinubi AA, Noronha CC, Okanlawon AO (2008) Antifertility potential of neem flower extract on adult female Sprague-Dawley rats. Afr Health Sci 8(3):168–173
121. Upadhyay SN, Kaushic C, Talwar GP (1990) Antifertility effects of neem (Azadirachta indica) oil by single intrauterine administration: a novel method for contraception. Proc Biol Sci 242(1305):175–179. https://doi.org/10.1098/rspb.1990.0121

GPSR Compliance

The European Union's (EU) General Product Safety Regulation (GPSR) is a set of rules that requires consumer products to be safe and our obligations to ensure this.

If you have any concerns about our products, you can contact us on ProductSafety@springernature.com

In case Publisher is established outside the EU, the EU authorized representative is:

Springer Nature Customer Service Center GmbH
Europaplatz 3
69115 Heidelberg, Germany

Batch number: 08024528

Printed by Printforce, the Netherlands